Eighth Edition

Production & Operations Analytics

Steven Nahmias
Santa Clara University

Tava Lennon Olsen
University of Auckland

WAVELAND PRESS, INC.
Long Grove, Illinois

For information about this book, contact:
Waveland Press, Inc.
4180 IL Route 83, Suite 101
Long Grove, IL 60047-9580
(847) 634-0081
info@waveland.com
www.waveland.com

10-digit ISBN 1-4786-3926-1
13-digit ISBN 978-1-4786-3926-8

Printed in the United States of America

7 6 5 4 3 2 1

Dedicated to the memory of my brother David.
Steven Nahmias

Dedicated to the memory of my parents.
Tava Lennon Olsen

About the Authors

Steven Nahmias is Professor of Analytics in the Department of Information Systems and Analytics in the Leavey School of Business Administration at Santa Clara University. He holds a B.A. in Mathematics and Sciences from Queens College, a B.S. in Operations Research from Columbia University, and the M.S. and Ph.D. degrees in Operations Research from Northwestern University. He previously served on the faculties of the University of Pittsburgh, Georgia Tech, and Stanford University.

He is widely known for his research on stochastic inventory models with an emphasis on perishables. He has authored or co-authored more than 50 scientific articles, which have appeared in a variety of national and international journals. He has served as Area Editor in Supply Chain Logistics for *Operations Research*, senior editor for *Manufacturing and Service Operations Management (M&SOM)*, associate editor for *Management Science*, and associate editor for *Naval Research Logistics*. He earned first prize in the Nicholson Student Paper competition (1971), second prize in the TIMS student paper competition (1972), and the University Award for Sustained Excellence in Scholarship from Santa Clara University (1998). In 2011 he was named Distinguished Fellow of the Manufacturing and Service Operations Management Society and in 2014 Distinguished Fellow of *INFORMS*.

In addition to his academic activities he has served as a consultant to a variety of companies and agencies. In his spare time he enjoys biking and golf. He is also a semi-professional jazz trumpet player and performs regularly with several bands in the Bay Area.

Tava Olsen holds the Ports of Auckland chair in Logistics and Supply Chain Management at the University of Auckland Business School. Prior to joining Auckland, she was Professor of Operations and Manufacturing Management in the Olin Business School at Washington University in St. Louis, which she joined after serving as an Assistant Professor in the Department of Industrial and Operations Engineering at the University of Michigan, Ann Arbor. She received her B.Sc. (honours) in Mathematics from the University of Auckland and earned both her M.S. in Statistics and her Ph.D. in Operations Research from Stanford University.

Her research interests include supply chain management; pricing and inventory control; and stochastic modeling of manufacturing, service, and healthcare systems. Among other journals, her publications have appeared in *Management Science*, *Operations Research*, *Manufacturing and Service Operations Management (M&SOM)*, *Production and Operations Management*, and the *Journal of Applied Probability*. She has twice received the Auckland Business School's Research Excellence Award.

She is the Director of the Centre for Supply Chain Management and Head of the Department of Information Systems and Operations Management. She is the Area Editor of the Operations and Supply Chain Area of *Operations Research* and is a senior editor of *Production and Operations Management*. She has previously served as an Associate Editor for *Management Science* and *M&SOM*. She is a past president of the Manufacturing and Service Operations (MSOM) society for which she received the society's distinguished service award. She enjoys hiking, biking, and spending time with her husband and two daughters.

Contents

Preface to the Eighth Edition

The seventh edition update focused on the first half of the text; whereas, this update has largely focused on the last half of the book. We have also taken the opportunity to reorganize some of the material in the book. The only material that has been lost is that which was so outdated that an update didn't seem relevant (e.g., some descriptions of facility layout software). We have also updated the title from Production and Operations Analysis to the more modern Production and Operations Analytics.

Chapter 1 has been significantly rewritten and updated. Some of the service strategy material that previously appeared in Chapter 7, material on the product/process matrix and flexible manufacturing systems that appeared in Chapter 11, and material on benchmarking that appeared in Chapter 12 of the prior edition can now be found in the first chapter of the new edition. In addition, there are now small sections on the productivity frontier and the innovation curve, which were missing from earlier editions. The focus is also now slightly more international. Finally, the detailed analysis of capacity growth planning, previously in the introductory chapter, is now discussed in Chapter 7 (Supply Chain Analytics).

Chapter 3 (Sales and Operations Planning) is partially revised. We had heard from instructors that they missed some of the deleted aggregate planning materials from the sixth edition. In particular, we again have a small section on aggregate units in Section 3.4; more details on the chase and level strategies are included in Appendix 3-A.

The chapter on supply chain management has been updated and divided into two chapters: Supply Chain Strategy (Chapter 6) and Supply Chain Analytics (Chapter 7). The discussion on risk pooling has been expanded within Chapter 6. The facility location material that was previously Chapter 11 is incorporated in Chapter 7. Service Operations Management is now Chapter 8.

Chapter 9 on production control systems has been significantly updated. Previous editions of this text used the term *just-in-time* (JIT) extensively. However, this term appears to have largely dropped out of favor, replaced by lean production. However, lean is much broader than a production control device. Indeed, lean production has become so pervasive that we do not think that devoting a chapter to it as a stand-alone concept is the most appropriate approach. Instead, we introduce lean as a strategy in Chapter 1. Then in Chapter 9 we discuss the mechanics behind lean production scheduling, namely kanban cards and pull production. Finally, we discuss the waste elimination side of lean in Chapter 10 on quality and assurance. Given how popular lean-six-sigma programs have become, this latter merging appears the most appropriate to us. However, we recognize that some instructors may wish to pull together the appropriate sections to cover the material together. The sections needed for this approach would be 1.7, 9.7, and 10.1.

Chapter 10 (Quality and Assurance) has also received a major rewrite. In addition to the incorporation of the quality management aspects of lean, the rest of the chapter has been updated and revised. The acceptance sampling portions of the chapter have been downplayed, with some sections moving to the appendix, and the quality management aspects have moved forward. We no longer use TQM (an outdated term) and instead discuss Six Sigma as a quality movement. A section on process capability has been added where we show how the 3.4 defects per million parts standard arises as the definition of Six Sigma quality.

Chapter 11 of the previous edition has been removed and the information on facilities layout and location redistributed through-out the text, as indicated in the table below.

7th Edition	8th Edition
11.1 The Facilities Layout Problem	1.4 Process Types and Layouts
11.2 Patterns of Flow	1.4 Process Types and Layouts and Appendix 1–A.
11.3 Types of Layouts	1.4 Process Types and Layouts
11.4 A Prototype Layout Problem and the Assignment Model	S1.10 A Prototype Layout Problem and the Assignment Model
11.5 More Advanced Mathematical Programming Formulations	S1.11 More Advanced Mathematical Programming Formulations
11.6 Computerized Layout Techniques	Removed except for centroids example
11.7 Flexible Manufacturing Systems	1.7 Strategic Initiatives
11.8 Locating New Facilities	7.2 Locating New Facilities
11.9 The Single-Facility Rectilinear Distance Location Problem	7.3 The Single-Facility Rectilinear Distance Location Problem
11.10 Euclidean Distance Problems	7.4 Euclidean Distance Problems
11.11 Other Location Models	7.5 Other Location Models
11.12 Historical Notes	Varies
11.13 Summary	Varies
APPENDIX 11–A FINDING CENTROIDS	APPENDIX 1–B USING CENTROIDS FOR LAYOUT DECISIONS
APPENDIX 11–B COMPUTING CONTOUR LINES	APPENDIX 7–B COMPUTING CONTOUR LINES

CHAPTER 1

Strategy and Competition

"However beautiful the strategy, you should occasionally look at the results."

—Winston Churchill

CHAPTER OVERVIEW

Purpose

The purpose of this chapter is to introduce the student to a variety of strategic issues that arise in the operations function of the firm.

KEY POINTS

1. *Strategy matters.* Firms that do not have a clear strategy risk trying to be all things to all people; this may result in them being no customer's first choice. Further, a firm's operations strategy, which is the means by which the firm deploys its resources to achieve its competitive goals, must be aligned with its business strategy.
2. *Manufacturing matters.* We contend that the loss of a manufacturing base in the United States, or any developed nation, is not healthy and will eventually lead to an overall loss in the standard of living and quality of life. This counters the argument that evolution into a service economy is a natural and healthy thing.
3. *Global competition.* How does one measure success and economic health on a global scale? One way is to examine classical measures of relative economic strength, which include: balance of trade, share of world exports, creation of jobs, and cost of labor. However, such macro measures do not adequately explain why certain countries dominate certain industries. National competitive advantage is a consequence of several factors (factor conditions, demand conditions, related and supporting industries, firm strategy structure, and rivalry), although productivity also plays an important role.
4. *Strategic dimensions.* Along with cost and/or product differentiation, other dimensions along which firms distinguish themselves include quality, delivery speed, delivery reliability, flexibility, and service. These dimensions require trade-offs and should be thought about explicitly. Within these dimensions, some attributes will be order winners, delighting the customer, and others will be order qualifiers, necessary for the product to even be considered.
5. *Layout Fundamentals.* Before deciding on the appropriate layout for a new facility, whether it be a factory, hospital, theme park, or anything else, one must first study the patterns of flow. The simplest flow pattern is straight-line flow, as might be encountered on an assembly line. Other patterns include U flow, L flow, serpentine flow, circular flow, and S flow. Another issue is desirability or undesirability of

locating operations near each other. For example, the emergency room in a hospital must be near the hospital entrance, and the maternity ward should be close to the neonatal care unit.

6. *Types of layouts.* In the factory setting, the appropriate type of layout depends on the manufacturing environment and the characteristics of the product. A product layout is where machines or workstations are organized around the sequence of operations required to produce the product. Product layouts are most typical for mass production. In the case of small- to medium-sized companies, a process layout makes more sense—grouping similar machines or similar processes together. Finally, layouts based on group technology might be appropriate. In this case, machines might be grouped into machine cells where each cell corresponds to a part family or group of part families. The product-process matrix can be used to determine the appropriate match of product type to process type.
7. *Product and process life cycles.* Most of us understand that products have natural life cycles: start-up, rapid growth, maturation, stabilization, or decline. However, it is rarely recognized that processes too have life cycles. Initially, new manufacturing processes have the characteristics of a job shop. As the process matures, automation is introduced. In the mature phases of a manufacturing process, most major operations are automated. A firm needs to match the phases of product and process life cycles to be the most successful in its arena.
8. *Learning and experience curves.* These curves are helpful in forecasting the decline in unit cost of a manufacturing process as one gains experience with the process. Learning curves are more appropriate when modeling the learning of an individual worker, and experience curves are more appropriate when considering an entire industry.
9. *Strategic initiatives.* We discuss several strategic initiatives that have allowed many companies to shine in their respective arenas, including business process reengineering, just-in-time/lean production, flexible manufacturing systems, competing on quality, benchmarking quality, time-based competition, servicization, and automation.

Strategy is a long-term plan of action designed to achieve a particular goal, most often winning. Its root is from the Greek *stratēgos,* which referred to a "military commander" during the age of Athenian democracy. Strategy was originally conceived in the military context. Two famous books dealing with military strategy are *The Prince* by Machiavelli and *The Art of War* by Sun Tzu.

Business strategy often relates closely to military strategy. Companies fight on an economic battlefield, and long-term strategies determine winners and losers. Business strategy is the highest level of corporate activity that bundles together the disparate functional area strategies. Business strategy sets the terms and goals for a company to follow.

Perhaps the reason that chief executive officers (CEOs) are compensated so highly is the realization that the strategic vision of the CEO is often the difference between the success and failure of a company. The strategic visions of industry giants such as Henry Ford, Jack Welch, and Bill Gates were central to the success of their companies that, at one time or another, dominated their competition.

Perhaps the most dramatic example is Apple Inc. With the introduction of the iPod in 2002 and the iPhone in 2007, Apple transformed itself from a failing computer company into a major force in portable computing and telecommunications. The fascinating transformation of the firm is described in the Snapshot Application.

Apple Adopts a New Business Strategy and Shifts Its Core Competency from Computers to Portable Electronics

Apple Computer was the kind of success story one sees in the movies. Two youngsters, Steve Wozniak and Steve Jobs, grew up with an interest in hobbyist computers. Working from a garage, they founded Apple Computer in April 1976 and soon introduced a build- your-own, hobbyist computer called the Apple I. The firm was incorporated a year later with the help of Mike Markula, and the Apple II was introduced in April 1977, ushering in the world of personal computing. Perhaps it was the fact that it was selected as the platform for Visicalc, the first spreadsheet program, that led to its success, as much as the superior capabilities of the hardware.

While personal computers have become a common part of our everyday lives, we forget that they are relatively new inventions. The nature of the personal computer marketplace was dramatically altered by the introduction of the first PC by IBM in 1981. IBM's open architecture allowed for inexpensive clones to enter the market and crowd out Apple's significantly more expensive products. By the turn of the century, Apple's future looked to be in doubt.

Apple's subsequent transformation and rebirth is a fascinating bit of business history. Around 2001, an independent consultant, Tony Fadell, was shopping around his concept of an MP3 music player linked to a music sales service. At that time, MP3 players were not new; one could "rip" music from one's CD collection and load the songs on the player. While no one else was interested in Fadell's idea, he was hired by Apple and assigned to a team of 30 people, including designers, programmers, and hardware engineers.

When Apple decided to go ahead with the MP3 player concept, they also decided that they needed a new design to separate themselves from the rest of the marketplace. Apple subcontracted much of the development and design work to PortalPlayer, who devoted all their resources to the project. Steve Jobs himself was intimately involved with the design and function of the new player. The iPod was a huge success and was redesigned over several models and generations, with sales estimated at 400 million units worldwide.

While the iPod was a huge success, Apple did not rest on its laurels. In 2007, Apple launched the first iPhone. Again, the concept of a smartphone was not new. Several companies (notably Motorola, Samsung, Palm, and Nokia) had marketed smartphones for several years. As with the iPod, Apple again produced an innovative product with unique features. Apple continues to improve the iPhone and introduces a new generation of the product virtually every year. As of this writing, 1.5 billion iphones have sold worldwide.

Apple's more recent product, the iPad, was also an instant success and essentially defined a new market category. This tablet computer, introduced in March 2010, is convenient for web surfing and reading e-books—and is the product against which the competition measures itself. Apple registered more than 1 million sales of the iPad in the first three months alone and has now sold 350 million units. In 2012, it had 58% of the tablet market (falling to about 26% in 2017). As a testament to Apple's phenomenal success in portable computing, the market capitalization of Apple surpassed that of the software behemoth Microsoft in 2010.

In 2015 Apple entered the smart watch category with its Apple Watch. While not a runaway success, it is a solid contributor to what is still a relatively small category (one estimate puts it at 50 percent of this market). It is particularly popular with professional women whose clothing often does not have pockets (delaying access to their phones). However, it is not clear that Apple has embraced this customer segment. It will be interesting to observe whether Apple dominates the category or whether a more nimble competitor might fill the gap.

Source: L. Kahney "Inside Look at the Birth of the iPod," 2004; Statista, Apple iPad Sales Worldwide, 2019; Statista, Apple Product Sales, 2019.

Success requires a **vision**, and visions must be articulated so that all employees share that vision. The **mission statement** is the formal articulation of the vision. A good mission statement should provide a clear description of the goals of the firm and is the first step toward formulating a coherent business strategy. Poor mission statements tend to be wordy and full of generalities. Good mission statements are direct, clear, and concise. Patricia Jones and Larry Kahaner (1995) list the 50 corporate mission statements that they perceive as the best. One example is the mission statement of the Gillette Corporation: "Our Mission is to achieve or enhance clear leadership, worldwide, in the existing or new core consumer product categories in which we choose to compete." They then list the areas they perceive as their core competencies. Intel defines their mission as: "Do a great job for our customers, employees, and stockholders by being the preeminent building block supplier to the computing industry." The statement then provides details on values and objectives. In many cases, their objectives are quite specific (e.g., "Lead in LAN products and Smart Network Services"). Certainly, the award for conciseness has to go to the General Electric Corporation, whose mission statement is three words: "Boundaryless . . . Speed . . . Stretch." The commentary following the statement provides an explanation of exactly what these words mean in the context of the corporation.

Having articulated a vision, the next step is to plan a strategy for achieving that vision. This is the firm's **business strategy**. The overall business strategy includes defining:

1. the market in which the enterprise competes;
2. the level of investment;
3. the means of allocating resources to and the integration of separate business units;
4. marketing strategy, financial strategy, and operations strategy.

Broadly defined, the **operations strategy** is the means by which the firm deploys its resources to achieve its competitive goals. For manufacturing firms, it is the sum total of all decisions concerning the production, storage, and distribution of goods. Important operations strategy decisions include where to locate new manufacturing facilities, how large these facilities should be, what processes to use for manufacturing and moving goods through the system, and what workers to employ. Service firms also require an operations strategy. The Disney theme park's continuing record of success is due in part to its careful attention to detail in every phase of its operations.

Examples of companies getting into trouble due to poorly defined or executed operations strategies abound. When the Apple Macintosh was introduced, the product was extremely successful. However, the company failed to keep up with consumer demand, resulting in constant backorders. According to Debbi Coleman, Apple's former director of worldwide manufacturing:

> Manufacturing lacked an overall strategy which created problems that took nine months to solve . . . we had extremely poor forecasting. Incoming materials weren't inspected for defects and we didn't have a mechanism for telling suppliers what was wrong, except angry phone calls. Forty percent of Mac materials were coming from overseas and no one from Apple was inspecting them before they were shipped. . . . One of the biggest tasks that high-tech manufacturers face is designing a manufacturing strategy that allows a company to be flexible so it can ride with the highs and lows of consumer and business buying cycles. (Fallon, 1985)

Recognizing the importance of strategy is not new. Every few years a hot new production control method or management technique comes along, often described by a three-letter acronym. While it is easy to be skeptical about what may be perceived as a fad, the new methods can have substantial value to corporations when implemented intelligently and in line with the corporate strategy. This chapter highlights some of these trends within operations strategy. It also covers key concepts necessary for formulating a strong operations strategy.

1.1 MACROECONOMIC TRENDS

Business and operations strategies should be considered within the broader context of the economy. Economists Allan Fisher (1935), Colin Clark (1940), and Jean Fourastié (1949) developed a three-sector framework of the economy. In their model, the **primary sector** is extractive—mining, agriculture, fishing, forestry, etc. The **secondary sector** is goods producing. The **tertiary sector** is all services (see Chapter 8).

In 1850 the primary sector in the United States made up over 60 percent of all employment, with the secondary and tertiary sectors accounting for under 20 percent each. In the twentieth century, the percentage of manufacturing jobs grew as the percentage of jobs in agriculture shrank. Later in the century, the percentage of service jobs grew as the percentage of manufacturing jobs decreased. In 2016, the primary sector was 1.9 percent of all employment in the United States and the secondary sector was 12.2 percent (down from a high of above 30 percent in the middle of the twentieth century). A pure service economy would use no labor to produce food or goods. The U.S. economy has a technologically advanced services sector accounting for about 80 percent of its output, compared to about 15 percent from manufacturing and less than 2 percent from agriculture (FocusEconomics, 2019). The services sector in other countries is also growing, with the following percentage shares of the economy: 81 percent in the United Kingdom, 79 percent in Canada, 78 percent in Australia, 77 percent in France, and 72 percent in Japan (World Bank, 2019). However, it seems very unlikely that service percentages will eventually reach 100 percent in large industrialized countries. (See problem 4 for whether 100 percent services might be practical for a small country.) As discussed next, there are strong arguments to support the idea that manufacturing matters for a country's economy.

Manufacturing Matters

A question that is being debated and has been debated by economists for several decades is the importance of a strong manufacturing base. As just discussed, there has been a shift in jobs from the manufacturing sector to the service sector in the United States and elsewhere. Because there are major disparities in labor costs in different parts of the world, there are strong incentives for U.S. firms to locate manufacturing facilities overseas to reduce labor costs. The growth of manufacturing overseas, and in China in particular, is well documented. Is a strong manufacturing base important for the health of the economy?

Some argue that the U.S. economy evolved from agriculture to manufacturing to services. However, this is far from clear. Although fewer U.S. workers are employed in the agricultural sector of the economy, agricultural production has not declined. The United States is the largest agricultural exporting country in the world because of large amounts of arable land and advanced farming technology (FocusEconomics, 2019). By utilizing new technologies, agriculture has been able to sustain growth while consuming fewer labor hours. Hence, there are problems with the argument that the U.S. economy has shifted away from agriculture.

Manufacturing has also had a dramatic rise in productivity. Average annual manufacturing productivity growth in the United States was 1.6 percent from 1987 to 1990 and 3.9 percent from 1990 to 2000. It increased to 4.3 percent between 2000 and 2007 before dropping off to 0.7 percent from 2007 to 2019 (U.S. Bureau of Labor Statistics, 2019). The steady improvement in manufacturing productivity partially accounted for the success of the U.S. economy prior to 2007. The poor productivity since the 2007/2008 global financial crisis has been noted and studied by economists. However, we have not found any of the posited explanations for the productivity drop particularly convincing, so will not repeat them here.

Service productivity has not risen in productivity like agriculture and manufacturing. In part this is because it is still very labor intensive. It is interesting to speculate whether the percentage employment in services will eventually reach an asymptote, and if so, at what value? What is the long-term stable mix between production and services going to be? More leisure time and/or wealth naturally lead to increased demand for services. However, whether we will continue to see increases in leisure time and/or wealth is beyond the scope of this text.

To sum up, the argument that the economy is undergoing natural stages of evolution is simply not borne out by the facts. We believe that all sectors of an economy—agricultural, manufacturing, and service—are important and that domestic economic well-being depends on properly linking the activities of these sectors.

Offshoring and Reshoring

Where to locate manufacturing as well as research and development (R&D) facilities is one of the key management decisions a firm must make. As just discussed, during the 1990s the United States saw an exodus of domestic manufacturing to China. The offshoring movement was rampant, not only in the United States but also in most developed countries. The primary driver for this exodus was wage rates, but other factors were relevant as well. Favorable tax treatments, proximity to natural resources, and proximity to markets are among the reasons why companies locate facilities offshore.

In recent years, the advantage of offshoring has been decreasing. For example, as the standard of living in China has improved, manufacturing wage rates have risen. When the disadvantages of offshoring are taken into account, the best course of action is no longer obvious. These disadvantages include longer lead times, infrastructure deficiencies, local politics, and quality problems.

An example cited in *The Economist* (2013) is the start-up company ET Water Systems. In 2005 the firm moved manufacturing operations to China in search of lower labor costs. However, the disadvantages of locating offshore became apparent as the company started suffering losses due to several factors, including the cost of funds tied up in goods in transit, the disconnect between manufacturing and design, and recurring quality problems. When the firm's chief executive, Mark Coopersmith, carefully looked at the total cost difference between manufacturing in China versus California he was amazed to discover that California was only 10 percent more expensive than China. He concluded that this cost difference was more than offset by the advantages of locating manufacturing domestically. ET Water Systems closed their plant in China and reshored the manufacturing function to General Electronics Assembly in San Jose.

To get some idea of the extent of reshoring compared to offshoring, Michael Porter and Jan Rivkin (2012) conducted an extensive survey of Harvard Business School alumni who made location decisions for their companies in 2011. They found that only 9 percent of the respondents were considering moving offshore activities back to the United States, 34 percent were planning on keeping their facilities where they were, and 57 percent were considering moving existing facilities out of the United States. Their results suggest that, for

U.S. manufacturing, offshoring still dominates both staying put and reshoring. The respondents' main reasons for offshoring were lower wages, proximity to customers, better access to skilled labor, higher labor productivity, lower taxes, proximity of suppliers, and proximity to other company operations. The respondents' main reasons for staying put or reshoring were proximity to the U.S. market, less corruption, greater safety, better intellectual property protection, similar language and culture, better infrastructure, and proximity to other company operations.

Has offshoring been a help or a hindrance to the U.S. economy? The answer is not simple. On the one hand, offshoring has resulted in loss of jobs domestically and lower average domestic wages, which in turn have yielded a lower tax base and a smaller domestic market. On the other hand, offshoring has improved the bottom line for many domestic firms and has resulted in lower costs of manufactured goods for the U.S. consumer.

1.2 COMPETING IN THE GLOBAL MARKETPLACE

In his excellent study of international competitiveness, Porter (1990) poses the following question: Why does one country become the home base for successful international competitors in an industry? There is no dispute that certain industries flourish in certain countries. The list below provides some examples.

1. Germany: printing presses, luxury cars, chemicals.
2. Switzerland: banking, pharmaceuticals, chocolate.
3. United States: software, movies, biotech.
4. Japan: automobiles, consumer electronics, robotics.
5. China: clothing, machinery and transportation equipment, consumer goods.

Several compelling explanations for why industries flourish in particular countries are listed below, along with counterexamples.

1. *Historical*. Some industries have thrived for years in some countries and are not easily displaced. Counterexample: The demise of the steel industry in the United States is one of many counterexamples.
2. *Tax structure*. Some countries, such as Germany, have no capital gains tax, thus providing a more fertile environment for industry. Counterexample: There is no reason that favorable tax treatment should favor certain industries over others.
3. *National character*. Many believe that workers from other countries, particularly from Pacific Rim countries, are better trained and more dedicated than U.S. workers. Counterexample: If this is true, why then do U.S. firms dominate in some industry segments? How does one explain the enormous success Japanese-based corporations have had running plants in the United States with a U.S. workforce?
4. *Natural resources*. There is no question that some industries are highly resource dependent, and these industries have a distinct advantage in some countries. One example is the forest products industry in the United States and Canada. Counterexample: Many industry sectors are essentially resource independent but still seem to flourish in certain countries.
5. *Government policies*. Some governments provide direct assistance to fledgling industries, such as MITI in Japan. The role of the U.S. government is primarily regulatory. For example, environmental standards in the United States are probably more stringent than

those of most Asian countries. Counterexample: This does not explain why some industries dominate in countries with strict environmental and regulatory standards such as Germany.

6. *Advantageous macroeconomic factors*. Exchange rates, interest rates, and government debt are some of the macroeconomic factors that provide nations with competitive advantage. For example, in the 1980s when interest rates were much higher in the United States than they were in Japan, it was much easier for Japanese firms to borrow for new projects. Counterexample: These factors do not explain why many nations have a rising standard of living despite rising deficits (Japan and Korea are two examples).
7. *Cheap, abundant labor*. Although cheap labor can attract new industry, most countries with cheap labor are very poor. On the other hand, many countries (Germany, Switzerland, and Sweden are examples) have a high standard of living, high wage rates, and shortages of qualified labor.
8. *Management practices*. There is evidence that Japanese management practices are more effective in general than Western-style practices. Counterexample: If U.S. management practices are so ineffective, why does it continue to dominate certain industries, such as software development and pharmaceuticals?

Talking about competitiveness is easier than measuring it. What are the appropriate ways to measure one country's success over another? Some possibilities are

- balance of trade,
- share of world exports,
- creation of jobs,
- low labor costs.

Arguments can be made against every one of these as a measure of international competitiveness. The United States and the United Kingdom have the largest trade deficits in the world, yet they have rising standards of living. Similar arguments can be made for other countries that import more than they export. The number of jobs created by an economy is a poor gauge of the health of that economy. More important is the quality of the jobs created. Finally, low labor costs correlate with a low standard of living. These counterexamples show that it is no easy task to develop an effective measure of international competitiveness.

Porter (1990) argues that the appropriate measure to compare national performance is the rate of productivity growth. Productivity is the value of output per unit of input of labor or capital. Porter argues that productivity growth in some industries appears to be stronger in certain countries, and that there are reasons for this. He suggests the following four determinants of national advantage.

1. *Factor conditions*. Porter's factor theory says all countries have access to the same technology (an assumption that is not strictly true) and that national advantages accrue from endowments of production factors such as land, labor, natural resources, and capital.
2. *Demand conditions*. If domestic consumers are sophisticated and demanding, they apply pressure on local industry to innovate faster, which gives firms an edge internationally. Consumers of electronics in Japan are very demanding, thus positioning this industry competitively in the international marketplace.

3. *Related and supporting industries*. Having world-class suppliers nearby is a strong advantage. For example, the Italian footwear industry is supported by a strong leather industry and a strong design industry.
4. *Firm strategy, structure, and rivalry*. The manner in which firms are organized and managed contributes to their international competitiveness. Japanese management style is distinctly different from U.S. management. In Germany, many senior executives possess a technical background, producing a strong inclination to product and process improvement. In Italy, there are many small family-owned companies, which encourages individualism.

SNAPSHOT APPLICATION

Global Manufacturing Strategies in the Automobile Industry

Consider the following four foreign automobile manufacturers: Honda, Toyota, BMW, and Mercedes Benz. Honda and Toyota are Japanese companies; BMW and Mercedes are German companies. The four account for the lion's share of foreign nameplates sold in the U.S. auto market. However, many assume that these cars are manufactured in their home countries. In fact, depending on the model, it could be more likely that a consumer buying a Honda, Toyota, BMW, or Mercedes is buying a car manufactured in the United States.

Honda was the first of the foreign automakers to commit to a significant investment in U.S.-based manufacturing facilities. Honda's first U.S. facility was its Marysville Motorcycle plant, which started production in 1979. Honda must have been pleased with the Ohio-based facility, since an automobile plant followed shortly. Automobile production in Marysville began in 1982. Today, Honda operates four plants in west-central Ohio, producing the Accord sedan and coupe, the Acura TL sedan, the Honda Civic line, and the Honda Element, with a manufacturing capacity of 440,000 vehicles annually.

Next to make a significant commitment in U.S. production facilities was Toyota. Toyota's plant in Georgetown, Kentucky, has been producing automobiles since 1986 and accounts for all Camrys sold in the domestic market. The Honda Accord and the Toyota Camry are two of the biggest- selling models in the United States—and they are produced here. They also top almost all reliability surveys.

The two German automakers were slower to commit to U.S.-based manufacturing facilities. BMW launched its Spartanburg, South Carolina, plant in March of 1995. BMW produces both the Z series sports cars and its SUV line in this plant. BMW's big sellers, its 3, 5, and 7 series sedans, are still manufactured in Germany.

Mercedes was the last of these four companies to make a significant commitment to production facilities here. The facility in Tuscaloosa, Alabama, is dedicated to producing the line of Mercedes SUVs. As with BMW, the more popular C, E, and S class sedans are still manufactured in Germany.

(One might ask why Volkswagen is not on this list. In fact, Volkswagen has 45 separate manufacturing facilities located in 18 countries around the world but only opened its first plant in 2011 in the United States.)

Sources: https://www.bmw.com/en/index.html; https://ohio.honda.com/our-operations;https://www.mbusi.com/[Mercedes Benz]; https://www.toyota.com/

Even though Porter makes a very convincing argument for national competitive advantage in some industries, there is a debate among economists as to whether the notion of international competitiveness makes any sense at all. Companies compete, not countries. The standard of living in a country depends on its own domestic economic performance and not on how it performs relative to other countries.

PROBLEMS FOR SECTIONS 1.1–1.2

1. What is a production and operations strategy? Discuss the common elements in marketing and financial strategies and the elements that differ.
2. Why is it undesirable for the United States to evolve into a service economy?
3. The statement was made in this chapter that "more leisure time and/or wealth naturally lead to increased demand for services." Explain why this is the case.
4. What would a pure services economy look like? Do you think this is practical for a small country with little land area to consider? Why or why not?
5. Consider the four determinants of national advantage suggested by Porter. Give examples of companies that have thrived as a result of each of these factors.
6. What factor advantage favors the aluminum industry in the United States over Japan and makes aluminum much cheaper to produce in the United States? (Hint: Aluminum production is very energy intensive. In what part of the United States is an inexpensive energy source available?)
7. Paul Krugman (1994) argues that because most of our domestic product is consumed domestically, we should not dwell on international competition. What industries in the United States have been hardest hit by foreign competition? What are the potential threats to the United States if these industries fail altogether?
8. Krugman points out some misguided government programs that have resulted from too much emphasis on international competitiveness. What risks are there from too little emphasis on international competitiveness?
9. Consider the Snapshot Application in this section concerning foreign automakers locating manufacturing facilities in the United States. Discuss the advantages and disadvantages of the strategy of locating manufacturing facilities where the product is consumed rather than where the company is located.
10. The North American Free Trade Agreement (NAFTA) was established in 1994.
    a. What was the purpose of NAFTA?
    b. At the time, political opponents characterized "the big sucking sound" as jobs would be lost as a result. Is there any evidence that this was, in fact, the case?
    c. Negotiations in 2018 resulted in the US-Mexico-Canada Agreement (USMCA). What factors contributed to replacing the 24-year-old NAFTA agreement?

1.3 A FRAMEWORK FOR OPERATIONS STRATEGY

Classical literature on competitiveness claims that firms position themselves strategically in the marketplace along one of two dimensions: lower cost or product differentiation (Porter, 1990).

Often new entrants to a market position themselves as the low-cost providers. Firms that have adopted this approach include the Korean automakers (Hyundai, Daewoo, Kia), discount outlets such as Costco, and retailers such as Walmart. While being the low-cost provider can be successful over the near term, it is a risky strategy. Consumers ultimately will abandon products that they perceive as poor quality regardless of cost. For example, many manufacturers of low-cost PC clones popular in the 1980s are long gone.

Most firms that have a long record of success in the marketplace have differentiated themselves from their competitors. By providing uniqueness to buyers, they are able to sustain high profit margins over time. One example is BMW, one of the most profitable auto firms in the world. BMW continues to produce high-performance, well-made cars that are often substantially more expensive than those of competitors in their class. Product differentiation within a firm has also been a successful strategy. Consider the success of General Motors in the early years compared to Ford. GM was able to successfully capture different market segments at the same time by forming five distinct divisions, while Henry Ford's insistence on providing only a single model almost led the company to bankruptcy (Womack, Jones, and Roos, 1990).

Strategic Dimensions

There are many dimensions to product differentiation. In particular, the following dimensions relate directly to the operations function:

- quality
- delivery speed
- delivery reliability
- flexibility
- service.

What does **quality** mean? It is a word often bandied about, but one that means different things in different contexts. A Honda Civic is a quality product and so is a Ferrari Testarosa. Consumers buying these products are both looking for quality cars but have fundamentally different objectives. We will explore these ideas in greater detail in Chapter 10 (Quality and Assurance).

Delivery speed can be an important competitive weapon in some contexts. Some firms base their primary competitive position on delivery speed, such as UPS and Federal Express. Online retailers also must be able to deliver products reliably and quickly to remain competitive. Amazon continually works to shorten delivery time. Building contractors that complete projects on time will have an edge.

Delivery reliability means being able to deliver products or services when promised. Online brokerages that execute trades reliably and quickly will retain customers. Contract manufacturers are measured on several dimensions, one being whether they can deliver on time. As third-party sourcing of manufacturing continues to grow, the successful contract manufacturers will be the ones that maintain a record of delivering high-quality products in a reliable fashion.

Flexibility means offering a wide range of products and/or being able to adjust to unexpected changes in the demand for the product mix offered. Successful manufacturers in the twenty-first century will be those that can respond the fastest to unpredictable changes in customer tastes. One of the authors was fortunate enough to tour Toyota's Motomachi Plant located in Toyoda City, Japan. What was particularly impressive was the ability to produce several different models in the same plant. In fact, each successive car on the assembly line was a different model. A right-hand drive Crown sedan, for the domestic market, was followed by a left-hand drive Lexus coupe, designated for shipment to the United States. Each car carried unique sets of instructions that could be read by both robot welders and human assemblers. This flexibility allowed Toyota to adjust the product mix on a real-time basis and to embark on a system in which customers could order custom-configured cars directly from terminals located in dealer showrooms (Port, 1999).

Service is a relatively new competitive dimension. It involves both the buying experience and product support. Servicization, as described later in this chapter, is one trend allowing companies to compete on the basis of service, even for manufactured products.

All these competitive dimensions are relevant for services as well as products. Delivery speed matters for tax preparation companies, while flexibility is important for consulting companies. Service, viewed as the buying experience, is particularly important for service companies, where often the customer is also the co-producer (see Chapter 8, Service Operations Management).

The competitive dimensions on which a firm chooses to focus are known as its **competitive priorities**. One way to think of operations strategy is the strategic positioning the firm chooses along each of the dimensions of cost, quality, delivery speed, delivery reliability, flexibility, and service. It achieves this positioning through all of the decisions it makes regarding both organizational structure and infrastructure. A firm that is competing on cost must find a way to achieve economies of scale; a firm competing on flexibility must have the correct technologies to achieve agility. Operations management is concerned with implementing the strategy to achieve leadership along the desired competitive priorities.

Order Qualifiers and Order Winners

Terry Hill has proposed an interesting way to look at competitive factors. He classifies them into two types: **order qualifiers** and **order winners**. A product not possessing a qualifying factor is eliminated from consideration. The order winner is the factor that determines who gets the sale among the field of qualifiers.

For example, everyone competes on quality to some extent. For this reason, Hill (1993) would often classify quality as an order qualifier, rather than an order winner. An option is immediately eliminated from consideration if it does not meet minimum quality standards. The particular aspect of quality on which a firm chooses to focus determines the nature of its competitive strategy and strategic positioning. For example, sound quality often functions as an order winner for Bose speakers, as their quality is clearly differentiated from the quality of cheaper speaker companies. Therefore, a specific aspect of quality may be an order winner, while basic performance and function will be quality aspects that are order qualifiers.

Over time, order winners often become order qualifiers. For example, electric windows when they were first introduced in cars were new and exciting and potential order-winners. However, many younger people today have never seen a car without electric windows, which now makes them an order qualifier.

Further, as a given product matures, order-winners increasingly become an operational task. It often becomes more important for a mature product to win based on price and delivery speed than it is for a newer, more innovative product that may compete on its features alone.

Measuring Performance

Along with a clear idea of a product's (or service's) competitive priorities is the need for clear metrics to measure performance. These then need to be aligned with performance metrics for the whole company (see also Chapter 3, Sales and Operations Planning).

Measuring a firm's success by the performance of its share price can result in shortsighted management practices. Boards of directors can become more concerned with the next quarterly report than with funding major long-term projects. In fact, Robert Hayes and Steven Wheelwright (1984) make a compelling argument that such factors led to a myopic management style in the United States, characterized by the following.

1. Managers' performance is measured on the basis of **return on investment** (ROI), which is simply the ratio of the profit realized by a particular operation or project over the investment made in that operation or project.
2. Performance is measured over short time horizons. There is little motivation for a manager to invest in a project that is not likely to bear fruit until after he or she has moved on to another position.

In order to improve ROI, a manager must either increase the numerator (profits) or decrease the denominator (investment). In the short term, decreasing the denominator by cutting back on the investment in new technologies or new facilities is easier than trying to increase profits by improving efficiency, the quality of the product, or the productivity of the operating unit. The long-term effects of decreasing investment are devastating. At some point, the capital costs required to modernize old factories become more than the firm can bear, and the firm loses its competitive position in the marketplace.

It would be encouraging if the problems of U.S. industries arising from overemphasis on short-term financial performance were decreasing; sadly, they appear to be worsening. Because of gross mismanagement and questionable auditing practices, two giants of U.S. industry were brought down in 2001: Enron and Arthur Andersen. "Enron went from the No. 7 company on the Fortune 500 to a penny stock in a stunning three weeks because it apparently lied on its financial statements," said Representative John D. Dingell, one-time member of the House Energy Committee. More recently, 19 major U.S. retailers filed for bankruptcy protection in 2017 including Toys"R"Us, The Limited, and Payless. Clearly, online shopping has not helped these retailers. In addition, private equity firms overburdened these businesses with debt—a classic example of short-term thinking (Dayen, 2017).

Measuring individual performance over the short term is a philosophy that seems to pervade life in the United States. Politicians are elected for two-, four-, or six-year terms. There is a strong incentive for them to show results in time for the next election. Even university professors are evaluated yearly on their professional performance in many institutions, even though most serious academic projects extend over many years. Setting appropriate metrics is a critical part of a good operations strategy.

The Productivity Frontier

Most manufacturing or service systems perform well for a limited set of tasks. A manufacturing (or service) environment cannot deliver on all competitive dimensions simultaneously. Therefore, trade-offs need to be made. It is important that these are made explicit so that the firm can align its operations strategy with its business strategy. One tool for visualizing such trade-offs is the **productivity frontier**. This frontier is depicted in Figure 1–1 for the competitive dimensions of price and quality.

Figure 1–1 shows a typical trade-off between quality and price. While ideally a firm would like to have both high quality and low price (the "North Star" in lean manufacturing terms, see Section 1.7), technology usually prevents this. For example, higher quality raw materials

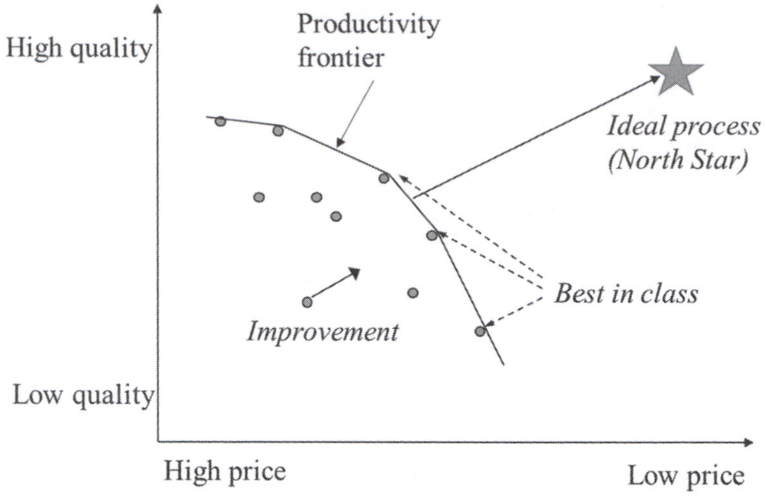

Figure 1–1 The productivity frontier (dots represent firms)

usually cost more than lower quality materials. Better trained and more knowledgeable workers require a higher rate of pay than less skilled workers. A firm can undertake process improvements that make improvements in both price and quality—but only if the firm is not on the productivity frontier.

The productivity frontier defines the current technological limit for best-in-class firms as they trade off between their competitive priorities. In a competitive marketplace we would expect to see firms positioned along this frontier, where some are competing on price, some on quality, and some on providing a reasonable standard for both. There is an analytic technique, known as data envelope analysis (DEA), that is based on linear programming techniques (see Supplement 1), which can help determine the productivity frontier in an industry. DEA is beyond the scope of this text (interested readers can reference Cooper, Seiford, and Tone, 2007).

Firms get into trouble if management does not recognize that trade-offs exist and does not set priorities for operations. It should not, for example, be left to the operations manager to decide the appropriate quality/cost trade-off. Instead, the operations strategy (e.g., high quality) needs to be aligned with the business strategy (e.g., competing on quality). Another example of short-term thinking is when firms cut costs to improve their bottom line. Doing this can erode their competitive advantage that relied on producing high quality products.

A study of three different plants making stamping dies in Michigan provides an excellent example of the existence of trade-offs (Pagell, Melnyk, and Handfield, 2000). The plants are named *Capital, Lifetime,* and *Overtime,* which refer to the key operations strategy pursued by each firm. In particular, capital had invested in equipment and was the most competitive in terms of lead time; lifetime had invested in its employees and was most competitive in quality and innovation; and overtime invested in neither and was most competitive in cost and volume flexibility. The article clearly demonstrates the existence of a productivity frontier.

PROBLEMS FOR SECTION 1.3

11. What disadvantages do you see if the chief executive officer (CEO) is primarily concerned with short-term ROI?

12. Can you think of companies that have gone out of business because they focused only on cost and were not able to achieve a minimum quality standard?
13. What are the different quality standards referred to in the example comparing the Honda Civic and the Ferrari?
14. Choose two competitive priorities that you believe would be difficult to compete on simultaneously. Explain your answer.
15. We stated that as products mature, order winners become increasingly an operational task. Give an example of a product, not discussed earlier, where this is the case. Justify your answer.
16. Choose an industry you are familiar with (e.g., cell phones) and approximately position the main firms in that industry on Figure 1–1.
17. Redraw Figure 1–1 with a different two competitive priorities and repeat the exercise in problem 16.

1.4 PROCESS TYPES AND LAYOUTS

One of the key questions in operations strategy is the best process to use for a particular product or service. The process involves the type of equipment and labor, the way they are used or scheduled, and how everything is physically laid out. All of these are important decisions that affect a product's competitive dimensions. Service industries also are faced with the problem of finding effective layouts. Achieving efficient patient flows in hospitals is one example. Another example is the obvious care used in laying out the many theme parks across the world. Anyone who has visited the San Diego Zoo or Disney World in Orlando, Florida, can appreciate the importance of effective layout and efficient management of people in service industries.

The Product–Process Matrix

Hayes and Wheelwright (1979) suggest answering the strategic question of matching products with appropriate processes using the **product–process matrix** pictured in Figure 1–2. Its purpose is to match the appropriate industry (assuming it is mature) with the appropriate process. The matrix is based on four types of manufacturing process: (1) jumbled flow, (2) disconnected line flow, (3) connected line flow, and (4) continuous flow.

Located in the upper left-hand corner of this matrix are companies that specialize in "one of a kind" jobs in which the manufacturing function has the characteristics of a jumbled flow shop. A commercial printer is an example of a jumbled flow shop. Production is in relatively small lots, and the shop is organized for maximum flexibility.

Farther down the diagonal are firms that still require a great deal of flexibility but produce a limited line of standardized items. Manufacturers of heavy equipment would fall into this category because they would produce in somewhat higher volumes. A disconnected line would provide enough flexibility to meet custom orders while still retaining economies of limited standardization.

The third category down the diagonal includes firms that produce a line of standard products for a large-volume market. Typical examples are producers of home appliances or electronic equipment, and automobile manufacturers. The assembly line or transfer line would be an appropriate process technology in this case.

Finally, the lower right-hand portion of the matrix would be appropriate for products involving continuous flow. Chemical processing, gasoline and oil refining, and sugar refining

Process structure Process life-cycle stage	I Low volume, low standardization, one of a kind	II Multiple products, low volume	III Few major products, higher volume	IV High volume, high standardization, commodity products
I Jumbled flow (job shop)	Commercial printer			Void
II Disconnected line flow (batch)		Heavy equipment		
III Connected line flow (assembly line)			Auto assembly	
IV Continuous flow	Void			Sugar refinery

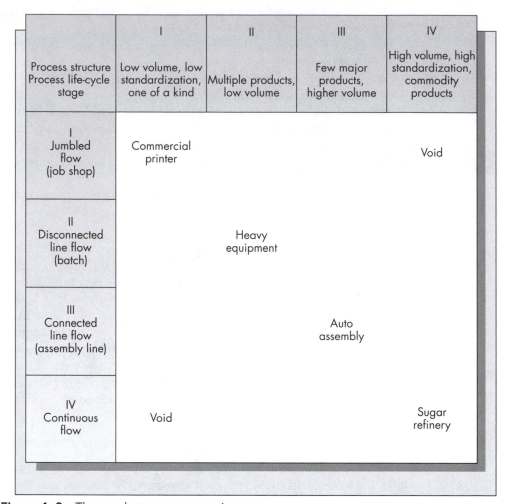

Figure 1–2 The product-process matrix

Source: Robert H. Hayes and Steven C. Wheelwright, "Link Manufacturing Process and Product Life Cycles" ©1979 by the President and Fellows of Harvard College; all rights reserved. Reprinted by permission.

are examples. Such processes are characterized by low unit costs, standardization of the product, high sales volume, and extreme inflexibility of the production process.

What is the point of this particular classification scheme? It provides a means of assessing whether a firm is operating in the proper portion of the matrix; that is, if the process is properly matched with the product structure. Firms choosing to operate off the diagonal should have a clear understanding of the reasons for doing so. One example of a successful firm that operates off the diagonal is Rolls-Royce. Another is a company producing handmade furniture. The manufacturing process in these cases would have the characteristics of a jumbled flow shop, but competitors might typically be located in the second or third position on the diagonal.

Firms that are located below the diagonal often suffer from large out-of-pocket costs. They may have invested in dedicated equipment but do not have the volume of product needed to cover the significant fixed costs associated with such equipment. Firms that are located above the diagonal often suffer from opportunity costs. This is because, as the equipment is less efficient, they may not be able to service the volume of product that is available in the marketplace.

This classification scheme also works for service operations, but we typically reduce the number of process types to only two: product and process layouts. These are covered later in this section.

Plant Layout

In addition to the type of process used, the physical plant layout also matters. Determining the best layout for a facility is a classical industrial engineering problem. The objectives in a plant layout study might include one or more of the following.

1. Minimize the investment required in new equipment.
2. Minimize the time required for production.
3. Utilize existing space most efficiently.
4. Provide for the convenience, safety, and comfort of the employees.
5. Maintain a flexible arrangement.
6. Minimize the materials handling cost.
7. Facilitate the manufacturing process.
8. Facilitate the organizational structure.

Thomas Vollmann and Elwood Buffa (1966) suggest nine steps as a guide for the analysis of layout problems.

1. Determine the compatibility of the materials handling layout models with the problem under study. Find all factors that can be modeled as materials flow.
2. Determine the basic subunits for analysis. Determine the appropriate definition of a department or subunit.
3. If a mathematical or computer model is to be used, determine the compatibility of the nature of costs in the problem and in the model. That is, if the model assumes that materials handling costs are linear and incremental (as most do), determine whether these assumptions are realistic.
4. How sensitive is the solution to the flow data assumptions? What is the impact of random changes in these data?
5. Recognize model idiosyncrasies and attempt to find improvements.
6. Examine the long-run issues associated with the problem and the long-run implications of the proposed solution.
7. Consider the layout problem as a systems problem.
8. Weigh the importance of qualitative factors.
9. Select the appropriate tools for analysis.

The objective used most frequently for quantitative analysis of layout problems is to minimize the materials handling cost. When minimizing the materials handling cost is the primary objective, a flow analysis of the facility is necessary. Flow patterns can be classified as horizontal or vertical. A horizontal flow pattern is appropriate when all operations are located on the same floor, and a vertical pattern is appropriate when operations are in multistory structures. Richard Francis and John White (1974) give six horizontal flow patterns and six vertical flow patterns. The six horizontal patterns appear in Figure 1–3.

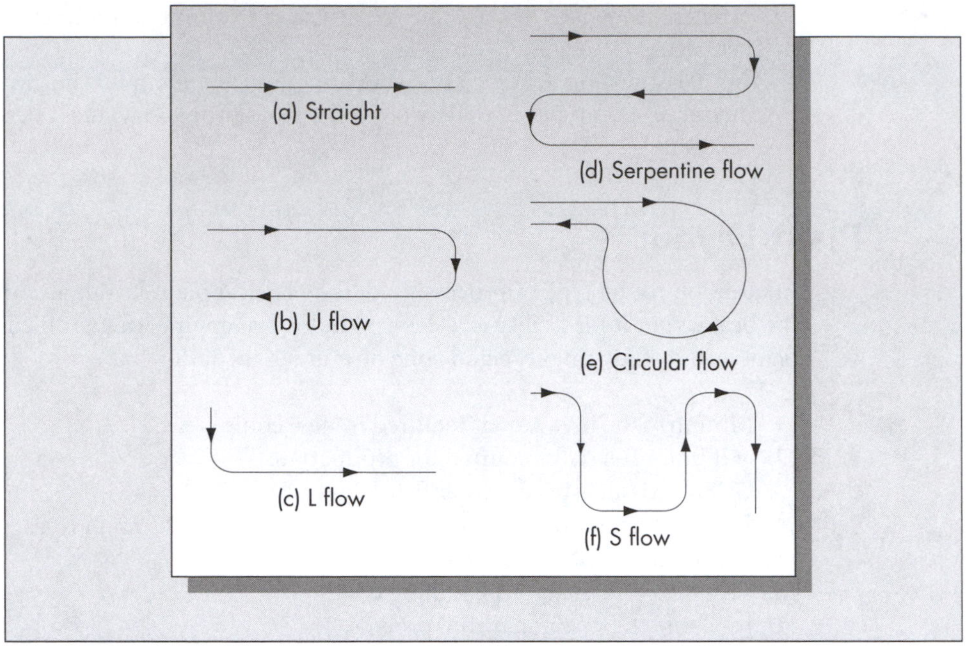

Figure 1–3 Six horizontal flow patterns

The simplest pattern is (a), which is straight-line flow. The main disadvantage of this pattern is that separate docks and personnel are required for receiving and shipping goods. The L shape is used to replace straight-line flow when the configuration of the building or the line requires it. The U shape has the advantage over the straight-line configuration of allowing shipping and receiving to be at the same location. The circular pattern is similar to the U shape. The remaining two patterns are used when the space required for production operations is too great to allow use of the other patterns.

Some products are too big to be moved, so the product remains fixed and the layout is based on the product size and shape. Examples of products requiring **fixed position** layouts are large airplanes, ships, and rockets. For such projects, once the basic frame is built, the various required functions would be located in fixed positions around the product. A project layout is similar in concept to the fixed position layout. This would be appropriate for large construction jobs such as commercial buildings or bridges. The required equipment is moved to the site and removed when the project is completed. A typical fixed position layout is shown in Figure 1–4.

Layout Strategies

Different philosophies of layout design are appropriate in different manufacturing environments. In a **product layout** (or product flow layout) machines are organized to conform to the sequence of operations required to produce the product. The product layout is typical of high-volume standardized production (the lower right-hand corner in Figure 1–2). An assembly line (or transfer line) is a product layout, because assembly facilities are organized according to the sequence of steps required to produce the item. Product layouts are desirable for flow-type mass production and provide the fastest cycle times in this environment. Transfer lines are expensive and inflexible, however, and become cumbersome when changes

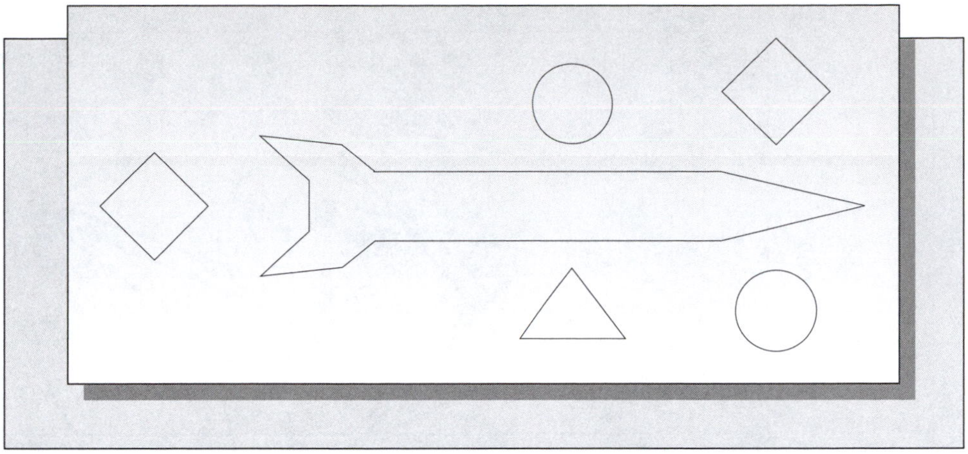

Figure 1–4 Fixed position layout

in the product flow are required. Furthermore, a transfer line can experience significant idle time. If one part of the line stops, the entire line may have to remain idle until the problem is corrected. Figure 1–5 shows a typical product layout.

Process layouts are the most common for small- to medium-volume manufacturers with processes found in the top left corner of the product-process matrix (Figure 1–2). A process layout groups similar machines having similar functions. A typical process layout would group lathes in one area, drills in one area, and so on. Process layouts are most effective when there is a wide variation in the product mix. Each product has a different routing sequence associated with it. In such an environment it would be difficult to organize the machines to conform with the production flow because flow patterns are highly variable. Process layouts have the advantage of minimizing machine idle time. Parts from multiple products or jobs queue up at each work center to facilitate high utilization of critical resources. Also, when design changes are common, parts routings will change frequently. In such an environment, a process layout affords minimal disruption. Figure 1–6 shows a typical process layout. The arrows correspond to part routings.

In services we can think of a process layout as equivalent to a departmental structure, where customer tasks may be served by anyone in a given department before being passed on to the next department. For example, most hospitals follow process layouts where a patient may be served by any nurse and scanner in radiology before returning to the emergency

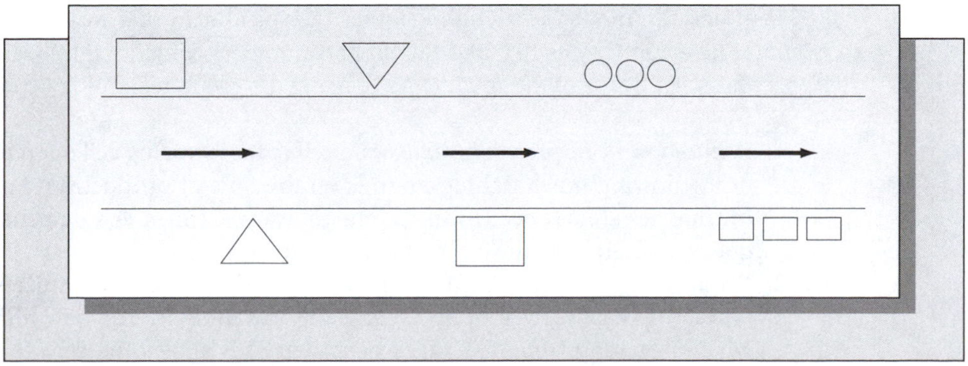

Figure 1–5 Product layout

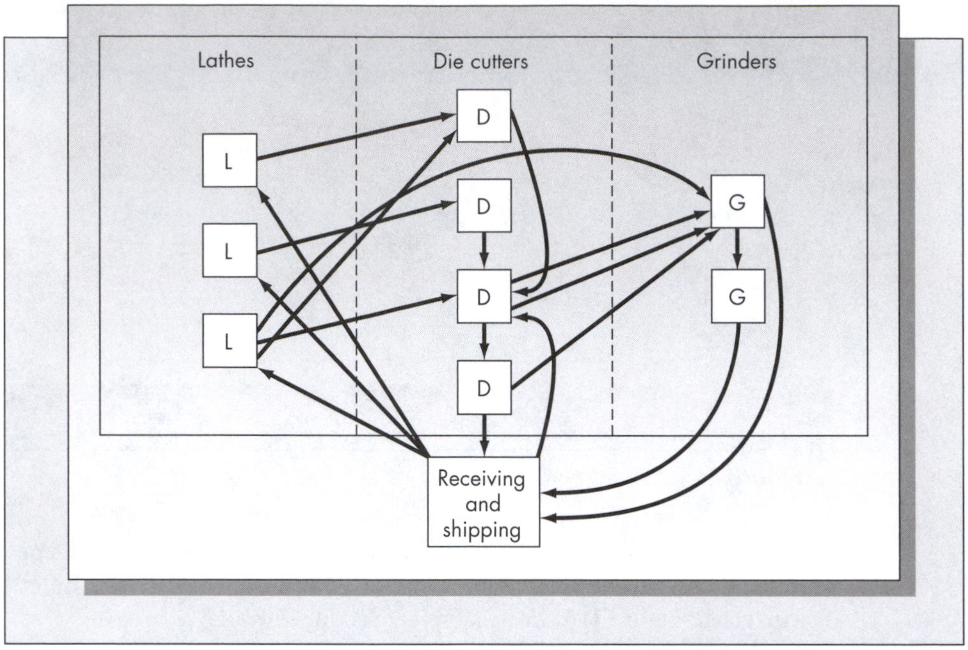

Figure 1–6 Process layout

department. In contrast, a product layout is where staff and equipment are organized by customer type. Hospitals with separate pediatric facilities are following a partial product layout approach.

Layouts Based on Group Technology

With increased emphasis on automated factories and flexible manufacturing systems, **group technology** layouts have received considerable attention. To implement a group technology layout, parts must be identified and grouped based on similarities in manufacturing function or design. Parts are organized into part families. Presumably, each family requires similar processing, which suggests a layout based on the needs of each family. In most cases, machines are grouped into machine cells where each cell corresponds to a particular part family or a small group of part families (see Figure 1–7).

The group technology concept seems best suited for large firms that produce a wide variety of parts in moderate to high volumes. A typical firm that would consider this approach might have as many as 10,000 different part numbers, which might be grouped into 50 or so part families. Some of the advantages of using the group technology concept are

1. *Reduced work-in-process inventories.* Each manufacturing cell operates as an independent unit, allowing much tighter controls on the flow of production. Large work-in-process inventories are not needed to maintain low cycle times. A side benefit of this is reduced queues of parts and the confusion that results.
2. *Reduced setup times.* Because manufacturing cells are organized according to part types, there should not be significant variation in the machine settings required when switching from one part to another. This allows the cells to operate much more efficiently.

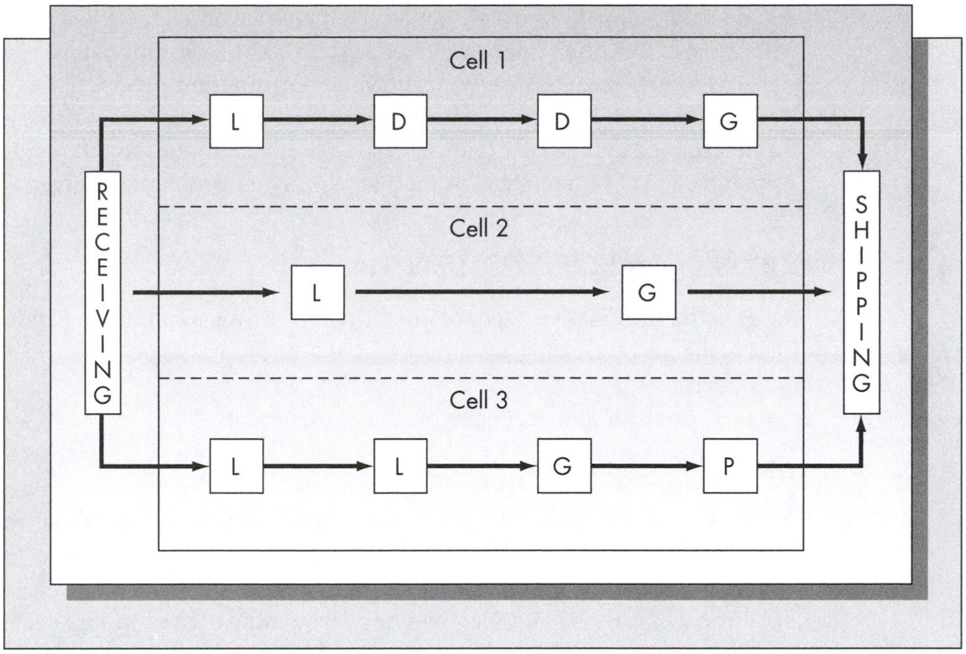

Figure 1–7 Group technology layout

3. *Reduced materials handling costs.* For a firm producing 10,000 parts, a process layout would require a dizzying variety of part routings. If volumes are large, process centers would necessarily have to be separated by large distances, thus requiring substantial materials handling costs. A layout based on group technology would overcome this problem.
4. *Better scheduling.* By isolating part groupings, it is much easier to keep track of the production flow within each cell. Reduction in cycle times and work-in-process queues results in more reliable due-date schedules.

There are several disadvantages of the group technology approach. One is that it can be difficult to determine suitable part families. Parts may be grouped by size and shape or by manufacturing process requirements. The first approach is easier but not as effective for developing layouts. How parts are grouped is a function, to a large degree, of the coding system used to classify these parts. (Mikell Groover and Emory Zimmers (1984) discuss several parts of coding systems and how they relate to the group technology concept.) Grouping part families according to the manufacturing flow requires a careful production flow analysis (Burbidge, 1975). This method is probably not feasible for a firm with a large number of parts, however.

Group technology layouts may require duplication of some machines. For a manufacturing cell to be self-contained, the cell must have all the machines necessary for the product being produced. Duplicating machines could be expensive and could result in greater overall idle time.

Under what circumstances would a group technology layout be preferred to a pure process or product layout? A simulation study provides some answers (Sassani, 1990). The researcher constructed a simulation of five manufacturing cells. Initially, when the products for each cell were well defined and the cells isolated, the system ran smoothly. However, as the product mix, product design, and demand patterns started to change, the simulation showed that the efficiency of the layout deteriorated.

The vast majority of existing manufacturing layouts are either process or product type, whereas services often follow more of a group technology approach. Firms producing a wide variety of parts may choose several layouts for different product lines or may choose some hybrid approach. Product variation and annual volume are the primary determining factors for making the appropriate choice. Chart-based techniques for choosing layouts are given in Appendix 1-A. Mathematical models for the layout problem are given in Supplement 1.

PROBLEMS FOR SECTION 1.4

18. Discuss the disadvantages of operating off the diagonal of the product-process matrix.
19. Locate the following operations in the appropriate position on the product–process matrix.
    a. A small shop that repairs musical instruments.
    b. An oil refinery.
    c. A manufacturer of office furniture.
    d. A manufacturer of major household appliances such as washers, dryers, and refrigerators.
    e. A manufacturing firm in the start-up phase.
20. For each of the eight objectives of a layout study listed in this section, give examples of situations in which that objective would be important and examples in which it would be unimportant.
21. Describe the differences among product, process, and group technology layouts. Describe the circumstances in which each type of layout is appropriate.

1.5 PRODUCT, PROCESS, AND INNOVATION LIFE CYCLES

The Product Life Cycle

The demand for new products typically undergoes cycles that can be identified and mapped over time. Understanding the nature of this evolution helps to identify appropriate strategies for production and operations at the various stages of the product cycle. A typical **product life cycle** is pictured in Figure 1–8. The product life cycle consists of four major segments:

1. start-up,
2. rapid growth,
3. maturation,
4. stabilization or decline.

During the start-up phase, the market for the product is developing, production and distribution costs are high, and competition is generally not a problem. During this phase the primary strategy concern is to apply the experiences of the marketplace and of manufacturing to improve the production and marketing functions. At this time, serious design flaws should be revealed and corrected.

The period of rapid growth sees the beginning of competition. The primary strategic goal during this period is to establish the product as firmly as possible in the marketplace. To do this, management should consider alternative pricing patterns that suit the various customer

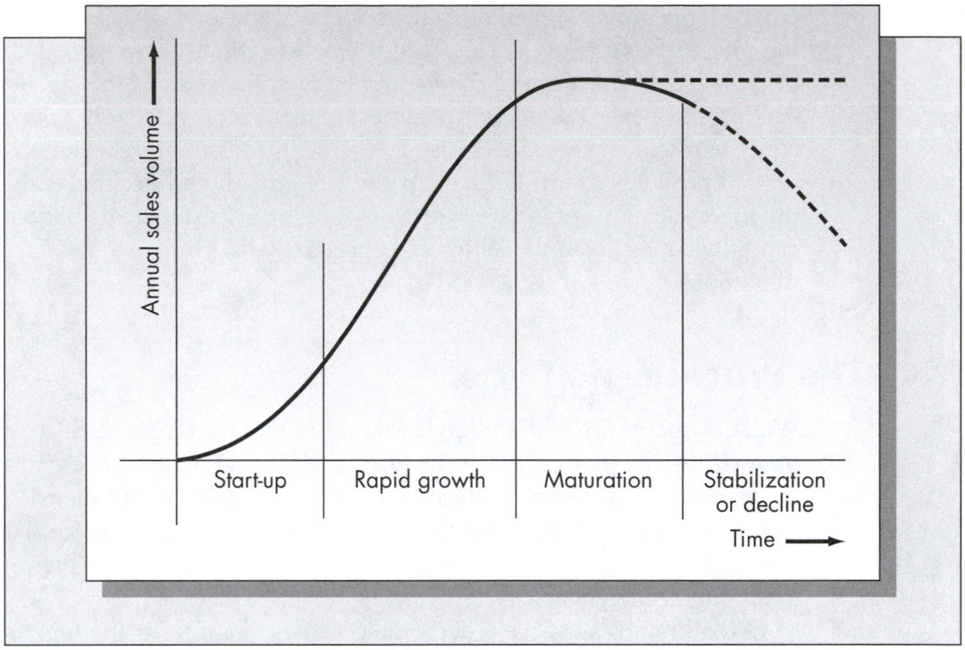

Figure 1–8 The product life-cycle curve

classes and should reinforce brand preference among suppliers and customers. The manufacturing process should be undergoing improvements and standardization as product volume increases. Flexibility and modularization of the manufacturing function are highly desirable at this stage.

During the maturation phase of the product life cycle, the objective should be to maintain and improve the brand loyalty that the firm cultivated in the growth phase. Management should seek to increase market share through competitive pricing. Cost savings should be realized through improved production control and product distribution. During this phase, the firm must listen to the messages of the marketplace. Most problems with product design and quality should have been corrected during the start-up and growth phases, but additional improvements should also be considered in the maturation phase.

The appropriate shape of the life-cycle curve in the final stage depends on the nature of the product. Many products will continue to sell, with the potential for annual growth continuing almost indefinitely—examples include commodities such as household goods, processed food, and automobiles. For such products, the company's primary goals in the final phase would be essentially the same as those described for the maturation phase. Other products will experience a natural decline in sales volume as the market for the product becomes saturated or as the product becomes obsolete. If this is the case, the company should adopt a strategy of squeezing out the most from the product or product line while minimizing investment in new manufacturing technology and media advertising.

Although a useful concept, the product life-cycle curve is not accurate in all circumstances. Marketing departments that base their strategies on the life-cycle curve may make poor decisions. One firm shifted advertising dollars from a successful stable product to a new product (Dhalla and Yuspeh, 1976). The assumption was that the new product was entering the growth phase of its life cycle, and the stable product was entering the declining phase of its life cycle. However, the new product never gained consumer acceptance; because of the drop in the advertising budget, the sales of the stable product went into a decline and never recovered. In some circumstances, it is more effective to build a model consistent with the product's history and with consumer behavior than to assume blindly that all products follow the same pattern of growth and decline. Although we believe that the life-cycle concept is a useful way of looking at customer demand patterns in general, a carefully constructed model for each product will ultimately be a far more effective planning tool.

The Process Life Cycle

William Abernathy and Phillip Townsend (1975) have classified three major stages of the **manufacturing process life cycle:** early, middle, and mature. These phases do not necessarily coincide exactly with the stages of the product life cycle, but they do provide a conceptual framework for planning improvements in the manufacturing process as the product matures. These phases correspond to the manufacturing processes given in Figure 1–2, which depicted the product/process matrix.

In the first phase of the process life cycle, the manufacturing function has the characteristics of a job shop. It must cope with a varied mix of relatively low-volume orders and be responsive to changes in the product design. The types and quality of the inputs may vary considerably, and the firm has little control over suppliers.

In the middle phase of the process life cycle, automation begins to play a greater role. The firm should be able to exert more control over suppliers as the volume of production increases. Unit production costs decline as a result of learning effects. The production process may involve batch processing and some transfer lines (assembly lines).

In the last phase of the process life cycle, most of the major operations are automated, the production process is standardized, and few manufacturing innovations are introduced. The production process may assume the characteristics of a continuous flow operation.

This particular evolutionary scenario is not appropriate for all new manufacturing ventures. Companies that thrive on small one-of-a-kind orders will maintain the characteristics of a job shop, for example. The process life-cycle concept applies to new products that eventually mature into high-volume items.

Thus, there is another way to look at the product–process matrix from Figure 1–2. It can be used to identify the proper match of the production process with the phases of the product life cycle. In the start-up phase of product development, the firm would typically be positioned in the upper left-hand corner of the matrix. As the market for the product matures, the firm would move down the diagonal to achieve economies of scale. Finally, the firm would settle at the position on the matrix that would be appropriate based on the characteristics of the product.

Experience curves (see the following section) show that unit production costs decline as the cumulative number of units produced increases. One may think of the experience curve in terms of the process life cycle shown in Figure 1–9. An accurate understanding of the relationship between the experience curve and the process life cycle can be very valuable. By matching the decline in unit cost with the various stages of the process life cycle, management can gain insight into the consequences of moving from one phase of the process life cycle into another. This insight will assist management in determining the proper timing of improvements in the manufacturing process.

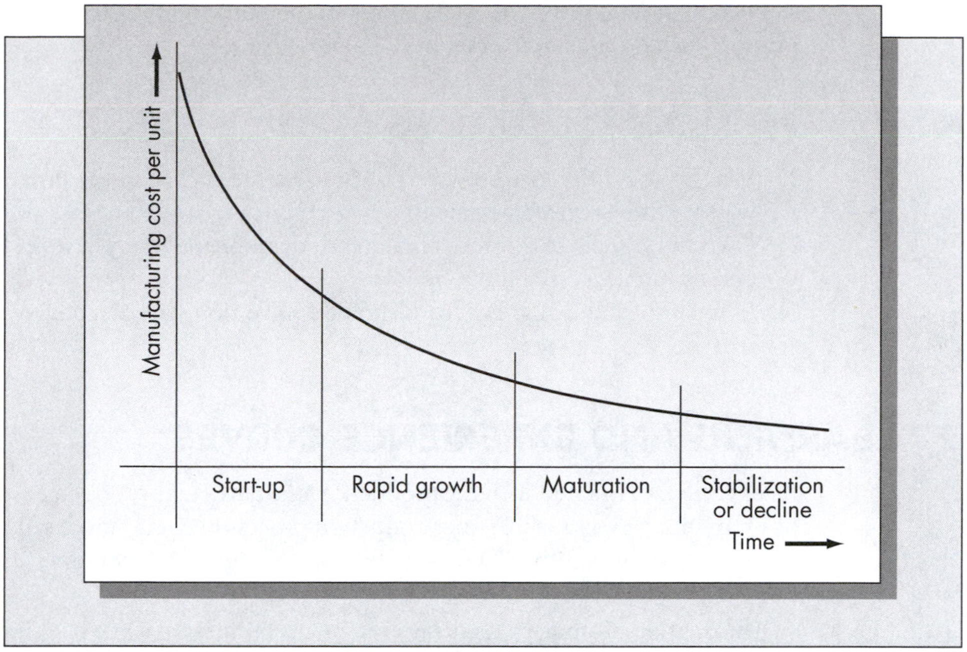

Figure 1–9 The process life cycle and the experience curve

1.6 THE INNOVATION LIFE CYCLE

A curve similar in shape to Figure 1–8 is the **innovation life cycle**, depicted in Figure 1–10. This shows the adoption of an innovative product over time. At the beginning of the product's lifetime, only **innovators** buy the product, followed by the **early adopters** (Rogers, 1962). After this the product volume increases significantly, and the **early majority** adopts

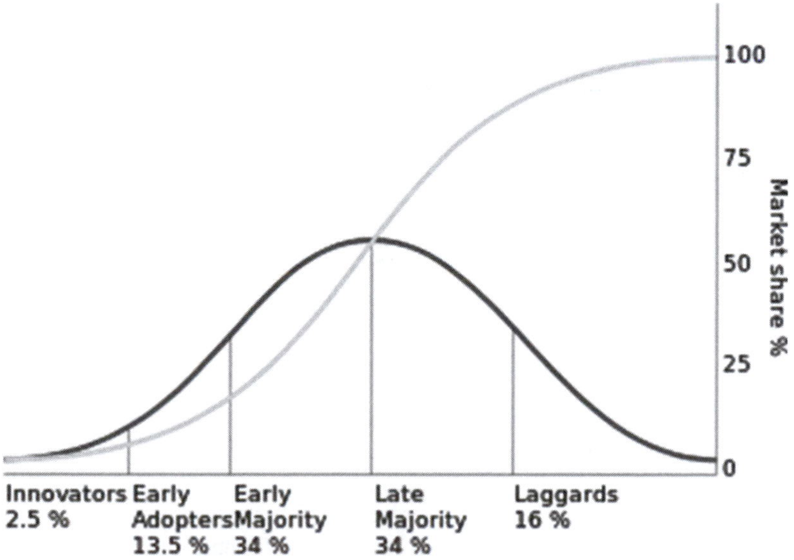

Figure 1–10 The process life cycle and the experience curve
Source: Wikimedia Commons, adapted from Rogers, 1962.

the product, followed by the **late majority**. Finally, the **laggards** come on board, and the product may be considered mature.

PROBLEMS FOR SECTIONS 1.5–1.6

22. What is the difference between the product life cycle and the process life cycle? In what way are these concepts related?
23. Give an example of a product that has undergone the four phases of the product life cycle and has achieved stability.
24. Discuss the following: "All firms should evolve along the diagonal of the product–process matrix."

1.7 LEARNING AND EXPERIENCE CURVES

As experience is gained with the production of a particular product, either by a single worker or by an industry as a whole, the production process becomes more efficient. As noted by the economist Adam Smith (1896) as far back as the eighteenth century:

> The division of labor, by reducing every man's business to some one simple operation, and by making this operation the sole employment of his life, necessarily increases very much the dexterity of the workman. (p. 9)

By quantifying the relationship that describes the gain in efficiency as the cumulative number of units produced increases, management can accurately predict the eventual capacity of existing facilities and the unit costs of production. Today we recognize that many other factors besides the improving skill of the individual worker contribute to this effect. Some of these factors include the following.

- Improvements in production methods.
- Improvements in the reliability and efficiency of the tools and machines used.
- Better product design.
- Improved production scheduling and inventory control.
- Better organization of the workplace.

Studies of the aircraft industry undertaken during the 1920s showed that the direct-labor hours required to produce a unit of output declined as the cumulative number of units produced increased. The term **learning curve** was adopted to explain this phenomenon. Similarly, it has been observed in many industries that marginal production costs also decline as the cumulative number of units produced increases. The term **experience curve** has been used to describe this second phenomenon.

Learning Curves

As workers gain more experience with the requirements of a particular process, or as the process is improved over time, the number of hours required to produce an additional unit declines. The **learning curve**, which models this relationship, is also a means of describing dynamic economies of scale. Experience has shown that these curves are accurately represented by an exponential relationship. Let $Y(u)$ be the number of labor hours required to produce the uth unit. Then the learning curve is of the form

$$Y(u) = au^{-b}$$

where a is the number of hours required to produce the first unit and b measures the rate at which the marginal production hours decline as the cumulative number of units produced increases.

Traditionally, learning curves are described by the percentage decline of the labor hours required to produce item $2n$ compared to the labor hours required to produce item n, and it is assumed that this percentage is independent of n. That is, an 80 percent learning curve means that the time required to produce unit $2n$ is 80 percent of the time required to produce unit n for any value of n. For an 80 percent learning curve

$$\frac{Y(2u)}{Y(u)} = \frac{a(2u)^{-b}}{au^{-b}} = 2^{-b} = .80.$$

It follows that

$$-b\ln(2) = \ln(.8)$$
$$\text{or } b = -\ln(.8)/\ln(2) = .3219. \text{ (ln is the natural logarithm).}$$

More generally, if the learning curve is a $100L$ percent learning curve, then

$$b = -\ln(L)/\ln(2).$$

Figure 1–11 shows an 80 percent learning curve. When graphed on double-log paper, the learning curve should be a straight line if the exponential relationship we have assumed is

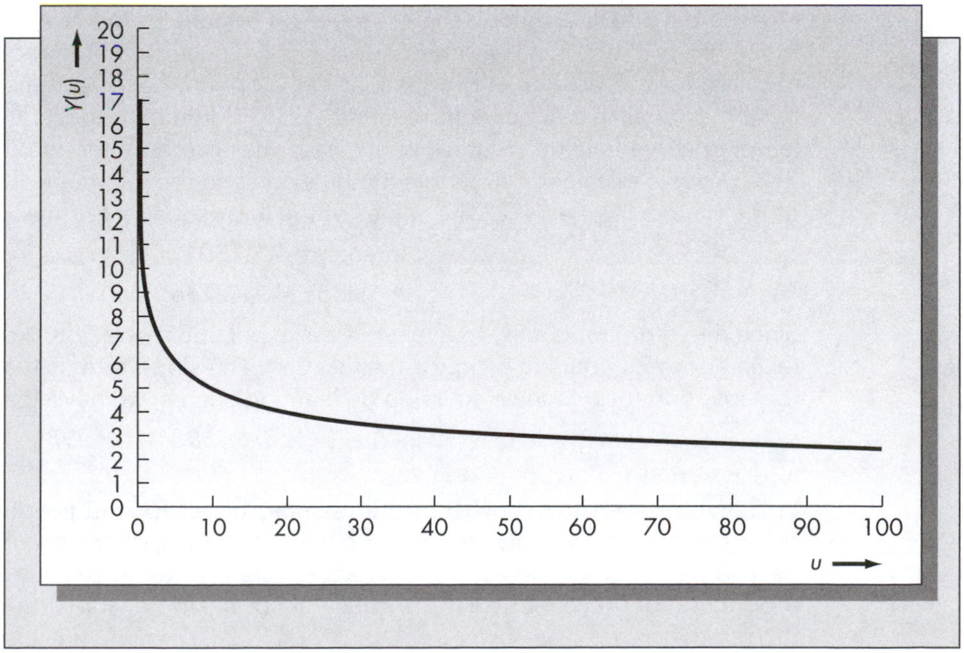

Figure 1–11 An 80 percent learning curve

accurate. If logarithms of both sides of the expression for $Y(u)$ are taken, a linear relationship results, since

$$\ln(Y(u)) = \ln(a) - b\ln(u).$$

Linear regression is used to fit the values of a and b to actual data after the logarithm transformation has been made. (General equations for finding least squares estimators in linear regression appear in Appendix 2–B.)

EXAMPLE 1.1

XYZ has kept careful records of the average number of labor hours required to produce one of its new products, a pressure transducer used in automobile fuel systems. These records are represented in the following table.

Cumulative Number of Units Produced (A)	Ln (Column A)	Hours Required for Next Unit (B)	Ln (Column B)
10.00	2.30	9.22	2.22
25.00	3.22	4.85	1.58
100.00	4.61	3.80	1.34
250.00	5.52	2.44	0.89
500.00	6.21	1.70	0.53
1,000.00	6.91	1.03	0.53
5,000.00	8.52	0.60	–0.51
10,000.00	9.21	0.50	–0.69

According to the theory, there should be a straight-line relationship between the logarithm of the cumulative number of units produced and the logarithm of the hours required for the last unit of production. The graph of the logarithms of these quantities for the data above appears in Figure 1–12. The figure suggests that the exponential learning curve is fairly accurate in this case. Using the methods outlined in Appendix 2–B, we have obtained estimators for the slope and the intercept of the least squares fit of the data in Figure 1–12. The values of the least squares estimators are

$$\text{Intercept} = 3.1301,$$
$$\text{Slope} = -.42276.$$

Since the intercept is $\ln(a)$, the value of a is $\exp(3.1301) = 22.88$. Hence, it should have taken about 23 hours to produce the first unit. The slope term is the constant $-b$. From the equation for b in the discussion on learning curves, we have that

$$\ln(L) = -b\ln(2) = (-.42276)(.6931) = -.293.$$

It follows that $L = \exp(-.293) = .746$.

Hence, these data show that the learning effect for the production of the transducers can be accurately described by a 75 percent learning curve. This curve can be used to predict the number of labor hours that will be required for continued production of these particular transducers. For example, substituting $u = 50{,}000$ into the relationship

$$Y(u) = 22.88u^{-.42276}$$

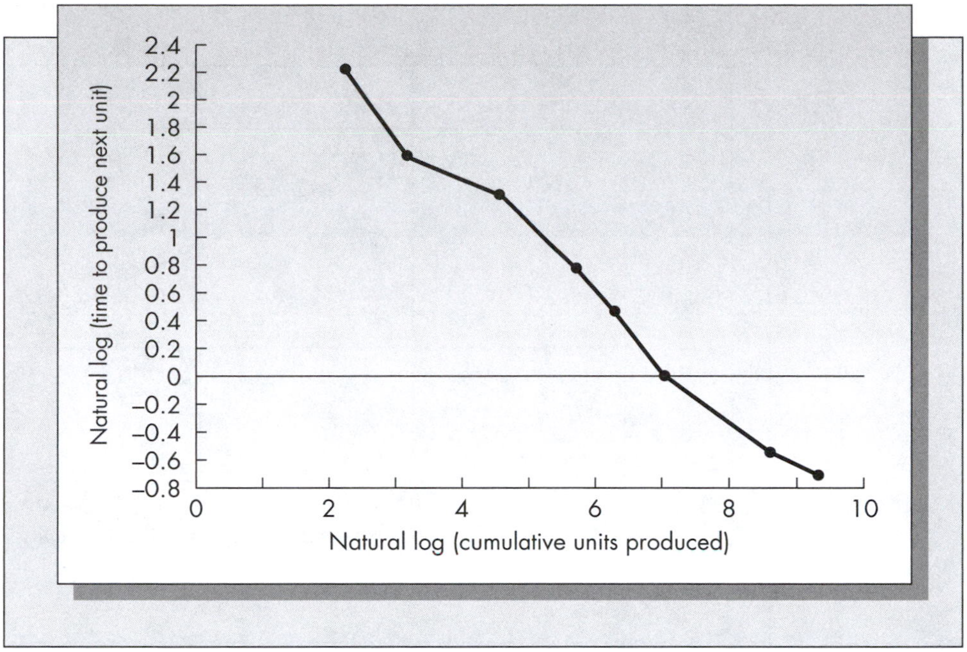

Figure 1–12 Log–log plot of XYZ data

gives a value of $Y(50{,}000) = .236$ hour. One must interpret such results with caution, however. A learning curve relationship may not be valid indefinitely. Eventually the product will reach the end of its natural life cycle, which could occur before 50,000 units have been produced in this example. Alternatively, there could be some absolute limit on the number of labor hours required to produce one unit that, because of the nature of the manufacturing process, can never be improved. Even with these limitations in mind, learning curves can be a valuable planning tool when properly used.

Experience Curves

Learning curves are a means of calibrating the decline in marginal labor hours as workers become more familiar with a particular task or as greater efficiency is introduced into the production process. Experience curves measure the effect that accumulated experience with production of a product or family of products has on overall cost and price. Experience curves are most valuable in industries that are undergoing major changes, such as the microelectronics industry, rather than very mature industries in which most radical changes have already been made, such as the automobile industry. The steady decline in the prices of integrated circuits (ICs) is a classic example of an experience curve. Figure 1–13 shows the average price per unit as a function of the industry's accumulated experience, in millions of units of production, during the period 1964 to 1972. This graph is shown on log–log scale, and the points fall very close to a straight line. This case represents a 72 percent experience curve. That is, the average price per unit declines to about 72 percent of its previous value for each doubling of the cumulative production of ICs throughout the industry.

Experience curves are generally measured in terms of cost per unit of production. In most circumstances, the price of a product or family of products closely tracks the cost of production. However, in some cases umbrella pricing occurs. That is, prices remain fairly

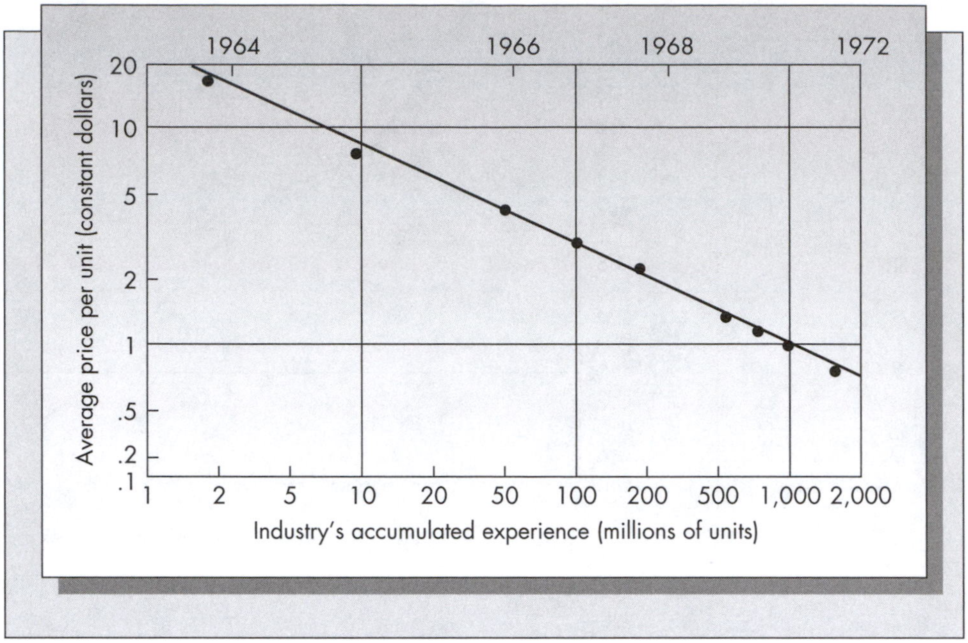

Figure 1–13 Prices of integrated circuits during the period 1964–1972

Source: Robert N. Noyce (September 1977), "Microelectronics," *Scientific American*. All rights reserved. Reprinted with permission of the publisher.

stable during a period in which production costs decline. Later, as competitive pressures of the marketplace take hold, prices decline more rapidly than costs until they catch up. This can cause a kink in the experience curve when price rather than cost is measured against cumulative volume. This type of phenomenon is pictured in Figure 1–14.

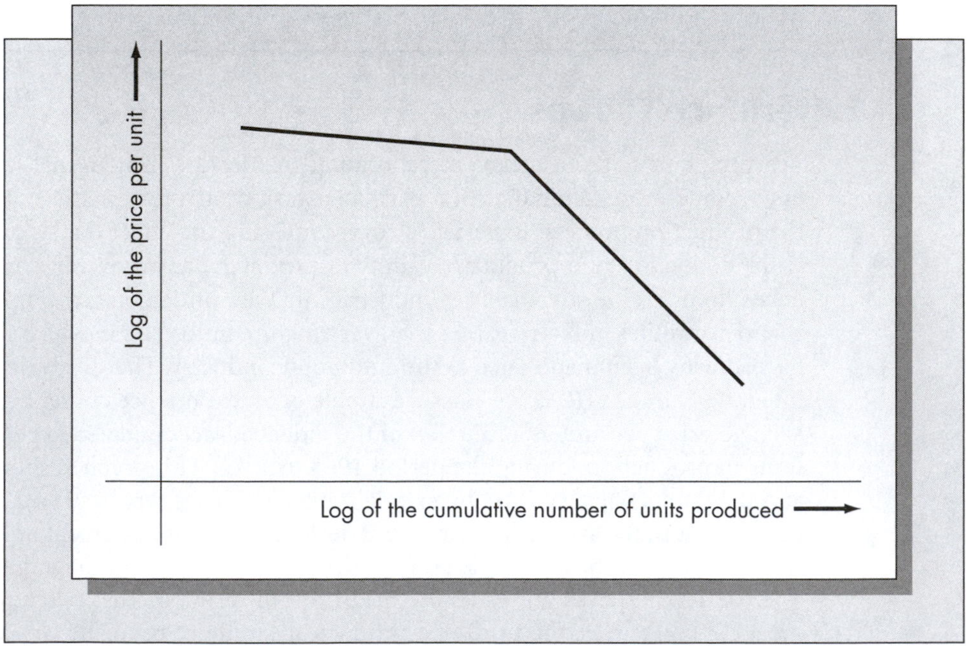

Figure 1–14 "Kinked" experience curve due to umbrella pricing

Learning curves have been the subject of criticism in the literature on a number of grounds: (a) they lack theoretical justification; (b) they confuse the effects of learning, economies of scale, and other technological improvements; and (c) they focus on cost rather than profit (Devinney, 1987). However, such curves are accurate descriptors of the way that marginal labor hours and costs decline as a function of the cumulative experience gained by the firm or industry.

Learning and Experience Curves and Manufacturing Strategy

We define a **learning curve strategy** as one in which the primary goal is to reduce costs of production along the lines predicted by the learning curve. Ford Motors adopted a learning curve strategy in seeking cost reductions in the Model T during the period 1909 to 1923. William Abernathy and Kenneth Wayne (1974) showed that the selling price of the Model T during this period closely followed an 85 percent experience curve. Ford's strategy during this time was clearly aimed at cost cutting; the firm acquired or built new facilities including blast furnaces, logging operations and sawmills, a railroad, weaving mills, coke ovens, a paper mill, a glass plant, and a cement plant. This allowed Ford to vertically integrate operations, resulting in reduced throughput time and inventory levels—a strategy similar in spirit to the lean philosophy discussed later in this chapter.

A learning curve strategy may not necessarily be the best choice over long-term planning horizons. Abernathy and Wayne (1974) make the argument that when manufacturing strategy is based on cost reduction, innovation is stifled. As consumer tastes changed in the 1920s, Ford's attention to cost cutting and standardization of the Model T manufacturing process resulted in its being slow to adapt to changing patterns of customer preferences. Ford's loss was General Motors's gain. GM was quick to respond to customer needs, recognizing that the open car design of the Model T would soon become obsolete. Ford thus found itself fighting for survival in the 1930s after having enjoyed almost complete domination of the market. Survival meant a break from the earlier rigid learning curve strategy to one based on innovation.

Another example of a firm that suffered from a learning curve strategy is Douglas Aircraft. The learning curve concept was deeply rooted in the airframe industry. Douglas made several commitments in the 1960s for delivery of jet aircraft based on extrapolation of costs down the learning curve. However, because of unforeseen changes in the product design, the costs were higher than anticipated, and commitments for delivery times could not be met. Douglas was forced into a merger with McDonnell Company as a result of the financial problems it experienced.

We are not implying by these examples that a learning curve strategy is wrong. Standardization and cost reduction based on volume production have been the keys to success for many companies. Failure to achieve quick time-to-volume can spell disaster in a highly competitive marketplace. What we are saying is that the learning curve strategy must be balanced with sufficient flexibility to respond to changes in the marketplace. Standardization must not stifle innovation and flexibility.

Economies of Scale and Economies of Scope

Economies of scale are generally considered the primary advantages of expanding existing capacity. As the scale of production increases, we move to the right in the product-process matrix (Figure 1–2). This allows a firm to move down the diagonal to product-type layouts such as assembly lines, which have much lower variable costs (but higher fixed costs) than job-shop production (see Section 7.1 for more details on this).

John Panzer and Robert Willig (1981) introduced the concept of **economies of scope**, which they defined as the cost savings realized from combining the production of two or more product lines at a single location. The idea is that the manufacturing processes for these product lines may share some of the same equipment and personnel so that the cost of production at one location could be less than at two or more different locations.

The notion of economies of scope extends beyond the direct cost savings that the firm can realize by combining the production of two or more products at a single location. It is often necessary to duplicate a variety of support functions at different locations. These functions include information storage and retrieval systems and clerical and support staff. Such activities are easier to coordinate if they reside at the same location. The firm also can realize economies of scope by locating different facilities in the same geographic region. In this way employees can, if necessary, call upon the talents of key personnel at a nearby location.

Joel Goldhar and Mariann Jelinek (1983) argue that considerations of economies of scope support investment in new manufacturing technology. Flexible manufacturing systems and computer-integrated manufacturing result in "efficiencies wrought by variety, not volume." These types of systems, argue the authors, allow the firm to produce multiple products in small lot sizes more cheaply using the same multipurpose equipment. (Flexible manufacturing systems are discussed in greater detail in Section 1.7.)

Management must weigh the benefits that the firm might realize by combining product lines at a single location against the disadvantages of lack of focus. Too many product lines produced at the same facility could cause the various manufacturing operations to interfere with each other. The proper sizing and diversity of the functions of a single plant must be balanced so that the firm can realize economies of scope without allowing the plant to lose its essential focus.

Another disadvantage of lack of focus is finding the right metrics to appropriately measure a diverse set of products or services. The Snapshot Application from Chapter 6 (Supply Chain Strategy) describes how Anheuser-Busch increased the focus of their breweries and then aligned the metrics used to the type of beer produced. The **focused factory** was introduced by Wickham Skinner (1974) who claimed, "A factory that focuses on a narrow product mix for a particular market niche will outperform the conventional plant, which attempts a broader mission."

PROBLEMS FOR SECTION 1.7

25. What are the factors that contribute to the learning curve/experience curve phenomenon?
26. What is a "learning curve strategy"? Describe how this strategy led to Ford's success up until the mid-1920s and Ford's problems after that time.
27. What are some of the pitfalls that can occur when using learning curves and experience curves to predict costs? Refer to the experience of Douglas Aircraft.
28. Consider the example of XYZ Corporation presented in this section. If the learning curve remains accurate, how long will it take to produce the 100,000th unit?
29. A start-up firm has kept careful records of the time required to manufacture its product, a shutoff valve used in gasoline pipelines.

Cumulative Number of Units Produced	Number of Hours Required for Next Unit
50	3.3
100	2.2
400	1.0
600	0.8
1,000	0.5
10,000	0.2

a. Compute the logarithms of the numbers in each column. (Use natural logs.)
b. Graph the ln(hours) against the ln(cumulative units) and eyeball a straight-line fit of the data. Using your approximate fit, estimate a and b.
c. Using the results of part (b), estimate the time required to produce the first unit and the appropriate percentage learning curve that fits these data.
d. Repeat parts (b) and (c), but use an exact least squares fit of the logarithms computed in part (a).

30. Consider the learning curve derived in problem 29. How much time will be required to produce the 100,000th unit, assuming the learning curve remains accurate?
31. Consider the experience curve plotted in Figure 1–15. What percentage experience curve does this represent?
32. Discuss the limitations of learning and experience curves.
33. An analyst predicts that an 80 percent experience curve should be an accurate predictor of the cost of producing a new product. Suppose that the cost of the first unit is \$1,000.

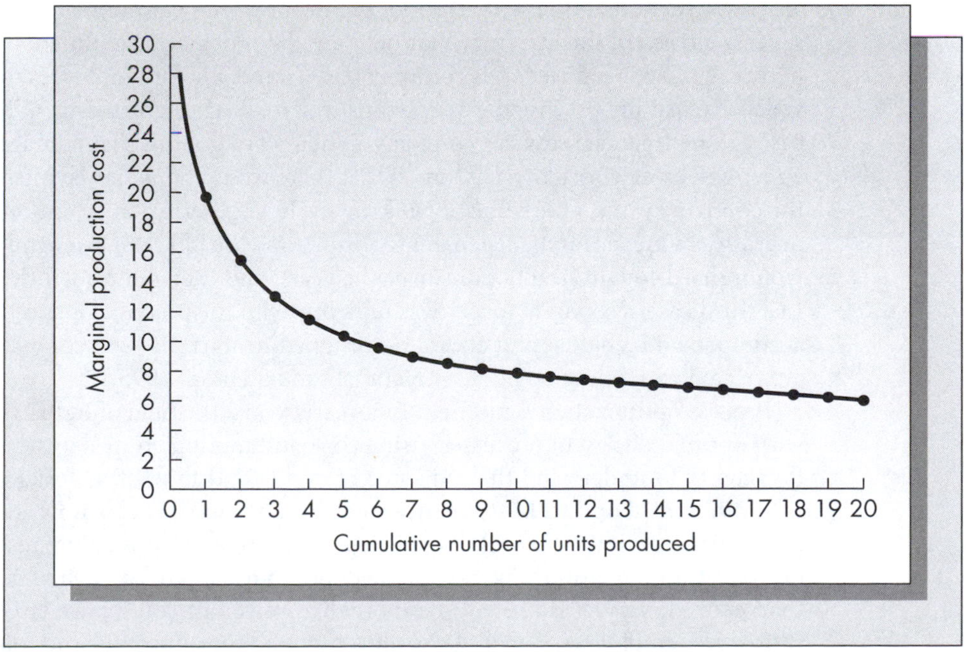

Figure 1–15 (for problem 31)

What would the analyst predict is the cost of producing the
a. 100th unit?
b. $10{,}000^{th}$ unit?

34. Suppose that the Mendenhall Corporation, a producer of women's handbags, has determined that a 73 percent experience curve accurately describes the evolution of its production costs for a new line. If the first unit costs $100 to produce, what should the 10,000th unit cost based on the experience curve?
35. Give some examples of economies of scale found in the building industry.
36. Give some examples of economies of scope found in the building industry.

1.8 STRATEGIC INITIATIVES

Operations management has been responsible for a variety of strategic initiatives over the years. Here we review some of the most important ones. Further details regarding many of these initiatives may be found in the relevant chapters of this text.

Business Process Reengineering

Business process reengineering (BPR) caught on after the publication of *Reengineering the Corporation: A Manifesto for Business Resolution* by Michael Hammer and James Champy (1993). BPR is not a specific technique, such as materials requirements planning or a production-planning concept like just-in-time. Rather, it is the idea that entrenched business processes can be changed and can be improved. The process is one of questioning why things are done a certain way—and not accepting the answer, "because that's the way we do it."

It is easier to list the steps one might consider in a reengineering effort than to actually implement one. In the real world, political realities cannot be ignored. For many of the success stories in the literature, not only are the processes simplified, but the headcount of personnel also is reduced. It is certainly understandable for employees to see BPR as a thinly veiled excuse for downsizing (euphemistically called "right-sizing"). This was exactly the case in one financial services company. When word got out that management was planning a reengineering effort, most assumed that there would be major layoffs. Some even thought the company was on the verge of bankruptcy. In another instance, union leaders saw reengineering as a means for management to throw away the job categories and work rules they had won in hard-fought negotiations over the years, and they persuaded the members to strike. In a third case, a senior manager was unhappy with the potential loss of authority that might accompany a reengineering effort. He resigned to start his own company. (These examples are related in a follow-up book by Hammer and Stanton, 1995.)

Process optimization is not new. In its early years, the field of industrial engineering dealt with optimal design of processes, using time and motion studies to set standards and using flowcharts to understand the sequence of events and flow of material in a factory. Why is BPR different? For one, BPR is concerned with business process flows rather than manufacturing process flows. Second, the concept is not one of optimizing an existing process but one of rethinking how things should be done from scratch. As such, it is more revolutionary than evolutionary. It is likely to be more disruptive but could have larger payoffs. To make BPR work, employees at every level must buy into the approach, and top management must champion it. Otherwise, the reengineering effort could be a costly failure.

Just-in-time, Lean Production, and the Toyota Production System

The **Toyota Production System** (TPS) is a manufacturing process on the one hand and a broad-based operations strategy on the other. It has been one of the key strategic developments for production and operations management in the past decades. The two quotations that follow compare the worst of U.S. mass production at General Motors' Framingham, Massachusetts, plant in 1986 with the best Japanese lean production.

> Next we looked at the line itself. Next to each workstation were piles—in some cases weeks' worth—of inventory. Littered about were discarded boxes and other temporary wrapping material. On the line itself the work was unevenly distributed with some workers running madly to keep up and others finding time to smoke and even read a newspaper. . . . At the end of the line we found what is perhaps the best evidence of old-fashioned mass production: an enormous work area full of finished cars riddled with defects. All these cars needed further repair before shipment, a task that can prove enormously time-consuming and often fails to fix fully the problems now buried under layers of parts and upholstery. (Womack et al., 1990)

Contrast this with the description below of Toyota's Takaoka plant in Toyoda City.

> The differences between Takaoka and Framingham are striking to anyone who understands the logic of lean production. For a start hardly anyone was in the aisles. The armies of indirect workers so visible at GM were missing, and practically every worker in sight was actually adding value to the car. . . . The final assembly line revealed further differences. Less than an hour's worth of inventory was next to each worker at Takaoka. The parts went on more smoothly and the work tasks were better balanced, so that every worker worked at about the same pace. . . . At the end of the line, the difference between lean and mass production was even more striking. At Takaoka we observed almost no rework area at all. Almost every car was driven directly from the line to the boat or the trucks taking cars to the buyer. (Womack et al., 1990)

These differences are dramatic. We should note that GM has since closed Framingham and that the plants run by the "big three" (namely, GM, Ford, and Chrysler LLC) are far more efficient and better managed than was Framingham.

The key stages of the development of TPS are detailed by Matthias Holweg (2007), who describes how many early treatments focused on the principles of **just-in-time** (JIT) production and **single minute exchange of dies** (SMED) that are core to TPS. Eventually, JIT became effectively synonymous with TPS. Earlier editions of this text used JIT to describe what is now known as **lean production**. The term "lean production" appears to have been first used by John Krafcik (1988) but popularized by *The Machine That Changed the World* (Womack et al., 1990).

The process elements of lean production (including JIT production through the use of **kanban** cards and SMEDs) will be discussed in detail in Chapter 9 (Production Control Systems: Push and Pull) as part of a complete analysis of production control systems. However, lean is a philosophy that includes treatment of inventory in the plant, relationships with suppliers, and distribution strategies. The core of the philosophy is to eliminate waste. This is accomplished by efficient scheduling of incoming orders, work-in-process inventories, and finished goods inventories. Tools for process improvement and elimination of waste will be discussed in Chapter 10 (Quality and Assurance).

Some key principles behind lean production are:

1. stable production,
2. careful product design,
3. low inventory,
4. excellent vendor relations, and
5. high quality production.

Lean production is part philosophy and part technique. The philosophical focus is on the elimination of all sources of waste. Lean consultants sometimes talk about the **North Star**, which is the goal the process should be aiming at (zero waste, zero inventories, and zero variability). The romantic version of lean is that it is the "one true path" to manufacturing. It promises to be easy to implement and to create a world where there are no "trade-offs." However, most practitioners of lean are far more pragmatic than this. They recognize that inventory is only one form of waste. The key idea is not to waste time figuring out the "optimal" amount of compensating inventory. Instead, lean seeks to eliminate the underlying problem that required inventory in the first place and to smooth out the flow of product through a plant. Inventory should not be eliminated without first eliminating the driving cause for it (where possible).

In summary, lean production requires a shift in mindset from the environment as a constraint to the environment as a control where all sources of waste and variability are examined. That is, inputs that are traditionally viewed as constraints—such as machine setup times, vendor deliveries, quality levels (scrap, rework), product designs, production schedules (e.g., customer due dates)—become controls. The result is that the system can be made much easier to manage by improving the environment.

1.9 FLEXIBLE MANUFACTURING SYSTEMS

A **flexible manufacturing systems** (**FMS**) is a collection of numerically controlled machines connected by a computer-controlled materials flow system. Typical flexible manufacturing systems are used for metal cutting and forming operations and certain assembly operations. Because the machines can be programmed, the same system can be used to produce a variety of different parts.

What advantages do such systems have over a conventional dedicated machine layout? They provide the opportunity to drastically slash the hidden costs of manufacturing. These include work-in-process inventory costs and overhead costs associated with indirect labor. They allow firms to quickly change tooling and product design with minimal additional investment in new equipment and personnel. However, they do require a substantial capital investment, which can be recouped only if their potential can be realized and they are used in the right environment.

Under what circumstances should a firm consider employing an FMS as part of its overall layout for manufacturing? As shown in Figure 1–16, a flexible manufacturing system is appropriate when the production volume and the variety of parts produced are moderate. For systems with low volume and high customization, stand-alone numerically controlled machines are appropriate. These can be programmed for each individual application. For high-volume production of standardized parts, fixed transfer lines are more appropriate.

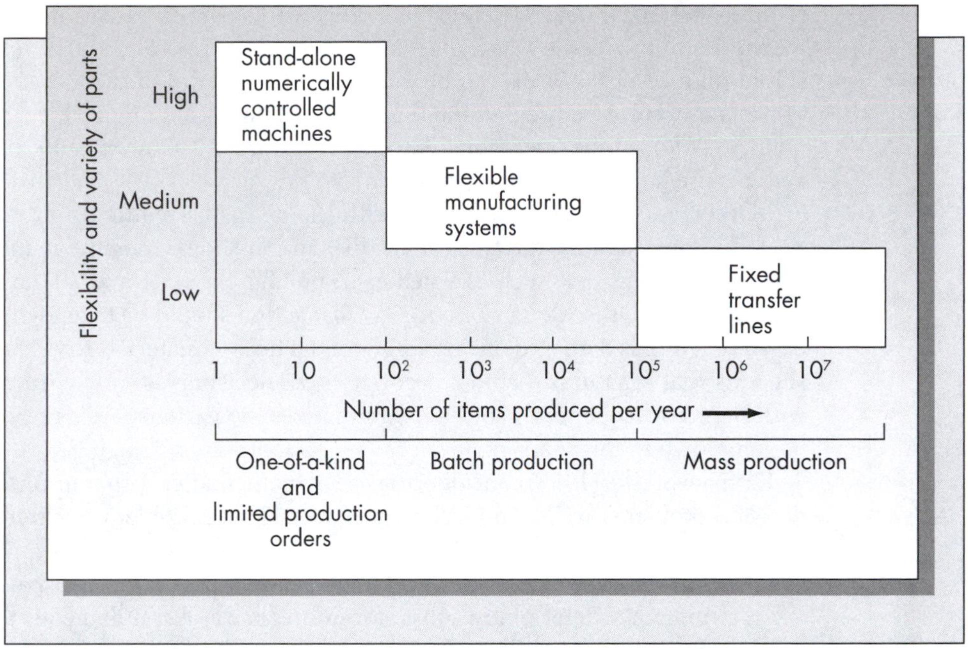

Figure 1–16 The position of FMS in the manufacturing hierarchy

(The similarity between Figure 1–16 and the product-process matrix of Figure 1–2 is not coincidental.)

The layout and structure of a typical FMS is pictured in Figure 1–17. Often the machines are controlled by a central computer, which also can be programmed for individual applications. Parts are typically loaded and unloaded at a single location along the materials handling path. The materials handling system consists of pallets, which are usually metal disks two

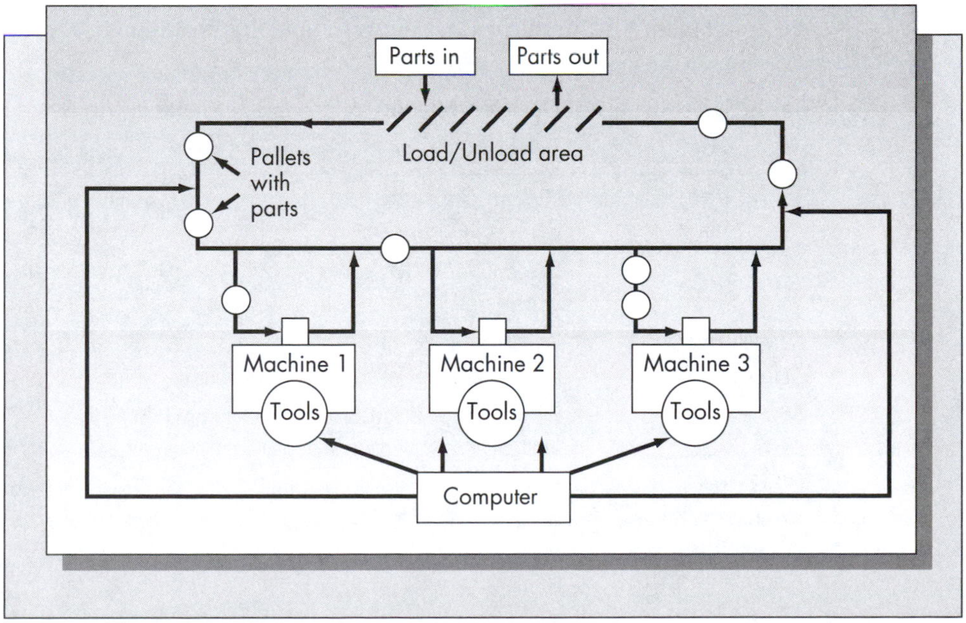

Figure 1–17 A typical flexible manufacturing system

to three feet in diameter. These pallets carry work-in-process inventory and queue up at the machines for processing. Each machine may contain from 5 to 100 different tools, which are stored on a carousel. Tools can be exchanged in a matter of seconds. Note that the routing of parts must also be programmed into the system, as not all products require the same sequencing of machine operations. Although Figure 1–17 shows a centralized computer, some systems use individually programmed machines.

Mathematical modeling can help with the decisions required to design and manage an FMS. Flexible manufacturing systems, like all job shops, must be managed on a continual basis. Some decisions, such as whether to purchase a system at all, have very far-reaching consequences. Other decisions, such as what action should be taken when a machine breaks down or a tool is damaged, may only affect the flow of materials for the next few hours. This suggests that a natural way to categorize these decisions is by the time horizon associated with their consequences. Table 1–1 summarizes the various levels of FMS decision making broken down in this manner.

Kathryn Stecke (1983) considers the use of mathematical programming for solving several decision problems arising in FMS management. These problems include the following.

1. *Part type selection problem.* From a given set of part types that have associated known requirements, determine a subset for immediate and simultaneous processing.
2. *Machine grouping problem.* Partition the machines into groups so that each machine in a group can perform the same set of operations.
3. *Production ratio problem.* Determine the relative ratios to produce the part types selected in step 1.
4. *Resource allocation problem.* Allocate the limited number of pallets and fixtures among the selected part types.
5. *Loading problem.* Allocate the operations and required tools of the selected part types among the machine groups subject to technological capacity constraints of the FMS.

We will not present the details of the mixed integer programming formulations (defined in Supplement 1). Our purpose is simply to note that mathematical programming is another

Table 1–1 Levels of Decision Making in Flexible Manufacturing Systems

Time Horizon	Major Issues	Modeling Methods
Long term (months–years)	Design of the system Parts mix changes	Queueing networks Simulation
	System modification and/or expansion	Optimization
Medium term (days–weeks)	Production batching Maximizing of machine utilization	Balancing methods Network simulation
	Planning for fluctuations in demand and/or availability of resources	
Short term (minutes–hours)	Work order scheduling and dispatching	Tool operation and allocation program
	Tool management	Work order dispatching algorithm
	Reaction to system failures	Simulation

means of helping with long-range and medium-range decisions affecting the FMS. Stecke (1983) shows how one would apply these models using actual data from an existing FMS. Section 9.9 also discusses FMS.

Competing on Quality

What competitive factors do U.S. managers believe to be important? Based on a survey of 217 industry participants, the following factors were deemed as the most important for gaining a competitive edge; they are listed in the order of importance.

1. Conformance quality
2. On-time delivery performance
3. Quality
4. Product flexibility
5. After-sale service
6. Price
7. Broad line (features)
8. Distribution
9. Volume flexibility
10. Promotion

In this list we see some important themes. **Quality** and **time management** emerge as leading factors. Quality control was brought to public attention with the Malcolm Baldrige National Quality Improvement Act of 1987 and the prestigious **Malcolm Baldrige National Quality Award** (modeled after the Japanese **Deming Prize**, which was first awarded in 1951). These awards are described in Chapter 10 (Quality and Assurance).

In the past decades, the Japanese and Germans gained a loyal following among U.S. consumers by producing quality products. However, U.S. firms have largely caught up on the quality dimension, and Chinese production is not far behind. If these trends continue, product quality will be assumed by the consumer. That is, quality may become an order qualifier rather than an order winner.

A U.S. firm known for its serious commitment to quality is Motorola (see the Snapshot Application in Chapter 10). Motorola, winner of the Baldrige Award in 1988, steadily drove down the rate of defects in its manufactured products. Defects were reported to be near 40 parts per million at the end of 1991, down from 6,000 parts per million in 1986. Motorola pioneered the goal known as **Six Sigma** (meaning six standard deviations away from the mean of a normal distribution), which translates to 3.4 parts per million (this will be discussed further in Chapter 10). Motorola felt that the process of applying for the Baldrige Award was so valuable that it started requiring all its suppliers to apply for the award as well. Six Sigma quality has become an industry standard, and a whole training industry has developed around it. Trainees receive **green belts** and **black belts** in the methods. Many of the tools are based on **total quality management (TQM)** programs that were developed in the 1980s.

Another trend in quality competition is **ISO 9000 certification**. ISO stands for *International Organization for Standardization*. That is, ISO 9000 provides a series of standards for quality management systems. It is possibly best described as "document what you do and do what you document." There are over one million organizations worldwide that have received this certification.

Benchmarking Quality

Benchmarking means measuring one's performance against that of one's competitors. Competitive benchmarking is gaining importance in light of increasing global competition. Setting one's priorities and being certain that those priorities are consistent with the needs of the marketplace are essential. Based on a database developed at Boston University, the competitive priorities in Europe, Japan, and the United States are summarized in Table 1–2. While this extensive benchmarking does not appear to have been repeated, we doubt that the priorities have changed significantly in the interim.

The table highlights several interesting points. First, consistent with our discussion, the Japanese firms surveyed rank product reliability as their top priority. The Japanese understand that product reliability could be their greatest competitive asset and plan to continue to stress this important dimension of quality. It is also interesting to see that price is mentioned only by the U.S. companies.

The authors list four types of benchmarking.

1. Product benchmarking.
2. Functional or process benchmarking.
3. Best practices benchmarking.
4. Strategic benchmarking.

Product benchmarking refers to the practice of tearing down a competitor's product to see what can be learned from its design and construction. It is said that when Toyota initiated its program to produce the Lexus to compete with cars such as Mercedes and BMW, it carefully examined the competitor's products to determine how and where welds were placed and how the cars were put together to achieve the look and feel of exceptional quality.

Functional benchmarking focuses on the process rather than on the product. Typical processes might be order entry, assembly, testing product development, and shipping. Functional benchmarking is possible only when companies are willing to cooperate and share information. It has the same goal as product benchmarking: to improve the process and ultimately the resultant product.

Best practices benchmarking is similar to functional benchmarking, except that it focuses on management practices rather than on specific processes. Best practices might consider factors such as the work environment and salary incentives for employees in firms with exceptional performance. General Electric is a strong advocate of best practices benchmarking (*Fortune*, 1991).

The goal of *strategic benchmarking* is to consider the results of other benchmarking comparisons in the light of the strategic focus of the firm. Specifically, what is the overall

Table 1–2 Top Five Competitive Priorities

Europe	Japan	United States
Conformance quality	Product reliability	Conformance quality
On-time delivery	On-time delivery	On-time delivery
Product reliability	Fast design change	Product reliability
Performance quality	Conformance quality	Performance quality
Delivery speed	Product customization	Price

Source: Miller, DeMeyer, and Nakane, 1992.

business strategy that has been articulated by the CEO, and are the results of other benchmarking studies consistent with this strategy?

Ultimately, what is the purpose of benchmarking? It is to ensure continuous improvement and is only one of the means of achieving this. Continuous improvement in product and process is the ultimate goal of any quality program (see Chapter 10). Competitive benchmarking provides a means of learning from one's competitors. Although benchmarking can be a useful tool, it is not a substitute for a clearly articulated business strategy and a vision for the firm.

Time-based Competition

Beyond quality another key competitive factor is **time to market.** What is **time-based competition**? It is not the time and motion studies popular in the 1930s that formed the basis of the industrial engineering discipline.

> Time-based competitors focus on the bigger picture, on the entire value-delivery system. They attempt to transform an entire organization into one focused on the total time required to deliver a product or service. Their goal is not to devise the best way to perform a task, but to either eliminate the task altogether or perform it in parallel with other tasks so that over-all system response time is reduced. Becoming a time-based competitor requires making revolutionary changes in the ways that processes are organized. (Blackburn, 1991)

Successful retailers understand time-based competition. The success of the fashion chains The Gap and The Limited was due largely to their ability to deliver the latest fashions to the customer in a timely manner. However, their speed has now been surpassed by Zara, a **fast fashion** leader. Zara can produce a garment from design to store shelf in ten to fifteen days (Petro, 2012). Such speed was previously unheard of in the fashion industry.

Walmart Inc. has been enormously successful in part because of its time-management strategy. Each stock item in a Walmart store is replenished twice a week, while the industry average is once every two weeks. This allows Walmart to achieve better inventory turnover rates than its competition and to respond more quickly to changes in customer demand. Walmart's strategies have enabled it to become the industry leader, with a growth rate three times the industry average and profits two times the industry average (Blackburn, 1991, chapter 3).

Time-based management is a more complex issue for manufacturers; in some industries, it is clearly the key factor leading to success or failure. The industry leaders in the dynamic random access memory (DRAM) industry changed four times between 1978 and 1987. In each case, the firm that was first to market with the next-generation DRAM dominated that market. However, now that DRAMs have become commoditized, time-based competition has become less important in that industry. None of the industry leaders from 1987 have significant market share today (although most were bought out in the intervening years).

Time-based competition has also been facilitated by FMS, as described earlier. FMS's ability to produce a wide variety of parts with minimal changeover times often allows for faster production than the use of dedicated production equipment. **Additive manufacturing** (or **3D printing**), discussed in Chapter 6, also allows for decreased production lead times.

Servicization

We will use the term "**servicization**" to describe the trend of manufacturing companies to bundle additional services with their products. Adding services is a means for firms to gain an edge over their competitors and to provide an alternative to inexpensive offshore labor. (Note that the Europeans coined the term "servitization," The term appears to have been first used by Sandra Vandermerwe and Juan Rada in 1988.)

An example of servicization is provided by Hyundai, which is a huge multinational company based in South Korea. Hyundai is known as a *chaebol*, meaning a collection of diverse

SNAPSHOT APPLICATION

The IBM Story

IBM has become almost synonymous with computing, but few realize that IBM had its roots in mechanical tabulating machines dating to the late 1800s. The company's start came in 1890 when the German immigrant, Herman Hollerith, developed a new process to track the U.S. census. Hollerith's concept involved the use of punched cards, which persisted into the 1960s. The original firm established by Hollerith was the Tabulating Machine Company. In 1924, 10 years after T.J. Watson joined the firm, the name was changed to International Business Machines, or IBM. IBM continued to innovate mechanical computing machines; it was a relatively late entry into the electronic computer business. In fact, the first commercial computer was produced by Engineering Research Associates of Minneapolis in 1950 and sold to the U.S. Navy. Remington Rand produced the Univac one year later, which was the first commercially viable machine. They sold 46 machines that year at a cost of over $1 million each. IBM entered the fray in 1953 when it shipped its first computer, the 701. During three years of production, IBM sold only 19 machines. In 1955, AT&T Bell Laboratories announced the first fully transistorized computer, the TRADIC, and a year later the Burroughs Corporation threw its hat into the ring and later became a major player in the computer business. Eventually, Burroughs merged with Sperry Rand to form Unisys.

In 1959, IBM produced its first transistorized-based mainframe computer, the 7000 series. However, it was not until 1964 that the firm became a leader in computer sales. That was the year that IBM announced the system 360. This was a family of six mutually compatible computers and 40 peripherals that worked seamlessly together. There is little question that the system 360 computers were state of the art at the time. Within two years, IBM was shipping 1000 systems per month and dominating the mainframe business. While all of us are familiar with personal computers of various types and configurations, the mainframe business has not gone away. Even today, IBM continues to be a presence in the mainframe business with their newest system z architecture.

While the quality of their hardware was an important factor in IBM's success, it was not the only factor. What really sealed IBM's domination of the business market was the total customer solution. IBM's industry-specific software and the 360 operating system were a large part of attracting customers away from competitors. IBM not only had one of the most successful sales forces in the business but it also was a master of after-sales service. Each client would have an SE (systems engineer) assigned to make sure that their needs were met. As much as anything else, it was IBM's commitment to after-sales service that locked in their position as market leader in mainframe computing. Clearly, the idea of edging out competitors by bundling services with products is not new; however, today the servicization concept is becoming an increasingly important means of gaining competitive advantage.

Source: https://www.ibm.com; www.computerhistory.org

companies under a single umbrella. Prior to spinning off several of its businesses as separate companies following the Asian financial crisis in 1997, Hyundai was the largest chaebol in South Korea. The Hyundai Motor Company was established in 1967 and first began selling cars in the United States in 1986. At that time, Japanese, American, and European carmakers were firmly entrenched in the lucrative U.S. market. One way in which Hyundai sought to differentiate itself from its competitors was by offering an exceptional warranty. Today Hyundai offers a comprehensive warranty package, including a seven year/100,000 mile powertrain warranty. Sales of Hyundai models have risen steadily since 1986, with the company now offering high-end luxury vehicles, along with its low-cost entry models. While there is no question that competitive pricing and improving reliability and performance of its products account for much of the company's success, one cannot deny that their exceptional warranties played a role as well.

Richard Wise and Peter Baumgartner (1999) have noted that many manufacturing companies have moved their energies downstream to remain competitive. As the installed base of products increase, and the demand for new products decrease, firms are changing their business model in order to remain competitive. The focus is no longer on the manufacturing function alone; it includes services required to operate and maintain products (e.g. financing, after sales parts and services, and training). The profit margins on service typically exceed those of manufacturing, thus providing an incentive for firms to move in this direction.

As firms became more specialized, they moved away from vertical integration. However, changing patterns of demands and profits are leading many companies back in this direction. As an example, the Boeing Company, the world's foremost manufacturer of commercial aircraft, has significantly broadened its view of the value chain. The company now offers financing, local parts supply, ground maintenance, logistics management, and even pilot training. Servicization can be the key to maintaining competitiveness. Leasing rather than buying is another type of servicization that will be discussed in Chapter 8 (Service Operations Management).

A final example of servicization is the **Internet of things (IoT)**. This is the network of devices, vehicles, and home appliances that are now connected to the internet. It is made possible by smart sensors embedded in the end-items. It allows for an exchange of information that was previously not possible. For example, we are close to the day where your refrigerator may order milk without you needing to do anything. Refrigerators are already equipped with touch screens for immediate food ordering. In some sense, buying a refrigerator in the future will be buying access to milk and other fresh groceries.

Servicization has been facilitated by **performance based contracting (PBC)**, which was designed to reduce excessive costs in government contracting. Costs were being driven up by unnecessary provisions that specified exactly how each contract was to be carried out. According to Jon Desenberg of the Washington-based Performance Institute, a think tank dedicated to improving government performance, the idea "is to let the contracted group come up with the best possible solution and only pay them based on solving the problem . . . not on the individual steps and minutia that we have for so many years required."

Government contracts are a "why," "how," and "what" proposition. The "why" is established by the funding agency. Under PBC the "how" is shifted from the government to the contractor to determine the best way to achieve the "what". This is not always a win-win for the contractor. The task of pricing a contract now becomes much more onerous. It can be difficult to estimate all costs in advance, and deciding who pays for contract changes can be problematic.

PBC's are typical for consumers seeking professional services. A plumber may quote a fixed price in advance for a simple job but might want to be compensated on a time plus materials basis if there is uncertainty upfront about the time required to complete the job. Anyone who has had a major remodeling of their homes is likely to have entered into a PBC with their contractor. It is rare that the final cost matches the quoted number for a variety of reasons: weather delays, poor estimation of material costs, difficulty in finding subcontractors—and more often than not, changes made by the homeowner to the original project.

While PBCs sound like a good solution for government contracting, it can lead to the wrong kind of behavior. As an example, Yujing Shen (2003) examined the result of PBCs in the Maine Addiction Treatment System. Because the system was being measured on the success of curing addicts, the center had a strong incentive to treat only the less severe cases. This, of course, runs counter to the purpose of a treatment center; it is the most severe cases of abuse that need attention. We can conclude that a PBC is not appropriate in all circumstances. For any contract it is important that incentives be properly aligned with desired outcomes.

Automation

Automation, namely the replacement of human labor by machines, has been an operations trend since the **Industrial Revolution** of the eighteenth century. It was at this time that mechanized looms were used to replace hand weavers. We are now said to be experiencing the **Fourth Industrial Revolution**. (The second industrial revolution was the technology revolution at the turn of the twentieth century, and the third is the digital revolution that we have be experiencing since the late 1950s.) The fourth industrial revolution is said to be the introduction of **cyber-physical** systems. These are automated systems that are also connected with both the internet and users.

Automation is now considered a natural part of a process's lifecycle and was discussed earlier in this chapter in this context. However, automation is continuing to improve significantly. Already there are some "lights out" factories where machines operate autonomously to produce items like printed circuit boards. Robots are able to do more and more tasks that were previously done by humans. Autonomous vehicles are likely to change supply chains substantially. It is likely that automation and cyber-physical systems will continue to have a considerable impact on operations decisions for the foreseeable future.

PROBLEMS FOR SECTIONS 1.8–1.9

37. What general features would you look for in a business process that would make that process a candidate for reengineering? Discuss a situation from your own experience in which it was clear that the business process could have been improved.
38. In what ways might the following techniques be useful as part of a reengineering effort?
    - Computer-based simulation.
    - Flowcharting.
    - Project management techniques.
    - Mathematical modeling.
    - Cross-functional teams.
39. What problems can you foresee arising in the following situations?
    a. Top management is interested in reengineering to cut costs, but the employees are skeptical.

   b. Line workers would like to see a reengineering effort undertaken to give them more say-so in what goes on, but management is uninterested.
40. Just-in-time/lean has been characterized as a system whose primary goal is to eliminate waste. Discuss how waste can be introduced in (a) relationships with vendors, (b) receipt of material into the plant, and (c) movement of material through the plant. How do JIT methods cut down on these forms of waste?
41. In what ways can JIT/lean systems improve product quality?
42. In each case listed, state whether the factor listed is an advantage or a disadvantage of FMS. Discuss the reasons for your choice.
    a. Cost.
    b. Ability to handle different parts requirements.
    c. Advances in manufacturing technology.
    d. Reliability.
43. For each of the case situations described, state which of the types of manufacturing systems would be most appropriate: (1) stand-alone numerically controlled machines, (2) FMS, (3) fixed transfer line, or (4) another system. Explain your choice in each case.
    a. A local machine shop that accepts custom orders from a variety of clients in the electronics industry.
    b. An international supplier of standard-sized metal containers.
    c. The metal-working division of a large aircraft manufacturer, which must serve the needs of a variety of divisions in the firm.
44. Studies have shown that the defect rates for many Japanese products are much lower than for their U.S.-made counterparts. Speculate on the reasons for these differences.
45. What does "time-based competition" mean? Give an example of a product that you purchased that was introduced to the marketplace ahead of its competitors.
46. Consider the old maxim, "Build a better mousetrap and the world will beat a path to your door." Discuss the meaning of this phrase in the context of time-based competition. In particular, is getting to the market first the only factor in a product's eventual success?
47. Define "servicization" and provide an example from your own experience of a case where services were the deciding factor in a purchase.
48. What are some of the services that IBM provided for its mainframe customers during its meteoric rise in sales in the 1960s?
49. How are improvements in automation likely to change how we view the product-process matrix from Section 1.4 (Figure 1–2)?

1.10 SUMMARY

This chapter discussed the importance of operations strategy and its relationship to the overall business strategy of the firm. Operations continues to grow in importance in the firm. While a larger portion of direct manufacturing continues to move offshore, the importance of the manufacturing function should not be underestimated. The success of the operations strategy can be measured along several dimensions. These include the obvious measures of cost and product characteristics but also include quality, delivery speed, delivery reliability, flexibility, and service. The productivity frontier depicts trade-offs among these characteristics.

We hear more and more frequently that we are part of a global community. When buying products today, we are less concerned with the country of origin than with the characteristics of the product. How many consumers of cell phones are even aware that Nokia is headquartered in Finland, Ericsson in Sweden, and Motorola in the United States? An interesting question explored by Porter is: Why do some industries seem to thrive in some countries? While the answer is complex, he suggests that the following four factors are most important: *factor conditions; demand conditions; related and supporting industries;* and *firm strategy, structure, and rivalry.*

Hayes and Wheelwright have developed the concept of the product–process matrix. This matrix depicts how processes relate to the type of products being produced. It is important to understand which types of processes are appropriate for which types of products and industries. However, it is also important to understand product and process life cycles. Both go through the four cycles of start-up, rapid growth, maturation, and stabilization or decline.

Very related to process-type is the type of layout. The two most common types are product layouts and process layouts. The product layout is usually a fixed transfer line arranged in the sequence of the manufacturing steps required. A process layout groups machines with similar functions. Part routings vary from product to product. The product layout is appropriate for high-volume production of a small number of products (at the bottom right of the product-process matrix). The process layout is appropriate for a low-volume job-shop environment (at the top left of the product-process matrix). Fixed position layouts are used for products that are too large to move. Further, there has been considerable interest in layouts based on group technology. Parts are grouped into families, and machine cells are developed consistent with this grouping. Group technology layouts are appropriate for automated factories.

Learning and experience curves are useful in modeling the decline in labor hours or the decline in product costs as experience is gained in the production of an item or family of items. These curves have been shown to obey an exponential law and can be useful predictors of the cost or time required for production. (Moore's Law, due to Gordon Moore, a founder of Intel, predicted the doubling of chip performance every 18 months. This is an example of an experience curve, and the prediction has continued to be accurate to the present day.)

Changing the way that one does things can be difficult. Even more difficult is changing the way that a company does things. For that reason, business process engineering (BPR) is a painful process, even when it works. The most dramatic successes of BPR have come in service functions, but the concept can be applied to any environment. It is the process of rethinking how and why things are done in a certain way. Intelligently done, BPR can lead to dramatic improvements. However, it can also be a time-consuming and costly process.

Just-in-time (JIT) or lean production is a philosophy that grew from the kanban system developed by Toyota. At the heart of the approach is the elimination of waste. Systems are put in place to reduce material flows to small batches to avoid large buildups of work-in-process inventories. While lean production developed on the factory floor, it is a concept that has been applied to the purchasing function as well. Successful application of lean purchasing requires the development of long-term relationships, and usually requires close proximity to suppliers. The mechanics of lean production are discussed in more detail in Chapter 9.

The dramatic successes of the Japanese during the 1970s and 1980s were to a large extent due to the outstanding quality of their manufactured products. Two Americans, W. Edwards Deming and Joseph Juran, visited Japan in the early 1950s and played an important role in making the Japanese aware of the importance of producing quality products. The quality movement in the United States has resulted in a much greater awareness of the importance of quality and initiation of important programs such as quality circles and the Six Sigma program at Motorola. (Both the statistical and the organizational issues concerning quality are discussed in detail in Chapter 10.) The Malcolm Baldrige National Quality Award recognizes performance excellence.

Being able to get to the market quickly with products that people want in the volumes that the marketplace requires is crucial if one wants to be a market leader. Time-based competition means that the time from product conception to its appearance in the marketplace must be reduced. To do so, one performs as many tasks concurrently as possible. In many instances, time to market is less important than time to volume. Being the first to the market may not mean much if one cannot meet product demand.

Servicization is the addition of service features to a manufactured product. For example, many airlines do not own the engines on their planes. Instead, they rent the engines from the manufacturer (e.g., Rolls-Royce) and pay per hour they are used to fly the plane. It is a way for manufacturing firms to make higher margins (i.e., be less of a commodity), but it also aligns incentives around the buyer's concerns.

Finally, automation continues to have a significant effect on how products are made. We are said to be entering the fourth industrial revolution, where cyber-physical systems are becoming more common. These systems integrate automation with both the internet and the user. An example is autonomous vehicles, which are poised to have a major effect on how we live our lives.

APPENDIX 1–A

LAYOUT DECISIONS

Two charts that supply useful information regarding flows are (1) the **activity relationship chart** and (2) the **from-to chart**.

Activity Relationship Chart

An activity relationship chart (also called a rel chart for short) is a graphical means of representing the desirability of locating pairs of operations near each other. The following letter codes have been suggested for determining a "closeness" rating.

A Absolutely necessary. Because two operations may use the same equipment or facilities, they must be located near each other.
E Especially important. The facilities may require the same personnel or records, for example.
I Important. The activities may be located in sequence in the normal work flow.
O Ordinary importance. It would be convenient to have the facilities near each other, but it is not essential.
U Unimportant. It does not matter whether the facilities are located near each other or not.
X Undesirable. Locating a welding department near one that uses flammable liquids would be an example of this category.

The closeness ratings are represented in an activity relationship chart that specifies the appropriate rating for each pair of departments. Consider the following example.

EXAMPLE 1.2

Meat Me, Inc. is a franchised chain of fast-food hamburger restaurants. A new franchise in a growing suburban community near Reston, Virginia, has the following departments.

1. Cooking burgers
2. Cooking fries

3. Packing and storing burgers
4. Drink dispensers
5. Counter servers
6. Drive-up server

The burgers are cooked on a large grill, and the fries are deep fried in hot oil. For safety reasons the company requires that these cooking areas not be located near each other. All hamburgers are individually wrapped after cooking and stored near the counter. The service counter can accommodate six servers, and the site has an area reserved for a drive-up window.

An activity relationship chart for this facility appears in Figure 1–18. In the chart, each pair of activities is given one of the letter designations A, E, I, O, U, or X. Once a final layout is determined, the proximity of the various departments can be compared to the closeness ratings in the chart. Note that Figure 1–18 gives only the closeness rating for each pair of departments. In the original conception of the chart, a number giving the reason for each closeness rating is also included in every cell. These numbers do not appear in our example.

From-To Chart

A from-to chart is similar to the mileage chart that appears at the bottom of many road maps and gives the mileage between selected pairs of cities. From-to charts are used to analyze the flow of materials between departments. The two most common forms are charts that show the distances between departments and charts that show the number of materials handling trips per day between departments. A from-to chart differs from an activity relationship chart in that the from-to chart is based on a specific layout. It is a convenient means of summarizing the flow data corresponding to a given layout.

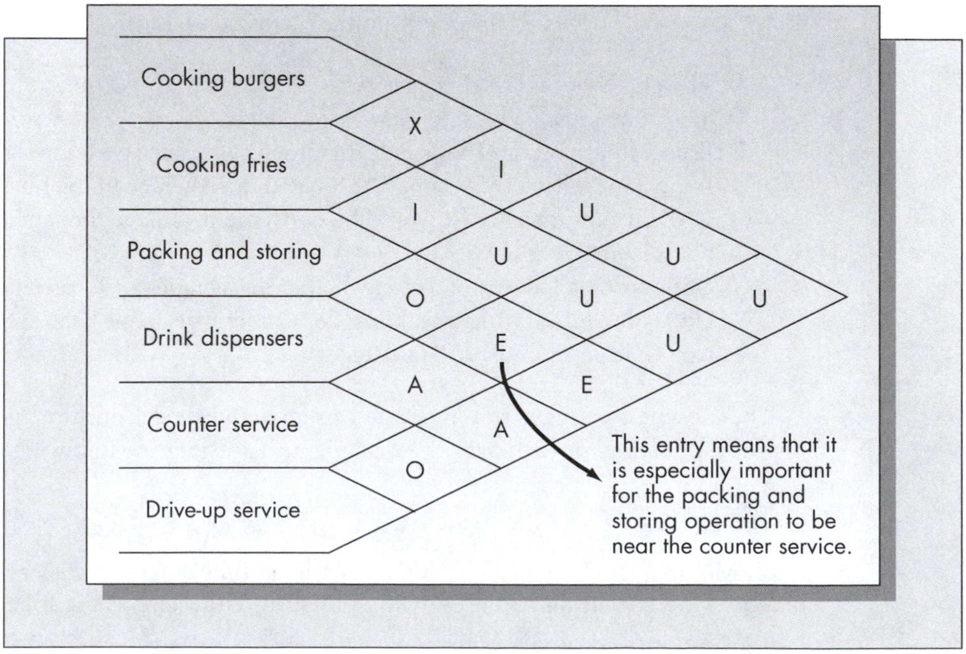

Figure 1–18 Activity relationship chart for Meat Me fast-food restaurant

EXAMPLE 1.3

A machine shop has six work centers. The work centers contain one or more of the following types of machines.

1. Saws
2. Milling machines
3. Punch presses
4. Drills
5. Lathes
6. Sanders

The from-to chart in Figure 1–19 shows the distance in feet between centers of the six departments. Note that the chart in the example is symmetric; that is, the travel distance between A and B is the same as the travel distance between B and A. This is not necessarily always the case, however. There may be one-way lanes, which allow material flow in one direction only or an automated materials handling system that moves pallets in one direction only.

Figure 1–20 shows a from-to chart that gives the number of materials handling trips per day. These figures could be based on a specific product mix produced in the shop or simply be representative of an average day.

Suppose that the firm's accounting department has estimated that the average cost of transporting material one foot in the machine shop is 20 cents. Using this fact, one can develop a third from-to chart that gives the average daily cost of materials handling from every department to every other department. For example, the distance separating saws and milling in the current layout is 18 feet (from Figure 1–19) and

From \ To	Saws	Milling	Punch press	Drills	Lathes	Sanders
Saws		18	40	30	65	24
Milling	18		38	75	16	30
Punch press	40	38		22	38	12
Drills	30	75	22		50	46
Lathes	65	16	38	50		60
Sanders	24	30	12	46	60	

Figure 1–19 From-to chart showing distances between six department centers (measured in feet)

From \ To	Saws	Milling	Punch press	Drills	Lathes	Sanders
Saws		43	26	14	40	
Milling			75	60		23
Punch press					45	16
Drills		22			28	
Lathes		45		30		60
Sanders		12				

Figure 1–20 From-to chart showing number of materials handling trips per day

there are an average of 43 materials handling trips between these two departments (from Figure 1–20). This translates to a total of (43)(18) = 774 feet traveled in a day or a total cost of (774)(0.2) = \$154.80 per day for materials handling between saws and milling. The materials handling costs for the other pairs of departments appear in Figure 1–21.

From \ To	Saws	Milling	Punch press	Drills	Lathes	Sanders
Saws		154.8	208	84	520	
Milling			570	900		138
Punch press					342	38.4
Drills		330			280	
Lathes		144		300		720
Sanders		72				

Figure 1–21 From-to chart showing materials handling cost per day (in \$)

From-to charts are not a means of determining layouts; they are a convenient way to express important flow characteristics of an existing layout. They can be useful in comparing the materials handling costs of a small number of alternatives. Because criteria other than the materials handling cost are relevant, the from-to chart should be supplemented with additional information, such as that contained in an activity relationship chart.

PROBLEMS FOR APPENDIX 1–A

50. A manufacturing facility consists of five departments: A, B, C, D, and E. At the current time, the facility is laid out as in Figure 1–22. Develop a from-to distance chart for this layout. Estimate the distance separating departments based on the flow pattern of the materials handling system in the figure. Assume that all departments are located at their centers as marked in Figure 1–22.
51. Four products are produced in the facility described in problem 50. The routing for these products and their forecasted production rates are

Product	Routing	Forecasted Production (units/week)
1	A–B–C–D–E	200
2	A–D–E	900
3	A–B–C–E	400
4	A–C–D–E	650

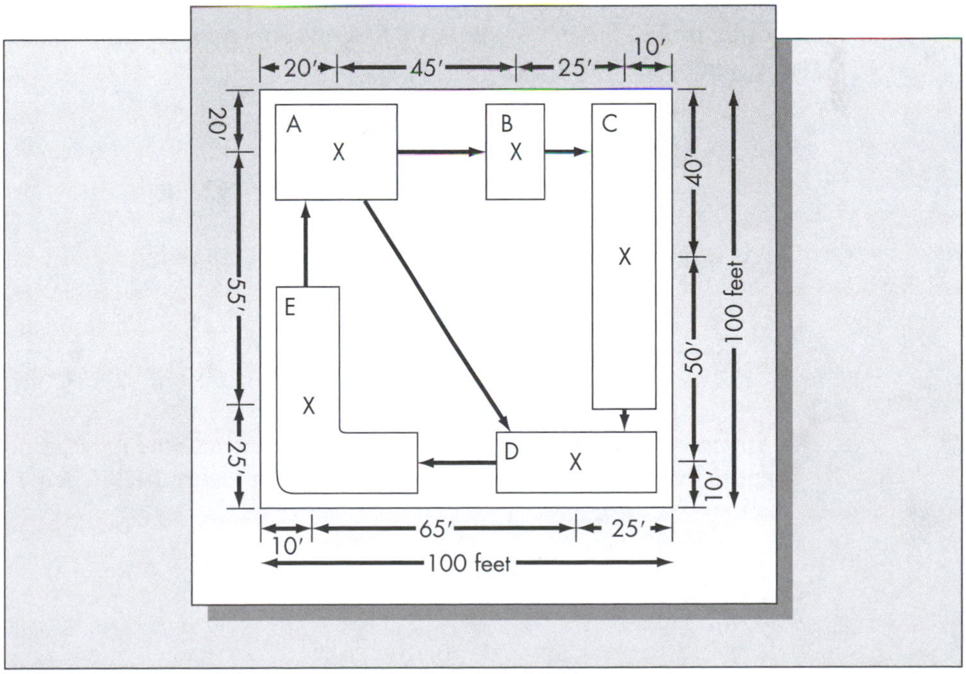

Figure 1–22 Layout of manufacturing facility (for problem 50)

Suppose that all four products are produced in batches of size 50.

a. Convert this information into a from-to chart similar to Figure 1–20 but giving the number of materials handling trips per week between departments.
b. Suppose that the cost of transporting goods 1 foot is estimated to be \$2. Using the from-to chart you determined in part (a), obtain a from-to chart giving materials handling cost per week between departments.
c. Develop an activity relationship chart for this facility based on the materials handling cost from-to chart you obtained in part (b). Use only the rankings A, E, I, O, U, with A being assigned to the highest cost and U the least.
d. Based on the results of parts (b) and (c), would you recommend a different layout for this facility?

52. Consider the example of the machine shop with the from-to charts given in Figures 1–19 through 1–21. Based on the results of Figure 1–21, in what ways might the current layout be improved?
53. Suggest a layout for the Meat Me fast-food restaurant, with the relationship chart shown in Figure 1–18. Assume that the facility is 50 feet square and half the restaurant will be for customer seating.

APPENDIX 1–B

USING CENTROIDS FOR LAYOUT DECISIONS

The **centroid** of any object is another term for the physical coordinates of the center of gravity. Centroids can be used for layout or location decisions if departments are assumed to be located at their centroids.

For a plate of uniform density, it would be the point at which the plate would balance exactly. Let R be any region in the plane. The centroid for R is defined by two points $\bar{x}$, $\bar{y}$. In order to find these two points, we first must obtain the moments of R, M_x, and M_y, which are given by the formulas

$$M_x = \int_R \int x \, dx \, dy,$$

$$M_y = \int_R \int y \, dx \, dy.$$

Let $A(R)$ be the area of R. Then the centroid of R is given by

$$\bar{x} = \frac{M_x}{A(R)}, \qquad \bar{y} = \frac{M_y}{A(R)}.$$

We now obtain explicit expressions for the moments when R is a finite sum of rectangles. Suppose that R is a simple rectangle as pictured in Figure 1–23. Then

$$M_x = \int_{y_1}^{y_2} dy \int_{x_1}^{x_2} x \, dx = \int_{y_1}^{y_2} dy \left. \frac{x^2}{2} \right|_{x_1}^{x_2}$$

$$= \int_{y_1}^{y_2} \frac{dy(x_2^2 - x_1^2)}{2} = \frac{x_2^2 - x_1^2}{2}(y_2 - y_1),$$

$$M_y = \int_{x_1}^{x_2} dx \int_{y_1}^{y_2} y\, dy = \int_{x_1}^{x_2} dx \left.\frac{y^2}{2}\right|_{y_1}^{y_2}$$

$$= \int_{x_1}^{x_2} \frac{dx(y_2^{\,2} - y_1^{\,2})}{2} = \frac{y_2^{\,2} - y_1^{\,2}}{2}(x_2 - x_1).$$

Note that because the area of the rectangle is $(x_2 - x_1)(y_2 - y_1)$, and $x_2^2 - x_1^2 = (x_2 - x_1)(x_1 + x_2)$ (and similarly for $y_1^2 - y_2^2$), we obtain

$$\bar{x} = \frac{x_1 + x_2}{2}, \qquad \bar{y} = \frac{y_1 + y_2}{2}.$$

The formulas for the moments of a rectangle may be used to find the centroid when R consists of a collection of rectangles as well. Suppose that R can be subdivided into k rectangles labeled $R_1, R_2, \ldots, R_k$ with respective boundaries defined by $[(x_{1i}, x_{2i}), (y_{1i}, y_{2i})]$ for $1 \leq i \leq k$. Since

$$M_x = \int_R \int x\, dx\, dy = \sum_{i=1}^{k} \int_{R_i} \int x\, dx\, dy,$$

it follows that

$$M_x = \sum_{i=1}^{k} \frac{x_{2i}^{\,2} - x_{1i}^{\,2}}{2}(y_{2i} - y_{1i}).$$

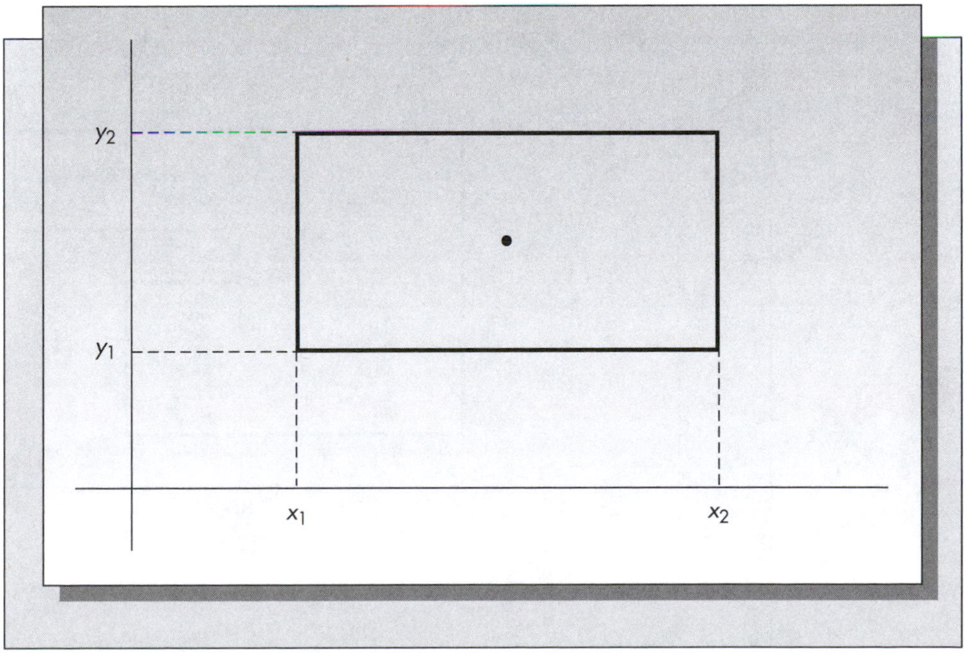

Figure 1–23 The centroid of a rectangle

Similarly,

$$M_y = \sum_{i=1}^{k} \frac{y_{2i}^2 - y_{1i}^2}{2}(x_{2i} - x_{1i}).$$

EXAMPLE 1.4

Consider the region R pictured in Figure 1–24. We will determine the centroid of the region using the given formulas. The region can be broken down into three rectangles in several different ways. For the one pictured in Figure 1–24, we have

$$\begin{aligned} x_{11} &= 10, & x_{21} &= 30, \\ y_{11} &= 5, & y_{21} &= 35, \\ x_{12} &= 30, & x_{22} &= 40, \\ y_{12} &= 5, & y_{22} &= 20, \\ x_{13} &= 30, & x_{23} &= 50, \\ y_{13} &= 25, & y_{23} &= 35. \end{aligned}$$

Substituting into the given formulas, we obtain

$$\begin{aligned} M_x &= \frac{30^2 - 10^2}{2}(35 - 5) + \frac{40^2 - 30^2}{2}(20 - 5) + \frac{50^2 - 30^2}{2}(35 - 25) \\ &= (400)(30) + (350)(15) + (800)(10) = 25{,}250, \\ M_y &= \frac{35^2 - 5^2}{2}(30 - 10) + \frac{20^2 - 5^2}{2}(40 - 30) + \frac{35^2 - 25^2}{2}(50 - 30) \\ &= (600)(20) + (187.5)(10) + (300)(20) = 19{,}875. \end{aligned}$$

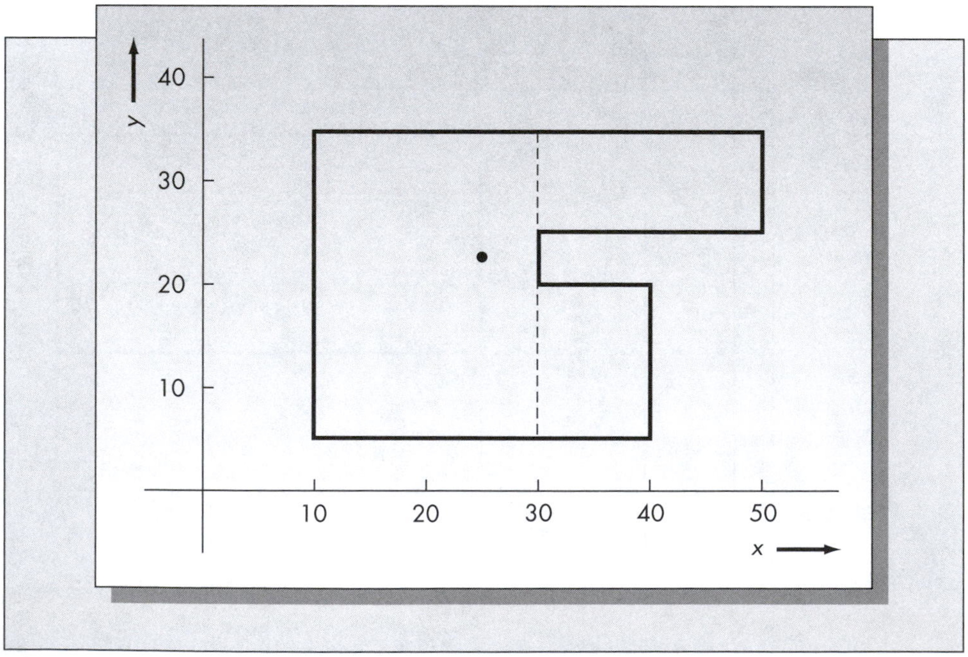

Figure 1–24 Centroid of a figure composed of rectangles

The total area of R is

$$A(R) = (20)(30) + (10)(15) + (20)(10) = 950.$$

It follows that the centroid is

$$\bar{x} = \frac{25,250}{950} = 26.579,$$

$$\bar{y} = \frac{19,875}{950} = 20.921.$$

The centroid is marked with a dot on Figure 1–24.

The accuracy of assuming that a department is located at its centroid depends upon the shape of the department. The assumption is most accurate when the shape of the department is square or rectangular but is less accurate for oddly shaped departments. Consider the following example.

EXAMPLE 1.5

A local manufacturing firm has recently completed construction of a new wing of an existing building to house four departments: A, B, C, and D. The wing is 100 feet by 50 feet. The plant manager has chosen an initial layout of the four departments. This layout appears in Figure 1–25. We have marked the centroid locations of the departments with a dot. From the figure we see that department A requires 1,800 square feet, B requires 1,200 square feet, C requires 800 square feet, and D requires 1,200 square feet.

One of the inputs required for the layout decision is the flow data—that is, the number of materials handling trips per unit time from every department to every other department. These data are given in the from-to chart appearing in Figure 1–26(a). The distance between departments is assumed to be the rectilinear distance between

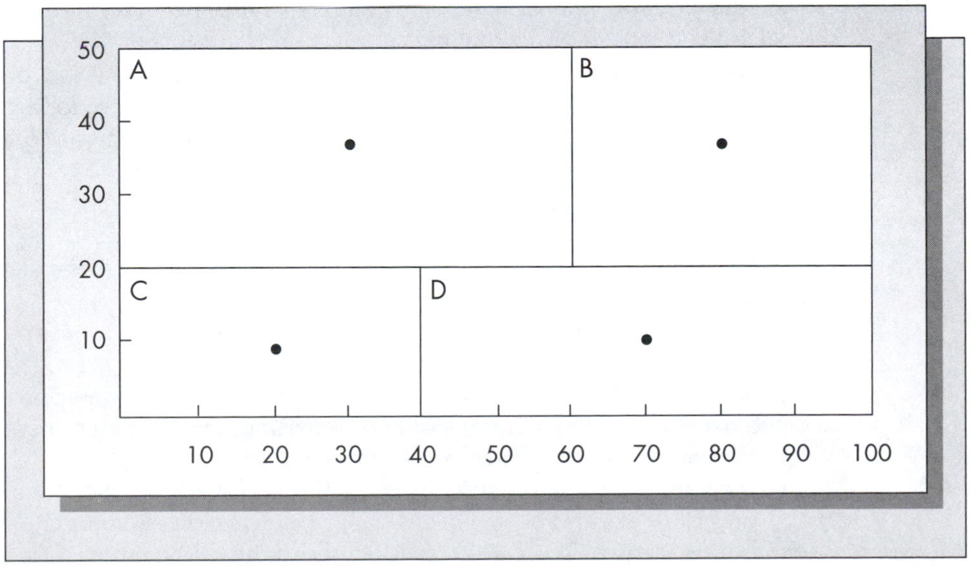

Figure 1–25 Initial layout (example 1.5)

From \ To	A	B	C	D
A		2	7	4
B	3		5	5
C	6	7		3
D	8	2	3	

(a) Number of materials handling trips per hour

From \ To	A	B	C	D
A		50	35	65
B	50		85	35
C	35	85		50
D	65	35	50	

(b) Rectilinear distance between departments in initial layout

From \ To	A	B	C	D	Totals
A		100	245	260	605
B	150		425	175	750
C	210	595		150	955
D	520	70	150		740
Totals:	880	765	820	585	3,050

(c) Total distance traveled per hour between departments in initial layout

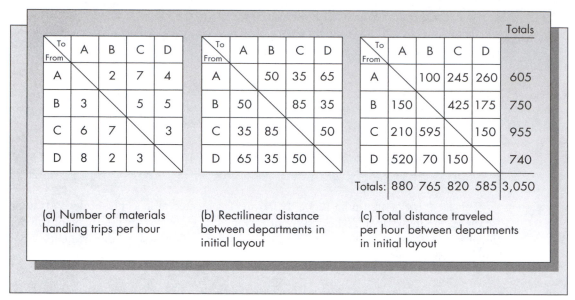

Figure 1–26 From-to charts for initial layout

centroid locations. From figure 1–25, we see that the centroid locations of the initial layout are

$$(x_A, y_A) = (30, 35), \qquad (x_C, y_C) = (20, 10),$$
$$(x_B, y_B) = (80, 35), \qquad (x_D, y_D) = (70, 10).$$

The rectilinear distance between A and B, for example, is given by the formula

$$|x_A - x_B| + |y_A - y_B| = |30 - 80| + |35 - 35| = 50.$$

One computes distances between other pairs of departments in a similar way. The calculations are summarized in the from-to chart in Figure 1–26(b). Notice that we have assumed that the distance from department i to department j is the same as the distance from department j to department i. This may not necessarily be true if material is transported in one direction only.

Normally, a third from-to chart also would be required. This would give the cost of transporting a unit load a unit distance from department i to department j. We assume we wish to minimize the product of flows and distances by assigning a value of 1 to these transport costs. Hence, one obtains the hourly cost of transporting materials to and from each of the departments by simply multiplying the entries in the from-to chart of Figure 1–26(a) by the entries in the from-to chart of Figure 1–26(b). These calculations are summarized in Figure 1–26(c). The total distance traveled per hour for the initial layout is 3,050 feet.

We now consider the effect of pairwise interchanges of two departments that either have adjacent borders or have the same area. The result of the exchange is determined by exchanging the location of the centroids. This is only an approximation, however, since exchanging the locations of two departments does *not* necessarily mean that the location of their centroids will be exchanged.

If we were to exchange the locations of A and B, for example, we would assume that $(x_A, y_A) = (80, 35)$ and $(x_B, y_B) = (30, 35)$, which would result in the new from-to distance chart appearing in Figure 1–27(a). Multiplying the original flow data in Figure 1–26(a) by this distance chart gives a new cost chart, which appears in Figure 1–27(b). Interchanging

(a)

From \ To	A	B	C	D
A		50	85	35
B	50		35	65
C	85	35		50
D	35	65	50	

(b)

From \ To	A	B	C	D	Totals
A		100	595	140	835
B	150		175	325	650
C	510	245		150	905
D	280	130	150		560
Totals	940	475	920	615	2,950

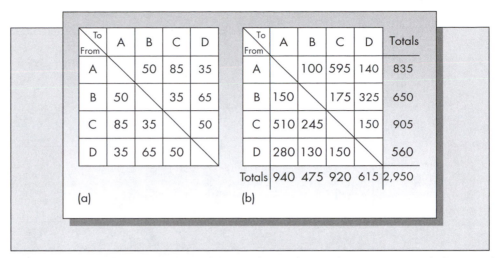

Figure 1–27 New distance and cost from-to charts after exchanging centroids for A and B

the centroids of A and B results in the predicted cost reduction from 3,050 to 2,950, or about 3 percent. (The actual cost of interchanging the locations of A and B would be slightly different because the centroids are not exactly exchanged.)

We can consider all pairwise interchanges of adjacent departments or departments with identical areas and pick the one that results in the largest decrease in the predicted cost. We will not present the details of the calculations but merely summarize the results. Exchanging the centroids of A and C results in a total predicted cost of 2,715, and exchanging the centroids of A and D results in a total predicted cost of 3,185. Two other exchanges must be considered as well: B and D, and C and D. Notice that exchanging B and C is not considered because they do not have equal areas and do not share a common border. Exchanging the centroids of B and D results in a total predicted cost of 2,735, and exchanging the centroids of C and D in a total predicted cost of 2,830.

The maximum predicted cost reduction is achieved by exchanging A and C. The new layout with A and C exchanged appears in Figure 1–28. Because C has the smaller area, it is placed in the upper left-hand corner of the space formerly occupied by A, so that the remaining space allows A to be contiguous. Notice that A is no longer rectangular. The actual cost of the new layout is not necessarily equal to the predicted value of 2,715. The centroid of A is computed using the method outlined above. It is determined by first finding the moments M_x and M_y given by

$$M_x = (40^2 - 0)(30 - 0)/2 + (60^2 - 40^2)(50 - 20)/2 = 54{,}000,$$
$$M_y = (30^2 - 0)(40 - 0)/2 + (50^2 - 20^2)(60 - 40)/2 = 39{,}000,$$

and dividing by the area of A to obtain

$$x_A = 54{,}000/1{,}800 = 30,$$
$$y_A = 39{,}000/1{,}800 = 21.66667.$$

The centroid of C is located at the center of symmetry, which is

$$(x_C, y_C) = (20, 40).$$

The centroids for B and D are unchanged. The cost of the new layout is actually 2,810, which is somewhat more than the 2,715 predicted at the last step but is still less than the original cost. The process is continued until predicted cost reductions are no

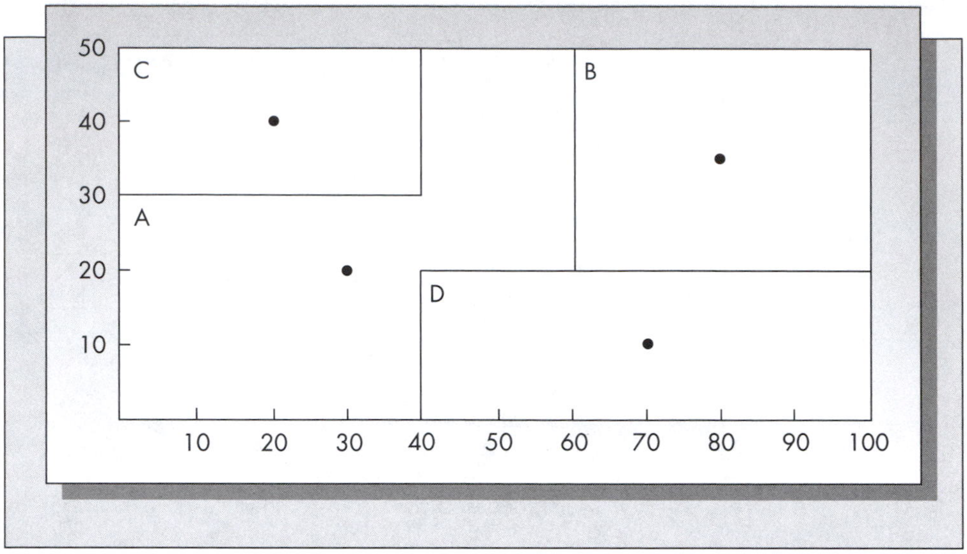

Figure 1–28 New layout with A and C interchanged

longer possible. At this point only three exchanges need to be considered: A and B, A and D, and B and D. Obviously, we need not consider exchanging A and C, and C and D may not be exchanged because they do not share a common border. The predicted cost resulting from exchanging the centroids of A and B is 2,763.33; of A and D, 3,641.33; and of B and D, 2,982. Clearly, the greatest predicted reduction is now achieved by exchanging the locations of A and B.

The new layout is pictured in Figure 1–29. The centroids for the respective departments are now

$$(x_A, y_A) = (70, 35), \qquad (x_C, y_C) = (20, 40),$$
$$(x_B, y_B) = (20, 15), \qquad (x_D, y_D) = (70, 10).$$

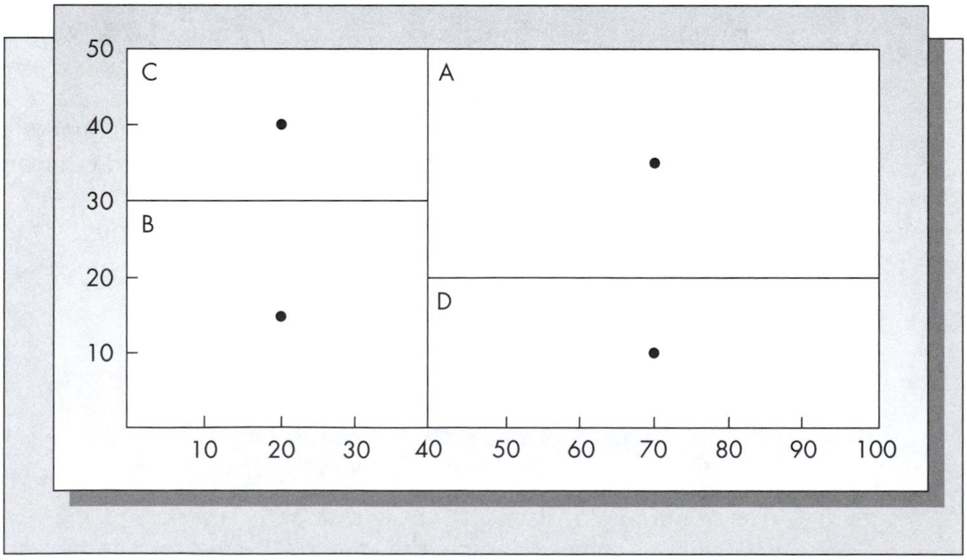

Figure 1–29 Second iteration, obtained from exchanging the locations of A and B

The actual cost of this layout is 2,530, which is considerably *less* than that predicted in the previous step. The process continues until no further reductions in the predicted costs can be achieved. Because A and B were exchanged at the previous step, that exchange need not be considered again. Exchanging A and C results in a predicted cost of 3,175; exchanging A and D, a predicted cost of 2,753; and exchanging B and D, a predicted cost of 3,325. (We do not consider exchanging C and D at this stage because they are not adjacent and do not have equal areas.) Since none of the predicted costs is less than the current cost, we terminate calculations. The final layout, pictured in Figure 1–29, required two iterations and resulted in a reduction of total distance traveled from 3,050 feet to 2,530 feet, or about 17 percent.

BIBLIOGRAPHY

Abernathy, W. J., and P. L. Townsend. "Technology, Productivity, and Process Change." *Technological Forecasting and Social Change* 7 (1975), pp. 379–96.

Abernathy, W. J., and K. Wayne. "Limits of the Learning Curve." *Harvard Business Review* 52 (September–October 1974), pp. 109–19.

Blackburn, J. D. *Time-Based Competition: The Next Battleground in American Manufacturing.* New York: McGraw-Hill/Irwin, 1991.

Burbidge, J. L. *The Introduction of Group Technology.* London: Heinemann, 1975.

Clark, C. *The Conditions of Economic Progress.* London: MacMillan & Co. Ltd., 1940 (revised and reprinted in 1951).

Cooper, W. W., L. M. Seiford, and K. Tone. *Data Envelopment Analysis*, New York: Springer 2007.

Davis, D. "Beating the Clock." *Electronic Business*, May 1989.

Dayen, D. "The Cause and Consequences of the Retail Apocalypse." *The New Republic,* November 15, 2017. Retrieved from https://newrepublic.com/article/145813/cause-consequences-retail-apocalypse

Devinney, T. M. "Entry and Learning." *Management Science* 33 (1987), pp. 102–12.

Dhalla, N. K., and S. Yuspeh. "Forget about the Product Life Cycle Concept." *Harvard Business Review* 54 (January–February 1976), pp. 102–12.

The Economist. "Reshoring Manufacturing: Coming Home." January 2013. Retrieved from http://www.economist.com/news/special-report/21569570-growing-number-americancompanies-are-moving-their-manufacturing-back-united

Fallon, M. "Lack of Strategy Hampered Macintosh, Executive Says." *San Jose Mercury News*, September 30, 1985, p. 11D.

Fisher, A. G. B. *The Clash of Progress and Security.* London: MacMillan & Co. Ltd.,1935.

FocusEconomics. "U.S. Economic Outlook." May 28, 2019. Retrieved from https://www.focus-economics.com/countries/united-states

Fortune. "How Jack Welch Keeps the Ideas Coming at GE." August 13, 1991.

Fourastié, J. *Le Grand Espoir du XXe Siècle.* Paris: Presses Universitaires de France, 1949.

Francis, R. L., and J. A. White. *Facility Layout and Location: An Analytical Approach.* Englewood Cliffs, NJ: Prentice Hall, 1974.

Goldhar, J. P., and M. Jelinek. "Plan for Economies of Scope." *Harvard Business Review* 61 (1983), pp. 141–48.

Groover, M. P., and E. W. Zimmers. *CAD/CAM: Computer Aided Design and Manufacturing.* Englewood Cliffs, NJ: Prentice Hall, 1984.

Hammer, M. S., and J. Champy. *Reengineering the Corporation: A Manifesto for Business Resolution.* New York: Harper Business, 1993.

Hammer, M. S., and S. A. Stanton. *The Reengineering Revolution.* New York: Harper Business, 1995.

Hayes, R. H., and S. Wheelwright. "Link Manufacturing Profess and Product Life Cycles." *Harvard Business Review* 57 (January–February 1979), pp. 133–40.

Hayes, R. H., and S. Wheelwright. *Restoring Our Competitive Edge: Competing through Manufacturing*. New York: John Wiley & Sons, 1984.

Hill, T. *Manufacturing Strategy: The Strategic Management of the Manufacturing Function*. Macmillan, 1993.

Holweg, M. "The Genealogy of Lean Production". *Journal of Operations Management* 25 (2) (2007), pp. 420–437.

Jones, P., and L. Kahaner. *Say It and Live It. The Fifty Corporate Mission Statements That Hit the Mark*. New York: Currency Doubleday, 1995.

Kahaney, L. "Inside Look at Birth of the iPod." *Wired*, July 21, 2004. Retrieved from https://www.wired.com/2004/07/inside-look-at-birth-of-the-ipod/

Krafcik, John F. "Triumph of the Lean Production System". *Sloan Management Review*. 30 (1) (1988), pp. 41–52.

Krugman, P. *Peddling Prosperity: Economic Sense and Nonsense in the Age of Diminished Expectations*. New York: W. W. Norton and Company, 1994.

Miller, J. G.; A. D. Meyer; and J. Nakane. *Benchmarking Global Manufacturing*. New York: McGraw-Hill/Irwin, 1992.

Noyce, R. N. "Microelectronics." *Scientific American*, September 1977.

Pagell, M., S. Melnyk, and R. Handfield, "Do Trade-offs Exist in Operations Strategy? Insights from the Stamping Die Industry." *Business Horizons* 43 (3) (2000), pp. 69-77.

Panzer, J. C., and R. O. Willig. "Economies of Scope." *American Economic Review*, 1981, pp. 268–72.

Petro, G.. "The Future of Fashion Retailing: The Zara Approach." *Forbes,* October 25, 2012.

Port, O. "Customers Move into the Driver's Seat." *Business Week*, October 4, 1999, pp. 103–06.

Porter, M. E. "The Competitive Advantage of Nations." *Harvard Business Review* (March-April 1990).

Porter, M. E. and J. W. Rivkin. "Choosing the United States." *Harvard Business Review* (March 2012), pp. 80–93.

Rogers, E. M. *Diffusion of innovations*. New York: Free Press of Glencoe. 1962.

Sassani, F. "A Simulation Study on Performance Improvement of Group Technology Cells." *International Journal of Production Research* 28 (1990), pp. 293–300.

Shen, Y. "Selection Incentives in a Performance-Based Contracting System." *Health Services Research* 38 (2003), pp. 535–552.

Skinner, W. "The Focused Factory." *Harvard Business Review* (May 1974).

Smith, A. *An Inquiry into the Nature and Causes of the Wealth of Nations*. London: George Bell & Sons, 1896, p. 9.

Statista, "Apple iPad Sales Worldwide, 2010–2018," 2019. Retrieved from https://www.statista.com/statistics/269915/global-apple-ipad-sales-since-q3-2010/

Statista, "Apple Product Sales (iPhone, iPad & iPod) 2009–2018," 2019. Retrieved from https://www.statista.com/statistics/253725/iphone-ipad-and-ipod-sales-comparison/

Stecke, K. E. "Formulation and Solution of Nonlinear Integer Production Planning Problems for Flexible Manufacturing Systems." *Management Science* 29 (1983), pp. 273–88.

U.S. Bureau of Labor Statistics. "Productivity Change in the Manufacturing Sector, 1987–2018." March 8, 2019. Retrieved from https://www.bls.gov/lpc/prodybar.htm

Vandermerwe, S. and J. Rada. "Servitization of Business: Adding Value by Adding Services." *European Management Journal* 6 (1988), pp. 314–324.

Vollman, T. E., and E. S. Buffa. "The Facilities Layout Problem in Perspective." *Management Science* 12 (1966), pp. 450–68.

Wise, R. and P. Baumgartner. "Go Downstream: The New Profit Imperative in Manufacturing." *Harvard Business Review* 77 (1999), pp. 133–141.

Womack, J. P., D. T. Jones, and D. Roos. *The Machine That Changed the World*. New York: Harper Perennial, 1990.

The World Bank. "Employment in Services." April 2019. Retrieved from https://data.worldbank.org/indicator/SL.SRV.EMPL.ZS

CHAPTER 2

Forecasting

"It's hard to make predictions, especially about the future."

—Neils Bohr

CHAPTER OVERVIEW

Purpose

To present and illustrate the most important methods for forecasting demand in the context of operations planning.

KEY POINTS

1. *Characteristics of forecasts.*
   - They are almost always going to be wrong.
   - A good forecast also gives some measure of error.
   - Forecasting aggregate units is generally easier than forecasting individual units.
   - Forecasts made further out into the future are less accurate.
   - A forecasting technique should not be used to the exclusion of known information.
2. *Subjective forecasting.* These methods measure either individual or group opinion. The better known subjective forecasting methods include sales force composites, customer surveys, jury of executive opinion, and the Delphi method.
3. *Objective forecasting methods (time series methods and regression).* Using *objective forecasting* methods, one makes forecasts based on past history. *Time series* forecasting uses only the past history of the series to be forecasted, while *regression models* often incorporate the past history of other series. In time series forecasting, the goal is to find predictable and repeatable patterns in past data. Based on the identified pattern, different methods are appropriate. Time series methods have the advantage of easily being incorporated into a computer program for automatic forecasting and updating. Repeatable patterns that we consider include increasing or decreasing linear trend, curvilinear trend (including exponential growth), and seasonal fluctuations. When using regression, one constructs a causal model that predicts one phenomenon (the dependent variable) based on the evolution of one or more other phenomenon (the independent variables). An example would be predicting the start or end of a recession based on housing starts (housing starts are a leading economic indicator of the health of the economy).
4. *Evaluation of forecasting methods.* The forecast error in any period, e_t, is the difference between the forecast for period t and the actual value of the series realized for period t ($e_t = F_t - D_t$). Three common measures of forecast error are MAD (average of the absolute errors over n periods), MSE (the average of the sum of the

squared errors over n periods), and MAPE (the average of the percentage errors over n periods).

5. *Methods for forecasting stationary time series.* We consider two forecasting methods when the underlying pattern of the series is stationary over time: moving averages and exponential smoothing. A *moving average* is simply the arithmetic average of the N most recent observations. *Exponential smoothing* forecasts rely on a weighted average of the most recent observation and the previous forecast. The weight applied to the most recent observation is α, where $0 < \alpha < 1$, and the weight applied to the last forecast is $1 - \alpha$. Both methods are commonly used in practice, but the exponential smoothing method is favored in inventory control applications—especially in large systems—because it requires much less data storage than does moving averages.
6. *Methods for forecasting series with trend.* When there is an upward or downward linear trend in the data, two common forecasting methods are *linear regression* and double exponential smoothing via *Holt's method.* Linear regression is used to fit a straight line to past data based on the method of least squares, and Holt's method uses separate exponential smoothing equations to forecast the intercept and the slope of the series each period.
7. *Methods for forecasting seasonal series.* A seasonal time series is one that has a regular repeating pattern over the same time frame. Typically, the time frame would be a year, and the periods would be weeks or months. The simplest approach for forecasting seasonal series is based on multiplicative seasonal factors. A multiplicative seasonal factor is a number that indicates the relative value of the series in any period compared to the average value over a year. Suppose a season consists of 12 months. A seasonal factor of 1.25 for a given month means that the demand in that month is 25 percent higher than the mean monthly demand. *Winters's method* is a more complex method based on triple exponential smoothing. Three distinct smoothing equations are used to forecast the intercept, the slope, and the seasonal factors each period.
8. *Box-Jenkins models.* George Box and Gwilym Jenkins developed forecasting methods based largely on a statistical analysis of the autoregressive function of a time series. Autoregression seeks to discover repeating patterns in data by considering the correlation of observations of the series with other observations separated by a fixed number of periods. These models have proven to be very powerful for forecasting some economic time series, but they require large data sets (at least 72 observations) and a knowledgeable user. We provide a brief review of these powerful methods.
9. *Other considerations.* When large data sets are available, filtering methods borrowed from electrical engineering can often provide excellent forecasts for economic time series (in addition to Box-Jenkins methods). Two of the better known filters are Kalman Filters and Wiener Filters. Neither of these methods are amenable to automatic forecasting. Monte Carlo simulation is another technique that can be useful for building a forecasting model. Finally, we discuss forecasting demand in the context of a lost sales inventory system.

As Charles Kettering eloquently stated, "My concern is with the future since I plan to spend the rest of my life there." But the future can never be known, so we make forecasts. We forecast traffic patterns and plan routes accordingly. We forecast which foods will be best in a particular restaurant, and order accordingly. We choose universities to attend based on forecasting our experiences there and the doors that a degree from that university will open. We make hundreds of forecasts every day, some carefully thought out, some made almost unconsciously. Forecasting plays a central role in all of our lives.

In the same way, forecasting plays a central role in the operations function of a firm. All business planning is based on forecasts. Sales of existing and new products, requirements and availabilities of raw materials, changing skills of workers, interest rates, capacity requirements, and international politics are only a few of the factors likely to affect the future success of a firm.

The functional areas of businesses that make the most use of forecasting methods are marketing and production. Marketing is responsible for forecasting sales of both new and existing product lines. Sales forecasts are the primary driver for sales and operations planning (S&OP), which will be discussed in detail in Chapter 3. In some circumstances, the forecasts prepared for marketing purposes may not be sufficient for operations planning. For example, to determine suitable stocking levels for spare parts, one must know schedules for planned replacements and be able to forecast unplanned replacements. Also, it could be that the S&OP planning function might be producing forecasts for aggregate units, while forecasts for individual SKUs (stock keeping units) might be required.

Firms benefit from good forecasting and pay the price for poor forecasting. Forecasting played a role in Ford Motor Company's early success and later demise. Henry Ford saw that the consumer wanted a simpler, less expensive car that was easier to maintain than most manufacturers were offering in the early 1900s. His Model T dominated the industry. However, Ford did not see that consumers would tire of the open Model T design. Ford's failure to forecast consumer desires for other designs nearly resulted in the end of a firm that had monopolized the industry only a few years before. During the 1960s, consumer tastes in automobiles slowly shifted from large, heavy gas guzzlers to smaller, more fuel-efficient automobiles. Detroit, slow to respond to this change, suffered when the OPEC oil embargo hit in the late 1970s and tastes shifted more dramatically to smaller cars. More recently, sales of hybrid vehicles also exceeded forecasts and in 2012 Toyota had a 120,000-car backlog of orders for its Prius. Tesla has also made headlines for it backlog but this has reportedly been caused by its problems with production capacity rather than a failure to forecast.

Seeing trends is the first step toward profiting from those trends. As an example, consider the trend towards greater use of renewable energy. Renewable energy sources include wind power, sun power, tidal power, geothermal power, etc. If energy can be generated and stored using renewable methods, this energy can be used to power electric cars, thus cutting down on gasoline consumption.

Some companies were able to see this trend and take advantage of it. In particular, the use of solar cells has grown dramatically in recent years. While Apple Corporation received a great deal of publicity for its fantastic successes in mobile computing, solar cell installations grew at a comparable rate. In the last decade, solar experienced an average annual growth rate in the United States of 50%. Globally, 89% of the world's solar power has been installed since 2011. China is the world's leader in solar power. Manufacturers that saw this trend developing are now reaping the rewards of their foresight.

Can all events be accurately forecasted? The answer is clearly no. Consider the experiment of tossing a coin. Assuming that it is a fair coin and the act of tossing does not introduce bias, the best you can say is that the probability of getting heads is 50 percent on any single toss. No one has been able to consistently top the 50 percent prediction rate for such an experiment over a long period of time. Many real phenomena are accurately described by a type of coin-flipping experiment. Games of chance played at casinos are random. By tipping the probabilities in its favor, the house is always guaranteed to win over the long term. There is evidence that daily prices of stocks follow a purely random process, much like a coin-flipping experiment. Studies have shown that professional money managers rarely outperform stock portfolios generated purely at random.

In production and operations management, we are primarily interested in forecasting product demand. Because demand is likely to be random in most circumstances, can forecasting methods provide any value? In most cases, the answer is yes. Although some portions of the demand process may be unpredictable, other portions may be predictable. Trends, cycles, and seasonal variation may be present, all of which give us an advantage over trying to predict the outcome of a coin toss. In this chapter we consider methods for predicting future values of a series based on past observations.

2.1 THE TIME HORIZON IN FORECASTING

We may classify forecasting problems along several dimensions. One is the time horizon. Figure 2–1 is a schematic showing the three time horizons associated with forecasting and typical forecasting problems encountered in operations planning associated with each. Short-term forecasting is crucial for day-to-day planning. Short-term forecasts, typically measured in days or weeks, are required for inventory management, production plans that may be derived from a materials requirements planning system (discussed in detail in Chapter 9), and resource requirements planning. Shift scheduling may require forecasts of workers' availabilities and preferences.

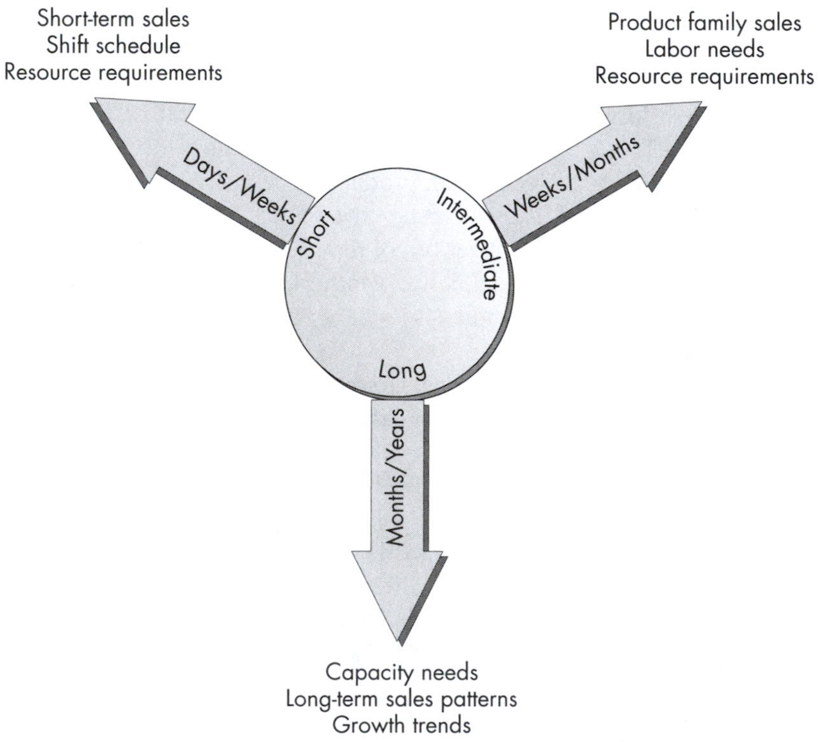

Figure 2–1 Forecast horizons in operation planning

The intermediate term is measured in weeks or months. Sales patterns for product families, requirements and availabilities of workers, and resource requirements are typical intermediate-term forecasting problems encountered in operations management.

Long-term production and manufacturing decisions, discussed in Chapter 1, are part of the overall manufacturing strategy of a firm. One example is long-term planning of capacity needs. When demands are expected to increase, the firm must plan for the construction of new facilities and/or the retrofitting of existing facilities with new technologies. Capacity planning decisions may require downsizing in some circumstances. For example, General Motors (GM) historically commanded about 45 percent of the U.S. car market. The percentage dropped to 28% in 2000 and 17% in 2018. As a result, GM was forced to significantly curtail its manufacturing operations to remain profitable.

2.2 CHARACTERISTICS OF FORECASTS

1. *They are usually wrong.* As strange as it may sound, this is probably the most ignored and most significant property of almost all forecasting methods. Forecasts, once determined, are often treated as known information. Resource requirements and production schedules may require modifications if the forecast of demand proves to be inaccurate. The planning system should be sufficiently robust to be able to react to unanticipated forecast errors.
2. *A good forecast* is more than a single number. Given that forecasts are generally wrong, a good forecast also includes some measure of the anticipated forecast error. This could be in the form of a range, or an error measure such as the variance of the distribution of the forecast error.
3. *Aggregate forecasts are more accurate.* Recall from statistics that the variance of the average of a collection of independent identically distributed random variables is lower than the variance of each of the random variables; that is, the variance of the sample mean is smaller than the population variance. This same phenomenon applies to forecasting. On a percentage basis, the error made in forecasting sales for an entire product line is generally less than the error made in forecasting sales for an individual item. This phenomenon, known as risk pooling, will be discussed in the inventory control context in Chapters 5 and 6.
4. *The longer the forecast horizon, the less accurate the forecast will be.* This property is quite intuitive. One can predict tomorrow's value of the Dow Jones Industrial Average more accurately than next year's value.
5. *Forecasts should not be used to the exclusion of known information.* A particular technique may result in reasonably accurate forecasts in most circumstances. However, there may be information available concerning the future demand that is not presented in the past history of the series. For example, the company may be planning a special promotional sale for a particular item so that the demand will probably be higher than normal. This information must be manually factored into the forecast.

2.3 SUBJECTIVE FORECASTING METHODS

We classify forecasting methods as either **subjective** or **objective**. A subjective forecasting method is based on human judgment. There are several techniques for soliciting opinions for forecasting purposes.

1. *Sales force composites*. In forecasting product demand, a good source of subjective information is the company sales force. The sales force has direct contact with consumers and is therefore in a good position to see changes in their preferences. To develop a sales force composite forecast, members of the sales force submit sales estimates of the products they will sell in the coming year. These estimates might be individual numbers, or there might be a range of numbers providing pessimistic, most likely, and optimistic estimates. Sales managers would then be responsible for aggregating individual estimates to arrive at overall forecasts for each geographic region or product group. Sales force composites may be inaccurate when compensation of sales personnel is based on meeting a quota. In that case, there is clearly an incentive for the sales force to lowball estimates.
2. *Customer surveys*. Customer surveys can signal future trends and shifting preference patterns. To be effective, however, surveys and sampling plans must be carefully designed to guarantee that the resulting data are statistically unbiased and representative of the customer base. Poorly designed questionnaires or an invalid sampling scheme may result in the wrong conclusions.
3. *Jury of executive opinion*. When there is no past history, as with new products, expert opinion may be the only source of information for preparing forecasts. The approach here is to combine the opinions of experts systematically to derive a forecast. For new product planning, opinions of personnel in the functional areas of marketing, finance, and operations should be solicited. Combining individual forecasts may be done in several ways. One is to have the individual responsible for preparing the forecast interview the executives directly and develop a forecast from the results of the interviews. Another is to require the executives to meet as a group and come to a consensus.
4. *The Delphi method*. The **Delphi method**, like the jury of executive opinion method, is based on soliciting the opinions of experts. The difference lies in the manner in which individual opinions are combined. (The method is named for the Delphic oracle of ancient Greece, who purportedly had the power to predict the future.) The Delphi method attempts to eliminate some of the inherent shortcomings of group dynamics, in which the personalities of some group members overshadow those of other members. The method requires a group of experts to express their opinions, preferably by individual sample survey. The opinions are then compiled and a summary of the results is returned to the experts, with special attention to those opinions that are significantly different from the group averages. The experts are asked if they wish to reconsider their original opinions in light of the group response. The process is repeated until (ideally) an overall group consensus is reached.

As with any particular technique, the Delphi method has advantages and disadvantages. Its primary advantage is that it provides a means of assessing individual opinion without the usual concerns of personal interactions. On the negative side, the method is highly sensitive to the care in the formulation of the questionnaire. Because discussions are intentionally excluded from the process, the experts have no mechanism for resolving ambiguous questions. Furthermore, it is not necessarily true that a group consensus will ever be reached. Olaf Helmer and Nicholas Rescher (1959) provide one of the earliest descriptions of the Delphi method.

2.4 OBJECTIVE FORECASTING METHODS

Objective forecasting methods are those in which the forecast is derived from an analysis of data. A **time series** method is one that uses only past values of the phenomenon we are predicting. **Causal models** are ones that use data from sources other than the series being predicted; that is, there may be other variables with values that are *linked* in some way to what is being forecasted. We discuss these first.

Causal Models

Let Y represent the phenomenon we wish to forecast and $X_1, X_2, \ldots, X_n$ be n variables that we believe to be related to Y. Then a causal model is one in which the forecast for Y is some function of these variables, say,

$$Y = f(X_1, X_2, \ldots, X_n).$$

Econometric models are special causal models in which the relationship between Y and $(X_1, X_2, \ldots, X_n)$ is linear. That is,

$$Y = \alpha_0 + \alpha_1 X_1 + \alpha_2 X_2 + \ldots + \alpha_n X_n$$

for some constants $(\alpha_1, \ldots, \alpha_n)$. The method of least squares is most commonly used to find estimators for the constants. (We discuss the method in Appendix 2–B for the case of one independent variable.)

Consider a simple example of a causal forecasting model. A realtor is trying to estimate his income for the succeeding year. In the past he has found that his income is close to being proportional to the total number of housing sales in his territory. He also has noticed that there has typically been a close relationship between housing sales and interest rates for home mortgages. He might construct a model of the form

$$Y_t = \alpha_0 + \alpha_1 X_{t-1},$$

where Y_t is the number of sales in year t and X_{t-1} is the interest rate in year $t - 1$. Based on past data he would then determine the least squares estimators for the constants α_0 and α_1. Suppose that the values of these estimators are currently $\alpha_0 = 385.7$ and $\alpha_1 = 1{,}878$. Hence, the estimated relationship between home sales and mortgage rates is

$$Y_t = 385.7 - 1{,}878X_{t-1},$$

where $X_{t-1,}$ the previous year's interest rate, is expressed as a decimal. Then if the current mortgage interest rate is 10 percent, the model would predict that the number of sales the following year in his territory would be 385.7 – 187.8 = 197.9, or about 198 houses sold.

Causal models of this type are common for predicting economic phenomena such as the gross national product (GNP) and the gross domestic product (GDP). Both MIT and the Wharton School of Business at the University of Pennsylvania have developed large-scale econometric models for making these predictions. Econometric prediction models are typically used by the economics and finance arms of the firm to forecast values of macroeconomic variables such as interest rates and currency exchange rates. Time series methods are more commonly used for operations planning applications.

Time Series Methods

Time series methods are often called naive methods, as they require no information other than the past values of the variable being predicted. *Time series* is just a fancy term for a collection of observations of some economic or physical phenomenon drawn at discrete points in time, usually equally spaced. The idea is that information can be inferred from the pattern of past observations and can be used to forecast future values of the series.

In time series analysis we attempt to isolate the patterns that arise most often, which include the following.

1. *Trend.* Trend refers to the tendency of a time series to exhibit a stable pattern of growth or decline. We distinguish between linear trend (the pattern described by a straight line) and nonlinear trend (the pattern described by a nonlinear function, such as a quadratic or exponential curve). When the pattern of trend is not specified, it is generally understood to be linear.
2. *Seasonality.* A seasonal pattern is one that repeats at fixed intervals. In time series we generally think of the pattern repeating every year, although daily, weekly, and monthly seasonal patterns are common as well. Fashion wear, ice cream, and heating oil exhibit a yearly seasonal pattern. Consumption of electricity exhibits a strong daily seasonal pattern.
3. *Cycles.* Cyclic variation is similar to seasonality, except that the length and the magnitude of the cycle may vary. One associates cycles with long-term economic variations (that is, business cycles) that may be present in addition to seasonal fluctuations.
4. *Randomness.* A pure random series is one in which there is no recognizable pattern to the data. One can generate patterns purely at random that often appear to have structure. An example of this is the methodology of stock market chartists who impose forms on random patterns of stock market price data. On the other side of the coin, data that appear to be random could have a very definite structure. Truly random data that fluctuate around a fixed mean form what is called a horizontal pattern.

Examples of time series exhibiting some of these patterns are given in Figure 2–2.

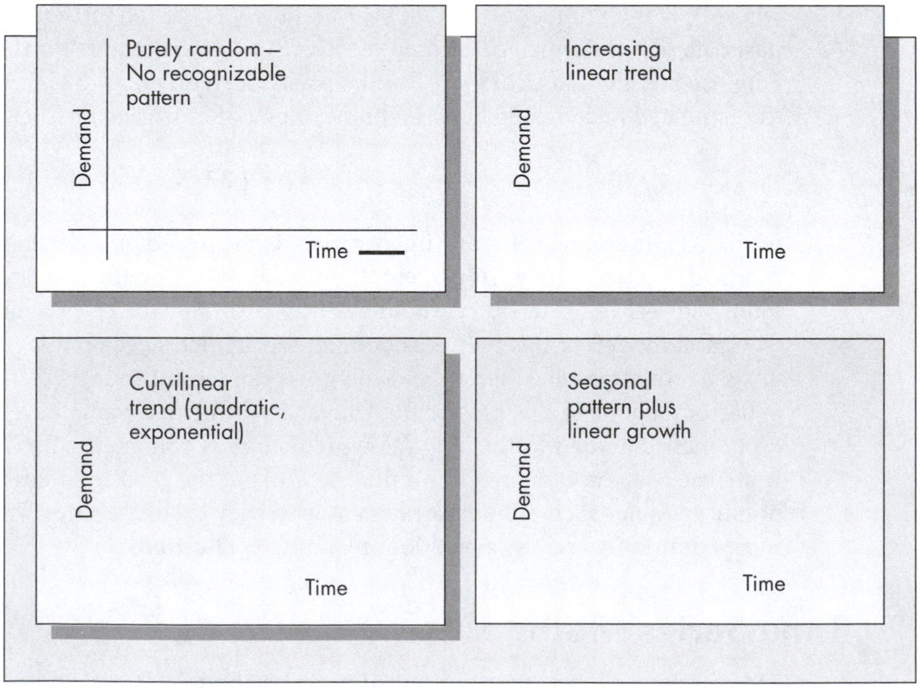

Figure 2–2 Time series patterns

PROBLEMS FOR SECTIONS 2.1–2.4

1. Name the four components of time series (i.e., the four distinct patterns exhibited by time series).
2. What distinguishes seasonality from cycles in time series analysis?
3. What is the appropriate type of forecasting method to use in each of the following scenarios?
   a. Ibis is attempting to predict the demand for hotel rooms next year, based on a history of demand observations.
   b. Sherwin-Williams has developed a new type of outdoor paint. The company wishes to forecast sales based on new housing starts.
   c. IBM is trying to ascertain the cost of being out of stock of a critical memory component. They email a survey to managers at various national spare parts centers. The surveys are returned to the managers for a reassessment, and the process is repeated until a consensus is reached.
4. Discuss the role of forecasting for the following functions of the firm.
   a. Marketing.
   b. Accounting.
   c. Finance.
   d. Production.
5. Distinguish among the following types of forecasts.
   a. Aggregate versus single item.
   b. Short-term versus long-term.
   c. Causal versus naive.
6. What is the advantage of the Delphi method over the jury of executive opinion method? What do these methods have in common?
7. What forecasting concerns did you have when, as a high school senior, you considered which college to attend? In particular, list the short-term, intermediate-term, and long-term forecasts you might have considered in making your final decision. What objective sources of data might you have used to provide you with better forecasts in each case?
8. Discuss the following quotation from an inventory control manager: "It's not my fault we ran out of those parts. The demand forecast was wrong."
9. Discuss the following statement: "Economists are predicting that mortgage interest rates in the United States will continue to be under 5 percent for at least 15 years."

2.5 NOTATION CONVENTIONS

The following discussion deals with time series methods. Define $D_1, D_2, \ldots, D_t, \ldots$ as the observed values of demand during periods $1, 2, \ldots, t, \ldots$. We will assume throughout that $\{D_t, t \geq 1\}$ is the time series we would like to predict. Furthermore, we will assume that if we are forecasting in period t, then we have observed $D_t, D_{t-1}, \ldots$ but have not observed D_{t+1}.

Define $F_{t-\tau,\, t}$ as the forecast made in period $t - \tau$ for the demand in period t, where $\tau = 1, 2, \ldots$. For the special case of $\tau = 1$, define $F_t = F_{t-1,\, t}$. That is, F_t is the forecast made in period $t - 1$ for the demand in period t, after having observed $D_{t-1}, D_{t-2} \ldots$, but before having observed D_t. For the time being we will assume that all forecasts are one-step-ahead forecasts; that is, they are made for the demand in the next period. Multiple-step-ahead forecasts will be discussed later.

Finally, note that a time series forecast is obtained by applying some set of weights to past data. That is,

$$F_t = \sum_{n=1}^{\infty} a_n D_{t-n} \quad \text{for some set of weights } a_1,\ a_2,\ \dots .$$

Most of the time series methods discussed in this chapter are distinguished only by the choice of weights.

2.6 EVALUATING FORECASTS

Define the **forecast error** in period t, e_t, as the difference between the forecast value for that period and the actual demand for that period. For multiple-step-ahead forecasts,

$$e_t = F_{t-\tau,\,t} - D_t,$$

and for one-step-ahead forecasts,

$$e_t = F_t - D_t.$$

Let $e_1, e_2, \dots, e_n$ be the forecast errors observed over n periods. Two common measures of forecast accuracy during these n periods are the **mean absolute deviation** (MAD) and the **mean squared error** (MSE) given by the following formulas:

$$\text{MAD} = (1/n)\sum_{i=1}^{n} |e_i|,$$

$$\text{MSE} = (1/n)\sum_{i=1}^{n} e_i^2.$$

Note that the MSE is similar to the variance of a random sample. The MAD is often the preferred method of measuring the forecast error because it does not require squaring. Furthermore, when forecast errors are normally distributed, as is generally assumed, an estimate of the standard deviation of the forecast error, σ_e, is given by 1.25 times the MAD.

Although the MAD and the MSE are the two most common measures of forecast accuracy, other measures are used as well. One that is not dependent on the magnitude of the values of demand is known as the **mean absolute percentage error** (MAPE) and is given by the formula

$$\text{MAPE} = \left[(1/n)\sum_{i=1}^{n} |e_i / D_i|\right] \times 100.$$

EXAMPLE 2.1

Artel, a manufacturer of static random access memories (SRAMs), has production plants in Austin, Texas, and Sacramento, California. The managers of these plants are asked to forecast production yields (measured in percent) one week ahead for their plants. Based on six weekly forecasts, the firm's management wishes to determine which manager is more successful at predicting his plant's yields. The results of their predictions are given in the following spreadsheet.

Week	P1	O1	|E1|	E1^2	|E1/O1|	P2	O2	|E2|	E2^2	|E2/O2|
1	92	88	4	16	0.04545	96	91	5	25	0.05495
2	87	88	1	1	0.01136	89	89	0	0	0
3	95	97	2	4	0.02062	92	90	2	4	0.02222
4	90	83	7	49	0.08434	93	90	3	9	0.03333
5	88	91	3	9	0.03297	90	86	4	16	0.04651
6	93	93	0	0	0	85	89	4	16	0.04494

Cell Formulas

Cell	Formula	Copied to
D2	=ABS(B2-C2)	D3:D7
E2	=ABS(D2/C2)	E3:E7

(Similar formulas and copies for cells H2 and I2)

Cell	Formula
B10	=AVERAGE(D2:D7)
B11	=AVERAGE(I2:I7)
B13	=AVERAGE(E2:E7)
B14	=AVERAGE(J2:J7)
B16	=AVERAGE(F2:F7)
B17	=AVERAGE(K2:K7)

Interpret P1 as the forecast made by the manager of plant 1 at the beginning of each week, O1 as the yield observed at the end of each week in plant 1, and E1 as the difference between the predicted and the observed yields. The same definitions apply to plant 2.

Compare the performance of these managers using the three measures MAD, MSE, and MAPE as defined previously. To compute the MAD we simply average the observed absolute errors:

$$MAD_1 = 17/6 = 2.83$$
$$MAD_2 = 18/6 = 3.00.$$

Based on the MADs, the first manager has a slight edge. To compute the MSE in each case, square the observed errors and average the results to obtain

$$MSE_1 = 79/6 = 13.17$$
$$MSE_2 = 70/6 = 11.67.$$

The second manager's forecasts have a lower MSE than the first, even though the MADs go the other way. Why the switch? The reason that the first manager now looks worse is that the MSE is more sensitive to one large error than is the MAD. Notice that the largest observed error of 7 was incurred by manager 1.

Compare performances based on the MAPE. To compute the MAPE we average the ratios of the errors and the observed yields:

$$MAPE_1 = .0325$$
$$MAPE_2 = .0336.$$

Using the MAD or the MAPE, the first manager has a very slight edge, but using the MSE, the second manager looks better. The forecasting abilities of the two managers would seem to be very similar. The method of evaluation chosen by management decides the "winner."

SNAPSHOT APPLICATION

Pfizer Bets Big on Forecasts of Drug Sales

One of Pfizer's most successful products was Lipitor, introduced in 1996. Lipitor, the first statin for reducing blood cholesterol, generated almost $13 billion in annual gross sales in its peak year of 2006. Not only was Lipitor the best-selling statin, but it was also the most profitable pharmaceutical ever sold until its patent expired in November 2011. At that point, generic versions of the drug that cost less entered the market. While annual gross sales approached $10 billion in 2011, they fell to about $4 billion in 2012 and reached a low of $1.8 billion in 2016 before increasing to $2.1 billion in 2018. Pfizer was forced to lower prices to compete with generics and saw their profit margin sink considerably.

To counteract this loss of revenue stream, a common strategy for drug companies is something called "evergreening." Evergreening means that the company plans to bring a new and more effective drug to the market to counter the anticipated losses when the patent expires. In Pfizer's case, this drug was torcetrapib. Torcetrapib not only decreased LDL (bad cholesterol) but also increased the levels of HDL (good cholesterol). In order to make the transition as smooth as possible, Pfizer would have to have sufficient production capacity online in 2011. Pfizer began production of torcetrapib as early as 2005 at a new $90 million plant in Loughberg, Ireland. This was reportedly a state-of-the-art facility.

Unfortunately, things did not go as planned for Pfizer. After investing $800 million in the development of torcetrapib, Pfizer informed the U.S. Food and Drug Administration that it was suspending Phase 3 clinical trials. It was at that point that Pfizer discovered serious side effects with the drug. They had to make the painful decision to discontinue their efforts after investing nearly a billion dollars. To Pfizer's credit, it is rare that a drug must be pulled from the market as late as Phase 3. Still, Pfizer's failure to accurately forecast the outcomes of the trials resulted in serious losses.

Source: Statista, "Worldwide Revenue of Pfizer's Lipitor from 2003 to 2018," April 24, 2019; Bala, R. S. Kunnumkal, and M. Sohoni. "Evergreening and Operational Risk Under Price Competition." SSRN (2015).

A desirable property of forecasts is that they should be unbiased. Mathematically, that means that $E(e_i) = 0$. One way of tracking a forecast method is to graph the values of the forecast error e_i over time. If the method is unbiased, forecast errors should fluctuate randomly above and below zero. An example is presented in Figure 2–3.

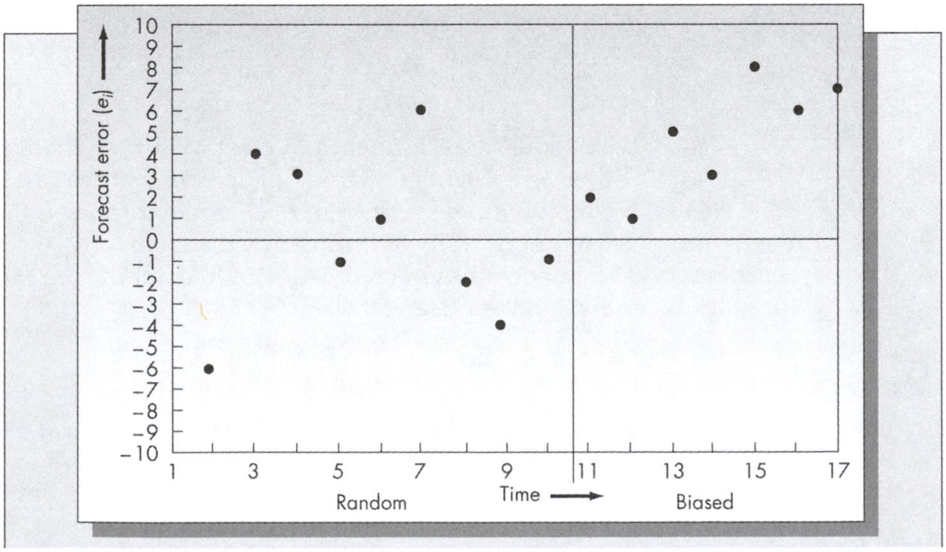

Figure 2–3 Forecast errors over time

An alternative to a graphical method is to compute the cumulative sum of the forecast errors, Σe_i. If the value of this sum deviates too far from zero (either above or below), it is an indication that the forecasting method is biased. At the end of this chapter we discuss a smoothing technique that also can be used to signal a bias in the forecasts. Statistical control charts also are used to identify unusually large values of the forecast error. (Statistical control charts are discussed in Chapter 10.)

PROBLEMS FOR SECTION 2.6

10. In his *Short Game Bible*, Dave Pelz attempts to characterize the skill of a golfer from a specific distance in terms of the ratio of the error in the shot and the intended shot distance. For example, a five iron that is hit 175 yards and is 20 yards off target would have an accuracy rating of 20/175 = .114, while a sand wedge hit 60 yards that is 10 yards off target would have an accuracy rating of 10/60 = .1667 (the lower the rating, the better). To what evaluation method discussed in this section is this most similar? Why does this evaluation method make more sense in golf than absolute or squared errors?
11. A forecasting method used to predict can opener sales applies the following set of weights to the last five periods of data: .1, .1, .2, .2, .4 (with .4 being applied to the most recent observation). Below are the observed values of can opener sales.

Period:	1	2	3	4	5	6	7	8
Observation:	18	22	26	33	14	28	30	52

Determine the following:

a. The one-step-ahead forecast for period 9.
b. The one-step-ahead forecast that was made for period 6.

12. A simple forecasting method for weekly sales of flash drives used by a local computer dealer is to form the average of the two most recent sales figures. Suppose sales for the drives for the past 12 weeks were as follows.

Week:	1	2	3	4	5	6	7	8	9	10	11	12
Sales:	86	75	72	83	132	65	110	90	67	92	98	73

a. Determine the one-step-ahead forecasts made for periods 3 through 12 using this method.
b. Determine the forecast errors for these periods.
c. Compute the MAD, the MSE, and the MAPE based on the forecast errors computed in part (b).

13. Two forecasting methods have been used to evaluate the same economic time series. The results are

Forecast from Method 1	Forecast from Method 2	Realized Value of the Series
223	210	256
289	320	340
430	390	375
134	112	110
190	150	225
550	490	525

Compare the effectiveness of these methods by computing the MSE, the MAD, and the MAPE. Do each of the measures of forecasting accuracy indicate that the same forecasting technique is best? If not, why?

14. What does the term *biased* mean in reference to a particular forecasting technique?
15. What is the estimate of the standard deviation of forecast error obtained from the data in problem 12?

2.7 METHODS FOR FORECASTING STATIONARY SERIES

In this section we will discuss two popular techniques, moving averages and exponential smoothing, for forecasting stationary time series. A **stationary** time series is one in which each observation can be represented by a constant plus a random fluctuation. In symbols,

$$D_t = \mu + \varepsilon_t,$$

where μ is an unknown constant corresponding to the mean of the series and ε_t is a random error with mean zero and variance σ^2.

The methods we consider in this section are more precisely known as single or simple exponential smoothing and single or simple moving averages. In addition, single moving averages also include weighted moving averages, which we do not discuss. For convenience, we will not use the modifiers single and simple in what follows. The meaning of the terms will be clear from the context.

Moving Averages

A simple but popular forecasting method is the method of **moving averages**. A moving average of order N is simply the arithmetic average of the most recent N observations. For the time being we restrict attention to one-step-ahead forecasts. Then F_t, the forecast made in period $t - 1$ for period t, is given by

$$F_t = (1/N)\sum_{i=t-N}^{t-1} D_i = (1/N)(D_{t-1} + D_{t-2} + \cdots + D_{t-N}).$$

In words, this says that the mean of the N most recent observations is used as the forecast for the next period. We will use the notation MA(N) for N-period moving averages.

EXAMPLE 2.2

Quarterly data for the failures of certain aircraft engines at a local military base during the last two years are 200, 250, 175, 186, 225, 285, 305, 190. Both three-quarter and six-quarter moving averages are used to forecast the numbers of engine failures. Determine the one-step-ahead forecasts for periods 4 through 8 using three-period moving averages, and the one-step-ahead forecasts for periods 7 and 8 using six-period moving averages.

Solution

The three-period moving-average forecast for period 4 is obtained by averaging the first three data points.

$$F_4 = (1/3)(200 + 250 + 175) = 208.$$

The three-period moving-average forecast for period 5 is

$$F_5 = (1/3)(250 + 175 + 186) = 204.$$

The six-period moving-average forecast for period 7 is

$$F_7 = (1/6)(200 + 250 + 175 + 186 + 225 + 285) = 220.$$

Quarter	Engine Failures	MA(3)	Error	MA(6)	Error
1	200				
2	250				
3	175				
4	186	208.33	22.33		
5	225	203.67	–21.33		
6	285	195.33	–89.67		
7	305	232.00	–73.00	220.17	–84.83
8	190	271.67	81.67	237.67	47.67

Other forecasts are computed in a similar fashion. Arranging the forecasts and the associated forecast errors in a spreadsheet, we obtain

Cell Formulas

Cell	Formula	Copied to
C4	=1/3*SUM(B2:B4)	C5:C8
E7	=1/6*SUM(B2:B7)	E8
D5	=C5-B5	D6:D8
F8	=E8-B8	F9

An interesting question is, how does one obtain multiple-step-ahead forecasts? For example, suppose in Example 2.2 that we are interested in using MA(3) in period 3 to forecast for period 6. Because the moving-average method is based on the assumption that the demand series is stationary, the forecast made in period 3 for *any* future period will be the same. That is, the multiple-step-ahead and the one-step-ahead forecasts are identical (although the one-step-ahead forecast will generally be more accurate). Hence, the MA(3) forecast made in period 3 for period 6 is 208. In fact, the MA(3) forecast made in period 3 for any period beyond period 3 is 208 as well.

An apparent disadvantage of the moving-average technique is that one must re compute the average of the last N observations each time a new demand observation becomes available. For large N this could be tedious. However, recalculation of the full N-period average is not necessary every period, since

$$\begin{aligned} F_{t+1} &= (1/N)\sum_{i=t-N+1}^{t} D_i = (1/N)\left[D_t + \sum_{i=t-N}^{t-1} D_i = D_{t-N}\right] \\ &= F_t + (1/N)[D_t - D_{t-N}]. \end{aligned}$$

This means that for one-step-ahead forecasting, we need only compute the difference between the most recent demand and the demand N periods old in order to update the forecast. However, we still need to keep track of all N past observations. Why?

Moving Average Lags behind the Trend

Consider a demand process in which there is a definite trend. For example, suppose that the observed demand is 2, 4, 6, 8, 10, 12, 14, 16, 18, 20, 22, 24. Consider the one-step-ahead MA(3) and MA(6) forecasts for this series.

Period	Demand	MA(3)	MA(6)
1	2		
2	4		
3	6		
4	8	4	
5	10	6	
6	12	8	
7	14	10	7
8	16	12	9
9	18	14	11
10	20	16	13
11	22	18	15
12	24	20	17

The demand and the forecasts for the respective periods are pictured in Figure 2–4. Notice that both the MA(3) and the MA(6) forecasts lag behind the trend. Furthermore, MA(6) has a greater lag. This implies that the use of simple moving averages is not an appropriate forecasting method when there is a trend in the series.

PROBLEMS ON MOVING AVERAGES

Problems 16 through 21 are based on the following observations of the demand for a certain part stocked at a parts supply depot during the calendar year.

Month	Demand	Month	Demand
January	89	July	223
February	57	August	286
March	144	September	212
April	221	October	275
May	177	November	188
June	280	December	312

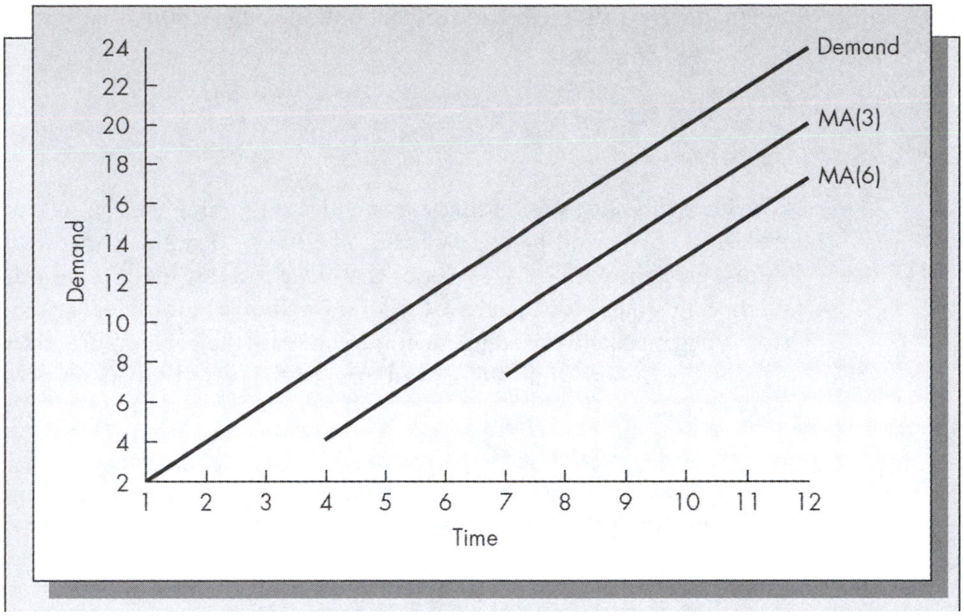

Figure 2–4 Moving-average forecasts log behind a trend

16. Determine the one-step-ahead forecasts for the demand for January of the following year using 3-, 6-, and 12-month moving averages.
17. Using a four-month moving average, determine the one-step-ahead forecasts for July through December.
18. Using a four-month moving average, determine the two-step-ahead forecast for July through December. (Hint: The two-step-ahead forecast for July is based on the observed demands in February through May.)
19. Compute the MAD for the forecasts obtained in problems 17 and 18. Which method gave better results? Based on forecasting theory, which method should have given better results?
20. Compute the one-step-ahead three-month and six-month moving-average forecasts for July through December. What effect does increasing N from 3 to 6 have on the forecasts?
21. What would an MA(1) forecasting method mean? Compare the accuracy of MA(1) and MA(4) forecasts for July through December.

Exponential Smoothing

Another very popular forecasting method for stationary time series is **exponential smoothing**. The current forecast is the weighted average of the last forecast and the current value of demand. That is,

$$\text{New forecast} = \alpha\,(\text{Current observation of demand}) + (1 - \alpha)(\text{Last forecast})$$

In symbols,

$$F_t = \alpha D_{t-1} + (1 - \alpha)F_{t-1},$$

where $0 < \alpha \le 1$ is the smoothing constant, which determines the relative weight placed on the current observation of demand. Interpret $(1 - \alpha)$ as the weight placed on past observations of demand. By a simple rearrangement of terms, the exponential smoothing equation for F_t can be written

$$\begin{aligned} F_t &= F_{t-1} - \alpha(F_{t-1} - D_{t-1}) \\ &= F_{t-1} - \alpha e_{t-1}. \end{aligned}$$

Written this way, we see that exponential smoothing can be interpreted as follows: the forecast in any period t is the forecast in period $t - 1$ minus some fraction of the observed forecast error in period $t - 1$. Notice that if we forecast high in period $t - 1$, e_{t-1} is positive and the adjustment is to decrease the forecast. Similarly, if we forecast low in period $t - 1$, the error is negative, and the adjustment is to increase the current forecast.

As before, F_t is the one-step-ahead forecast for period t made in period $t - 1$. Notice that since

$$F_{t-1} = \alpha D_{t-2} + (1 - \alpha)F_{t-2},$$

we can substitute above to obtain

$$F_t = \alpha D_{t-1} + \alpha(1 - \alpha)D_{t-2} + (1 - \alpha)^2 F_{t-2}.$$

We can now substitute for F_{t-2} in the same fashion. If we continue in this way, we obtain the infinite expansion for F_t,

$$F_t = \sum_{i=0}^{\infty} \alpha(1-\alpha)^i D_{t-i-1} = \sum_{i=0}^{\infty} a_i D_{t-i-1},$$

where the weights are $a_0 > a_1 > a_2 > \ldots > a_i = \alpha(1 - \alpha)^i$, and

$$\sum_{i=0}^{\infty} a_i = \sum_{i=0}^{\infty} \alpha(1-\alpha)^i = \alpha \sum_{i=0}^{\infty} (1-\alpha)^i = \alpha \times 1/[1-(1-\alpha)] = 1.$$

Hence, exponential smoothing applies a declining set of weights to all past data. The weights are graphed as a function of i in Figure 2–5.

In fact, we could fit the continuous exponential curve $g(i) = \alpha \exp(-\alpha i)$ to these weights, which is why the method is called exponential smoothing. The smoothing constant α plays essentially the same role here as the value of N does in moving averages. If α is large, more weight is placed on the current observation of demand and less weight on past observations, which results in forecasts that will react quickly to changes in the demand pattern but may have much greater variation from period to period. If α is small, then more weight is placed on past data and the forecasts are more stable.

When using an automatic forecasting technique to predict demand for a production application, stable forecasts (that is, forecasts that do not vary a great deal from period to period) are very desirable. Demand forecasts are used as the starting point for production planning and scheduling. Substantial revision in these forecasts can wreak havoc with employee work schedules, component bills of materials, and external purchase orders. For this reason, a value of α between .1 and .2 is generally recommended for production applications. (See, for example, Brown, 1962.)

Multiple-step-ahead forecasts are handled the same way for simple exponential smoothing as for moving averages; that is, the one-step-ahead and the multiple-step-ahead forecasts are the same.

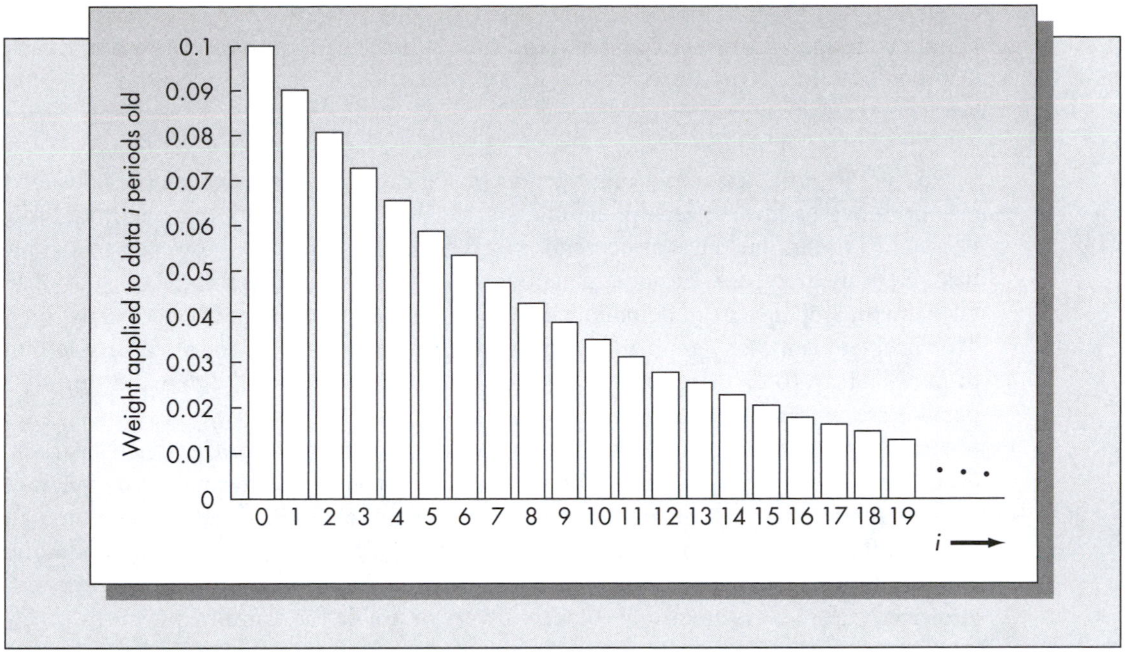

Figure 2–5 Weights in exponential smoothing

EXAMPLE 2.3

Consider Example 2.2 in which moving averages were used to predict aircraft engine failures. The observed numbers of failures over a two-year period were 200, 250, 175, 186, 225, 285, 305, 190. We will now forecast using exponential smoothing. In order to get the method started, assume that the forecast for period 1 was 200. Suppose that $\alpha = .1$. The one-step-ahead forecast for period 2 is

$$F_2 = \alpha D_1 + (1 - \alpha)F_1 = (.1)(200) + (.9)(200) = 200.$$

Similarly,

$$F_3 = \alpha D_2 + (1 - \alpha)F_2 = (.1)(250) + (.9)(200) = 205.$$

Other one-step-ahead forecasts are computed in the same fashion. The observed numbers of failures and the one-step-ahead forecasts for each quarter are listed below.

Quarter	Failures	Forecast
1	200	200 (by assumption)
2	250	200
3	175	205
4	186	202
5	225	201
6	285	203
7	305	211
8	190	220

Notice the effect of the smoothing constant. Although the original series shows high variance, the forecasts are quite stable. Repeat the calculations with a value of $\alpha = .4$. There will be much greater variation in the forecasts.

Because exponential smoothing requires that at each stage we have the previous forecast, it is not obvious how to get the method started. We could assume that the initial forecast is equal to the initial value of demand, as we did in Example 2.3. However, this approach has a serious drawback. Exponential smoothing puts substantial weight on past observations, so the initial value of demand will have an unreasonably large effect on early forecasts. This problem can be overcome by allowing the process to evolve for a reasonable number of periods (say 10 or more) and using the arithmetic average of the demand during those periods as the initial forecast.

In order to appreciate the effect of the smoothing constant, we have graphed a particularly erratic series in Figure 2–6, along with the resulting forecasts using values of $\alpha = .1$ and $\alpha = .4$. Notice that for $\alpha = .1$ the predicted value of demand results in a relatively smooth pattern, whereas for $\alpha = .4$, the predicted value exhibits significantly greater variation. Although smoothing with the larger value of α does a better job of tracking the series, the stability afforded by a smaller smoothing constant is very desirable for planning purposes.

EXAMPLE 2.3 (CONTINUED)

Consider again the problem of forecasting aircraft engine failures. Suppose that we were interested in comparing the performance of MA(3) with the exponential smoothing forecasts obtained earlier (ES(.1)). The first period for which we have a forecast using MA(3) is period 4, so we will make the comparison for periods 4 through 8 only.

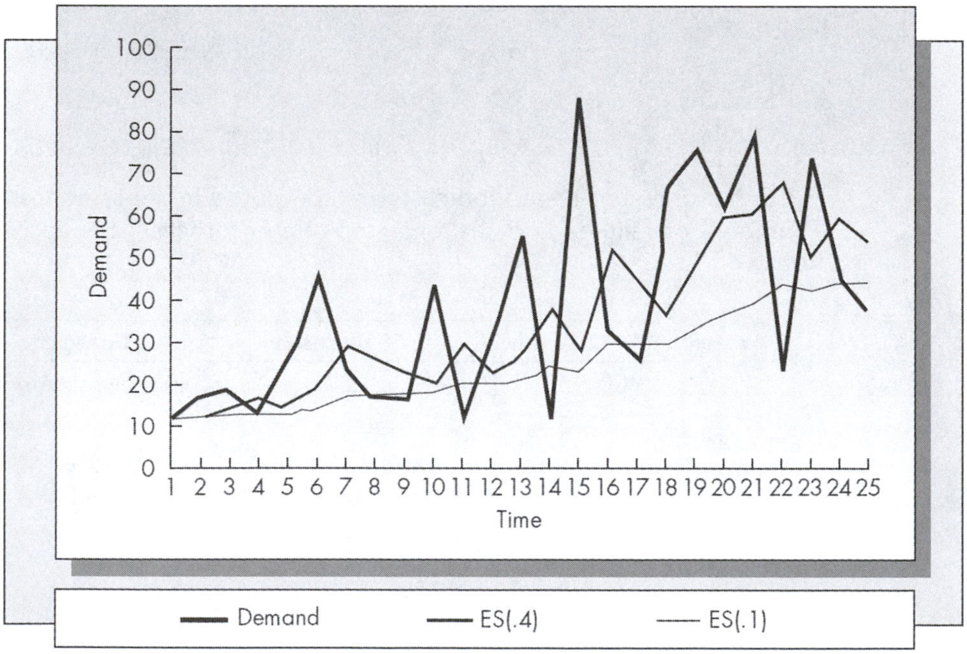

Figure 2–6 Exponential smoothing for different values of alpha

Quarter	Failures	Forecast
1	200	200.00
2	250	200.00 (by assumption)
3	175	205.00
4	186	202.00
5	225	200.40
6	285	202.86
7	305	211.07
8	190	220.47

Cell Formulas

Cell	Formula	Copied to
C3	=SUM(B$12*B2,(1-B$12)*C2)	C4:C9

The arithmetic average of the absolute errors, the MAD, is 57.6 for the three-period moving average and 49.2 for exponential smoothing. The respective values of the MSE are 4,215.6 and 3,458.4. Based on this comparison only, one might conclude that exponential smoothing is a superior method for this series. This is not necessarily true, however.

In Example 2.3 we compared exponential smoothing with $\alpha = .1$ and a moving average with $N = 3$. How do we know that those parameter settings are consistent? The MA(3) forecasts exhibit much greater variability than the ES(.1) forecasts do, suggesting that $\alpha = .1$ and $N = 3$ are not consistent values of these parameters.

Determining consistent values of α and N can be done in two ways. One is to equate the average age of the data used in making the forecast. A moving-average forecast consists of equal weights of $1/N$ applied to the last N observations. Multiplying the weight placed on each observation by its "age," we get the average age of data for moving averages as

$$\begin{aligned}\text{Average age} &= (1/N)(1+2+3+\cdots N) = (1/N)\big(N\big)(N+1)/2 \\ &= (N+1)/2.\end{aligned}$$

For exponential smoothing, the weight applied to data i periods old is $\alpha(1-\alpha)^{i-1}$. Assume that we have an infinite history of past observations of demand. Hence, the average age of data in an exponential smoothing forecast is

$$\text{Average age} = \sum_{i=1}^{\infty} i\alpha(1-\alpha)^{i-1} = 1/\alpha.$$

We omit the details of this calculation.

Equating the average age of data for the two methods, we obtain

$$\frac{N+1}{2} = \frac{1}{\alpha},$$

which is equivalent to

$$\alpha = 2/(N+1) \quad \text{or} \quad N = \frac{2-\alpha}{\alpha}.$$

Hence, we see that we would have needed a value of $N = 19$ for $\alpha = .1$ or a value of $\alpha = .5$ for $N = 3$ in order for the methods to be consistent in the sense of average age of data.

In Appendix 2–A of this chapter, we derive the mean and variance of the forecast error for both moving averages and exponential smoothing in terms of the variance of each individual observation, assuming that the underlying demand process is stationary. We show that both methods are unbiased; that is, the expected value of the forecast error is zero. Furthermore, by equating the expressions for the variances of the forecast error, one obtains the same relationship between α and N as by equating the average age of data. This means that if both exponential smoothing and moving averages are used to predict the same stationary demand pattern, forecast errors are normally distributed, and $\alpha = 2/(N + 1)$, then both methods will have exactly the same distribution of forecast errors. (However, this does *not* mean that the forecasts obtained by the two methods are the same.)

Multiple-Step-Ahead Forecasts

Thus far, we have talked only about one-step-ahead forecasts. That is, we have assumed that a forecast in period t is for the demand in period $t + 1$. However, there are cases where we are interested in making a forecast for more than one step ahead. For example, a retailer planning for the Christmas season might need to make a forecast for December sales in June in order to have enough time to prepare. Since the underlying model assumed for both moving averages and exponential smoothing is stationary (i.e., not changing in time), the one-step-ahead and multiple-step-ahead forecasts for moving averages and exponential smoothing are the same. That is, a forecast made in June for July sales is the same as a forecast made in June for December sales. (In the case of the retailer, the assumption of stationarity would probably be wrong, since December sales would likely be greater than a typical month's sales. That would suggest that these methods would *not* be appropriate in this case.)

Comparison of Exponential Smoothing and Moving Averages

There are several similarities and several differences between exponential smoothing and moving averages.

Similarities

1. Both methods are derived with the assumption that the underlying demand process is stationary (that is, can be represented by a constant plus a random fluctuation with zero mean). However, we should keep in mind that although the methods are appropriate for stationary time series, we don't necessarily believe that the series are stationary forever. By adjusting the values of N and α we can make the two methods more or less responsive to shifts in the underlying pattern of the data.
2. Both methods depend on the specification of a single parameter. For moving averages the parameter is N, the number of periods in the moving average, and for exponential smoothing the parameter is α, the smoothing constant. Small values of N or large values of α result in forecasts that put greater weight on current data, and large values of N and small values of α put greater weight on past data. Small N and large α may be more responsive to changes in the demand process but will result in forecast errors with higher variance.
3. Both methods will lag behind a trend if one exists.

4. When $\alpha = 2/(N + 1)$, both methods have the same distribution of forecast error. This means that they should have roughly the same level of accuracy, but it does *not* mean that they will give the same forecasts.

Differences

1. The exponential smoothing forecast is a weighted average of *all* past data points (as long as the smoothing constant is strictly less than 1). The moving-average forecast is a weighted average of only the last N periods of data. This can be an important advantage for moving averages. An outlier (an observation that is not representative of the sample population) is washed out of the moving-average forecast after N periods but remains forever in the exponential smoothing forecast.
2. In order to use moving averages, one must save all N past data points. In order to use exponential smoothing, one need only save the last forecast. This is the most significant advantage of the exponential smoothing method and one reason for its popularity in practice. In order to appreciate the consequence of this difference, consider a system in which the demand for 300,000 inventory items is forecasted each month using a 12-month moving average. The forecasting module alone requires saving 300,000 × 12 = 3,600,000 pieces of information. If exponential smoothing were used, only 300,000 pieces of information need to be saved. This issue is less important today than it was in the past because the cost of information storage has decreased enormously in recent years. However, it is still easier to manage a system that requires less data. It is primarily for this reason that exponential smoothing appears to be more popular than moving averages for production-planning applications.

PROBLEMS FOR SECTION 2.7

22. Handy, Inc., produces a solar-powered electronic calculator that has experienced the following monthly sales history for the first four months of the year, in thousands of units.

January	23.3	March	30.3
February	72.3	April	15.5

   a. If the forecast for January was 25, determine the one-step-ahead forecasts for February through May using exponential smoothing with a smoothing constant of $\alpha = .15$.
   b. Repeat the calculation in part (a) for a value of $\alpha = .40$. What difference in the forecasts do you observe?
   c. Compute the MSEs for the forecasts you obtained in parts (a) and (b) for February through April. Which value of α gave more accurate forecasts, based on the MSE?
23. Compare and contrast exponential smoothing when α is small (near zero) and when α is large (near 1).
24. Observed weekly sales of ball peen hammers at the town hardware store over an eight-week period have been 14, 9, 30, 22, 34, 12, 19, 23.
   a. Suppose that three-week moving averages are used to forecast sales. Determine the one-step-ahead forecasts for weeks 4 through 8.
   b. Suppose that exponential smoothing is used with a smoothing constant of $\alpha = .15$. Find the exponential smoothing forecasts for weeks 4 through 8. [To get the method started, use the same forecast for week 4 as you used in part (a).]

c. Based on the MAD, which method did better?
d. What is the exponential smoothing forecast made at the end of week 6 for the sales in week 12?

25. Determine the following:
a. The value of α consistent with $N = 6$ in moving averages.
b. The value of N consistent with $\alpha = .05$.
c. The value of α that results in a variance of forecast error, σ_e^2, 10 percent higher than the variance of each observation, σ^2 (refer to the formulas derived in Appendix 2–A).
26. Referring to the data in problem 22, what is the exponential smoothing forecast made at the end of March for the sales in July? Assume $\alpha = .15$.
27. For the data for problems 16 through 21, use the arithmetic average of the first six months of data as a baseline to initialize the exponential smoothing.
a. Determine the one-step-ahead exponential smoothing forecasts for August through December, assuming $\alpha = .20$.
b. Compare the accuracy of the forecasts obtained in part (a) with the one-step-ahead six-month moving-average forecasts determined in problem 20.
c. Comment on the reasons for the result you obtained in part (b).

2.8 TREND-BASED METHODS

Both exponential smoothing and moving-average forecasts will lag behind a trend if one exists. We will consider two forecasting methods that specifically account for a trend in the data regression analysis and Holt's method. **Regression analysis** is a method that fits a straight line to a set of data. **Holt's method** is a type of double exponential smoothing that allows for simultaneous smoothing on the series and on the trend.

Regression Analysis

Let $(x_1, y_1), (x_2, y_2), \dots, (x_n, y_n)$ be n paired data points for the two variables X and Y. Assume that y_i is the observed value of Y when x_i is the observed value of X. Refer to Y as the dependent variable and X as the independent variable. We believe that a relationship exists between X and Y that can be represented by the straight line

$$\hat{Y} = a + bX.$$

Interpret $\hat{Y}$ as the predicted value of Y. The goal is to find the values of a and b so that the line $\hat{Y} = a + bX$ gives the best fit of the data. The values of a and b are chosen so that the sum of the squared distances between the regression line and the data points is minimized (see Figure 2–7). In Appendix 2–B we derive the optimal values of a and b in terms of the given data.

When applying regression analysis to the forecasting problem, the independent variable often corresponds to time and the dependent variable to the series to be forecast. Assume that $D_1, D_2, \dots, D_n$ are the values of the demand at times $1, 2, \dots, n$. Then it is shown in Appendix 2–B that the optimal values of a and b are given by

$$b = \frac{S_{xy}}{S_{xx}}$$

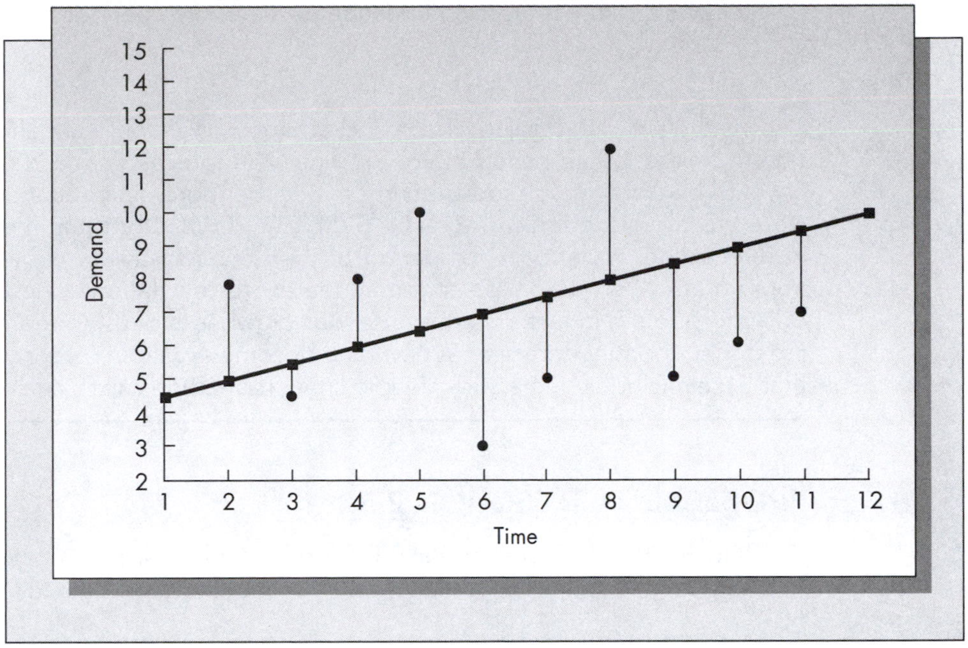

Figure 2–7 An example of a regression line

and

$$a = \bar{D} - b(n+1)/2,$$

where

$$S_{xy} = n\sum_{i=1}^{n} iD_i - \frac{n(n+1)}{2}\sum_{i=1}^{n} D_i,$$
$$S_{xx} = \frac{n^2(n+1)(2n+1)}{6} - \frac{n^2(n+1)^2}{4},$$

and $\bar{D}$ is the arithmetic average of the observed demands during periods 1, 2, … , n.

EXAMPLE 2.4

We will apply regression analysis to the problem of predicting aircraft engine failures (Examples 2.2 and 2.3). Recall that the demand for aircraft engines during the last eight quarters was 200, 250, 175, 186, 225, 285, 305, 190. Suppose that we use the first five periods as a baseline in order to estimate our regression parameters. Then

$$S_{xy} = 5\,[200 + (250)(2) + (175)(3) + (186)(4) + (225)(5)] - [(5)(6)/2]$$
$$[200 + 250 + 175 + 186 + 225] = -70.$$

$$S_{xx} = (25)(6)(11)/6 - (25)(36)/4 = 50.$$

Then

$$b = S_{xy}/S_{xx} = -70/50 = -7/5,$$
$$a = 207.2 - (-7/5)(3) = 211.4.$$

It follows that the regression equation based on five periods of data is

$$\hat{D}_t = 211.4 - (75)t.$$

$\hat{D}_t$ is the predicted value of demand at time t. We would use this regression equation to forecast from period 5. For example, the forecast made in period 5 for period 8 would be obtained by substituting $t = 8$ in the regression equation just given, which would result in the forecast 211.4 – (7/5)(8) = 200.2. Note that if we were interested in forecasting in period 7 for period 8, then this regression equation would not be appropriate. We would have to repeat the entire calculation using the data from periods 1 through 7. In fact, one of the serious drawbacks of using regression for forecasting is that updating forecasts as new data become available is very cumbersome. (Note that Excel includes single and multiple linear regression capabilities.)

PROBLEMS FOR SECTION 2.8

28. Shoreline Park in Mountain View, California, keeps close tabs on the number of patrons using the park, recording the following figures for the first six months of the calendar year.

Month	Number of Patrons	Month	Number of Patrons
January	133	April	640
February	183	May	1,876
March	285	June	2,550

  a. Draw a graph of these six data points. Assume that January = period 1, February = period 2, and so on. Using a ruler, "eyeball" the best straight-line fit of the data. Estimate the slope and intercept from your graph.
  b. Compute the exact values of the intercept a and the slope b from the regression equations.
  c. What are the forecasts obtained for July through December from the regression equation determined in part (b)?
  d. Comment on the results you obtained in part (c). Specifically, how confident would you be about the accuracy of the forecasts that you obtained?

29. The Mountain View Department of Parks and Recreation must project the total use of Shoreline Park for the next calendar year.
  a. Forecast the total number of people who will visit the park based on the regression equation.
  b. Determine the forecast for the same quantity using a six-month moving average.
  c. Draw a graph of the most likely shape of the curve describing the park's usage by month during this next calendar year, and predict the same quantity using your graph. Is your prediction closer to the answer you obtained for part (a) or part (b)? Discuss your results.

Double Exponential Smoothing Using Holt's Method

Holt's method is a type of double exponential smoothing designed to track time series with linear trend. The method requires the specification of two smoothing constants, α and β, and uses two smoothing equations: one for the value of the series (the intercept) and one for the trend (the slope). The equations are

$$S_t = \alpha D_t + (1 - \alpha)(S_{t-1} + G_{t-1}),$$
$$G_t = \beta(S_t - S_{t-1}) + (1 - \beta)G_{t-1}.$$

Interpret S_t as the value of the intercept at time t and G_t as the value of the slope at time t. The first equation is very similar to that used for simple exponential smoothing. When the most current observation of demand, D_t, becomes available, it is averaged with the prior forecast of the current demand, which is the previous intercept, $S_{t-1,}$ plus 1 times the previous slope, $G_{t-1.}$ The second equation can be explained as follows. Our new estimate of the intercept, S_t, causes us to revise our estimate of the slope to $S_t - S_{t-1}$. This value is then averaged with the previous estimate of the slope, G_{t-1}. The smoothing constants may be the same, but for most applications more stability is given to the slope estimate (implying $\beta \leq \alpha$).

The τ-step-ahead forecast made in period t, which is denoted by $F_{t,\, t+\tau}$, is given by

$$F_{t,\, t+\tau} = S_t + \tau G_t.$$

EXAMPLE 2.5

Apply Holt's method to the problem of developing one-step-ahead forecasts for the aircraft engine failure data. Recall that the original series was 200, 250, 175, 186, 225, 285, 305, 190. Assume that both α and β are equal to .1. In order to get the method started, we need estimates of both the intercept and the slope at time zero. Suppose that these are $S_0 = 200$ and $G_0 = 10$. Then we obtain

$$S_1 = (.1)(200) + (.9)(200+ 10) = 209.0$$
$$G_1 = (.1)(209 - 200) + (.9)(10) = 9.9$$
$$S_2 = (.1)(250) + (.9)(209 + 9.9) = 222.0$$
$$G_2 = (.1)(222 - 209) + (.9)(9.9) = 10.2$$
$$S_3 = (.1)(175) + (.9)(222 + 10.2) = 226.5$$
$$G_3 = (.1)(226.5 - 222) + (.9)(10.2) = 9.6$$

and so on.

Comparing the one-step-ahead forecasts to the actual numbers of failures for periods 4 through 8, we obtain the following.

Period	Actual	S	G	Forecast	|Error|
1	200	200.00	0.00	200.00	0.00
2	250	209.00	9.90	218.90	31.10
3	175	222.01	10.21	232.22	57.22
4	186	226.50	9.64	236.14	50.14
5	225	231.12	9.14	240.26	15.26
6	285	238.74	8.98	247.72	37.28
7	305	251.45	9.36	260.81	44.19
8	190	265.23	9.80	275.02	85.02

$\alpha = 0.1$
$\beta = 0.1$
$S_0 = 200$
$G_0 = 10$

Cell Formulas

Cell	Formula	Copied to
C3	SUM(B$12*B2,(1-B$12)*(C2+D2))	C4:C9
D3	SUM(B$13*(C3-C2),(1-B$13)*(D2))	D4:D9
E3	C3+D3	E4:E9
F3	ABS(B3-E3)	F4:F9

Averaging the numbers in the final column, we obtain a MAD of 46.4. Notice that this is lower than that for simple exponential smoothing or moving averages. Holt's method does better for this series because it is explicitly designed to track the trend in the data, whereas simple exponential smoothing and moving averages are not. Note that the forecasts in the given table are one-step-ahead forecasts. Suppose you needed to forecast the demand in period 2 for period 5. This forecast is $F_{2,5} = S_2 + (3)G_2 = 222 + (3)(10.2) = 252.6$.

The initialization problem also arises in getting Holt's method started. The best approach is to establish some set of initial periods as a baseline and use regression analysis to determine estimates of the slope and intercept using the baseline data.

Both Holt's method and regression are designed to handle series that exhibit trend. However, with Holt's method it is far easier to update forecasts as new observations become available.

MORE PROBLEMS FOR SECTION 2.8

30. For the data in problem 28, use the results of the regression equation to estimate the slope and intercept of the series at the end of June. Use these numbers as the initial values of slope and intercept required in Holt's method. Assume that $\alpha = .15$, $\beta = .10$ for all calculations.
    a. Suppose that the actual number of visitors using the park in July was 2,150 and the number in August was 2,660. Use Holt's method to update the estimates of the slope and intercept based on these observations.
    b. What are the one-step-ahead and two-step-ahead forecasts that Holt's method gives for the number of park visitors in September and October?
    c. What is the forecast made at the end of July for the number of park attendees in December?
31. Continuing with problem 30, suppose that because of serious flooding, the park was closed for most of December. During that time only 53 people visited. Comment on the effect this observation would have on predictions of future use of the park. If you were in charge of forecasting the park's usage, how would you deal with this data point?
32. Discuss some of the problems that could arise when using either regression analysis or Holt's method for obtaining multiple-step-ahead forecasts.

2.9 METHODS FOR SEASONAL SERIES

This section considers forecasting methods for seasonal problems. A **seasonal series** is one that has a pattern that repeats every N periods for some value of N (which is at least 3). A typical seasonal series is pictured in Figure 2–8.

We refer to the number of periods before the pattern begins to repeat as the length of the season (N in the picture). Note that this is different from the popular usage of the word *season* as a time of year. In order to use a seasonal model, one must be able to specify the length of the season.

There are several ways to represent seasonality. The most common is to assume that there exists a set of multipliers c_t, for $1 \leq t \leq N$, with the property that $\Sigma c_t = N$. The multiplier c_t represents the average amount that the demand in the tth period of the season is above or below the overall average. For example, if $c_3 = 1.25$ and $c_5 = .60$, then, on average, the demand in the third period of the season is 25 percent above the average demand and the demand in the fifth period of the season is 40 percent below the average demand. These multipliers are known as **seasonal factors**.

Seasonal Factors for Stationary Series

The following is a simple method of computing seasonal factors for a time series with seasonal variation and no trend. It requires a minimum of two seasons of data—as will most methods to account for seasonality.

The method is as follows:

1. Compute the sample mean of all the data.
2. Divide each observation by the sample mean. This gives seasonal factors for each period of observed data.

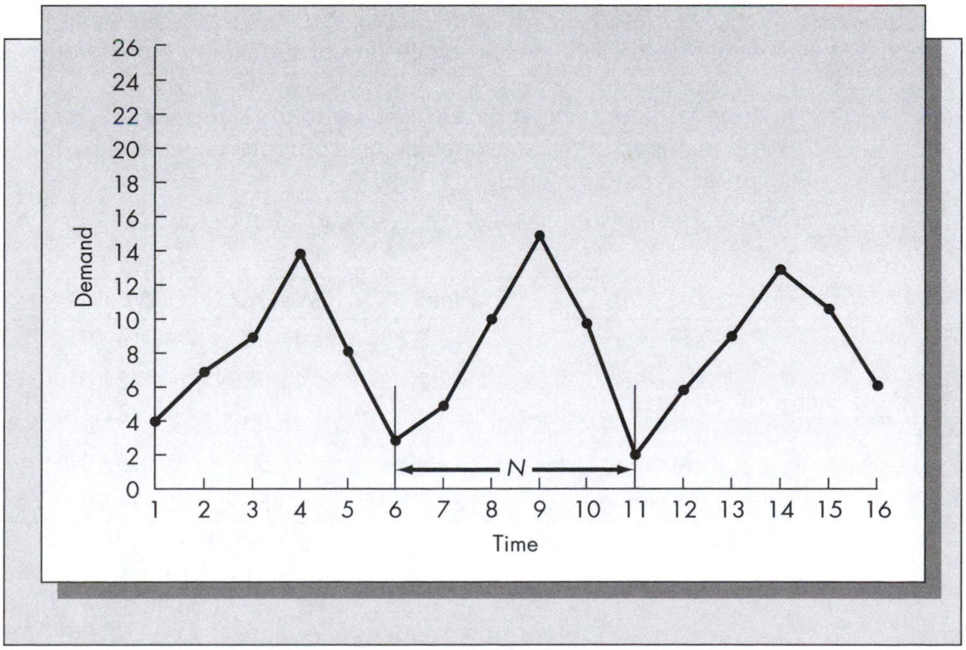

Figure 2–8 A seasonal demand series

3. Average the factors for like periods within each season. That is, average all the factors corresponding to the first period of a season, all the factors corresponding to the second period of a season, and so on. The resulting averages are the N seasonal factors. They will always add to exactly N.

EXAMPLE 2.6

The County Transportation Department needs to determine usage factors by day of the week for a toll bridge connecting different parts of the city. In the current study, they are considering only working days. Suppose that the numbers of cars using the bridge each working day over the past four weeks were (in thousands of cars)

	Week 1	Week 2	Week 3	Week 4
Monday	16.2	17.3	14.6	16.1
Tuesday	12.2	11.5	13.1	11.8
Wednesday	14.2	15.0	13.0	12.9
Thursday	17.3	17.6	16.9	16.6
Friday	22.5	23.5	21.9	24.3

Find the seasonal factors corresponding to daily usage of the bridge.

Solution

To solve this problem:

1. Compute the arithmetic average of all of the observations (20 in this case).
2. Divide each observation by the average computed in step 1. This will give 20 factors.
3. Average factors corresponding to the same period of each season. That is, average all factors for Mondays, all factors for Tuesdays, and so on. This will give five seasonal factors: one for each day of the week. Note that these factors will sum to five.
4. Forecasts for the numbers of cars using the bridge by day of the week are obtained by multiplying the seasonal factors computed in step 3 by the average value computed in step 1.

These steps are summarized in the spreadsheet below.

	Week 1	Week 2	Week 3	Week 4
Monday	16.20	17.30	14.60	16.10
Tuesday	12.20	11.50	13.10	11.80
Wednesday	14.20	15.00	13.00	12.90
Thursday	17.30	17.60	16.90	16.60
Friday	22.50	23.50	21.90	24.30

Step 1: Compute the overall average of all of the observations

Average = 16.43 Formula = AVERAGE(B2:E6)

Step 2: Divide each observation by the Mean

	Week 1	Week 2	Week 3	Week 4
Monday	0.99	1.05	0.89	0.98
Tuesday	0.74	0.70	0.80	0.72
Wednesday	0.86	0.91	0.79	0.79
Thursday	1.05	1.07	1.03	1.01
Friday	1.37	1.43	1.33	1.48

Step 3: Average Factors corresponding to the same day of the week

Seasonal Factor	
Monday	0.98
Tuesday	0.74
Wednesday	0.84
Thursday	1.04
Friday	1.40

Note that these factors sum to exactly five.

Step 4: Build the forecasts by multiplying the mean, 16.425, by the appropriate factor

Forecast	
Monday	16.05
Tuesday	12.15
Wednesday	13.78
Thursday	17.10
Friday	23.05

Cell Formulas

Cell	Formula	Copied to
B10	AVERAGE	(B2:E6)
B14	B2/$B$10	B15:B18
C14	C2/$C$10	C15:C18
(Similar formulas and copies for cells D14 and E14)		
B22	AVERAGE(B14:E14)	B23:B26
B30	$B$10*B22	B31:B34

Note: Example 2.6

Determining the Deseasonalized Series

In cases where there is both seasonal variation and trend, a useful technique is to form the **deseasonalized** series by removing the seasonal variation from the original series. To illustrate this, consider the following simple example that consists of two seasons of data. Note that this method is an approximation that won't be as accurate as Winters's method (to follow), especially for series with a strong trend and weak seasonality.

EXAMPLE 2.7

Period	Demand
1	10
2	20
3	26
4	17
5	12
6	23
7	30
8	22

Following the steps described above, the reader should verify that we obtain the following four seasonal factors.

0.550
1.075
1.400
0.975

To obtain the deseasonalized series, one simply divides each observation by the appropriate seasonal factor. For this example, the divisions would be 10/.550, 20/1.075, etc., resulting in the following.

Period	Demand	Deseasonalized Demand
1	10	18.182
2	20	18.605
3	26	18.571
4	17	17.436
5	12	21.818
6	23	21.395
7	30	21.429
8	22	22.564

Notice that the deseasonalized demand shows a clear trend. To forecast the deseasonalized series one could use any of the trend-based methods discussed earlier in this chapter. Suppose that we fit a simple linear regression to this data where time is the independent variable as described in Section 2.8. Doing so one obtains the regression fit of this data as $y_t = 16.91 + 0.686t$. To forecast, one first applies the regression to the deseasonalized series and then re-seasonalizes by multiplying by the appropriate factor. For example, if one wishes to forecast for periods 9 through 12, the regression equation gives the following forecasts for the deseasonalized series: 23.08, 23.77, 24.46, 25.14. The final forecasts are obtained by multiplying by the appropriate seasonal factors, giving the final forecasts for periods 9 through 12 as 12.70, 25.55, 34.24, and 25.51.

PROBLEMS FOR SECTION 2.9

33. Sales of walking shorts at Hugo's Department Store appear to exhibit a seasonal pattern. The proprietor of the store, Wally Hugo, has kept careful records of his sales. During the past two years there have been the following monthly sales of the shorts.

	Year 1	Year 2		Year 1	Year 2
Jan.	12	16	July	112	130
Feb.	18	14	Aug.	90	83
March	36	46	Sep.	66	52
April	53	48	Oct.	45	49
May	79	88	Nov.	23	14
June	134	160	Dec.	21	26

Assuming no trend in shorts sales over the two years, obtain estimates for the monthly seasonal factors.

34. A popular brand of tennis shoe has had the following demand history by quarters over a three-year period.

Year 1	Demand	Year 2	Demand	Year 3	Demand
1	12	1	16	1	14
2	25	2	32	2	45
3	76	3	71	3	84
4	52	4	62	4	47

a. Determine the seasonal factors for each quarter.
b. Based on the result of part (a), determine the deseasonalized demand series.
c. Predict the demand for each quarter of year 4 for the deseasonalized series from a six-quarter moving average.
d. Using the results from parts (a) and (c), predict the demand for the shoes for each quarter of year 4.

Winters's Method for Seasonal Problems

The moving-average method just described can be used to predict a seasonal series with or without a trend. However, as new data become available, the method requires that all seasonal factors be recalculated from scratch. **Winters's method** is a type of triple exponential smoothing, which has the important advantage of being easy to update as new data become available.

We assume a model of the form

$$D_t = (\mu + G_t)c_t + \varepsilon_t.$$

Interpret μ as the base signal or intercept at time $t = 0$ excluding seasonality, G_t as the trend or slope component, c_t as the multiplicative seasonal component in period t, and finally ε_t as the error term. Because the seasonal factor multiplies both the base level and the trend term, we are assuming that the underlying series has a form similar to that pictured in Figure 2–9.

Again, assume that the length of the season is exactly N periods and that the seasonal factors are the same each season and have the property that $\Sigma c_t = N$. Three exponential smoothing equations are used each period to update estimates of deseasonalized series, the seasonal factors, and the trend. These equations may have different smoothing constants, which we will label α, β, and γ.

1. *The series.* The current level of the deseasonalized series, S_t, is given by

$$S_t = \alpha(D_t/c_{t-N}) + (1 - \alpha)(S_{t-1} + G_{t-1}).$$

   Notice what this equation does. By dividing by the appropriate seasonal factor, we are deseasonalizing the newest demand observation. This is then averaged with the current forecast for the deseasonalized series, as in Holt's method.
2. *The trend.* The trend is updated in a fashion similar to Holt's method.

$$G_t = \beta[S_t - S_{t-1}] + (1 - \beta)G_{t-1}.$$

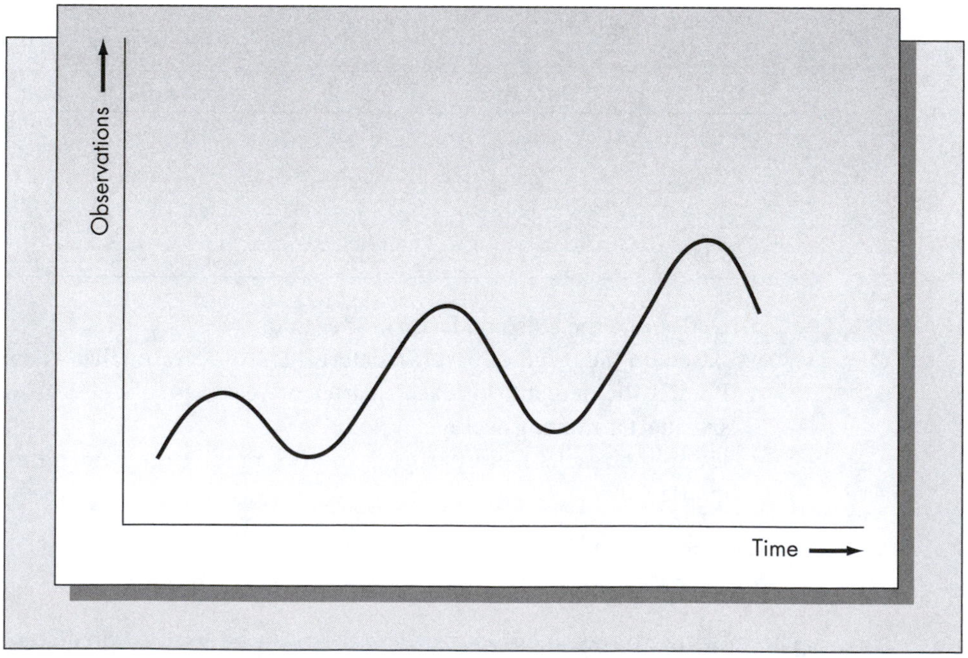

Figure 2–9 Seasonal series with increasing trend

3. *The seasonal factors.*

$$c_t = \gamma(D_t/S_t) + (1 - \gamma)c_{t-N}.$$

The ratio of the most recent demand observation over the current estimate of the deseasonalized demand gives the current estimate of the seasonal factor. This is then averaged with the previous best estimate of the seasonal factor, c_{t-N}. Each time that a seasonal factor is updated, it is necessary to norm the most recent N factors to add to N.

Finally, the forecast made in period t for any future period $t + \tau$ is given by

$$F_{t,\,t+\tau} = (S_t + \tau G_t)c_{t+\tau-2N}.$$

Note that this forecasting equation assumes that $\tau \le N$. If $N < \tau \le 2N$, the appropriate seasonal factor would be $c_{t+\tau-2N}$; if $2N < \tau \le 3N$, the appropriate seasonal factor would be $c_{t+\tau-3N}$; and so on.

Initialization Procedure

In order to get the method started, we need to obtain initial estimates for the series, the slope, and the seasonal factors. Winters suggests that a minimum of two seasons of data be available for initialization. Assume that exactly two seasons of data are available; that is, $2N$ data points. Suppose that the current period is $t = 0$, so that the past observations are labeled $D_{-2N+1}, D_{-2N+2}, \ldots, D_0$.

1. Calculate the sample means for the two separate seasons of data:

$$V_1 = \frac{1}{N}\sum_{j=-2N+1}^{-N} D_j,$$

$$V_2 = \frac{1}{N}\sum_{j=-N+1}^{0} D_j.$$

2. Define $G_0 = (V_2 - V_1)/N$ as the initial slope estimate. If $m > 2$ seasons of data are available for initialization, then compute $V_1, \ldots, V_m$ as in step 1 and define $G_0 = (V_m - V_1)/[(m - 1)N]$. If we locate V_1 at the center of the first season of data [at period $(-3N + 1)/2$] and V_2 at the center of the second season of data [at period $(-N + 1)/2$], then G_0 is simply the slope of the line connecting V_1 and V_2 (refer to Figure 2–10).
3. Set $S_0 = V_2 + G_0[(N - 1)/2]$. This estimates the value of the series at time $t = 0$. Note that S_0 is the value assumed by the line connecting V_1 and V_2 at $t = 0$ (see Figure 2–10).
4. a. The initial seasonal factors are computed for each period in which data are available and then averaged to obtain one set of seasonal factors. The initial seasonal factors are obtained by dividing each of the initial observations by the corresponding point along the line connecting V_1 and V_2. This can be done graphically or by using the following formula:

$$c_t = \frac{D_t}{V_i - [(N+1)/2 - j]G_0} \quad \text{for } -2N+1 \le t \le 0,$$

where $i = 1$ for the first season, $i = 2$ for the second season, and j is the period of the season. That is, $j = 1$ for $t = -2N + 1$ and $t = -N + 1$; $j = 2$ for $t = -2N + 2$ and $t = -N + 2$; and so on.

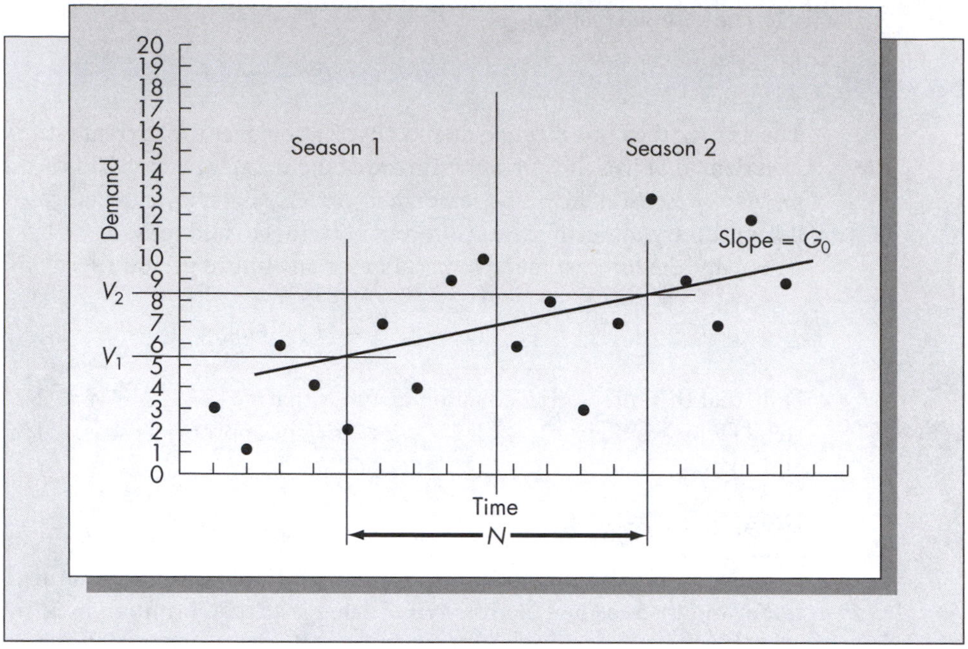

Figure 2–10 Initialization for Winters's method

b. Average the seasonal factors. Assuming exactly two seasons of initial data, we obtain

$$c_{-N+1} = \frac{c_{-2N+1} + c_{-N+1}}{2}, \ldots, c_0 = \frac{c_{-N} + c_0}{2}.$$

c. Normalize the seasonal factors.

$$c_j = \left[\frac{c_j}{\sum_{i=0}^{-N+1} c_i} \right] N \quad \text{for } -N+1 \leq j \leq 0.$$

This initialization procedure just discussed is the one suggested by Winters. It is not the only means of initializing the system. Seasonal factors could be determined by the method of moving averages discussed in the first part of this section. Another alternative would be to fit a linear regression to the baseline data and use the resulting slope and intercept values, as was done in Holt's method, to obtain S_0 and G_0. The seasonal factors would be obtained by dividing each demand observation in the baseline period by the corresponding value on the regression line, averaging like periods, and norming. The actual values of the initial estimates of the intercept, the slope, and the seasonal factors will be similar no matter which initialization scheme is employed.

EXAMPLE 2.8

Assume that the initial data set is the same as that of Example 2.7, in which centered moving averages were used to find the seasonal factors. Recall that we have two seasons of data: 10, 20, 26, 17, 12, 23, 30, 22. Then

$$V_1 = (10 + 20 + 26 + 17)/4 = 18.25,$$
$$V_2 = (12 + 23 + 30 + 22)/4 = 21.75,$$
$$G_0 = (21.75 - 18.25)/4 = 0.875,$$
$$S_0 = 21.75 + (0.875)(1.5) = 23.06.$$

The initial seasonal factors are computed as follows:

$$c_{-7} = \frac{10}{18.25 - (5/2 - 1)(0.875)} = 0.5904,$$
$$c_{-6} = \frac{20}{18.25 = (5/2 - 2)(0.875)} = 1.123.$$

The other factors are computed in a similar fashion. They are

$$c_{-5} = 1.391, \qquad c_{-4} = 0.869, \qquad c_{-3} = 0.5872,$$
$$c_{-2} = 1.079, \qquad c_{-1} = 1.352, \qquad c_0 = 0.9539.$$

We then average c_{-7} and c_{-3}, c_{-6} and c_{-2}, and so on, to obtain the four seasonal factors:

$$c_{-3} = 0.5888, \qquad c_{-2} = 1.1010, \qquad c_{-1} = 1.3720, \qquad c_0 = 0.9115.$$

Finally, norming the factors to ensure that the sum is 4 results in

$$c_{-3} = 0.5900, \qquad c_{-2} = 1.1100, \qquad c_{-1} = 1.3800, \qquad c_0 = 0.9200.$$

Notice how closely these factors agree with those obtained from the moving-average method.

Suppose that we wish to forecast the following year's demand at time $t = 0$. The forecasting equation is

$$F_{t,t+\tau} = (St + \tau G_t)c_{t+\tau-N},$$

which results in

$$F_{0,1} = (S_0 + G_0)c_{-3} = (23.06 + 0.875)(0.59) = 14.12,$$
$$F_{0,2} = (S_0 + 2G_0)c_{-2} = [23.06 + (2)(0.875)](1.11) = 27.54,$$
$$F_{0,3} = 35.44,$$
$$F_{0,4} = 24.38.$$

Now, suppose that at time $t = 1$ we observe a demand of $D_1 = 16$. We now need to update our equations. Assume that $\alpha = .2$, $\beta = .1$, and $\gamma = .1$. Then

$$S_1 = \alpha(D_1/c_{-3}) + (1 - \alpha)(S_0 + G_0) = (0.2)(16/.59) + (0.8)(23.06 + 0.875) = 24.57,$$
$$G_1 = \beta(S_1 - S_0) + (1 - \beta)G_0 = (0.1)(24.57 - 23.06) + (0.9)(0.875) = 0.9385,$$
$$c_1 = \gamma(D_1/S_1) + (1 - \gamma)c_{-3} = (0.1)(16/24.57) + (0.9)(0.59) = 0.5961.$$

At this point, we would renorm c_{-2}, c_{-1}, c_0, and the new value of c_1 to add to 4. Because the sum is 4.0061, it is close enough (the norming would result in rounding c_1 down to.59 once again).

Forecasting from period 1, we obtain

$$F_{1,2} = (S_1 + G_1)c_{-2} = (24.57 + 0.9385)(1.11) = 28.3144,$$
$$F_{1,3} = (S_1 + 2G_1)c_{-1} = [24.57 + (2)(0.9385)](1.38) = 36.4969,$$

and so on.

Now suppose that we have observed one full year of demand, given by $D_1 = 16$, $D_2 = 33$, $D_3 = 34$, and $D_4 = 26$. Each time a new observation becomes available, the intercept, slope, and most current seasonal factor estimates are updated. One obtains

$$S_2 = 26.35, \quad S_3 = 26.83, \quad S_4 = 27.89,$$
$$G_2 = 1.0227, \quad G_3 = 0.9678, \quad G_4 = 0.9770,$$
$$c_2 = 1.124, \quad c_3 = 1.369, \quad c_4 = 0.9212.$$

As c_1, c_2, c_3, and c_4 sum to 4.01, normalization is not necessary. Suppose that we were interested in the forecast made in period 4 for period 10. The forecasting equation is

$$F_{t,t+\tau} = (S_t + \tau G_t)c_{t+\tau-2N},$$

which results in

$$F_{4,10} = (S_4 + 6G_4)c_2 = [27.89 + 6(0.9770)](1.124) = 37.94.$$

An important consideration is the choice of the smoothing constants α, β, and γ to be used in Winters's method. The issues here are the same as those discussed for simple exponential smoothing and Holt's method. Large values of the smoothing constants will result in more responsive but less stable forecasts. One method for setting α, β, and γ is to experiment with various values of the parameters that retrospectively give the best fit of previous forecasts to the observed history of the series. Because one must test many combinations of the three constants, the calculations are tedious. Furthermore, there is no guarantee that the best values of the smoothing constants based on past data will be the best values for future forecasts. The most conservative approach is to guarantee stable forecasts by choosing the smoothing constants to be between 0.1 and 0.2.

MORE PROBLEMS FOR SECTION 2.9

35. Consider the data for problem 34.
    a. Using the data from years 2 and 3, determine initial values of the intercept, slope, and seasonal factors for Winters's method.
    b. Assume that the observed demand for the first quarter of year 4 was 18. Using $\alpha = .2$, $\beta = .15$, and $\gamma = .10$, update the estimates of the series, the slope, and the seasonal factors.
    c. What are the forecasts made at the end of the first quarter of year 4 for the remaining three quarters of year 4?
36. Suppose the observed quarterly demand for year 4 was 18, 51, 86, 66. Compare the accuracy of the forecasts obtained for the last three quarters of year 4 in problems 34(d) and 35(c) by computing both the MAD and the MSE.
37. Determine updated estimates of the slope, the intercept, and the seasonal factors for the end of year 4 based on the observations given in problem 36. Using these updated estimates, determine the forecasts that Winters's method gives for all of year 6 made at the end of year 4. Use the values of the smoothing constants given in problem 35(b).

2.10 BOX-JENKINS MODELS

The forecasting models introduced in this section are significantly more sophisticated than those previously discussed in this chapter. The goal is to present the basic concepts of **Box-Jenkins** analysis so that the reader can appreciate the power of these methods. However, an in-depth coverage is beyond the scope of this book. The methods are named for two well-known statisticians, George Box and Gwilym Jenkins formerly from the University of Wisconsin and University of Lancaster, respectively. The approach they developed is based

SNAPSHOT APPLICATION

Sport Obermeyer Slashes Costs with Improved Forecasting

Sport Obermeyer is a leading supplier in the U.S. fashion ski apparel market. The firm was founded in 1950 by engineer/ski instructor Klaus Obermeyer. Virtually all the firm's offerings are redesigned annually to incorporate changes in style, fabrics, and colors. For 40 years, the firm was able to meet demand by producing during the summer months after receiving firm orders from customers.

Things changed during the 1990s, and problems developed. First, volumes increased. There was insufficient capacity among suppliers to produce the required volume in the summer. Second, the firm developed a complex global supply chain strategy (see Section 6.1) to reduce costs. A parka sold in the United States might be sewn in China from fabrics and parts from Japan, South Korea, and Germany. Together these changes lengthened the production lead time, thus requiring the firm to commit to production before orders were placed by customers.

The firm undertook several "quick response" initiatives to reduce lead times. These included encouraging some customers to place orders earlier, locating raw materials near the Chinese production facility, and instituting an air freight system to expedite delivery to its Denver distribution center. Even with these changes in place, the problem of stockouts and markdowns due to oversupply were not solved. The company still had to commit about half the production based on forecasts. In the fashion industry, there is often no statistical history on which to base forecasts, and forecast errors can be huge. Products that outsell original forecasts by a factor of 2 or undersell original forecasts by a factor of 10 are common.

Sport Obermeyer needed some help with forecasting to avoid expensive miscalculations. The customary procedure was to base the forecasts on a consensus of members of the buying committee. The problem with consensus forecasting is that the dominant personalities in a group carry more weight. A forecast obtained in this way might represent only the opinion of one person. To overcome this problem, the research team (Fisher, Hammand, Obermeyer, and Raman, 1994) recommended that members of the committee supply *individual* forecasts.

The dispersion among individual forecasts turned out to be a reliable indicator of forecast accuracy. When committee members' forecasts were close, forecasts were more accurate. This provided a mechanism for signaling the products whose sales were likely to be poorly forecast. This did not solve the problem of poorly forecast items, but it allowed the firm to commit first to production of items whose forecasts were likely to be accurate. By the time production had to begin on the problem items, information on early sales patterns would be available.

The team noticed that retailers were remarkably similar. That meant that even if only the first 20 percent of orders for a product were in, that information could dramatically improve forecasts. Production plans for these "trouble" items could now be committed with greater confidence. In this way, the firm could separate products into two categories: reactive and nonreactive. The nonreactive items are those for which the forecast is likely to be accurate. These are produced early in the season. The reactive items are those whose forecasts are updated later in the season from early sales figures. The firm's experience was that stockout and markdown costs were reduced from 10.2 percent of sales to 1.8 percent of sales on items that could be produced reactively. Sport Obermeyer was able to produce 30 percent of its season's volume reactively and experienced a cost reduction of about 5 percent of sales.

What are the lessons here? One is that even in cases where there is no statistical history, statistical methodology can be successfully applied to improve forecasting accuracy. Another is not to assume that things should be done a certain way. Sport Obermeyer assumed that consensus forecasting was the best approach. In fact, by requiring the buying committee to reach a consensus, valuable information was being ignored. The *differences* among individual forecasts proved to be important.

Source: This application is based on the work of a team from the Wharton School and the Harvard Business School. The results are reported in Fisher et al. (1994).

on exploiting the autocorrelation structure of a time series. While Box-Jenkins methods are based on statistical relationships in a time series, much of the basic theory goes back to Norbert Wiener's (1949) book, and before.

Box-Jenkins models are also known as **ARIMA** models. ARIMA is an acronym for **autoregressive integrated moving average**. The autocorrelation function plays a central role in the development of these models, a feature that distinguishes ARIMA models from the other methods discussed in this chapter. As we have assumed throughout this chapter, denote the time series of interest as $D_1, D_2, \ldots$. We will assume initially that the series is stationary. That is, $E(D_i) = \mu$ and $\mathrm{Var}(D_i) = \sigma^2$ for all $i = 1, 2, \ldots$. Practically speaking, *stationarity* means that there is no growth or decline in the series, and variation remains relatively constant. It is important to note that stationarity does not imply independence. Hence, it is possible that values of D_i and D_j are dependent random variables when $i \neq j$ even though their marginal density functions are the same. It is this dependence we wish to exploit. (Note: A more precise way to characterize stationarity is that the joint distribution of $D_t, D_{t+1}, \ldots, D_{t+k}$ is the same as the joint distribution of $D_{t+m}, D_{t+m+1}, \ldots, D_{t+m+k}$ at any time t and pair of positive integers m and k.)

The assumption of stationarity implies that the marginal distributions of any two observations separated by the same time interval are the same. That is, D_t and D_{t+1} have the same joint distribution as D_{t+m} and D_{t+m+1} for any $m \geq 1$. This implies that the covariance of D_t and D_{t+1} is exactly the same as the covariance of D_{t+m} and D_{t+m+1}. Hence, the covariance of any two observations depends only on the number of periods separating them. In this context, the covariance is also known as the autocovariance, since we are comparing two values of the same series separated by a fixed lag.

Let Cov (D_{t+m}, D_{t+m+k}) be the covariance of D_{t+m} and D_{t+m+k} given by $\mathrm{Cov}(D_{t+m}, D_{t+m+k}) = E(D_{t+m}D_{t+m+k}) - E(D_{t+m})E(D_{t+m+k})$ for any integer $k \geq 1$. The correlation coefficient of these two random variables is given by

$$\rho_k = \frac{\mathrm{Cov}(D_{t+m}, D_{t+m+k})}{\sqrt{\mathrm{Var}(D_{t+m})}\sqrt{\mathrm{Var}(D_{t+m+k})}}.$$

This is often referred to as the autocorrelation coefficient of lag k, since it refers to the correlation between all values of the series separated by k periods. These autocorrelation coefficients are typically computed for several values of k. It is these autocorrelation coefficients that will play the key role in building ARIMA models.

The autocorrelation coefficients are estimated from a history of the series. In order to guarantee reliable estimators, Box and Jenkins (1970) suggest that one have at least 72 data points of past history of the series. Hence, these models are only meaningful when one has a substantial and reliable history of the series being studied.

Estimating the Autocorrelation Function

Let $D_1, D_2, \ldots D_n$ be a history of observations of a time series. The **autocorrelation coefficient** of lag k is estimated from the following formula:

$$r_k = \frac{\sum_{t=k+1}^{n} (D_t - \bar{D})(D_{t-k} - \bar{D})}{\sum_{t=1}^{n} (D_t - \bar{D})^2},$$

where $\bar{D}$ is the sample mean (that is, the average) of the observed values of the series. Refer to r_k as sample autocorrelation coefficients. This calculation is typically done for 10 or 15 values of k. For most of the time series discussed earlier in the chapter, one identifies the appropriate patterns by just looking at a graph of the data. This is not the case here, however.

EXAMPLE 2.9

If observations are completely random (i.e., form a white noise process), then we would expect that there would be no significant autocorrelations among the observed values of the series. To test this, we generated a time series using the random number generator built into Excel. This series appears in Table 2–1. Each value is 100 times the RAND function. The reader can check that the sample autocorrelations for lags of 1 to 10 periods for these 36 observations are

$$
\begin{aligned}
r_1 &= 0.098 \\
r_2 &= -0.118 \\
r_3 &= 0.018 \\
r_4 &= -0.080 \\
r_5 &= 0.0752 \\
r_6 &= 0.006 \\
r_7 &= -0.270 \\
r_8 &= -0.207 \\
r_9 &= 0.117 \\
r_{10} &= 0.136.
\end{aligned}
$$

Table 2–1 Time Series with 36 Values Generated by a Random Number Generator (White Noise Series)

Period	Value	Period	Value	Period	Value
1	42	13	47	25	88
2	93	14	52	26	73
3	17	15	28	27	60
4	5	16	58	28	56
5	38	17	41	29	49
6	2	18	47	30	51
7	67	19	48	31	59
8	66	20	50	32	80
9	11	21	81	33	40
10	65	22	93	34	60
11	88	23	45	35	20
12	91	24	24	36	35

If we had an infinite number of observations and a perfect white noise series, we would expect that all of the autocorrelations would be zero. However, since we only have a finite series, there will be statistical variation resulting in nonzero values of the autocorrelations. The question is whether these values are significantly different from zero. (The data from Table 2–1 appear in Figure 2–11 and the autocorrelations in Figure 2–12.)

Several statistical tests have been proposed to answer this question. One is the Box-Pierce Q statistic, computed from the formula

$$Q = n\sum_{k=1}^{h} r_k^2,$$

where h is the maximum length of the lag being considered, and n is the number of observations in the series. Under the null hypothesis that the series is white noise, the Q statistic has the chi-square distribution with $(h - m)$ degrees of freedom, where m is the number of parameters in the model that has been fitted to the data. This test can be applied to any set of data not necessarily fitted to a specific model by setting $m = 0$.

Applying the formula for the Q statistic to the preceding autocorrelations, we obtain a value of $Q = 6.62$. Comparing this to the critical values of the chi-square statistic in Table 2–2 with 10 degrees of freedom (df), we see that this value is substantially smaller than any value in the table (for example, the value for a right tail probability of .1 at 10 degrees of freedom is 15.99). Hence, we could not reject the null hypothesis that the data form a white noise process, and we conclude that the autocorrelations are not significant.

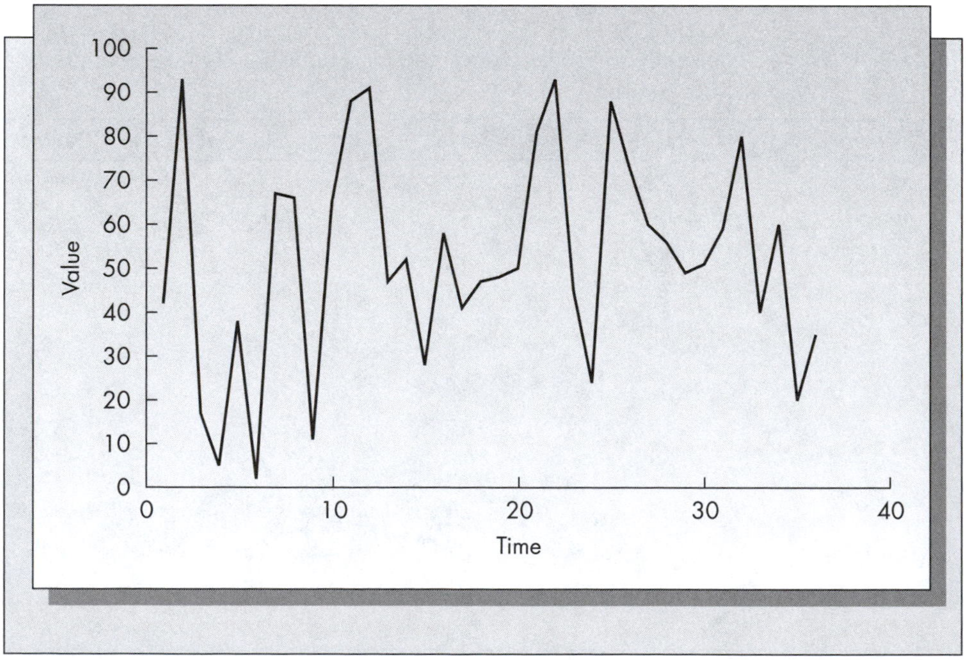

Figure 2–11 Plot of data in Table 2–1

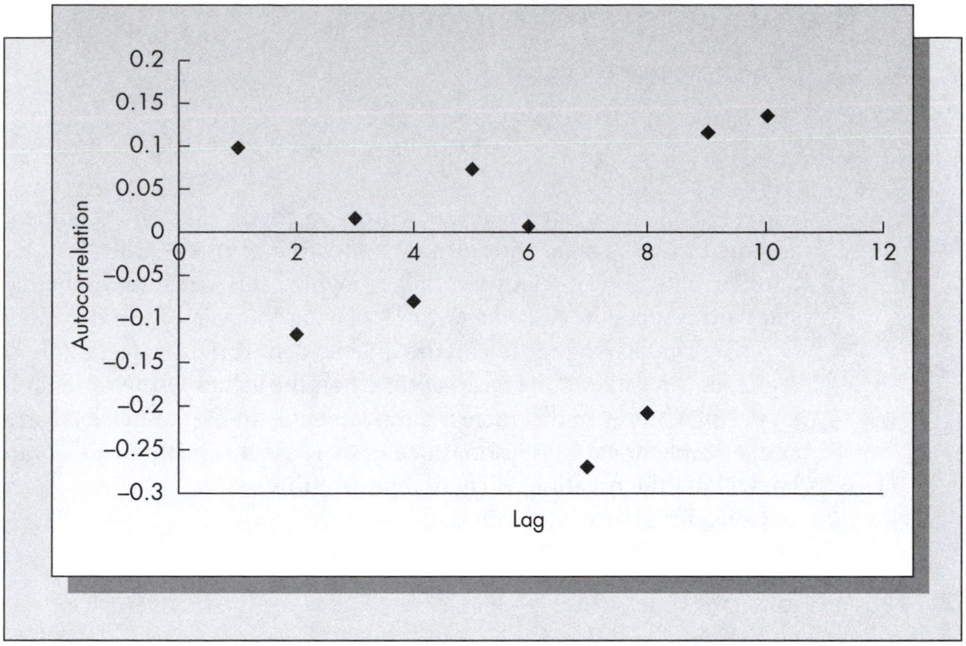

Figure 2–12 Plot of autocorrelations of series in Figure 2–11

Of course, the real interest is in those cases where the autocorrelations are significant. The basic idea behind the method is to compare the graph of the autocorrelation function (known as the correlogram) to those of known processes to identify the appropriate model. (Note: In addition to considering autocorrelations, most texts on Box-Jenkins analysis also recommend examining the partial autocorrelations. We will not discuss partial autocorrelations here, but the reader should be aware that these also provide information about the underlying structure of the process.)

Table 2–2 Partial Table of Critical Values of the Chi-Square Statistic

	Tail Value Probability		
df	**0.1**	**0.05**	**0.01**
1	2.70	3.84	6.63
2	4.60	5.99	9.21
3	6.25	7.81	11.34
4	7.77	9.48	13.27
5	9.23	11.07	15.08
6	10.64	12.59	16.81
7	12.01	14.06	18.47
8	13.36	15.50	20.09
9	14.68	16.91	21.66
10	15.99	18.30	23.20

The Autoregressive Process

The **autoregressive model** is

$$D_t = a_0 + a_1 D_{t-1} + a_2 D_{t-2} + \ldots + a_p D_{t-p} + \varepsilon_t,$$

where $a_0, a_1, \ldots, a_p$ are the linear regression coefficients and ε_t is the error term (generally assumed to be normal with mean 0 and variance σ^2 as earlier in the chapter). The reader familiar with linear regression will recognize this equation as being very similar to the standard regression equation with D_t playing the role of the dependent variable and D_{t-1}, $D_{t-2}, \ldots, D_{t-p}$ playing the role of the independent variables. Hence, the autoregressive model regresses the value of the series at time t on the values of the series at times $t - 1, t - 2, \ldots, t - p$. Note, however, that there is a fundamental difference between an autoregressive model and a simple linear regression, since in this case it is likely that the variables are correlated. We will use the notation AR(p) to represent this model.

Consider a basic AR(1) model,

$$D_t = a_0 + a_1 D_{t-1} + \varepsilon_t.$$

In order for the process to be stable, we require $|a_1| < 1$. If $a_1 > 0$, it means that successive values of the series are positively correlated—that is, large values tend to be followed by large values, and small values tend to be followed by small values. This means that the series will be relatively smooth. If $a_1 < 0$, then the opposite is true, so the series will appear much spikier. The difference is illustrated in a realization of two AR(1) processes in Figure 2–13. (Figure 2–13 was generated in Excel using the built-in RAND function and the normal variate generator given in Chapter 5 problem 34 with $a_0 = 10$ and $\sigma = 30$.)

Of course, it is unlikely one can recognize an AR(1) process by simply examining a graph of the raw data. Rather, one would examine the autocorrelation function. It is easy to show that the autocorrelation function for an AR(1) process is

$$\rho_j = a_1{}^j.$$

The autocorrelation functions for the two cases illustrated in Figure 2–13 are given in Figure 2–14. If the sample autocorrelation function of a series has a pattern resembling one of those in Figure 2–14, it would suggest that an AR(1) process is appropriate. The theoretical autocorrelation functions for higher-order AR processes can be more complex. [In the case of AR(2), the patterns are either similar to one of the two pictured in Figure 2–14 or follow a damped sine wave.] In practice, one would rarely include more than one or two AR terms in the model. Determining the autocorrelation structure for higher-order AR processes is not difficult. One must solve a series of linear equations known as the Yule-Walker equations. We will not elaborate further here but refer the interested reader to Box and Jenkins (1970).

The Moving-Average Process

The **moving-average process** provides another means of describing a stationary stochastic process used to model time series. The term *moving average* as used here should not be confused with the moving average discussed earlier in this chapter in Section 2.7. In that case, the moving average was an average of past values of the series. In this case, the moving average is a weighted average of past forecast errors.

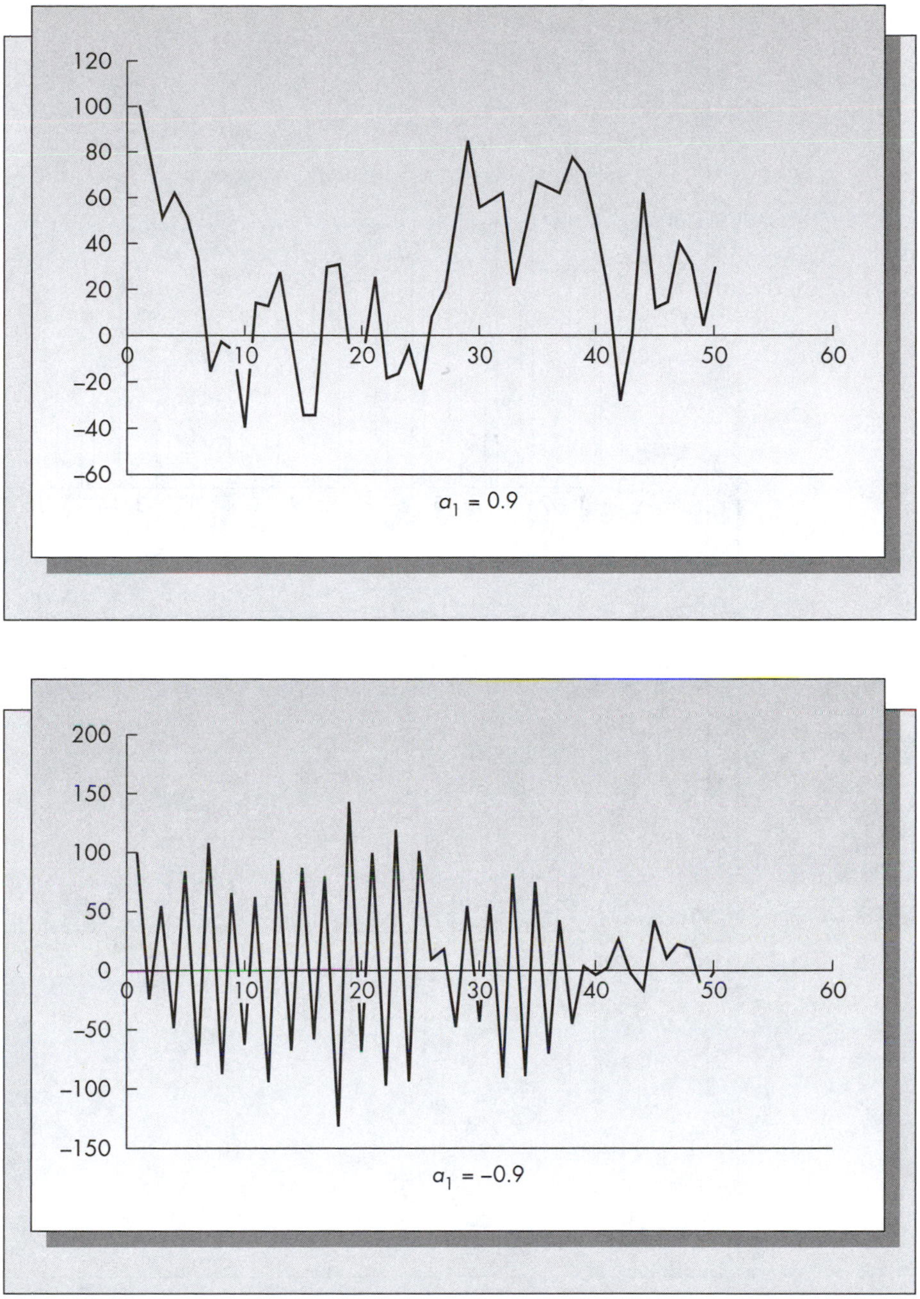

Figure 2–13 Realizations of an AR(1) process with $a_1 = 0.9$ and $a_1 = -0.9$, respectively

The general moving-average process has the form

$$D_t = b_0 - b_1\varepsilon_{t-1} - b_2\varepsilon_{t-2} - \dots - b_q\varepsilon_{t-q} + \varepsilon_t.$$

[The weights $b_1, b_2, \dots, b_q$ are shown with negative signs by convention.] We will denote this model as MA(q). The intuition behind the moving-average process is not as straightforward as the autoregressive process, but the two are related. Consider the first-order AR(1) process,

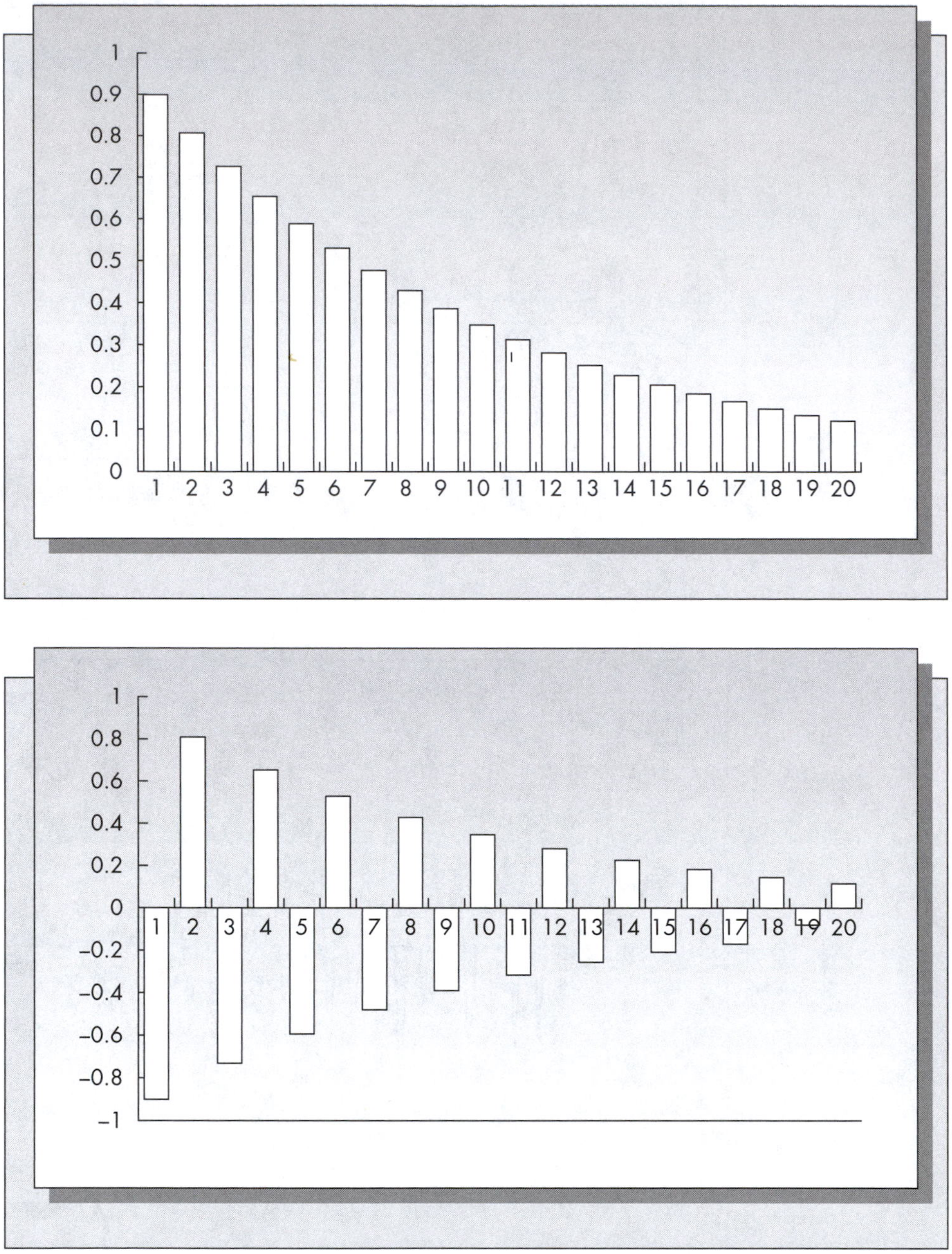

Figure 2–14 Theoretical autocorrelation functions for AR(1) processes pictured in Figure 2–13

$D_t = a_0 + a_1 D_{t-1} + \varepsilon_t$. By back-substituting for D_{t-1}, D_{t-2}, … , we see that the AR(1) process can also be written as

$$D_t = a_0 \sum_{i=0}^{\infty} a_1^i + a_1 \varepsilon_{t-1} + a_1^2 \varepsilon_{t-2} + \cdots + \varepsilon_t,$$

which is easily recognized as an MA(∞) process.

The reader will better understand the power of MA processes when we examine the autocorrelation function. Consider first the simplest MA process, namely, an MA(1) process. In this case,

$$D_t = b_0 - b_1\varepsilon_{t-1} + \varepsilon_t.$$

The autocorrelation structure for this case is very simple:

$$\rho_1 = \frac{-b_1}{1+b_1^2},$$
$$\rho_2 = \rho_3 = \cdots = 0.$$

Hence, an MA(1) process has only one significant autocorrelation at lag 1. In general, MA(1) processes tend to have spiky patterns independent of the sign of b_1, because successive errors are assumed to be uncorrelated. Figure 2–15 shows realizations of an MA(1) process with b_1 equal to −0.9 and 0.9, respectively. Finding the autocorrelation structure for higher-order MA processes is a challenging mathematical problem, requiring the solution of a collection of nonlinear equations. Again, we refer the interested reader to Box and Jenkins (1970). The main characteristic identifying an MA(q) process is that only the first q autocorrelations are nonzero.

Mixtures: ARMA Models

The real power in Box-Jenkins methodology comes in being able to mix both AR and MA terms. Any model that contains one or more AR terms and one or more MA terms is known as an **ARMA** model, for **autoregressive moving average**. An ARMA(p, q) model contains p autoregressive terms and q moving-average terms. For example, one would write an ARMA(1,1) model as

$$D_t = c + a_1 D_{t-1} - b_1\varepsilon_{t-1} + \varepsilon_t.$$

The ARMA(1,1) model is quite powerful and can describe many real processes accurately. It requires identification of the two parameters a_1 and b_1. The ARMA(1, 1) process is equivalent to an MA(∞) process and also equivalent to an AR(∞) process, thus showing the power that can be achieved with a parsimonious model. The autocorrelation function of the ARMA(1,1) process has characteristics of both the MA(1) and AR(1) processes. The autocorrelation at lag 1 is determined primarily by the MA(1) term, while autocorrelations at lags greater than 1 are determined by the AR(1) term.

ARIMA Models

Thus far we have assumed that the underlying stochastic process generating the time series is stationary. However, in many real problems, patterns such as trend or seasonality are present, which imply nonstationarity. The Box-Jenkins approach would be of limited utility if it were unable to address such situations. Fortunately, there is a simple technique for converting many nonstationary processes into stationary processes.

Consider first a process with a linear trend, such as the one pictured in Figure 2–7. Simple methods for dealing with linear trends were discussed in Section 2.8 of this chapter. How

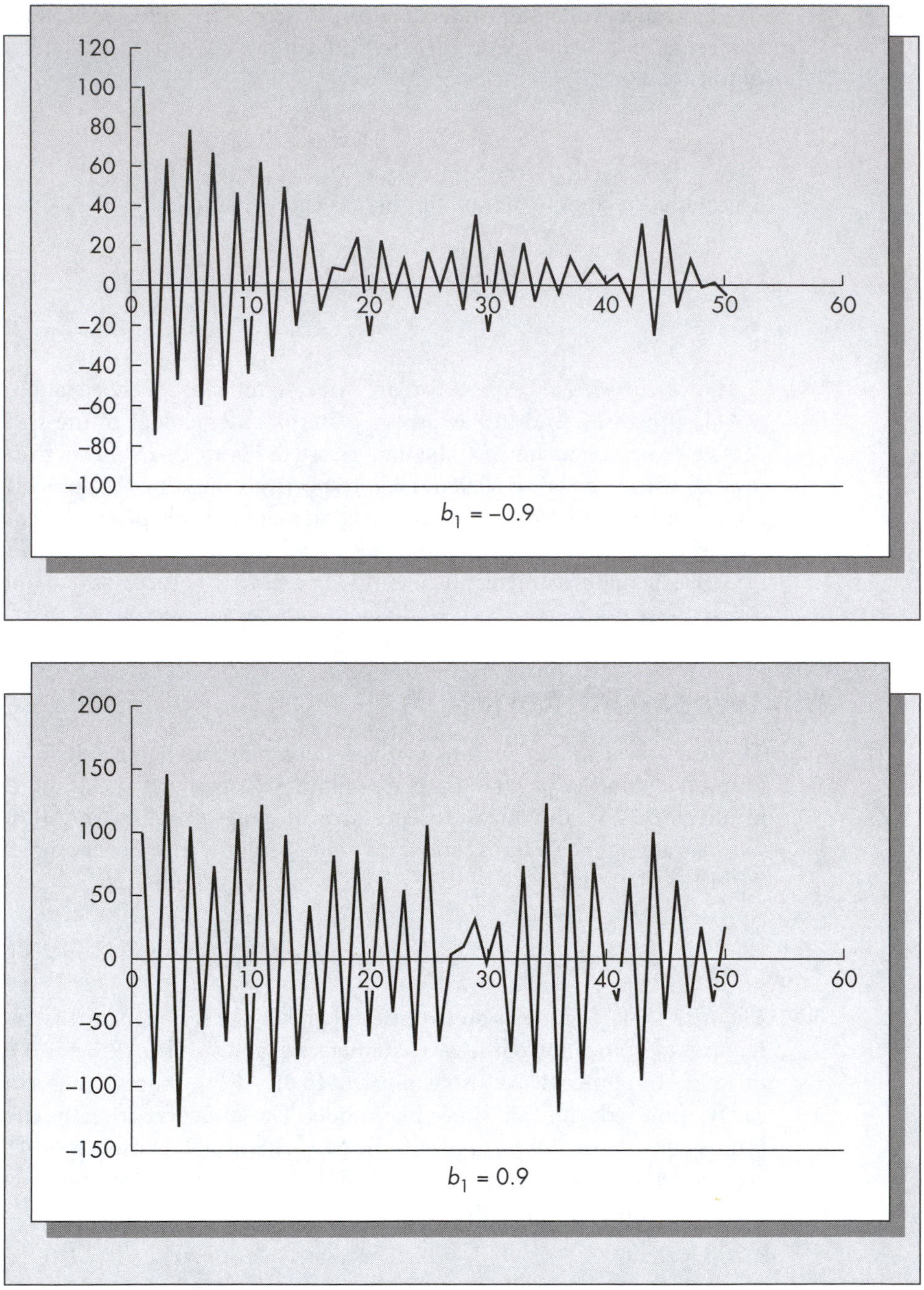

Figure 2–15 Realization of an MA(1) process with $b_1 = -0.9$ and $_b1 = 0.9$, respectively

can a process with a linear trend be converted to one with no trend? The answer turns out to be surprisingly simple. Suppose our original process D_t has a linear trend. Consider the new process U_t given by

$$U_t = D_t - D_{t-1}.$$

The process U_t tracks the slope of the original process. If the original process had a linear trend, then the slope should be relatively constant, implying that U_t would be stationary. In the same way, if the original process increased or decreased according to a quadratic function, differencing the first difference process (forming a second difference) will result in a stationary process. Differencing is the discrete analogue of a derivative. Going from the process U_t back to the original process D_t requires summing values of U_t, which is the discrete analogue of integration. For that reason, when differencing is introduced, we use the term **integration** to describe it. An ARMA model based on data derived from differencing is denoted **ARIMA**, which stands for **autoregressive integrated moving average**.

A common notation is $U_t = \nabla D_t$. If two levels of differencing were required to achieve stationarity, then one would need a double summation to retrieve the original series. In the case of two levels of differencing,

$$\nabla^2 D_t = D_t - D_{t-1} - (D_{t-1} - D_{t-2}) = D_t - 2D_{t-1} + D_{t-2}.$$

Differencing can also be used to remove seasonality from a time series. Suppose that a seasonal pattern repeats every 12 months. Then defining

$$U_t = \nabla^{12} D_t = D_t - D_{t-12}$$

would result in a process with no seasonality.

An ARIMA process has three constants associated with it: p for the number of autoregressive terms, d for the order of differencing, and q for the number of moving-average terms. The general ARIMA process would be denoted ARIMA(p, d, q). Thus, for example, ARMA(1,1) can also be denoted ARIMA(1,0,1). While these parameters can be any nonnegative integers, it is very rare that any of the values of p, d, or q would exceed 2. Thus, virtually all the ARIMA models one finds in practice correspond to values of 0, 1, or 2 for the parameters p, d, and q. While this might seem limiting, these few cases cover an enormous range of practical forecasting scenarios.

It is important to note that observations are lost when differencing. For example, if one uses a single level of differencing ($U_t = D_t - D_{t-1}$), the first difference process, U_t, will have one less observation than the original series. Similarly, each level of differencing will reduce the sample size by 1. Seasonal differencing effectively reduces the data set by the length of the season. Another way to represent the differencing operation is via the backshift operator. That is,

$$BD_t = D_{t-1},$$

which means that one would represent the first difference process as

$$U_t = D_t - D_{t-1} = D_t - BD_t = (1 - B)D_t.$$

The backshift operator can also be used to simplify the notation for AR, MA, and ARMA models. Consider the simple AR(1) process given by $D_t = a_0 + a_1 D_{t-1} + \varepsilon_t$. Writing this in the form $D_t - a_1 D_{t-1} = a_0 + \varepsilon_t$ reduces the process to the alternate notation $(1 - a_1 B)D_t = a_0 + \varepsilon_t$. Similarly, the reader should check that the MA(1) model using the backshift operator is $D_t = b_0 + (1 - b_1 B)\, \varepsilon_t$.

Using ARIMA Models for Forecasting

Given an ARIMA model, how does one go about using it to provide forecasts of future values of the series? The approach is similar to that discussed earlier in this chapter for the simpler methods. For example, a forecast based on a simple AR(p) model is a weighted average of past p observations of the series. A forecast based on an MA(q) model is a weighted average of the past q forecast errors. And finally, one must take into account the level of differencing and any transformations made on the original data.

As an example, consider an ARIMA (1, 1, 1) model (that is, a model having one level of differencing, one moving-average term, and one autoregressive term). This can be represented using the backshift notation as

$$(1 - B)(1 + a_1 B)D_t = c + (1 - b_1 B)\,\varepsilon_t,$$

or as

$$(1 + a_1 B)\nabla D_t = c + (1 - b_1 B)\,\varepsilon_t.$$

Writing out the model without backshift notation we have

$$D_t = c + (1 + a_1)D_{t-1} - a_1 D_{t-2} + \varepsilon_t - b_1 \varepsilon_{t-1}.$$

Suppose this model has been fitted to a time series with the result that $c = 15$, $a_1 = 0.24$, and $b_1 = 0.70$. Suppose that the last five values of the time series used to fit this model were 31.68, 29.10, 43.15, 56.74, and 62.44 based on a total of 76 observations. The first period in which one can forecast is period 77. The one-step-ahead forecast for period 77 made at the end of period 76 is

$$\widehat{D}_{77} = 15 + (1 + 0.24)D_{76} - 0.24D_{75} - 0.70\varepsilon_{76}.$$

$\widehat{D}_{77}$ is the conditional expected value of D_{77} having observed the demand in periods 1, 2, … , 76. Since this is the first forecast made for this series, there is no observed previous value of the error, and the final term drops out. Hence, the one-step-ahead forecast made in period 76 for the demand in period 77 is $15 + (1.24)(62.44) - (0.24)(56.74) = 78.81$. Now, suppose we observed a value of 70 for the series in period 77. That means that the forecast error observed in period 77 is $\varepsilon_{77} = 78.81 - 70 = 8.81$. The one-step-ahead forecast for period 78 made in period 77 would be $15 + (1.24)(70) - 0.24(62.44) - 0.70(8.81) = 86.25$.

When using an ARIMA model for multiple-step-ahead forecasts, the operative rule is to use the forecasts for the unobserved demand and use zero for the unobserved errors. Thus in the preceding example, a two-step-ahead forecast made at the end of period 76 for demand in period 78 would be based on the assumption that the observed demand in period 77 was the one-step-ahead forecast, 86.25. The observed forecast error for period 77 would be assumed to be zero.

Summary of the Steps Required for Building ARIMA Models

There are four major steps required for building Box-Jenkins forecasting models.

1. *Data transformations.* The Box-Jenkins methodology is predicated on starting with a stationary time series. To be certain that the series is indeed stationary, several preliminary steps might be required. We know that differencing eliminates trend and seasonality.

However, if the mean of the series is relatively fixed, it still may be the case that the variance is not constant, thus possibly requiring a transformation of the data (for example, stock market data are often transformed by the logarithm function).

2. *Model identification.* This step refers to determining exactly which ARIMA model seems to be most appropriate. Proper model identification is both art and science. It is difficult, if not impossible, to identify the appropriate model by only examining the series itself. It is far more effective to study the sample autocorrelations and partial autocorrelations to discern patterns that match those of known processes. In some cases, the autocorrelation structure will point to a simple AR or MA process, but it is more common that some mixture of these terms would be required to get the best fit. However, one must not add terms willy-nilly. The operative concept is parsimony—using the most economical model that adequately describes the data.
3. *Parameter estimation.* Once the appropriate model has been identified, the optimal values of the model parameters (i.e., $a_0, a_1, \ldots, a_p$ and $b_0, b_1, \ldots, b_q$) must be determined. Typically, this is done via either least squares fitting methods or the method of maximum likelihood. In either case, this step is done by a computer program.
4. *Forecasting.* Once the model has been identified and the optimal parameter values determined, the model provides forecasts of future values of the series. Box-Jenkins models are most effective in providing one-step-ahead forecasts, but they can also provide multiple-step-ahead forecasts.
5. *Evaluation.* The pattern of residuals (forecast errors) can provide useful information regarding the quality of the model. The residuals should form a white noise (i.e., random) process with zero mean. Residuals should be normally distributed as well. When there are patterns in the residuals, it suggests that the forecasting model can be improved.

CASE STUDY Using Box-Jenkins Methodology to Predict Monthly International Airline Passenger Totals

This study is based on data (Brown, 1962) analyzed by Box and Jenkins (1970) using ARIMA methods. It illustrates the basic steps in transforming data and building ARIMA models. The data represent the monthly international airline sales from the period January 1949 to December 1960. The raw data appear in Table 2–3 and are pictured in Figure 2–16. From the figure, it is clear that there are several nonstationarities in this data. First, there is clearly an increasing linear trend. Second, there is seasonality, with a pattern repeating yearly. Third, there is increasing variance over time. In cases where the mean and variance increase at a comparable rate (which would occur if the series is increasing by a fixed percentage), a logarithmic transformation will usually eliminate the nonstationarity due to increasing variance. (Changing variance is known as heteroscedasticity, and constant variance is known as homoscedasticity.) Applying a natural log transformation to the data yields a homoscedastic series as shown in Figure 2–17.

Next, we need to apply two levels of differencing to eliminate both trend and seasonality. The trend is eliminated by applying a single level of differencing and the seasonality by applying 12 periods of differencing. After these three transformations are applied to the original data, the resulting data appear in Figure 2–18. Note that the transformed data now appear to form a random white noise process centered at zero showing neither trend nor seasonality.

It is to this set of data that we wish to fit an ARMA model. To do so, we determine the sample autocorrelations. While this can be accomplished with many of the software programs available

Table 2–3 International Airline Passengers: Monthly Totals (Thousands of Passengers), January 1949–December 1960*

	Jan.	Feb.	Mar.	Apr.	May	June	July	Aug.	Sept.	Oct.	Nov.	Dec.
1949	112	118	132	129	121	135	148	148	136	119	104	118
1950	115	126	141	135	125	149	170	170	158	133	114	140
1951	145	150	178	163	172	178	199	199	184	162	146	166
1952	171	180	193	181	183	218	230	242	209	191	172	194
1953	196	196	236	235	229	243	264	272	237	211	180	201
1954	204	188	235	227	234	264	302	293	259	229	203	229
1955	242	233	267	269	270	315	364	347	312	274	237	278
1956	284	277	317	313	318	374	413	405	355	306	271	306
1957	315	301	356	348	355	422	465	467	404	347	305	336
1958	340	318	362	348	363	435	491	550	404	359	310	337
1959	360	342	406	396	420	472	548	559	463	407	362	405
1960	417	391	419	461	472	535	622	606	508	461	390	432

*144 observations.
Source: From Box and Jenkins (1970), p. 531.

for forecasting (including general statistical packages, such as SAS), we have done the calculations directly in Excel using the formulas for sample autocorrelations given earlier. The autocorrelations for lags of 1 to 12 periods for the series pictured in Figure 2–18 appear in Table 2–4.

Although not explicitly discussed in this section, when both seasonal differencing and period-to-period differencing are applied simultaneously, one must determine an ARMA model for each level of differencing. That is, one would like to find an ARMA model corresponding

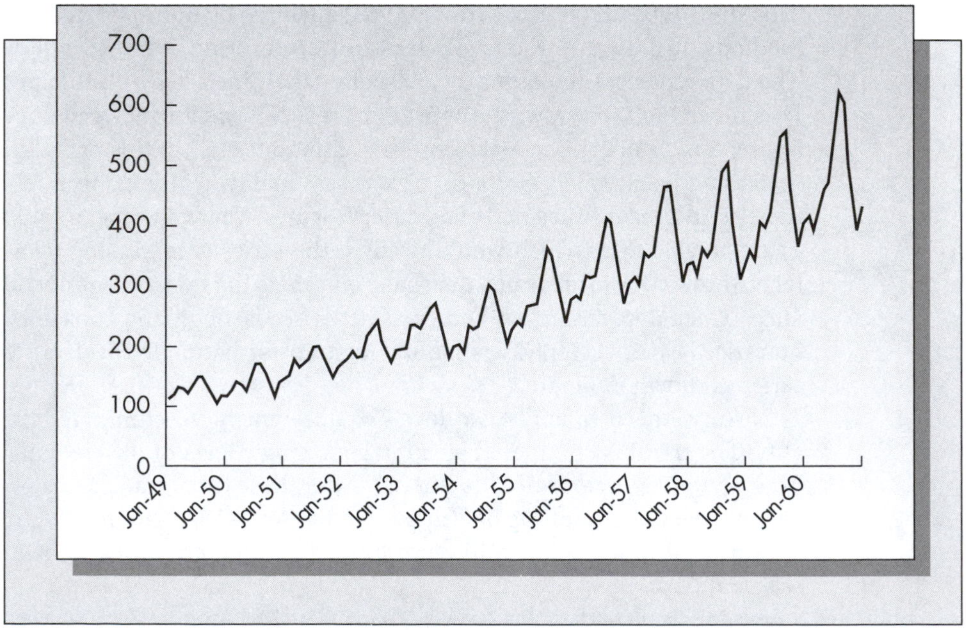

Figure 2–16 International airline passengers (thousands)

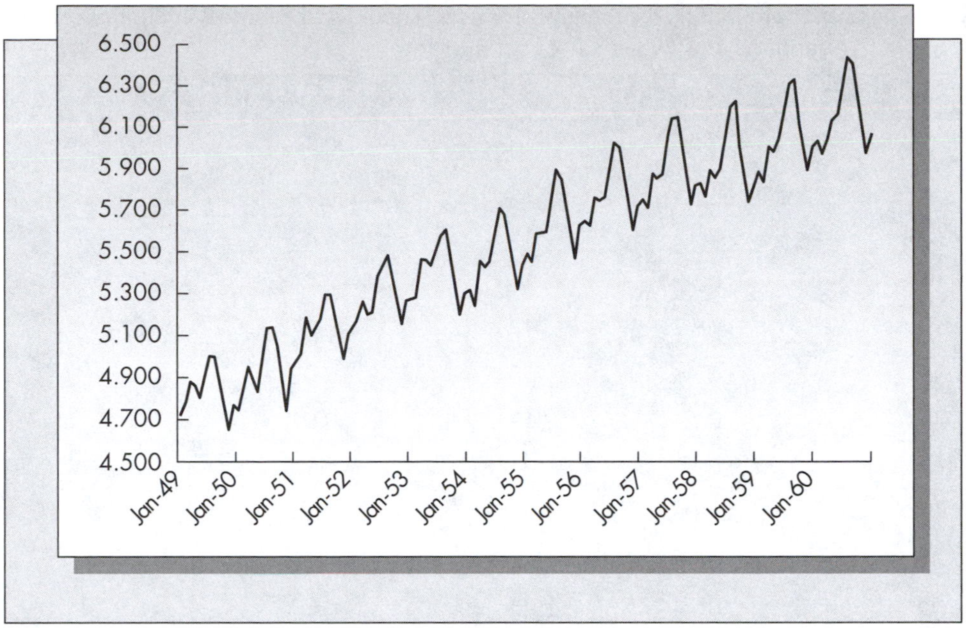

Figure 2–17 Natural log of international airline passengers

to the single level of differencing and to the seasonal level of differencing. Thus, when examining the autocorrelation function, we look for patterns at lags of 1 period and patterns at lags of 12 periods. From Table 2–4, it is clear that there are significant autocorrelations at lags of exactly 1 and 12 periods. This suggests that MA(1) models are appropriate for both differencing levels.

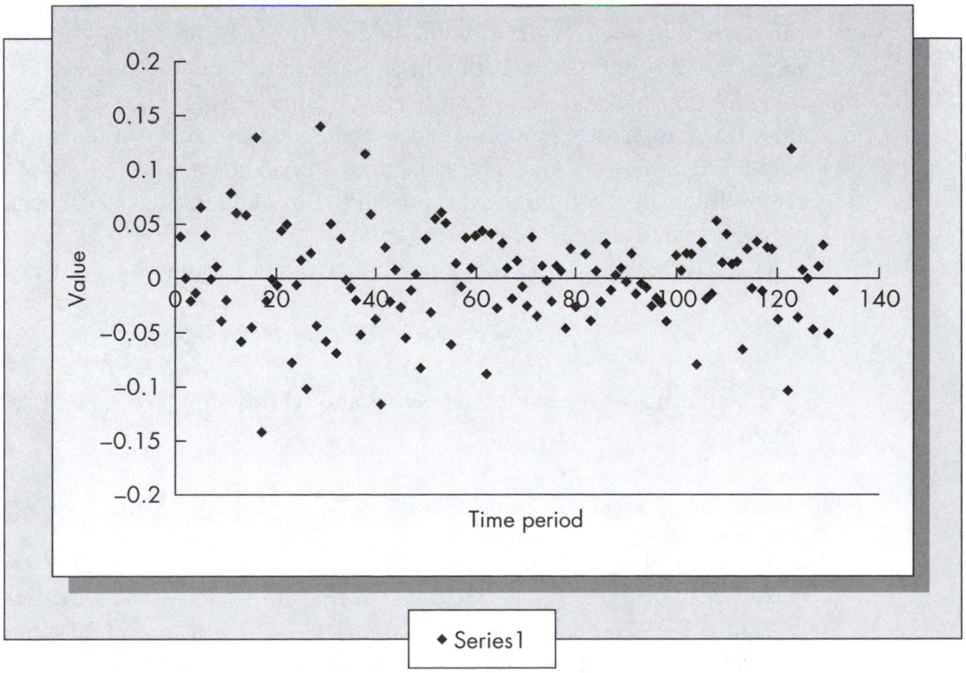

Figure 2–18 Natural log of airline data with single and seasonal differencing

Table 2–4 Autocorrelations for the Transformed Airline Data Pictured in Figure 2–18 (after taking logarithms and two levels of differencing)

Lag	Autocorrelation	Lag	Autocorrelation	Lag	Autocorrelation
1	–0.34	13	0.15	25	–0.10
2	0.11	14	–0.06	26	0.05
3	–0.20	15	0.15	27	–0.03
4	0.02	16	–0.14	28	0.05
5	0.06	17	0.07	29	–0.02
6	0.03	18	0.02	30	–0.05
7	–0.06	19	–0.01	31	–0.05
8	0.00	20	–0.12	32	0.20
9	0.18	21	0.04	33	–0.12
10	–0.08	22	–0.09	34	0.08
11	0.06	23	0.22	35	–0.15
12	–0.39	24	–0.02	36	–0.01

Note that if we let z_t represent the log-transformed series, we would denote the series pictured in Figure 2–18 as $(1 - B)(1 - B^{12})\, z_t$ or as $\nabla\nabla^{12}z_t$ to indicate that both first-order and 12th-order differencing were applied. Since we are assuming an MA(1) model for both the first difference process and the 12th-order difference process, the model we wish to fit can be denoted $\nabla\nabla^{12}z_t = c + (1 - b_1B)(1 - b_2B^{12})e_t$, where the parameters b_1 and b_2 are to be determined based on one of several fitting criteria. The exact values we obtain for the parameters will depend on the optimization method we use, but generally all methods will yield similar values. Least squares is probably the most common method used, but maximum likelihood and Bayesian methods have also been suggested. Using XLSTAT, a program that contains an ARIMA forecasting module and is embedded in Excel, we obtain the parameter values $b_1 = 0.333$ and $b_2 = 0.544$. The value of the constant c is small enough to be ignored. [These values differ slightly from those reported in Box and Jenkins (1970) because XLSTAT used a different search algorithm.]

When forecasting using this model, it is convenient to write it out explicitly in the form

$$z_t - z_{t-1} - (z_{t-12} - z_{t-13}) = \varepsilon_t - b_1\, \varepsilon_{t-1} - b_2(\varepsilon_{t-12} - b_1\, \varepsilon_{t-13}).$$

Substituting for the parameter values and rearranging terms, the forecasting equation we obtain for the log series is

$$z_t = z_{t-1} + z_{t-12} - z_{t-13} + \varepsilon_t - 0.333\varepsilon_{t-1} - 0.544\varepsilon_{t-2} + 0.181\varepsilon_{t-13}.$$

To forecast the original series, we apply the antilog to namely, $D_t = \exp(z_t)$. Because of the two levels of differencing, period 14 is the first period for which we can determine a forecast. In Figure 2–19, we show the original series starting at period 14 and the one-step-ahead forecast using the preceding ARIMA model. Note how closely the ARIMA forecast tracks the original series.

A Simple Arima Model Predicts the Performance of the U.S. Economy

In years past, a very complex and very large regression model known as the FRB-MIT-PENN (FMP) model (for Federal Reserve Bank—Massachusetts Institute of Technology—University of Pennsylvania) was used to predict several basic measures of the U.S. economy. This model required massive amounts of data and past history. Nelson (1972) employed the ARIMA methodology outlined in this section to obtain predictors for many of the same fundamental measures of the U.S. economy considered in the FMP model. These include gross national product, consumer's expenditures on durable goods, nonfarm inventory investment, and several others.

Perhaps the most interesting case is the prediction of the gross national product. The ARIMA model he obtained is surprisingly simple:

$$z_t = z_{t-1} + 0.615(z_{t-1} - z_{t-2}) + 2.76 + \varepsilon_t,$$

which is easily seen to be an AR(1) model with one level of differencing. What is most impressive and surprising is that the forecast errors obtained from this and Nelson's other ARIMA models had lower forecast errors in predicting future values of these measures than did the complex FMP model. Again, this points to the power of these methods in providing accurate forecasts in a variety of scenarios.

Box-Jenkins Modeling—A Critique

The preceding case study highlights the power of ARIMA models. However, one must be aware that there are only a limited number of situations in which one would or could use them. An important requirement is that one must have a substantial history of observations. Typical recommendations vary between 50 and 100 observations of past history, and that is for a nonseasonal series. When seasonality is present, the requirement is more severe. In some sense, each season of data is comparable to a single observation.

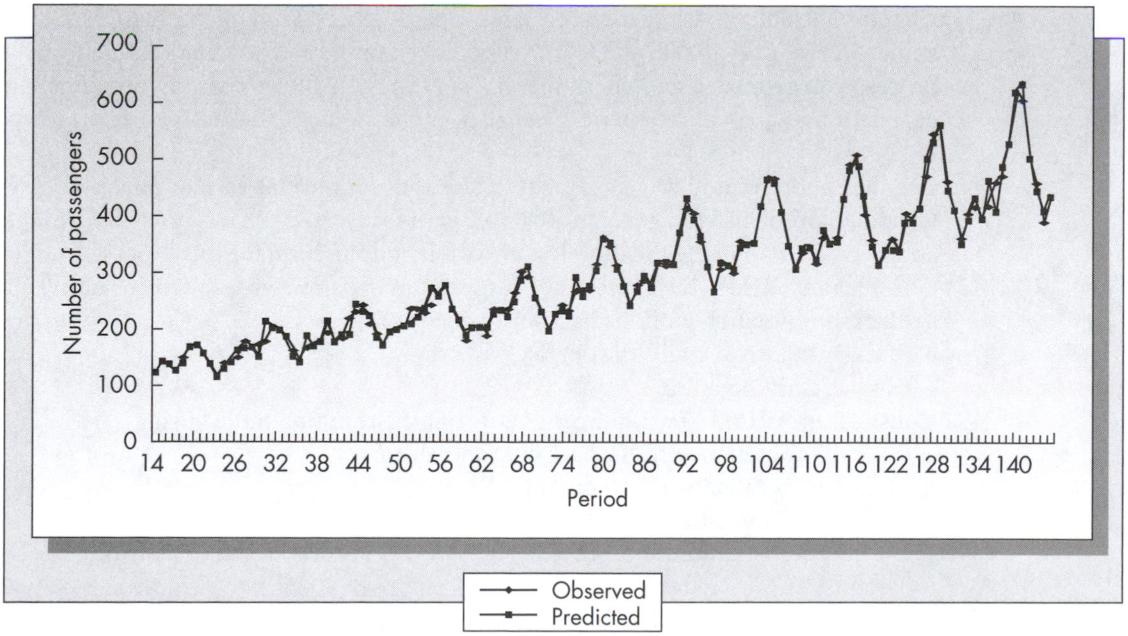

Figure 2–19 Observed versus predicted number of airline sales

In the operations context, one would be hard-pressed to find many applications where that much history is available. Style goods, for example, are typically managed with a very small amount of data, if any (see the Sport Obermeyer Snapshot Application). Even when enough data are available, it may not make sense to invest the amount of time and energy required to develop an ARIMA model. Consider forecasting usage rates in a Walmart store, for example, where one must manage tens of thousands of SKUs. ARIMA models are most useful for forecasting economic series with substantial history (such as GNP) or continuous processes (such as chemical processes or stock market prices). In the latter case, one can conceivably choose a small enough time increment size to generate any number of observations.

Another shortcoming of ARIMA models is that they are not easily updated. If the problem parameters change, one must redo the analysis to find a new model, or at least rerun the software to find new optimal values of the parameters.

Even with all of these shortcomings, Box-Jenkins methods are very powerful. In the right situation, they can result in much more accurate short-term forecasts than any of the other methods discussed in this chapter.

PROBLEMS FOR SECTION 2.10

38. Consider the white noise process data in Table 2–1. Enter this data into Excel and generate the sample autocorrelations for lags of 1 to 10 periods, using the formula for the sample autocorrelations (r_k). Alternatively, if you have access to an ARIMA computing module, generate these autocorrelations using your software.
39. Use the random number generator in Excel and the normal variate generator given in in Chapter 5 problem 34 to generate a sample path for an AR(1) process such as the ones shown in Figure 2–13 for $a_1 = 0.7$. Compute the sample autocorrelation function for the process you generated and check that the correlogram has the same structure as shown in Figure 2–14a.
40. Use the random number generator in Excel and the normal variate generator given in Chapter 5 problem 34 to generate a sample path for an MA(1) process such as the ones shown in Figure 2–15 for $b_1 = 0.7$. Compute the sample autocorrelation function for the process you generated and check that the correlogram shows only one significant autocorrelation at a lag of one period. What does the value of this auto-correlation appear to be?
41. Use the random number generator in Excel and the normal variate generator given in Chapter 5 problem 34 to generate a sample path for an ARMA(1,1) process. Use $a_1 = 0.8$ and $b_1 = -0.6$. Compute the sample autocorrelation function for this process.
42. Consider an ARIMA(2,1,1) process. For the notation below, write out the resulting model.
    a. Backshift notation with the backshift operator B.
    b. Backshift notation with the operator ∇.
    c. No backshift notation.
43. Consider an ARIMA(0,2,2) process. Write out the resulting model using
    a. Backshift notation with the backshift operator B.
    b. Backshift notation with the operator ∇.
    c. No backshift notation.
44. Using back-substitution, show that an ARMA(1, 1) model may be written as either an AR (∞) or an MA (∞) model.
45. Consider the seasonal time series pictured in Figure 2–8. What level of differencing would be required to make this series stationary?

46. The U.S. Federal Reserve in St. Louis stores a host of economic data on its website (https://fred.stlouisfed.org/categories/106). Download the following time series.
    a. U.S. GNP.
    b. Annual Federal Funds rate.
    c. Consumer Price Index.
    In each case, use a minimum of 50 data points over a period not including the most recent 25 years. Graph the data points; determine the level of differencing that appears to be required. Once you have obtained a stationary process, use the methods outlined in this section to arrive at an appropriate ARIMA model for each case. Compare the most recent 25 years of observations with the predictions obtained from your model using both one-step-ahead and two-step-ahead forecasts.

2.11 PRACTICAL CONSIDERATIONS

Model Identification and Monitoring

Determining the proper model depends both on the characteristics of the history of observations and on the context in which the forecasts are required. When historical data are available, they should be examined carefully in order to determine if obvious patterns exist, such as trend or seasonal fluctuations. Usually, these patterns can be spotted by graphing the data. Statistical tests, such as significance of regression, can be used to verify the existence of a trend. Identifying complex relationships requires more sophisticated methods. The **sample autocorrelation function** can reveal intricate relationships that simple graphical methods cannot as we saw in Section 2.10.

Once a model has been chosen, forecasts should be monitored regularly to see if the model is appropriate or if some unforeseen change has occurred in the series. As we indicated, a forecasting method should not be biased. That is, the expected value of the forecast error should be zero. In addition to the methods mentioned in Section 2.6, one means of monitoring the bias is the **tracking signal** (Trigg, 1964). Following earlier notation, let e_t be the observed error in period t and $|e_t|$ the absolute value of the observed error. The smoothed values of the error and the absolute error are given by

$$E_t = \beta e_t + (1 - \beta)E_{t-1},$$
$$M_t = \beta|e_t| + (1 - \beta)M_{t-1}.$$

The tracking signal is the ratio

$$T_t = \left|\frac{E_t}{M_t}\right|.$$

If forecasts are unbiased, the smoothed error E_t should be small compared to the smoothed absolute error M_t. Hence, a large value of the tracking signal indicates biased forecasts, which suggest that the forecasting model is inappropriate. The value of T_t that signals a significant bias depends on the smoothing constant β. For example, a value of T_t exceeding 0.51 indicates nonrandom errors for a β of .1 (Trigg, 1964). The tracking signal also can be used directly as a variable smoothing constant. This is considered in problem 55.

Simple versus Complex Time Series Methods

The literature on forecasting is voluminous. In this chapter we have touched only on a number of fairly simple techniques. The reader is undoubtedly asking, do these methods actually work? The results from the literature suggest that the simplest methods are often as accurate as sophisticated ones. J. Scott Armstrong (1984) reviews 25 years of forecasting case studies with the goal of ascertaining whether or not sophisticated methods work better. In comparing the results of 39 case studies, he found that in 20 cases sophisticated methods performed about as well as simple ones, in 11 cases they outperformed simple methods, and in 7 cases they performed significantly worse.

SNAPSHOT APPLICATION

Nate Silver Presidential Election Forecasts

In a stunning victory, President Barack Obama won re-election in 2012 with 332 electoral votes versus Governor Romney's 206. This was in stark contrast to the Romney victory confidently predicted by Republican pundits just days before the election. Perhaps this was a result of a Gallup poll weeks before the election that showed Romney with a lead in the popular vote. However, American presidents are not elected by popular vote; they are elected on a state by state basis through the Electoral College.

A statistician named Nate Silver forecasted the outcome of the election exactly. He based his forecasts on careful analysis of state by state (and even county by county) polls. His methodology had previously proven very successful in baseball. He also accurately predicted the outcome of the 2008 presidential election, successfully forecasting the results for 49 out of 50 states. In 2012 he was correct in all fifty states, predicting not only an Obama victory with probability exceeding 90 percent but also predicting the number of electoral votes for each candidate exactly.

In the 2016 election, almost every poll predicted Hillary Clinton would win the presidency (Silver, 2017). When Trump received 306 Electoral College votes, political pundits proclaimed the death of data—blaming pollsters and forecasters. Silver, however, said that the data analytics did not fail; rather, major media outlets did not understand probability and relied on shopworn conventional wisdom.

Silver believes the pundits ignored and or misinterpreted the available evidence. Instead, they chose a narrative that they failed to adjust as information became available. Silver cautioned that predictions are more difficult in a general election without an incumbent and with an average economy. He also pointed to the number of undecided voters, saying it only required 1 in every 50 voters to change their minds to have a 7-point lead fall to a 3-point lead. At the end of the campaign, the polls showed only a 3 percent separation between the candidates. Despite polls showing a highly competitive race in the Electoral College, many people were convinced the Electoral College would be an advantage to Clinton. It was an enormous disadvantage, with more Trump voters in swing states.

There were errors in interpretation. Using the same data, one model estimated Trump's chances about 30 times higher than another. Rather than perceiving the discrepancies as an indication of the challenges of poll interpretation, all models were criticized uniformly for inaccurately forecasting a Clinton victory. Many predicted that Trump had no chance of winning, but the Silver model put his chances at almost 30 percent. Silver believes his general election model was sound; it correctly captured that the Electoral College was a Clinton vulnerability. He thinks that was a successful application of modeling but acknowledges that the average person would rank a 30 percent forecast as a poor prediction.

A more sophisticated forecasting method is one that requires the estimation of a larger number of parameters from the data. Trouble can arise when these parameters are estimated incorrectly. To give some idea of the nature of this problem, consider a comparison of simple moving averages and regression analysis for the following series: 7, 12, 9, 23, 27. Suppose that we are interested in forecasting at the end of period 5 for the demand in period 15 (that is, we require $F_{5,15}$). The five-period moving-average forecast made at the end of period 5 is 15.6, and this would be the forecast for period 15. The least squares fit of the data is $\widehat{D}_t = 0.3 + 5.1t$. Substituting $t = 15$, we obtain the regression forecast of 76.8. In Figure 2–20 we picture the realization of the demand through period 15. Notice what has happened. The apparent trend that existed in the first five periods was extrapolated to period 15 by the regression equation. However, there really was no significant trend in this particular case. The more complex model gave *significantly* poorer results for the long-term forecast.

There is some evidence that the arithmetic average of forecasts obtained from different methods is more accurate than a single method (see Makridakis and Winkler, 1983). This is perhaps because often a single method is unable to capture the underlying signal in the data and different models capture different aspects of the signal. (For a discussion of this phenomenon, see Armstrong, 1984.)

What do these observations tell us about the application of forecasting techniques to production planning? At the aggregate level of planning, forecast accuracy is extremely important and multiple-step-ahead forecasts play an integral role in the planning of workforce and production levels. For that reason, blind reliance on time series methods is not advised at this level. At a lower level in the system, such as routine inventory management for spare parts, the use of simple time series methods such as moving averages or exponential smoothing makes a great deal of sense. At the individual item level, short-term forecasts for a large number of items are required, and monitoring the forecast for each item is impractical at best. The risk of severe errors is minimized if simple methods are used.

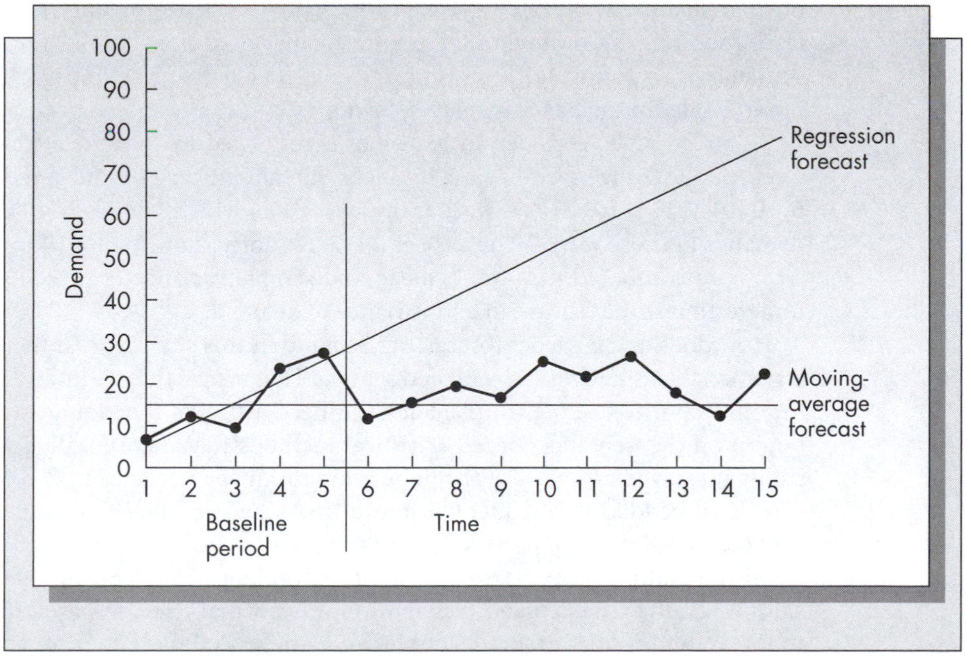

Figure 2–20 The difficulty with long-term forecasts

2.12 OVERVIEW OF ADVANCED TOPICS IN FORECASTING

Simulation as a Forecasting Tool

Computer simulation is a powerful technique for tackling complex problems. A computer simulation is a description of a problem reduced to a computer program. The program is designed to recreate the key aspects of the dynamics of a real situation. When a problem is too complex to model mathematically, simulation is a popular alternative. By rerunning the program under different starting conditions and/or different scenarios, one can, by a trial-and-error process, discover the best strategy for managing a system.

Simulation is a common tool for modeling manufacturing planning problems such as complex material flow problems in the plant. It is less commonly used as a forecasting tool. As one notable example, Compaq Computer experimented with a powerful simulation-based forecasting tool for assisting with the process of new product introductions (McWilliams, 1995). The program recommended the optimal timing and pricing of new product introductions by incorporating forecasts of component availability and price changes, fluctuating demand for a given feature or price, and the impact of rival models. The company credited the tool with some highly profitable strategic decisions.

Forecasting Demand in the Presence of Lost Sales

Retailers rely heavily on forecasting. Basic items (items that don't change appreciably from season to season, such as men's dress shirts) generally have substantial sales history, arguing for the use of time series methods to forecast demand. However, there is an important difference between what is observed and what one wants to forecast. The goal is to forecast *demand,* but one only observes *sales*. What's the difference? Suppose a customer wants to buy a blouse in a certain size and color and finds it's not available on the shelf? What will she do? Perhaps she will place a special order with a salesperson, but, more likely, she will just leave the store and try to find the product somewhere else. This is known as a lost sale. The difficulty is that most brick-and-mortar retailers have no way to track lost sales. Thus, they observe sales but need to estimate demand.

As an example, consider an item that is restocked to 10 units at the beginning of each week. Suppose that over the past 15 weeks the sales history for the item was 7, 5, 10, 10, 8, 3, 6, 10, 10, 9, 5, 0, 10, 10, 4. Consider those weeks in which sales were 10 units. What were the demands in those weeks? The answer is that we don't know. We only know that it was *at least* 10. If you computed the sample mean and sample variance of these numbers, they would underestimate the true mean and variance of demand.

How does one go about forecasting demand in this situation? In the parlance of classical statistics, this is known as a censored sample. That means that we know the values of demand for only a portion of the sample. For the other portion of the sample, we know only a lower bound on the demand. Special statistical methods that incorporate censoring give significantly improved estimates of the population mean and variance in this case. These methods can be embedded into sequential forecasting schemes, such as exponential smoothing, to provide significantly improved forecasts.

Steven Nahmias (1994) considered the problem of forecasting in the presence of lost sales when the true demand distribution was normal. He compared the method of maximum likelihood for censored samples and a new method, either of which could be incorporated into exponential smoothing routines. He also showed that both of these methods would result in substantially improved forecasts of both the mean and the variation of the demand.

To see how dramatic this difference can be, consider a situation in which the true weekly demand for a product is a normal random variable with mean 100 and standard deviation 30. Suppose that items are stocked up to 110 units at the start of each week. Exponential smoothing is used to obtain two sets of forecasts: the first accounts for lost sales (includes censoring) and the second does not (does not including censoring). Figures 2–21 and 2–22 show the estimators for the mean and the standard deviation, respectively, with and without

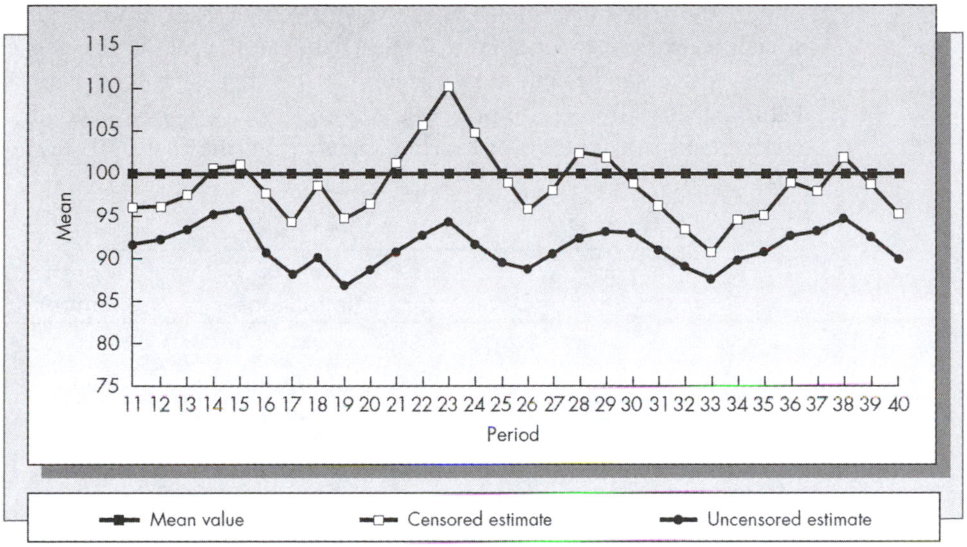

Figure 2–21 Tracking the mean when lost sales are present

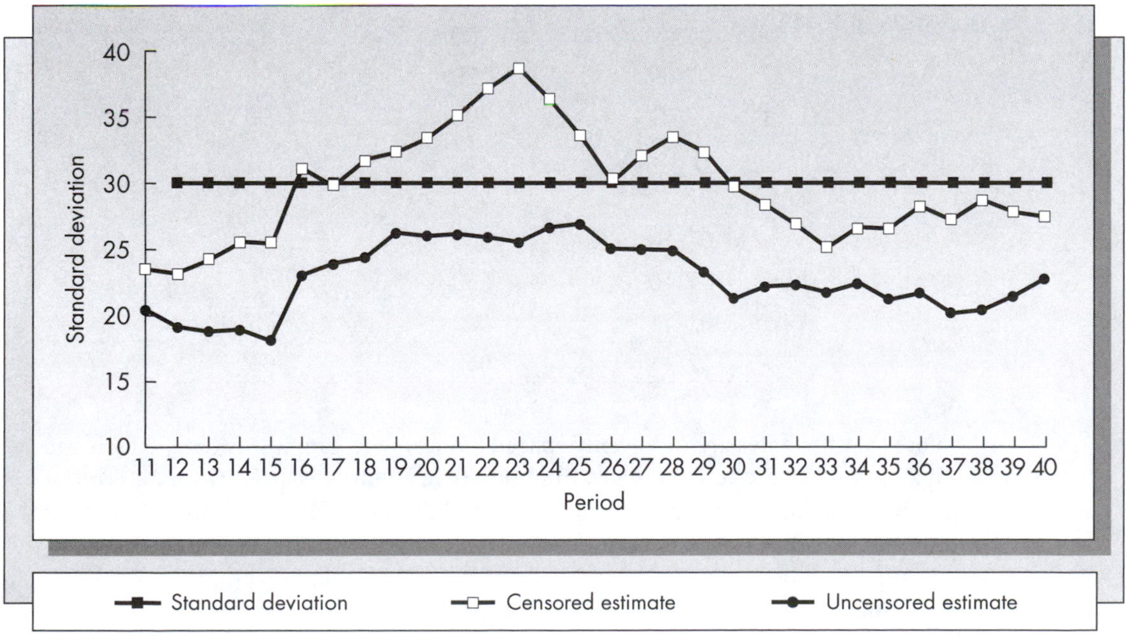

Figure 2–22 Tracking the standard deviation when lost sales are present

censoring. Notice the severe low bias when lost sales are ignored in both cases. That means that by not correctly accounting for the difference between sales and demand, one underestimates both the mean and the variance of the demand. Since both the mean and the variance of demand are inputs for determining optimal stocking levels, these levels could be severely underestimated.

CASE STUDY Predicting the Growth of Facebook

Facebook, founded by Mark Zuckerberg in 2004, is the quintessential social network and one of the true phenomena of the early twenty-first century. While Myspace, another social network, was founded a year earlier than Facebook, it never enjoyed the extraordinary success experienced by Facebook. Consider what forecasters would have faced in February 2011 if asked to project the number of Facebook users in 2013. The following chart tracks the numbers of Facebook users from the year of its founding in 2004 to early 2011.

Date	Users (in millions)	Month Number
12/04	1	12
12/05	5.5	24
12/06	12	36
4/07	20	40
10/07	50	46
08/08	100	56
01/09	150	61
02/09	175	62
04/09	200	64
07/09	250	67
09/09	300	69
12/09	350	72
02/10	450	74
07/10	500	79
09/10	550	81
01/11	600	85
02/11	650	86

Note that the dates are not evenly spaced. One way to graph this data, which is useful in analyzing it, is to convert the dates to numbers of months elapsed from an arbitrary starting point. If we define month 1 as January 2004, then the dates in column 1 translate to the number of months elapsed in column 3.

Treating numbers of months as the independent variable, the graph of numbers of users versus months elapsed appears in the graph on page 123.

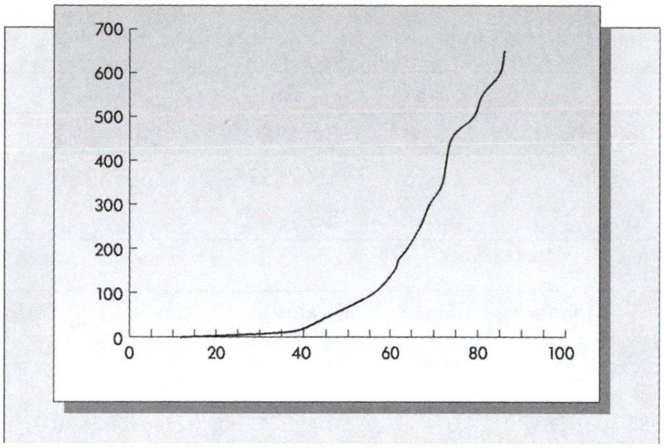

The graph seems to show exponential growth. To test this hypothesis, we consider graphing the natural logarithm of the numbers of users versus elapsed numbers of months. Doing so results in the following graph.

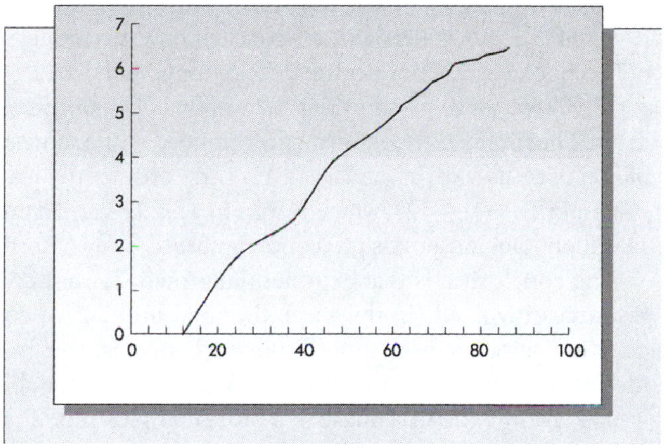

The fact that the graph of the natural logs is almost linear means that the assumption of exponential growth is an accurate one. Fitting a simple linear regression to this curve results in the following.

SUMMARY OUTPUT

Regression Statistics	
Multiple R	0.987306
R Square	0.974772
Adjusted R Square	0.973091
Standard Error	0.307178
Observations	17

ANOVA

	df	SS	MS	F	Significance F
Regression	1	54.68893	54.68893	579.5877578	2.11649E-13
Residual	15	1.415375	0.094358		
Total	16	56.10431			

	Coefficients	Standard Error	*t*-Stat	P-value	Lower 95%	Upper 95%	Lower 95.0%	Upper 95.0%
Intercept	–0.4425	0.22592	–1.95867	0.069013361	–0.924037859	0.039035397	–0.924037859	0.039035397
X Variable 1	0.086085	0.00357	24.07463	2.11649E-13	0.078463009	0.093706009	0.078463009	0.093706009

This Excel output shows that the logs follow a straight-line relationship very closely. In fact, it's rare to see an R square value of more than.97, so this fit is extremely strong. Let's see what numbers of users would be projected by this model.

The regression model is $\hat{y} = a + bx$ where $\hat{y}$ represents the logarithm of the number of users, a is the intercept and b is the slope. The independent variable, x, is the number of months elapsed since January 2004. The regression output indicates that the least squares estimators are $a = -.4425$ and $b = .086085$. Let's consider how this model would be used for forecasting numbers of Facebook users. Consider the forecast for January 2012, which is month 97. Setting $x = 97$ gives $\hat{y} = 7.90785$. Since $\hat{y} = ln(y)$, it follows that $y = \exp(\hat{y}) = \exp(7.90785) = 2718.53$ millions of users. That is, if Facebook's growth continued at the same rate, there would be approximately 2.7 billion users as soon as January 2012. Let's consider the prediction for January 2014. This month corresponds to $x = 121$, which results in $\hat{y} = 9.34$ and a predicted number of users of more than 11 billion! Obviously, this prediction is absurd, as it is more than the number of people on earth.

The conclusion is that exponential growth cannot continue indefinitely if there are finite resources involved (in this case, the total number of computer users on the Earth). So the question remains, how should one have forecast the growth of Facebook users? Clearly the model of continued exponential growth was unreasonable. The observed exponential growth was likely only the first phase of a more complex model. But what would have been an appropriate model in this case? Based on past experience with company growth curves, it would have been reasonable to postulate a logistic curve. The mathematical form of the standard logistic curve is $P(t) = \frac{1}{1+e^{-t}}$. This results in the curve pictured below.

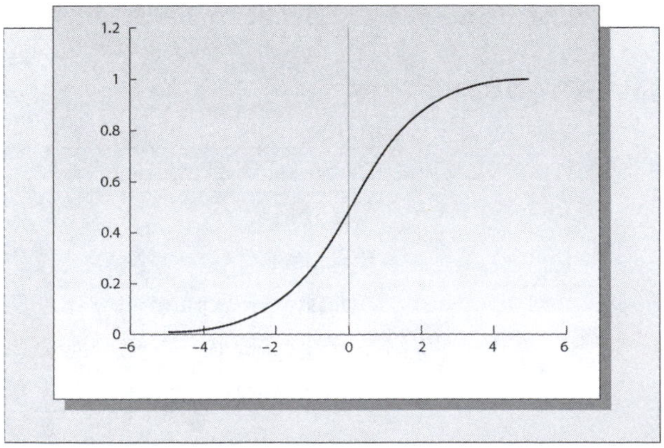

The goal is to try to provide the best fit of this curve to the original Facebook data. Since we're not sure where the point of inflection occurs, our estimate at this stage is probably not going to be very accurate. However, we can do a lot better than simply extrapolating from an exponential growth curve. In order to fit a logistic curve to a set of data, we need to know a few things. First, where in the data set is the inflection point? Second, what is likely to be the value of the asymptote (which is one for the standard curve)? And finally, what is the appropriate time scale?

Consider estimating the asymptote. This is probably not going to be very accurate since we don't know where the inflection point is. From the original graph, it does appear that the inflection point occurred somewhere near month 80. The number of users in month 80 was approximately 525 million. This would imply a value of the asymptote of approximately (2)(525 million) = 1.05 billion, which would be the estimate of the peak value of Facebook users.

We also need to transform the time scale. If we assume in our original curve that the first half of the logistic curve corresponds to months 20 to 80, this would be the same as 5 to 0 in the standard logistic curve above. The reader can verify that this would correspond to the transformation of the time scale following the equation $u = 12^*t + 80$. (Check the values of u and t at –5, 0, and 5 to verify this transformation.) Furthermore, if the asymptote is K rather than 1, we scale the function by multiplying by K. Applying these two transformations, the appropriate logistic curve to fit our original data would be $P(u) = \dfrac{K}{1+\exp(-(u-80)/12))}$ where u corresponds to number of elapsed months, and $K = 1.05 \times 10^9$. Consider the forecast for January 2012 based on this model. Substituting $u = 97$ gives a forecast of $P(u)= 0.845$ billion users. According to the logistic model, the expected number of users would be close to the postulated maximum of 1.05 billion by month 140, which corresponds to August of 2015.

Hindsight informs us that the January 2012 estimate was very accurate. At December 31, 2011, the number of Facebook users was.845 billion (Statista, 2019). However, Facebook reached 1.05 billion users by December 31, 2012. In the third quarter of 2015, the number exceeded 1.5 billion. Further growth continued through the first quarter of 2019 when there were 2.38 billion monthly active users. The interested reader might repeat the above exercise using recent data to forecast the number of users five years into the future (Statista, for example, compiles the number of monthly active Facebook users worldwide).

2.13 LINKING FORECASTING AND INVENTORY MANAGEMENT

Inventory control under demand uncertainty will be treated in detail in Chapter 5. In practice, inventory management and demand forecasting are closely linked. The forecasting method could be any one of the methods discussed in this chapter. One of the inputs required for inventory control models is the distribution of the demand over a period or over an order replenishment lead time. Is there a link between the distribution of forecast errors and the distribution of demand? The answer is that there is such a link. The forecast error distribution plays a key role in the correct application of inventory models in practice.

The first issue is to decide on the appropriate form of the distribution of demand. Most commercial systems assume that the demand distribution is normal. That means we need only estimate the mean μ and the standard deviation σ to specify the entire distribution. (Statistical methods, known as goodness-of-fit techniques, can be applied to test the accuracy of the normality assumption.) Whether or not the inventory system is linked to a forecasting system, we must have a history of observations of demand to obtain statistical estimates of the mean and variance. (When no history of demand exists, personal judgment must be substituted for statistical estimation. Methods for aggregating subjective judgment are discussed in Section 2.3.)

In the beginning of Chapter 5, we discuss how one would estimate the mean and the variance of demand directly from a history of demand observations. In practice, however, we don't generally use a long history of observations, because we believe that the underlying demand distribution does not remain constant indefinitely. That is why we adjust the N for moving averages or the α for exponential smoothing to balance stability and responsiveness. If that is the case, what is the appropriate variance estimator one should use?

In Appendix 2–A, we show that for simple moving averages of order N, the variance of forecast error, σ_e^2, is given by

$$\sigma_e^2 = \sigma^2\left(\frac{N+1}{N}\right)$$

and for exponential smoothing the variance of forecast error is

$$\sigma_e^2 = \sigma^2\left(\frac{2}{2-\alpha}\right).$$

Notice that in both cases, the value of σ_e^2 *exceeds* the value of σ^2. Also notice that as N gets large and as α gets small, the values of σ_e and σ grow close. These cases occur when one uses the entire history of demand observations to make a forecast. In Chapter 5 one of the inputs needed to determine safety stocks for inventory is the distribution of demand. The problem is, if we have estimators for both σ_e and σ, which should be used as the standard deviation estimator for setting safety stocks?

The seemingly obvious answer is that we should use the estimator for σ, since it represents the standard deviation of demand. The correct answer, however, is that we should use σ_e. The reason is that the process of forecasting introduces sampling error into the estimation process, and this sampling error is accounted for in the value of σ_e. The forecasting error variance is higher than the demand variance because the forecast is based on only a limited portion of the demand history.

We also can provide an intuitive explanation. If a forecast is used to estimate the mean demand, we keep safety stocks in order to protect against the error in this forecast. Hence, the distribution of forecast errors is more relevant than the distribution of demands. This is an important practical point that has been the source of much confusion in the literature. We will return to this issue in Chapter 5 and discuss its relevance in the context of safety stock calculations.

Most inventory control systems use the method suggested by Robert Brown (1959 and 1962) to estimate the value of σ_e. (In fact, it appears that Brown was the first to recognize the importance of the distribution of forecast errors in inventory management applications.) The method requires estimating the MAD of forecast errors using exponential smoothing. This is accomplished using the smoothing equation

$$\text{MAD}_t = \alpha|F_t - D_t| + (1 - \alpha)\text{MAD}_{t-1}.$$

The MAD is converted to an estimate of the standard deviation of forecast error by multiplying by 1.25. That is, the estimator for σ_e obtained at time t is

$$\hat{\sigma}_e = 1.25\ \text{MAD}_t.$$

A small value of α, generally between 0.1 and 0.2, is used to ensure stability in the MAD estimator. This approach to estimating the MAD works for any of the forecasting methods discussed in this chapter. Safety stocks are then computed using this estimator for σ_e.

2.14 HISTORICAL NOTES AND ADDITIONAL TOPICS

Forecasting is a rich area of research. Its importance in business applications cannot be overstated. The simple time series methods discussed in this chapter have their roots in basic statistics and probability theory. The method of exponential smoothing is generally credited to Brown (1959 and 1962), who worked as a consultant to A. D. Little (a pioneering professional services firm) at the time. Although not grounded in basic theory, exponential smoothing is probably one of the most popular forecasting methods used today. Brown was also the first to recognize the importance of the forecast error distribution and its implications for inventory management. However, interest in using statistical methods for forecasting goes back a century or more. (See, for example, Yule, 1926.)

Not discussed in this chapter is forecasting of time series using spectral analysis and state space methods. These methods are highly sophisticated and require substantial structure to exist in the data. They often rely on the use of the autocorrelation function; in that sense, they are similar conceptually to Box-Jenkins methods discussed briefly in Section 2.10. The groundbreaking work in this area is credited to Norbert Wiener (1949) and Rudolph Kalman (1960). However, there have been few applications of these methods to forecasting economic time series. Most applications have been in the area of signal processing in electrical engineering. [Wilbur Davenport and William Root (1987) provide a good summary of the fundamental concepts in this area.]

The **Kalman filter** is a type of exponential smoothing technique in which the value of the smoothing constant changes with time and is chosen using optimization. The idea of adjusting the value of the smoothing constant based on some measure of prior performance of the model has been used in several ad hoc ways. A typical approach requires the calculation of the tracking signal (Trigg and Leach, 1967). The tracking signal is then used as the value of the smoothing constant for the next forecast. The idea is that when the tracking signal is large, it suggests that the time series has undergone a shift; a larger value of the smoothing constant should be more responsive to a sudden shift in the underlying signal. Other methods have also been suggested. We have intentionally not included an explicit discussion of adaptive response rate methods for one reason: there is little evidence that they work in the context of predicting economic time series. Most studies that compare the effectiveness of different forecasting methods for many different series show no advantage for adaptive response rate models. (See, for example, Armstrong, 1984.) The Trigg-Leach method is discussed in more detail in problem 57.

We have included a reasonably self-contained discussion of Box-Jenkins ARIMA models. The basic ideas behind these methods, such as the auto-correlation structure of a process, go back many years. However, Box and Jenkins (1970) were the first to put these ideas together into a comprehensive step-by-step approach for building ARIMA models for short-term forecasting. Readers may find their book daunting, as they focus on issues of mathematical concern. For the reader seeking a more comprehensive coverage of ARIMA models at a level consistent with ours, a good starting point is the text by Spyros Makridakis, Steven Wheelright, and Rob Hyndman (1998).

2.15 SUMMARY

This chapter provided an introduction to a number of the more popular time series forecasting techniques as well as a brief discussion of other methods, including the Delphi method and causal models. A moving-average forecast is obtained by computing the arithmetic average of the N most recent observations of demand. An exponential smoothing forecast is obtained by computing the weighted average of the current observation of demand and the most recent forecast of that demand. The weight applied to the current observation is α and the weight applied to the last forecast (that is, the past observations) is $1 - \alpha$. Small N and large α result in responsive forecasts, and large N and small α result in stable forecasts. Although the two methods have similar properties, exponential smoothing is generally preferred because it requires storing only the previous forecast, whereas moving averages requires storing the last N demand observations.

When there is a trend in the series, both moving averages and exponential smoothing lag behind the trend. We discussed two time series techniques that are designed to track the trend. One is regression analysis, which uses least squares to fit a straight line to the data, and the other is Holt's method, which is a type of double exponential smoothing. Holt's method has the advantage that forecasts are easier to update as new demand observations become available.

We also discussed techniques for seasonal series. We employed classical decomposition to show how simple moving averages could be used to estimate seasonal factors and obtain the deseasonalized series when there is a trend and showed how seasonal factors could be estimated quickly when there is no trend. The extension of Holt's method to deal with seasonal problems, called Winters's method, is a type of triple exponential smoothing technique.

Section 2.12 provided a brief overview of advanced methods that are beyond the scope of this text. Section 2.13 discussed the relationship between forecasting and inventory control. The key point is that the standard deviation of forecast error is the appropriate measure of variation for computing safety stocks (see Chapter 5 for further elaboration).

ADDITIONAL PROBLEMS ON FORECASTING

47. John Kittle, an independent insurance agent, uses a five-year moving average to forecast the number of claims made in a single year for one of the large insurance companies he sells for. He has just discovered that a clerk in his employ incorrectly entered the number of claims made four years ago as 1,400 when it should have been 1,200.
    a. What adjustment should Mr. Kittle make in next year's forecast to take into account the corrected value of the number of claims four years ago?
    b. Suppose that Mr. Kittle used simple exponential smoothing with $\alpha = .2$ instead of moving averages to determine his forecast. What adjustment is now required in next year's forecast? (Note that you do not need to know the value of the forecast for next year in order to solve this problem.)

48. A method of estimating the MAD discussed in Section 2.13 recomputes it each time a new demand is observed according to the following formula:

$$\text{MAD}_t = \alpha|e_t| + (1 - \alpha)\ \text{MAD}_{t-1}.$$

Consider the one-step-ahead forecasts for aircraft engine failures for quarters 2 through 8 obtained in Example 2.3. Assume an initial value of the MAD = 50 in period 1. Using the same value of α, what values of the MAD does this method give for periods 2 through 8? Discuss the advantages and disadvantages of this approach vis-à-vis direct computation of the MAD.

49. Herman Hahn is attempting to set up an integrated forecasting and inventory control system for his hardware store. When Herman indicates that outdoor lights are a seasonal item on the computer, he is prompted by the program to input the seasonal factors by quarter.

Unfortunately, Herman has not kept any historical data, but he estimates that first-quarter demand for the lights is about 30 percent below average, the second-quarter demand about 20 percent below average, third-quarter demand about average, and fourth-quarter demand about 50 percent above average. What should he input for the seasonal factors?

50. Irwin Richards, a publisher of business textbooks, publishes in the areas of management, marketing, accounting, production, finance, and economics. The president of the firm is interested in getting a relative measure of the sizes of books in the various fields. Over the past three years, the average numbers of pages of books published in each area were

	Year 1	Year 2	Year 3
Management	835	956	774
Marketing	620	540	575
Accounting	440	490	525
Production	695	680	624
Finance	380	425	410
Economics	1,220	1,040	1,312

Using the quick and dirty methods discussed in Section 2.9, find multiplicative factors for each area giving the percentage above or below the mean number of pages.

51. Over a two-year period, the Topper Company sold the following numbers of lawn mowers.

Month:	1	2	3	4	5	6	7	8	9	10	11	12
Sales:	238	220	195	245	345	380	270	220	280	120	110	85
Month:	13	14	15	16	17	18	19	20	21	22	23	24
Sales:	135	145	185	219	240	420	520	410	380	320	290	240

a. In column A input the numbers 1 to 24 representing the months and in column B the observed monthly sales. Compute the three-month moving-average forecast and place this in the third column. Be sure to align your forecast with the period *for* which you are forecasting (the average of sales for months 1, 2, and 3 should be placed in row 4; the average of sales for months 2, 3, and 4 in row 5; and so on.) In the fourth column, compute the forecast error for each month in which you have obtained a forecast.

b. In columns 5, 6, and 7 compute the absolute error, the squared error, and the absolute percentage error. Using these results, find the MAD, MSE, and MAPE for the MA(3) forecasts for months 4 through 24.

c. Repeat parts (a) and (b) for six-month moving averages. (These calculations should appear in columns 8 through 12.) Which method, MA(3) or MA(6), was more accurate for these data?

52. Repeat the calculations in problem 51 using simple exponential smoothing and allow the smoothing constant α to be a variable. That is, the smoothing constant should be a cell location. By experimenting with different values of α, determine the value that appears to minimize the
    a. MAD
    b. MSE
    c. MAPE

    Assume that the forecast for month 1 is 225.

53. Baby It's You, a maker of baby foods, has found a high correlation between the aggregate company sales (in \$100,000) and the number of births nationally the preceding year. Suppose that the sales and the birth figures during the past eight years are

	Year							
	1	2	3	4	5	6	7	8
Sales (in \$100,000)	6.1	6.4	8.3	8.8	5.1	9.2	7.3	12.5
U.S. births (in millions)	2.9	3.4	3.5	3.1	3.8	2.8	4.2	3.7

a. Assuming that U.S. births represent the independent variable and sales the dependent variable, determine a regression equation for predicting sales based on births. Use years 2 through 8 as your baseline. (Hint: You will require the general regression formulas appearing in Appendix 2–B to solve this problem.)

b. Suppose that births are forecasted to be 3.3 million in year 9. What forecast for sales revenue in year 10 do you obtain using the results of part (a)?

c. Suppose that simple exponential smoothing with $\alpha = .15$ is used to predict the number of births. Use the average of years 1 to 4 as your initial forecast for period 5; determine an exponentially smoothed forecast for U.S. births in year 9.

d. Combine the results in parts (a), (b), and (c) to obtain a forecast for the sum of total aggregate sales in years 9 and 10.

54. Hy and Murray are planning to set up an ice cream stand in Shoreline Park. After six months of operation, the observed sales of ice cream (in dollars) and the number of park attendees are

	Month					
	1	2	3	4	5	6
Ice cream sales	325	335	172	645	770	950
Park attendees	880	976	440	1,823	1,885	2,436

a. Determine a regression equation treating ice cream sales as the dependent variable and time as the independent variable. Based on this regression equation, what should the dollar sales of ice cream be in two years (month 30)? How confident are you about this forecast? Explain your answer.
b. Determine a regression equation treating ice cream sales as the dependent variable and park attendees as the independent variable. (Hint: You will require the general regression equations in Appendix 2–B in order to solve this part.)
c. Suppose that park attendance is expected to follow a logistic curve (see Figure 2–23). The park department expects the attendance to peak out at about 6,000 attendees per month. Plot the data of park attendees by month and "eyeball" a logistics curve fit of the data using 6,000 as your maximum value. Based on your curve and the regression equation determined in part (b), predict ice cream sales for months 12 through 18.

55. A suggested method for determining the "right" value of the smoothing constant α in exponential smoothing is to retrospectively determine the α value that results in the minimum forecast error for some set of historical data. Comment on the appropriateness of this method and some of the potential problems that could result.
56. Lakeroad, a manufacturer of hard disks for personal computers, has sold the following numbers of disks.

Year	Number Sold (in 000s)	Year	Number Sold (in 000s)
1	0.2	5	34.5
2	4.3	6	68.2
3	8.8	7	85.0
4	18.6	8	58.0

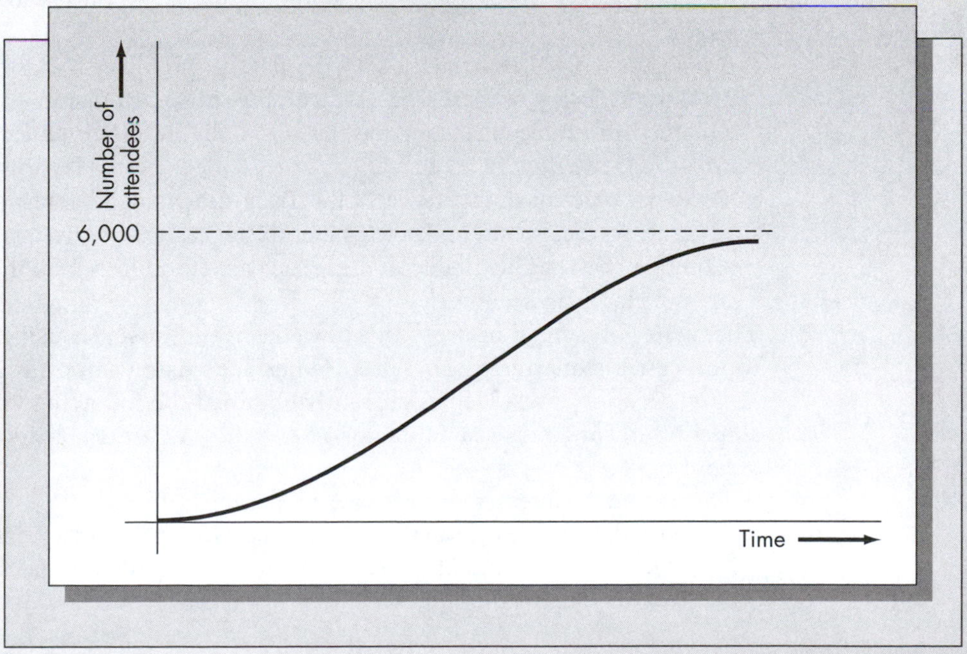

Figure 2–23 Logistic curve (for problem 52)

a. Suppose the firm uses Holt's method for forecasting sales. Assume $S_0 = 0$ and $G_0 = 8$. Using $\alpha = .2$ and $\beta = .2$, find one-step-ahead forecasts for years 2 through 9 and compute the MAD and MSE for the forecasts during this period. What is the sales forecast for year 20 made at the end of year 8? Based on the results of year 8, why might this forecast be very inaccurate?

b. By experimenting with various values of α and β, determine the values of the smoothing constants that appear to give the most accurate forecasts.

57. Along with smoothing the original series, the Trigg-Leach adaptive response-rate exponential smoothing method also smooths the error e_t and the absolute error $|e_t|$ according to the equations

$$E_t = \beta e_t + (1 - \beta)E_{t-1},$$
$$M_t = \beta|e_t| + (1 - \beta)M_{t-1}$$

and defines the smoothing constant to be used in forecasting the series in period t as

$$\alpha_t = \left|\frac{E_t}{M_t}\right|.$$

The forecast made in period t for period $t + 1$ is obtained by the usual exponential smoothing equation, using α_t as the smoothing constant. That is,

$$F_{t+1} = \alpha_t D_t + (1 - \alpha_t)F_t.$$

The idea behind the approach is that when E_t is close in magnitude to M_t, it suggests that the forecasts are biased. In that case, a larger value of the smoothing constant results that makes the exponential smoothing more responsive to sudden changes in the series.

a. Apply the Trigg-Leach method to the data in problem 22. Using the MAD, compare the accuracy of these forecasts with the simple exponential smoothing forecasts obtained in problem 22. Assume that $E_1 = e_1$ (the observed error in January) and $M_1 = |e_1|$. Use $\beta = .1$.

b. For what types of time series will the Trigg-Leach adaptive response rate give more accurate forecasts, and under what circumstances will it give less accurate forecasts? Comment on the advisability of using such a method for situations in which forecasts are not closely monitored.

58. The owner of a small brewery in Milwaukee, Wisconsin, is using Winters's method to forecast his quarterly beer sales. He has been using smoothing constants of $\alpha = .2$, $\beta = .2$, and $\gamma = .2$. He has currently obtained the following values of the various slope, intercept, and seasonal factors: $S_{10} = 120$, $G_{10} = 14$, $c_{10} = 1.2$, $c_9 = 1.1$, $c_8 = .8$, and $c_7 = .9$.

a. Determine the forecast for beer sales in quarter 11.

b. Suppose that the actual sales turn out to be 128 in quarter 11. Find S_{11} and G_{11}, and find the updated values of the seasonal factors. Also determine the forecast made at the end of quarter 11 for quarter 13.

59. The U.S. gross national product (GNP) in billions of dollars during the period 1964 to 1984 was as follows.

Year	GNP	Year	GNP
1964	649.8	1975	1,598.4
1965	705.1	1976	1,782.8
1966	772.0	1977	1,990.5
1967	816.4	1978	2,249.7
1968	892.7	1979	2,508.2
1969	963.9	1980	2,732.0
1970	1,015.5	1981	3,052.6
1971	1,102.7	1982	3,166.0
1972	1,212.8	1983	3,401.6
1973	1,359.3	1984	3,774.7
1974	1,472.8		

Source: Economic Report of the President, February 1986.

a. Use Holt's method to predict the GNP. Determine a regression fit of the data for the period 1964 to 1974 to estimate the initial values of the slope and intercept. (Hint: If you are doing the regression by hand, transform the years by subtracting 1963 from each value to make the calculations less cumbersome.) Using Holt's method, determine forecasts for 1975 to 1984. Assume that $\alpha = .2$ and $\beta = .1$. Compute the MAD and the MSE of the one-step-ahead forecasts for the period 1975 to 1984.

b. Determine the percentage increase in GNP from 1964 to 1984 and graph the resulting series. Use a six-year moving average and simple exponential smoothing with $\alpha = .2$ to obtain one-step-ahead forecasts of this series for the period 1975 to 1984. (Use the arithmetic average of the observations from 1964 to 1974 to initialize the exponential smoothing.)

In both cases (i.e., MA and ES forecasts), convert your forecasts of the percentage increase for the following year to a forecast of the GNP itself and compute the MAD and the MSE of the resulting forecasts. Compare the accuracy of these methods with that of part (a).

c. Discuss the problem of predicting GNP. What methods other than the ones used in parts (a) and (b) might give better predictions of this series?

APPENDIX 2–A

Forecast Errors for Moving Averages and Exponential Smoothing

The forecast error e_t is the difference between the forecast for period t and the actual demand for that period. In this appendix we will derive the distribution of the forecast error for both moving averages and exponential smoothing.

The demand is assumed to be generated by the process

$$D_t = \mu + \varepsilon_t,$$

where ε_t is normal with mean zero and variance σ^2.

Case 1. Moving Averages

Consider first the case in which forecasts are generated by moving averages. Then the forecast error is $e_t = F_t - D_t$, where F_t is given by

$$F_t = \frac{1}{N}\sum_{i=t-N}^{t-1} D_i.$$

It follows that

$$E(F_t - D_t) = (1/N)\sum_{i=t-N}^{t-1} E(D_i) - E(D_t) = (1/N)(N\mu) - \mu = 0.$$

This proves that when demand is stationary, moving-average forecasts are unbiased.

Also,

$$\begin{aligned}\mathrm{Var}(F_t - D_t) &= \mathrm{Var}(F_t) + \mathrm{Var}(D_t)\\ &= (1/N^2)\sum_{i=t-N}^{t-1} \mathrm{Var}(D_i) + \mathrm{Var}(D_t)\\ &= (1/N^2)(N\sigma^2) + \sigma^2\\ &= \sigma^2(1 + 1/N)\\ &= \sigma^2[(N+1)/N].\end{aligned}$$

It follows that the standard deviation of the forecast error, σ_e, is

$$\sigma_e = \sigma\sqrt{\frac{N+1}{N}}.$$

This is the standard deviation of the forecast error for simple moving averages in terms of the standard deviation of each observation.

Having derived the mean and the variance of the forecast error, we still need to specify the *form* of the forecast error distribution. By assumption, the values of D_t form a sequence of independent, identically distributed, normal random variables. Since F_t is a linear combination of $D_{t-1}, D_{t-2}, \ldots, D_{t-N}$, it follows that F_t is normally distributed and independent of D_t. It now follows that e_t is normal as well. Hence, the distribution of e_t is completely specified by its mean and variance.

As the expected value of the forecast error is zero, we say the method is unbiased. Notice that this is a result of the assumption that the demand process is stationary. Consider the variance of the forecast error. The value of N that minimizes σ_e is $N = +\infty$. This means that the variance is minimized if the forecast is the average of all the past data. However, our intuition tells us that we can do better if we use more recent data to make our forecast. The discrepancy arises because we really do not believe our assumption that the demand process is stationary for all time. A smaller value of N will allow the moving-average method to react more quickly to unforeseen changes in the demand process.

Case 2. Exponential Smoothing

Now consider the case in which forecasts are generated by exponential smoothing. In this case F_t may be represented by the weighted infinite sum of past values of demand.

$$F_t = \alpha D_{t-1} + \alpha(1-\alpha)D_{t-2} + \alpha(1-\alpha)^2 D_{t-3} + \ldots,$$
$$E(F_t) = \mu[\alpha + \alpha(1-\alpha) + \alpha(1-\alpha)^2 + \ldots] = \mu.$$

Notice that this means that $E(e_t) = 0$, so that both exponential smoothing and moving averages are unbiased forecasting methods when the underlying demand process is a constant plus a random term.

$$\begin{aligned}\mathrm{Var}(F_t) &= \alpha^2\sigma^2 + (1-\alpha)^2\alpha^2\sigma^2 + \cdots \\ &= \sigma^2\alpha^2\sum_{n=0}^{\infty}(1-\alpha)^{2n}.\end{aligned}$$

It can be shown that

$$\sum_{n=0}^{\infty}(1-\alpha)^{2n} = \frac{1}{1-(1-\alpha)^2}$$

so that

$$\mathrm{Var}(F_t) = \frac{\sigma^2\alpha^2}{1-(1-\alpha)^2} = \frac{\sigma^2\alpha}{2-\alpha}.$$

Since

$$\mathrm{Var}(e_t) = \mathrm{Var}(F_t) + \mathrm{Var}(D_t),$$
$$\mathrm{Var}(e_t) = \sigma^2[\alpha/(2-\alpha) + 1] = \sigma^2[2/(2-\alpha)],$$

or

$$\sigma_e = \sigma\sqrt{\frac{2}{2-\alpha}}.$$

This is the standard deviation of the forecast error for simple exponential smoothing in terms of the standard deviation of each observation. The distribution of the forecast error for exponential smoothing is normal for essentially the same reasons as stated above for moving averages.

Notice that if we equate the variances of the forecast error for exponential smoothing and moving averages, we obtain

$$2/(2-\alpha) = (N+1)/N$$

or $\alpha = 2/(N+1)$, which is exactly the same result as we obtained by equating the average age of data for the two methods.

APPENDIX 2–B

Derivation of the Equations for the Slope and Intercept for Regression Analysis

In this appendix we derive the equations for the optimal values of a and b for the regression model. Assume that the data are $(x_1, y_1), (x_2, y_2), \dots, (x_n, y_n)$, and the regression model to be fitted is $Y = a + bX$. Define

$$g(a,b) = \sum_{i=1}^{n} [y_i - (a + bx_i)]^2.$$

Interpret $g(a, b)$ as the sum of the squares of the distances from the line $a + bx$ to the data points y_i. The object of the analysis is to choose a and b to minimize $g(a, b)$. This is accomplished where

$$\frac{\partial g}{\partial a} = \frac{\partial g}{\partial b} = 0.$$

That is,

$$\frac{\partial g}{\partial a} = -\sum_{i=1}^{n} 2[y_i - (a + bx_i)] = 0,$$

$$\frac{\partial g}{\partial b} = -\sum_{i=1}^{n} 2x_i[y_i - (a + bx_i)] = 0,$$

which results in the two equations

$$an + b\sum_{i=1}^{n} x_i = \sum_{i=1}^{n} y_i, \tag{1}$$

$$a\sum_{i=1}^{n} x_i + b\sum_{i=1}^{n} x_i^2 = \sum_{i=1}^{n} x_i y_i. \tag{2}$$

These are two linear equations in the unknowns a and b. Multiplying Equation (1) by Σx_i and Equation (2) by n gives

$$an\sum_{i=1}^{n} x_i + b\left(\sum_{i=1}^{n} x_i\right)^2 = \left(\sum_{i=1}^{n} x_i\right)\left(\sum_{i=1}^{n} y_i\right), \tag{3}$$

$$an\sum_{i=1}^{n} x_i + bn\sum_{i=1}^{n} x_i^2 = n\sum_{i=1}^{n} x_i y_i. \tag{4}$$

Subtracting Equation (3) from Equation (4) results in

$$b\left[n\sum_{i=1}^{n} x_i^2 - \left(\sum_{i=1}^{n} x_i\right)^2\right] = n\sum_{i=1}^{n} x_i y_i - \left(\sum_{i=1}^{n} x_i\right)\left(\sum_{i=1}^{n} y_i\right). \tag{5}$$

Define $S_{xy} = n\,\Sigma\, x_i y_i - (\Sigma x_i)(\Sigma y_i)$ and $S_{xx} = n\Sigma x^2_i - (\Sigma x_i)^2$. It follows that Equation (5) may be written $bS_{xx} = S_{xy}$, which gives

$$b = \frac{S_{xy}}{S_{xx}}. \tag{6}$$

From Equation (1) we have

$$an = \sum_{i=1}^{n} y_i - b\sum_{i=1}^{n} x_i$$

or

$$a = \overline{y} - b\overline{x}, \tag{7}$$

where $\overline{y} = (1/n)\sum y_i$ and $\overline{x} = (1/n)\sum x_i$.

These formulas can be specialized to the forecasting problem when the independent variable is assumed to be time. In that case, the data are of the form $(1, D_1), (2, D_2), \ldots, (n, D_n)$, and the forecasting equation is of the form $\check{D} = a + bt$. The various formulas can be simplified as follows:

$$\sum x_i = 1 + 2 + 3 + \cdots + n = \frac{n(n+1)}{2},$$

$$\sum x_i^2 = 1 + 4 + 9 + \cdots + n^2 = \frac{n(n+1)(2n+1)}{6}.$$

Hence, we can write

$$S_{xy} = n\sum_{i=1}^{n} iD_i - n(n+1)/2 \sum_{i=1}^{n} D_i,$$

$$S_{xx} = \frac{n^2(n+1)(2n+1)}{6} - \frac{n^2(n+1)^2}{4},$$

$$b = \frac{S_{xy}}{S_{xx}},$$

$$a = \overline{D} - \frac{b(n+1)}{2}.$$

APPENDIX 2–C

Glossary of Notation for Chapter 2

a = Estimate of the intercept in regression analysis.

α = Smoothing constant used for single exponential smoothing. One of the smoothing constants used for Holt's method or one of the smoothing constants used for Winters's method.

b = Estimate of the slope in regression analysis.

β = Second smoothing constant used for either Holt's method or Winters's method.

c_t = Seasonal factor for the tth period of a season.

γ = Third smoothing constant used for Winters's method.

D_t = Demand in period t. Refers to the series whose values are to be forecasted.

e_t = $F_t - D_t$ = (Observed) forecasting error in period t.
ε_t = Random variable representing the random component of the demand.
F_t = One-step-ahead forecast made in period $t - 1$ for demand in period t.
$F_{t,t+\tau}$ = τ-step-ahead forecast made in period t for the demand in period $t + \tau$.
G_t = Smoothed value of the slope for Holt's and Winters's methods.
μ = Mean of the demand process.

MAD = Mean absolute deviation $= (1/n)\sum_{i=1}^{n}|e_i|$.

MAPE = Mean absolute percentage error $= (1/n)\sum_{i=1}^{n}|e_i / D_i| \times 100$.

MSE = Mean squared error $= (1/n)\sum_{i=1}^{n} e_i^2$.

S_t = Smoothed value of the series (intercept) for Holt's and Winters's methods.
σ^2 = Variance of the demand process.
T_t = Value of the tracking signal in period t (refer to problem 57).

BIBLIOGRAPHY

Armstrong, J. S. "Forecasting by Extrapolation: Conclusions from Twenty-Five Years of Research." *Interfaces* 14 (1984), pp. 52–66.

Box, G. E. P., and G. M. Jenkins. *Time Series Analysis, Forecasting, and Control*. San Francisco: Holden Day, 1970.

Brown, R. G. *Statistical Forecasting for Inventory Control*. New York: McGraw-Hill, 1959.

Brown, R. G. *Smoothing, Forecasting, and Prediction of Discrete Time Series*. Englewood Cliffs, NJ: Prentice Hall, 1962.

Davenport, W. B., and W. L. Root. *An Introduction to the Theory of Random Signals and Noise* (original copyright 1958). New York: IEEE Press, 1987.

Fisher, M. L.; J. H. Hammond; W. R. Obermeyer; and A. Raman, "Making Supply Meet Demand in an Uncertain World." *Harvard Business Review* (May–June 1994), pp. 221–40.

Helmer, O., and N. Rescher. "On the Epistemology of the Inexact Sciences." *Management Science* 6 (1959), pp. 25–52.

Kalman, R. E. "A New Approach to Linear Filtering and Prediction Problems." *Journal of Basic Engineering* 82 (1960), pp. 35–44.

Makridakis, S.; S. C. Wheelwright; and R. J. Hyndman. *Forecasting: Methods and Applications*, 3rd ed. New York: John Wiley & Sons, 1998.

Makridakis, S., and R. L. Winkler. "Averages of Forecasts." *Management Science* 29 (1983), pp. 987–96.

McWilliams, G. "At Compaq, a Desktop Crystal Ball." *Business Week*, March 20, 1995, pp. 96–97.

Nahmias, S. "Demand Estimation in Lost Sales Inventory Systems." *Naval Research Logistics* 41 (1994), pp. 739–57.

Nelson, C. R. "The Prediction Performance of the FRB-MIT-PENN Model of the U.S. Economy." *The American Economic Review* 62, no. 5 (December 1972), pp. 902–917.

Silver, N. "The Real Story of 2016." January 19, 2017. Retrieved from https://fivethirtyeight.com/features/the-real-story-of-2016/

Statista. "Number of Monthly Active Facebook Users Worldwide." 2019. Retrieved from https://www.statista.com/statistics/264810/number-of-monthly-active-facebook-users-worldwide/

Trigg, D. W. "Monitoring a Forecasting System." *Operational Research Quarterly* 15 (1964), pp. 271–74.

Trigg, D. W., and A. G. Leach. "Exponential Smoothing with Adaptive Response Rate." *Operational Research Quarterly* 18 (1967), pp. 53–59.

Wiener, N. *Extrapolation, Interpolation, and Smoothing of Stationary Time Series*. Cambridge, MA: MIT Press, 1949.

Yule, G. U. "Why Do We Sometimes Get Nonsense Correlations between Time Series? A Study of Sampling and the Nature of Time Series." *Journal of the Royal Statistical Society* 89 (1926), pp. 1–64.

CHAPTER 3

Sales and Operations Planning

"In preparing for battle I have always found that plans are useless, but planning is indispensable."

—Dwight D. Eisenhower

CHAPTER OVERVIEW

Purpose

To present the process by which companies go from technical forecasts to aggregate level sales and operations plans.

KEY POINTS

1. *The sales and operations planning (S&OP) process.* This chapter could also be called Macro Planning, since the purpose of an S&OP process is to develop top-down sales and operations plans for the entire firm. The key goals of the process are (1) to make aggregate level plans that all divisions as well as suppliers can use; (2) resolve the inherent tensions between sales and operations divisions; and (3) anticipate and react to strategic challenges in matching supply with demand for the firm.
2. *Key Performance Indicators.* Inherent in any S&OP process are a set of metrics—key performance indicators (KPIs)—that the firm uses to judge the performance of the different divisions. Effective KPIs measure important factors, are relatively easy to compute, and are actionable, in the sense that those being measured by a given KPI can also effect its change. Operational KPIs may be efficiency or effectiveness focused and must be aligned with the strategic goals of the firm.
3. *The role of uncertainty.* It is important to recognize explicitly the role of uncertainty in planning. Different types of uncertainty require different types of management responses. The S&OP process needs to carefully go over possible major sources of uncertainty and plan for them appropriately.
4. *Costs in aggregate operations plans.* The following are key costs to be considered in developing aggregate operations plans.
   - *Smoothing costs.* The cost of changing production and/or workforce levels.
   - *Holding costs.* The opportunity cost of dollars invested in inventory.
   - *Shortage costs.* The costs associated with lost demand or back orders.
   - *Labor costs.* These include direct labor costs on regular time, overtime, subcontracting costs, and idle time costs.
5. *Solving aggregate planning problems.* Approximate solutions to aggregate planning problems can be found graphically and exact solutions via linear programming. A level plan has constant production or workforce levels over the planning

horizon, while a chase strategy keeps zero inventory and minimizes holding and shortage costs. A linear programming formulation assumes that all costs are linear and typically does not take into account management policy, such as avoiding hiring and firing as much as possible.

6. *Disaggregating plans.* While aggregate planning is useful for providing approximate solutions for macro planning at the firm level, the question is whether these aggregate plans provide any guidance for planning at the lower levels of the firm. A disaggregation scheme is a mean of taking an aggregate plan and breaking it down to get more detailed plans at lower levels of the firm.

As we go through life, we make both micro and macro decisions. Micro decisions might be what to eat for breakfast, what route to take to work, what auto service to use, or which movie to rent. Macro decisions are the kind that change the course of one's life: where to live, what to major in, which job to take, whom to marry. A company must also make both micro and macro decisions every day. Some macro decisions are highly strategic in nature, such as process technology or market entry choices (see Chapter 1). Other macro decisions are more tactical, such as planning companywide workforce and production levels or setting sales target. In this chapter we explore tactical decisions made at the macro level in the context of a process known as **sales and operations planning** (**S&OP**).

S&OP begins with demand forecasts and turns them into targets for both sales and operations (techniques for demand forecasting were presented in Chapter 2). A firm may want to produce more than is forecast if stock-outs are unacceptable or if they are planning a major promotion; they may want to produce less than is forecast if overstocks are costly or they see the product winding down in its lifecycle. Such decisions must be made at a high strategic level and must involve both the sales and the operations staff.

One of the key goals in S&OP is to resolve the fundamental tension between sales and operations divisions in an organization. Sales divisions are usually measured on revenue; they want the product 100 percent available with as many different varieties as possible. Meanwhile, operations divisions are frequently measured on cost; they want to keep both capacity and inventory costs low, which means limiting overproduction and product varieties. The best way to resolve these inherent tensions is through a formal S&OP process in which the heads of sales and operations meet with other high level executives. This chapter explores what such processes look like.

Core to the S&OP process is a review of divisional Key Performance Indicators (KPIs). Section 3.2 reviews key challenges in KPI selection and key types of operational KPIs. One of the fundamental challenges in KPI selection is insuring that the KPI is both sufficiently high-level to be well aligned with corporate strategy and sufficiently low-level to provide a meaningful guide to behavior for people being evaluated by it.

The S&OP process would be a lot simpler if demand were certain. As noted in Chapter 2, however, it is not, and forecasts are generally wrong. In fact, there are a range of uncertainties that must be considered in the planning process from known unknowns to unknown unknowns, terms which will be formally defined in Section 3.3. Part of an S&OP process must include decisions around how to plan for and mitigate risk or uncertainty.

An important part of S&OP is **aggregate planning**, which might also be called macro production planning. It addresses the problem of deciding how many employees the firm should retain and, for a manufacturing firm, the quantity and the mix of products to be produced. Macro planning is not limited to manufacturing firms. Service organizations must determine employee staffing needs as well. For example, airlines must plan staffing levels for flight attendants and pilots, and hospitals must plan staffing levels for nurses and doctors. Macro planning strategies are a fundamental part of the firm's overall business strategy. Some firms operate on the philosophy that costs can be controlled only by making frequent changes in the size and/or composition of the workforce. Firms with seasonal demand are particularly likely to operate in this manner. Other firms have a reputation for retaining employees, even in bad times; Southwest Airlines and Toyota are two well-known examples.

Aggregate planning methodology is designed to translate demand forecasts into a blueprint for planning staffing and production levels for the firm over a predetermined planning horizon. Aggregate planning methodology is not limited to top-level planning. Although generally considered to be a macro planning tool for determining overall workforce and production levels, large companies may find aggregate planning useful at the plant level as well. Production planning may be viewed as a hierarchical process in which purchasing, production, and staffing decisions must be made at several levels in the firm. Aggregate planning methods may be applied at almost any level, although the concept is one of managing groups of items rather than single items.

Determining optimal production levels for all products produced by a large firm can be an enormous undertaking. Aggregate planning addresses this problem by assuming that individual items can be grouped together. However, finding an effective aggregating scheme can be difficult, and often revenue dollars are used for simplicity. One particular aggregating scheme that has been suggested is *items*, *families*, and *types*. **Items** (or **stock keeping units**, **SKU**s), represent the finest level of detail and are identified by separate part numbers, bar codes, and/or radio frequency ID (RFID) tags when appropriate. **Families** are groups of items that share a common manufacturing setup, and **types** are natural groups of families. This particular aggregation scheme is fairly general; there is no guarantee that it will work in every application.

This chapter outlines the S&OP process. Key to the process are KPIs. Thus, this chapter briefly reviews principles involved in setting KPIs. The chapter also describes the types of uncertainty that must be planned for in the process and how they are best addressed. Core to the operations component of any S&OP process is determining an aggregate production plan for capacity and inventory. Some aggregate production plans are heuristic (i.e., approximate), and some are optimal. We hope to convey to the reader an understanding of the issues involved in S&OP, a knowledge of the basic tools available for providing production plans, and an appreciation of the difficulties associated with such planning in the real world.

3.1 THE S&OP PROCESS

The S&OP process is designed to produce a plan that is used by all divisions within in the organization, as well as suppliers to the organization. The process is also sometimes referred to as **sales, inventory, and operations planning** (SIOP) to emphasize the important role that inventory can play as a buffer between sales planning and operations planning. It is also sometimes called **Integrated Business Planning** (IBP) to emphasize that it is a holistic approach across all divisions of an organization.

In any organization, demand is a function of sales effort and pricing, and supply is a function of operations effort and capacity. Therefore, in order to best balance supply with demand, a strategic approach must be applied. As defined by J. Andrew Grimson and David Pyke (2007), "S&OP is a business process that links the corporate strategic plan to daily operations plans and enables companies to balance demand and supply for their products" (p. 323).

SNAPSHOT APPLICATION

Heineken International was founded in 1864 by Gerard Adriaan Heineken in Amsterdam. It has a global network of distributors and over 170 breweries in more than 70 countries. Supply chain planning within the beer supply chain is complicated by the fact that most beers, including Heineken, are produced in batches. In fact, Heineken has a highly prescribed process for its brewing that all producers globally must follow to ensure taste consistency. It also has a robust S&OP process, which it implements in all of its subsidiaries. The key mantra for this process, pictured in Figure 3–1, is "one single plan."

> The global beer market is a dynamic environment with changing markets, strong competition, and changing customer preferences. Heineken realized that a global Sales and Operations Planning (S&OP) program would become a key enabler in supporting aggressive expansion targets and would become necessary to support a retail globalization landscape, which is applying increasing pressures on costs and service. Heineken's S&OP process integrates finance, marketing, sales, and supply chain departments with the objective of aligning the organization towards a synchronized operating plan globally. This program is supported by a very strong project management approach which has been designed to provide enough consistency across regions yet provide enough flexibility to embrace and benefit from local cultural differences. (Smits & English, 2010)

Notice how this description matches our general description of S&OP yet leaves some flexibility in terms of the actual running of meetings to allow for local customs to apply. The benefits Heineken has realized from their S&OP process are better cross-functional decisions, enabled growth; higher capacity utilization; lower supply chain costs; and reduced working capital.

Source: Heineken N.V. "2018 Annual Report."

A key phrase in Grimson and Pike's definition of S&OP is *business process*. That is, it a consistent set of steps that a company follows on a regular basis. Each business will apply its process differently but some common elements include the following.

1. *Strategically focused.* As discussed above, finding the appropriate balance for supply and demand requires knowledge of the company's strategic plan. C-suite executives are involved in S&OP planning to ensure that the company's strategy is appropriately represented.
2. *Cross-functional team.* Because the S&OP process must balance the interests of different parts of the organization, the team performing the process must be both cross-functional and balanced.
3. *Aggregated.* It is usually not possible to forecast at the individual stock keeping unit (SKU) level. Thus, S&OP processes work with aggregated units of production, such as product families. One natural unit of aggregation is sales dollars, but of course there is not a one-to-one mapping from sales dollars to units produced.
4. *Time fences.* The agreed sales and operations plan can only be fixed for a relatively short period (e.g., a week or month). Forecasts for future time periods are assumed to be flexible within a given range. Oftentimes, fences are used to show when the level of certainty changes; the fences can be referred to as frozen (i.e., fixed), slushy (somewhat fixed), and liquid (highly flexible).

There are a number of key inputs required for the S&OP process.

1. *Technical demand forecast.* Advanced forecasting techniques such as those discussed in Chapter 2 are used to understand raw demand.

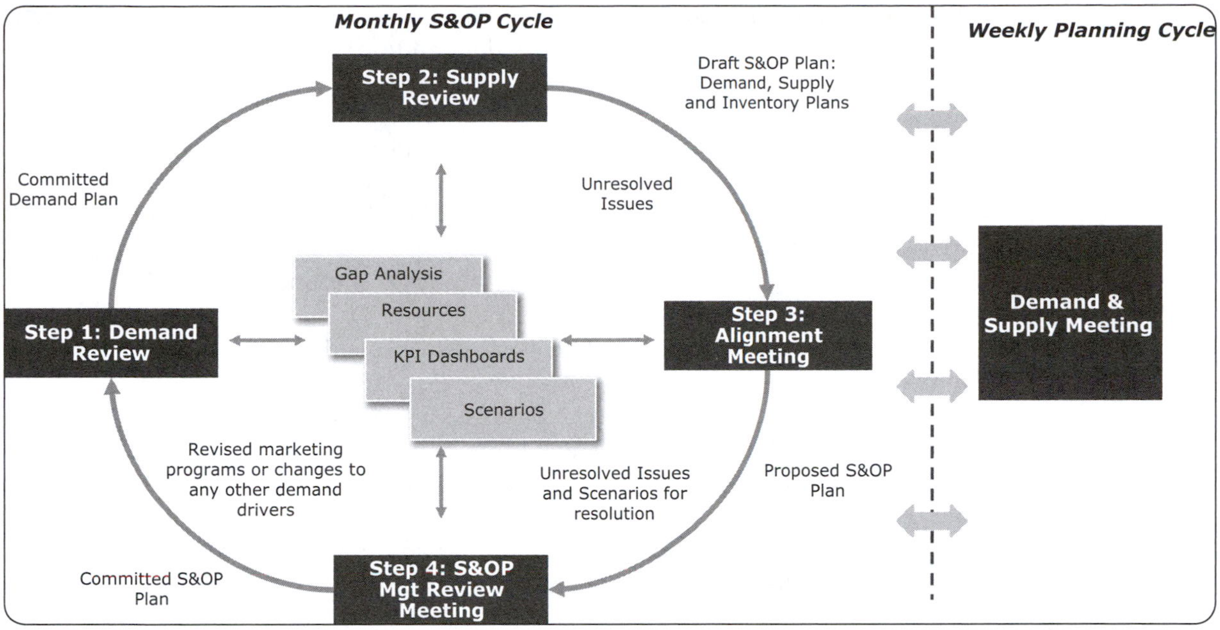

Figure 3–1 S&OP overview

2. *Sales plans*. The marketing and sales divisions will use the technical forecasts and their own promotion and marketing plans to provide a sales forecast.
3. *Operations plans*. Operations will produce a production plan that includes plans for capacity, supply, and inventory.
4. *Innovation plans*. The planned roll out of new products and future product innovation must be considered in the context of sales and operations plans.
5. *Financial plans*. Any public company must announce earnings forecasts, commonly with a breakdown of how such earnings are to be achieved. Because negative market surprises can affect a company's ability to raise capital, the sales and operations plans must also consider the financial forecasts. Further, cash flow constraints in many smaller companies must also be considered in the planning process.

The S&OP process must iterate between these five key inputs to resolve tensions and unforeseen constraints (e.g., a supplier who cannot deliver until six weeks from today). Plans are revised throughout this iterative process until they become finalized for the entire company. It is then left to the sales and operations divisions of the company to execute these plans, such as disaggregating sales dollars into actual production units.

One sample iterative process is shown in Figure 3–1 (produced for Heineken by Johan Smits and Michael English, 2010). Notice how the outcome of the S&OP meeting is a committed S&OP plan for the following month. Key uncertainties are shown as "scenarios."

The chief executive officer (CEO) typically leads the S&OP meeting and makes the key decisions with respect to trade-offs (Sheldon, 2006). A well run meeting is focused on decision making, rather than information dissemination; all participants are expected to have studied the materials that are prepared ahead of time by the different divisions.

The vice president of sales and/or marketing will be present and is the process owner for the demand plan. The vice president of operations and/or supply chain management will also be present and is the process owner for the operations plan. The chief financial officer (CFO) will attend and prepares the financial plan, which measures performance for the master business plan. The master scheduler prepares documents and is the process owner for weekly schedules (ordinarily not part of the S&OP process). In a larger organization this will take place at the divisional level, although the CEO may attend divisional S&OP meetings at least semi-regularly.

The following is a standard agenda for a monthly S&OP meeting (Sheldon, 2006).

- Review of last 30 days–accuracy of financial, demand, and operations plans (usually by product family).
- Review of 30- to 60-day plan expectations–risks for each plan.
- Review 90- to 120-day risks as required.
- Review balance of 12-month horizon for exceptions.

Typically, an S&OP plan is considered fixed (with agreed windows of flexibility for later time periods); therefore, the operations division must plan for uncertainty within its own execution plan. The following are common strategies for dealing with uncertainty.

1. *Buffering*: Maintaining excess resources (inventory, capacity, or time) to cover for fluctuations in supply or demand. Such buffering is an explicit component of planning under uncertainty, such as the inventory models considered in Chapter 5.
2. *Pooling*: Sharing buffers to cover multiple sources of variability (e.g., demand from different markets). This will be covered in more detail in Section 6.5.
3. *Contingency planning*: Establishing a preset course of action for an anticipated scenario. This may involve strategies such as bringing on a backup supplier or sourcing from the spot market. This is best done explicitly in the S&OP process.

Figure 3–2 illustrates how Heineken Netherlands Supply positions the S&OP process within their planning framework (Every Angle, 2010).

Notice how in Figure 3–2, S&OP straddles the strategic and tactical portions of the planning framework. Many of the other planning topics, such as the master production schedule and materials requirements planning, will be covered in Chapter 9.

PROBLEMS FOR SECTION 3.1

1. Why don't the head of sales or the head of operations chair the S&OP meeting?
2. Why do you think it is important that the CFO attends the S&OP meeting?
3. Describe the likely key tensions between the sales and operations divisions of a small manufacturer that mostly produces for the local market. Now describe them for a global organization that sells in many regions and outsources most of its production to China.
4. Why is S&OP described as an iterative process?
5. What are the trade-offs between having long fixed or frozen periods versus short fixed or frozen periods as outputs of the S&OP process?
6. What are the trade-offs between using revenue dollars versus sales quantities as the units in the aggregate S&OP plan?

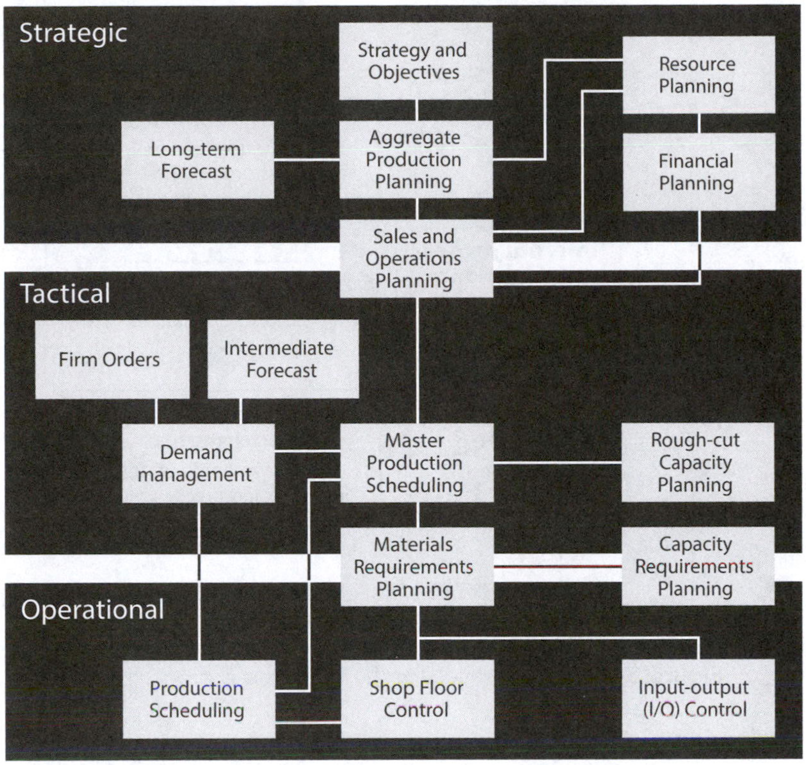

Figure 3–2 Planning and control framework for Heineken Netherlands Supply

3.2 KEY PERFORMANCE INDICATORS

One of the quantitative inputs to the S&OP process will be the **key performance indicators** (**KPIs**) for the divisions and the organization. The CFO is present at the meeting to present key financial performance measurements such as net profit and return on investment. However, as discussed in Chapter 1, such financial measures tend to emphasize short-term goals. This issue is often counteracted by the **balanced score card approach**, which provides a more well-rounded performance measurement approach covering a wider scope of metrics than pure finance measures. The balanced score card approach was developed by Robert Kaplan and David Norton (1992); it consists of performance measures designed to reflect strategy and communicate a vision to the organization. There are four standard perspectives for the approach, although many organizations have adapted these for their own use; these perspectives are (1) customer; (2) internal; (3) innovation and learning; and (4) financial. Figure 3–3 depicts these perspectives.

Operational metrics commonly sit within internal process measures. There are two key types of operational KPIs: (1) **efficiency**-related KPIs measure resource utilization (e.g., time, cost, materials, etc.); (2) **effectiveness**-related KPIs measure how well the process meets or exceeds customer requirements (e.g., defects, complaints, satisfaction scores, etc.). Which of these are most important will depend on the product-line strategy (see Chapter 1). Efficiency is most important for products that compete on cost. Effectiveness is most important for high-margin products that compete on innovation or fashion dimensions.

Some evaluation criteria for the merits of a KPI (ABB, 2010) are: (a) *Importance*—Are you measuring the things that really matter? (b) *Ease*—Does the measurement "flow" from the

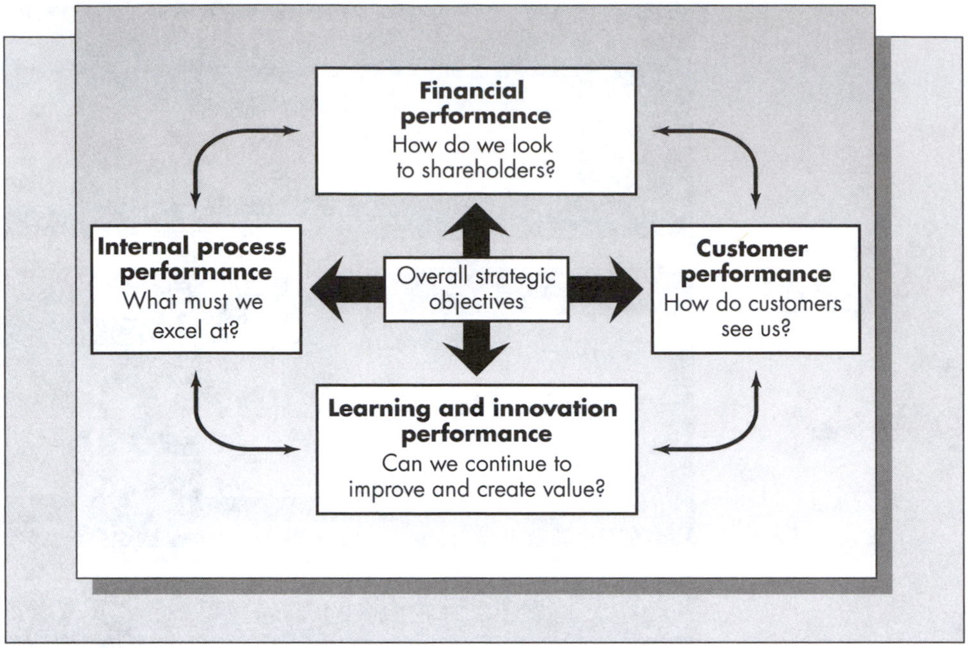

Figure 3–3 Perspectives for the balanced scorecard approach

activity being monitored? and (c) *Actionable*—Can the metric initiate appropriate actions? To both measure accurately and to incentivize behavior correctly, the KPI must be well understood by employees. One of the key challenges in choosing KPIs is in achieving organizational alignment. KPIs must be **aligned** to the strategic plan in order to monitor the essential performance results implied by the vision and mission statements and to monitor current strategic goals or objectives. To maintain alignment, the KPIs must cascade down through the business processes. Finally, the KPIs may be assigned to business units once they have been associated with business processes. Going the opposite direction starting from business units is more likely to result in misaligned KPIs. In general, a high level KPI (e.g., monthly profit) is easier to align, but employees may disengage if they don't feel that they can control the KPI. This is why KPIs must be **actionable**.

In their book *The Goal,* Eliyahu Goldratt and Jeff Cox (2014) state "The goal of a firm is to make money." Many of the issues described in the book arise from misaligned KPIs to this goal. Goldratt and Cox suggest that there are really only three operational KPIs—throughput, inventory, and operational expenses; however, their definitions of these terms are nonstandard. Throughput is defined as the rate at which sales generate money in the system; inventory is defined as all the money the system has invested in purchasing items it intends to sell; and operating expenses are defined as all the money the system spends to convert inventory into throughput. The beauty of these metrics is that they indeed align incentives with the goal of making money. The challenge is that firms have found them difficult to implement in practice. They have not been widely adopted by the field of managerial accounting.

A final challenge in KPI use is **gaming** of the KPI by employees. If the KPI definition leaves room for interpretation, then those being judged by it ordinarily will interpret it in their favor. One of the most common operational KPIs is **DIFOT**, which stands for **Delivery In-Full and On-Time**. However, the definition of "on-time" needs to be spelled out carefully—does it relate to when the order is shipped or when the customer receives the order? Of course, the customer only cares about the latter, but many firms measure the former—which is both easier to measure and more fully under operations control (i.e., more actionable). Clearly there are trade-offs when choosing between the two perspectives.

PROBLEMS FOR SECTION 3.2

7. In some S&OP processes, all divisions are measured by the same set of KPIs. What are the advantages and disadvantages to this approach?
8. Give two efficiency-related KPIs and two effectiveness-related KPIs that you believe Walmart is likely to use to evaluate its suppliers? Its in-store managers?
9. What types of organizations are likely to face the largest challenges in achieving alignment between operational KPIs and strategic priorities?
10. Evaluate the following metrics along the three dimensions of importance, ease, and actionability: DIFOT, ROI, machine utilization, and number of customer complaints.
11. Do you think efficiency-related KPIs or effectiveness-related KPIs will be easier for employees to "game" in practice? Explain your answer.
12. Why do you think that the KPIs proposed by Goldratt and Cox (2014) have not been widely implemented?

3.3 THE ROLE OF UNCERTAINTY

One of the key challenges in sales and operations planning is the effective management of uncertainty or risk. In a survey by IBM of 400 supply chain executives, risk was identified as the second most important key challenge (IBM, 2010). Sixty percent of respondents ranked risk as important "to a very great extent" or "to a significant extent" (the challenge that was ranked highest was supply chain visibility). S&OP is the highest level of tactical planning; the organization must acknowledge major sources of uncertainty explicitly and work to manage and/or mitigate risk.

Arnoud De Meyer, Christoph Loch, and Michael Pich (2002) highlight four different types of uncertainty. While they particularly focus on project management, the categorization also applies to S&OP processes and operational risks.

1. *Variation.* **Variation** is anything that causes the system to depart from regular, predictable behavior that may be described by a probability distribution (see Appendix 5–A). Typical sources for variation include variability in customer orders, differential worker skill levels, and variability in quality or lead times.
2. *Foreseen uncertainty.* **Foreseen uncertainty** ("**known unknowns**" as they are often called) are risks that can be anticipated and for which plans can be made. They are ordinarily described by a probability of occurrence and have a larger impact than variation as described in the first uncertainty. Typical sources include supply breakdowns, design changes, natural disasters, labor strikes, etc.
3. *Unforeseen uncertainty.* **Unforeseen uncertainty** ("**unknown unknowns**") are risks that could not be predicted ahead of time or that are considered too unlikely to make contingency plans for. Typical sources include natural disasters of unusual scale or in regions where they do not occur customarily (e.g., an earthquake on the eastern coast of the United States) or competitor innovations that were not anticipated. These are termed "black swan" events by Nassim Taleb (2010).

4. *Chaos*. **Chaos** is unforeseen uncertainty where the event not only affects the operations but also the fundamental goal of the project or company. For example, in the wake of the New Zealand city of Christchurch's 2010 earthquake, a number of local beverage manufacturers offered to switch to bottling water instead of regular products. Not only was the earthquake unforeseen because the rupture occurred along a previously unknown fault line that is thought not to have moved for at least 16,000 years, but the goal for the beverage manufacturers shifted from profit from beverage sales to humanitarian aid.

Each of the above forms of uncertainty must be dealt with and planned for in different ways. However, all require a firm to have effective management and communication processes.

Variation is the type of uncertainty most commonly dealt with by operations analysts and managers and will be described in more detail in Section 5.1. Because variation is considered routine and within an operations manager's purview to anticipate and plan for, it is not usually much discussed during the S&OP process. Strategies used by operations to mitigate variation include holding inventory, deliberately maintaining excess capacity, or incurring deliberate delays in order fulfillment.

Foreseen uncertainty is best dealt with through specific contingency plans, risk mitigation processes, and clear ownership responsibility for processes. It is foreseen uncertainty that is best dealt with explicitly within the S&OP process. For example, if it is known that a supplier is currently struggling with labor issues, then the S&OP meeting discussions should include plans to be put in place if the supplier's workforce goes on strike.

Unforeseen uncertainty may be mitigated by generic contingency plans and good leadership, while chaos requires strong leadership and crisis management processes. By definition, neither unforeseen uncertainty nor chaos may be explicitly planned for in the S&OP process. However, an organization with strong cross-functional ties, which good S&OP processes foster, is going to manage such occurrences better than a siloed organization with little cross-functional planning.

PROBLEMS FOR SECTION 3.3

13. Why does the S&OP process typically not explicitly recognize variation even though it is a fact of life for both operations and sales divisions?
14. Classify the following risks into variation, foreseen uncertainty, unforeseen uncertainty, and chaos.
    a. A hurricane on the East Coast floods a regional warehouse destroying a large amount of stock.
    b. A machine on the plant floor breaks down for an hour.
    c. Bad weather on the weekend causes an increase in demand for umbrellas.
    d. A cool summer causes a decrease in demand for air conditioners for that season.
    e. The excavation process for a new manufacturing plant in the Midwest uncovers an archaeological find of such significance that no building can take place on that site, and a new site for the plant must be found.
    f. Competitors to the iPad launch smaller tablet computers before the iPad mini is ready to launch, thus negatively affecting demand for the iPad.
    g. The Second World War caused auto manufacturers to switch to producing military vehicles.
    h. A drug is found to have dangerous side effects following its launch.

i. The transportation disruptions, including the grounding of all airplanes, following the attacks on September 11, 2001 severed many supply chains.

15. List two examples each of variation, foreseen uncertainty, and unforeseen uncertainty that you have personally experienced in your studies.
16. Give an example of chaos, either from your own experience or from others, within the educational domain.

3.4 AGGREGATE PLANNING OF CAPACITY

The operations division is responsible for determining an **aggregate plan** for capacity usage throughout the planning horizon. This plan uses the sales forecasts determined by collaboration at the S&OP meeting. The forecast is expressed in terms of aggregate production units or dollars. The operations division must then determine aggregate production quantities and the levels of resources required to achieve these production goals. In practice, this translates to finding the number of workers that should be employed and the number of aggregate units to be produced in each of the planning periods 1, 2, … , T. The objective of such aggregate planning is to balance the advantages of producing to meet demand as closely as possible against the disturbance caused by changing the levels of production and/or the workforce levels.

As just noted, aggregate planning methodology requires the assumption that demand is known with certainty. This is simultaneously a weakness and a strength of the approach. It is a weakness because it does not provide any buffer against unanticipated forecast errors. However, most inventory models that allow for random demand require that the average demand be constant over time. Aggregate planning allows the manager to focus on the systematic changes that are generally not present in models that assume random demand. By assuming deterministic demand, the effects of seasonal fluctuations and business cycles can be incorporated into the planning function. As discussed in Sections 3.1 and 3.3, variation caused by demand uncertainty may be buffered using inventory, capacity, or customer delays.

Aggregate Units

The aggregate planning approach is predicated on the existence of an aggregate unit of production. When the types of items produced are similar, an aggregate production unit can correspond to an "average" item, but if many different types of items are produced, it would be more appropriate to consider aggregate units in terms of weight (tons of steel), volume (gallons of gasoline), amount of work required (worker-years of programming time), or dollar value (value of inventory in dollars). What the appropriate aggregation scheme should be is not always obvious. It depends on the context of the particular planning problem and the level of aggregation required.

EXAMPLE 3.1

A plant manager working for a large national appliance firm is considering implementing an aggregate planning system to determine the workforce and production levels in his plant. This particular plant produces six models of washing machines. The characteristics of the machines are listed below.

Model Number	Number of Worker-Hours Required to Produce	Selling Price ($)
A5532	4.2	285
K4242	4.9	345
L9898	5.1	395
L3800	5.2	425
M2624	5.4	525
M3880	5.8	725

The plant manager must decide on the particular aggregation scheme to use. One possibility is to define an aggregate unit as one dollar of output. Unfortunately, the selling prices of the various models of washing machines are not consistent with worker-hours required to produce them. The ratio of the selling price divided by the worker-hours is $67.86 for A5532 and $125.00 for M3880. (The company bases its pricing on the fact that the less expensive models have a higher sales volume.) The manager notices that the percentages of the total number of sales for these six models have been fairly constant, with values of 32 percent for A5532, 21 percent for K4242, 17 percent for L9898, 14 percent for L3800, 10 percent for M2624, and 6 percent for M3880. He decides to define an aggregate unit of production as a fictitious washing machine requiring $(.32)(4.2) + (.21)(4.9) + (.17)(5.1) + (.14)(5.2) + (.10)(5.4) + (.06)(5.8) = 4.856$ hours of labor time. He can obtain sales forecasts for aggregate production units in essentially the same way by multiplying the appropriate fractions by the forecasts for unit sales of each type of machine.

The approach used by the plant manager in Example 3.1 was possible because of the relative similarity of the products produced. However, defining an aggregate unit of production at a higher level of the firm is more difficult. In cases in which the firm produces a large variety of products, a natural aggregate unit is sales dollars. Although, as we saw in the example, this will not necessarily translate to the same number of units of production for each item, it will generally provide a good approximation for planning at the highest level of a firm that produces a diverse product line. In other words, it is appropriate as input to the S&OP process.

Costs in Aggregate Capacity Planning

As with most of the optimization problems considered in production management, the goal of the analysis is to choose the aggregate plan that minimizes cost. It is important to identify and measure those specific costs that are affected by the planning decision.

1. *Smoothing costs.* **Smoothing costs** are those costs that accrue as a result of changing the production levels from one period to the next. In the aggregate planning context, the most salient smoothing cost is the cost of changing the size of the workforce. Increasing the size of the workforce requires time and expense to advertise positions, interview prospective employees, and train new hires. Decreasing the size of the workforces means that workers must be laid off. Severance pay is thus one cost of decreasing the size of the

workforce. However, there are other costs associated with firing workers that may be harder to measure.

Firing workers could have far-reaching consequences. Firms that hire and fire frequently develop a poor public image. This could adversely affect sales and discourage potential employees from joining the company. It may adversely affect employee morale. Furthermore, workers who are laid off might not wait for business to pick up; they could seek employment elsewhere. Firing workers can have a detrimental effect on the future size of the labor force in the area if those workers obtain employment in other industries. That is, if the workers retrain for new and different jobs and do not want to return to their original industry, then there may not be workers available to hire when demand picks up. Finally, most companies are not at liberty to hire and fire at will. Labor agreements restrict the freedom of management to freely alter workforce levels.

Most of the models that we consider assume that the costs of increasing and decreasing the size of the workforce are linear functions of the number of employees that are hired or fired. That is, there is a constant dollar amount charged for each employee hired or fired. The assumption of linearity is probably reasonable up to a point. As the supply of labor becomes scarce, there may be additional costs required to hire more workers, and the costs of laying off workers may go up substantially if the number of workers laid off is too large. A typical cost function for changing the size of the workforce appears in Figure 3–4.

2. *Holding costs.* **Holding costs** are the costs that accrue as a result of having capital tied up in inventory. If the firm can decrease its inventory, the money saved could be invested elsewhere with a return that will vary with the industry and with the specific company. (See Chapter 4 for a more complete discussion of holding costs.) Holding costs

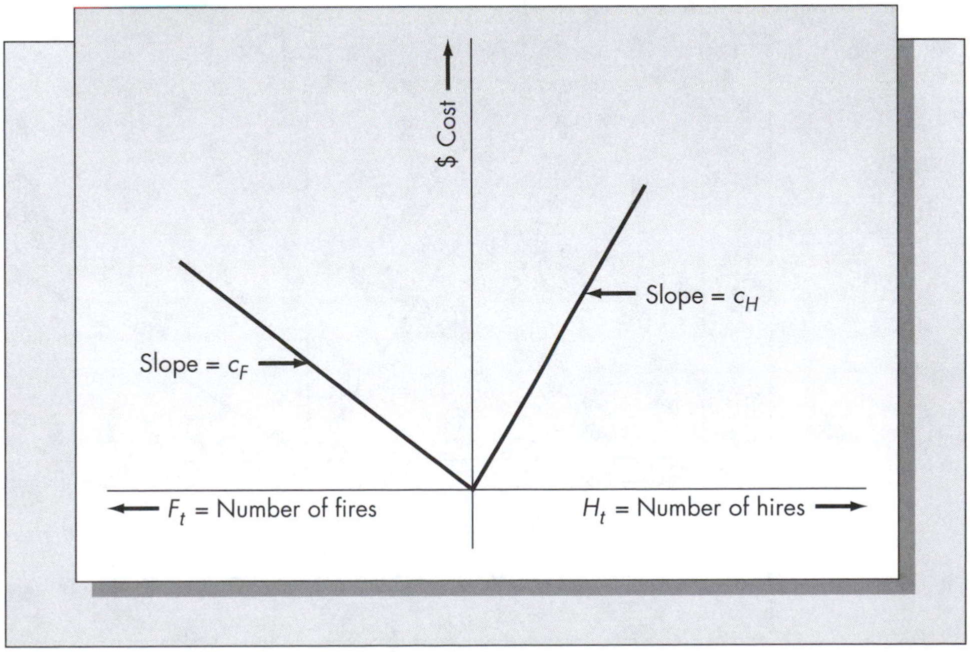

Figure 3–4 Cost of changing the size of the workforce

are almost always assumed to be linear in the number of units being held at a particular point in time. We will assume for the purposes of the aggregate planning analysis that the holding cost is expressed in terms of dollars per unit held per planning period. We also will assume that holding costs are charged against the inventory remaining on hand at the *end* of the planning period. This assumption is made for convenience only. Holding costs could be charged against starting inventory or average inventory as well.

3. *Shortage costs.* Holding costs are charged against the aggregate inventory as long as it is positive. In some situations, it may be necessary to incur shortages, which are represented by a negative level of inventory. Shortages can occur when forecasted demand exceeds the capacity of the production facility or when demands are higher than anticipated. For the purposes of aggregate planning, it is generally assumed that excess demand is backlogged and filled in a future period. In a highly competitive situation, however, it is possible that excess demand is lost, and the customer goes elsewhere. This case, which is known as lost sales, is more appropriate in the management of single items and is more common in a retail than in a manufacturing context.

   As with holding costs, **shortage costs** are generally assumed to be linear. Convex functions also can accurately describe shortage costs, but linear functions seem to be the most common. Figure 3–5 shows a typical holding/shortage cost function.
4. *Regular time costs.* **Regular time costs** involve the cost of producing one unit of output during regular working hours. Included in this category are the actual payroll costs of regular employees working on regular time, the direct and indirect costs of materials, and other manufacturing expenses. When all production is carried out on regular time, regular payroll costs become a "sunk cost," because the number of units produced must equal the number of units demanded over any planning horizon of sufficient length. If there is no overtime or worker idle time, regular payroll costs do not have to be included in the evaluation of different strategies.

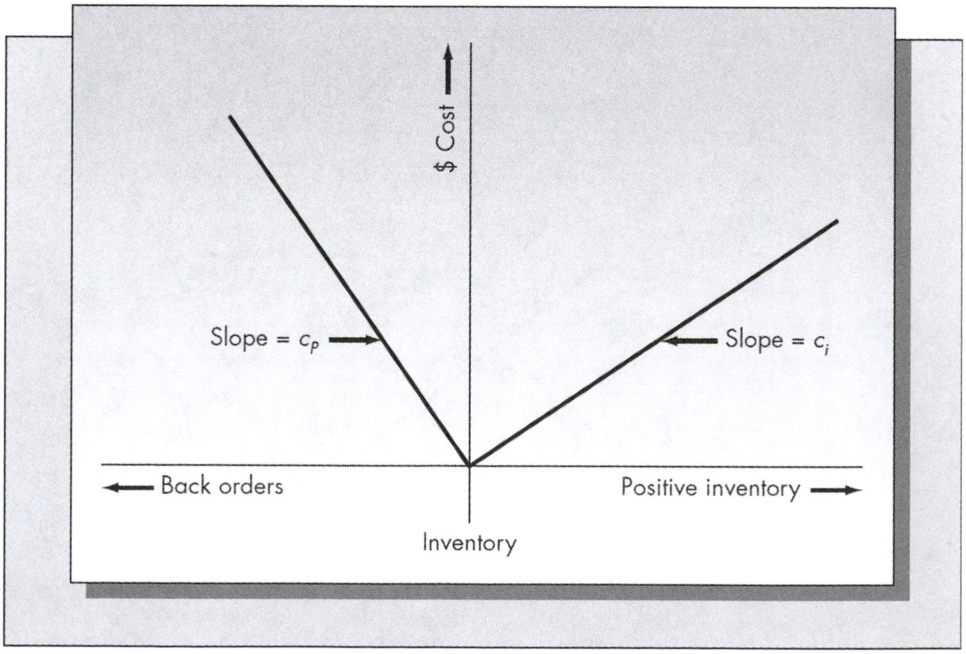

Figure 3–5 Holding and backorder costs

SNAPSHOT APPLICATION

HP Enterprise Services Uses Optimization for Workforce Planning

The S&OP process discussed in this chapter includes optimization of several of the firm's functions, including production planning and workforce planning. In practice, the workforce planning problem is much more difficult, since both the supply and demand for workers is often uncertain. This means that both demand and supply are likely to be random variables, thus making the problem potentially very complex. Hewlett Packard Enterprise (HPE) focuses on global business and technology. These services are provided by more than 60,000 employees located in over 60 countries. Prior to 2015, the company operated an Enterprise Services business, which it spun off in 2017. After a merger with Computer Sciences Corporation, it became DXC Technology.

Providing an optimal solution to the planning problem of enterprise services is probably not realistic due to the scale of the problem. Traditional operations research methods such as Markov Decision Process modeling or large scale mixed integer programming (two methods that have been proposed in the literature) quickly become computationally unwieldy.

A group of researchers (Santos et al. 2013) suggested a two stage approach for solving this problem. The first stage (supply and demand consolidation) indicates those jobs for which an employee is fully qualified. For those employees who are partially qualified, a transition table provides scores that indicate the degree of qualification. Supply uncertainty in this context is primarily due to employee attrition, which can be 30 percent or more in some locations. Under ideal circumstances, the firm would like to be able to match job requirements with those employees who are 100 percent qualified. However, such a stringent rule results in poor demand fulfilment levels. Rather, the research team designed a flexible matching scheme based on concepts developed for the analytical hierarchy process, a tool that allows one to determine appropriate weights in a multi-criteria decision problem.

The second stage of the analysis is to build a mixed integer programming (MIP) module to allocate workers to jobs. This is done in stages: first, available employees are allocated to jobs. Second, employees who are currently committed to jobs, but will be freed up at some future time, are allocated. If neither of these schemes covers the job requirements, then new hires are recommended. Finally, if positions are still unfilled, a "gap" is declared. As noted above, mixed integer optimization can be very demanding computationally, so the researchers employ a heuristic for this phase. Gaps are quite common, so procedures are recommended to deal with them.

Hewlett Packard first implemented this approach at its facility in Bangalore, India. Resource utilization rates improved from approximately 75 percent to approximately 90 percent as a result of this planning tool. HP then proceeded to implement this system on a worldwide scale.

5. *Overtime and subcontracting costs.* **Overtime** and **subcontracting costs** are the costs of production of units not produced on regular time. Overtime refers to production by regular-time employees beyond the normal workday, and subcontracting refers to the production of items by an outside supplier. Again, it is generally assumed that both of these costs are linear.
6. *Idle time costs.* The complete formulation of the aggregate planning problem also includes a cost for underutilization of the workforce, or idle time. In most contexts, the **idle time cost** is zero, as the direct costs of idle time would be taken into account in labor costs and lower production levels. However, idle time could have other consequences for the firm. For example, if the aggregate units are input to another process, idle time on the line could result in higher costs to the subsequent process. In such cases, one would explicitly include a positive idle cost.

When planning is done at a relatively high level of the firm, the effects of intangible factors are more pronounced. Any solution to the aggregate planning problem obtained from a cost-based model must be considered carefully in the context of company policy. An optimal solution to a mathematical model might result in a policy that requires frequent hiring and firing of personnel. Such a policy may be infeasible because of prior contract agreements or undesirable because of the potential negative effects on the firm's public image.

A Prototype Problem

We will illustrate the various techniques for solving aggregate planning problems with the following example.

EXAMPLE 3.2

Densepack is to plan workforce and production levels for the six-month period January to June. The firm produces a line of disk drives for mainframe computers that are plug compatible with several computers produced by major manufacturers. Forecast demands over the next six months for a particular line of drives produced in the Milpitas, California, plant are 1,280, 640, 900, 1,200, 2,000, and 1,400. There are currently (end of December) 300 workers employed in the Milpitas plant. Ending inventory in December is expected to be 500 units, and the firm would like to have 600 units on hand at the end of June.

There are several ways to incorporate the starting and the ending inventory constraints into the formulation. The most convenient is simply to modify the values of the predicted demand by the netting procedure described next. This method assumes that starting inventory is less than the first period demand. It is not necessary if the aggregate plan is produced using linear programming (see Section 3.5).

Define net predicted demand in period 1 as the predicted demand minus initial inventory. If there is a minimum ending inventory constraint, then this amount should be added to the demand in period T. Minimum buffer inventories also can be handled by modifying the predicted demand. If there is a minimum buffer inventory in every period, this amount should be added to the first period's demand. If there is a minimum buffer inventory in only one period, this amount should be added to that period's demand and subtracted from the next period's demand. Actual ending inventories should be computed using the original demand pattern, however.

Returning to our example, we define the net predicted demand for January as 780 (1,280 – 500) and the net predicted demand for June as 2,000 (1,400 + 600). By considering net demand, we may make the simplifying assumption that starting and ending inventories are both zero. The net predicted demand and the net cumulative demand for the six months January to June are as follows.

Month	Net Predicted Demand	Net Cumulative Demand
January	780	780
February	640	1,420
March	900	2,320
April	1,200	3,520
May	2,000	5,520
June	2,000	7,520

The cumulative net demand is pictured in Figure 3–6. A production plan is the specification of the production levels for each month. If shortages are not permitted, then cumulative production must be at least as great as cumulative demand each period. In addition to the cumulative net demand, Figure 3–6 also shows one feasible production plan.

In order to illustrate the cost trade-offs of various production plans, we will assume in the example that there are only three costs to be considered: cost of hiring workers, cost of firing workers, and cost of holding inventory. Define

$$c_H = \text{Cost of hiring one worker} = \$500,$$

$$c_F = \text{Cost of firing one worker} = \$1{,}000,$$

$$c_I = \text{Cost of holding one unit of inventory for one month} = \$80.$$

We require a means of translating aggregate production in units to workforce levels. Because not all months have an equal number of working days, we will use a day as an indivisible unit of measure and define

$$K = \text{Number of aggregate units produced by one worker in one day.}$$

In the past, the plant manager observed that over 22 working days, with the workforce level constant at 76 workers, the firm produced 245 disk drives. That means that on average the production rate was 245/22 = 11.1364 drives per day when there were 76 workers employed at the plant. It follows that one worker produced an average of 11.1364/76 = 0.14653 drive in one day. Hence, $K = 0.14653$ for this example.

The final data needed to evaluate a plan for this example is the number of working days per month. In what follows we assume this to be 20, 24, 18, 26, 22, and 15 for January to June, respectively.

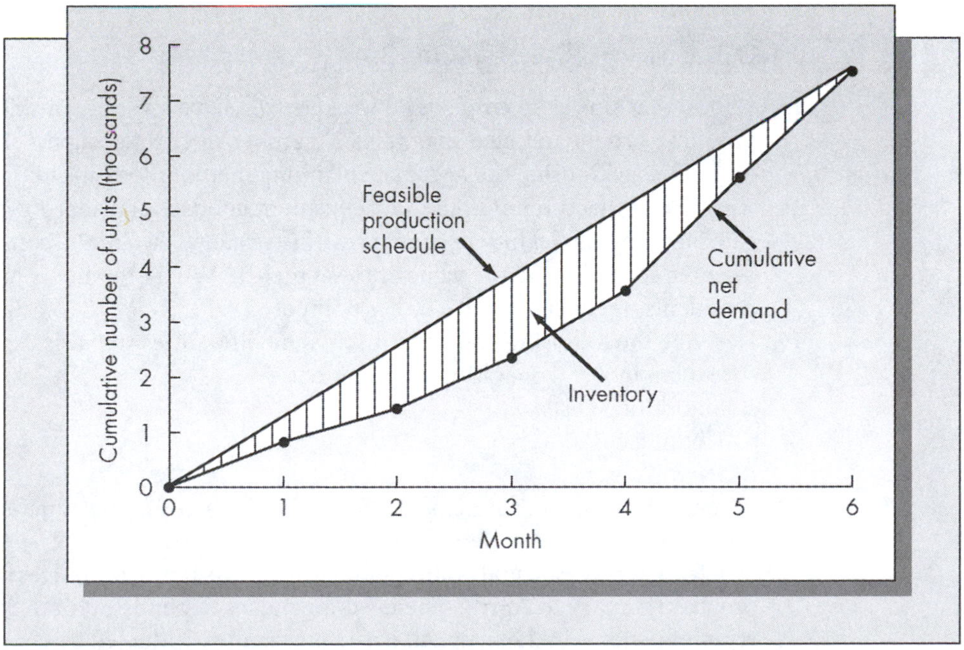

Figure 3–6 A feasible aggregate plan for Densepack

Chase, Level, and Mixed Strategies

Two extremes in capacity planning are the **zero inventory plan**, also known as a **chase strategy**, and the **constant workforce plan**, also known as a **level strategy**. Under the zero inventory plan, the workforce is changed each month in order to produce enough units to most closely match the demand pattern. Capacity is adjusted up and down (i.e., workers are hired and fired) to achieve this matching. Under the constant workforce plan, capacity is kept constant during the planning period (i.e., no workers are hired or fired) and instead inventory is kept between periods; capacity is set to the minimum possible to ensure no shortages in any period.

For Example 3.2, consider a zero inventory plan that uses all available inventory in January and produces the required 600 units of ending inventory in June. This plan hires a total of 755 workers, fires a total 145 workers, and achieves a total cost of $572,900 (calculations are shown in Appendix 3-A). The best constant workforce plan sets capacity to 411 workers each month (hiring 111 at the beginning of January) with no further hiring or firing, has a total inventory cost of $524,960, and a total cost of $580,460 once the initial workers are considered (calculations again shown in Appendix 3-A). While the zero inventory plan has slightly lower costs, it is unlikely to be practical because there may be constraints on the total capacity of the plant or on the maximum change that is possible from one month to the next.

The zero inventory plan and constant workforce strategies are pure strategies: they are designed to achieve one objective. They are useful in enhancing intuition and for ballpark calculations. However, with more flexibility, small modifications can result in dramatically lower costs. Such plans are described as **mixed strategies**. Optimal mixed strategies are typically found using linear programming formulations, which can incorporate a variety of additional practical constraints. Linear programming formulations are the subject of the next section.

PROBLEMS FOR SECTION 3.4

17. A local machine shop employs 60 workers who have a variety of skills. The shop accepts one-time orders and also maintains a number of regular clients. Discuss some of the difficulties with using the aggregate planning methodology in this context.
18. A large manufacturer of household consumer goods is considering integrating an aggregate planning model into its manufacturing strategy. Two of the company vice presidents disagree strongly as to the value of the approach. What arguments might each of the vice presidents use to support his or her point of view?
19. Describe the following costs and discuss the difficulties that arise in attempting to measure them in a real operating environment.
    a. Smoothing costs
    b. Holding costs
    c. Payroll costs
20. Discuss the following statement: "Since we use a rolling production schedule, I really don't need to know the demand beyond next month."
21. St. Clair County Hospital is attempting to assess its needs for nurses over the coming four months (January to April). The need for nurses depends on both the numbers and the types of patients in the hospital. Based on a study conducted by consultants, the hospital has determined that the following ratios of nurses to patients are required.

Patient Type	Numbers of Nurses Required per Patient	Patient Forecasts Jan.	Feb.	Mar.	Apr.
Major surgery	0.4	28	21	16	18
Minor surgery	0.1	12	25	45	32
Maternity	0.5	22	43	90	26
Critical care	0.6	75	45	60	30
Other	0.3	80	94	73	77

a. How many nurses should be working each month to most closely match patient forecasts?
b. Suppose the hospital does not want to change its policy of not increasing the nursing staff size by more than 10 percent in any month. Suggest a schedule of nurse staffing over the four months that meets this requirement and also meets the need for nurses each month.

22. Give an example of an environment where a chase strategy would be highly disruptive. Now give one where a level strategy would be highly disruptive.
23. Is a chase or level strategy more appropriate for aggregate planning in an air conditioning manufacturing plant where demand is highly seasonal and the workforce is relatively skilled? Explain your answer.

3.5 SOLVING AGGREGATE PLANNING PROBLEMS

Linear programming is a term used to describe a general class of optimization problems. The objective is to determine values of n decision variables in order to maximize or minimize a linear function of these variables that is subject to m linear constraints on these variables. Supplement 1, which follows this chapter, provides an overview of linear programming. We recommend that readers who have not seen linear programming before begin by reading the supplement before reading this section. The primary advantage in formulating a problem as a linear program is that the simplex method yields optimal solutions very efficiently.

When all cost functions are linear, there is a linear programming formulation of the general aggregate planning problem. Because of the efficiency of commercial linear programming codes, this means that (essentially) optimal solutions can be obtained for very large problems (we include the qualifier because rounding may give suboptimal solutions, a point discussed in more detail later in this section).

Cost Parameters and Given Information

The following values are assumed to be known:

c_H = Cost of hiring one worker,
c_F = Cost of firing one worker,
c_I = Cost of holding one unit of stock for one period,
c_R = Cost of producing one unit on regular time,
c_O = Incremental cost of producing one unit on overtime,
c_U = Idle cost per unit of production,
c_S = Cost to subcontract one unit of production,

n_t = Number of production days in period t,
K = Number of aggregate units produced by one worker in one day,
I_0 = Initial inventory on hand at the start of the planning horizon,
W_0 = Initial workforce at the start of the planning horizon,
D_t = Forecast of demand in period t.

The cost parameters also may be time dependent; that is, they may change with t. Time-dependent cost parameters could be useful for modeling changes in the costs of hiring or firing due, for example, to shortages in the labor pool, or changes in the costs of production and/or storage due to shortages in the supply of resources, or changes in interest rates. It is also possible to include costs associated with salaried workers if these cannot be rolled into the cost of regular time production.

Problem Decision Variables

The following are the problem's decision variables:

W_t = Workforce level in period t,
P_t = Production level in period t,
I_t = Inventory level in period t,
H_t = Number of workers hired in period t,
F_t = Number of workers fired in period t,
O_t = Overtime production in units,
U_t = Worker idle time in units ("undertime"),
S_t = Number of units subcontracted from outside.

The overtime and idle time variables are determined in the following way. The term Kn_t represents the number of units produced by one worker in period t, so that Kn_tW_t would be the number of units produced by the entire workforce in period t. However, we do not require that $Kn_tW_t = P_t$. If $P_t > Kn_tW_t$, then the number of units produced exceeds what the workforce can produce on regular time. This means that the difference is being produced on overtime, so that the number of units produced on overtime is exactly $O_t = P_t - Kn_tW_t$. If $P_t < Kn_tW_t$, then the workforce is producing less than it should be on regular time, which means that there is worker idle time. The idle time is measured in units of production rather than in time, and is given by $U_t = Kn_tW_t - P_t$.

Problem Constraints

Three sets of constraints are required for the linear programming formulation. They are included to ensure that conservation of labor and conservation of units are satisfied.

1. Conservation of workforce constraints.

W_t	=	W_{t-1}	+	H_t	–	F_t	for $1 \le t \le T$.
Number of workers in t	=	Number of workers in $t - 1$	+	Number hired in t	–	Number fired in t	

2. Conservation of units constraints.

I_t	=	I_{t-1}	+	P_t	+	S_t	−	D_t	for $1 \leq t \leq T$.
Inventory in t	=	Inventory in $t-1$	+	Number of units produced in t	+	Number of units subcontracted in t	−	Demand in t	

3. Constraints relating production levels to workforce levels.

P_t	=	Kn_tW_t	+	O_t	−	U_t	for $1 \leq t \leq T$.
Number of units produced in t	=	Number of units produced by regular workforce in t	+	Number of units produced on overtime in t	−	Number of units of idle production in t	

In addition to these constraints, all problem variables should be nonnegative. These constraints and the nonnegativity constraints are the minimum that must be present in any formulation. Notice that (1), (2), and (3) constitute $3T$ constraints, rather than 3 constraints, where T is the length of the forecast horizon.

The formulation also requires specification of the initial inventory, I_0, and the initial workforce, W_0, and may include specification of the ending inventory in the final period, I_T. The inclusion of these variables means that the netting procedure described in Example 3.2 does not need to be applied.

The objective function includes all the costs defined earlier. The linear programming formulation is to choose values of the decision variables W_t, P_t, I_t, H_t, F_t, O_t, U_t, and S_t to

$$\text{minimize} \quad \sum_{t=1}^{T}(c_H H_t + c_F F_t + c_I I_t + c_R P_t + c_O O_t + c_U U_t + c_S S_t)$$

subject to

$$W_t = W_{t-1} + H_t - F_t \qquad \text{for } 1 \leq t \leq T \quad \text{(conservation of workforce)} \tag{A}$$

$$P_t = Kn_tW_t + O_t - U_t \qquad \text{for } 1 \leq t \leq T \quad \text{(production and workforce)} \tag{B}$$

$$I_t = I_{t-1} + P_t + S_t - D_t \qquad \text{for } 1 \leq t \leq T \quad \text{(inventory balance)} \tag{C}$$

$$H_t, F_t, I_t, O_t, U_t, S_t, W_t, P_t \geq 0 \quad \text{(nonnegativity)} \tag{D}$$

plus any additional constraints that define the values of starting inventory, starting workforce, ending inventory, or any other variables with values that are fixed in advance.

Rounding the Variables

In general, the optimal values of the problem variables will not be integers. However, fractional values for many of the variables do not make sense. These variables include the size of the workforce, the number of workers hired each period, and the number of workers fired each period, and also may include the number of units produced each period. (It is possible that fractional numbers of units could be produced in some applications.) One way to deal with this problem is to require in advance that some or all of the problem variables assume only integer values. Unfortunately, this makes the solution algorithm considerably more complex. The resulting problem, known as an integer linear programming problem, requires much more computational effort to solve than does ordinary linear programming. For a moderate-sized problem, solving the problem as an integer linear program is a reasonable alternative.

If an integer programming code is unavailable or if the problem is simply too large to solve by integer programming, linear programming still provides a workable solution. However, after the linear programming solution is obtained, some of the problem variables must be rounded to integer values. Simply rounding off each variable to the closest integer may lead to an infeasible solution and/or one in which production and workforce levels are inconsistent. It is not obvious what is the best way to round the variables. We recommend the following conservative approach: round the values of the numbers of workers in each period t to W_t, the next larger integer. Once the values of W_t are determined, the values of the other variables, H_t, F_t, and P_t, can be found along with the cost of the resulting plan. Conservative rounding will always result in a feasible solution but will rarely give the optimal solution. The conservative solution generally can be improved by trial-and-error experimentation.

There is no guarantee that if a problem can be formulated as a linear program, the final solution makes sense in the context of the problem. In the aggregate planning problem, it does not make sense that there should be both overtime production and idle time in the same period, and it does not make sense that workers should be hired and fired in the same period. This means that either one or both of the variables O_t and U_t must be zero, and either one or both of the variables H_t and F_t must be zero for each t, $1 \leq t \leq T$. This requirement can be included explicitly in the problem formulation by adding the constraints

$$O_t U_t = 0 \qquad \text{for } 1 \leq t \leq T,$$
$$H_t F_t = 0 \qquad \text{for } 1 \leq t \leq T,$$

since if the product of two variables is zero it means that at least one must be zero. Unfortunately, these constraints are not linear, as they involve a product of problem variables. However, it turns out that it is not necessary to explicitly include these constraints, because the optimal solution to a linear programming problem always occurs at an extreme point of the feasible region. It can be shown that every extreme point solution automatically has this property. If this were not the case, the linear programming solution would be meaningless.

Extensions

Linear programming also can be used to solve somewhat more general versions of the aggregate planning problem. Uncertainty of demand can be accounted for indirectly by assuming that there is a minimum buffer inventory B_t each period. In that case, we would include the constraints

$$I_t \geq B_t \qquad \text{for } 1 \leq t \leq T.$$

The constants B_t would have to be specified in advance. Upper bounds on the number of workers hired and the number of workers fired each period could be included in a similar way. Capacity constraints on the amount of production each period could easily be represented by the set of constraints:

$$P_t \leq C_t \qquad \text{for } 1 \leq t \leq T.$$

The linear programming formulation introduced in this section assumed that inventory levels would never go negative. However, in some cases it might be desirable or even necessary to allow demand to exceed supply, for example, if forecast demand exceeded production capacity over some set of planning periods. In order to treat backlogging of excess demand, the inventory level I_t must be expressed as the difference between two nonnegative variables, say I_t+ and I_t^-, satisfying

$$I_t = I_t + - I_t^-,$$
$$I_t^+ \geq 0, \qquad I_t^- \geq 0.$$

The holding cost would now be charged against I_t+ and the penalty cost for back orders (say c_P) against I_t^-. However, notice that for the solution to be sensible, it must be true that I_t^+ and I_t^- are not both positive in the same period t. As with the overtime and idle time and the hiring and firing variables, the properties of linear programming will guarantee that this holds without having to explicitly include the constraint $I_t^+ I_t^- = 0$ in the formulation.

In the development of the linear programming model, we stated the requirement that all the cost functions must be linear. This is not strictly correct. Linear programming also can be used when the cost functions are **convex piecewise-linear functions**.

A **convex function** is one with an increasing slope. A **piecewise-linear function** is one that is composed of straight-line segments. Hence, a convex piecewise-linear function is a function composed of straight lines that have increasing slopes. A typical example is presented in Figure 3–7.

In practice, it is likely that some or all of the cost functions for aggregate planning are convex. For example, if Figure 3–7 represents the cost of hiring workers, then the marginal cost of hiring one additional worker increases with the number of workers that have already been hired. This is probably more accurate than assuming that the cost of hiring one additional worker is a constant independent of the number of workers previously hired. As more workers are hired, the available labor pool shrinks, and more effort must be expended to hire the remaining available workers.

In order to see exactly how convex piecewise-linear functions would be incorporated into the linear programming formulation, we will consider a very simple case. Suppose that the cost of hiring new workers is represented by the function pictured in Figure 3–8. According to the figure, it costs c_{H1} to hire each worker until H^* workers are hired, and it costs c_{H2} for each worker hired beyond H^* workers, with $c_{H1} < c_{H2}$. The variable H_t, the number of workers hired in period t, must be expressed as the sum of two variables:

$$H_t = H_{1t} + H_{2t}.$$

Interpret H_{1t} as the number of workers hired up to H^* and H_{2t} as the number of workers hired beyond H^* in period t. The cost of hiring is now represented in the objective function as

$$\sum_{t=1}^{T} (c_{H1} H_{1t} + c_{H2} H_{2t}),$$

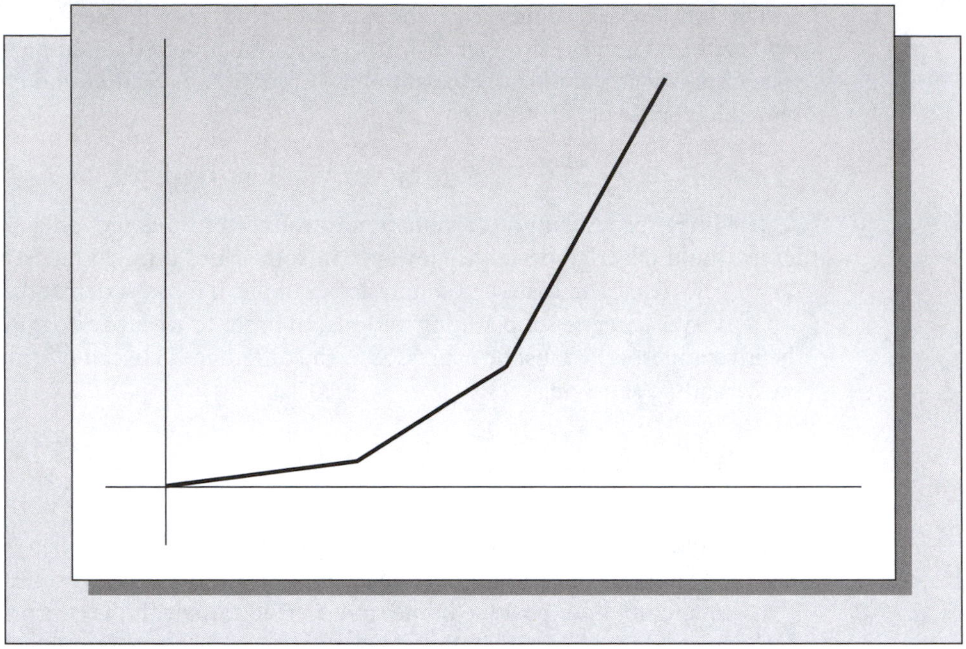

Figure 3–7 A convex piecewise linear function

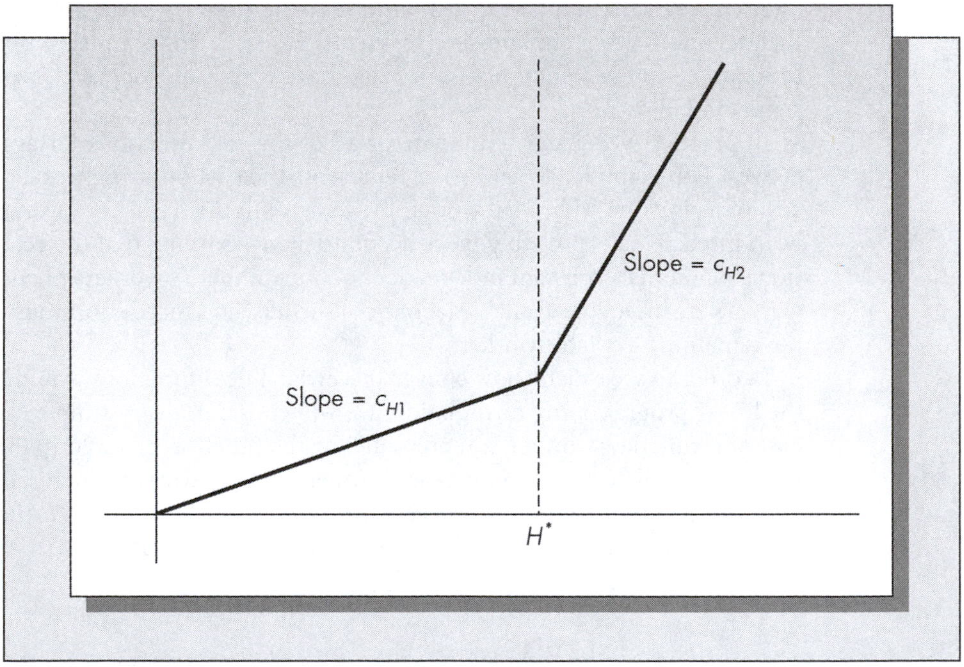

Figure 3–8 Convex piecewise-linear hiring cost function

and the additional constraints

$$H_t = H_{1t} + H_{2t}$$
$$0 \le H_{1t} \le H^*$$
$$0 \le H_{2t}$$

must also be included.

In order for the final solution to make sense, it can never be the case that $H_{1t} < H^*$ and $H_{2t} > 0$ for some t. (Why?) However, because linear programming searches for the minimum cost solution, it will force H_{1t} to its maximum value before allowing H_{2t} to become positive, since $c_{H1} < c_{H2}$. This is the reason that the cost functions must be convex. This approach can easily be extended to more than two linear segments and to any of the other cost functions present in the objective function. The technique is known as separable convex programming and is discussed in greater detail in Frederick Hillier and Gerald Lieberman (2015).

We will demonstrate the use of linear programming by finding the optimal solution to the example presented in Section 3.4. As there is no subcontracting, overtime, or idle time allowed, and the cost coefficients are constant with respect to time, the objective function is simply

$$\text{minimize} \quad \left(500\sum_{t=1}^{6} H_t + 1{,}000\sum_{t=1}^{6} F_t + 80\sum_{t=1}^{6} I_t \right).$$

The boundary conditions comprise the specifications of the initial inventory of 500 units, the initial workforce of 300 workers, and the ending inventory of 600 units. These are best handled by including a separate additional constraint for each boundary condition. We then use the original values for demand (i.e., 1,280 in January) rather than the netted amount (i.e., 780).

The constraints are obtained by substituting $t = 1, \ldots, 6$ into Equations (A), (B), and (C). The full set of constraints expressed in standard linear programming format (with all problem variables on the left-hand side and nonnegative constants on the right-hand side) is as follows:

$$\begin{aligned} W_1 - W_0 - H_1 + F_1 &= 0,\\ W_2 - W_1 - H_2 + F_2 &= 0,\\ W_3 - W_2 - H_3 + F_3 &= 0,\\ W_4 - W_3 - H_4 + F_4 &= 0,\\ W_5 - W_4 - H_5 + F_5 &= 0,\\ W_6 - W_5 - H_6 + F_6 &= 0; \end{aligned} \tag{A}$$

$$\begin{aligned} P_1 - I_1 + I_0 &= 1{,}280,\\ P_2 - I_2 + I_1 &= 640,\\ P_3 - I_3 + I_2 &= 900,\\ P_4 - I_4 + I_3 &= 1{,}200,\\ P_5 - I_5 + I_4 &= 2{,}000,\\ P_6 - I_6 + I_5 &= 1{,}400; \end{aligned} \tag{B}$$

$$P_1 - 2.931W_1 = 0,$$
$$P_2 - 3.517W_2 = 0,$$
$$P_3 - 2.638W_3 = 0, \qquad \text{(C)}$$
$$P_4 - 3.810W_4 = 0,$$
$$P_5 - 3.224W_5 = 0,$$
$$P_6 - 2.198W_6 = 0;$$

$$W_1, \ldots, W_6, P_1, \ldots, P_6, I_1, \ldots, I_6, F_1, \ldots, F_6, H_1, \ldots, H_6 \geq 0; \qquad \text{(D)}$$

$$W_0 = 300,$$
$$I_0 = 500, \qquad \text{(E)}$$
$$I_6 = 600.$$

The values in equations (C) come from multiplying the value of K, number of aggregate units per worker, found earlier by the number of working days in the month. The complete formulation in Excel is therefore as follows.

	A	B	C	D	E	F	G	H	I	J
1	**Cost of hiring**	$500								
2	**Cost of firing**	$1,000								
3	**Holding cost**	$80								
4	**K**	0.14653								
5	**Ending inv**	600								
6										
7	**Month**	**Hired**	**Fired**	**Inventory**	**Workers**	**Production**		**Worker**	**Inventory**	**Adjusted**
8		(H_t)	(F_t)	(I_t)	(W_t)	(P_t)		**balance**	**Balance**	**Demand**
9	Start			500	300					
10	January	0.00	27.02	20.00	272.98	800.00		0	1280	1280
11	February	0.00	0.00	340.00	272.98	960.00		0	640	640
12	March	0.00	0.00	160.00	272.98	720.00		0	900	900
13	April	0.00	0.00	0.00	272.98	1040.00		0	1200	1200
14	May	464.80	0.00	378.38	737.78	2378.38		0	2000	2000
15	June	0.00	0.00	600.00	737.78	1621.62		0	1400	1400
16	Totals	464.80	27.02	1498.38						
17										
18	**Month**	**Days**	**Units per**	**Production**						
19		(n_t)	**worker**	**balance**						
20	January	20	2.931	0						
21	February	24	3.517	0						
22	March	18	2.638	0						
23	April	26	3.810	0						
24	May	22	3.224	0						
25	June	15	2.198	0						
26										
27	**Total cost**	$379,292.22								

Cell Formulas

Cell	Formula	Copied to
B4	=245/22/76	
H9	=E9-E8-B9+C9	H10:H14
I9	=F9-D9+D8	I10:I14
C20	=B20*$B$4	C21:C25
D20	=F10-C20*E10	D21:D25
B16	=SUM(B10:B15)	C16:D16
B27	=$B$1*$B$16+$B$2*$C$16+$B$3*$D$16	

The Solver window for this problem follows.

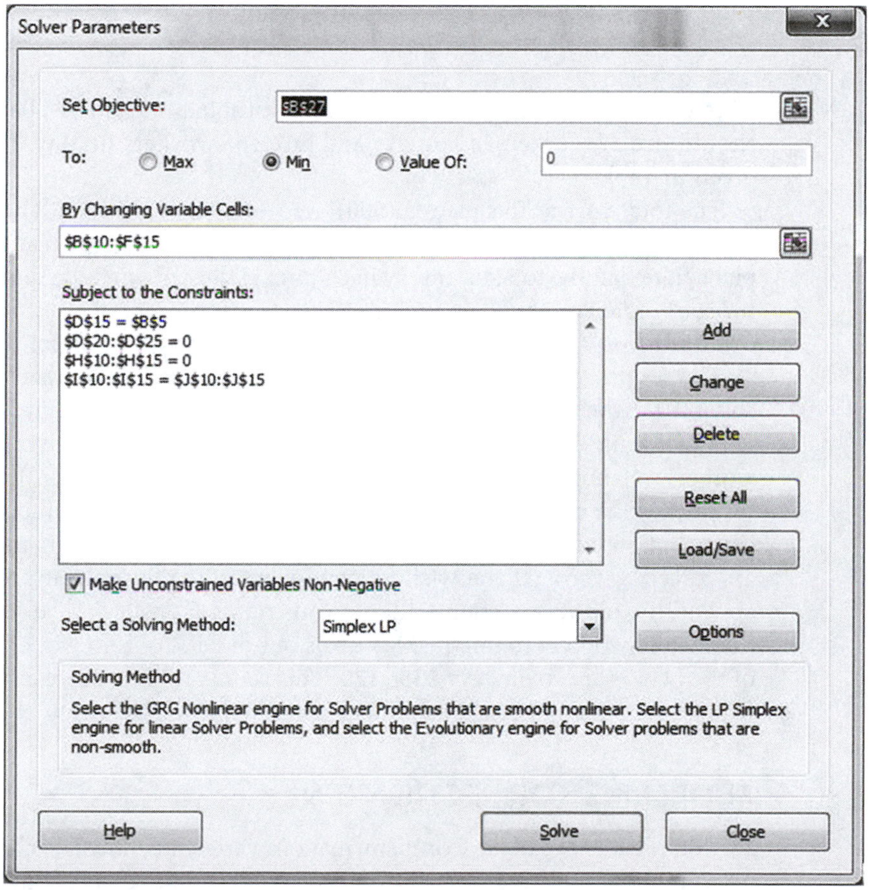

The value of the objective function at the optimal solution is 379,292.22,which is *considerably* less than that achieved with either the zero inventory plan or the constant workforce plan. However, this cost is based on fractional values of the variables. The actual cost will be slightly higher after rounding.

Following the rounding procedure recommended earlier, we will round all the values of W_t to the next higher integer. That gives $W_1 = \ldots = W_4 = 273$ and $W_5 = W_6 = 738$.

Table 3–1 Aggregate Plan for Densepack Obtained from Rounding the Linear Programming Solution

A	B	C	D	E	F	G	H
Month	Number of Workers	Number Hired	Number Fired	Number of Units per Worker	Number of Units Produced (B × E)	Demand	Ending Inventory (H + F – G)
							500
January	273		27	2.931	800	1,280	20
February	273			3.517	960	640	340
March	273			2.638	720	900	160
April	273			3.810	1,040	1,200	0
May	738	465		3.224	2,379	2,000	379
June	738			2.198	1,622	1,400	601
Totals		465	27				1500

This determines the values of the other problem variables. This means that the firm should fire 27 workers in January and hire 465 workers in May. The complete solution is given in Table 3–1.

The total cost of this plan is (500)(465) + (1,000)(27) + (80)(1,500) = \$379,500, which represents a substantial savings over both the zero inventory plan and the constant workforce plan. Note that the total inventory in column H does not include the 500 units of inventory at the start of January but does include the ending inventory of 601 units. The cost of the initial 500 units would have been accounted for in the aggregate plan for the previous six months.

The results of the linear programming analysis suggest another plan that might be more suitable for the company. Because the optimal strategy is to decrease the workforce in January and build it back up again in May, a reasonable alternative might be to not fire the 27 workers in January and to hire fewer workers in May. In this case, the most efficient method for finding the correct number of workers to hire in May is to simply re-solve the linear program, but without the variables $F_1, \ldots, F_6$, as no firing of workers means that these variables are forced to zero. (If you wish to avoid reentering the problem into the computer, simply append the old formulation with the constraints $F_1 = 0, F_2 = 0, \ldots, F_6 = 0$.) The optimal number of workers to hire in May turns out to be 374 if no workers are fired, and the cost of the plan is approximately \$386,120. This is only slightly more expensive than the optimal plan and has the important advantage of not requiring the firing of any workers.

PROBLEMS FOR SECTION 3.5

24. Mr. Meadows Cookie Company makes a variety of chocolate chip cookies in the plant in Albion, Michigan. Based on orders received and forecasts of buying habits, it is estimated that the demand for the next four months is 850, 1,260, 510, and 980, expressed in thousands of cookies. During a 46-day period when there were 120 workers, the company produced 1.7 million cookies. Assume that the number of workdays over the four months are respectively 26, 24, 20, and 16. There are currently 100 workers employed, and there is no starting inventory of cookies.
    a. What is the minimum constant workforce required to meet demand over the next four months?

b. Assume that $c_I = 10$ cents per cookie per month, $c_H = \$100$, and $c_F = \$200$. Evaluate the cost of the plan derived in part (a).
c. Formulate as a linear program. Be sure to define all variables and include the required constraints.
d. Solve for the optimal solution.

25. Harold Grey owns a small farm in the Salinas Valley that grows apricots. The apricots are dried on the premises and sold to a number of large supermarket chains. Based on past experience and committed contracts, he estimates that sales over the next five years in thousands of packages will be as follows.

Year	Forecasted Demand (thousands of packages)
1	300
2	120
3	200
4	110
5	135

Assume that each worker stays on the job for at least one year and that Grey currently has three workers on the payroll. He estimates that he will have 20,000 packages on hand at the end of the current year. Assume that, on the average, each worker is paid \$25,000 per year and is responsible for producing 30,000 packages. Inventory costs have been estimated to be 4 cents per package per year, and shortages are not allowed.

Based on the effort of interviewing and training new workers, Farmer Grey estimates that it costs \$500 for each worker hired. Severance pay amounts to \$1,000 per worker.

a. Assuming that shortages are not allowed, determine the minimum constant workforce that he will need over the next five years.
b. Evaluate the cost of the plan found in part (a).
c. Formulate this as a linear program.
d. Solve the problem and round-off the solution and determine the cost of the resulting plan.

26. A local semiconductor firm, Superchip, is planning its workforce and production levels over the next year. The firm makes a variety of microprocessors and uses sales dollars as its aggregate production measure. Based on orders received and sales forecasts provided by the marketing department, the estimate of dollar sales for the next year by month is on the next page.

Inventory holding costs are based on a 25 percent annual interest charge. It is anticipated that there will be 675 workers on the payroll at the end of the current year, and inventories will amount to \$120,000. The firm would like to have at least \$100,000 of inventory at the end of December next year. It is estimated that each worker accounts for an average of \$60,000 of production per year (assume that one year consists of 250 working days). The cost of hiring a new worker is \$200, and the cost of laying off a worker is \$400.

a. Formulate this as a linear program.

b. Solve the problem. Round the variables in the resulting solution and determine the cost of the plan you obtain.

Month	Production Days	Predicted Demand (in $10,000)
January	22	340
February	16	380
March	21	220
April	19	100
May	23	490
June	20	625
July	24	375
August	12	310
September	19	175
October	22	145
November	20	120
December	16	165

27. Revisit Mr. Meadows Cookie Company from problem 24. Suppose that the cost of hiring workers each period is $100 for each worker until 20 workers are hired, $400 for each worker when between 21 and 50 workers are hired, and $700 for each worker hired beyond 50.
    a. Write down the complete linear programming formulation of the revised problem.
    b. Solve the revised problem for the optimal solution. What difference does the new hiring cost function make in the solution?
28. Leather-All produces a line of handmade leather products. At the present time, the company is producing only belts, handbags, and attaché cases. The predicted demand for these three types of items over a six-month planning horizon is as follows.

Month	Number of Working Days	Belts	Handbags	Attaché Cases
1	22	2,500	1,250	240
2	20	2,800	680	380
3	19	2,000	1,625	110
4	24	3,400	745	75
5	21	3,000	835	126
6	17	1,600	375	45

The belts require an average of two hours to produce, the handbags three hours, and the attaché cases six hours. All the workers have the skill to work on any item. Leather-All has 46 employees, all of whom have a share in the firm and cannot be fired. There are an additional 30 locals who are available and can be hired for short periods at higher cost. Regular employees earn $8.50 per hour on regular time and $14.00 per hour on overtime. Regular time comprises a seven-hour workday, and the regular employees will work as much overtime as is available. The additional workers are hired for $11.00 per hour and are kept on the payroll for at least one full month. Costs of hiring and firing are negligible.

Because of the competitive nature of the industry, Leather-All does not want to incur any demand backorders.

a. Using worker hours as an aggregate measure of production, convert the forecasted demands to demands in terms of aggregate units.
b. What would be the size of the workforce needed to satisfy the demand for the coming six months on regular time only? Would it be to the company's advantage to bring the permanent workforce up to this level? Why or why not?
c. Formulate the problem of optimizing Leather-All's hiring schedule as a linear program. Define all problem variables and include whatever constraints are necessary.
d. Solve the problem formulated in part (c) for the optimal solution. Round all the relevant variables and determine the cost of the resulting plan.

3.6 DISAGGREGATING PLANS

As we saw earlier in this chapter, aggregate planning may be done at several levels of the firm. Aggregate planning might be done for a single plant, a product family, a group of families, or for the firm as a whole.

There are two views of production planning: bottom-up or top-down. The bottom-up approach means that one would start with individual item production plans. These plans could then be aggregated up the chain of products to produce aggregate plans. The top-down approach, which is the one treated in this chapter, is to start with an aggregate plan at a high level. These plans would then have to be "**disaggregated**" to produce detailed production plans at the plant and individual item levels.

It is not clear that disaggregation is an issue in all circumstances. If the aggregate plan is used only for macro planning purposes, and not for planning at the detail level, then one need not worry about disaggregation. However, if it is important that individual item production plans and aggregate plans be consistent, then it might be necessary to consider disaggregation schemes.

The disaggregation problem is similar to the classic problem of resource allocation. Consider how resources are allocated in a university, for example. A university receives revenues from tuition, gifts, interest on the endowment, and research grants. Costs include salaries, maintenance, and capital expenditures. Once an annual budget is determined, each school (arts and science, engineering, business, law, etc.) and each budget center (staff, maintenance, buildings and grounds, etc.) would have to be allocated its share. Budget centers would have to allocate funds to each of their subgroups. For example, each school would allocate funds to individual departments, and departments would allocate funds to faculty and staff in that department.

In the manufacturing context, a disaggregation scheme is just a means of allocating aggregate units to individual items. Just as funds are allocated on several levels in the university, aggregate units might have to be disaggregated at several levels of the firm. This is the idea behind hierarchical production planning championed by several researchers at MIT and reported in detail in Gabrial Bitran and Arnoldo Hax (1981) and Hax and Dan Candea (1984).

We will discuss one possible approach to the disaggregation problem from Bitran and Hax (1981). Suppose that X^* represents the number of aggregate units of production for a particular planning period. Further, suppose that X^* represents an aggregation of n different items

SNAPSHOT APPLICATION

Welch's Uses Aggregate Planning for Production Scheduling

Welch's is a food manufacturer based in Concord, Massachusetts. They are probably best known for grape jelly and grape juices, but they produce a wide variety of processed foods. Stuart Allen and Edmund Schuster (1994) created an aggregate planning model for Welch's primary production facility, accounting for the following characteristics.

- Dynamic, uncertain demand, resulting in changing buffer stock requirements.
- Make-to-stock environment.
- Dedicated production lines.
- Production lines that each produces two or more families of products.
- Large setup times and setup costs for families, as opposed to low setup times and costs for individual items.

The two primary objectives of the model were to smooth peak demands through time so as not to exceed production capacity and to allocate production requirements among the families of products to balance family holding and setup costs. The planning is done with a six-month time horizon for demand forecasting. The six-month period is divided into two portions: the next four weeks and the remaining five months. Detailed plans are developed for the near term, including regular and overtime production allocation.

The model has two primary components: a family planning model, which finds the optimal timing and sizing of family production runs, and a disaggregation planning model, which takes the results of family planning and determines lot sizes for individual items within families.

There were several implementation issues specific to Welch's environment. Product run lengths must be tied to the existing eight-hour shift structure. To do so, the recommendation was that production run lengths be expressed as multiples of one-quarter shift (two hours).

The model was implemented on a personal computer. Computing times are very moderate. Solution techniques include a mixed integer mathematical program and a linear programming formulation with relaxation (that is, rounding of variables to integer values).

This case demonstrates that the concepts discussed in this chapter can be useful in a real production planning environment. Although the system described here is not based on any of the specific models discussed in this chapter, this application shows that aggregation and disaggregation are useful concepts. Hierarchical aggregation for production scheduling is a valuable planning tool.

$(Y_1, Y_2, \ldots, Y_n)$. The question is how to divide up (i.e., disaggregate) X^* among the n items. We know that holding costs are already included in the determination of X^*, so we need not include them again in the disaggregation scheme. Suppose that K_j represents the fixed cost of setting up for production of Y_j, and λ_j is the annual usage rate for item j. A reasonable optimization criterion in this context is to choose $Y_1, Y_2, \ldots, Y_n$ to minimize the average annual cost of setting up for production. As we will see in Chapter 4, the average annual setup cost for item j is $K_j \lambda_j / Y_j$. Hence, disaggregation requires solving the following mathematical programming problem:

$$\text{minimize} \quad \sum_{j=\lambda}^{J} \frac{K_j \lambda_j}{Y_j}$$

subject to

$$\sum_{j=1}^{J} Y_j = X^*$$

and

$$a_j \le Y_j \le b_j \qquad \text{for } 1 \le j \le J.$$

The upper and lower bounds on Y_j account for possible side constraints on the production level for item j.

A number of feasibility issues need to be addressed before the family run sizes Y_j are further disaggregated into lots for individual items. The objective is to schedule the lots for individual items within a family so that they run out at the scheduled setup time for the family. In this way, items within the same family can be produced within the same production setup.

The concept of disaggregating the aggregate plan along organizational lines in a fashion that is consistent with the aggregation scheme is an appealing one. Whether or not the methods discussed in this section provide a workable link between aggregate plans and detailed item schedules remains to be seen.

Another approach to the disaggregation problem was explored by Chen-Hua Chung and Lee Krajewski (1984). They developed a mathematical programming formulation of the problem. Inputs to the program include aggregate plans for each product family. This includes setup time, setup status, total production level for the family, inventory level, workforce level, overtime, and regular time availability. The goal of the analysis is to specify lot sizes and timing of production runs for each individual item, consistent with the aggregate information for the product family. Although such a formulation provides a potential link between the aggregate plan and the master production schedule, the resulting mathematical program requires many inputs and can result in a very large mixed integer problem that could be very time-consuming to solve.

PROBLEMS FOR SECTION 3.6

29. What does "disaggregation of aggregate plans" mean?
30. Discuss the following quotation from a production manager: "Aggregate planning is useless to me because the results have nothing to do with my master production schedule."

3.7 SALES AND OPERATION PLANNING ON A GLOBAL SCALE

Globalization of manufacturing operations is commonplace. Many major corporations are now classified as multinationals; manufacturing and distribution activities routinely cross international borders. With the globalization of both sources of production and markets, firms must rethink production planning strategies. One issue explored in this chapter was smoothing of production plans over time; costs of increasing or decreasing workforce levels (and, hence, production levels) play a major role in the optimization of any aggregate plan. When formulating global production strategies, other smoothing issues arise. Exchange rates, costs of direct labor, and tax structure are just some of the differences among countries that must be factored into a global strategy.

Why the increased interest in global operations? Cost and competitiveness is the short answer to the question. According to Michael McGrath and Roberto Bequillard (1989):

> The benefits of a properly executed international manufacturing strategy can be very substantial. A well-developed strategy can have a direct impact on the financial performance and ultimately be reflected in increased profitability. In the electronics industry, there are examples of companies attributing 5% to 15% reduction in cost of goods sold, 10% to 20% increase in sales, 50% to 150% improvement in asset utilization, and 30% to 100% increase in inventory turnover to their internationalization of manufacturing. (p. 23)

Morris Cohen and Hau Lee (1989) outline some of the issues that a firm must consider when planning production levels on a worldwide basis.

- In order to achieve the kinds of economies of scale required to be competitive today, multinational plants and vendors must be managed as a global system.
- Duties and tariffs are based on material flows. Their impact must be factored into decisions regarding shipments of raw material, intermediate product, and finished product across national boundaries.
- Exchange rates fluctuate randomly and affect production costs and pricing decisions in countries where the product is produced and sold.
- Corporate tax rates vary widely from one country to another.
- Global sourcing must take into account longer lead times, lower unit costs, and access to new technologies.
- Strategies for market penetration, local content rules, and quotas constrain product flow across borders.
- Product designs may vary by national market.
- Centralized control of multinational enterprises creates difficulties for several reasons, and decentralized control requires coordination.
- Cultural, language, and skill differences can be significant.

Determining optimal globalized manufacturing strategies is clearly a daunting problem for any multinational firm. One can formulate and solve mathematical models similar to the linear programming formulations of the aggregate planning models presented in this chapter, but the results of these models must always be weighed against judgment and experience. The Cohen-Lee model assumed multiple products, plants, markets, raw materials, vendors, vendor supply contract alternatives, time periods, and countries. Their formulation was a large-scale mixed integer, nonlinear program.

One issue not treated in their model is that of exchange rate fluctuations and their effect on both pricing and production planning. Pricing, in particular, is traditionally done by adding a markup to unit costs in the home market. This completely ignores the issue of exchange rate fluctuations and can lead to unreasonable prices in some countries. When the dollar is strong relative to other currencies, retail prices charged to overseas customers are not competitive in local markets. Many firms therefore use locally competitive pricing strategies to counteract this problem.

Several researchers have explored the possibility of using manufacturing capacity as a hedge against exchange rate fluctuations. Bruce Kogut and Nalin Kulatilaka (1994), for example, developed a mathematical model for determining when it is optimal to switch production from one location to another. Since the cost of switching is assumed to be positive, there must be a sufficiently large difference in exchange rates before switching is recommended. As an example, they considered a situation where a firm could produce its product in either the United States or Germany. If production is currently being done in one location, the model provides a means of determining if it is economical to switch locations based on the relative strengths of the euro and dollar. Such models provide a means of rationalizing international production planning strategies (for exploration of similar topics, see Huchzermeier and Cohen, 1996; Simchi-Levi, Kaminski, and Simchi-Levi, 2008).

3.8 HISTORICAL NOTES

The aggregate planning problem was conceived in an important series of papers that appeared in the mid-1950s. In 1955, Charles Holt, Franco Modigliani, and Herbert Simon discussed the structure of the problem and introduced the quadratic cost approach. In 1956, Holt, Modigliani, and John Muth concentrated on the computational aspects of the model. In a textbook, the authors presented a complete description of the method and its application to production planning for a paint company (Holt, Modigliani, Muth, and Simon, 1960).

The textbook represented a landmark work in the application of quantitative methods to production planning problems. The authors developed a set of formulas that were easy to implement, and they undertook the implementation of the method. The work (Holt et al, 1960) details the application of the approach to a large manufacturer of household paints in the Pittsburgh area. The analysis was implemented in the company, but a subsequent visit to the firm indicated that serious problems arose when the linear decision rule was followed, primarily because of the firm's policy of not firing workers when the model indicated that they should be fired.

It appears to have been known in the early 1950s that production planning problems could be formulated as linear programs. Edward Bowman (1956) discussed the use of a transportation model for production planning. The particular linear programming formulation of the aggregate planning problem discussed in Section 3.5 is essentially the same as the one developed by Fred Hansmann and Sidney Hess (1960). Other linear programming formulations of the production planning problem generally involved multiple products or more complex cost structures (see, for example, Newson, 1975a and 1975b).

Other work on the aggregate planning problem focused on aggregation and disaggregation issues (Axsater, 1981; Bitran and Hax, 1981; and Zoller, 1971), the incorporation of learning curves into linear decision rules (Ebert, 1976), extensions to allow for multiple products (Bergstrom and Smith, 1970), and inclusion of marketing and/or financial variables (Damon and Schramm, 1972, and Leitch, 1974).

William Taubert (1968) considered a technique he referred to as the search decision rule. The method required developing a computer simulation model of the system and searching the response surface using standard search techniques to obtain a (not necessarily optimal) solution. Applications of Taubert's approach (see Buffa and Taubert,1972) yielded results comparable to those in the paint company case (Holt et al., 1960).

Morton Kamien and Lode Li (1990) developed a mathematical model to examine the effects of subcontracting on aggregate planning decisions. Under certain circumstances subcontracting was preferred to producing in-house and provided an additional means of smoothing production and workforce levels.

3.9 SUMMARY

Modern firms take a much less siloed approach to planning than firms of the past, realizing that large benefits can accrue from a collaborative approach among the different divisions including operations, sales and marketing, and finance. A well-designed S&OP process can manage the inherent tensions between these divisions and make trade-offs from a strategic perspective. The key output of this process from an operational perspective is a fixed sales plan to guide the operations division.

A common component of the S&OP process will be reviewing divisional KPIs. Because KPIs have such a large impact on employee incentives, they need to be chosen carefully. The primary challenge in KPI selection is to ensure that the KPI is aligned with the strategic imperatives of the firm while still remaining actionable as far as the person being measured by it is concerned. It is also important to try to mitigate opportunities for gaming of KPIs by employees.

The key output from the S&OP process is a fixed forecast; therefore, planning for routine uncertainty in these numbers is customarily left to the operations division. While risk pooling (see Chapter 6) can be used to mitigate some of the uncertainty, there will always be natural variation that must be buffered. Such buffering can take the form of inventory, spare capacity, or time in the form of customer lead times. This sort of buffering typically takes place outside the S&OP process and will be discussed in later chapters in this text.

At a higher level than routine variation are more major risks that are discussed within the S&OP process. Such known unknowns should have contingency plans and/or mitigation strategies associated with them. They may be demand-side risks, such as new product introductions by competitors or demand shocks caused by extreme weather, or they may be supply-side risks, such as supplier failure or major quality issues. An effective S&OP process can anticipate such risks and set strategies for dealing with them that are in line with the company's goals. This may include the explicit relaxing of certain KPIs that become less relevant when working under exceptional circumstances.

Once the strategies for dealing with uncertainty have been determined, most firms assume determinist demand in designing an aggregate production plan. Indeed, fixed demand forecasts over a specified planning horizon are a required input. This assumption is not made for the sake of realism but to allow the analysis to focus on the changes in the demand that are systematic rather than random. The goal of the analysis is to determine for each period the number of workers who should be employed, the production that should occur, and the inventory that should be carried over for each period.

The objective of an aggregate production plan is to minimize costs of production, payroll, holding, and changing the size of the workforce or capacity. The costs of making changes are generally referred to as smoothing costs. The aggregate planning models discussed in this chapter assume that all the costs are linear functions. This assumption is probably a reasonable approximation for most real systems within a given reasonable range of values. It is unlikely that the primary problem with applying a linear programming formulation to a real situation will be that the shape of the cost function is incorrect; it is more likely that the primary difficulty will be in estimating the costs correctly and other required input information.

Aggregate production plans will be of little use to the firm if they cannot be coordinated with detailed item schedules (i.e., the master production schedule). The problem of disaggregating aggregate plans is a difficult one, but one that must be addressed if the aggregate plans are to have value to the firm. There have been some mathematical programming formulations of the disaggregating problem suggested in the literature.

ADDITIONAL PROBLEMS ON AGGREGATE PLANNING

31. An aggregate planning model is being considered for the following applications. Suggest an aggregate measure of production and discuss the difficulties of applying aggregate planning in each case.
    a. Planning for the size of the faculty in a university.
    b. Determining workforce requirements for a travel agency.
    c. Planning workforce and production levels in a fish-processing plant that produces canned sardines, anchovies, kippers, and smoked oysters.
32. A local firm manufactures children's toys. The projected demand over the next four months for one particular model of toy robot is

Month	Workdays	Forecasted Demand (in aggregate units)
July	23	3,825
August	16	7,245
September	20	2,770
October	22	4,440

Assume that a normal workday is eight hours. Hiring costs are $350 per worker and firing costs (including severance pay) are $850 per worker. Holding costs are $4.00 per aggregate unit held per month. Assume that it requires an average of 1 hour and 40 minutes for one worker to assemble one toy. Shortages are not permitted. Assume that the ending inventory for June was 600 toys and the manager wishes to have at least 800 units on hand at the end of October. Assume that the current workforce level is 35 workers.

a. Ignoring all production costs and assuming that all employed workers are producing 8 hours per day (i.e., there are no variables in the model to account for worker idling), find the optimal plan by formulating as a linear program.
b. Now assume that each worker is paid a salary of $14.40 per hour regardless of whether they are producing toys and that how many of the 8 hours in the normal workday workers will produce toys is a decision. Modify your model (it will be slightly different than the example model of this chapter). Does this change the answer from (a)? Why or why not?
c. Suppose we didn't include worker salaries in (b) but still allowed workers to produce toys for less than 8 hours per day (i.e., suppose idling workers is costless in the model in part (a)). What would the solution be? Compare this solution to parts (a) and (b).

33. The Paris Paint Company is in the process of planning labor force requirements and production levels for the next four quarters. The marketing department has provided production with the following forecasts of demand for Paris Paint over the next year.

Quarter	Demand Forecast (in thousands of gallons)
1	380
2	630
3	220
4	160

Assume that there are currently 280 employees with the company. Employees are hired for at least one full quarter. Hiring costs amount to $1,200 per employee, and firing costs are $2,500 per employee. Inventory costs are $1 per gallon per quarter. It is estimated that one worker produces 1,000 gallons of paint each quarter.

Assume that Paris currently has 80,000 gallons of paint in inventory and would like to end the year with an inventory of at least 20,000 gallons.

a. Determine the minimum constant workforce plan for Paris Paint and the cost of the plan. Assume that stock-outs are not allowed.
b. If Paris were able to backorder excess demand at a cost of $2 per gallon per quarter, determine a minimum constant workforce plan that holds less inventory than the plan you found in part (a) but incurs stock-outs in quarter 2. Determine the cost of the new plan.
c. Formulate this as a linear program. Assume that stock-outs are not allowed.
d. Solve the linear program. Round the variables and determine the cost of the resulting plan.

34. Suppose that Paris Paint has the capacity to employ a maximum of 370 workers. Suppose that regular-time employee costs are $12.50 per hour. Assume seven-hour days, five-day weeks, and four-week months. Overtime is paid on a time-and-a-half basis. Subcontracting is available at a cost of $7 per gallon of paint produced. Overtime is limited to three hours per employee per day, and no more than 100,000 gallons can be subcontracted in any quarter.
    a. Formulate as a linear program.
    b. Solve the linear program. Round the variables and determine the cost of the resulting plan.
35. The Mr. Meadows Cookie Company can obtain accurate forecasts for 12 months based on firm orders. These forecasts and the number of workdays per month are as follows.

Month	Demand Forecast (in thousands of cookies)	Workdays
1	850	26
2	1,260	24
3	510	20
4	980	18
5	770	22
6	850	23
7	1,050	14
8	1,550	21
9	1,350	23
10	1,000	24
11	970	21
12	680	13

During a 46-day period when there were 120 workers, the firm produced 1,700,000 cookies. Assume that there are 100 workers employed at the beginning of month 1 and zero starting inventory.

a. Find the minimum constant workforce needed to meet monthly demand.
b. Assume c_I = \$0.10 per cookie per month, c_H = \$100, and c_F = \$200. Add columns that give the cumulative on-hand inventory and inventory cost. What is the total cost of the constant workforce plan?
c. Solve for the optimal plan using linear programming. Compare your solution to *b*.

36. The Yeasty Brewing Company produces a popular local beer known as Iron Stomach. Beer sales are somewhat seasonal, and Yeasty is planning its production and workforce levels on March 31 for the next six months. The demand forecasts are as follows.

Month	Production Days	Forecasted Demand (in hundreds of cases)
April	11	85
May	22	93
June	20	122
July	23	176
August	16	140
September	20	63

As of March 31, Yeasty had 86 workers on the payroll. Over a period of 26 working days when there were 100 workers on the payroll, Yeasty produced 12,000 cases of beer. The cost to hire each worker is \$125, and the cost of laying off each worker is \$300. Holding costs amount to 75 cents per case per month.

As of March 31, Yeasty expects to have 4,500 cases of beer in stock, and it wants to maintain a minimum buffer inventory of 1,000 cases each month. It plans to start October with 3,000 cases on hand.

a. Based on this information, find the minimum constant workforce plan for Yeasty over the six months, and determine hiring, firing, and holding costs associated with that plan.
b. Suppose that it takes one month to train a new worker. How will that affect your solution?
c. Suppose that the maximum number of workers that the company can expect to be able to hire in one month is 10. How will that affect your solution to part (a)?
d. Formulate the problem levels as a linear program. [You may ignore the conditions in parts (b) and (c).]
e. Solve the resulting linear program. Round the appropriate variables and determine the cost of your solution.
f. Suppose Yeasty does not wish to fire any workers. What is the optimal plan subject to this constraint?

37. A local canning company sells canned vegetables to a supermarket chain in the Minneapolis area. A typical case of canned vegetables requires an average of 0.2 day of labor to produce. The aggregate inventory on hand at the end of June is 800 cases. The demand for the vegetables can be accurately predicted for about 18 months based on orders received by the firm. The predicted demands for the next 18 months are as follows.

Month	Forecasted Demand (hundreds of cases)	Workdays	Month	Forecasted Demand (hundreds of cases)	Workdays
July	23	21	April	29	20
August	28	14	May	33	22
September	42	20	June	31	21
October	26	23	July	20	18
November	29	18	August	16	14
December	58	10	September	33	20
January	19	20	October	35	23
February	17	14	November	28	18
March	25	20	December	50	10

The firm currently has 25 workers. The cost of hiring and training a new worker is \$1,000, and the cost to lay off one worker is \$1,500. The firm estimates a cost of \$2.80 to store a case of vegetables for a month. They would like to have 1,500 cases in inventory at the end of the 18-month planning horizon.

a. Develop a spreadsheet to find the minimum constant workforce aggregate plan and determine the total cost of that plan.
b. Develop a spreadsheet to find a plan that hires and fires workers monthly in order to minimize inventory costs. Determine the total cost of that plan as well.

APPENDIX 3–A

AGGREGATE PLANNING HEURISTICS

Section 3.4 introduced two extremes in capacity planning, namely the **zero inventory plan**, also known as a **chase strategy**, and the **constant workforce plan**, also known as a **level strategy**. This appendix shows how to compute these plans using Example 3.2 (Densepack).

Evaluation of a Chase Strategy (Zero Inventory Plan)

Here, we will develop a production plan for Densepack that minimizes the levels of inventory the firm must hold during the six-month planning horizon. Table 3–2 summarizes the input information for the calculations and shows the minimum number of workers required in each month.

One obtains the entries in the final column of Table 3–2, the minimum number of workers required each month, by dividing the forecasted net demand by the number of units produced per worker. The value of this ratio is then rounded *upward* to the next higher integer. We must round upward to guarantee that shortages do not occur.

Recall that the number of workers employed at the end of December is 300. Hiring and firing workers each month to match forecast demand as closely as possible results in the aggregate plan given in Table 3–3.

Table 3–2 Initial Calculations for Zero Inventory Plan for Densepack

A	B	C	D	E
Month	Number of Working Days	Number of Units Produced per Worker (B × 0.14653)	Forecast Net Demand	Minimum Number of Workers Required (D/C rounded up)
January	20	2.931	780	267
February	24	3.517	640	182
March	18	2.638	900	342
April	26	3.810	1,200	315
May	22	3.224	2,000	621
June	15	2.198	2,000	910

Table 3–3 Zero Inventory Aggregate Plan for Densepack

A	B	C	D	E	F	G	H	I
Month	Number of Workers	Number Hired	Number Fired	Number of Units per Worker	Number of Units Produced (B × E)	Cumulative Production	Cumulative Net Demand	Ending Inventory (G – H)
January	267		33	2.931	783	783	780	3
February	182		85	3.517	640	1,423	1,420	3
March	342	160		2.638	902	2,325	2,320	5
April	315		27	3.810	1,200	3,525	3,520	5
May	621	306		3.224	2,002	5,527	5,520	7
June	910	289		2.198	2,000	7.527	7,520	7
Totals		755	145					30

The number of units produced each month (column F in Table 3–3) is obtained by the following formula:

Number of units produced	=	Number of workers	×	Average number of aggregate units produced in a month by a single worker

rounded to the nearest integer.

The total cost of this production plan is obtained by multiplying the totals at the bottom of Table 3–3 by the appropriate costs. For this example, the total cost of hiring, firing, and holding is (755)(500) + (145)(1,000) + (30)(80) = \$524,900. This cost must now be adjusted to include the cost of holding for the ending inventory of 600 units, which was netted out of the demand for June. Hence, the total cost of this plan is 524,900 + (600)(80) = \$572,900. Note that the initial inventory of 500 units does not enter into the calculation because its cost will have been accounted for in the previous year's aggregate plan and then it is used in lieu of production in January.

It is usually impossible to achieve zero inventory at the end of each planning period because it is not possible to employ a fractional number of workers. For this reason, there will almost always be some inventory remaining at the end of each period in addition to the inventory required to be on hand at the end of the planning horizon.

It is possible that the ending inventory in one or more periods could build up to a point where the size of the workforce could be reduced by one or more workers. In this example, there is sufficient inventory on hand to reduce the workforce by one worker in the months of both March and May. The reader should check that the resulting plan hires a total of 753 workers and fires a total of 144 workers and has a total of only 13 units of inventory. The cost of this modified plan comes to $569,540.

Evaluation of a Constant Workforce Plan

Now assume that the goal is to eliminate completely the need for hiring and firing during the planning horizon. In order to guarantee that shortages do not occur in any period, it is necessary to compute the minimum workforce required for *every* month in the planning horizon. For January, the net cumulative demand is 780 and there are 2,931 units produced per worker, resulting in a minimum workforce of 267 in January. There are exactly $2.931 + 3.517 = 6.448$ units produced per worker in January and February combined, which have a cumulative demand of 1,420. Hence $1{,}420/6.448 = 220.22 \approx 221$ workers are required to cover both January and February. Continuing to form the ratios of the cumulative net demand and the cumulative number of units produced per worker for each month in the horizon results in Table 3–4.

The maximum number of workers required for the entire six-month planning period is the maximum entry in column D in Table 3–4, which is 411 workers. It is only a coincidence that the maximum ratio occurred in the final period.

Because there are 300 workers employed at the end of December, the constant workforce plan requires hiring 111 workers at the beginning of January. No further hiring and firing of workers are required. The inventory levels that result from a constant workforce of 411 workers appear in Table 3–5. The monthly production levels in column C of the table are obtained by multiplying the number of units produced per worker each month by the fixed workforce size of 411 workers. The total of the ending inventory level is $5{,}962 + 600 = 6{,}562$. (Recall the 600 units that were netted out of the demand for June.) Hence, the total inventory cost of this plan is $(6{,}562)(80) = \$524{,}960$. To this we add the cost of increasing the workforce from 300 to 411 in January, which is $(111)(500) = \$55{,}500$, giving a total cost of this plan of $580,460.

Table 3–4 Initial Calculations for Zero Inventory Plan for Densepack

A	B	C	D
Month	Cumulative Net Demand	Cumulative Number of Units Produced per Worker	Ratio B/C (rounded up)
January	780	2.931	267
February	1,420	6.448	221
March	2,320	9.086	256
April	3,520	12.896	273
May	5,520	16.120	343
June	7,520	18.318	411

Table 3–5 Inventory Levels for Constant Workforce Schedule

A	B	C	D	E	F
Month	Number of Units per Worker	Number of Units Produced (B × 411)	Cumulative Production	Cumulative Net Demand	Ending Inventory (D – E)
January	2.931	1,205	1,205	780	425
February	3.517	1,445	2,650	1,420	1,230
March	2.638	1,084	3,734	2,320	1,414
April	3.810	1,566	5,300	3,520	1,780
May	3.224	1,325	6,625	5,520	1,105
June	2.198	903	7,528	7,520	8
Total					5,962

This is somewhat higher than the cost of the zero inventory plan, which was \$569,540 but not by a lot.

Note that neither plan considered the cost of producing one unit on regular time, c_R. This is because we are always producing almost the same number of units under any plan (cumulative demand), so they are not important for planning the workforce. Given only these costs, it is likely that the company would prefer the constant workforce plan in order to avoid any unaccounted for costs of making frequent changes in the workforce. However, if there is a cost to idling workers, then they may prefer the zero inventory plan.

Appendix 3–B

Glossary of Notation for Chapter 3

α = Smoothing constant for production and demand used in Bowman's model.
β = Smoothing constant for inventory used in Bowman's model.
c_F = Cost of firing one worker.
c_H = Cost of hiring one worker.
c_I = Cost of holding one unit of stock for one period.
c_O = Cost of producing one unit on overtime.
c_P = Penalty cost for back orders.
c_R = Cost of producing one unit on regular time.
c_S = Cost to subcontract one unit of production.
c_U = Idle cost per unit of production.
D_t = Forecast of demand in period t.
F_t = Number of workers fired in period t.
H_t = Number of workers hired in period t.
I_t = Inventory level in period t.
K = Number of aggregate units produced by one worker in one day.
λ_j = Annual demand for family j (refer to Section 3.6).
n_t = Number of production days in period t.
O_t = Overtime production in units.
P_t = Production level in period t.

S_t = Number of units subcontracted from outside.
T = Number of periods in the planning horizon.
U_t = Worker idle time in units ("undertime").
W_t = Workforce level in period t.

BIBLIOGRAPHY

ABB Group. "Key Performance Indicators: Identifying and Using Key Metrics for Performance." Retrieved from https://slideus.org/philosophy-money.html?utm_source=kpi-G7d39Wu

Allen, S. J., and E. W. Schuster. "Practical Production Scheduling with Capacity Constraints and Dynamic Demand: Family Planning and Disaggregation." *Production and Inventory Management Journal* 35 (1994), pp. 15–20.

Axsater, S. "Aggregation of Product Data for Hierarchical Production Planning." *Operations Research* 29 (1981), pp. 744–56.

Bergstron, G. L., and B. E. Smith. "Multi-Item Production Planning–An Extension of the HMMS Rules." *Management Science* 16 (1970), pp. 100–103.

Bitran, G. R., and A. Hax. "Disaggregation and Resource Allocation Using Convex Knapsack Problems with Bounded Variables." *Management Science* 27 (1981), pp. 431–41.

Bowman, E. "Production Scheduling by the Transportation Method of Linear Programming." *Operations Research* 4 (1956) pp. 100–103.

Buffa, E. S., and W. H. Taubert. *Production-Inventory Systems: Planning and Control* (Rev. ed.). New York: McGraw-Hill/Irwin, 1972.

Chung, C., and L. J. Krajewski. "Planning Horizons for Master Production Scheduling." *Journal of Operations Management* (1984), pp. 389–406.

Cohen, M., and H. L. Lee. "Resource Deployment Analysis of Global Manufacturing and Distribution Networks." *Journal of Manufacturing and Operations Management* 2 (1989), pp. 81–104.

De Meyer, A., C. H. Loch, and M. T. Pich. "Managing Project Uncertainty: From Variation to Chaos." *MIT Sloan Management Review* (2002).

Damon, W. W., and R. Schramm. "A Simultaneous Decision Model for Production, Marketing, and Finance." *Management Science* 19 (1972), pp. 16 –72.

Ebert, R. J. "Aggregate Planning with Learning Curve Productivity." *Management Science* 23 (1976), pp. 171–82.

Every Angle. "Customer Case Heineken Netherlands." 2010. Retrieved from https://www.slideshare.net/Every_Angle/heineken-every-angle-customer-case-2010-english

Goldratt, E. M. and J. Cox. *The Goal: A Process of Ongoing Improvement.* 4th Revised Edition. Great Barrington, MA: North River Press, 2014.

Grimson, J. A., and D. F. Pyke. "Sales and Operations Planning: An Exploratory Study and Framework." *International Journal of Logistics Management* 18 (2007), pp.322–346.

Hansmann, F., and S. W. Hess. "A Linear Programming Approach to Production and Employment Scheduling." *Management Technology* 1 (1960), pp. 46–51.

Hax, A. C., and D. Candea. *Production and Inventory Management.* Englewood Cliffs, NJ: Prentice Hall, 1984.

Hillier, F. S., and G. J. Lieberman. *Introduction to Operations Research* (10th ed.). New York: McGraw Hill, 2015

Holt, C. C., F. Modigliani, and J. F. Muth. "Derivation of a Linear Decision Rule for Production and Employment." *Management Science* 2 (1956), pp. 159–77.

Holt, C. C., F. Modigliani, J. F. Muth, and H. A. Simon. *Planning Production, Inventories, and Workforce.* Englewood Cliffs, NJ: Prentice Hall, 1960.

Holt, C. C., F. Modigliani; and H. A. Simon. "A Linear Decision Rule for Employment and Production Scheduling." *Management Science* 2 (1955), pp. 1–30.

Huchzermeier, A., and M. Cohen. "Valuing Operational Flexibility under Exchange Rate Risk." *Operations Research* 44 (1996), pp. 100-113.

IBM. "The Smarter Supply Chain of the Future: Insights from the Global Chief Supply Chain Officer Study." 2010. Retrieved from https://www.ibm.com/downloads/cas/AN4AE4QB

Kaplan, R. S., and D. P. Norton. "The Balanced Scorecard: Measures that Drive Performance." *Harvard Business Review* (1992), pp. 71–80.

Kamien, M. I., and L. Li. "Subcontracting, Coordination, Flexibility, and Production Smoothing in Aggregate Planning." *Management Science* 36 (1990), pp. 1352–1363.

Kogut, B., and N. Kulatilaka. "Operating Flexibility, Global Manufacturing, and the Option Value of a Multinational Network." *Management Science* 40 (1994), pp. 123–139.

Leitch, R. A. "Marketing Strategy and Optimal Production Schedule." *Management Science* 20 (1974), pp. 903–11.

McGrath, M.E., and R. B. Bequillard. "International Manufacturing Strategies." In K. Ferdows (Ed.), *Managing International Manufacturing* (pp. 23–40). New York: North-Holland, 1989.

Newson, E. F. P. "Multi-Item Lost Size Scheduling by Heuristic, Part 1: With Fixed Resources." *Management Science* 21 (1975a), pp. 1186–93.

Newson, E. F. P. "Multi-Item Lost Size Scheduling by Heuristic, Part 2: With Fixed Resources." *Management Science* 21 (1975b), pp. 1194–1205.

Santos, C., et al. "HP Enterprise Services Uses Optimization for Resource Planning." *Interfaces* 43 (2013), pp. 152–169.

Sheldon, D. H. *World Class Sales & Operations Planning.* Ft. Lauderdale, FL: J Ross Publishing and APICS, 2006.

Simchi-Levi, D., P. Kaminski, and E. Simchi-Levi. *Designing and Managing the Supply Chain: Concepts, Strategies, and Case Studies* (3rd ed.). New York: McGraw-Hill/Irwin, 2008.

Smits, J. and M. English. "The Journey to Worldclass S&OP at Heineken." 2010. Retrieved from https://www.yumpu.com/en/document/read/29387059/the-journey-to-world-class-sop-at-heineken-supply-chain-

Taleb, N. N. *The Black Swan: The Impact of the Highly Improbable* (2nd ed.). New York: Random House, 2010.

Taubert, W. H. "A Search Decision Rule for the Aggregate Scheduling Problem." *Management Science* 14 (1968), pp. B343–53.

Zoller, K. "Optimal Disaggregation of Aggregate Production Plans." *Management Science* 17 (1971), pp. B53–49.

SUPPLEMENT 1

Linear Programming

S1.1 INTRODUCTION

Linear programming is a mathematical technique for solving a broad class of optimization problems. These problems require maximizing or minimizing a linear function of n real variables subject to m linear constraints. One can formulate and solve a large number of real problems with linear programming, such as those shown in the partial list below.

1. Scheduling of personnel.
2. Several varieties of blending problems including cattle feed, sausages, ice cream, and steel making.
3. Inventory control and production planning.
4. Distribution and logistics problems.
5. Assignment problems.

Problems with thousands of decision variables and thousands of constraints are easily solvable on computers today. Linear programming was developed to solve logistics problems during World War II. George Dantzig, a mathematician employed by the RAND Corporation at the time, developed a solution procedure he labeled the simplex method. At the time, the term "program" referred to logistics schedules and similar plans (Dantzig, 1963); thus, the field was known as linear programming. Today, it is often just called linear optimization.

That the simplex method turned out to be so efficient for solving large problems quickly was a surprise even to Dantzig. That fact, coupled with the simultaneous development of the electronic computer, established linear programming as an important tool in logistics management. The success of linear programming in industry spawned the development of the disciplines of operations research and management science. The simplex method has withstood the test of time. There are some more efficient methods for solving specially structured linear programs (e.g., Karmarkar's algorithm), but the simplex method remains the standard for general problems.

In Section S1.2 we consider a typical manufacturing problem that we formulate and solve using linear programming. Later we explore how one solves small problems (that is, having exactly two decision variables) graphically and how one solves large problems using a computer.

Linear programming (or, more generally, mathematical programming or mathematical optimization) has been an important tool for logistics planning for a wide variety of operations management problems. Today, Microsoft publishes Solver, an Excel add-in that makes optimization accessible to a much wider audience. The Solver tool uses the simplex algorithm for linear programming problems. It also has algorithms for problems with integer variables and non-linear objectives and constraints. Such problems are briefly discussed in the final two sections of this supplement.

S1.2 A PROTOTYPE LINEAR PROGRAMMING PROBLEM

EXAMPLE S1.1

Sidneyville manufactures household and commercial furnishings. The Office Division produces two desks, rolltop and regular. Sidneyville constructs the desks in its plant outside Medford, Oregon, from a selection of woods. The woods are cut to a uniform thickness of 1 inch. For this reason, one measures the wood in units of square feet. One rolltop desk requires 10 square feet of pine, 4 square feet of cedar, and 15 square feet of maple. One regular desk requires 20 square feet of pine, 16 square feet of cedar, and 10 square feet of maple. The desks yield respectively \$115 and \$90 profit per sale. At the current time the firm has available 200 square feet of pine, 128 square feet of cedar, and 220 square feet of maple. The firm has backlogged orders for both desks and would like to produce the number of rolltop and regular desks that would maximize profit. How many of each should it produce?

Solution

The first step in formulating a problem as a linear program is to identify the decision variables. In this case there are two decisions required: the number of rolltop desks to produce and the number of regular desks to produce. We must assign symbol names to each of these decision variables. Let

$$x_1 = \text{Number of rolltop desks to be produced; and}$$

$$x_2 = \text{Number of regular desks to be produced.}$$

Now that we have identified the decision variables, the next step is to identify the **objective** function and the constraints. The objective function is the quantity we wish to maximize or minimize. The objective is to maximize profits, so the objective function equals the total profit when producing x_1 rolltop desks and x_2 regular desks. Each rolltop desk contributes \$115 to profit, so the total contribution to profit from all rolltop desks is $115x_1$. Similarly, the contribution to profit from all regular desks is $90x_2$. Hence, the total profit is $115x_1 + 90x_2$. This is known as the objective function.

The next step is to identify the constraints. The number of desks Sidneyville can produce is limited by the amount of wood available. The three types of wood constitute the critical resources. To obtain the constraints, we need to find expressions for the amount of each type of wood consumed by construction of x_1 rolltop desks and x_2 regular desks. Those expressions are then bounded by the amount of each type of wood available.

The number of square feet of pine used to make x_1 rolltop desks is $10x_1$. The number of square feet of pine used to make x_2 regular desks is $20x_2$. It follows that the total amount of pine consumed in square feet is $10x_1 + 20x_2$. This quantity cannot exceed the number of square feet of pine available, which is 200. Hence we obtain the first constraint:

$$10x_1 + 20x_2 \leq 200.$$

The other two constraints are similar. The second constraint is to ensure that the firm does not exceed the available supply of cedar. Each rolltop desk requires 4 square feet of cedar, so x_1 rolltop desks require $4x_1$ square feet of cedar. Each regular desk requires 16 square feet of cedar, so x_2 regular desks require $16x_2$ square feet of cedar. It follows that the constraint on the supply of cedar is

$$4x_1 + 16x_2 \leq 128.$$

In the same way, the final constraint ensuring that we do not exceed the supply of maple is

$$15x_1 + 10x_2 = 220.$$

Because we cannot produce a negative number of desks, we also include nonnegativity constraints:

$$x_1 \geq 0,$$
$$x_2 \geq 0.$$

We have now constructed the complete linear programming formulation of the Sidneyville problem. The goal is to find values of x_1 and x_2 to maximize $115x_1 + 90x_2$, subject to the constraints

$$\begin{aligned} 10x_1 + 20x_2 &\leq 200, \\ 4x_1 + 16x_2 &\leq 128, \\ 15x_1 + 10x_2 &\leq 220, \\ x_1, x_2 &\geq 0. \end{aligned}$$

We next consider how such a problem is solved. We will briefly outline the theory behind the simplex method. However, as linear programming problems are almost never solved by hand any longer, you will not need to understand the mechanics of the simplex method in order to use linear programming. You will need to know how to formulate problems as linear programs, enter the formulations into the computer, recognize special problems, and analyze the computer's output.

At the optimal solution to the Sidneyville problem, the value of x_1 is 12 and the value of x_2 is 4. That is, Sidneyville should produce exactly 12 rolltop desks and 4 regular desks. The value of the objective function at the optimal solution is \$1,740.00. That this is the (unique) optimal solution means that every other pair of values of x_1 and x_2 will result in either a lower profit or not meeting the constraints, or both.

Sidneyville's manager of production planning is very skeptical about mathematics. When presented with this solution, his response was, "There's only one problem with this production plan. We have a specialist make the rolltop portion of the rolltop desk. She can only do four desks a day, and we want to be ready to ship out in two days. There is no way we can produce twelve of those desks in two days. I knew this math stuff was a lot of hooey!"

The manager was wrong. The trouble is not that the formulation is incorrect but that it does not include all relevant constraints, as labor hours turned out to be a critical resource. The lesson here is that for the final solution to be meaningful, the model must include *all* relevant constraints.

S1.3 STATEMENT OF THE GENERAL PROBLEM

The Sidneyville problem is an example of a **linear program** in which there are two **decision variables** and three **constraints**. Linear programming problems may have any number of decision variables and any number of constraints. Suppose that there are n decision variables, labeled $x_1, x_2, \ldots, x_n$, subject to m resource constraints. Then we may write the problem of maximizing the **objective** subject to the constraints as

$$\begin{aligned} \text{maximize} \quad & c_1x_1 + c_2x_2 + \ldots + c_nx_n, \\ \text{subject to} \quad & a_{11}x_1 + a_{12}x_2 + \ldots + a_{1n}x_n \leq b_1, \\ & a_{21}x_1 + a_{22}x_2 + \ldots + a_{2n}x_n \leq b_2, \\ & a_{m1}x_1 + a_{m2}x_2 + \ldots + a_{mn}x_n \leq b_m, \\ & x_1, x_2, \ldots, x_n \geq 0. \end{aligned}$$

Interpret $c_1, c_2, \ldots, c_n$ as the profit coefficients per unit of output of the activities $x_1, x_2, \ldots, x_n$; a_{ij} as the amount of resource i consumed by one unit of activity j; and b_i as the amount of resource i available, for $i = 1, \ldots, m$ and $j = 1, \ldots, n$. We usually require that the constants $b_1, \ldots, b_m$ be nonnegative. This particular formulation includes problems in which we want to maximize profit subject to constraints on the available resources. However, linear programming can be used to solve a much larger variety of problems. Other possible formulations will be discussed later.

Definitions of Commonly Used Terms

1. *Objective function.* This is the quantity we wish to maximize or minimize. In the given formulation, the objective function is the term $c_1x_1 + c_2x_2 + \ldots + c_nx_n$. In business applications, one typically minimizes cost or maximizes profit. We use the abbreviations "min" for a minimization problem and "max" for a maximization problem.
2. *Constraints.* Each constraint is a linear inequality or equation, that is, a linear combination of the problem variables followed by a relational operator ($\leq$ or $=$ or $\geq$) followed by a nonnegative constant. Although the given formulation shows all $\leq$ constraints, $\geq$ and $=$ type constraints are also common. For example, suppose there is a contractual agreement that requires a minimum number of labor hours daily. This would result in a $\geq$ constraint.
3. *Right-hand side (RHS).* The right-hand side is the constant following the relational operator in a constraint. In the given constraint, the constants $b_1, b_2, \ldots, b_m$ are the right-hand sides. These constants are sometimes required to be nonnegative numbers. However, we do *not* require that the constants a_{ij} be nonnegative. This means that any constraint can be written with a nonnegative right-hand side by multiplying through by the constant -1 whenever the right-hand side is negative. Consider the following simple example. Suppose that when formulating a problem as a linear program we obtain the constraint

$$4x_1 - 2x_2 \leq -5.$$

Because the right-hand side is negative, this is not a constraint with a nonnegative right-hand side. However, if we multiply through by —1, this constraint becomes

$$-4x_1 + 2x_2 \geq 5,$$

which is then more standard.
4. *Feasible region.* The feasible region is the set of values of the decision variables, $x_1, x_2, \ldots, x_n$, that satisfy the constraints. Because each of the constraints is generated by a linear equation or linear inequality, the feasible region has a particular structure. The technical term for this structure is **convex polytope**. In two dimensions, a convex polytope is a convex set with boundaries that are straight lines. In three dimensions, the boundaries are formed by planes. A convex set is characterized as follows: pick any two points in the set and connect them with a straight line; the line lies entirely within the set.
5. *Extreme points.* Because of the structure of the feasible region, there will be a finite number of feasible points, with the property that they cannot be expressed as a linear combination of any other set of feasible points. These points are known as extreme points or corner points, and they play an important role in linear programming. The concept of extreme points will become clearer when we discuss graphical solutions.

6. *Feasible solution.* A feasible solution is one particular set of values of the decision variables that satisfies the constraints. A feasible solution is also one point in the feasible region. It may be an extreme point or an interior point.
7. *Optimal solution.* The optimal solution is the feasible solution that maximizes or minimizes the objective function. In some cases, the optimal solution may not be unique. When this is the case, there will be an infinite number of optimal solutions.

Features of Linear Programs

Linear programming is a very powerful tool. Many real problems have been successfully formulated and solved using this technique. However, to use the method correctly, one must be aware of its limitations. Two important features of linear programs are linearity and continuity. Many problems that may appear to be solvable by linear programming fail one or both of these two crucial tests.

Linearity

Optimization problems can be formulated as linear programs only when (a) the objective can be expressed as a linear function of the decision variables and (b) all constraints can be expressed as linear functions of the decision variables.

Linearity implies that quantities change in fixed proportions. For example, if it costs \$10 to produce one unit, then it costs \$20 to produce two units, and \$100 to produce ten units. If one ounce of orange juice supplies 30 mg of vitamin C, then three ounces must supply 90 mg. Linearity must hold in the objective function and the constraints. In the objective function, this means that the profit or cost per unit must be the same independent of the number of units. In the constraints, linearity means that the amount of each resource consumed is the same per unit whether one produces a single unit or many units.

However, one often observes nonlinear relationships in the real world. Economies of scale in production mean that the marginal cost of producing a unit decreases as the number of units produced increases. When this occurs, the cost of production is a nonlinear function of the number of units produced. An example of scale economies is a fixed setup cost for production. The formula for the EOQ discussed in Chapter 4 says that the lot size increases as the square root of the demand rate. Hence, EOQ is a nonlinear function of the demand. When either the objective function or a constraint is a nonlinear function of the decision variables, the problem is a nonlinear programming problem and cannot be solved by linear programming. (In some circumstances, convex programming problems can be solved by linear programming by approximating the objective function with a piecewise-linear function. An example of this approach appears at the end of Section 3.5.)

Continuity

Decision variables should be continuous (that is, able to assume any nonnegative value) as opposed to discrete or integer valued. This can be a serious restriction. The solution to many problems makes sense only if the decision variables are integer valued. In particular, Example S1.1 is, strictly speaking, not a linear programming problem because the number of desks produced must be integer valued. (We were lucky that the optimal solution turned out to be integer valued in the example.) One might think that the optimal integer solution is equal to the continuous solution rounded off to the nearest integer. Unfortunately, this is not always the case. First, rounding may lead to infeasibility; that is, the rounded-off solution may lie outside the feasible region. Second, even if the rounded solution is feasible, it may not be optimal. It can happen that the optimal integer solution is in an entirely different portion of the feasible region than the rounded-off linear programming solution!

To give the reader some idea of the difficulties that can arise when the solution must be integer valued, consider Example S1.1. Suppose that the profit from selling rolltop desks was \$150 rather than \$115. Then the objective would be to maximize $150x_1 + 90x_2$ subject to the same set of constraints. The optimal linear programming solution is

$$x_1 = 14.666666\ldots$$
$$x_2 = 0.$$

Rounding the solution to the nearest integer gives $x_1 = 15$ and $x_2 = 0$, which is infeasible. Substituting these values into the final constraint results in a requirement of 225 square feet of maple. However, only 220 square feet are available. Rounding x_1 down to 14 results in a feasible but suboptimal solution. At $x_1 = 14$ and $x_2 = 0$, there are 10 feet of maple still available, which is enough to make one regular desk. The optimal integer solution in this case is $x_1 = 14$ and $x_2 = 1$.

When the decision variables must be integer valued, we say that the problem is an integer linear programming problem. Finding optimal integer solutions to linear programs can be time-consuming, even for modest-sized problems. However, Excel does offer an option for defining some or all of the problem variables as integer valued. Excel does a fine job solving small integer linear programming problems. For larger problems, one would use a computer program designed to solve integer programming problems. In some cases, and especially when the values of the variables are relatively large, careful rounding of the linear programming solution should give acceptable results.

S1.4 SOLVING LINEAR PROGRAMMING PROBLEMS GRAPHICALLY

Graphing Linear Inequalities

In this section we will solve two-variable linear programming problems graphically. Although most real problems have more than two variables, understanding the procedure for solving two-variable linear programs will improve your grasp of the concepts underlying the simplex method.

The first step is to graph the linear inequalities represented by the constraints. A linear inequality corresponds to all points in the plane on one side of a straight line. There are thus two steps to graphing linear inequalities.

1. Draw the straight line representing the boundary of the region corresponding to the constraint expressed as an equation.
2. Determine which side of the line corresponds to the inequality.

To illustrate the method, consider the first constraint in Example S1.1:

$$10x_1 + 20x_2 \leq 200.$$

The boundary of the region represented by this inequality is the straight line

$$10x_1 + 20x_2 = 200.$$

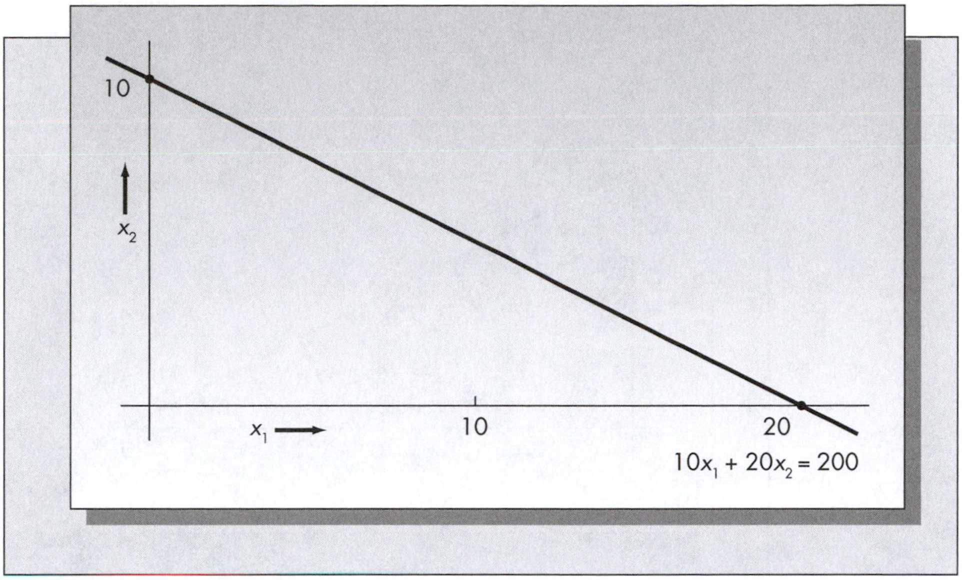

Figure S1–1 Graphing a constraint boundary

The easiest way to graph a straight line is to determine the two intercepts. These are found by setting x_2 to zero and solving for x_1, and then setting x_1 to zero and solving for x_2. First setting x_2 to zero gives

$$10x_1 = 200 \qquad \text{or} \qquad x_1 = 20.$$

Similarly, setting x_1 to zero and solving for x_2 gives

$$20x_2 = 200 \qquad \text{or} \qquad x_2 = 10.$$

Hence, the line $10x_1 + 20x_2 = 200$ must pass through the points (20, 0) (the x_1 intercept) and (0, 10) (the x_2 intercept). A graph of this line is shown in Figure S1–1. Now that we have graphed the line defining the boundary of the half space, we must determine which side of the line corresponds to the inequality. To do so, we pick any point *not* on the line, substitute the values of x_1 and x_2, and see if the inequality is satisfied or not. If the inequality is satisfied, then that point belongs in the half space; if it is not, then that point does not belong in the half space. If the boundary line does not go through the origin [that is, the point (0, 0)], then the most straightforward approach is to use the origin as the point to be tested.

Substituting (0, 0) into the inequality, we obtain

$$(10)(0) + (20)(0) = 0 \leq 200.$$

Because 0 is less than 200, the test confirms that the origin lies within the region represented by the inequality. This means that the graph of the inequality includes all the points below the boundary line in Figure S1–1, as shown in Figure S1–2.

The origin test to determine which side of the line is appropriate only works when the boundary line itself does not go through the origin. If it does, then some other point not on the line must be used for the test. For example, consider the constraint

$$4x_1 - 3x_2 \geq 0.$$

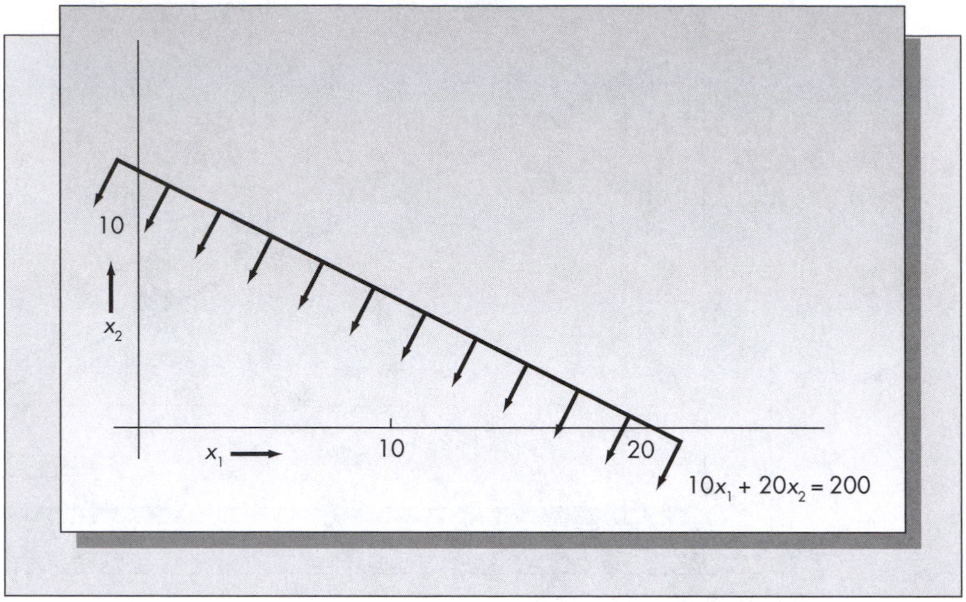

Figure S1–2 Half space representing the inequality $10x_1 + 20x_2 \leq 200$

When we try to graph the line $4x_1 - 3x_2 = 0$, we see that substituting $x_1 = 0$ gives $x_2 = 0$ and substituting $x_2 = 0$ gives $x_1 = 0$. This means that the line passes through the origin. In order to graph the line, we must determine another point that lies on it. Just pick any value of x_1 and solve for the corresponding value of x_2. For example, substituting $x_1 = 3$ gives $x_2 = 4$, meaning that the point (3, 4) lies on the line as well as the point (0, 0). The boundary line is graphed by connecting these points, as shown in Figure S1–3.

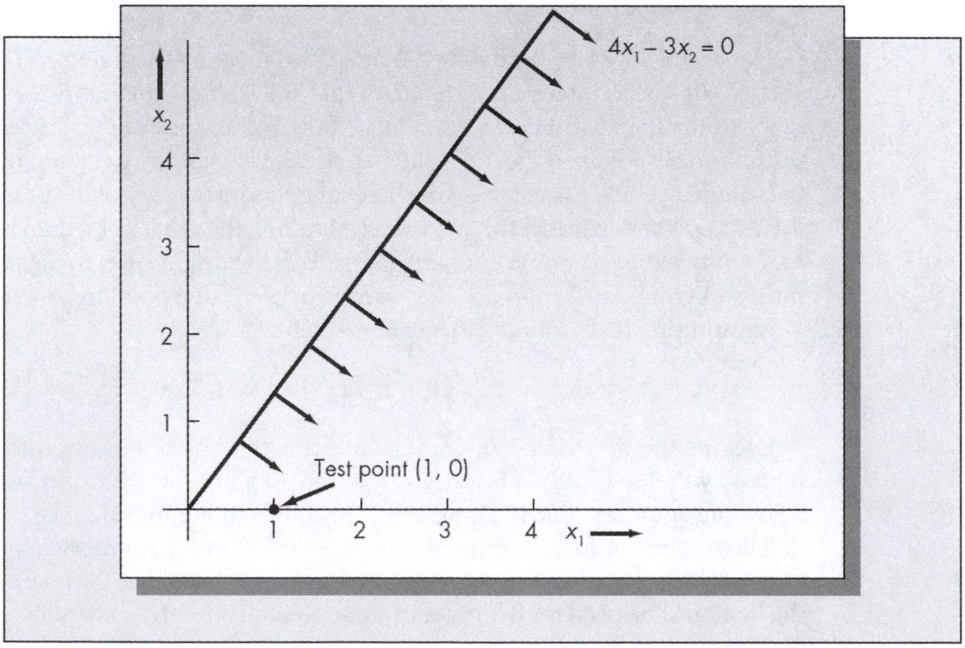

Figure S1–3 Graphing a constraint boundary that passes through the origin

Next, we determine which side of the inequality corresponds to the region of interest. As noted above, the origin test does not work when the line passes through the origin. We pick any point *not* on the line to do the test. In this case, one point that does not lie on the line is $x_1 = 1$ and $x_2 = 0$. Substituting these values into the inequality gives

$$(4)(1) - 0 = 4 > 0.$$

The inequality is satisfied, so the point (1, 0) lies in the region. The inequality corresponds to the points below the line as pictured in Figure S1–3.

Graphing the Feasible Region

The graph of the feasible region is found by graphing the linear inequalities represented by the constraints and determining the region of intersection of the corresponding half spaces. We will determine a graph of the feasible region for the Sidneyville problem in this way.

We have graphed the feasible region corresponding to the first constraint in Figure S1–2. Consider the other two constraints:

$$4x_1 + 16x_2 \leq 128,$$
$$15x_1 + 10x_2 \leq 220.$$

The half spaces corresponding to these constraints are found in the same way. First we graph the straight lines corresponding to the region boundaries. In the first case the intercepts are $x_1 = 32$ and $x_2 = 8$, and in the second case they are $x_1 = 14.6667$ and $x_2 = 22$. Using the origin test, we see that the appropriate half spaces are the points lying below both lines. In addition, we must also include the nonnegativity constraints, $x_1 \geq 0$ and $x_2 \geq 0$. The resulting feasible region is pictured in Figure S1–4.

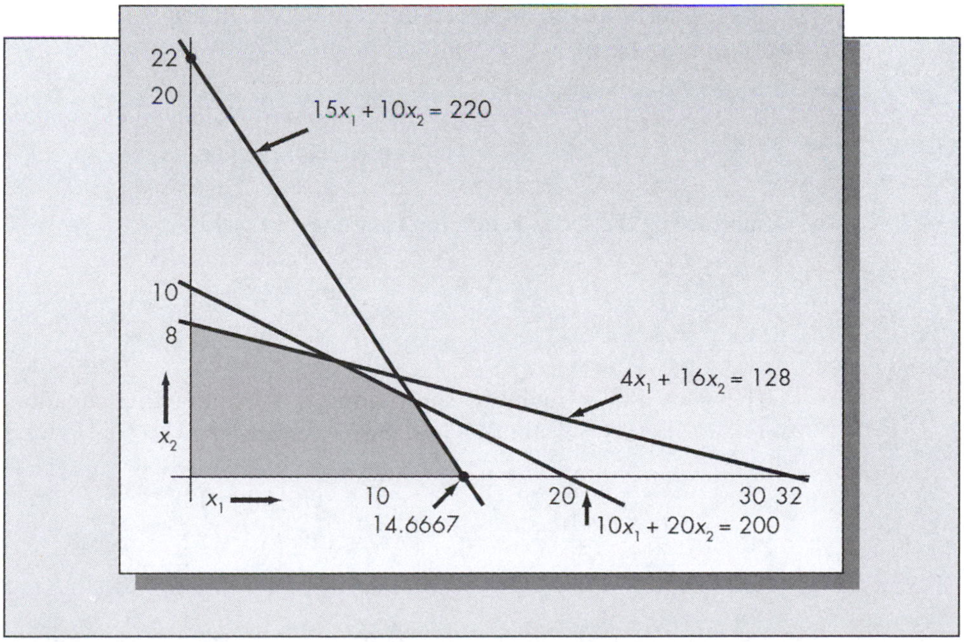

Figure S1–4 Feasible region for Sidneyville problem (Example S1.1)

Finding the Optimal Solution

The feasible region pictured in Figure S1–4 has several interesting properties that are common to all linear programming problems. Pick any two feasible solutions (that is, points in the region) and connect these points with a straight line. The resulting line lies completely in the region, meaning that the feasible region is a convex set. The region boundaries are straight lines. These lines intersect at points known as the extreme points. There are a total of five extreme points in the feasible region pictured in Figure S1–4.

An important property of linear programs is that the optimal solution always occurs at an extreme point of the feasible region. For the Sidneyville problem, this means that among the infinite number of solutions (i.e., points) in the feasible region, the optimal solution will be one of only five points! (It can happen that two extreme points are optimal, in which case all the points on the line connecting them are optimal as well.) This is true no matter what objective function we assume. This means that we can find the optimal solution to the Sidneyville problem by identifying the five extreme points, substituting the (x_1, x_2) coordinates of these points into the objective function, and determining the point that results in the maximum profit. We will consider this method first, even though there is a more efficient graphical solution procedure.

From Figure S1–4 we see that one of the extreme points is the origin (0, 0). Another is the x_2 intercept corresponding to the constraint $4x_1 + 16x_2 = 128$, which is (0, 8). A third is the x_1 intercept corresponding to the constraint $15x_1 + 10x_2 = 220$, which is (14.6667, 0). The other two extreme points correspond to the intersections of pairs of boundary lines. They are found by simultaneously solving the equations corresponding to the boundaries.

First, we simultaneously solve the equations:

$$4x_1 + 16x_2 = 128,$$
$$10x_1 + 20x_2 = 200.$$

These equations can be solved in several ways. Multiplying the first equation by 10 and the second by 4 gives

$$40x_1 + 160x_2 = 1{,}280,$$
$$40x_1 + 80x_2 = 800.$$

Subtracting the second equation from the first yields

$$80x_2 = 480$$
$$x_2 = 6.$$

The value of x_1 is found by substituting $x_2 = 6$ into either equation (both will yield the same value for x_1). Substituting into the first equation gives

$$4x_1 + (16)(6) = 128$$
$$4x_1 = 128 - 96 = 32$$
$$x_1 = 8.$$

Check that substituting $x_2 = 6$ into the equation $10x_1 + 20x_2 = 200$ gives the same result.

The last extreme point is found by solving

$$15x_1 + 10x_2 = 220,$$
$$10x_1 + 20x_2 = 200$$

simultaneously. We will not present the details of this calculation. The reader should be able to show by the same methods just used that the simultaneous solution in this case is

$$x_1 = 12,$$
$$x_2 = 4.$$

We have now identified all five extreme points. The next step is to substitute the corresponding values of x_1 and x_2 into the objective function and see which gives the largest profit. The objective function is $115x_1 + 90x_2$.

Extreme Point	Value of Objective Function
(0, 0)	(115)(0) + (90)(0) = 0
(0, 8)	(115)(0) + (90)(8) = 720
(14.666... , 0)	(115)(14.666...) + (90)(0) = 1,686.67
(8, 6)	(115)(8) + (90)(6) = 1,460
(12, 4)	(115)(12) + (90)(4) = 1,740

The maximum value of the objective function is 1,740 and is achieved at the point (12, 4). Hence, we have shown that one method of finding the optimal solution to a linear programming problem is to find all the extreme points, substitute their values into the objective function, and pick the one that gives the largest objective function value for maximization problems or the smallest objective function value for minimization problems. Next we show how one can quickly identify the optimal extreme point graphically.

Identifying the Optimal Solution Directly by Graphical Means

One identifies the optimal solution directly in the following way. The objective function is a linear combination of the decision variables. In our example the objective function is $115x_1 + 90x_2$. Consider the family of straight lines defined by the equation

$$Z = 115x_1 + 90x_2.$$

As Z is varied, one generates a family of parallel lines. The variable Z is the profit obtained when producing x_1 rolltop desks and x_2 regular desks such that (x_1, x_2) lies on the line $Z = 115x_1 + 90x_2$. As an example, consider $Z = 4{,}140$. Figure S1–5 shows the line $4{,}140 = 115x_1 + 90x_2$. Notice that it lies completely outside the feasible region, meaning that no feasible combination of x_1 and x_2 results in a profit of \$4,140. Reducing Z to 2,070 drops the Z line closer to the feasible region, also pictured in Figure S1–5.

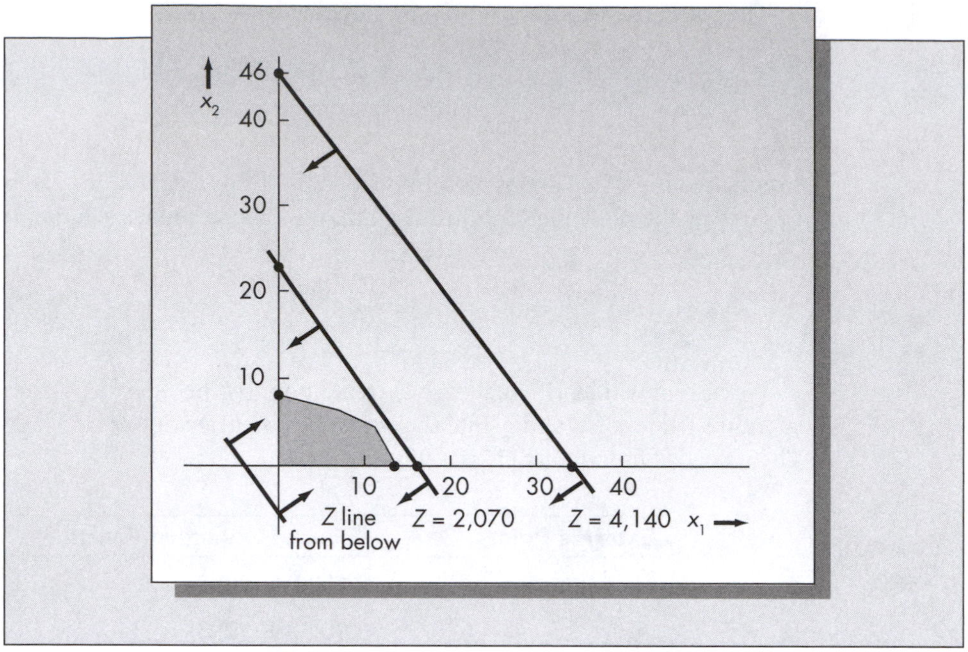

Figure S1–5 Approaching the feasible region with the *Z* line

The graphical method of identifying the optimal solution is to pick one value of *Z*, such as $Z = 3{,}000$, that takes us beyond the feasible region, place a ruler on the *Z* line, and move the ruler parallel to the *Z* line toward the feasible region. The extreme point that is hit first is the optimal solution. Once one graphically determines which extreme point is optimal, the coordinates of that point are found by solving the appropriate equations as shown above. This approach avoids having to identify all the extreme points of the feasible region.

One problem can arise. If we pick a small starting value of *Z* and move toward the feasible region from below, this method will identify a *different* extreme point. In Figure S1–5, suppose that we chose $Z = -3{,}000$. Then the *Z* line would be located under the feasible region. As we moved the *Z* line toward the feasible region, the first extreme point encountered is the origin. This means that the origin solves Example S1.1 with a minimization objective rather than a maximization objective (i.e., if your goal is to *minimize* profit, the best strategy is not to produce any desks).

Hence, this approach identifies two extreme points. One corresponds to the maximum solution and one to the minimum solution. In this case it is obvious which is which. When it is not obvious, the (x_1, x_2) values for each extreme point should be found and substituted into the objective function to be certain which is the minimum solution and which is the maximum solution.

S1.5 THE SIMPLEX METHOD: AN OVERVIEW

The **simplex method** is an algorithm that moves sequentially from one extreme point to another until it reaches the optimal solution. If the origin (that is, all problem variables set to zero) is a feasible solution, it will serve as the initial extreme point. (When the origin is not feasible, there are techniques available for determining an initial feasible solution to get the method started.) At each iteration, the method considers all adjacent extreme points (those that can be reached by moving along an edge) and moves to the one that gives the best improvement in the objective function. The algorithm continues to move from one adjacent extreme point to another, finally terminating when the optimal solution is reached.

For the problem of Example S1.1 the origin is feasible, so that is the initial feasible solution. The two extreme points adjacent to the origin are $(x_1, x_2) = (14.6667, 0)$ and $(x_1, x_2) = (0, 8)$. The largest improvement in the objective function is obtained by moving to the point (14.6667, 0). There are two extreme points adjacent to (14.6667, 0). They are (0, 0) and (12, 4). Clearly, a greater improvement is obtained by moving to (12, 4). At this point the method recognizes that a movement to another adjacent extreme point lowers the profit and therefore the current solution is optimal. (One reason this method is so efficient is because it is indeed the case that if no improvement is possible from the current extreme point, then that point can be proven to be optimal.)

In the worst case, the simplex method could conceivably need to search all the extreme points of the feasible region before identifying the optimal solution. If this were common, the simplex method would not be a practical solution method for solving linear programming problems. Let us consider why.

Suppose that a linear program had 25 variables and 25 less than or equal to constraints. For each constraint, one adds a slack variable converting the problem to one with only equality constraints. This is known as standard form. Hence, in standard form we have a problem of 50 variables and 25 constraints. Each basic solution (extreme point) corresponds to setting 25 variables to zero and solving the resulting system of 25 linear equations in 25 unknowns. It follows that the number of such solutions (that is, extreme points) equals the number of combinations of 50 things taken 25 at a time. This turns out to be about 1.264×10^{14} (about 126 trillion). To give the reader some idea of the magnitude of this number, suppose that we had a computer program that could identify 100 extreme points every second. At that rate, it would take about 40,000 years to find all the extreme points for a problem of this size!

It is indeed fortunate that the simplex method rarely needs to search all the extreme points of the feasible region to discover the optimal solution. In fact, for a problem of this size, on average it would need to evaluate only about 25 extreme points. (The reason for this was understood only very recently. The proof requires very sophisticated mathematics.) Hence, the simplex method turns out to be a very efficient solution procedure.

We will not explore the mechanics of the method or additional theory underpinning its concepts. With today's easy access to computing and the wide availability of excellent software, it is unlikely that anyone with a real problem would solve it manually. There are many excellent texts that delve more deeply into the theoretical and computational aspects of linear programming. Frederick Hillier and Gerald Lieberman (2015) provide a good starting point for the interested reader (see Hadley, 1962 for a more detailed treatment of the theory).

S1.6 SOLVING LINEAR PROGRAMMING PROBLEMS WITH EXCEL

Excel has become the standard for spreadsheet programs. One of the useful features of Excel is the availability of the add-in called Solver that solves both linear and nonlinear programming problems. Excel continues to improve the algorithms behind Solver, and the linear programming part of Solver is excellent.

Because Solver is part of a spreadsheet program, problems are not entered algebraically. Consider Example S1.1. The algebraic representation is

$$\text{maximize } 115x_1 + 90x_2$$

subject to

$$10x_1 + 20x_2 \leq 200,$$
$$4x_1 + 16x_2 \leq 128,$$
$$15x_1 + 10x_2 \leq 220,$$
$$x_1, x_2 \geq 0.$$

It is convenient to write the problem in a matrix format before entering the information into the Excel spreadsheet. The matrix representation for this problem is

Variable names:	x_1	x_2	Operator	RHS
Objective function:	115	90	max	
subject to				
Constraint 1	10	20	≥	200
Constraint 2	4	16	≥	128
Constraint 3	15	10	≥	220

The spreadsheet will look very much like this. The only differences are that one must specify the locations of the variable values (which we recommend be a row located directly under the variable names) and the algebraic formulas for the objective function and the constraints. These will be located in a column between "Operator" and "RHS (right-hand side)." We will label this column "Value" in the spreadsheet.

Note that the column labeled "Operator" is not required. It is a convenience for the user to help keep track of the direction of the objective function and constraints. Excel requires this information, and it is entered manually when Solver is invoked.

At this point the spreadsheet should look similar to Figure S1–6. Notice the additional row labeled "Var Values" and the additional column labeled "Value." It is in the column labeled "Value" that one enters the algebraic form of the linear programming problem. The locations of the variable values are cells C4 and D4. The algebraic form of the objective function will be entered in cell F5 and the constraints in cells F7, F8, and F9. The formula for cell F5 is =C5*$C$4+D5*$D$4. The formula can be typed in or entered using the mouse to point and click cell locations.

Notice that we have used absolute addressing for the variable values (C4 and D4). This allows us to copy the formula from cell F5 to cells F7, F8, and F9 without having to retype the algebraic form for the constraints. You may wish to assign name labels to these cells so that they can later be referred to by name rather than by cell location. (This is done most conveniently by invoking the formula bar, accessing the label area, and typing in a label of your choice. Note that the label "profit" appears just above column A. This was the name assigned to cell F5.)

Check that the algebraic form of the constraints is correct after copying cell F5 to cells F7, F8, and F9. For example, the formula corresponding to cell F7 should be: =C7*$C$4 + D7*$D$4.

Place 0s in the cells corresponding to the variable values (C4 and C5). After Solver has completed its search, the optimal values of these variables will appear in these cell locations.

The problem is now completely defined. To use Solver to obtain the solution, click on data and choose Solver from the ribbon that appears (as mentioned earlier, you must have first installed the Solver add-in). The Solver dialog box will appear as in Figure S1–7. The first

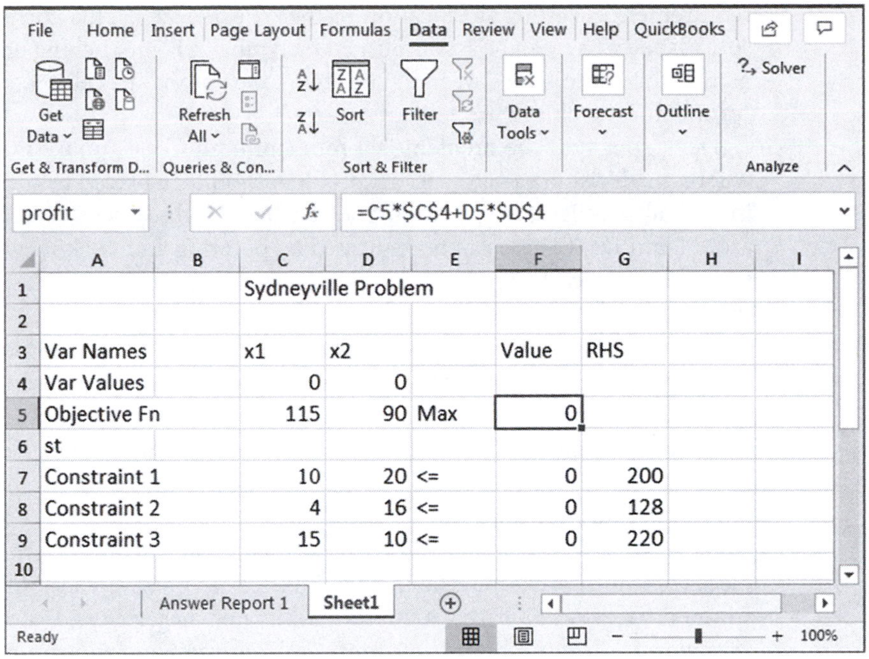

Figure S1–6 Excel spreadsheet for Sidneyville problem

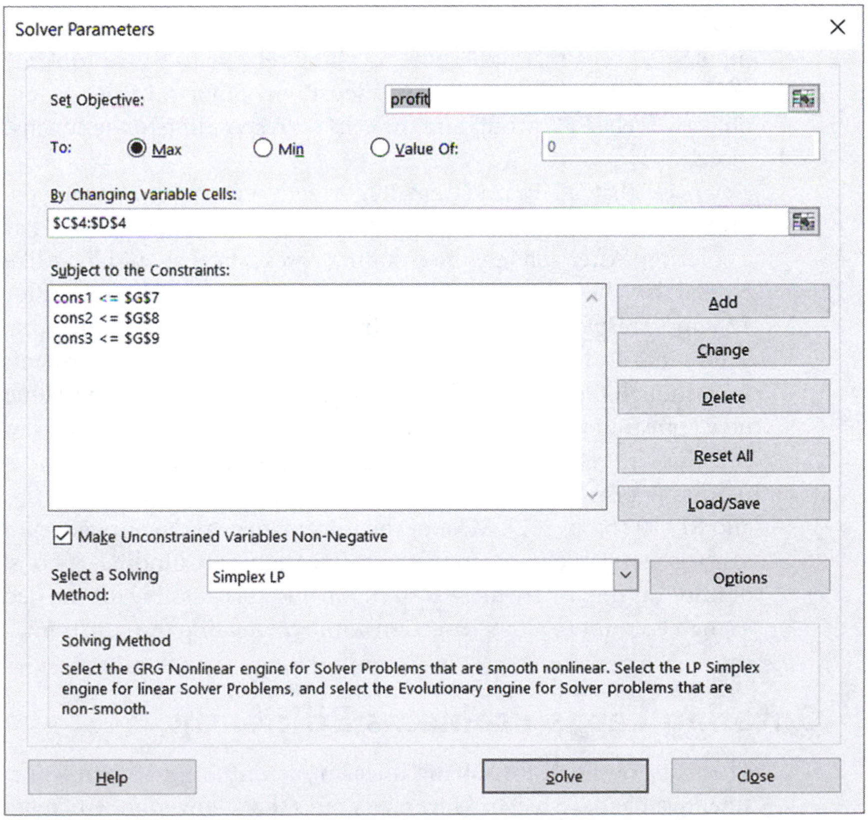

Figure S1–7 Solver dialog box

requirement is setting the location of the objective cell. This corresponds to cell F5 in our spreadsheet. This can be accomplished by typing in F5 or clicking on the cell with the mouse. Notice that in this case we assigned the name "profit" to cell F5, so we can simply type in "profit" for the objective.

Next, specify if the problem is a min (minimum) or a max (maximum). Excel refers to where the decision variables in the problem should be placed by saying these are the "changing variable cells." You must tell Solver where to find these cells. In this spreadsheet they are C4 and D4. (These can be indicated by pointing and clicking with the mouse, manually typing in the cell locations, or entering preassigned cell names as we did for the objective function.)

Next, tell Excel where to find the algebraic definitions for the constraints. The constraints can be entered into the system one at a time by clicking on the Add button (a more efficient method for entering groups of constraints is described later). For each constraint you first tell the system where to find the algebraic form for the left-hand side of the constraint (F7, F8, and F9 in our case), the logical operator for the constraint ($\leq$, $=$, or $\geq$), and the location of the RHS value (G7, G8, and G9 in our case).

Because Solver is a general-purpose mathematical programming tool, two additional pieces of information must be included. First, you need to check the box for "Make Unconstrained Variables Non-Negative" so that the non-negativity constraints is included Second, "Simplex LP" should be chosen as the solving method (appears in the dropdown box). If your output includes values of Lagrange variables, you'll know that you forgot to specify this option.

At this point your dialog box should look like Figure S1–7. Notice that we have named some of the cells and the names appear in the dialog box rather than the cell locations. Using named cells that have some meaning relative to your problem will be very helpful later when you obtain the solution and sensitivity reports. In the dialog box, we have named the objective cell (F5) "profit" and the cells corresponding to the constraints "cons1," "cons2," and "cons3."

Check that all information is correctly entered and that you have specified a linear problem. Now simply click the Solve button and Excel will whirl away and quickly produce a solution. After solving, the resulting spreadsheet should look like the one in Figure S1–8. Notice that the values of the variables in cells C4 and D4 now reflect the optimal solution of 12 and 4, respectively. The value in cell F5 is the value of the optimal profit of $1,740. The values in cells F7, F8, and F9 are the values of the left-hand sides of the constraints.

Although the optimal values of the variables and the optimal value of the objective function now appear in your spreadsheet, Excel has the option of printing several types of reports. The two that are relevant for linear programming are the answer and sensitivity reports. These reports appear on different sheets of the Excel workbook and are shown in Figures S1–9 and S1–10 (on p. 202). Most of the information in the answer report appears in the original spreadsheet. We also are told which constraints are binding. A constraint that is labeled "not binding" is one where there is slack. That is, it is possible for the decision variables to change a small amount in either direction without violating the constraint.

Entering Large Problems Efficiently

The steps outlined for solving linear programming problems on a spreadsheet are fine for small problems. Excel has several features that allow more efficient entry of larger problems, however.

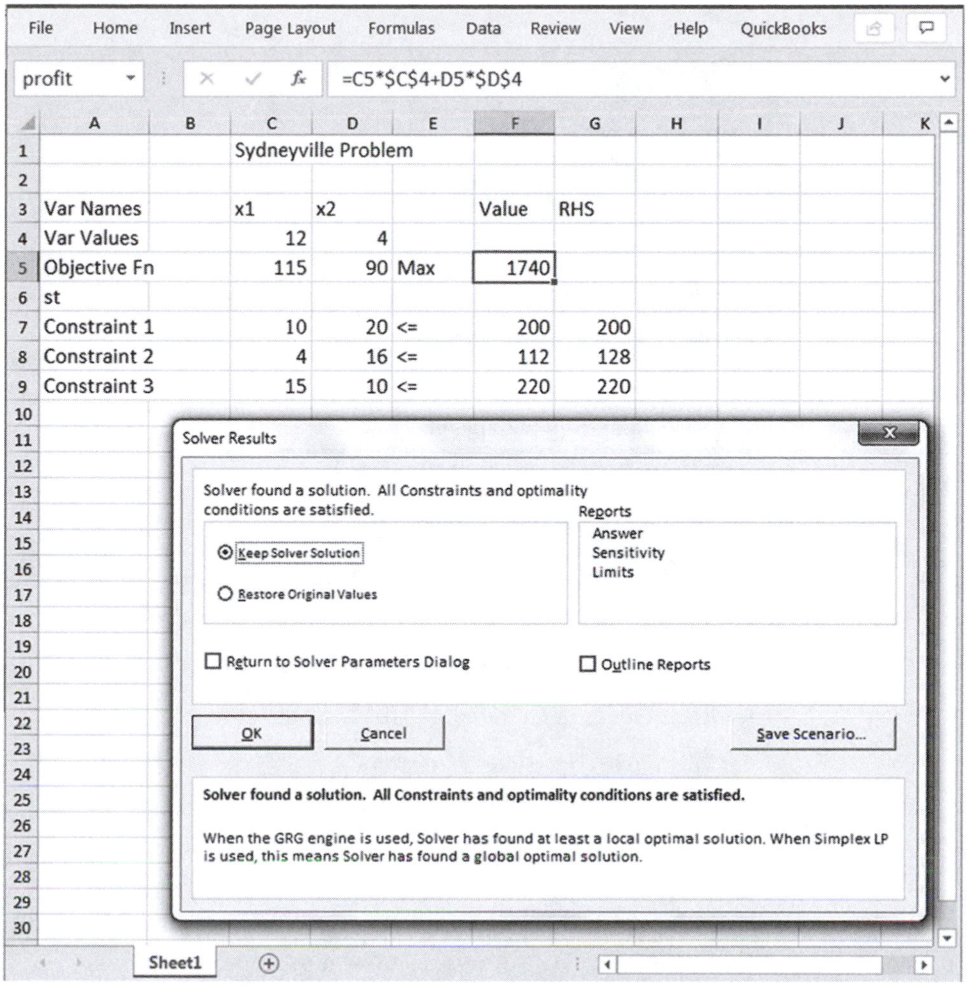

Figure S1–8 The Excel spreadsheet displaying the optimal solution

One such feature is the SUMPRODUCT function. SUMPRODUCT is a vector or array product, which multiplies the elements of one array times another array term by term and adds the results. This means that the algebraic form for the objective function and for the constraints can be entered with this function. Recall that for cell F5 we used the formula =C5*$C$4+D5*$D$4. This formula also could have been entered as =SUMPRODUCT($C$4:$D$4, C5:D5). While this may not appear to be much of an improvement here, it saves a lot of typing when entering large problems.

The second shortcut for entering large problems is to group constraints by type and enter all constraints in a group at one time. In the case of the Sidneyville problem, there are three $\leq$ constraints. These constraints could be entered with one command, similar to the single command for entering the nonnegativity constraints. The appropriate formulas for the constraints appear in cells F7, F8, and F9. The single command for entering these three constraints into Solver is: $F$7:$F$9 $\leq$ $G$7:$G$9. This can be typed in directly or entered by pointing to the appropriate cells in the spreadsheet. The Solver dialog box for the Sidneyville problem using this approach is shown in Figure S1–11 (p. 203).

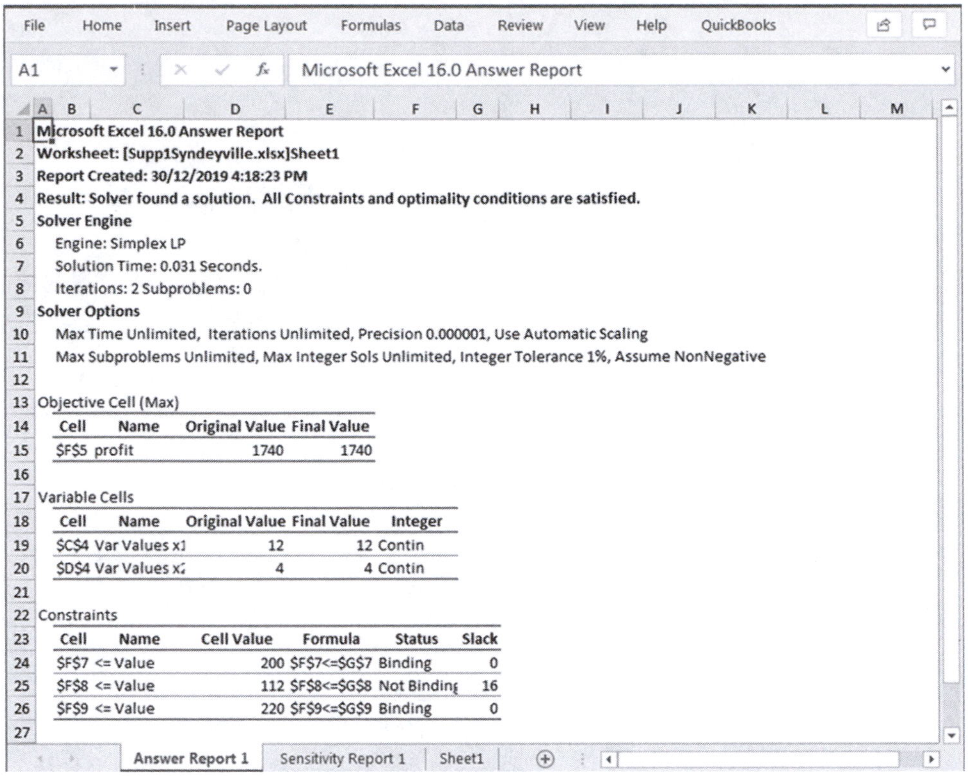

File Home Insert Page Layout Formulas Data Review View Help QuickBooks

A1 Microsoft Excel 16.0 Answer Report

Microsoft Excel 16.0 Answer Report
Worksheet: [Supp1Syndeyville.xlsx]Sheet1
Report Created: 30/12/2019 4:18:23 PM
Result: Solver found a solution. All Constraints and optimality conditions are satisfied.
Solver Engine
Engine: Simplex LP
Solution Time: 0.031 Seconds.
Iterations: 2 Subproblems: 0
Solver Options
Max Time Unlimited, Iterations Unlimited, Precision 0.000001, Use Automatic Scaling
Max Subproblems Unlimited, Max Integer Sols Unlimited, Integer Tolerance 1%, Assume NonNegative

Objective Cell (Max)

Cell	Name	Original Value	Final Value
$F$5	profit	1740	1740

Variable Cells

Cell	Name	Original Value	Final Value	Integer
$C$4	Var Values x1	12	12	Contin
$D$4	Var Values x2	4	4	Contin

Constraints

Cell	Name	Cell Value	Formula	Status	Slack
$F$7	<= Value	200	$F$7<=$G$7	Binding	0
$F$8	<= Value	112	$F$8<=$G$8	Not Binding	16
$F$9	<= Value	220	$F$9<=$G$9	Binding	0

Answer Report 1 | Sensitivity Report 1 | Sheet1

Figure S1–9 Answer report for Sidneyville problem

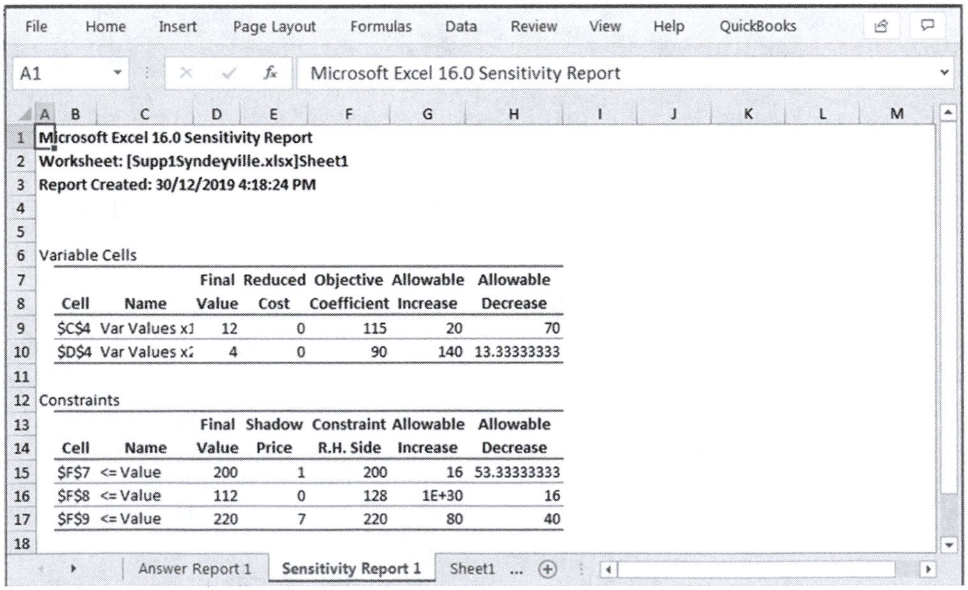

File Home Insert Page Layout Formulas Data Review View Help QuickBooks

A1 Microsoft Excel 16.0 Sensitivity Report

Microsoft Excel 16.0 Sensitivity Report
Worksheet: [Supp1Syndeyville.xlsx]Sheet1
Report Created: 30/12/2019 4:18:24 PM

Variable Cells

Cell	Name	Final Value	Reduced Cost	Objective Coefficient	Allowable Increase	Allowable Decrease
$C$4	Var Values x1	12	0	115	20	70
$D$4	Var Values x2	4	0	90	140	13.33333333

Constraints

Cell	Name	Final Value	Shadow Price	Constraint R.H. Side	Allowable Increase	Allowable Decrease
$F$7	<= Value	200	1	200	16	53.33333333
$F$8	<= Value	112	0	128	1E+30	16
$F$9	<= Value	220	7	220	80	40

Answer Report 1 | Sensitivity Report 1 | Sheet1 ...

Figure S1–10 Sensitivity report for Sidneyville problem

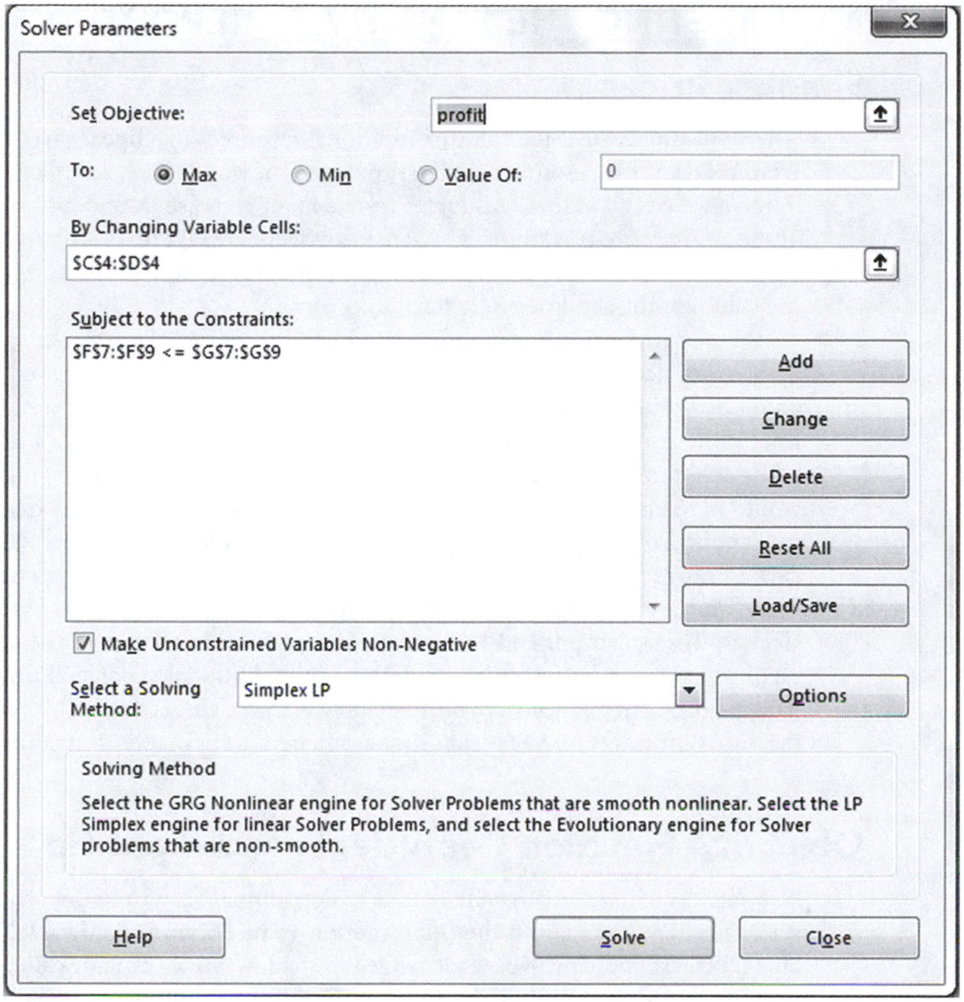

Figure S1–11 Solver dialog box with efficient data entry

Using SUMPRODUCT and entering constraints in groups can save a lot of time for large problems. However, the Solver dialog box is not as informative, since only cell locations and not cell names appear. For problems of fewer than 10 constraints total, entering the constraints one at a time is fine.

We noted earlier that an advantage of the spreadsheet for solving linear programming is that one can construct a general-purpose template. A template might have up to 10 variables and 10 constraints. Variable names could be x_1 through x_{10} and constraint names const_1 through const_{10}. The SUMPRODUCT functions for the objective function and the constraints would be programmed in advance. One would then simply enter the coefficients of the problem to be solved and save as a new file name. (This is much faster than typing in the full algebraic representation for every new problem as one must do with other linear programming systems.)

S1.7 INTERPRETING THE SENSITIVITY REPORT

Shadow Prices

An interesting issue is the value of additional resources in a linear programming problem. In Example S1.1, the resources are the three types of wood: pine, cedar, and maple. An increase in the level of any of these resources results in an increase in the value of the right-hand side of the appropriate constraint. The **sensitivity report** (Figure S1–10) gives information about the value to the objective function of additional resources. This information is contained in the values of the shadow prices for the constraints.

The **shadow price** is defined as the improvement in the objective function realized by adding one additional unit of a resource. For Example S1.1, the first constraint refers to the amount of pine needed. Since this constraint is binding at the optimal solution (since the final value and the constraint right-hand-side values are the same), it is likely that if we had additional pine we could increase revenue. The shadow price tells us just how beneficial this would be. Since the shadow price for this constraint is \$1, it means that for each additional unit square foot of pine, the objective (profit) increases by \$1. (This will hold only within a certain range as discussed subsequently.) Consider the second constraint (cedar). Since the final value and the right-hand side for this constraint are different (112 versus 120), there is slack in this constraint and it is non-binding. That means we are not consuming the entire quantity of cedar at the optimal solution, and additional cedar will not improve the profit. This is borne out by the fact that the shadow price for this constraint is zero. The final shadow price of 7 indicates that every additional square foot of maple contributes \$7 to the profit.

Objective Function Coefficients and Right-Hand Sides

The shadow prices remain valid as long as the optimal basis does not change. (The basis is the set of positive variables in the final solution.) The columns Allowable Increase and Allowable Decrease indicate over what range the shadow prices remain valid. This means that we can determine the effect on the objective function of changes in constraint right-hand sides without re-solving the problem.

The first part of the sensitivity report (Figure S1–10) gives the values of the objective function coefficients for which the shadow prices remain valid. The current values of the profits are \$115 and \$90, respectively. The shadow prices are valid as long as the first objective function coefficient does not increase more than 20 or decrease more than 70 (i.e., the objective function coefficient for x_1 is between 45 and 135). Similarly, the allowable range for the objective function coefficient for x_2 is 76⅔ to 230.

The second part of the sensitivity report provides the ranges on the right-hand sides of the constraints for which the shadow prices remain valid. Hence, the shadow price of \$1 for pine is valid for an increase of 16 or less and a decrease of 53⅓ or less in the right-hand side of the first constraint. That is, the right-hand side of the first constraint could be any value between 146 and 216. The shadow price of 0 for cedar is valid for any increase (1E+30 should be interpreted as infinity) and a decrease of 16 or less, and the shadow price for maple is valid for an increase of 80 or less and a decrease of 40 or less.

If either the objective function coefficients or the right-hand sides increase or decrease beyond the allowable ranges, the shadow prices no longer remain valid, and the problem would have to be re-solved to determine the effect. Note that Excel uses the convention that a positive shadow price means an increase in the objective function per unit increase in the right-hand side of a constraint, and a negative shadow price means a decrease in the objective function per unit increase in the right-hand side, irrespective of whether the problem is a max or a min. (Other linear programming systems may have other conventions.) Also note that changes to objective function coefficients or right-hand sides can be only one at a time in these ranges. The rules for simultaneous changes in right-hand sides or objective function coefficients are much more complex and will not be discussed here.

Adding a New Variable

We can use the results of the sensitivity analysis to determine whether it is profitable to add a new activity (variable) without re-solving the problem. For small problems, such as Example S1.1, simply inputting and solving the new problem on the computer is quick and easy. However, in real applications, the number of decision variables and constraints could be in the hundreds or even in the thousands. Reentering a problem of this magnitude is a major task, to be avoided whenever possible.

Suppose the firm is considering producing a third product, a vanity table, which would require the same woods used in making the desks. Each vanity table would contribute $75 to profit, but each would require 8 square feet of pine, 6 square feet of cedar, and 10 square feet of maple. We could determine if it would be worth producing vanity tables in addition to the desks by solving the problem with three activities and comparing the values of the objective functions at the optimal solutions.

There is a faster way. The dual prices in Figure S1–10 tell us the value of each unit of resource at the current solution. The decrease in profit resulting from reducing the supply of pine is $1 per square foot, which translates to $8.00 for 8 square feet of pine. There is no cost for decreasing the supply of cedar. The cost of decreasing the supply of maple by 10 square feet is (10)(7) = $70. Hence the total decrease in profit from the consumption of resources required to produce one vanity table is $78. The contribution to profit is only $75. We conclude that it is not optimal to produce vanity tables in addition to the desks with the current resources. Had we determined, however, that it was profitable to produce the vanity table, we would have had to re-solve the problem with three activities to find the optimal numbers of desks and vanity tables to produce.

Using Sensitivity Analysis

To cement your understanding of the information in Figure S1–10, consider the following questions.

EXAMPLE S1.1 *(CONTINUED)*

a. Sidneyville's sales manager has renegotiated the contract for regular desks and now expects to make a profit of $125 on each. He excitedly conveys this information to the firm's production manager, expecting that the optimal mix of rolltop and regular desks will change as a result. Does it?
b. Suppose that the new contract also has a higher profit for the rolltop desks. If the new profit for the rolltop desks is $140, how will this change the optimal solution?
c. A logging company has offered to sell Sidneyville an additional 50 square feet of maple for $5.00 per square foot. Based on the original objective function, would you recommend that it accept the offer?

d. Assuming that Sidneyville purchases the 50 square feet of maple, how is the optimal solution affected?
e. The firm is considering a pine desk that would require 25 square feet of pine and no other wood. What profit for pine desks would be required to make its production worthwhile, assuming current levels of resources and original profits on regular and rolltop desks?
f. During inspection, the quality department discovered that 50 square feet of pine had water damage and could not be used. Will it be optimal to produce both desks under these circumstances? Will the product mix change?

Solution

a. According to Figure S1–10, the allowable increase in the coefficient of the objective function for variable x_2, the regular desks, is 140. Because the increase to $125 is still within the allowable range, the optimal mix of rolltop and regular desks will remain the same: namely, $x_1 = 12$ and $x_2 = 4$.
b. The allowable increase in the objective function for the rolltop desks (x_1) is 20, or to a maximum value of 135. As 140 is outside the allowable range, it is possible that the basis will change. However, the allowable ranges in Figure S1–10 are only valid if the profit for regular desks is $90. The allowable ranges will change when the profit for regular desks is changed to $125, even though the optimal solution does not. The output for part (a) (that is, with profits of $115 and $125) follows.

OBJ Coefficient Ranges			
Variable	**Current Coefficient**	**Allowable Increase**	**Allowable Decrease**
x_1	115.000000	72.500000	52.500000
x_2	**125.000000**	**105.000000**	**48.333330**

This shows that the allowable increase in the coefficient for x_1 is now 72.5. Because 140 is within the allowable range, the solution for parts (a) and (b) will be the same, which is also the same as our original solution of $x_1 = 12$ and $x_2 = 4$.
c. Since the dual price for the third constraint corresponding to maple is 7, it is profitable to purchase the maple for $5.00 a square foot. The allowable increase of the right-hand side over which this dual price applies is 80, so it is worth purchasing the full 50 additional square feet.
d. Because the increase of 50 is within the allowable right-hand-side range, we know that the basis will not change. That is, it still will be optimal to produce both the rolltop and the regular desks. However, if the right-hand side changes, the values of the basic variables *will* change. We must re-solve the problem with the new right-hand-side value to determine the updated solution. The solution follows.

Objective Function Value: 2090.00		
Variable	**Value**	**Reduced Cost**
x_1	17.00	.00
x_2	**1.50**	**.00**

To retain feasibility we round x_2 to 1. (This is *not* the optimal integer solution, however. The optimal integer solution is $x_1 = 18$ and $x_2 = 0$ with a profit of $2,070, which is obtained from Excel by identifying both x_1 and x_2 as integer variables. The suboptimal solution of $x_1 = 17$ and $x_2 = 1$ results in a profit of $2,045.)

e. The dual price for pine is \$1 per square foot. As each desk consumes 25 square feet of pine, the profit for each pine desk must exceed \$25 for pine desks to be profitable to produce.
f. The right-hand side of the first constraint can decrease as much as 53.33333 and the current basis will remain optimal. That means that a decrease of 50 square feet will not change the basis; it will still be profitable to produce both desks. However, the production quantities will decrease. We must re-solve the problem to determine the correct levels of the new quantities.

Objective Function Value: 1690.00		
Variable	**Value**	**Reduced Cost**
x_1	14.50	.00
x_2	.25	.00

Again, we need to round these variables. Rounding both x_1 and x_2 down guarantees feasibility. If we produce 14 rolltop desks and 0 regular desks, we require 140 square feet of pine (150 are available), 56 square feet of cedar (128 are available), and 210 square feet of maple (220 are available). There does not appear to be enough wood to produce an additional desk of either type, so we leave the solution at $x_1 = 14$ and $x_2 = 0$. (This is the optimal integer solution.)

S1.8 RECOGNIZING SPECIAL PROBLEMS

Several problems can occur when solving linear programming problems. In this section, we will discuss the causes of these problems and how one recognizes them when using Excel.

Unbounded Solutions

The feasible region of a linear program is not necessarily bounded. The feasible region for the Sidneyville problem pictured in Figure S1–4 is bounded. However, consider the following linear programming problem.

EXAMPLE S1.2

Maximize

$$2x_1 + 3x_2$$

subject to

$$\begin{aligned} x_1 + 4x_2 &\geq 8, \\ x_1 + x_2 &\geq 5, \\ 2x_1 + x_2 &\geq 7, \\ x_1, x_2 &\geq 0. \end{aligned}$$

Figure S1–12 shows the feasible region. Notice that it is unbounded. Because we can make x_1 and x_2 as large as we like, there is no limit to the size of the objective function. When this occurs, the problem is unbounded and there is no optimal solution.

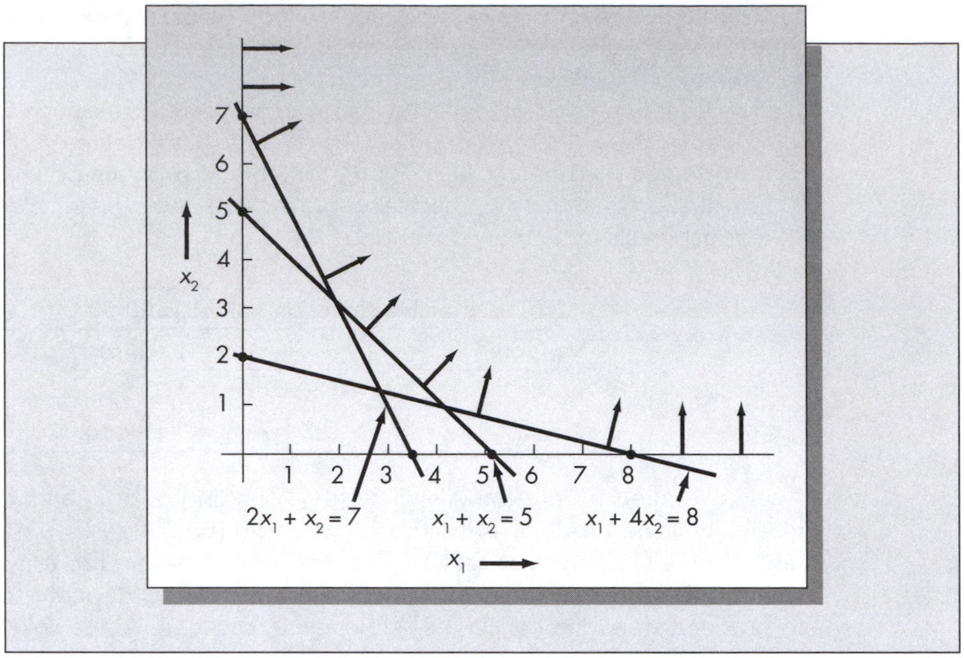

Figure S1–12 Feasible region for Example S1.2

When this problem is inputted in Excel, Solver writes in somewhat arbitrary feasible values for the problem variables and displays a message that set the objective cell values do not converge. The Excel output for this problem appears in Figure S1–13.

Empty Feasible Region

It is possible for two or more constraints to be inconsistent. When that occurs, there will be no feasible solution. Consider the following problem.

EXAMPLE S1.3

Maximize

$$2x_1 + 3x_2$$

subject to

$$x_1 + 4x_2 \leq 8,$$
$$x_1 + x_2 \geq 10,$$
$$x_1, x_2 \geq 0.$$

The feasible region for this example appears in Figure S1–14. Notice that there is no intersection of the half spaces defined by the two constraints in the positive quadrant. In this case, the feasible region is empty, and we say that the problem is infeasible. The Excel output appears in Figure S1–15 (p. 210). Note that the solution $x_1 = 8$ and $x_2 = 0$ shown is not feasible because it results in negative slack in the first constraint.

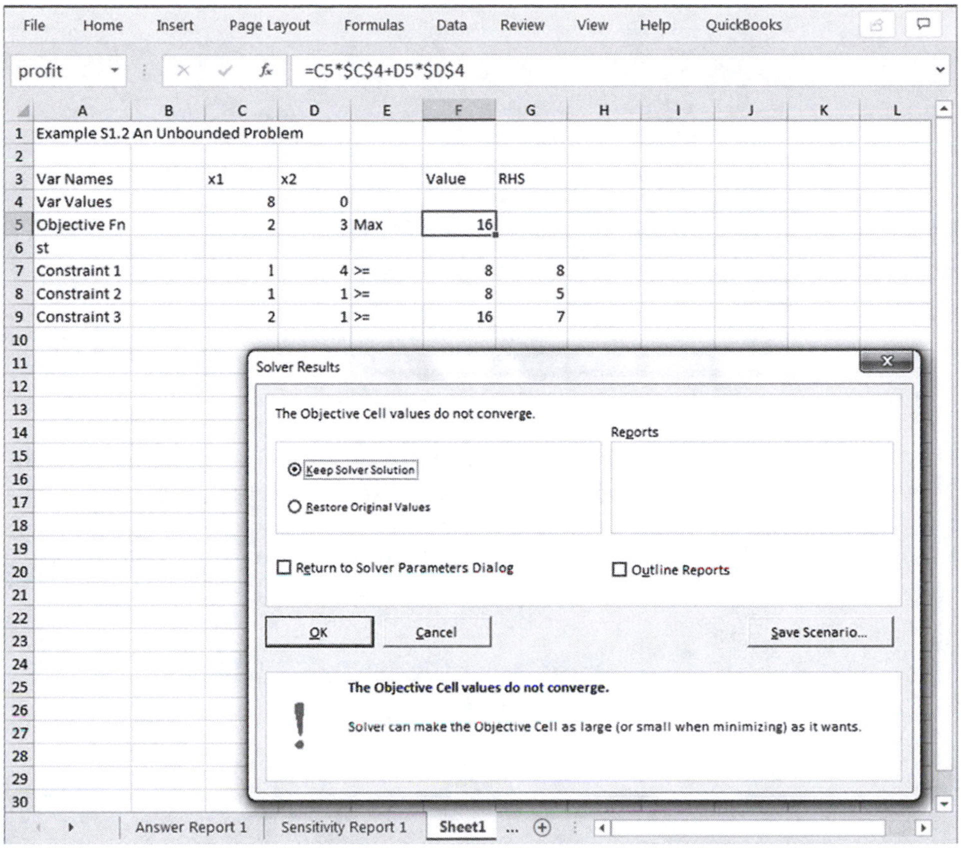

Figure S1–13 Excel output for Example S1.2

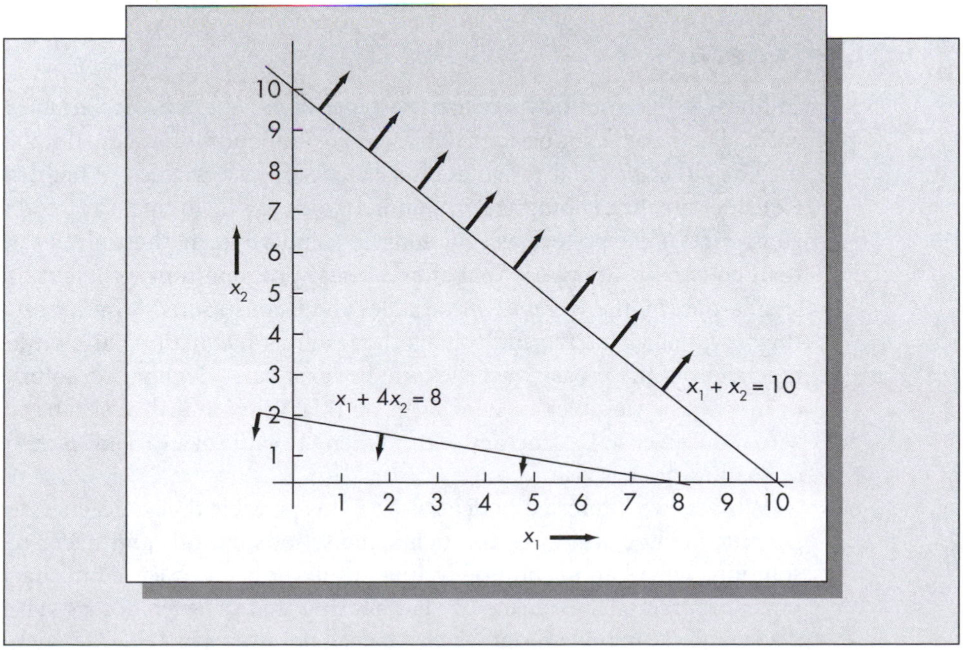

Figure S1–14 Feasible region for Example S1.3

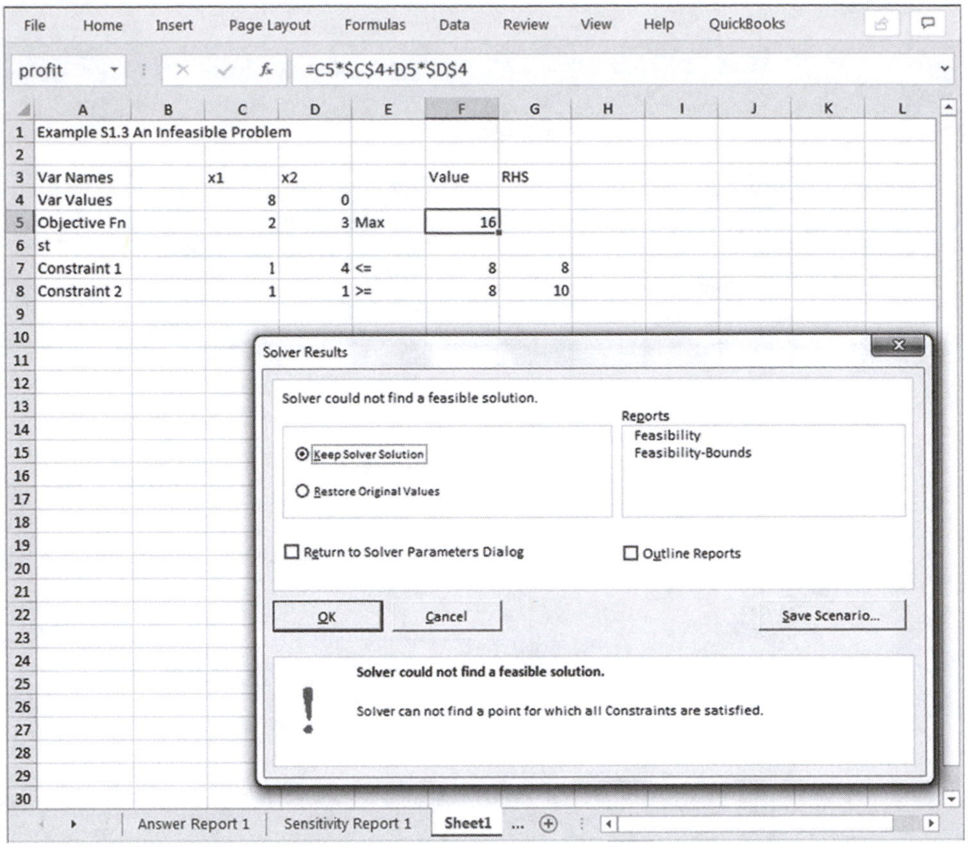

Figure S1–15 Excel output for Example S1.3

Degeneracy

In linear programming there are two types of variables: **basic** variables and **nonbasic** variables. The number of basic variables equals the number of constraints, and basic variables may be either original variables or slack or surplus variables. What defines basic variables? Consider any linear program in standard form. By including slack and surplus variables, all constraints are expressed as equations. In standard form, there always will be more variables than constraints. Suppose that after a linear programming problem has been expressed in standard form, there are $n + m$ variables and n constraints. A **basic solution** is found by setting m variables to zero and solving the resulting n equations in n unknowns. In most cases the values of the n basic variables will be positive. A **degenerate solution** occurs when one or more basic variables are zero at the optimal solution. In Excel, a basic variable is one with zero reduced cost. Degeneracy occurs when the value of a variable is zero and its reduced cost or shadow price is also zero.

Why are we interested in degeneracy? It is possible that if degenerate solutions occur, the simplex method will cycle through some set of solutions and never recognize the optimal solution. The phenomenon of cycling has never been observed in practice, and most computer programs have means of guaranteeing that it never occurs. The bottom line is that degeneracy is an issue about which we need not worry, but one of which we should be aware.

Multiple Optimal Solutions

The optimal solution to a linear program is not always unique. There are cases in which there are multiple optimal solutions. In Section S1.4 we saw that two-variable problems could be solved by graphical means by approaching the feasible region with the Z line. Assuming that we approach the feasible region from the correct side, the first feasible point with which the Z line comes into contact is the optimal solution.

However, suppose that the Z line is parallel to one of the constraints. In that case it does not contact a single point first, but an entire edge.

EXAMPLE S1.4

Consider the feasible region pictured in Figure S1–12 corresponding to the constraints of Example S1.2. Suppose that the objective function is min $3x_1 + 3x_2$. Then the Z line has slope –1 and is parallel to the constraint boundary $x_1 + x_2 = 5$. As the Z line approaches the feasible region, it meets the edge defined by this constraint and both extreme points along this edge. This means that both extreme points *and* all points along the edge are optimal.

Unfortunately, Excel does not indicate that there are multiple optimal solutions to this problem. Our only clue is that the solution is degenerate; the surplus variable for the third constraint has both zero dual price and zero value. Our graphical solution tells us that both extreme points (2, 3) and (4, 1) are optimal, and so are all points along the edge connecting these extreme points. [The points along the edge can be written in the form

$$x_1 = \alpha(2) + (1 - \alpha)(4),$$
$$x_2 = \alpha(3) + (1 - \alpha)(1),$$

where α is a number between zero and one. This is known as a convex combination of these two extreme points.]

Redundant Constraints

It is possible for one or more constraints to be redundant. That means that these constraints can be eliminated from the formulation without affecting the solution. In simple two-variable problems, redundant constraints can be recognized graphically because they lie outside the feasible region. Excel does not recognize or signal if one or more constraints are redundant. Sometimes redundant constraints can cause degeneracy, but degeneracy can result when constraints are not redundant as well.

EXAMPLE S1.5

Consider Example S1.1. Suppose that we add the following additional constraint:

$$x_1 + x_2 \le 20.$$

Figure S1–16 shows the resulting feasible region. It is exactly the same as the feasible region pictured in Figure S1–4. The additional constraint has no effect, as it lies completely outside the feasible region defined by the first three constraints. The optimal solution will, of course, be exactly the same.

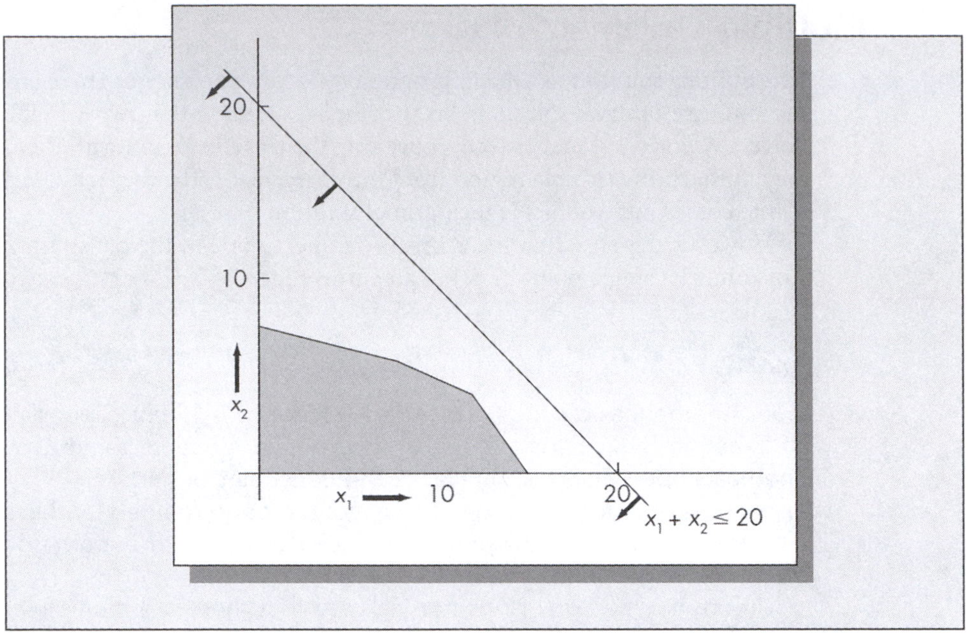

Figure S1–16 Feasible region for Example S1.5 showing a redundant constraint

If we had originally formulated the problem with the four constraints

$$10x_1 + 20x_2 \leq 200,$$
$$4x_1 + 16x_2 \leq 128,$$
$$15x_1 + 10x_2 \leq 220,$$
$$x_1 + x_2 \leq 20,$$

and solved, the output indicates that the optimal solution is $x_1 = 12$ and $x_2 = 4$ as before and gives us no clue that the last constraint is redundant. The only way to see that the final constraint is redundant is to graph the feasible region as we did in Figure S1–16. Graphing is possible, however, in two-variable problems only.

Does redundancy cause a problem? Not really. We would certainly like to be able to write our linear program as economically as possible, but if one or more constraints are redundant, the optimal solution is unaffected.

S1.9 THE APPLICATION OF LINEAR PROGRAMMING TO PRODUCTION AND OPERATIONS ANALYSIS

In Chapter 3 we showed how linear programming could be used to find optimal solutions (subject to rounding errors) for aggregate planning problems. In Chapter 7 we will use linear programming to solve transportation and transshipment problems. Although these are the only explicit uses of linear programming in this book, there have been successful linear programming applications for many other operations management problems. (In this section we interpret linear programming in the broad sense to include integer linear programming.) Scheduling and distribution are perhaps two areas in which applications are most common.

Researchers applied linear programming to the problem of providing a coordinated vehicle scheduling and routing system for delivery of consumable products to customers of the Du Pont Company (Fisher, Greenfield, Jaikumar, and Uster,1982). The primary issue was to determine delivery routes (loops) for the trucks to the company's clients in various regions of the country. A refrigerated truck drives a weekly loop that includes several dozen customers. The largest region considered, Chicago, had 16 loops and several hundred cities, whereas the Houston region, the smallest, had 4 loops and less than 80 cities.

The basic mathematical formulation of the problem was a generalized assignment problem. The **assignment problem** is a linear programming problem in which the decision variables are restricted to zeros and ones. The mathematical formulation used in this study follows.

1. Given data
   d_{ik} = Cost of including customer i in loop k,
   a_i = Demand from customer i.
2. Problem variables

$$y_{ik} = \begin{cases} 1 \text{ if customer } i \text{ is assigned to loop } k, \\ 0 \text{ if customer } i \text{ is not assigned to loop } k. \end{cases}$$

3. Generalized assignment problem

$$\text{Min} \sum_{k=1}^{K} \sum_{i=1}^{n} d_{ik} y_{ik}$$

subject to

$$\sum_{k=1}^{K} y_{ik} = 1, \text{ for } i = 1,\ldots,n,$$

$$\sum_{i=1}^{n} a_i y_{ik} \leq b_k, \text{ for } k = 1,\ldots,K,$$

$$y_{ik} = 0 \text{ or } 1, \text{ for all } i \text{ and } k,$$

where K is the total number of loops in the region and n is the number of customers.

The implementation of this model was reported to have saved Du Pont over $200 million. A more complex mathematical model solving a similar problem for Air Products Corporation won the Institute of Management Sciences Practice Award in 1983 (Bell et al., 1983).

S1.10 A PROTOTYPE LAYOUT PROBLEM AND THE ASSIGNMENT MODEL

An important application area for optimization is for facility layout problems. In some cases, layout problems can be formulated as assignment problems, as described in the previous section. In particular, the assignment model is appropriate when only a discrete set of alternative locations is being considered and there are no interactions among departments. Although moderately sized assignment problems can be solved by hand, *extremely* large problems can be solved only with the aid of computers.

EXAMPLE S1.6

Because of an increase in the volume of sales, the owner of a small manufacturing firm located in Muncie, Indiana, has decided to expand production capacity. A new wing has been added to the plant that will house four machines: (1) a punch press, (2) a grinder, (3) a lathe, and (4) a welding machine. There are only four possible locations for these machines (A, B, C, and D). However, the welding machine, which is the largest machine, will not fit in location B.

The plant foreman has made estimates of the impact in terms of materials handling costs of locating each of the machines in each of the possible locations. These costs, expressed in terms of dollars per hour, are represented in the following table.

		Location			
		A	B	C	D
Machines	**1**	94	13	62	71
	2	62	19	84	96
	3	75	88	18	80
	4	11	M	81	21

The entry *M* stands for a very large cost. It is used to indicate that machine 4 is not permitted in location B. As the objective is to assign the four machines to locations in order to minimize the total materials handling cost, an optimal solution will never assign machine 4 to location B.

Minimum cost assignments can rarely be found by inspection. For example, one reasonable approach might be to assign the machines to the lowest cost locations in sequence. The lowest cost in the matrix is 11, so machine 4 would be located in location A. Eliminating the last row and the first column, we see that the next remaining lowest cost is 13, so machine 1 would be assigned to location B. Eliminating the first row and the second column, the smallest remaining cost is 18, so machine 3 would be assigned to location C. Finally, machine 2 must be assigned to location D. The total cost of this solution is 11 + 13 + 18 + 96 = \$138. As we will see, this solution is suboptimal. (The optimal cost is \$114. Can you find the optimal solution?)

A simple approach such as this will rarely result in an optimal solution. Linear programming (with integer variables) can be used, but it turns out there is a more efficient solution algorithm from which one can obtain an optimal solution to the assignment problem by hand for moderately sized problems.

The Assignment Algorithm

The solution algorithm presented in this section is based on the following observation.

Result: If a constant is added to, or subtracted from, all entries in a row or column of an assignment cost matrix, the optimal assignment is unchanged.

In applying this result, the objective is to continue subtracting constants from rows and columns in the cost matrix until a zero cost assignment can be made. The zero cost assignment for the modified matrix is optimal for the original problem. This leads to the following algorithm for solving assignment problems.

Solution Procedure for Assignment Problems

1. Locate the smallest number in row 1 and subtract it from all the entries in row 1. Repeat for all rows in the cost matrix.
2. Locate the smallest number in column 1 and subtract it from all the entries in column 1. Repeat for all columns in the cost matrix.
3. At this point each row and each column will have at least one zero. If it is possible to make a zero cost assignment, then do it. That will be the optimal solution. If not, go to step 4.
4. Determine the maximum number of zero cost assignments. This will equal the smallest number of lines required to cover all zeros. The lines are found by inspection and are not necessarily unique. The important point is that the number of lines drawn not exceed the maximum number of zero cost assignments.
5. Choose the smallest uncovered number and do the following.
   a. Subtract it from all other uncovered numbers.
   b. Add it to the numbers where the lines cross.
   c. Return to step 3.

The process is continued until one can make a zero cost assignment. Notice that step 5 is merely an application of the result in the following way: If the smallest uncovered number is subtracted from every element in the matrix and then added to every covered element, it will be subtracted once and added twice where lines cross.

EXAMPLE S1.6 *(CONTINUED)*

Assume that the original cost matrix in Example S1.6 is

		Location			
		A	B	C	D
Machine	**1**	94	13	62	71
	2	62	19	84	96
	3	75	88	18	80
	4	11	M	81	21

Step 1. Subtracting the smallest number from each row gives

81	0	49	58
43	0	65	77
57	70	0	62
0	*M*	70	10

Because *M* is very large relative to the other costs, subtracting 11 from *M* still leaves a very large number, which we again denote as *M* for convenience. At this point at least one zero appears in each row, and in each column except the last.

Step 2. Subtracting 10 from every number in the final column gives

81	0	49	48
43	0	65	67
57	70	0	52
0	M	70	0

At this point we have at least one zero in every row and every column (step 3). However, that does not necessarily mean that a zero cost assignment is possible. In fact, it is possible to make at most three zero cost assignments at this stage, as shown in step 4.

Step 4.

81	0	49	48
43	0	65	67
57	70	0	52
0	M	70	0

The three assignments shown in step 4 are 1–B, 3–C, and 4–A. There are other ways of assigning three locations at zero cost as well. It does not matter which we choose at this stage, only that we know three are possible. The next step is to find three lines that cover all the zeros. These are shown here.

81	0	49	48
43	0	65	67
57	70	0	52
0	M	70	0

Again, the choice of lines is not unique. However, it is important that no more than three lines be used. Finding three lines to cover all zeros is done by trial and error.

Step 5. The smallest uncovered number, 43, is subtracted from all other uncovered numbers and added to the numbers where the lines cross. The resulting matrix is

		Location			
		A	B	C	D
Machine	**1**	38	0	6	5
	2	0	0	22	24
	3	57	113	0	52
	4	0	M	70	0

It is now possible to make a zero cost assignment, as shown in the matrix. The optimal assignment is machine 1, the punch press, to location B; machine 2, the grinder, to location A; machine 3, the lathe, to location C; and machine 4, the welder, to location D. The total materials handling cost per hour of the optimal solution is obtained by referring to the original assignment cost matrix. It is $13 + 62 + 18 + 21 = \$114$ per hour.

The assignment algorithm also can be used when the number of sites is larger than the number of machines. For example, suppose that there were six potential sites for locating the four machines, and the original cost matrix was

		Location					
		A	B	C	D	E	F
Machine	**1**	94	13	62	71	82	25
	2	62	19	84	96	24	29
	3	75	88	18	80	16	78
	4	11	*M*	81	21	45	14

The procedure is to add two dummy machines, 5 and 6, with zero costs. The problem is then solved using the assignment algorithm as if there were six machines and six locations. The locations to which the dummy machines are assigned are the ones that are not used.

S1.11 MORE ADVANCED MATHEMATICAL PROGRAMMING FORMULATIONS

The assignment model described in Section S1.10 can be useful for determining optimal layouts for a limited number of real problems. The primary limitation of the simple assignment model is that, in most cases, the number of materials handling trips and the associated materials handling cost are assumed to be independent of the location of the other facilities. In Example S1.6, the cost of assigning the punch press to location A was assumed to be $94 per hour. However, in most cases this cost would depend upon the location of the other machines as well.

A formulation of the problem that takes this feature into account is considerably more complex. In order to avoid confusion, we will assume that the problem is to assign machines to locations. The problem could be to assign other types of subfacilities to locations, of course, but we will retain this terminology for convenience. Define the following quantities:

n = Number of machines;

c_{ij} = Cost per time period of assigning machine i to location j. This cost could be a one-time relocation cost that is converted to an annual equivalent;

d_{jr} = Cost of making a single materials handling trip from location j to location r;

f_{ik} = Mean number of trips per time period from machine i to machine k;

S_i = The set of locations to which machine i could feasibly be assigned;

$$A_{ijkr} = \begin{cases} f_{ik} d_{jr} & \text{if } i \neq k \text{ or } j \neq r, \\ c_{ij} & \text{if } i \neq k \text{ and } j \neq r; \end{cases}$$

$$x_{ij} = \begin{cases} 1 & \text{if machine } i \text{ is assigned to location } j, \\ 0 & \text{otherwise.} \end{cases}$$

Interpret a_{ijkr} as the materials handling cost per unit time when machine i is assigned to location j and machine k is assigned to location r. This cost is incurred only if both x_{ij} and

x_{kr} are equal to 1. Hence, it follows that the total cost of assigning machines to locations is given by

$$\frac{1}{2}\sum_{i=1}^{n}\sum_{j=1}^{n}\sum_{k=1}^{n}\sum_{r=1}^{n} a_{ijkr}\, x_{ij}\, x_{kr} \tag{1}$$

As all indices are summed from 1 to n, each assignment will be counted twice; hence the need to multiply by ½. Constraints are included to ensure that each machine gets assigned to exactly one location, and each location is assigned exactly one machine. These are

$$\sum_{i=1}^{n} x_{ij} = 1, \qquad j = 1, \dots, n; \tag{2}$$

$$\sum_{j=1}^{n} x_{ij} = 1, \qquad i = 1, \dots, n; \tag{3}$$

$$x_{ij} = 0 \text{ or } 1, \qquad i = 1, \dots, n \text{ and } j = 1, \dots, n \tag{4}$$

$$x_{ij} = 0, \qquad i = 1, \dots, n \text{ and } j \notin S(i). \tag{5}$$

The mathematical programming formulation is to minimize the total materials handling cost of the assignment, (1), subject to the constraints (2), (3), (4), and (5). This formulation is known as the quadratic assignment problem. Because this problem is no longer linear, the simplex method for linear problems cannot be used, and convex programming techniques are needed. Excel Solver's "GRG non-linear" algorithm will solve this type of problem so long as it is not too large.

BIBLIOGRAPHY

Bell, W. J., L. M. Dalberto, M. L. Fisher, A. J. Greenfield, R. Jaikumar, P. Kedia, R. G. Mack, and P. J. Prvtzman. "Improving the Distribution of Industrial Gases with an On-Line Computerized Routing and Scheduling Optimizer." *Interfaces* 13 (1983), pp. 4–23.

Dantzig, G. B. *Linear Programming and Extensions.* Princeton: Princeton University Press, 1963.

Fisher, M., A. J. Greenfield, R. Jaikumar, and J. T. Uster III. "A Computerized Vehicle Routing Application." *Interfaces* 12 (1982), pp. 42–52.

Hadley, G. *Linear Programming.* Reading, MA: Addison-Wesley, 1962.

Hillier, F. S., and G. J. Lieberman. *Introduction to Operations Research* (10th ed.). New York: McGraw Hill, 2015.

CHAPTER

4

Inventory Control Subject to Known Demand

"We want to turn our inventory faster than our people."

—James Sinegal

CHAPTER OVERVIEW

Purpose

To consider methods for controlling individual item inventories when product demand is assumed to follow a known pattern (that is, demand forecast error is zero).

KEY POINTS

1. *Classification of inventories.*
   - *Raw materials.* These are resources required for production or processing.
   - *Components.* These could be raw materials or subassemblies that will later be included in a final product.
   - *Work-in-process (WIP).* These are inventories that are in the plant waiting for processing.
   - *Finished goods.* These are items that have completed the production process and are waiting to be shipped out.
2. *Why hold inventory?*
   - *Economies of scale.* It is often cheaper to order or produce in large batches than in small batches.
   - *Uncertainties.* Demand uncertainty, lead time uncertainty, and supply uncertainty all provide reasons for holding inventory.
   - *Speculation.* Inventories may be held in anticipation of a rise in their value or cost.
   - *Transportation.* Transporting goods from one location to another requires having pipeline inventories.
   - *Smoothing.* As noted in Chapter 3, inventories provide a means of smoothing out an irregular demand pattern.
   - *Logistics.* System constraints may require holding inventories.
   - *Control costs.* Holding inventory can lower the costs necessary to monitor a system. (For example, it may be less expensive to order monthly and hold the units than to order weekly and closely monitor orders and deliveries.)
3. *Characteristics of inventory systems.*
   - *Patterns of demand.* The two patterns are (a) constant versus variable, and (b) known versus uncertain.

- *Replenishment lead times.* The time that elapses between placement of an order (or initiation of production) until the order arrives (or is completed).
- *Review times.* Current inventory levels are checked at these points in time.
- *Treatment of excess demand.* When demand exceeds supply, excess demand may be either backlogged or lost.

4. *Relevant costs.*
   - *Holding costs.* The opportunity cost of lost investment revenue, physical storage costs, insurance, breakage and pilferage, and obsolescence are all costs of holding inventory.
   - *Order costs.* These generally consist of two components: a fixed component and a variable component. The fixed component is incurred whenever a positive order is placed (or a production run is initiated), and the variable component is a unit cost paid for each unit ordered or produced.
   - *Penalty costs.* These are incurred when demand exceeds supply. In this case, excess demand may be back ordered (to be filled at a later time) or lost. Lost demand results in lost profit, and back orders require record keeping. In both cases, one risks losing customer goodwill.
5. *The basic EOQ model.* The EOQ model dates back to 1915 and forms the basis for all the inventory control models developed subsequently. It treats the basic trade-off between the fixed cost of ordering and the variable cost of holding. If h represents the holding cost per unit time and K the fixed cost of setup, then the order quantity that minimizes costs per unit time is $Q = \sqrt{2K\lambda / h}$, where λ is the rate of demand. This formula is very robust.
6. *The EOQ with finite production rate.* This is an extension of the basic EOQ model to account for items produced internally rather than ordered from an outside supplier—in which case the rate of production is finite rather than infinite as would be required in the simple EOQ model. We show that the optimal batch size for production now follows the formula $Q = \sqrt{2K\lambda / h'}$, where $h' = h(1 - \lambda/P)$ and P is the rate of production ($P > \lambda$).
7. *Quantity discounts.* We consider two types of quantity discounts: all-units and incremental discounts. In the case of all-units discounts, the discount is applied to all the units in the order. In the case of incremental discounts, the discount is applied to only the units above the break point. The all-units case is by far the most common in practice, but one does encounter incremental discounts in industry. In the case of all-units discounts, the optimization procedure requires searching for the lowest point on a broken annual cost curve. In the incremental discounts case, the annual cost curve is continuous but has discontinuous derivatives.
8. *Resource-constrained multiple product systems.* Consider a retail store that orders many different items but cannot exceed a fixed budget. If we optimize the order quantity of each item separately, then each item should be ordered according to its EOQ value. However, suppose doing so exceeds the budget. In Section 4.8, a model is developed that explicitly takes into account the budget constraint and adjusts the EOQ values accordingly. In most cases, the optimal solution subject to the budget constraint requires an iterative search of the Lagrange multiplier. However, when the condition $c_1/h_1 = c_2/h_2 = \ldots = c_n/h_n$ is met, the optimal order quantities are a simple scaling of the optimal EOQ values. Note that this problem is mathematically identical to one in which the constraint is on available space rather than available budget.
9. *EOQ models for production planning.* Suppose that n distinct products are produced on a single production line or machine. Assume we know the holding costs, order costs, demand rates, and production rates for each of the items. The goal is to determine the optimal sequence to produce the items and the optimal batch size for each of the items to meet the demand and minimize costs. Simply setting a batch size for each item equal to its EOQ value (that is, optimal lot size with a finite production rate), is likely to be suboptimal because it is likely to result in stock-outs.

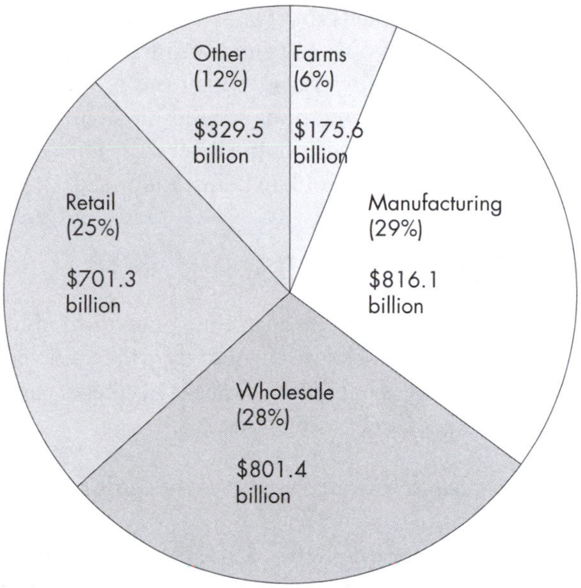

Figure 4–1 Total investment in inventories in the U.S. economy (first quarter 2019)

The investment in inventories in the United States is enormous. In the first quarter of 2019, the investment in inventories was almost $2.83 trillion (Bureau of Economic Analysis, 2019). Figure 4–1 shows investment in inventories by sectors of the economy. The inventory models we will be discussing in this chapter and in Chapter 5 can be applied to all the sectors of the economy shown in Figure 4–1 but are most applicable to the manufacturing, wholesale, and retail sectors, which compose approximately 82 percent of the total. The investment in inventories accounts for between 20 and 25 percent of the total annual GNP. Clearly there is enormous potential for improving the efficiency of our economy by intelligently controlling inventories. Companies that use scientific inventory control methods have a significant competitive advantage in the marketplace.

A major portion of this text is devoted to presenting and analyzing several mathematical models that can assist with controlling the replenishment of inventories. Both Chapters 4 and 5 assume that the demand for the item is external to the system. In most cases, this means that the inventory is being acquired or produced to meet the needs of a customer. In a manufacturing environment, however, demands for certain parts are the result of production schedules for higher-level assemblies; the production-lot-sizing decisions at one level of the system result in the demand patterns at other levels. The interaction of components, subassemblies, and final products plays an important role in determining future demand. Systems of this type are referred to as **materials requirements planning** (MRP) systems or dependent demand systems. MRP is treated in detail in Chapter 9.

Two questions succinctly address the issue of inventory management (1) When should an order be placed? And (2) How much should be ordered? The complexity of the resulting model depends upon the assumptions one makes about the various parameters of the system. The major distinction is between models that assume known demand (this chapter) and those that assume random demand (Chapter 5), although, as we will see, the form of the cost functions and the assumptions one makes about physical characteristics of the system also play an important role in determining the complexity of the resulting model.

In general, the models that we discuss can be used interchangeably to describe either replenishment from an outside vendor or internal production. This means that from the point of view of the model, inventory control and production planning are often synonymous. For example, the lot-sizing methods treated in Chapter 9 could just as well have been included in this chapter. The issue is not the label that is placed on a technique but whether it is being correctly applied to the problem being addressed.

4.1 TYPES OF INVENTORIES

When we consider inventories in the context of manufacturing and distribution, there is a natural classification scheme suggested by the value added from manufacturing or processing. (This certainly is not the only means of categorizing inventories, but it is the most natural one for manufacturing applications.)

1. *Raw materials.* **Raw materials** are the resources required in the production or processing activity of the firm.
2. *Components.* **Components** correspond to items that have not yet reached completion in the production process. Components are sometimes referred to as **subassemblies**.
3. *Work-in-process.* **Work-in-process** (WIP) is inventory either waiting in the system for processing or being processed. Work-in-process inventories include component inventories and may include some raw materials inventories as well. The level of work-in-process inventory is often used as a measure of the efficiency of a production scheduling system. The lean production approach, discussed in detail in Chapter 9, is aimed at reducing WIP to a minimum.
4. *Finished goods.* Also known as end items, **finished goods** are the final products of the production process. During production, value is added to the inventory at each level of the manufacturing operation, culminating with finished goods.

The appropriate label to place on inventory depends upon the context. For example, components for some operations might be the end products for others.

4.2 MOTIVATION FOR HOLDING INVENTORIES

The key motivations for holding inventories are as follows.

1. *Economies of scale.* Consider a company that produces a line of similar items, such as air filters for automobiles. Each production run of a particular size of filter requires that the production line be reconfigured and the machines recalibrated. Because the company must invest substantial time and money in setting up to produce each filter size, enough filters should be produced at each setup to justify this cost. This means that it could be economical to produce a relatively large number of items in each production run and store them for future use. This allows the firm to amortize fixed setup costs over a larger number of units. (This argument assumes that the setup cost is a fixed constant. In some circumstances it can be reduced, thus justifying smaller lot sizes. This forms the basis of the lean philosophy discussed in detail in Chapter 9.) This is one of many examples of **economies of scale**, which was described in more detail in Chapter 1.
2. *Uncertainties.* Uncertainty often plays a major role in motivating a firm to store inventories. Uncertainty of external demand is the most important factor. For example, a retailer

stocks different items so that he or she can be responsive to consumer preferences. If a customer requests an item that is not available immediately, it is likely that the customer will go elsewhere. Worse, the customer may never return. Inventory provides a buffer against the uncertainty of demand.

Other uncertainties provide a motivation for holding inventories as well. One is the uncertainty of the lead time. Lead time is defined as the amount of time that elapses from the point that an order is placed until it arrives. In the production planning context, interpret the lead time as the time required to produce the item. Even when future demand can be predicted accurately, the company needs to hold buffer stocks to ensure a smooth flow of production or continued sales when replenishment lead times are uncertain.

A third significant source of uncertainty is the supply. From October 1973 to March 1974, OPEC placed an embargo on oil for some nations; for others, prices quadrupled. Two industries that relied (and continue to rely) heavily on oil and gasoline are the electric utilities and the airlines. Firms in these and other industries risked having to curtail operations because of fuel shortages.

Additional uncertainties that could motivate a firm to store inventory include the uncertainty in the supply of labor, the price of resources, and the cost of capital.

3. *Speculation.* If the value of an item or natural resource is expected to increase, it may be more economical to purchase large quantities at current prices and store the items for future use than to pay the higher prices at a future date. In the early 1970s, for example, the Westinghouse Corporation sustained severe losses on its contracts to build nuclear plants for several electric utility companies because it guaranteed to supply the uranium necessary to operate the plants at a fixed price. Unfortunately for Westinghouse, the price of the uranium skyrocketed between the time the contracts were signed and the time the plants were built.

   Other industries require large quantities of costly commodities that have experienced considerable fluctuation in price. The escalating trade war with China in 2019 endangered the shipment of rare earth materials produced primarily in China. Those raw materials are crucial for electronics—everything from smartphones to electric cars to wind turbines to missiles. Fearing a ban on shipments to the United States, some companies chose to purchase and store large quantities of the materials to insure availability and/or to realize substantial savings if prices skyrocketed.

   The speculative motive also can be a factor for a firm facing the possibility of a labor strike. The cost of production could increase significantly when there is a severe shortage of labor.
4. *Transportation.* In-transit or **pipeline** inventories exist because transportation times are positive. When transportation times are long, as is the case when transporting oil from the Middle East to the United States, the investment in pipeline inventories can be substantial. One of the disadvantages of producing overseas is the increased transportation time—hence the increase in pipeline inventories. This factor has been instrumental in motivating some firms to establish production operations domestically.
5. *Smoothing.* Changes in the demand pattern for a product can be deterministic or random. Seasonality is an example of a deterministic variation, while unanticipated changes in economic conditions can result in random variation. Producing and storing inventory in anticipation of peak demand can help to alleviate the disruptions caused by changing production rates and workforce levels. Smoothing costs and planning for anticipated swings in the demand were considered in the aggregate planning models in Chapter 3.

6. *Logistics.* We use the term **logistics** to describe reasons for holding inventory that differ from those already outlined. Certain constraints can arise in the purchasing, production, or distribution of items that force the system to maintain inventory. One such case is an item that must be purchased in minimum quantities. Another is the logistics of manufacture; it is virtually impossible to reduce all inventories to zero and expect any continuity in a manufacturing process.
7. *Control costs.* An important issue, and one that often is overlooked, is the cost of maintaining the inventory control system. A system in which more inventory is carried does not require the same level of control as one in which inventory levels are kept to a bare minimum. It can be less costly to the firm in the long run to maintain large inventories of inexpensive items than to expend worker time to keep detailed records for these items. Even though control costs could be a major factor in determining the suitability of a particular technique or system, they are rarely factored into the types of inventory models we will be discussing.

4.3 CHARACTERISTICS OF INVENTORY SYSTEMS

The key characteristics of any inventory system are as follows.

1. *Demand.* The assumptions one makes about the patterns and characteristics of demand are often the most significant factors in determining the complexity of the resulting control model.
   a. *Constant versus variable.* The simplest inventory models assume that the rate of demand is a constant. The economic order quantity (EOQ) model and its extensions are based on this assumption. Variable demand arises in a variety of contexts, including aggregate planning (Chapter 3) and materials requirements planning (Chapter 9).
   b. *Known versus random.* It is possible for demand to be constant in expectation but still be random. Synonyms for random are **uncertain** and **stochastic**. Virtually all stochastic demand models assume that the average demand rate is constant. Random demand models are generally both more realistic and more complex than their deterministic counterparts.
2. *Lead time.* If items are ordered from the outside, the **lead time** is defined as the amount of time that elapses from the instant that an order is placed until it arrives. If items are produced internally, lead time is the amount of time required to produce a batch of items. We will use the Greek letter τ to represent lead time, which is expressed in the same units of time as demand. That is, if demand is expressed in units per year, then lead time should be expressed in years.
3. *Review time.* In some systems the current level of inventory is known at all times. This is an accurate assumption when demand transactions are recorded as they occur. One example of a system in which inventory levels are known at all times is a modern supermarket. As an item is passed through the checkout scanner, the transaction is recorded in the storewide inventory database, and the inventory level is decreased by one unit. We will refer to this case as **continuous review**. In the other case, referred to as **periodic review**, inventory levels are known only at discrete points in time. An example of periodic review is a small kiosk in which a physical count of stock on hand is required to determine the current levels of inventory.
4. *Excess demand.* Another important distinguishing characteristic is how the system reacts to excess demand (that is, demand that cannot be filled immediately from stock). The

two most common assumptions are that excess demand is either **back ordered** (held over to be satisfied at a future time) or **lost** (generally satisfied from outside the system). Other possibilities include partial back ordering (part of the demand is back ordered, and part of the demand is lost) or customer impatience (if the customer's order is not filled within a fixed amount of time, he or she cancels). The vast majority of inventory models, especially the ones that are used in practice, assume full back ordering of excess demand.

5. *Changing inventory.* In some cases, the inventory undergoes changes over time that may affect its utility. Some items have a limited shelf life, such as food; others may become obsolete, such as automotive spare parts. Mathematical models that incorporate the effects of perishability or obsolescence are generally quite complex and beyond the scope of this text. A brief discussion can be found in Section 5.8.

4.4 RELEVANT COSTS

Because we are interested in optimizing the inventory system, we must determine an appropriate optimization or performance criterion. Virtually all inventory models use cost minimization as the optimization criterion. An alternative performance criterion might be profit maximization. However, cost minimization and profit maximization are essentially equivalent criteria for most inventory control problems. Although different systems have different characteristics, virtually all inventory costs can be placed into one of three categories: holding cost, order cost, or penalty cost. We discuss each in turn.

Holding Cost

The **holding cost,** also known as the carrying cost or the inventory cost, is the sum of all costs that are proportional to the amount of inventory physically on hand at any point in time. The components of the holding cost include a variety of seemingly unrelated items. Some of these are:

- cost of providing the physical space to store the items;
- taxes and insurance;
- breakage, spoilage, deterioration, and obsolescence;
- opportunity cost of alternative investment.

The last item often turns out to be the most significant in computing holding costs for most applications. Inventory and cash are in some sense equivalent. Capital must be invested to either purchase or produce inventory; decreasing inventory levels results in increased capital. This capital could be invested by the company either internally (in its own operation) or externally.

What is the interest rate that could be earned on this capital? You and I can place our money in a simple savings account with an interest rate of 1 percent, or possibly a long-term certificate of deposit with a return of maybe 2.5 percent. We could earn somewhat more by investing in high-yield bond funds or buying short-term industrial paper or second deeds of trust.

In general, however, most companies must earn higher rates of return on their investments than do individuals in order to remain profitable. The value of the interest rate that corresponds to the opportunity cost of alternative investment is related to (but not the same as) a number of standard accounting measures, including the internal rate of return, the return on assets, and the hurdle rate (the minimum rate that would make an investment attractive to the firm). Finding the right interest rate for the opportunity cost of alternative investment is very difficult. Its value is estimated by the firm's accounting department and is usually an amalgam of the accounting measures listed earlier. For convenience, we will use the term **cost of capital** to refer to this component of the holding cost. We may think of the holding cost as an aggregated interest rate comprised of the four components we listed. For example,

$$\begin{aligned} 28\% &= \text{cost of capital} \\ 2\% &= \text{taxes and insurance} \\ 6\% &= \text{cost of storage} \\ 1\% &= \text{breakage and spoilage} \\ \hline 37\% &= \text{total interest charge.} \end{aligned}$$

This would be interpreted as follows: We would assess a charge of 37 cents for every dollar that we have invested in inventory during a one-year period. However, as we generally measure inventory in units rather than in dollars, it is convenient to express the holding cost in terms of dollars per unit per year rather than dollars per dollar per year. Let c be the dollar value of one unit of inventory, I be the annual interest rate, and h be the holding cost in terms of dollars per unit per year. Then we have the relationship

$$h = Ic.$$

Hence, in this example, an item valued at \$180 would have an annual holding cost of $h = (0.37)(\$180) = \$66.60$. If we held 300 of these items for five years, the total holding cost over the five years would be

$$(5)(300)(66.60) = \$99{,}900$$

This example raises an interesting question. Suppose that during the five-year period the inventory level did not stay fixed at 300 but varied on a continuous basis. We would expect inventory levels to change over time. Inventory levels decrease when items are used to satisfy demand and increase when units are produced or new orders arrive. How would the holding cost be computed in such a case? In particular, suppose the inventory level $I(t)$ during some interval (t_1, t_2) behaves as in Figure 4–2.

The holding cost incurred at any point in time is proportional to the inventory level at that point in time. In general, the total holding cost incurred from a time t_1 to a time t_2 is h multiplied by the area under the curve described by $I(t)$. The *average* inventory level during the period (t_1, t_2) is the area under the curve divided by $t_2 - t_1$. For the cases considered in this chapter, simple geometry can be used to find the area under the inventory level curve. When $I(t)$ is described by a straight line, its average value is obvious. In cases such as that pictured in Figure 4–2, in which the curve of $I(t)$ is complex, the average inventory level would be determined by computing the integral of $I(t)$ over the interval (t_1, t_2) and dividing by $t_2 - t_1$.

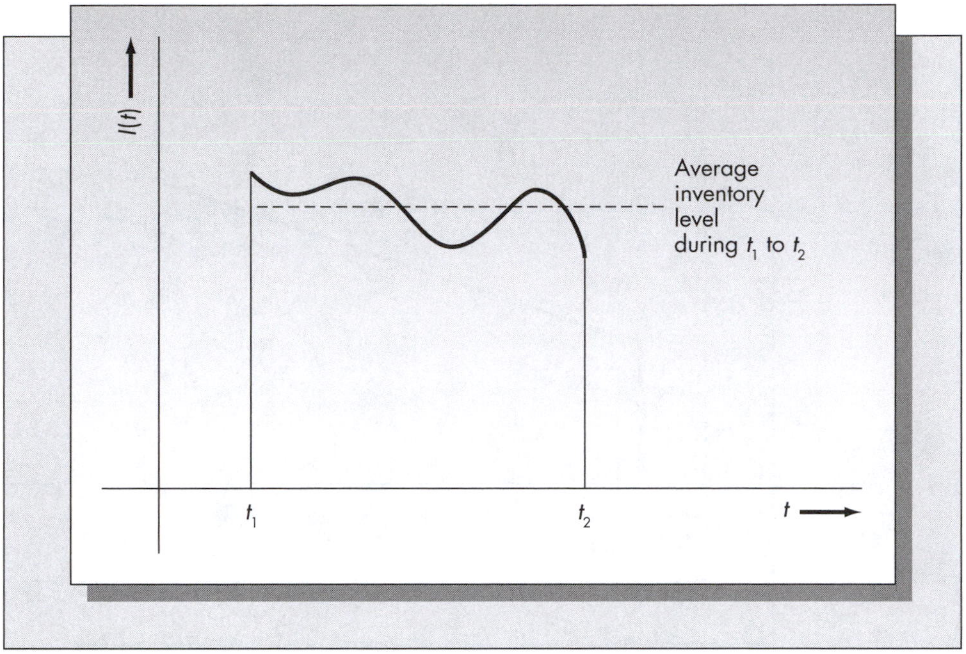

Figure 4–2 Inventory as a function of time

Order Cost

The holding cost includes all those costs that are proportional to the amount of inventory on hand, whereas the **order cost** depends on the amount of inventory that is ordered or produced.

In most applications, the order cost has two components: a fixed and a variable component. The fixed cost, K, is incurred independent of the size of the order as long as it is not zero. The variable cost, c, is incurred on a per-unit basis. We also refer to K as the setup cost and c as the proportional order cost. Define $C(x)$ as the cost of ordering (or producing) x units. It follows that

$$C(x) = \begin{cases} 0 & \text{if} \quad x = 0, \\ K + cx & \text{if} \quad x > 0. \end{cases}$$

The order cost function is pictured in Figure 4–3.

When estimating the setup cost, one should include *only* those costs that are relevant to the current ordering decision. For example, the cost of maintaining the purchasing department of the company is *not* relevant to daily ordering decisions and should not be factored into the estimation of the setup cost. This is an overhead cost that is independent of the decision of whether or not an order should be placed. The appropriate costs comprising K would be the bookkeeping expense associated with the order, the fixed costs independent of the size of the order that might be required by the vendor, costs of order generation and receiving, and shipping and handling costs.

Penalty Cost

The **penalty cost,** also known as the shortage cost or the stock-out cost, is the cost of not having sufficient stock on hand to satisfy a demand *when it occurs.* This cost has a different

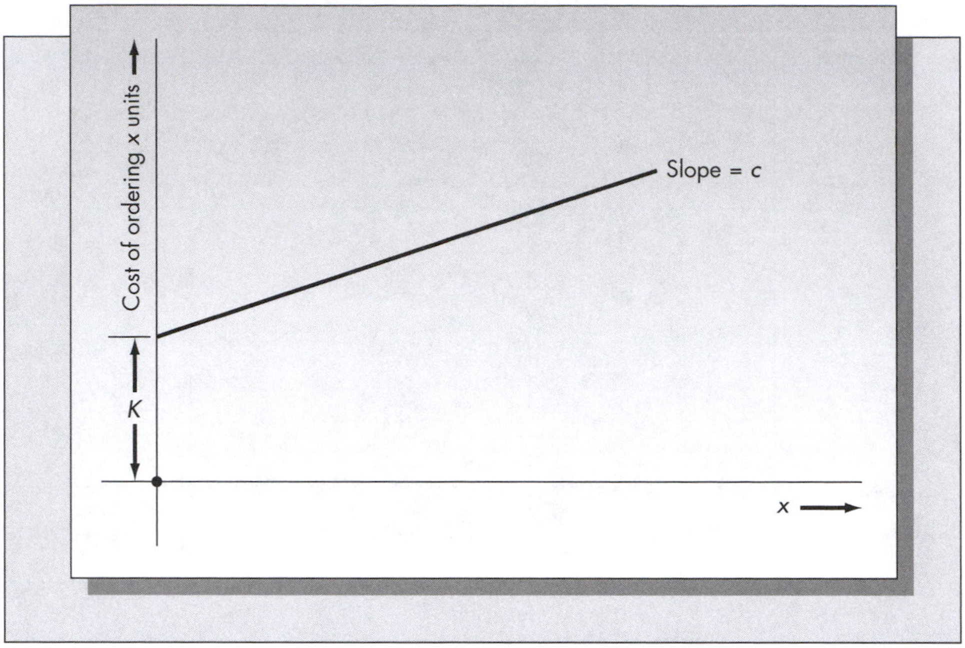

Figure 4–3 Order cost function

interpretation depending on whether excess demand is back ordered (orders that cannot be filled immediately are held on the books until the next shipment arrives) or lost (known as lost sales). In the back-order case, the penalty cost includes whatever bookkeeping and/or delay costs might be involved. In the lost-sales case, it includes the lost profit that would have been made from the sale. In either case, it would also include the **loss-of-goodwill** cost, which is a measure of customer satisfaction. Estimating the loss-of-goodwill component of the penalty cost can be very difficult in practice.

We use the symbol p to denote penalty cost and assume that p is charged on a per-unit basis. That is, each time a demand occurs that cannot be satisfied immediately, a cost p is incurred independent of how long it takes to eventually fill the demand. An alternative means of accounting for shortages is to charge the penalty cost on a per-unit-per-unit-time basis (as we did with the holding cost). This approach is appropriate if the time that a back order stays on the books is important—for example, if a back order results in stopping a production line because of the unavailability of a part. The models considered in this chapter assume that penalty costs are charged on a per-unit basis only. Penalty cost models are not considered in this chapter. Penalty costs are included in Chapter 5, but models incorporating a time-weighted penalty cost are not.

We present those inventory models that have had the greatest impact in the user community. Many of the techniques discussed in both this chapter and in Chapter 5 form the basis for commercial inventory control systems or in-house systems. In most cases, the models are simple enough that optimal operating policies can be calculated by hand, but they are often complex enough to capture the essential trade-offs in inventory management.

PROBLEMS FOR SECTIONS 4.1–4.4

1. What are the two questions that inventory control addresses?
2. Discuss the cost penalties incurred by a firm that holds too much inventory and one that holds too little inventory.

3. ABC, Inc., produces a line of touring bicycles. Specifically, what are the four types of inventories (raw materials, components, work-in-process, and finished goods) that would arise in the production of this item?
4. I Carry rents trucks for moving and hauling. Each truck costs the company an average of $8,000, and the inventory of trucks varies monthly depending on the number that are rented out. During the first eight months of last year, I Carry had the following ending inventory of trucks on hand.

Month	Number of Trucks	Month	Number of Trucks
January	26	May	13
February	38	June	9
March	31	July	16
April	22	August	5

I Carry uses a 20 percent annual interest rate to represent the cost of capital. Yearly costs of storage amount to 3 percent of the value of each truck, and the cost of liability insurance is 2 percent.

a. Determine the total handling cost incurred by I Carry during the period January to August. Assume for the purposes of your calculation that the holding cost incurred in a month is proportional to the inventory on hand at the end of the month.
b. Assuming that these eight months are representative, estimate the average annual cost of holding trucks.

5. Stationery Supplies is considering installing an inventory control system in its store in Provo, Utah. The store carries about 1,400 different inventory items and has annual gross sales of about $80,000. The inventory control system would cost $12,500 to install and about $2,000 per year in additional supplies, time, and maintenance. If the savings to the store from the system can be represented as a fixed percentage of annual sales, what would that percentage have to be in order for the system to pay for itself in five years or less?
6. List and discuss all the uncertainties that would motivate Stationery Supplies to maintain inventories of its 1,400 items.
7. Stationery Supplies orders plastic erasers from a company in Nürnberg, Germany. It takes six weeks to ship the erasers from Germany to Utah. Stationery Supplies maintains a standing order of 200 erasers every six months (shipped on the first of January and the first of July).

a. Assuming the ordering policy the store is using does not result in large buildups of inventory or long-term stock-outs, what is the annual demand for erasers?
b. Draw a graph of the pipeline inventory (that is, the inventory ordered but not received) of the erasers during one year. What is the average pipeline inventory of erasers during the year?
c. Express the replenishment lead time in years and multiply the annual demand you obtained in part (a) by the lead time. What do you notice about the result that you obtain?

8. Penalty costs can be assessed only against the number of units of demand that cannot be satisfied or against the number of units weighted by the amount of time that an order stays on the books. Consider the following history of supply and demand transactions for a particular part.

Month	Number of Items Received	Demand during Month
January	200	520
February	175	1,640
March	750	670
April	950	425
May	500	280
June	2,050	550

Assume that starting inventory at the beginning of January is 480 units.

a. Determine the ending inventory each month. Assume that excess demands are back ordered.

b. Assume that each time a unit is demanded that cannot be supplied immediately, a one-time charge of $10 is made. Determine the stock-out cost incurred during the six months (1) if excess demand at the end of each month is lost, and (2) if excess demand at the end of each month is back ordered.

c. Suppose that each stock-out costs $10 per unit per month that the demand remains unfilled. If demands are filled on a first-come, first-served basis, what is the total stock-out cost incurred during the six months using this type of cost criterion? (Assume that the demand occurs at the beginning of the month for purposes of your calculation.) Notice that you must assume that excess demands are back ordered for this analysis to make any sense.

d. Discuss under what circumstances the cost criterion used in part (b) might be appropriate and under what circumstances the cost criterion used in part (c) might be appropriate.

9. HAL Ltd. produces a line of high-capacity disk drives for mainframe computers. The housings for the drives are produced in Hamilton, Ontario, and shipped to the main plant in Toronto. HAL uses the drive housings at a fairly steady rate of 720 per year. Suppose that the housings are shipped in trucks that can hold 40 housings at one time. It is estimated that the fixed cost of loading the housings onto the truck and unloading them on the other end is $300 for shipments of 120 or fewer housings (i.e., three or fewer truckloads). Each trip made by a single truck costs the company $160 in driver time, gasoline, oil, insurance, and wear and tear on the truck.

   a. Compute the annual costs of transportation and loading and unloading the housings for the following policies: (1) shipping one truck per week, (2) shipping one full truckload as often as needed, and (3) shipping three full truckloads as often as needed.

   b. For what reasons might the option with the highest annual cost be more desirable from a systems point of view than the policy having the lowest annual cost?

4.5 THE EOQ MODEL

The **EOQ model (economic order quantity model)** is the simplest and most fundamental of all inventory models. It describes the important trade-off between fixed order costs and holding costs; it is the basis for the analysis of more complex systems.

The Basic Model

The basic EOQ model has four key assumptions as follows.

1. The demand rate is known and is a constant λ units per unit time. (The unit of time may be days, weeks, months, etc. In what follows we assume that the default unit of time is a year. However, the analysis is valid for other time units as long as all relevant variables are expressed in the same units.)
2. Shortages are not permitted.
3. There is no order lead time. (This assumption will be relaxed.)
4. The costs include
   a. Setup cost at K per positive order placed.
   b. Proportional order cost at c per unit ordered.
   c. Holding cost at h per unit held per unit time.

Assume with no loss in generality that the on-hand inventory at time zero is zero. Shortages are not allowed, so we must place an order at time zero. Let Q be the size of the order. It follows that the on-hand inventory level increases instantaneously from zero to Q at time $t = 0$.

Consider the next time an order is to be placed. At this time, either the inventory is positive, or it is again zero. A little reflection shows that we can reduce the holding costs by waiting until the inventory level drops to zero before ordering again. At the instant that on-hand inventory equals zero, the situation looks exactly the same as it did at time $t = 0$. If it was optimal to place an order for Q units at that time, then it is still optimal to order Q units. It follows that the function that describes the changes in stock levels over time is the familiar sawtooth pattern of Figure 4–4.

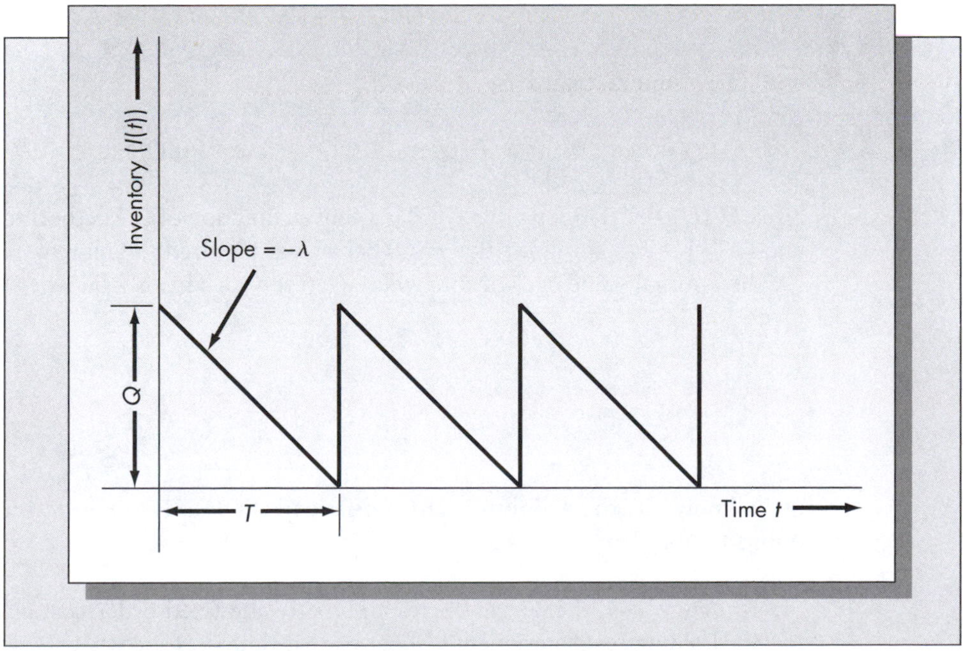

Figure 4–4 Inventory levels for the EOQ model

The objective is to choose Q to minimize the average cost per unit time. Unless otherwise stated, we will assume that a unit of time is a year, so that we minimize the average annual cost. Other units of time, such as days, weeks, or months, are also acceptable, as long as all time-related variables are expressed in the same units. One might think that the appropriate optimization criterion would be to minimize the *total* cost in a cycle. However, this ignores the fact that the cycle length itself is a function of Q and must be explicitly included in the formulation.

Next, we derive an expression for the average annual cost as a function of the lot size Q. In each cycle, the total fixed plus proportional order cost is $C(Q) = K + cQ$. In order to obtain the order cost per unit time, we divide by the cycle length T. As Q units are consumed each cycle at a rate λ, it follows that $T = Q/\lambda$. This result also can be obtained by noting that the slope of the inventory curve pictured in Figure 4–4, $-\lambda$, equals the ratio $-Q/T$.

Consider the holding cost. Because the inventory level decreases linearly from Q to 0 each cycle, the average inventory level during one order cycle is $Q/2$. Because all cycles are identical, the average inventory level over a time horizon composed of many cycles is also $Q/2$. It follows that the average annual cost, say $G(Q)$, is given by

$$\begin{aligned} G(Q) &= \frac{K + cQ}{T} + \frac{hQ}{2} = \frac{K + cQ}{Q/\lambda} + \frac{hQ}{2} \\ &= \frac{K\lambda}{Q} + \lambda c + \frac{hQ}{2}. \end{aligned}$$

The three terms composing $G(Q)$ are annual setup cost, annual purchase cost, and annual holding cost, respectively.

We now wish to find Q to minimize $G(Q)$. Consider the shape of the curve $G(Q)$. We have that the first derivative is

$$G'(Q) = -K\lambda/Q^2 + h/2,$$

and the second derivative is

$$G''(Q) = 2K\lambda/Q^3 > 0 \qquad \text{for } Q > 0.$$

Since $G''(Q) > 0$, it follows that $G(Q)$ is a convex function of Q. Furthermore, since $G'(0) = -\infty$ and $G'(\infty) = h/2$, it follows that $G(Q)$ behaves as pictured in Figure 4–5.

The optimal value of Q occurs where $G'(Q) = 0$. This is true when $Q^2 = 2K\lambda/h$, which gives

$$Q^* = \sqrt{\frac{2K\lambda}{h}}.$$

Q^* is known as the **economic order quantity** (EOQ). There are a number of interesting points to note.

1. In Figure 4–5, the curves corresponding to the fixed order cost component $K\lambda/Q$ and the holding cost component $hQ/2$ also are included. Notice that Q^* is the value of Q where the two curves intersect. (If you equate $hQ/2$ and $K\lambda/Q$ and solve for Q, you will obtain the

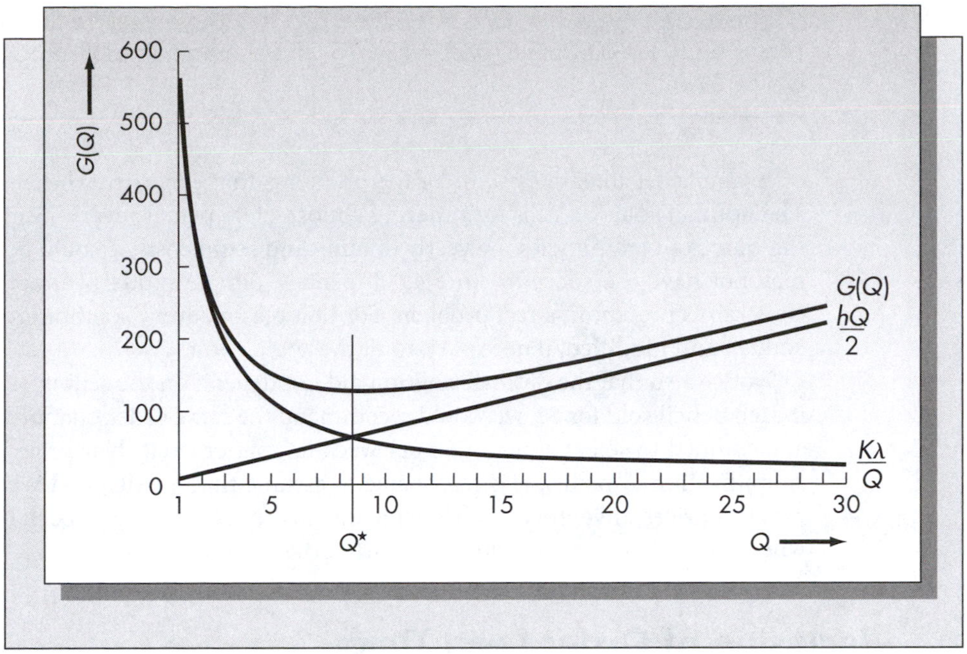

Figure 4–5 The average annual cost function $G(Q)$

EOQ formula.) In general, the minimum of the sum of two functions will *not* occur at the intersection of the two functions. It is an interesting coincidence that it does in this case.

2. Notice that the proportional order cost component, c, does not appear explicitly in the expression for Q^*. This is because the term λc appearing in the definition of $G(Q)$ is independent of Q. As all feasible policies must replenish inventory at the rate of demand, the proportional order cost incurred per unit time is λc independent of Q. Because λc is a constant, we generally ignore it when computing average costs. Notice that c *does* affect the value of Q^* indirectly, as h appears in the EOQ formula and $h = Ic$.

EXAMPLE 4.1

Number 2 pencils at the campus bookstore are sold at a fairly steady rate of 60 per week. The pencils cost the bookstore 2 cents each and sell for 15 cents each. It costs the bookstore $12 to initiate an order, and holding costs are based on an annual interest rate of 25 percent. Determine the optimal number of pencils for the bookstore to purchase and the time between placement of orders. What are the yearly holding and setup costs for this item?

Solution

First, we convert the demand to a yearly rate so that it is consistent with the interest charge, which is given on an annual basis. (Alternatively, we could have converted the annual interest rate to a weekly interest rate.) The annual demand rate is $\lambda = (60)(52) = 3{,}120$. The holding cost h is the product of the annual interest rate and the variable cost of the item. Hence, $h = (0.25)(0.02) = 0.005$. Substituting into the EOQ formula, we obtain

$$Q^* = \sqrt{\frac{2K\lambda}{h}} = \sqrt{\frac{(2)(12)(3{,}120)}{0.005}} = 3{,}870.$$

The cycle time is $T = Q/\lambda = 3{,}870/3{,}120 = 1.24$ years. The average annual holding cost is $h(Q/2) = 0.005(3{,}870/2) = \$9.675$. The average annual setup cost is $K\lambda/Q$, which is also \$9.675.

Example 4.1 illustrates some of the problems that can arise when using simple models. The optimal solution calls for ordering almost 4,000 pencils every 15 months. Even though this value of Q minimizes the yearly holding and setup costs, it could be infeasible: the store may not have the space to store 4,000 pencils. Simple models cannot account for all the constraints present in a real problem. For that reason, every solution must be considered in context and modified, if necessary, to fit the application.

Notice also that the optimal solution did not depend on the selling price of 15 cents. Even if each pencil sold for \$2, we would recommend the same order quantity, because the pencils are assumed to sell at a rate of 60 per week no matter what their price. This is, of course, a simplification of reality. It is reasonable to assume that the demand is relatively stable for a range of prices. Inventory models explicitly incorporate selling price in the formulation only when the selling price is included as part of the optimization.

Inclusion of Order Lead Time

One of the assumptions made in our derivation of the EOQ model was that there was no order lead time. We now relax that assumption. Suppose in Example 4.1 that the pencils had to be ordered four months in advance. If we were to place the order exactly four months before the end of the cycle, the order would arrive at exactly the same point in time as in the zero lead time case. The optimal timing of order placement for Example 4.1 is shown in Figure 4–6.

Rather than say that an order should be placed so far in advance of the end of a cycle, it is more convenient to indicate reordering in terms of the on-hand inventory. Define

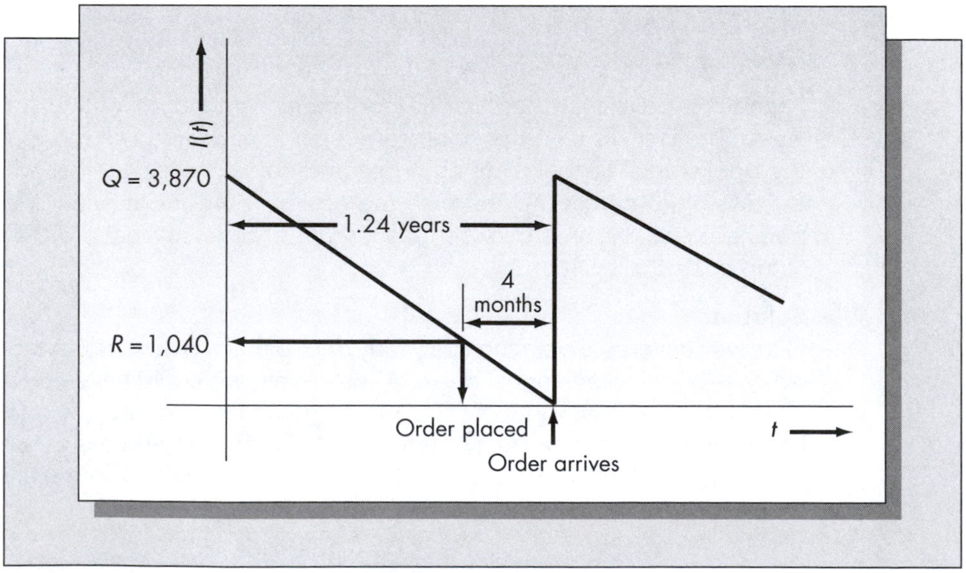

Figure 4–6 Reorder point calculation for Example 4.1

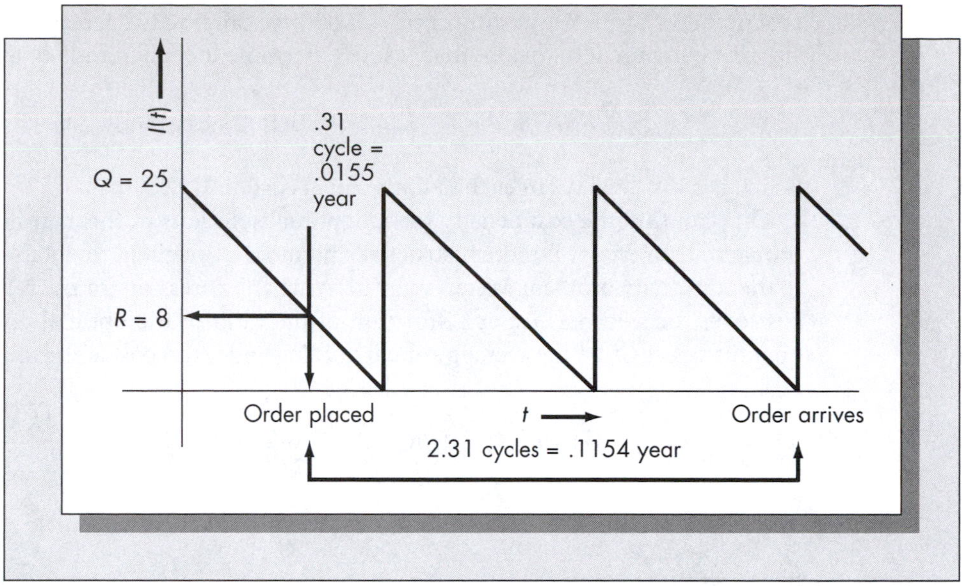

Figure 4–7 Reorder point calculation for lead times exceeding one cycle

R, the reorder point, as the level of on-hand inventory at the instant an order should be placed. From Figure 4–6, we see that R is the product of the lead time and the demand rate ($R = \lambda\tau$). For the example, $R = (3{,}120)(0.3333) = 1{,}040$. Notice that we converted the lead time to years before multiplying. *Always express all relevant variables in the same units of time.*

Determining the reorder point is more difficult when the lead time exceeds a cycle. Consider an item with an EOQ of 25, a demand rate of 500 units per year, and a lead time of six weeks. The cycle time is $T = 25/500 = 0.05$ year, or 2.6 weeks. Forming the ratio of τ/T, we obtain 2.31. This means that there are exactly 2.31 cycles in the lead time. Every order must be placed 2.31 cycles in advance (see Figure 4–7).

Notice that for the purpose of computing the reorder point, this is exactly the same as placing the order *0.31 cycle in advance.* This is true because the level of on-hand inventory is the same whether we are at a point 2.31 or 0.31 cycle before the arrival of an order. In this case, 0.31 cycle is 0.0155 year, thus giving a reorder point of $R = (0.0155)(500) = 7.75 \approx 8$. In general, when $\tau > T$, use the following procedures.

a. Form the ratio τ/T.
b. Consider only the fractional remainder of the ratio. Multiply this fractional remainder by the cycle length to convert back to years.
c. Multiply the result of step (b) by the demand rate to obtain the reorder point.

Sensitivity

In this part we examine the issue of how sensitive the annual cost function is to errors in the calculation of Q. Consider Example 4.1. Suppose that the bookstore orders pencils in batches of 1,000, rather than 3,870 as the optimal solution indicates. What additional cost is it incurring by using a suboptimal solution? To answer the question, we consider the average annual cost function $G(Q)$. By substituting $Q = 1{,}000$, we can find the average annual cost for this lot size and compare it to the optimal cost to determine the magnitude of the penalty. We have

$$G(Q) = K\lambda/Q + hQ/2 = (12)(3{,}120)/1{,}000 + (0.005)(1{,}000)/2 = \$39.94,$$

which is considerably larger than the optimal cost of \$19.35.

One can find the cost penalty for suboptimal solutions in this manner for any particular problem. However, it is more instructive and more convenient to obtain a universal solution to the sensitivity problem. We do so by deriving an expression for the ratio of the suboptimal cost over the optimal cost as a function of the ratio of the optimal and suboptimal order quantities. Let G^* be the average annual holding and setup cost at the optimal solution. Then

$$\begin{aligned} G^* = K\lambda / Q^* + hQ^*/2 &= \frac{K\lambda}{\sqrt{2k\lambda/h}} + \frac{h}{2}\sqrt{\frac{2K\lambda}{h}} = 2\sqrt{\frac{K\lambda h}{2}} \\ &= \sqrt{2K\lambda h}. \end{aligned}$$

It follows that for any Q,

$$\begin{aligned} \frac{G(Q)}{G^*} &= \frac{K\lambda/Q + hQ/2}{\sqrt{2K\lambda h}} \\ &= \frac{1}{2Q}\sqrt{\frac{2K\lambda}{h}} + \frac{Q}{2}\sqrt{\frac{h}{2K\lambda}} \\ &= \frac{Q^*}{2Q} + \frac{Q}{2Q^*} \\ &= \frac{1}{2}\left[\frac{Q^*}{Q} + \frac{Q}{Q^*}\right]. \end{aligned}$$

To see how one would use this result, consider using a suboptimal lot size in Example 4.1. The optimal solution was $Q^* = 3{,}870$, and we wished to evaluate the cost error of using $Q = 1{,}000$. Forming the ratio Q^*/Q gives 3.87. Hence, $G(Q)/G^* = (0.5)(3.87 + 1/3.87) = (0.5)(4.128) = 2.06$. This says that the average annual holding and setup cost with $Q = 1{,}000$ is 2.06 times the optimal average annual holding and setup cost.

In general, the cost function $G(Q)$ is relatively insensitive to errors in Q. For example, if Q is twice as large as it should be, Q/Q^* is 2 and G/G^* is 1.25. Hence, an error of 100 percent in the value of Q results in an error of only 25 percent in the annual holding and setup cost. Notice that you obtain the same result if Q is half Q^*, since $Q^*/Q = 2$. However, this does *not* imply that the average annual cost function is symmetric. In fact, suppose that the order quantity differed from the optimal by ΔQ units. A value of $Q = Q^* + \Delta Q$ would result in a *lower* average annual cost than a value of $Q = Q^* - \Delta Q$.

EOQ and Lean

Largely as a result of the success of Toyota's kanban system, a new philosophy has emerged about the role and importance of inventories in manufacturing environments. This philosophy, known as **lean**, says that excess work-in-process inventories are not desirable, and inventories should be reduced to the bare essentials. (We discuss the lean philosophy in more detail in Chapter 1 and the mechanics of kanban in Chapter 9.) EOQ is the result of traditional thinking about inventories and scale economies in economics. Are the EOQ and lean approaches at odds with each other?

Proponents argue that an essential part of implementing lean is reducing setup times, and hence setup costs. As setup costs decrease, traditional EOQ theory says that lot sizes should be reduced. In this sense, the two ways of thinking are compatible. However, there are times when they may not be. We believe that there is substantial value to the lean approach that may not be incorporated easily into a mathematical model. Quality problems can be identified and rectified before inventories of defective parts accumulate. Plants can be more flexible if they are not burdened with excess in-process inventories. Toyota's success with the lean approach, as evidenced by substantially lower inventory costs per car than are typical for U.S. auto manufacturers, is a testament to the value of lean.

However, we believe that every new approach must be incorporated carefully into the firm's business and not adopted blindly without evaluating its consequences and appropriateness. Lean, although an important development in material management, is not always the best approach. The principles underlying EOQ (and MRP, discussed in Chapter 9) are sound and should not be ignored. The following example illustrates this point.

EXAMPLE 4.2

The Rahway, New Jersey, plant of Metalcase, a manufacturer of office furniture, produces metal desks at a rate of 200 per month. Each desk requires 40 Phillips head metal screws purchased from a supplier in North Carolina. The screws cost 3 cents each. Fixed delivery charges and costs of receiving and storing shipments of the screws amount to about $100 per shipment, independent of the size of the shipment. The firm uses a 25 percent interest rate to determine holding costs. Metalcase would like to establish a standing order with the supplier and is considering several alternatives. What standing order size should they use?

Solution

First, we compute the EOQ. The annual demand for screws is

$$(200)(12)(40) = 96{,}000.$$

The annual holding cost per screw is (0.25)(0.03) = 0.0075. From the EOQ formula, the optimal lot size is

$$Q^* = \sqrt{\frac{(2)(100)(96{,}000)}{0.0075}} = 50{,}597.$$

Note that the cycle time is $T = Q/\lambda = 50{,}597/96{,}000 = 0.53$ year or about once every six months. Hence the optimal policy calls for replenishment of the screws about twice a year. A lean approach would be to order the screws as frequently as possible to minimize the inventory held at the plant. Implementing such an approach might suggest a policy of weekly deliveries. Such a policy makes little sense in this context, however. This policy would require 52 deliveries per year, incurring setup costs of $5,200 annually. The EOQ solution gives a total annual cost of both setups and holding of less than $400. For a low-value item such as this with high fixed-order costs, small lot sizes in accordance with lean are inappropriate. The point is that no single approach should be blindly adopted for all situations. The success of a method in one context does not ensure its appropriateness in all other contexts.

PROBLEMS FOR SECTION 4.5

10. A specialty coffeehouse sells Colombian coffee at a fairly steady rate of 280 pounds annually. The beans are purchased from a local supplier for $2.40 per pound. The coffeehouse estimates that it costs $45 in paperwork and labor to place an order for the coffee, and holding costs are based on a 20 percent annual interest rate.
    a. Determine the optimal order quantity for Colombian coffee.
    b. What is the time between placement of orders?
    c. What is the average annual cost of holding and setup due to this item?
    d. If replenishment lead time is three weeks, determine the reorder level based on the on-hand inventory.
11. For the situation described in problem 10, draw a graph of the amount of inventory on order. Using your graph, determine the average amount of inventory on order. Also compute the demand during the replenishment lead time. How do these two quantities differ?
12. A large automobile repair shop installs about 1,250 mufflers per year, 18 percent of which are for imported cars. All the imported-car mufflers are purchased from a single local supplier at a cost of $18.50 each. The shop uses a holding cost based on a 25 percent annual interest rate. The setup cost for placing an order is estimated to be $28.
    a. Determine the optimal number of imported-car mufflers the shop should purchase each time an order is placed and the time between placement of orders.
    b. If the replenishment lead time is six weeks, what is the reorder point based on the level of on-hand inventory?
    c. The current reorder policy is to buy imported-car mufflers only once a year. What are the additional holding and setup costs incurred by this policy?
13. Consider the coffeehouse discussed in problem 10. Suppose that its setup cost for ordering was really only $15. Determine the error made in calculating the annual cost of holding and setup incurred as a result of using the wrong value of *K*. (Note that this implies that its current order policy is suboptimal.)
14. A local machine shop buys hex nuts and molly screws from the same supplier. The hex nuts cost 15 cents each, and the molly screws cost 38 cents each. A setup cost of $100 is assumed for all orders. This includes the cost of tracking and receiving the orders. Holding costs are based on a 25 percent annual interest rate. The shop uses an average of 20,000 hex nuts and 14,000 molly screws annually.
    a. Determine the optimal size of the orders of hex nuts and molly screws and the optimal time between placement of orders of these two items.
    b. If both items are ordered and received simultaneously, the setup cost of $100 applies to the combined order. Compare the average annual cost of holding and setup if these items are ordered separately; if they are both ordered when the hex nuts would normally be ordered; and if they are both ordered when the molly screws would normally be ordered.
15. David's Delicatessen purchases Hebrew National salamis regularly to satisfy a growing demand for the salamis in Silicon Valley. The owner, David Gold, estimates that the demand for the salamis holds fairly steady at 175 per month. The salamis cost Gold $1.85 each. The fixed cost of calling his brother in New York and having the salamis flown in is $200. It takes three weeks to receive an order. Gold's accountant, Irving Wu, recommends an annual cost of capital of 22 percent, a cost of shelf space of 3 percent of the value of the item, and a cost of 2 percent of the value for taxes and insurance.
    a. How many salamis should Gold purchase and how often should he order them?
    b. How many salamis should Gold have on hand when he phones his brother to send another shipment?

c. Suppose that the salamis sell for \$3 each. Are these salamis a profitable item for Gold? If so, what annual profit can he expect to realize from this item? (Assume that he operates the system optimally.)
d. If the salamis have a shelf life of only 4 weeks, what is the trouble with the policy that you derived in part (a)? What policy would Gold have to use in that case? Is the item still profitable?

16. In view of the results derived in the section on sensitivity analysis, discuss the following quotation of an inventory control manager: "If my lot sizes are going to be off the mark, I'd rather miss on the high side than on the low side."

4.6 EXTENSION TO A FINITE PRODUCTION RATE

An implicit assumption of the simple EOQ model is that the items are obtained from an outside supplier. When that is the case, it is reasonable to assume that the entire lot is delivered at the same time. However, if we wish to use the EOQ formula when the units are produced internally, then we are effectively assuming that the production rate is infinite. When the production rate is much larger than the demand rate, this assumption is probably satisfactory as an approximation. However, if the rate of production is comparable to the rate of demand, the simple EOQ formula will lead to incorrect results.

Assume that items are produced at a rate P during a production run. We require that $P > \lambda$ for feasibility. All other assumptions will be identical to those made in the derivation of the simple EOQ. When units are produced internally, the curve describing inventory levels as a function of time is slightly different from the sawtooth pattern of Figure 4–4. The change in the inventory level over time for the finite production rate case is shown in Figure 4–8.

Let Q be the size of each production run. Let T, the cycle length, be the time between successive production startups. Write $T = T_1 + T_2$, where T_1 is uptime (production time) and T_2 is downtime. Note that the maximum level of on-hand inventory during a cycle is *not* Q.

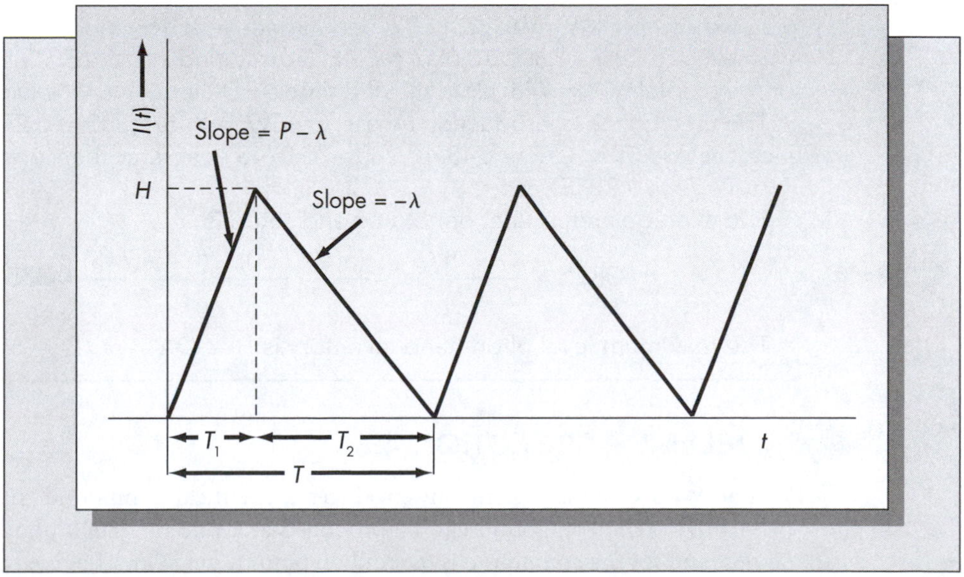

Figure 4–8 Inventory levels for finite production rate model

The number of units consumed each cycle is λT, which must be the same as the number of units produced each cycle, which is simply Q. It follows that $Q = \lambda T$, or $T = Q/\lambda$. Define H as the maximum level of on-hand inventory. As items are produced at a rate P for a time T_1, it follows that $Q = PT_1$, or $T_1 = Q/P$. From Figure 4–8 we see that $H/T_1 = P - \lambda$. This follows from the definition of the slope as the rise over the run. Substituting $T_1 = Q/P$ and solving for H gives $H = Q(1 - \lambda/P)$.

We now determine an expression for the average annual cost function. Because the average inventory level is $H/2$, it follows that

$$G(Q) = \frac{K}{T} + \frac{hH}{2} = \frac{K\lambda}{Q} + \frac{hQ}{2}(1 - \lambda / P).$$

Notice that if we define $h' = h(1 - \lambda/P)$, then this $G(Q)$ is identical to that of the infinite production rate case with h' substituted for h. It follows that

$$Q^* = \sqrt{\frac{2K\lambda}{h'}}.$$

EXAMPLE 4.3

A local company produces an erasable programmable read-only memory (EPROM) for several industrial clients. It has experienced a relatively flat demand of 2,500 units per year for the product. The EPROM is produced at a rate of 10,000 units per year. The accounting department has estimated that it costs \$50 to initiate a production run, each unit costs the company \$2 to manufacture, and the cost of holding is based on a 30 percent annual interest rate. Determine the optimal size of a production run, the length of each production run, and the average annual cost of holding and setup. What is the maximum level of the on-hand inventory of the EPROMs?

Solution

First, we compute $h = (0.3)(2) = 0.6$ per unit per year. The modified holding cost is $h' = h(1 - \lambda/P) = (0.6)(1 - 2{,}500/10{,}000) = 0.45$. Substituting into the EOQ formula and using h' for h, we obtain $Q^* = 745$. Note that the simple EOQ equals 645, which is 14 percent less.

The time between production runs is $T = Q/\lambda = 745/2{,}500 = 0.298$ year. The uptime each cycle is $T_1 = Q/P = 745/10{,}000 = 0.0745$ year, and the downtime each cycle is $T_2 = T - T_1 = 0.2235$ year.

The average annual cost of holding and setup is

$$G(Q^*) = \frac{K\lambda}{Q^*} + \frac{h'Q^*}{2} = \frac{(50)(2{,}500)}{745} + \frac{(0.45)(745)}{2} = 335.41.$$

The maximum level of on-hand inventory is $H = Q^*(1 - \lambda / P) = 559$ units.

PROBLEMS FOR SECTION 4.6

17. The Wod Chemical Company produces a chemical compound that is used as a lawn fertilizer. The compound can be produced at a rate of 10,000 pounds per day. Annual demand for the compound is 0.6 million pounds per year. The fixed cost of setting up for a production run of the chemical is \$1,500, and the variable cost of production is \$3.50

per pound. The company uses an interest rate of 22 percent to account for the cost of capital, and the costs of storage and handling of the chemical amount to 12 percent of the value. Assume that there are 250 working days in a year.

a. What is the optimal size of the production run for this particular compound?
b. What proportion of each production cycle consists of uptime, and what proportion consists of downtime?
c. What is the average annual cost of holding and setup attributed to this item? If the compound sells for $3.90 per pound, what is the annual profit the company is realizing from this item?

18. Determine the batch size that would result in problem 17 if you assumed that the production rate was infinite. What is the additional average annual cost that would be incurred using this batch size?
19. HAL Ltd., discussed in problem 9, can produce the disk drive housings in the Hamilton, Ontario, plant at a rate of 150 housings per month. The housings cost HAL $85 each to produce, and the setup cost for beginning a production run is $700. Assume an annual interest rate of 28 percent for determining the holding cost.
    a. What is the optimal number of housings for HAL to produce in each production run?
    b. Find the time between initiation of production runs, the time devoted to production, and the downtime each production cycle.
    c. What is the maximum dollar investment in housings that HAL has at any point in time?
20. Filter Systems produces air filters for domestic and foreign cars. One filter, part number JJ39877, is supplied on an exclusive contract basis to Oil Changers at a constant 200 units monthly. Filter Systems can produce this filter at a rate of 50 per hour. Setup time to change the settings on the equipment is 1.5 hours. Worker time (including overhead) is charged at the rate of $55 per hour, and plant idle time during setups is estimated to cost the firm $100 per hour in lost profit.

    Filter Systems has established a 22 percent annual interest charge for determining holding cost. Each filter costs the company $2.50 to produce; they are sold for $5.50 each to Oil Changers. Assume 6-hour days, 20 working days per month, and 12 months per year for your calculations.
    a. How many JJ39877 filters should Filter Systems produce in each production run of this particular part to minimize annual holding and setup costs?
    b. Assuming that it produces the optimal number of filters in each run, what is the maximum level of on-hand inventory of these filters that the firm has at any point in time?
    c. What percentage of the working time does the company produce these particular filters, assuming that the policy in part (a) is used?

4.7 QUANTITY DISCOUNT MODELS

We have assumed up until this point that the cost c of each unit is independent of the size of the order. Often, however, the supplier is willing to charge less per unit for larger orders. The purpose of the discount is to encourage the customer to buy the product in larger batches. Such quantity discounts are common for many consumer goods.

Although many different types of discount schedules exist, there are two that seem to be the most popular: all-units and incremental. In each case we assume that there are one or more breakpoints defining changes in the unit cost. However, there are two possibilities: either the discount is applied to all the units in an order (all-units), or it is applied only to the additional units beyond the breakpoint (incremental). The all-units case is more common.

EXAMPLE 4.4

The Weighty Trash Bag Company has the following price schedule for its large trash can liners.

For orders of less than 500 bags, the company charges 30 cents per bag; for orders of 500 or more but fewer than 1,000 bags, it charges 29 cents per bag; and for orders of 1,000 or more, it charges 28 cents per bag. In this case the breakpoints occur at 500 and 1,000. The discount schedule is all-units because the discount is applied to all of the units in an order. The order cost function $C(Q)$ is defined as

$$C(Q) = \begin{cases} 0.30Q & \text{for} \quad 0 \le Q < 500, \\ 0.29Q & \text{for} \quad 500 \le Q < 1{,}000, \\ 0.28Q & \text{for} \quad 1{,}000 \le Q. \end{cases}$$

The function $C(Q)$ is pictured in Figure 4–9. In Figure 4–10, we consider the same breakpoints, but assume an incremental quantity discount schedule.

Note that the average cost per unit with an all-units schedule will be less than the average cost per unit with the corresponding incremental schedule.

The all-units schedule appears irrational in some respects. In Example 4.4, 499 bags would cost \$149.70, whereas 500 bags would cost \$145.00. Why would Weighty charge less for a larger order? One reason would be to provide an incentive for the purchaser to buy more. If you were considering buying 400 bags, you might choose to move up to the breakpoint to obtain the discount. Furthermore, it is possible that Weighty has stored its bags in lots of 100, so that its savings in handling costs might more than compensate for the lower total cost.

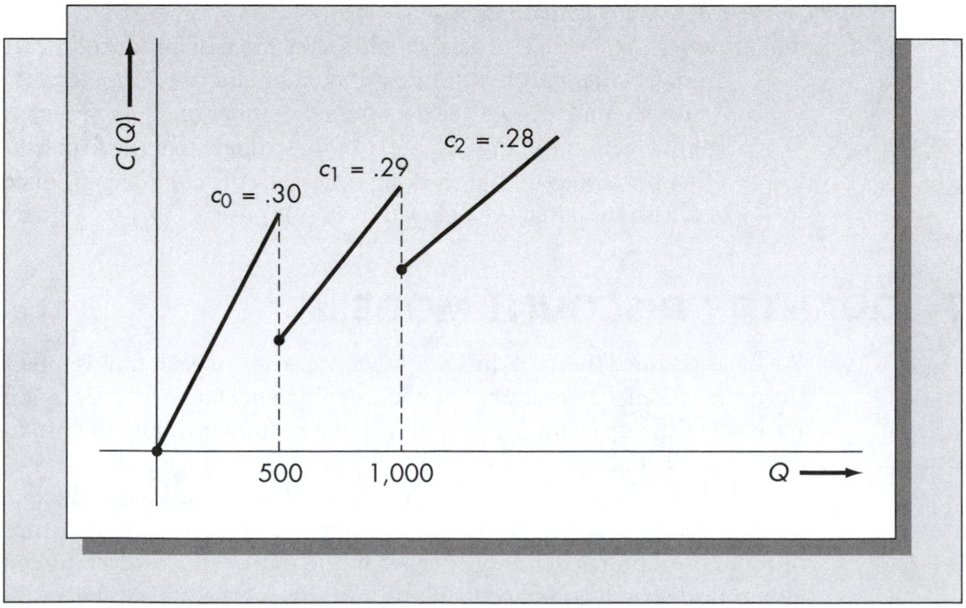

Figure 4–9 All-units discount order cost function

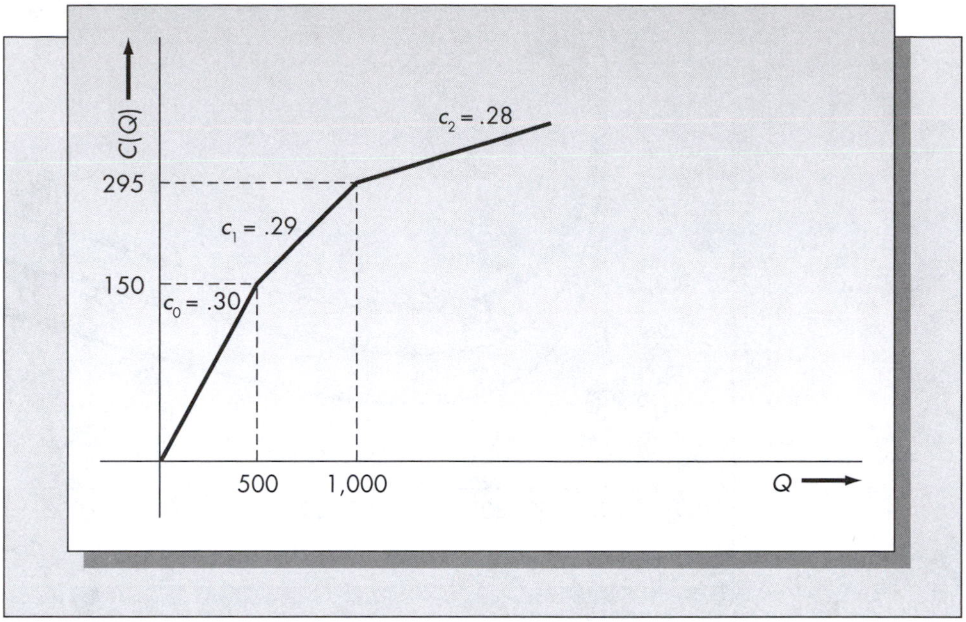

Figure 4–10 Incremental discount order cost function

Optimal Policy for All-Units Discount Schedule

We will illustrate the solution technique using Example 4.4. Assume that a company deciding how many bags to order from Weighty uses trash bags at a fairly constant rate of 600 per year. The accounting department estimates that the fixed cost of placing an order is \$8, and holding costs are based on a 20 percent annual interest rate. From Example 4.4, the respective unit costs are $c_0 = 0.30$, $c_1 = 0.29$, and $c_2 = 0.28$.

The first step toward finding a solution is to compute the EOQ values corresponding to each of the unit costs, which we will label $Q^{(0)}$, $Q^{(1)}$, and $Q^{(2)}$, respectively.

$$Q^{(0)} = \sqrt{\frac{2K\lambda}{Ic_0}} = \sqrt{\frac{(2)(8)(600)}{(0.2)(0.30)}} = 400,$$

$$Q^{(1)} = \sqrt{\frac{2K\lambda}{Ic_1}} = \sqrt{\frac{(2)(8)(600)}{(0.2)(0.29)}} = 406,$$

$$Q^{(2)} = \sqrt{\frac{2K\lambda}{Ic_2}} = \sqrt{\frac{(2)(8)(600)}{(0.2)(0.28)}} = 414.$$

We say that the EOQ value is realizable if it falls within the interval that corresponds to the unit cost used to compute it. Since $0 \leq 400 < 500$, $Q^{(0)}$ is realizable. However, neither $Q^{(1)}$ nor $Q^{(2)}$ is realizable; ($Q^{(1)}$ would have to have been between 500 and 1,000, and $Q^{(2)}$ would have to have been 1,000 or more). Each EOQ value corresponds to the minimum of a different annual cost curve. In this example, if $Q^{(2)}$ were realizable, it would necessarily have to have been the optimal solution, as it corresponds to the lowest point on the lowest curve. The three average annual cost curves for this example appear in Figure 4–11. Because each curve is valid only for certain values of Q, the average annual cost function is given by the

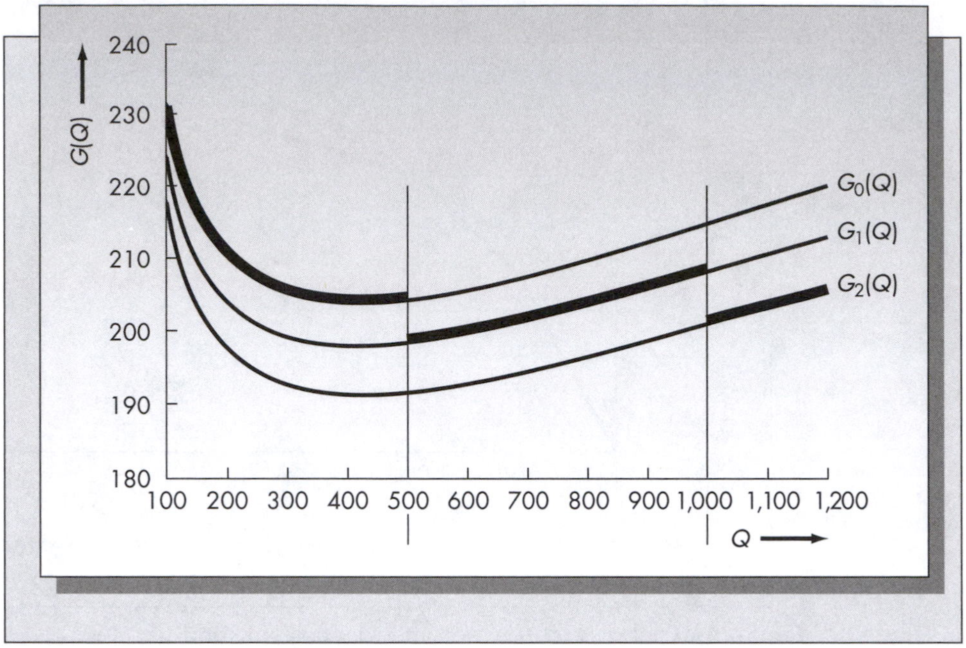

Figure 4–11 All-units discount average annual cost function

discontinuous curve shown in heavy shading. The goal of the analysis is to find the minimum of this discontinuous curve.

There are three candidates for the optimal solution: 400, 500, and 1,000. In general, the optimal solution will be either the largest realizable EOQ or one of the breakpoints that exceeds it. The optimal solution is the lot size with the lowest average annual cost.

The average annual cost functions are given by

$$G_j(Q) = \lambda c_j + \lambda K/Q + Ic_j\,Q/2 \quad \text{for } j = 0, 1, \text{and } 2.$$

The broken curve pictured in Figure 4–11, $G(Q)$, is defined as

$$G(Q) = \begin{cases} G_0(Q) & \text{for} \quad 0 \le Q < 500, \\ G_1(Q) & \text{for} \quad 500 \le Q < 1{,}000, \\ G_2(Q) & \text{for} \quad 1{,}000 \le Q. \end{cases}$$

Substituting Q equals 400, 500, and 1,000, and using the appropriate values of c_j, we obtain

$$\begin{aligned} G(400) &= G_0(400) \\ &= (600)(0.30) + (600)(8)/400 + (0.2)(0.30)(400)/2 = \$204.00 \\ G(500) &= G_1(500) \\ &= (600)(0.29) + (600)(8)/500 + (0.2)(0.29)(500)/2 = \$198.10 \\ G(1{,}000) &= G_2(1{,}000) \\ &= (600)(0.28) + (600)(8)/1{,}000 + (0.2)(0.28)(1{,}000)/2 = \$200.80. \end{aligned}$$

Hence, we conclude that the optimal solution is to place a standing order with Weighty for 500 units at an average annual cost of $198.10.

SmartOps Assists in Designing Caterpillar's Inventory Control System

Caterpillar Inc. is the world's leading manufacturer of construction and mining equipment, diesel and natural gas engines, industrial turbines, and diesel-electric locomotives. It is a Fortune 100 company with sales of $13.5 billion In the first quarter of 2019.

In 1997, management embarked on a project aimed at improving product availability and inventory turns. In 2001, the project team was expanded to include both internal Caterpillar personnel and two analysts from SmartOps, David Alberti and SmartOps founder Sridhar Tayur. The team was charged with examining the following questions: (1) What product availability is possible at what cost and what inventory levels? (2) How much inventory reduction is possible? (3) What mix and deployment of inventory is necessary to achieve the firm's service objectives?

The team focused its attention on backhoe loaders produced by the company's Clayton, North Carolina, facility. A key part of their focus was on lead times both for products being shipped from the plant and for materials and equipment shipped into the plant from suppliers. While lead times coming out of the plant were relatively short (typically one week), the lead times of products coming from suppliers could be quite long and tended to have high variance. The team developed a comprehensive model of the entire inventory supply chain. Inputs to the model included inventory stocking locations, historical forecast errors for different seasons and different products, subassemblies and bill of materials for each product, lead times, and inventory review times. The model took into account the multi-echelon (that is, multilevel) nature of the system.

In order to provide an accurate picture of the system, the analysts differentiated three types of orders placed by dealers: (a) sold orders based on firm sales to customers, (b) orders by dealers placed for the purpose of replenishing inventory, and (c) orders placed by dealers to replenish stock of machines rented to customers. Note that replenishment lead times for each of these demand types are different: four to six weeks for sold orders, six weeks for inventory replenishment, and fourteen weeks for replenishment of rental equipment.

The model provided a comprehensive picture of the important service versus inventory trade-offs Caterpillar could expect from the Clayton plant. For example, the team found that the Clayton plant could achieve a consistent 45- and 30-day product availability to dealers with 50 percent less finished goods inventory. However, to achieve a 14- or 10-day availability the finished goods inventory would have to be increased by approximately 90 percent. They also found that current service levels could be maintained with a total supply chain inventory reduction of 30–50 percent by repositioning inventory from finished goods to components.

After careful consideration of the trade-offs involved, management made the following changes to the system: (1) total inventory in the supply chain was reduced by 16 percent, (2) the mean time for orders was reduced by 20 percent and the variance reduced by 50 percent, and (3) order fulfillment was increased 2 percent.

The analysis provided by the team demonstrated to management that there were no simple answers to their inventory management problems. The Caterpillar supply chain was very complex with interaction among the multiple levels. Seeing what trade-offs were possible allowed management to set priorities that ultimately resulted in lower inventory costs and higher levels of service to the customer.

Source: Keene, S., D. Alberti, G. Henby, A. Brohinsky, and S. Tayur. "Caterpillar's Building Construction Products Division Improves and Stabilizes Product Availability" *Interfaces* 36 (2006), pp. 283–295.

Summary of the Solution Technique for All-Units Discounts

1. Determine the largest realizable EOQ value. The most efficient way to do this is to compute the EOQ for the lowest price first and then continue with the next higher price. Stop when the first EOQ value is realizable (that is, within the correct interval).
2. Compare the value of the average annual cost at the largest realizable EOQ and at all the price breakpoints that are greater than the largest realizable EOQ. The optimal Q is the point at which the average annual cost is a minimum.

Incremental Quantity Discounts

Reconsider Example 4.4 assuming incremental quantity discounts. That is, the trash bags cost 30 cents each for quantities of 500 or fewer; for quantities between 500 and 1,000, the first 500 cost 30 cents each and the remaining amount cost 29 cents each; for quantities of 1,000 and over the first 500 cost 30 cents each, the next 500 cost 29 cents each, and the remaining amount cost 28 cents each. We need to determine a mathematical expression for the function $C(Q)$ pictured in Figure 4–10. From the figure, we see that the first price break corresponds to $C(Q) = (500)(0.30) = \$150$ and the second price break corresponds to $C(Q) = 150 + (0.29)(500) = \$295$. It follows that

$$C(Q) = \begin{cases} 0.30Q & \text{for} \quad 0 \le Q < 500, \\ 150 + 0.29(Q - 500) = 5 + 0.29Q & \text{for} \quad 500 \le Q < 1{,}000, \\ 295 + 0.28(Q - 1{,}000) = 15 + 0.28Q & \text{for} \quad 1{,}000 \le Q, \end{cases}$$

so that

$$C(Q)/Q = \begin{cases} 0.30 & \text{for} \quad 0 \le Q < 500, \\ 0.29 + 5/Q & \text{for} \quad 500 \le Q < 1{,}000, \\ 0.28 + 15/Q & \text{for} \quad 1{,}000 \le Q. \end{cases}$$

The average annual cost function, $G(Q)$, is

$$G(Q) = \lambda C(Q)/Q + K\lambda/Q + I[C(Q)/Q]Q/2.$$

In this example, $G(Q)$ will have three different algebraic representations [$G_0(Q)$, $G_1(Q)$, and $G_2(Q)$] depending on the interval in which Q falls. Because $C(Q)$ is continuous, $G(Q)$ also will be continuous. The function $G(Q)$ appears in Figure 4–12.

The optimal solution occurs at the minimum of one of the average annual cost curves. The solution is obtained by substituting the three expressions for $C(Q)/Q$ in the defining equation for $G(Q)$, computing the three minima of the curves, determining which of these minima fall into the correct interval, and, finally, comparing the average annual costs at the realizable values. We have that

$$G_0(Q) = (600)(0.30) + (8)(600)/Q + (0.20)(0.30)Q/2,$$

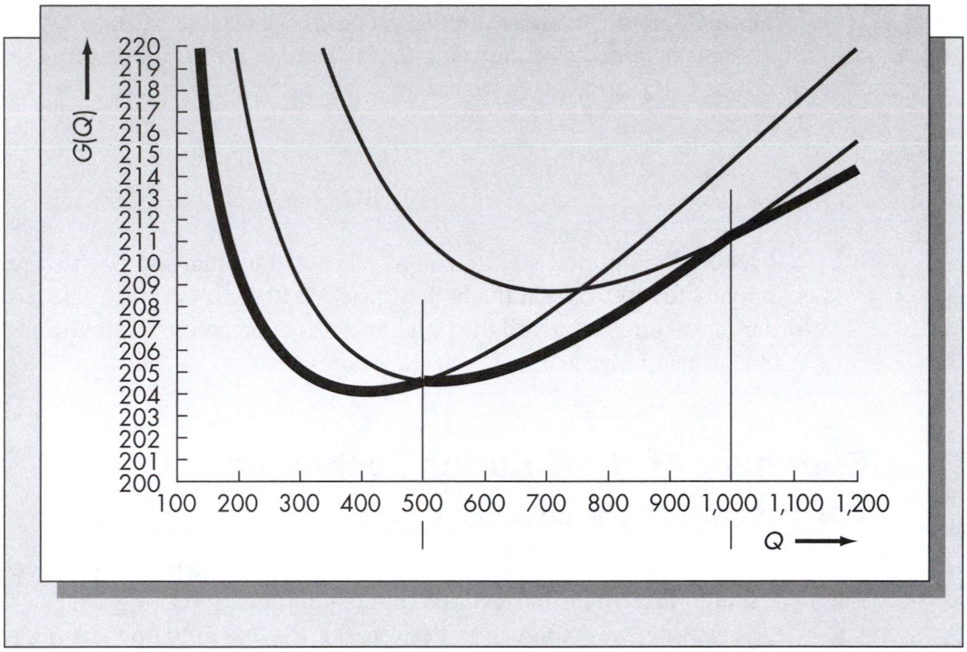

Figure 4–12 Average annual cost function for incremental discount schedule

which is minimized at

$$Q^{(0)} = \sqrt{\frac{2K\lambda}{Ic_0}} = \sqrt{\frac{(2)(8)(600)}{(0.20)(0.30)}} = 400;$$

$$\begin{aligned} G_1(Q) &= (600)(0.29 + 5Q) + (8)(600)/Q + (0.20)(0.29 + 5/Q)(Q/2) \\ &= (0.29)(600) + (13)(600)/Q + (0.20)(0.29)Q/2 + (0.20)(5)/2, \end{aligned}$$

which is minimized at

$$Q^{(1)} = \sqrt{\frac{(2)(13)(600)}{(0.20)(0.29)}} = 519;$$

and finally

$$\begin{aligned} G_2(Q) &= (600)(0.28 + 15/Q) + (8)(600)/Q + (0.20)(0.28 + 15/Q)Q/2 \\ &= (0.28)(600) + (23)(600)/Q + (0.20)(0.28)Q/2 + (0.20)(15)/2, \end{aligned}$$

which is minimized at

$$Q^{(2)} = \sqrt{\frac{(2)(23)(600)}{(0.20)(0.28)}} = 702.$$

Both $Q^{(0)}$ and $Q^{(1)}$ are realizable. $Q^{(2)}$ is not realizable because $Q^{(2)} < 1{,}000$. The optimal solution is obtained by comparing $G_0(Q^{(0)})$ and $G_1(Q^{(1)})$. Substituting into the earlier expressions for $G_0(Q)$ and $G_1(Q)$, we obtain

$$G_0(Q^{(0)}) = \$204.00,$$
$$G_1(Q^{(1)}) = \$204.58.$$

Hence, the optimal solution is to place a standing order with the Weighty Trash Bag Company for 400 units at the highest price of 30 cents per unit. The cost of using a standard order of 519 units is only slightly higher. Notice that compared to the all-units case, we obtain a smaller batch size at a higher average annual cost.

Summary of the Solution Technique for Incremental Discounts

1. Determine an algebraic expression for $C(Q)$ corresponding to each price interval. Use that to determine an algebraic expression for $C(Q)/Q$.
2. Substitute the expressions derived for $C(Q)/Q$ into the defining equation for $G(Q)$. Compute the minimum value of Q corresponding to each price interval separately.
3. Determine which minima computed in (2) are realizable (that is, fall into the correct interval). Compare the values of the average annual costs at the realizable EOQ values and pick the lowest.

Other Discount Schedules

Although all-units and incremental discount schedules are the most common, there are a variety of other discount schedules as well. One example is the carload discount schedule pictured in Figure 4–13.

The rationale behind the carload discount schedule is the following. A carload consists of M units. The supplier charges a constant c per unit up until you have paid for the cost of a full carload, at which point there is no charge for the remaining units in that carload. Once the first carload is full, you again pay c per unit until the second carload is full, and so forth.

Determining optimal policies for the carload or other discount schedules could be extremely difficult. Each discount schedule has a unique solution procedure. Some can be extremely complex.

PROBLEMS FOR SECTION 4.7

21. Your local grocery store stocks facial tissue in single packages and in more economical 12-packs. You are trying to decide which to buy. The single package costs 45 cents, and the 12-pack costs \$5. You consume tissues at a fairly steady rate of one pack every three months. Your opportunity cost of money is computed assuming an interest rate of 25 percent and a fixed cost of \$1 for the additional time it takes you to buy bathroom tissue when you go shopping. (We are assuming that you shop often enough so that you don't require a special trip when you run out.)

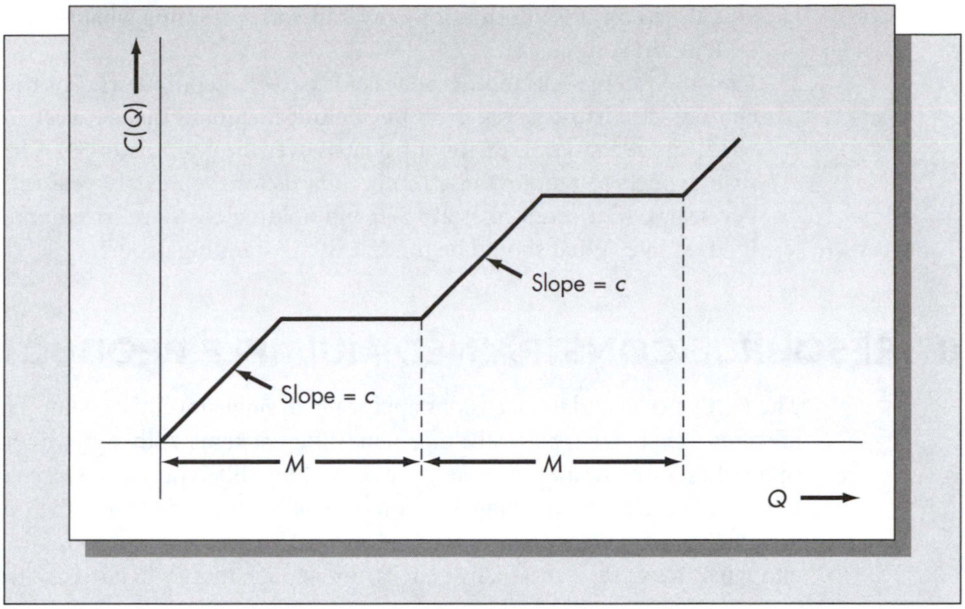

Figure 4–13 Order cost function for carload discount schedule

a. How many single packages should you be buying in order to minimize the annual holding and setup costs of purchasing tissues?
b. Determine if it is more economical to purchase the tissues in 12-packs.
c. Are there reasons other than those discussed in the problem that would motivate you not to buy the tissues in 12-packs?

22. A purchasing agent for a particular type of silicon wafer used in the production of semiconductors must decide among three sources. Source A will sell the silicon wafers for $2.50 per wafer, independently of the number of wafers ordered. Source B will sell the wafers for $2.40 each but will not consider an order for fewer than 3,000 wafers, and Source C will sell the wafers for $2.30 each but will not accept an order for fewer than 4,000 wafers. Assume an order setup cost of $100 and an annual requirement of 20,000 wafers. Assume a 20 percent annual interest rate for holding cost calculations.
    a. Which source should be used, and what is the size of the standing order?
    b. What is the optimal value of the holding and setup costs for wafers when the optimal source is used?
    c. If the replenishment lead time for wafers is three months, determine the reorder point based on the on-hand level of inventory of wafers.
23. Assume that two years have passed, and the purchasing agent mentioned in problem 22 must recompute the optimal number of wafers to purchase and from which source to purchase them. Source B has decided to accept any size offer but sells the wafers for $2.55 each for orders of up to 3,000 wafers and $2.25 each for the incremental amount ordered over 3,000 wafers. Source A still has the same price schedule, and Source C went out of business. Now which source should be used?
24. In the calculation of an optimal policy for an all-units discount schedule, you first compute the EOQ values for each of the three order costs, and you obtain: $Q^{(0)} = 800$, $Q^{(1)} = 875$, and $Q^{(2)} = 925$. The all-units discount schedule has breakpoints at 750 and

900. Based on this information only, can you determine what the optimal order quantity is? Explain your answer.

25. Parasol Systems sells motherboards for personal computers. For the first 25 boards purchased, the firm charges \$350 per board; for quantities between 26 and 50, it charges \$315; and it charges \$285 for any boards over the quantity of 50. A large communications firm expects to require these motherboards for the next 10 years at a rate of at least 140 per year. Order setup costs are \$30 and holding costs are based on an 18 percent annual interest rate. What should be the size of the standing order?

4.8 RESOURCE-CONSTRAINED MULTIPLE PRODUCT SYSTEMS

The EOQ model and its extensions apply only to single inventory items. However, these models are often used in companies stocking many different items. Although we could certainly compute optimal order quantities separately for each of the different items, there could exist constraints that would make the resulting solution infeasible. In Example 4.1, the optimal solution called for purchasing 3,870 pencils every 1.24 years. The bookstore, however, may not have allocated enough space to store that many pencils, nor enough money to purchase that many at one time.

EXAMPLE 4.5

Three items are produced in a small fabrication shop. The shop management has established the requirement that the shop never have more than \$30,000 invested in the inventory of these items at one time. The management uses a 25 percent annual interest charge to compute the holding cost. The relevant cost and demand parameters are given in the following table. What lot sizes should the shop be producing so that they do not exceed the budget?

	Item		
	1	2	3
Demand rate λ_j	1,850	1,150	800
Variable cost c_j	50	350	85
Setup cost K_j	100	150	50

Solution

If the budget is not exceeded when using the EOQ values of these three items, then the EOQs are optimal. Hence, the first step is to compute the EOQ values for all items to determine whether or not the constraint is active.

$$\text{EOQ}_1 = \sqrt{\frac{(2)(100)(1{,}850)}{(0.25)(50)}} = 172,$$

$$\text{EOQ}_2 = \sqrt{\frac{(2)(150)(1{,}150)}{(0.25)(350)}} = 63,$$

$$\text{EOQ}_3 = \sqrt{\frac{(20)(50)(800)}{(0.25)(85)}} = 61.$$

If the EOQ value for each item is used, the maximum investment in inventory would be

$$(172)(50) + (63)(350) + (61)(85) = \$35{,}835.$$

Note that we are assuming for the purposes of this section that the three products are ordered simultaneously. By staggering order cycles, it is possible to meet the constraint with larger lot sizes. We will not consider that case here.

Because the EOQ solution violates the constraint, we need to reduce these lot sizes. But how?

The optimal solution turns out to be very easy to find in this case. We merely multiply each EOQ value by the ratio (30,000)/(35,835) = 0.8372. In order to guarantee that we do not exceed the \$30,000 budget, we round each value *down* (adjustments can be made subsequently). Letting Q^*_1, Q^*_2, and Q^*_3 be the optimal values, we obtain

$$Q^*_1 = (172)(0.8372) \approx 144,$$
$$Q^*_2 = (63)(0.8372)0 \approx 52,$$
$$Q^*_3 = (61)(0.8372) \approx 51,$$

where $\approx$ should be interpreted as rounding to the next lower integer in this case.

The total budget required for these lot sizes is \$29,735. The remaining \$265 can now be used to increase the lot sizes of products 1 and 3 slightly. (For example, $Q^*_1 = 147$, $Q^*_3 = 52$ results in a budget of \$29,970.)

In general, budget- or space-constrained problems are not solved so easily. Suppose that n items have unit costs of $c_1, c_2, \ldots, c_n$, respectively, and the total budget available for them is C. Then the budget constraint can be written

$$c_1Q_1 + c_2Q_2 + \cdots + c_nQ_n \le C.$$

Let

$$\text{EOQ}_i = \sqrt{\frac{2K_i\lambda_i}{h_i}} \qquad \text{for } i = 1, \ldots, n,$$

where K_i, h_i, and λ_i are the respective cost and demand parameters.

There are two possibilities: either the constraint is active, or it is not. If the constraint is not active, then

$$\sum_{i=1}^{n} c_i \text{EOQ}_i \le C,$$

and the optimal solution is $Q_i = \text{EOQ}_i$. If the constraint is active, then

$$\sum_{i=1}^{n} c_i \text{EOQ}_i > C,$$

and the EOQ solution is no longer feasible. If we include the following assumption, the solution to the active case is relatively easy to find:

$$\textit{assumption: } c_1/h_1 = c_2/h_2 = \ldots = c_n/h_n.$$

If this assumption holds and the constraint is active, we prove in Appendix 4–A that the optimal solution is

$$Q_i^* = m\text{EOQ}_i,$$

where the multiplier m solves

$$m = C \bigg/ \left[\sum_{i=1}^{n} (c_i \text{EOQ}_i) \right].$$

Since $c_i/h_i = c_i/(I_i c_i) = 1/I_i$, the condition that the ratios be equal is equivalent to the requirement that the same interest rate be used to compute the holding cost for each item, which is reasonable in most circumstances.

Suppose that the constraint is on the available space. Let w_i be the space consumed by one unit of product i for $i = 1, 2, \ldots, n$ (this could be floor space measured, say, in square feet, or volume measured in cubic feet), and let W be the total space available. Then the space constraint is of the form

$$w_1 Q_1 + w_2 Q_2 + \cdots + w_n Q_n \leq W.$$

This is mathematically of the same form as the budget constraint, so the same analysis applies. However, our condition for a simple solution now is that the ratios w_i/h_i are equal. That is, the space consumed by an item should be proportional to its holding cost. When the interest rate is fixed, this is equivalent to the requirement that the space consumed should be proportional to the value of the item. This requirement would probably be too restrictive in most cases. For example, flash drives take up far less space than legal pads but are more expensive.

Now consider the problem in which the constraint is active but the proportionality assumption is not met. This problem is far more complex than the one just solved. It requires formulating the **Lagrangian** function. The details of the formulation of this problem can be found in Appendix 4–A. As we show, the optimal lot sizes are now of the form

$$Q_i^* = \sqrt{\frac{2K_i\lambda_i}{h_i + 2\theta w_i}},$$

where θ is a constant chosen so that

$$\sum_{i=1}^{n} w_i Q_i^* = W.$$

The constant θ, known as the **Lagrange multiplier**, reduces the lot sizes by increasing the effective holding cost. The correct value of θ can be found by trial and error or by a search technique such as interval bisection or Goal Seek in Excel. Note that $\theta > 0$, so that the search can be limited to positive numbers only. The value of θ can be interpreted as the decrease in the average annual cost that would result from adding an additional unit of resource. In this case, it would represent the marginal benefit of an additional square foot of space.

EXAMPLE 4.6

Revisiting the fabrication shop of Example 4.5, suppose that there are only 2,000 square feet of floor space available, in addition to the budget constraint. Assume that the three products consume respectively 9, 12, and 18 square feet per unit.

First, we check to see if the EOQ solution is feasible. Setting $w_1 = 9$, $w_2 = 12$, and $w_3 = 18$, we find that

$$\sum_{i=1}^{n} \text{EOQ}_i w_i = (172)(9) + (63)(12) + (61)(18) = 3{,}402,$$

which is obviously infeasible. Next, we check if the budget-constrained solution provides a feasible solution to the space-constrained problem. The budget-constrained solution requires (147)(9) + (52)(12) + (52)(18) = 2,883 square feet of space, which is also infeasible.

The next step is to compute the ratios w_i/h_i for $1 \le i \le 3$. These ratios turn out to be 0.72, 0.14, and 0.85, respectively. Because their values are different, the simple solution obtained by a proportional scaling of the EOQ values will not be optimal. Hence, we must determine the value of the Lagrange multiplier θ.

We can determine upper and lower bounds on the optimal value of θ by assuming equal ratios. If the ratios were equal, the multiplier, m, would be

$$m = W / \sum_{i=1}^{n} (\text{EOQ}_i w_i) = 2{,}000 / 3{,}402 = 0.5879,$$

which would give the three lot sizes as 101, 37, and 36, respectively. The three values of θ that result in these lot sizes are respectively $\theta = 1.32$, $\theta = 6.86$, and $\theta = 2.33$. (These values were obtained by setting the given expression for Q^*_1 equal to the lot sizes 101, 37, and 36, and solving for θ.)

The true value of θ will be between 1.32 and 6.86. If we start with $\theta = 3.5$, we obtain $Q_1^* = 70$, $Q_2^* = 45$, and $Q_3^* = 23$, and $\Sigma w_i Q_i^* = (70)(9) + (45)(12) + (23)(18) = 1{,}584$, which implies $\theta < 3.5$. After considerable experimentation, we finally find that the optimal value of $\theta = 1.75$, and $Q_1^* = 92$, $Q_2^* = 51$, and $Q_3^* = 31$, giving $\Sigma w_i Q_i^* = 1{,}998$. Notice that these lot sizes are in very different proportions from the ones obtained assuming a constant multiplier. Searching for the optimal value of the Lagrange multiplier, although tedious to do by hand, can be accomplished very efficiently using a computer. Spreadsheets are a useful means for effecting such calculations.

We have only touched on some of the complications that could arise when applying EOQ-type models to a realistic system with multiple products. Often, real problems are far too complex to be expressed accurately as a solvable mathematical model. For this reason, simple models such as the ones presented in this and subsequent chapters are used by practitioners. However, any solution recommended by a mathematical model must be considered in the context of the system in which it is to be used.

PROBLEMS FOR SECTION 4.8

26. A local outdoor vegetable stand has exactly 1,000 square feet of space to display three vegetables: tomatoes, lettuce, and zucchini. The appropriate data for these items are given in the following table.

	Item		
	Tomatoes	Lettuce	Zucchini
Annual demand (in pounds)	850	1,280	630
Cost per pound	\$0.29	\$0.45	\$0.25

The setup cost for replenishment of the vegetables is \$100 in each case, and the space consumed by each vegetable is proportional to its costs, with tomatoes requiring 0.5 square foot per pound. The annual interest rate used for computing holding costs is 25 percent. What are the optimal quantities that should be purchased of these three vegetables?

27. Suppose that the vegetables discussed in problem 26 are purchased at different times. In what way could that have an effect on the order policy that the stand owner should use?
28. Suppose that in problem 26 the space consumed by each vegetable is not proportional to its cost. In particular, suppose that one pound of lettuce required 0.4 square foot of space, and one pound of zucchini required 1 square foot of space. Determine upper and lower bounds on the optimal values of the order quantities in this case. Test different values of the Lagrange multiplier to find the optimal values of the order quantities. (A spreadsheet is ideally suited for this kind of calculation. If you solve the problem using a spreadsheet, place the Lagrange multiplier in a cell so that its value can be changed easily.)

4.9 EOQ MODELS FOR PRODUCTION PLANNING

Simple lot-sizing models have been successfully applied to a variety of manufacturing problems. In this section we consider an extension of the EOQ model with a finite production rate (discussed in Section 4.6) to the problem of producing n products on a single machine. Following the notation used in this chapter, let

λ_j = demand rate for product j,
P_j = production rate for product j,
h_j = holding cost per unit per unit time for product j, and
K_j = cost of setting up the production facility to produce product j.

The goal is to determine the optimal procedure for producing n products on the machine to minimize the cost of holding and setups and to guarantee that no stock-outs occur during the production cycle.

We require the assumption that

$$\sum_{j=1}^{n} \lambda_j / P_j \leq 1.$$

This assumption is needed to ensure that the facility has sufficient capacity to satisfy the demand for all products. Notice that this is stronger than the assumption made in Section 4.6 that $\lambda_j < P_j$ for each j. To see why this assumption is necessary, consider the case of two products with identical demand and production rates. In each cycle, one would produce product 1 first, then product 2. Clearly $P_1 \geq 2\lambda_1$ and $P_2 \geq 2\lambda_2$, so that enough could be produced to meet the total demand for both products each cycle. This translates to $\lambda_1/P_1 \leq 0.5$

and $\lambda_2/P_2 \leq 0.5$, giving a value of the sum less than or equal to one. Similar reasoning holds for more than two products with nonidentical demand and production rates.

We also will assume that the policy used is a **rotation cycle policy**. That means that in each cycle there is exactly one setup for each product, and products are produced in the same sequence in each production cycle. The importance of this assumption will be discussed further at the end of the section.

At first, one might think that the optimal solution is to sequentially produce lot sizes for each product optimized by treating each product in isolation. From Section 4.6, this would result in a lot size for product j of

$$Q_j = \sqrt{\frac{2K_j\lambda_j}{h_j'}},$$

where $h_j' = h_j(1 - \lambda_j/P_j)$. The problem with this approach is that because we have only a single production facility, it is likely that some of the lot sizes Q_j will not be large enough to meet the demand between production runs for product j, thus resulting in stock-outs.

Let T be the cycle time. During time T we assume that exactly one lot of each product is produced. In order that the lot for product j will be large enough to meet the demand occurring during time T, it follows that the lot size must be

$$Q_j = \lambda_j T.$$

From Section 4.6, the average annual cost associated with product j can be written in the form

$$G(Q_j) = K_j\lambda_j/Q_j + h_j' Q_j/2.$$

The average annual cost for all products is the sum

$$\sum_{j=1}^{n} G(Q_j) = \sum_{j=1}^{n} (K_j\lambda_j / Q_j + h_j' Q_j / 2).$$

Substituting $T = Q_j/\lambda_j$, we obtain the average annual cost associated with the n products in terms of the cycle time T as

$$G(T) = \sum_{j=1}^{n} (K_j / T + h_j'\lambda_j T / 2).$$

The goal is to find T to minimize $G(T)$. The necessary condition for an optimal T is

$$\frac{dG(T)}{dT} = 0.$$

Setting the first derivative with respect to T to zero gives

$$\sum_{j=1}^{n} (-K_j / T^2 + h_j'\lambda_j / 2) = 0.$$

Solving for T, we obtain the optimal cycle time T^* as

$$T^* = \sqrt{\frac{2\sum_{j=1}^{n} K_j}{\sum_{j=1}^{n} h'_j \lambda_j}}.$$

If setup times are a factor, we must check that there is enough time each cycle to account for both setup times and production of the n products. Let s_j be the setup time for product j. Ensuring that the total time required for setups and production each cycle does not exceed T leads to the constraint

$$\sum_{j=1}^{n} (s_j + Q_j / P_j) \leq T.$$

Using the fact that $Q_j = \lambda_j T$, this condition translates to

$$\sum_{j=1}^{n} (s_j + \lambda_j T / P_j) \leq T,$$

which gives, after rearranging terms,

$$T \geq \frac{\sum_{j=1}^{n} s_j}{1 - \sum_{j=1}^{n} (\lambda_j / P_j)} = T_{\min}.$$

Because $T_{\min}$ cannot be exceeded without compromising feasibility, the optimal solution is to choose the cycle time T equal to the *larger* of T^* and $T_{\min}$.

EXAMPLE 4.7

Bally produces several styles of men's and women's shoes at a single facility near Bergamo, Italy. The leather for both the uppers and the soles of the shoes is cut on a single machine. This Bergamo plant is responsible for seven styles and several colors in each style. (The colors are not considered different products for our purposes, because no setup is required when switching colors.) Bally would like to schedule cutting for the shoes using a rotation policy that meets all demand and minimizes setup and holding costs. Setup costs are proportional to setup times.

The firm estimates that setup costs amount to an average of \$110 per hour, based on the cost of worker time and the cost of forced machine idle time during setups. Holding costs are based on a 22 percent annual interest charge.

The relevant data for this problem appear in Table 4–1.

Solution

The first step is to verify that the problem is feasible. To do so we compute $\Sigma \lambda_j/P_j$. The reader should verify that this sum is equal to 0.69355. Because this is less than one,

Table 4–1 Relevant Data for Example 4.7

Style	Annual Demand (units/year)	Production Rate (units/year)	Setup Time (hours)	Variable Cost ($/unit)
Women's pump	4,520	35,800	3.2	$40
Women's loafer	6,600	62,600	2.5	26
Women's boot	2,340	41,000	4.4	52
Women's sandal	2,600	71,000	1.8	18
Men's wingtip	8,800	46,800	5.1	38
Men's loafer	6,200	71,200	3.1	28
Men's oxford	5,200	56,000	4.4	31

there will be a feasible solution. Next we compute the value of T^*, but to do so we need to do several intermediate calculations.

First, we compute setup costs. Setup costs are assumed to be $110 times setup times. Second, we compute modified holding costs (h'_j). This is done by multiplying the cost of each product by the annual interest rate (0.22) times the factor $1 - \lambda_j/P_j$. These calculations give

Setup Costs (K_j)	Modified Holding Costs (h'_j)
352	7.69
275	5.12
484	10.79
198	3.81
561	6.79
341	5.62
484	6.19

The sum of the setup costs is 2,695, and the sum of the products of the modified holding costs and the annual demands is 230,458.4. Substituting these figures into the formula for T^* gives the optimal cycle time as 0.1529 year. Assuming a 250-day work year, this means that the rotation cycle should repeat roughly every 38 working days. The optimal lot size for each of the shoes is found by multiplying the cycle time by the demand rate for each item. The reader should check that the following lot sizes result.

Style	Optimal Lot Sizes for Each Production Run
Women's pump	691
Women's loafer	1,009
Women's boot	358
Women's sandal	398
Men's wingtip	1,346
Men's loafer	948
Men's oxford	795

The plant would cut the soles and uppers in these lot sizes in sequence (although the sequence does not necessarily have to be this one) and would repeat the rotation cycle roughly every 38 days (0.1529 year). However, this solution can be implemented only if T^* is at least T_{min}. To determine T_{min} we must express the setup times in years. Assuming 8 working hours per day and 250 working days per year, one would divide the setup times given in hours by 2,000 (250 times 8). The reader should check that the resulting value of T_{min} is 0.04, thus making T^* feasible and, hence, optimal.

The total average annual cost of holding and setups at an optimal policy can be found by computing the value of $G(T)$ when $T = T^*$. It is $35,244.44. It is interesting to note that if the plant manager chooses to implement this policy, the facility will be idle for a substantial portion of each rotation cycle. The total uptime each rotation cycle is found by dividing the lot sizes by the production rates for each style and summing the results. It turns out to be 0.106 year. Hence, the optimal rotation policy that minimizes total holding and setup costs results in the cutting operation remaining idle about one-third of the time.

The reader should be aware that the relatively simple solution to this problem was the result of two assumptions. One was that the setup costs were not sequence dependent. In Example 4.7, it is possible that the time required to change over from one shoe style to another could depend on the styles. For example, a changeover from a woman's style to a man's style probably takes longer than from one woman's style to another. A second assumption was that the plant used a rotation cycle policy. That is, in each cycle Bally does a single production run of each style. When demand rates and setup costs differ widely, it might be advantageous to do two or more production runs of a product in a cycle. This is a challenging problem to solve. If setups are not sequence dependent, then the problem is usually solved by first finding an optimal frequency of production and then trying to sequence the products to meet this frequency. However, if setups do depend on the production sequence, then a series of so-called traveling salesman problems must be solved, which is a non-trivial exercise.

PROBLEMS FOR SECTION 4.9

29. A metal fabrication shop has a single punch press. There are currently three parts that the shop has agreed to produce that require the press, and it appears that they will be supplying these parts well into the future. You may assume that the press is the critical resource for these parts, so that we need not worry about the interaction of the press with the other machines in the shop. The relevant information is listed below.

Part Number	Annual Contracted Amount (demand)	Setup Cost	Cost (per unit)	Production Rate (per year)
1	2,500	$80	$16	45,000
2	5,500	120	18	40,000
3	1,450	60	22	26,000

Holding costs are based on an 18 percent annual interest rate, and the products are to be produced in sequence on a rotation cycle. Setup times can be considered negligible.

a. What is the optimal time between setups for part number 1?

b. What percentage of the time is the punch press idle, assuming an optimal rotation cycle policy?

c. What are the optimal lot sizes of each part put through the press at an optimal solution?
d. What is the total annual cost of holding and setup for these items on the punch press, assuming an optimal rotation cycle?

30. Tomlinson Furniture has a single lathe for turning the wood for various furniture pieces, including bedposts, rounded table legs, and other items. Four forms are turned on the lathe and produced in lots for inventory. To simplify scheduling, one lot of each type will be produced in a cycle, which may include idle time. The four products and the relevant information concerning them appears in the following table.

Piece	Monthly Requirements	Setup Time (hours)	Unit Cost	Production Rate (units/day)
J–55R	125	1.2	$20	140
H–223	140	0.8	35	220
K–18R	45	2.2	12	100
Z–344	240	3.1	45	165

Worker time for setups is valued at $85 per hour, and holding costs are based on a 20 percent annual interest charge. Assume 20 working days per month and 12 months per year for your calculations.

a. Determine the optimal length of the rotation cycle.
b. What are the optimal lot sizes for each product?
c. What are the percentages of uptime and downtime for the lathe, assuming that it is not used for any other purpose?
d. Draw a graph showing the change in the inventory level over a typical cycle for each product.
e. Discuss why the solution you obtained might not be feasible for the firm or why it might not be desirable even when it is feasible.

4.10 POWER-OF-TWO POLICIES

The inventory models treated in this chapter form the basis of more complex cases. In almost every case treated here, we were able to find relatively straightforward algebraic solutions. However, even when demand is assumed known, there exist several extensions of these basic models whose optimal solutions may be difficult or impossible to find. In those cases, effective approximations are very valuable. Here, we discuss an approach that has proven to be successful in a variety of deterministic environments. The idea is based on choosing the best replenishment interval from a set of possible intervals proportional to powers of two. While the analysis of power-of-two policies in complex environments is beyond the scope of this book, we can illustrate the idea in the context of the basic EOQ model.

From Section 4.5, we know that the order quantity that minimizes average annual holding and setup costs when demand is fixed at λ units per unit time is the EOQ given by

$$Q^* = \sqrt{\frac{2K\lambda}{h}},$$

and the optimal time between placement of orders, say T^*, is given by

$$T^* = Q^*/\lambda = \sqrt{\frac{2K}{\lambda h}}.$$

It is possible, and even likely, that optimal order intervals are inconvenient. For example, the optimal solution might call for ordering every 3.393 weeks. However, one might only want to place orders at the beginning of a day or a week. To account for this, suppose that we impose the constraint that ordering must occur in some multiple of a base time, T_L. To find the optimal solution under the constraint that the order interval must be a multiple of T_L, one would simply compare the costs at the two closest multiples of T_L to T^* and pick the order interval with the lower cost. [That is, find k for which $kT_L \le T^* \le (k+1)T_L$ and choose the reorder interval to be either kT_L or $(k+1)T_L$ depending on which results in a lower average annual cost.]

Now suppose we add the additional restriction that the order interval must be a power of two times T_L. That is, the order interval must be of the form 2^kT_L for some integer $k \ge 0$. While it is unlikely one would impose such a restriction in the context of the simple EOQ problem, the ultimate goal is to explore these policies for more complex problems whose optimal solutions are hard to find. The question we wish to address is: Under such a restriction (known as a power-of-two policy), what is the worst cost error we will incur relative to that of the optimal reorder interval T^*? On the surface, it appears that such a restriction would result in serious errors. As k increases, the distance between successive powers of two grows rapidly. For example, if $k = 12$, $2^k = 4{,}096$ and $2^{k+1} = 8{,}192$, a very wide interval. If $T^* = 6{,}000$ and $T_L = 1$, for example, it seems that forcing the order interval to be either 4,096 or 8,192 would result in a large cost error (as the error in T is nearly 30 percent in either direction). However, this turns out not to be the case. In fact, we can prove that the cost error in every case is bounded by slightly more than 6 percent. While the result seems unintuitive at first, recall that the average annual cost function is relatively insensitive to errors in Q, as we saw in Section 4.5. Since Q and T are proportional, a similar cost insensitivity holds with respect to T.

We know from Section 4.5 that the average annual holding and setup cost as a function of the order quantity, Q, is given by

$$G(Q) = \frac{K\lambda}{Q} + \frac{hQ}{2}.$$

Since $Q = \lambda T$, the average annual cost can also be expressed in terms of the reorder interval, T, as

$$G(T) = \frac{K}{T} + \frac{h\lambda T}{2}.$$

In the case of a power-of-two policy, we wish to find the value of k that minimizes $G(2^kT_L)$. Since $G(T)$ is a continuous convex function of T, it follows that $G(2^kT_L)$ is a discrete convex function of k. That means that the optimal value of k, say k^*, satisfies

$$k^* = \min\{k : G(2^{k+1}T_L) \ge G(2^kT_L)\}.$$

Substituting for $G(T)$, the optimality condition becomes

$$\frac{K}{2^{k+1}T_L} - \frac{K}{2^k T_L} \geq \frac{h\lambda 2^k T_L}{2} - \frac{h\lambda 2^{k+1} T_L}{2},$$

which reduces to

$$\frac{K/2}{2^k T_L} \leq h\lambda 2^{k-1} T_L.$$

Rearranging terms gives

$$2^k \geq \frac{1}{T_L}\sqrt{\frac{K}{h\lambda}},$$

or

$$2^k T_L \geq \frac{1}{\sqrt{2}} T^*.$$

We assume that $T^* \geq T_L$, which means that $\sqrt{2}T^* > T_L$. Hence, to summarize, we seek the smallest value of k that satisfies the simultaneous inequalities

$$\frac{1}{\sqrt{2}} T^* \,''\, 2^k T_L \,''\, \sqrt{2} T^*.$$

Note that this implies that as long as $T^* \geq T_L$, the optimal power-of-two solution will always lie between $.707T^*$ and $1.41T^*$. The next question is: What is the worst-case cost error of this policy? Since $Q = \lambda T$, if $T = T^*/\sqrt{2}$, then $Q = Q^*/\sqrt{2}$. Similarly, if $T = \sqrt{2}\,T^*$, then $Q = \sqrt{2}\,Q^*$. It follows that the worst-case cost error of the power-of-two policy is given by

$$\frac{G(Q)}{G(Q^*)} = \frac{1}{2}\left(\frac{Q^*}{Q} + \frac{Q}{Q^*}\right) = \frac{1}{2}\left(\frac{1}{\sqrt{2}} + \sqrt{2}\right) = 1.0607,$$

or slightly more than 6 percent. (Because of symmetry, we obtain the same result when $Q = Q^*/\sqrt{2}$ or when $Q = \sqrt{2}\,Q^*$.)

The real "power" of power-of-two policies occurs when trying to solve more complex problems whose optimal policies are difficult to find. Consider the following scenario. A single warehouse is the sole supplier of N retailers. The demand rate experienced by each retailer is known and constant. As with the simple EOQ problem, assume that shortages at both the warehouse and the retailers are not permitted. There are fixed costs for ordering at both the warehouse and the retailers plus holding costs at these locations. These costs do not necessarily need to be the same. It is assumed that there is no lead time for placement or arrival of orders.

Unlike the simple EOQ problem, determining an optimal policy (that is, one that minimizes long-run average costs) for this problem could be extremely difficult, or even close to impossible. In many cases, even the form of an optimal policy may not be known. Clearly, effective approximations are very important.

One approximation for this problem is a so-called nested policy. In this case, a retailer would automatically order whenever the warehouse does, and possibly at other times as well. Although nested policies can have arbitrarily large cost errors, various adaptations of power-of-two policies can be shown to have 94 percent or even 98 percent guaranteed effectiveness. Power-of-two approximations have also been shown to be equally effective in serial production systems and more complex arborescent assembly systems (see Muckstadt and Roundy, 1993; Roundy, 1985).

CASE STUDY Mia Buys a Business

Mia Robinson decided it was time for a change. She had been a social worker for 10 years. Although she found the work rewarding, it was time for something else. When she saw that Herbie's Hut was for sale, she made some inquiries.

Herbie's Hut was a small gift shop and toy store in her old neighborhood of Skokie, a Chicago suburb. Herb Gold had been running the business for 40 years, but it was a demanding job and he decided it was time to retire. The business was pretty steady, so Herb figured that there would be buyers. He was delighted to hear from Mia, whom he had known when she was a child growing up in the neighborhood. The money was less important to Herb than making sure the store was in good hands. He had a good feeling about Mia, so he agreed to let the business go for a modest price.

After Mia purchased the business, she decided to remodel the store. Herbie's Hut would reopen as Mia's Marvels. Skylights, new wallpaper, and a fresh layout gave the store a sprightlier look that reflected Mia's personality. During this time, Mia took a careful look at the stock that came with the purchase. She noticed that several items were severely understocked. Two of those were ID bracelets and Lego sets. She spoke with Herb about it, and he apologized. Herb had not had a chance to reorder these popular items.

Mia was concerned that she might run out of these items quickly and that it would hurt her reputation. When she called the suppliers, she realized that she had no idea how much to order. She asked Herb, and he said his rule of thumb was to order about one month's supply. This sounded reasonable to Mia; she could figure out what a month's supply was from the store's sales records, which Herb had given her.

The store stocked hundreds of different items. It occurred to Mia that if she reordered every item monthly, she would be spending a lot of time processing orders. Mia thought that perhaps there was a better way. She consulted her boyfriend Bob, who had an MBA degree. Bob found his old class notes and showed Mia the EOQ formula, $\sqrt{\frac{2K\lambda}{h}}$, used for determining order sizes. "What do these symbols represent?" asked Mia. Bob explained that K is the fixed cost of each order, the Greek letter (λ) is the sales rate, and h is the cost of holding. Mia still wasn't sure what it all meant.

Let's take an item you need to order now, and see how this works out," suggested Bob. So they considered the bracelets.

"First, what's the process you would go through to place a new order?" asked Bob. "Well," replied Mia, "I would call the supplier to place the order, and when it arrived I would unpack the shipment, possibly put some of the items on display, and store the rest." "How long does that take?" inquired Bob. "I don't know," she responded, "maybe a couple of hours. Also, that supplier charges a fixed cost of $50 per order in addition to the cost of each bracelet." Figuring Mia's time at \$50 per hour Bob computed a total fixed cost of (2)(50) + 50 = \$150.

"Ok. Let's look at the sales rate," said Bob. Based on Herb's records, Mia estimated that the store sold an average of about 75 bracelets a month. That would give the value of λ. Finally, they needed to estimate the holding cost. Bob had spent some time thinking about this. "Holding cost can be thought of in two ways. The symbol h refers to the cost of holding a single unit of inventory for some fixed time, typically one year. It can also be thought of as an interest rate, which is then multiplied by the unit cost of the item. I think the most appropriate value of the interest rate is the rate of return you expect to earn in this business. I did a little research on the web and it seems that for businesses of this type, a 15 percent annual return on investment seems to be about right. We'll use that to figure the holding cost." Since Mia's cost of each bracelet was \$30, this resulted in an annual holding cost of $h = (.15)(30) = \$4.50$ annually.

Since the holding cost was based on an annual interest rate, the sales rate had to be yearly as well. This translated to an annual sales rate of (12)(75) = 900. Plugging these numbers into the formula gave a lot size of $\sqrt{\frac{(2)(150)(900)}{4.5}} = 245$. This made more sense to Mia than ordering 75 bracelets every month—the results translated to a little more than a three-month supply.

Once Mia got the idea, it didn't take her long to set up a spreadsheet for all the items in the store. For example, a similar analysis of the Lego sets resulted in an EOQ value of 80 Lego sets. That seemed all well and good until her first order arrived. Uh oh, she thought, these things are pretty bulky. They filled up almost half of her backroom storage area. She realized that there was a little more to figuring out the right lot sizes than just applying the EOQ formula.

Mia asked Bob what she should do about the fact that the Lego order took up so much space. "Well, I'm not sure," answered Bob. "I guess you shouldn't order so many Lego sets." "Well thanks for nothing, genius." Mia responded. "I know that now, but just how many should I order?"

So Bob did his research and found out this was a tougher problem than figuring the EOQ. As a first cut, Bob read that if the value of each item were proportional to the space it consumed, the solution would be pretty simple. Realizing that this was only an approximation, he figured it would be better than nothing and hoped it would satisfy Mia.

"Ok," Bob said, "let's consider your spreadsheet that computes the EOQ values for all of the items in the store. Now let's add a column to the spreadsheet that indicates the cubic feet of space consumed by each item. Multiply the two columns for each item and add up the total."

This was an easy calculation once Mia was able to approximate the space requirements for each of the items. Multiplying the two columns and adding resulted in a total space requirement of 120,000 cubic feet. Mia had only 50,000 cubic feet of storage space, so Bob suggested that each of the order quantities be reduced by multiplying by the constant 50,000/120,000 = .4167. "If you reduce all of your order quantities by around 60 percent, you shouldn't run of the space," Bob recommended.

"Bob, that's great! That will be a big help," said Mia. She added another column to her spreadsheet that reduced all of the lot sizes by 60 percent. But as she started looking at the numbers, something bothered her.

"You know, Bob, this will solve my space problem, but there's something about it that doesn't make sense to me. Reducing the lot size of the Lego sets from 80 to 32 sounds about right, but why should I reduce the lot size for the bracelets? They hardly take up any room at all."

She's absolutely right on the money about that, thought Bob. You really wanted to reduce the lot sizes of the bulky items a lot more than the small items. But how do you do that? Bob consulted his friend Cathy, who has a PhD in operations research. Cathy explained that this was a nontrivial optimization problem that involved finding the value of something called the Lagrange multiplier. Given Mia's spreadsheet, this was a piece of cake for Cathy. Once the calculation was completed, the lot sizes for the bulky items were indeed reduced much more than those of the smaller items.

By applying a few basic concepts from inventory control, Mia was able to get a handle on the problem of managing her stock. Her willingness to apply scientific principles meant that Mia was well on her way to creating a successful business.

4.11 HISTORICAL NOTES AND ADDITIONAL TOPICS

The interest in using mathematical models to control the replenishment of inventories dates back to the early part of the twentieth century. Ford Harris (1915) is generally credited with the development of the simple EOQ model. R. H. Wilson (1934) is also recognized for his analysis of the model. The procedure we suggest for the all-units quantity discount model is attributed to C. West Churchman, Russell Ackoff, and E. Leonard Arnoff (1957), while George Hadley and Thomson Whitin (1963) contributed the incremental discount model. Kenneth Arrow (1958) provides an excellent discussion of the economic motives for holding inventories.

Hadley and Whitin (1963) may have been the first to consider budget- and space-constrained problems, although several researchers have studied the problem subsequently. Meir Rosenblatt (1981) derives results similar to those of Section 4.8 and considers several extensions not treated in this section.

Academic interest in inventory management problems took a sudden upturn in the late 1950s and early 1960s with the publication of a number of texts in addition to those just mentioned (Bowman and Fetter, 1961; Fetter and Dalleck, 1961; Hanssmann, 1961; Magee and Boodman, 1967; Scarf, Gilford, and Shelly, 1963; Star and Miller, 1962; Wagner, 1962; and Whitin, 1957).

This chapter assumed a rotation cycle policy for systems with setups. In general, finding non-rotation policies is not an easy problem. An early discussion on this is provided by William Maxwell (1964). John Magee and David Boodman (1967) provide some heuristic ways of dealing with more general setup cost problems.

By and large, the vast majority of the published research on inventory systems since the 1960s, including inventory models with setup times and costs, has been on stochastic models, which will be the subject of Chapter 5. Lineu Barbosa and Moshe Friedman (1978), as well as Leroy Schwarz and Linus Schrage (1971) extended the EOQ model.

4.12 SUMMARY

This chapter presented several popular models used to control inventories when the demand is known. We discussed the various economic motives for holding inventories, which include economies of scale, uncertainties, speculation, and logistics. In addition, we mentioned some of the physical characteristics of inventory systems that are important in determining the complexity and applicability of the models to real problems. These include demand, lead time, review time, back-order/lost-sales assumptions, and the changes that take place in the inventory over time. There are three significant classes of costs in inventory management: holding or carrying costs, order or setup costs, and penalty costs for not meeting demand. The holding cost is usually expressed as the product of an interest rate and the cost of the item.

The grandfather of all inventory control models is the simple EOQ model, in which demand is assumed to be constant, no stock-outs are permitted, and only holding and order costs are present. The optimal batch size is given by the classic square root formula. The first extension of the EOQ model we considered was for items produced internally at a finite production rate. We showed that the optimal batch size in this case could be obtained from the EOQ formula with a modified holding cost.

Two types of quantity discounts were considered: all-units, in which the discounted price is valid on all the units in an order, and incremental, in which the discount is applied only to the additional units beyond the breakpoint. The optimal solution to the all-units case will often occur at a breakpoint, whereas the optimal solution to the incremental case will almost never occur at a breakpoint.

In most real systems, the inventory manager cannot ignore the interactions that exist between products. These interactions impose constraints on the system. Constraints might arise because of limitations in the space available to store inventory or in the budget available to purchase items. We considered both cases and showed that when the ratio of the item value or space consumed by the item over the holding cost is the same for all items, a solution to the constrained problem can be obtained easily. When this condition is not met, the formulation requires introducing a Lagrange multiplier. The correct value of the Lagrangian multiplier can be found by trial and error or by some type of search. The presence of the multiplier reduces lot sizes by effectively increasing holding costs.

The finite production rate model of Section 4.6 was extended to multiple products under the assumption that all the products are produced on a single machine. Assuming a rotation cycle policy in which one lot of each product is produced each cycle, we showed how to determine the cycle time minimizing the sum of annual setup and holding costs for all products. Rotation cycles provide a straightforward way to schedule production on one machine, but they may be suboptimal when demand or production rates differ widely or setup costs are sequence dependent.

ADDITIONAL PROBLEMS ON DETERMINISTIC INVENTORY MODELS

31. Peet's Coffee in Menlo Park, California, sells Melitta Number 4 coffee filters at a fairly steady rate of about 60 boxes of filters monthly. The filters are ordered from a supplier in Trenton, New Jersey. Peet's manager is interested in applying some inventory theory to determine the best replenishment strategy for the filters.

    Peet's pays \$2.80 per box of filters and estimates that fixed costs of employee time for placing and receiving orders amount to about \$20. Peet's uses a 22 percent annual interest rate to compute holding costs.

    a. How large a standing order should Peet's have with its supplier in Trenton, and how often should these orders be placed?
    b. Suppose that it takes three weeks to receive a shipment. What inventory of filters should be on hand when an order is placed?

c. What are the average annual holding and fixed costs associated with these filters, assuming they adopt an optimal policy?
d. The Peet's store in Menlo Park is rather small. In what way might this affect the solution you recommended in part (a)?

32. A local supermarket sells a popular brand of shampoo at a fairly steady rate of 380 bottles per month. The cost of each bottle to the supermarket is 45 cents, and the cost of placing an order has been estimated at $8.50. Assume that holding costs are based on a 25 percent annual interest rate. Stock-outs of the shampoo are not allowed.
a. Determine the optimal lot size the supermarket should order and the time between placements of orders for this product.
b. If the procurement lead time is two months, find the reorder point based on the on-hand inventory.
c. If the item sells for 99 cents, what is the annual profit (exclusive of overhead and labor costs) from this item?

33. Diskup produces a variety of personal computer products. Flash drives are produced at a rate of 1,800 per day and are shipped out at a rate of 800 per day. The drives are produced in batches. Each drive costs the company 20 cents, and the holding costs are based on an 18 percent annual interest rate. Shortages are not permitted. Each production run of a drive type requires recalibration of the equipment. The company estimates that this step costs $180.
a. Find the optimal size of each production run and the time between runs.
b. What fraction of the time is the company producing flash drives?
c. What is the maximum dollar investment in inventory that the company has in these drives?

34. Berry Computer is considering moving some of its operations overseas in order to reduce labor costs. In the United States, its main circuit board costs Berry $75 per unit to produce, while overseas it costs only $65 to produce. Holding costs are based on a 20 percent annual interest rate, and the demand has been a fairly steady 200 units per week. Assume that setup costs are $200 both locally and overseas. Production lead times are one month locally and six months overseas.
a. Determine the average annual costs of production, holding, and setup at each location, assuming that an optimal solution is employed in each case. Based on these results only, which location is preferable?
b. Determine the value of the pipeline inventory in each case. (The pipeline inventory is the inventory on order.) Does comparison of the pipeline inventories alter the conclusion reached in part (a)?
c. Might considerations other than cost favor local over overseas production?

35. A large producer of household products purchases a glyceride used in one of its deodorant soaps from outside of the company. It uses the glyceride at a fairly steady rate of 40 pounds per month, and the company uses a 23 percent annual interest rate to compute holding costs. The chemical can be purchased from two suppliers, A and B. A offers the following all-units discount schedule

Order Size	Price per Pound
$0 \le Q < 500$	$1.30
$500 \le Q < 1{,}000$	1.20
$1{,}000 \le Q$	1.10

whereas B offers the following incremental discount schedule: $1.25 per pound for all orders less than or equal to 700 pounds, and $1.05 per pound for all incremental amounts over 700 pounds. Assume that the cost of order processing for each case is $150. Which supplier should be used?

36. The president of Value Filters became very enthusiastic about using EOQs to plan the sizes of her production runs and instituted lot sizing based on EOQ values before she could properly estimate costs. For one particular filter line, which had an annual demand of 1,800 units per year and which was valued at $2.40 per unit, she assumed a holding cost based on a 30 percent annual interest rate and a setup cost of $100. Sometime later, after the cost accounting department had time to perform an analysis, it found that the appropriate value of the interest rate was closer to 20 percent, and the setup cost was about $40. What was the additional average annual cost of holding and setup incurred from using the original cost estimates?
37. Consider the carload discount schedule pictured in Figure 4–13. Suppose $M = 500$ units, $c = \$10$ per unit, and a full carload of 500 units costs $3,000.
    a. Develop a graph of the average cost per unit, $C(Q)/Q$, assuming this schedule.
    b. Suppose that the units are consumed at a rate of 800 per week, order setup cost is $2,500, and holding costs are based on an annual interest charge of 22 percent. Graph the function $G(Q) = \lambda C(Q)/Q + K\lambda/Q + I(C(Q)/Q)Q/2$ and find the optimal value of Q. (Assume that 1 year = 50 weeks.)
    c. Repeat part (b) for $\lambda = 1{,}000$ per week and $K = \$1{,}500$.
38. Harold Gwynne is considering starting a sandwich-making business from his dormitory room in order to earn some extra income. However, he has only a limited budget of $100 to make his initial purchases. Harold divides his needs into three areas: breads, meats and cheeses, and condiments. He estimates that he will be able to use all the products he purchases before they spoil, so perishability is not a relevant issue. The demand and cost parameters are as follows.

	Breads	Meats and Cheeses	Condiments
Weekly demand	6 packages	12 packages	2 pounds
Cost per unit	$0.85	$3.50	$1.25
Fixed order cost	$12	$8	$10

    The choice of these fixed costs is based on the fact that these items are purchased at different locations in town. They include the cost of Harold's time in making the purchase. Assume that holding costs are based on an annual interest rate of 25 percent.
    a. Find the optimal quantities that Harold should purchase of each type of product so that he does not exceed his budget.
    b. If Harold could purchase all the items at the same location, would that alter your solution? Why?
39. Mike's Garage, a local automotive service and repair shop, uses oil filters at a fairly steady rate of 2,400 per year. Mike estimates that the cost of his time to make and process an order is about $50. It takes one month for the supplier to deliver the oil filters to the garage, and each one costs Mike $5. Mike uses an annual interest rate of 25 percent to compute his holding cost.
    a. Determine the optimal number of oil filters that Mike should purchase and the optimal time between placement of orders.

b. Determine the level of on-hand inventory at the time a reorder should be placed.
c. Assuming that Mike uses an optimal inventory control policy for oil filters, what is the annual cost of holding and order setup for this item?

40. An import/export firm has leased 20,000 cubic feet of storage space to store six items that it imports from the Far East. The relevant data for the six items it plans to store in the warehouse follow.

Item	Annual Demand (units)	Cost per Unit ($)	Space per Unit (feet^3)
DVD	800	$200.00	12
32-inch flat screen TV	1,600	150.00	18
Blank DVDs (box of 10)	8,000	30.00	3
Blank CDs (box of 50)	12,000	18.00	2
Compact stereo	400	250.00	24
Telephone	1,200	12.50	3

Setup costs for each product amount to $2,000 per order, and holding costs are based on a 25 percent annual interest rate. Find the optimal order quantities for these six items so that the storage space is never exceeded. (Hint: Use a cell location for the Lagrange multiplier and search over different values until the storage constraint is satisfied as closely as possible.)

41. A manufacturer of greeting cards must determine the size of production runs for a popular line of cards. The demand for these cards has been a fairly steady 2 million per year, and the manufacturer is currently producing the cards in batch sizes of 50,000. The cost of setting up for each production run is $400.

Assume that for each card the material cost is 35 cents, the labor cost is 15 cents, and the distribution cost is 5 cents. The accounting department of the firm has established an interest rate to represent the opportunity cost of alternative investment and storage costs at 20 percent of the value of each card.

a. What is the optimal value of the EOQ for this line of greeting cards?
b. Determine the additional annual cost resulting from using the wrong production lot size.

42. Suppose that in problem 41 the firm decides to account for the fact that the production rate of the cards is not infinite. Determine the optimal size of each production run assuming that cards are produced at the rate of 75,000 per week.

43. Pies 'R' Us bakes its own pies on the premises in a large oven that holds 100 pies. They sell the pies at a fairly steady rate of 86 per month. The pies cost $2 each to make. Prior to each baking, the oven must be cleaned out, which requires one hour's time for four workers, each of whom is paid $8 per hour. Inventory costs are based on an 18 percent annual interest rate. The pies have a shelf life of three months.

a. How many pies should be baked for each production run? What is the annual cost of setup and holding for the pies?
b. The owner of Pies 'R' Us is thinking about buying a new oven that requires one-half the cleaning time of the old oven and has a capacity twice as large as the old one. What is the optimal number of pies to be baked each time in the new oven?

c. The net cost of the new oven (after trading in the old oven) is $350. How many years would it take for the new oven to pay for itself?

44. The Kirei-Hana Japanese Steak House in San Francisco consumes 3,000 pounds of sirloin steak each month. Yubi Hirai, the new restaurant manager, recently completed an MBA degree. She learned that the steak was replenished using an EOQ value of 2,000 pounds. The EOQ value was computed assuming an interest rate of 36 percent per year. Assume that the current cost of the sirloin steak to the steak house is $4 per pound.
    a. What is the setup cost used in determining the EOQ value?

    A meat wholesaler offers a 5 percent discount on orders in quantities of 3,000 pounds or more.

    b. Should Ms. Hirai accept the offer from the wholesaler? If so, how much can be saved?
    c. Ms. Hirai misunderstood the offer as an all-units discount. In fact, the 5 percent discount was incremental and applied only to the amounts ordered over 3,000 pounds. Should this offer be accepted, and if so, how much is now saved?

45. Green's Buttons of Rolla, Missouri, supplies all the New Jersey Fabrics stores with eight different styles of buttons for men's dress shirts. The plastic injection molding machine can produce only one button style at a time, and substantial time is required to reconfigure the machine for different button styles. As Green's has contracted to supply fixed quantities of buttons for the next three years, its demand can be treated as fixed and known. The relevant data for this problem appear in the following table.

Button Type	Annual Sales	Production Rate (units/day)	Setup Time (hours)	Variable Cost
A	25,900	4,500	6	$0.003
B	42,000	5,500	4	0.002
C	14,400	3,300	8	0.008
D	46,000	3,200	4	0.002
E	12,500	1,800	3	0.010
F	75,000	3,900	6	0.005
G	30,000	2,900	1	0.004
H	18,900	1,200	3	0.007

Assume 250 working days per year. Green's accounting department has established 1 percent annual interest rate for the cost of capital and a 3 percent annual interest rate to account for storage space. Setup costs are $20 per hour required to reconfigure the equipment for a new style. Suppose that the firm decides to use a rotation cycle policy for production of the buttons.

    a. What is the optimal rotation cycle time?
    b. How large should the lots be?
    c. What is the average annual cost of holding and setups at the optimal solution?
    d. What contractual obligations might Green's have with New Jersey Fabrics that would prevent them from implementing the policy you determined in parts (a) and (b)? More specifically, if Green's agreed to make three shipments per year for each button style, what production policy would you recommend?

APPENDIX 4–A

MATHEMATICAL DERIVATIONS FOR MULTIPRODUCT CONSTRAINED EOQ SYSTEMS

Consider the standard multiproduct EOQ system with a budget constraint that was discussed in Section 4.8. Mathematically, the problem is to find values of the variables Q_1, Q_2, ... , Q_n to

$$\text{minimize} \quad \sum_{i=1}^{n}\left[\frac{h_i Q_i}{2}+\frac{K_i \lambda_i}{Q_i}\right]$$

subject to

$$\sum_{i=1}^{n} c_i Q_i \le C.$$

Let EOQ_i be the respective unconstrained EOQ values. Then there are two possibilities:

$$\sum_{i=1}^{n} c_i \text{EOQ}_i \le C, \tag{1}$$

$$\sum_{i=1}^{n} c_i \text{EOQ}_i > C. \tag{2}$$

If Equation (1) holds, the optimal solution is the trivial solution; namely, set $Q_i = \text{EOQ}_i$. If Equation (2) holds, then we are guaranteed that the constraint is binding at the optimal solution. That means that the constraint may be written

$$\sum_{i=1}^{n} c_i Q_i = C.$$

In this case, we introduce the Lagrange multiplier θ, and the problem is now to find Q_1, Q_2, ... , Q_n, and θ to solve the unconstrained problem:

$$\text{minimize } G(Q_1, Q_2, \ldots, Q_n, \theta) = \sum_{i=1}^{n}\left(\frac{h_i Q_i}{2}+\frac{K_i \lambda_i}{Q_i}\right) + \theta \sum_{i=1}^{n}(c_i Q_i - C).$$

Necessary conditions for optimality are that

$$\frac{\partial G}{\partial Q_i} = 0 \qquad \text{for } i = 1, \ldots, n$$

and

$$\frac{\partial G}{\partial \theta} = 0.$$

The first n conditions give

$$\frac{h_i}{2} - \frac{K_i \lambda_i}{Q_i^2} + \theta c_i = 0 \qquad \text{for } i = 1, \ldots, n.$$

Rearranging terms, we get

$$Q_i = \sqrt{\frac{2K_i \lambda_i}{h_i + 2\theta c_i}} \qquad \text{for } i = 1, \ldots, n,$$

and we also have the final condition

$$\sum_{i=1}^{n} c_i Q_i = C.$$

Now consider the case where $c_i/h_i = c/h$ independent of i. By dividing the numerator and the denominator by h_i, we may write

$$\begin{aligned} Q_i &= \sqrt{\frac{2K_i \lambda_i}{h_i}} \sqrt{\frac{1}{1 + 2\theta c / h}} \\ &= \text{EOQ}_i \sqrt{\frac{1}{1 + 2\theta c / h}} \\ &= \text{EOQ}_i m, \end{aligned}$$

where

$$m = \sqrt{\frac{1}{1 + 2\theta c / h}}.$$

Substituting this expression for Q_i into the constraint gives

$$\sum_{i=1}^{n} c_i \text{EOQ}_i m = C$$

or

$$m = \frac{C}{\sum_{i=1}^{n} c_i \text{EOQ}_i}.$$

APPENDIX 4–B

Glossary of Notation for Chapter 4

c = Proportional order cost
EOQ = Economic order quantity (optimal lot size)
$G(Q)$ = Average annual cost associated with lot size Q
h = Holding cost per unit time
h' = Modified holding cost for finite production rate model
I = Annual interest rate used to compute holding cost
K = Setup cost or fixed order cost
λ = Demand rate (units per unit time)
P = Production rate for finite production rate model
Q = Lot size or size of the order
s_i = Setup time for product i (refer to Section 4.9)
T = Cycle time; time between placement of successive orders
τ = Order lead time
θ = Lagrange multiplier for space-constrained model (refer to Section 4.8)
w_i = Space consumed by one unit of product i (refer to Section 4.8)
W = Total space available (refer to Section 4.8)

BIBLIOGRAPHY

Arrow, K. A. Chapter 1. In S. Karlin, and H. Scarf (Eds.), *Studies in the Mathematical Theory of Inventory Production.* Stanford, CA: Stanford University Press, 1958.

Barbosa, L. C., and M. Friedman. "Deterministic Inventory Lot Size Models—A General Root Law." *Management Science* 23 (1978), pp. 820–29.

Bureau of Economic Analysis, Table 5.8.5B. Private Inventories and Domestic Final Sales by Industry. May 30, 2019. Retrieved from https://apps.bea.gov/scb/2019/06-june/pdf/0619-selected-nipa-tables.pdf

Bowman, E. H., and R. B. Fetter. *Analysis for Production Management.* New York: McGraw-Hill/Irwin, 1961.

Churchman, C. W., R. L. Ackoff, and E. L. Arnoff. *Introduction to Operations Research.* New York: John Wiley & Sons, 1957.

Fetter, R. B., and W. C. Dalleck. *Decision Models for Inventory Management.* New York: McGraw-Hill/Irwin, 1961.

Hadley, G. J., and T. M. Whitin. *Analysis of Inventory Systems.* Englewood Cliffs, NJ: Prentice Hall, 1963.

Hanssmann, F. "A Survey of Inventory Theory from the Operations Research Viewpoint." In R. L. Ackoff (Ed.), *Progress in Operations Research.* New York: John Wiley & Sons, 1961.

Harris, F. W. *Operations and Cost* (Factory Management Series). Chicago: Shaw, 1915.

Magee, J. F., and D. M. Boodman. *Production Planning and Inventory Control* (2nd ed.). New York: McGraw-Hill, 1967.

Maxwell, W. L. "The Scheduling of Economic Lot Sizes." *Naval Research Logistics Quarterly 11* (1964), pp. 89–124.

Muckstadt, J. A., and R. O. Roundy. "Analysis of Multi-Stage Production Systems." In S. C. Graves, A. H. G. Rinnooy Kan, and P. H. Zipkin (Eds.), *Logistics of Production and Inventory* (59–132). Volume 4 of. *Handbooks in Operations Research and Management Science.* Amsterdam: North-Holland, 1993.

Rosenblatt, M. J. "Multi-Item Inventory System with Budgetary Constraint: A Comparison between the Lagrangian and the Fixed Cycle Approach." *International Journal of Production Research* 19 (1981), pp. 331–39.

Roundy, R. 0. "98%-Effective Integer-Ratio Lot-Sizing for One Warehouse Multi-Retailer Systems." *Management Science* 31 (1985), pp. 1416–30.

Scarf, H. E., D. M. Gilford, and M. W. Shelly. *Multistage Inventory Models and Techniques.* Stanford, CA: Stanford University Press, 1963.

Schwarz, L. B., and L. Schrage. "Optimal and Systems Myopic Policies for Multiechelon Production/ Inventory Assembly Systems." *Management Science* 21 (1971), pp. 1285–94.

Starr, M. K., and D. W. Miller. *Inventory Control: Theory and Practice.* Englewood Cliffs, NJ: Prentice Hall, 1962.

Wagner, H. M. *Statistical Management of Inventory Systems.* New York: John Wiley & Sons, 1962.

Whitin, T. M. *The Theory of Inventory Management* (2nd ed.). Princeton, NJ: Princeton University Press, 1957.

Wilson, R. H. "A Scientific Routine for Stock Control." *Harvard Business Review* 13 (1934), pp. 116–28.

CHAPTER

5

Inventory Control Subject to Uncertain Demand

"Knowing what you've got, knowing what you need, knowing what you don't—that's inventory control."

—Frank Wheeler [protagonist in *Revolutionary Road*]

CHAPTER OVERVIEW

Purpose

To understand how one deals with uncertainty (randomness) in the demand when computing replenishment policies for a single inventory item.

KEY POINTS

1. *What is uncertainty and when should it be assumed?* Uncertainty means that demand is a random variable. A random variable is defined by its probability distribution, which is generally estimated from a past history of demands. In practice, it is common to assume that demand follows a normal distribution. When demand is assumed normal, one only needs to estimate the mean, μ, and variance, σ^2. Clearly, demand is uncertain to a greater or lesser extent in all real-world applications. What value, then, does the analysis of Chapters 3 and 4 have, where demand was assumed known? Chapter 3 focused on *systematic* or predictable changes in the demand pattern, such as peaks and valleys. The constant demand models in Chapter 4 results are useful if the variance of demand is low relative to the mean. In this chapter we consider items whose primary variation is due to uncertainty rather than predictable causes.

   If demand is described by a random variable, it is unclear what the optimization criterion should be, since the cost function is a random variable as well. To handle this, we assume that the objective is to minimize *expected* costs. The use of the expectation operator is justified by the law of large numbers from probability, since an inventory control problem invariably spans many planning periods. The law of large numbers guarantees that the arithmetic average of the incurred costs and the expected costs grow close as the number of planning periods gets large.
2. *The newsvendor model.* Consider a news vendor who decides each morning how many papers to buy to sell during the day. Since daily demand is highly variable, it is modeled with a random variable, *D*. Suppose that *Q* is the number of papers he or she purchases. If *Q* is too large, there are unsold papers; if *Q* is too small, some demands go unfilled. If we let c_o be the unit overage cost, and c_u be the unit

underage cost, then we show that the optimal number of papers to purchase at the start of a day, say Q^*, satisfies $F(Q^*) = c_u/(c_u + c_o)$, where $F(Q^*)$ is the cumulative distribution function of D evaluated at Q^* (which is the same as the probability that demand is less than or equal to Q^*).

3. *Lot size–reorder point systems.* The newsvendor model is appropriate for a problem that essentially restarts from scratch every period. Yesterday's newspaper has no value in the market, save for the possible scrap value of the paper itself. However, most inventory control situations that one encounters in the real world are not like this. Unsold items continue to have value in the marketplace for many periods. For these cases we use an approach that is essentially an extension of the EOQ model of Chapter 4.

   The lot size–reorder point system relies on the assumption that inventories are reviewed continuously rather than periodically. That is, the state of the system is known at all times. The system consists of two decision variables: Q and R. Q is the order size and R is the reorder point. That is, when the inventory of stock on hand reaches R, an order for Q units is placed. The model also allows for a positive order lead time, τ. It is the demand over the lead time that is the key uncertainty in the problem, since the lead time is the response time of the system.

4. *Service levels in* (Q, R) *systems.* We assume two types of service. Type 1 service is the probability of not stocking out in the lead time and is represented by the symbol α. Type 2 service is the proportion of demands that are filled from stock (also known as the fill rate) and is represented by the symbol β. Finding the optimal (Q, R) subject to a Type 1 service objective is very easy. One merely finds R from $F(R) = \alpha$ and sets $Q = \text{EOQ}$. Unfortunately, what one generally means by service is the Type 2 criterion, and finding (Q, R) in that case is more difficult. For Type 2 service, we only consider the normal distribution. The solution can be obtained using standardized loss tables, $L(z)$, which are supplied in the back of the book. As with the cost model, setting $Q = \text{EOQ}$ and solving for R will usually give good results if one does not want to bother with an iterative procedure.

5. *Periodic review systems under uncertainty.* The newsvendor model treats a product that perishes quickly (after one period). However, periodic review models also make sense when unsold product can be used in future periods. In this case the form of the optimal policy is known as an (s, S) policy. Let u be the starting inventory in any period. Then the (s, S) policy is: If $u \le s$, order to S (that is, order $S - u$). If $u > s$, don't order. Unfortunately, finding the optimal values of (s, S) each period is much more difficult than finding the optimal (Q, R) policy, and is beyond the scope of this book. We also briefly discuss service levels in periodic review systems.

6. *Multiproduct systems.* Virtually all inventory control problems occurring in the operations planning context involve multiple products. One issue that arises in multiproduct systems is determining the amount of effort one should expend managing each item. Clearly, some items are more valuable to the business than others. The ABC classification system is one means of ranking items. Items are sequenced in decreasing order of annual dollar volume of sales or usage. Ordering the items in this way and graphing the cumulative dollar volume gives an exponentially increasing curve known as a Pareto curve. Typically, 20 percent of the items account for 80 percent of the annual dollar volume (A items), the next 30 percent of the items account for the next 15 percent of the dollar volume (B items), and the final 50 percent of the items account for the final 5 percent of the dollar volume (C items). A items should receive the most attention. Their inventory levels should be reviewed often, and they should carry a high service level. B items do not need such close scrutiny, and C items are typically ordered infrequently in large quantities.

7. *Other issues.* The discussion of stochastic inventory models in this chapter barely reveals the tip of the iceberg in terms of the vast quantity of research done on this topic. Two important areas of research are multi-echelon inventory systems and perishable inventory systems. A multi-echelon inventory system is one in which items are stored at multiple locations linked by a network. Supply chains, discussed in detail in Chapter 6, are such a system. Another important area of research are items that change during storage, thus affecting their useful lifetime. One class of such items are perishable items. Perishable items have a fixed lifetime known in advance (i.e., food, and pharmaceuticals). A related problem is managing items subject to obsolescence. Obsolescence differs from perishability in that the useful lifetime of an item subject to obsolescence cannot be predicted in advance. Mathematical models for analyzing such problems are quite complex and well beyond the scope of this book.

The management of uncertainty plays an important role in the success of any firm. What are the sources of uncertainty that affect a firm? A partial list includes uncertainty in consumer preferences and trends in the market, uncertainty in the availability and cost of labor and resources, uncertainty in vendor resupply times, uncertainty in weather and its ramifications on operations logistics, uncertainty of financial variables such as stock prices and interest rates, and uncertainty of demand for products and services.

Before the terrible tragedy of September 11, 2001, passengers could arrive at the airport no more than 30 or 40 minutes before the scheduled departure time of their flight. Now we might arrive two hours before the flight. Increased airport security has not only increased the average time required but it has also increased the uncertainty of this time. To compensate for this increased uncertainty, we arrive far in advance of our scheduled departure time to provide a larger buffer time. This same principle will apply when managing inventories.

The uncertainty of demand and its effect on inventory management strategies are the subjects of this chapter. When a quantity is uncertain, it means that we cannot predict its value exactly in advance. For example, a department store cannot exactly predict the sales of a particular item on any given day. An airline cannot exactly predict the number of people who will choose to fly on any given flight. How, then, can these firms choose the number of items to keep in inventory or the number of flights to schedule on any given route?

Although exact sales of an item or numbers of seats booked on a plane cannot be predicted in advance, one's past experience can provide useful information for planning. As shown in Section 5.1, previous observations of any random phenomenon can be used to estimate its *probability distribution*. By properly quantifying the consequences of incorrect decisions, a well-thought-out mathematical model of the system being studied will result in intelligent strategies. When uncertainty is present, the objective is almost always to minimize expected cost or maximize expected profits.

Demand uncertainty plays a key role in many industries, but some are more susceptible to business cycles than others. The world economy saw one of the worst recessions in recent times in 2008. The stock market eventually dropped to about half of its high in 2007, and unemployment levels soared. The retailing industry in particular is very sensitive to the vicissitudes of consumer demand. Low cost providers such as Costco and Walmart fared well, but high-end retailers such as Nordstrom and Bloomingdales suffered. Matching supply and demand becomes even more critical in times of recession.

As some level of demand uncertainty seems to characterize almost all inventory management problems in practice, one might question the value of the deterministic inventory control models discussed in Chapter 4. There are two reasons for studying deterministic models. One is that they provide a basis for understanding the fundamental trade-offs encountered in inventory management. Another is that they may be good approximations depending on the degree of uncertainty in the demand.

To better understand the second point, let D be the demand for an item over a given period of time. We express D as the sum of two parts, D_{Det} and D_{Ran}. That is,

$$D = D_{\text{Det}} + D_{\text{Ran}},$$

where

$$D_{\text{Det}} = \text{Deterministic component of demand,}$$
$$D_{\text{Ran}} = \text{Random component of demand.}$$

There are a number of circumstances under which it would be appropriate to treat D as being deterministic even though D_{Ran} is not zero. Some of these are:

1. when the variance of the random component, D_{Ran}, is small relative to the magnitude of D,
2. when the predictable variation is more important than the random variation,
3. when the problem structure is too complex to include an explicit representation of randomness in the model.

An example of circumstance 2 occurs in the aggregate planning problem from Chapter 3. Although the forecast error of the aggregate demands over the planning horizon may not be zero, we are more concerned with planning for the anticipated changes in the demand than for the unanticipated changes. An example of circumstance 3 occurs in material requirements planning (treated in detail in Chapter 9). The intricacies of the relationships among various component levels and end items make it difficult to incorporate demand uncertainty into the analysis.

However, for many items, the random component of the demand is too significant to ignore. As long as the expected demand per unit time is relatively constant and the problem structure not too complex, explicit treatment of demand uncertainty is desirable. In this chapter we will examine several of the most important stochastic inventory models and the key issues surrounding uncertainty. As discussed in Chapter 4, the word **stochastic** is a synonym for random.

As discussed in Chapter 4, inventory control models come in two key types: (1) periodic review and (2) continuous review. **Periodic review** means that the inventory level is known at discrete points in time only; **continuous review** means that the inventory level is known at all times. Periodic review models may be for one planning period or for multiple planning periods. For one-period models, the objective is to balance the costs of overage (ordering too much) and underage (ordering too little). Single-period models are useful in several contexts: planning for initial shipment sizes for high-fashion items, ordering policies for food products that perish quickly, or determining run sizes for items with short useful lifetimes, such as newspapers. Because of this last application, the single-period stochastic inventory model has come to be known as the **newsvendor model.** The newsvendor model will be the first one considered in this chapter.

The vast majority of computer-based inventory control systems on the market use some variant of the continuous review models treated in the remainder of this chapter. They are, in a sense, extensions of the EOQ model that incorporate uncertainty. Their popularity in practice is attributable to several factors. First, the policies are easy to compute and easy to implement. Second, the models accurately describe most systems in which there is ongoing replenishment of inventory items under uncertainty. A detailed discussion of service levels is included as well. Because estimating penalty costs is difficult in practice, service level approaches are more frequently implemented than penalty cost approaches.

Multiperiod stochastic inventory models dominate the professional literature on inventory theory. There are enough results in this fascinating research area to compose an entire volume on the topic. However, the level of mathematical sophistication they require is beyond that of this book.

5.1 THE NATURE OF RANDOMNESS

In order to clarify what the terms **randomness** and **uncertainty** mean in the context of inventory control, we begin with an example.

EXAMPLE 5.1

On consecutive Sundays, Mac, the owner of a local newsstand, purchases a number of copies of *The Computer Journal,* a popular weekly magazine. He pays 25 cents for each copy and sells each for 75 cents. Copies he has not sold during the week can be returned to his supplier for 10 cents each. The supplier is able to salvage the paper for printing future issues. Mac has kept careful records of the demand each week for the *Journal*. (This includes the number of copies actually sold plus the number of customer requests that could not be satisfied.) The observed demands during each of the last 52 weeks were

15	19	9	12	9	22	4	7	8	11
14	11	6	11	9	18	10	0	14	12
8	9	5	4	4	17	18	14	15	8
6	7	12	15	15	19	9	10	9	16
8	11	11	18	15	17	19	14	14	17
13	12								

There is no discernible pattern to these data, so it is difficult to predict the demand for the *Journal* in any given week. However, we can represent the demand experience of this item as a frequency histogram, which gives the number of times each weekly demand occurrence was observed during the year. The histogram for this demand pattern appears in Figure 5–1.

One uses the frequency histogram to estimate the probability that the number of copies of the *Journal* sold in any week is a specific value. These probability estimates are obtained by dividing the number of times that each demand occurrence was observed during the year by 52. For example, the probability that demand is 10 is estimated to be 2/52 = .0385, and the probability that the demand is 15 is 5/52 = .0962. The collection of all the probabilities is known as the **empirical probability distribution**.

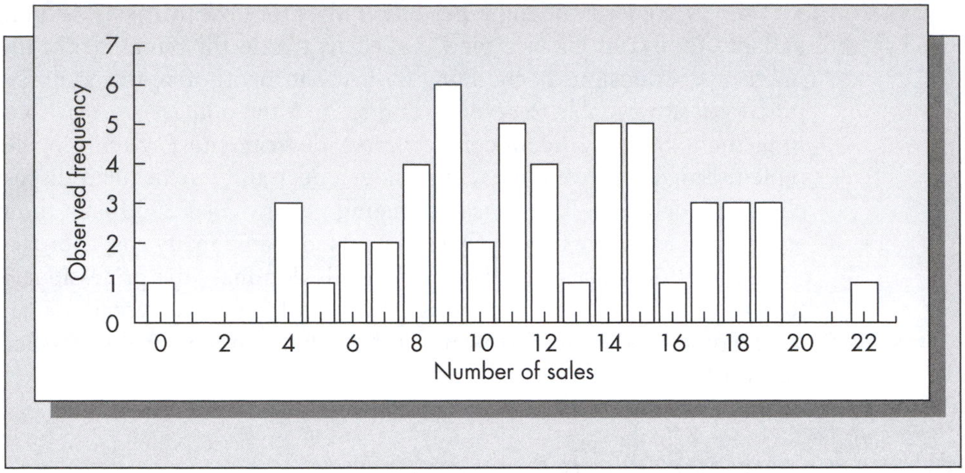

Figure 5–1 Frequency histogram for a 52-week history of sales of *The Computer Journal* at Mac's

Cumulative probabilities also can be estimated in a similar way. For example, the probability that there are nine or fewer copies of the *Journal* sold in any week is (1 + 0 + 0 + 0 + 3 + 1 + 2 + 2 + 4 + 6) = 19/52 = .3654.

Although empirical probabilities can be used in subsequent analyses, they are inconvenient for a number of reasons. First, they require maintaining a record of the demand history for every item. This can be costly and cumbersome. Second, the distribution must be expressed (in this case) as 23 different probabilities. Other items may have an even wider range of past values. Finally, it is more difficult to compute optimal inventory policies with empirical distributions.

For these reasons, we generally approximate the demand history using a continuous distribution. The form of the distribution chosen depends upon the history of past demand and its ease of use. By far the most popular distribution for inventory applications is the normal distribution. One reason is the frequency with which it models demand fluctuations accurately. Another is its convenience. The normal model of demand must be used with care, however, as it admits the possibility of negative values. When using the normal distribution to describe a nonnegative phenomenon such as demand, the likelihood of a negative observation should be small (less than .01 suffices for most applications).

A normal distribution is determined by two parameters: the mean μ and the variance σ^2. These can be estimated from a history of demand by the sample mean $\bar{D}$ and the sample variance s^2. Let $D_1, D_2, \ldots, D_n$ be n past observations of demand. Then

$$\bar{D} = \frac{1}{n}\sum_{i=1}^{n} D_i,$$
$$s^2 = \frac{1}{n-1}\sum_{i=1}^{n}(D_i - \bar{D})^2.$$

For the data pictured in Figure 5–1 we obtain

$$\bar{D} = 11.73,$$
$$s = 4.74.$$

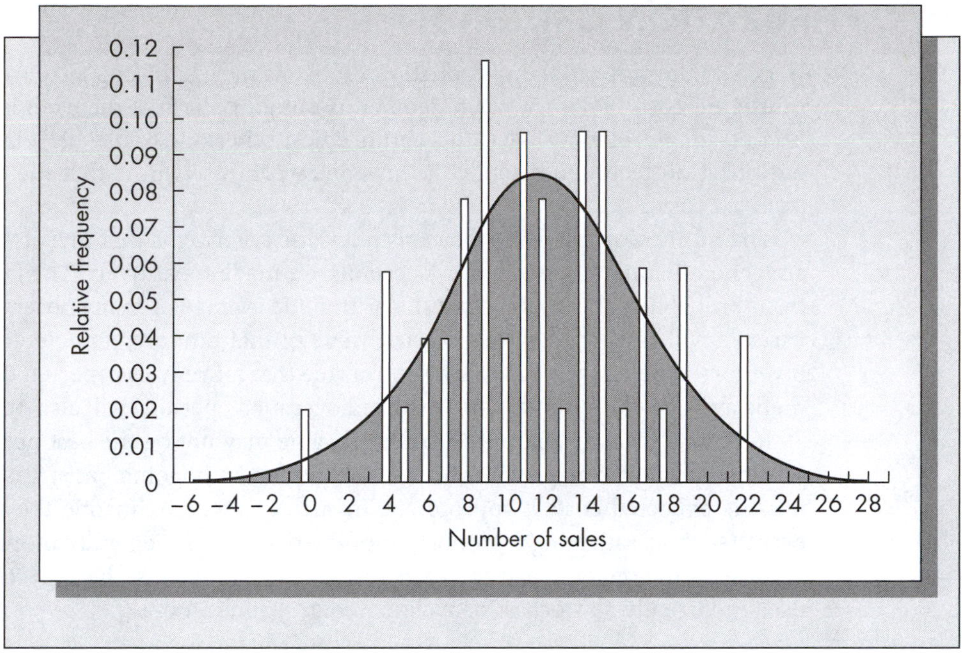

Figure 5–2 Frequency histogram and normal approximation

The normal density function, $f(x)$, is given by the formula

$$f(x) = \frac{1}{\sigma\sqrt{2\pi}} \exp\left[-\frac{1}{2}\left(\frac{x-\mu}{\sigma}\right)^2\right] \qquad \text{for } -\infty < x < +\infty.$$

We substitute $\bar{D}$ as the estimator for μ and s as the estimator for σ.

The **relative frequency histogram** is the same as the frequency histogram pictured in Figure 5–1, except that the y-axis entries are divided by 52. In Figure 5–2 we show the normal density function that results from the substitution we made, superimposed on the relative frequency histogram.

In practice, exponential smoothing is used to recursively update the estimates of the mean and the standard deviation of demand. The standard deviation is estimated using the mean absolute deviation (MAD). Both exponential smoothing and MAD are discussed in detail in Chapter 2. Let $\bar{D}_t$ be the estimate of the mean after observing demand D_t, and let MAD_t be the estimate of the MAD. Then

$$\bar{D}_t = \alpha D_t + (1-\alpha)\bar{D}_{t-1},$$
$$\text{MAD}_t = \alpha\left|D_t - \bar{D}_{t-1}\right| + (1-\alpha)\text{MAD}_{t-1},$$

where $0 < \alpha < 1$ is the smoothing constant. For normally distributed demand

$$\sigma \approx 1.25 * \text{MAD}.$$

A smoothing constant of $\alpha \approx .1$ is generally used to ensure stability in the estimates (see Chapter 2 for a more complete discussion of the issues surrounding the choice of the smoothing constant).

5.2 OPTIMIZATION CRITERION

In general, optimization in production problems means finding a control rule that achieves minimum cost. However, when demand is random, the cost incurred is itself random, and it is no longer obvious what the optimization criterion should be. Virtually all stochastic optimization techniques applied to inventory control assume that the goal is to minimize *expected* costs.

The motivation for using the expected value criterion is that inventory control problems are generally ongoing problems. Decisions are made repetitively. The law of large numbers from probability theory says that the arithmetic average of many observations of a random variable will converge to the expected value of that random variable. In the context of the inventory problem, if we follow a control rule that minimizes expected costs, then the arithmetic average of the actual costs incurred over many periods will also be a minimum.

In certain circumstances, the expected value may not be the best optimization criterion. When a product is acquired only once and not on an ongoing basis, it is not clear that minimizing expected costs is appropriate. In such a case, maximizing the probability of some event (such as satisfying a proportion of the demand) is generally more suitable. However, because of the ongoing nature of most production problems, the expected value criterion is used in virtually all stochastic inventory control applications.

PROBLEMS FOR SECTIONS 5.1 AND 5.2

1. Suppose that Mac only kept track of the number of magazines sold. Would this give an accurate representation of the demand for the *Journal*? Under what circumstances would the actual demand and the number sold be close, and under what circumstances would they differ by a substantial amount?
2. What is the difference between deterministic and random variations in the pattern of demands? Provide an example of a real problem in which predictable variation would be important and an example in which random variation would be important.
3. Oakdale Furniture Company uses a special type of woodworking glue in the assembly of its furniture. During the past 36 weeks, the following amounts of glue (in gallons) were used by the company.

25	38	26	31	21	46	29	19	35	39	24	21
17	42	46	19	50	40	43	34	31	51	36	32
18	29	22	21	24	39	46	31	33	34	30	30

a. Compute the mean and the standard deviation of this sample.

b. Consider the following class intervals for the number of gallons used each week:

Less than 20.
20–27.
28–33.
34–37.
38–43.
More than 43.

Determine the proportion of data points that fall into each of these intervals. Compare these proportions to the probabilities that a normal variate with the mean and the standard deviation you computed in part (a) falls into each of these intervals. Based on the comparison of the observed proportions and those obtained from assuming a normal distribution, would you conclude that the normal distribution provides an adequate fit of these data? (This procedure is essentially the same as a chi-square goodness-of-fit test.)

c. Assume that the numbers of gallons of glue used each week are independent random variables, having the normal distribution with mean and standard deviation computed in part (a). What is the probability that the total number of gallons used in six weeks does not exceed 200 gallons? (Hint: The mean of a sum of random variables is the sum of the means, and the *variance* of a sum of *independent* random variables is the sum of the variances.)

4. In problem 3, what other probability distributions might accurately describe Oakdale's weekly usage of glue?
5. Rather than keeping track of each demand observation, Maria Sucasas, a member of the marketing staff with a large company that produces a line of switches, has kept only grouped data. For switch C9660Q, used in small power supplies, she has observed the following numbers of units of the switch shipped over the last year.

Units Shipped	Number of Weeks
0–2,000	3
2,001–5,000	6
5,001–9,000	12
9,001–12,000	17
12,001–18,000	10
18,001–20,000	4

Based on these observations, estimate the mean and the standard deviation of the weekly shipments. (Hint: This is known as grouped data. For the purposes of your calculation, assume that all observations occur at the midpoint of each interval.)

6. a. Returning to the Oakdale Furniture Company, under what circumstances might the major portion of the usage of the glue be predictable?

b. If the demand were predictable, would you want to use a probability law to describe it? Under what circumstances might the use of a probability model of demand be justified even if the demand could be predicted exactly?

5.3 THE NEWSVENDOR MODEL

Recall from Example 5.1 that Mac wishes to determine the number of copies of *The Computer Journal* he should purchase each Sunday. A study of the historical data showed that the demand during any week is a random variable that is approximately normally distributed, with mean 11.73 and standard deviation 4.74. Each copy is purchased for 25 cents and sold for 75 cents, and he is paid 10 cents for each unsold copy by his supplier. One obvious solution is that he should buy enough to meet the mean demand, which is approximately 12 copies. There is something wrong with this solution. Suppose Mac purchases a copy that he does not sell. His out-of-pocket expense is 25 cents − 10 cents = 15 cents. Suppose, on the other hand, that he is unable to meet the demand of a customer. In that case, he loses 75 cents − 25 cents = 50 cents profit. Hence, there is a significantly greater penalty for not having enough than there is for having too much. If he only buys enough to satisfy mean demand, he will stock-out with the same frequency that he has an oversupply. Our intuition tells us that he should buy more than the mean, but how much more? This question is answered in this section.

Notation

This problem is an example of the newsvendor model, in which a single product is to be ordered at the beginning of a period and can be used only to satisfy demand during that period. Assume that all relevant costs can be determined on the basis of ending inventory. Define

c_o = Cost per unit of positive inventory remaining at the end of the period (known as the **overage cost**).

c_u = Cost per unit of unsatisfied demand. This can be thought of as a cost per unit of negative ending inventory (known as the **underage cost**).

In the development of the model, we will assume that the demand D is a continuous nonnegative random variable with density function $f(x)$ and cumulative distribution function $F(x)$. [A brief review of probability theory is given in Appendix 5–A. In particular, both $F(x)$ and $f(x)$ are defined there.]

The **decision variable** Q is the number of units to be purchased at the beginning of the period. The goal of the analysis is to determine Q to minimize the expected costs incurred at the end of the period.

Development of the Cost Function

A general outline for analyzing most stochastic inventory problems follows.

1. Develop an expression for the cost incurred as a function of both the random variable D and the decision variable Q.
2. Determine the expected value of this expression with respect to the density function or probability function of demand.
3. Determine the value of Q that minimizes the expected cost function.

Define $G(Q, D)$ as the total overage and underage cost incurred at the end of the period when Q units are ordered at the start of the period and D is the demand. If Q units are purchased and D is the demand, $Q - D$ units are left at the end of the period as long as $Q \geq D$. If $Q < D$, then $Q - D$ is negative, and the number of units remaining on hand at the end of the period is 0. Notice that

$$\max\{Q-D,0\} = \begin{cases} Q-D & \text{if } Q \geq D, \\ 0 & \text{if } Q \leq D. \end{cases}$$

In the same way, $\max\{D - Q, 0\}$ represents the excess demand over the supply, or the unsatisfied demand remaining at the end of the period. For any realization of the random variable D, either one or the other of these terms will be zero.

Hence, it now follows that

$$G(Q, D) = c_o \max(0, Q - D) + c_u \max(0, D - Q).$$

Next, we derive the expected cost function. Define

$$G(Q) = E(G(Q, D)).$$

Using the rules outlined in Appendix 5–A for taking the expected value of a function of a random variable, we obtain

$$\begin{aligned} G(Q) &= c_o \int_0^\infty \max(0, Q-x) f(x)dx + c_u \int_0^\infty \max(0, x-Q) f(x)dx \\ &= c_o \int_0^Q (Q-x) f(x)dx + c_u \int_Q^\infty (x-Q) f(x)dx. \end{aligned}$$

Determining the Optimal Policy

We would like to determine the value of Q that minimizes the expected cost $G(Q)$. In order to do so, it is necessary to obtain an accurate description of the function $G(Q)$. We have that

$$\frac{dG(Q)}{dQ} = c_o \int_0^Q 1 f(x)dx + c_u \int_Q^\infty (-1) f(x)dx = c_o F(Q) - c_u(1 - F(Q)).$$

(This is a result of Leibniz's rule, which indicates how one differentiates integrals. Leibniz's rule is stated in Appendix 5–A.)

It follows that

$$\frac{d^2G(Q)}{dQ^2} = (c_o + c_u) f(Q) \geq 0 \qquad \text{for all } Q \geq 0.$$

Because the second derivative is nonnegative, the function $G(Q)$ is said to be **convex** (bowl shaped). We can obtain additional insight into the shape of $G(Q)$ by further analysis. Note that

$$\left.\frac{dG(Q)}{dQ}\right|_{Q=0} = c_o F(0) - c_u(1 - F(0)) = -c_u < 0 \qquad \text{since } F(0) = 0.$$

Since the slope is negative at $Q = 0$, $G(Q)$ is decreasing at $Q = 0$. The function $G(Q)$ is pictured in Figure 5–3.

It follows that the optimal solution, say Q^*, occurs where the first derivative of $G(Q)$ equals zero. That is,

$$G'(Q^*) = (c_o + c_u)F(Q^*) - c_u = 0.$$

Rearranging terms gives

$$F(Q^*) = c_u/(c_o + c_u).$$

We refer to the right-hand side of the last equation as the **critical ratio**. Because c_u and c_o are positive numbers, the critical ratio is strictly between zero and one. This implies that for a continuous demand distribution this equation is always solvable.

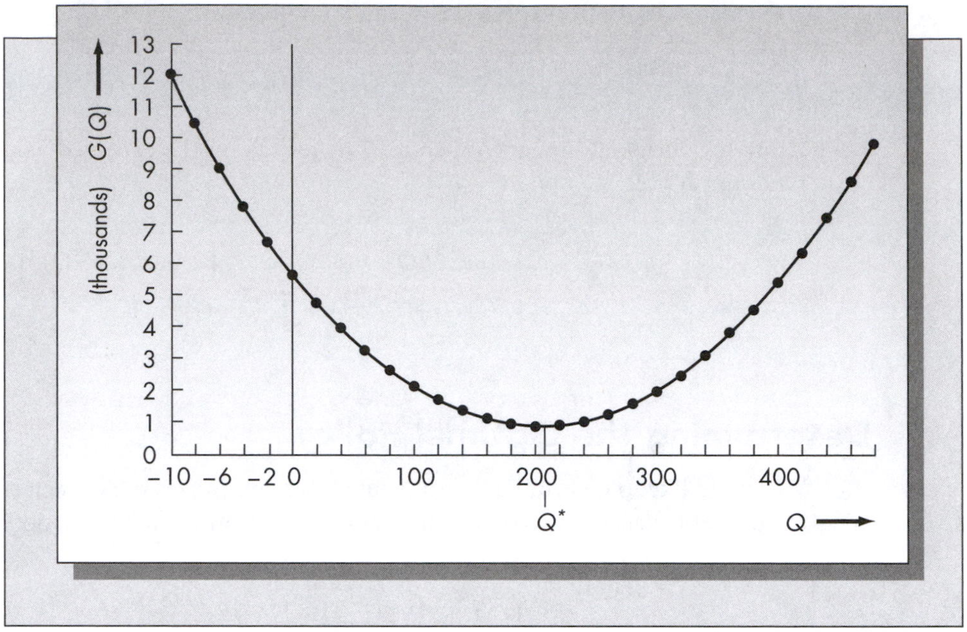

Figure 5–3 Expected cost function for newsvendor model

As $F(Q^*)$ is defined as the probability that the demand does not exceed Q^*, the critical ratio is the probability of satisfying all the demand during the period if Q^* units are purchased at the start of the period. It is important to understand that this is *not* the same as the proportion of satisfied demands. When underage and overage costs are equal, the critical ratio is exactly one-half. In that case Q^* corresponds to the *median* of the demand distribution. When the demand density is symmetric (such as the normal density), the mean and the median are the same.

EXAMPLE 5.1 (CONTINUED)

Consider the example of Mac's newsstand. From past experience, we saw that the weekly demand for the *Journal* is approximately normally distributed with mean $\mu = 11.73$ and standard deviation $\sigma = 4.74$. Because Mac purchases the magazines for 25 cents and can salvage unsold copies for 10 cents, his overage cost is $c_o = 25 - 10 = 15$ cents. His underage cost is the profit on each sale, so that $c_u = 75 - 25 = 50$ cents. The critical ratio is $c_u/(c_o + c_u) = 0.50/0.65 = .77$. Hence, he should purchase enough copies to satisfy all the weekly demand with probability .77. The optimal Q^* is the 77th percentile of the demand distribution (see Figure 5–4).

Using either Table A–1 or Table A–4 at the back of the book, we obtain a standardized value of $z = 0.74$. The optimal Q is

$$Q^* = \sigma z + \mu = (4.74)(0.74) + 11.73 = 15.24 \approx 15.$$

Hence, he should purchase 15 copies every week. This value may also be obtained by the Excel formula =ROUND(NORM.INV(0.77, 11.73, 4.74),0).

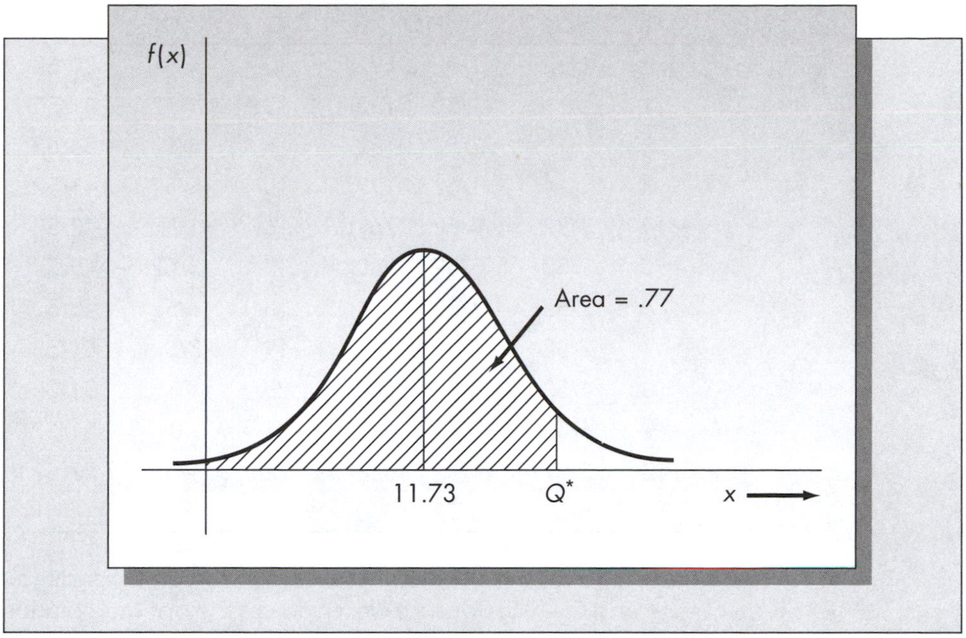

Figure 5–4 Determination of the optimal order quantity for the newsvendor example

Optimal Policy for Discrete Demand

Our derivation of the newsvendor formula assumed that the demand in the period was described by a continuous probability distribution. We noted a number of reasons for the desirability of working with continuous distributions. However, in some cases, and particularly when the mean demand is small, it may not be possible to obtain an accurate representation of the observed pattern of demand using a continuous distribution. For example, the normal approximation of the 52-week history of the demand for the *Journal* pictured in Figure 5–2 may not be considered sufficiently accurate for our purposes.

The procedure for finding the optimal solution to the newsvendor problem when the demand is assumed to be discrete is a natural generalization of the continuous case. In the continuous case, the optimal solution is the value of Q that makes the distribution function equal to the critical ratio $c_u/(c_u + c_o)$. In the discrete case, the distribution function increases by jumps; it is unlikely that any of its values exactly equal the critical ratio. The critical ratio will generally fall between two values of $F(Q)$. The optimal solution procedure is to locate the critical ratio between two values of $F(Q)$ and choose the Q corresponding to the *higher* value. (The fact that you always round Q up rather than simply round it off can be easily proven mathematically. This is different from assuming that units are ordered in discrete amounts but demand is continuous, in which case Q is rounded to the closest integer, as we did earlier.)

EXAMPLE 5.2

We will solve the problem faced by Mac's newsstand using the empirical distribution derived from a one-year history of the demand, rather than the normal approximation of that demand history. The empirical probabilities are obtained from Figure 5–1 by dividing each of the heights by 52. We obtain

Q	f(Q)	F(Q)	Q	f(Q)	F(Q)
0	1/52	1/52 (.0192)	12	4/52	30/52 (.5769)
1	0	1/52 (.0192)	13	1/52	31/52 (.5962)
2	0	1/52 (.0192)	14	5/52	36/52 (.6923)
3	0	1/52 (.0192)	15	5/52	41/52 (.7885)
4	3/52	4/52 (.0769)	16	1/52	42/52 (.8077)
5	1/52	5/52 (.0962)	17	3/52	45/52 (.8654)
6	2/52	7/52 (.1346)	18	3/52	48/52 (.9231)
7	2/52	9/52 (.1731)	19	3/52	51/52 (.9808)
8	4/52	13/52 (.2500)	20	0	51/52 (.9808)
9	6/52	19/52 (.3654)	21	0	51/52 (.9808)
10	2/52	21/52 (.4038)	22	1/52	52/52 (1.0000)
11	5/52	26/52 (.5000)			

The critical ratio for this problem was .77, which corresponds to a value of $F(Q)$ between $Q = 14$ and $Q = 15$. Because we round up, the optimal solution is $Q^* = 15$. Notice that this is exactly the same order quantity obtained using the normal approximation.

Extension to Include Starting Inventory

In the derivation of the newsvendor model, we assumed that the starting inventory at the beginning of the period was zero. Suppose now that the starting inventory is some value u and $u > 0$. The optimal policy for this case is a simple modification of the case $u = 0$. Note that this extension would not apply to newspapers but would be appropriate for a product with a shelf life that exceeds one period.

Consider the expected cost function $G(Q)$, pictured in Figure 5–3. If $u > 0$, it just means that we are starting at some point other than 0. We still want to be at Q^* after ordering, as that is still the lowest point on the cost curve. If $u < Q^*$, this is accomplished by ordering $Q^* - u$. If $u > Q^*$, we are past where we want to be on the curve. Ordering any additional inventory merely moves us up the cost curve, which results in higher costs. In this case it is optimal to simply not order.

Hence the optimal policy when there is a starting inventory of $u > 0$ is

Order $Q^* - u$ if $u < Q^*$.
Do not order if $u \geq Q^*$.

Note that Q^* should be interpreted as the **order-up-to point** rather than the order quantity when $u > 0$. It is also known as a target or **base stock** level.

EXAMPLE 5.2 (CONTINUED)

Suppose that Mac has received 6 copies of the *Computer Journal* at the beginning of the week from another supplier. The optimal policy still calls for having 15 copies on hand after ordering, so now he would order the difference 15 – 6 = 9 copies. (Set $Q^* = 15$ and $u = 6$ to get the order quantity of $Q^* - u = 9$.)

SNAPSHOT APPLICATION

Using Inventory Models to Manage the Seed-Corn Supply Chain at Syngenta

Each year farmers plant tens of thousands of acres of corn worldwide to meet the demands of a growing population. Companies such as Minnesota-based Syngenta Seeds supply seed to these farmers. In the United States alone, the market for corn seed is approximately $2.3 billion annually. Syngenta is one of eight firms that accounts for 73 percent of this market. Where do Syngenta and its competitors obtain this seed? The answer is that it is produced by growing corn and harvesting the seeds. Corn grown for the purpose of harvesting seed is known as seed-corn.

The problem of determining how much seed-corn to plant is complicated by several factors. One is that there are hundreds of different seed hybrids. Some hybrids do better in warmer, more humid climates, while others do better in cooler, dryer climates. The color, texture, sugar content, and so forth, of the corn produced by different hybrids varies as well. Farmers will not reuse a hybrid that yielded disappointing results. Hence, annual demand is hard to predict. In addition to facing uncertain demand, seed-corn producers also face uncertain yields. Their seed-corn plantings are subject to the same set of risks faced by all farmers: frost, draught, and heat spells.

Syngenta must decide each season how much seed-corn to plant. Since demand for the seeds is uncertain, the problem sounds like a straightforward application of the newsvendor model with uncertain demand and uncertain supply. However, Syngenta's decision problem has an additional feature that makes it more complicated than an ordinary news vendor problem. Syngenta, along with many of its competitors, plants seed-corn in both northern and southern hemispheres. Since the hemispheric seasons are the inverse of each other, the plantings are done at different times of the year. In particular, the seed-corn is planted in the spring season in each hemisphere, so that the South American planting occurs about six months after the North American planting. This gives the company a second chance: to increase production levels in South America to make up for shortfalls in North America or to decrease production levels in South America when there are surpluses in North America.

The problem of planning the size of the seed-corn planting was tackled by a team of researchers from the University of Iowa in collaboration with a vice president in charge of supply at Syngenta (Jones, Kegler, Lowe, and Traub, 2003).

Using discrete approximations of the demand and yield distributions, they were able to formulate the planning problem as a linear program, so that it could be solved on a firmwide scale. A retrospective analysis showed that the company could have saved upwards of $5 million using the model. Also, the analysts were able to identify a systematic bias in the forecasts for seed generated by the firm that resulted in consistent overproduction. The mathematical model is now used to help guide the firm on its planting decisions each year.

Extension to Multiple Planning Periods

The underlying assumption made in the derivation of the newsvendor model was that the item "perished" quickly and could not be used to satisfy demand in subsequent periods. In most industrial and retail environments, however, products are durable. Inventory left at the end of a period can be stored and used to satisfy future demand.

This means that the ending inventory in any period becomes the starting inventory in the next period. Previously, we indicated how the optimal policy is modified when starting inventory is present. However, when the number of periods remaining exceeds one, the value of Q^* also must be modified. In particular, the interpretation of both c_o and c_u will be different. We consider only the case in which there are infinitely many periods remaining. The optimal value of the order-up-to point when a finite number of periods remain will fall between the one-period and the infinite-period solutions.

In our derivation and subsequent analysis of the EOQ formula in Chapter 4, we saw that the variable order cost c only entered into the optimization to determine the holding cost ($h = Ic$). In addition, we saw that all feasible operating policies incurred the same average annual cost of replenishment, λc. It turns out that essentially the same thing applies in the infinite horizon newsvendor problem. As long as excess demand is back ordered, all feasible policies will order the demand over any long period of time. Similarly, as long as excess demand is back ordered, the number of units sold will be equal to the demand over any long period of time. Hence, both c_u and c_o will be independent of both the proportional order cost c and the selling price of the item. Interpret c_u as the loss-of-goodwill cost and c_o as the holding cost in this case. (Appendix 5-B rigorously establishes that this is the correct interpretation of the underage and overage costs.)

EXAMPLE 5.3

Suppose that Mac (from Examples 5.1 and 5.2) is considering how to replenish the inventory of a very popular paperback thesaurus that is ordered monthly. Copies of the thesaurus unsold at the end of a month are still kept on the shelves for future sales. Assume that customers who request copies of the thesaurus when they are out of stock will wait until the following month. Mac buys the thesaurus for \$1.15 and sells it for \$2.75. Mac estimates a loss-of-goodwill cost of 50 cents each time he cannot fill an order for a thesaurus. Monthly demand for the book is fairly closely approximated by a normal distribution with mean 18 and standard deviation 6. Mac uses a 20 percent annual interest rate to determine his holding cost. How many copies of the thesaurus should he purchase at the beginning of each month?

Solution

The overage cost in this case is the cost of holding, which is $(1.15)(0.20)/12 = 0.0192$. The underage cost is the loss-of-goodwill cost, which is assumed to be 50 cents. Hence, the critical ratio is $0.5/(0.5 + 0.0192) = .9630$. From Table A–1 at the back of this book, this corresponds to a z value of 1.79. The optimal value of the order-up-to point $Q^* = \sigma z + \mu = (6)(1.79) + 18 = 28.74 \approx 29$ = ROUND(NORM.INV(0.9630,18,6),0) in Excel.

EXAMPLE 5.3 (CONTINUED)

Assume that a local bookstore also stocks the thesaurus and that customers will purchase the thesaurus there if Mac is out of stock. In this case excess demands are lost rather than back ordered. The order-up-to point will be different from that obtained assuming full back ordering of demand. In Appendix 5–B we show that in the lost sales case the underage cost should be interpreted as the loss-of-goodwill cost plus the lost profit. The overage cost should still be interpreted as the holding cost only. Hence, the lost sales solution for this example gives $c_u = 0.5 + 1.6 = 2.1$. The critical ratio is $2.1/(2.1 + 0.0192) = .9909$, giving a z value of 2.36. The optimal value of Q in the lost sales case is $Q^* = \sigma z + \mu = (6)(2.36) + 18 = 32.16 \approx 32$ = ROUND(NORM.INV(0.9909,18,6),0) in Excel.

Although the multiperiod solution appears to be sufficiently general to cover many types of real problems, it suffers from one serious limitation: there is no fixed cost of ordering. This means that the optimal policy, which is to order up to Q^*, requires that ordering take place in every period. In most real systems, however, there are fixed costs associated with ordering, and it is not optimal to place orders each period. Unfortunately, if we include a fixed charge for placing an order, it becomes extremely difficult to determine optimal operating policies. For this reason, we approach the problem of random demand when a fixed charge for ordering is present in a different way. We will assume that inventory levels are reviewed continuously and develop a generalization of the EOQ analysis presented in Chapter 4. This analysis is presented in Section 5.4.

PROBLEMS FOR SECTION 5.3

7. A newsvendor keeps careful records of the number of papers he sells each day and the various costs that are relevant to his decision regarding the optimal number of newspapers to purchase. For what reason might his results be inaccurate? What would he need to do in order to measure the daily demand for newspapers accurately?
8. Billy's Bakery bakes fresh bagels each morning. The daily demand for bagels is a random variable with a distribution estimated from prior experience.

Number of Bagels Sold in One Day	Probability
0	.05
5	.10
10	.10
15	.20
20	.25
25	.15
30	.10
35	.05

   The bagels cost Billy's 8 cents to make, and they are sold for 35 cents each. Bagels unsold at the end of the day are purchased by a nearby charity soup kitchen for 3 cents each.
   a. Based on the given discrete distribution, how many bagels should Billy's bake at the start of each day? (Your answer should be a multiple of 5.)
   b. If you were to approximate the discrete distribution with a normal distribution, would you expect the resulting solution to be close to the answer that you obtained in part (a)? Why or why not?
   c. Determine the optimal number of bagels to bake each day using a normal approximation. (Hint: You must compute the mean μ and the variance σ^2 of the demand from the given discrete distribution.)
9. The Crestview Printing Company prints a popular Christmas card once a year and distributes the cards to stationery and gift shops throughout the United States. It costs Crestview 50 cents to print each card, and the company receives 65 cents for each card sold.

   Because the cards have the current year printed on them, those cards that are not sold are generally discarded. Based on past experience and forecasts of current buying patterns, the probability distribution of the number of cards to be sold nationwide for the next Christmas season is estimated to be

Quantity Sold	Probability
100,000–150,000	.10
150,001–200,000	.15
200,001–250,000	.25
250,001–300,000	.20
300,001–350,000	.15
350,001–400,000	.10
400,001–450,000	.05

Determine the number of cards that Crestview should print this year.

10. Happy Henry's car dealer sells an imported car called the EX123. Once every three months, a shipment of the cars arrives at Happy Henry's. Emergency shipments can be made between these three-month intervals to resupply the cars when inventory falls short of demand. The emergency shipments require two weeks; buyers are willing to wait this long for the cars but will generally go elsewhere if they must wait for the next three-month shipment.

    From experience, it appears that the demand for the EX123 over a three-month interval is normally distributed with a mean of 60 and a variance of 36. The cost of holding an EX123 for one year is $500. Emergency shipments cost $250 per car over and above normal shipping costs.

    a. How many cars should Happy Henry's be purchasing every three months?
    b. Repeat the calculations assuming that excess demands are back ordered from one three-month period to the next. Assume a loss-of-goodwill cost of $100 for customers having to wait until the next three-month period and a cost of $50 per customer for bookkeeping expenses.
    c. Repeat the calculations assuming that when Happy Henry's is out of stock of EX123s, the customer will purchase the car elsewhere. In this case, assume that the cars cost Henry an average of $10,000 and sell for an average of $13,500. Ignore loss-of-goodwill costs for this calculation.

11. Irwin's sells a particular model of fan, with most of the sales being made in the summer months. Irwin's makes a one-time purchase of the fans prior to each summer season at a cost of $40 each and sells each fan for $60. Any fans unsold at the end of the summer season are marked down to $29 and sold in a special fall sale. Virtually all marked-down fans are sold. The following is the number of sales of fans during the past 10 summers: 30, 50, 30, 60, 10, 40, 30, 30, 20, 40.

    a. Estimate the mean and the variance of the demand for fans each summer.
    b. Assume that the demand for fans each summer follows a normal distribution, with mean and variance given by what you obtained in part (a). Determine the optimal number of fans for Irwin's to buy prior to each summer season.
    c. Based on the observed 10 values of the prior demand, construct an empirical probability distribution of summer demand and determine the optimal number of fans for Irwin's to buy based on the empirical distribution.
    d. Based on your results for parts (b) and (c), would you say that the normal distribution provides an adequate approximation?

12. The buyer for Needless Markup, a famous "high end" department store, will be placing an order for handbags made in Italy for the Christmas season. The unit cost of the handbag

to the store is \$28.50, and it will sell for \$150.00. Any handbags not sold by the end of the season are purchased by a discount firm for \$20.00. In addition, store accountants estimate that there is a cost of \$0.40 for each dollar tied up in inventory, since the dollar invested elsewhere could have yielded a profit. Assume that this cost is attached to unsold bags only.

a. Suppose that it is equally likely that there will be anywhere from 50 to 250 sales of this handbag during the holiday season. Based on this, how many bags should the buyer purchase? (Hint: This means that the correct distribution of demand is uniform. You may solve this problem assuming either a discrete or a continuous uniform distribution.)
b. A detailed analysis of past data shows that the number of handbags sold is better described by a normal distribution, with mean 150 and standard deviation 20. Now what is the optimal number of bags to be purchased?
c. The expected demand was the same in parts (a) and (b), but the optimal order quantities should have been different. What accounted for this difference?

5.4 LOT SIZE–REORDER POINT SYSTEMS

The form of the optimal solution for the simple EOQ model with a positive lead time analyzed in Chapter 4 is: When the level of on-hand inventory hits R, place an order for Q units. In that model the only independent decision variable was Q, the order quantity. The value of R was determined from Q, λ, and τ, where λ and τ are demand rate and lead time, respectively. In what follows, we also assume that the operating policy is of the (Q, R) form. However, when generalizing the EOQ analysis to allow for random demand, we treat Q and R as independent decision variables.

The multiperiod newsvendor model was unrealistic for two reasons: it did not include a setup cost for placing an order, and it did not allow for a positive lead time. In most real systems, however, both a setup cost and a lead time are present. For these reasons, the kinds of models discussed in this section are used much more often in practice and, in fact, form the basis for the policies used in many commercial inventory systems.

Note that Q in this section is the amount ordered, whereas Q in Section 5.3 was the order-up-to point.

We make the following assumptions.

1. The system is continuous review. That is, demands are recorded as they occur, and the level of on-hand inventory is known at all times.
2. Demand is random and stationary. Although we cannot predict the value of demand, the *expected* value of demand over any time interval of fixed length is constant. Assume that the expected demand rate is λ units per year.
3. There is a fixed positive lead time τ for placing an order.
4. The following costs are assumed: setup cost at $\$K$ per order; holding cost at $\$h$ per unit held per year; proportional order cost of $\$c$ per item; stock-out cost of $\$p$ per unit of unsatisfied demand (this is also called the shortage cost or the penalty cost).

Describing Demand

In the newsvendor problem, the appropriate random variable is the demand during the period. One period is the amount of time required to effect a change in the on-hand inventory level. This is known as the response time of the system. In the context of our current problem, the response time is the reorder lead time τ. Hence, the random variable of interest is the demand during the lead time. We will assume that the demand during the lead time is a continuous random variable D with probability density function (or pdf) $f(x)$ and cumulative distribution function (or cdf) $F(x)$. Let $\mu = E(D)$ and $\sigma = \sqrt{\text{var}(D)}$ be the mean and the standard deviation of demand during the lead time.

Decision Variables

There are two decision variables for this problem, Q and R, where Q is the lot size or order quantity and R is the reorder level in units of inventory. [Authors' note: When lead times are very long, it may happen that an order should be placed again before a prior order arrives. In that case, the reorder decision variable R should be interpreted as the inventory position (on-hand plus on-order) when a reorder is placed, rather than the inventory level.] Unlike the EOQ model, this problem treats Q and R as independent decision variables. The policy is implemented as follows: when the level of on-hand inventory reaches R, an order for Q units is placed that will arrive in τ units of time. The operation of this system is pictured in Figure 5–5.

Derivation of the Expected Cost Function

The analytical approach we will use to solve this problem is, in principle, the same as that used in the derivation of the newsvendor model. Namely, we will derive an expression for the expected average annual cost in terms of the decision variables (Q, R) and search for the optimal values of (Q, R) to minimize this cost.

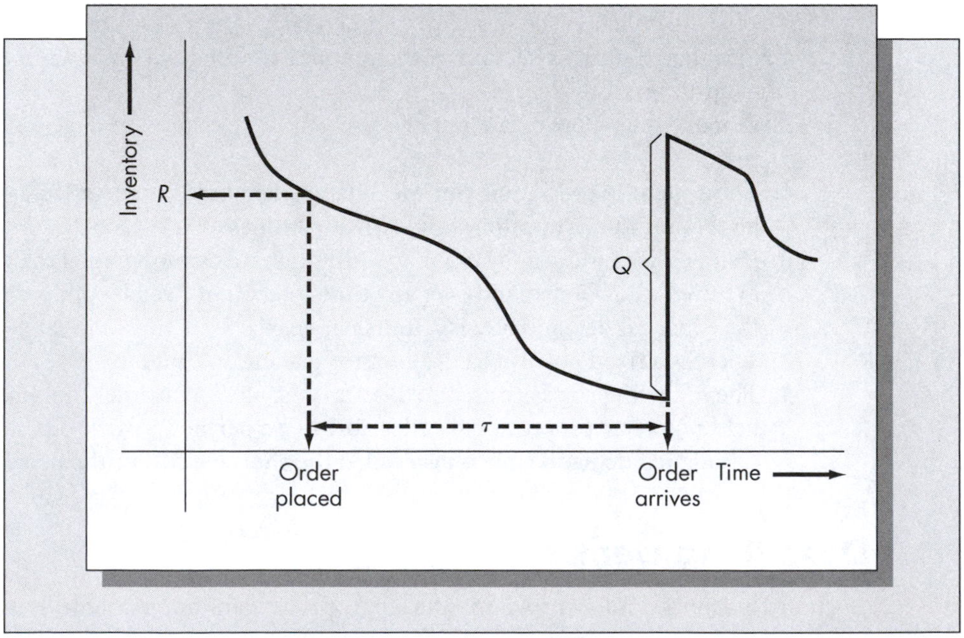

Figure 5–5 Changes in inventory over time for continuous-review (Q, R) system

The Holding Cost

We assume that the mean rate of demand is λ units per year. The expected inventory level varies linearly between s and $Q + s$. We call s the **safety stock**; it is defined as the expected level of on-hand inventory just before an order arrives; it is given by the formula $s = R - \lambda\tau$. The expected inventory level curve appears in Figure 5–6.

We estimate the holding cost from the average of the expected inventory curve. The average of the function pictured in Figure 5–6 is $s + Q/2 = R - \lambda\tau + Q/2$. An important point to note here is that this computation is only an approximation. When computing the average inventory level, we include both the cases when inventory is positive and the cases when it is negative. However, the holding cost should *not* be charged against the inventory level when it is negative, so that we are underestimating the true value of expected holding cost. An exact expression for the true average inventory is quite complex and has been derived only for certain specific demand distributions. In most real systems, however, the proportion of time spent out of stock is generally small, so this approximation should be reasonably accurate.

Setup Cost

A cycle is defined as the time between the arrival of successive orders of size Q. Consistent with the notation used in Chapter 4, let T represent the expected cycle length. Because the setup cost is incurred exactly once each cycle, we need to obtain an expression for the average length of a cycle in order to accurately estimate the setup cost per unit of time.

There are a number of ways to derive an expression for the expected cycle length. From Figure 5–6, we see that the distance between successive arrivals of orders is Q/λ. Another argument is that the expected demand during T is λT. However, because the number of units that are entering inventory each cycle is Q and there is conservation of units, the number of units demanded each cycle on average also must be Q. Setting $Q = \lambda T$ and solving for T gives the same result.

It follows, therefore, that the average setup cost incurred per unit time is $K/T = K\lambda/Q$.

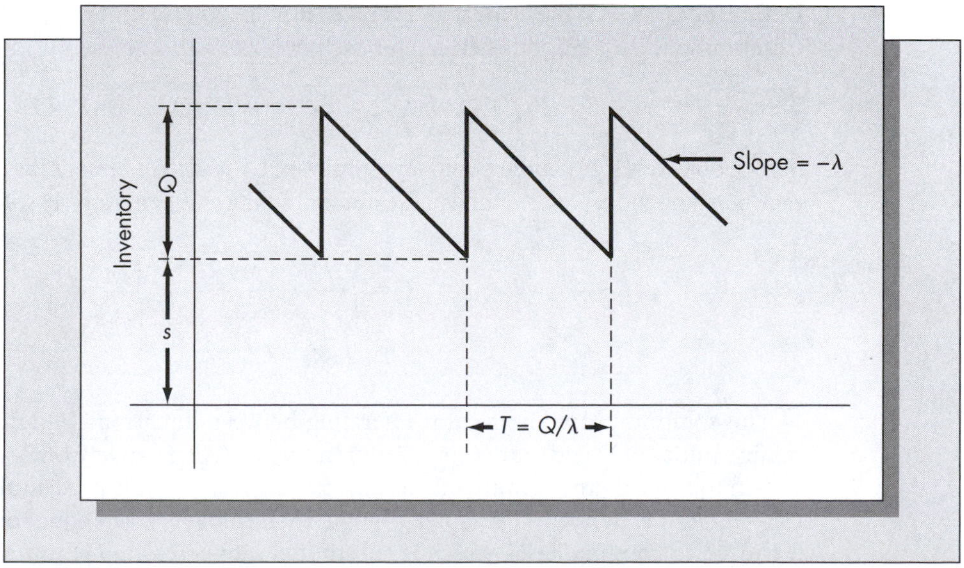

Figure 5–6 Expected inventory level for (Q, R) inventory model

Penalty Cost

From Figure 5–5 we see that the only portion of the cycle during which the system is exposed to shortages is between the time that an order is placed and the time that it arrives (the lead time). The number of units of excess demand is simply the amount by which the demand over the lead time, D, exceeds the reorder level, R. It follows that the expected number of shortages that occur in one cycle is given by the expression

$$E(\max(D-R,0)) = \int_R^\infty (x-R)f(x)dx,$$

which is defined as $n(R)$.

Note that this is essentially the same expression that we derived for the expected number of stock-outs in the newsvendor model. As $n(R)$ represents the expected number of stock-outs incurred in a cycle, it follows that the expected number of stock-outs incurred per unit of time is $n(R)/T = \lambda n(R)/Q$.

Proportional Ordering Cost Component

Over a long period of time, the number of units that enter inventory and the number that leave inventory must be the same. This means that every feasible policy will necessarily order a number of units equal to the demand over any long interval of time. That is, every feasible policy, on average, will replenish inventory at the rate of demand. It follows that the expected proportional order cost per unit of time is λc. Because this term is independent of the decision variables Q and R, it does not affect the optimization. We will henceforth ignore it.

It should be pointed out, however, that the proportional order cost will generally be part of the optimization in an indirect way. The holding cost h is usually computed by multiplying an appropriate value of the annual interest rate I by the value of the item c. For convenience we use the symbol h to represent the holding cost, but keep in mind that it could also be written in the form Ic.

The Cost Function

Define $G(Q, R)$ as the expected average annual cost of holding, setup, and shortages. Combining the expressions derived for each of these terms gives

$$G(Q, R) = h(Q/2 + R - \lambda\tau) + K\lambda/Q + p\lambda n(R)/Q.$$

The objective is to choose Q and R to minimize $G(Q, R)$. We present the details of the optimization in Appendix 5–C. As shown, the optimal solution is to iteratively solve the two equations

$$Q = \sqrt{\frac{2\lambda[K + pn(R)]}{h}} \tag{1}$$

$$1 - F(R) = Qh/p\lambda. \tag{2}$$

The solution procedure requires iterating between equations (1) and (2) until two successive values of Q and R are (essentially) the same. (Alternatively one can use Excel's Solver add-in to solve them as simultaneous equations.) The iterative procedure is started by using Q_0 = EOQ (as defined in Chapter 4). One then finds R_0 from equation (2). That value of R is used to compute $n(R)$, which is substituted into equation (1) to find Q_1, which is then substituted into equation (2) to find R_1, and so on. Convergence generally occurs within two or three iterations. When units are integral, the computations should be continued until successive values of both Q and R are within a single unit of their previous values. When units are continuous, a convergence requirement of less than one unit may be required depending upon the level of accuracy desired.

When the demand is normally distributed, $n(R)$ is computed by using the standardized loss function. The standardized loss function $L(z)$ is defined as

$$L(z) = \int_z^\infty (t - z)\phi(t)dt,$$

where $\varphi(t)$ is the standard normal density. The standardized loss function, $L(z)$, can be computed in Excel or through Table A–4 at the back of this book. To use Excel, write $L(z)$ in the following way:

$$L(z) = \int_z^\infty (t - z)\phi(t)\,dt = \int_z^\infty t\phi(t)\,dt - z(1 - \Phi(z)) = \phi(z) - z(1 - \Phi(z)).$$

The first equality results from the definition of the cumulative distribution function, and the second equality is a consequence of a well-known property of the standard normal distribution (see, for example, Hadley and Whitin, 1963, p. 444). To put this formula into Excel, we write =NORM.S.DIST(z,FALSE)-z* NORM.S.DIST(z,TRUE), where z needs to be a number.

If lead time demand is normal with mean μ and standard deviation σ, then it can be shown that

$$n(R) = \sigma L\left(\frac{R - \mu}{\sigma}\right) = \sigma L(z).$$

The standardized variate z is equal to $(R - \mu)/\sigma$. Calculations of the optimal policy are carried out using Table A–4 at the back of this book or Excel. [Authors' note: In rare cases equations (1) and (2) may not be solvable. This can occur when the penalty cost p is small compared to the holding cost h. The result is either diverging values of Q and R or the right side of equation (2) having a value exceeding 1 at some point in the calculation. In this case, the recommended solution is to set Q = EOQ and $R = R_0$ as long as the right side of equation (2) is less than 1. This often leads to negative safety stock, discussed later in this chapter.]

EXAMPLE 5.4

Harvey's Specialty Shop is well known for stocking international gourmet foods. One of the items that Harvey sells is a popular mustard that he purchases from an English company. The mustard costs Harvey \$10 a jar and requires a six-month lead time for replenishment of stock. Harvey uses a 20 percent annual interest rate to compute holding costs and estimates that if a customer requests the mustard when he is out of stock, the loss-of-goodwill cost is \$25 a jar. Bookkeeping expenses for placing an order amount to about \$50. During the six-month replenishment lead time, Harvey estimates that he sells an average of 100 jars, but there is substantial variation from one six-month period to the next. He estimates that the standard deviation of demand during each six-month period is 25. Assume that demand is described by a normal distribution. How should Harvey control the replenishment of the mustard?

Solution

We wish to find the optimal values of the reorder point R and the lot size Q. In order to get the calculation started, we need to find the EOQ. However, this requires knowledge of the annual rate of demand, which does not seem to be specified. But notice that if the order lead time is six months and the mean lead time demand is 100, that implies that the mean yearly demand is 200, giving a value of $\lambda = 200$. It follows that the

$$\text{EOQ} = \sqrt{2K\lambda / h} = \sqrt{(2)(50)(200) / (0.2)(10)} = 100.$$

The next step is to find R_0 from equation (2). Substituting $Q = 100$, we obtain

$$1 - F(R_0) = Q_0 h/p\lambda = (100)(2)/(25)(200) = .04.$$

From Table A–4 we find that the z value corresponding to a right tail of .04 is $z = 1.75$. Solving $R = \sigma z + \mu$ gives $R = (25)(1.75) + 100 = 144$. Furthermore, $z = 1.75$ results in $L(z) = 0.0162$. Hence, $n(R) = \sigma L(z) = (25)(0.0162) = 0.405$.

We can now find Q_1 from equation (1):

$$Q_1 = \sqrt{\frac{(2)(200)}{2}[50 + (25)(0.405)]} = 110.$$

This value of Q is compared with the previous one, which is 100. They are not close enough to stop. Substituting $Q = 110$ into equation (2) results in $1 - F(R_1) = (110)(2)/(25)(200) = .044$. Table A–4 now gives $z = 1.70$ and $L(z) = 0.0183$. Furthermore, $R_1 = (25)(1.70) + 100 = 143$. We now obtain $n(R_1) = (25)(0.0183) = 0.4575$, and $Q_2 = \sqrt{(200)[50 + (25)(0.4575)]} = 110.85 \approx 111$. Substituting $Q_2 = 111$ into equation (2) gives $1 - F(R2) = .0444$, $z = 1.70$, and $R_2 = R_1 = 143$. Because both Q_2 and R_2 are within one unit of Q_1 and R_1, we may terminate computations.

We conclude that the optimal values of Q and R are $(Q, R) = (111, 143)$. Hence, each time that Harvey's inventory of this type of mustard hits 143 jars, he should place an order for 111 jars.

EXAMPLE 5.4 (CONTINUED)

Determine the following for Harvey's Specialty Shop.

1. Safety stock.
2. The average annual holding, setup, and penalty costs associated with the inventory control of the mustard.
3. The average time between placement of orders.
4. The proportion of order cycles in which no stock-outs occur.
5. The proportion of demands that are not met.

Solution

1. The safety stock is $s = R - \mu = 143 - 100 = 43$ jars.
2. We will compute average annual holding, setup, and penalty costs separately.

   The holding cost is $h[Q/2 + R - \mu] = 2[111/2 + 143 - 100] = \$197$ per year.

   The setup cost is $K\lambda/Q = (50)(200)/111 = \$90.09$ per year.

   The stock-out cost is $p\lambda n(R)/Q = (25)(200)(0.4575)/111 = \$20.61$ per year.

   Hence, the total average annual cost associated with the inventory control of the mustard, assuming an optimal control policy, is \$307.70 per year.
3. $T = Q/\lambda = 111/200 = 0.556$ year $= 6.7$ months.
4. Here we need to compute the probability that no stock-out occurs in the lead time. This is the same as the probability that the lead time demand does not exceed the

reorder point. We have $P\{D \le R\} = F(R) = 1 - .044 = .956$. We conclude that there will be no stock-outs in 95.6 percent of the order cycles.

5. The expected demand per cycle must be Q (see the argument in the derivation of the expected setup cost). The expected number of stock-outs per cycle is $n(R)$. Hence, the proportion of demands that stock out is $n(R)/Q = 0.4575/111 = .004$. Another way of stating this result is that on average 99.6 percent of the demands are satisfied as they occur.

Before ending this section, it should be noted that equations (1) and (2) are derived under the assumption that all excess demand is back ordered. That is, when an item is demanded that is not immediately available, the demand is filled at a later time. However, in many competitive situations, such as retailing, a more accurate assumption is that excess demand is lost. This case is known as **lost sales**. As long as the likelihood of being out of stock is relatively small, equations (1) and (2) will give adequate solutions to both lost sales and back-order situations. If this is not the case, a slight modification to equation (2) is required. The lost-sales version of equation (2) is

$$1 - F(R) = Qh/(Qh + p\lambda). \qquad (2')$$

The effect of solving equations (1) and (2′) simultaneously rather than equations (1) and (2) will be to increase the value of R slightly and decrease the value of Q slightly.

Inventory Level versus Inventory Position

An implicit assumption required in the analysis of (Q, R) policies was that each time an order of Q units arrives, it increases the inventory level to a value greater than the reorder point R. If this were not the case, one would never reach R, and one would never order again. This will happen if the demand during the lead time exceeds Q, which is certainly possible. On the surface, this seems like a serious flaw in the approach, but it isn't. To avoid this problem, one bases the reorder decision on the inventory position rather than on the inventory level. The inventory position is defined as the total stock on hand plus on order. The inventory position varies from R to $R + Q$ as shown in Figure 5–7.

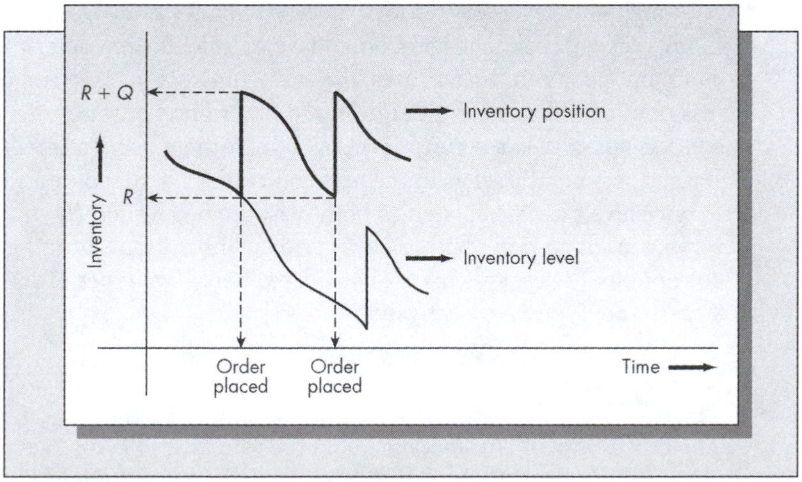

Figure 5–7 Inventory level versus inventory position

SNAPSHOT APPLICATION

Inventory Management Software for the Small Business

The software for PC-based inventory management continues to grow at an increasing rate. As personal computers have become more powerful, the inventory management needs of larger businesses can be handled by software designed to run on PCs. One example is an inventory system called inFlow, which can handle inventories stored at multiple locations. Its monthly costs range from $69 to $599. A less expensive alternative is Inventoria (the downloadable software program is $79.99), which might be more appropriate for single location businesses, such as one-off retail stores. Some other choices include Goldenseal Accounting, Info Plus, iMagic Inventory, Inventory Tracker Plus, NCR Counterpoint, Odero, and many others. Each of the packages have certain advantages, including compatibility with other software systems, additional features such as financial functions, single versus multiple locations, industry specific applications, etc. Of course, inventory management modules are also available in large ERP systems, such as those sold by United States-based Oracle Corporation and German-based SAS.

An application that makes use of the most modern hardware and software is ShopKeep POS. ShopKeep is designed to run on the Apple iPad, and all of the records are stored in the cloud. This means that inventory data can be accessed from any iPad in the system at any time. One customer, The Bean, a small chain of coffee shops based in New York City, changed their inventory control function from a Windows-based system to ShopKeep, running on iPads and iPhones. The system gives them the opportunity to run reports from any iPad connected to the system. According to owner Ike Estavo, one of the most useful functions of ShopKeep is identifying the most profitable items per square foot of display space.

The personal computer has become ubiquitous since the first edition of this book was published in 1989. The huge growth in personal computers has been accompanied by an explosion in software choices for almost every conceivable application. Inventory control is no exception. A challenge for operations managers is to choose wisely among the multitude of applications, only a few of which are mentioned here. One must be able to pick the level of generality and complexity appropriate for the desired application. No matter what software package is chosen, it is vitally important to understand the basic principles of inventory management discussed here. Hoping that the software will solve one's problem without really understanding the key elements of that problem is a big mistake.

Source: The information discussed here was obtained from the website of the vendors mentioned.

5.5 SERVICE LEVELS IN (Q, R) SYSTEMS

Although the inventory model described in Section 5.4 is quite realistic for describing many real systems, managers often have a difficult time determining an exact value for the stock-out cost p. In many cases, the stock-out cost includes intangible components such as loss of goodwill and potential delays to other parts of the system. A common substitute for a stock-out cost is a service level. Although there are a number of different definitions of service, it generally refers to the probability that a demand or a collection of demands is met. Service levels can be applied in both periodic review and (Q, R) systems. The application of service levels to periodic-review systems will be discussed in Section 5.6. Service levels for continuous-review systems are considered here. Two types of service are considered, labeled Type 1 and Type 2, respectively.

Type 1 Service

The probability of not stocking out in the lead time is **Type 1 service**. We will use the symbol α to represent this probability. As specification of α completely determines the value of R, the computation of R and Q can be decoupled. The computation of the optimal (Q, R) values subject to a Type 1 service constraint is very straightforward.

a. Determine R to satisfy the equation $F(R) = \alpha$.
b. Set $Q = \text{EOQ}$.

Interpret α as the proportion of cycles in which no stock-out occurs. A Type 1 service objective is appropriate when a shortage occurrence has the same consequence independent of its time or amount. One example would be where a production line is stopped whether 1 unit or 100 units are short. However, Type 1 service is not how service is interpreted in most applications. Usually when we say we would like to provide 95 percent service, we mean that we would like to be able to fill 95 percent of the demands when they occur, not fill all the demands in 95 percent of the order cycles. Also, as different items have different cycle lengths, this measure will not be consistent among different products, making the proper choice of α difficult.

Type 2 Service

Type 2 service measures the proportion of demands that are met from stock. We will use the symbol β to represent this proportion. As we saw in part 5 of Example 5.4, $n(R)/Q$ is the average fraction of demands that stock out each cycle. Hence, specification of β results in thc constraint $n(R)/Q = 1 - \beta$.

This constraint is more complex than the one arising from Type 1 service, as it involves both Q and R. It turns out that although the EOQ is not optimal in this case, it usually gives pretty good results. If we use the EOQ to estimate the lot size, then we would find R to solve $n(R) = \text{EOQ}(1 - \beta)$.

EXAMPLE 5.5

Let's revisit Harvey's Specialty Shop. Harvey feels uncomfortable with the assumption that the goodwill cost of a stock-out is \$25; he decides to use a service level criterion instead. Suppose that he chooses to use a 98 percent service objective.

1. *Type 1 service.* If we assume an α of .98, then we find R to solve $F(R) = 0.98$. From Table A–1 or A–4, we obtain $z = 2.05$. Setting $R = \sigma z + \mu$ gives $R = 151$.
2. *Type 2 service.* Here $\beta = 0.98$. We are required to solve the equation

$$n(R) = \text{EOQ}(1 - \beta),$$

which is equivalent to

$$L(z) = \text{EOQ}(1 - \beta)/\,\sigma.$$

Substituting EOQ $= 100$ and $\beta = .98$, we obtain

$$L(z) = (100)(0.02)/25 = 0.08.$$

From Table A–4 or Excel, we obtain $z = 1.02$. Setting $R = \sigma z + \mu$ gives $R = 126$. Notice that the same values of α and β give *considerably* different values of R.

In order to understand more clearly the difference between these two measures of service, consider the following example. Suppose that we have tracked the demands and stock-outs over 10 consecutive order cycles with the following results.

Order Cycle	Demand	Stock-Outs
1	180	0
2	75	0
3	235	45
4	140	0
5	180	0
6	200	10
7	150	0
8	90	0
9	160	0
10	40	0

Based on a Type 1 measure of service, we find that the fraction of periods in which there is no stock-out is 8/10 = 80 percent. That is, the probability that all the demands are met in a single order cycle is 0.8, based on these observations. However, the Type 2 service provided here is considerably better. In this example, the total number of demands over the 10 periods is 1,450 (the sum of the numbers in the second column), and the total number of demands that result in a stock-out is 55. Hence, the number of satisfied demands is 1,450 – 55 = 1,395. The proportion of satisfied demands is 1,395/1,450 = .9621, or roughly 96 percent.

The term **fill rate** is often used to describe Type 2 service; it is generally what most managers mean by service. (The fill rate in this example is 96 percent.) We saw in the example that there is a significant difference between the proportion of cycles in which all demands are satisfied (Type 1 service) and the fill rate (Type 2 service). *Even though it is easier to determine the best operating policy satisfying a Type 1 service objective, this policy will not accurately approximate a Type 2 service objective and should not be used in place of it.*

Optimal (*Q*, *R*) Policies Subject to Type 2 Constraint

Using the EOQ value to estimate the lot size gives reasonably accurate results when using a fill rate constraint, but the EOQ value is only an approximation of the optimal lot size. A more accurate value of Q can be obtained as follows. Consider the pair of equations (1) and (2) we solved for the optimal values of Q and R when a stock-out cost was present. Solving for p in equation (2) gives

$$p = Qh \,/\, [(1 - F(R))\lambda],$$

which now can be substituted for p in equation (1), resulting in

$$Q = \sqrt{\frac{2\lambda\{K + Qhn(R)/[(1 - F(R))\lambda]\}}{h}},$$

which is a quadratic equation in Q. It can be shown that the positive root of this equation is

$$Q = \frac{n(R)}{1 - F(R)} + \sqrt{\frac{2K\lambda}{h} + \left(\frac{n(R)}{1 - F(R)}\right)^2}. \tag{3}$$

Equation (3) will be called the SOQ formula (for service level order quantity). [Authors' note: The SOQ formula also could have been derived by more conventional Lagrange multiplier techniques. We include this derivation to demonstrate the relationship between the fill rate objective and the stock-out cost model.] This equation is solved simultaneously with

$$n(R) = (1 - \beta)Q \tag{4}$$

to obtain optimal values of (Q, R) satisfying a Type 2 service constraint.

The reader should note that the version of equation (4) used in the calculations is in terms of the standardized variate z and is given by

$$L(z) = (1 - \beta)Q/\sigma.$$

The solution procedure is essentially the same as that required to solve equations (1) and (2) simultaneously. Either use Excel's Solver add-in, or start with Q_0 = EOQ, find R_0 from (4), use R_0 in (3) to find Q_1, and so on, and stop when two successive values of Q and R are sufficiently close (within one unit is sufficient for most problems).

EXAMPLE 5.5 (CONTINUED)

Returning to Example 5.5, $Q_o = 100$ and $R_o = 126$. Furthermore, $n(R_o) = (0.02)(100) = 2$. Using $z = 1.02$ gives $1 - F(R_o) = 0.154$. Continuing with the calculations,

$$Q_1 = \frac{2}{0.154} + \sqrt{(100)^2 + \left(\frac{2}{0.154}\right)^2} = 114.$$

Solving equation (4) gives $n(R_1) = (114)(0.02)$, which is equivalent to

$$L(z) = (114)(0.02)/25 = 0.0912.$$

From Table A–4, $z = 0.95$, so that

$$1 - F(R_1) = 0.171$$

and

$$R_1 = \sigma z + \mu = 124.$$

Carrying the computation one more step gives $Q_2 = 114$ and $R_2 = 124$. As both Q and R are within one unit of their previous values, we terminate computations. Hence, we conclude that the optimal values of Q and R satisfying a 98 percent fill rate constraint are $(Q, R) = (114, 124)$.

Consider the cost error resulting from the EOQ substituted for the SOQ. In order to compare these policies, we compute the average annual holding and setup costs (notice that there is no stock-out cost) for the policies $(Q, R) = (100, 126)$ and $(Q, R) = (114, 124)$.

Recall the formulas for average annual holding and setup costs.

$$\text{Holding cost} = h(Q/2 + R - \mu).$$

$$\text{Setup cost} = K\lambda/Q.$$

For (100, 126):

$$\left.\begin{array}{lll} \text{Holding cost} & = 2(100/2 + 126 - 100) & = \$152 \\ \text{Setup cost} & = (50)(200)/100 & = \$100 \end{array}\right\} \text{Total} = \$252.$$

For (114, 124)

$$\left.\begin{array}{lll} \text{Holding cost} & = 2(114/2 + 124 - 100) & = \$162 \\ \text{Setup cost} & = (50)(200)/114 & = \$88 \end{array}\right\} \text{Total} = \$250.$$

We see that the EOQ approximation gives costs close to the optimal in this case.

Imputed Shortage Cost

Consider the solutions that we obtained for (Q, R) in Example 5.5 when we used a service level criterion rather than a shortage cost. For a Type 2 service of $\beta = 0.98$ we obtained the solution (114, 124). Although no shortage cost was specified, this solution clearly corresponds to *some* value of p. That is, there is some value of p such that the policy (114, 124) satisfies equations (1) and (2). This particular value of p is known as the imputed shortage cost.

The imputed shortage cost is easy to find. One solves for p in equation (2) to obtain $p = Qh/[(1 - F(R))\lambda]$. The imputed shortage cost is a useful way to determine whether the value chosen for the service level is appropriate.

EXAMPLE 5.5 (*CONTINUED*)

Consider Harvey's Specialty Shop again. Using a value of $\alpha = .98$ (Type 1 service), we obtained the policy (100, 151). The imputed shortage cost is $p = (100)(2)/[(0.02)(200)] = \$50$.

Using a value of $\beta = 0.98$ (Type 2 service) we obtained the policy (114, 124). In this case the imputed cost of shortage is $p = (114)(2)/[(0.171)(200)] = \$6.67$.

Scaling of Lead Time Demand

In all previous examples the demand during the lead time was given. However, in most applications demand would be forecast on a periodic basis, such as monthly. In such cases one would need to convert the demand distribution to correspond to the lead time.

Assume that demands follow a normal distribution. Because sums of independent normal random variables are also normally distributed, the form of the distribution of lead time demand is normal. Hence, all that remains is to determine the mean and the standard deviation. Let the periodic demand have mean λ and standard deviation υ, and let τ be the lead time in periods. As both the means and the *variances* (not standard deviations) are additive, the mean demand during lead time is $\mu = \lambda\tau$ and the variance of demand during lead time is $\upsilon^2\tau$. Hence, the standard deviation of demand during lead time is $\sigma = \upsilon\sqrt{\tau}$ (although the square root may not always be appropriate). [Authors' note: Often in practice it turns out that there is more variation in the demand process than is described by a pure normal distribution. For that reason, the standard deviation of demand is generally expressed in the form $\upsilon\tau^q$ where the correct value of q, generally between 0.5 and 1, must be determined for each item or group of items by an analysis of historical data.]

EXAMPLE 5.6

Weekly demand for a certain type of automotive spark plug in a local repair shop is normally distributed with mean 34 and standard deviation 12. Procurement lead time is six weeks. Determine the lead time demand distribution.

Solution

The demand over the lead time is also normally distributed with mean (34)(6) = 204 and standard deviation $(12)\sqrt{6} = 29.39$. These would be the values of μ and σ that one would use for all remaining calculations.

Estimating Sigma When Inventory Control and Forecasting Are Linked

Thus far in this chapter we have assumed that the distribution of demand over the lead time is known. In practice, one assumes a *form* for the distribution, but its parameters must be estimated from real data. Assuming a normal distribution for lead time demand (which is the most common assumption), one needs to estimate the mean and the standard deviation. When a complete history of past data is available, the standard statistical estimates for the mean and the standard deviation (i.e., those suggested in Section 5.1) are fine. However, most forecasting schemes do *not* use all past data. Moving averages use only the past N data values and exponential smoothing places declining weights on past data.

In these cases, it is unclear exactly what are the right estimators for the mean and the standard deviation of demand. This issue was discussed in Section 2.13. The best estimate of the mean is simply the forecast of demand for the next period. For the variance, one should use the estimator for the variance of forecast error. The rationale for this is rarely understood. The variance of forecast error and the variance of demand are *not* the same thing. This was established rigorously in Appendix 2–A.

Why is it appropriate to use the standard deviation of forecast error to estimate σ? The reason is that it is the forecast that we are using to estimate demand. Safety stock is held to protect against errors in forecasting demand. In general, the variance of forecast error will be higher than the variance of demand. This is the result of the additional sampling error introduced by a forecasting scheme that uses only a portion of past data. In cases where the underlying demand process is nonstationary (that is, where there is trend or seasonality), the variance of demand due to *systematic* changes could be higher than the variance of forecast error. All of our analysis assumes stationary (constant mean and variance) demand patterns.

Robert Brown (1959) was apparently the first to recommend using the standard deviation of forecast error in safety stock calculations. The method he recommended, which is still in widespread use today, is to track the MAD (mean absolute deviation) of forecast error using the formula

$$\text{MAD}_t = \alpha\,\text{MAD}_{t-1} + (1 - \alpha)|F_t - D_t|,$$

where F_t is the forecast of demand at time t and D_t is the actual observed demand at time t. The estimator for the standard deviation of forecast error at time t is 1.25 MAD_t. While this method is very popular in commercial inventory control systems, apparently few realize that by using this approach they are estimating not demand variance but forecast error variance, and that these are not the same quantities.

Lead Time Variability

Thus far, we have assumed that the lead time τ is a known constant. However, lead time uncertainty is common in practice. For example, the time required to transport commodities, such as oil, that are shipped by sea depends upon weather conditions. In general, it is very difficult to incorporate the variability of lead time into the calculation of optimal inventory policies. The problem is that if we assume that successive lead times are independent random variables, then it is possible for lead times to cross; that is, two successive orders would not necessarily be received in the same sequence that they were placed.

Order crossing is unlikely when a single supplier is used. If we are willing to make the simultaneous assumptions that orders do not cross and that successive lead times are independent, the variability of lead time can be easily incorporated into the analysis. Suppose that the lead time τ is a random variable with mean μ_τ, and variance σ_τ^2. Furthermore, suppose that demand in any time t has mean λt and variance $\upsilon^2 t$. Then it can be shown that the demand during lead time has mean and variance (see, for example, Hadley and Whitin, 1963, p. 153)

$$\mu = \lambda\mu_\tau,$$
$$\sigma^2 = \mu_\tau v^2 + \lambda^2\sigma_\tau^2.$$

EXAMPLE 5.7

Harvey's Specialty Shop orders an unusual olive from the island of Santorini, off the Greek coast. Over the years, Harvey has noticed considerable variability in the time it takes to receive orders of these olives. On average, the order lead time is four months, and the standard deviation is six weeks (1.5 months). Monthly demand for the olives is normally distributed with mean 15 (jars) and standard deviation 6.

Setting $\mu_\tau = 4$, $\sigma_\tau = 1.5$, $\lambda = 15$, and $\upsilon = 6$, we obtain

$$\mu = \mu_\tau\lambda = (4)(15) = 60,$$
$$\sigma^2 = \mu_\tau\upsilon^2 + \lambda^2\sigma_\tau^2 = (4)(36) + (225)(2.25) = 650.25$$

One would proceed with the calculations of optimal inventory policies using $\mu = 60$ and $\sigma^2 = 650.25$ as the mean and the variance of lead time demand.

Negative Safety Stock

When the shortage cost or service level is relatively low, it is possible for a negative safety stock situation to arise. Recall that safety stock is the expected inventory on hand at the arrival of an order. If the safety stock is negative, there would be a backorder situation in more than 50 percent of the order cycles (that is, the Type 1 service would be under 50 percent). When this occurs, the optimal z value is negative, and the optimal $L(z)$ value would exceed $L(0) = .3989$. We illustrate with another example from Harvey's Specialty Shop.

EXAMPLE 5.8

Harvey sells a high-end espresso machine, an expensive and bulky item. For this reason, Harvey attaches a very high holding cost of \$50 per year to each machine. Harvey sells about 20 of these yearly and estimates the variance of yearly demand to be 50.

Annual demand follows the normal distribution. Since customers are willing to wait for the machine when Harry is out of stock, the shortage penalty is low. He estimates it to be \$25 per unit. Order lead time is six months. The fixed cost of ordering is \$80. Assume that the lot size used for reordering is the EOQ value. Determine the following:

1. optimal reorder level,
2. safety stock,
3. resulting Type 1 service level,
4. resulting Type 2 service level.

Solution

1. We have that $K = \$80$, $p = \$25$, $h = \$50$, and $\lambda = 20$. Scaling lead time demand to 0.5 year, we obtain $\mu = 10$ and $\sigma = \sqrt{(50)(0.5)} = 5$. The reader can check that the EOQ rounded to the nearest integer is $Q = 8$. The equation for determining R_0 is

$$1 - F(R_0) = \frac{Q_0 h}{p\lambda} = \frac{(8)(50)}{(25)(20)} = 0.8.$$

   Since $1 - F(R_0)$ exceeds 0.5, the resulting z value is negative. In this case $z = -0.84$, $L(z) = 0.9520$, and $R_0 = \sigma z + \mu = (5)(-.84) + 10 = 5.8$, which we round to 6. Hence, the optimal solution based on the EOQ is to order eight units when the on-hand inventory falls to six units.
2. The safety stock is $S = R - \mu = 6 - 10 = -4$.
3. The Type 1 service level is $\alpha = F(R) = .20$ (or 20 percent).
4. The Type 2 service level is $\beta = 1 - n(R)/Q = 1 - \sigma L(z)/Q = 1 - (5)(.9520)/8 = 0.405$ (or 41 percent).

Note: Had we tried to solve this problem for the optimal (Q, R) iteratively using equations (1) and (2), the equations would have diverged and not yielded a solution. This is a result of having a very low shortage cost compared to the holding cost. Furthermore, if Q were larger, the problem could be unsolvable. For example, if one assumes the setup cost $K = 150$, then EOQ = 11 and $1 - F(R_0) = Q_0 h/p\lambda = (11)(50)/(25)(20) = 1.1$, which is obviously not solvable. Such circumstances are very rare in practice, but, as we see from this example, it is possible for the model to fail. This is a consequence of the fact that the model is not exact. Exact (Q, R) models are beyond the scope of this book and are known only for certain demand distributions.

PROBLEMS FOR SECTIONS 5.4 AND 5.5

13. An automotive warehouse stocks a variety of parts that are sold at neighborhood stores. One particular part, a popular brand of oil filter, is purchased by the warehouse for \$1.50 each. It is estimated that the cost of order processing and receipt is \$100 per order. The company uses an inventory carrying charge based on a 28 percent annual interest rate.

    The monthly demand for the filter follows a normal distribution with mean 280 and standard deviation 77. Order lead time is assumed to be five months.

    Assume that if a filter is demanded when the warehouse is out of stock, then the demand is back ordered, and the cost assessed for each back-ordered demand is \$12.80. Determine the following quantities.

    a. The optimal values of the order quantity and the reorder level.
    b. The average annual cost of holding, setup, and stock-out associated with this item assuming that an optimal policy is used.

c. Evaluate the cost of uncertainty for this process. That is, compare the average annual cost you obtained in part (b) with the average annual cost that would be incurred if the lead time demand had zero variance.

14. Weiss's paint store uses a (Q, R) inventory system to control its stock levels. For a particularly popular white latex paint, historical data show that the distribution of monthly demand is approximately normal, with mean 28 and standard deviation 8. Replenishment lead time for this paint is about 14 weeks. Each can of paint costs the store \$6. Although excess demands are back ordered, the store owner estimates that unfilled demands cost about \$10 each in bookkeeping and loss-of-goodwill costs. Fixed costs of replenishment are \$15 per order, and holding costs are based on a 30 percent annual rate of interest.
    a. What are the optimal lot sizes and reorder points for this brand of paint?
    b. What is the optimal safety stock for this paint?
15. After taking a production seminar, Al Weiss, the owner of Weiss's paint store, decides that his stock-out cost of \$10 may not be very accurate and switches to a service level model. He decides to set his lot size by the EOQ formula and determines his reorder point so that there is *no stock-out* in 90 percent of the order cycles.
    a. Find the resulting (Q, R) values.
    b. Suppose that, unfortunately, he really wanted to satisfy 90 percent of his demands (that is, achieve a 90 percent fill rate). What fill rate did he actually achieve from the policy determined in part (a)?
16. Suppose that in problem 13 the stock-out cost is replaced with a Type 1 service objective of 95 percent. Find the optimal values of (Q, R) in this case.
17. Suppose that in problem 13 a Type 2 service objective of 95 percent is substituted for the stock-out cost of \$12.80. Find the resulting values of Q and R. Also, what is the imputed cost of shortage for this case?
18. Suppose that the warehouse from problem 13 mistakenly used a Type 1 service objective when it really meant to use a Type 2 service objective (see problems 16 and 17). What is the additional holding cost incurred each year for this item because of this mistake?
19. IBM introduced the Winchester drive in 1973—the precursor of today's terabyte drives. It remained the standard until 2011. Disk Drives Limited (DDL) produced a line of internal Winchester disks for microcomputers. The drives used a 3.5-inch platter that DDL purchased from an outside supplier. Demand data and sales forecasts indicated that the weekly demand for the platters closely approximated a normal distribution with mean 38 and variance 130. The platters required a three-week lead time for receipt. DDL had been using a 40 percent annual interest charge to compute holding costs. The platters cost \$18.80 each, order cost was \$75.00 per order, and the company used a stock-out cost of \$400.00 per platter. (Because the industry was so competitive, stock-outs were very costly.)
    a. Because of a prior contractual agreement with the supplier, DDL purchased the platters in lots of 500. What would have been the reorder point?
    b. If DDL renegotiated its contract with the supplier, what lot size should it have written into the agreement?
    c. How much of a penalty in terms of setup, holding, and stock-out cost was DDL paying for contracting to buy too large a lot?
    d. DDL's president was uncomfortable with the \$400 stock-out cost and decided to substitute a 99 percent fill rate criterion. If DDL used a lot size equal to the EOQ, what would its reorder point have been? Also, find the imputed cost of shortage.

20. Bobbi's Restaurant in Boise, Idaho, is a popular place for weekend brunch. The restaurant serves real maple syrup with French toast and pancakes. Bobbi buys the maple syrup from a company in Maine that requires three weeks for delivery. The syrup costs Bobbi \$4 a bottle and may be purchased in any quantity. Fixed costs of ordering amount to about \$75 for bookkeeping expenses, and holding costs are based on a 20 percent annual rate. Bobbi estimates that the loss of customer goodwill for not being able to serve the syrup when requested amounts to \$25. Based on past experience, the weekly demand for the syrup is normal with mean 12 and variance 16 (in bottles). For the purposes of your calculations, you may assume that there are 52 weeks in a year and that all excess demand is back ordered.
    a. How large an order should Bobbi be placing with her supplier for the maple syrup, and when should those orders be placed?
    b. What level of Type 1 service is being provided by the policy you found in part (a)?
    c. What level of Type 2 service is being provided by the policy you found in part (a)?
    d. What policy should Bobbi use if the stock-out cost is replaced with a Type 1 service objective of 95 percent?
    e. What policy should Bobbi use if the stock-out cost is replaced with a Type 2 service objective of 95 percent? (You may assume an EOQ lot size.)
    f. Suppose that Bobbi's supplier requires a minimum order size of 500 bottles. Find the reorder level that Bobbi should use if she wishes to satisfy 99 percent of her customer demands for the syrup.

5.6 ADDITIONAL DISCUSSION OF PERIODIC-REVIEW SYSTEMS

(s, S) Policies

In our analysis of the newsvendor problem, we noted that a severe limitation of the model from a practical viewpoint is that there is no setup cost included in the formulation. The (Q, R) model treated in the preceding sections included an order setup cost but assumed that the inventory levels were reviewed continuously; that is, known at all times. How should the system be managed when there is a setup cost for ordering but inventory levels are known only at discrete points in time?

The difficulty that arises from trying to implement a continuous-review solution in a periodic-review environment is that the inventory level is likely to overshoot the reorder point R during a period, making it impossible to place an order the instant the inventory reaches R. To overcome this problem, the operating policy is modified slightly. Define two numbers, s and S, to be used as follows: When the level of on-hand inventory is *less than or equal to s*, an order for the difference between the inventory and S is placed. If u is the starting inventory in any period, then the (s, S) policy is

If $u \leq s$, order $S - u$.

If $u > s$, do not order.

Determining optimal values of (s, S) requires a numerical algorithm (e.g., Zheng, Y.-S., and A. Federgruen, 1991). Several explicit approximations have been suggested. One such approximation is to compute a (Q, R) policy using the methods described earlier, and set $s = R$ and $S = R + Q$. This approximation will give reasonable results in many cases, and is probably the most commonly used (for a comprehensive comparison of several approximate (s, S) policies, see Porteus, 1985).

Tropicana Uses Sophisticated Modeling for Inventory Management

Tropicana is one of the world's largest suppliers of citrus-based juice products. The company was founded by an Italian immigrant, Anthony Rossi, in 1947; it was acquired by PepsiCo, Inc. in 1998. From its production facilities in Bradenton, Florida, Tropicana makes daily rail shipments to its regional distribution centers (DCs).

The focus of this application is the largest DC located in Jersey City, New Jersey, which services the northeast United States and Canada. Shipments from Bradenton to Jersey City require four days: one day for loading, two days in transit, and one day for unloading. This is the order lead time as seen from the DC. Lead time variability is not considered to be a significant issue.

Based on a statistical analysis of past data, Tropicana planners have determined that daily demands on the DC are closely approximated by a normal distribution for each of their products. Product classes are assumed to be independent.

Trains are sent from Florida five times per week; arrivals coincide with each business day at the Jersey City DC. Demand data and inventory levels are reviewed very frequently, a reorder point R triggers replenishment, and lot size Q is defined by the user. Hence, they operate a standard (Q, R) continuous-review review system with a positive lead time. The state variable is the inventory position defined as the total amount of stock on hand at, and in transit to, the DC.

Let μ_D and σ_D be the mean and standard deviation of daily demand for a particular product line. According to the theory outlined in this chapter, the mean and standard deviation of demand over the four-day lead time should be

$$\mu = \mu_D \tau = 4\mu_D,$$
$$\sigma = \sigma_D \sqrt{\tau} = 2\sigma_D.$$

However, analysis of Tropicana data showed that the standard deviation of lead time demand is closer to $\sigma_D \tau^{0.7} = 2.64\sigma_D$.

The firm's objective is to maintain a 99.5 percent Type 2 service level at the DC. Values of (Q, R) are computed using the methodology of this chapter. Planners check the inventory on hand and in transit and place an order equal to the EOQ, when this value falls below the reorder level, R. This analysis is carried out for a wide range of the company's products and individually determined for each regional DC.

Source: Based on joint work of Steven Nahmias and Tim Rowell of Tropicana.

Service Levels in Periodic-Review Systems

Service levels also may be used when inventory levels are reviewed periodically. Consider first a Type 1 service objective. That is, we wish to find the order-up-to point Q so that all the demand is satisfied in a given percentage of the periods. Suppose that the value of Type 1 service is α. Then Q should solve the equation

$$F(Q) = \alpha.$$

This follows because $F(Q)$ is the probability that the demand during the period does not exceed Q. Notice that one simply substitutes a for the critical ratio in the newsvendor model. To find Q to satisfy a Type 2 service objective of β, it is necessary to obtain an expression for the fraction of demands that stock out each period. Using essentially the same notation as that used for (Q, R) systems, define

$$n(Q) = \int_Q^\infty (x - Q) f(x)dx.$$

Note that $n(Q)$, which represents the expected number of demands that stock out at the end of the period, is the same as the term multiplying c_u in the expression for the expected cost function for the newsvendor model discussed in Section 5.3. As the demand per period is μ, it follows that the proportion of demands that stock out each period is $n(Q)/\mu$. Hence, the value of Q that meets a fill rate objective of β solves

$$n(Q) = (1 - \beta)\mu.$$

The specification of either a Type 1 or Type 2 service objective completely determines the order quantity, independent of the cost parameters.

EXAMPLE 5.9

Mac, the owner of the newsstand described in Example 5.1, wishes to use a Type 1 service level of 90 percent to control his replenishment of *The Computer Journal*. The z value corresponding to the 90th percentile of the unit normal is $z = 1.28$. Hence,

$$Q^* = \sigma z + \mu = (4.74)(1.28) + 11.73 = 17.8 \approx 18.$$

Using a Type 2 service of 90 percent, we obtain

$$n(Q) = (1 - \beta)\mu = (0.1)(11.73) = 1.173.$$

It follows that $L(z) = n(Q)/\sigma = 1.173/4.74 = 0.2475$. From Table A–4 at the back of this book, we find

$$z \approx 0.35;$$

then

$$Q^* = \sigma z + \mu = (4.74)(0.35) + 11.73 = 13.4 \approx 13.$$

As with (Q, R) models, notice the striking difference between the resulting values of Q^* for the same levels of Type 1 and Type 2 service.

Fixed Order Size Model

If a positive order lead time is included, only a slight modification of these equations is required. In particular, the response time of the system is now the order lead time plus one period. Hence, we would now use the distribution of demand over $\tau + T$, where T is the time between inventory reviews.

This periodic review service level model is very useful in retail settings, in particular. It is common in retailing to place orders at fixed points in time to take advantage of bundling multiple orders together.

EXAMPLE 5.10

Stroheim's is a dry goods store located in downtown Milwaukee, Wisconsin. Stroheim's places orders weekly with their suppliers for all of their reorders. The lead time for men's briefs is 4 days. Stroheim's uses a 95 percent service level. Assuming a Type 1 service, what is the order up to level for the briefs? Assume demands for briefs are uncertain with daily mean demand of 20 and daily standard deviation of 12.

Solution

The total response time (review time plus lead time) is 7 + 4 = 11 days. No matter what the form of the daily demand distribution, the Central Limit Theorem indicates that the demand over 11 days should be close to normal. The parameters are $\mu = (11)$

(20) = 220 and $\sigma = 12\sqrt{11} = 39.80$. Hence it follows that for a Type 1 service objective of 95 percent, the order-up-to-point should be $Q = \sigma z + \mu = (39.80)(1.645) + 220 = 286$.

If Stroheim's was interested in a Type 2 service objective, the z value would be lower, and the corresponding order up to point would be lower as well. In particular, $n(Q) = (1 - \beta)\mu = (.05)(220) = 11$, and $L(z) = n(Q)/\sigma = 11/39.80 = .2764$, which gives a z value of 0.27 and a corresponding order up to level of 231.

PROBLEMS FOR SECTION 5.6

21. Consider the Crestview Printing Company mentioned in problem 9. Suppose that Crestview wishes to produce enough cards to satisfy all Christmas demand with probability 90 percent. How many cards should they print? Suppose the probability is 97 percent. What would you recommend in this case? (Your answer will depend upon the assumption you make concerning the shape of the cumulative distribution function.)
22. Consider Happy Henry's car dealer described in problem 10.
    a. How many EX123s should Happy Henry's purchase to satisfy all the demand over a three-month interval with probability .95?
    b. How many cars should be purchased if the goal is to satisfy 95 percent of the demands?
23. For the problem of controlling the inventory of white latex paint at Weiss's paint store, suppose that the paint is reordered on a monthly basis rather than on a continuous basis.
    a. Using the (Q, R) solution you obtained in part (a) of problem 14, determine appropriate values of (s, S).
    b. Suppose that the demands during the months of January to June were

Month	Demand	Month	Demand
January	37	April	31
February	33	May	14
March	26	June	40

If the starting inventory in January was 26 cans of paint, determine the number of units of paint ordered in each of the months January to June following the (s, S) policy you found in part (a).

5.7 MULTIPRODUCT SYSTEMS

ABC Analysis

One issue that we have not discussed is the cost of implementing an inventory control system and the trade-offs between the cost of controlling the system and the potential benefits that accrue from that control. In multiproduct inventory systems, not all products are equally profitable. Control costs may be justified in some cases and not in others. For example, spending \$200 annually to monitor an item that contributes only \$100 a year to profits is clearly not economical.

For this reason, it is important to differentiate profitable from unprofitable items. To do so, we borrow a concept from economics. Vilfredo Pareto, who studied the distribution of

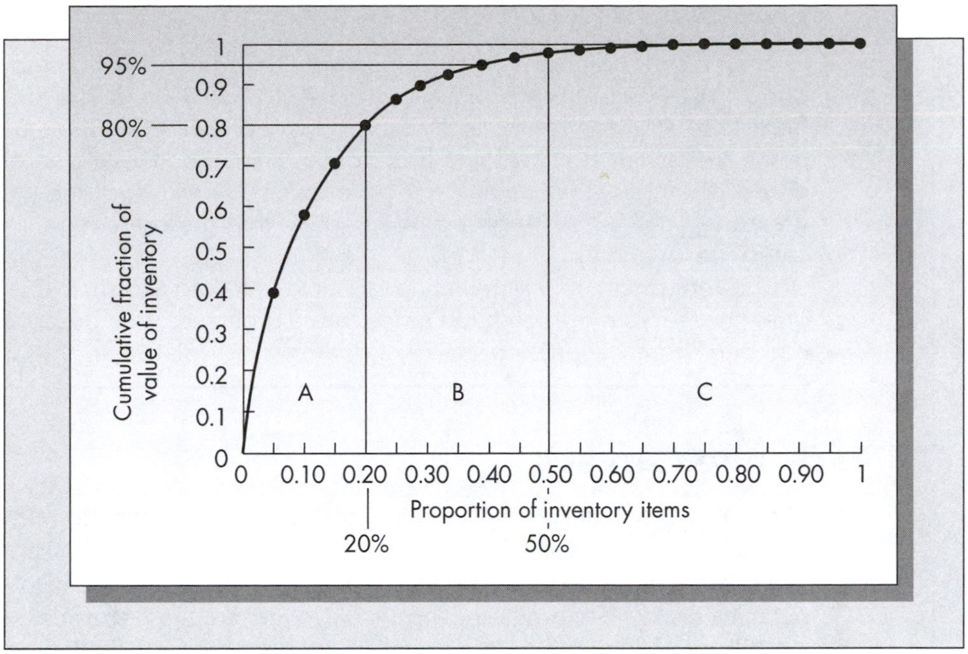

Figure 5–8 Pareto curve: Distribution of inventory by value

wealth in the nineteenth century, noted that a large portion of the wealth was owned by a small segment of the population. This *Pareto effect* also applies to inventory systems: a large portion of the total dollar volume of sales is often accounted for by a small number of inventory items. Assume that items are ranked in decreasing order of the dollar value of annual sales. The *cumulative* value of sales generally results in a curve much like the one pictured in Figure 5–8.

Typically, the top 20 percent of the items account for about 80 percent of the annual dollar volume of sales, the next 30 percent of the items for the next 15 percent of sales, and the remaining 50 percent for the last 5 percent of dollar volume. These figures are only approximate and will vary slightly from one system to another. The three item groups are labeled A, B, and C, respectively. When a finer distinction is needed, four or five categories could be used. Even when using only three categories, the percentages used in defining A, B, and C items could be different from the 80 percent, 15 percent, and 5 percent recommended.

Because A items account for the lion's share of the yearly revenue, these items should be watched most closely. Inventory levels for A items should be monitored continuously. More sophisticated forecasting procedures might be used, and more care would be taken in the estimation of the various cost parameters to determine operating policies. Inventories for B items could be reviewed periodically; items could be ordered in groups rather than individually; and somewhat less sophisticated forecasting methods could be used. The minimum degree of control would be applied to C items. For very inexpensive C items with moderate levels of demand, large lot sizes are recommended to minimize the frequency that these items are ordered. For expensive C items with very low demand, the best policy is generally not to hold any inventory. One would simply order these items as they are demanded.

EXAMPLE 5.10 (CONTINUED)

A sample of 20 different stock items from Harvey's Specialty Shop is selected at random. These items vary in price from $0.25 to $24.99 and in average yearly demand from 12 to 786. The results of the sampling are presented in Table 5–1. In Table 5–2 the items are ranked in decreasing order of the annual dollar volume of sales. Notice that only 4 of the 20 stock items account for over 80 percent of the annual dollar volume generated by the entire group. Also notice that there are high-priced items in both categories A and C.

This report was very illuminating to Harvey. He had assumed that R077, a packaged goat cheese from the south of France, was a profitable item because of its cost, and he had been virtually ignoring TTR77, a domestic chocolate bar.

Exchange Curves

Much of our analysis assumes a single item in isolation and that the relevant cost parameters K, h, and p (or just K and h in the case of service levels) are constants with "correct" values that can be determined. However, it may be more appropriate to think of one or all of the cost parameters as policy variables. The correct values are those that result in a control system with characteristics that meet the needs of the firm and the goals of management. In a typical

Table 5–1 Performance of 20 Stock Items Selected at Random

Part Number	Price	Yearly Demand	Dollar Volume
5497J	$2.25	260	$585.00
3K62	2.85	43	122.55
88450	1.50	21	31.50
P001	0.77	388	298.76
2M993	4.45	612	2,723.40
4040	6.10	220	1,342.00
W76	3.10	110	341.00
JJ335	1.32	786	1,037.52
R077	12.80	14	179.20
70779	24.99	334	8,346.66
4J65E	7.75	24	186.00
334Y	0.68	77	52.36
8ST4	0.25	56	14.00
16113	3.89	89	346.21
45000	7.70	675	5,197.50
7878	6.22	66	410.52
6193L	0.85	148	125.80
TTR77	0.77	690	531.30
39SS5	1.23	52	63.96
93939	4.05	12	48.60

Table 5–2 Twenty Stock Items Ranked in Decreasing Order of Annual Dollar Volume

Part Number	Price	Yearly Demand	Dollar Volume	Cumulative Dollar Volume	
70779	$24.99	334	$8,346.66	$8,346.66	A items: 20% of items account for 80.1% of total value.
45000	7.70	675	5,197.50	13,544.16	
2M993	4.45	612	2,723.40	16,267.56	
4040	6.10	220	1,342.00	17,609.56	
JJ335	1.32	786	1,037.52	18,647.08	B items: 30% of items account for 14.8% of total value.
5497J	2.25	260	585.00	19,232.08	
TTR77	0.77	690	531.30	19,763.38	
7878	6.22	66	410.52	20,173.90	
16113	3.89	89	346.21	20,520.11	
W76	3.10	110	341.00	20,861.11	
P001	0.77	388	298.76	21,159.87	C items: 50% of items account for 5.1% of total value.
4J65E	7.75	24	186.00	21,345.87	
R077	12.80	14	179.20	21,525.07	
6193L	0.85	148	125.80	21,650.87	
3K62	2.85	43	122.55	21,773.42	
39SS5	1.23	52	63.96	21,837.38	
334Y	0.68	77	52.36	21,889.74	
93939	4.05	12	48.60	21,938.34	
88450	1.50	21	31.50	21,969.84	
8ST4	0.25	56	14.00	21,983.84	

multiproduct system, the same values of setup cost K and interest rate I are used for all items. We can treat the ratio K/I as a policy variable; if this ratio is large, lot sizes will be larger and the average investment in inventory will be greater. If this ratio is small, the number of annual replenishments will increase.

To see exactly how a typical exchange curve is derived, consider a deterministic system consisting of n products with varying demand rates $\lambda_1, \ldots, \lambda_n$ and item values $c_1, \ldots, c_n$. If EOQ values are used to replenish stock for each item, then

$$Q_i = \sqrt{\frac{2K\lambda_i}{Ic_i}} \quad \text{for } 1 \le i \le n.$$

For item i, the cycle time is Q_i/λ_i, so that λ_i/Q_i is the number of replenishments in one year. The total number of replenishments for the entire system is $\Sigma\lambda_i/Q_i$. The average on-hand inventory of item i is $Q_i/2$, and the value of this inventory in dollars is $c_i Q_i/2$. Hence, the total value of the inventory is $\Sigma c_i Q_i/2$.

Each choice of the ratio K/I will result in a different value of the number of replenishments per year and the dollar value of inventory. As K/I is varied, one traces out a curve such as the one pictured in Figure 5–9. An exchange curve such as this one allows management to

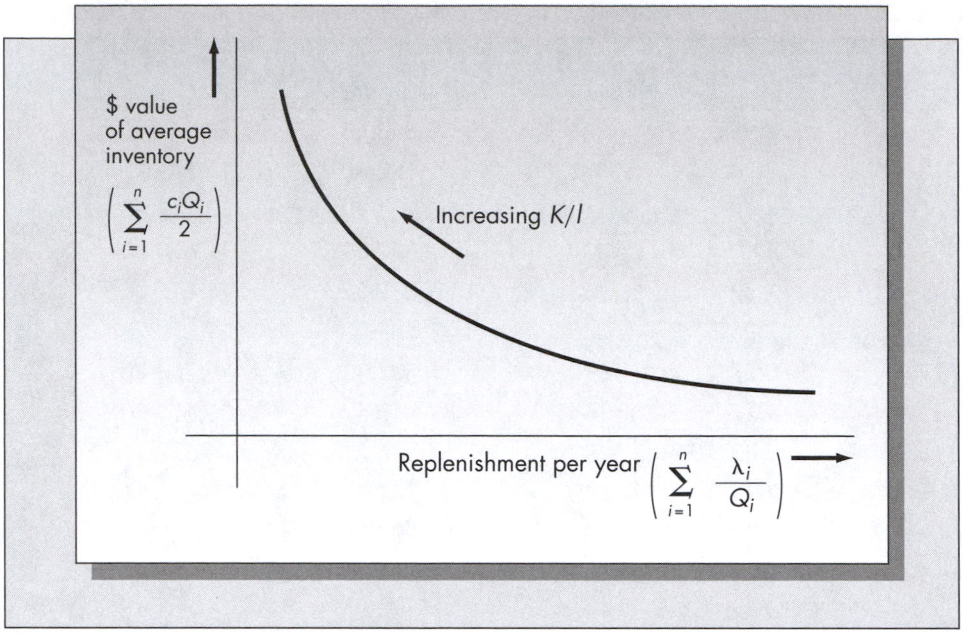

Figure 5–9 Exchange curve of replenishment frequency and inventory value

easily see the trade-off between the dollar investment in inventory and the frequency of stock replenishment.

Exchange curves also can be used to compare various safety stock and service level strategies. As an example, consider a system in which a fill rate constraint is used (i.e., Type 2 service) for all items. Furthermore, suppose that the lead time demand distribution for all items is normal, and each item gets equal service. The dollar value of the safety stock is $\Sigma c_i (R_i - \mu_i)$, and the annual value of back-ordered demand is $\Sigma c_i \lambda_i n(R_i)/Q_i$. A fixed value of the fill rate β will result in a set of values of the control variables $(Q_1, R_1), \ldots, (Q_n, R_n)$, which can be computed by the methods discussed earlier in the chapter. Each set of (Q, R) values yields a pair of values for the safety stock and the back-ordered demand. As the fill rate is increased, the investment in safety stock increases, and the value of back-ordered demand decreases. The exchange curve one would obtain is pictured in Figure 5–10. Such an exchange curve is a useful way for management to assess the dollar impact of various service levels.

EXAMPLE 5.11

Consider the 20 stock items listed in Tables 5–1 and 5–2. Suppose that Harvey, the owner of Harvey's Specialty Shop, is reconsidering his choices of the setup cost of $50 and interest charge of 20 percent. Harvey uses the EOQ formula to compute lot sizes for the 20 items for a range of values of K/I from 50 to 500. The resulting exchange curve appears in Figure 5–11.

Harvey is currently operating at $K/I = 50/0.2 = 250$, which results in approximately 22 orders per year and an average inventory cost of $5,447 annually. By reducing K/I to 100, the inventory cost for these 20 items is reduced to $3,445 and the order frequency is increased to 34 orders a year. After some thought, Harvey decides that the $2,000 savings in inventory cost is definitely worth the additional time and bookkeeping expenses required to track the additional 12 orders annually. (He is fairly comfortable

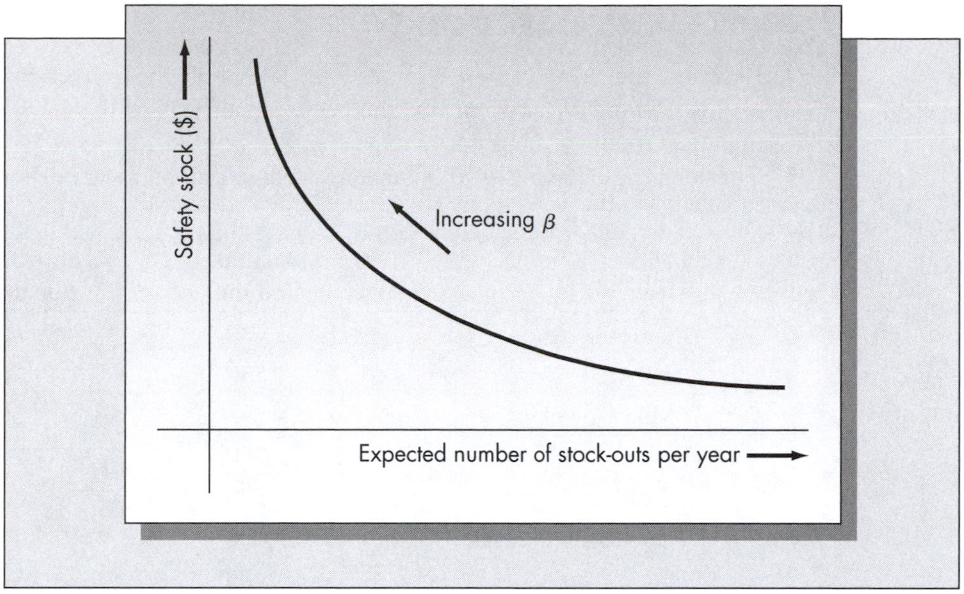

Figure 5–10 Exchange curve of the investment in safety stock and β

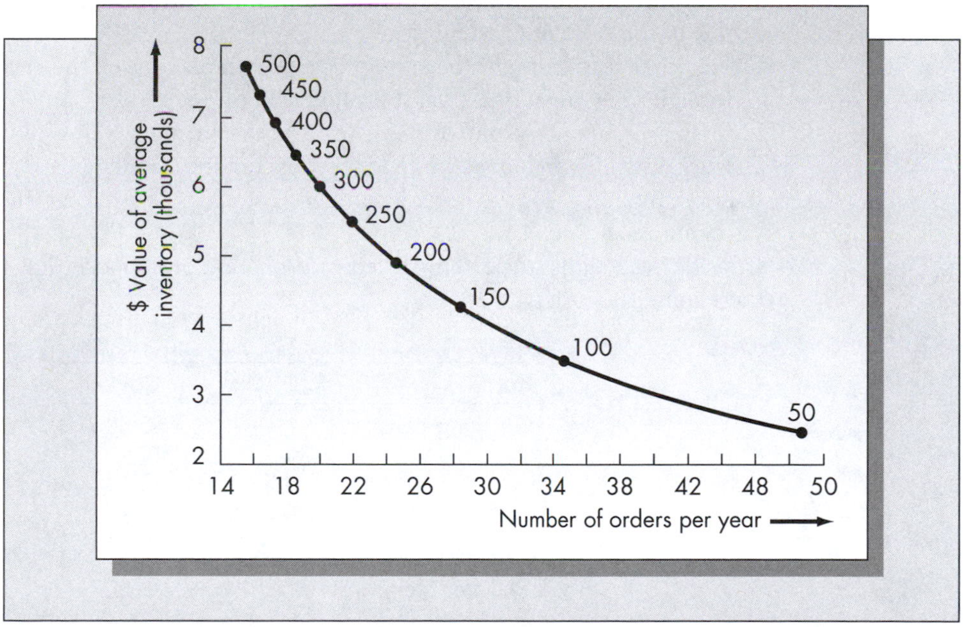

Figure 5–11 Exchange curve: Harvey's Specialty Shop

with the 20 percent interest rate, which means that the true value of his setup cost for ordering is closer to \$20 than \$50. In this way the exchange curve can assist in determining the correct value of cost parameters that may otherwise be difficult to estimate.) He also considers moving to $K/I = 50$ but decides that the additional savings of about \$1,000 are not worth having to process almost an additional 50 orders a year.

PROBLEMS FOR SECTION 5.7

24. Describe the ABC classification system. What is the purpose of classifying items in this fashion? What would be the primary value of ABC analysis to a retailer? To a manufacturer?
25. Consider the following list of retail items sold in a small neighborhood gift shop.

Item	Annual Volume	Average Profit per Item
Greeting cards	3,870	$0.40
T-shirts	1,550	1.25
Men's jewelry	875	4.50
Novelty gifts	2,050	12.25
Children's clothes	575	6.85
Chocolate cookies	7,000	0.10
Earrings	1,285	3.50
Other costume jewelry	1,900	15.00

a. Rank the item categories in decreasing order of the annual profit. Classify each into one of the A, B, or C categories.
b. For what reason might the store proprietor choose to sell the chocolate cookies even though they might be her least profitable item?

26. From management's point of view, what is the primary value of an exchange curve? Discuss both the exchange curve for replenishment frequency and inventory value and the exchange curve for expected number of stock-outs per year and the investment in safety stock.
27. Consider the eight stock items listed in problem 25. Suppose that the average costs of these item categories are

Item	Cost
Greeting cards	$0.50
T-shirts	3.00
Men's jewelry	8.00
Novelty gifts	12.50
Children's clothes	8.80
Chocolate cookies	0.40
Earrings	4.80
Other costume jewelry	12.00

Compare the total number of replenishments per year and the dollar value of the inventory of these items for the following values of the ratio of K/I: 100, 200, 500, 1,000. From the four points obtained, estimate the exchange curve of replenishment frequency and inventory value.

5.8 OVERVIEW OF ADVANCED TOPICS

This chapter has treated only a small portion of inventory models available. Considerable research has been devoted to analyzing far more complex stochastic inventory control problems, but most of this research is beyond the scope of our coverage. This section presents a brief overview of two areas not discussed in this chapter that account for a large portion of the research on inventory management.

Multi-echelon Systems

Firms involved in the manufacture and distribution of consumer products must take into account the interactions of the various levels in the distribution chain. Typically, items are produced at one or more factories, shipped to warehouses for intermediate storage, and subsequently shipped to retail outlets to be sold to the consumer. We refer to the levels of the system as **echelons**. For example, the factory–warehouse–retailer three-echelon system is pictured in Figure 5–12.

A system of the type pictured in Figure 5–12 is generally referred to as a *distribution* system. In such a system, the demand arises at the lowest echelon (the retailer in this case) and is transmitted up to the higher echelons. Production plans at the factory must be coordinated with orders placed at the warehouse by the retailers. Another type of multi-echelon system arising in manufacturing contexts is known as an *assembly* system. In an assembly system, components are combined to form subassemblies, which are eventually combined to form end items (that is, final products). A typical assembly system is pictured in Figure 5–13.

In the assembly system, external demand originates at the end-item level only. Demand for the components arises only when "orders" are placed at higher levels of the system, which are the consequence of end-item production schedules. Materials requirements planning methodology (see Chapter 9) is designed to model systems of this type but does not consider the effect of uncertainty on the final demand. Although there has been some research on this problem, precisely how the uncertainty in the final demand affects the optimal production and replenishment policies of components remains an open question.

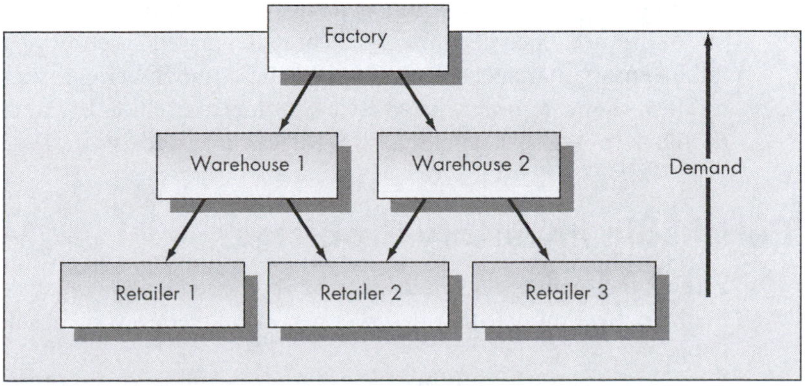

Figure 5–12 Typical three-level distribution system

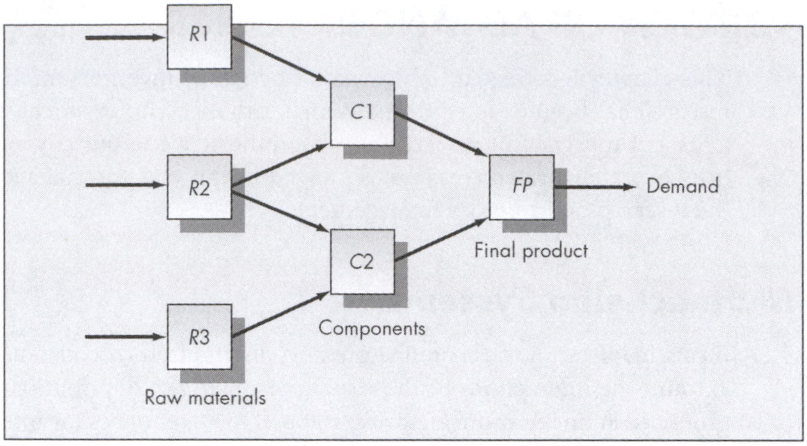

Figure 5–13 Typical three-level assembly system

One multi-echelon system that has received a great deal of attention in the literature (Sherbrooke, 1968) was designed for an application in military logistics. Consider the replenishment problem that arises in maintaining a supply of spare parts to support jet aircraft engine repair. When an engine (or engine component) fails, it is sent to a central depot for repair. The depot maintains an inventory of engines so that it can ship a replacement to the base without waiting for the repair to be completed. The problem is to determine the correct level of spare engines at the base and at the depot given the conflicting goals of budget limitations and the desirability of a high system fill rate.

The U.S. military's investment in spare parts is enormous. As far back as 1968 it was estimated that the U.S. Air Force alone had $10 billion invested in repairable item inventories. Considering the effects of inflation and the sizes of the other armed services, the current investment in repairable items in the U.S. military probably exceeds $100 billion. Mathematical inventory models have made a significant impact on the management of repairable item inventories, both in military and nonmilitary environments.

In the private sector, retailing is an area in which multi-echelon distribution systems are common. Most major retail chains utilize a distribution center (DC) as an intermediate storage point between the manufacturer and the retail store. Determining how to allocate inventory between the DC and the stores is an important strategic issue. DC inventory allows "risk pooling" among the stores and facilitates redistribution of store inventories that might grow out of balance. Several commercial computer-based inventory packages are available for retail inventory management, perhaps the best known being Inforem, designed and marketed by IBM. (For a comprehensive review of the research on the retailer inventory management problem, see Agrawal and Smith, 2015; Nahmias and Smith, 1993.)

Perishable Inventory Problems

Although certain types of decay can be accounted for by adjustments to the holding cost, the mathematical models that describe perishability, decay, or obsolescence are generally quite complex.

Typical examples of fixed-life perishable inventories include food, blood, drugs, and fresh flowers. In order to keep track of the age of units in stock, it is necessary to know the amount

SNAPSHOT APPLICATION

Intel Uses Multiechelon Inventory Modelling to Manage the Supply Chain for Boxed CPUs

Intel Corporation, headquartered in Santa Clara, California, is the largest producer of integrated circuits in the world. Intel microprocessors have been the "guts" of most personal computers ever since IBM first adopted the Intel 8088 as the CPU in the original IBM PC in 1981. In 2005, a group at Intel and one outside consultant was charged with the problem of managing the supply chain of its branded boxed CPUs. The division typically sold 100–150 configurations of these products.

The group chose to model the problem as a three echelon inventory system: (1) global supply, (2) boxing sites, and (3) boxed CPU warehouses. The group quickly realized that there were many factors affecting the supply chain, but the primary driver of system effectiveness was demand uncertainty. Predicting demand uncertainty was difficult since product life cycles were often relatively short, resulting in very little historical data for some products.

Another complicating factor was that the three echelons were far apart geographically, resulting in a very long cumulative lead time for the system. Hence, a make-to-order business model was simply not feasible. For that reason, proper positioning of inventory along the entire supply chain was critical. The mathematical model developed for this application was a nonlinear mathematical programming model that could be solved by standard dynamic programming techniques.

A key issue was metrics. That is, how should one evaluate the quality of the model's results? One metric was the total inventory in the system. This translated to the total dollar investment required by Intel. A second metric was the quality of the service provided to the customer. Both Type 1 and Type 2 service level metrics (as discussed in detail in this chapter) were considered. Intel management chose a modified version of Type 2 service (that is, the fill rate). Their metric was the percentage of an order filled on the date the customer asked for delivery of the product.

The team decided that it was not worth trying to apply their optimization tool to all 150 products, so they embarked on an ABC analysis, and ultimately chose to apply the model only to products with more than 3,000 sales per month and 12,000 sales per quarter. Also, it became clear that in order to be responsive to customer demand, the vast majority of the inventory would be held at the third echelon, namely the boxed CPU warehouses. This inventory had the most value added and, therefore, was the most expensive. That expense was offset by the fact that keeping the inventory closer to the customer significantly improved service levels. The system was implemented at Intel in late 2005. Comparisons with pre- and post-implementation metrics showed that the model resulted in a modest decline in system-wide inventory levels but a significantly improved level of customer service.

The priorities set by the project team were an important outcome of this study. These priorities provided a good blueprint on how to go about a project of this type to maximize the likelihood of implementation. The four goals set were: (1) keep models and processes simple, (2) make things better now, (3) implement changes in a phased manner, and (4) be clear about what constitutes success.

Source: Wieland, B., P. Mastrantonio, S. Willems, and K. Kempf. "Optimizing Inventory Levels within Intel's Channel Supply Demand Operations." *Interfaces* 42 (2012), pp. 517–527.

of each age level of inventory on hand. This results in a complicated multidimensional system state space. Perishability adds an additional penalty for holding inventory, thus reducing the size of orders placed. The most notable application of the theory of perishable inventory models has been to the management of blood banks.

A somewhat simpler but related problem is that of exponential decay, in which it is assumed that a fixed fraction of the inventory in stock is lost each period. Exponential decay is an accurate description of the changes that take place in volatile liquids such as alcohol and gasoline and is an exact model of decline in radioactivity of radioactive pharmaceuticals and nuclear fuels.

Obsolescence has been modeled by assuming that the length of the planning horizon is a random variable with a known probability distribution. In practice, such models are valuable only if the distribution of the useful lifetime of the item can be accurately estimated in advance.

A related problem is the management of style goods, such as fashion items in the garment industry. The style goods problem differs from the inventory models considered in this chapter—not only is the demand itself uncertain but the distribution of demand is uncertain as well. This is typical of problems for which there is no prior demand history. The procedures suggested to deal with the style goods problem generally involve some type of Bayesian updating scheme to combine current observations of demand with prior estimates of the demand distribution.

5.9 HISTORICAL NOTES AND ADDITIONAL READINGS

Research on stochastic inventory models was sparked by the national war effort and appears to date back to the 1940s, although the first published material appeared in the early 1950s (see Arrow, Harris, and Marschak, 1951; Dvorestsky, Kiefer, and Wolfowitz, 1952). The important text by Kenneth Arrow, Samuel Karlin, and Herbert Scarf (1958) produced considerable interest in the field.

The model of a lot size–reorder point system with stock-out cost criterion discussed in this chapter is attributed to Thomson Whitin (1957). The extensions to service levels are attributed to Brown (1967). Brown also treats a variety of other topics of practical importance, including the relationship between forecasting models and inventory control. Edward Silver, David Pyke, and Douglas Thomas (2017) cover exchange curves in detail and also provide more in-depth coverage of a number of the topics we discuss in this chapter. Stephen Love (1979) provides an excellent summary of many of the issues we treat. George Hadley and Whitin (1963) also deal with these topics at a more sophisticated mathematical level.

There was considerable interest in the development of approximate (s, S) policies. Scarf (1959), and later Donald Iglehart (1963), proved the optimality of (s, S) policies; Arthur Veinott and Harvey Wagner (1965) considered methods for computing optimal (s, S) policies. A variety of approximation techniques have been suggested. Richard Ehrhardt (1979) considered using regression analysis to fit a grid of optimal policies, whereas James Freeland and Evan Porteus (1980) adapted dynamic programming methods to the problem. Porteus (1985) numerically compared the effectiveness of various approximation techniques. Robert Kaplan (1970) and Steven Nahmias (1979), among others, examined some of the issues regarding lead time uncertainty.

Andrew Clark and Scarf (1960) offered the original formulation of the standard multi-echelon inventory problem. Other researchers developed extensions (see, for example, Bessler and Veinott, 1966; and Federgruen and Zipkin, 1984). Bryan Deuermeyer and Leroy Schwarz (1979), as well as Eppen and Schrage (1981), studied retailer-warehouse systems. Charles Schmidt and Nahmias (1985) analyzed an assembly system when demand for the final product was random.

Craig Sherbrooke (1968) wrote the classic work on the type of multi-echelon model that has become the basis for the inventory control systems implemented in the military. Stephen

Graves (1985) extended Sherbrooke's analysis. John Muckstadt and Joseph Thomas (1980) discussed the advantages of implementing multi-echelon models in an industrial environment (see Nahmias, 1981 for a review of models for managing repairables).

Interest in perishable inventory control models appears to stem from the problem of blood bank inventory control, although food management has a far greater economic impact. Most of the mathematical models for perishables assume that the inventory is issued from stock on an oldest-first basis (a notable exception is the study by Cohen and Pekelman, 1978; see Nahmias, 1982 for a review of the research on perishable inventory control).

A number of researchers have studied the style goods problem (for example, Hausman and Peterson, 1972; Hartung, 1973; Murray and Silver, 1966). Those interested in additional readings on stochastic inventory models should refer to the various review articles in the field (Graves, Kan, and Zipkin, 1993; Nahmias, 1978; Scarf, 1963; and Veinott, 1966).

5.10 SUMMARY

This chapter presented an overview of several inventory control methods when the demand for the item is random. The newsvendor model is based on the assumption that the product has a useful life of exactly one planning period. We assumed that the demand during the period is a continuous random variable with cumulative distribution function $F(x)$ and that there are specified overage and underage costs of c_o and c_u charged against the inventory remaining on hand at the end of the period or the excess demand, respectively. The optimal order quantity Q^* solves the equation

$$F(Q^*) = \frac{c_u}{c_u + c_o}.$$

Extensions to discrete demand and multiple planning periods also were considered.

From a practical standpoint, the newsvendor model has a serious limitation: it does not allow for a positive order setup cost. For that reason, we considered an extension of the EOQ model known as a lot size–reorder point model. The key random variable for this case was the demand during the lead time. We showed how optimal values of the decision variables Q and R could be obtained by the iterative solution of two equations. The system operates in the following manner. When the level of on-hand inventory hits R, an order for Q units is placed (which will arrive after the lead time τ). This policy is the basis of many commercial inventory control systems.

Service levels provide an alternative to stock-out costs. Two service levels were considered: Type 1 and Type 2. Type 1 service is the probability of not stocking out in any order cycle, and Type 2 service is the probability of being able to meet a demand when it occurs. Type 2 service, also known as the fill rate, is the more natural definition of service for most applications.

Several additional topics for (Q, R) systems also were considered. The imputed shortage cost is the effective shortage cost resulting from specification of the service level. Assuming normality, we showed how one transforms the distribution of periodic demand to lead time demand. Finally, we considered the effects of lead time variability.

If a setup cost is included in the multiperiod version of the newsvendor problem, the optimal form of the control policy is known as an (s, S) policy. This means that if the starting inventory in a period, u, is less than or equal to s, an order for $S - u$ units is placed. An effective approximation for the optimal (s, S) policy can be obtained by solving the problem as if it were continuous review to obtain the corresponding (Q, R) policy, and setting $s = R$ and $S = Q + R$. We also discussed the application of service levels to periodic-review systems.

Most real inventory systems involve the management of more than a single product. We discussed several issues that arise when managing a multiproduct inventory system. One of these is the amount of time and expense that should be allocated to the control of each item. The ABC system is a method of classifying items by their annual volume of sales. Another issue concerns the correct choice of the cost parameters used in computing inventory control policies. Because many of the cost parameters used in inventory analysis involve managerial judgment and are not easily measured, it would be useful if management could compare the effects of various parameter settings on the performance of the system. A convenient technique for making these comparisons is via exchange curves. We discussed two of the most popular exchange curves: (1) the trade-off between the investment in inventory and the frequency of stock replenishment and (2) the trade-off between the investment in safety stock and service levels.

ADDITIONAL PROBLEMS ON STOCHASTIC INVENTORY MODELS

28. An artist's supply shop stocks a variety of different items to satisfy the needs of both amateur and professional artists. In each case described, what is the appropriate inventory control model that the store should use to manage the replenishment of the item described? Choose your answer from the following list and be sure to explain your answer in each case.

Simple EOQ	Newsvendor model with service level
Finite production rate	(Q, R) model with stock-out cost
EOQ with quantity discounts	(Q, R) model with Type 1 service level
Resource-constrained EOQ	(Q, R) model with Type 2 service level
Newsvendor model	Other type of model

a. A highly volatile paint thinner is ordered once every three months. Cans not sold during the three-month period are discarded. The demand for the paint thinner exhibits considerable variation from one three-month period to the next.
b. A white oil-based paint sells at a fairly regular rate of 600 tubes per month and requires a six-week order lead time. The paint store buys the paint for \$1.20 per tube.
c. Burnt sienna oil paint does not sell as regularly or as heavily as the white. Sales of the burnt sienna vary considerably from one month to the next. The useful lifetime of the paint is about two years, but the store sells almost all the paint prior to the two-year limit.
d. Synthetic paint brushes are purchased from an East Coast supplier who charges \$1.60 for each brush in orders of under 100 and \$1.30 for each brush in orders of 100 or greater. The store sells the brushes at a fairly steady rate of 40 per month for \$2.80 each.

e. Camel hair brushes are purchased from the supplier of part (d), who offers a discount schedule similar to the one for the synthetic brushes. The camel hair brushes, however, exhibit considerable sales variation from month to month.

29. Annual demand for number 2 pencils at the campus store is normally distributed with mean 1,000 and standard deviation 250. The store purchases the pencils for 6 cents each and sells them for 20 cents each. There is a two-month lead time from the initiation to the receipt of an order. The store accountant estimates that the cost in employee time for performing the necessary paperwork to initiate and receive an order is $20 and recommends a 22 percent annual interest rate for determining holding cost. The cost of a stock-out is the cost of lost profit plus an additional 20 cents per pencil, which represents the cost of loss of goodwill.
   a. Find the optimal value of the reorder point R assuming that the lot size used is the EOQ.
   b. Find the simultaneous optimal values of Q and R.
   c. Compare the average annual holding, setup, and stock-out costs of the policies determined in parts (a) and (b).
   d. What is the safety stock for this item at the optimal solution?
30. Consider the problem of satisfying the demand for number 2 pencils faced by the campus store mentioned in problem 29.
   a. Re-solve the problem, substituting a Type 1 service level criterion of 95 percent for the stock-out cost.
   b. Re-solve the problem, substituting a Type 2 service level criterion of 95 percent for the stock-out cost. Assume that Q is given by the EOQ.
   c. Find the simultaneous optimal values of Q and R assuming a Type 2 service level of 95 percent.
31. Answer the following true or false.
   a. The lead time is always less than the cycle time.
   b. The optimal lot size for a Type 1 service objective of X percent is always less than the optimal lot size for a Type 2 service objective of X percent for the same item.
   c. The newsvendor model does not include a fixed order cost.
   d. ABC analysis ranks items according to the annual value of their demand.
   e. For a finite production rate model, the optimal lot size to produce each cycle is equal to the maximum inventory each cycle.
32. One of the products stocked at Weiss's paint store (problem 14) is a certain type of highly volatile paint thinner that, due to chemical changes in the product, has a shelf life of exactly one year. Al Weiss purchases the paint thinner for $20 a gallon can and sells it for $50 a can. The supplier buys back cans not sold during the year for $8 for reprocessing. The demand for this thinner generally varies from 20 to 70 cans a year. Al assumes a holding cost for unsold cans at a 30 percent annual interest rate.
   a. Assuming that all values of the demand from 20 to 70 are equally likely, what is the optimal number of cans of paint thinner for Al to buy each year?
   b. More accurate analysis of the demand shows that a normal distribution gives a better fit of the data. The distribution mean is identical to that used in part (a), and the standard deviation estimator turns out to be 7. What policy do you now obtain?
33. Semicon is a start-up company that produces semiconductors for a variety of applications. The process of burning in the circuits requires large amounts of nitric acid, which has a shelf life of only three months. Semicon estimates that it will need between 1,000 and 3,000 gallons of acid for the next three-month period and assumes that all values in this interval are equally likely. The acid costs them $150 per gallon. The company assumes a

30 percent annual interest rate for the money it has invested in inventory, and the acid costs the company \$35 a gallon to store. (Assume that all inventory costs are attached to the end of the three-month period.) The disposal cost of acid that is left over at the end of the three-month period is \$75 per gallon. If the company runs out of acid during the three-month period, it can purchase emergency supplies quickly at a price of \$600 per gallon.

a. How many gallons of nitric acid should Semicon purchase? Experience with the marketplace later shows that the demand is closer to a normal distribution, with mean 1,800 and standard deviation 480.

b. Suppose that now Semicon switches to a 94 percent fill rate criterion. How many gallons should now be purchased at the start of each three-month period?

34. *Newsvendor simulator.* In order to solve this problem, your spreadsheet program will need to have a function that produces random numbers [=RAND() in Excel]. The purpose of this exercise is to construct a simulation of a periodic-review inventory system with random demand. We assume that the reader is familiar with the fundamentals of Monte Carlo simulation.

Your spreadsheet should allow for cell locations for storing values of the holding cost, the penalty cost, the proportional order cost, the order-up-to point, the initial inventory, and the mean and the standard deviation of periodic demand.

An efficient means of generating an observation from a standard normal variate is the formula

$$Z = [-2\ln(U_1)]^{0.5}\cos(2\pi U_2),$$

where U_1 and U_2 are two independent draws from a (0, 1) uniform distribution. (See, for example, Fishman, 1973.) Note that two independent calls of =RAND() are required. Since Z is approximately standard normal, the demand X is given by

$$X = \sigma Z + \mu$$

where μ and σ, are the mean and the standard deviation of one period's demand. Alternatively, a less efficient but simpler method of generating a normal random variable in Excel is =NORM.INV(RAND(), μ, σ). Below is a suggested layout of the spreadsheet.

NEWSVENDOR SIMULATOR

Holding cost = Mean demand =

Order cost = Std. dev. demand =

Penalty cost = Initial inventory =

Order-up-to point =

Period	Starting Inventory	Order Quantity	Demand	Ending Inventory	Holding Cost	Penalty Cost	Order Cost
1							
2							
3							
.							
.							
.							
20							
Totals							

Each time you recalculate, the simulator will generate a different sequence of demands. A set of suggested parameters is

$$\begin{aligned} h &= 2 \\ c &= 5 \\ p &= 20 \\ \mu &= 100 \\ \sigma &= 20 \\ I_0 &= 50 \\ \text{Order-up-to point} &= 150 \end{aligned}$$

35. *Using the simulator for optimization.* You will be able to do this problem only if the program you are using does not restrict the size of the spreadsheet. Assume the parameters given in problem 34 but use $\sigma = 10$. Extend the spreadsheet from problem 34 to 1,000 rows or more and compute the average cost per period. If the cost changes substantially as you recalculate the spreadsheet, you need to add additional rows.

    Now, experiment with different values of the order-up-to point to find the one that minimizes the average cost per period. Compare your results to the theoretical optimal solution.
36. *Newsvendor calculator.* Design a spreadsheet to calculate the optimal order-up-to point for a newsvendor problem with normally distributed demands. In order to avoid table look-ups, use the approximation formula

$$Z = 5.0633[F^{0.135} - (1 - F)^{0.135}]$$

    for the inverse of the standard normal distribution function. (This formula is from Ramberg and Schmieser, 1972.) The optimal order-up-to point, y^*, is of the form $y^* = \sigma Z + \mu$.
    a. Assume that all parameters are as given in problem 34. Graph y^* as a function of p, the penalty cost, for $p = 10, 15, \ldots, 100$. By what percentage does y^* increase if p increases from 10 to 20? from 50 to 100?
    b. Repeat part (a) with $\sigma = 35$. Comment on the effect that the variance in demand has on the sensitivity of y^* to p.
37. A large national producer of canned foods plans to purchase 100 combines that are to be customized for its needs. One of the parts used in the combine is a replaceable blade for harvesting corn. Spare blades can be purchased at the time the order is placed for \$100 each but will cost \$1,000 each if purchased at a later time because a special production run will be required.

    It is estimated that the number of replacement blades required by a combine over its useful lifetime can be closely approximated by a normal distribution with mean 18 and standard deviation 5.2. The combine maker agrees to buy back unused blades for \$20 each. How many spare blades should the company purchase with the combines?
38. Crazy Charlie's, a discount electronics store, uses simple exponential smoothing with a smoothing constant of $\alpha = .2$ to track the mean and the MAD of monthly item demand. One particular item, a wifi receiver, has experienced the following sales over the last three months: 126, 138, 94.

    Three months ago the receiver had stored values of mean = 135 and MAD = 18.5.
    a. Using the exponential smoothing equations given in Section 5.1, determine the current values for the mean and the MAD of monthly demand. (Assume that the stored values were computed *prior* to observing the demand of 126.)

b. Suppose that the order lead time for this particular item is 10 weeks (2.5 months). Assuming a normal distribution for monthly demand, determine the current estimates for the mean and the standard deviation of lead time demand.
c. This particular receiver is ordered directly from Japan; as a result, there has been considerable variation in the replenishment lead time from one order cycle to the next. Based on an analysis of past order cycles, it is estimated that the standard deviation of the lead time is 3.3 weeks. All other relevant figures are given in parts (a) and (b). Find the mean and the standard deviation of lead time demand in this case.
d. If Crazy Charlie's uses a Type 1 service objective of 98 percent to control the replenishment of this item, what is the value of the reorder level? (Assume that the lead time demand has a normal distribution.)

39. The home appliance department of a large department store is using a lot size–reorder point system to control the replenishment of a particular model of coffee machine. The store sells an average of 10 machines each week. Weekly demand follows a normal distribution with variance 26.

    The store pays $20 for each coffee machine, which it sells for $75. Fixed costs of replenishment amount to $28.The accounting department recommends a 20 percent interest rate for the cost of capital. Storage costs amount to 3 percent and breakage to 2 percent of the value of each item.

    If a customer demands the machine when it is out of stock, the customer will generally go elsewhere. Loss-of-goodwill costs are estimated to be about $25 per coffee machine. Replenishment lead time is three months.

    a. If lot sizes are based on the EOQ formula, what reorder level should be used for the machines?
    b. Find the optimal values of (Q, R).
    c. Compare the average annual costs of holding, ordering, and stock-out for the policies that you found in parts (a) and (b).
    d. Re-solve the problem using equations (1) and (2′) rather than (1) and (2). What is the effect of including lost sales explicitly?

40. Re-solve problem 39 replacing the stock-out cost with a 96 percent Type 1 service level.
41. Re-solve problem 39 replacing the stock-out cost with a 96 percent Type 2 service level. What is the imputed shortage cost?
42. Consider the equation giving the expected average annual cost of the policy (Q, R) in a continuous-review inventory control system from Section 5.4:

$$G(Q,R)=h\left(\frac{Q}{2}+R-\lambda T\right)+\frac{K\lambda}{Q}+\frac{p\lambda n(R)}{Q}.$$

    Design a spreadsheet to compute $G(Q, R)$ for a range of values of $Q \geq$ EOQ and $R \geq \mu$. Use the following approximation formula for $L(z)$ to avoid table look-ups:

$$L(z) = \exp(-0.92 - 1.19z - 0.37z^2).$$

    (This formula is from Parr, 1972.) Store the problem parameters c, h, p, μ, σ, and λ in cell locations. Visually search through the tabled values of $G(Q, R)$ to discover the minimum value and estimate the optimal (Q, R) values in this manner. Compare your results to the true optimal found from manual calculation.

    a. Solve Example 5.4.
    b. Solve problem 13 in this manner.
    c. Solve problem 14 in this manner.

43. The daily demand for a spare engine part is a random variable with a distribution, based on past experience, as follows.

Number of Demands per Day	Probability
0	.21
1	.38
2	.19
3	.14
4	.08

The part is expected to be obsolete after 400 days. Assume that demands from one day to the next are independent. The parts cost \$1,500 each when acquired in advance of the 400-day period and \$5,000 each when purchased on an emergency basis during the 400-day period. Holding costs for unused parts are based on a daily interest rate of 0.08 percent. Unused parts can be scrapped for 10 percent of their purchase price. How many parts should be acquired in advance of the 400-day period? (Hint: Let $D_1, D_2, \ldots, D_{400}$ represent the daily demand for the part. Assume each D_i has mean μ and variance σ^2. The central limit theorem says that the total demand for the 400-day period, ΣD_i, is approximately normally distributed with mean 400μ and variance $400\sigma^2$.)

44. Cassorla's Clothes sells a large number of white dress shirts. The shirts, which bear the store label, are shipped from a manufacturer in New York City. Hy Cassorla, the proprietor, says, "I want to be sure that I never run out of dress shirts. I always try to keep at least a two months' supply in stock. When my inventory drops below that level, I order another two-month supply. I've been using that method for 20 years, and it works."

    The shirts cost \$6 each and sell for \$15 each. The cost of processing an order and receiving new goods amounts to \$80, and it takes three weeks to receive a shipment. Monthly demand is approximately normally distributed with mean 120 and standard deviation 32. Assume a 20 percent annual interest rate for computing the holding cost.

    a. What value of Q and R is Hy Cassorla using to control the inventory of white dress shirts?
    b. What fill rate (Type 2 service level) is being achieved with the current policy?
    c. Based on a 99 percent fill rate criterion, determine the optimal values of Q and R that he should be using. (Assume four weeks in a month for your calculations.)
    d. Determine the difference in the average annual holding and setup costs between the policies in parts (b) and (c).
    e. Estimate how much time would be required to pay for a \$25,000 inventory control system, assuming that the dress shirts represent 5 percent of Hy's annual business and that similar savings could be realized on the other items as well.

45. The Poisson distribution is discussed in Appendix 5–D at the end of this chapter. Assume that the distribution of bagels sold daily at Billy's Bakery in problem 8 follows a Poisson distribution with mean 16 per day. Using Table A–3 in the back of the book or the Poisson distribution function built into Excel, determine the optimal number of bagels for Billy's to bake each day.

46. Consider the Crestview Printing Company described in problem 9. Suppose that sales of cards (in units of 50,000) follow a Poisson distribution with mean 6.3. Using Table A–3 in the back of the book or the Poisson distribution function built into Excel find the optimal number of cards for Crestview to print for the next Christmas season.

47. The Laplace distribution is discussed in Appendix 5–D. As noted there, the Laplace distribution could be a good choice for describing demand for slow-moving items and for fast-moving items with high variation. The cdf of the Laplace distribution is given by

$$F(x) = 0.5[1 + \text{sgn}\,(x - \mu)](1 - \exp(-|x - \mu|/\theta),$$

and the inverse of the cdf is given by

$$F^{-1}(p) = \mu - \theta\,\text{sgn}\,(p - 0.5)\ln(1 - 2|p - 0.5|),$$

where sgn(x) is the sign of x. The mean is μ and the variance is $2\theta^2$. Since the inverse distribution function can be written down explicitly, one does not have to resort to tables to solve newsvendor problems when demand follows the Laplace distribution.

Solve problem 10, part (a), assuming the demand for the EX123 follows a Laplace distribution with parameters $\mu = 60$ and $\theta = 3\sqrt{2}$ (which will give exactly the same mean and variance).

48. Solve problem 11 assuming the demand for fans over the selling season follows a Laplace distribution with the same mean and variance as you computed in problem 11(a).
49. Solve problem 12(b) assuming the demand for handbags over the selling season follows a Laplace distribution with mean 150 and standard deviation 20.
50. Solve problem 13 assuming that the lead time demand for oil filters follows a Laplace distribution with mean and variance equal to that which was computed in problem 13. What difference do you see in the (Q, R) values as compared to those for the normal case?
51. Solve problem 14 assuming that the lead time demand for white latex paint follows a Laplace distribution with mean and variance equal to the mean and variance of lead time demand you computed for problem 14. What difference do you see in the (Q, R) values as compared to those for the normal case?

APPENDIX 5–A

NOTATIONAL CONVENTIONS AND PROBABILITY REVIEW

Demand will be denoted by D, which is assumed to be a random variable. The cumulative distribution function (cdf) of demand is $F(x)$ and is defined by

$$F(x) = P\{D \le x\} \qquad \text{for } -\infty < x < +\infty.$$

When D is continuous, the probability density function (pdf) of demand, $f(x)$, is defined by

$$f(x) = \frac{dF(x)}{dx}.$$

When D is discrete, $f(x)$ is the probability function (pf) defined by

$$f(x) = P\{X = x\} = F(x) - F(x - 1).$$

Note that in the continuous case the density function is not a probability and the value of $f(x)$ is not necessarily less than 1, although it is always true that $f(x) \geq 0$ for all x.

The expected value of demand, $E(D)$, is defined as

$$E(D) = \int_{-\infty}^{+\infty} xf(x)dx$$

in the continuous case, and

$$E(D) = \sum_{x=-\infty}^{+\infty} xf(x)$$

in the discrete case.

We use the symbol μ to represent the expected value of demand $[E(D) = \mu]$. In what follows we assume that D is continuous; similar formulas hold in the discrete case. Let $g(x)$ be any real-valued function of the real variable x. Then

$$E(g(D)) = \int_{-\infty}^{+\infty} g(x)f(x)dx.$$

In particular, let $g(D) = \max(0, Q - D)$. Then

$$E(g(D)) = \int_{-\infty}^{+\infty} \max(0,\ Q - x)f(x)dx.$$

Because demand is nonnegative, it must be true that $f(x) = 0$ for $x < 0$. Furthermore, when $x > Q$, $\max(0, Q - x) = 0$, so we may write

$$E(g(D)) = \int_{0}^{Q} (Q - x)f(x)dx.$$

In the analysis of the newsvendor model, Leibniz's rule is used to determine the derivative of $G(Q)$. According to Leibniz's rule:

$$\frac{d}{dy}\int_{a_1(y)}^{a_2(y)} h(x, y)dx = \int_{a_1(y)}^{a_2(y)} [\partial h(x, y)/\partial y]dx + h(a_2(y), y)a_2'(y) - h(a_1(y), y)a_1'(y).$$

APPENDIX 5–B

ADDITIONAL RESULTS AND EXTENSIONS FOR THE NEWSVENDOR MODEL

1. Interpretation of the overage and underage costs for the single period problem

Define

S = Selling price of the item.
c = Variable cost of the item.

h = Holding cost per unit of inventory remaining in stock at the end of the period.
p = Loss-of-goodwill cost plus bookkeeping expense (charged against the number of back orders on the books at the end of the period).

We show how c_u and c_o should be interpreted in terms of these parameters. As earlier, let Q be the order quantity and D the demand during the period. Assume without loss of generality that starting inventory is zero. Then the cost incurred at the end of the period is

$$cQ + h\max(Q - D, 0) + p\max(D - Q, 0) - S\min(Q, D).$$

The expected cost is

$$G(Q) = cQ + h\int_0^Q (Q-x)f(x)dx + p\int_Q^\infty (x-Q)f(x)dx - S\int_0^Q xf(x)dx - SQ\int_Q^\infty f(x)dx.$$

Using

$$\int_0^Q xf(x)dx = \int_0^\infty xf(x)dx - \int_Q^\infty xf(x)dx = \mu - \int_Q^\infty xf(x)dx,$$

the expected cost may be written

$$G(Q) = cQ + h\int_0^Q (Q-x)f(x)dx + (p+S)\int_Q^\infty (x-Q)f(x)dx - S\mu.$$

The optimal order quantity satisfies

$$G'(Q) = 0$$

or

$$c + hF(Q) - (p + S)(1 - F(Q)) = 0,$$

which results in

$$F(Q) = \frac{p+S-c}{p+S+h}.$$

Setting $c_u = p + S - c$ and $c_o = h + c$ gives the critical ratio in the form $c_u/(c_u + c_o)$.

2 The Newsvendor Cost When Demand is Normal

When the one period demand follows a normal distribution, we can obtain an explicit expression for the optimal one period cost for the newsvendor model from Section 5.3. We know that the expected single period cost function is

$$G(Q) = c_o\int_0^Q (Q-t)f(t)dt + c_u\int_Q^\infty (t-Q)f(t)dt,$$

and the optimal value of Q satisfies

$$F(Q^*) = \frac{c_u}{c_u + c_o}.$$

It is convenient to express $G(Q)$ in a slightly different form. Note that

$$\int_0^Q (Q-t)f(t)dt = \int_0^\infty (Q-t)f(t)dt - \int_Q^\infty (Q-t)f(t)dt = Q - \mu + n(Q) \text{ where}$$
$$n(Q) = \int_Q^\infty (t-Q)f(t)dt.$$

It follows that $G(Q)$ can be written in the form

$$G(Q) = c_o(Q - \mu) + (c_u + c_o)n(Q),$$

where μ is the expected demand.

We know from Section 5.4 that one can write an explicit expression for $n(Q)$ when demand is normal. In particular, we showed that $n(Q) = \sigma L(z)$ where $L(z)$ is the standard loss integral. As noted in Section 5.4,

$$L(z) = \varphi(z) - z[1 - \Phi(z)],$$

where $\varphi(z)$ is the standard normal density, $\Phi(z)$ the standard normal distribution function, and $z = (Q - \mu)/\sigma$ is the standardized value of the order quantity Q.

It follows that in the normal case we can write

$$G(Q) = c_o(Q - \mu) + (c_u + c_o)\phi\,[\varphi(z) - z(1 - \Phi(z)].$$

The standard normal density function can be computed directly, and the cumulative distribution function is available through tables and is a built-in function in Excel.

It might also be of interest to know the minimum value of the expected cost. This is the value of the function G when $Q = Q^*$. Note that at $Q = Q^*$ the cumulative distribution function $\Phi(z^*)$ is equal to the critical ratio $c_u/(c_u + c_o)$, which means the complementary cumulative distribution function $1 - \Phi(z^*)$ is equal to the ratio $c_o/(c_u + c_o)$. Also, since $z = (Q - \mu)/\sigma$, it follows that $Q - \mu = \sigma z$. Making these two substitutions in the expression above for $G(Q)$ we obtain

$$G(Q^*) = c_o\sigma z^* + (c_u + c_o)\sigma\left[\phi(z^*) - z^*\left(\frac{c_o}{c_o + c_u}\right)\right] = (c_u + c_o)\sigma\phi(z^*).$$

[Porteus, 2002, p. 13 derived essentially the same expression; Gerard Cachon provided helpful discussions on this point.]

Note that if one knows z^* no table look up is required to compute this expression since we know that

$$\phi(z) = \frac{1}{\sqrt{2\pi}}e^{-0.5z^2}.$$

The expression for $G(Q^*)$ shows that the optimal newsvendor cost increases linearly in both the underage and overage costs as well as in the standard deviation of demand.

3 Extension to Infinite Horizon Assuming Full Backordering of Demand

Let $D_1, D_2, \ldots$ be an infinite sequence of demands. Assume that the demands are independent identically distributed random variables having common distribution function $F(x)$ and density $f(x)$. The policy is to order up to Q each period. As all excess demand is back ordered, the order quantity in any period is exactly the previous period's demand. The number of units sold will also equal the demand. In order to see this, consider the number of units sold in successive periods:

number of units sold in period 1 $= \min(Q, D_1)$,
number of units sold in period 2 $= \max(D_1 - Q, 0) + \min(Q, D_2)$,
number of units sold in period 3 $= \max(D_2 - Q, 0) + \min(Q, D_3)$,

and so on.

These relationships follow because back ordering of excess demand means that the sales are made in the subsequent period. Now, notice that

$$\min(Q, D_i) + \max(D_i - Q, 0) = D_i \qquad \text{for } i = 1, 2, \ldots,$$

which follows from considering the cases $Q < D_i$ and $Q \geq D_i$.

Hence, the expected cost over n periods is

$$\begin{aligned} &cQ + (c - S)E(D_1 + D_2 + \ldots + D_{n-1}) - (S)E[\min(Q, D_n)] + nL(Q) \\ &= cQ + (c - S)(n - 1)\mu - (S)E[\min(Q, D_n)] + nL(Q) \end{aligned}$$

where

$$L(Q) = h\int_0^Q (Q - x)f(x)dx + p\int_Q^\infty (x - Q)f(x)dx.$$

Dividing by n and letting $n \to \infty$ gives the average cost over infinitely many periods as

$$(c - S)\mu + L(Q).$$

The optimal value of Q occurs where $L'(Q) = 0$, which results in

$$F(Q) = p / (p + h).$$

4 Extension to Infinite Horizon Assuming Lost Sales

If excess demand is lost rather than back ordered, then the previous argument is no longer valid. The number of units sold in period 1 is $\min(Q, D_1)$, which is also the number of units ordered in period 2; the number of units sold in period 2 is $\min(Q, D_2)$ (since there is no back-ordering of excess demand), which is also the number of units ordered in period 3; and so on. As shown in Section 1 of this appendix,

$$E[\min(Q,D)] = \mu - \int_Q^\infty (x-Q)f(x)dx.$$

Hence, it follows that the expected cost over n periods is given by

$$cQ + [(n-1)c - nS]\left[\mu - \int_Q^\infty (x-Q)f(x)\,dx\right] + nL(Q).$$

If we divide by n and let $n \to \infty$, we obtain the following expression for the average cost per period:

$$(c-S)\left[\mu - \int_Q^\infty (x-Q)f(x)\,dx\right] + L(Q).$$

Differentiating with respect to Q and setting the result equal to zero gives the following condition for the optimal Q:

$$F(Q) = \frac{p+S-c}{p+S+h-c}.$$

Setting $c_u = p + S - c$ and $c_o = h$ gives the critical ratio in the form $c_u/(c_u + c_o)$. Hence, we interpret c_u as the cost of the loss of goodwill plus the lost profit per sale, and c_o as the cost of holding only.

APPENDIX 5–C

DERIVATION OF THE OPTIMAL (Q, R) POLICY

From Section 5.4, the objective is to find values of the variables Q and R to minimize the function

$$G(Q, R) = h(Q/2 + R - \lambda\tau) + K\lambda/Q + p\lambda n(R)/Q. \quad (1)$$

Because this function is to be minimized with respect to the two variables (Q, R), a necessary condition for optimality is that $\partial G/\partial Q = \partial G/\partial R = 0$. The two resulting equations are

$$\frac{\partial G}{\partial Q} = \frac{h}{2} - \frac{K\lambda}{Q^2} - \frac{p\lambda n(R)}{Q^2} = 0, \quad (2)$$

$$\frac{\partial G}{\partial R} = h + p\lambda n'(R)/Q = 0. \quad (3)$$

Note that since $n(R) = \int_R^\infty (x-R)f(x)dx,$ one can show that

$$n'(R) = -(1 - F(R)).$$

From equation (2) we obtain

$$\frac{1}{Q^2}[K\lambda + p\lambda n(R)] = \frac{h}{2}$$

or

$$Q^2 = \frac{2K\lambda + 2p\lambda n(R)}{h},$$

which gives

$$Q = \sqrt{\frac{2\lambda[K + pn(R)]}{h}}. \tag{4}$$

From equation (3) we obtain

$$h + p\lambda[-(1 - F(R))]/Q = 0,$$

which gives

$$1 - F(R) = Qh/p\lambda. \tag{5}$$

Authors' note: We use the term *optimal* somewhat loosely here. Technically speaking, the model that gives rise to these two equations is only approximate for several reasons. For one, the use of the average expected inventory is an approximation, since this is not the same as the expected average inventory (because one should not charge holding costs when inventory goes negative). As we saw in the discussion of negative safety stock in this chapter, there are cases where the right-hand side of equation (5) can exceed 1, and the model "blows up." This would not be the case for a truly exact model, which is beyond the scope of this book.

APPENDIX 5–D

PROBABILITY DISTRIBUTIONS FOR INVENTORY MANAGEMENT

In this chapter we have made frequent reference to the normal distribution as a model for demand uncertainty. Although the normal distribution certainly dominates applications, it is not the only choice available. In fact, it could be a poor choice in some circumstances. In this appendix we discuss other distributions for modeling demand uncertainty.

1. The Poisson Distribution as a Model of Demand Uncertainty

One situation in which the normal distribution may be inappropriate is for slow-moving items—that is, ones with small demand rates. Because the normal is an infinite distribution, when the mean is small it is possible that a substantial portion of the density curve extends into the negative half line. This could give poor results for safety stock calculations. A common choice for modeling slow movers is the Poisson distribution. The Poisson is a discrete distribution defined on the positive half line only. Let X have the Poisson distribution with parameter μ. Then

$$f(x)=\frac{e^{-\mu}\mu^x}{x!} \qquad \text{for } x=0,\ 1,\ 2,\ \ldots$$

(The derivation of the Poisson distribution and its relationship to the exponential distribution are discussed in detail in Section 13.3).

An important feature of the Poisson distribution is that both the mean and the variance are equal to μ (giving $\sigma=\sqrt{\mu}$). Hence, it should be true that the observed standard deviation of periodic demand (or lead time demand) is close to the square root of the mean periodic demand (or lead time demand) for the Poisson to be an appropriate choice.

Table A–3 at the back of the book is a table of the complementary cumulative Poisson distribution. This table allows one to compute optimal policies for the newsvendor and the (Q, R) models assuming that demand follows the Poisson distribution. For the newsvendor case, one simply applies the method outlined in Section 5.3 for discrete demand and obtains the probabilities from Table A–3. For the (Q, R) model, the process requires an iterative solution similar to that for the normal distribution described in Section 5.5. The example that follows illustrates the procedures.

EXAMPLE 5.12

Returning to Mac's newsstand, a few regular customers have asked Mac to stock a particular music magazine. Mac has agreed even though sales have been slow. Based on past data, Mac has found that a Poisson distribution with mean 5 closely fits the weekly sales pattern for the magazine. Mac purchases the magazines for \$0.50 and sells them for \$1.50. He returns unsold copies to his supplier, who pays Mac \$0.10 for each. Find the number of these magazines he should purchase from his supplier each week.

Solution

The overage cost is c_o = \$0.50 – \$0.10 = \$0.40 and the underage cost is c_u = \$1.50 – \$0.50 = \$1.00. It follows that the critical ratio is $c_u/(c_o + c_u) = 1/1.4 = .7143$. As Table A–3 gives the complementary cumulative distribution, we subtract table entries from 1 to obtain values of the cumulative distribution function. From the table we see that

$$F(5) = 1 - .5595 = .4405,$$
$$F(6) = 1 - .3840 = .6160,$$
$$F(7) = 1 - .2378 = .7622.$$

Because the critical ratio is between $F(6)$ and $F(7)$, we move to the larger value, thus giving an optimal order quantity of 7 magazines.

It is not difficult to calculate optimal (Q, R) policies when the demand follows a Poisson distribution. Define $P(x)$ as the complementary cumulative probability of the Poisson. That is,

$$P(x)=\sum_{k=x}^{\infty} f(k).$$

Table A–3 gives values of $P(x)$. It can then be shown that for the Poisson distribution

$$n(R) = \mu P(R) - RP(R + 1),$$

which means that all the information required to compute optimal policies appears in Table A–3 (see Hadley and Whitin, 1963, p. 441). This relationship allows one to compute the (Q, R) policy using the pair of equations (1) and (2) from Section 5.4.

EXAMPLE 5.13

A department store uses a (Q, R) policy to control its inventories. Sales of a particular fitness tracker average 1.4 trackers per week. The tracker costs the store \$50.00, and fixed costs of replenishment amount to \$20.00. Holding costs are based on a 20 percent annual interest charge, and the store manager estimates a \$12.50 cost of lost profit and loss of goodwill if the tracker is demanded when out of stock. Reorder lead time is 2 weeks, and statistical studies have shown that demand over the lead time is accurately described by a Poisson distribution. Find the optimal lot sizes and reorder points for this item.

Solution

The relevant parameters for this problem are

$$\lambda = (1.4)(52) = 72.8.$$
$$h = (50)(0.2) = 10.0.$$
$$K = \$20.$$
$$\mu = (1.4)(2) = 2.8.$$
$$p = \$12.50.$$

To start the solution process, we compute the EOQ. It is

$$\text{EOQ} = \sqrt{\frac{2K\lambda}{h}} = \sqrt{\frac{(2)(20)(72.8)}{10}} = 17.1.$$

The next step is to find R solving

$$P(R) = Qh/p\lambda.$$

Substituting the EOQ for Q and solving gives $P(R_0) = .1868$. From Table A–3, we see that $R = 4$ results in $P(R) = .3081$ and $R = 5$ results in $P(R) = .1523$. Assuming the conservative strategy of rounding to the *larger* R, we would choose $R = 5$ and $P(R) = .1523$. It now follows that

$$n(R_0) = (2.8)(.1523) - (5)(.0651) = 0.1009.$$

Hence Q_1 is

$$Q_1 = \sqrt{\frac{(2)(72.8)[20 + (12.5)(1.1009)]}{10}} = 17.6.$$

It follows that $P(R_1) = .1934$, giving $R_1 = R_0 = 5$. Hence the solution has converged, because successive R (and, hence, Q) values are the same. The optimal solution is $(Q, R) = (18, 5)$.

2. The Laplace Distribution

A continuous distribution, which has been suggested for modeling slow-moving items or ones with more variance in the tails than the normal, is the Laplace distribution.

The Laplace distribution has been called the pseudo-exponential distribution, as it is mathematically an exponential distribution with a symmetric mirror image. (The exponential distribution is discussed at length in Chapter 13 in the context of reliability management.)

The mathematical form of the Laplace pdf is

$$f(x) = \frac{1}{2\theta}\exp(-|x-\mu|/\theta) \qquad \text{for } -\infty < x < +\infty.$$

Because the pdf is symmetric around μ, the mean is μ. The variance is $2\theta^2$. The Laplace distribution is also a reasonable model for slow-moving items and is an alternative to the normal distribution for fast-moving items when there is more spread in the tails of the distribution than the normal distribution gives. As far as inventory applications are concerned, the Laplace distribution significantly simplified the calculation of the optimal policy for the (Q, R) model (Presutti and Trepp, 1970).

One can show that for any value of $R > \mu$, the complementary cumulative distribution $(P(R))$ and the loss integral $(n(R))$ are given by

$$P(R) = 0.5\exp(-[(R-\mu)/\theta])$$
$$n(R) = 0.5\theta\exp(-[(R-\mu)/\theta])$$

so that the ratio $n(R)/P(R) = \theta$, independent of R. This fact results in a very simple solution for the (Q, R) model. Recall that the two equations defining the optimal policy were

$$Q = \sqrt{\frac{2\lambda[K + pn(R)]}{h}}$$
$$P(R) = Qh/p\lambda \qquad [\text{where } P(R) = 1 - F(R)].$$

The simplification is achieved by using the SOQ formula presented in Section 5.5. The SOQ representation is an alternative representation for Q that does not include the stock-out cost, p. Using $P(R) = 1 - F(R)$, the SOQ formula is

$$Q = \frac{n(R)}{P(R)} + \sqrt{\frac{2K\lambda}{h} + \left(\frac{n(R)}{P(R)}\right)^2} = \theta + \sqrt{\frac{2K\lambda}{h} + \theta^2}$$

independently of R! Hence, the optimal Q and R can be found in a simple one-step calculation. When using a cost model, find the value of $P(R)$ from $P(R) = Qh/p\lambda$. Then, using the representation $P(R) = \exp[-(R-\mu)/\theta)]/2$, it follows that $R = -\theta\ln(2P(R)) + \mu$. Using a service level model, one simply uses the formulas for R given in Section 5.5. We illustrate with an example.

EXAMPLE 5.14

Returning to Example 5.13, suppose that we wish to use the Laplace distribution to compute the optimal policy rather than the Poisson distribution. As both the mean and the variance of the lead time demand are 2.8, we set $\mu = 2.8$ and $2\theta^2 = 2.8$, giving $\theta = 1.1832$. It follows that

$$Q = 1.1832 + \sqrt{\frac{(2)(20)(72.8)}{10} + (1.1832)^2} = 18.3.$$

As with Example 5.13, we obtain $P(R) = .1934$. Using $R = -\theta \ln(2P(R)) + \mu$ and substituting $P(R) = .1934$ gives $R = 3.92$, which we round to 4. Notice that this solution differs slightly from the one we obtained assuming Poisson demand. However, recall that using the Poisson distribution we found that the optimal R was between 4 and 5, which we chose to round to 5 to be conservative.

3. Other Lead Time Demand Distributions

Many other probability distributions have been recommended for modeling lead time demand. Some of these include the negative binomial distribution, the gamma distribution, the logarithmic distribution, and the Pearson distribution. (See Silver et al, 2017, p. 275 for articles that discuss these distributions.) The normal distribution probably accounts for the lion's share of applications, with the Poisson accounting for almost all the rest. We included the Laplace distribution because of the interesting property that optimal (Q, R) policies can be found without an iterative solution and because it could be a good alternative to the Poisson for low-demand items.

APPENDIX 5–E

Glossary of Notation for Chapter 5

α = Desired level of Type 1 service.

β = Desired level of Type 2 service.

c_o = Unit overage cost for newsvendor model.

c_u = Unit underage cost for newsvendor model.

D = Random variable corresponding to demand. It is the one-period demand for the newsvendor model and the lead time demand for the (Q, R) model.

EOQ = Economic order quantity.

$F(t)$ = Cumulative distribution function of demand. Values of the standard normal CDF appear in Table A–4 at the end of the book.

$f(t)$ = Probability density function of demand.

$G(Q)$ = Expected one-period cost associated with lot size Q (newsvendor model).

$G(Q, R)$ = Expected average annual cost for the (Q, R) model.

h = Holding cost per unit per unit time.

I = Annual interest rate used to compute holding cost.

K = Setup cost or fixed order cost.

λ = Expected demand rate (units per unit time).

$L(z)$ = Normalized loss function. Used to compute $n(R) = \sigma L(z)$. Tabled values of $L(z)$ appear in Table A–4 at the end of the book.

μ = Mean demand [lead time demand for (Q, R) model].

$n(R)$ = Expected number of stock-outs in the lead time for (Q, R) model.

p = Penalty cost per unit for not satisfying demand.

Q = Lot size or size of the order.

S = Safety stock; $S = R - \lambda\tau$ for (Q, R) model.
SOQ = Service order quantity.
T = Expected cycle time; mean time between placement of successive orders.
τ = Order lead time.
Type 1 service = Proportion of cycles in which all demand is satisfied.
Type 2 service = Proportion of demands satisfied.

BIBLIOGRAPHY

Agrawal, N., and S. A. Smith (Eds.). *Retail Supply Chain Management* (rev. 2nd ed.). New York: Springer, 2015.

Arrow, K. A., T. E. Harris, and T. Marschak. "Optimal Inventory Policy." *Econometrica* 19 (1951), pp. 250–72.

Arrow, K. A., S. Karlin, and H. E. Scarf, Eds. *Studies in the Mathematical Theory of Inventory and Production.* Palo Alto, CA: Stanford University Press, 1958.

Bessler, S. A., and A. F. Veinott, Jr. "Optimal Policy for a Dynamic Multiechelon Inventory Model." *Naval Research Logistics Quarterly* 13 (1966), pp. 355–89.

Brown, R. G. *Statistical Forecasting for Inventory Control.* New York: McGraw-Hill, 1959.

Brown, R. G. *Decision Rules for Inventory Management.* Hinsdale, IL: Dryden Press, 1967.

Clark, A., and H. E. Scarf. "Optimal Policies for a Multiechelon Inventory Problem." *Management Science* 6 (1960), pp. 475–90. PDF available at http://dido.econ.yale.edu/~hes/pub/echelon1.pdf

Cohen, M. A., and D. Pekelman. "LIFO Inventory Systems." *Management Science* 24 (1978), pp. 1150–62.

Deuermeyer, B. L., and L. B. Schwarz. "A Model for the Analysis of System Service Level in Warehouse-Retailer Distribution Systems: Identical Retailer Case." Institute for Research in the Behavioral, Economic, and Management Sciences, Krannert Graduate School of Management, Purdue University, 1979

Dvoretsky, A., J. Kiefer, and J. Wolfowitz. "The Inventory Problem: I. Case of Known Distributions of Demand." *Econometrica* 20 (1952), pp. 187–222.

Ehrhardt, R. "The Power Approximation for Computing (*s*, *S*) Inventory Policies." *Management Science* 25 (1979), pp. 777–86.

Eppen, G., and L. Schrage. "Centralized Ordering Policies in a Multi-Warehouse System with Lead Times and Random Demand." In L. B. Schwarz (Ed.), *Multilevel Production/Inventory Control Systems: Theory and Practice* (pp. 51–68). Amsterdam: North Holland, 1981.

Federgruen, A., and P. Zipkin. "Computational Issues in an Infinite Horizon Multi-Echelon Inventory Model." *Operations Research* 32 (1984), pp. 818–36.

Fishman, G. S. *Concepts and Methods in Discrete Event Digital Simulation.* New York: John Wiley & Sons, 1973.

Freeland, J. R., and E. L. Porteus. "Evaluating the Effectiveness of a New Method of Computing Approximately Optimal (*s*, *S*) Inventory Policies." *Operations Research* 28 (1980), pp. 353–64.

Graves, S. C. "A Multiechelon Inventory Model for a Repairable Item with One for One Replenishment." *Management Science* 31 (1985), pp. 1247–56.

Graves, S. C., A. H. G. Rinnooy Kan, and P. Zipkin (Eds.). *Handbooks in Operations Research and Management Science.* Volume 4, *Logistics of Production and Inventory.* Amsterdam: North Holland, 1993.

Hadley, G. J., and T. M. Whitin. *Analysis of Inventory Systems.* Englewood Cliffs, NJ: Prentice Hall, 1963. Retrieved from https://babel.hathitrust.org/cgi/pt?id=mdp.39015000461452&view=1up&seq=9

Hartung, P. "A Simple Style Goods Inventory Model." *Management Science* 19 (1973), pp. 1452–58.

Hausman, W. H., and R. Peterson. "Multiproduct Production Scheduling for Style Goods with Limited Capacity, Forecast Revisions, and Terminal Delivery." *Management Science* 18 (1972), pp. 370–83.

Iglehart, D. L., "Optimality of (*s*, *S*) Inventory Policies in the Infinite Horizon Dynamic Inventory Problem." *Management Science* 9 (1963), pp. 259–67.

Jones, P. C., G. Kegler, T. J. Lowe, and R. D Traub. "Managing the Seed-Corn Supply Chain at Syngenta." *Interfaces* 33 (1), January-February 2003, pp. 80–90.

Kaplan, R. "A Dynamic Inventory Model with Stochastic Lead Times." *Management Science* 16 (1970), pp. 491–507.

Love, S. F. *Inventory Control.* New York: McGraw-Hill, 1979.

Muckstadt, J. M., and L. J. Thomas. "Are Multiechelon Inventory Models Worth Implementing in Systems with Low Demand Rate Items?" *Management Science* 26 (1980), pp. 483–94.

Murray, G. R., and E. A. Silver. "A Bayesian Analysis of the Style Goods Inventory Problem." *Management Science* 12 (1966), pp. 785–97.

Nahmias. S. "Inventory Models." In J. Belzer, A. G. Holzman, and A. Kent (Eds.), *The Encyclopedia of Computer Science and Technology, Volume* 9 (pp. 447–83). New York: Marcel Dekker, 1978.

Nahmias, S. "Simple Approximations for a Variety of Dynamic Lead Time Lost-Sales Inventory Models." *Operations Research* 27 (1979), pp. 904–24.

Nahmias, S. "Managing Reparable Item Inventory Systems: A Review." In L. B. Schwarz (Ed.), *Multilevel Production/Inventory Control Systems: Theory and Practice* (pp. 253–77). Amsterdam: North Holland, 1981.

Nahmias, S. "Perishable Inventory Theory: A Review." *Operations Research* 30 (1982), pp. 680–708.

Nahmias, S., and S. Smith. "Mathematical Models of Retailer Inventory Systems: A Review." In R. K. Sarin (Ed.), *Perspectives in Operations Management: Essays in Honor of Elwood S. Buffa* (pp. 249–78). Norwell, MA: Kluwer, 1993.

Parr, J. O. "Formula Approximations to Brown's Service Function." *Production and Inventory Management* 13 (1972), pp. 84–86.

Porteus, E. L. *Foundations of Stochastic Inventory Theory*. Redwood City, CA: Stanford Business Books, 2002.

Porteus, E. L. "Numerical Comparisons of Inventory Policies for Periodic Review Systems." *Operations Research* 33 (1985), pp. 134–52.

Presutti, V., and R. Trepp. "More Ado about EOQ." *Naval Research Logistics Quarterly* 17 (1970), pp. 243–51.

Ramberg, J. S., and B. W. Schmeiser. "An Approximate Method for Generating Symmetric Random Variables." *Communications of the ACM* 15 (1972), pp. 987–89.

Scarf, H. E. "The Optimality of (*s*, *S*) Policies in the Dynamic Inventory Problem." Technical Report No. 11, April 10, 1959. Retrieved from https://statistics.stanford.edu/sites/g/files/sbiybj6031/f/KAR%20ONR%2011.pdf

Scarf, H. E. "A Survey of Analytical Techniques in Inventory Theory." In H. E. Scarf, D. M. Gilford, and M. W. Shelly (Eds.), *Multi-Stage Inventory Models and Techniques* (185–225). Palo Alto, CA: Stanford University Press, 1963. PDFs available at http://dido.econ.yale.edu/~hes/pub/survey.pdf

Schmidt, C. P., and S. Nahmias. "Optimal Policy for a Single Stage Assembly System with Stochastic Demand." *Operations Research* 33 (1985), pp. 1130–45.

Sherbrooke, C. C. "METRIC: Multiechelon Technique for Recoverable Item Control." *Operations Research* 16 (1968), pp. 122–41.

Silver, E. A., D. F. Pike, and D. J Thomas. *Inventory and Production Management in Supply Chains* (4th ed.). Boca Raton, FL, CRC Press, 2017.

Veinott, A. F. "The Status of Mathematical Inventory Theory." *Management Science* 12 (1966), pp. 745–77.

Veinott, A. F., and H. M. Wagner. "Computing Optimal (*s*, *S*) Inventory Policies." *Management Science* 11 (1965), pp. 525–52.

Whitin, T. M. *The Theory of Inventory Management*. Rev. ed. Princeton, NJ: Princeton University Press, 1957.

Zheng, Y.-S., and A. Federgruen. "Finding Optimal (*s*, *S*) Policies Is About as Simple as Evaluating A Single Policy." *Operations Research* 39 (1991), pp. 654–665.

CHAPTER 6

Supply Chain Strategy

"Strategic partnering used to mean stealing revenue or pushing cost onto someone else in the supply chain. You are a pig at the trough if you view it that way. With technology, there are so many efficiencies that can be shared."

—William M. Gibson

CHAPTER OVERVIEW

Purpose

To understand what a modern supply chain is, how supply chains are organized and managed, and to review the newest developments in this important area.

KEY POINTS

1. *What is a supply chain and how should it be strategically positioned?* Managing the flow of goods has been an issue since the industrial revolution and was traditionally called logistics. The term "supply chain management" (SCM) emerged in the 1980s to describe the activities of a firm linking the entire network of suppliers, factories, warehouses, stores, and customers. It requires management of goods, money, and information among all the relevant players. For most products, it is not possible to have a supply chain that is both low cost and highly responsive. A supply chain's strategic positioning must therefore align with the product's positioning in the marketplace. A product that competes on price must be delivered through a low cost supply chain. For an innovative or high fashion item, the supply chain must be able to respond quickly to changes in customer demand.
2. *The role of information and technology in supply chains.* A supply chain involves the transfer of goods, money, and information. Modern supply chain management seeks to eliminate the inefficiencies that arise from poor information flows. One way to ameliorate this problem is through vendor-managed inventories, where vendors, rather than retailers, are responsible for keeping inventory on the shelves. Advances in technology have also improved the availability of information in supply chains and are changing how supply chains are managed.
3. *The bullwhip effect.* This effect implies that demand variability tends to increase when moving up the supply chain away from the customer. That is, orders at the factory tend to be more variable than end-customer demand. The four primary causes of bullwhip are: demand forecast updating, order batching, price fluctuations, and shortage gaming. The bullwhip effect can be ameliorated by information sharing, channel alignment, price stabilization, and the discouragement of shortage gaming.

4. *Supply chain contracts.* One key method for aligning incentives in the supply chain is through the use of contracts. The simplest types of contracts are wholesale price contracts, where the retailer pays the supplier a fixed price per item. However, these contracts tend to lead to misaligned incentives, including understocking by the retailer. More sophisticated contracts include buy-back and revenue-sharing contracts. In addition to helping with information sharing, vendor managed inventory can also help align incentives in the supply chain.
5. *Risk pooling.* One key technique for mitigating uncertainty and improving planning is risk pooling. This principle states that the sum of a number of uncertain variables is inherently less variable than each individual variable. Decreased variability makes it easier to plan for, schedule, and manage pooled systems. There are a variety of ways to operationalize risk pooling, including product postponement, regional warehouses, aggregate capacity plans, and flexible capacity. Postponement is also called delayed differentiation. A classic example is Benetton's change to dyeing their garments after they were knitted (instead of before), to allow for a pooled inventory of "gray stock."
6. *Multilevel distribution systems.* Typically in large systems, stock is stored at multiple locations. Distribution centers (DCs) receive stock from plants and factories and then ship to either smaller local DCs or directly to stores. Some of the advantages of employing DCs include economies of scale, tailoring the mix of product to a particular region or culture, and safety stock reduction via risk pooling.
7. *Designing products for supply chain efficiency.* "Thinking outside the box" has become a cliché. It means looking at a problem in a new way, often not taking constraints at face value. Postponement, which is a way to increase pooling, is one example of thinking outside the box. Ikea provides another example with its design of flatpack furniture for supply chain efficiency. In some cases, with enough thought to product design, a decoupling point can be used to separate end customer demand from intermediate inventory, allowing for greater customization options.
8. *Global supply chain management.* Today, most firms are multinational. Products are designed for, and shipped to, a wide variety of markets around the world. As an example, consider the market for automobiles. Fifty years ago, virtually all the automobiles sold in the United States were produced domestically. Today, that number is probably closer to 50 percent. Global market forces are shaping the new economy. Vast markets, such as China, are now emerging, and the major industrial powers are vying for a share. Technology, cost considerations, and political and macroeconomic forces have driven globalization. Selling in diverse markets presents special problems for supply chain management.

At 10pm we realize we need a jar of peanut butter for our child's lunch the next day, so we take a trip to the grocery store. Not only is the store open but there are also a large variety of brands, styles, and sizes of peanut butter available. It is easy to take such things for granted. However, there is a complex myriad of activities that had to be carefully coordinated to ensure that the peanut butter was there when we needed it. And this goes for clothes, hardware, and all the other consumer goods we purchase. Do we appreciate the fact that many of these goods are produced and shipped all over the world before they make it to our homes? The logistics of coordinating all the activities that afford us this convenience is the essence of supply chain management.

The term **supply chain management** (SCM) seems to have emerged in the late 1980s and continues to gain interest at an increasing rate. The trade literature abounds with book titles and articles relating to some aspect of SCM. Software and consulting firms specializing in SCM solutions are now commonplace. These companies have grown at remarkable rates and include giants SAP and Oracle, which offer SCM solutions as part of comprehensive enterprise systems. Although the SCM label is somewhat recent, the problems considered are not. The topics in Chapters 2 to 5 are all relevant for managing supply chains. So what is different about SCM? The simple answer is that SCM looks at the problem of managing the flow of goods as an integrated system.

Many definitions of SCM have been proposed, and it is instructive to examine some of them. The simplest and most straightforward, attributed to Hau Lee from Stanford, defines supply chain management as the management of materials, information, and financial flows in a network consisting of suppliers, manufacturers, distributors, and customers. While short, this definition is fairly complete. It indicates that it is not only the flow of goods that is important but also the flow of information and money.

Another definition focuses on only the flow of goods.

> Supply chain management is a set of approaches utilized to efficiently integrate suppliers, manufacturers, warehouses, and stores, so that merchandise is produced and distributed at the right quantities, to the right locations, and at the right time, in order to minimize system wide costs while satisfying service level requirements. (Simchi-Levi, Kaminsky, and, Simchi-Levi, 2008).

Since this definition focuses on the flow of goods, it implies that SCM is relevant only to manufacturing firms. Is SCM also relevant to service organizations? Today it is well recognized that many aspects of SCM do indeed apply to service organizations, but the term used is typically **value chain** rather than supply chain. Indeed, Stanford University renamed its Global Supply Chain Management Forum the "Value Chain Innovation Initiative," probably to make it more inclusive of services.

Most writers agree that "logistics" deals with essentially the same issues as supply chain management. The term logistics originated in the military and concerned problems of moving personnel and material in critical times. It was later adopted by business and became a common moniker for professional societies and academic programs. One might argue that although classical logistics treated the same set of problems as SCM, it did not consider the supply chain as an integrated system (Copacino, 1997). A summary of the "umbrella" of activities composing SCM appears in Figure 6–1.

Taking the broad view of the field depicted in Figure 6–1, one might consider this entire text to deal with some aspect of SCM. However, most of the chapters focus on specific topics, like inventory control or scheduling, as discrete topics. SCM, on the other hand, treats the entire product delivery system from suppliers of raw materials to distribution channels of finished goods. While important and useful in many contexts, simple formulas such as the EOQ are unlikely to shed much light on effective management of complex, integrated supply chains.

Where does the supply chain fit into the overall business strategy of the firm? In Chapter 1, it was noted that an important part of the firm's competitive edge is its strategic positioning in the marketplace. Examples of strategic positioning include being the low-cost provider (such as Hyundai automobiles) or the high-quality provider (such as Mercedes-Benz) or exploiting

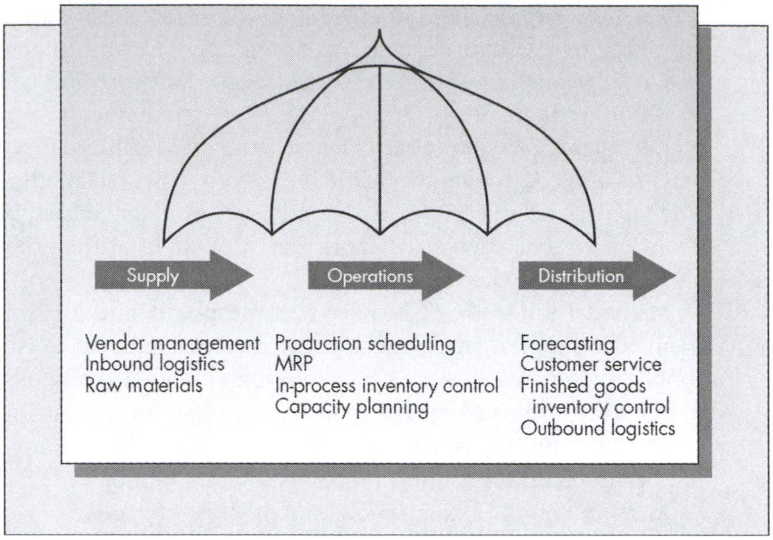

Figure 6–1 The supply chain umbrella

market segmentation to achieve shares of both markets (as does General Motors with different brands aimed at different market segments). The design of the supply chain also reflects a firm's strategic positioning.

In a supply chain, the primary trade-off is between cost and response time. Ground transportation (boat, truck, or rail) is less expensive but airfreight is faster. Will deliveries be more reliable if the product is moved via the firm's internal system, or would it be better to subcontract the logistics operation to a third party? Third-party logistics (abbreviated 3PL) is becoming more common for the same reason that third-party manufacturing has become so widespread. Companies such as Foxconn Technology, a Taiwanese multinational electronics manufacturer, are able to achieve significant economies of scale by providing manufacturing services to large numbers of firms producing similar products. In this way, it can be less expensive to subcontract manufacturing than to do it in-house. Cost is not the only issue. For example, Apple Inc. outsources to Foxconn because they have long realized that their core competencies lie in aesthetics and innovation and not in manufacturing. Cisco Systems, Inc. also outsources to Foxconn. They have transitioned from a manufacturing-focused company to a design-focused firm. They have also embraced a servicization strategy (see Chapter 1).

Key elements of effective supply chain management include the use of **information**, **analytics**, and **incentives**. The Snapshot Application below highlights Walmart, whose extraordinary success was achieved to some extent because of its sophisticated supply chain strategies. Supply chain analytics will be the focus of Chapter 7, while this chapter focuses on supply chain strategy decisions, including the use of information and incentives.

One of the key challenges in supply chain management is aligning incentives—in particular, how to design appropriate contracts to align incentives. Modern technologies make this easier. In fact, it has been over 20 years since William Gibson, then chairman of Manugistics Inc., provided the quotation that begins this chapter in a *Wall Street Journal* article (*WSJ*, April 12, 1996). Technologies and opportunities for supply chain collaboration have only continued to increase since he made his pronouncement.

SNAPSHOT APPLICATION

Walmart Wins with Solid Supply Chain Management

Although there are many examples of companies winning or losing because of good or bad supply chain strategies, perhaps the most dramatic are the stories of Walmart and Kmart. Both companies were founded in 1962. In only a few years, Kmart had become a household term, while Walmart was largely unknown except for a few communities in the South. Walmart stores were typically located in rural areas, so they rarely competed head-on with chains like Sears and Kmart located in large cities and surrounding suburbs. In 1987 Kmart's sales were almost twice those of Walmart (about $25 billion annually versus about $15 billion annually). By 1990 Walmart sales overtook those of Kmart, and in 1994 Walmart's annual sales almost tripled Kmart's (about $80 billion versus about $27 billion)! Walmart is now the largest discount retailer in the United States, surpassing Sears as well as Kmart (the companies merged in 2005 to form Sears Holding Corporation, which filed for bankruptcy in 2018 before restructuring in 2019). What accounted for this dramatic turn-around?

While one can point to several factors leading to Walmart's success, there is no question that the firm's emphasis on solid supply chain management was one of the most important. Rather than using capital to improve logistics, Kmart diversified its business, acquiring Borders, builders Square, Sports Authority, and Office Max. Sam Walton, in contrast, was obsessed with operations (Duff and Ortega, 1995). He invested tens of millions of dollars in a companywide computer system linking cash registers to headquarters to track inventory and to restock popular items. He also invested heavily in trucks and DCs, which improved delivery and reduced costs sharply. Sophisticated scanner systems reduced price-check delays at the cash register and kept shelves stocked. Meanwhile, Kmart scanners were outdated and did not provide purchasing information to headquarters (Schoenberger, 2002). DCs were late delivering merchandise to stores 11 percent of the time compared to a retail average of 5 percent. The supply chain system mishandled store orders 15 percent of the time compared to an industry average of less than one-half percent. The percentage of goods lost to mishandling and theft was three times that of competitors.

In 2019 Walmart again topped the Fortune Global 500 list, which ranks the world's largest corporations by revenue. It has dominated this list for over two decades. It is the biggest private employer in the world with over two million employees, and the largest retailer in the world (over 11,000 stores in 28 countries). It has moved into e-commerce and has begun testing of last-mile grocery delivery. It will be interesting to observe how it competes against Amazon in the upcoming years.

6.1 SUPPLY CHAIN STRATEGY

As just discussed, one of the key trade-offs in supply chain design is between cost and speed. It is important to recognize that the best choices on these dimensions depend on the product being produced. Marshall Fisher (1997) presented a simple framework for this choice. He categorized products as **functional** or **innovative,** and he categorized supply chains as **efficient** versus **responsive**. He argued that functional products should be produced using efficient supply chains, while innovative products require responsive supply chains.

A functional product is a commodity-type product that competes primarily on cost. A canonical functional product is Campbell's Soup. Only 5 percent of the products they produce are new each year, and most products have been in the market for many years. The soup has highly predictable demand and long life cycles (particularly relative to lead times). This type of product is best produced using an efficient supply chain, where the primary focus is on costs. Such costs include inventory holding, transportation, handling, and manufacturing costs. Large manufacturing batches achieve scale economies, and minimum order quantities reduce handling/order processing costs. Supply chain performance is evaluated using traditional cost-based metrics, such as inventory turns, factory utilization (extent to which the business uses its available production capacity), and transportation utilization (extent to which the business maximizes the use of its fleet or third-party contracts).

An innovative product is one that typically has a high margin but highly variable demand and a short life cycle. An example is Sport Obermeyer's fashion skiwear. Each year, 95 percent of the products are completely new designs, and demand forecasts err by as much as 200 percent. There is a short retail season, and cost concerns focus around inventory obsolescence and lost sales (see the Snapshot Application in Chapter 2). This type of product is best produced using a market responsive supply chain, where a premium is paid for flexibility. Such flexibility can be achieved by using faster transportation, lower transportation and factory utilization, and smaller batches. Supply chain performance is best measured not with traditional cost measures but by opportunity costs of lost sales and poor service.

Cost and speed are both key measures of supply chain effectiveness. However, Hau Lee (2004) argues that they are not the whole story. He proposes the "Triple A" supply chain, which is **agile**, **adaptable**, and **aligned**. Agility is defined as the ability to respond to short-term changes in demand or supply quickly. Adaptability is defined as the ability to adjust the supply chain design to accommodate market changes. Finally, alignment relates to the existence of incentives for supply chain partners to improve performance of the entire chain. While desirable traits for a supply chain, the three features come with costs, and the cost to implement must be balanced against strategic necessity.

SNAPSHOT APPLICATION

Anheuser-Busch Re-Engineers Their Supply Chain

Anheuser-Busch merged with InBev in 2008 to become the largest beer company in the world with sales of $56.4 billion in 2017 (Statista, 2018). Its share of the U.S. beer market was 45.6 percent. The company sells iconic brands including Budweiser and Bud Light. In the late 1990s Anheuser-Busch undertook a major reorganization of its supply chain (John & Willis, 1998). They found that less than 10 percent of their volume accounted for around 80 percent of the brand and package combinations. Management decided to switch to focused production. Some breweries were dedicated for the efficient production of large volume items, such as Bud and Bud Light. Others were responsible for producing lower volume niche products, such as wheat beers or products targeted at holidays. (Refer to the discussion of the focused factory in Chapter 1.)

By aligning supply chain metrics to the type of product being produced, decisions were in line with Fisher's recommendations for matching product type with supply chain type. In particular, Anheuser-Busch made sure that their high-volume breweries were measured by efficiency metrics, such as utilization and volume of output; whereas, their niche product breweries were measured by how effectively they met demand. Previously, benchmarking applied across all breweries, which did not create the correct incentives for either the high volume or niche products.

In addition to focused facilities and supply chains, Anheuser-Busch also undertook a number of initiatives in line with topics covered later in this chapter and the next. In particular, a mixed integer programming model (i.e., supply chain analytics) was used to identify the optimal number and location of distribution points in the supply chain. Contracts with a thousand different trucking companies (across the whole network) were largely replaced by a single contract with a single dedicated carrier, decreasing the complexity of the network. Inventory pooling was used to reduce both risk and cost. Finally, replenishment agreements were renegotiated, and vendor-managed inventory agreements were put in place, which improved the alignment of the incentives in the supply chain.

6.2 THE ROLE OF INFORMATION AND TECHNOLOGY IN THE SUPPLY CHAIN

We continually hear that we are living in the "information age". The availability of information is increasing at an exponential rate. New sources of information in the form of academic and trade journals, magazines, newsletters, and so on are introduced every day. The explosion of information available on the internet has been truly phenomenal. Web searches are now the first choice of many people for information on almost anything.

Knowledge is power. In supply chains, information is power. It provides the decision maker the power to get ahead of the competition, the power to run a business smoothly and efficiently, and the power to succeed in an ever more complex environment. Information plays a key role in the management of the supply chain. As we saw in the earlier chapters of this book, many aspects of operations planning start with the forecast of sales and build a plan for manufacturing or inventory replenishment from that forecast. Forecasts, of course, are based on information.

Jan Hammond (2006) wrote a case about Barilla, an Italian food company, to illustrate the role of information in supply chains. In the late 1980s, Barilla's head of logistics tried to introduce a new approach to dealing with distributors, which he called **just-in-time distribution** (JITD). In brief, the idea was to obtain sales data directly from its distributors (Barilla's customers) and use this data to allow Barilla to determine when and how large the deliveries should be. At that time, Barilla was operating in the traditional fashion. Distributors would independently place weekly orders based on standard reorder point methods. This led to wide swings in the demand on Barilla's factories due to the "bullwhip effect" (discussed in detail in the next section). The JITD idea met with staunch resistance, both within and outside Barilla. The Barilla sales and marketing organizations, in particular, were most threatened by the proposed changes. Distributors were concerned that their prerogatives would be compromised. Without going into the details covered in the case, management eventually did prevail, and the JITD system was implemented with several of Barilla's largest distributors. The results were striking. Delivery reliability improved, and the variance of orders placed on the factory were substantially reduced. The program proved a win-win for Barilla and its customers.

Barilla's success with its JITD program is one example of what today is known as **vendor-managed inventory** (VMI). VMI programs have been the mainstay of many successful retailers in the United States and abroad. For example, Procter & Gamble has assumed responsibility for keeping track of inventories for several of its major clients, such as Walmart. Years ago, grocery store managements were responsible for keeping track of their inventories. Today, it is commonplace that employees of the manufacturers, rather than store employees, check inventories in local stores. With VMI programs, it becomes the manufacturer's responsibility to keep the shelves stocked. While stock-outs certainly hurt the retailer, they hurt the manufacturer even more. When a product stocks out, customers typically will substitute another, so the store makes the sale anyway. The manufacturer suffers the penalty of lost sales, so the manufacturer has a strong incentive to keep shelves stocked. We explore incentive ideas further in Section 6.4.

Electronic Commerce

Electronic commerce, or simply e-commerce, is a catch-all term for a wide range of methods for effecting business transactions. It includes electronic data interchange (EDI), email, electronic funds transfers, electronic publishing, image processing, electronic bulletin boards, shared databases, blockchain, and all manner of internet-based business systems.

The internet has fundamentally changed supply chain management. First, it allows easy transmission of information to users within firms and to customers and trading partners. This includes point-of-sale demand information, purchase orders, and inventory status information. Originally, such information sharing was through EDI systems that allowed the transmission of standard business documents in a predetermined format between computers. Dedicated EDI systems have now mostly been replaced by internet-based systems. Cloud computing has also facilitated information sharing and enterprise systems. For example, SAP and Oracle, two of the largest enterprise system providers (see Chapter 9), offer cloud computing options for their systems. A further change on the horizon is blockchain, described further below. One of the key advantages of such systems from a supply chain management perspective is speed. By transmitting information quickly, lead times are reduced. As noted in Chapter 5, it is the uncertainty of demand over the replenishment lead time that results in the need to carry safety stock.

In addition to facilitating information sharing, the internet has also provided both **business to consumer** (B2C) and **business to business** (B2B) opportunities. The rise of B2C commerce was rocky with the now infamous dot-com bust in the early twenty-first century. Many internet-only companies failed at that time; however, Amazon.com, which is now the world's largest online retailer, has thrived. Amazon started as a discount bookseller. By selling from a single location, it could reap the benefits of stock centralization (to be discussed in Section 6.5) and avoid the costly capital investment in brick and mortar stores. However, Amazon has grown so dominant that it is expanding into brick and mortar stores.

The model for internet-based retailing is essentially the same as that for catalog shopping. Catalog retailers have been around since the nineteenth century. National Mail Order Catalog day is August 18 because on that date in 1872 Aaron Montgomery Ward produced the first mail order catalog for general merchandise; Sears followed with its own catalog in 1894. Today, Lands' End and L.L.Bean are two successful examples. However, catalogs need to be developed, printed, and mailed regularly, which is quite expensive, especially considering that most catalogs are thrown away. For this reason, the internet has a significant advantage over the traditional mail-order catalog business, and both Lands' End and L.L.Bean have put significant investment into their websites. Further, the move away from paper catalogs has allowed them to reach consumers around the world.

Along with the growth of B2C commerce, there has also been a less visible growth of B2B commerce. Hundreds, if not thousands, of firms offer internet-based supply chain solutions. Many tailor their products to specific industry segments, such as PrintingForLess.com (printing) and ShowMeTheParts.com (automotive aftermarket). There are also intermediaries, such as Li & Fung Limited, who specialize in third-party sourcing. While Li & Fung started as a traditional trading company, today they provide access to a network of over 15,000 suppliers in more than 40 economies. This would not be possible without e-commerce.

It is interesting to speculate what effect **3D printing** will have on e-commerce. 3D printing, or **additive manufacturing**, allows the construction of three-dimensional objects by a "printer," just as regular printers allow for the production of two-dimensional text. Such printers are currently widely used in the production of manufacturing prototypes but are less widely used in consumer goods. The range of materials that may be printed is limited but increasing. Examples of products produced by 3D printing include custom jewelry and home decor, such as lampshades. In time, instead of buying items from the internet (or store) we may simply buy a design and print it up at home or in our local print shop. Such a development could drastically simplify supply chain management!

Barcodes and QR Codes

The point-of-sale **barcode** system, now ubiquitous in supermarkets and retailers, has greatly facilitated e-commerce. Barcodes were first introduced as a way to speed up supermarket checkouts. The first retail scanning of a barcode appears to have been a pack of Wrigley chewing gum in June 1974 (Varchaver, 2004). A powerful advantage of barcodes is their role in information gathering. Assuming the check-out operator scans the products correctly, the retailer now has accurate electronic information on sales.

The reason for the caveat on correct scanning is that a major source of data error is operators who scan incorrectly. For example, an operator may scan one can of mushroom soup and hit "times three," rather than separately scanning each of the mushroom, tomato, and cream of asparagus soups that the customer actually purchased. Retailers have put significant educational effort into ensuring correct scanning techniques by employees. Yet, one of the authors recently saw a till operator take the faster route of multiplying a single purchase using the till, rather than the more laborious process of individual scanning.

With the introduction of customer loyalty cards in addition to barcodes, retailers now have information on individual purchasing behavior. This means they can target promotions to the individual customer. This can be done immediately at the point of sale or through mailings or emails. While marketing strategies are outside the scope of this text, it is easy to see that the availability of such information has radically increased the potential strategies for retailers.

Another semi-recent development is the two-dimensional barcode or **QR (Quick Response) code**, as it is more commonly known. These codes allow for significantly more information than the traditional barcode. They are not commonly used at check-out but rather to provide additional information to customers, who scan them using an app on their smartphone.

RFID Technology

Radio frequency identification (RFID) tags are an emerging technology that change the way information is stored and transmitted in a supply chain. Unlike bar-codes, which are the same across all items of the same stock-keeping unit (SKU), RFID tags can contain item specific information, such as farm of origin (for produce) or date of production.

RFID tags were invented in 1973 but are only now becoming commercially viable. They are microchips powered by batteries (active tags) or radio signals (passive tags). The passive device is smaller and cheaper and is likely to emerge as the device of choice for inventory management applications. Passive tags receive power from the readers (at which time they are "woken up") and transmit a unique code. Passive tags can only be read at close distances, but they provide a simple means of electronic identification.

Passive RFID tags cost between $.07 and $.15 each, which makes them too expensive for small grocery items (such as gum) but practical for larger items (such as jeans). JCPenney announced in 2012 that they were moving to 100 percent RFID tagging of all items in their stores, but then later backed away from full deployment. One of their key hurdles was that the radio signals interfered with their existing anti-theft sensors (Bjork, 2014). Fast-fashion retailer Zara managed to overcome this technological hurdle and has placed the chips inside their plastic security tags, which also allows them to be reused. All Zara's in-store items are now tagged with RFID.

One of the main advantages of such tagging is in drastically speeding up and improving accuracy in stock-taking, thereby reducing stock-outs and increasing sales. Retailers that had fully adopted RFID in 2018 achieved a 9.2 percent return on investment, and the financial benefits increase over time (Sain and Wong, 2018). Improved inventory accuracy and increased returns on investment facilitate more advanced uses, including multichannel fulfillment and an improved customer experience. Burberry customers use their smartphones to interact with the RFID tags to learn, for example, about the origins and/or construction of items (Busby, 2018).

Beyond retail item tagging, other applications of RFID technology include (1) EZ Pass for paying bridge or highway tolls, (2) tagging of luggage on flights, and (3) tagging of cargo containers at most of the world's ports. As the cost of RFID tags declines, we will see a much wider range of applications in the context of supply chain management. Reconciling shipments against bills-of-lading (i.e., cargo records) or packing and pick lists can be performed quickly and accurately electronically, eliminating the need to perform these functions manually.

Of course, RFID technology has far broader implications than its application to supply chains. For example, thousands of people in Sweden have had microchips about the size of a grain of rice inserted into the skin just above the thumb by a syringe similar to one used for vaccinations (Savage, 2018). Sweden has a long history of embracing new technologies and is moving toward becoming a cashless society. The chip can be used as an e-ticket to events, to board trains, for access to homes and offices (no need for keys), and for sharing social media profiles by touching smartphones to transfer information. While implants of microchips in humans is not yet widespread, the microchipping of pets has become commonplace. In Australia, RFID ear tags are mandatory to register farm and ranch livestock; the tags are widely used to register livestock with the U.S. National Animal Identification System (Weiss, 2018). RFID technology applications potentially have enormous benefits but have also raised privacy concerns.

RFID technology has made inroads into many industries, including the following (Schuster, Allen, and Brock, 2007).

- *Warehousing*. By tagging inventory stored in a warehouse, improved tracking of items and order fulfillment significantly affects several measures of customer service.
- *Maintenance*. Maintenance programs require keeping track of the location, suitability, and condition of spare parts, and RFID tags provide instant, accurate monitoring of parts and components.
- *Pharmaceuticals*. Tracking and tracing pharmaceuticals can help ameliorate the problems of counterfeit drugs and theft in the industry. These problems have been estimated to run to $30 billion annually worldwide.
- *Medical devices*. RFID technology can provide continuous access to the identity, location, and state of medical devices—bolstering patient care.
- *Animal tracking*. The appearance of mad cow disease (BSE) in the United Kingdom, Canada, and the United States raised an alarm worldwide about food safety. Being able to track individual livestock can be invaluable when trying to trace the source of problems.

- *Shelf-life tracking*. We are all familiar with expiration dates on foods such as dairy products, packaged meats, fish, and canned goods. RFID technology provides the opportunity for "dynamic" shelf-life tracking—updating the shelf-life indicator to take into account environmental conditions such as temperature.
- *Retailing*. It is common for expensive retail items, such as leather jackets, to be tagged with a transmitter that sounds an alarm if removed from the store. Inexpensive RFID chips provide the opportunity to tag almost all retail items, virtually eliminating theft altogether.
- *Defense*. Logistic support has always been a key to securing victory in battle, in both modern and ancient times. It is currently estimated that the U.S. Department of Defense manages an inventory valued at $67 billion. Keeping track of this inventory is obviously a top priority, and RFID technology helps accomplish this goal.

Blockchain Technology

Blockchain is another emerging technology that may lead to significant supply chain change. Many people are familiar with Bitcoin (a cryptocurrency sent from user to user without the use of banks) but the real SCM potential lies with other aspects of the technology. In brief, a blockchain is a distributed and secure ledger. Such a ledger allows information to be kept on the entire history of a product as it travels along the supply chain.

A common way to access blockchain information is through QR codes placed on the item. The customer can then use a QR reader on their smartphone to access detailed information about the product. One emerging application is in agriculture, where consumers can use the technology to learn exactly which farm the steak they are purchasing came from.

Blockchain also allows for so-called **smart contracts**, where the contract is automatically triggered based on some externally verified event. For example, payment could be processed once the product's RFID tag passes some specific reader, with no need for human intervention. We will discuss supply chain contracting in Section 6.4, but it is interesting to imagine what more complex contracts might be possible under blockchain.

6.3 THE BULLWHIP EFFECT

As described in the previous section, Barilla's experience of wide swings in demand at its factories, prior to the implementation of their VMI program, is one example of the **bullwhip effect**. Both practitioners and academics have taken a considerable interest in this effect of increasing demand variability moving up the supply chain. In the early 1990s, executives at Procter & Gamble (P&G) were studying replenishment patterns for Pampers disposable diapers. They were surprised to see that the orders placed by distributors had much more variation than did sales at retail stores. Furthermore, orders of materials to suppliers had even greater variability. Demand for diapers is pretty steady, so one would assume that variance would be low in the entire supply chain. However, this was clearly not the case. P&G coined the term "bullwhip" effect for this phenomenon. It also has been referred to as the "whiplash" or "whipsaw" effect.

This phenomenon was observed by other firms as well. HP experienced the bullwhip effect in patterns of sales of its printers. Orders placed by a reseller exhibited wider swings than retail sales, and orders placed by the printer division to the company's integrated circuit

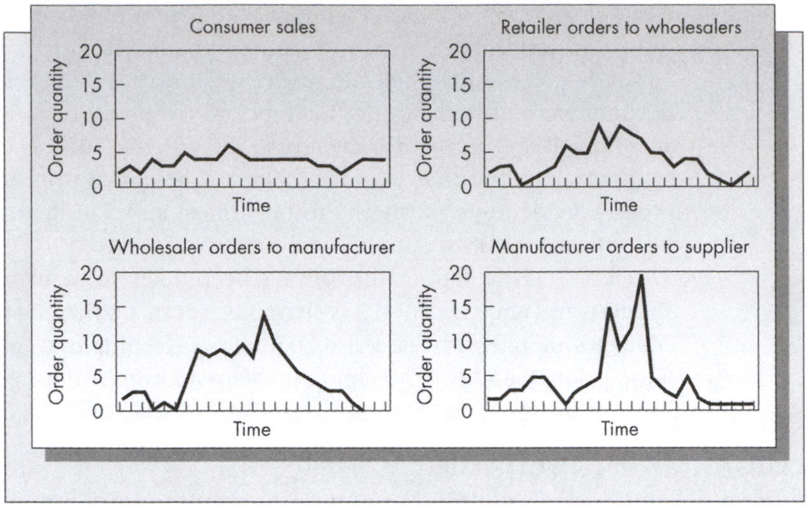

Figure 6–2 Increasing variability of orders up the supply chain
Source: Lee, Padmanabhan, and Whang, 1997.

division had even wider swings. Figure 6–2 shows how variance increases as one moves up the supply chain.

Another example of the bullwhip effect is the simulated beer distribution game (Sterman, 1989). Participants play the roles of retailers, wholesalers, distributors, and manufacturers of beer. Communication among participants is prohibited: each player must make ordering decisions based on what is demanded from the downstream player only. What one observes are wild swings in orders placed downstream even when the original demands are fairly stable. This is a consequence of the bullwhip effect.

The telescoping variation in the demand patterns in a supply chain results in a planning nightmare for many industries. What can be done to alleviate these effects? There are four primary causes of the bullwhip effect (Lee, Padmanabhan and Whang, 1997).

- Demand forecast updating.
- Order batching.
- Price fluctuations.
- Shortage gaming.

We consider each of these effects in turn. *Demand forecasts* at each stage of the supply chain are the result of demands observed one level downstream (as in the beer distribution game). Only at the final stage of the chain (the retailer) are consumer demands observed directly. When each individual in a serial supply chain determines his or her demand forecasts individually, the bullwhip effect results. The retailer builds in safety stocks to protect against uncertainty in consumer demand. These safety stocks cause the retailers' orders to have greater variance than the consumer demand. The distributor observes these swings in the orders of the retailer and builds in even larger safety stocks, and so on.

Order batching can result in smooth demand patterns being translated to spiky demand patterns at lower levels of the product structure. The natural tendency to save fixed costs by ordering less frequently (which is the idea behind the EOQ formula) gives rise to order batching. Given the cost structure at each level, this is certainly a reasonable response. This is particularly present in material requirements planning (MRP) systems, which will be discussed in Chapter 9. Consider a basic two-level MRP system in which the pattern of final demand is fixed and constant. Since items are produced in batches of size Q, the demand pattern one level down is much spikier.

When prices *fluctuate*, there is a speculative motive for holding inventories. (This was first discussed in the inventory management context by Kenneth Arrow (1958) and relates to the motivations for holding cash postulated by the economist John Maynard Keynes.) In the food industry, the majority of transactions from the manufacturer to distributors are made under a "forward buy" arrangement. This refers to the practice of buying in advance of need because of an attractive price offered by manufacturers. Such practices also contribute to the bullwhip effect. Large orders are placed when promotions are offered.

Shortage gaming occurs when product is in short supply, and manufacturers place customers on allocation. When customers find out that they may not get all the units they request, they inflate orders to make up for the anticipated shortfall. For example, if a computer manufacturer expects to receive only half the requested units of a CPU in short supply, it can simply double the size of its order. If anticipated demands don't materialize, orders can be canceled. The result is that the manufacturer gets an inflated picture of the real demand for the product. This could have dire consequences for a manufacturer placing large amounts of capital into capacity expansion based on "phantom" demands.

Clearly, the bullwhip effect is not the result of poor planning or irrational behavior on the part of players in the supply chain. Each individual acts to optimize his or her position. What, then, can be done to alleviate the situation? There are several potential remedies; however, these remedies must take into account that people behave selfishly. The motivation for bullwhip-inducing behavior must be eliminated.

Researchers recommend four initiatives (Lee et al., 1997).

- Information sharing.
- Channel alignment.
- Price stabilization.
- Discouragement of shortage gaming.

Information sharing means that all parties involved share information on point-of-sale (POS) data and base forecasts on these data only. Information sharing can be accomplished by several techniques, as discussed earlier. The trend toward information sharing is increasing. Computer manufacturers, for example, are now requiring sell-through data from resellers (these are data on the withdrawal of stocks from the central warehouse). In some cases, manufacturers are linked directly to sources of POS data.

Channel alignment is the coordination of efforts in the form of pricing, transportation, inventory planning, and ownership between upstream and downstream sites in the supply chain. VMI, described earlier, is one method for improving alignment. One of the tendencies that defeats channel alignment is order batching. As we noted earlier, fixed costs motivate order batching. Reducing fixed costs will result in smaller order sizes. One of the important components of fixed costs is the paperwork required to process an order. With new information technologies, these costs are substantially reduced. Another factor motivating large batches is transportation economies of scale. It is cheaper on a per-unit basis to order a full truckload than a partial one. If manufacturers allow customers to order an assortment of items in a single truckload, the motivation to order large batches of single items is reduced. Another trend encouraging small batch ordering is the outsourcing of logistics to third parties. Logistics companies can consolidate loads from multiple suppliers. As mentioned earlier, outsourcing of logistics (like outsourcing of manufacturing) is expanding rapidly.

Pricing promotions motivate customers to buy in large batches and store items for future use; this is very common in the grocery industry. By *stabilizing pricing*, sales patterns will have less variation. In retailing, the effect of stable pricing is evident by comparing sales patterns at retailers such as Macy's that run frequent promotions and warehouse stores such as Costco that offer everyday low pricing. The warehouse stores experience steadier sales than do the department stores. In fact, for many department stores, promotional sales account for most of their business. Major grocery manufacturers such as P&G, Kraft, and Pillsbury are moving toward a value-pricing strategy and away from promotional pricing for the same reason.

Finally, one way to minimize excessive orders as a result of shortage gaming is to allocate based on past sales records rather than on orders. This will reduce the tendency of customers to exaggerate orders. Several companies (including General Motors) are moving in this direction.

The bullwhip effect has been observed in a variety of settings and is theoretically predicted to occur by so-called *systems dynamics* models (Sterman, 1989). However, a well-designed production system should work to smooth variability, particularly that associated with seasonality. Gérard Cachon, Taylor Randall, and Glen Schmidt (2007) performed an empirical study on demand variability across a wide range of U.S. firms. They found that industries with seasonality tended to smooth production relative to demand, whereas industries without seasonality tended to amplify. Retailers, in particular, tended to perform a smoothing rather than amplifying effect on demand variability.

The study by Cachon et al. was primarily at the industry level, whereas Robert Bray and Haim Mendelson (2012) performed a firm-level study that found 65 percent of firms exhibited a positive overall bullwhip. The 2012 study confirmed the findings of the 2007 study that in highly seasonal settings the production smoothing effect outweighs the inherent uncertainty amplification caused by demand shocks. Interestingly, the bullwhip effect is significantly reduced for data after 1995. While much of this can be explained by improved information systems, perhaps the increasing awareness of the bullwhip effect is also a contributing factor. Bray and Mendelson empirically verified the impressive bullwhip reduction achieved by Caterpillar Inc., which focused its efforts on initiatives similar to the four outlined above for bullwhip reduction.

In summary, the bullwhip effect is the consequence of individual agents in the supply chain acting in their own best interests. To mitigate bullwhip effects, several changes must be made. Incentives must be put in place to reduce demand forecast errors, to reduce excessive order sizes in allocation situations, and to encourage information sharing and system alignment. As these initiatives become policy, everyone, especially the consumer, will benefit from reduced costs and improved supply chain efficiency.

PROBLEMS FOR SECTIONS 6.1, 6.2, AND 6.3

1. Name two products you are familiar with that are *not* discussed in this chapter, one of which is best classed as functional and the other as innovative.

2. What product characteristics would be necessary for a product to be able to have a supply chain that is simultaneously both efficient and responsive?
3. Until recently, Amazon.com was a purely e-commerce company, yet Barnes & Noble maintains both retail bookstores and a significant internet business. What are the advantages and disadvantages of Barnes and Noble's strategy and why do you think Amazon is moving into "bricks and mortar" businesses?
4. What are the key benefits that the internet has provided in aiding efficiency or effectiveness in supply chain management?
5. Give three examples of revenue or cost benefits made possible by RFID tags when contrasted with barcodes.
6. Give a supply chain application where you think the distributed ledger provided by blockchain might be particularly useful.
7. What is the bullwhip effect? What is the origin of the term?
8. Do you think that eliminating intermediaries would always get rid of the bullwhip effect? Under what circumstances might it not work?
9. Discuss the four initiatives recommended for ameliorating the bullwhip effect in supply chains.

6.4 INCENTIVES IN THE SUPPLY CHAIN

One of the key challenges in SCM is that a supply chain is rarely owned by any one firm. It generally involves different players, often with competing objectives. Aligning incentives along the chain can considerably increase supply chain profitability. Consider the following example.

EXAMPLE 6.1

Suppose that the retailer SnowInc buys ski jackets from the Chinese manufacturer JacketCo. Due to the short selling season and long delivery lead time for the ski jacket, this is a one-time purchase decision. SnowInc estimates the season's demand for one type of jacket to be normally distributed with mean 500 and standard deviation 100. Retail price for the jacket is $100. If not sold at the end of the season SnowInc unloads them for $10 per jacket. JacketCo charges a wholesale price of $30 per jacket; its per unit manufacturing cost is $12. How many jackets should SnowInc order, and what are the two firms' profits? If both firms were owned by the same company, how would these answers change?

Solution

We first consider the decision from SnowInc's perspective. From the newsvendor model (see Section 5.3), they have an overage cost of $30 – $10 = $20 and an underage cost of $100 – $30 = $70. The critical ratio is therefore 70/(20 + 70) = 7/9. The spreadsheet below shows that SnowInc should order 576 jackets. [Note: not all significant figures are displayed in the spreadsheet.] The expected profit for SnowInc would be approximately $32,320. The calculations for JacketCo are 576*($30 – $12) = $10,368. The total channel profit is $42,688.

If SnowInc and JacketCo are owned by the same company, the jacket cost is $12 rather than $30. Using the lower number in the spreadsheet yields an order of 701 jackets for a total channel profit of $43,524 ($30,906 to SnowInc and $12,618 to JacketCo), which is a 2 percent increase in total profit.

	A	B
1	Price	$100
2	Cost	$30
3	Salvage Value	$10
4	Cost of Inventory Left (overage cost)	$20
5	Cost of Unsatisfied Demand (underage cost)	$70
6	Mean Demand	$500
7	Standard Deviation of Demand	$100
8		
9	Critical Fractile	0.7778
10	z	0.7647
11	Amount to Stock	576
12		
13	L(z)--loss function	0.1279
14	Expected Lost Sales	13
15	Expected Sales	487
16	Expected Left Over Inventory	89
17		
18	Expected Cost	$2,680.31
19	Expected Profit	$32,319.79
20		

Cell Formulas

Cell	Formula
B4	=B2−B3
B5	=B1−B2
B9	=B5/(B4+B5)
B10	=NORM.S.INV(B9)
B11	=B6+B7*B10
B13	=NORM.S.DIST(B10,0) =B10*(1−NORM.S.DIST(B10,1))
B14	=B7*B13
B15	=B6−B14
B16	=B11−B15
B18	=(B4+B5)*B7*NORM.S.DIST(B10,0)
B19	=B5*B15−(B4*B16)

In Example 6.1, if 701 jackets are ordered then the supply chain creates more profit and availability improves than if 576 jackets are ordered. In fact, there is only a 2 percent chance that the retailer will stock out (one minus the critical ratio). The reason that so many jackets are ordered (relative to mean demand of 500 jackets) is because the margins are so high relative to the overstock costs.

Given that the supply chain makes more profit and customers are happier if the retailer orders 701 jackets, why doesn't the retailer do this on their own initiative? The answer is clear from the numbers above. Ordering 576 jackets produces an expected profit of $32,320 for the retailer whereas ordering 701 jackets produces an expected profit of $30,906. There is no incentive for the retailer to make this change.

In general, the incentives in Example 6.1 can be written as follows. Suppose the manufacturer produces an item for c, sells it to the retailer for w, and the retailer sells it for p. Then the underage cost for the retailer is $p - w$, for the manufacturer $w - c$, and for the supply-chain as a whole $p - c$. Because $(p - w) > (p - c)$, the retailer will under stock from the standpoint of maximizing profit to the whole supply chain. In economics, this is known as **double marginalization**. Because each firm naturally considers its own margin, the decisions are not what a centralized decision maker would do.

The issue of double marginalization arises because the firms are following a standard wholesale price contract agreement. With different contracts, it is often possible to align incentives of each party so that the optimal supply chain decision is reached; in this case, the supply chain is said to be **coordinated**. For example, many book publishers offer **buy-back agreements** to bookstores that allow the stores to return unsold copies for a full refund. In this way, the overage cost to retailers is reduced, and they are encouraged to stock more. Barry Pasternack (1985) showed how the right buy-back contract can coordinate the supply chain, resulting in system optimal profit; the question then becomes how to split the profits. Assuming that a win-win (both parties benefit) arrangement can be found, both parties will enter into such an agreement. However, buy-back contracts can be problematic in environments where the supplier really does not want to take the goods back or the goods may be damaged in transit.

Revenue sharing contracts are another type of contract that, with the right parameter choices, can coordinate the supply chain. In this case, the retailer pays much less for the goods up front but gives the supplier a portion of all revenue earned. Because there is less risk of overstock, the retailer is again incentivized to order more than they would with a simple wholesale price contract. Such contracts were used particularly effectively in the movie rental industry before it was displaced by Netflix and similar firms. Movie studios charged low upfront DVD costs to incentivize stores to stock more. This made for both happier customers and more rentals. The movie studios then took a cut from every movie rented.

VMI, as discussed in Section 6.2, is another way to mitigate double marginalization and coordinate the supply chain. Because the supplier is deciding inventory at the retailer's facility, it need not consider the retailer's margins. Of course, restrictions such as shelf-space limitations need to be in place for such contracts to ensure the supplier doesn't overstock on the retailer's behalf (Fry, Kapuscinski, and Olsen, 2001).

There are many more kinds of contacts than those described here that, in the right situations, work to coordinate the supply chain. Further, as discussed earlier, blockchain is likely to open the door to even more sophisticated contract types. However, there has also been significant behavioral research to evaluate how such contacts perform in practice, when the parties may have considerations beyond expected profit, such as perceptions of fairness. Somewhat surprisingly, behavioral considerations often work against supply chain coordination. In lab studies, wholesale price contracts can do better than profit-maximizing theory would predict (e.g., Loch and Wu, 2008). This combined with their simplicity probably explains their continued widespread use.

PROBLEMS FOR SECTION 6.4

10. How is double marginalization affected by the profit margin of the retailer $(p - w)$ relative to the profit margin of the whole supply chain $(p - c)$? Is its effect greatest when most of the profit margin sits with the retailer or with the supplier?
11. Why might it not be practical for the supplier in Example 6.1 to simply tell the retailer to order 701 jackets and give the supplier a kick-back for the profit differential? Can you think of different supply chains where such an arrangement might be effective?
12. What types of industries, beyond book publishing, are likely to find buy-back agreements effective?
13. What types of industries, beyond movie rentals, are likely to find revenue sharing agreements effective? When are they least likely to be practical?
14. Why might the supplier be likely to overstock (relative to what is supply chain optimal) in a VMI arrangement if shelf-space restrictions or similar are not in place by the retailer?

6.5 RISK POOLING

A key strategy for increasing efficiency and mitigating uncertainty in supply chains is **risk pooling**. Put simply, risk pooling means that when one adds multiple sources of variability together, the whole is inherently less variable. After a brief review of the theory of pooling, this section discusses three key versions of risk pooling: (1) inventory/location pooling, (2) product pooling and postponement, and (3) capacity pooling. First, we explain the mathematics behind pooling, which is common to all three versions of pooling.

The Theory of Pooling

A common measure of relative variability is known as the **coefficient of variation** (CV). The coefficient of variation of a random variable is the ratio of the standard deviation over the mean. In symbols, if X is a random variable with mean μ and standard deviation σ, then $CV(X) = \alpha/\mu$.

Now, consider n independent sources of variation, represented by the random variables $X_1, X_2, \ldots, X_n$. Assume that these are independent and identically distributed and each have mean μ and standard deviation σ. These might represent demands at n stores of a single retailer. Now consider $CV(W)$ where $W = \sum_{i=1}^{n} X_i$ (the sum of the demands at the n stores). Since the random variables are assumed to be independent, it follows that the variance of W is $n\sigma^2$ and hence the standard deviation of W is $\sigma\sqrt{n}$. Since the expected value (that is, the mean) of W is $n\mu$, it follows that the coefficient of variation of W is $CV(W) = \frac{\sigma\sqrt{n}}{n\mu} = \frac{\sigma}{\mu\sqrt{n}} = \frac{CV(X_1)}{\sqrt{n}}$. Clearly, this is decreasing in n. We can see that the higher the CV of an individual demand, X_1, the greater the benefit is likely to be from pooling (i.e., the operational value of decreasing an already small CV will be minimal). Also, the greater the level of pooling (the larger the n) the greater the benefit.

Even if the random variables $X_1, X_2, \ldots, X_n$ do not have the same distribution, a similar phenomenon will hold. Correlation between the variables affects the strength of the effect, with positive correlation decreasing the effects of pooling and negative correlation increasing the effects.

Pooling is analogous to **portfolio diversification** in finance. In that case, one seeks financial products (e.g., stocks and bonds) that are negatively correlated and that are of sufficient variety (i.e., large enough n) such that the risk of the total portfolio is decreased. In operations management, the goal is to aggregate sources of uncertainty (ideally those that are negatively correlated) so that the whole is easier to manage.

Inventory/Location Pooling

If two geographic sources of demand can be served by the same supply, then pooling will imply that the aggregate demand will have a lower coefficient of variation. This in turn implies that the inventory needed to achieve a target service level will be less. Of course, there are declining marginal returns to pooling, so if the coefficient of variation is already low it will have little benefit.

Inventory pooling can be achieved either by a centralized warehouse, which has the disadvantage of moving inventory away from some customers, or by **virtual pooling** (where items are transshipped from one location to another for resupply as needed), which may increase transportation costs.

EXAMPLE 6.2

Consider a toy retailer with warehouses in St. Louis and Kansas City, Missouri. The warehouses stock an identical popular toy delivered to stores in the two cities. Assume a warehouse serves only stores in its city. Weekly demand for St. Louis is normally distributed with mean 2,000 and standard deviation 400 (that is N(μ = 2,000, σ = 400)). Weekly demand for Kansas City is distributed N(μ = 2,000, σ = 300). We assume that the two cities are far enough apart so that demand in the two cities is independent.

The following parameters are seen by both warehouses:

Replenishment lead time in weeks ~ N(μ =2, σ = 0.1);
Fixed shipping cost of replenishment: $500;
Cost per toy: $10; and
Holding cost per toy: 20 percent of toy's value per year.

How many toys should each location order at a time, and when should they reorder if they want a 99 percent cycle service level? What if they pool the two locations?

Solution

From the EOQ equation of Section 4.5, shown in the spreadsheet below, both warehouses should order 7,221 toys at a time. Notice how weekly demand has been converted to yearly demand so it is in the same time units as the holding cost.

	A	B	C	D	E	F	G
1	λ	104280	Demand rate of the item, in units/unit time				
2	K	$500	Fixed cost incurred with each replenishment, in dollars				
3	h	2	Holding cost per unit per unit time				
4	EOQ	7220.8	Economic order quantity				
5	EOQ rounded up	7221					

Cell Formulas

Cell	Formula
B1	=2000*52.14
B3	=0.2*10
B4	=SQRT(2*B2*B1/B3)

From the Type 1 service reorder point model of Section 5.5, shown in the spreadsheet below, St. Louis should keep a safety stock of 1,396 toys, and Kansas City should keep a safety stock of 1,092 toys (only calculations for St Louis are shown but Kansas City is similar with cell B3 equal to 300 instead of 400).

	A	B	C	D	E	F	G
1	Service level = (1-α)	0.01	Probability of a stockout in a reorder cycle				
2	λ	2000	Demand rate of the item, in units/unit time (E[D])				
3	σ(λ)	400	Standard deviation of the demand per unit time				
4	μ(t)	2	Expected lead time, in unit time				
5	σ(t)	0.1	Standard deviation of the leadtime				
6							
7							
8	Solve for safety stock and reorder point						
9	σ(LTD)	600.00	Standard deviation of demand during lead time, in units				
10	z	2.33	Safety factor				
11	Safety Stock	1395.81	zσ(LTD)				
12	Reorder Point	5395.81	Safety stock + λμ(t)				

Cell Formulas

Cell	Formula
B9	=SQRT(B4*B3*B3+B2*B2*B5*B5
B10	=NORMS.S.INV(1−B1)
B11	=B10*B9
B12	=B11+B2*B4

If the two warehouses are pooled (either physically or virtually through an information system), then weekly demand becomes 4,000 (= 2*2000) and the standard deviation of weekly demand is 500 (= $\sqrt{300^2 + 400^2}$). Substituting these numbers into the above spreadsheets shows that the company should order 10,199 toys at a time (a 29 percent reduction) and keep a safety stock of 1,890 toys (a 24 percent reduction).

Example 6.2 shows that pooling inventory can result in a significant reduction of inventory in the system. Notice how centralizing inventory has reduced both safety stock and average inventory. This is generally true with pooled inventories.

As another illustration, consider the following simple scenario. Assume n independent retail locations stock similar items. For example, these could be Macy's department stores located in different cities. Further, assume that the stock level of a particular item is determined from the newsvendor model. Referring to the discussion in Section 5.5, there are known values of the unit overage cost, c_o, and the unit underage cost, c_u. In Section 5.3 it was proven that the optimal stocking level is the $c_u/(c_u + c_o)$ fractile of the demand distribution. (If we assume that stocking levels are determined based on service levels instead of costs, then the critical ratio *is* the service level and c_o and c_u need not be known.)

To simplify the analysis, suppose that the demand for this item follows the same normal distribution at each store with mean μ and standard deviation σ, and the demands are independent from store to store. Let z^* be the value of the standard normal variate that corresponds to a left-tail probability equal to the critical ratio. Then, as is shown in Section 5.3, the optimal policy is to order up to $Q = \sigma z^* + \mu$ at each location. The safety stock held at each location is σz^*, so the total safety stock in the system is $n\sigma z^*$. Refer to this case as the decentralized system.

Alternatively, suppose that all inventory for this item is held at a single distribution center (DC) and shipped overnight on an as-needed basis to the stores. Let's consider the amount of safety stock needed in this case to provide the same level of service as with the decentralized system. Since store demands are independent normal random variables with mean μ and variance σ^2, the aggregated demand from all n stores is also normal, but with mean $n\mu$ and variance $n\sigma^2$. The standard deviation of the aggregated demand therefore has standard deviation $\sigma\sqrt{n}$. This means that to achieve the same level of service, the warehouse needs to stock up to the level Q_w given by $Q_w = n\mu + z\sigma\sqrt{n}$. The total safety stock is now $z\sigma\sqrt{n}$. This corresponds to a centralized system.

Forming the ratio of the safety stock in the decentralized system over the safety stock in the centralized system gives $z\sigma n/(z\sigma\sqrt{n}) = \sqrt{n}$. Hence, the decentralized system will have to hold $\sqrt{n}$ times more safety stock than the centralized system to achieve the same level of service. Even for small values of n, this difference is significant, and for large retailers such as Walmart that have thousands of stores, this difference is enormous.

Of course, this example is a simplification of reality. For most products, it is impractical for every sale to come from a DC, and it is impractical for a single DC to serve the entire country. Furthermore, the assumption that demands at different stores for the same item are independent is also unlikely to be true. Demands for some items, such as desirable fashion items, are likely to be positively correlated, while others, such as seasonal items like bathing suits, might be negatively correlated. However, even with these caveats, the advantages of centralization are substantial and account for the fact that multilevel distribution systems are widespread in many industries, especially retailing. These results are based on the work of Gary Eppen (1979) (who allowed for correlated demands). This model was extended by Eppen and Linus Schrage (1981) and by Nesim Erkip, Warren Hausman, and Steven Nahmias (1990) to more general settings.

Several authors have examined the issue of how item characteristics affect the optimal breakdown of DC versus in-store inventory. John Muckstadt and Joseph Thomas (1980) showed that high-cost, low-demand items derived the greatest benefit from centralized stocking, while Steven Nahmias and Stephen Smith (1994) showed how other factors, such as the probability of a lost sale and the frequency of shipments from the DC to the stores, also affect the optimal breakdown between store and DC inventory. Nahmias and Smith (1993) provide a comprehensive review of inventory control models for retailing, and Leroy Schwarz (1981) assembled an excellent collection of articles on general multi-echelon inventory models.

In general, we get the most benefit from aggregating products that have a high coefficient of variation. This can occur either because they move slowly (have a low μ) or because they have high uncertainty (a large σ), or both. Products that have high demand (a large μ) and accurate forecasts (a small σ) are probably best left disaggregated and stocked close to demand. Warehousing and storage requirements can also inform aggregation decisions, where products needing a temperature-controlled environment (for example) may be best stored together in a limited number of facilities. The following section elaborates on multi-level distribution systems.

Product Pooling and Postponement

A universal design may be used to pool demand between two product categories. We see the reverse of this effect when products are tailored to a specific market (e.g., boys and girls diapers). The likely negative effect on variability, and hence inventory and production planning, is often overlooked when making such product decisions. One way to mitigate such effects is using **delayed differentiation**, or **postponement**, where the configuration of the final product is delayed as long as possible.

The first application of this principle known to the authors was implemented by Benetton (Signorelli and Heskett, 1984). The Benetton Group is a well-known Italian fashion company. Many of their knitted products were made of wool, which traditionally was dyed before it was knitted. In 1972 Benetton implemented a unique strategy—dying the garments *after* they were knitted. One might well question the strategy, since labor and production costs for garments dyed after manufacture are about 10 percent higher than for garments knitted from dyed thread.

The advantage of reversing the order of the dying and the knitting operations was that it provided Benetton a chance to gain additional data on consumer preferences for colors before committing to the final mix of colors. Benetton's knitwear included nearly 500 color and style combinations. Undyed garments are referred to as **gray stock**. Keeping inventories of gray stock, rather than dyed garments, had several advantages. First, if a specific color became more popular than anticipated, Benetton could meet the demand for that color. Second, the company would run less risk of having large unsold stockpiles of garments in unpopular colors. These advantages more than offset the higher costs of dying the knitted garments rather than the raw wool.

How does postponement correlate with inventory management theory? As we saw in Chapter 5, safety stock is retained to protect against demand uncertainty over the replenishment lead time. The lead time for garments of a specific color was reduced by postponing the dying operation. It follows that the uncertainty was reduced as well, thus achieving comparable service levels with less safety stock. Further, because intermediate inventories for different end items were pooled (e.g., all sweaters had the same base garment at Benetton) the inherent demand uncertainty was reduced.

Postponement has become a key strategy in many diverse industries. Hewlett-Packard (HP) provided one example (Lee, Billington, and Carter, 1993). HP is one of the world's leading producers of inkjet and laser printer. Printers are sold worldwide. While the basic mechanisms of the printers sold overseas are the same as the U.S. versions, subassemblies, such as power supplies, must be customized for local markets. HP's original strategy was to configure printers for local requirements (i.e., localize them) at the factory. That is, printers with the correct power supplies, plugs, and manuals would be produced at the factory, sorted, and shipped overseas as final products. The result was that HP needed to carry large safety stocks of all printer configurations.

In order to reduce inventories and improve the service provided by DCs to retail customers, HP adopted a strategy similar to Benetton's. Printers sent from the factory would be generic or gray stock. Localization would be done at the DC level rather than at the factory.

As with Benetton, this had the ultimate effect of reducing necessary safety stocks while improving service. Printers shipped to overseas DCs by boat required one-month transit times. Generic printers that would be localized at the DC were less bulky and could be shipped in bulk pallets and larger lots. In addition, the final packaging materials could be handled at the DCs. With local customization of the product, the replenishment lead time for configured printers was dramatically reduced. Also, the demand on the factory was now the aggregate of the demands at the DCs, which had relatively smaller variation due to pooling. DC localization for HP Deskjet-Plus printers resulted in an 18 percent reduction in inventories with no reduction in service levels.

Harold E. Edmondson, an HP vice president at the time, stated:

> The results of this model analysis confirmed the effectiveness of the strategy for localizing the printers at remote distribution centers. Such a design strategy has significant benefits in terms of increased flexibility to meet customer demands, as well as savings in both inventory and transportation costs … I should add that the design for localization concept is now part of our manufacturing and distribution strategy. (quoted in Lee et al., 1993, p. 11)

Firms in many industries are becoming aware of the risk-pooling and lead time reduction benefits from postponement and local customization of products. **Modularization** strategies are another side of this same coin, where demand aggregation can occur due to common components.

Capacity Pooling

Almost all production systems require capacity that is larger than expected demand to allow for fluctuations in either demand or supply. This excess capacity is called **safety capacity** and forms a buffer against variability, similar to how inventory buffers variation. If capacity can be shifted among products, the different product demands are effectively one pool. In this case, less safety capacity is needed. However, flexible capacity is usually either more expensive or less efficient than dedicated capacity, so such pooling needs to be done carefully. There has been significant research on the "right" level of flexibility for various types of systems.

For manufacturing systems, capacity pooling typically involves investing in more flexible machines. Setup times for changing between product types need to be carefully managed in any non-dedicated system. Toyota has led the way in designing equipment to produce multiple products with minimal setup times between different types of products. (Toyota's single minute exchange of dies, SMED, were described in Chapter 1.) For assembly or service systems, capacity pooling typically involves cross-training workers, and workers may be more expensive because they need higher skill sets. Also, there are cognitive limits on how much cross-training one person can absorb.

The good news on capacity pooling is that its benefits can be achieved without a fully flexible system. As shown originally by William Jordan and Stephen Graves (1995), a little flexibility goes a long way. Consider Figure 6–3. On the right is a fully flexible system where each of 10 plants can produce any of 10 products. Not shown is the fully dedicated system where each product is only produced by one plant. On the left is a strategy known as **chaining** where each product is connected to two plants in such a fashion that the whole system becomes linked. Such a chain is almost as effective as full flexibility.

Of course, there are many more options for capacity pooling than just the two described above. This is particularly true when capacity corresponds to people rather than machines.

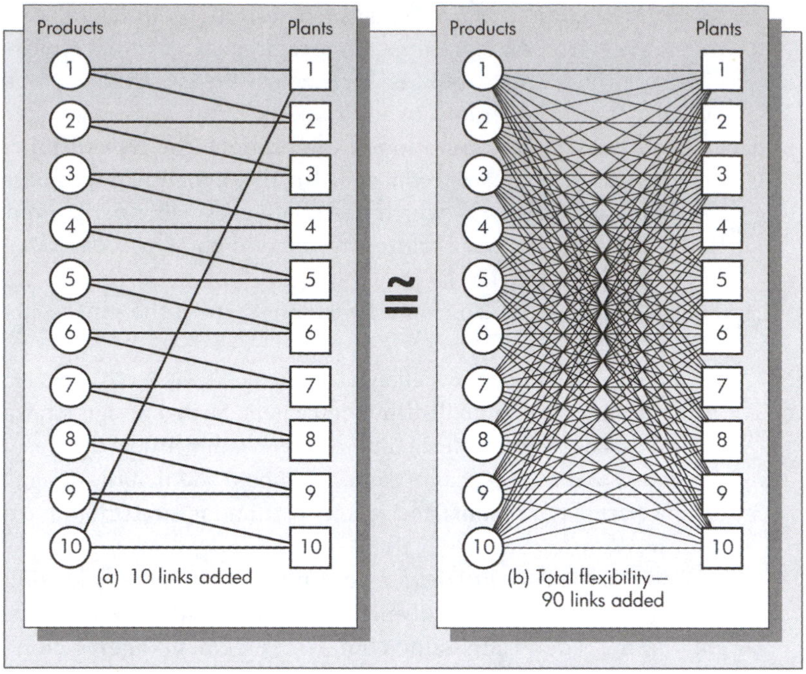

Figure 6–3 Flexibility Configurations
Source: Jordan and Graves (1995).

For example, instead of training all staff to work either on products (or customer types) 1 and 10 or 1 and 2, as shown in Figure 6–3, some staff may be trained on types 1 and 3, 1 and 4, etc. In addition, some staff may be dedicated to a product or customer type, while others are cross-trained. Such a strategy is called **tailored pairing** and is close to optimal (Bassamboo, Randhawa, and Van Mieghem, 2012).

The key finding in most capacity pooling research is that more pooling is better, but there are decreasing returns to scope (i.e., number of types of product pooled). Furthermore, it is important to try to create a circuit with the pooling that encompasses as broad a range of capacity and products as possible. In practice, pooling is constrained by the day-to-day realities of running a production system. As we saw above, a little pooling can go a long way.

6.6 MULTILEVEL DISTRIBUTION SYSTEMS

In many large commercial and military inventory systems, it is common for stock to be stored at multiple locations. For example, inventory may flow from manufacturer to regional warehouses, from regional warehouses to local warehouses, and from local warehouses to stores or other point-of-use locations. Determining the allocation of inventory among these multiple locations is an important strategic consideration that can have a significant impact on the bottom line. A typical multilevel distribution system is pictured in Figure 6–4.

In the parlance of traditional inventory theory, such systems are referred to as multi-echelon inventory systems. As mentioned in Chapter 5, the interest in this area was originally sparked by a seminal paper by Andrew Clark and Herbert Scarf (1960), which was part of a major initiative at the Rand Corporation to study logistics and inventory management problems arising in the military. This paper laid the theoretical framework for what later became a

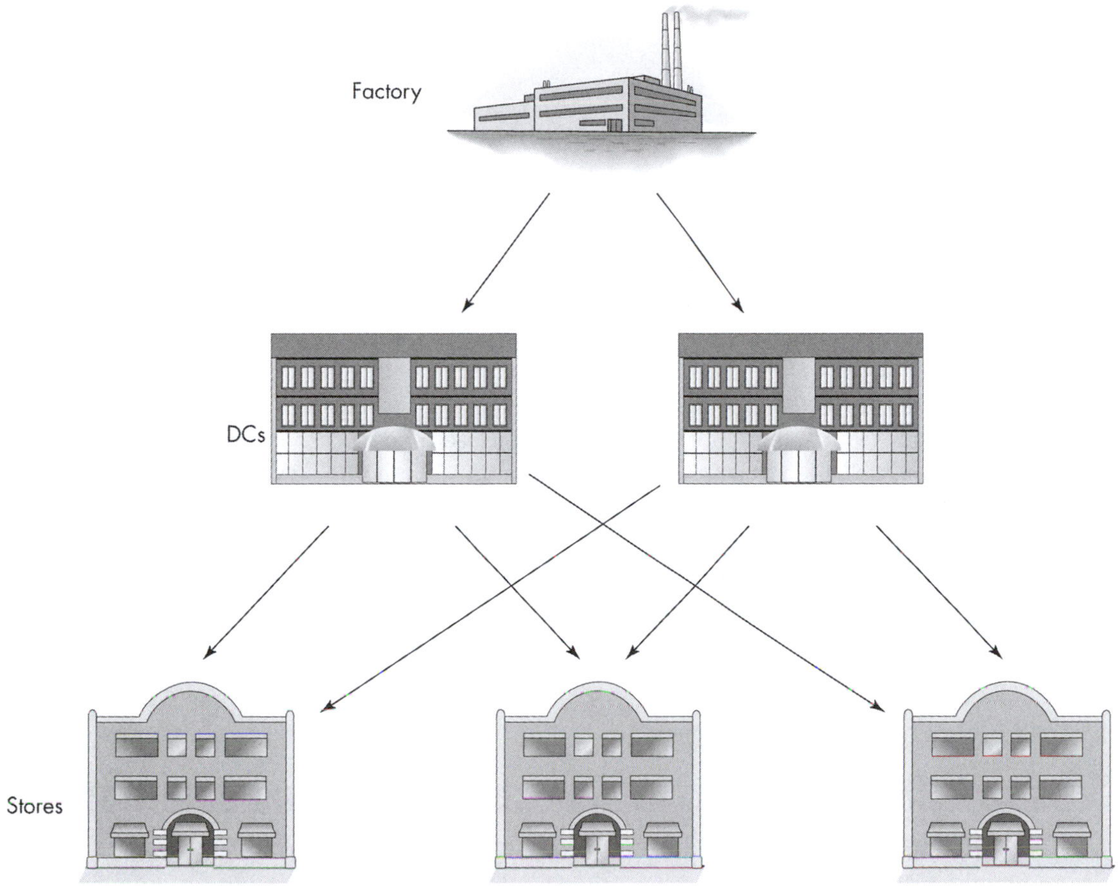

Figure 6–4 Typical multilevel distribution system

very large body of research in multi-echelon inventory theory. (Chapter 5 provides a brief overview of multi-echelon inventory models.)

There are a number of advantages of intermediate storage locations in a large-scale distribution system.

1. *Risk pooling.* Suppose a DC serves 50 retail outlets. As seen earlier, by holding the majority of the stock at the DC rather than at the stores, pooling implies that the same system-wide service level could be achieved with less total inventory.
2. *DCs can be designed to meet local needs.* In retailing, the mix of products sold depends on the location of the store. For example, one sells more short-sleeve shirts in Arizona than in Maine. Hence, DCs located in the Southwest would stock a different assortment of products than DCs located in the Northeast. DCs could be designed to take these differences into account.
3. *Economies of scale in storage and movement of goods.* DCs allow for **economies of scale** in the following way. If all product were shipped directly from the factory to the local outlet, shipment sizes would be relatively small. Large bulk shipments from factories located overseas could be made to DCs at a lower unit cost. In that way, the smaller shipments would be made over shorter distances from DCs to stores.
4. *Faster response time.* Because DCs can be located closer to the customer than can factories, demands can be met more quickly. Retailers try to locate DCs in the United States

within one day's drive away from any store. (This is why Ohio is a popular place for retail DCs, since this is the geographic center of the country.) In this way, store inventories can be replenished overnight if necessary.

Multilevel distribution systems also could have several disadvantages.

1. *More inventory than simpler distribution systems.* Depending on the number of distinct locations used, it is likely that a multilevel system will require more overall total inventory than a single-level system. This is because the transportation lead times will mean that safety stock will be built in at each level. This, in turn, means that more money is tied up in the pipeline.
2. *Increased total order lead times from the plant to the customer.* When DCs experience stock-outs, the response time of the system from manufacturer to point of sale could actually be worse in a multilevel system. This is a consequence of the fact that the total lead time from plant to store is the sum of the lead times at each level of the system. (However, in a well-managed system, stock-outs at the DC should be very rare occurrences.)
3. *Greater costs for storage and movement of goods.* One must not ignore the cost of the DC itself. Building a modern, large-scale facility could cost upward of $500 million. In addition, there are recurring costs such as rent and labor. As a result, building and maintaining a multilocation distribution system is expensive.
4. *Possible contributions to the bullwhip effect.* The bullwhip effect discussed in Section 6.3 is the propensity of the variance of orders to increase as one moves up the supply chain. Adding additional levels to a distribution system could cause a bullwhip effect.

6.7 DESIGNING PRODUCTS FOR SUPPLY CHAIN EFFICIENCY

Product design was traditionally a function totally divorced from operations management issues. Designers were concerned primarily with aesthetics and marketability, paying little attention to manufacturing and logistics. As quality and reliability advanced to the forefront, it became clear that product reliability and product design were closely linked. The **design for manufacturability** (DFM) movement developed out of a need to know why products failed and how those failures could be minimized. Once it was understood that reliability could be factored in at the design stage, the link between design and manufacturing was forged. Another term for DFM is **concurrent engineering**.

In recent years, firms have realized that the logistics of managing the supply chain can have as much impact on the bottom line as product design and manufacturing. More than ever, we are seeing innovative designs that take supply chain considerations into account. One way of describing this concept is **design for logistics** (DFL). Another is three-dimensional concurrent engineering (3-DCE), a term adopted by Charles Fine (1998). The three dimensions are: product, process, and supply chains. 3-DCE carries the concept of concurrent engineering one step further. Concurrent engineering means that product-related issues (functionality, marketability) and process-related issues (how the product is produced, reliability, and quality of the final product) are joint considerations in the design phase. Three-dimensional concurrent engineering means that supply chain logistics are also considered in the product design phase.

There are two significant ways that logistics considerations enter into the product design phase.

1. Product design for efficient transportation and shipment.
2. Postponement of final product configuration (as discussed earlier).

Products that can be packed, moved, and stored easily streamline both manufacturing and logistics. Buyers prefer products that are easy to store and easy to move. Some products tend to be large and bulky and present a special challenge. One example is furniture. Swedish-based Ikea expertly designs products that are modular and easily stored. Furniture is sold in simple-to-assemble kits that allow Ikea retailers to store furniture in the same warehouse-like locations where the items are displayed.

Additional Issues in Supply Chain Design

While products can be better designed for efficient supply chain operation, there are several important issues to consider in the design of the supply chain itself. Fine (1998) refers to supply chain design as the "ultimate core competency." The following are four important relevant issues.

- The configuration of the supplier base.
- Outsourcing arrangements.
- Channels of distribution.
- The decoupling point.

Configuration of the Supplier Base

The number of suppliers, their locations, and their sizes are important considerations for efficient supply chain design. The trend has been to reduce the number of suppliers and to develop long-term arrangements with the existing supplier base. One example is the Xerox Corporation, which underwent an impressive turnaround in the 1980s (Jacobson and Hillkirk, 1986). One of the company's strategies was to streamline the supply chain on the procurement side by reducing the number of its suppliers from 5,000 to only about 400. Those 400 suppliers included a mix of both overseas and local suppliers. The overseas suppliers were chosen primarily on a cost basis, while the local suppliers could provide more timely deliveries when necessary. [Four decades later, Xerox has become another example of a corporation that dominated its industry, as did Kmart, but could not maintain its business model.]

Cooperative efforts between manufacturers and suppliers (as well as manufacturers and retailers on the other end of the chain) have been gaining popularity. Traditionally, this relationship was an adversarial one. Today, with the advent of arrangements like vendor-managed inventories, suppliers and manufacturers work closely and often share what was once proprietary data to improve both party's performance.

Outsourcing Arrangements

Outsourcing of manufacturing has become a popular trend in recent years. Successful contract manufacturers such as Foxconn experienced rapid growth in the past decades. Many firms are outsourcing supply chain functions to third parties as well; 3PL is becoming big business. For example, Walmart now outsources a major portion of its logistics operations.

A 3PL agreement might only outsource transport operations, or it might also outsource warehousing, purchasing, and inventory management. A new term, 4PL, has been coined to reflect 3PL providers that offer end-to-end logistics support. Originally, the term was only applied to firms that manage external 3PL providers (as well as providing full logistics support). However, more common usage allows the 3PL activities to take place internally to the 4PL. In some sense, Amazon.com is a 4PL provider as it sells products for other companies providing both the web interface as well as the pick-and-pack and dispatch operations.

Consider the following example of a smaller firm that decided to outsource purchasing (Dreifus, 1992). The Singer Furniture Company contracted with the Florida-based IMX Corporation to handle its procurements. In the agreement, IMX handled all negotiations of consolidated shipments, took care of the paperwork, conducted quality-control inspection, and searched for new suppliers in exchange for a fixed commission. In a trial run, IMX suggested a switch from a Taiwanese supplier of bedposts to one based in the Caribbean, which resulted in a 50 percent reduction in price to Singer. From Singer's point of view, the following were the primary advantages of entering into the arrangement.

- *Buying economies of scale.* IMX bought for several clients (including other furniture makers), which allowed it to negotiate lower unit costs on larger orders. There were also economies of scale in shipping costs from overseas. IMX claimed that these economies saved at least 20 percent for clients.
- *Reduced management overhead.* Singer was able to eliminate its international purchasing unit and several related management functions.
- *Lower warehousing costs.* IMX took much of the risk on several lines of product by absorbing local storage costs.
- *Above-board accounting.* The firms agreed on open procedures and disclosures to be sure that the quality programs were functioning and charges were in line with costs.

Channels of Distribution

Failure to establish solid channels of distribution can spell the end for many firms. The design of the distribution channel includes the number and the configuration of DCs; arrangements with third-party wholesalers; licensing and other agreements with retailers, including VMI programs; and establishment of the means to move product quickly through the channel. Direct selling to consumers has been an enormously successful strategy for some retailers, especially for products that become obsolete quickly. By eliminating intermediaries, manufacturers reduce the risk of product dying in the supply channel. This has been a successful strategy for computer maker Dell and for Amazon.com. Dell Computer's original success can be attributed partially to its outstanding supply chain design, highlighted in the Snapshot Application.

The Decoupling Point

One idea in supply chain design is to introduce a **decoupling point**, also known as the **push-pull boundary**. As we will see in Chapter 9, a push system is based on forecasts, while a pull system reacts to demand (either end-customer demand or demand from the next manufacturing stage). The decoupling point is where the supply chain transitions from make-to-stock to make-to-order. Some systems are purely make-to-stock and have no decoupling point. However, a postponement strategy coupled with make-to-order end-customer demand can be an effective strategy for some supply-chains. An inventory of intermediate items can be kept at the decoupling point; customization then can occur after the decoupling point. [Note, the authors prefer the term *decoupling point* over the term *push-pull boundary* because pull systems can also be make-to-stock in terms of final customer demand (see Chapter 9).] A firm that effectively used this strategy was Raychem Corporation, whose varieties of heat-shrinkable tubing were so great that it made sense to keep an inventory of unexpanded tubing in stock (Lennon, 1994).

SNAPSHOT APPLICATION

Dell Computer Designs the Ultimate Supply Chain

Dell Computer was one of the biggest success stories of the 1990s. How successful? The stock price increased a whopping 27,000 percent during the decade. There is no question that a sleek, efficient supply chain design was a contributing factor to Dell's phenomenal success.

Michael Dell, the firm's founder, built customized computers from stock components and sold them from his dorm room at the University of Texas. In 1984, he formed PC's Limited. The company offered mail-order clones of the IBM XT, which was selling for several thousand dollars. PC's Limited sold a basic box consisting of a motherboard with an Intel 8088 chip, a power supply, a floppy drive, and a controller for $795. The buyer had to add a graphics card, a monitor, and a hard drive to make the system functional, but the final product cost less than half as much as an equivalent IBM XT and ran faster. (This was, in fact, the first computer purchased by the senior author.) In 1988, the company changed its name to Dell Computer Corporation. By 1992, Dell was the youngest CEO of a Fortune 500 company. In 2013, Dell took the company private and merged it with EMC and VMware in 2016 (the largest technology deal in history) to form Dell Technologies. From very modest beginnings, Dell Technologies became one of the world's largest IT companies serving global corporations, governments, small businesses, and consumers.

To understand Dell's success, one must understand the nature of the personal computer marketplace. The central processor is really the computer, and its power determines the power of the PC. In the PC marketplace, Intel has been the leader in designing every new generation of processor chip. A few companies, such as AMD and Texas Instruments, have had limited success in this market, but the relentless new-product introduction by Intel has resulted in competitors' processors becoming obsolete before they can garner significant market share. Each new generation of microprocessors renders old technology obsolete. For that reason, computers and computer components have a relatively short life span. As new chips are developed at an ever-increasing pace, the obsolescence problem becomes even more critical.

How does one avoid getting stuck with obsolete inventory? Dell's solution was simple and elegant. Don't keep *any* inventory! For a long time, all PCs purchased from Dell were made to order. Dell did not build to stock. They did and do store components, however. Although Dell can't guarantee all components are used, their market strategy is designed to move as much of the component inventory as possible. Dell focuses only on the latest technology, and both systems and components are priced to sell quickly. Furthermore, because of their high volume, they can demand quantity discounts from suppliers. As the so-called clockspeed of the industry (a term coined by Fine, 1998) increases, Dell's advantage over manufacturers with more traditional supply chain designs also increases.

Dell's supply chain strategy is not designed to produce the least expensive computer. Less expensive brands, such as Asus and Acer, manufacture in low cost labor markets and in large production quantities. Dell's more agile supply chain allows them to design and market products with the latest technology quickly. However, it is true that personal computers have evolved into a much more commoditized market. For this reason, Dell has moved away from their configure-to-order strategy for many of its products. Also, this has prompted Dell and other U.S.-based manufacturers to expand into other areas, such as peripherals, servers, software, and services. Dell frequently contracts with large organizations (e.g., universities) to provide all of their computing needs. This is another example of servicization, which was discussed in Chapter 1.

PROBLEMS FOR SECTIONS 6.5–6.7

15. Describe how inventory pooling works in a multi-echelon inventory system. Under what circumstances does one derive the greatest benefit from pooling? The least benefit?
16. Cross training workers is one way to achieve capacity pooling in a manufacturing environment. What are the advantages and disadvantages to such cross-training?
17. Describe the concept of postponement in supply chains. If you were planning a trip to a distant place, what decisions would you want to postpone as long as possible?
18. Many automobiles can be ordered in one of two engine sizes (e.g., 2 liter or 2.5 liter) but are virtually identical in every other way. How might these automakers use the concept of postponement in their production planning?
19. Why might a retailer want to consider developing a three-level multilevel system? (The three levels might be labeled National Distribution Center, Regional Distribution Center, and Store.)
20. What are the characteristics of items from which one derives the greatest benefit from centralized storage? the least benefit?
21. Discuss why having too many suppliers can be troublesome. Can having too few suppliers also be a problem?
22. Many large companies that have their own manufacturing facilities and logistics organizations outsource a portion of their production or their supply chain operations. Why do you suppose this is?
23. A 3PL provider may have only transport services or may also provide warehouse and inventory management services (often called 4PL services). What are the advantages and disadvantages to a firm in outsourcing its warehouse and inventory management operations along with transportation?

6.8 GLOBAL SUPPLY CHAIN MANAGEMENT

Economic barriers between countries have come down dramatically in the last few decades. China, a longtime holdout, signed a free trade agreement with the Association of Southeast Asian Nations (ASEAN) in 1997. In addition to those ten nations, China has free trade agreements with Australia, Chile, Costa Rica, Georgia, Iceland, Korea, Maldives, New Zealand, Pakistan, Peru, and Switzerland. At the time of this writing, there is a populist pushback against frictionless trade. Nevertheless, few industries only produce in and serve their home markets.

One example of a dramatic shift in the marketplace occurred in the automobile industry. In the 1950s, there was no question that the family car would be a U.S. nameplate because that was what almost *everybody* in the United States purchased in those years. Foreign automakers began to make inroads into the U.S. market during the late fifties and early sixties, with market share of U.S. companies falling 20 percent by 1970 (Womack, Jones, and Roos, 1990). After that date, domestic market share eroded more quickly. While U.S. firms clung to old designs and old ways of doing business, foreign manufacturers took advantage of changing tastes. The big winners were Japanese automobile manufacturers who saw their share of the world market soar from almost zero in the mid-fifties to about 30 percent by 1990. In 2007 Toyota surpassed General Motors as the leading automobile manufacturer (Chozick & Shirouzui, 2007). In the first quarter of 2019, Toyota had the largest market share of sales worldwide—11.54 percent compared to 8.33 percent for General Motors (CSI Market, 2019).

In addition to the fact that U.S. consumers are buying more foreign cars, it is also true that U.S. firms are producing more cars in other countries, and foreign competitors are producing more cars in the United States. Mexico, and to a lesser extent Canada, have also become key manufacturing locations for U.S.-headquartered companies' automobile production. However, as of the time of this writing, China leads the world in automobile production, accounting for one-third of the 70 million cars produced worldwide in 2018. Almost all of the bestselling cars in China were produced as joint ventures with foreign manufacturers; Chinese law requires foreign carmakers to form joint ventures with Chinese brands (Statista, 2019). General Motors has a joint venture with the Shanghai Automotive Industry Corporation, and Hyundai formed a joint venture with the Beijing automotive group.

Automobile supply chains are simply one example of supply chains that have undergone major changes in the last few decades. In general, supply chains have become both more global and more complex. Between 1995 and 2007, the number of transnational companies more than doubled from 38,000 to 79,000 and foreign subsidiaries nearly tripled from 265,000 to 790,000 (IBM, 2010). The heterogeneity of government and local regulations and cultures across different countries makes managing global supply chains more challenging.

A further complexity in managing global supply chains is managing volatile exchange rates. If revenue is reported in U.S. dollars but earned in a foreign currency, then any changes in exchange rates can directly affect reported earnings. Similarly, if suppliers are paid in their local currency, then exchange rate changes will affect the home country's reported costs. Companies often use **exchange rate hedging** to mitigate the risk from exchange rate movements. Currency options are purchased so that if the exchange rate moves in a favorable direction for the firm, then there is no payout; if it moves in an unfavorable direction, then the options pay out. The cost of the options depend in part on the **strike price** for the option, which is the point at which they begin to pay out.

Fuel costs are a key issue in supply chain design. Oil prices peaked at $147 a barrel in 2008 but fell to 13-year low of $26 in February 2016 (Carey, 2016). Geopolitical crises contribute to the volatility of fuel prices. Fuel costs are the second largest expense for airlines after payroll and account for 20 to 25 percent of airline expenses. Southwest Airlines used fuel price options to hedge against fuel price rises and saved $2 billion between 2001 and 2015—but it lost $1 billion in 2016. American Airlines discontinued hedges in 2014. In March 2018, oil was approximately $62 a barrel (Leff, 2018). An option to hedge fuel at $70 a barrel was $10. For the option to pay off, the price of oil would have to rise to about $82 per barrel.

One effect of rising fuel costs was the decision by shipping companies to use **slow steaming**, where vessels designed to travel at 25 knots are deliberately slowed to between 18 and 20 knots to reduce the usage of fuel. In fact, some companies use super slow steaming (12 to 14 knots) to reduce costs even further. Clearly, this has significant negative implications with respect to supply chain responsiveness and is particularly problematic for supply chains of perishable items.

One response to increasing labor costs in China and increasing shipping costs due to fuel is the rise of **onshoring** or **reshoring.** As explained in Chapter 1, this is the practice of returning offshored manufacturing operations to the home country. This is most common in the United States where labor unions have become more flexible, and automation has made routine tasks more cost effective (*The Economist*, 2013). Another factor in this decision is to decrease risk, particularly in the face of increasing global tensions. In food supply chains, concerns over food miles, food safety, and sustainability have also led to changing supply chain configurations. Supply chain designs must be evaluated regularly and rethought as global trends change the key trade-offs that were the incentives to choose one location over another.

PROBLEMS FOR SECTION 6.8

24. What do you see as the new emerging markets in the world in the next 20 years? For what reason has Africa been slow to develop as a new market and as a desirable location for manufacturing?
25. As noted earlier in this chapter, the share of U.S. sales accounted for by multinational firms is increasing. What events might reverse this trend?
26. What difficulties for supply chain management are created by the growth of globalization?
27. What industries do you think are particularly likely to onshore manufacturing operations in the next few years?
28. What is the downside to using currency option hedging?

6.9 SUMMARY

The rise of the information age brought increased interest in supply chain management. When firms can see data on supply chain costs, they can work to reduce them. There is anecdotal evidence to suggest that firms first worked to push costs out of manufacturing and then out of their supply chains. Because of this, supply chain software has become big business. Most of the biggest names in enterprise systems (see Chapter 9), including Oracle and SAP, now offer supply chain modules as part of their total system solutions.

Both information systems and new technologies have allowed information to be gathered and shared in ways that were not previously possible. Supply chain partners can share data on market trends or even manage each other's inventory through vendor-managed inventory agreements. Point-of-sale entry systems based on barcoding are now ubiquitous, and RFID technology solutions are becoming increasingly popular as well. Blockchain is on the horizon as an important new technology for supply chains.

Delivering the product to the consumer was traditionally a secondary consideration for manufacturing firms. Today, however, supply chain considerations are taken into account even at the product design level. Products that are easily stacked can be shipped and stored more easily and less expensively. The strategy of postponement has turned out to be a fundamental design principle that has proven to be highly cost effective. By designing products whose final configuration can be postponed, firms can delay product differentiation. This allows them to gain valuable time in determining demand and also to pool intermediate inventory for a variety of end products.

The bullwhip effect is a phenomenon observed in the late 1980s. The variance of orders increases dramatically as one moves up the supply chain. Most people agree that the bullwhip effect is the result of different agents acting to optimize their own positions; finding solutions is challenging. Information sharing, reduced order batching, and decreased order lead times help, but they do not solve the issue of each party operating in their own interests.

Misaligned incentives in the supply chain are a major source of supply chain inefficiency. About the only way to mitigate these issues are through a change in approach to partnerships and contracts. When different parties in the supply chain consider themselves to be partners, they are more likely to work together to find win-win solutions—solutions that both increase the performance of the supply chain and also increase each party's individual profit. Moving beyond wholesale price contracts to agreements such as buy-back contracts, revenue sharing contracts, and vendor-managed inventory agreements help to coordinate the supply chain if operated under appropriate parameters.

The problem of misaligned incentives is exacerbated in global supply chains because the parties are often located in different geographic areas, may have different cultural backgrounds, and are affected by governments providing different local incentives. Variable exchange rates, rising fuel costs, and increased global tensions also challenge global supply chain management. Sustainability and risk management are important considerations for many global corporations.

BIBLIOGRAPHY

Arrow, K. J. "Historical Background." In K. J. Arrow, S. Karlin, and H. Scarf (Eds.), *Studies in the Mathematical Theory of Inventory and Production* (chapter 1). Palo Alto, CA: Stanford University Press, 1958.

Bassamboo , A., R. S. Randhawa, and J. A. Van Mieghem. "A Little Flexibility Is All You Need: On the Asymptotic Value of Flexible Capacity in Parallel Queuing Systems." *Operations Research* 60 (2012), pp. 1423–1435.

Bjork, C. "Zara Builds Its Business Around RFID." *Wall Street Journal,* September 16, 2014.

Bray, R. L., and H. Mendelson. "Information Transmission and the Bullwhip Effect: An Empirical Investigation." *Management Science* 58, no. 5 (2012), pp. 860–875.

Busby, A. "Why the Time Is Now for the Forgotten Technology of Retail." *Forbes*, November 12, 2018. Retrieved from https://www.forbes.com/sites/andrewbusby/2018/11/12/why-the-time-is-now-for-the-forgotten-technology-of-retail/amp/

Cachon, G. P., T. Randall, and G. M. Schmidt. "In Search of the Bullwhip Effect." *Manufacturing and Service Operations Management* 9, no. 4 (2007), pp. 457–479.

Carey, S. "Airlines Retreat on Fuel Hedging." *Wall Street Journal*, March 21, 2016. Retrieved from https://www.wsj.com/articles/airlines-pull-back-on-hedging-fuel-costs-1458514901

Clark, A. J., and H. E. Scarf. "Optimal Policies for a Multiechelon Inventory Problem." *Management Science* 6 (1960), pp. 475–90.

CSI Market. "World Wide Market Share in the First Quarter 2019." (2019). Retrieved from https://csimarket.com/economy/Vehicle_Unit_Sales_glance.php

Copacino, W. C. *Supply Chain Management: The Basics and Beyond*. Boca Raton, FL: CRC Press, 1997.

Dreifus, S. B., ed. *Business International's Global Desk Reference*. New York: McGraw-Hill, 1992.

Duff, C., and R. Ortega. "How Wal-Mart Outdid a Once-Touted Kmart in Discount Store Race." *The Wall Street Journal*, March 24, 1995.

The Economist. "Special Report: Outsourcing and Offshoring." January 19, 2013. Retrieved from https://www.economist.com/sites/default/files/20130119_offshoring_davos.pdf

Eppen, G. D. "Effects of Centralization on Expected Costs in a Multi-Location Newsboy Problem." *Management Science* 25 (1979), pp. 498–501.

Eppen, G. D., and L. Schrage. "Centralized Ordering Policies in a Multi-Warehouse System with Leadtimes and Random Demand." In L. B. Schwarz (Ed.), *Multi-Level Production/Inventory Systems: Theory and Practice* (pp. 51–67). New York: North Holland, 1981.

Erkip, N.; W. H. Hausman; and S. Nahmias. "Optimal Centralized Ordering Policies in Multi-Echelon Inventory Systems with Correlated Demands." *Management Science* 36 (1990), pp. 381–392.

Fine, C. H. *Clockspeed: Winning Industry Control in the Age of Temporary Advantage*. Reading, MA: Perseus Books, 1998.

Fisher, M. L. "What Is the Right Supply Chain for Your Product?" *Harvard Business Review* (1997), pp. 105–106.

Fry, M.J.; Kapuscinski, R; and T. L. Olsen. "Coordinating Production and Delivery Under a (z, Z)-Type Vendor-Managed Inventory Contract." *Manufacturing & Service Operations Management* 3 (2001), pp. 89–173.

Hammond, J. H. "Barilla SpA (B)." Harvard Business School Supplement 695-064 (Revised July 2006.)

IBM. "The Smarter Supply Chain of the Future." *Global Chief Supply Chain Officer Study,* 2010. Retrieved from https://www.ibm.com/downloads/cas/AN4AE4QB

Jacobson, G., and J. Hillkirk. *Xerox, American Samurai*. New York: Macmillan, 1986.

John, C. G. and M. Willis. "Supply Chain Re-Engineering at Anheuser-Busch." *Supply Chain Management Review* 2 (1998), pp. 28–36.

Jordan, W. C., and S. C. Graves. "Principles on the Benefits of Manufacturing Process Flexibility." *Management Science* 41 (1995), pp. 577–594.

Lee, H. L. "The Triple-A Supply Chain," *Harvard Business Review,* October (2004).

Lee, H. L., C. Billington, and B. Carter. "Hewlett-Packard Gains Control of Inventory and Service through Design for Localization." *Interfaces* 23, no. 4 (1993), pp. 1–11.

Lee, H. L., P. Padmanabhan, and S. Whang. "The Bullwhip Effect in Supply Chains." *Sloan Management Review* Spring (1997), pp. 93–102.

Leff, G. "Why American Airlines Is Brilliant Not to Hedge Fuel." *View from the Wing,* March 2, 2018. Retrieved from https://viewfromthewing.boardingarea.com/2018/03/02/american-airlines-brilliant-not-hedge-fuel/

Lennon, T. M. *Response-time approximations for multi-server polling models, with manufacturing applications.* Stanford University Ph.D. Thesis, (1994).

Loch, C. and Y. Wu. (2008). "Social Preferences and Supply Chain Performance: An Experimental Study." *Management Science* 54, no. 11 (2008), pp. 1835–1849.

Muckstadt, J. A., and L. J. Thomas. "Are Multi-Echelon Inventory Methods Worth Implementing in Systems with Low Demand Rate Items?" *Management Science* 26 (1980), pp. 483–94.

Nahmias, S., and S. A. Smith. "Mathematical Models of Retailer Inventory Systems: A Review." In R. K. Sarin (Ed.), *Perspectives in Operations Management* (pp. 249–78). Norwell, MA: Kluwer, 1993.

Nahmias, S., and S. A. Smith. "Optimizing Inventory Levels in a Two-Echelon Retailer System with Partial Lost Sales." *Management Science* 40 (1994), pp. 582–96.

Pasternack, B. "Optimal Pricing and Returns Policies for Perishable Commodities." *Marketing Science* 4 (1985), pp. 166–176.

Sain, J., and A. Wong. "Transforming Modern Retail: Findings of the 2018 RFID in Retail Study." Accenture, 2018. Retrieved from https://www.accenture.com/_acnmedia/PDF-84/Accenture-Transforming-Modern-Retail-RFID-in-Retail-Study-POV-2018.pdf#zoom=50

Savage, M. "Thousands of Swedes Are Inserting Microchips Under Their Skin." National Public Radio, October 22, 2018.

Schoenberger, C. "How Kmart Blew It." *Forbes*, January 18, 2002. Retrieved from https://www.forbes.com/2002/01/18/0118kmart.html#4def8fb55818

Schuster, E. W.; S. J. Allen; and D. L. Brock. *Global RFID: The Value of the EPCglobal Network for Supply Chain Management*. Berlin: Springer, 2007.

Schwarz, L. B. (Ed.). *Multi-Level Production/Inventory Systems: Theory and Practice*. New York: North Holland, 1981.

Signorelli, S., and H. Heskett. "Benneton." Harvard Business School Case 9-685-014, 1984.

Simchi-Levi, D.; P. Kaminski; and E. Simchi-Levi. *Designing and Managing the Supply Chain: Concepts, Strategies, and Case Studies* (3rd ed.). New York: McGraw-Hill/ Irwin, 2008.

Statista Research Department. "Anheuser-Busch InBev Statistics & Facts." Statista, September 19, 2018. Retrieved from https://www.statista.com/topics/1904/anheuser-busch-inbev-ab-inbev/

Statista Research Department. "Production of Cars in China from 2009 to 2019." Statista, March 20, 2019. Retrieved from https://www.statista.com/statistics/281133/car-production-in-china/

Sterman, R. "Modeling Managerial Behavior: Misperception of Feedback in a Dynamic Decision Making Experiment." *Management Science* 35, no. 3 (1989), pp. 321–39.

Varchaver, N. "Scanning the Globe: The Humble Bar Code Began as an Object of Suspicion and Grew into a Cultural Icon. Today It's Deep in the Heart of the Fortune 500." *Fortune Magazine,* May 31, 2004. Retrieved from http://money.cnn.com/magazines/fortune/fortune_archive/2004/05/31/370719/index.htm

Weiss, H. "Why You're Probably Getting a Microchip Implant Someday." *The Atlantic*, September 21, 2018.

Womack, J. P., D. T. Jones, and D. Roos. *The Machine That Changed the World*. New York: Harper Perennial, 1990.

CHAPTER

7

Supply Chain Analytics

"Supply chains cannot tolerate even 24 hours of disruption. So if you lose your place in the supply chain because of wild behavior, you could lose a lot. It would be like pouring cement down one of your oil wells."

—Thomas Friedman

CHAPTER OVERVIEW

Purpose

To understand how analytics can help with supply chain design and to provide mathematical formulations for problems involving facility location and transportation and routing.

KEY POINTS

1. *Supply chain analytics.* Data has become ubiquitous and computers have become powerful. Only a foolish company would not take advantage of these factors in order to improve decision making. Supply chain analytics is the science of using data and computer algorithms for improved decision making.
2. *Capacity growth planning.* An important strategic issue in supply chain management is determining the timing and sizing of new capacity additions. Simple models (i.e., the make or buy problem) and more complex exponential growth models are explored in this chapter.
3. *Locating new facilities.* Where to locate a new facility is a complex and strategically important problem. Hospitals need to be close to high-density population centers, and airports need to be near large cities—but not too near because of noise pollution. In cases where the primary objective is to locate a facility close to its customer base, quantitative methods can be very useful. In these cases, one must specify how distance is measured. Straight-line distance (also known as Euclidean distance) measures the shortest distance between two points. However, straight-line distance is not always the most appropriate measure. For example, when locating a firehouse, one must consider the layout of streets. Using rectilinear distance (as measured by only horizontal and vertical movements) would make more sense in this context. Another consideration is that not all customers are of equal size. For example, a bakery would make larger deliveries to a supermarket or warehouse store than to a convenience store, in which case a weighted distance criterion is useful. In this chapter, we review several quantitative techniques for finding the best location of a single facility under various objectives.

4. *The transportation problem.* The transportation problem is one of the early applications of linear programming. Assume *m* production facilities (sources) and *n* demand points (sinks). The unit cost of shipping from every source to every sink is known; the objective is to determine a shipping plan that satisfies the supply and demand constraints at minimum cost. The linear programming formulation of the transportation problem has been successfully solved with hundreds of thousands of variables and constraints. A generalization of the transportation problem is the transshipment problem, where intermediate nodes can be used for storage as well as being demand or supply points. Transshipment problems can also be solved with linear programming.
5. *Routing in supply chains.* Consider a delivery truck that must make deliveries to several customers. The objective is to find the optimal sequence of deliveries that minimizes the total distance travelled. This problem is known as the traveling salesman problem and is very difficult to solve optimally. The calculations required to find the optimal solution grow exponentially with the problem size (known mathematically as an NP hard problem). We present a simple heuristic, known as the savings method, for obtaining approximate solutions.

There are a staggering 2.5 quintillion bytes of data created each day (Marr, 2018). In 2018, 90 percent of the world's data was generated in just the two prior years. This is what **big data** looks like. Companies are investing in information systems and decision tools to harness the power of this information. Yet, many of the decision tools for effective supply chain design have been known for decades. What is new is the ease of access to data. This chapter describes key tools, which when combined with good data, can form an important part of a supply chain manager's toolkit.

The capacity of a plant is the number of units that the plant can produce in a given time. Capacity policy plays a key role in determining the firm's competitive position in the marketplace. Factors to consider when planning a capacity strategy include:

- predicted patterns of demand,
- costs of constructing and operating new facilities,
- new process technology, and
- competitors' strategies.

Thus, capacity planning is an important piece of any effective supply chain design.

Where to locate facilities and the efficient design of those facilities are important strategic issues for businesses, nonprofit institutions, the military, and government. During the 1980s, Northern California's Silicon Valley experienced tremendous growth in microelectronics and related industries. Most of these firms were "start-ups," adapting a new technology to a specialized segment of the marketplace. With the surge in demand for microcomputers came support industries producing hard disk drives, floppy disks (ask your parent), semiconductor manufacturing equipment, local area networks, and a host of other related products. In many cases, large investments in capital equipment were required before the first unit could be sold. The venture capital investors bore the risk that the firm would survive. Today's Silicon Valley start-ups tend to be less capital intensive, yet they are still located in the same region for strategic reasons.

However, the importance of strategy in decision making does not imply that mathematical and computer models are not useful for solving location problems. What it does mean is that quantitative solutions must not be taken on blind faith. They must be carefully considered in the context of the problem. Used properly, the results of mathematical and computer models can significantly reduce the number of alternatives that the analyst must consider.

Quantitative techniques are most useful when the goal is to minimize or maximize a single dimensional objective such as cost or profit. The objective function used in location problems generally involves either Euclidean or rectilinear distance (these terms will be defined in Sections 7.3 and 7.4). However, minimizing total distance traveled may not make sense in all cases. As an extreme example, consider the problem of locating a school. A location that requires 100 students to travel 10 miles each is clearly not equally desirable to one that requires 99 students to travel 1 mile and one student to travel 901 miles. A problem such as this requires a different objective to total distance.

Transportation route planning is particularly well suited to quantitative modeling and will be extensively covered in this chapter. The objective is usually to minimize distance or cost subject to constraints on what points are visited. Problems occur at the firmwide level, where the question is to determine which plants to use to service the sources of demand. But routing can also be considered at the individual customer level in order to come up with a specific route for a specific truck.

More generally, mathematical modeling has been used successfully in many supply chain applications. Sophisticated models lie at the heart of several commercial software products such as those offered by JDA software. The Snapshot Application in this section discusses a successful application of advanced mathematical models to IBM's supply chain for semiconductors.

7.1 CAPACITY GROWTH PLANNING

Capacity planning is an extremely complex issue. Each time a company considers expanding existing productive capacity, it must sift through a myriad of possibilities. First, the decision must be made whether to increase capacity by modifying existing facilities. From an overhead point of view, this is an attractive alternative. It is cheaper to effect major changes in existing processes and plants than to construct new facilities. However, such a strategy ultimately could be penny wise and pound foolish. Diminishing returns quickly set in if the firm tries to push the productive capacity of a single location beyond its optimal value.

If a decision is made to construct a new plant, additional decisions remain.

1. *When.* The timing of construction of new facilities is an important consideration. Lead times for construction and changing patterns of demand are two factors that affect timing.
2. *Where.* Locating new facilities is a complex issue. Consideration of the logistics of material flows suggests that new facilities be located near suppliers of raw materials and market outlets. If labor costs were the key issue, overseas locations might be preferred. Tax incentives are sometimes given by states and municipalities trying to attract new industry. Cost of living and geographical desirability are factors that would affect the company's ability to hire and keep qualified employees.

SNAPSHOT APPLICATION

IBM Streamlines Its Semiconductor Supply Chain Using Sophisticated Mathematical Models

An important part of IBM's success is its focus on servicization (see Chapter 1). IBM has also been a leader in the use of optimization and other business analytics tools, both selling them as software and services and applying them to their own supply chains. In the mid-1980s they applied such tools to their supply chain for spare parts (Cohen, Kamesam, Kleindorfer, Lee, and Tekerian, 1990), realizing a 10 percent improvement in parts availability and savings of approximately $20 million annually. More recently, they have applied a combination of optimization and heuristics to improve the planning of their semiconductor supply chain (Degbotse et al., 2013).

IBM has been in the semiconductor business since 1957 and has manufacturing and contract manufacturing facilities in Asia and North America. These facilities make products that range from silicon wafers to complex semiconductor devices. Until the 1990s, IBM facilities were separated by regions. A North American facility would supply component parts to local assembly plants in North America, for example. The regional supply chains were managed and planned independently, in part because enterprise supply chain optimization was not feasible.

Aided by more powerful computers and algorithms, IBM developed a central planning engine to coordinate planning across the extended supply chain (Degbotse et al., 2013). Its purpose is "to determine a production and shipment plan for the enterprise by using limited material inventories and capacity availability to satisfy a prioritized demand statement." Because such a problem is still too large-scale to solve optimally, they used heuristic decomposition and mixed integer programming. They state that the result is "a unified production, shipping, and distribution plan with no evidence of the original decomposition."

They found the following benefits: (a) on-time deliveries to commit date increased by 15 percent, (b) asset utilization increased by 2–4 percent of costs, and (c) inventory decreased by 25–30 percent. By coordinating and planning the supply chain as a whole, they effectively pooled the different regions into one centralized system. Further, the benefits include both increased service and decreased costs, rather than a trade-off between the two. This is the ideal outcome for any process improvement.

What is the lesson learned from this case? As supply chains get larger and more complex, more and more sophisticated methods will be required to manage those systems. Generic software products may not be able to provide sufficient power and customization to be effective in such environments. IBM's experience is only one example of how the management of complex supply chain structures can be improved with the aid of the modeling methods discussed in this text.

3. *How much.* Once management has decided when and where to add new capacity, it must decide on the size of the new facility. Adding too much capacity means that the capacity will be underutilized. This is an especially serious problem when capital is scarce. On the other hand, adding too little capacity means that the firm will soon be faced with the problem of increasing capacity again.

Make or Buy: A Prototype Capacity Expansion Problem

A classic problem faced by the firm is known as the **make-or-buy decision.** The firm can purchase the product from an outside source for c_1 per unit but can produce it internally for a lower unit price, $c_2 < c_1$. However, in order to produce the product internally, the

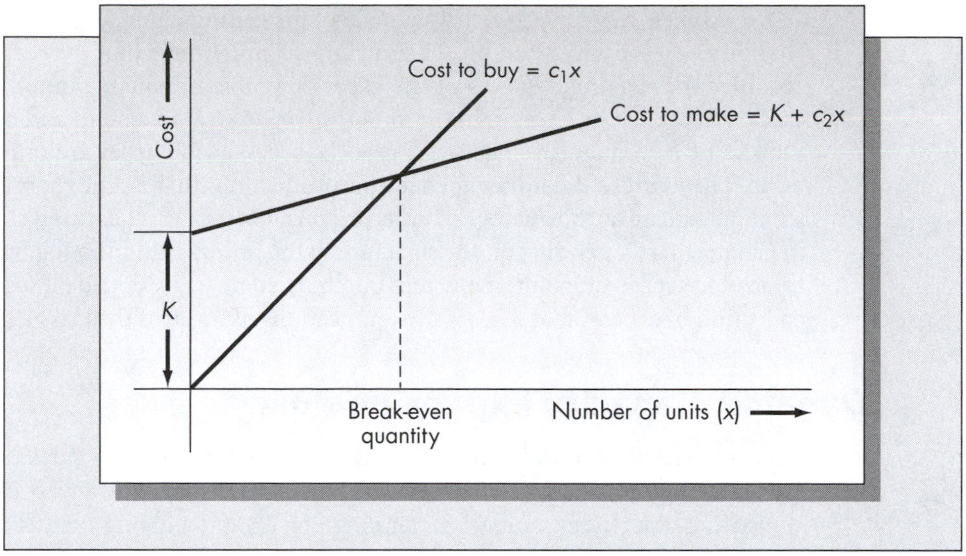

Figure 7–1 Break-even curves

company must invest $\$K$ to expand production capacity. Which strategy should the firm adopt?

The make-or-buy problem contains many of the elements of the general **capacity expansion** problem. It clarifies the essential trade-off of investment and economies of scale. The total cost of the firm to produce x units is $K + c_2x$. This is equivalent to $K/x + c_2$ per unit. As x increases, the cost per unit of production decreases, since K/x is a decreasing function of x. The cost to purchase outside is c_1 per unit, independent of the quantity ordered. By graphing the total costs of both internal production and external purchasing, we can find the point at which the costs are equal. This is known as the **break-even quantity**. The break-even curves are pictured in Figure 7–1.

The break-even quantity solves

$$K + c_2x = c_1x,$$

giving $x = K/(c_1 - c_2)$.

EXAMPLE 7.1

A large international computer manufacturer is designing a new model of personal computer and must decide whether to produce the keyboards internally or to purchase them from an outside supplier. The supplier is willing to sell the keyboards for $50 each, but the manufacturer estimates that the firm can produce the keyboards for $35 each. Management estimates that expanding the current plant and purchasing the necessary equipment to make the keyboards would cost $8 million. Should they undertake the expansion?

The break-even quantity is $x = 8{,}000{,}000/(50 - 35) = 533{,}333$. Hence, the firm would have to sell at least 533,333 keyboards in order to justify the $8 million investment required for the expansion.

Break-even curves such as this are useful for getting a quick ballpark estimate of the desirability of a capacity addition. Their primary limitation is that they are static. They do not consider the dynamic aspects of the capacity problem, which cannot be ignored in most cases. These include changes in the anticipated pattern of demand and considerations of the time value of money. Even as static models, break-even curves are only rough approximations. They ignore the learning effects of production; that is, the marginal production cost should decrease as the number of units produced increases. (Learning curves were discussed in Chapter 1.) Depending on the structure of the production function, it may be economical to produce some units internally and purchase some units outside. (See Manne, 1967, for a collection of articles that discuss the implications of some of these issues.)

Dynamic Capacity Expansion Policy

Capacity decisions must be made in a dynamic environment. In particular, the dynamics of the changing demand pattern determine when the firm should invest in new capacity. Two competing objectives in capacity planning are: (1) maximizing market share and (2) maximizing capacity utilization.

A firm that bases its long-term strategy on maximization of capacity utilization runs the risk of incurring shortages in periods of higher-than-anticipated demand. An alternative strategy to increasing productive capacity is to produce to inventory and let the inventory absorb demand fluctuations. However, this can be very risky. Inventories can become obsolete, and holding costs can become a financial burden.

Alternatively, a firm may assume the strategy of maintaining a "capacity cushion." This capacity cushion is excess capacity that the firm can use to respond to sudden demand surges; it puts the firm in a position to capture a larger portion of the marketplace if the opportunity arises.

Consider the case where the demand exhibits an increasing linear trend. Two policies, (a) and (b), are represented in Figure 7–2. In both cases the firm is acquiring new capacity at

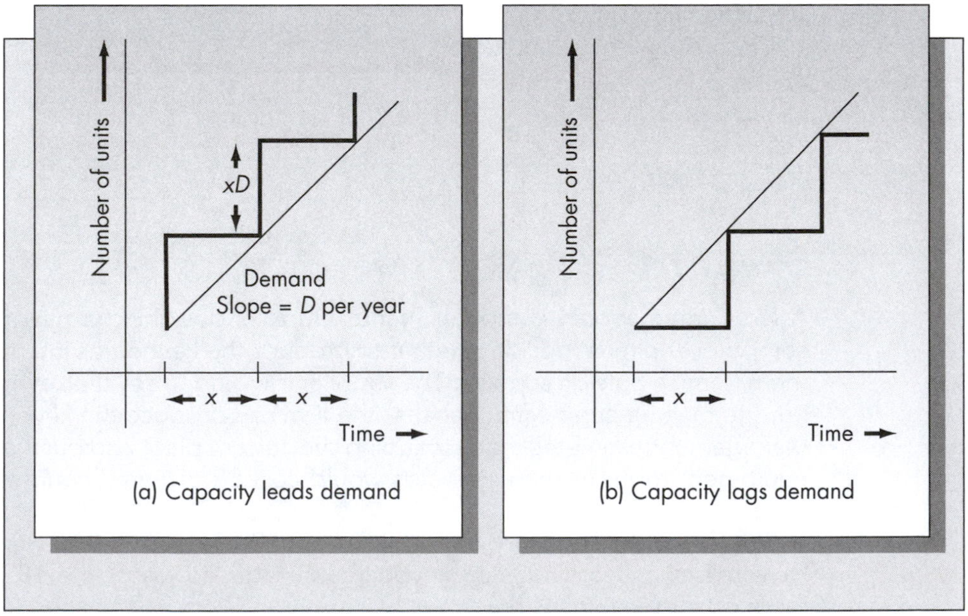

Figure 7–2 Capacity planning strategies

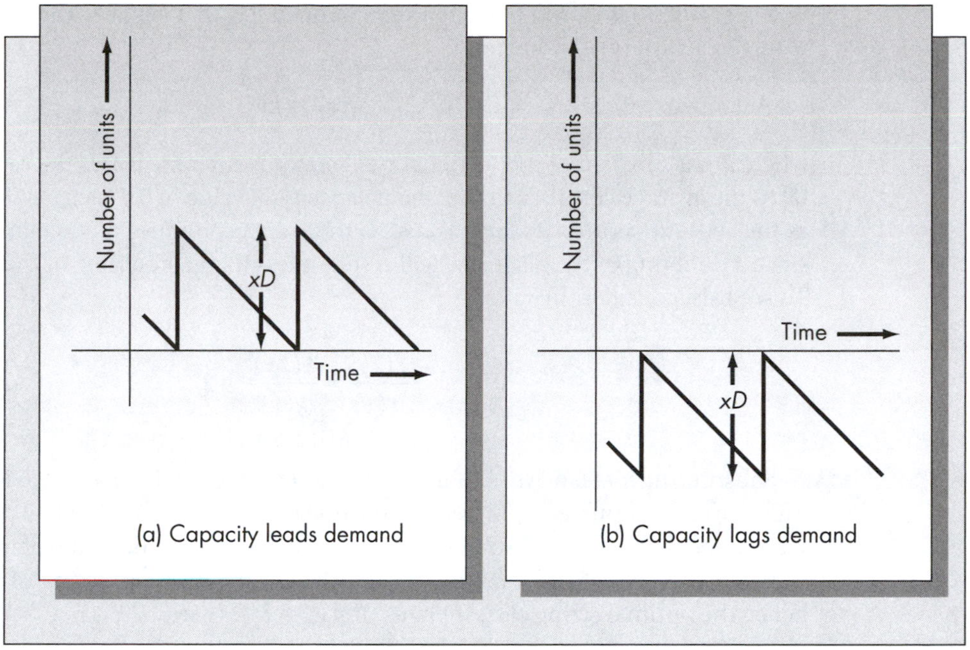

Figure 7–3 Time path of excess or deficient capacity

equally spaced intervals x, $2x$, $3x$, ... and increasing the capacity by the same amount at each of these times. However, in case (a) capacity leads demand, meaning that the firm maintains excess capacity at all times; whereas in case (b), the capacity lags the demand, meaning that the existing capacity is fully utilized at all times. Following policy (a) or (b) results in the time path of excess capacity or a capacity shortfall, as shown in Figure 7–3.

Consider the following specific model (Manne, 1967). Define

D = Annual increase in demand.

x = Time interval between introduction of successive plants.

r = Annual discount rate, compounded continuously.

$f(y)$ = Cost of opening a plant of capacity y.

From Figure 7–2a (which is the strategy assumed in the model), we see that if the time interval for plant replacement is x, it must be true that the plant size at each replacement is xD. Furthermore, the present value of a cost of \$1 incurred t years into the future is given by e^{-rt}. (A discussion of discounting and the time value of money appears in Appendix 7–A.)

Define $C(x)$ as the sum of discounted costs for an infinite horizon given a plant opening at time zero. It follows that

$$\begin{aligned} C(x) &= f(xD) + e^{-rx} f(xD) + e^{-2rx} f(xD) + \cdots \\ &= f(xD)[1 + e^{-rx} + (e^{-rx})^2 + (e^{-rx})^3 + \cdots] \\ &= \frac{f(xD)}{1 - e^{-rx}}. \end{aligned}$$

Experience has shown that a representation of $f(y)$ that explains the economies of scale for plants in a variety of industries is

$$f(y) = ky^a,$$

where k is a constant of proportionality. The exponent a measures the ratio of the incremental to the average costs of a unit of plant capacity. A value of 0.6 seems to be common (known as the six-tenths rule). As long as $a < 1$, there are economies of scale in plant construction, since a doubling of the plant size will result in less than a doubling of the construction costs. To see this, consider the ratio

$$\frac{f(2y)}{f(y)} = \frac{k(2y)^a}{k(y)^a} = 2^a.$$

Substituting $a = 0.6$, we obtain $2^a = 1.516$. This means that if $a = 0.6$ is accurate, the plant capacity can be doubled by increasing the dollar investment by about 52 percent. Henceforth, we assume that $0 < a < 1$ so that there are economies of scale in the plant sizing.

Given a specific form for $f(y)$, we can solve for the optimal timing of plant additions and hence the optimal sizing of new plants. If $f(y) = ky^a$, then

$$C(x) = \frac{k(xD)^a}{1 - e^{-rx}}.$$

Consider the logarithm of $C(x)$:

$$\begin{aligned} \log[C(x)] &= \log[k(xD)^a] - \log[1 - e^{-rx}] \\ &= \log(k) + a\log(xD) - \log[1 - e^{-rx}]. \end{aligned}$$

It can be shown that the cost function $C(x)$ has a unique minimum with respect to x and furthermore that the value of x for which the derivative of $\log[C(x)]$ is zero is the value of x that minimizes $C(x)$. The optimal solution satisfies

$$\frac{rx}{e^{rx} - 1} = a.$$

[As shown by $\dfrac{d\log[C(x)]}{dx} = \dfrac{aD}{xD} - \dfrac{(-e^{-rx})(-r)}{1 - e^{-rx}} = \dfrac{a}{x} - \dfrac{r}{e^{rx} - 1} = 0$, which gives $\dfrac{rx}{e^{rx} - 1} = a$.]

The function $f(u) = u/(e^u - 1)$ appears in Figure 7–4, where $u = rx$. By locating the value of a on the ordinate axis, one can find the optimal value of u on the abscissa axis.

EXAMPLE 7.2

A chemicals firm is planning for an increase of production capacity. The firm has estimated that the cost of adding new capacity obeys the law $f(y) = .0107y^{.62}$, where cost is measured in millions of dollars and capacity is measured in tons per year.

For example, substituting $y = 20{,}000$ tons gives $f(y) = \$4.97$ million plant cost. Furthermore, suppose that the demand is growing at a constant rate of 5,000 tons per

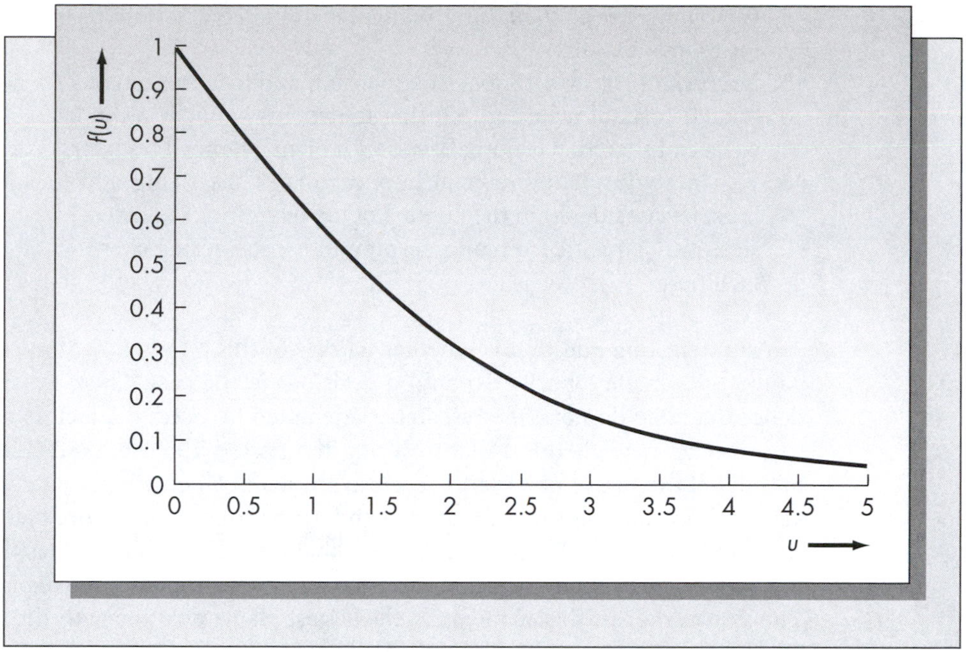

Figure 7–4 The function $u/(e^u - 1)$

year and future costs are discounted using a 16 percent interest rate. From Figure 7–4 we see that, if $a = .62$, the value of u is approximately 0.9. Solving for x, we obtain the optimal timing of new plant openings: $x = u/r = .9/.16 = 5.625$ years.

The optimal value of the plant capacity should be $xD = (5.625)(5{,}000) = 28{,}125$ tons. Substituting $y = 28{,}125$ into the equation for $f(y)$ gives the cost of each plant at the optimal solution as $6.135 million.

Much of the research into the capacity expansion problem consists of extensions of models of this type. This particular model could be helpful in some circumstances, but it ignores a number of fundamental features one would expect to find in the real world.

1. *Finite plant lifetime.* The assumption of the model is that, once constructed, a plant has an infinite lifetime. However, companies close plants for a variety of reasons. Equipment becomes obsolete or unreliable and cannot be replaced easily. Labor costs or requirements may dictate moving either to less expensive locations domestically or to overseas locations. Major changes in the process technology may not be easily adaptable to existing facilities.
2. *Demand patterns.* We have assumed that demand grows at a constant rate per year. Models have been proposed to account for more complex growth patterns of demand. In truth, demand uncertainty is a key factor. In many industries, foreign competition has made significant inroads into established markets, thus forcing rethinking of earlier strategies.
3. *Technological developments.* The model assumes that the capacity of all new plants constructed remains constant and that the cost of building a plant of given size remains constant as well. This is obviously unreasonable. Major changes in process technology occur on a regular basis, changing both the maximum size of new plants and the costs associated with a fixed plant size.

4. *Government regulation.* Environmental and safety restrictions may limit choices of plant location and scale.
5. *Overhead costs.* Most capacity expansion and location models do not explicitly account for the costs of overhead. During the energy crunch in the late 1970s, costs of energy overhead soared, wreaking havoc with plant overhead budgets.
6. *Tax incentives.* The financial implications of the sizing and location of new facilities must be considered in the context of tax planning. Tax incentives are offered by local or state municipalities to major corporations considering sites for the construction of new facilities.

An interesting question is whether models of this type really capture the way that companies have made capacity expansion decisions in the past. There is some preliminary evidence that they do not. One researcher attempted to assess the factors that motivated firms to construct new chemical plants during the period 1957 to 1982 (Lieberman, 1987). He found that the size of new plants increased by about 8 percent per year independent of market conditions. In periods of high demand, firms constructed more plants. The preliminary study indicated that the rational thinking that leads to models such as the one developed in this section does not accurately reflect the way that companies make plant-sizing decisions. The results suggest that firms build the largest plants possible with the existing technology. (Given sufficient economies of scale, however, this policy may theoretically be optimal.)

PROBLEMS FOR SECTION 7.1

1. A start-up company, Macrotech, plans to produce a device to translate Morse code to a written message on a home computer and to send written messages in Morse code over the airwaves. The device is primarily of interest to ham radio enthusiasts. The president, Ron Lodel, estimates that it would require a $30,000 initial investment. Each unit costs him $20 to produce and each sells for $85.
   a. How many units must be sold in order for the firm to recover its initial investment?
   b. What is the total revenue at the break-even volume?
   c. If the price were increased to $100 each, find the break-even volume.
2. For problem 1, suppose that sales are expected to be 100 units in the first year and increase at a rate of 40 percent per year. How many years will it take to recoup the $30,000 initial investment? Assume that each unit sells for $85.
3. A domestic producer of baby carriages, Pramble, buys the wheels from a company in the north of England. Currently the wheels cost $4 each, but for a number of reasons the price will double. In order to produce the wheels themselves, Pramble would have to add to existing facilities at a cost of $800,000. It estimates that its unit cost of production would be $3.50. At the current time, the company sells 10,000 carriages annually. (Assume that there are four wheels per carriage.)
   a. At the current sales rate, how long would it take to pay back the investment required for the expansion?
   b. If sales are expected to increase at a rate of 15 percent per year, how long will it take to pay back the expansion?
4. Based on past experience, a chemicals firm estimates that the cost of new capacity additions obeys the law $f(y) = .0205y^{.58}$ where y is measured in tons per year and $f(y)$ in millions of dollars. Demand is growing at the rate of 3,000 tons per year, and the accounting department recommends a rate of 12 percent per year for discounting future costs.

a. Determine the optimal timing of plant additions and the optimal size of each addition.
b. What is the cost of each addition?
c. What is the present value of the cost of the next four additions? Assume an addition has just been made for the purposes of your calculation. (Refer to Appendix 7–A for a discussion of cost discounting.)

5. A major oil company is considering the optimal timing for the construction of new refineries. From past experience, each doubling of the size of a refinery at a single location results in an increase in the construction costs of about 68 percent. Furthermore, a plant size of 10,000 barrels per day costs \$6 million. Assume that the demand for the oil is increasing at a constant rate of two million barrels yearly, and the discount rate for future costs is 15 percent.
   a. Find the values of k and a assuming a relationship of the form $f(y) = ky^a$. Assume that y is in units of barrels per day.
   b. Determine the optimal timing of plant additions and the optimal size of each plant.
   c. Suppose that the largest single refinery that can be built with current technology is 15,000 barrels per day. Determine the optimal timing of plant additions and the optimal size of each plant in this case. (Assume 365 days per year for your calculations.)
6. A Japanese steel manufacturer is considering expanding operations. From experience, it estimates that new capacity additions obey the law $F(y) = .00345y^{.51}$, where the cost $f(y)$ is measured in millions of dollars and y is measured in tons of steel produced. If the demand for steel is assumed to grow at the constant rate of 8,000 tons per year and future costs are discounted using a 10 percent discount rate, what is the optimal number of years between new plant openings?

The following problems are designed to be solved by spreadsheet.

7. Consider the following break-even problem: the cost of producing Q units, $c(Q)$, is described by the curve $c(Q) = 48Q[1 - \exp(-.08Q)]$, where Q is in hundreds of units of items produced and $c(Q)$ is in thousands of dollars.
   a. Graph the function $c(Q)$. What is its shape? What economic phenomenon gives rise to a cumulative cost curve of this shape?
   b. At what production level does the cumulative production cost equal \$1,000,000?
   c. Suppose that these units can be purchased from an outside supplier at a cost of \$800 each, but the firm must invest \$850,000 to build a facility that would be able to produce these units at a cost $c(Q)$. At what cumulative volume of production does it make sense to invest in the facility?
8. Maintenance costs for a new facility are expected to be \$112,000 for the first year of operation. It is anticipated that these costs will increase at a rate of 8 percent per year. Assuming a rate of return of 10 percent, what is the present value of the stream of maintenance costs over the next 30 years?
9. Suppose the supplier of keyboards described in Example 7.1 is willing to offer the following incremental quantity discount schedule.

Cost per Keyboard	Order Quantity
\$50	$Q \leq 100{,}000$
\$45	$100{,}000 < Q \leq 500{,}000$
\$40	$500{,}000 < Q$

Determine the cost to the firm for order quantities in increments of 20,000 for $Q = 200{,}000$ to $Q = 1{,}000{,}000$, and compare that to the cost to the firm of producing internally for these same values of Q. What is the break-even order quantity?

10. Delon's Department Store sells several of its own brands of clothes plus several well-known designer brands. Delon's is considering building a plant in Malaysia to produce silk ties. The plant will cost the firm $5.5 million. The plant will be able to produce the ties for $1.20 each. On the other hand, Delon's can subcontract to have the ties produced and pay $3.00 each. How many ties will Delon's have to sell worldwide to break even on its investment in the new plant?

7.2 LOCATING NEW FACILITIES

The previous section discussed capacity expansion decisions—specifically, determining the amount and timing of new capacity additions. A related issue is the **location of the new facility.** Deciding where to locate a plant is a complex problem. Many factors must be carefully considered by management before making the final choice.

The following information about the plant itself is relevant to the location decision.

1. *Size of the plant.* This includes the required acreage, the number of square feet of space needed for the building structure, and constraints that might arise as a result of special needs.
2. *Product lines to be produced.*
3. *Process technology to be used.*
4. *Labor force requirements.* These include both the number of workers required and the specific skills needed.
5. *Transportation needs.* Depending on the nature of the product produced and the requirements for raw materials, the plant may have to be located near major interstate highways or rail lines.
6. *Utilities requirements.* These include special needs for power, water, sewage, or fossil fuels such as natural gas. Plants that have unusual power needs should be located in areas where energy is less expensive or near sources of hydroelectric power.
7. *Environmental issues.* Because of government regulations, there will be few allowable locations if the plant produces significant waste products.
8. *Interaction with other plants.* If the plant is a satellite of existing facilities, it is likely that management would want to locate the new plant near the others.
9. *International considerations.* Whether to locate a new facility domestically or overseas is a very sensitive issue. Although labor costs may be lower in some locations, such as China, tariffs, import quotas, inventory pipeline costs, and market responsiveness also must be considered.
10. *Tax treatment.* Tax consideration is an important variable in the location decision. Favorable tax treatment is given by some countries, such as Ireland, to encourage new industry. There are also significant differences in state tax laws designed to attract domestic manufacturers.

Mathematical models are useful for assisting with many operational decisions. However, they generally are of only limited value for determining a suitable location for a new plant. Because so many factors and constraints enter into the decision process, such decisions are generally made based on the inputs of one or more of the company's divisions, and the decision process can span several years. Roger Schmenner (1982) examined the decision process at a number of *Fortune* 500 companies. His results showed that in most firms the decision of where to locate a new facility was made either by the corporate staff or by the CEO, even though the request for new facilities might have originated at the division level. The degree of decision-making autonomy enjoyed at the division level depended on the firm's management style.

Based on a sample survey, Schmenner (1982) found that decisions about the location of new facilities were influenced by the following major factors.

1. *Labor costs.* For industries such as apparel, leather, furniture, and consumer electronics, labor costs are a primary concern. The concern diminishes for capital-intensive industries.
2. *Unionization.* A motivating factor for a firm considering expanding an existing facility, as opposed to building a new facility, is the potential for eliminating union influence in the new facility. A fresh labor force may be more difficult to organize.
3. *Proximity to markets.* When transportation costs account for a major portion of the cost of goods sold, locating new plants near existing markets is essential.
4. *Proximity to supplies and resources.* In certain industries, the decision about where to locate plants is based on the location of resources. For example, firms producing wood or paper products must be located near forests, and firms producing processed food must be in proximity to farms.
5. *Proximity to other facilities.* Many companies tend to place manufacturing divisions and corporate facilities in the same geographic area. For example, IBM originated in Westchester County in New York, and located many of its divisions in that state. By locating key personnel near each other, the firm has been able to realize economies of scope.
6. *Quality of life in the region.* When other issues do not dictate the choice of a location, choosing a site that will be attractive to employees may help in recruiting key personnel. This is especially true in high-tech industries that must compete for workers with particular skills.

We have just covered important qualitative considerations when deciding on the best location for a new plant or other facility. The remainder of this section will consider quantitative techniques called **location models** that can help with location decisions. There are many circumstances in which the objective can be quantified easily, and analytical solution methods are most useful in these cases. Usually, management will balance the quantitative considerations against the qualitative considerations listed above. For example, while quality of life in the region is an important consideration, it will not outweigh a high cost for that location.

In some sense, the plant layout problem, considered in Supplement 1, is a special case of the location problem. However, the problems considered in this chapter differ from layout problems in the following way: we will assume that there are one or more existing facilities already in place, and we wish to find the optimal location of a new facility. We will concentrate on the problem of locating a single new facility, although there are versions of the problem in which one must determine the locations of multiple facilities.

Listed below are some examples of one-facility location problems that can be solved by the methods discussed in this chapter.

1. Location of a new storage warehouse for a company with an existing n production and distribution centers.
2. Determining the best location for a new machine in a job shop.
3. Locating the student center in a university.
4. Finding the best location for a new hospital in a metropolitan area.

5. Finding the most suitable location for a power generating plant designed to serve a geographic region.
6. Determining the placement of an ATM in a neighborhood.
7. Locating a new police station in a community.

Undoubtedly, the reader can think of many other examples of location problems. Analytical methods assume that the objective is to locate the new facility to minimize some function of the distance of the new location from existing locations. For example, when locating a new warehouse, an appropriate objective would be to minimize the total distance traveled to the warehouse from production facilities and from the warehouse to retail outlets. A hospital would be located so that it is easily accessible to the largest proportion of the population in the area it serves. A machine would be placed to minimize the weighted sum of materials handling trips to and from the machine. Clearly, choosing the location of a facility to minimize some function of the distance separating the new facility from existing facilities is appropriate for many real location problems.

Measures of Distance

Two measures of distance are most common: **Euclidean distance** and **rectilinear distance**. Euclidean distance is also known as straight-line distance. The Euclidean distance separating two points is simply the length of the straight line connecting the points. Suppose that an existing facility is located at the point (a, b) and let (x, y) be the location of the new facility. Then the Euclidean distance between (a, b) and (x, y) is

$$\sqrt{(x-a)^2+(y-b)^2}.$$

The rectilinear distance (also known as metropolitan distance, recognizing the fact that streets usually run in a crisscross pattern) is given by the formula

$$|x-a|+|y-b|.$$

Figure 7–5 illustrates the difference between these two distance measures.

Rectilinear distance is appropriate for many location problems. Distances in metropolitan areas tend to be more closely approximated by rectilinear distances than by Euclidean distances even when the street pattern is not a perfect grid. In many manufacturing environments, material is transported across aisles arranged in regular patterns. It is a fortunate coincidence that rectilinear distance is more common than Euclidean distance, because the rectilinear distance problem is easier to solve.

PROBLEMS FOR SECTION 7.2

11. Consider the problem of locating a new hospital in a metropolitan area. List the factors that can be quantified and those that cannot. Comment on the usefulness of quantitative methods in the decision-making process.
12. A coordinate system is superimposed on a map. Three existing facilities are located at (5, 15), (10, 20), and (6, 9). Compute both the rectilinear and the Euclidean distances separating each facility from a new facility located at $(x, y) = (8, 8)$.

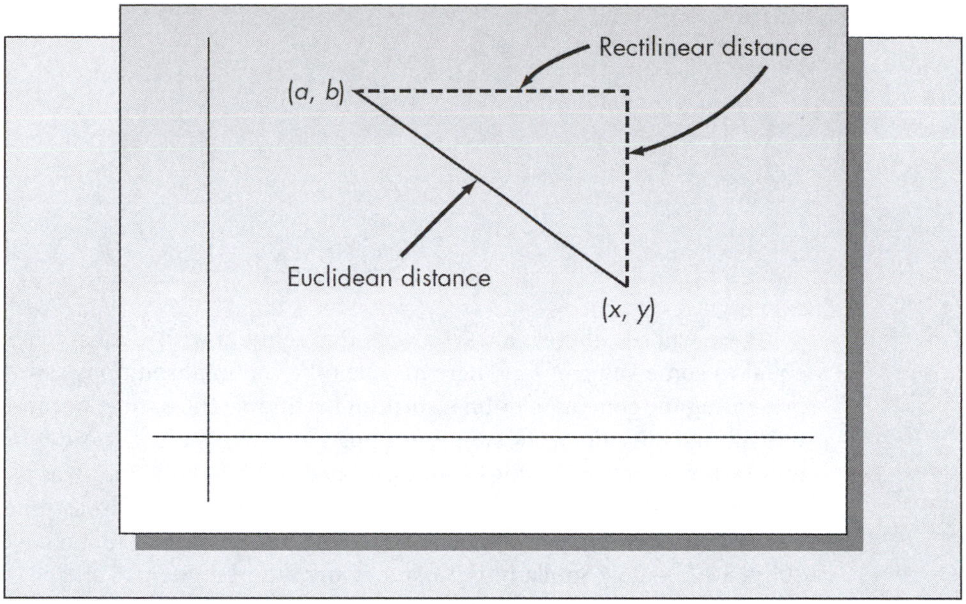

Figure 7–5 Euclidean and rectilinear distances

13. For the situation described in problem 12, suppose that there are only three feasible locations for the new facility: (8, 16), (6, 15), and (4, 18).
    a. What is the optimal location if the objective is to minimize the total rectilinear distance to the three existing facilities?
    b. What is the optimal location if the objective is to minimize the total Euclidean distance to the three existing facilities?
14. For each of the seven examples of location problems listed in this section, indicate which distance measure, Euclidean or rectilinear, would be more appropriate. (Discuss how, in some cases, one or the other objective could be appropriate for the same problem, depending on the optimization criterion used.)

7.3 THE SINGLE-FACILITY RECTILINEAR DISTANCE LOCATION PROBLEM

In this section we will present a solution to the general problem of locating a new facility among n existing facilities. The objective is to locate the new facility to minimize a weighted sum of the rectilinear distances from the new facility to existing facilities. Assume that the existing facilities are located at points $(a_1, b_1), (a_2, b_2), \ldots, (a_n, b_n)$. Then the goal is to find values of x and y to minimize

$$f(x,y)=\sum_{i=1}^{n} w_i(|x-a_i|+|y-b_i|).$$

The weights are included to allow for different traffic rates between the new facility and the existing facilities. A simplifying property of the problem is that the optimal values of x and y may be determined separately, as

$$f(x, y) = g_1(x) + g_2(y),$$

where

$$g_1(x) = \sum_{i=1}^{n} w_i |x - a_i|$$

and

$$g_2(y) = \sum_{i=1}^{n} w_i |y - b_i|.$$

As we will see, there is always an optimal solution with x equal to some value of a_i and y equal to some value of b_i. (There may be other optimal solutions as well.) However, before presenting the general solution algorithm for finding the optimal location of the new facility, we will first consider a few simple examples as background. Consider first the case in which there are exactly two existing facilities, located at (5, 10) and (20, 30) as pictured in Figure 7–6. Assume that the weight applied to each of these facilities is 1. If x assumes any value between 5 and 20, the value of $g_1(x)$ is equal to 15. (For example, if $x = 13$, then $g_1(x) = |5 - 13| + |13 - 20| = 8 + 7 = 15$.) Similarly, if y assumes any value between 10 and 30, then $g_2(y) = 20$. Any value of x outside the closed interval [5, 20] and any value of y outside the closed interval [10, 30] results in larger values of $g_1(x)$ and $g_2(y)$. Hence, the optimal solution is (x, y) with $5 \leq x \leq 20$ and $10 \leq y \leq 30$. All locations in the shaded region pictured in Figure 7–6 are optimal.

As in the example, there always will be an optimal location of the new facility with coordinates coming from the set of coordinates of the existing facilities. Suppose that the existing facilities have locations (3, 3), (6, 9), (12, 8), and (12, 10). Assume again that the weight applied to these locations is 1. Ranking the x locations in increasing order gives 3, 6, 12, 12. A *median* value is such that half of the x values lie above it and half of the x values lie below it. Any value of x between 6 and 12 is a median location and is optimal for this problem. The optimal value of $g_1(x)$ is 15. (The reader should experiment with a number of different values

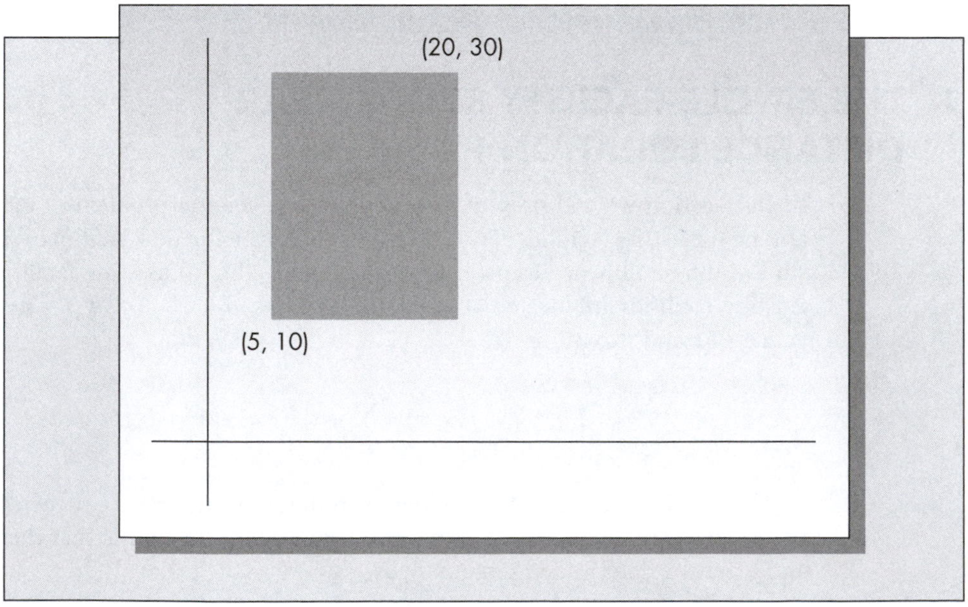

Figure 7–6 Optimal locations of the new facility for a rectilinear distance measure

of x between 6 and 12 to confirm that this is the case.) Ranking the y values in increasing order gives 3, 8, 9, 10. The median value of y is between 8 and 9, and the optimal value of $g_2(y) = 8$. The optimal solution is to locate (x, y) at the median of the existing facilities. This result carries over to the case in which there are weights different from 1.

When the weights are large, this approach is inconvenient. A quicker method of finding the optimal location of the new facility is to compute the accumulated weights and determine the location or locations corresponding to half of the accumulated weight. The procedure is best illustrated by example.

EXAMPLE 7.3

University of the Far West has purchased audiovisual equipment that permits faculty to prepare professional-looking recordings of their presentations. The equipment will be used by faculty from six schools on campus: business, education, engineering, humanities, law, and science. The locations of the buildings on the campus are pictured in Figure 7–7. The coordinates of the locations and the numbers of faculty that are anticipated to use the equipment are as follows.

School	Campus Location	Number of Faculty
Business	(5, 13)	31
Education	(8, 18)	28
Engineering	(0, 0)	19
Humanities	(6, 3)	53
Law	(14, 20)	32
Science	(10, 12)	41

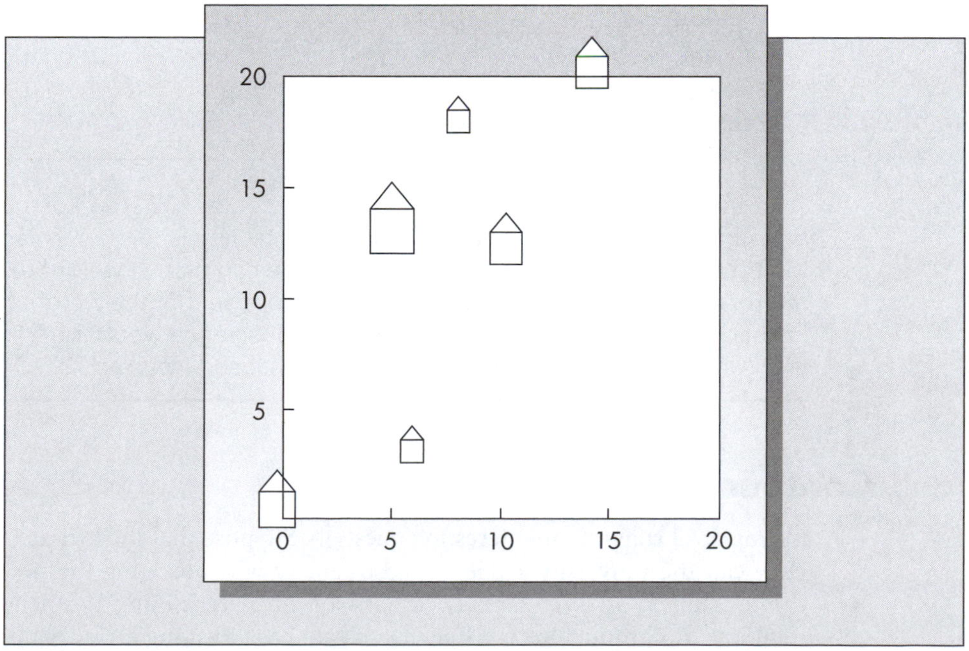

Figure 7–7 Location of six campus buildings (Example 7.3)

The campus is laid out with large grassy areas separating the buildings, and walkways are mainly east–west or north–south, so that distances between buildings are rectilinear. The university planner would like to locate the new facility to minimize the total travel time of all faculty planning to use it.

We will find the optimal values of the x and y coordinates separately. Consider the optimal x coordinate value. We first rank x coordinates in increasing value and accumulate the weights.

School	x Coordinate	Weight	Cumulative Weight
Engineering	0	19	19
Business	5	31	50
Humanities	6	53	103
Education	8	28	131
Science	10	41	172
Law	14	32	204

The optimal value of the x coordinate is found by dividing the total cumulative weight by 2 and identifying the first location at which the cumulative weight exceeds this value. The first time that the cumulative weight exceeds $204/2 = 102$ occurs at $x = 6$, when the cumulative weight is 103. Hence, the optimal $x = 6$.

We use the same procedure to find the optimal value of the y coordinate. The rankings are given here.

School	y Coordinate	Weight	Cumulative Weight
Engineering	0	19	19
Humanities	3	53	72
Science	12	41	113
Business	13	31	144
Education	18	28	172
Law	20	32	204

In this case the cumulative weight first exceeds 102 when $y = 12$. Hence, $y = 12$ is optimal. The optimal location of the new facility is (6, 12). This solution is unique. Multiple optima will occur when a value of the cumulative weight exactly equals half of the total cumulative weight. For example, suppose that the weight for science were 19 instead of 41. Then the cumulative weight for science would be 91, exactly half of the total weight of 182, and all values of y on the closed interval [12, 13] would be optimal.

Contour Lines

Example 7.3 suggests an interesting question. Suppose that the location (6, 12) is infeasible. How can the university gauge the cost penalty when locating the new facility elsewhere? Contour lines, or isocost lines, can assist with determining the penalty of non-optimal solutions. A **contour line** is a line of constant cost; locating a new facility anywhere along a contour line results in exactly the same cost.

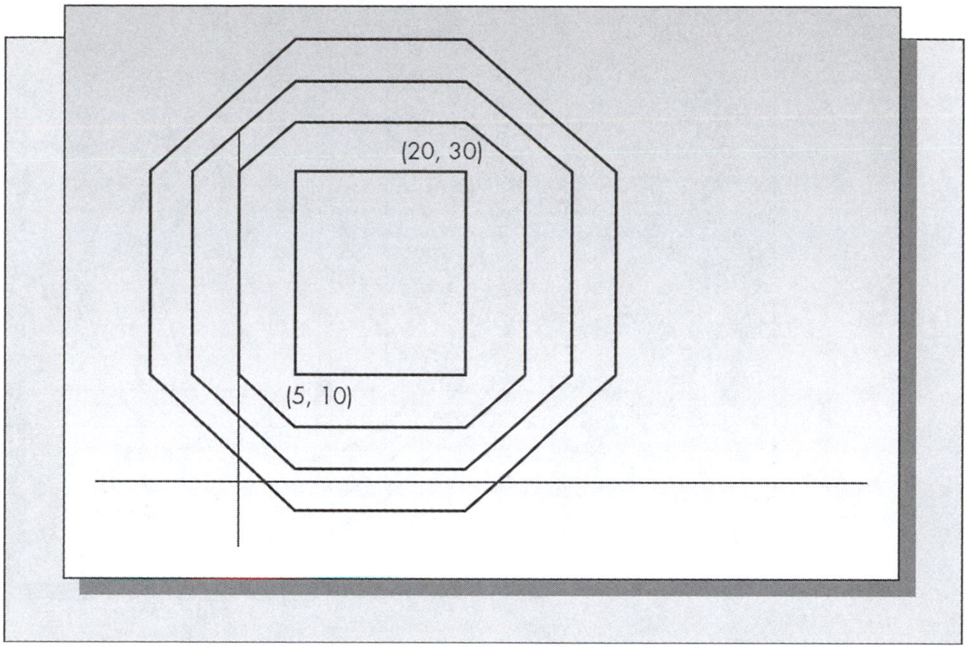

Figure 7–8 Contour lines for the two-facility problem pictured in Figure 7–6

Figure 7–8 pictures the contour lines for the simple example depicted in Figure 7–6, in which there are only two existing facilities and the weights are equal. The total rectilinear distance from any point along a contour line to the two points (5,10) and (20, 30) is the same. Determining contour lines involves computing the appropriate slope for each of the regions obtained by drawing vertical and horizontal lines through each of the points (a_i, b_i). The procedure for determining contour lines is outlined in Appendix 7–B at the end of this chapter.

In Figure 7–9 we have pictured contour lines for the Example 7.3 problem in which the university must locate an audiovisual center. If the optimal location (6, 12) is infeasible, the university administration could use this map to see the penalties associated with alternative sites.

Minimax Problems

We have assumed thus far that the new facility should be placed to minimize the sum of the weighted distances to all existing facilities. There are circumstances in which this objective is inappropriate, however. Consider the following example. The city is considering locations for a paramedic facility. The paramedics should be able to respond to emergency calls anywhere in the city. Certain conditions, such as a severe heart attack, must be treated quickly if the patient is to have any chance of surviving. Hence, the facility should be located so that *all* locations in the city can be reached in a given amount of time.

In such a case the objective would be to determine the location of the new facility to minimize the maximum distance to the existing facilities rather than the total distance. Let $f(x, y)$ be the maximum distance from the new facility to the existing facilities. Then

$$f(x,y) = \max_{1 \le i \le n} \left(|x - a_i| + |y - b_i|\right).$$

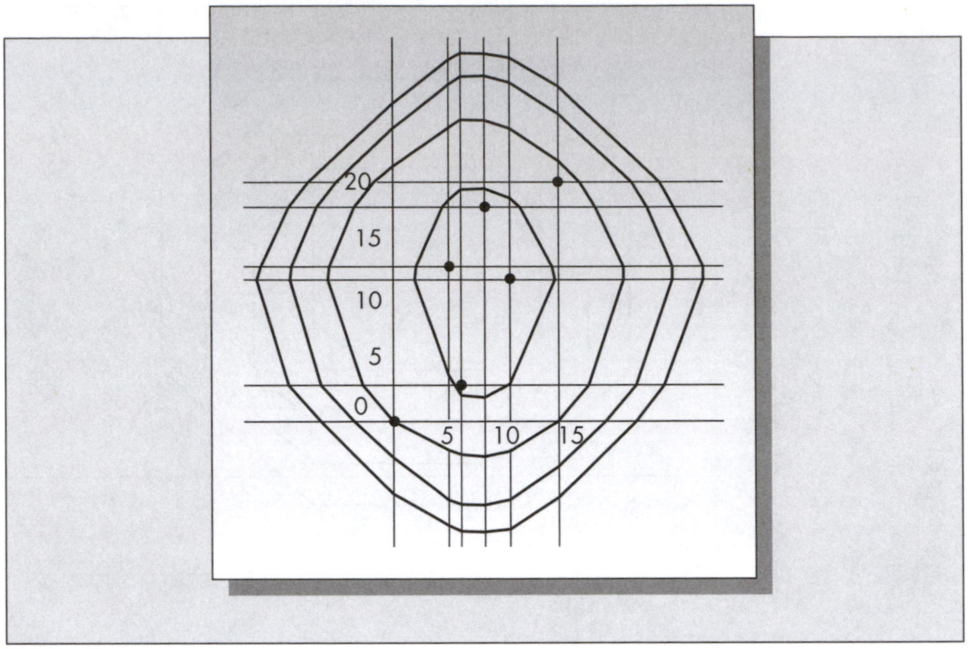

Figure 7–9 Contour lines for the university location in Example 7.3

The objective is to find (x^*, y^*) that satisfies

$$f(x^*, y^*) = \min_{x,y} f(x, y).$$

The procedure for finding the optimal minimax location is straightforward. Linear programming can be used to show that the procedure we will outline is optimal. (We will not present the details here but refer the interested reader to Francis, McGinnis, and White, 1992.) Define the numbers c_1, c_2, c_3, c_4, and c_5:

$$\begin{aligned} c_1 &= \min_{1 \le i \le n} (a_i + b_i), \\ c_2 &= \max_{1 \le i \le n} (a_i + b_i), \\ c_3 &= \min_{1 \le i \le n} (-a_i + b_i), \\ c_4 &= \max_{1 \le i \le n} (-a_i + b_i), \\ c_5 &= \max (c_2 - c_1, c_4 - c_3). \end{aligned}$$

Let

$$\begin{aligned} x_1 &= (c_1 - c_3)/2, \\ y_1 &= (c_1 + c_3 + c_5)/2, \end{aligned}$$

and

$$\begin{aligned} x_2 &= (c_2 - c_4)/2, \\ y_2 &= (c_2 + c_4 - c_5)/2. \end{aligned}$$

Then all points that lie along the line connecting (x_1, y_1) and (x_2, y_2) are optimal. That is, every optimal solution to the minimax problem, (x^*, y^*), can be expressed in the form

$$x^* = \lambda x_1 + (1 - \lambda)x_2,$$
$$y^* = \lambda y_1 + (1 - \lambda)y_2,$$

where λ is a constant satisfying $0 \leq \lambda \leq 1$. The optimal value of the objective function is $c_5/2$.

EXAMPLE 7.4

Consider Example 7.3 of the University of the Far West. As some faculty members have disabilities, the president has decided to locate the audiovisual facility to minimize the maximum distance from the facility to the six schools on campus. Recall that the locations of the schools are: (5, 13), (8, 18), (0, 0), (6, 3), (14, 20), (10, 12).

The values of the constants $c_1, \ldots, c_5$ are

$$c_1 = \min_{1 \leq i \leq n}(a_i + b_i) = \min(18, 26, 0, 9, 34, 22) = 0,$$
$$c_2 = \max_{1 \leq i \leq n}(a_i + b_i) = \max(18, 26, 0, 9, 34, 22) = 34,$$
$$c_3 = \min_{1 \leq i \leq n}(-a_i + b_i) = \min(8, 10, 0, -3, 6, 2) = -3,$$
$$c_4 = \max_{1 \leq i \leq n}(-a_i + b_i) = \max(8, 10, 0, -3, 6, 2) = 10,$$
$$c_5 = \max(c_2 - c_1, c_4 - c_3) = \max(34, 13) = 34.$$

Hence, it follows that

$$x_1 = (c_1 - c_3)/2 = [0 - (-3)]/2 = 1.5,$$
$$y_1 = (c_1 + c_3 + c_5)/2 = (0 - 3 + 34)/2 = 15.5,$$

and

$$x_2 = (c_2 - c_4)/2 = (34 - 10)/2 = 12,$$
$$y_2 = (c_2 + c_4 - c_5)/2 = (34 + 10 - 34)/2 = 5.$$

All points on the line connecting (x_1, y_1) and (x_2, y_2) are optimal. The value of the objective function at the optimal solution(s) is 34/2 = 17. The optimal locations for the minimax problem are pictured in Figure 7–10. Recall that the optimal solution when we used a weighted objective was (6, 12). It is interesting to note that one optimal solution to this problem, (6, 11), is quite close to that solution.

PROBLEMS FOR SECTION 7.3

15. A machine shop has five machines, located at (3, 3), (3, 7), (8, 4), (12, 3), and (14, 6), respectively. A new machine is to be located in the shop with the following expected numbers of loads per hour transported to the existing machines: $\frac{1}{8}$, $\frac{1}{8}$, $\frac{1}{4}$, 1, and $\frac{1}{6}$. Material is transported along parallel aisles, so a rectilinear distance measure is appropriate. Find the coordinates of the optimal location of the new machine to minimize the weighted sum of the rectilinear distances from the new machine to the existing machines.
16. Solve the problem of locating the new audiovisual center at the University of the Far West described in Example 7.3, assuming that the weights are respectively 80, 12, 56, 104, 42, and 17 for the schools of business, education, engineering, humanities, law, and science.

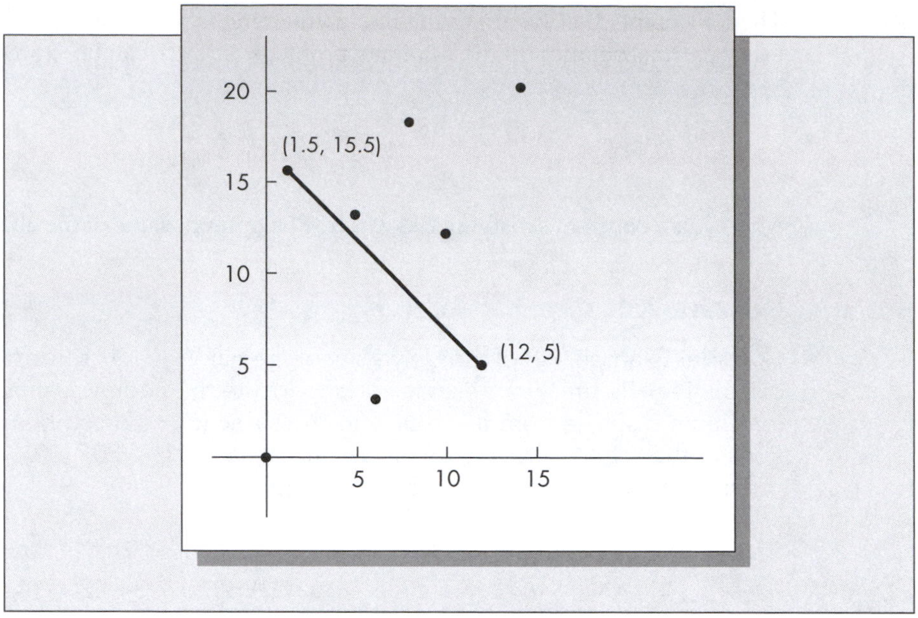

Figure 7–10 Optimal solutions for minimax location objective

17. Armand Bender plans to visit six customers in Manhattan. Three are located in a building at 34th Street and 7th Avenue. The remaining customers are at 48th and 8th, 38th and 3rd, and 42nd and 5th. Streets are separated by 200 feet and avenues by 400 feet. He plans to park once and walk to all of the customers. Assume that he must return to his car after each visit to pick up different samples. At what location should he park in order to minimize the total distance traveled to the clients? (Hint: In designing your grid, be sure to account for the fact that avenues are twice as far apart as streets.)
18. In problem 17, suppose that Mr. Bender can park only in lots located at (1) 40th Street and 8th Avenue, (2) 46th Street and 6th Avenue, and (3) 33rd Street and 5th Avenue. Where should he plan to park?
19. An industrial park consists of 16 buildings. The corporations in the park are sharing the cost of construction and maintenance for a new first-aid center. Because of the park's layout, distances between buildings are most closely approximated by a rectilinear distance measure. Weights for the buildings are determined based on the frequency of accidents. Find the optimal location of the first-aid center to minimize the weighted sum of the rectilinear distances to the 16 buildings.

Building	a_i	b_i	w_i	Building	a_i	b_i	w_i
1	0	0	9	9	14	6	11
2	10	3	7	10	19	0	17
3	8	8	4	11	20	4	14
4	12	20	3	12	14	25	6
5	4	9	2	13	3	14	5
6	18	16	12	14	6	6	8
7	4	1	4	15	9	21	15
8	5	3	5	16	10	10	4

20. Draw contour lines for the problem of locating a machine shop described in problem 15. (Refer to Appendix 7–B.)
21. Draw contour lines for the location problem described in problem 17. (Refer to Appendix 7–B.)
22. Two facilities, located at (0, 0) and (0, 10), have respective weights 2 and 1. Draw contour lines for this problem. (Refer to Appendix 7–B.)
23. Solve problem 15 assuming a minimax rectilinear objective.
24. Solve problem 17 assuming a minimax rectilinear objective.
25. Solve problem 19 assuming a minimax rectilinear objective.

7.4 EUCLIDEAN DISTANCE PROBLEMS

Although the rectilinear distance measure is appropriate for many real problems, there are applications in which the appropriate measure of distance is the straight-line measure. An example is locating power-generating facilities in order to minimize the total amount of electrical cable that must be laid to connect the plant to the customers. This section will consider the Euclidean problem and a variant of it known as the gravity problem, which has a far simpler solution.

The Gravity Problem

The **gravity problem** corresponds to the case of an objective equal to the square of the Euclidean distance. Hence, the objective is to find values of (x, y) to minimize

$$f(x,y) = \sum_{i=1}^{n} w_i\left[\left(x-a_i\right)^2 + \left(y-b_i\right)^2\right].$$

This objective is appropriate when the cost of locating new facilities increases as a function of the square of the distance of the new facility to the existing facilities. Although such an objective is not common, the solution to this problem is straightforward and often has been used as an approximation to the more common straight-line distance problem.

The optimal values of (x, y) are easily determined by differentiation. The partial derivatives of the objective function with respect to x and y are

$$\frac{\partial f(x,y)}{\partial x} = 2\sum_{i=1}^{n} w_i(x-a_i),$$
$$\frac{\partial f(x,y)}{\partial y} = 2\sum_{i=1}^{n} w_i(y-b_i).$$

Setting these partial derivatives equal to zero and solving for x and y gives the optimal solution

$$x^* = \frac{\sum_{i=1}^{n} w_i a_i}{\sum_{i=1}^{n} w_i}$$
$$y^* = \frac{\sum_{i=1}^{n} w_i b_i}{\sum_{i=1}^{n} w_i}.$$

The term *gravity problem* arises for the following reason. Suppose that one places a map of the area in which the facility is to be located on a heavy piece of cardboard. Weights proportional to the numbers w_i are placed at the locations of the existing facilities. The gravity solution is the point on the map at everything is in balance (Keefer, 1934). Although one could certainly solve the gravity problem this way, it is so easy to find (x^*, y^*) using the given formulas that there seems little reason to employ the physical model.

EXAMPLE 7.5

We will find the solution to the problem of locating the audiovisual center for the University of the Far West assuming a squared Euclidean distance location measure. Substituting the values of the weights and the building locations into the given formulas, we obtain

$$x^* = 1{,}555/204 = 7.6,$$
$$y^* = 2{,}198/204 = 10.8.$$

which is somewhat different from the rectilinear solution (6, 12).

The Straight-Line Distance Problem

The **straight-line distance measure** arises much more frequently than does the squared-distance measure. The objective in this case is to find (x, y) to minimize

$$f(x,y)=\sum_{i=1}^{n} w_i\sqrt{(x-a_i)^2+(y-b_i)^2}.$$

Unfortunately, it is not as easy to find the optimal solution mathematically when using a Euclidean distance measure as it is when using squared Euclidean distance. As with the gravity problem, there is also a physical model that one could construct to find the optimal solution. In this case one places a map of the area on a tabletop. Holes are punched in the tabletop at the locations of existing facilities, and weights are suspended on strings through the holes. The size of the weights should be proportional to the relative weights of the locations. The strings are attached to a ring. If there is no friction, the ring will come to rest at the location of the optimal solution (see Figure 7–11). Although this method can be used to find a solution, it does have drawbacks. In particular, the friction between the string and the table must be negligible to be sure that the ring comes to rest in the correct position.

Determining the optimal solution mathematically is more difficult for Euclidean distance than for either rectilinear or squared Euclidean distance. There are no known simple algebraic solutions; all existing methods require an iterative procedure. The procedure below (from Francis et al., 1992) will yield an optimal solution as long as the location of the new facility does not overlap with the location of an existing facility.

Define

$$g_i(x,y)=\frac{w_i}{\sqrt{(x-a_i)^2+(y-b_i)^2}}.$$

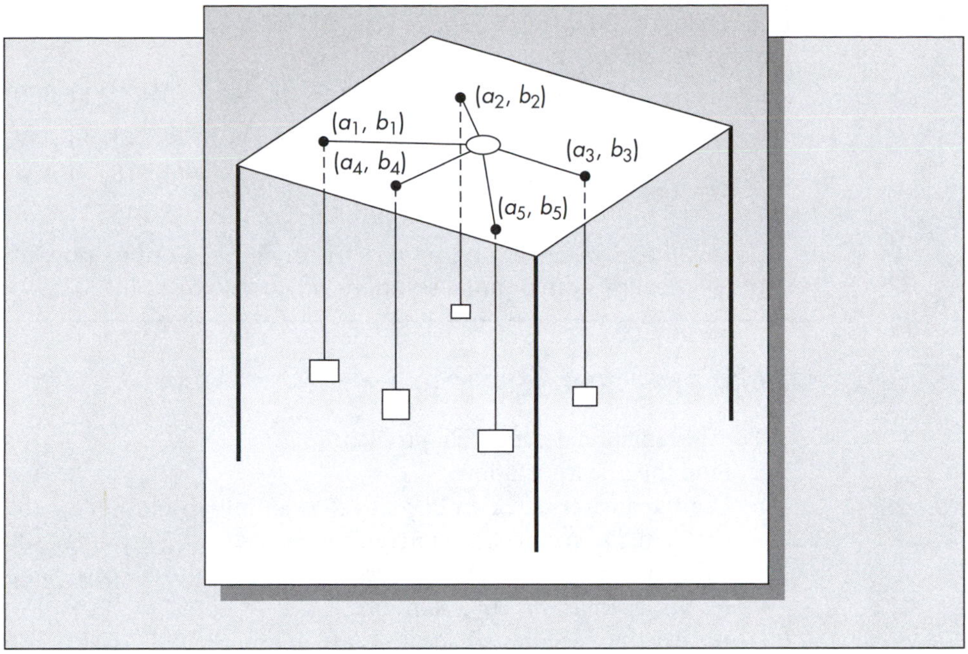

Figure 7–11 Solution of the Euclidean distance problem using a physical model

Let

$$x = \frac{\sum_{i=1}^{n} a_i g_i(x, y)}{\sum_{i=1}^{n} g_i(x, y)},$$

$$y = \frac{\sum_{i=1}^{n} b_i g_i(x, y)}{\sum_{i=1}^{n} g_i(x, y)}.$$

The procedure is as follows. Begin the process with an initial solution (x_0, y_0). The gravity solution is generally recommended to get the method started. Compute the values of $g_i(x,y)$ at this solution and determine new values of x and y from the given formulas. Then, recompute $g_i(x,y)$ using the new values of x and y, giving rise to yet another pair of (x, y) values. Continue to iterate in this fashion until the values of the coordinates converge. This procedure will give an optimal solution as long as the values of (x, y) at each iteration do not correspond to some existing location. The problem is that if we substitute $x = a_i$ and $y = b_i$, $g_i(x,y)$ is undefined, because the denominator is zero. It is unlikely that this will happen, but if it does, a modification of this method is required (see Francis et al., 1992).

EXAMPLE 7.6

We will solve the university's location problem assuming a Euclidean distance measure. To get the procedure started we use the solution to the gravity problem, which, to three decimals, is $(x_0, y_0) = (7.622, 10.775)$. The sequence of (x, y) values obtained from iterating the given equations for x and y is

$$(x_1, y_1) = (7.895, 11.432), (x_5, y_5) = (8.429, 11.880),$$
$$(x_2, y_2) = (8.102, 11.715), (x_6, y_6) = (8.481, 11.889),$$
$$(x_3, y_3) = (8.251, 11.822), (x_7, y_7) = (8.518, 11.894),$$
$$(x_4, y_4) = (8.356, 11.863), (x_8, y_8) = (8.545, 11.899).$$

Clearly, these values are beginning to converge. Continuing with the iterations one eventually reaches the optimal solution, which is (8.621, 11.907).

PROBLEMS FOR SECTION 7.4

26. For the situation described in problem 15,
    a. Find the gravity solution.
    b. Use the answer obtained in part (a) as an initial solution for the Euclidean distance problem. Determine (x_i, y_i) iteratively for $1 \le i \le 5$ if you are solving the problem by hand, or for $1 \le i \le 20$ if you are solving with the aid of a computer. Based on your results, estimate the final solution.
27. Three existing facilities are located at (0, 0), (5, 5), and (10, 10). The weights applied to these facilities are 1, 2, and 3, respectively. Find the location of a new facility that minimizes the weighted Euclidean distance to the existing facilities.
28. A telecommunications system is going to be installed at a company site. The switching center that governs the system must be located in one of the five buildings to be served. The goal is to minimize the amount of underground cabling required. In which building should the switching center be located? Assume a straight-line distance measure.

Building	Location
1	(10, 10)
2	(0, 4)
3	(6, 15)
4	(8, 20)
5	(15, 0)

7.5 OTHER LOCATION MODELS

The particular models discussed in Sections 7.3 and 7.4 have been referred to as "planar location models" (Francis et al., 1992). Seven assumptions are associated with these models of location problems.

1. A plane is an adequate approximation of a sphere.
2. Any point on the plane is a valid location for a facility.
3. Facilities may be idealized as points.
4. Distances between facilities are adequately represented by planar distance.
5. Travel costs are proportional to distance.
6. Fixed costs are ignored.
7. Issues of distribution can be ignored.

Unless distances separating facilities are extremely large, the first assumption should be a reasonable approximation of reality. The second assumption can be very restrictive. In some circumstances, only a small number of feasible locations exist, in which case the simplest approach is to evaluate the cost of locating the new facility at each of these locations and choose the one with the lowest cost. When the number of feasible alternatives is large, one could construct contour lines and consider only those locations along a given contour line with an acceptably low cost.

Depending upon the nature of the problem, assumption three may or may not be problematic. For example, if one uses location models to solve factory layout problems, then the size of the facilities becomes an issue. However, for a problem such as locating warehouses nationwide, idealizing facilities as points is reasonable. Assumption four requires that one use a particular measure of distance to compare various configurations. However, one distance measure may not be sufficient to explain all facility interactions. For example, rectilinear distances are typically used in problems in which facilities are to be located in cities. However, closed roads, rivers, or unusual street patterns could make this assumption inaccurate.

The final three assumptions deal with issues that typically arise in distribution problems. Transportation costs depend on the terrain. It is more expensive to transport goods over mountains than on flat interstate highways, for example. Fixed costs of transporting goods can be substantial depending on the mode of transportation. The fixed cost of transporting goods by air, for example, is very high.

More complex location models exist that deal with several of these shortcomings. We briefly review these in the following section.

Locating Multiple Facilities

Section 7.4 treated the problem of locating a single facility among n existing facilities. However, applications exist in which the goal is to locate multiple facilities among n existing facilities. For example, a nationwide consumer products producer might be considering where to locate five new regional warehouses.

In some circumstances, multifacility location problems can be solved as a sequence of single location problems. That is, the optimal locations of the new facilities can be determined one at a time. However, when there is any interaction among the new facilities, this approach will not work.

This section will show how linear programming (see Supplement 1) can be used to solve the multiple-facility rectilinear distance location problem. Assume that existing facilities have locations at points $(a_1, b_1), \ldots, (a_n, b_n)$ as in Section 7.4. Suppose that m new facilities are to be located at $(x_1, y_1), \ldots, (x_m, y_m)$. Then the objective function to be minimized may be written in the form

$$\text{minimize} \quad f_1(\mathbf{x}) + f_2(\mathbf{y}),$$

where

$$f_1(\mathbf{x}) = \sum_{1 \le j < k \le m} v_{jk} \, |x_j - x_k| \; + \; \sum_{j=1}^{m} \sum_{i=1}^{n} w_{ij} \, |x_j - a_i|$$

and

$$f_2(\mathbf{y}) = \sum_{1 \le j < k \le m} v_{jk} \, |y_j - y_k| \quad \sum_{j=1}^{m} \sum_{i=1}^{n} w_{ij} \, |y_j - b_i|.$$

The interaction of new facilities j and k is measured by v_{jk}. The presence of these terms prevents us from solving the multifacility problem as a sequence of one-facility problems. However, as with the single-facility problem, the optimal x and y locations may be determined independently. We present the linear programming formulation for finding the optimal x coordinates. The optimal y coordinates may be found in the same way.

The trick in transforming the problem of finding $\mathbf{x}$ to minimize $f_1(\mathbf{x})$ is to eliminate the absolute value function from the objective, as this is not a strictly linear function. The means for doing so is a standard trick in linear programming. (A similar technique was applied in the linear programming formulation of the aggregate planning problem in Section 3.5.)

For any constants a and b write $|a - b| = c + d$, but require that $cd = 0$ (either c or d or both must be zero). If $a > b$, then $|a - b| = c$, and if $a < b$, then $|a - b| = d$. Hence, we may think of c as the positive part of $a - b$ and d as the negative part of $a - b$. Substituting $|x_j - x_k| = c_{jk} + d_{jk}$ and $|x_j - a_i| = e_{ij} + f_{ij}$, we obtain the linear programming formulation of the problem of determining $\mathbf{x}$:

$$\text{minimize} \quad \sum_{1 \le j < k \le m} v_{jk}(c_{jk} + d_{jk}) + \sum_{j=1}^{m} \sum_{i=1}^{n} w_{ij}(e_{ij} + f_{ij})$$

subject to

$$\begin{array}{lll} x_j - x_k - c_{jk} + d_{jk} = 0, & 1 \le j < k \le n; & \\ x_j - a_i - e_{ij} + f_{ij} = 0, & 1 \le i \le n, & 1 \le j \le n; \\ c_{jk} \ge 0, \quad d_{jk} \ge 0, & 1 \le j < k \le n; & \\ e_{ij} \ge 0, \quad f_{ij} \ge 0, & 1 \le i \le n, & 1 \le j \le n; \\ x_j \text{ unrestricted in sign.} & & \end{array}$$

We do not need to explicitly include the constraints $c_{jk}d_{jk} = 0$ and $e_{ij}f_{ij} = 0$. One can show that at the minimum cost solution these relationships always hold (which is certainly fortunate as these are not linear relationships). One additional substitution is common prior to solving the problem. Since some linear programming codes require that all variables be nonnegative, we can substitute $x_j = x_j^+ - x_j^-$, where $x_j^+ \ge 0$ and $x_j^- \ge 0$. Commercial linear programming codes are based on the simplex method, which is extremely efficient. One can solve realistically sized problems easily even on a personal computer.

Multifacility gravity problems require the solution of a system of linear equations, so that gravity problems involving large numbers of facilities are easily solved as well. Multifacility Euclidean problems are solved by utilizing a multidimensional version of the iterative solution method described in Section 7.4. (We will not review these methods here; the interested reader should refer to Francis et al., 1992.)

Further Extensions

Facilities Having Positive Areas

All previous models assumed that facilities are approximated by points in the plane. When the area of the facilities is small compared to the area covered by the available locations, this assumption is reasonable. However, in certain applications the areas of the facilities cannot be ignored. For example, when finding locations for machines in a job shop, the machines must be far enough apart for them to be able to operate efficiently. Good algorithms exist for this problem (see, e.g., Tompkins, White, Bozer, and Tanchoco, 2010).

Location–Allocation Problems

Often the decision of where to locate new facilities must be accompanied by the decision of which of the existing locations will be served by each new facility. For example, a firm may be considering where to locate several regional warehouses. In addition to determining the location and the number of these new warehouses, the firm also must decide which of the retail outlets will be serviced by which warehouses.

Location–allocation problems are difficult to solve owing to the large number of decision variables. The mathematical programming formulation of the problem, assuming that a rectangular distance measure is used, is

$$\text{minimize} \quad \sum_{j=1}^{m} \sum_{i=1}^{n} z_{ij} w_{ij} \left[\left| x_j - a_i \right| + \left| y_j - b_i \right| \right] + g(m)$$

subject to

$$\sum_{j=1}^{m} z_{ij} = 1 \qquad \text{for } 1 \le i \le n,$$

where

w_{ij} = cost per unit time per unit distance if the existing facility i is serviced by new facility j;

$z_i = \begin{cases} 1 & \text{if existing facility } i \text{ is serviced by new facility } j, \\ 0 & \text{otherwise;} \end{cases}$

m = total number of new facilities, $1 \le m \le n$;

(x_j, y_j) = coordinates of new facility j, $1 \le j \le m$;

(a_i, b_i) = coordinates of existing facility i, $1 \le i \le n$;

$g(m)$ = cost per unit time of providing m new facilities.

This problem formulation has decision variables: m, the number of new facilities; z_{ij}, the specification of which of the existing locations i will be serviced by facility j; and (x_j, y_j), the location of the new facilities. The optimization is difficult due to the presence of the zero-one variables z_{ij} and the inclusion of m as a decision variable. The problem can be solved by considering successive values of $m = 1, 2, \dots$, and enumerating all combinations of z_{ij} for each value of m. Given a fixed m and set of z_{ij} values, the solution can be obtained using the methods for locating multiple facilities discussed earlier in this section. However, the number of different z_{ij} values grows quickly as a function of m, so only moderately sized problems can be solved in this fashion. Integer programming algorithms can also be used effectively on this problem.

Discrete Location Problems

The models considered in this chapter for location of new facilities assumed that the new facilities could be located anywhere in the plane. This is not the case for most applications.

Contour lines assist with evaluating alternative locations but cannot be constructed for problems in which one must locate multiple facilities. An alternative approach is to restrict a priori the possible locations to some discrete set of possibilities. When there is only a single facility and the number of possible locations is small, the easiest approach is to evaluate the cost of each location and pick the smallest.

When there are multiple facilities, the assignment model discussed in Section S1.10 can be used to determine the optimal locations of the new facilities. In certain types of warehouse-layout problems, new facilities can take up more than one potential site. For example, suppose that we must determine which locations in a warehouse should store k items. Suppose that the appropriate storage area in the warehouse is composed of n grid squares and each item stored takes up more than a single square. Each square would be numbered, and the storage location of an item specified by the numbers of the grid squares covered by the item. The resulting model (discussed in Francis et al., 1992) is a generalization of the simple assignment model appearing in Section S1.10. We will not present the details of the model here.

Network Location Models

Planar location models assume that the goal is to locate one or more new facilities in order to minimize some function of the distance separating the new and the existing facilities. A rectilinear, Euclidean, or other distance measure is generally assumed. In certain applications, the distances should be measured over an existing network and are not accurately approximated by standard measures. Overland transport must follow road networks; water transport must follow shipping lanes and sea routes; and air transport is confined to predetermined air corridors. In other applications, the network may correspond to a network of power cables or telephone wires. In many of these applications the new facility or facilities must be placed on or very near to a location on the network, and distances can be measured only in terms of the network. Network location models are beyond the scope of our coverage. (For information on network location models, see Francis et al., 1983 and Tansel, Francis, and Lowe, 1983.)

International Issues

The problem of locating facilities is a part of the larger issue of global supply chain management. No longer is the issue isolated to industry giants; globalization now plays a greater role than ever before for many businesses, and its importance will continue to grow. Lower wage rates were traditionally the primary reason for firms based in developed countries to locate plants in less developed countries. Virtually all the large semiconductor manufacturers have fabrication facilities overseas, typically in places like Malaysia and the Philippines. As outlined in Chapter 3, Morris Cohen and Hau Lee (1989) provide a good overview of some of the issues that one must take into account when locating facilities in other countries.

Only recently have we begun to understand the complexity of the problem of locating new facilities and their effect on global supply chain operations. Well-constructed and well-thought-out mathematical models will continue to assist us in managing these increasingly complex networks. However, many of the issues alluded to in this section are difficult to quantify, thus making good judgment crucial.

PROBLEMS FOR SECTION 7.5

29. For each of the location problems described, discuss which of the seven assumptions listed in this section are likely to be violated.

a. Locating three new machines in a machine shop.
b. Locating an international network of telecommunications facilities.
c. Locating a hospital in a sparsely populated area.
d. Locating spare parts depots to support a field repair organization.

30. Consider problem 15 of this chapter. Suppose that two new machines, A and B, are to be located in the shop. Machine A has $\frac{1}{8}, \frac{1}{8}, \frac{1}{4}$, 1, and $\frac{1}{6}$ as the expected numbers of loads transported to the existing five machines, respectively, and machine B has $\frac{1}{4}, \frac{1}{6}$, 3, $\frac{1}{5}$, and $\frac{1}{2}$ as the expected numbers of loads transported, respectively. Furthermore, suppose that there are two loads per hour on average transported between the new machines. Assume a rectilinear distance measure.
a. Formulate the problem of determining the optimal locations of the new machines as a linear program.
b. Use Excel Solver to solve the problem formulated in part (a).

31. Consider the University of the Far West described in Example 7.3. Suppose that the university administration has decided that two audiovisual centers are needed. Each center would have different facilities. The anticipated numbers of faculty members using each center are

School	Faculty Members Using Center A	Faculty Members Using Center B
Business	13	18
Education	40	23
Engineering	24	17
Humanities	20	23
Law	30	9
Science	16	21

Furthermore, there will be a total of 16 staff persons at centers A and B. They will need to interact frequently.
a. Formulate the problem of determining the optimal locations of the two audiovisual centers as a linear program. Assume that rectilinear distances are used throughout.
b. Use Excel Solver to solve the problem formulated in part (a).

32. A real estate firm wishes to open four new offices in the Boston area. There are six potential sites available. Based on the number of employees in each office and the location of the properties that each employee will manage, the firm estimated the total travel time in hours per day for each office and each location. Find the optimal assignment of offices to sites to minimize employee travel time.

		Offices			
		A	B	C	D
Sites	1	10	3	3	8
	2	13	5	2	6
	3	12	9	9	4
	4	14	2	7	7
	5	17	7	4	3
	6	12	8	5	5

33. A large supermarket chain in the Southeast requires five additional warehouses in the Atlanta area. It has identified five sites for these warehouses. The annual transportation costs (in $000) for each warehouse at each site are given in the following table. Find the assignment of warehouses to sites to minimize the total annual transportation costs.

		Warehouses				
		A	B	C	D	E
Sites	**1**	41	47	38	46	50
	2	39	37	42	36	45
	3	43	46	45	42	46
	4	51	54	47	58	56
	5	44	40	42	41	45

EXTRA PROBLEMS FOR SECTIONS 7.3–7.5

34. Felicity Green, an independent TV repairer, is considering purchasing a home in Ames, Iowa, that she will use as a base of operations for her repair business. Felicity's primary sources of business are 10 industrial accounts located throughout the Ames area. She has overlaid a grid on a map of the city and determined the following locations for these clients as well as the expected number of calls per month she receives.

Client	Grid Location	Expected Calls per Month
1	(5, 8)	2
2	(10, 3)	1
3	(14, 14)	1
4	(2, 2)	3
5	(1, 17)	1
6	(18, 25)	$\frac{1}{2}$
7	(14, 3)	$\frac{1}{4}$
8	(25, 4)	4
9	(35, 1)	3
10	(16, 21)	$\frac{1}{6}$

Find the optimal location of her house, assuming

a. a weighted rectilinear distance measure.
b. a squared Euclidean distance measure.
c. the goal is to minimize the maximum rectilinear distance to any client.

35. An electronics firm located near Phoenix, Arizona, is considering where to locate a new phone switch that will link five buildings. The buildings are located at (0, 0), (2, 6), (10, 2), (3, 9), and (0, 4). The objective is to locate the switch to minimize the cabling required to those five buildings.
   a. Determine the gravity solution.
   b. Determine the optimal location assuming a straight-line distance measure. (If you are solving this problem by hand, iterate the appropriate equations at least five times and estimate the optimal solution.)
36. Consider three locations at (0, 0), (0, 6), and (3, 3) with equal weights. Using the methods described in Appendix 7–B, find contour lines for the rectilinear location problem.
37. A company is considering where to locate its cafeteria to service six buildings. The locations of the buildings and the fraction of the company's employees working at these locations are

Building	a_i	b_i	Fraction of Workforce
A	2	6	$\frac{1}{12}$
B	1	0	$\frac{1}{12}$
C	3	3	$\frac{1}{6}$
D	5	9	$\frac{1}{4}$
E	4	2	$\frac{1}{4}$
F	10	7	$\frac{1}{6}$

   a. Find the optimal location of the cafeteria to minimize the weighted rectilinear distance to all the buildings.
   b. Find the optimal location of the cafeteria to minimize the maximum rectilinear distance to all the buildings.
   c. Find the gravity solution.
   d. Suppose that the cafeteria must be located in one of the buildings. In which building should it be located if the goal is to minimize weighted rectilinear distance?
   e. Solve part (d) assuming weighted Euclidean distance.
38. Design a spreadsheet to compute the total rectilinear distance from a set of up to 10 existing locations to any other location. Assume that existing locations are placed in Columns A and B and the new location in cells D1 and E1. Initialize column A with the value of cell D1 and column B with the value of cell E1, so that the total distance will be computed correctly when there are fewer than 10 locations.
   a. Suppose that existing facilities are located at (0, 0), (5, 15), (110, 120), (35, 25), (80, 10), (75, 20), (8, 38), (50, 65), (22, 95), and (44, 70), and the new facility is to be located at (50, 50). Determine the total rectilinear distance of the new facility to the existing facilities.

b. Suppose the new facility in part (a) can be located only at $x = 0, 5, 10, \ldots, 100$ and $y = 0, 10, 20, \ldots, 100$. By systematically varying the x and y coordinates, find the optimal location of the new facility.

39. Solve problem 38, assuming a Euclidean distance measure.
40. Solve problem 19 using a spreadsheet. To do so, enter the building numbers in column A, the x coordinates in column B, and the associated weights in column C. Sort columns A, B, and C in ascending order by using column B as the primary sort key. Now accumulate the weights (column C) in column D using the sum function. Divide the total accumulated weight by 2 and visually identify the optimal x coordinate value. It will be where the cumulative weight first exceeds half the total cumulative weight. Repeat the process for the y coordinates.
41. Design a spreadsheet to compute the optimal solution to the gravity problem. Allow for up to 20 locations. Let column A be the location number, column B the x coordinates $(a_1, \ldots, a_n)$ of existing locations, and column C the y coordinates $(b_1, \ldots, b_n)$ of existing locations. Store the optimal solution in cell D1. Find the gravity solution to problem 19 using your spreadsheet.
42. Extend the results of problem 41 to find the optimal location of a new facility among a set of existing facilities assuming a straight-line Euclidean distance measure. Let column E correspond to $g_i(x, y)$, where the initial (x, y) values appear in F1 and F2. In locations G1 and G2, store

$$x = \frac{\sum_{i=1}^{n} a_i g_i(x, y)}{\sum_{i=1}^{n} g_i(x, y)}.$$

$$y = \frac{\sum_{i=1}^{n} b_i g_i(x, y)}{\sum_{i=1}^{n} g_i(x, y)}.$$

Start with the gravity solution in cells F1 and F2. After calculation, replace the values in cells F1 and F2 with the values in cells G1 and G2. Continue in this manner until the solution converges. Using your spreadsheet,

a. solve problem 19.
b. solve problem 28.
c. solve problem 34 assuming a Euclidean distance measure.
d. solve problem 37 assuming a Euclidean distance measure.

7.6 THE TRANSPORTATION PROBLEM

The **transportation problem** is a mathematical model for optimally scheduling the flow of goods from production facilities to distribution centers. Assume that a fixed amount of product must be transported from a group of **sources** (plants) to a group of **sinks** (warehouses). The unit cost of transporting from each source to each sink is assumed to be known. The goal is to find the optimal flow paths and the amounts to be shipped on those paths to minimize the total cost of all shipments.

The transportation problem can be viewed as a prototype supply chain problem. Although most real-world problems involving the shipment of goods are more complex, the model provides an illustration of the issues and methods one would encounter in practice.

EXAMPLE 7.7

The Pear Tablet Corporation produces several types of tablet computers. Their most popular product is the 64 gigabytes (GB) tablet with a 10-inch display. Pear produces these tablets in three plants located in Sunnyvale, California; Dublin, Ireland; and Bangkok, Thailand. Periodically, shipments are made from these three production facilities to four distribution warehouses located in the United States in Amarillo, Texas; Teaneck, New Jersey; Chicago, Illinois; and Sioux Falls, South Dakota. Over the next month, it has been determined that these warehouses should receive the following proportions of the company's total production of the 64 GB tablets.

Warehouse	Percentage of Total Production
Amarillo	31
Teaneck	30
Chicago	18
Sioux Falls	21

The production quantities at the factories in the next month are expected to be (in thousands of units)

Plant	Anticipated Production (in 1,000s of units)
Sunnyvale	45
Dublin	120
Bangkok	95

Since the total production at the three plants is 260 units, the amounts shipped to the four warehouses will be (rounded to the nearest unit)

Warehouse	Total Shipment Quantity (1,000s)
Amarillo	80
Teaneck	78
Chicago	47
Sioux Falls	55

While the shipping cost may be lower between certain plants and distribution centers, Pear has established shipping routes between every plant and every warehouse. This is in case of unforeseen problems such as a forced shutdown at a plant, unanticipated swings in regional demands, or poor weather along some routes. The unit costs for shipping 1,000 units from each plant to each warehouse is given in the following table.

Shipping Costs per 1,000 Units in $					
		TO			
		Amarillo	Teaneck	Chicago	Sioux Falls
F	Sunnyvale	250	420	380	280
R	Dublin	1,280	990	1,440	1,520
O	Bangkok	1,550	1,420	1,660	1,730
M					

The goal is to determine a pattern of shipping that minimizes the total transportation cost from plants to warehouses. The network representation of Pear's distribution problem appears in Figure 7–12.

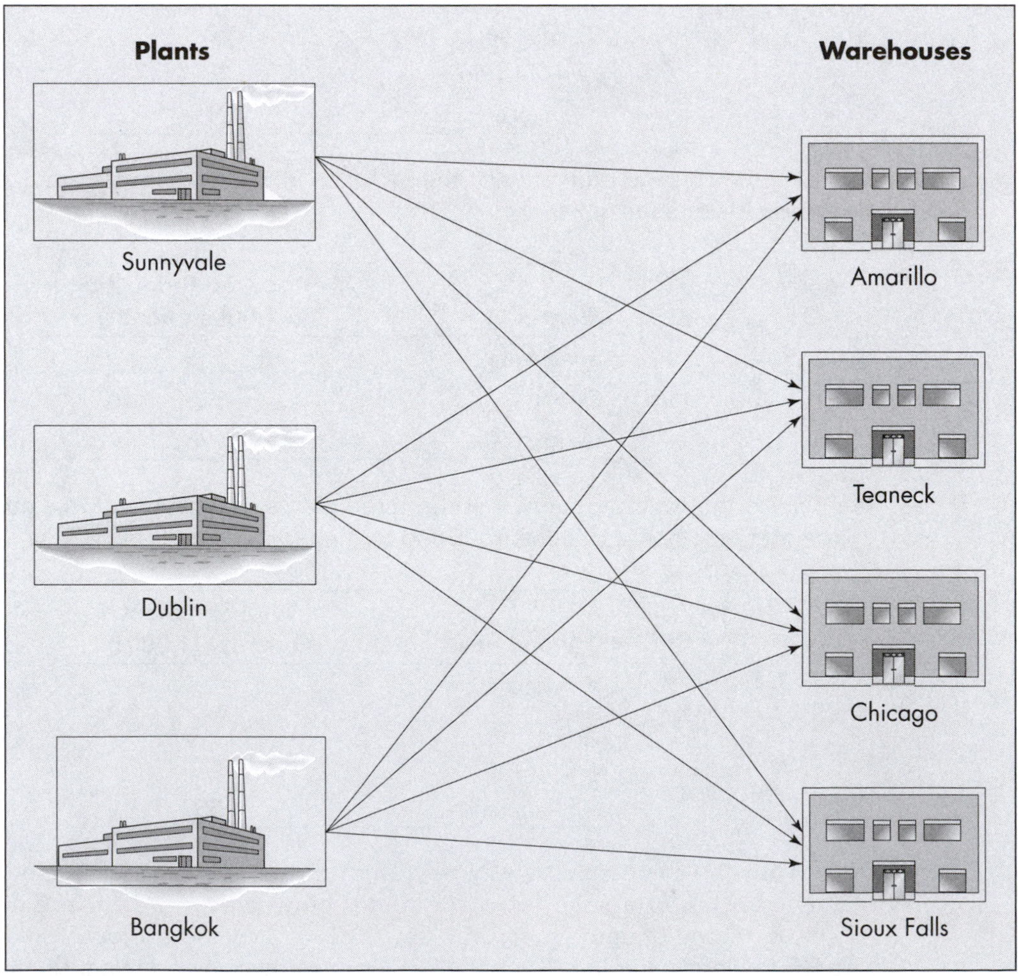

Figure 7–12 Pear Tablet transportation problem

Several heuristics for solving transportation problems have been proposed, such as a greedy heuristic that allocates capacity first to the cheapest options. However, it is unlikely that anyone with a real problem would use a heuristic, since optimal solutions can be found efficiently by linear programming. In fact, because of the special structure of the transportation problem, today's specialized codes can solve problems with millions of variables.

Let m be the number of sources and n the number of sinks. (In Example 7.7, $m = 3$ and $n = 4$.) Let

$$x_{ij} = \text{flow from source } i \text{ to sink } j \quad \text{for} \quad 1 \le i \le m \quad \text{and} \quad 1 \le j \le n$$

and define c_{ij} as the cost of shipping one unit from i to j. It follows that the total cost of making all shipments is

$$\sum_{i=j}^{m} \sum_{j=1}^{n} c_{ij} x_{ij}.$$

For the case of Pear Company, described in Example 7.7, the objective function is

$$250x_{11} + 420x_{12} + 380x_{13} + 280x_{14} + \ldots + 1{,}730x_{34}.$$

Since many routes are obviously not economical, it is likely that many of the decision variables will equal zero at the optimal solution.

The constraints are designed to ensure that the total amount shipped out of each source equals the amount available at that source, and the amount shipped into any sink equals the amount required at that sink. Since there are m sources and n sinks, there are a total of $m + n$ constraints (excluding nonnegativity constraints). Let a_i be the total amount to be shipped out of source i and b_j the total amount to be shipped into sink j. Then linear programming constraints may be written:

$$\sum_{j=1}^{n} x_{ij} = a_i \qquad \text{for } 1 \le i \le m$$

$$\sum_{i=1}^{m} x_{ij} = b_j \qquad \text{for } 1 \le j \le n$$

$$x_{ij} \ge 0 \qquad \text{for } 1 \le i \le m \text{ and } 1 \le j \le n.$$

For the Pear Tablet Company problem, we obtain the following seven constraints:

$$\begin{aligned}
x_{11} + x_{12} + x_{13} + x_{14} &= 45 && \text{(shipments out of Sunnyvale)}\\
x_{21} + x_{22} + x_{23} + x_{24} &= 120 && \text{(shipments out of Dublin)}\\
x_{31} + x_{32} + x_{33} + x_{34} &= 95 && \text{(shipments out of Bangkok)}\\
x_{11} + x_{21} + x_{31} &= 80 && \text{(shipments into Amarillo)}\\
x_{12} + x_{22} + x_{32} &= 78 && \text{(shipments into Teaneck)}\\
x_{13} + x_{23} + x_{33} &= 47 && \text{(shipments into Chicago)}\\
x_{14} + x_{24} + x_{34} &= 55 && \text{(shipments into Sioux Falls)}
\end{aligned}$$

and the required nonnegativity constraints are

$$x_{ij} \ge 0 \quad \text{for} \quad 1 \le i \le 3 \quad \text{and} \quad 1 \le j \le 4.$$

Solution of Pear's transportation problem using Excel Solver

	A	B	C	D	E	F	G	H	I	J	K	L	M	N	O	P
1			**Solution of Example 7.7 (Pear's Transportation Problem)**													
2																
3	Variables	x11	x12	x13	x14	x21	x22	x23	x24	x31	x32	x33	x34	Operator	Value	RHS
4																
5	Values	0	0	0	45	42	78	0	0	38	0	47	10			
6																
7	Objective Coeff	250	420	380	280	1280	990	1440	1520	1550	1420	1660	1730	Min	297800	
8	st															
9	Constraint 1	1	1	1	1									=	45	45
10	Constraint 2					1	1	1	1					=	120	120
11	Constraint 3									1	1	1	1	=	95	95
12	Constraint 4	1				1				1				=	80	80
13	Constraint 5		1				1				1			=	78	78
14	Constraint 6			1				1				1		=	47	47
15	Constraint 7				1				1				1	=	55	55
16																
17																
18		Notes:	Formula for Cell O9: =SUMPRODUCT (B9:M9,$B$5:$M$5). Copied to O10 to O15.													
19			Changing cells for Solver are $B$5:$M$5.													
20																

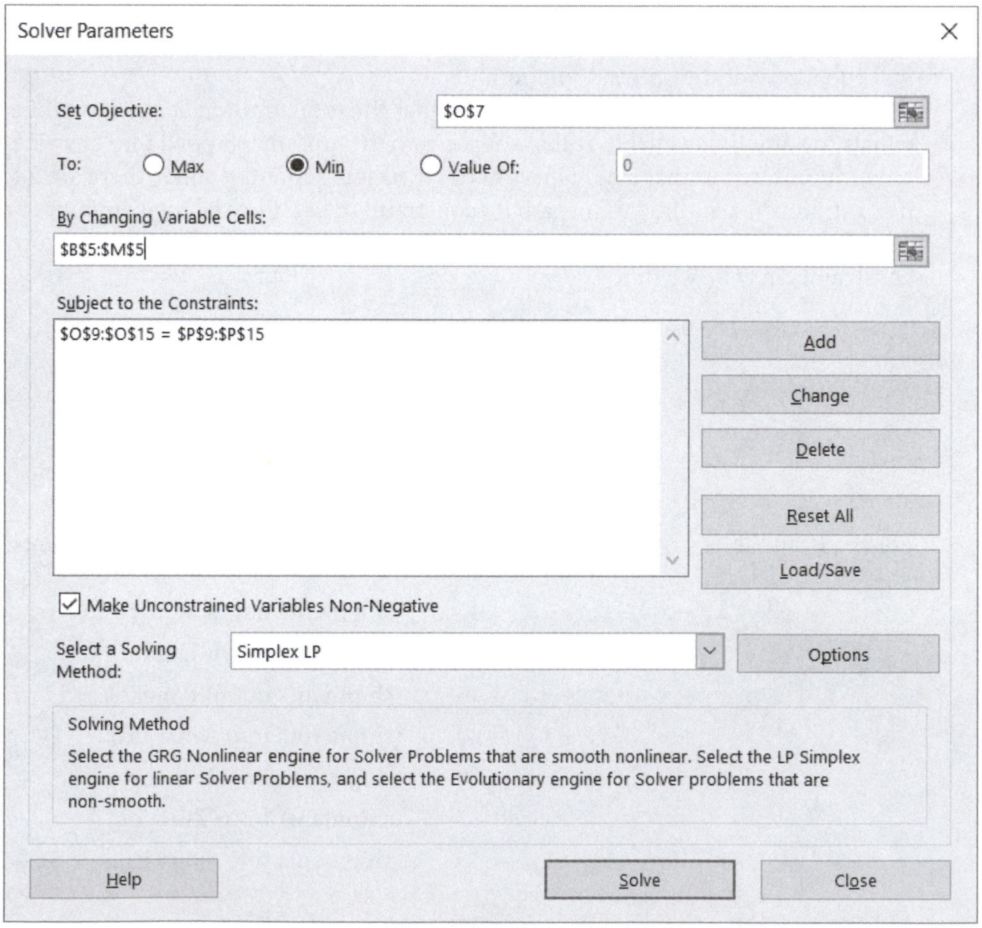

Figure 7–13 Solution of Pear's transportation problem using Excel Solver [The solver parameters screen will vary depending on the version of Excel used.]

The problem was entered in Excel Solver. The spreadsheet used and the solution appear in Figure 7–13. The solution obtained is

$$x_{14} = 45, \quad x_{21} = 42, \quad x_{22} = 78, \quad x_{31} = 38, \quad x_{33} = 47, \quad \text{and} \quad x_{34} = 10,$$

with all other values equaling zero. The total cost of this solution is $297,800.

7.7 GENERALIZATIONS OF THE TRANSPORTATION PROBLEM

The Pear Company example is the simplest type of transportation problem. Every link from source to sink is feasible, and the total amount available from the sources exactly equals the total demand at the sinks. Several of these requirements can be relaxed without making the problem significantly more complicated.

Infeasible Routes

Suppose in the example that the firm has decided to eliminate the routes from Dublin to Chicago and from Bangkok to Sioux Falls. This would be accounted for by placing very high costs on these arcs in the network. Traditionally, uppercase M has been used to signify a very high cost. In practice, of course, one would have to assign a number to these locations. As long as that number is much larger than the other costs, an optimal solution will never assign flow to these routes. For the example, suppose we assign costs of $1,000,000 to each of these routes and re-solve the problem. The reader can check that one now obtains the following solution:

$$x_{14} = 45, \quad x_{21} = 32, \quad x_{22} = 78, \quad x_{31} = 48, \quad \text{and} \quad x_{33} = 47,$$

with all other values equaling zero. The cost of the new solution is $298,400, only slightly larger than the cost obtained when all the routes were feasible.

Unbalanced Problems

An unbalanced transportation problem is one in which the total amount shipped from the sources is not equal to the total amount required at the sinks. This can arise if the demand exceeds the supply or vice versa. There are two ways of handling unbalanced problems. One is to add either a dummy row or a dummy column to absorb the excess supply or demand. A second method for solving unbalanced problems is to alter the appropriate set of constraints to either $\leq$ or $\geq$ form. Both methods will be illustrated.

Suppose in Example 7.7 that the demand for the tablets was higher than anticipated. Suppose that the respective requirements at the four warehouses are now Amarillo, 90; Teaneck, 78; Chicago, 55; and Sioux Falls, 55. This means that the total demand is 278 and the total supply is 260. To turn this into a balanced problem, we add an additional fictitious factory to account for the 18-unit shortfall. This can be labeled as a dummy row in the transportation tableau and all entries for that row assigned an arbitrarily large unit cost. Note that when supply exceeds demand and one adds a dummy column, the costs in the dummy column do *not* have to be very large numbers, but they do all have to be the same. (In fact, one can assign zero to all costs in the dummy column.) In the example, we assigned a cost of 10^6 to each cell in the dummy row. The resulting Excel spreadsheet and Solver solution (shown in the row labeled "Values") appear in Figure 7–14. The optimal solution calls for assigning the shortfall to two warehouses: 8 units to Chicago and 10 units to Sioux Falls.

Solution of Example 7.7 with excess demand and dummy row

				Solution of Example 7.7 with Excess Demand and a Dummy Row														
Variables	x11	x12	x13	x14	x21	x22	x23	x24	x31	x32	x33	x34	x41	x42	x43	x44	Oper	Value	RHS
Values	0	0	0	45	42	78	0	0	48	0	47	0	0	0	8	10			
Obj Func	250	420	380	280	1280	990	1440	1520	1550	1420	1660	1730	1.E+06	1.E+06	1.E+06	1.E+06	Min	2E+07	
st																			
Constraint 1	1	1	1	1													=	45	60
Constraint 2					1	1	1	1									=	120	130
Constraint 3									1	1	1	1					=	95	95
Constraint 4													1	1	1	1	=	18	18
Constraint 5	1				1				1				1				=	90	90
Constraint 6		1				1				1				1			=	78	78
Constraint 7			1				1				1				1		=	55	55
Constraint 8				1				1				1				1	=	55	55

Figure 7–14 Solution of Example 7.7 with excess demand and dummy row

Unbalanced transportation problems also can be formulated as linear programs by using inequality constraints. In the previous example, where there is excess demand, one would use equality for the first three constraints, to be sure that all the supply is shipped, and less than or equal to constraints for the last four, with the slack accounting for the shortfall. The reader should check that one obtains the same solution by doing so as was obtained by adding a dummy row. This method has the advantage of giving an accurate value for the objective function. (In the case where the supply exceeds the demand, the principle is the same, but the details differ. The first three supply constraints are converted to *greater than or equal to* form, while the last four demand constraints are still equality constraints. The slack in the first three constraints corresponds to the excess supply.)

7.8 MORE GENERAL NETWORK FORMULATIONS

The transportation problem is a special type of network where all nodes are either **supply nodes** (also called sources) or **demand nodes** (also called sinks). Linear programming also can be used to solve more complex network distribution problems. One example is the **transshipment problem**. In this case, one or more of the nodes in the network are transshipment points rather than supply or demand points. Note that a transshipment node can be either a supply or a demand node (but no node is both a supply and a demand node).

For general network flow problems, we use the following balance of flow rules.

If	Apply the Following Rule at Each Node
1. Total supply > total demand	Inflow – outflow ≥ supply or demand
2. Total supply < total demand	Inflow – outflow ≤ supply or demand
3. Total supply = total demand	Inflow – outflow = supply or demand

The decision variables are defined in the same way as with the simple transportation problem. That is, x_{ij} represents the total flow from node i to node j. For general network flow problems, we represent the supply as a negative number attached to that node and the demand as a positive number attached to that node. This convention along with the flow rules will result in the correct balance-of-flow equations.

EXAMPLE 7.8

Consider the example of Pear Tablets. The company has decided to place a warehouse in Sacramento to be used as a transshipment node and has expanded the Chicago facility to also allow for transshipments. Suppose that in addition to being transshipment nodes, both Chicago and Sacramento are also demand nodes. The new network is pictured in Figure 7–15. Note that several of the old routes have been eliminated in the new configuration.

We define a decision variable for each arc in the network. In this case, there are a total of 10 decision variables. The objective function is to minimize

$$250x_{16} + 76x_{14} + 380x_{15} + 1{,}440x_{25} + 1{,}660x_{35} + 110x_{46} + 95x_{48} + 180x_{56} + 120x_{57} + 195x_{58}.$$

The total supply available is still 260 units, but the demand is 285 units (due to the additional 25 units demanded at Sacramento). Hence, this corresponds to case 2 of the flow rules in which total demand exceeds total supply. Applying rule 2 to each node gives the following eight constraints for this problem:

Node 1: $-x_{14} - x_{15} - x_{16} \le -45,$

Node 2: $-x_{25} \le -120,$

Node 3: $-x_{35} \le -95,$

Node 4: $x_{14} - x_{46} - x_{48} \le 25,$

Node 5: $x_{16} + x_{46} + x_{56} - x_{56} - x_{57} - x_{58} \le 47,$

Node 6: $x_{16} + x_{46} + x_{56} \le 80,$

Node 7: $x_{57} \le 78,$

Node 8: $x_{48} + x_{58} \le 55.$

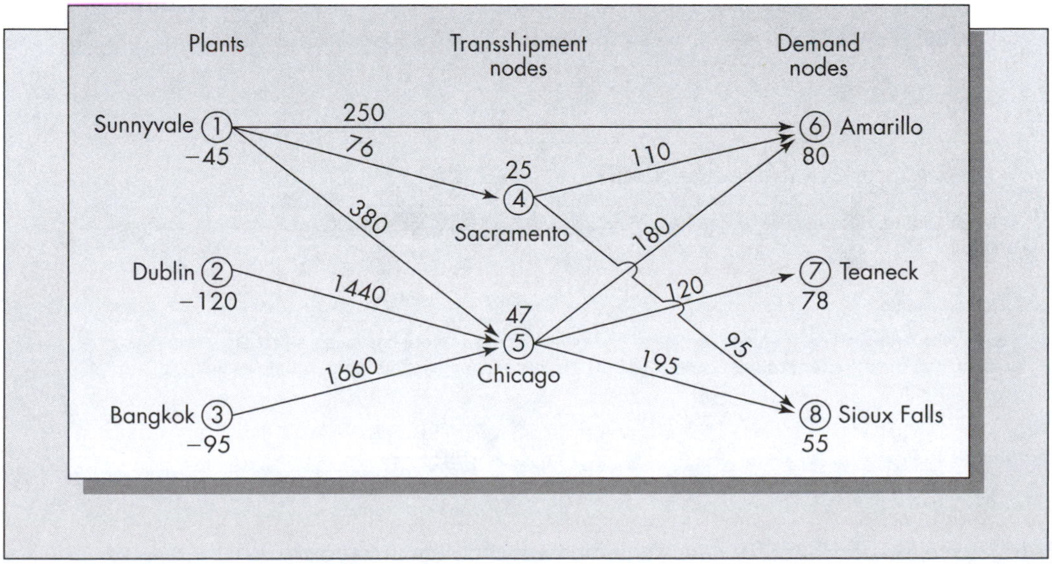

Figure 7–15 Pear Tablets problem with transshipment nodes

	A	B	C	D	E	F	G	H	I	J	K	L	M	N
1			**Excel Spreadsheet for Pear Transshipment Problem in Example 7.8**											
2														
3	Variables	×14	×15	×16	×25	×35	×46	×48	×56	×57	×58	Operator	Value	RHS
4														
5	Values	45	0	0	120	95	0	20	80	78	10			
6														
7	Obj Funct	76	380	250	1440	1660	110	95	180	120	195	Min	361530	
8	st													
9	Node 1	−1	−1	−1								<=	−45	−45
10	Node 2				−1							<=	−120	−120
11	Node 3					−1						<=	−95	−95
12	Node 4	1					−1	−1				<=	25	25
13	Node 5		1		1	1			−1	−1	−1	<=	47	47
14	Node 6			1			1		1			<=	80	80
15	Node 7									1		<=	78	78
16	Node 8							1			1	<=	30	55

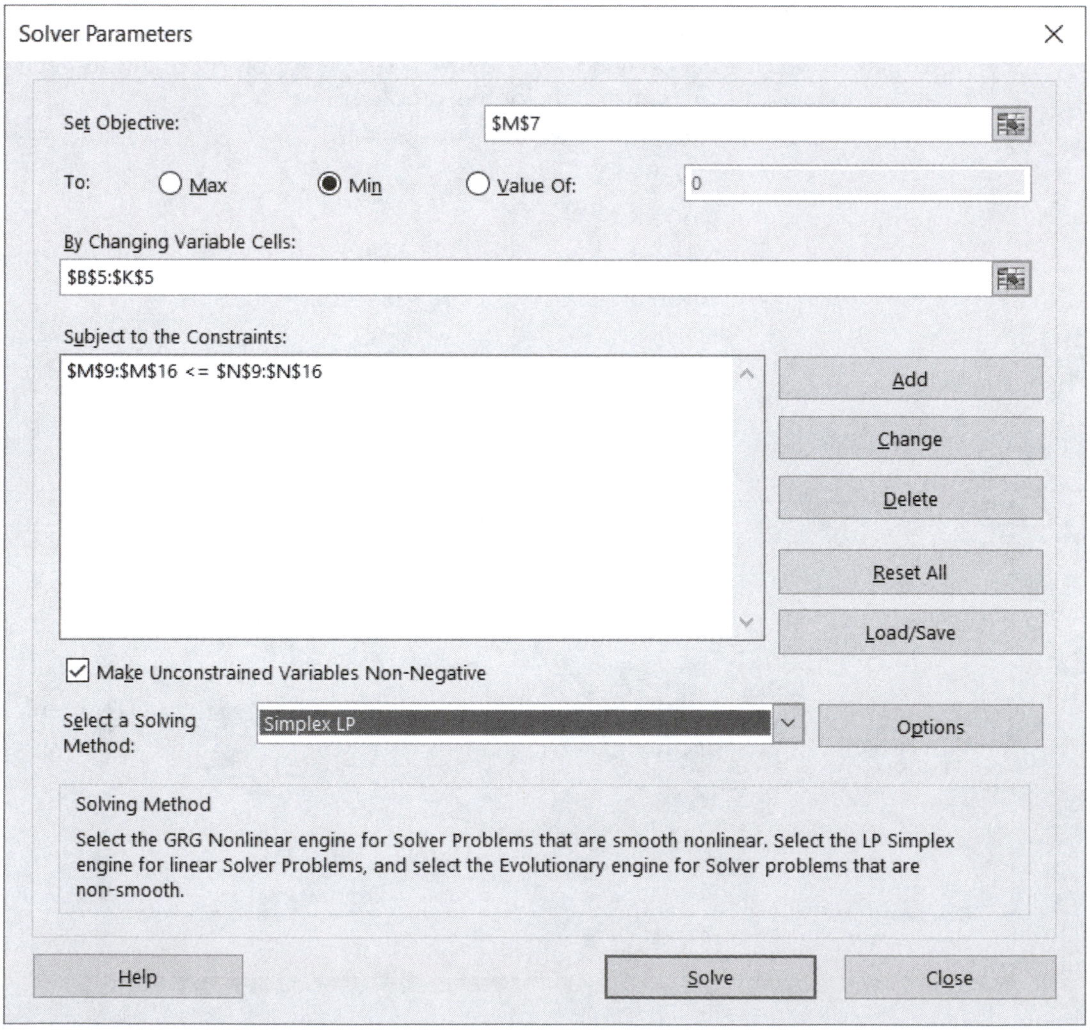

Figure 7–16 Excel spreadsheet for Pear transshipment problem in Example 7.8

For some linear programming codes, all right-hand sides would have to be nonnegative. In those cases, one would multiply the first three constraints by –1. However, this is not required in Excel, so we can enter the constraints just as they appear.

The Excel spreadsheet and solution for Pear's transshipment problem appears in Figure 7–16. Note that the solution calls for shipping all units from the sources. Since this problem had less supply than demand, it is interesting to see where the shortfall occurs. The amount shipped into Sacramento is 45 units (all from Sunnyvale), and the total shipped out of Sacramento is 20 units. Thus, all the demand (25 units) is satisfied in Sacramento. For the other transshipment point, Chicago, the shipments into Chicago are 120 units from Dublin and 95 units from Bangkok, and the shipments out of Chicago are 48 units to Amarillo and 120 units to Teaneck. The difference is 120 + 95 – (48 + 120) = 47 units. Hence, there is also no shortage in Chicago. Total shipments into the demand nodes at Amarillo, Teaneck, and Sioux Falls are respectively 80, 78, and 30 units. Hence, all the shortage (25 units) is absorbed at the Sioux Falls location at the optimal solution.

Real networks can be extremely complex by virtue of their sheer magnitude. A typical soft drink distribution system could involve anywhere between 10,000 and 120,000 accounts (Simchi-Levi, Kaminsky, and Simchi-Levi, 2008). Some retailers have thousands of stores and hundreds of thousands of products. For this reason, efficient data aggregation may be required to solve problems of this magnitude. Customer aggregation is generally accomplished by combining accounts that are nearby. Combining customers with like or similar zip codes is a common means of geographic aggregation. Product aggregation is discussed in detail in Chapter 3 in the context of manufacturing. Product aggregation rules for a supply chain are likely to be different from those discussed in Chapter 3. For example, one might aggregate products according to where they are picked up or where they are delivered. (For a more comprehensive discussion of the practical issues surrounding implementation of supply chain networks, see Simchi-Levi et al., 2008.)

PROBLEMS FOR SECTIONS 7.6–7.8

43. Consider Example 7.7 of the Pear Company assuming the following supplies and demands.

Plant	Production	Warehouse	Requirement
Sunnyvale	60	Amarillo	100
Dublin	145	Teaneck	84
Bangkok	125	Chicago	77
		Sioux Falls	69

Use Excel's Solver (or other linear programming code) to determine the optimal solution.

44. Resolve Pear's transportation problem assuming that the following routes are eliminated: Sunnyvale to Teaneck, Dublin to Chicago, and Bangkok to Amarillo. What is the percentage increase in total shipping costs at the optimal solution due to the elimination of these routes?

45. Resolve Pear's transshipment problem assuming that an additional transshipment point is located at Oklahoma City. Assume that the unit costs of shipping from the three plants to Oklahoma City are Sunnyvale, $170; Dublin, $1,200; and Bangkok, $1,600; and the respective costs of shipping from Oklahoma City to the demand nodes are Amarillo, $35; Teaneck, $245; and Sioux Falls, $145. Assume that Oklahoma City is only a transshipment point and has no demand of its own. Find the new shipping pattern with the addition of the new transshipment point and the savings, if any, of introducing this additional node.
46. Major Motors produces its Trans National model in three plants: (1) Flint, Michigan, (2) Fresno, California, and (3) Monterrey, Mexico. Dealers receive cars from three regional distribution centers: (1) Phoenix, Arizona, (2) Davenport, Iowa, and (3) Columbia, South Carolina. Anticipated production at the plants over the next month (in 100s of cars) is 43 at Flint, 26 at Fresno, and 31 at Monterrey. Based on firm orders and other requests from dealers, Major Motors has decided that it needs to have the following numbers of cars at the regional distribution centers at month's end: Phoenix, 26; Davenport, 28; and Columbia, 30. Suppose that the cost of shipping 100 cars from each plant to each distribution center is given in the following matrix (in $1,000s).

		TO		
		Phoenix	Davenport	Columbia
F	Flint	12	8	17
R	Fresno	7	14	21
O	Monterrey	18	22	31
M				

a. Convert the problem to a balanced problem by adding an appropriate row or column and find the optimal solution using Solver.
b. Now find the optimal solution using Solver and inequality constraints.
c. Do the solutions in (a) and (b) match? Why or why not?
d. Suppose that the route between Monterrey and Columbia is no longer available due to a landslide on a key road in Mexico. Modify the model in (b) and resolve to find the optimal solution. Has the objective value increased or decreased? Explain why.

47. Consider the problem of Major Motors described in problem 46. In order to be able to deal more effectively with unforeseen events (such as the road closing), Major Motors has established two transshipment points between the factories and the regional distribution centers at Santa Fe, New Mexico, and Jefferson City, Missouri. The cost of shipping 100 cars to the transshipment points is (in $1,000s),

		TO	
		Santa Fe	Jefferson City
F	Flint	8	6
R	Fresno	6	9
O	Monterrey	9	14
M			

while the cost of shipping from the transshipment points to the distribution centers is

		TO		
		Phoenix	Davenport	Columbia
F	Santa Fe	3	8	10
R	Jefferson City	5	5	9
O				
M				

Assuming that none of the direct routes between the factories and the distribution centers is available, find the optimal flow of cars through the transshipment points that minimizes the total shipping costs.

48. Toyco produces a line of Bonnie dolls and accessories at its plants in New York and Baltimore that must be shipped to distribution centers in Chicago and Los Angeles. The company uses Air Freight, Inc., to make its shipments. Suppose that it can ship directly or through Pittsburgh and Denver. The daily production rates at the plants are respectively 5,000 and 7,000 units daily, and the demands at the distribution centers are respectively 3,500 and 8,500 units daily. The costs of shipping 1,000 units are given in the following table. Find the optimal shipping routes and the associated cost.

		TO			
		Pittsburgh	Denver	Chicago	Los Angeles
F	New York	$182	$375	$285	$460
R	Baltimore	77	290	245	575
O	Pittsburgh	—	275	125	380
M	Denver	—	—	90	110

49. Reconsider problem 48 if there is a drop in the demand for dolls to 3,000 at Chicago and 7,000 at Los Angeles. Find the optimal shipping pattern in this case. How much of the total decrease in demand of 2,000 units is absorbed at each factory at the optimal solution?
50. Reconsider problem 48 assuming that the maximum amount that can be shipped from either New York or Baltimore through Pittsburgh is 2,000 units due to the size of the plane available for this route.

7.9 DETERMINING DELIVERY ROUTES IN SUPPLY CHAINS

An important aspect of supply chain logistics is moving product from one place to another efficiently. The transportation and transshipment problems discussed earlier in this chapter deal with this problem at a macro or firmwide level. At a micro level, deliveries to customers also must be planned efficiently. Because of the scale of the problem, efficient delivery schedules can have a very significant impact on the bottom line. As a result, they become an important part of designing the entire supply chain.

Determining optimal delivery schedules turns out to be a very difficult problem in general, rivaling the complexity of job shop scheduling problems discussed in detail in Chapter 11. Vehicle scheduling is closely related to a classical operations research problem known as the **traveling salesman problem**. The problem is described in the following way. A salesman starts at his home base, labeled city 1. He must then make stops at $n - 1$ other cities, visiting each city exactly once. The problem is to determine the optimal sequence in which to visit the cities to minimize the total distance traveled. Although this problem is easy to state, it turns out to be very hard to solve. If the number of cities is small, it is possible to enumerate all the possible tours. There are $n!$ orderings of n objects ($n!$ is equal to n times $(n - 1)$ times $(n - 2) \ldots$ times 1). For modest values of n, one can enumerate all the tours and compute their distances directly. For example, for $n = 5$ there are 120 sequences. This number grows very fast, however. For $n = 10$, the number of sequences grows to over 3 million, and for $n = 25$ it grows to more than 1.55×10^{25}. To get some idea of how large this number is, suppose that we could evaluate 1 trillion sequences per second on a supercomputer. Then, for a 25-city problem, it would take nearly 500,000 years to evaluate all the sequences!

Total enumeration is hopeless for solving all but the smallest traveling salesman problems. Problems such as this are known in mathematics as **NP hard**. NP hardness means that there are no known polynomial time solutions, meaning that the time required to solve such problems is an exponential function of the number of cities rather than a polynomial function. We will not dwell on the traveling salesman problem here except to note that methods of solution have been proposed that are vast improvements over total enumeration. However, finding optimal solutions to even moderate-sized problems is still difficult.

Finding optimal routes in vehicle scheduling is a similar, but more complex, problem. Assume that there is a central depot with one or more delivery vehicles and n customer locations, each having a known requirement. The question is how to assign vehicles to customer locations to meet customer demand and satisfy whatever constraints there might be at minimum cost. More real vehicle scheduling problems are too large and complex to solve optimally.

Because optimality may be impossible to achieve, methods for determining "good" solutions are important. We will discuss a simple technique for finding good routes, known as the savings method (Clarke and Wright, 1964).

Suppose that there is a single depot from which all vehicles depart and return. Customer locations and needs are known. Identify the depot as location 0 and customers as locations 1, 2, ... , n. We assume that there are known costs of traveling from the depot to each customer location, given by

$$c_{0j} = \text{Cost of making one trip from the depot to customer } j.$$

To implement the method, we will also need to know the costs of trips between customers. This means that we will also assume that the following constants are known:

$$c_{ij} = \text{Cost of making a trip from customer location } i \text{ to customer location } j.$$

For our purposes we consider only the case in which $c_{ij} = c_{ji}$ for all $1 \le i, j \le n$. This does not necessarily hold in all situations, however. For example, if there are one-way streets, the distance from i to j may be different from the distance from j to i. The method proceeds as follows. Suppose initially that there is a separate vehicle assigned to each customer location. Then the initial solution consists of n separate routes from the depot to each customer location and back. It follows that the total cost of all round-trips for the initial solution is

$$2\sum_{j=1}^{n} c_{0j}.$$

Now, suppose that we link customers i and j. That is, we go from the depot to i to j and back to the depot again. In doing so, we would save one trip between the depot and location i and one trip between the depot and location j. However, there would be an added cost of c_{ij} for the trip from i to j (or vice versa). Hence, the savings realized by linking i and j is

$$s_{ij} = c_{0i} + c_{0j} - c_{ij}.$$

The method is to compute s_{ij} for all possible pairs of customer locations i and j, and then rank the s_{ij} in decreasing order. One then considers each of the links in descending order of savings and includes link (i, j) in a route if it does not violate feasibility constraints. If including the current link violates feasibility, one goes to the next link on the list and considers including that on a single route. One continues in this manner until the list is exhausted. Whenever link (i, j) is included on a route, the cost savings is s_{ij}.

The total number of calculations of s_{ij} required is

$$\binom{n}{2} = \frac{n!}{2!(n-2)!} = \frac{n(n-1)}{2}.$$

(When c_{ij} and c_{ji} are not equal, twice as many savings terms must be computed.)

The savings method is feasible to solve by hand for only small values of n. For example, for $n = 10$ there are 45 terms, and for $n = 100$ there are nearly 5,000 terms. However, as long as the constraints are not too complex, the method can be implemented easily on a computer.

We illustrate the method with the following example.

EXAMPLE 7.9

Whole Grains is a small bakery that supplies five major customers with bread each morning. If we locate the bakery at the origin of a grid [i.e., at the point (0, 0)], then the five customer locations and their daily requirements are

Customer	Location	Daily Requirements (loaves)
1	(15, 30)	85
2	(5, 30)	162
3	(10, 20)	26
4	(5, 5)	140
5	(20, 10)	110

The relative locations of the Whole Grains bakery and its five customers are shown in Figure 7–17.The bakery has several delivery trucks, each having a capacity of 300 loaves. We shall assume that the cost of traveling between any two locations is simply the straight-line or Euclidean distance between the points. Recall that the formula for the straight-line distance separating the points (x_1, y_1) and (x_2, y_2) is

$$\sqrt{(x_1 - x_2)^2 + (y_1 - y_2)^2}.$$

The goal is to find a delivery pattern that both meets customer demand and minimizes delivery costs, subject to not exceeding the capacity constraint on the size of the delivery trucks.

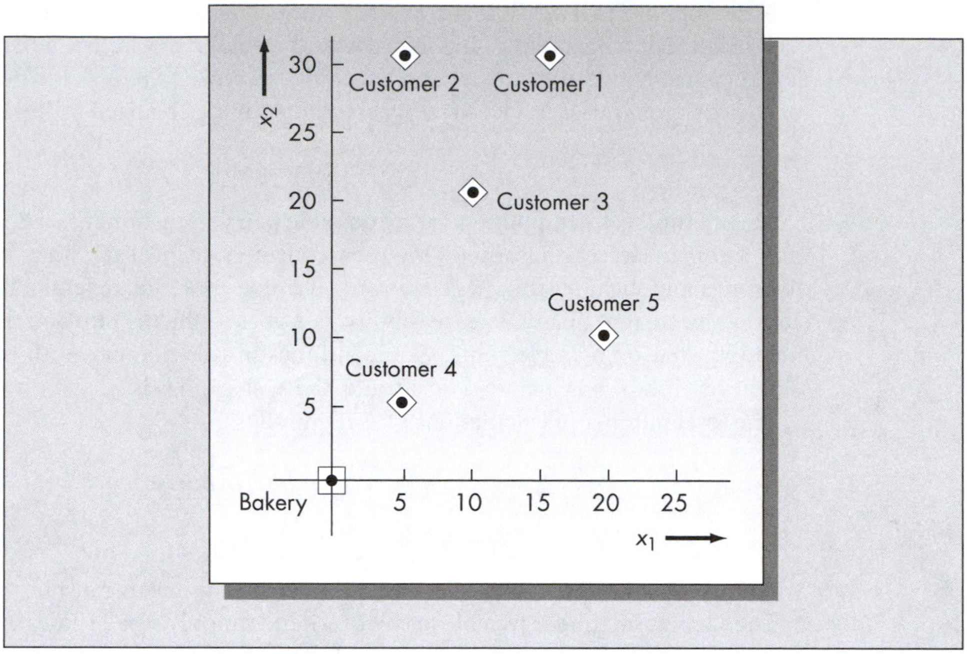

Figure 7–17 Customer locations in Example 7.9

Solution

The first step is to compute the cost for each pair (i, j) where i and j vary from 0 to 5. We are assuming that this cost is the straight-line distance between the points representing customer locations. The straight-line distances are given in the following matrix.

Cost Matrix (c_{ij})		TO					
		0	1	2	3	4	5
F	0		33.5	30.4	22.4	7.1	22.4
R	1			10.0	11.2	26.9	20.6
O	2				11.2	25.0	25.0
M	3					15.8	14.1
	4						15.8

Next, we compute the savings for all pairs (i, j), $1 \le i < j \le 5$. There are a total of 10 savings terms to compute for this example:

$$s_{12} = c_{01} + c_{02} - c_{12} = 33.5 + 30.4 - 10 = 53.9,$$
$$s_{13} = c_{01} + c_{03} - c_{13} = 33.5 + 22.4 - 11.2 = 44.7.$$

The remaining terms are computed in the same way, with the results

$$s_{14} = 13.7, \quad s_{25} = 27.8,$$
$$s_{15} = 35.3, \quad s_{34} = 13.7,$$
$$s_{23} = 41.6, \quad s_{35} = 30.7,$$
$$s_{24} = 12.5, \quad s_{45} = 13.7.$$

The next step is to rank the customer pairs in decreasing order of their savings values. This results in the ranking

(1, 2), (1, 3), (2, 3), (1, 5), (3, 5), (2, 5), (1, 4), (3, 4), (4, 5), and (2, 4).

Note that (1, 4), (3, 4), and (4, 5) have the same savings. Ties are broken arbitrarily, so these three pairs could have been ranked differently. We now begin combining customers and creating vehicle routes by considering the pairs in ranked order, checking each time that we do not violate the problem constraints. Because (1, 2) is first on the list, we first try linking customers 1 and 2 on the same route. Doing so results in a load of 85 + 162 = 247 loaves. Next, we consider combining 1 and 3, which means including 3 on the same route. This results in a load of 247 + 26 = 273, which is still feasible. Hence, we have now constructed a route consisting of customers 1, 2, and 3. The next pair on the list is (2, 3). However, 2 and 3 are already on the same route. Next on the list is (1, 5). Linking customer 5 to the current route is infeasible, however. As the demand at location 5 is 110 loaves, adding location 5 to the current route would exceed the truck's capacity. The next feasible pair on the list is (4, 5) which we make into a new route. The solution recommended by the savings method consists of two routes, as shown in Figure 7–18.

We should point out that the savings method is only a heuristic. It does not necessarily produce an optimal routing. The problem is that forcing the choice of a highly ranked link may preclude other links that might have slightly lower savings but might be better choices in a global sense by allowing other links to be chosen downstream. There have been suggested modifications of the savings method to attempt to overcome this difficulty (see Eilon, Watson-Gandy, and Christofides,1971). However, the authors point out that these modifications do not always result in a more cost-effective solution.

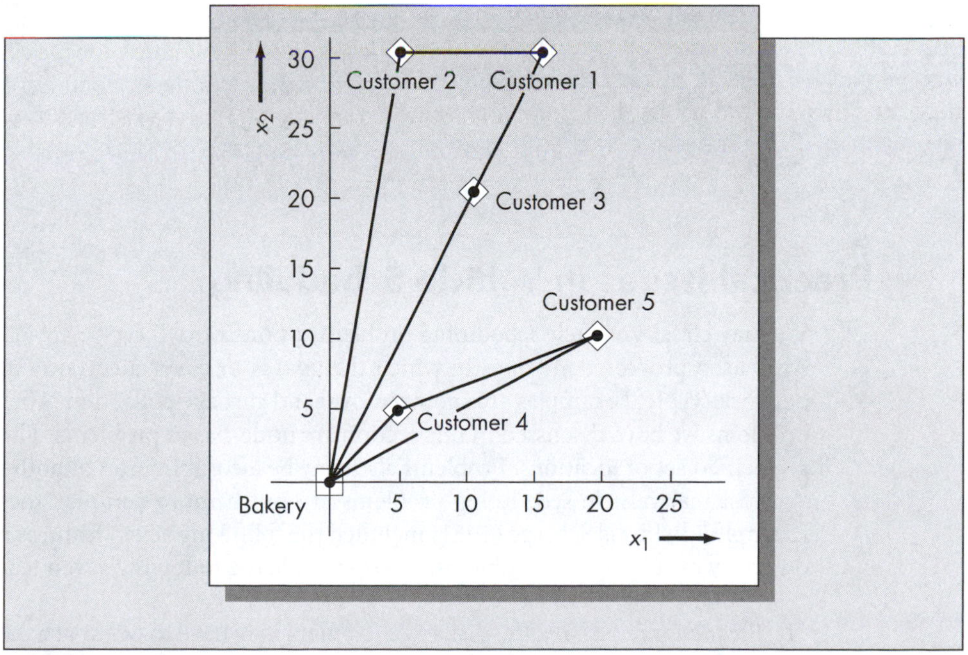

Figure 7–18 Vehicle routing found from the savings method for Example 7.9

SNAPSHOT APPLICATION

J. B. HUNT Saves Big with Routing and Scheduling Algorithm

J.B. Hunt Transport Services, Inc., is one of the largest transportation logistics companies in North America. A significant portion of their business lies in *drayage*, which is the transport of goods from an origin to a destination within the same urban area (i.e., it does not include long-haul operations). The term originated from the days when transport was by dray horses. One of the most common types of drayage is the transport of containerized cargo between and among rail ramps and shipping docks.

Jennifer Pazour and Lucas Neubert (2013) describe a routing and scheduling project done for J.B. Hunt Transport to determine cross-town drayage moves between rail ramps. While this scheduling was originally done manually, the size of the fleet and scale of movements had grown too large for manual calculations to remain practical. The combination of fleet and movements made for 4.13×10^{32} routes, although not all would be feasible because they ignored availability constraints for ramps and drivers and geographical considerations. Because of the scale of the problem, the authors produced a heuristic to decide the routing of trucks and the scheduling of drivers on the routes.

J.B. Hunt retains a fleet of drivers and also hires third-party drivers to cover overloads. Since third-party drivers are paid by the load, the primary objective of the heuristic was to maximize the number of loads covered by company drivers. However, there could be many solutions with identical numbers of company-driver loads; therefore, the heuristic attempted to minimize empty travel miles, which are costly in terms of fuel usage and vehicle wear and tear.

The constraints in the optimization are: (1) every load must be covered, either by a company driver or a third-party contractor; and (2) every company driver must be assigned a route. The heuristic generates feasible routes that consist of combinations of legs that either move a container across town or move the empty truck to where it is needed. A truck may do twelve or more loads in a day. Third-party contractors are not assigned to routes. They are assumed to do a single container load from an origin to a destination, which is a conservative assumption. The heuristic also considers operational constraints, including the number of loads per driver schedule, driver start times, driver start and end locations, hourly traffic patterns, load-time windows, and required driver service hours.

The implementation of the cross-town application positively impacted J.B. Hunt's intermodal drayage operation by automating and enhancing planning work flow for dispatchers, allowing the fleet size to grow without making planning impossible, reducing the number of costly outsourced loads, and significantly improving operational efficiency. J.B. Hunt has documented the annualized cost savings of the cross-town heuristic implementation at $581,000.

Practical Issues in Vehicle Scheduling

We may classify vehicle scheduling problems as one of two types: arc-based or node-based. Arc-based problems are ones in which the goal is to cover a certain collection of arcs in a network. Typical examples are snow removal and garbage collection. The type of distribution problems we have discussed in this section are node-based problems. The objective is to visit a specified set of locations. Problems also may be a combination of both of these.

Most real vehicle scheduling problems are much more complex than that described in Example 7.9. Linus Schrage (1981) included the following seven features in his analysis of the difficulty of solving real problems. These include the following seven features.

1. *Frequency requirements.* Visits to customers may have to occur at a certain frequency and that frequency may vary from customer to customer. In our example, bread deliveries are made daily to each customer, so frequency is not an issue. Consider, however, the

problem of delivering oil or gas for residential use. The frequency of delivery depends on the usage rate, so delivery frequency will vary from customer to customer.

2. *Time windows.* Visits to customer locations sometimes must be made at specific times. Ride booking systems and postal and bank pickups and deliveries are typical examples.
3. *Time-dependent travel time.* When deliveries are made in urban centers, rush-hour congestion can be an important factor. This is an example of the case in which travel time (and hence the cost associated with a link in the network) depends on the time of day.
4. *Multidimensional capacity constraints.* There may be constraints on weight as well as on volume. This can be a thorny issue, especially when the same vehicles are used to transport a variety of different products.
5. *Vehicle types.* Large firms may have several vehicle types from which to choose. Vehicle types may differ according to capacity, the cost of operation, and whether the vehicle is constrained to closed trips only (where it must return to the depot after making deliveries). When several types of vehicles are available, the number of feasible alternatives increases dramatically.
6. *Split deliveries.* If one customer has a particular requirement, it could make sense to have more than one vehicle assigned to that customer.
7. *Uncertainty.* Routing algorithms invariably assume that all the information is known in advance. In truth, however, the time required to cross certain portions of a network could be highly variable, depending on factors such as traffic conditions, weather, and vehicle breakdowns.

PROBLEMS FOR SECTION 7.9

51. Re-solve Example 7.9 assuming that the capacity of the vehicles is only 250 loaves of bread.
52. Re-solve Example 7.9 assuming that the distance between any two locations is the rectangular distance rather than the Euclidean distance. (See Section 7.5 for a definition of rectangular distance.)
53. Add the following customer locations and requirements to Example 7.9 and re-solve.

Customer	Location	Daily Requirement
6	(12, 12)	78
7	(23, 3)	126

54. Suppose that one wishes to schedule vehicles from a central depot to five customer locations. The cost of making trips between each pair of locations is given in the following matrix. (Assume that the depot is location 0.)

Cost Matrix (c_{ij})		TO 0	1	2	3	4	5
	0		20	75	33	10	30
F	1			35	5	20	15
R	2				18	58	42
O	3					40	20
M	4						25

Assume that these costs correspond to distances between locations and that each vehicle is constrained to travel no more than 50 miles on each route. Find the routing suggested by the savings method.

55. All-Weather Oil and Gas Company is planning delivery routes to six natural gas customers. The customer locations and gas requirements (in gallons) are given in the following table.

Customer	Location	Requirements (gallons)
1	(5, 14)	550
2	(10, 25)	400
3	(3, 30)	650
4	(35, 12)	250
5	(10, 7)	300

Assume that the depot is located at the origin of the grid and that the delivery trucks have a capacity of 1,200 gallons. Also assume that the cost of travel between any two locations is the straight-line (Euclidean) distance between them. Find the route schedule obtained from the savings method.

7.10 SUMMARY

Business analytics arose with increased information (or "big data"). Mathematical modeling can play an important role in efficient supply chain management. This chapter provided a number of analytic methods for supply chain design.

We first discussed two methods for assisting with capacity expansion decisions. Break-even curves provide a means of determining the sales volume necessary to justify investing in new or existing facilities. A simple model for a dynamic expansion policy is presented that gives the optimal timing and sizing of new facilities assuming constant demand growth and discounting of future costs.

We also discussed issues that arise in trying to decide where to locate new facilities. This problem is very complex in that there are many factors that relate to the decision of where to locate production, design, and management facilities. Location models are appropriate when locating one or more new facilities within a specified area already containing a finite number of existing facilities. The objective is to locate new facilities to minimize some function of the distance separating new and existing facilities. Three distance measures were considered: rectilinear, Euclidean, and squared Euclidean. The first two are the most common for describing real problems and depend on whether movement occurs according to a crisscross street pattern (rectilinear) or is measured by straight-line distances (Euclidean).

Some of the location problems discussed in this chapter go back hundreds of years. The problem of finding the location of a single new facility to minimize the sum of the Euclidean distances to the existing facilities has been referred to as the Steiner–Weber problem or the general Fermat problem. Richard Francis, Leon McGinnis, and John White (1992) state that the problem with exactly three facilities was posed by Pierre de Fermat and solved by the mathematician Evangelista Torricelli prior to 1640. The work on the rectilinear distance problem is more recent (Hakimi, 1964).

The optimal solution to the weighted rectilinear distance problem is to locate the (x, y) coordinates at the median of the existing coordinates. When using a squared Euclidean distance measure, the optimal location of the new facility is at the center of gravity of the existing coordinates. No simple algebraic solution for the Euclidean distance problem is known, but iterative solution techniques exist. The chapter also included a brief discussion of several more complex location problems, including location of multiple facilities, location of facilities having nonzero areas, location–allocation problems, discrete location problems, and network location models. The transportation and transshipment problems, discussed in this chapter, are further examples of mathematical optimization models. They can assist firms with determining efficient schedules for moving product from the factory to the market. There are also mathematically based techniques for efficient scheduling of delivery vehicles.

APPENDIX 7–A

PRESENT WORTH CALCULATIONS

Including the time value of money in the decision process is common when considering alternative investment strategies. The idea is that a dollar received today has greater value than one received a year from now. For example, if a dollar were placed in a savings account paying 2 percent, it would be worth \$1.02 in one year. More generally, if it were invested at a rate of return of r (expressed as a decimal), it would be worth $1 + r$ in a year, $(1 + r)^2$ in two years, and so on. (Note, in a deflationary environment a dollar today is worth less than one in the future so r becomes negative; but most companies invest assuming r to be positive.)

In the same way, a cost of \$1 incurred in a future year has a present value of less than \$1 today. For example, at 2 percent interest, how much would one need to place in an account today so that the total principal plus interest would equal \$1 in a year? The answer is $1/(1.02) = 0.9804$. Similarly, the present value of a \$1 cost incurred in two years at 2 percent is $1/(1.02)^2 = 0.9612$. In general, the present value of a cost of \$1 incurred in t years assuming a rate of return r is $(1 + r)^{-t}$.

These calculations assume that there is no compounding. Compounding means that one earns interest on the interest, so to speak. For example, 2 percent compounded semiannually means that one earns 1 percent on \$1 after six months and 1 percent on the original \$1 plus interest earned in the first six months. Hence, the total return is

$$(1.01)(1.01) = \$1.0201$$

after one year, or slightly more than 2 percent. If the interest were compounded quarterly, the dollar would be worth

$$(1 + .02/4)^4 = 1.020151$$

at the end of the year. The logical extension of this idea is continuous compounding. One dollar invested at 2 percent compounded continuously would be worth

$$\lim_{n \to \infty}\left(1 + 0.02/n\right)^n = e^{0.02} = 1.0202$$

at the end of a year. The number $e = 2.7172818 \ldots$ is defined as

$$e = \lim_{n \to \infty} \left(1 + 1/n\right)^n .$$

Notice that continuous compounding only increases the effective simple interest rate from 2 percent to 2.02 percent.

More generally, C invested at a rate of r for t years compounded continuously is worth Ce^{rt} at the end of t years.

Reversing the argument, the present value of a cost of C incurred in t years assuming continuous compounding at a discount rate r is Ce^{-rt}. A stream of costs $C_1, C_2, \ldots, C_n$ incurred at times $t_1, t_2, \ldots, t_n$ has present value

$$\sum_{i=1}^{n} c_i e^{-rt_i} .$$

John Freidenfelds (1981) presents a comprehensive treatment of discounting and its relationship to the capacity expansion problem.

APPENDIX 7–B

COMPUTING CONTOUR LINES

This appendix outlines the procedure for computing contour lines, or isocost lines, such as those pictured in Figures 7–8 and 7–9. (See Francis et al., 1992 for the theoretical justification for this procedure).

1. Plot the points $(a_1, b_1), (a_2, b_2), \ldots, (a_n, b_n)$ on graph paper. Draw a horizontal line (parallel to the x axis) and a vertical line (parallel to the y axis) through each point.
2. Number the horizontal and vertical lines in sequence from left to right and from top to bottom. (If none of the original points is collinear, there will be exactly n horizontal and n vertical lines.)
3. Let C_j be the sum of the weights associated with the points along vertical line j, and D_i the sum of the weights associated with the points along horizontal line i.
4. Compute the following numbers:

$$M_0 = -\sum_{i=1}^{n} w_i, \qquad N_0 = M_0 = \sum_{i=1}^{n} w_i,$$
$$M_1 = M_0 + 2C_1, \qquad N_1 = N_0 + 2D_1,$$
$$M_2 = M_1 + 2C_2, \qquad N_2 = N_1 + 2D_2,$$

and so on.

(The final values of M_i and N_j will both be $\sum_{i=1}^{n} w_i$.)

5. Define the region (i, j) as the region bounded by the ith and $(i + 1)$th vertical lines and the jth and $(j + 1)$th horizontal lines. The regions to the left of the first vertical line are

labeled $(0, j)$, and those below the first horizontal line are labeled $(i, 0)$. The slope of any contour line passing through region (i, j) is given by

$$S_{i,j} = -M_i/N_j.$$

Once the slopes are determined, a contour line is constructed by starting at any point and moving through each region at the angle determined by the slope computed in step 5. We will present a simple example to illustrate the method.

EXAMPLE 7.10

Assume $(a_1, b_1) = (1, 1)$, $(a_2, b_2) = (5, 3)$, and $(a_3, b_3) = (3, 8)$, and $w_1 = 2$, $w_2 = 3$, and $w_3 = 6$. The first step is to plot these points on a grid as we have done in Figure 7–19. Drawing vertical and horizontal lines through each of the three points yields three vertical lines labeled 1, 2, and 3 and three horizontal lines labeled 1, 2, 3 as well. Regions are labeled from (0, 0) to (3, 3) as in the figure.

Next we compute $M_0, \ldots, M_3$ and $N_0, \ldots, N_3$.

$$M_0 = -(2 + 3 + 6) = -11 = N_0,$$

$$M_1 = -11 + (2)(2) = -7, \qquad N_1 = -11 + 2(2) = -7,$$

$$M_2 = -7 + (2)(6) = +5, \qquad N_2 = -7 + (2)(3) = -1,$$

$$M_3 = -5 + (2)(3 = +11, \qquad N_3 = -1 + (2)(6) = +11.$$

Next the ratios are computed to find the slope for each region.

$$S_{0,0} = -(-11)/(-11) = -1, \qquad S_{0,2} = -(-11)/(-1) = -11,$$

$$S_{1,0} = -(-7)/(-11) = -0.64, \qquad S_{1,2} = -(-7)/(-1) = -7,$$

$$S_{2,0} = -(5)/(-11) = +0.45, \qquad S_{2,2} = -(5)/(-1) = +5,$$

$$S_{3,0} = -(11)/(-11) = +1, \qquad S_{3,2} = -(11)/(-1) = +11,$$

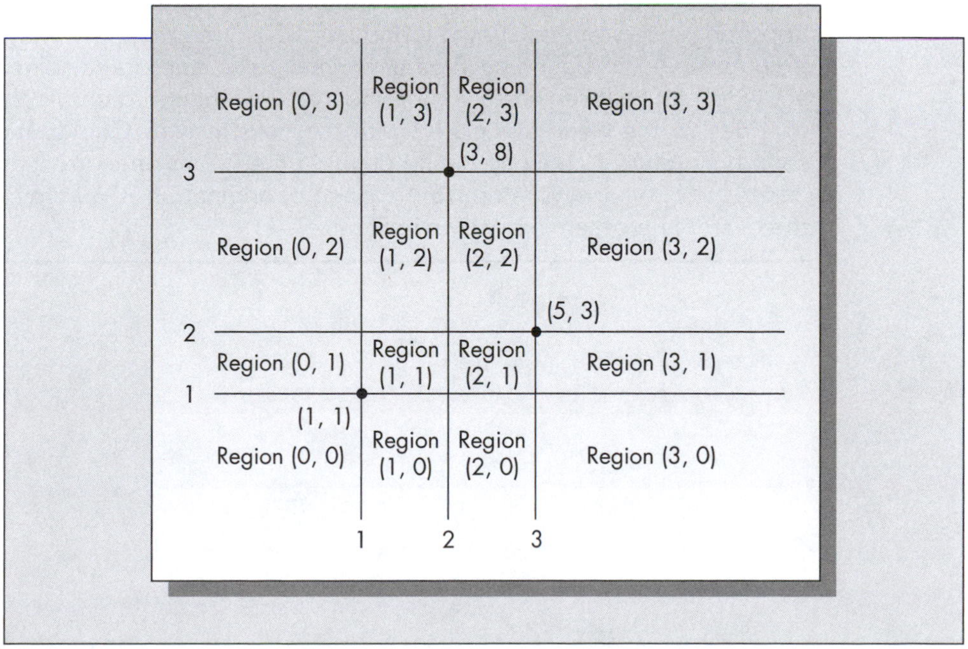

Figure 7–19 Regions for Example 7.10

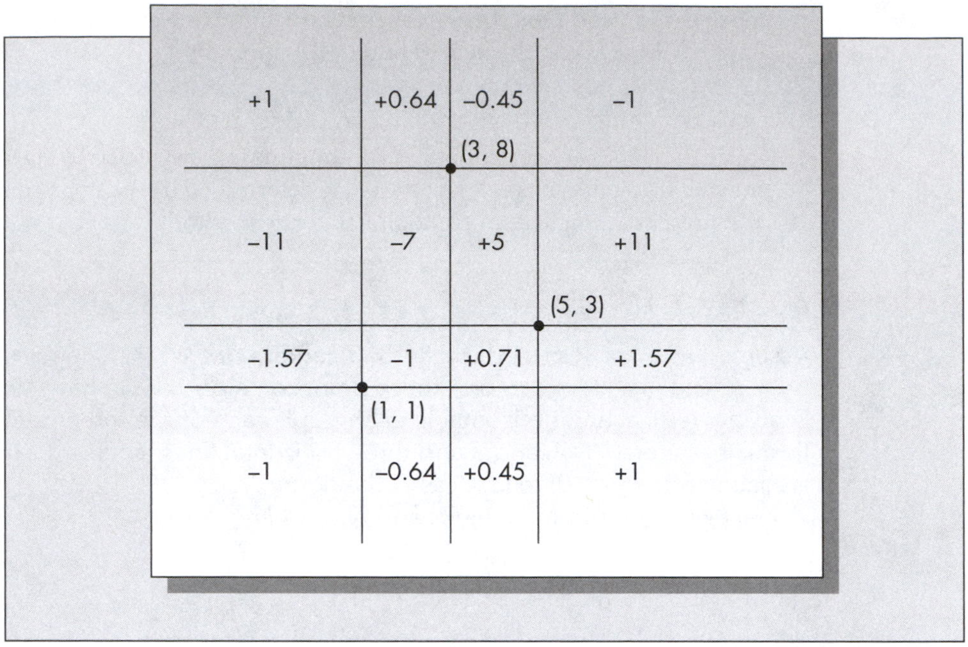

Figure 7–20 Slopes for Example 7.10

$$S_{0,1} = -(-11)/(-7) = -1.57, \qquad S_{0,3} = -(-11)/(11) = +1,$$
$$S_{1,1} = -(-7)/(-7) = -1, \qquad S_{1,3} = -(-7)/(11) = +0.64,$$
$$S_{2,1} = -(5)/(-7) = +0.71, \qquad S_{2,3} = -(5)/(11) = -0.45,$$
$$S_{3,1} = -(11)/(-7) = +1.57, \qquad S_{3,3} = -(11)/(11) = -1.$$

Before constructing the contour lines, it is convenient to place the slopes in the appropriate regions, as shown in Figure 7–20. A contour line may be started at any point on a region boundary. From the initial point, one draws a line with the appropriate slope for that region to the boundary of the next region. At that point the slope changes to the value associated with the next region. One continues until the line segments return to the originating point. (If the slopes are correct and the drawing accurate, one will always return to the point of origination.) Two typical contour lines for the example problem are shown in Figure 7–21.

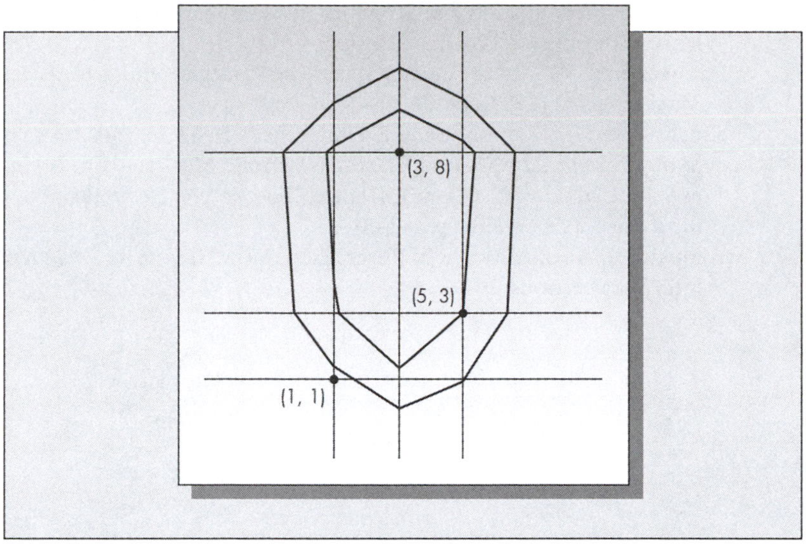

Figure 7–21 Sample contour lines for Example 7.10

BIBLIOGRAPHY

Arntzen, B. C., G. G. Brown, T. P. Harrison, and L. L. Trafton. "Global Supply Chain Management at Digital Equipment Corporation." *Interfaces* 25 (1995), pp. 69–93.

Clarke, G., and G. W. Wright. "Scheduling of Vehicles from a Central Depot to a Number of Delivery Points." *Operations Research* 12 (1964), pp. 568–81.

Cohen, M., P. V. Kamesam, P. Kleindorfer, H. Lee, and A. Tekerian. "Optimizer: IBM's Multi-Echelon Inventory System for Managing Service Logistics." *Interfaces* 20, no. 1 (1990), pp. 65–82.

Cohen, M., and H. L. Lee. "Resource Deployment Analysis of Global Manufacturing and Distribution Networks." *Journal of Manufacturing and Operations Management* 2 (1989), pp. 81–104.

Degbotse, A., B. T. Denton, K. Fordyce, R. J. Milne, R. Orzell, and C. T. Wang. "IBM Blends Heuristics and Optimization to Plan Its Semiconductor Supply Chain." *Interfaces* 43, no. 2 (2013), pp. 130–141.

Eilon, S., C. D. T. Watson-Gandy, and N. Christofides. *Distribution Management: Mathematical Modeling and Practical Analysis*. London: Griffin, 1971.

Francis, R. L., L. F. McGinnis, and J. A. White. "Locational Analysis." *European Journal of Operational Research* 12 (1983), pp. 220–52.

Francis, R. L., L. F. McGinnis, and J. A. White. *Facility Layout and Location: An Analytical Approach* (2nd ed.). Englewood Cliffs, NJ: Prentice Hall, 1992.

Freidenfelds, J. *Capacity Expansion: Analysis of Simple Models with Applications*. New York: Elsevier North Holland, 1981.

Hakimi, S. L. "Optimum Location of Switching Centers and the Absolute Centers and Medians of a Graph." *Operations Research* 12 (1964), pp. 450–59.

Keefer, K. B. "Easy Way to Determine the Center of Distribution." *Food Industries* 6 (1934), pp. 450–51.

Lieberman, M. "Scale Economies, Factory Focus, and Optimal Plant Size." Paper present at Stanford Graduate School of Business, July 22, 1987.

Manne, A. S. "Calculations for a Single Producing Area." In A. S. Manne, (Ed.), *Investments for Capacity Expansion: Size, Location, and Time Phasing,* pp. 28–48. Cambridge, MA: MIT Press, 1967.

Marr, B., "How Much Data Do We Create Every Day? The Mind-Blowing Stats Everyone Should Read." *Forbes*, M ay 21, 2018.

Pazour, J. A., and L. C. Neubert. "Routing and Scheduling of Cross-Town Drayage Operations at J.B. Hunt Transport." *Interfaces* 43, no. 2 (2013), pp. 117–129.

Schrage, L. "Formulation and Structure of More Complex/ Realistic Routing and Scheduling Problems" *Networks* 11 (1981), pp. 229–32.

Schmenner, R. W. "Multiplant Manufacturing Strategies among the Fortune 500." *Journal of Operations Management* 2 (1982), pp. 77–86.

Simchi-Levi, D. P. Kaminski, and E. Simchi-Levi. *Designing and Managing the Supply Chain: Concepts, Strategies, and Case Studies* (3rd ed.). New York: McGraw-Hill/ Irwin, 2008.

Tansel, B., C. R. L. Francis, and T. J. Lowe. "Location on Networks: A Survey (Parts 1 and 2)." *Management Science* 29 (1983), pp. 482–511.

Tompkins, J. A., J. A. White, Y. Bozer, and J. M. A. Tanchoco. *Facilities Planning* (4th ed.). New York: John Wiley & Sons, 2010.

CHAPTER

8

Service Operations Management

"The goal as a company is to have customer service that is not just the best, but legendary."

—Sam Walton, Founder of Walmart

CHAPTER OVERVIEW

Purpose

To understand the challenges unique to managing service operations and to learn key tools for matching supply with demand in services.

KEY POINTS

1. *What is a service?* To be considered a service, there must be an intangible portion to the offering. It involves something that is not a good (or more informally, something that cannot be dropped on your foot). A service is also usually time-perishable; it cannot be stored. Further, it frequently involves the customer as co-producer; the service cannot take place without the customer's involvement.
2. *Service operations strategy.* While strategy within service operations is fundamentally similar to general operations strategy, as discussed in Chapter 1, there are two particular features that make it more challenging for a service firm. First, in part, because capacity in services tends to correspond to people rather than machines, there is a "fear of focus" that develops for many firms as they try to be all things to all customers. Second, defining and measuring quality is more difficult for services than for goods because it tends to be more subjective.
3. *Bottleneck analysis.* A system's capacity is the capacity of its bottleneck—that is, the step or resource in the system that can process the fewest customers per hour. In order to find the bottleneck, the capacity of all resources must be calculated. The utilization of a resource is the rate at which customers arrive divided by the capacity (i.e., the rate at which customers can be processed). Thus, the bottleneck resource will have the highest percentage utilization.
4. *Poisson arrivals.* Arrivals to many service systems are both unscheduled and highly variable. The Poisson process is often a good model for these types of systems, particularly over the short term, and can be used to make system predictions. The model is based on an assumption of independent behavior by a large number of potential customers.
5. *Pooling.* As in Chapter 6, pooling is a key technique for mitigating uncertainty and improving planning. In service systems, pooling strategies involve either combining

variable streams of arrivals into one larger, inherently less variable, stream or cross-training staff so that they may serve multiple classes of customer.

6. *Queueing systems.* A queue (waiting line) represents customers waiting for service. The structural aspects of queueing models include arrivals, service, queue discipline, capacity of the queue, number of servers, and the network structure. Measures of performance of queueing systems can be determined analytically for simpler cases and by simulation for more complex cases.
7. *The M/M/1 queue.* One of the simplest queueing models is known as the M/M/1 queue. It assumes Poisson arrivals and exponentially distributed service times. Although rarely completely accurate, it yields simple intuitive formulas for various system measures.
8. *Little's law.* This "law" states that the average number of customers in the system is equal to the average customer's time in the system multiplied by the customer arrival rate. Because it is an equation, any one of the values can be calculated if the other two are known. For example, average time in system can be calculated by dividing the average number of customers in the system by the arrival rate.
9. *Incentives in services.* Like any operational system, metrics drive performance and incentivize behavior in service systems. A common contract in services is a Service Level Agreement (SLA) where the firm agrees to meet a specified goal for service a certain percentage of the time. A widely used type of SLA is a delay percentile contract, which puts an upper bound on waiting time for a certain percentage of customers. Unfortunately, this type of contract can lead to perverse incentives for the service provider; there are more effective contracts available.
10. *The human element.* Human behavior is more important to consider in the design of service systems than it is for most production systems because, as mentioned above, services frequently involve the customer as coproducer. For example, a self-service system must be intuitive and pleasant for customers who have never encountered the system before, while being sufficiently fast for use by existing customers.
11. *Revenue management.* An important technique for managing variable customer arrivals and perishable capacity, as are typical for service systems, is to use differentiated prices, or revenue management. For example, airlines will price seats higher if the plane has little spare capacity and lower if there are a large number of unsold seats. Further, they will try to segment the customers by willingness to pay, so that business travelers, who need more flexibility and perhaps a more comfortable trip, will typically pay more for a ticket than price-sensitive leisure travelers.

One of the benefits of social media is instant access to the opinions of others. A popular site for finding out what others think about consumer services is Yelp. If you find yourself in a new city and want to find a restaurant, Yelp provides you with feedback from other patrons about the quality of the food and the service. Previously relegated to word of mouth, opinions are now available electronically. Yelp reviews can be misleading. One of the authors recently went to Yelp to learn about local pool services. Influenced by five glowing reviews, he hired a company. During the month the firm was treating the pool, the algae became so bad that the pool turned an emerald green, and workers broke some equipment. It is likely that the owners or family members and friends wrote the reviews. (And yes, the author did his civic duty and posted a very negative review.)

The previous example illustrates the difficulty of measuring quality in service systems. In part, this is because the fundamental principle that differentiates a service from a good is an element of **intangibility**, which means something that cannot be perceived by the sense of touch. This intangibility also means that services typically cannot be inventoried in advance of consumption and hence are **time-perishable**. For example, a movie shown in a theater is of no use to a customer if they are not there at the same time as it is being shown.

Another distinctive feature of services is that they typically involve the customer as **coproducer**. Whether it is the customer selecting and consuming the food in a restaurant or riding the roller coaster in a theme park, the service does not happen without the customer helping to "produce" the service.

The line between goods and services is not solid, especially in light of the trend toward servicization, as discussed in Chapter 1. Restaurants and retailers are usually classed as service businesses, despite the fact that the former needs food for the transaction and the latter sells goods. A typical representation of this idea is that there is a continuum of industries from commodity goods producers (such as wheat farmers) on the left to pure experience services (such as movies) on the right. Retailers fall around the middle of the continuum and are to the left of most services but to the right of most production industries.

In the marketing literature, a popular trend is known as **service dominant logic** (SDL). It states that "all firms are service firms" because all firms provide "the applications of competences (knowledge and skills) for the benefit of a party" (Lusch and Vargo, 2014). In some sense, this idea is similar to the process view of a firm, often taken in operations, which is that firms use process competencies to convert inputs to outputs. The term **services** (as opposed to simply "service") is used to reflect service operations as defined in this chapter.

As mentioned above, services typically cannot be inventoried. Therefore, when there is the inevitable mismatch between supply and demand, customers either wait or leave without service. The presence of variability exacerbates this mismatch. There are five key types of variability that must be planned for in most service systems (Frei, 2006).

1. *Arrival variability*. There is variability in the timing of customer arrivals to the service by day of the week, time of day, number of customers in a group, and even from minute to minute.
2. *Request variability*. There is variability in customer expectations and the type of service they wish to consume.
3. *Capability variability*. There is variability both in the ability of staff to provide the service and in customer abilities when they serve as coproducer.
4. *Effort variability*. There is variability in the effort put into the service both by staff and by customers.
5. *Subjective preference variability*. Even if customers receive identical services, there is still variability in how they perceive the service due to individual and subjective preferences. For example, the moods of customers affect their perceptions, as do individual tastes.

Improving the performance of a service system typically involves either reducing or better accommodating at least one of these types of variability. This chapter considers both possibilities. It provides tools for mitigating the mismatch between supply and demand that adds cost to any service system.

The two key goals for the chapter are:

- to foster readers' abilities to analyze services in terms of the potential to deliver the services promised; and
- to provide readers with tools that they can apply to the design and improvement of service systems.

The chapter begins with a discussion of strategy applied to service industries before moving on to the more technical side of service system analysis. Service systems can be modeled as flow systems with stochastic (random) arrivals and service times. Thus, queueing theory is a helpful tool for the analysis of service systems. However, the human element cannot be forgotten in service system design, which is covered in Section 8.6. The chapter concludes with sections on call and contact centers and revenue management.

8.1 SERVICE OPERATIONS STRATEGY

As discussed in Chapters 1 and 6, operations strategy involves trade-offs; managing service operations is no different. The most common trade-off in services is between quality and cost. However, quality is more difficult to measure in services, and customers may not be willing to pay for it. For example, Bose can charge a premium for their headphones because their superior fidelity can be measured. In contrast, McDonald's cannot charge a premium for their friendly service (although one author had a friend who was fired from McDonald's for failing to smile at an undercover McDonald's quality assessment employee posing as a customer!).

In this section, we first discuss where services fit in the economic landscape. Then, we consider what is meant by service quality. We discuss the key decisions that must be made in positioning a service in the marketplace and how best to plan for the five key types of variability. Servicization, introduced in Chapter 1, is revisited here in the context of decisions about leasing versus buying. We conclude by discussing service competition and how it differs from competition among goods-producing firms.

The Service Economy

Section 1.1 described the three-sector framework of the economy consisting of the primary sector (extractive), the secondary sector (goods producing), and the tertiary sector (services). The tertiary sector is sometimes divided further into **domestic services** (including restaurants and hotels, barber and beauty shops, laundry and dry cleaning, and maintenance and repair), **trade and commerce services** (including transportation, retailing, real estate, communication, and finance and insurance), **refining and extending human capacities** (including health, education, research, recreation, and the arts), and the **experience economy** (which in its purest form provides entertainment value only, such as theme parks).

Over 25 years ago, Fortune magazine predicted that "in the new U.S. economy, service—bold, fast, unexpected, innovative, and customized—is the ultimate strategic imperative. … Everyone has become better at developing products. The one place you can differentiate yourself is in the service you provide" (Henkoff, 1994). While these predictions were a little premature, they do emphasize the importance of service as a competitive strategy. While novel products will always have their place in the marketplace, service quality is indeed an important piece of a firm's competitive strategy.

Service Quality

As mentioned earlier, the primary trade-off in most service operations is between cost and quality; however, how quality is defined differs from application to application. In a fast food restaurant, quality is speed of service and consistency of food and experience. In a romantic restaurant, quality includes atmosphere and attentiveness of staff. Therefore, quality in services must be defined relative to customers' expectations.

There are many key elements that define service quality. There is the *consistency* of the service, and the *delay* incurred before service. Also important is the ability of staff to perform the promised service *dependably*, *accurately*, *promptly*, *courteously*, and with a *friendly demeanor*. The *appearance* of physical facilities, equipment, personnel, and communication materials help set the *atmosphere* for the service. The appropriate level of *communication* with the customer is a key element. The ease with which the service is *accessed* includes hours of operation, location, and availability of the appropriate server. Another element is the level of *personalization* of the service, as well as the *pleasure* or "fun" level of the service (where appropriate). Also key to service quality are the *credibility* and *technical level* of the service plus the *safety* and *security* of the service (Metters, King-Metters, Pullman, and Walton, 2008).

The elements of service quality that are most important will obviously depend on the industry, but all will be important to some extent. For example, many people will choose a bank depending on its convenience, the services offered, and its reputation for security. Yet they may switch banks if treated poorly by staff who fail to be sufficiently friendly. Who among us has not been put off by an unhelpful or unfriendly staff member in some service environment? As humans, we tend to avoid unnecessary sources of conflict or stress and will simply switch to another firm rather than risk another unpleasant encounter. Most problematic for the firm is that we are unlikely to tell them what occurred, preventing them from taking corrective action.

Measuring Quality

Because there are so many elements that define quality, it is typically more difficult to measure quality for services than for goods. Even defining a "defect" can be problematic. For example, if the service was performed as prescribed but the customer was not happy because he or she misunderstood part of the experience or simply entered into the experience in a negative mood, should that be considered faulty service? The answer to this question will depend on what the service provider is going to do with this information. Clearly, the staff member involved should not be penalized. However, the information should be collected so that the process can be redesigned to be more "foolproof."

Quality in services is often measured by customer surveys, yet this can be a problem if unhappy customers disengage and therefore do not fill in the survey at the same rate that satisfied customers do (or if the reverse occurs and only dissatisfied customers give feedback). In addition, if there is separation between the time the service is consumed and the time of the survey, customers may not accurately remember perceptions at the time of the service. One tool for mitigating this issue is to place push-button perception collectors at the end of the service. For example, London's Heathrow Airport has small kiosks with four buttons ranging from a very sad face to a very happy face. Arriving international travelers are encouraged to push one of the buttons to indicate how their customs and immigration experience was. It is not clear how this information is used but Heathrow has addressed the immediacy issue.

Controlling Quality

Another complication in service quality is that controlling defects when customers are involved in coproduction can be highly problematic. Even if the issue is the customer's fault, they will quite naturally blame the process—and hence the service provider. Some service providers try to "train" customers in order to speed up service, which will improve the quality of the experience. For example, Starbucks coffee shops have the servers repeat the customer's order in the preferred sequence (e.g., size before coffee type) and jargon (skinny rather than skim milk) in the deliberate desire to make regular customers follow this "language." The practice both increases process efficiency and grows brand loyalty.

Paying for Quality

High quality services are usually more costly than lower quality services and therefore must have a revenue source. While most people expect to pay more at a romantic restaurant than at McDonald's, most customers also expect servers from any business to have a pleasant demeanor and sufficient skills to provide the service properly. Many service providers struggle with the issue of whether to hire staff based on ability, personality, or both (which will cost more). If they have both able and personable staff, they must work out how to retain them through either higher salary or increased job benefits, and also how to fund this. If customers are not willing to pay for both ability and personality, then the firm needs to decide where to compromise. For example, many banks by necessity hire for ability, but Commerce Bank set out to distinguish itself by simplifying its account offerings and therefore being able to hire for personality (Frei and Hajim, 2002). Banks are an example of a service industry where it is difficult to extract revenue for high quality service.

A commonly cited statistic, used to emphasize the importance of quality service, is that it costs five times more money to acquire a new customer than to retain a current customer (e.g., Hart, Heskett, and Sasser, 1990). Further, Frederick Reichheld (1996) argues that it is more profitable to serve long-time customers because they purchase more frequently and can be served more efficiently. Service businesses must therefore work hard to understand customers' preferences and to retain current customers.

Key Decisions for Service Businesses

There are four key decisions that any service business must make (Frei, 2008).

1. *The offering.* What precisely is the service provider going to offer customers? Will there be a variety of options for customers to choose among or one standard service offering? Can customers customize their experience? What types of customers will be served? Is the goal to design for a long-term customer relationship or a more transactional service?
2. *The funding mechanism.* There are usually more ways to charge customers than simply fee-for-service. For example, banks charge account fees, have flexibility in the interest they pay out and charge, and choose the transaction fees they apply. Airlines charge passengers for the ticket but also often for bags or seat booking preferences.
3. *The employee management system.* What management structure will be used in the organization? How will employees be trained in the various processes? How will the processes be structured to accommodate employee variation and to foolproof the service? What type of environment or atmosphere will be created in the workplace?

4. *The customer management system.* What type of environment will the customer experience? In how much of the service will the customer be expected to participate? How will the customer be communicated with, and how will customers communicate their needs? Is this to be a long-term relationship with the customer; if so, how will this be promoted?

These decisions define how the service provider competes in the marketplace. There are a couple of key points for managers to keep in mind. First, a firm's culture is not happenstance—it is the result of the company's deliberate decisions. As described in the Snapshot Application, Southwest Airlines made a deliberate decision to be a "fun" place to work and has aligned its policies to promote this. Second, firms can design customer policies in ways that help to improve the work experience of their employees. For example, most airlines have moved to self-service check-in for domestic travel. This reduces the workload for ticket agents. It can also lead to more satisfied customers because such terminals allow customers to select their preferred seating.

Managing Variability

The introduction to this chapter gave five key types of variability that must be planned for: arrival, request, capability, effort, and subjective preference variability. Table 8–1 (p. 443) outlines how such variability can be accommodated or reduced in a variety of situations. Notice how many of the uncompromised reductions of variability in the table relate to the targeting of specific types of customers. For service firms to have an effective strategy, they cannot be all things to all people, even if their employees are, in theory, capable of such flexibility. For any firm, and service firms are no exception, a well-designed strategy is key for long-term success.

Servicization and Leasing

As introduced in Chapter 1, servicization is when companies bundle additional services with their products. One very old form of servicization is leasing. Leases may be viewed as a service provided to the consumer by the seller. In most cases, leasing is simply another way for the consumer to finance a purchase. It can also be viewed as a means for the consumer to reduce risk. The trend toward leasing of goods and services has increased substantially in recent years for several reasons.

Car leasing has long been an option for consumers. Leases are more popular when financing is difficult or expensive to obtain. The prospect of a low monthly payment attracts consumers, who may ultimately pay more in the long run. Consider the buyer who likes to drive a relatively new car and trades in his or her automobile every three years. This buyer faces the risk of not being able to predict the trade-in value of the car accurately three years down the road. In a lease situation, this residual risk is assumed by the seller. The terms of the lease are predicated on an assumption about the residual value of the car at the end of the lease. If the manufacturer overestimates the residual value of the car, the consumer benefits by simply turning the car in at the end of the lease period. If the manufacturer underestimates the residual value, the lessee wins by purchasing the car at the end of the lease and selling it for a higher market price. Hence, the manufacturer absorbs the risk of estimating the depreciation, which provides an incentive to the consumer to lease rather than buy. Of course, since

Southwest Airlines Competes with Service

Southwest Airlines is the darling of operations management texts for a number of reasons. First, it has been successful partly due to effective operations management. Second, it is one of the very few U.S. airlines that have never filed for bankruptcy. Finally, its very well-thought-out corporate and business strategy matches the firm's operational strategy.

Southwest's effective operations are based on their overarching goal of short turnaround times at airports. Because planes are very capital intensive, and only planes in the air earn money, being able to have short turnaround times has led to extra flights and hence extra revenue. Many of their decisions are guided by this goal. They only operate Boeing 737s, which (a) reduces the complexity of their maintenance and spare-parts systems; (b) reduces the training needed for their pilots; and (c) makes it easier for one plane to be substituted for another should there be a problem. They also typically fly in and out of less congested secondary airports (e.g., Chicago's Midway airport rather than O'Hare). They do not charge for checked baggage. This is due in part to the company's desire to avoid departure delays caused by congestion in the passenger compartment.

The mission of Southwest Airlines is "dedication to the highest quality of customer service delivered with a sense of warmth, friendliness, individual pride, and Company Spirit." Further, Southwest states: "We are committed to provide our employees a stable work environment with equal opportunity for learning and personal growth. Creativity and innovation are encouraged for improving the effectiveness of Southwest Airlines. Above all, employees will be provided the same concern, respect, and caring attitude within the organization that they are expected to share externally with every Southwest Customer."

The company's golden rule is "treat others the way you want to be treated," which is posted on a wall at the entrance to the headquarters of the company. Southwest hires employees with servant's hearts—employees who follow the golden rule, treat others with respect and embrace the Southwest family (Brian, 2018). Posted on another wall at the headquarters is "Customers come second." Southwest focuses on internal customer service as well as external customer service. The philosophy is that satisfied employees create satisfied customers, and satisfied customers create satisfied shareholders.

Southwest has realized that high service quality can only come through empowered, happy, and loyal employees. Southwest works hard to achieve such a workforce. They state that they "hire for attitude and train for skill." Notice they do not pretend that they can hire for both attitude and skill and still be a low-cost airline. Instead, they hire for attitude and make sure they have processes in place to ensure employees can succeed. Their CEO, Gary Kelly, has stated that "our people are our single greatest strength and most enduring long-term competitive advantage." They have innovative hiring processes, such as asking applicants to share their most embarrassing moment and then observing the empathy of the other participants, to ensure that they are indeed hiring for attitude.

Southwest's definition of "high-quality service" is, of course, informed by it being a low-cost airline. While they do still offer free in-flight nonalcoholic beverages and snacks, they do not offer any bells and whistles. Their employees dress informally and do not try to coddle customers. Instead, they are friendly and often quirky. Safety announcements sometimes involve singing, and games are often played during flights. One of the authors was on a flight in which there was a competition to see which section of the aircraft could have the most passengers pack their snack boxes into their peanut bags. Not only was it an amusing game but it also significantly decreased the trash to be collected—which had a small impact on decreasing turnaround time because the trash to be removed had already been compacted.

It is easier to rally employees around the goal of short turnaround times than the goal of lowering costs. This is another example of the alignment of their culture, their metrics, and their processes. Southwest is also very focused on what they do and do not offer. They are very deliberate about not servicing all cities. Boarding is another innovative process at Southwest. Seats are not preassigned. Southwest typically flies point-to-point, rather than using the hub and spoke system followed by most airlines.

Former president Colleen Barrett said of Southwest: "We are not an airline with great customer service. We are a great customer service organization that happens to be in the airline business." Southwest tied with Jet Blue as number one in customer satisfaction in 2018 (Gilbertson, 2019).

Source: Frei and Hajim (2001) and www.southwest.com.

Table 8–1 Strategies for Managing Customer-Introduced Variability

	Classic Accommodation	Low-Cost Accommodation	Classic Reduction	Uncompromised Reduction
Arrival	• Make sure plenty of employees are on hand	• Hire lower-cost labor • Automate tasks • Outsource customer contact • Create self-service options	• Require reservations • Provide off-peak pricing • Limit service availability	• Create complementary demand to smooth arrivals without requiring customers to change their behavior
Request	• Make sure many employees with specialized skills are on hand • Train employees to handle many kinds of requests	• Hire lower-cost specialized labor • Automate tasks • Create self-service options	• Require customers to make reservations for specific types of service • Persuade customers to compromise their requests • Limit service breadth	• Limit service breadth • Target customers on the basis of their requests
Capability	• Make sure employees are on hand who can adapt to customers' varied skill levels • Do work for customers	• Hire lower-cost labor • Create self-service options that require no special skills	• Require customers to increase their level of capability before they use the service	• Target customers on the basis of their capability
Effort	• Make sure employees are on hand who can compensate for customers' lack of effort • Do work for customers	• Hire lower-cost labor • Create self-service options with extensive automation	• Use rewards and penalties to get customers to increase their effort	• Target customers on the basis of motivation • Use a normative approach to get customers to increase their effort
Subjective Preference	• Make sure employees are on hand who can diagnose differences in expectations and adapt accordingly	• Create self-service options that permit customization	• Persuade customers to adjust their expectations to match the value proposition	• Target customers on the basis of their subjective preferences

(Frei, 2006. With permission from Harvard Business Publishing.)

automobiles tend to depreciate most in the first several years, buying and holding a car will be a less expensive alternative in the long run for most vehicles—especially those that are more durable and more reliable. Since auto leases are typically two to four years, the buyer who keeps cars for a long time will not choose to lease.

Leasing (that is, renting) versus buying is also an important choice for the consumer when it comes to choosing how and where to live. While owning a home was long touted as the "American Dream," demand for rental apartments soared in 2019 (Olick, 2019). A Freddie Mac survey found that 82% of renters said renting is more affordable than owning (compared to 67% one year earlier). In the United States, most homeowners have a mortgage on their primary property. A mortgage is a loan provided to homeowner, with the home itself as the collateral. In periods when housing prices rise faster than inflation, homeowners have done very well. However, rising housing prices are not a certainty. There have been several periods in which housing prices have dropped precipitously, including the Great Depression of the 1930s and the Global Financial Crisis of 2007. Renting is a sensible choice for many. Landlords must absorb the risks of repairs and price fluctuations. In other words, the renter purchases the service (somewhere to live) rather than buying the product (the house). This is the essence of servicization.

Service Competition

Competition in service environments is complicated by several factors. In his classic article, Michael Porter (1979) provided insights that changed the strategy field. His five forces model of strategy continues to shape business practice and academic thinking today.

- *Relatively low entry barriers.* It is usually much easier to set up a service business than a production system, which requires equipment and specialized materials. This means that it may be easy for new entrants to enter the market. Thus, the existing service business will need to find a way to grow brand loyalty. Starbucks is a service company that has had unusual success along these lines.
- *Minimal opportunities for economies of scale.* Most services are highly labor dependent. If more capacity is needed, then more staff must be hired. Therefore, there are few opportunities for economies of scale in such proportional scaling of capacity. Like the issue of relatively low entry barriers, this factor means that incumbent firms are constantly under threat from new entrants in a way that firms with economies of scale, who can underprice the competition, are not.
- *Product substitution.* Services are often highly substitutable—customers can easily find an alternative for meeting their needs. For example, there is often very little difference between banking providers. Customers can easily switch banks with few negative consequences (other than the hassle involved). Further, innovations can sometimes replace the need for the service. For example, the internet has meant that in many cases customers can perform the service themselves (e.g., search for travel information) rather than rely on a service provider (e.g., a travel agent).
- *Exit barriers.* Some who start a service-based business do so because of their passion for the work rather than because they have a particularly innovative service offering. For example, boutique owners, bed and breakfast operators, and art gallery owners may all have had a dream of owning their own operation. This makes them less likely to exit the service marketplace even if the firm is not very successful financially.

A firm positioning itself in the services marketplace should consider and evaluate all these factors. Note that, just as for production firms, it is important for firms to choose a strategic position in the marketplace deliberately. In service firms, there is more customer participation and staff are usually more adaptable than in firms that manufacture equipment. Some service firms develop a **fear of focus**—trying to avoid a specific competitive position within the marketplace. This rarely works well. By trying to appeal to everyone, firms can end up appealing to no one.

PROBLEMS FOR SECTION 8.1

1. Give three examples of services you are familiar with that include all of the following: there is an intangible portion to the offering, it is time-perishable, and it involves the customer as coproducer. Explain your answers.
2. Choose five of the dimensions of service quality. For each, name a service firm that you believe competes effectively on this dimension. Explain your answers.
3. Choose a service firm you are familiar with and describe how it has made the four key decisions of: offering, funding mechanism, employee management system, and customer management system.
4. Choose a service firm you are familiar with and describe how each of the five key types of variability from Table 8–1 apply in its setting (e.g., make concrete its source of arrival variability). For each of the five types, describe what sort of accommodation and/or reduction is typically applied.
5. Give an example of a service firm, outside the coffee shop industry, that has managed to compete in an industry with low entry barriers using brand loyalty.
6. Why can car leasing be viewed as a service? What are the advantages and disadvantages of car leasing from the buyer's point of view? Why do manufacturers offer leases?

8.2 FLOW SYSTEMS

The first step in analyzing a process within a service system is to consider the aggregate flow of customers or their orders. This is often done using a process flow diagram to identify the flow and then calculating the capacity of the system to meet the demand for the service. System **utilization**, which at a broad level is demand divided by capacity, provides a measure of system efficiency. It is useful in both services and goods-producing systems.

Process Flow Diagrams

There are a variety of types of **process flow diagrams** from the very high level to the very detailed. Lean production systems apply a process known as **value stream mapping**; a similar process can be used for service systems. Section 1.4 describes patterns of flow within facilities. (Readers interested in learning more about such tools can consult Rother and Shook, 2003).

Figure 8–1 depicts a simple flow diagram for a doctor's office. Patients check in with the receptionist and then wait to be called by a nurse. At the nurse's station, weight and various vital statistics are recorded before patients are shown to an examination room. The doctor sees the patient in the examination room. After treatment, some proportion of patients must revisit the receptionist to schedule a follow-up visit. Finally, the patient leaves the office.

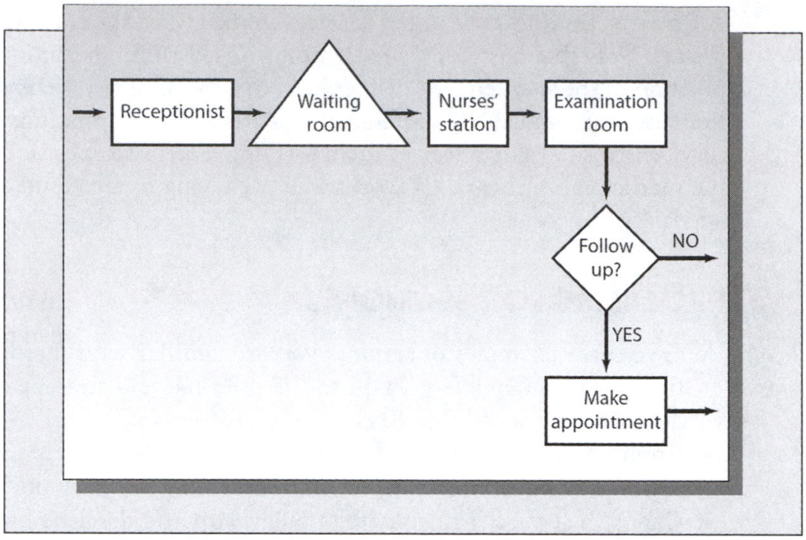

Figure 8–1 Patient flow at a doctor's office

A more sophisticated flow diagram will also show the **resources** used for a customer to complete service. Resources may correspond to staff who perform the service or equipment that is used to process the customer's order. For example, in a doctor's office resources include the receptionist, the nurses, the doctor (who may treat multiple patients in different examination rooms), and the equipment used. In a restaurant, resources include tables, wait staff, the maître d', chefs, ovens, and other cooking equipment.

Capacity

Key to determining the performance of any system is the **capacity** of the relevant resources. This is defined as the number of customers per hour (or per any relevant time unit) that may be processed by the resource. If the resource is only available some fraction of the time, this should be taken into account when calculating capacity. Similarly, if the resource, or set of resources, can process customers in parallel (i.e., in batches), then the capacity calculation should also consider this factor. Capacity is therefore defined as:

$$\text{Number of parallel resources} \times \frac{\text{Units per batch}}{\text{Time per batch}} \times \text{Fraction of time available.}$$

In many cases, both the batch size and the fraction of time available are equal to one. For example, if a single server takes 5 minutes to process a customer then its capacity is 0.2 customers per minute, or 12 customers per hour. For two servers, the capacity is doubled.

EXAMPLE 8.1

Suppose a theme park roller coaster has a train with six carriages. Each carriage takes eight customers. A train arrives at the loading point every 30 seconds. However, each hour, on the hour, the roller coaster is shut down for 3 minutes to check the system. What is the capacity of the roller coaster?

Solution
We will calculate capacity as customers per hour. Therefore, the time per batch is calculated as 30/(60 × 60) hours, rather than just 30 seconds. Total capacity is as follows:

$$6\times\frac{8}{30/(60\times 60)}\times\frac{57}{60}=\frac{6\times 8\times 60\times 57}{30}=5472 \text{ customers per hour.}$$

The system **bottleneck** is defined as the resource group with the smallest capacity (assuming that all customers go through all resources). If some group of required resources can only process 10 customers per hour, then there is no way to get more than 10 customers per hour through the system as a whole. Therefore, the **system capacity** is defined as the capacity of the bottleneck resource group.

Calculations are more complex if customers follow different paths through the system or have different service requirements. Also, if there are multiple paths that must be completed in parallel for the customer's service to be complete (e.g., if a patient must receive the results from both x-rays and blood tests before moving on to treatment), all paths must be considered separately. Finally, the customer mix will determine the precise resource capacity when different types of customers have differing processing requirements. Similar principles to those discussed here may be used for such systems, although the calculations are more complicated. (for more detail, see Anupindi, Chopra, Deshmukh, Van Mieghem, and Zemel, 2012).

Flow Rates and Utilization

In most flow analysis, we assume that the **arrival rate** of customers to the system is less than the system capacity. Let λ denote the number of customers per unit time that arrive to the system. While it is possible to do transient (that is, short-term) calculations with an arrival rate greater than the system capacity, one usually assumes **steady state**. In steady state, the system is assumed to have been running long enough, under a similar set of conditions, so that the effect of the starting state disappears. Further, the conditions under which the system is running are assumed to be relatively constant with no large effect from seasonality.

A common cause of confusion is the calculation of the arrival rate to a set of resources within the process (e.g., the arrival rate to the examination room in Figure 8–1). Unless customers are leaving without service or being created somehow (e.g., a maternity ward), the arrival rate to any resource group in the process must be the same as the arrival rate to the start of the process, which must also equal (on average) the rate at which customers depart the system. Customers do not depart at the processing rate of the final server because, so long as arrivals do not exceed its capacity (the usual assumption), the server will be idle for portions of time when there are no customers to process.

If there are s resources that can process μ customers per unit time, then the resources have capacity $s\mu$ and the **utilization** of this set of resources is given by

$$\rho=\frac{\lambda}{s\mu}.$$

Notice that utilization ρ is always nonnegative and is less than one so long as the arrival rate is less than the capacity. It is a dimensionless measure that represents the fraction of time the set of resources is busy. It can also be calculated as

$$\frac{\text{Amount of work that needs to be done}}{\text{Time available to work}}.$$

EXAMPLE 8.2

Consider the roller coaster in Example 8.1 and suppose that customers arrive to the ride at a rate of 90 customers per minute and all customers wait for service. What is the average utilization of the roller coaster?

Solution

Using the capacity calculated in Example 8.1, the service rate $s\mu = 5472$ customers per hour. We must convert the arrival rate, λ, to the same time units, namely $90 \times 60 = 5400$ customers per hour. Utilization is therefore

$$\rho = \frac{\lambda}{s\mu} = \frac{5400}{5472} = 0.9868.$$

Thus, each seat in the roller coaster will be full 98.68 percent of the time and empty 1.32 percent of the time.

PROBLEMS FOR SECTION 8.2

7. Consider the flow in Figure 8–1 but assume that no patients require follow-up appointments. All patients first check in with the single receptionist, which takes an average of 6 minutes. They are then seen by one of two nurses who take weight, blood pressure, etc., which takes an average of 10 minutes. Finally, they are seen by a GP in an examination room, which takes an average of 16 minutes. There are three GPs. The table below summarizes this data. Assume the clinic has no waiting room or examination room constraints.

   Assume that there are eight patients arriving an hour during working hours.

Resource	Minutes per patient	Number of resources
Receptionist	6	1
Nurse	10	2
Doctor	16	3

   a. What is the capacity of each resource?
   b. Which resource is the bottleneck and hence what is the capacity of the system?
   c. Suppose they wish to increase system capacity, what actions do you recommend?
   d. What are the utilizations of the receptionist, nurses, and GPs, respectively (during working hours)?

e. Now suppose that 20 percent of patients require follow-up appointments, which take 2 minutes to book on average. This means that, on average, the receptionist spends $6 + 0.2 \times 2 = 6.4$ minutes per patient. What is the new capacity and utilization of the receptionist?

8. Does increasing the capacity of a resource ever increase the arrival rate to the resource group that follows it in the process flow? Explain your answer.
9. What does a utilization of greater than one mean in practice?

8.3 MODELING UNSCHEDULED ARRIVALS

One of the key features of a service system is that unscheduled arrivals must be buffered using waiting time, rather than inventory. Who has not experienced hours of waiting in line for service? We line up to wait our turn at banks, supermarkets, hair stylists, and restaurants. Estimates are that the average time a person spends in line in a lifetime is six months (Horton, 2015)—that is a significant amount of time waiting in line, or **queueing**, as it is more formally known. As we will see in Section 8.5, queueing is exacerbated by variability. This section examines arrival process variability.

Figure 8–2 shows the arrivals of calls to a call center. The variations seen in the first three boxes from month to month, day to day, and hour to hour are relatively predictable seasonal variations. Usually, staffing can account for seasonal variation of a predictable nature. However, the fourth box shows an extraordinary amount of variation in the number of calls from

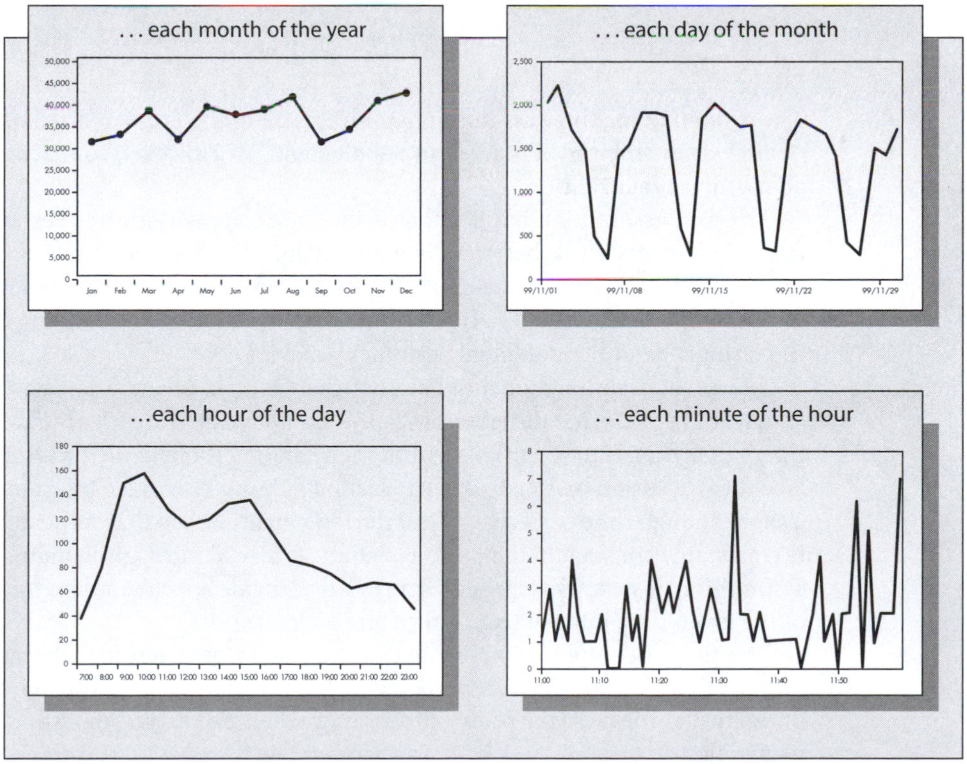

Figure 8–2 Unscheduled arrivals to a call center

Source: Gans, Koole, and Mandelbaum, 2003.

minute to minute. Such variability is impossible to predict, leading to difficulties in determining suitable staffing levels (even if it could be predicted). It causes delays in call centers and queueing in other service systems. A good model for arrivals of this type is the Poisson process.

Poisson Process

Let $N(t)$ be the number of arrivals to some service facility between time zero and t. Then $\{N(t): 0 \leq t < \infty\}$ is called a **stochastic process** because it is random (i.e., stochastic) and it evolves over time (so can be called a process). Arrivals to service systems can follow many different types of stochastic processes, but a common model to use for unscheduled arrivals is the **Poisson process**. The Poisson process is an arrival process that satisfies the following three key assumptions.

1. The number of arrivals in disjoint intervals are independent.
2. The number of arrivals in an interval depends only on the interval's length.
3. For a very short interval (of duration h): the probability of one arrival is approximately λh; and the probability of more than one arrival is negligible.

A more formal statement of these three assumptions is made in Supplement 2.1. Together, they imply that for any $s, t \geq 0$, the distribution of the number of arrivals in an interval $(s, s + t]$ (i.e., $N(t + s) - N(s)$) is **Poisson** with mean λt, which means that

$$P\{N(t)=n\}=\frac{(\lambda t)^n e^{-\lambda t}}{n!} \quad \text{for } n=0,1,2,\ldots$$

The proof that the three assumptions above result in the Poisson distribution for the number of arrivals in an interval is given in Supplement 2.1. However, some comments are in order about their practicality.

The first assumption implies that if the service provider observes an unusually high or low number of arrivals between 9am and 10am, this does not affect the distribution of the number of customers likely to arrive between 10am and 11am. This assumption is reasonable in a system where customer arrivals are not driven by some underlying force but instead are the result of random individual customer behavior.

The second assumption implies that the number of arrivals between 9am and 10am should have the same distribution as the number between 10am and 11am because both intervals are an hour long. This assumption can be problematic for systems with seasonality (see the discussion on steady state in Section 8.2) but is avoided by studying the system over a short enough time period (e.g., just during lunchtime) so that the arrival rate to the system is reasonably constant in the period of study. If arrival rates are changing slowly enough, the assumption of steady state may still form a reasonable approximation for an accurate analysis of the given time interval (e.g., Green and Kolesar, 1991).

The final assumption implies that a very short time interval should have at most one arrival. Thus, if the system experiences batch arrivals, then each batch must be counted as one arrival, otherwise the probability of more than one arrival in a short interval will not be negligible.

Other than the assumptions above, there are two further models of arrivals that result in a Poisson process. First, Martin Lariviere and Jan Van Mieghem (2004) showed that Poisson arrivals can also occur as the result of strategic customers trying to avoid congestion if the population is large and the time horizon is long. Second, as discussed next, if the time between arrivals has an exponential distribution, then the number of arrivals will follow a Poisson process.

Exponential Interarrival Times

Let $N(t)$ be a Poisson process with rate λ, and let $T_1, T_2, \ldots$ be successive interarrival times; that is, the first customer arrives at time T_1, the second at time $T_1 + T_2$, the third at $T_1 + T_2 + T_3$, and so on. If arrivals follow a Poisson process then the interarrival times $T_1, T_2, \ldots$ have an **exponential distribution**. That is, if X is a random variable representing the time between successive arrivals, then

$$P\{X > t\} = \exp(-\lambda t).$$

Note that $E[X] = 1/\lambda$ and $\text{var}[X] = 1/\lambda^2$ so that the coefficient of variation (see Section 6.5) $CV[X] = 1$.

Not only does a Poisson process result in exponential interarrival times, but the relationship also goes the other direction. That is, if the time between any two consecutive arrivals is exponential with mean $1/\lambda$ (and independent of all other interarrival times), then the number of arrivals in any interval of length t must be Poisson with mean λt. This equivalence is proven in Chapter 13 (see also Figure 13–5 for further understanding).

In Chapter 13 (on reliability modeling) we discuss the **memoryless property** of the exponential distribution and its relationship to the Poisson process. Both the exponential and the Poisson distribution play a key role in queueing theory, just as they do in reliability theory. When we talk about purely random arrivals in queueing, we mean that the arrival process is a Poisson process. A purely random service process means that service times have the exponential distribution. We use the term "purely random" because of the memoryless property of the exponential distribution.

The memoryless property states that no matter how much time has passed, the distribution of the time until the next arrival is the same as the distribution of the time of first arrival, T_1. (This idea is formalized in Supplement 2.1.) Such a property may make little intuitive sense until one considers it in the light of coin flips. Even if one has thrown 20 heads in a row, the probability of a head on the next throw (from a fair coin, of course) is still 50/50. Alternatively, think of the property with regard to forgetfulness around the home. Even though you have not forgotten to turn a light off recently, this has little effect on the likelihood that you will be careless and forget to turn one off today (family members leaving lights on, at least in our experience, follow a highly random process that appears Poisson to an observer).

Because the exponential distribution has the memoryless property, it is the continuous equivalent of the geometric (coin flip) distribution (as will be discussed further in Chapter 13). This is equivalent to describing arrivals as "random" in that knowing something about one period of time tells nothing about the following time interval (see the first Poisson assumption given earlier). As a final example, suppose that cabs arrive at a cabstand according to a Poisson process. You arrive at the stand at some random time and wait for the next cab. Your waiting time is exponential, with exactly the same distribution as the time between two successive arrivals of cabs. However, time has already passed, so if you add the time that has passed to your expected wait it will be longer than the original exponential distribution. This is what is termed the **inspection paradox**, described further in Supplement 2.1.

EXAMPLE 8.3

Suppose customers arrive at a 7–11 convenience store according to a Poisson process with a rate of 10 per hour. What are the probabilities of (a) no customers in an hour; (b) exactly 5 customers in an hour; (c) exactly 10 customers in two hours; and (d) at least two customers in half an hour?

Solution

Here $\lambda = 10$ per hour and we must compute:

a. $P\{N(1) = 0\} = e^{-10\times 1} = 0.00005;$

b. $P\{N(1) = 5\} = \dfrac{(10)^5 e^{-10\times 1}}{5!} = 0.038;$

c. $P\{N(2) = 10\} = \dfrac{(10\times 2)^{10} e^{-10\times 2}}{10!} = 0.0058;$

d. $P\{N(0.5) \geq 2\} = 1 - P\{N(0.5) = 0\} - P\{N(0.5) = 1\} = 1 - e^{-5} - \dfrac{(5)^1 e^{-5}}{1!} = 0.96.$

Notice that the probability of 10 arrivals in two hours is not twice the probability of 5 arrivals in one hour; in fact, it is less than the probability of 5 arrivals in one hour. Explaining why this is the case is left as an exercise for the reader (see problem 12).

General Arrival Processes

If arrivals are not purely random (Poisson) then an important metric is the arrival process variation. Let c_a^2 be the squared coefficient of variation associated with the arrival process (see also Section 6.5). That is, $c_a^2 = \lambda^2 \sigma_a^2$, where σ_a^2 is the variance of the times between arrivals and λ is the arrival rate. For a Poisson process, $\sigma_a^2 = \lambda^2$ and $c_a^2 = 1$. Notice that if the average time between arrivals stays the same but the variance increases, then c_a^2 increases. Also, c_a^2 is dimensionless, in that it is just a number that says something about how variable the system is.

Thus, $c_a^2 = 1$ when arrivals are independent, such as customers walking into a store. However, when there is a schedule, we expect to see $c_a^2 < 1$ For example, patients arriving for doctor's appointments should be much less variable than unscheduled emergency department patients. It is possible to have $c_a^2 > 1$ when the system is highly variable, which can occur when there are batch arrivals. Examples of batch arrivals include the arrival of people from tour buses to an attraction or students arriving at the local coffee shop right before or after classes, when arrivals are much more common than when classes are in session.

If there is no data on the exact interarrival times, a reasonably good estimate can be calculated as follows. Let N be the number of arrivals per time period (e.g., per hour). Calculate the mean $E[N]$ and variance $\mathrm{var}[N]$. Then $c_a^2 \approx \mathrm{var}[N]/E[N]$. This estimate is exact for the Poisson process because if N is Poisson then $\mathrm{var}[N] = E[N] = \lambda t$, where t is the time period considered. The lack of a square on the denominator of this estimate is correct and follows from the central limit theorem for renewal processes (e.g., Wolff, 1989). Although this result is an approximation, it is often easier to calculate than the exact form. It relies on the mean and variance of the number of arrivals per time period, rather than the mean and variance of the time between two arrivals, which requires an accurate stopwatch.

Pooling in Services

We saw uses of pooling in Chapter 6, but it can also be effective in service systems. For example, suppose we have only one bed and unscheduled patient arrivals. Then, if we want to be 95 percent sure the bed is available for a new patient, its utilization can be at most 5 percent. However, suppose we have thousands of beds and we want to be 95 percent sure a bed is available when a new patient arrives. We should be able to keep each bed at above 99 percent utilization and still be reasonably sure at least one will be available for the new arrival. Thus, if we can pool streams of patients into one set of bed resources, then we will be able to keep the bed utilizations higher than if each type of patient gets a specialized bed type.

This question is not just academic. Hospitals often wish to separate patients by type, and emergency departments often separate the arrival streams of low and high priority patients. Such service lines are usually more efficient and provide higher quality of care than the more general pooled layouts. However, their lack of pooling can mean that they will have either low utilizations or poor service levels. Therefore, service environments should treat specialization with extreme caution. Even a relatively small amount of cross training or cross sharing of resources can help mitigate utilization problems caused by dedicated service lines. These ideas are related to decisions of product versus process layouts as discussed in Section 1.4.

Probability of Delay

A useful service metric is the probability that a customer is delayed. If there is only one server and arrivals are Poisson, the probability a customer is delayed is simply ρ, where ρ is the utilization rate. However, if arrivals are not Poisson then this is only an approximation. As an extreme example, suppose customers arrive in *exactly* one-minute intervals and each takes *exactly* 45 seconds to serve. Then $\rho = 0.75$ but no customer is ever delayed. The server is busy 75 percent of the time, but an arriving customer never sees the server busy because he arrived exactly 15 seconds after the previous customer left the system.

When arrivals are Poisson, the probability that a customer is delayed is the same as the probability all servers are busy at any random instant. If arrivals are truly random (Poisson), an arriving customer will see the system in its usual or typical state. If arrivals are not truly random, the probability that all servers are busy at the instant the customer arrives, so that the customer is delayed, is not exactly the same as the probability all servers are busy at a randomly chosen instant in time (although the difference is likely to be small). The so-called **PASTA**—Poisson Arrivals See Time Averages—property guarantees that customers arriving according to a Poisson process indeed see the system in its usual or "time-averaged" state (e.g., see Wolff, 1989).

Hirotaka Sakasegawa (1977) developed a useful and simple approximation for the probability of delay, p^d, which is

$$p^d \approx \rho^{-1+\sqrt{2(s+1)}},$$

where ρ is the utilization and s is the number of servers. This approximation is used for both Poisson and non-Poisson arrivals, although it is usually more accurate for Poisson arrivals. If the number of servers $s = 1$ then the right-hand-side equals ρ, which is the exact probability of delay for Poisson arrivals. Further, the right-hand-side approaches zero as s grows to infinity; with a large enough number of servers, customers will not be delayed. The formula is based on a least squares fit to the exact formula for the so-called M/M/s queue (see

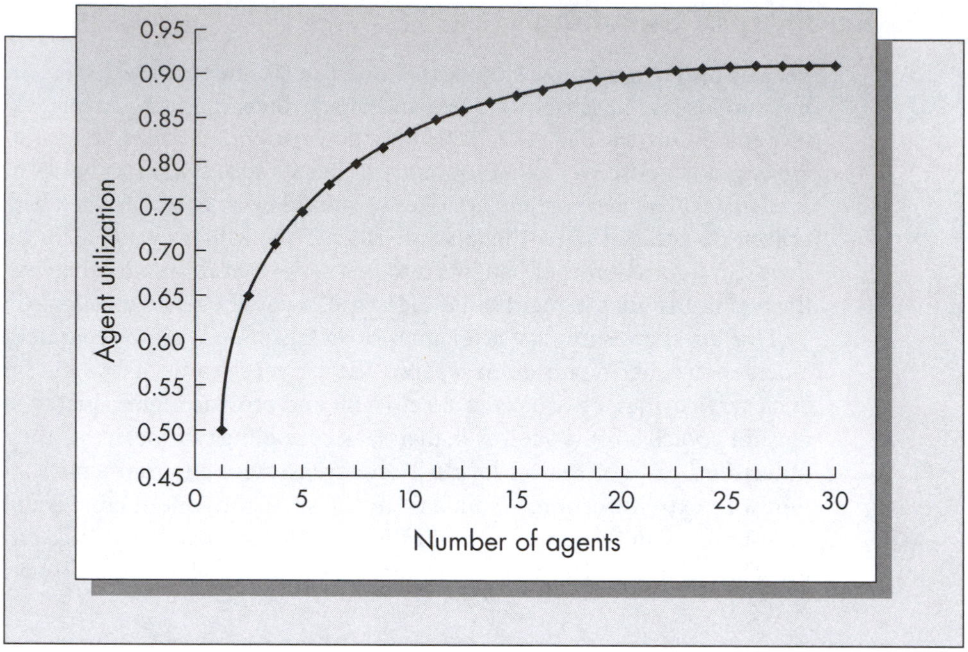

Figure 8–3 Scaling of utilization in number of agents

Section 8.4). Although the derivation of this formula is outside the scope of this text, the intuition behind it should be clear to the reader.

Figure 8–3 uses the Sakasegawa formula to derive the number of agents needed to achieve a 50 percent no hold rate (i.e., no delay) in a call center. If there is only one agent, the utilization is only 50 percent. However, as the scale grows so can agent utilization.

While the Sakasegawa approximation is simple and intuitive, the actual probability of delay depends on both the staffing decisions made by the firm and the arrival and service processes. Ward Whitt (2004) has defined an **efficiency-driven** regime as one where p^d tends to one as both arrival rate and s grow to infinity, whereas a **quality-driven** regime has p^d tending to zero (as in the Sakasegawa formula). The so-called **quality-and-efficiency** (QED) regime is one where staffing is matched to arrival rate growth so that p^d tends to a value strictly between 0 and 1. An approximation useful for this scenario is given in Supplement 2.4.

EXAMPLE 8.4

Consider the roller coaster example of Examples 8.1 and 8.2. What is the probability of delay?

Solution

In Example 8.1 we saw that the number of seats (servers) in the roller coaster train/carriages was $s = 6 \times 8 = 48$; in Example 8.2 we calculated $\rho = 0.9868$. Under the Sakasegawa formula, $p^d = \rho^{-1+\sqrt{2(s+1)}} = 0.9869^{-1+\sqrt{2(48+1)}} = 0.889$. This probability is much less than $\rho = 0.9868$ due to the pooling effects of the servers. Therefore, an arriving customer has only an 89 percent chance of being delayed even though the seats in the ride will have an occupancy of over 98 percent.

PROBLEMS FOR SECTION 8.3

10. Suppose that arrivals to a coffee shop follow a Poisson process with average rate of one customer every 5 minutes. What are the probabilities of (a) no customers in 5 minutes; (b) exactly one customer in a minute; (c) exactly two customers in 2 minutes; and (d) at least two customers in 10 minutes?
11. Suppose that arrivals to an emergency room follow a Poisson process with mean two patients every 5 minutes. Calculate the probabilities of (a) no customers in 2 minutes; (b) exactly 2 customers in a minute; (c) exactly 5 customers in 3 minutes; and (d) no more than 2 customers in 5 minutes.
12. In Example 8.3 we saw that the probability of 10 arrivals in two hours is not double the probability of 5 arrivals in one hour; in fact, it is less than the probability of 5 arrivals in one hour. Explain why this is the case.
13. The following data has been collected on the interarrival times of calls to a call center: 3.33, 4.21, 5.12, 1.24, 0.11, 10.23, 7.65, 1.23, 4.44, 2.89, 4.92, 3.87, 2.67, 3.51, and 5.90 minutes. Estimate the squared coefficient of variation of the arrival process. Is the arrival process likely to be Poisson? Why or why not?
14. The following data has been collected on the number of customers seen to arrive at a museum in a succession of 5-minute intervals: 5, 1, 3, 7, 5, 5, 6, 7, 5, 7, 4, 8, 1, 5, 2, 3, and 5. Estimate the squared coefficient of variation of the arrival process. If this data was known to come from a Poisson process, what would be your estimate of λ, the rate of customer arrivals?
15. Recreate Figure 8–3 for a call center that wants to achieve a 40 percent no hold rate. Suppose that agents only become cost effective (where the revenue they generate exceeds their cost) if they have 90 percent utilization. How large, in terms of number of agents, does the call center need to be to have agents who generate a positive profit? If mean call time is 5 minutes, what would the arrival rate need to be for 90 percent agent utilization of that number of agents?
16. What does problem 15 imply about whether an online and catalog retailer, such as L.L.Bean, should have regional or centralized call centers? What other considerations might affect this decision?
17. Suppose a bank has three tellers who are each busy 80 percent of the time. Estimate the probability of delay for a randomly arriving customer.

8.4 QUEUEING SYSTEMS

Queueing theory is the science of waiting-line processes. Virtually all the results in queueing theory assume that both the arrival and service processes are random. The interaction between these two processes makes queueing an interesting and challenging area. Figure 8–4 shows a typical queueing system. Customers arrive at one or more service facilities. If other

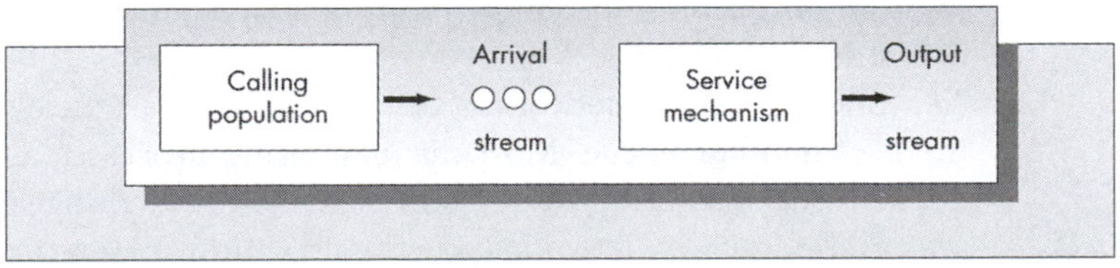

Figure 8–4 Typical single-server queueing system

customers are already waiting for service, depending on the service discipline, newly arriving customers would wait their turn for the next available server and then exit the system when service is completed.

Queueing problems are common in operations management. In the context of manufacturing, most complex systems can be thought of as networks of queues. However, queueing problems occur most frequently in service systems. Call or contact centers are a good example of complex queueing systems. Telephone calls are routed through switching systems, where they queue up until they are either switched to the next switching station or routed to their final destination. Abner Erlang, a Danish telephone engineer, studied a village telephone exchange in the early 1900s, determining the number of callers attempting to call someone outside the village who had to wait because all of the lines were in use. He developed formulas to calculate the number of lines and the number of operators necessary for optimum service. His formulas remain the most widely used today (Prisco, 2019).

Structural Aspects of Queueing Models

Queueing systems share a number of common structural elements.

1. *Arrival process*. The process describing arrivals of customers to the system was discussed in Section 8.3.
2. *Service process*. The service process is characterized by the distribution of the time required to serve a customer. The easiest case to analyze is when the distribution of service times is exponential; other more general service distributions can also provide queueing results.
3. *Service discipline*. This is the rule by which customers in the queue are served. Most queueing problems occurring in service systems are first-come, first-served (FCFS). This is the rule we usually think of as "fair." However, other service disciplines are also common. When we buy milk, we may check the dates of the bottles and buy the one with the latest expiration date. Thinking of the milk as the queue, this means that the service discipline is last-come, first-served (LCFS). Hospital emergency rooms will give priority to patients with a life-threatening condition (such as trauma from an automobile accident) over patients with less severe problems. This is referred to as a priority service discipline.
4. *Capacity of the queue*. In some cases, the size of the queue might be limited. For example, restaurants and movie theaters can accommodate only a limited number of customers. From a mathematical point of view, the simplest assumption is that the queue size is unlimited. Even where there is a finite capacity, it is reasonable to ignore the capacity constraint if the queue is unlikely to fill.
5. *Number of servers*. Queues may be either single-server or multiserver. A bank is the most common example of a multiserver queue. Customers form a single line and are served by the next available server. By contrast, the checkout area of a typical supermarket is *not* a multiserver queue. Because a shopper must commit to a specific checkout line, this is a parallel system of (possibly dependent) single-server queues. Another example of a multiserver queue is the landing area of a large airport; planes may take off or land on one of several runways.
6. *Network structure*. A network of queues results when the output of one queue forms the input of another queue. Most manufacturing processes are generally some form of a queueing network. Highway systems, telephone switching systems, and medical facilities are other examples. Network queueing structures are often too complex to analyze mathematically and are therefore simulated (see Section 8.5).

Notation

David Kendall's (1953) shorthand notation for single-station queueing systems takes the form:

$$\text{Label 1/ Label 2/ Number},$$

where Label 1 is an abbreviation for the arrival process, Label 2 is an abbreviation for the service process, and Number indicates the number of servers. [More complex notations exist that include capacity restrictions and specification of the queueing discipline (see, for example, Shortie, Thompson, Gross and Harris, 2018). The letter "M" is used to denote pure random arrivals or pure random service. This means that interarrival times are exponential (i.e., the arrival process is Poisson) or service times are exponential. The "M" stands for "Markovian," a reference to the memoryless property of the exponential distribution. The simplest queueing problem is the one labeled M/M/1. Another symbol that is commonly used is "G," for general distribution. Hence, G/G/s would correspond to a queueing problem in which the interarrival distribution is general, the service distribution is general, and there are s servers. There are other labels for other distributions, but we do not consider them here. Some useful notation, some of which has already been covered, follows.

λ = Arrival rate to system.

μ = Service rate per server.

s = Number of servers.

ρ = Utilization rate $= \lambda/(s\mu)$.

w = Expected time a customer spends in the system in steady state.

w_q = Expected time a customer spends in the queue in steady state.

l = Expected number of customers in the system in steady state.

l_q = Expected number of customers in the queue in steady state.

W = Variable representing time in system for an arbitrary customer in steady state; $E[W] = w$.

L = Variable representing number of customers in system in steady state; $E[L] = l$.

p_n = Steady-state probability of n customers in the system; $p_n = P\{L = n\}$.

p^d = Steady-state probability an arbitrary customer in steady state is delayed; $p^d = P\{W > 0\}$.

Little's Law

In this section we show some useful relationships between the steady state expected values l, l_q, w, and w_q. Because w_q is the expected time in the queue only, whereas w is the expected time in the queue plus the expected time in service, it follows that w_q and w differ by the expected time in service. That is,

$$w = w_q + 1/\mu.$$

(If the service rate is μ, it follows that the mean service time is $1/\mu$.)

Little's law is named for John Little of the Massachusetts Institute of Technology, who proved that it holds under very general circumstances. It is a simple but very useful relationship between the l's and the w's. The basic result is

$$l = \lambda w.$$

We will not present a formal proof of this result, providing only the following intuitive explanation. Consider a customer who joins the queue in steady state. At the instant the customer is about to complete service, he looks over his shoulder at the customers who have arrived behind him. There will be, on average, l customers in the system. The expected amount of time that has elapsed since he joined the queue is, by definition, w. Because customers arrive at a constant rate λ, it follows that during a time w, on average, there will have been λw arrivals, giving $l = \lambda w$. For example, if customers arrive at the rate of 2 per minute and each spends an average of 5 minutes in the system, then there will be 10 customers in the system on average. Another version of Little's law is

$$l_q = \lambda w_q.$$

The argument here is essentially the same, except that the customer looks over his shoulder as he enters service, rather than when completing service.

The M/M/1 Queue

The **M/M/1 queue** assumes Poisson arrivals, exponential service times, and a single server serving customers in a FCFS fashion. As discussed above, Poisson arrivals are a reasonably good assumption for unscheduled systems. Further, if there is a mix of many different types of jobs, the exponential distribution can be realistic for service times. Otherwise, it tends to be a too variable distribution. However, it will often provide a reasonable upper bound because its extra variability leads to the overestimation of most system statistics.

Supplement 2.2 shows that the steady state probability of n customers in the M/M/1 queue is given by

$$p_n = \rho^n(1 - \rho) \qquad \text{for } n = 0, 1, 2, \ldots$$

This distribution, known as the geometric distribution, is pictured in Figure 8–5. Several aspects of this result are both interesting and surprising. First, the geometric distribution is the discrete analog of the exponential distribution. Second, the probability of state n is a decreasing function of n, so long as $\rho < 1$. As ρ gets close to one, the variance increases and the distribution "spreads out" (large values become more likely). As ρ gets close to zero, the probabilities associated with larger values drop off to zero more quickly. This means that the most likely state is *always* state 0 (as long as $\rho < 1$). This is an extremely surprising result! As ρ approaches one, the queues get longer and longer. One would have thought that the probability of n in the system for some large value of n would be higher than the probability of zero in the system when ρ is near one. This turns out not to be the case. What is true is that, for ρ close to one, the probability that the system is in state zero is close to the probability that the system is in state 1 or state 2; whereas, for ρ close to zero, the probability that the system is in

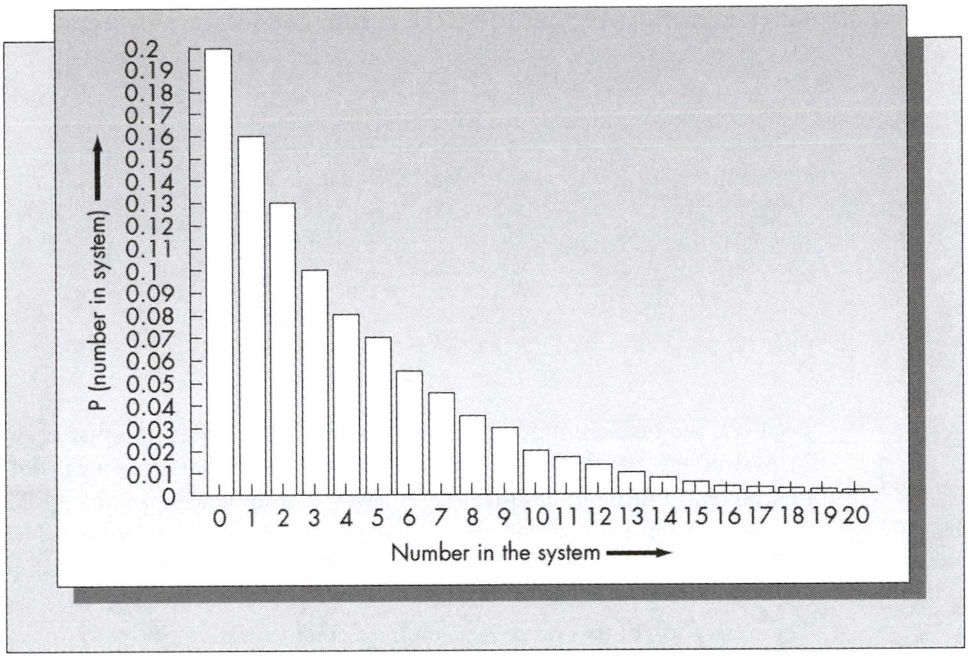

Figure 8–5 Geometric distribution of number in system for M/M/1 queue ($\rho = .8$)

state zero is much larger than the probability that it is state 1 or 2. This phenomenon holds *only* for Poisson arrivals and exponential services, however.

The distribution of p_n may be used to calculate l, l_q, w, and w_q. The expected value of a random variable is the sum of its outcomes weighted by the probabilities of those outcomes. It follows that the average, or expected, number of customers in the system in steady state, l, is:

$$l = \sum_{i=0}^{\infty} ip_i = \sum_{i=0}^{\infty} i(1-\rho)\rho^i = (1-\rho)\rho\sum_{i=0}^{\infty} i\rho^{i-1}.$$

To complete the calculation, we use the fact that

$$\sum_{i=0}^{\infty} i\rho^{i-1} = \frac{d}{d\rho}\left(\sum_{i=0}^{\infty} \rho^i\right) = \frac{d}{d\rho}\left(\frac{1}{1-\rho}\right) = \frac{1}{(1-\rho)^2}.$$

It follows that

$$l = \frac{(1-\rho)\rho}{(1-\rho)^2} = \frac{\rho}{(1-\rho)}.$$

For the case of l_q, we note that the number in the queue is exactly one less than the number in the system as long as there is at least one in the system. It follows that

$$l_q = \sum_{i=1}^{\infty}(i-1)p_i = \sum_{i=1}^{\infty} ip_i - \sum_{i=1}^{\infty} p_i = l - (1-p_0) = l - \rho = \rho^2/(1-\rho).$$

Given knowledge of l and l_q, we can obtain w and w_q directly from Little's law. From Little's law, $w = l/\lambda$, giving

$$w = \frac{\rho}{\lambda(1-\rho)} = \frac{1/\mu}{(1-\rho)} = \frac{1}{(\mu-\lambda)}.$$

Similarly, $w_q = l_q/\lambda$, which gives

$$w_q = \frac{\rho^2}{\lambda(1-\rho)}.$$

Let W be the random time a customer spends in the system, so that $E[W] = w$. Then, for the M/M/1 queue, the distribution of W is known and remarkably turns out also to be an exponential distribution; it has mean $w = 1/(\mu - \lambda)$ and hence rate $(\mu - \lambda)$. That is,

$$P\{W \le t\} = 1 - e^{-(\mu-\lambda)t} \qquad \text{for all } t \ge 0.$$

The derivation of this result is given in Supplement 2.2.

EXAMPLE 8.5

Customers arrive one at a time, completely at random, to an ATM at the rate of six per hour. Customers take an average of 4 minutes to complete their transactions. However, ATM tasks are highly variable ranging from simple withdrawals to complex deposits; thus, service times may be considered truly random. Customers queue up on a first-come, first-served basis and no customers leave without service. Assume there is only one ATM.

a. Find the following expected measures of performance for this system: the expected number of customers in the system, the expected number of customers waiting for service, the expected time in the system, and the expected time in the queue.
b. What is the probability that there are more than five people in the system at a random point in time?
c. What is the probability that the waiting time in the system exceeds 10 minutes?
d. Given these results, do you think that management should consider adding another ATM?

Solution

The statement that customers arrive one at a time completely at random implies that the input process is a Poisson process. The arrival rate is $\lambda = 6$ per hour. The statement that service times are also truly random implies service times may be modeled by an exponential distribution. The mean service time is 4 minutes = 1/15 hour, so that the service rate is $\mu = 15$ per hour. Thus, this is an M/M/1 queue with utilization rate $\rho = \lambda/\mu = 6/15 = 2/5 = 0.4$.

a. $l = \rho(1 - \rho) = (2/5)/(3/5) = 2/3$ (= 0.6667 customers)
   $l_q = \rho L = (2/5)(2/3) = 4/15$ (= 0.2667 customers)
   $w = l/\lambda = (2/3)/6 = 2/18 = 1/9$ hour (= 6.6667 minutes)
   $w_q = l_q/\lambda = (4/15)/6 = 4/90 = 2/45$ hour (5 2.6667 minutes)

b. Here we are interested in $P\{L > 5\}$. In general,

$$P\{L > k\} = \sum_{i=k+1}^{\infty} p_i = \sum_{i=k+1}^{\infty} (1-\rho)\rho^i = (1-\rho)\sum_{i=k+1}^{\infty} \rho^i$$
$$= (1-\rho)\rho^{k+1}\sum_{i=0}^{\infty} \rho^i = (1-\rho)\rho^{k+1}(1/(1-\rho)) = \rho^{k+1}.$$

Hence, $P\{L > 5\} = \rho^6 = (0.4)^6 = 0.0041$.

c. Here we are interested in $P\{W > 1/6\}$.

$$P\{W > t\} = e^{-(\mu-\lambda)t} = e^{-(15-6)/6} = e^{-1.5} = 0.223.$$

d. The answer is not obvious. Looking at the expected measures of performance, it would appear that the service provided is reasonable. The expected number of customers in the system is fewer than one, and the average waiting time in the queue is less than 3 minutes. However, from part (c) we see that the proportion of customers who have to spend more than 10 minutes in the system is more than 20 percent. This means that there are probably plenty of irate customers, even though, on average, the system looks good. This illustrates a pitfall of only considering expected values when evaluating queueing service systems.

PROBLEMS FOR SECTION 8.4

18. A supermarket manager notices that there are 20 customers at the checkouts and also knows that arrivals to the checkout at that time of day are at a rate of about two per minute. About how long are customers spending in the checkout process (queueing and being served) on average?
19. Suppose that the billing cycle for a firm is 60 days and they invoice on average \$5000 per day. What is the average total dollar amount of outstanding invoices that they carry?
20. Which of the three variables in Little's law do you think is generally the most difficult to estimate (and why)?
21. A teller works at a rural bank. Customers arrive to complete their banking transactions on average one every 10 minutes; their arrivals follow a Poisson arrival process. Because of the range of possible transactions, the time taken to serve each customer may be assumed to follow an exponential distribution with a mean time of 7 minutes. Customers wait in a single queue to get their banking done, and no customer leaves without service.
    a. Calculate the average utilization of the teller.
    b. Calculate how long customers spend on average to complete their transactions at the bank (time in queue plus service time). What percentage of that time is spent queueing?
    c. How many customers are in the bank on average?
    d. Calculate the probability a customer will spend less than 30 minutes at the bank (time in queue plus service time).
    e. Calculate the probability that there are more than two customers in the bank.
    f. What do a–e imply about customer service at the bank?
22. Customers arrive at a local bakery with an average time between arrivals of 5 minutes. However, there is quite a lot of variability in the customers' arrivals, as one would expect in an unscheduled system. The single bakery server requires an amount of time having the exponential distribution with mean 4.5 minutes to serve customers (in the order in which they arrive). No customers leave without service.

a. Calculate the average utilization of the bakery server.
b. Calculate how long customers spend on average to complete their transactions at the bakery (time in queue plus service time). What percentage of that time is spent queueing?
c. How many customers are in the bakery on average?
d. Calculate the probability a customer will spend more than an hour at the bakery (time in queue plus service time).
e. What is the probability that there are fewer than two customers in the bakery?
f. Why are the estimated waits in this system so long? Are the assumptions behind them reasonable? Why or why not?

8.5 GENERAL QUEUEING MODELS

This section considers results for G/G/s queues and other general queueing models. We also discuss simulation as a tool if the model is too complex for queueing analysis. Supplement 2 (*Queueing Techniques*) contains further queueing theory results and covers some of the more technical results that are too detailed for this chapter.

In the M/M/1 model of the previous section, the distribution of the service time is exponential. In many cases, this assumption is unwarranted. One would expect that service times would rarely be exponential because the exponential distribution has the memoryless property; the amount of time remaining in service would have to be independent of the time already spent. One would think that a modal distribution (such as the normal or Erlang) would be a more accurate model of service times in most circumstances. For that reason, models with general service times are of great interest.

Define c_s^2 to be the **squared coefficient of variation** associated with the service process. As for the arrival process (see Section 8.3), we can compute it with two alternate formulas. First, $c_s^2 = \mu^2\sigma_s^2$ where σ_s^2 is the variance of the service times. Second, $c_s^2 \approx \text{var}[S]/E[S]$ where S is the number of services per unit time (e.g., per hour), not including idle time. This second value is less convenient to calculate than for arrivals because of the need to exclude idle time. If we observe that a server has served a certain number of customers in an hour, we must ensure that the entire hour was used for processing for the calculation to be correct. Then, $c_s^2 < 1$ when there is some uniformity in customer service times, $c_s^2 \approx 1$ when the service tasks are very customer specific (with $c_s^2 = 1$ for exponential service distributions), and $c_s^2 > 1$ when the customers have unusually variable service requirements.

Expected Time in System for a Single Server System

For the G/G/1 system, an approximation for the expected time in system is the following.

$$w \approx \frac{1}{\mu}\left(\frac{c_a^2 + c_s^2}{2}\right)\frac{\rho}{1-\rho} + \frac{1}{\mu}.$$

The expected time in queue, w_q, is the first term of this equation. (Recall that $w = w_q + 1/\mu$.) If arrivals follow a Poisson process (i.e., when $c_a^2 = 1$) then this formula is exact and called the **Pollaczek-Khintchine** (P-K) formula, If service times are also exponentially distributed, then the formula is the same as the one for the M/M/1 queue. If arrivals are not Poisson then it is an approximation; it becomes proportionately more accurate as utilization, ρ, tends to 1.

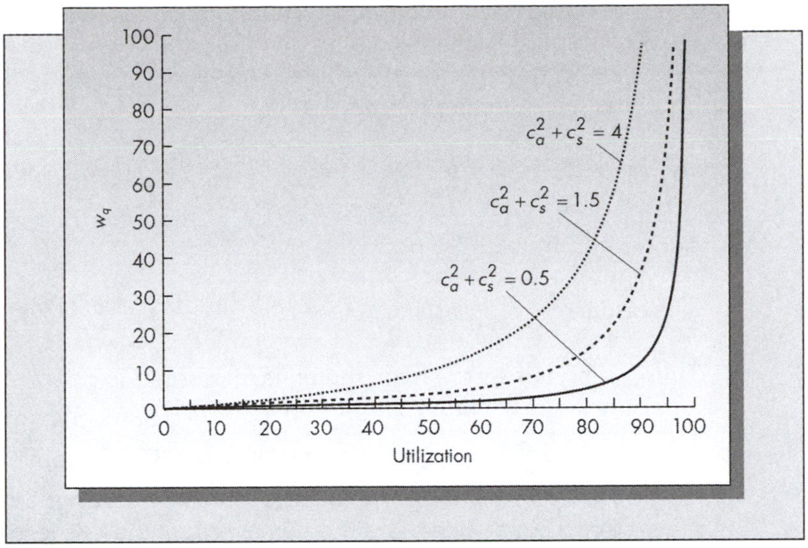

Figure 8–6 Growth in time in queue for a G/G/1 queue

This formula implies that even if a server is not fully utilized and even if a server is after (or before) the system bottleneck (as defined in Section 8.2), significant waiting (queueing) can occur. In service systems where the customer's experience is influenced by waiting time, a bottleneck analysis is likely not sufficient for good system design. The formula shows that waits increase linearly with variability and exponentially with utilization. Figure 8–6 shows the growth of expected time in queue for a range of variabilities as utilization grows.

The impact of not understanding the relationships shown in Figure 8–6 has huge implications with respect to staffing decisions. Many managers consider utilizations of less than 100 percent inefficient, yet such utilizations are only feasible in a system with no variability (which almost never occurs in the real world). Many hospital emergency departments target utilizations of 80 percent, because they understand that high utilizations result in long waits. However, as can be seen from the formula, it is not really possible to have a one-size-fits-all target for utilization (because results also depend on variability).

EXAMPLE 8.6

A large discount warehouse is assessing the number of checkout stands it needs. During a period of the day when the arrival rate of customers is about one every 12 minutes, there is only one checkout stand open. It takes an average of 8 minutes to check out one customer. The checkout time follows a normal distribution with standard deviation 1.3 minutes. The arrival process may be assumed to be a Poisson process. Find l, l_q, w, and w_q for this system. How far off would your calculations be if you assumed that the service distribution was exponential?

Solution

The arrival rate is $\lambda = 5$ per hour and the service rate is $\mu = 60/8 = 7.5$ per hour, giving $\rho = 5/7.5 = 2/3$. Notice how we put λ and μ into consistent units to compute ρ. We also must use consistent units when computing the squared coefficient of variation associated with the service process, using $c_s^2 = \mu^2\sigma_s^2$. The standard deviation of the service time is 1.3 minutes, or $1.3/60 = 0.02167$ hour. It follows that the variance of the service time is $0.02167^2 = 4.6944 \times 10^{-4}$ hours^2. Therefore $c_s^2 = 4.6944 \times 10^{-4} \times (7.5)^2 = 0.0264$. (If we had computed both μ and σ_s in minutes, then we would have the same answer for c_s^2, which is unitless.) Since arrivals are Poisson, $c_a^2 = 1$ and the results will be exact, not approximate, it therefore follows that

$$w_q = \frac{1}{\mu}\left(\frac{c_a^2 + c_s^2}{2}\right)\frac{\rho}{1-\rho} = \frac{1}{7.5}\left(\frac{1+0.0264}{2}\right)\frac{2/3}{1-2/3} = 0.13686 \text{ hour} = 8.21 \text{ minutes}.$$

Therefore,

$$w = w_q + \frac{1}{\mu} = 8.21 + 8 \text{ minutes} = 16.21 \text{ minutes}.$$

Then, using Little's law (with the units expressed in hours),

$$l_q = \lambda w_q = 5 \times 0.1368 = 0.6843 \text{ customers},$$
$$l = \lambda w = 5 \times 0.2702 = 1.351 \text{ customers}.$$

Hence, each customer should expect to wait in line about 8 minutes and expect to be in the system about twice that time. On average, there is less than one customer in the queue.

Now, suppose that we had assumed that the service distribution was exponential. We can either use the same method substituting $c_s^2 = 1$ or more simply, use the performance measures for an M/M/1 queue as follows.

$$l = \rho/(1-\rho) = (2/3)/(1/3) = 2 \text{ customers}.$$
$$l_q = l - \rho = 2 - 2/3 = 4/3 \text{ customers}.$$
$$w = l/\lambda = 2/5 = 0.4 \text{ hours (24 minutes)}.$$
$$w_q = l_q/\lambda = 1.3333/5 = 0.2667 \text{ hours (16 minutes)}.$$

We see that assuming that the service distribution is exponential results in substantial errors in the system performance measures. In particular, w_q is too high by 100 percent! The M/M/1 model is very sensitive to the assumption that the service time is exponential and typically overestimates the performance measures for most real systems (i.e., it is rare for a system to be more variable than an M/M/1 system, although it is possible). If the results of a queueing analysis are to be accurate, it is vitally important that any assumptions regarding the form of the service or the arrival process be verified by direct observation of the system.

One interesting special case of the M/G/1 queue is when the service distribution is deterministic (labeled M/D/1). In this case $\sigma_s^2 = c_s^2 = 0$.

As discussed in the next subsection, when the number of servers exceeds one, exact algebraic expressions are not known in general. (For more complete coverage of queueing, see Shortie et al., 2018 or Kleinrock, 1975.) Explicit results for many versions of the M/M queueing model are available, however. Examples include systems with priority services, jockeying (switching between queues), and impatient customers, just to mention a few.

Multiple Parallel Servers

Consider a queue with s servers in parallel as shown in Figure 8–7. When customers arrive, they queue up in a single line. The next customer in the line is served by the next available server. This is the G/G/s queue.

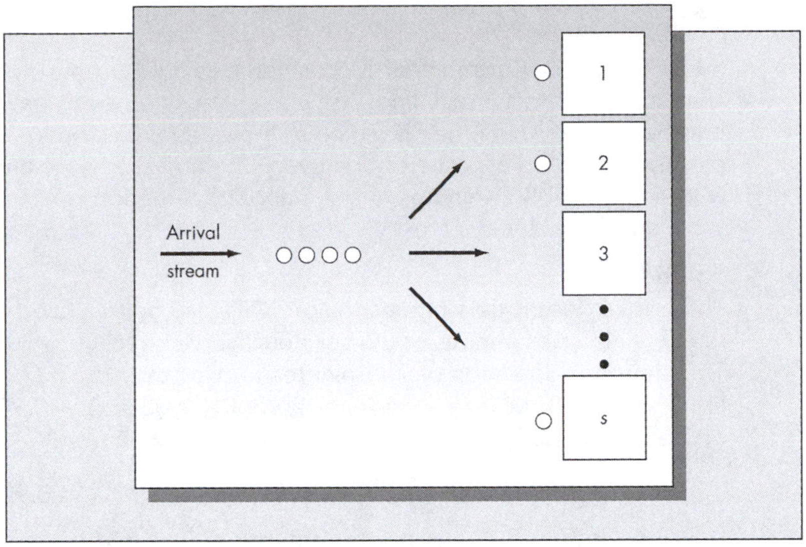

Figure 8–7 *s* servers in parallel

For the G/G/s system, an approximation for expected time in system is as follows.

$$w \approx \frac{1}{s\mu}\left(\frac{c_a^2 + c_s^2}{2}\right)\frac{p^d}{1-\rho} + \frac{1}{\mu},$$

where p^d is the probability of delay and may be approximated using the formula in Section 8.3 or the one in Supplement 2.4. As in the G/G/1 system, this approximation for w becomes proportionately more accurate as utilization, ρ, tends to 1. However, it is no longer exact even for Poisson arrivals. Exact results do exist for M/M/s queues and may be found in Supplement 2.3.

Some comments are in order on this formula. First, w scales linearly in $1/\mu$. If we double the service time and scale arrivals to match, the waiting time will double. If we transform waits from hours to minutes by multiplying $1/\mu$ by sixty, w will also become sixty times larger and be transformed from hours to minutes. Second, the delay in queue scales down by the number of servers, s, while the final $1/\mu$ does not. This is because no matter how many servers there are, the customer will always require an average of $1/\mu$ time to be served; however, the pooling of the servers cuts down the queueing delay. Third, as for the G/G/1 queue, although variability does not affect average utilization it does have a large impact on delays. The key measure of variability is $c_a^2 + c_s^2$, so both arrival and service time variability contribute equally to delays. Finally, waiting time depends on both utilization rates and variability; when utilization gets close to 1, waits (and hence lines by Little's law) get long.

This formula provides insight into the circumstances that would give rise to exceptionally long queues. First, anywhere capacity is expensive (e.g., theme parks) or the service is measured primarily in terms of cost (e.g., government departments), we expect to see high utilizations and hence long waits. Second, any system with high variability either in arrivals (e.g., tourist attractions receiving tour buses) or service times (e.g., emergency departments) will also experience long waits. Third, small scale systems (with a small s) will have longer waits than larger scale systems, even at similar utilization levels (this is a pooling effect). A final caveat is that this formula assumes that no one leaves because the line is too long. Systems with customers departing without service are described in the next subsection.

EXAMPLE 8.7

Suppose the Department of Motor Vehicles (DMV) employs three servers that serve customers in one (virtual) line. Assume customers never leave once they have taken a number, which holds their position in the queue. Customers arrive one at a time completely at random at a rate of one every 3 minutes. Service times are also quite variable with a mean of 8.5 minutes and a standard deviation of 7.5 minutes. Approximate w, w_q, l, and l_q for this system.

Solution

The above description implies that arrivals may be modeled by a Poisson process with rate $\lambda = 1/3$ per minute or 20 per hour. Service rate $\mu = 60/8.5 = 7.05882$ per hour. The standard deviation of 7.5 minutes implies that $c_s^2 = (7.5)^2/(8.5)^2 = 0.7785$. There are 3 servers. Thus, $s = 3$ and $\rho = 20/3(3 \times 7.05882) = 0.9444$. We approximate the probability of delay as

$$p^d \approx \rho^{-1+\sqrt{2(s+1)}} = (0.94444)^{-1+\sqrt{2\times(3+1)}} = 0.9008.$$

Notice how this is significantly smaller than ρ due to the pooling effects of having three servers. We can then calculate

$$w_q \approx \frac{1}{s\mu}\left(\frac{c_a^2 + c_s^2}{2}\right)\frac{p^d}{1-\rho} = \frac{8.5}{3}\left(\frac{1+0.7785}{2}\right)\frac{0.9008}{1-0.9444} = 40.85 \text{ minutes} = 0.681 \text{ hour}.$$

Therefore,

$$w = w_q + \frac{1}{\mu} \approx 40.85 + 8.5 \text{ minutes} = 49.35 \text{ minutes}.$$

Then, using Little's law (with the units expressed in minutes),

$$l_q = \lambda w_q \approx 1/3 \times 40.85 = 13.62 \text{ customers},$$
$$l = \lambda w \approx 1/3 \times 49.35 = 16.45 \text{ customers}.$$

Hence, each customer should expect to spend almost 50 minutes in the system, and on average there are more than 13 customers in the queue. Even though there are three servers and each will be idle more than 5 percent of the time, there is significant waiting in this system. Given the cost emphasis placed on government departments, utilizations of 94 percent are not unreasonable. Therefore, spending close to an hour in the DMV (and if the average time is 50 minutes, then many customers actually spend more than an hour) should not be a surprise. As discussed in the next subsection, waits are also exacerbated by the fact that in this system we assume no one leaves without service. Indeed, for many government services people do wait indefinitely because the service is not discretionary.

Systems with Abandonment

The queueing models above all show that the expected wait grows toward infinity as utilization approaches 100 percent. Further, the formulas assume $\rho < 1$. What happens in reality when inflow is greater than outflow ($\rho > 1$)? In most service systems, customers will not actually wait indefinitely but will leave without the service. This is known as **abandonment** or **customer impatience**. If the customer leaves after observing the queue length without joining, this is termed **balking**; if the customer leaves after having joined the queue and has waited some (usually random) length of time, this is termed **reneging**. Either way, the abandonment represents lost revenue to the firm and lost customer goodwill.

In most systems abandonment is not observed so a manager may make the assumption that if $\rho < 1$ then abandonment is not occurring. This is a very poor assumption for two reasons. First, abandonment can occur in any system even when queues are short. Second, even when utilization is less than one, long queues can still form, leading to significant abandonment. Service firms can find themselves in a downward cycle of service if managers place too much importance on high server utilization and do not recognize the need for buffer capacity to make sure the lines do not get too long. In such a scenario, the manager may cut staffing because seemingly there is slack in the system. This results in longer waits and more abandonment, resulting in more slack, and so the downward cycle continues.

There are some limited analytic models of abandonment (e.g., see Wang, Li, and Zhigan, 2010). Most models typically assume that customers have a "patience" distribution; when their patience tolerance has been exceeded, they leave if they have not yet been placed in service. M/M/1 queueing models with exponential abandonment are particularly tractable. In practice, customer behavior is actually more complex than a simple distribution and may depend on factors like the length of the queue (e.g., Bolandifar, Dehoratius, Olsen, and Wiler, 2019). Section 8.6 discusses the psychology of queueing.

There has also been work on systems with balking. Here, the customer will join the system if the value of service minus the combined cost of service and waiting is nonnegative. The earliest work in this area showed that, at equilibrium, customers join the queue only if the length of the queue is below a threshold (Naor, 1969). Refael Hassin and Moshe Haviv (2003) extensively review game theoretic results of this sort.

Priorities

While first-come, first-served is considered fair, it is not always the order of service. Some customers may simply be higher priority than others (e.g., severe trauma cases in the emergency department versus minor ailments). Furthermore, it may not be efficient to treat all customers equally if they have different service requirements, as illustrated in the following example.

EXAMPLE 8.8

Suppose we have a system with two types of Poisson arrival streams. Type 1 customers arrive at a rate of 9 arrivals per hour, and each requires exactly 6 minutes of work. Type 2 customers arrive at a rate of 1.5 arrivals per hour, and each requires exactly 36 minutes of work. What is the expected waiting time in the system if each type has a line to itself? Then, what is the expected time in system if there is only a single FCFS line with two servers? Finally, suppose the system gives priority to Type 1 arrivals but still shares the two servers. It can be shown (e.g., Kleinrock, 1976) that the wait in system for type 1, w_1, equals 18.0 minutes and the time in system for type 2, w_2, equals 117.8 minutes. How do these waits compare to the ones previously computed without priorities?

Solution

For type 1 customers, the arrival rate is $\lambda = 9$ per hour, and the service rate is $\mu = 60/6 = 10$ per hour, giving $p = 9/10 = 0.9$. Poisson arrivals and exact service times imply that

Using Queueing to Make Staffing Decisions Saves Lives

Emergency departments (EDs) are complex queueing systems where too much waiting time can have severe negative health consequences. Delays in treatment have been shown to increase mortality rates and to cause patients to abandon the queue, known in the medical literature as leaving without being seen (LWBS). LWBS by itself is also associated with negative consequences because some patients leave because they feel too ill to wait—despite needing treatment. It is therefore not a stretch to say that improved staffing of EDs, if it results in lower waits and lower LWBS rates, saves lives.

One way to improve waits and LWBS rates is to add more staff. Unfortunately, because capacity is expensive and budgets are limited, this is usually not possible. One queueing model for ED staffing decreased LWBS rates in an urban hospital even for periods *without* increased staffing levels (Green, Soares, Giulio, and Green, 2006). The researchers modeled the nonlinear queueing effects—which are not accounted for in the simple nurse to patient ratios frequently used by hospitals.

Arrivals to EDs are, of course, highly nonstationary. To get around this, the researchers used a stationary independent period by period (SIPP) model where each independent period of time (e.g., each hour) is assumed to be stationary. They also made a so-called "lag" adjustment, which accounts for the fact that in nonstationary queueing models the arrival rate peaks before the waiting time.

A simple M/M/s queueing model was used to estimate the staff needed during each staffing interval. This estimate was based on the requirement that no more than 20 percent of patients wait more than one hour (i.e., a delay percentile metric). The model showed that staffing on weekdays should be increased from 55 to 58 provider hours, but on weekends 53 provider hours were sufficient. However, the model also indicated a significant shift in when the hours were provided, from the middle of the night to much earlier in the day. Unfortunately, the ED did not have the budget to add all the additional staffing. However, they did shift hours around and added a few hours. Further, there were practical considerations to be considered. For example, it was not deemed possible to have different daily schedules. Hence, there were only two final schedules—one for weekdays and one for weekends.

Even though not all of the recommended hours were followed, the queueing model did result in better placement of the limited resources. Four hours fewer were provided on both Saturdays and Sundays, with those extra hours going to the weekday schedule. Further, there was a four-day subset for which there were no more and no fewer hours, simply a rearrangement of schedules guided by the queueing model. In this time interval, LWBS events declined from 9.2 percent to 7.2 percent even though the number of arriving patients increased by 5.5 percent (548 patients) from the initial study period to the one with the new schedule. The schedule was a success!

Source: Green et al. (2006) and Bolandifar et al., (2019).

$c_a^2 = 1$ and $c_s^2 = 0$ respectively. Using the formula for an M/G/1 queue, the wait in system for type 1, w_1, equals $\frac{1}{\mu}\left(\frac{c_a^2+c_s^2}{2}\right)\frac{\rho}{1-\rho}+\frac{1}{\mu}=\frac{60}{10}\left(\frac{1+0}{2}\right)\frac{0.9}{1-0.9}+6=33$ minutes.

For type 2 customers, the arrival rate is $\lambda = 1.5$ per hour and the service rate is $\mu = 60/36 = 1.667$ per hour, giving $p = 1.5/1.667 = 0.9$. Again, $c_a^2 = 1$ and $c_s^2 = 0$. Then, w_2 equals $\frac{1}{\mu}\left(\frac{c_a^2+c_s^2}{2}\right)\frac{\rho}{1-\rho}+\frac{1}{\mu}=36\left(\frac{1+0}{2}\right)\frac{0.9}{1-0.9}+36=198$ minutes. Finally, the average customer wait, w, is computed as $9/10.5 \times 33 + 1.5/10.5 \times 198 = 56.6$ minutes, where 9/10.5 is the fraction of type 1 customers and 1.5/10.5 is the fraction of type 2 customers.

If the customers are pooled into one queue, we must first calculate c_s^2, because the presence of two types of customers implies that service is no longer deterministic. In particular, the expected service time is $E[S] = 9/10.5 \times 6 + 1.5/10.5 \times 36 = 10.286$ and the second moment of service time $E[S^2] = 9/10.5 \times 6^2 + 1.5/10.5 \times 36^2 = 216$. Therefore, $c_s^2 = \mathrm{var}[S]/(E[S])^2 = (216 - 10.286^2)/(10.286)^2 = 1.042$. It is known that the combination of two streams of Poisson arrivals is also Poisson (e.g., Wolff, 1989), so that $c_a^2 = 1$. Then, putting this into the M/G/2 system with $p^d = \rho^{-1+\sqrt{2(s+1)}} = 0.9^{-1+\sqrt{6}} = 0.858$, we have that

$$w_1 = w_2 = w = \frac{1}{s\mu}\left(\frac{c_a^2 + c_s^2}{2}\right)\frac{p^d}{1-\rho} + \frac{1}{\mu} = \frac{10.286}{2}\left(\frac{1+1.042}{2}\right)\frac{0.858}{1-0.9} + 10.286 = 55.4$$

minutes. Thus, total expected wait has decreased very slightly at significant expense to type 1 customers. Pooling servers improves overall system performance but at a significant cost to some customers. However, pooling the servers *significantly* increased the variability of service time from zero to above 1. Therefore, there is only a very modest decrease in overall wait obtained by the pooling. It is even possible to construct examples where pooling increases overall wait, if the gap between the two types of service is large enough and enough variability has been added to the system by such pooling (see problem 25).

Finally, suppose the system gives priority to type 1 arrivals but still shares the two servers. As noted (the formulas are beyond the scope of this text), $w_1 = 18.0$ minutes and $w_2 = 117.8$ minutes. Therefore, the time in system averaged across both types, w, equals $9/10.5 \times 18 + 1.5/10.5 \times 117.8 = 32.2$ minutes. Compared to two separate lines (no pooling) both type 1 and type 2 are better off. Compared to the FCFS system type 1 is significantly better off but type 2 is not. Further, notice how much the overall average wait w has decreased by using this priority system. We explain how this is possible next.

The order of service does not affect server utilization (unless there is abandonment or some similar effect); however, it can have a large effect on waiting times. If the goal is to minimize the average wait in the system, short jobs should be served first. We all intuitively know this when we let the person with one sheet of paper at the photocopier ahead of us when we have a large job. We understand that the effect on our own wait is minimal but the effect on his is large. In general, giving priority to customers with the shortest expected service time will minimize w when it is computed across all customer classes. Chapter 11 contains more discussion on the optimal scheduling of jobs or customers.

In call centers it is easy to give priority based on customer specific information, because customers do not observe the queue. However, in other service environments, deviating from FCFS can be problematic because customers see it as unfair. Further, customers have a good reason to be upset because FCFS minimizes delay variance (across all customers) and so indeed is the most "fair" system (see Section 11.9). Hayriye Ayhan and Tava Olsen (2000) proposed a rule to serve the next customer with the largest value of $a\mu$, where a (i.e., age) is how long the customer has waited thus far and $1/\mu$ is the customer's expected service time. This is shown to minimize the second moment of delay as utilization approaches 1. It provides a balance between efficiency (a small w) and fairness.

In short, queueing theory tells us it is better to pool customers and to give priority to customers with short service times. However, there are a variety of other considerations that may make this unattractive or impractical. Regardless, customer prioritization can have a significant impact on service system performance. Optimization of queueing systems is discussed further in Supplement 2.6.

Other Queueing Extensions

Many more variants of queueing models are possible than those that have been discussed thus far. In some cases, there is a finite limit K on the number of customers that can be present in the system at any time. Supplement 2.3 presents results for the M/M/1 queue with finite capacity K. (for a comprehensive treatment of traditional queueing models, see Shortie et al., 2018).

Queueing networks are not generally very tractable, but there are a number of notable exceptions discussed in Supplement 2.5. Approximate analytical results are available for nonexponential cases (see Suri and Hildebrant, 1984, for example). Also, the exponential assumption is not required for some simple network configurations. (See Buzacott and Yao, 1986, for a review of the analytical results for network queueing models of flexible manufacturing systems.) Beyond these, other studied queueing networks include closed queueing networks and polling models. In a closed network, customers never leave but instead are recycled back to the beginning (this can represent the interaction between computer users and the core). Polling models have a single server that cycles through multiple classes of customers. Setup times between classes can be accommodated as can time for the server to walk between the different classes. In general, the availability of results depends on the restrictiveness of the assumptions.

Simulation

Many real queueing problems are not amenable to the type of mathematical analysis discussed in this section. Some of the reasons are listed below.

1. The system is a queueing network. Queueing networks are common in manufacturing systems where the output of one process is the input to one or more other processes. They are also common in complex service systems such as emergency departments. Some systems are very complex with feedback loops and other unusual features. Such systems are generally too complicated to be analyzed mathematically.
2. The interest is in transient rather than steady state behavior. The results discussed here are for steady state only. As stated earlier, steady state means that the system has evolved for a sufficiently long period so that initial conditions do not affect the probability distribution of system states. In many real problems, short-term behavior is an important part of proper systems management.
3. Interarrival times and/or service times are not exponential. Only approximations are available even for G/G/s queueing systems. Additional features, such as a finite waiting room, priority service, abandonment from the queue, and so on, can make such systems too complex to analyze mathematically.
4. Human behavior needs to be considered. The behavior of customers and/or servers may not obey a simple mathematical model. For example, in many service systems it has been

observed that servers will speed up if the line is long. Also, customer behavior in switching/jockeying between parallel queues creates complications. Such behavior is usually too complex for queueing analysis.

One way to deal with a complex queueing system is simulation. A simulation (in our context) is a computer program that recreates the essential steps of a service process. In some sense, the computer "experiences" the process. In this way, one can see how the system responds to different parameter settings available to the system designer. Inputs that would take place over multiple years can be simulated in the space of a few seconds on a computer.

Problems amenable to simulation almost always involve some element of randomness. Computer-based simulators that incorporate randomness are called **Monte Carlo simulators**. At the heart of a Monte Carlo simulation are random numbers. See Appendix 8–A for more details on how such random numbers are generated.

Simulation can be used to analyze many types of complex problems where uncertainty affects outcomes, but it probably has been used most often for queueing problems. Simulators can be written in a general-purpose programming language (such as Java or C#), or constructed using special-purpose simulation packages, such as ProModel, ARENA, or GPSS. Many packages are specifically designed for certain types of applications, such as queueing or network problems. MedModel is a package designed specifically for simulating medical facilities.

More recently, two approaches for developing simulations have become popular, opening up the use of simulation to a much wider audience. Spreadsheet programs have gained a great deal of popularity in recent years. Because many spreadsheets have a random number generator built in, they can be used to construct simulators. Excel, in particular, can generate observations from many distributions. The add-ins @Risk, Crystal Ball, or Risk Solver Platform facilitate building simulations in Excel; they are all very similar. These packages include a much wider array of distributions and convenient report generation (e.g., see Powell and Baker, 2017; Winston and Albright, 2019).

For complex queueing networks, graphics-based simulation packages are far easier to use than spreadsheets. These programs allow the user to construct a model of the system using graphical icons. Icons represent service facilities and waiting areas, and arrows represent the direction of flow. Random arrivals and random service can be easily incorporated into the model. Both ProModel and ARENA, mentioned above, are graphics based and allow experienced users to build simulations of complex systems very quickly. Such programs employ live animation to show the simulation in action, and summary statistics are collected after the simulation has run its course. Further details on the type of simulation engine used in these packages may be found in Appendix 8–A.

Improving a Service Process

The queueing principles in this section yield insights into how to improve the performance of a service system. These are easiest seen using the following elements of the process.

1. *The arrival process.* The focus should be on reducing variability, possibly by scheduling appointments or using peak load pricing. Further, the arrival rate can be decreased by eliminating the need for the service, possibly by moving work offline.
2. *The service process.* If process rates can be increased, performance improves. Often lean programs can be used to improve efficiency. Frequently forgotten is the improvement

from reducing service variability, which can be achieved by improved process design or further standardization of options.

3. *The scheduling rules and process flow*. As discussed above, priorities make a large difference to the service experience. Giving priority to customers with short service times improves overall expected waits. One way to achieve this and still appear fair is to have dedicated servers for small jobs, which is the approach taken by supermarkets. Using these servers for larger tasks when there are no small tasks to address eliminates any inefficiency caused by decreasing pooling.
4. *Pooling for added efficiencies*. Pooling allows servers not to sit idle while other servers are overworked. Even a small number of cross-trained servers can work to pool work effectively without the need for all servers to be trained in all tasks.
5. *Transform customers into servers*. If customers can serve themselves, the work content for paid servers decreases. Examples include salad bars and self-check-in kiosks at airports.
6. *Make the wait "feel" less long*. Sometimes a firm does not actually need to decrease the waiting time if it can make it feel less long to the customer. Techniques for this are discussed in the next section.

PROBLEMS FOR SECTION 8.5

23. Customers send emails to a help desk that has three employees who answer the emails (and this is their only responsibility). Customer requests arrive according to a Poisson process at a rate of 30 per hour. It takes on average 4 minutes to write a response email; the standard deviation of service times is 2 minutes.
    a. What is the utilization of the employees?
    b. What is the average time an email spends waiting before an employee starts working on it? What is the average time to complete an email request (from the time the email is sent to when it is answered)?
    c. How many emails on average are in the system queue waiting to be answered?
    d. If the manager wants to decrease customer waiting, what options are available to him/her? Which do you recommend?
24. At the SuperSpeedy drive-through, the time between consecutive customer arrivals has a mean of 50 seconds and a standard deviation of 30 seconds. There are two servers whose service time averages 80 seconds with a standard deviation of 20 seconds. Assume that no customers leave the drive-through after entry.
    a. What is the utilization of the employees?
    b. What is the average time a customer spends at the drive-through? What fraction of that is waiting in the queue?
    c. How many cars on average are in the drive-through lane (including those in service)?
    d. What suggestions would you have for the drive-through to improve customer satisfaction?
25. Modify Example 8.8 to find an example where pooling increases average waiting time across the two customer types.
26. What are the advantages and disadvantages of queueing analysis versus simulation?

8.6 THE HUMAN ELEMENT IN SERVICE SYSTEMS

As discussed earlier, one key characteristic of service processes is that they tend to involve the customer as coproducer. This means that human psychology and decision biases should be considered when designing service systems. This section outlines the psychology of queueing, guidelines for introducing technology into services, and principles for giving service guarantees and refunds.

The Psychology of Queueing

David Maister (1985) has detailed a number of psychological principles for the design of queueing systems.

1. *Unoccupied time feels longer than occupied time.* If a customer is not doing anything while waiting, the wait feels longer. Mirrors at elevators were introduced so customers could check their appearance for any adjustments, and the elevator wait then seemed shorter. An airport moved arrival gates farther from baggage claim to make the walk to retrieve luggage longer, which cut down on complaints about delayed bags.
2. *Pre-process waits feel longer than in-process waits.* Customers who feel they are making progress are more tolerant of delay. For example, the wait in the doctor's waiting room is usually more frustrating than the wait in the examination room after having seen a nurse.
3. *Anxiety makes waits seem longer.* Worry makes customers more sensitive to their surroundings and less tolerant of waiting. For example, emergency room waits are generally long and are more intolerable because of being anxious and probably in pain.
4. *Uncertain waits feel longer than known, finite waits.* Uncertainty is a form of anxiety and will make the waiting less tolerable. Waits for a reservation are not bothersome when the customer arrives early but become increasingly unpleasant once the reservation time passes. Many call centers have understood this principle and now provide estimates of time on hold when the customer calls.
5. *Unexplained waits are longer than explained waits.* This principle is also based on the anxiety associated with uncertainty. Many airlines are getting better at understanding this principle and will provide customers with an explanation of why their flight is delayed, rather than simply telling them the estimated time of departure.
6. *Unfair waits are longer than equitable waits.* This principle relates to the mood of the customer. A negative mood, whether it is anxiety or anger, will make the wait less tolerable. Seeing someone cut in line makes most people angry!
7. *The more valuable the service, the longer I will wait.* This is natural. People will wait a long time for needed medical treatment but much less time for coffee. Further, a customer's tolerance for waiting in queue is proportional to the complexity or quality of service anticipated by that customer.
8. *Solo waiting feels longer than group waiting.* This is a corollary to the first principle about unoccupied time. Waiting in a group is simply more pleasant than solo waiting.

All eight principles highlight that it is not the length of the delay that matters; rather, what matters most is the customer experience, particularly relative to expectations. Sensible firms consider these principles when they design their service. Often this can be done quite simply. For example, putting in ropes to delineate the line prevents customers from jumping the queue and keeps the queue fair.

SNAPSHOT APPLICATION

Disney Uses Both the Science and the Psychology of Queueing

A company that is a world leader in the science of queueing is Disney Corporation. They provide a service that is only of value for the experience it provides. It is therefore critical that customers leave Disney's parks feeling like they have had a great day.

According to Disney, they employ more than 75 industrial engineers who help with queue management at their parks around the world. They measure the capacity of the rides, optimize the flow within the rides, set the capacity of the queues, etc. However, Disney is also very aware of the psychology of queueing. Knowing that uncertain waits feel longer, they will post the expected waiting time at the start of each line. Further, knowing that unoccupied time also feels longer, they provide significant entertainment for customers in line for each ride. They introduced interactive entertainment in some of their queues.

In 1999, Disney implemented a technological innovation called FastPass. The reservation system for rides (experiences) was available without charge to anyone with a pass to the theme park. Customers would insert their passes into FastPass machines and would receive a paper ticket with the next available reservation time. Because the time windows advance 5 minutes at a time, customers in groups will always have overlapping times (assuming they scan their tickets at approximately the same time). The return times are displayed on a board and may be anywhere from an hour to many hours after the current time. At some point during the day, the system will run out of reservations, and no more FastPass tickets will be available for that ride. Disney arranged separate entries at rides for customers with FastPasses. Customers would hand the paper passes to cast members, who strictly enforced no entry before the allotted time. FastPass reservations were available only one at a time on the day of visits to the park.

In 2014, Disney introduced FastPass+ at Disney World. Customers with park passes could reserve 3 experiences per day 30 days in advance of a visit (60 days in advance if staying on Disney property)—again at no additional charge. The 3 reservations are stored on MagicBand bracelets or credit-card-like passes, which have RFIC chips. Visitors tap the bracelets or cards on Mickey-shaped readers to gain entry to the experiences. Entrants can change reservation times or experiences using the MyDisney Experience site or app. Disney doubled the number of experiences for which one can use FastPass+; character greetings, parades, and nighttime shows are available in addition to rides.

There is a lot of science behind the experiences for which Disney offers the FastPass and FastPass+ options and the capacity allocated to reservations for those experiences. Most of that science is, unfortunately, proprietary. One likely effect of the system is that many customers spend less time queueing and more time wandering the park and probably buying consumables (i.e., food and souvenirs), which is clearly good for Disney's bottom line. Some customers will simply stand in line for other rides while waiting for their reservation, and those customers will probably end up riding more rides during a day than without the system. It is likely that the system has therefore made the lines for less attractive rides longer than they were before. Fully understanding the queueing implications of the FastPass options is an interesting open research question.

Source: Disney.com and Pawlowski, 2008.

Introducing Technology into Services

With the increasing role of technology in services (e.g., self-service checkouts, online ordering, helpful applications, etc.) companies must consider how to introduce technology carefully. Frances Frei and Hanna Rodriguez-Farrar (2008) recommend the following three key principles.

1. *Be helpful before being intrusive.* If customers can see the obvious value of the technology, then they will be much more tolerant of data gathering than if they view the technology

as solely operating for the firm's benefit. For example, Facebook has been slowly increasing its intrusiveness; it is unlikely it would have reached today's popularity if it had launched with its current level of intrusiveness. As a second example, customer loyalty cards have provided customers with benefits even though they are used by firms to collect and analyze data on customer behavior. If the technology makes the service easier or more pleasant, then customers are often willing to share data, which can be very helpful to firms looking to optimize their offerings.

2. *Roll out functions at a pace consumers can absorb.* Customers can get overwhelmed by too much technology. Firms need to consider what the customers can successfully navigate as they introduce technology. The firm may need to deliberately hold back functionality if it feels customers will not be able to absorb it. They should also consider the demographics of the customer base when introducing technology. For example, BMW overwhelmed customers of its 7-series automobiles with too many new features (e.g., joystick control and keyless entry) and received significant customer backlash; Audi rolled out similar technology successfully.
3. *Framing matters.* If consumers feel the technology is simply being used for cost savings, then they may revolt or turn away from the service provider. However, if the technology is presented to the customer as something that will improve their service experience, then they will be much more likely to view it positively. For example, self-checkout machines are clearly designed to save on labor costs but, so long as customers see improved waiting time with these machines, they will be tolerant of the technology.

Guidelines for Service Guarantees and Refunds

Because the definition of good service is usually relative, firms often advertise service guarantees. Sanjeev Bordoloi, James Fitzsimmons, and Mona Fitzsimmons (2019) provide the following guidelines for such service guarantees.

1. *Focus on customers.* The customer does not care about firm-centric metrics; service guarantees should be relative to customer-centric metrics, such as delay.
2. *Set clear standards.* The guarantee should not be vague about what constitutes service within the guidelines.
3. *Guarantee feedback.* If customers are rewarded for reporting service failures that do not meet the guarantee, the firm has implemented an effective method of quality control (as well as keeping customers happy).
4. *Promote an understanding of the service delivery system.* A firm cannot introduce a service guarantee without knowing what is achievable. A thorough understanding of the service process is necessary before introducing a service guarantee.
5. *Build customer loyalty.* Happy customers are usually loyal customers. Service guarantees properly implemented ensure that complaints are resolved quickly and effectively, which builds customer loyalty.

Unfortunately, service failures are all too frequent. In such cases, firms need to remedy the failure. Christopher Hart (1988) provided key guidelines for such remedies.

1. *Unconditional.* Customer satisfaction should not come with exceptions; refunds should not come with strings attached.

2. *Easy to understand and communicate.* The customer should not be confused by offers or misunderstand what is guaranteed.
3. *Meaningful.* If the offer is too trivial to have any real meaning to the customer, then it may be better not to offer anything.
4. *Easy to invoke.* Fulfilling or initiating offers should not consume large amounts of firm or customer resources.
5. *Easy to collect.* It should not be a headache for the customer to receive a discount or other remedy.

Of course, offering refunds that are conditional or difficult to collect will save money, but usually such savings are shortsighted. Recall the statistic that it costs five times more money to acquire a new customer than to retain a current customer. Handling service refunds properly is vitally important in maintaining customer loyalty.

8.7 CALL AND CONTACT CENTERS

An important class of service system is the **call** or **contact center**. There are two types. The first is **outgoing**, where agents sell products and services to potential clients. The second is **incoming** where customers call for service, the focus of this section. A call center is entirely phone call focused, whereas a contact center will answer emails and likely participate in web chats.

Call Center Basics

In a typical call center, calls come in to one of a number of parallel trunk lines. If no trunk lines are free, the caller will receive a busy signal; otherwise, they will reach **the interactive voice response** (IVR) system. Firms pay telecommunication companies for the number of trunk lines they reserve. Thus, they may want to limit the number of lines both to save expense and because they prefer a customer to receive a busy signal rather to experience a very long wait.

The IVR system interacts with the Automatic Call Distributor (ACD), which routes the call based on customer selections. This system may or may not be linked to the central data server. Have you ever had the experience of typing in your customer ID number to the IVR system only to find that you have to tell your ID to the agent answering your call? If so, the two computer systems were not connected. The first input of your customer ID was to route your call based on your priority as a customer (priority routing will be discussed later). Appropriate design of IVR systems is very important but outside the scope of this text.

In *best practice* call centers, many hundreds of agents would cater to many thousands of phone callers per hour (Gans, Koole, and Mandelbaum, 2003). Agent utilization levels would average between 90 percent to 95 percent, with no customer encountering a busy signal and about half of the customers receiving an answer immediately. The waiting time of those delayed would be measured in seconds. The number of customers who would decide to hang up while waiting would vary from the negligible to a mere 1 to 2 percent. Unfortunately, many of the call centers we interact with are not best practice!

Just as nurse staffing decisions can benefit from queueing theory, so can call center staffing. In fact, call centers tend to have more access to data than hospitals; therefore, call center staffing is much more a science than an art. (For an excellent practical overview of using queueing models for staffing decisions, see Green, Kolesar, and Whitt, 2007).

Metrics

Because there so much data is available and there is often a pressing requirement to lower costs, metrics matter even more in call centers than in other operations systems. Researchers have detailed how agents would hang up on customers, which made their average talk-time metric look better (Gans et al., 2007). Clearly, there were not the proper systems in place (e.g., random call monitoring) to catch such undesirable behavior—yet this is not a one-off occurrence. The authors and their students have experienced similar behavior; there is even a Dilbert comic based on this phenomenon (with Dogbert being the agent hanging up on customers). Further, while such undesirable behavior can happen within a firm, it is even more likely when the metrics are driven by misaligned incentives across firms.

As discussed in Section 6.4, contracts drive incentives. Call centers are typically contracted by some form of **service-level agreement** (SLA) through which firms agree to meet a specified level of service a certain percentage of the time. For example, firms may contract to have fewer than a given percentage of customers abandon their calls or that average delays across the day do not exceed a given number 99 percent of the time. A very common contract for call centers is the **delay percentile contract** where the service level requirement is based on customer delay.

Under a delay percentile contract, the provider agrees to meet a certain fraction of the calls, say 80 percent, within a certain window of time, say 20 seconds. However, because the contract leaves unspecified what should happen to the other 20 percent, the provider has an incentive to ignore calls that have exceeded the 20 second threshold, assigning them the lowest possible priority in the system. Joseph Milner and Tava Olsen (2010) studied this phenomenon and demonstrated that a better contract is one where there is a convex increasing cost to delays; of course, such contracts are less common in practice in part because they are more difficult to implement.

Delay percentile contracts have also been documented as leading to undesirable behavior in the British health system (BBC, 2011). Hospitals in the British system are required to treat patients within 18 weeks. To avoid additional penalties, Royal Cornwell Hospital treated newer referrals first. Patients who had waited longer than 18 weeks were downgraded in priority (rather than upgraded as one would want) relative to those who had not yet exceeded the 18-week requirement. If the hospital had been measured on the total cumulative days that patients had waited beyond 18 weeks (a convex increasing cost), there would have been no incentive for this undesirable behavior. It should come as no surprise to the reader that metrics matter in service system design.

Call Routing

As mentioned earlier, calls to call centers are frequently routed by the priority of the customer to the firm. High priority customers receive both little to no wait and better trained agents than low priority customers. Diane Brady (2000) found different treatment for the best customers versus the least valuable customers. If a strict segmentation strategy is followed, where high-priority customer agents are reserved only for high-priority customers and never used for lower priority customers, then this decreased pooling may result in very long delays for the low priority customers. It may also result in very low utilizations for the high-priority agents (to ensure good service). Therefore, most call centers will allow more flexibility in assignments of customers to agents.

While preferential treatment for priority clients is not a new phenomenon it does appear that asking customers to pay for priority is increasing in popularity. For example, United

Airlines has introduced a portfolio of Visa cards that gives customers the benefits of preferred status without them needing to fly any miles. In some sense, paying for priority is a type of revenue management, the subject of the next section.

8.8 REVENUE MANAGEMENT

Andy Boyd (2002) defines **revenue management** as "the science of maximizing profits through market demand forecasting and the mathematical optimization of pricing and inventory." There are a number of key terms to pull out of this definition. First, the decisions involved in revenue management surround both inventory and pricing. Here, "inventory," in the sense of revenue management, may mean seats of a certain class on an airplane or tickets for a certain area of a sports arena. Second, revenue management is a science, in that it uses data more than gut instinct in order to make decisions. Finally, revenue management's key goal is to maximize the firm's profit, not the customers' welfare or what is "socially optimal." This section contains a number of examples where, indeed, revenue management may not be good for the consumer.

Since having been developed in the early 1980s as a result of U.S. airline deregulation, revenue management tools have spilled over to many other industries. Many companies find it particularly attractive because revenue management addresses the revenue side of the balance sheet rather than the cost side. Hence, it is less painful than downsizing or process reengineering. Other application areas for revenue management include hotels, car rentals, rail tickets, tour operators, cargo freight, energy, entertainment, and restaurants. In fact, any service industry that faces finite perishable capacity and uncertain demand may benefit from revenue management tools.

Airline Revenue Management Overview

In the airline industry, the fundamental revenue management question is, at any point of time, how many seats of a certain class should be made available at what price? Seat classes may mean different types of seats in the airplane (e.g., business versus economy) but they also refer to the type of restrictions attached to the seat. For example, an economy seat that can be purchased at any time with no restrictions and that is fully refundable will be more expensive than an economy seat that requires purchase three weeks in advance with penalties for changing the ticket after purchase.

Boyd (2002) provides the following example. A mid-size carrier might have 1000 daily departures with an average of 200 seats per flight leg, which results in $200 \times 1000 = 200{,}000$ seats per network day. If there are 365 network days maintained in inventory, then there are $365 \times 200{,}000 = 73$ million seats in inventory at any given time. Therefore, the mechanics of managing final inventory represents a challenge because of volume. Airline revenue management therefore often treats price for a class as given and then optimizes inventory availability for that price. Price for each class is then set at a higher level (although this changes as computers become more powerful). Even setting inventory availability requires sophisticated forecasting techniques that can determine customer demand and willingness-to-pay for each class.

Effective revenue management can have a significant impact but requires sophisticated tools to be done correctly. One of the key firms operating in this space is *PROS Revenue Management.* It was founded in 1985. The website lists the company's mission as helping people and companies outperform by enabling smarter selling in the digital economy. It describes itself as developing predictive and prescriptive guidance to help companies in more than

30 industries around the globe. Each day, *PROS* drives more than 400 million prices through 1.7 billion forecasts. In 2018 the company had 1,000 employees and revenues of $197 million.

Revenue Management Basics

Revenue management involves dividing customers by their willingness to pay (i.e., **market segmentation**). Consider the following simple example. Suppose that there are 1000 potential customers for some service. The customers differ in their valuation of the service and are willing to pay between $0 and $100 for it, with all values in between equally likely (i.e., customer valuation follows a uniform distribution). If the firm charges $0, then all 1000 customers will buy the service—but the firm will make no money. If the firm charges $100, then no one will buy the service; again the firm makes no money. If they charge $$x$ ($\$0 < x < \$100$), then they make $$x\,(100 - x)1000$. Thus, it is easy to show that, if they have no capacity constraints and costs are fixed, the optimal amount to charge is $50 so that half the customers buy service and they make revenue of $25,000.

Now, suppose the firm is able to introduce a service-plus option for which they charge $75. The standard service still costs $50, and they offer service-minus for $25. Further, suppose that service-plus is designed to appeal to all customers with high valuations, so that all 250 customers with valuations over $75 buy it. Customers with valuations between $50 and $75 buy standard service. Finally, service minus is bought by customers with valuations between $25 and $50. (This is not very realistic but is intended as an example only.) Now the firm makes revenue of $\$25 \times 250 + \$50 \times 250 + \$75 \times 250 = \$27{,}500$ (a 10 percent revenue increase). The difficulty is setting the conditions for service-minus so that they are not attractive to any (or too many) customers willing to pay at least $50 and conditions for regular service so that they are not attractive to any (or too many) customers willing to pay at least $75. Otherwise, all customers will simply buy service-minus, and the firm will only make $\$25 \times 750 = \$18{,}750$ revenue.

Revenue management involves finding offerings that appeal to price-sensitive customers and more expensive offerings that will appeal to high-value customers. Hotels use premium offerings such as executive lounges to extract revenue from nonprice-sensitive customers; they use discounting sites such as priceline.com to sell unbooked hotel rooms to price-sensitive customers.

Thus, revenue management is both a science and an art. The art lies in appropriately defining product or service offerings that will appeal to different customer segments. The science is in forecasting the best prices and quantity of offerings for the different segments at different points in time. The optimization algorithms behind revenue management are beyond the scope of this text (for a good overview, see Talluri and van Ryzin, 2004).

Lead Time Pricing

Another potential application for revenue management tools is dynamic lead time pricing. Amazon has already worked out how to segment customers by urgency of delivery; their "super saver" shipping is free but if you want the item urgently, it will be expensive. However, Amazon has close to unlimited capacity for delivery. In smaller service firms, or make-to-order manufacturing, capacity is more limited. Thus, it would be dangerous for a small firm to guarantee a short lead time for a single fixed price as Amazon does.

As a concrete example, many higher-end U.S. furniture stores offer customers a choice of fabrics for sofa and chair purchases. The customer places their order and then frequently waits three months or more for delivery. We have yet to see a store where customers are offered the option to pay more to get their custom furniture sooner, yet shouldn't this be an option, at least for furniture made domestically? Further, shouldn't the price paid depend on the firm's current order backlog (which should be easily assessable through the enterprise resource planning (ERP) system)?

As another example, consider the automobile industry. As Internet commerce grew, automakers such as Ford and GM in 2000 attempted to change U.S. purchase habits to a more efficient, build-to-order system common in Europe (Vellequette, 2016). The companies hoped consumers would order cars online and have them delivered to their doors within a week. The goal was aggressive, never realized, and the plan withered. GM did cut dealer order times from 78 days to 33. Less than 5 percent of new vehicles in the United States are custom ordered compared to 50 percent in Europe. Most high-end cars in Europe are made-to-order, yet lead time dependent pricing does not seem to be much practiced yet. Revenue management does not appear to have yet made it to the automobile industry.

Dynamic lead time pricing allows the quotation to depend on current congestion. With static quotation/pricing the firm will need to work to a worst case bound, which may be very long. For example, consider an M/M/1 queue at 90 percent utilization. Using the formulas from Section 8.4, the expected lead time is 10 times the expected service time and the 95th percentile of lead time is around 30 times the expected service time. Yet, 10 percent of customers wait no time, and 39 percent of customers wait five times the expected service time or less. Clearly, there is value to be had by offering short lead times to customers willing to pay for them, especially when congestion is light. Conversely, when congestion is heavy, offering a discount to customers willing to wait may alleviate the need for overtime costs. Mustafa Akan, Barış Ata, and Tava Olsen (2013) suggest the following for implementing lead time dependent pricing. First, decide on a time unit to quote lead time in (e.g., days or weeks). Then predict what fraction of customers will pay for "premium" delivery, which is the fewest number of time units that is practical to produce and deliver the product to the customer if the system were empty. Next, reserve approximately that fraction of capacity per time unit (e.g., one couch per day). However, the firm should never waste the reserved capacity, but rather use it for backlogged nonpremium customers. They can either deliver early or store until it is time to ship, depending on whether they think early delivery will cannibalize demand for premium delivery. The firm should not offer premium delivery if the capacity has already been committed. Further, they should consider a "super saver" discount for long lead times when they are busy, which will act as a tool for smoothing demand. We believe that true revenue building opportunities are available for firms willing to consider dynamic pricing techniques for lead times.

Nontraditional Applications for Revenue Management

There is a current trend toward more nontraditional applications for revenue management. Lead time pricing, as described above, is one such application but there are many others. For example, ticket pricing for baseball games has become quite sophisticated (e.g., Biderman, 2010) as has ticketing for other events such as movies, concerts, and other sporting events. Promoters often use revenue management principles to decide on their pricing strategies.

Of course, airlines have long used revenue management to price their seats, but they are now getting creative about using it in other ways as well. Offers are made at the time of check-in for upgrades. Air New Zealand runs a type of auction for upgrades where passengers bid what they are willing to pay for a one-class upgrade, and the highest bids win the upgrades. They have also run a reverse auction system where fixed seats on a pair of flights are offered for two fixed dates (with the gap between the flights being sufficient for a vacation). The auction runs from a high value down to the reserve value and stops when a single customer has bid on, and therefore bought, the flight pair (this is known as a Dutch auction).

It is interesting to speculate on future revenue management trends. "Big data" and business analytics are likely to increase opportunities, not all of which will be good for the consumer. The wise manager will evaluate potential opportunities but should also be wary of alienating existing customers.

PROBLEMS FOR SECTIONS 8.6–8.8

27. Give an example (other than Disney) of a firm that you think is good at using queueing psychology. Explain your answer.
28. Consider your most recent unpleasant waiting experience. What went wrong and how could it have been improved?
29. Give an example of a firm that has recently introduced or increased the use of technology in its service offering. How did they perform relative to the three guidelines given in Section 8.6?
30. Give an example of a service guarantee with which you are familiar. How does it rate relative to the guidelines in Section 8.6?
31. Give an example of a service failure you have experienced. Was there any remedy given? What could the firm have done better?
32. Write (and send) a letter or email to a firm detailing a recent service failure you have personally experienced. Analyze the reply (if any) with respect to the guidelines in Section 8.6.
33. Disney's FastPass+ system is free to all ticket holders. Other theme parks, such as Universal Studios, charge a fee for priority queue access. Discuss the advantages and disadvantages of charging fees for priority access.
34. List the possible metrics that may be used by call centers to measure performance. Which of these are easiest to measure? Which are most important from the customer's perspective?
35. Give an example of an application of revenue management that you have seen or heard of outside of airlines and hotels. Describe how it is offered.
36. Give an example of an industry to which revenue management has not been applied but one which you believe would benefit from its application. Why do you think it has not been used?

8.9 HISTORICAL NOTES AND ADDITIONAL READINGS

Service operations management is a significantly younger research field than either manufacturing or inventory management. It was not recognized as a discipline within the *Decision Science Institute* until the late 1980s (Bordoloi et al., 2019). The special interest group on service management (within the society of *Manufacturing and Services Operations Management*) and a new section on service science were established in 2007 within the *Institute for Operations Research and Management Science*. It is now a thriving area of research encompassing important subfields such as healthcare management, call center management, and revenue management.

There are a number of comprehensive textbooks on service management (i.e., Bordoloi et al., 2019; Haksever and Render, 2018, and Metters et al., 2008). Such texts cover the material in this chapter in further detail. They also typically cover other operations topics found in this text, such as forecasting or quality management, but from a services slant. It should be noted that most of the building blocks for successful operations management are not fundamentally different whether one is applying them to service or production systems.

Queueing theory as a discipline is much older than service operations management. As mentioned earlier, Erlang (1909) was responsible for many of the early theoretical developments in the area of queueing. His work modeled the number of calls with the Poisson distribution and solved for the mean delay in an M/D/1 queue. However, the Poisson distribution itself is about a century older than Erlang's work. Queueing theory became a thriving research field in the 1960s and 1970s, although often from a purely mathematical standpoint. Leonard Kleinrock (1975; 1976) wrote two early texts on the subject, covering theory and applications. Many textbooks are devoted entirely to queueing (e.g., Shortie et al., 2018).

Revenue management was originally called yield management. It was used primarily in the airline industry to fill seats. American Airlines was an early adopter; its revenue management program was generating close to $1 billion annually in the 1990s (Cook, 1998). Revenue management has also been called "revenue optimization," "demand management," and "demand chain management." As a research discipline it did not really take off until the 2000s. Kalyan Talluri and Garrett van Ryzin (2004) wrote one of the earliest and most comprehensive texts on the field.

8.10 SUMMARY

Services are an increasing proportion of the economies of developed nations. Traditionally, services have lagged production systems in operational efficiency. Two of the key reasons for this are their need to include the customer as a coproducer and the perishable nature of most service capacity. This chapter examined tools for analyzing and managing service systems. Listed below are some key tools for mitigating the mismatch between supply and demand in service systems.

1. Turn customers into servers (i.e., no supply/demand mismatch).
2. Apply better predictive models (i.e., better forecasting).
3. Implement pooling (unless it increases service variability significantly).
4. Make use of queueing models to predict capacity to meet acceptable wait standards.
5. Utilize queueing models and simulation to guide improvement.
6. Schedule different customer types differently.
7. Decrease or accommodate variability.
8. Make system performance visible to employees and align incentives accordingly.
9. Engage tools for "lean" operations (see Chapter 9).
10. Don't forget the psychological aspects.
11. Employ revenue management tools.

One of the most often forgotten tools above is the need to decrease or accommodate variability in order to improve performance. Variability has a large impact on delays, both on their mean and their distribution. In most service systems, long delays lead to customer frustration and/or abandonment—and need to be avoided. A frequently ignored benefit of lean improvement programs is their drive toward standardization and hence the reduction of variation. However, many other tools, such as the careful use of pricing, can also help reduce variability.

ADDITIONAL PROBLEMS FOR CHAPTER 8

37. Consider the following commercial bread-making process. First, the dough is mixed in batches in the single mixer, and it takes 15 minutes for a batch of dough that will produce 100 loaves. The batch of dough is then proofed, which takes an hour but has no effective capacity constraint. Then the dough is baked in the single oven, which takes half an hour, and is done in batches of 100 loaves. Finally, the loaves are sliced and packed one at a time into bags, which takes 30 seconds per loaf on one of two slicing and bagging machines. What is the capacity of the mixer, oven, and the bagging machines? What is the bottleneck step? What is the capacity of the system?
38. Suppose that arrivals to a hairdresser follow a Poisson process with mean 12 customers per hour. Calculate the probabilities of (a) no customers in 20 minutes; (b) exactly one customer in 5 minutes; (c) exactly 12 customers in an hour; and (d) fewer than three customers in 10 minutes.
39. The following data has been collected on the interarrival times of patients to an emergency department: 3.772, 1.761, 0.743, 15.988, 0.412, 7.541, 6.900, 3.447, 7.024, 1.061, 5.449, 0.309, 0.766, 4.807, 8.143, 0.093, 9.524, 0.012, 4.634, and 0.195 minutes. Estimate the squared coefficient of variation of the arrival process. Is the arrival process likely to be Poisson? Why or why not? Estimate the arrival rate?
40. The following data has been collected on the *number* of customers seen to arrive to a doctor's office in a succession of 15 minute intervals: 4, 5, 4, 3, 3, 3, 1, 2, 1, 3, 2, 4, 1, 2, and 4. Estimate the squared coefficient of variation of the arrival process. Is the arrival process likely to be Poisson? Why or why not? Estimate the arrival rate.
41. Suppose that on average we observe 50 people at the beach and a rate of arrivals of one person every 3 minutes. How long, on average, do people spend at the beach?
42. Patients arrive to a small hospital emergency room according to a Poisson process with an average rate of 1.5 per hour. Because service times for these patients vary considerably, the service times are accurately described by an exponential distribution. Suppose that the average service time is 26 minutes. If there is only a single doctor working at any point in time, find the following measures of service for this emergency room.
    a. The expected total time in the system.
    b. The expected time each customer has to wait.
    c. The expected number of patients waiting for service.
    d. If the emergency room has a triage nurse who gives priority to more serious conditions, explain qualitatively how that would change your results to parts (a), (b), and (c). What additional information would one need to know to analyze a system like this?
43. Students arrive one at a time, completely at random, to an advice clinic at a rate of 10 per hour. Students take on average 5 minutes of advice, but there is wide variation in the time they need. This variation can be modeled by the exponential distribution.
    a. Assume there is only one advisor serving in the clinic. Find the following expected measures of performance for this system: the expected time in the clinic, the expected time in the queue for advice, the expected number of students in the clinic, and the expected number of students waiting for advice.
    b. Again assuming one advisor, what is the probability that there are more than ten students in the clinic at a random point in time? What is the probability that the time spent in the clinic exceeds 30 minutes?
    c. Now suppose that a second advisor is hired. Repeat your answer for (a). What are the advisors' utilizations? Would you recommend a second advisor be hired?

d. What are some ways that the advice clinic could improve students' experiences at the clinic without hiring more staff?

44. An ice cream truck is parked at a local beach, and customers queue up to buy ice cream at a rate of one per minute. The arrival pattern of people buying ice cream is essentially random. It takes 40 seconds on average to serve a customer ice cream, with a standard deviation of 20 seconds. Find the following expected measures of performance for this system: the expected time in the queue for ice cream, the expected total time to get ice cream, and the expected number of customers waiting for ice cream.
45. Cars traveling the George Washington Bridge from New Jersey to New York City must pay a toll. About 38 percent of the commuters use E-ZPass, which registers the toll to their account electronically. E-ZPass customers go quickly through the toll area averaging a wait time of 30 seconds because of the need to slow down. However, paying customers must queue up at the cash booths. They require an average service time of twenty seconds each with a standard deviation of 10 seconds. If cars are arriving to the toll area at an average rate of 8 per minute and there are 3 cash toll booths, what is the ratio of total time in the system for E-ZPass commuters versus cash commuters?
46. A local café has a single cash register, with a single assistant to work it, and three servers working to fill the customer orders. Customers arrive with exponential interarrival time an average of one every 2 minutes. The time to place their order and pay at the register is normally distributed with mean 90 seconds and standard deviation 20 seconds. Each customer's order is then passed to one of the servers who take on average 5 minutes with standard deviation 1.5 minutes, also normally distributed, to fill the order.
    a. Calculate the capacity of the register and the servers. What is the bottleneck in this system?
    b. Calculate the average utilizations of the register and the servers
    c. What is the probability a customer is delayed at the register?
    d. What is the expected time from a customer's arrival to the order being passed on to the servers (including any queueing time)?
    e. Estimate the probability that there is a delay between a customer placing his order and a server beginning to work on the order.
    f. Using the formula in S2.5 in Supplement 2, estimate the squared coefficient of variation of arrivals of orders to the servers.
    g. Estimate the expected time from the servers receiving an order to it being ready for the customer (including any queueing time).
    h. If we add (d) and (g) we get the total time from a customer walking in to receiving their order. What assumptions have been made to compute this time? Which *one* of these assumptions is the least realistic for this system? Explain your answer.
    i. List as many ways you can think of that would decrease the average time a customer spends waiting from placing their order to receiving their food.

APPENDIX 8–A

SIMULATION IMPLEMENTATION

This appendix discusses how simulations are implemented by computers. In particular, it discusses random number generation and entity-driven logic for process simulations.

Random Number Generation

At the heart of any Monte Carlo simulation are random numbers, which in this context are drawn from a uniform (0,1) distribution. That is, they are numbers between zero and one with the property that every number drawn has an equal likelihood of being selected. Random number generators are algorithms that produce what appear to be independent realizations of uniform variates. The algorithms used do not cycle for a very large number of steps, thus producing number sequences that appear random. However, because the recursive algorithms used are deterministic, the resulting string of numbers is referred to as "pseudorandom" numbers (e.g., see Fishman, 1973).

From uniform variates, one obtains observations of random variables having virtually any distribution using results from probability theory. For example, the central limit theorem says that the sums of independent random variables are approximately normally distributed. (This is a *very* loose statement of the central limit theorem.) Therefore, the sum of a reasonable number of independent draws from a uniform (0,1) distribution will be approximately normal. (Convergence occurs very quickly so the number does not have to be very large.) For example, if we let U_1, U_2, ... be successive draws from a uniform (0,1) distribution, then

$$Z = \sum_{i=1}^{12} U_i - 6$$

is approximately standard normal. (This is meant for illustrative purposes only. In practice, there are more efficient ways to generate normal variates.)

Entity Driven Logic

The interface for most graphics-based, high-level simulation packages (such as Pro-Model or ARENA) is known as **entity-based logic**. An entity is any object or customer that moves through the processes in the simulation. Entities are the "brains" in the simulation. If the modeler needs anything to happen that is not a predefined function in the package, an entity needs to do it. Therefore, an entity may serve as a breakdown demon or a lunchtime angel, if such events need to depend on more than just time elapsed.

In general, entities move from one station or location in the simulation to the next. Locations may represent servers, queues, decision points, transportation, etc. Most simulation packages will try to move an entity as far as possible through the process before the entity encounters a delay. At that point, the entity is placed in a queue, and the next entity ready for movement is picked up. This continues until there is nothing more that can occur at the current time, and the simulation clock is advanced to the earliest next event (which may be an entity arrival, an entity completing service, or any other time-flagged event). This is the reason such simulation models are often called **discrete event simulation**.

The implication of this type of logic is that some innocuous looking processes can be remarkably difficult to simulate if they are not well modeled by entity-driven logic. For example, consider customers queueing at supermarket checkouts who jockey between queues if the line next to them gets shorter. At any given point in time, there will be customer entities queueing at each of the server resources. If there were no jockeying allowed, the server resources would simply pick the next customer in line whenever a departure occurs from their station, which is very easy to implement in any major simulation package. With jockeying, all entities need to re-evaluate whenever a departure occurs—if someone switched queues, the departure decreased the number in one queue and increased the number in another. Entity-driven logic involves releasing all waiting entities from some sort of gate process, routing them through a decision node, and then routing them back to the gate process. The gate process needs to be set up by the user, rather than using built-in resource queues. In any commercial simulation package the authors have used, there is no simple way to put customers in line for a checkout server resource and then release them to jockey. Thus, if the process consists of more than simple flows of entities through locations, simulations can get complicated quite quickly.

BIBLIOGRAPHY

Anupindi, R., S. Chopra, S. D. Deshmukh, J. A. Van Mieghem, and E. Zemel. *Managing Business Process* Flows, 3rd ed. Englewood Cliffs, NJ: Prentice Hall, 2012.

Akan, M., B. Ata, and T. L. Olsen. "Congestion-Based Leadtime Quotation for Heterogenous Customers with Convex-Concave Delay Costs: Optimality of a Cost-Balancing Policy Based on Convex Hull Functions." *Operations Research,* Vol. 60 (6) (2013), pp. 1505-1519.

Ayhan, H., and T. L. Olsen. "Scheduling of a Multi-Class Single Server Queue Under Non-Traditional Performance Measures." *Operations Research* 48 (2000), pp. 482–489.

BBC News. "Royal Cornwall Hospital Patients 'Jumping Waiting List.'" August 27, 2011. Accessed from http://www. bbc.co.uk/news/uk-england-cornwall-14686568

Biderman, D. "When Did Buying Tickets Get So Complicated?" *Wall Street Journal*. January 4, 2010. Retrieved from https://www.wsj.com/articles/SB100014240527487040654045746366622642639610

Bolandifar, E., N. Dehoratius, T. L. Olsen, and J. Wiler. "An Empirical Study of the Behavior of Patients Who Leave the Emergency Department without Being Seen." Working paper, (2019).

Bordoloi, S., J. A. Fitzsimmons, and M. J. Fitzsimmons. *Service Management: Operations, Strategy, and Information Technology,* 9th ed. New York: McGraw-Hill, 2019.

Boyd, E. A. *Revenue Management and Dynamic Pricing: Part I.* 2002. https://www.ima.umn.edu/2002-2003/W9.23-27.02/19236

Brady, D. "Why Service Stinks." *Businessweek*, October 23, 2000.

Brian, O. "Southwest Airlines Has a Legendary Culture Because of This One Rule." Linkedin, August 26, 2018. Retrieved from https://www.linkedin.com/pulse/southwest-airlines-has-legendary-culture-because-one-rule-ori-brian

Buzacott, J. A., and D. D. Yao. "On Queueing Network Models of Flexible Manufacturing Systems." *Queueing Systems* 1 (1986), pp. 5–27.

Cook, T. S. "Sabre Soars." *OR/MS Today* (1998), pp. 26–31.

Erlang, A. K. "The Theory of Probabilities and Telephone Conversations." *Nyt Tidskrift for Matematik B* 20 (1909), p. 33.

Fishman, G. *Concepts and Methods in Discrete Event Digital Simulation.* New York: John Wiley & Sons, 1973.

Frei, F. X. "Breaking the Trade-Off Between Efficiency and Service." *HBR Magazine*, November 2006.

Frei, F. X. "The Four Things a Service Business Must Get Right." *HBR Articles*, April 2008.

Frei, F. X. and C. B. Hajim. "Rapid Rewards at Southwest Airlines." *HBR Cases*, August 2001.

Frei, F. X. and C. B. Hajim. "Commerce Bank." *HBR Cases*, December 2002.

Frei, F. X., and H. Rodriguez-Farrar. "Innovation at Progressive (A): Pay-As-You-Go Insurance." *HBS Case Teaching Note* 5-608-044. April 2008.

Gans, N., G. Koole, and A. Mandelbaum. "Telephone Call Centers: Tutorial, Review, and Research Prospects." *Manufacturing & Service Operations Management (M&SOM)* 5 (2003), pp. 79–141.

Gilbertson, D. "Southwest, JetBlue Top J.D. Power Airline Rankings." *USA Today*, May 29, 2019.

Green, L., and P.J. Kolesar. "The Pointwise Stationary Approximation for Queues with Nonstationary Arrivals." *Management Science* 37 (1991) pp. 84–97

Green, L., P. J. Kolesar, and W. Whitt. "Coping with Time-Varying Demand when Setting Staffing Requirements for a Service Systems." *Production and Operations Management* 16 (2007), pp. 13–39.

Green, L., J. Soares, J. Giulio, and R. Green. "Using Queueing Theory to Increase the Effectiveness of Physician Staffing in the Emergency Department." *Academic Emergency Medicine* 13 (2006), pp. 61–68.

Haksever, C., and B. Render. *Service Management and Operations*. Hackensak, NJ: World Scientific, 2018.

Hart, C. W. L. "The Power of Unconditional Service Guarantees." *HBR Articles*, July 1988.

Hart, C. W. L., J. L. Heskett, and W. E. Sasser Jr. "The Profitable Art of Service Recovery." *HBR Articles*, July 1990.

Hassin, R., and M. Haviv. *To Queue or Not to Queue: Equilibrium Behavior in Queueing Systems*. Boston: Kluwer Academic Publishers, 2003.

Henkoff, R. "Service is Everybody's Business." *FORTUNE Magazine*, June 27, 1994.

Horton, H. "The Last Person in the Queue Should Be Served First, Scientific Research Suggests." *The Telegraph*, September 9, 2015.

Kendall, D. G. "Stochastic Processes Occurring in the Theory of Queues and their Analysis by the Method of the Imbedded Markov Chain." *The Annals of Mathematical Statistics* 24 (1953), pp. 338–354.

Kleinrock, L. *Queueing Systems. Vol. I: Theory*. New York: Wiley Interscience, 1975.

Kleinrock, L. *Queueing Systems. Vol. II: Computer Applications*. New York: Wiley Interscience, 1976.

Lariviere, M. and J. A. Van Mieghem. "Strategically Seeking Service: How Competition Can Generate Poisson Arrivals." *Manufacturing & Service Operations Management* 6 (2004), pp. 23–40.

Lusch, R. F., and S. L. Vargo. *Service-Dominant Logic: Premises, Perspectives, Possibilities*. Cambridge, U.K.: Cambridge University Press, 2014.

Maister, D. H. "The Psychology of Waiting Lines." *Technical Report* (1985). Accessed from http://davidmaister. com/articles/the-psychology-of-waiting-lines/

Metters, R. D., K. H. King-Metters, M. Pullman, and S. Walton. *Successful Service Operations Management*, 2nd ed. Cincinnati, OH: South-Western Publishing, 2008.

Milner, J. M., and T. L. Olsen. "Service-Level Agreements in Call Centers: Perils and Prescriptions." *Management Science* 54 (2010), pp. 238–252.

Naor, P. "The Regulation of Queue Size by Levying Tolls." *Econometrica* 37 (1969) pp. 15–24.

Olick, D. "Apartment Rental Demand Soars as More Millennials Believe It's Cheaper Than Owning a Home." CNBC, July 19, 2019.

Pawlowski, A. "Queueing Psychology: Can Waiting in Line Be fun?" *CNN*, November 20, 2008.

Porter, M. E. "How Competitive Forces Shape Strategy." *Harvard Business Review* (March/April 1979).

Powell, S. G., and K. R. Baker. *Business Analytics: The Art of Modeling with Spreadsheets*, 5th ed. New York: Wiley, 2017.

Prisco, J. "Waiting Game: An Extended Look at How We Queue." CNN Style, February 15, 2019.

Reichheld, F. F. *The Loyalty Effect*. Watertown, MA: Harvard Business School Press, 1996.

Rother, M. and J. Shook. *Learning to See: Value-Stream Mapping to Create Value and Eliminate Muda*. Cambridge, MA: Lean Enterprise Institute, 2003.

Sakasegawa, H. "An Approximation Formula $L_q = \alpha \cdot \rho^\beta/(1 - \rho)$," *Annals of the Institute of Statistical Mathematics* 29 (1977), pp. 67–75.

Shortie, J. F., J. M. Thompson, D. Gross, and C. M. Harris. *Fundamentals of Queueing Theory*, 5th ed. New York: John Wiley & Sons, 2018.

Suri, R., and R. R. Hildebrant. "Modeling Flexible Manufacturing Systems Using Mean Value Analysis." *SME Journal of Manufacturing Systems* 3 (1984), pp. 27–38.

Talluri, K, and G. van Ryzin. *The Theory and Practice of Revenue Management*. New York: Springer, 2004.

Vellequette, L. "Why Americans Reject Build-to-Order Cars." Automotive News, June 6, 2016.

Wang, K., N. Li, and Z. Zhiang. "Queueing Systems with Impatient Customers: A Review." *Proceedings of the IEEE international conference on service operations and logistics and informatics (SOLI)*, 2010.

Winston, W. L., and S. C. Albright. *Practical Management Science,* 6th ed. Boston: Cengage, 2019.

Whitt, W. "Efficiency-Driven Heavy-Traffic Approximations for Many-Server Queues with Abandonments." *Management Science* 50 (2004), pp. 1449–1461.

Wolff, R. W. *Stochastic Modeling and the Theory of Queues.* Englewood Cliffs, NJ: Prentice Hall, 1989.

SUPPLEMENT

2 Queueing Techniques

Chapter 8 described those aspects of queueing theory that are most critical for managing service systems. However, queueing theory is a very large field and is the subject of numerous stand-alone textbooks. This supplement contains other important queueing theory results and covers some of the more technical details omitted from Chapter 8. In particular, it covers additional details on the Poisson process and exponential distribution, derives the results given for the M/M/1 queue, covers further M/M queue results, gives some infinite server queueing results, briefly covers queueing networks, and touches on the optimization of queueing systems.

S2.1 DETAILS OF THE POISSON PROCESS AND EXPONENTIAL DISTRIBUTION

This section details some of the more technical details around Poisson arrival processes. In particular, as discussed in Section 8.3, the following are the key assumptions for a Poisson process $\{N(t): t \geq 0\}$.

1. The number of arrivals in disjoint intervals are independent.
2. The number of arrivals in an interval depends only on the interval's length.
3. For a very short interval (of duration h):
   a. the probability of one arrival is approximately λh; and
   b. the probability of more than one arrival is negligible.

These assumptions can be used to derive the Poisson distribution. Mathematically we write them as follows.

a. $P\{N(t+s) - N(t) = n; N(t) - N(0) = m\} = P\{N(t+s) - N(t) = n\}P\{N(t) - N(0) = m\}$ for any $s, t \geq 0$ and integers $m, n \geq 0$.
b. $P\{N(t+h) - N(t) = 1\} = \lambda h + o(h)$.
c. $P\{N(t+h) - N(t) > 1\} = o(h)$.

Where $o(h)$ is a function such that $\lim_{h \to 0} o(h)/h = 0$. We can then write

$$\begin{aligned} P\{N(t+h)=n\} &= \sum_{m=0}^{n} P\{N(t)=m\}P\{N(t+h)-N(t)=n-m\} \\ &= P\{N(t)=n\}(1-\lambda h+o(h))+P\{N(t)=n-1\}(\lambda h+o(h))+o(h) \\ &= P\{N(t)=n\}(1-\lambda h)+P\{N(t)=n-1\}\lambda h+o(h). \end{aligned}$$

Defining $p_n(t) = P\{N(t) = n\}$, and letting $n = 0$ we have that

$$\frac{p_0(t+h)-p_0(t)}{h} = -\lambda p_0(t) + o(h)/h.$$

Taking the limit as $h \downarrow 0$ gives (through using differential equations and the fact that $p_0(0) = 1$)

$$p_0(t) = e^{-\lambda t}.$$

Further,

$$\frac{p_n(t+h)-p_n(t)}{h} = -\lambda(p_n(t) - p_{n-1}(t)) + o(h)/h.$$

Again taking the limit as $h \downarrow 0$, this equation can be used recursively to find that (as desired)

$$P\{N(t)=n\} = p_n(t) = \frac{e^{-\lambda t}(\lambda t)^n}{n!} \text{ for } n = 0,1,2\ldots.$$

As noted in Section 8.3 and shown in Section 13.3, the Poisson process can also be derived from an assumption of exponential interarrival times. The exponential distribution has a property related to the memoryless property that is particularly useful in queueing analysis. It has to do with what are known as forward and backward recurrence times. Let $N(t)$ be a Poisson process with rate λ, and T_1, T_2, ... be successive interarrival times. Consider some deterministic time t that falls between the two successive interarrival times, say T_{i-1}, and T_i. The forward recurrence time is the random variable $T_i - t$, or the time that elapses from t until the next arrival. The exponential distribution is the only distribution that has the property that the distribution of the forward recurrence time also has the exponential distribution with rate λ *independent* of t. In queueing, this means that if a server is busy when a customer arrives, the amount of time that elapses until the completion of service is still exponential with rate μ. This leads to an apparent paradox.

It turns out that the backward recurrence time of a Poisson process with rate λ, $t - T_{i-1}$, also has the exponential distribution with rate λ. The astute reader will sense something wrong here. Adding $t - T_{i-1}$ and $T_i - t$ gives $T_i - T_{i-1}$, which is just one interarrival time. However, if $t - T_{i-1}$ has an exponential distribution with rate λ and $T_i - t$ is also exponential with rate λ, it should follow that $E(t - T_{i-1} + T_i - t) = E(T_i - T_{i-1}) = 2/\lambda$, contradicting the assumption that interarrival times are exponential with rate λ! This apparent contradiction is known as the **waiting time paradox** or the **inspection paradox**. It has to do with the fact that we picked a point in time at random and found the interval that included that point rather than picking an interval at random. Intervals covering a random point are twice as long, on average. We will not dwell on this point here but note that it perplexed mathematicians (as it is probably perplexing the reader) for many years. We hope that the interested reader will follow up on their own. William Feller (1966, p. 11) provides a good starting point; Leonard Kleinrock (1975, p. 169) also discusses the paradox in the context of queueing.

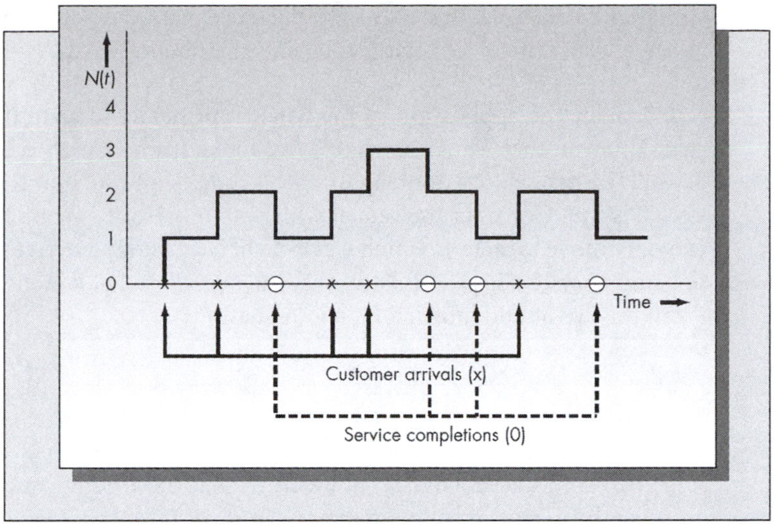

Figure S2–1 The process $L(t)$

S2.2 ANALYSIS OF THE M/M/1 QUEUE

This section derives the results given in Section 8.2 for the M/M/1 queue using what is known as a **birth and death analysis.** The process $N(t)$, the number of arrivals up until time t, is a pure birth process. It increases by one at each arrival. The process $L(t)$, the number of customers in the system at time t, is known as a birth and death process because it both increases and decreases. It increases by one at each arrival and decreases by one at each completion of a service. A realization of $L(t)$ is shown in Figure S2–1.

Notice that the state of the system either increases by one or decreases by one. The intensity or rate at which the system state increases is λ and the intensity at which the system state decreases is μ. [At this point we consider only the case in which the arrival and the service rates are fixed and independent of the state of the system; the extension to the more general case will be considered in Supplement 2.3.] This means that we can represent the rate at which the system changes state by the diagram in Figure S2–2.

Suppose that the system has evolved to a steady-state condition. That means that the state of the system is independent of the starting state. Because we are in steady state, we consider only the stationary probabilities p_n. The following derivation is based on the balance principle.

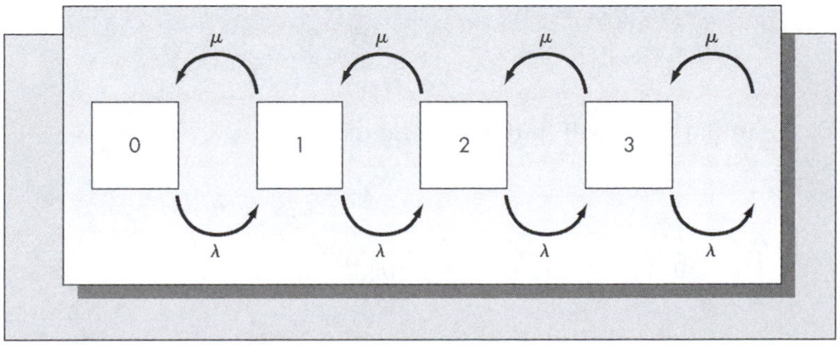

Figure S2–2 State changes for the M/M/1 queue

Balance Principle: In steady state, the rate of entry into a state must equal the rate of entry out of a state if a steady state probability distribution exists.

Consider the application of the balance principle to state 0. We enter state 0 only from state 1. Given that we are in state 1, we move from state 1 to state 0 at a rate μ (see Figure S2–2). The probability of being in state 1 is p_1. It follows that the rate at which we move into state 0 is μp_1. Consider the rate at which we move out of state 0. When we are in state 0, we can only move to state 1, which we do (when a customer arrives) at rate λ. As the probability of being in state 0 is p_0, it follows that the overall rate at which we move out of state 0 is λp_0. From this we obtain our first balance equation:

$$\mu p_1 = \lambda p_0.$$

Consider state 1. From Figure S2–2, we see that we can enter state 1 in two ways: from state 0 or from state 2. Given that we are in state 0, we enter state 1 at rate λ, and given that we are in state 2, we enter state 1 at rate μ. It follows that the rate at which we enter state 1 is $\lambda p_0 + \mu p_2$. We can leave state 1 by going either to state 0 if a service completion occurs or state 2 if an arrival occurs. Hence, the rate at which we leave state 1 is $\lambda p_1 + \mu p_1 = (\lambda + \mu)p_1$. It follows that the second balance equation is

$$\mu p_2 + \lambda p_0 = (\lambda + \mu)\, p_1.$$

The form of the remaining balance equations is essentially the same as that of the second balance equation. In general,

$$\mu p_{i+1} + \lambda p_{i-1} = (\lambda + \mu) p_i \qquad \text{for } 1 \le i < \infty.$$

These equations, along with one other condition, allow us to obtain an explicit solution for the steady-state probabilities. The method of solution is to first express each p_i in terms of its p_0. From the first balance equation we have

$$p_1 = (\lambda/\mu) p_0.$$

The second balance equation gives

$$\mu p_2 = (\lambda + \mu)\, p_1 - \lambda p_0 = (\lambda + \mu)(\lambda/\mu) p_0 - \lambda p_0 = (\lambda^2/\mu) p_0 + \lambda\, p_0 - \lambda\, p_0 = (\lambda^2/\mu)\, p_0.$$

Dividing both sides by μ gives

$$p_2 = (\lambda/\mu)^2 p_0.$$

Similarly, we will find in general that

$$p_i = (\lambda/\mu)^i p_0.$$

The solution is obtained by using the condition that

$$\sum_{i=0}^{\infty} p_i = 1,$$

since $p_0, p_1, p_2, \ldots$ forms a probability distribution on the states of the system. Substituting for each p_i, we have

$$\sum_{i=0}^{\infty} (\lambda/\mu)^i \, p_0 = 1.$$

Let $\rho = \lambda/\mu$ be the utilization rate. For a solution to exist, it must be true that $\rho < 1$. In that case

$$\sum_{i=0}^{\infty} \rho^i = 1/(1-\rho),$$

known as the geometric series, from which we obtain

$$p_0 = (1 - \rho)$$

and

$$p_i = \rho^i (1 - \rho) \qquad \text{for } i = 1, 2, 3 \ldots$$

as given in Section 8.4. (This formula is also valid when $i = 0$.)

Waiting Time Distribution

We now derive the distribution of the waiting time W for a random customer joining the queue in steady state. To derive this distribution, we condition on the number of customers in the system at steady state, n, and uncondition by multiplying by the probability p_n. Suppose that a customer joining the queue at a random point in time finds n customers already in the system. The customer must then wait for n service completions before entering service. As W is the total time in the system, this means that in this case W will be the sum of $n + 1$ service completions. Let $S_1, S_2, \ldots$ be the times of the successive services. By assumption, these random variables are mutually independent and exponentially distributed with common mean $1/\mu$. The time for $n + 1$ service completions is $S_1 + S_2, + \ldots + S_{n+1}$, which we know has the Erlang distribution with parameters μ and $n + 1$ (see Section 13.3).

That is,

$$P\{W > t \mid n \text{ in the system}\} = \sum_{k=0}^{n} \frac{e^{-\mu t} (\mu t)^k}{k!}.$$

We know from the previous subsection that the unconditional probability of n in the system in the steady state, p_n, has the geometric distribution. Substituting $\rho = \lambda/\mu$, we may write p_n in the form

$$p_n = \left(\frac{\mu - \lambda}{\mu}\right)\left(\frac{\lambda}{\mu}\right)^n.$$

Unconditioning on p_n gives

$$P\{W > t\} = \sum_{n=0}^{\infty}\sum_{k=0}^{n} \frac{e^{-\mu t}(\mu t)^k}{k!}\left(\frac{\mu-\lambda}{\mu}\right)\left(\frac{\lambda}{\mu}\right)^n$$
$$= \sum_{k=0}^{\infty}\sum_{n=k}^{\infty} \frac{e^{-\mu t}(\mu t)^k}{k!}\left(\frac{\mu-\lambda}{\mu}\right)\left(\frac{\lambda}{\mu}\right)^n$$
$$= \left(\frac{\mu-\lambda}{\mu}\right)e^{-\mu t}\sum_{k=0}^{\infty}\frac{(\mu t)^k}{k!}\sum_{n=k}^{\infty}\left(\frac{\lambda}{\mu}\right)^n.$$

Using the fact that

$$\sum_{n=k}^{\infty}\left(\frac{\lambda}{\mu}\right)^n = \left(\frac{\lambda}{\mu}\right)^k \frac{1}{1-\lambda/\mu},$$

and substituting this into the earlier equation gives, after simplifying,

$$P\{W > t\} = e^{-\mu t}\sum_{k=0}^{\infty}\frac{(\lambda t)^k}{k!} = e^{-\mu t}e^{+\lambda t} = e^{-(\mu-\lambda)t}.$$

The summation term equals $e^{+\lambda t}$ because it is the Taylor series expansion for e (and because Poisson probabilities sum to one). What we have shown is the surprising result that W has the exponential distribution with parameter $\mu - \lambda$. This implies that W has the memoryless property. That is, suppose that a customer has already been waiting for s units of time. The probability of waiting at least an additional t units of time is the same as the probability that a newly joining customer waits at least t units of time. This result is not intuitive and is rather depressing for the poor customer, who has already spent a substantial amount of time waiting for service!

We will not present the derivation (it is similar to the one previously given) but state that the distribution of W_q is essentially exponential with the complementary cumulative distribution function

$$P\{W_q > t\} = \rho e^{-(\mu-\lambda)t} \qquad \text{for all } t \geq 0.$$

Note that the probability that the waiting time in the queue is zero (i.e., there is no delay) is positive. It is equal to the probability that the system is empty, p_0. That is

$$P\{W_q = 0\} = p_0 = 1 - \rho = 1 - p^d.$$

S2.3 FURTHER RESULTS FOR M/M QUEUES

This section gives further known results for M/M queues. In particular, we derive results for when transitions are state dependent, when there are multiple servers, and when there is a finite system capacity.

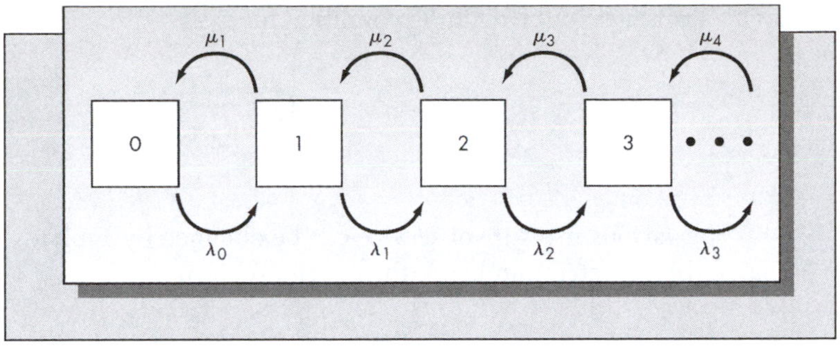

Figure S2–3 State changes for the M/M/1 queue with state-dependent service and arrival rates

Consider first the case where *both* the arrival and the service rates depend on the state. Several versions of the M/M/1 model are special cases of these circumstances. The transition diagram is the same as that pictured in Figure S2–2, except that both λ and μ are state dependent (see Figure S2–3). The balance equation principle applied to this system yields

$$\mu_1 p_1 = \lambda_0 p_0,$$
$$\lambda_0 p_0 + \mu_2 p_2 = (\lambda_1 + \mu_1) p_1,$$
$$\lambda_1 p_1 + \mu_3 p_3 = (\lambda_2 + \mu_2) p_2,$$

and so on.

Expressing each of the state probabilities in terms of p_0, as we did earlier, results in the following:

$$p_1 = \frac{\lambda_0}{\mu_1} p_0,$$
$$p_2 = \frac{\lambda_0 \lambda_1}{\mu_2 \mu_1} p_0,$$
$$p_3 = \frac{\lambda_0 \lambda_1 \lambda_2}{\mu_3 \mu_2 \mu_1} p_0,$$

and so on.

Define

$$a_n = \frac{\lambda_{n-1} \lambda_{n-2} \ldots \lambda_o}{\mu_n \mu_{n-1} \ldots \mu_1}$$

so that

$$p_n = a_n p_0 \qquad \text{for } n = 1, 2, 3 \ldots.$$

Again using the fact that $p_0, p_1, \ldots$ is a probability distribution, we have that

$$\sum_{n=0}^{\infty} p_n = 1.$$

This translates to the defining condition for p_0 as

$$p_0 = \frac{1}{1 + \sum_{n=1}^{\infty} a_n}.$$

The various measure of service can be obtained by applying their definitions. In particular, l, the expected number in the system, is given by

$$l = \sum_{n=0}^{\infty} n p_n.$$

If there are assumed to be s servers then l_q, the expected number in the queue, is given by

$$l_q = \sum_{n=s}^{\infty} (n - s) p_n.$$

Little's law still applies and can be used to find the expected waiting times given the expected number in the system and the expected number in the queue. However, to apply Little's law when the arrival rate is state dependent, we must determine the *overall* average expected arrival rate, or the effective arrival rate, which we call λ_{eff}. Because the arrival rate is λ_n, when the system is in state n, it follows that the effective arrival rate is

$$\lambda_{eff} = \sum_{n=0}^{\infty} \lambda_n p_n.$$

We can use these general results for a variety of configurations of the queue with exponential interarrival times and exponential service times. In particular, they can be used to find results for the M/M/s queue and queues with finite capacity, as shown next.

The M/M/s Queue

Consider the M/M/s queue; that is, the case in which there are s servers in parallel. This case is pictured in Figure 8–7 from Section 8.5. In order to apply the results from the state-dependent model we need to establish the following.

Suppose that m servers are busy at a random point in time and also suppose that the distribution of each of the servers is exponential with rate μ. What is the distribution of the time until the next service completion? Let $T_1, T_2, \ldots, T_m$ be the service times of the customers who are currently in service. By assumption, these are independent exponential random variables. Furthermore, if t is a random point in time, the remaining time in service from t to the end of the service completion for each of the customers is also exponential with the same distribution. (This is a consequence of the properties of the exponential distribution discussed earlier.)

It follows that the time until the next service completion, say T, is distributed as the *minimum* of $T_1, T_2, \ldots, T_m$.

Result: Let $T_1, T_2, \ldots, T_m$ be independent exponential random variables with common exponential distribution with rate μ, and define $T = \min(T_1, T_2, \ldots, T_m)$. Then T is also exponentially distributed with rate $m\mu$.

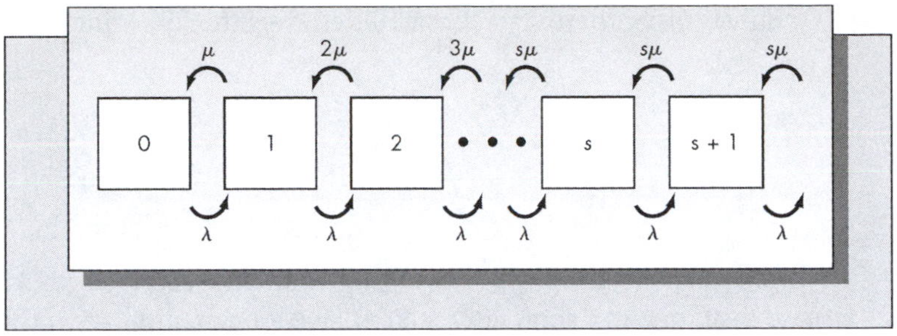

Figure S2–4 Transition rate diagram when there are *s* servers in parallel

(This result is proven in Chapter 13 in the context of series systems of components subject to exponential failure.)

Why is this result important? It means that the distribution between customer departures is still exponential and that the methods just derived for state-dependent queues still apply. If there are s servers, then it follows that the transition rate diagram is as pictured in Figure S2–4.

Comparing Figures S2–3 and S2–4, we see that

$$
\begin{aligned}
\mu_1 &= \mu \\
\mu_2 &= 2\mu \\
&\vdots \\
\mu_s &= s\mu \\
\mu_{s+1} &= s\mu \\
\mu_{s+2} &= s\mu
\end{aligned}
$$

and $\lambda_i = \lambda$ for all $i = 0, 1, 2, \ldots$.

Substituting, it follows that

$$
\begin{aligned}
P_1 &= \frac{\lambda}{\mu} P_0, \\
P_2 &= \frac{1}{2}\left(\frac{\lambda}{\mu}\right)^2 P_0, \\
P_3 &= \frac{1}{(3)(2)}\left(\frac{\lambda}{\mu}\right)^3 P_0, \\
&\vdots \\
P_n &= \frac{1}{n!}\left(\frac{\lambda}{\mu}\right)^n P_0 \qquad \text{for } 0 \le n \le s.
\end{aligned}
$$

When $n > s$, we obtain

$$
P_n = \frac{1}{s!s^{n-s}}\left(\frac{\lambda}{\mu}\right)^n P_0 \qquad \text{for } n > s.
$$

Substituting these state probabilities gives the following for p_0:

$$p_0 = \left\{ \sum_{n=0}^{s-1} \frac{1}{n!} \left(\frac{\lambda}{\mu} \right)^n + \sum_{n=s}^{\infty} \frac{1}{s!s^{n-s}} \left(\frac{\lambda}{\mu} \right)^n \right\}^{-1}.$$

This can be simplified by noting that the second term is a geometric series. Letting $\rho = \lambda/s\mu$, one can show after a bit of algebraic manipulation that

$$p_0 = \left\{ \sum_{n=0}^{s-1} \frac{(s\rho)^n}{n!} + \frac{(s\rho)^s}{s!} (1-\rho)^{-1} \right\}^{-1}.$$

When computing the standard performance measures, it turns out that l_q has the simplest form. Again, we present only the results, not the details. The derivations are similar to those presented in earlier sections.

$$l_q = \frac{s^s \rho^{s+1}}{s!(1-\rho)^2} p_0.$$
$$l = l_q + s\rho.$$
$$w_q = l_q / \lambda.$$
$$w = w_q + 1/\mu.$$

As with the single-server queue, the condition that $\rho < 1$ is required to guarantee that the queue does not grow without bound.

EXAMPLE S2.1

Tony's Barbershop is run, owned, and operated by Anthony Jones, who has been cutting hair for more than 20 years. Anthony does not take appointments, so the arrival pattern of customers is essentially random. Traditionally, the arrival rate had been about one customer every 50 minutes. Two months ago, the local paper ran an article about Anthony that improved business substantially. Currently, the arrival rate is closer to one customer every 35 minutes. Haircuts require an average of 25 minutes, but the times vary considerably depending on customer needs. A trim might require as little as 5 minutes, but a shampoo and styling could take as long as an hour or more. For this reason, the exponential distribution seems to provide a reasonably good fit for the service time distribution.

Anthony's customers had always been patient. After business picked up, some have complained that the wait is too long. Anthony is considering taking his cousin Marvin into the business to improve customer service. Assume that Marvin cuts hair at the same rate as Anthony.

a. How much has the quality of service declined since more customers have started using the shop?
b. How much improvement in the performance of the system are customers likely to see with an additional barber in the shop?

Solution

a. First, we will determine the various performance measures for the system prior to the appearance of the newspaper article. The average time between arrivals was one every 50 minutes, which gives an arrival rate of

$$\lambda = 60/50 = 1.2 \text{ arrivals per hour.}$$

Each haircut requires an average of 25 minutes, which translates to a service rate of

$$\mu = 60/25 = 2.4 \text{ haircuts per hour.}$$

It follows that $\rho = \lambda/\mu = 1.2/2.4 = 0.5$. (That is, Tony was busy half the time.) The values of the performance measures are

$$l = \rho / (1 - \rho) = 0.5/0.5 = 1.$$
$$l_q = \rho l = 0.5.$$
$$w = l/\lambda = 1 / 1.2 = 0.8333 \text{ hour.}$$
$$w_q = l_q/\lambda = 0.5/1.2 = 0.4167 \text{ hour.}$$

This means that originally customers waited (0.4167)(60) = 25 minutes for a haircut on average.

After the article appeared, the arrival rate increased to one customer every 35 minutes. This means that λ became 60/35 = 1.7143 and ρ = 0.7143. The performance measures are now

$$l = 0.7143/(1 - 0.7143) = 2.5.$$
$$l_q = \rho l = (0.7143)(2.5) = 1.7857.$$
$$w = l/\lambda = 2.5 / 1.7143 = 1.458 \text{ hours.}$$
$$w_q = l_q/\lambda = 1.78 / 1.7143 = 1.0383 \text{ hours.}$$

The customers clearly have a valid gripe. A customer has to wait an average of more than an hour before getting a haircut. In fact, because the distribution of w_q is exponential, many would have to wait quite a bit longer than this.

b. Adding an additional barber improves the system performance dramatically. With two barbers, we have

$$\rho = \lambda/(s\mu) = 1.7143/(2)(2.4) = 0.3571 \qquad (s\rho = 0.7143).$$

$$P_o = \left\{1 + 0.7143 + \frac{(0.7143)^2}{2!}\frac{1}{1 - 0.3571}\right\}^{-1} = (2.111)^{-1} = 0.4737.$$

It follows that

$$l_q = \frac{(2)^2(0.3571)^{2+1}}{2!(1 - 0.3571)^2}(0.4737) = 0.1044.$$
$$l = l_q + s\rho = 0.1044 + 0.7143 = 0.8187.$$
$$w_q = l_q/\lambda = 0.1044/1.7143 = 0.0609 \text{ hour (3.65 minutes).}$$
$$w = w_q + 1/\mu = 0.0609 + 0.4167 = 0.4776 \text{ hour (about 28.7 minutes).}$$

With only a single barber, customers could expect to wait more than an hour for a haircut. With the addition of another barber, time is reduced to less than 4 minutes on average.

The M/M/1 Queue with a Finite Capacity

Another special version of the general M/M/1 queue with state-dependent service and arrival rates is the case in which there is a finite waiting area. If arrivals occur when the waiting area is full, they are turned away. Problems of this type are common in service systems such as restaurants, movie theaters, and concert halls. They can also occur in manufacturing systems in which buffers between work centers have a finite capacity. This is the case, for example, with lean systems. (See the discussion of lean in Chapter 9.)

Suppose that the maximum number of customers permitted in the system is K. The transition rate diagram for this case is exactly the same as that pictured in Figure S2–2 except that the transitions do not occur beyond state K. Because the transition rate diagram is the same up to state K, the balance equations will yield the same relationship between p_n and p_0 for $n = 1, 2, \ldots, K$. That is,

$$p_n = \rho^n p_0 \qquad \text{for } n = 1, 2, 3, \ldots, K.$$

Then p_0 is found from

$$\sum_{n=0}^{K} p_n = 1,$$

which gives

$$p_0 = \left(\sum_{n=0}^{K} \rho^n\right)^{-1}.$$

An explicit expression for the finite geometric series is obtained in the following way:

$$\sum_{n=0}^{K} \rho^n = \sum_{n=0}^{\infty} \rho^n - \sum_{n=K+1}^{\infty} \rho^n = \frac{1}{1-\rho} - \frac{\rho^{K+1}}{1-\rho} = \frac{1-\rho^{K+1}}{1-\rho}.$$

It follows that

$$p_0 = \frac{1-\rho}{1-\rho^{K+1}},$$

from which we obtain

$$p_n = \frac{(1-\rho)\rho^n}{1-\rho^{K+1}} \qquad \text{for } n = 0, 1, 2, \ldots, K.$$

For the case of the finite waiting room, it is not necessary that $\rho < 1$. In fact, p_n has this value for *all* values of $\rho \neq 1$. When $\rho = 1$, it turns out that all states are equally likely, so that

$$p_n = 1/(K+1) \qquad \text{for } 0 \le n \le K \qquad (\text{when } \rho = 1 \text{ only}).$$

Little's law still applies, but we must use a modified value for the arrival rate because not all arriving customers are permitted to enter the system. When there are K or more in the system, the arrival rate is zero, so the overall arrival rate is less than λ. The effective arrival rate, λ_{eff}, is computed as follows:

$$\begin{aligned}\lambda_{eff} &= \lambda P\{\text{Number in the system} < K\} + 0P\{\text{Number in the system} = K\}\\ &= \lambda(1 - P\{\text{number in the system} = K\})\\ &= \lambda\,(1 - p_K).\end{aligned}$$

The measures of performance are obtained from l, the expected number in the system in steady state, which is found from

$$l = \sum_{n=0}^{K} np_n = \sum_{n=0}^{K} \frac{1-\rho}{1-\rho^{K+1}} n\rho^n.$$

The calculation proceeds by noting that

$$\sum_{n=0}^{K} n\rho^{n-1} = \frac{d}{d\rho}\sum_{n=0}^{K}\rho^n.$$

Using the earlier expression for the finite geometric sum, we eventually obtain

$$l = \frac{\rho}{1-\rho} - \frac{(K+1)\rho^{K+1}}{1-\rho^{K+1}}.$$

The remaining measures of performance are found from

$$\begin{aligned}l_q &= l - (1 - p_0).\\ w &= l / \lambda_{eff}.\\ w_q &= l_q / \lambda_{eff}.\end{aligned}$$

Similar formulas can be derived for the case of a finite capacity queue and multiple parallel servers. (See, for example, Hillier and Lieberman, 1990).

EXAMPLE S2.2

A popular attraction at the New Jersey Shore is a street artist who will paint a caricature in about five minutes. However, because the times required for each drawing vary considerably, they are accurately described by an exponential distribution. People are willing to wait their turn, but when there are more than 10 waiting for a picture, customers are turned away and asked to return at a later time. At peak times one can expect as many as 20 customers per hour. Assume that customers arrive completely at random at the peak arrival rate.

a. What proportion of the time is the queue at maximum capacity?
b. How many customers are being turned away on average? Determine the measures of performance for this queueing system.
c. If the waiting area were doubled in size, how would that affect your answers to parts (a) and (b)?

Solution

The arrival rate is $\lambda = 20$ per hour and the service rate is $\mu = 12$ per hour, so that $\rho = 20/12 = 1.667$. The maximum number in the system is $K = 11$ (10 in the queue plus the customer being served).

The probability that the system is full is p_K, which is given by

$$p_K = \frac{(1-\rho)\rho^K}{1-\rho^{K+1}} = \frac{(1-1.667)(1.667)^{11}}{1-1.667^{12}} = \frac{-184.25}{-459.5} = 0.40.$$

b. As the arrival rate is 20 per hour and 40 percent of the time the system is full, there are, during peak periods, (20)(0.40) = 8 customers per hour being turned away. This gives λ_{eff} = 12 per hour.

$$l = \frac{\rho}{1-\rho} - \frac{(K+1)\rho^{K+1}}{1-\rho^{K+1}} = \frac{1.667}{1-1.667} - \frac{(12)(1.667)^{12}}{1-1.667^{12}} = -2.5-(-12.03) = 9.53.$$

We need to determine p_0 to compute l_q.

$$p_0 = \frac{(1-\rho)}{1-\rho^{K+1}} = \frac{1-1.667}{1-1.667^{12}} = 0.00145,$$

which gives $l_q = l - (1 - p_0) = 9.53 - (1 - 0.00145) = 8.53$.

Notice that the small value of p_0 means that the system is rarely empty. In particular, the artist is idle only 0.145 percent of the time!

We showed earlier that $\lambda_{eff} = 12$, so that

$$w = l / \lambda_{eff} = 9.53/12 = 0.7942 \text{ hour (about 48 minutes).}$$

$$w_q = l_q / \lambda_{eff} = 8.53/12 = 0.7108 \text{ hour (about 43 minutes).}$$

c. If the waiting area were doubled in size, then $K = 21$. In that case p_K is given by

$$p_k = \frac{(1-1.667)(1.667)^{21}}{1-1.667^{22}} = \frac{-30{,}533.28}{-76{,}309.3} = 0.40.$$

Interestingly, doubling the capacity of the queue makes no difference relative to the probability that the system is full. The reason is that because the arrival rate exceeds the service rate, the system reaches capacity quickly in either case. In both cases the effective arrival rate λ_{eff} is approximately equal to the service rate (although it will always be true that $\lambda_{eff} < \mu$). Even for much larger values of K, p_K is 0.4 in this example.

S2.4 INFINITE SERVER RESULTS

This section considers results for queues with an infinite number of servers. It gives exact results for the M/G/∞ queue and also limiting results as the number of servers, s, approaches ∞. These limiting results provide an approximation for the probability of delay that is typically more accurate (although a little more complicated) than the Sakasegawa (1977) formula given in Section 8.3.

The M/G/∞ queue

Another version of the queueing problem with general service distribution for which there are explicit results is the case in which there are an infinite number of servers. Customers arrive at the system completely at random according to a Poisson process with rate λ. The service time distribution is arbitrary with service rate μ. At the instant of arrival, the customer enters service. An infinite number of servers means that there is always a server available, no matter how many customers are in the system. Although this might seem unrealistic, many real problems can be modeled in this way. Because there is no queue of customers waiting for service, there is no waiting time for service. Hence, both measures of performance, l_q and

w_q, are zero. However, the number of customers in the system, L, is not zero. Note that the number of customers in the system is equal to the number of busy servers.

Result: For the M/G/∞ queue with arrival rate λ and service rate μ, the distribution of the number of customers in the system (or the number of busy servers) in steady state is Poisson with rate λ/μ. That is

$$P\{L=k\}=\frac{e^{-(\lambda/\mu)}(\lambda/\mu)^k}{k!} \qquad \text{for } k=0,1,2,\ldots.$$

This is a powerful result. It follows that both the mean and the variance of the number of customers in the system in steady state is λ / μ.

EXAMPLE S2.3

A common inventory control policy for high-value items is the one-for-one policy, also known as the $(S - 1, S)$ policy. This means that the target stock is S and at each occurrence of a demand, a reorder for one unit is placed. Suppose that demands are generated by a stationary Poisson process with rate λ and that the lead time required for replenishment is a random variable with arbitrary distribution with mean $1/\mu$. This problem is exactly an M/G/∞ queue. The number of customers in the system is equivalent to the number of outstanding orders, which by the earlier result has the Poisson distribution with mean λ/μ. If the lead time is fixed at τ, the expected number of units on order is just $\lambda\tau$. It follows that the expected number of units in stock is $S - \lambda\tau$. This result can be used to determine an expression for the expected holding, stock-out, and replenishment costs, which can then be optimized with a choice of S. (See Hadley and Whitin, 1963, p. 212, for example.)

EXAMPLE S2.4

Personnel planning is an important function for many firms. Consider a company's department with a desired headcount of 100 positions. Suppose that employees leave their positions at a rate of 3.4 per month and that it requires an average of 4 months for the firm to fill open positions. Analysis of past data shows that the number of employees leaving the firm per month has the Poisson distribution and that the time required to fill positions follows the Weibull distribution. What is the probability that there are more than 15 positions unfilled at any point in time? How many jobs within the department are filled on average? How many positions should the firm have in order for the head count of working employees to be 100 on average?

Solution

To determine the distribution of filled positions, we model the problem as an M/G/∞ queue. Each time that an employee leaves, his or her position enters the queue of unfilled positions. Assuming that the search for a replacement starts immediately, the correct model is an infinite number of servers. According to the theory, the expected number of unfilled positions is independent of the time required to replace each employee. The expected number of unfilled positions is λ/μ. For this application, λ corresponds to the rate at which employees leave their positions, which is 3.4 per month, and μ the rate that jobs are filled, which is 1/4 per month. Thus, the mean number of unfilled jobs is $\lambda/\mu = 3.4/(1/4) = (3.4)(4) = 13.6$. Hence, there are $100 - 13.6 = 86.4$ positions filled on average.

The probability that there are more than 15 unfilled positions is the probability that a Poisson random variable with mean 13.6 exceeds 15. Interpolating from Table A–3, this probability is approximately 0.29.

It also follows that if the department were allotted 114 positions rather than 100, there would be, on average, 100 positions filled at any point in time (although the actual number is a random variable). [John Peterson of Smith-Kline-Beecham brought this application to our attention.]

Infinite Server Limits

Useful queueing approximations can be developed by taking the limit as the number of servers tends towards infinity. However, as the number of servers increases, we must also scale either the arrival rate or service rate; otherwise delays would go to zero. Turning this around, we must decide the rate at which the number of servers are increased, relative to the arrival rate scaling up. As mentioned in Section 8.3, one useful scenario is where staffing is matched to arrival rate growth so that the probability of delay p^d tends to a value strictly between 0 and 1. This occurs if the scaling in the number of servers, s, occurs such that

$$\sqrt{s}(1-\rho_s)\to\beta \qquad \text{for } 0<\beta<1,$$

where ρ_s is the utilization when there are s servers. This is the so-called quality-and-efficiency (QED) driven regime, or Halfin-Whitt scaling. This scaling results in the following approximation for the probability of delay p^d (Whitt, 2004). Define

$$\beta=(1-\rho)\sqrt{s},$$

and a measure of peakedness

$$z=1+(c_a^2-1)(1-(c_s^2/2)) \qquad \text{for } 0\le c_s^2\le 1.$$

Note that the above equation for z is only valid for $c_s^2 \le 1$. (See Whitt, 2004, equation 1.6 for a more accurate (but more complicated) estimate of this value, which is not limited by the range of c_s^2.) The estimate for probability of delay is then given by

$$\rho^d=\frac{1}{1+\beta\ \Phi(\beta/\sqrt{z})/(\sqrt{z}\,\phi(\beta/\sqrt{z}))}.$$

where $\Phi(.)$ is the cumulative standard normal distribution and $\phi(.)$ is the standard normal density (see Section 5.1). Numerical tests show this approximation to be reasonably accurate across a wide range of G/G/s systems. It can also be extended to the case of a finite capacity waiting room (see Whitt, 2004).

EXAMPLE S2.5

In Example 8.4, suppose that customer arrivals are Poisson and service is deterministic. What is the probability of delay using the Halfin-Whitt estimate? How does it compare to the Sakasegawa approximation? Now repeat this for Example 8.6.

Solution

For Example 8.4 given the above description, $c_a^2 = 1$ and $c_s^2 = 0$. Therefore, $z = 1$. Further, $\beta = (1 - \rho)\sqrt{s} = (1 - 0.9868)\sqrt{48} = 0.09145$. Therefore, $p^d = \dfrac{1}{1+0.09145\Phi(0.09145)/\phi(0.09145)} = 0.8901$.

Under the Sakasegawa formula, $p^d = 0.889$, so the two values are very close. For Example 8.6, $c_a^2 = 1$ and $c_s^2 = 0.7785$, which imply $z = 1$. Further, $\beta = (1 - \rho)\sqrt{s} = (1 - 0.94444)\sqrt{3} = 0.09623$. Therefore, $p^d = \dfrac{1}{1+0.09623\,\Phi(0.09623)/\phi(0.09623)} = 0.8846$.

Under the Sakasegawa formula, $p^d = 0.9008$, which is 1.8 percent larger. It is encouraging that these probabilities are not entirely different.

S2.5 QUEUEING NETWORKS

As previously discussed, analyzing queueing networks exactly is difficult; therefore, simulation is often used to obtain an accurate estimation of system statistics. However, some notable exceptions to this statement are discussed below.

An interesting result for M/M/s queues is that the departure stream of customers from the system forms a Poisson process (of rate λ if no customers are created or lost by the server). This means that if another station is downstream from this queue, then it receives a stream of Poisson customers arriving, which can be analyzed exactly.

A **Jackson network** is a network of J M/M/s queueing nodes. Each node i receives a Poisson stream of external arrivals at rate a_i, $1 \le i \le J$. Jobs that are completed at node i are routed to node j with probability p_{ij} and out of the system with probability $1-\sum_{j=1}^{J} p_{ij}$. The flow balance equations for the aggregate arrival rates, λ_i, are as follows

$$\lambda_i = \alpha_i + \sum_{j=1}^{J} p_{ji}\,\lambda_j \qquad \text{for } 1 \le i \le J.$$

This implies that each node i forms its own M/M/s queueing system with arrival rate λ_i. James Jackson (1957) showed that the distribution of customers in the system is simply the product of the probabilities for each queueing node. That is, the queues act as if they are independent of each other, even though they are clearly not.

In general, the departure stream from general queueing models is not Poisson, which of course means that there are typically not exact results for performance of downstream stations. However, an approximation technique was developed by Ward Whitt (1983), called the **queueing network analyzer** (QNA), for quite general networks of G/G/s queues. It assumes an open network (where customers eventually leave the system), no capacity constraints, FCFS service—but that customers can be created or destroyed at stations, and the routing can be quite general.

One of the useful approximations in the QNA is an expression for the variability of departures from a G/G/1 queue. This variability is given by

$$\rho^2 c_s^2 + (1-\rho^2)c_a^2.$$

Notice how if $\rho = 1$ then this is equal to c_s^2, the variability of the service process; whereas, if $\rho = 0$ then it equals c_a^2, the variability of the arrival process. This makes intuitive sense because if $\rho = 1$ then the server is consistently busy and the customers flowing out look like the service process, whereas if $\rho = 0$ (and there are arrivals) then the service time must be negligible and arrivals are just passed straight through. This expression can then be used as the variability of arrivals to a downstream station if the queues are in series.

S2.6 OPTIMIZATION OF QUEUEING SYSTEMS

Classical queueing analysis is descriptive rather than prescriptive. In practice, this means that given the various input and service distributions, one determines the measures of performance. These measures of performance do not directly translate to optimal decisions concerning the design of the system. This section shows how one would use queueing theory to develop models for determining optimal system configurations (based on Hillier and Lieberman, 1990, chapter 17).

Typical Service System Design Problems

1. The State Highway Board must determine the number of tollbooths to have available on a new interstate toll road. The more tollbooths open at any point in time, the less wait commuters will experience. However, there is a one-time cost to build additional tollbooths and ongoing costs of salaries for additional toll takers.
2. A plant is being built by a major manufacturer of solid-state (memory) drives. The company management is considering several options for the manufacturing equipment. A new machine for the drives has double the throughput of the conventional equipment but at more than triple the cost. Is the investment justified?
3. A translation service is considering how large of a client base to develop. The company wishes to have a large enough number of clients to make it busy, but not so many that it cannot provide reasonable turnaround times.

Modeling Framework

1. Consider the example of the state highway board. The more time commuters spend on the highway, the less time they spend working and contributing to society. If we view the goal as societal optimization, then there is clearly a direct economic benefit to reducing commute time. Suppose that an economic analysis of the highway problem resulted in an estimate of the cost incurred when a commuter spends w units of time in the system as the function $h(w)$. A typical case is pictured in Figure S2–5.

Let W be the time in system of a customer chosen at random. Then W is a random variable. For the M/M/1 queue, we showed that W has the exponential distribution with parameter $\mu - \lambda$. Given the distribution of W, it follows that the expected waiting cost of a customer chosen at random is

$$E(h(W)) = \int_0^\infty h(w)(\mu - \lambda)e^{-(\mu-\lambda)w}\,dw.$$

Because the arrival rate of customers is λ units per unit time, the overall waiting cost per unit time is $\lambda E(h(W))$. In the case of the tollbooths, one would determine the distribution of W for each number of servers being considered, say W_s. If the cost per unit time for

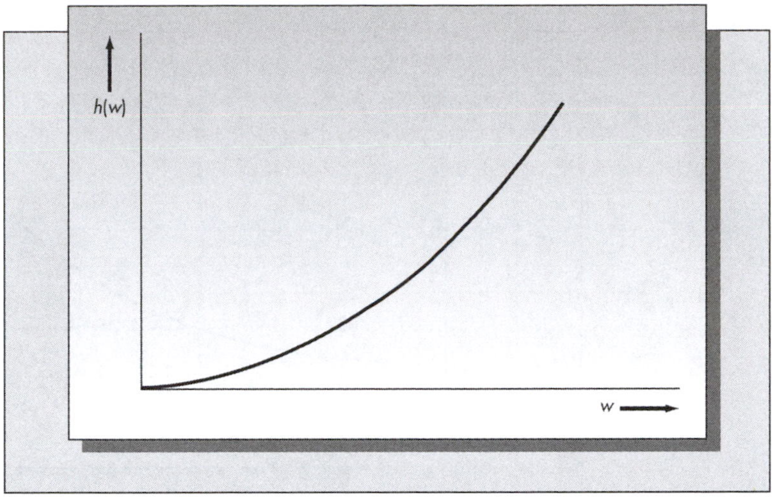

Figure S2–5 A typical waiting cost function

maintaining each server is c, then the objective would be to find the optimal value of s to minimize

$$sc + \lambda E(W_s).$$

2. Consider the example of the firm producing solid-state drives. The firm can purchase a larger μ (service rate), but only for an increased cost. To determine the best decision in this case, the firm would have to be able to quantify the costs associated with various levels of service. Suppose that the annual cost of the manufacturing operation when the throughput rate of the process is μ, is given by the function $f(\mu)$. Because the cost decreases as μ increases, this would be a monotonically decreasing function of μ. Furthermore, suppose that the one-time cost of purchasing equipment with service rate μ is $C(\mu)$. We would expect that $C(\mu)$ would be a monotonically increasing function of μ. Let I be the annual interest rate of alternative investments. Then the total annual cost is

$$IC(\mu) + f(\mu).$$

This function will be convex in μ (see Figure S2–6), so an optimal minimizing μ will exist and can be found easily. When there are only several possible values of μ, the objective function can be evaluated at these values, and the choice yielding the lowest cost can be made.

3. Consider the example of the translation service. In this case, the decision variable is the arrival rate λ. The larger the client base, the more jobs the firm will receive and the larger the value of λ. There are several possible formulations of this problem. One would be to determine the value of the expected number in the system that allows the firm to meet its obligations. In that case, we would assume that s and μ are given and that the objective is to determine λ so that l is equal to a target value. Another approach would be to find λ so that the probability that the number of customers in the system does not exceed a target level is at least some specified probability (such as .95).

To illustrate this case, consider the following specific example. Mary Worth runs her own translation service. She requires an average of 1.2 hours to complete a job, but the size of jobs

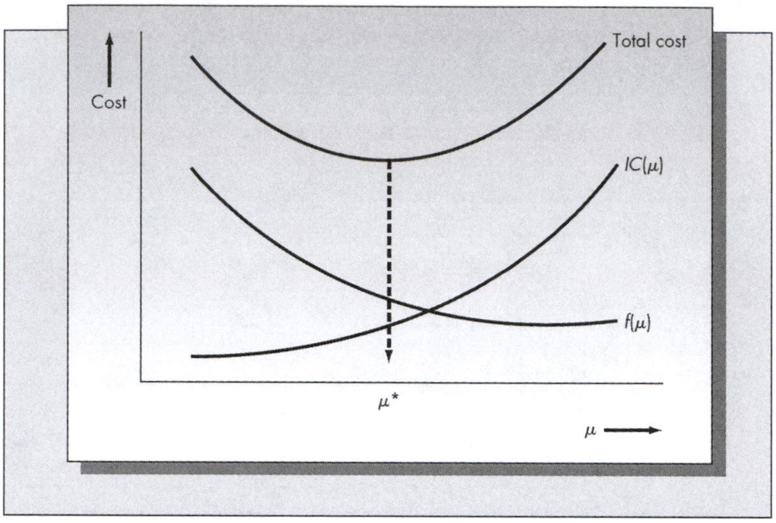

Figure S2–6 Optimizing the service rate, μ

varies considerably and can be described by an exponential distribution. Furthermore, the pattern of arrivals of jobs appears to be completely random. Mary works eight-hour days and does not want to have more than two days of work piled up at any point in time. Suppose that she wants the likelihood of this occurring to be no more than 5 percent. This means that the number of customers in the queue should not exceed 16/1.2 = 13.33 5 percent of the time. Hence, she wishes to have an arrival rate of jobs so that

$$P\{L > 13\} \le 0.05.$$

We showed in the solution to Example 8.5 that

$$P\{L > k\} = \rho^{k+1}.$$

Hence, we wish to find λ to solve

$$(\lambda/\mu)^{13} = 0.05.$$

Using $\mu = 1/1.2 = 0.8333$ per hour, and taking natural logarithms of both sides, we obtain

$$13 \ln(\lambda / 0.8333) = \ln(0.05)$$
$$\ln(\lambda / 0.8333) = -0.23044$$
$$\lambda / 0.8333 = \exp(-0.23044) = 0.7942,$$

giving $\lambda = 0.662$.

The solution, then, is that she should plan to have enough clients to generate about 0.662 jobs each working hour, or about 5.3 jobs per day.

Assigning service territories to repairmen, or deciding which region a blood bank is to cover, are other examples of problems in which the objective is to find the optimal value of the arrival rate, λ.

BIBLIOGRAPHY

Feller, W. *An Introduction to Probability Theory and Its Applications*. Vol. 2. New York: John Wiley & Sons, 1966.

Hadley, G., and T. M. Whitin. *Analysis of Inventory Systems*. Englewood Cliffs, NJ: Prentice Hall, 1963.

Hillier, F.S., and G.J. Lieberman (1990). *Introduction to Operations Research*, 5th ed. New York: McGraw-Hill, 1990.

Jackson, J. R. "Networks of Waiting Lines." *Operations Research* 5 (1957), pp. 518–521.

Kleinrock, L. *Queuing Systems*. Vol. 1, Theory. New York: Wiley Interscience, 1975.

Whitt, W. "A Diffusion Approximation for the G/GI/n/m Queue." *Operations Research* 52 (2004), pp. 922–941.

Whitt, W. "The Queuing Network Analyzer." *Bell System Technical Journal* 62 (1983), pp. 2779–2815.

CHAPTER

9

Production Control Systems: Push and Pull

"The most dangerous kind of waste is the waste we do not recognize."

—Shigeo Shingo

CHAPTER OVERVIEW

Purpose

To understand the push and pull philosophies in production planning and to compare MRP and kanban-type methods for scheduling the flow of goods in a factory.

KEY POINTS

1. *Enterprise Systems.* An enterprise system is an information system designed to manage all processes within a company and its supply chain. It is typically modular in nature and includes modules for operations, sales, finance, etc. Here our focus will be on the operational components of such systems that are used for production control.
2. *Push versus pull.* There are two fundamental philosophies for moving material through the factory. A push system is one in which production planning is done for all levels in advance. Once production is completed, units are pushed to the next level. A pull system is one in which items are moved from one level to the next only when requested. Each of the methods has particular advantages and disadvantages. Materials requirements planning (MRP) is the basic push system. Based on forecasts for end items over a specified planning horizon, the MRP planning system determines production quantities for each level of the system. The earliest of the pull systems is kanban developed by Toyota, which has exploded into the just-in-time (JIT) and lean production movements. Here the fundamental goal is to reduce work-in-process to a bare minimum. To do so, items are only moved when requested by the next higher level in the production process.
3. *MRP basics.* The MRP explosion calculus is a set of rules for converting a master production schedule (MPS) to a build schedule for all the components comprising the end product. The MPS is a production plan for the end item or final product by period. It is derived from the forecasts of demand adjusted for returns, on-hand inventory, and the like. At each stage in the process, one computes the production amounts required at each level of the production process by doing two basic operations: (1) offsetting the time when production begins by the lead time required at the current level and (2) multiplying the higher-level requirement by the gozinto

factor (i.e., how many of part A are required for part B). The simplest production schedule at each level is lot for lot (L4L), which means one produces the number of units required each period. However, if one knows the holding and setup cost for production, it is possible to construct a more cost efficient lot-sizing plan. Some available heuristics for this include: (1) EOQ lot sizing, (2) the Silver–Meal heuristic, (3) the least unit cost heuristic, and (4) part-period balancing. Optimal lot sizing requires dynamic programming and can be achieved via the efficient Wagner–Whitin algorithm. Lot sizing when capacity constraints are explicitly accounted for is difficult to solve optimally but can be approximated efficiently.

4. *MRP trade-offs.* MRP as a planning system has advantages and disadvantages over other planning systems. Some of the disadvantages include (1) forecast uncertainty is ignored; (2) capacity constraints are largely ignored; (3) the choice of the planning horizon can have a significant effect on the recommended lot sizes; (4) lead times are assumed fixed, but they should depend on the lot sizes; (5) MRP ignores the losses due to defectives or machine downtime; (6) data integrity can be a serious problem; and (7) in systems where components are used in multiple products, it is necessary to peg each order to a specific higher-level item. Advantages of MRP include (1) the ability to react to changes in demand, since demand forecasts are an integral part of the system; (2) allowance for lot sizing at the various levels of the system, thus affording the opportunity to reduce setups and setup costs; and (3) planning of production levels at all levels of the firm for several periods into the future, thus affording the firm the opportunity to look ahead to better schedule shifts and adjust workforce levels in the face of changing demand.
5. *Kanban basics.* Kanban is the Japanese word for card or ticket. Kanban controls the flow of goods in the plant by using a variety of different kinds of cards. Each card is attached to a palette or container of goods. Production cannot commence until production ordering kanbans are available. This guarantees that production at one level will not begin unless there is demand at the next level. This prevents work-in-process inventories from building up between work centers when a problem arises anywhere in the system. Part of what made kanban so successful at Toyota was the development of single minute exchange of dies (SMED), which reduced changeover times for certain operations from several hours to several minutes. Kanban is not the only way to implement a pull system. Information flows can often be controlled more efficiently with a central information processor (e.g., an enterprise system) than with cards. In this case, production authorization is given on screens rather than by physical cards. Note that pull control is not synonymous with a make-to-order system because it is possible to run pull production control within a plant that is producing for a make-to-stock inventory of finished goods (indeed, that is what Toyota does).
6. *Kanban trade-offs.* Kanban flow, and pull systems more generally, have several advantages and disadvantages when compared with MRP as a production planning system. Some of the advantages include (1) reduced work-in-process inventories, thus decreasing inventory costs and waste; (2) it is easy to quickly identify quality problems before large inventories of defective parts build up; and (3) when coordinated with a pull purchasing program, it ensures the smooth flow of materials throughout the entire supply chain. Some disadvantages include (1) a lack of forward planning; (2) an implicit assumption of a stable system (pull systems are particularly inappropriate for seasonal products that need to be produced well ahead of demand); and (3) a lack of visibility outside of the plant (if implemented through kanban cards rather than an enterprise system).

7. *Other control systems.* There are a number of hybrid systems in addition to MRP and kanban (or pure pull) control. Some key examples include CONWIP, POLCA, and synchronous manufacturing, all of which are briefly described in this chapter. In addition, flexible manufacturing systems, where each workstation can produce a variety of parts, require a different type of production control because the flexibility means that a part's route through the system is also a production decision.

The supply chain (see Chapter 6) is the set of all activities that convert raw material to the final product, and then deliver that product to the customer. One of the key activities in the supply chain is the actual production process. How well things are managed in the factory plays a fundamental role in the reliability and quality of the final product. There are two fundamentally different philosophies for managing the flow of goods in the factory. As we will see in this chapter, the methods developed in Chapters 4 and 5 for managing inventories are not always appropriate in the factory context.

The two approaches we consider are "**push**" and "**pull**" control systems. To appreciate exactly what distinguishes push and pull systems will require an understanding of exactly how these methods work, which will be covered in detail in this chapter. The simplest definition that we have seen is that "a pull system initiates production as a reaction to present demand, while a push system initiates production in anticipation of future demand" (Karmarkar, 1989, p. 123). A key tool for push system control is **materials requirements planning (MRP)**. A key tool for pull system control is kanban control. MRP incorporates forecasts of future demand while kanban does not.

To better understand the difference between push and pull, consider the following simple example. Garden spades are produced by a plant in Muncie, Indiana. Each spade consists of two parts: the metal digger and the wooden handle. The parts are connected by two screws. The plant produces spades at an average rate of 100 per week. The metal digger is produced in batches of 400 on the first two days of each month, and the handles are ordered from an outside supplier. The assembly of spades takes place during the first week of each month.

Consider now the demand pattern for the screws. Exactly 800 screws are needed during the first week of each month. Assuming four weeks per month, the weekly demand pattern for the screws is 800, 0, 0, 0, 800, 0, 0, 0, 800, 0, 0, 0, and so on. Using a weekly demand rate of 200 and appropriate holding and setup costs, suppose that the EOQ formula gives an order quantity of 1,400. A little reflection shows that ordering the screws in lots of 1,400 doesn't make much sense. If we schedule a delivery of 1,400 screws at the beginning of a month, 800 are used immediately and 600 are stored for later use. At the beginning of the next month another order for 1,400 has to be made, since the 600 screws stored are insufficient to meet the next month's requirement. It makes more sense to either order 800 screws at the beginning of each month or some multiple of 800 every several months.

The EOQ solution was clearly inappropriate here. Why? Recall that in deriving the EOQ formula, we assumed that demand was known and constant. The demand pattern in this example is known, but it is certainly not constant. In fact, it is very spiky. If we were to apply the methods of Chapter 5, we would assume that the demand was random. It is easy to show that over a one-year period the weekly demand has mean 200 and standard deviation 350. These values could be used to generate (Q, R) values assuming some form of a distribution for weekly demand. But this solution would not make any sense either. The demand pattern for the screws is not random; it is predictable, since it is a consequence of the production plan for the spades, which is known. The demand is *variable,* but it is not *random.*

We still have not solved the problem of how many screws to buy and when they should be delivered. One approach might be to just order once at the beginning of the year to meet the demand for an entire year. This would entail a one-time delivery of 10,400 screws at the start of each year (assuming 200 per week). What would be the advantage of this approach? Screws are very inexpensive items. By purchasing enough for an entire year's production, we would incur the fixed delivery costs only once.

There is a completely different way to approach this problem. One could simply decide to schedule deliveries of screws at the beginning of every month. This approach might be more expensive than the once-a-year delivery strategy, since fixed costs would be 12 times higher. However, it could have other advantages that more than compensate for the higher fixed costs. Monthly deliveries eliminate the need to store screws in the plant. If usage rates vary, delivery sizes could be adjusted to match need. Also, if a problem arose with the screws caused by either a defect in production or a design change in the spades, the company would not be stuck with a large inventory of useless items.

These two policies illustrate the basic difference between push and pull (although, as we will see, there is much more to these production control philosophies than this). In a push system, we determine lot sizes based on forecasts of future demands and possibly on cost considerations. In a pull system, we try to reduce lot sizes to their minimum to eliminate waste and unnecessary buildups of inventory.

Push systems are commonly implemented through MRP, which is usually embedded in an enterprise system or enterprise resource planning (ERP) system. MRP may be considered to be a top-down planning system in that all production quantity decisions are derived from demand forecasts. Lot-sizing decisions are found for every level of the productive system. Items are produced based on this plan and *pushed* to the next level.

In pull systems, requests for goods originate at a higher level of the system and are *pulled* through the various levels of production. Pull is often implemented using a **kanban system** introduced by Toyota. Kanban is a Japanese word meaning card or ticket. Originally, kanban cards were the only means of implementing pull control. However, as we will see in this chapter, it is only one of a number of systems for pull control. The kanban system was introduced by Toyota to reduce excess work-in-process (WIP) inventories. Of course, the **Toyota production system** (as introduced in Chapter 1) is much broader than just pull production control, which is the focus of this chapter.

Earlier editions of this textbook used the term **just-in-time (JIT)** to cover the concept of pull production. However, that term has largely fallen out of favor, possibly because it tended to be misunderstood. Some authors use the term **lean production** to cover the pull production control that we describe here. However, like the Toyota production system (TPS), lean is much broader than simply production control.

One might wonder why this chapter is about pull and not lean production. Pull is a set of principles for moving material through the factory that can be compared directly to MRP. Lean production encompasses all of the concepts of pull control elaborated on in this chapter but is much more than just a production control system. The broad ideas behind lean production were introduced in Chapter 1. The core principles behind these methods relate to the elimination of waste and variability and hence will be described in more detail within the context of quality management (see Chapter 11). In this chapter, we focus on the principles that relate to production control.

When one reads descriptions of lean production systems from practitioners (i.e., Nicholas and Soni, 2006), one is struck by a few things. First, practitioners view lean production in a very broad sense as noted above. Second, there appear to be many success stories of lean production concepts implemented in the United States. We know that the EOQ and EPQ formulas developed in Chapter 4 can be used to determine appropriate run sizes in a factory. Are not these concepts fundamentally at odds with those of lean production? The answer is yes, they are. Run sizes recommended by these formulas could be large depending on costs and usage rates, which is verboten in a lean production system. Which is the better approach?

The answer is that it depends. If the objective is to set run sizes to balance holding and setup costs, the models of Chapter 4 (and Chapter 5 when uncertainty is included) are fine. However, there are many costs and benefits that these models ignore because they are difficult to quantify. How does one quantify the cost of having to rework a large batch of items because a setting on a machine was wrong? How does one quantify the chaos that results from having large stockpiles of work-in-process inventory all over the plant? Thus, many of the benefits of lean production are hard to incorporate into a model. Simple economic trade-offs tell only a small part of the story. This suggests that modeling the true benefits of lean production is an area of opportunity for researchers. When we have models that take into account all of the costs and benefits of these disparate approaches to running the factory, we can make more intelligent comparisons among different production planning philosophies.

This chapter begins with a brief discussion on enterprise systems. It then provides the basic explosion calculus of MRP and how lot-sizing strategies other than lot for lot are incorporated into a basic single-level MRP solution. It finishes the discussion of push systems by considering capacity constraints and the advantages and disadvantages of MRP. Switching to pull systems, the basic kanban mechanism is presented, followed by a discussion of how it fits into the broader Toyota production system. Advantages and disadvantages are also presented. Finally, other control systems (beyond MRP and kanban) are discussed.

9.1 ENTERPRISE SYSTEMS

An **enterprise system** is a computer system designed to manage the processes of a business. Particularly when considered in the context of production control, it is often also called an **enterprise resource planning** (**ERP**) system. Such systems are multi-functional in scope. That is, they track a range of activities including financial results, procurement, sales, operations, and human resources. They are also, by definition, integrated, usually around a central database. However, they are also usually modular in structure and usable in any combination of modules.

Two of the largest enterprise systems vendors are SAP and Oracle. Within the category of supply chain management software, Gartner (2018) reports that in 2017 SAP held a 26.6 percent market share, received revenue of US$3.26 billion, and had a growth rate of 11.2 percent. For the same year, Oracle held a 13.7 percent market share, received revenue of US$1.68 billion, and had a growth rate of 8.1 percent. The next largest player in this market was JDA with only a 4.4 percent market share. However, there are also a vast array of small to medium companies making up about half the market share. The total market grew 13.9 percent in 2017 reaching US$12.2 billion, illustrating the popularity of such systems.

The three key reasons that companies adopt such systems are to (1) simplify and standardize IT systems; (2) have access to accurate information so that interactions and communications with customers and suppliers can be improved; and (3) improve the availability and quality of data (Mabert, Soni, and Venkataramanan, 2001). Implicit in this third item is how such data can be used to improve business processes. We suspect that were the survey repeated today, improved business analytics and decision making might be listed more prominently alongside the need for high quality data.

Implementing an enterprise system is no easy undertaking. It can take many years to complete, particularly for large companies and companies that need excessive reengineering and customization. The spending needed for software, hardware, training, implementation teams, and consulting can be very costly. The media commonly publish stories of implementation failures. For example, Nike's $400 million upgrade to its supply chain and ERP systems in 2000 caused $100 million in lost sales, a 20 percent stock dip, and a collection of class-action lawsuits (Koch, 2004). The following six key elements were common in successful ERP implementation (Mabert et al., 2001).

1. Senior management was thoroughly involved from the outset and established clear priorities.
2. A cross-functional implementation team with a more senior management leader was established.
3. The teams spent more time up front to define in great detail exactly how the implementation would be carried out.
4. Clear guidelines were laid out on performance measurement. These metrics were not just technical ones but also included business operations.
5. Clear guidelines were established on how to use outside consultants. In addition, there was a knowledge transfer process from consultants to in-house experts for both system configuration information and long-run maintenance.
6. Detailed plans were developed for training users.

However, even with these elements in place, companies may want to pause before moving ahead with purchasing an enterprise system. Thomas Davenport (1998) advises "SAP isn't a software package, it's a way of doing business" (p. 125). He highlights the strategic issues around a company changing its processes to conform with SAP or similar systems. Such conformance may mean that the company's processes are easily replicable by anyone else in the industry. He provides six key questions that should be answered before any purchasing decisions are made.

1. How might an Enterprise System (ES) strengthen our competitive advantages?
2. How might it erode them?
3. What will be the system's effect on our organization and culture?
4. Do we need to extend the system across all our functions, or should we implement only certain modules?
5. Would it be better to roll the system out globally or to restrict it to certain regional units?
6. Are there other alternatives for information management that might actually suit us better than an ES?

However, even given these cautions, it appears that the installation of a basic enterprise system is becoming a requirement for doing business in today's digital age. Such systems are often in charge of the production control for a company. Yet, it is still important for operations managers to understand the trade-offs between different production control mechanisms so that they can effectively use such systems. Such trade-offs are a key focus of this chapter.

9.2 MRP BASICS

In general, a production plan is a complete specification of the amounts of each end item or final product and subassembly produced, the exact timing of the production lot sizes, and the final schedule of completion. The production plan may be broken down into several

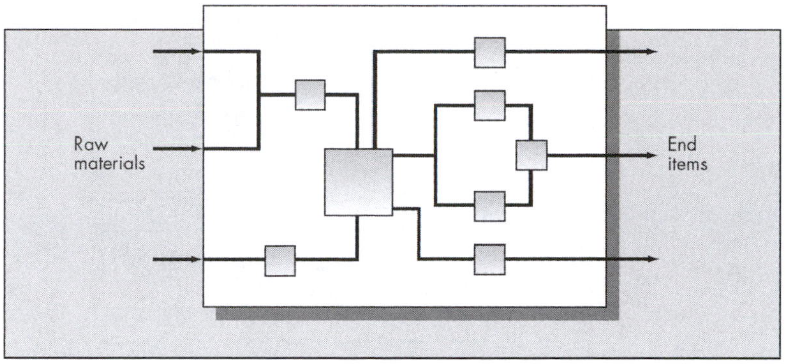

Figure 9–1 Schematic of the productive system

component parts: (1) the master production schedule (MPS), (2) the materials requirements planning (MRP) system, and (3) the detailed job shop schedule. Each of these parts can represent a large and complex subsystem of the entire plan.

At the heart of the production plan are the forecasts of demand for the end items produced over the planning horizon. An end item is the output of the productive system; that is, products shipped out the door. Components are items in intermediate stages of production, and raw materials are resources that enter the system. A schematic of the productive system appears in Figure 9–1. It is important to bear in mind that raw materials, components, and end items are defined in a relative and not an absolute sense. Hence, we may wish to isolate a portion of a company's operation as a productive system. End items associated with one portion of the company may be raw materials for another portion. A single productive system may be the entire manufacturing operation of the firm or only a small part of it.

The **master production schedule (MPS)** is a specification of the exact amounts and timing of production of each of the end items in a productive system. The MPS refers to *unaggregated* items. As such, the inputs for determining the MPS are forecasts for future demand by item rather than by aggregate items, as discussed in Chapter 3. The MPS is then broken down into a detailed schedule of production for each of the components that comprise an end item. The materials requirements planning (MRP) system is the means by which this is accomplished. Finally, the results of the MRP are translated into specific shop floor schedules (using methods such as those discussed in Chapter 11) and requirements for raw materials.

The data sources for determining the MPS include the following.

1. Firm customer orders.
2. Forecasts of future demand by item.
3. Safety stock requirements.
4. Seasonal plans.
5. Internal orders from other parts of the organization.

An important part of the success of MRP is the integrity and timeliness of the data. The information/enterprise system that supports the MRP receives inputs from the production, marketing, and finance departments of the firm. These are often provided using an S&OP process, as described in Chapter 3. A smooth flow of information among these three functional areas is a key ingredient to a successful production planning system.

We may consider the control of the productive system to be composed of three major phases. Phase 1 is the gathering and coordinating of the information required to develop the

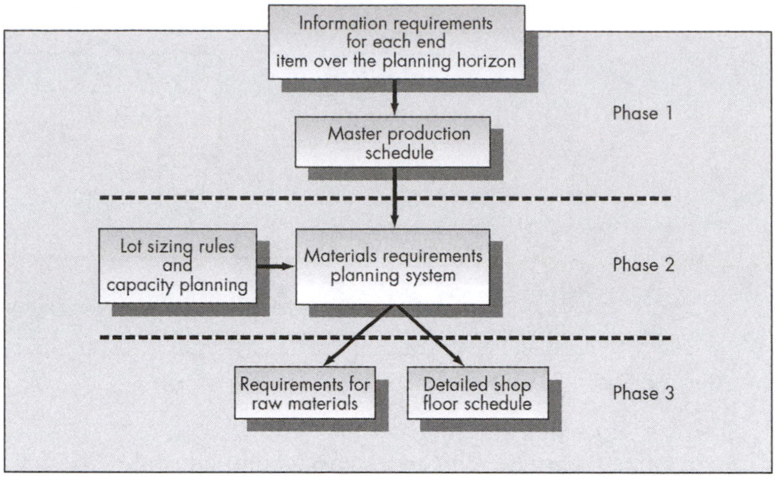

Figure 9–2 The three major control phases of the productive system

master production schedule (the S&OP process). Phase 2 is the determination of planned order releases using MRP. Finally, phase 3 is the development of detailed shop floor schedules and resource requirements from the MRP planned order releases. Figure 9–2 is a schematic of these three control phases of the productive system.

This chapter is concerned with the way that the MPS is used as input to the MRP system. We will show in detail exactly how the MRP calculus works; that is, how product structures are converted to parent–child relationships between levels of the productive system, how production lead times are used to obtain time-phased requirements, and how lot-sizing methods result in specific schedules. In the lot-sizing sections we will consider both optimal and heuristic lot-sizing techniques for uncapacitated systems and a straightforward heuristic technique for capacitated lot sizing.

9.3 THE EXPLOSION CALCULUS

Explosion calculus is a term that refers to the set of rules by which gross requirements at one level of the product structure are translated into a production schedule at that level and requirements at lower levels. At the heart of any MRP system is the product structure. The product structure refers to the relationship between the components at adjacent levels of the system. The **product structure diagram** details the parent–child relationship of components and end items at each level, the number of periods required for production of each component, and the number of components required at the child level to produce one unit at the parent level (the **gozinto** factor).

A typical product structure appears in Figure 9–3. In order to produce one unit of the end item, two units of A and one unit of B are required. Assembly of A requires one week, and assembly of B requires two weeks. A and B are "children" of the end item. In order to produce A, one unit of C and two units of D are required. In order to produce B, two units of C and three units of E are required. The respective production lead times also appear on the product structure diagram. Product structure diagrams can be quite complex, with as many as 15 or more levels in some industries.

The explosion calculus (also known as the **bill-of-materials** explosion) follows a set of rules that translate the planned order releases for end items and components into production

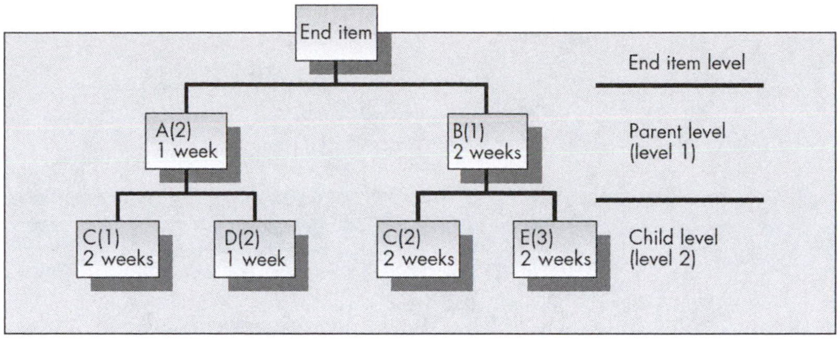

Figure 9–3 Typical product structure diagram

schedules for lower-level components. The method involves properly phasing requirements in time and accounting for the number of components required at the child level to produce a single parent item. The method is best illustrated by example.

EXAMPLE 9.1

The Harmon Music Company produces a variety of wind instruments at its plant in Joliet, Illinois. Because the company is relatively small, it would like to minimize the amount of money tied up in inventory. For that reason, production levels are set to match predicted demand as closely as possible. In order to achieve this goal, the company has adopted an MRP system to determine production quantities.

One of the instruments produced is the model 85C trumpet. The trumpet retails for $800 and has been a reasonably, if not spectacularly, profitable item for the company. Based on orders from music stores around the country, the production manager receives predictions of future demand for about four months into the future.

Figure 9–4 shows the trumpet and its various subassemblies. Figure 9–5 gives the product structure diagram for the construction of the trumpet. The bell section and the lead pipe and valve sections are welded together in final assembly. Before the welding, three slide assemblies and three valves are produced and fitted to the valve casing assembly. [The astute reader will know that the valves and the slides are not identical. Hence, each valve and each slide should be treated as a separate item. However, if we agree that slides and valves correspond to matching groups of three, our approach is valid. This allows us to demonstrate the multiplier effect when multiple components are needed for a single end item.] The forming and shaping of the bell section require two

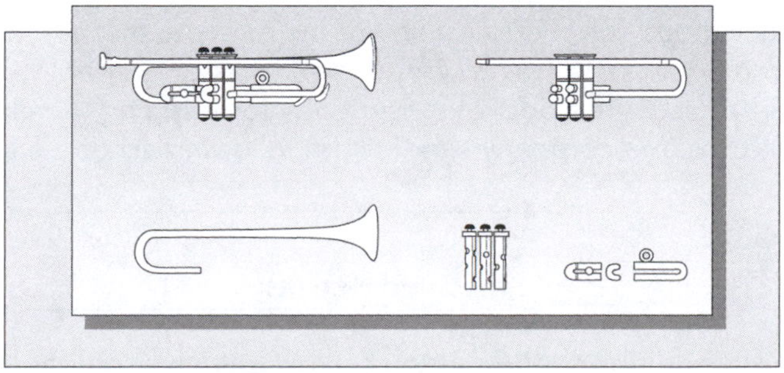

Figure 9–4 Trumpet and subassemblies

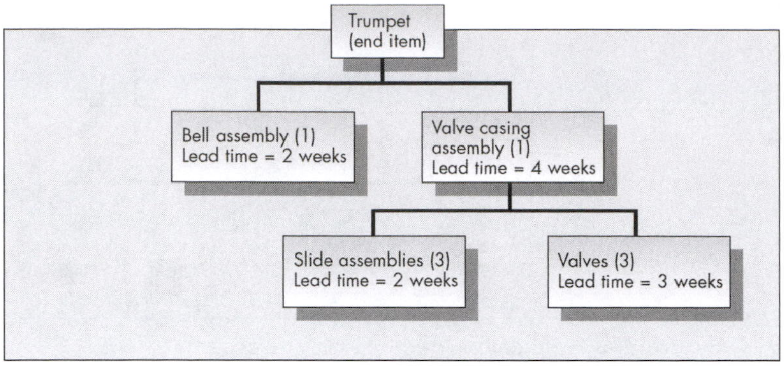

Figure 9–5 Product structure diagram for Harmon trumpet

weeks, and the forming and shaping of the lead pipe and valve sections require four weeks. The valves require three weeks to produce, and the slide assemblies two weeks.

The trumpet assembly problem is a three-level MRP system. Level 0 corresponds to the final product or end item, which is the completed trumpet. Level 1, the child level relative to the trumpet, corresponds to the bell and valve casing assemblies. Level 2 corresponds to the slide and valve assemblies. The information in the product structure diagram is often represented as an **indented bill of materials (BOM)**, which is a more convenient representation for preparation of computer input. The indented BOM for the trumpet is

1 Trumpet
  1 Bell assembly
  1 Valve assembly
    3 Slide assemblies
    3 Valves

It takes seven weeks to produce a trumpet. Hence, the company must begin production now on trumpets to be shipped in seven weeks. For that reason we will consider only forecasts for demands that are at least seven weeks into the future. If we label the current week as week 1, then Harmon requires forecasts for the sales of trumpets for weeks 8 to 17. Suppose that the predicted demands for those weeks are

Week	8	9	10	11	12	13	14	15	16	17
Demand	77	42	38	21	26	112	45	14	76	38

These forecasts represent the numbers of trumpets that the firm would like to have ready to ship in the specified weeks. Harmon periodically receives returns from its various suppliers. These are instruments that are defective for some reason or are damaged in shipping. Once the necessary repairs are completed, the trumpets are returned to the pool of those ready for shipping. Based on the current and anticipated returns, the company expects the following schedule of receipts to the inventory.

Week	8	9	10	11
Scheduled receipts	12		6	9

In addition to the scheduled receipts, the company expects to have 23 trumpets in inventory at the end of week 7. The MPS for the trumpets is now obtained by netting out the inventory on hand at the end of week 7 and the scheduled receipts, in order to obtain the net predicted demand.

Week	8	9	10	11	12	13	14	15	16	17
Net predicted demand	42	42	32	12	26	112	45	14	76	38

Having determined the MPS for the end product, we must translate it into a production schedule for the components at the next level of the product structure. These are the bell assembly and the valve casing assembly. Consider first the bell assembly. The first step is to translate the MPS for trumpets into a set of gross requirements by week for the bell assembly. Because there is exactly one bell assembly used for each trumpet, this is the same as the MPS. The next step is to subtract any on-hand inventory or scheduled receipts to obtain the net requirements (here there are none). The net requirements are then translated back in time by the production lead time, which is two weeks for the bell assembly, to obtain the time-phased requirements. Finally, the lot-sizing algorithm is applied to the time-phased requirements to obtain the planned order release by period. **Lot for lot** (L4L) means that the production quantity each week is just the time-phased net requirement. A lot-for-lot production rule means that no inventory is carried from one period to another.

Assuming a lot-for-lot production rule, we obtain the following MRP calculations for the bell assembly.

Week	6	7	8	9	10	11	12	13	14	15	16	17
Gross requirements			42	42	32	12	26	112	45	14	76	38
Net requirements			42	42	32	12	26	112	45	14	76	38
Time-phased net requirements	42	42	32	12	26	112	45	14	76	38		
Planned order release (lot for lot)	42	42	32	12	26	112	45	14	76	38		

As we will see later, lot for lot is rarely an optimal production rule. Optimal and heuristic production scheduling rules will be examined in Section 9.4.

The calculation is essentially the same for the valve casing assembly, except that the production lead time is four weeks rather than two weeks. The calculations for the valve casing assembly follow.

Week	4	5	6	7	8	9	10	11	12	13	14	15	16	17
Gross requirements					42	42	32	12	26	112	45	14	76	38
Net requirements					42	42	32	12	26	112	45	14	76	38
Time-phased net requirements	42	42	32	12	26	112	45	14	76	38				
Planned order release (lot for lot)	42	42	32	12	26	112	45	14	76	38				

Now consider the MRP calculations for the valves. Assume that the company expects an on-hand inventory of 186 valves at the end of week 3 and a receipt from an outside supplier of 96 valves at the start of week 5. There are three valves required for each trumpet. One obtains gross requirements for the valves by multiplying the production schedule for the valve casing assembly by 3. Net requirements are obtained by subtracting on-hand inventory and scheduled receipts. The MRP calculations for the valves are

Week	2	3	4	5	6	7	8	9	10	11	12	13
Gross requirements			126	126	96	36	78	336	135	42	228	114
Scheduled receipts				96								
On-hand inventory		186	60	30								
Net requirements			0	0	66	36	78	336	135	42	228	114
Time-phased net requirements		66	36	78	336	135	42	228	114			
Planned order release (lot for lot)		66	36	78	336	135	42	228	114			

Net requirements are obtained by subtracting on-hand inventory and scheduled receipts from gross requirements. Because the on-hand inventory of 186 in period 3 exceeds the gross requirement in period 4, the net requirements for period 4 are 0. The remaining 60 units (186 – 126) are carried into period 5. In period 5 the scheduled receipt of 96 is added to the starting inventory of 60 to give 156 units. The gross requirements for period 5 are 126, so the net requirements for period 5 are 0, and the additional 30 units are carried over to period 6. Hence, the resulting net requirements for period 6 are 96 – 30 = 66.

The net requirements are phased back three periods in order to obtain the time-phased net requirements and the production schedule. Note that the valves are produced internally. The scheduled receipt of 96 corresponds to defectives that were sent out for rework. A similar calculation is required for the slide assemblies.

Example 9.1 represents the essential elements of the explosion calculus. Note that we have assumed for the sake of the example that the production scheduling rule is lot for lot. That is, in each period the production quantity is equal to the net requirements for that period. However, such a policy may be suboptimal and even infeasible. For example, the schedule requires the delivery of 336 valves in week 9. However, suppose that the plant can produce only 200 valves in one week. If that is the case, a lot-for-lot scheduling rule is infeasible. We discuss this case later.

PROBLEMS FOR SECTION 9.3

1. The inventory control models discussed in Chapters 4 and 5 are often labeled *independent* demand models, and MRP is often labeled a *dependent* demand system. What do the terms *independent* and *dependent* mean in this context?
2. What information is contained in a product structure diagram?
3. For the example of Harmon Music presented in this section, determine the planned order release for the slide assemblies. Assume lot-for-lot scheduling.

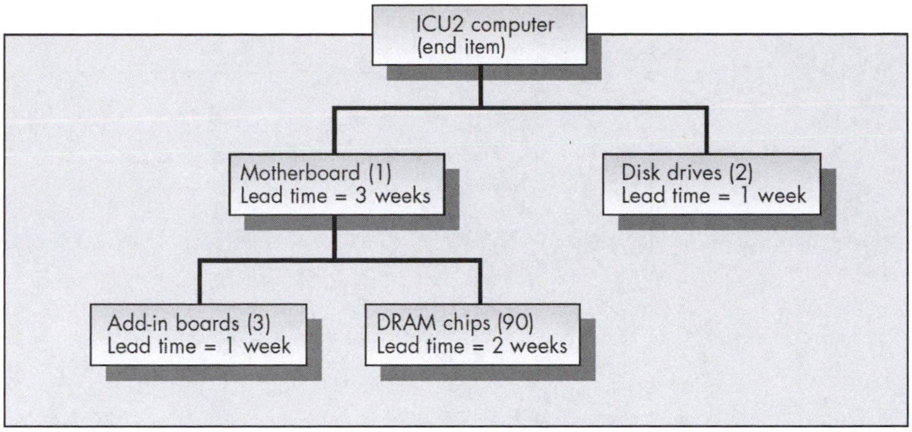

Figure 9–6 Product structure diagram for ICU2 computer (for problem 4)

4. The Noname Computer Company builds a gaming computer designated model ICU2. It imports the motherboard of the computer from Taiwan, but the company inserts the sockets for the chips and boards in its plant in Lubbock, Texas. Each computer requires a total of ninety dynamic random access memory (DRAM) chips. Noname sells the computers with three add-in boards and two disk drives. The company purchases both the DRAM chips and the disk drives from an outside supplier. The product structure diagram for the ICU2 computer is given in Figure 9–6. Suppose that the forecasted demands for the computer for weeks 6 to 11 are 220, 165, 180, 120, 75, 300. The starting inventory of assembled computers in week 6 will be 75, and the production manager anticipates returns of 30 in week 8 and 10 in week 10.
   a. Determine the MPS for the computers.
   b. Determine the planned order release for the motherboards assuming a lot-for-lot scheduling rule.
   c. Determine the schedule of outside orders for the disk drives.
5. For problem 4, suppose that Noname has 23,000 DRAM chips in inventory. It anticipates receiving a lot of 3,000 chips in week 3 from another firm that has gone out of business. At the current time, Noname purchases the chips from two vendors, A and B. A sells the chips for less but will not fill an order exceeding 10,000 chips per week.
   a. If Noname has established a policy of inventorying as few chips as possible, what order should it be placing with vendors A and B over the next six weeks?
   b. Noname has found that not all the DRAM chips purchased function properly. From past experience it estimates an 8 percent failure rate for the chips purchased from vendor A and a 4 percent failure rate for the chips purchased from vendor B. What modification in the order schedule would you recommend to compensate for this problem?
6. Consider the product structure diagram given in Figure 9–3. Assume that the MPS for the end item for weeks 10 through 17 is

Week	10	11	12	13	14	15	16	17
Net requirements	100	100	40	40	100	200	200	200

Assume that lot-for-lot scheduling is used throughout. Also assume that there is no entering inventory in period 10 and no scheduled receipts.

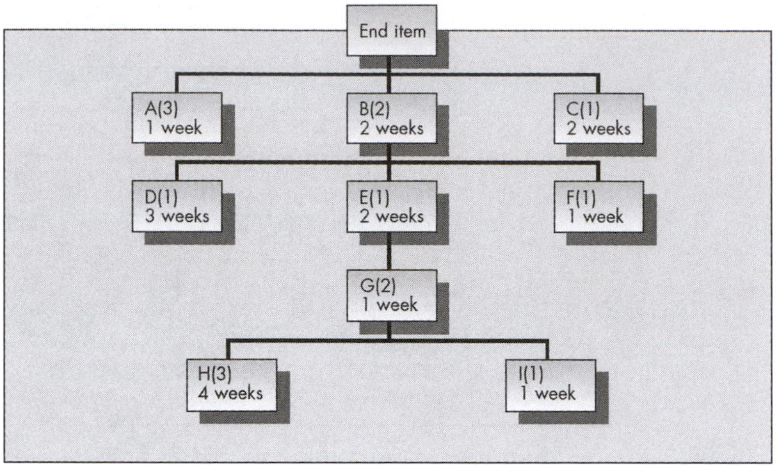

Figure 9–7 Product structure diagram (for problem 9)

a. Determine the planned order release for component A.
b. Determine the planned order release for component B.
c. Determine the planned order release for component C. (Hint: Note that C is required for both A and B.)

7. What alternatives are there to lot-for-lot scheduling at each level? Discuss the potential advantages and disadvantages of other lot-sizing techniques.
8. One of the inputs to the MRP system is the forecast of demand for the end item over the planning horizon. From the point of view of production, what advantages are there to a forecasting system that smooths the demand (that is, provides forecasts that are relatively constant) versus one that achieves greater accuracy but gives "spiky" forecasts that change significantly from one period to the next?
9. An end item has the product structure diagram given in Figure 9–7.
   a. Write the product structure diagram as an indented bill-of-materials list.
   b. Suppose that the MPS for the end item is

Week	30	31	32	33	34	35
MPS	165	180	300	220	200	240

   If production is scheduled on a lot-for-lot basis, find the planned order release for component F.
   c. Using the data in part (b), find the planned order release for component I.
   d. Using the data in part (b), find the planned order release for component H.

9.4 ALTERNATIVE LOT-SIZING SCHEMES

In Example 9.1 we assumed that the production scheduling rule was lot for lot. That is, the number of units scheduled for production each period was the same as the net requirements for that period. In fact, this policy is assumed for convenience and ease of use only. It is, in general, not optimal. The problem of finding the best (or near best) production plan can be characterized as follows. We have a known set of time-varying demands and costs of setup and holding. What production quantities will minimize the total holding and setup costs over the planning horizon? Note that neither the methods of Chapter 4 (which assumes known but constant demands) nor those of Chapter 5 (which assumes random demands) are appropriate.

In this section we will discuss several popular heuristic (i.e., approximate) lot-sizing methods that easily can be incorporated into the MRP calculus. We first consider using the EOQ formula (from Chapter 4) to do the lot sizing. While this is highly intuitive, it is rarely optimal. The Silver–Meal heuristic, which we consider next, usually provides better results. Finally, two other popular heuristics are the least unit cost (LUC) heuristic and the part-period balancing heuristic. Details of these latter two heuristics are relegated to Appendix 9–A as are details of the Wagner–Whitin algorithm, which provides optimal lot sizes.

EOQ Lot Sizing

To apply the EOQ formula, we need three inputs: the average demand rate, λ; the holding cost rate, h; and the setup cost, K. Consider the valve casing assembly in Example 9.1. Suppose that the setup operation for the machinery used in this assembly operation takes two workers about three hours. The workers average \$22 per hour. That translates to a setup cost of $(22)(2)(3) = \$132$.

The company uses a holding cost based on a 22 percent annual interest rate. Each valve casing assembly costs the company \$141.82 in materials and value added for labor. Hence, the holding cost amounts to $(141.82)(0.22)/52 = \$0.60$ per valve casing assembly per week.

The planned order release resulting from a lot-for-lot policy requires production in each week. Consider the total holding and setup cost incurred from weeks 6 through 15 when using this policy. If we adopt the convention that the holding cost is charged against the inventory each week, then the total holding cost over the 10-week horizon is zero. As there is one setup incurred each week, the total setup cost incurred over the planning horizon is $(132)(10) = \$1{,}320$.

This cost can be reduced significantly by producing larger amounts less often. As a "first cut" we can use the EOQ formula to determine an alternative production policy. The total of the time-phased net requirements over weeks 8 through 17 is 439, for an average of 43.9 per week. Using $\lambda = 43.9$, $h = 0.60$, and $K = 132$, the EOQ formula gives

$$Q = \sqrt{\frac{2K\lambda}{h}} = \sqrt{\frac{(2)(132)(43.9)}{0.6}} = 139.$$

If we schedule the production in lot sizes of 139 while guaranteeing that all net requirements are filled, the resulting MRP calculations for the valve casing assembly are listed below.

Week	4	5	6	7	8	9	10	11	12	13	14	15	16	17
Net requirements					42	42	32	12	26	112	45	14	76	38
Time-phased net requirements	42	42	32	12	26	112	45	14	76	38				
Planned order release (EOQ)	139	0	0	0	139	0	139	0	0	139				
Planned deliveries					139	0	0	0	139	0	139	0	0	139
Ending inventory					97	55	23	11	124	12	106	92	16	117

One finds the ending inventory each period from the formula

$$\text{Ending inventory} = \text{Beginning inventory} + \text{Planned deliveries} - \text{Net requirements}.$$

Consider the cost of using EOQ lot sizing rather than lot for lot. During periods 8 through 17 there are a total of four setups, resulting in a total setup cost of (132)(4) = \$528. The most direct way to compute the holding cost is to simply accumulate the ending inventories for the 10 periods and multiply by h. The cumulative ending inventory is 97 + 55 + 23 + ... + 117 = 653. Hence, the total holding cost incurred over the 10 periods is (0.6)(653) = \$391.80. The total holding and setup cost when lot sizes are computed from the EOQ formula is \$528 + \$391.80 = \$919.80. This is a considerable improvement over the cost of \$1,320 obtained when using lot-for-lot production scheduling. (However, this savings does not consider the cost impact that lot sizing at this level may have upon lower levels in the product tree. It is possible, though unlikely, that in a global sense lot for lot could be more cost effective than EOQ.) Note that the use of the EOQ to set production quantities results in an entirely different pattern of gross requirements for the valve and slide assemblies one level down. In particular, the gross requirements for the valves are now

Week	4	5	6	7	8	9	10	11	12	13
Gross requirements	417	0	0	0	417	0	417	0	0	417

In the remainder of this section, we discuss other popular lot-sizing schemes when demand is known and time is varying. It should be pointed out that the problem of determining lot sizes subject to time-varying demand occurs in contexts other than MRP. We have included it here to illustrate how these methods can be linked to the MRP explosion calculus.

The Silver–Meal Heuristic

The Silver–Meal heuristic (named for Harlan Meal and Edward Silver) is a forward method that requires determining the average cost per period as a function of the number of periods the current order should span and stopping the computation when this function first increases.

Define $C(T)$ as the average holding and setup cost per period if the current order spans the next T periods. As above, let $(r_1, \ldots, r_n)$ be the requirements over the n-period horizon. Consider period 1. If we produce just enough in period 1 to meet the demand in period 1, then we just incur the order cost K. Hence,

$$C(1) = K.$$

If we order enough in period 1 to satisfy the demand in both periods 1 and 2, then we must hold r_2 for one period. Hence,

$$C(2) = (K + hr_2)/2.$$

Similarly,

$$C(3) = (K + hr_2 + 2hr_3)/3,$$

and, in general,

$$C(j) = (K + hr_2 + 2hr_3 + \ldots + (j-1)\,hr_j)\,/\,j.$$

Once $C(j) > C(j-1)$, we stop and set $y_1 = r_1 + r_2 + \ldots + r_{j-1}$, and begin the process again starting at period j.

EXAMPLE 9.2

A machine shop uses the Silver–Meal heuristic to schedule production lot sizes for computer casings. Over the next five weeks the demands for the casings are $r = (18, 30, 42, 5, 20)$. The holding cost is \$2 per case per week, and the production setup cost is \$80. Find the recommended lot sizing.

Solution

Starting in period 1:

$C(1) = 80,$

$C(2) = [80 + (2)(30)]/2 = 70,$

$C(3) = [80 + (2)(30) + (2)(2)(42)]/3 = 102.67$. Stop because $C(3) > C(2)$.

Set $y_1 = r_1 + r_2 = 18 + 30 = 48$.

Starting in period 3:

$C(1) = 80,$

$C(2) = [80 + (2)(5)]/2 = 45,$

$C(3) = [80 + (2)(5) + (2)(2)(20)]/3 = 56.67$. Stop.

Set $y_3 = r_3 + r_4 = 42 + 5 = 47$.

Because period 5 is the final period in the horizon, we do not need to start the process again. We set $y_5 = r_5 = 20$. Hence, the Silver–Meal heuristic results in the policy $\mathbf{y} = (48, 0, 47, 0, 20)$. (Hint: You can streamline calculations by noting that $C(j+1) = [j/(j+1)]\,[C(j) + hr_{j+1}]$.)

To show that the Silver–Meal heuristic will not always result in an optimal solution, consider the following counterexample.

EXAMPLE 9.3

Let $r = (10, 40, 30)$, $K = 50$, and $h = 1$. The Silver–Meal heuristic gives the solution $\mathbf{y} = (50, 0, 30)$, but the optimal solution is $(10, 70, 0)$.

In closing this section, we note that Edward Silver, David Pyke, and Douglas Thomas (2017, p. 219) recommend conditions under which the Silver–Meal heuristic should be used instead of the EOQ. The condition is based on the variance of periodic demand; the higher the variance, the better the improvement the heuristic gives. However, our feeling is that given today's computing technology and the ease with which the heuristic solution can be found, the additional computational costs of using Silver–Meal (or one of the following methods described in the next subsection) instead of EOQ are minimal and not an important consideration.

Other Lot-Sizing Methods

The Silver–Meal heuristic just provided is easy to use and gives lot sizes with costs that are generally near the true optimal. However, there are another two heuristics that are popular, namely the **least unit cost** (**LUC**) heuristic and the **part-period balancing** heuristic. We provide details of these methods in Appendix 9–A. The key idea behind the LUC heuristic is to choose the order horizon that minimizes the cost per unit of demand rather than the cost per period. For the part-period balancing heuristic the order horizon is set equal to the number of periods that most closely matches the total holding cost with the setup cost over that period. All three heuristics are reasonable methods based on the structure of the problem but don't necessarily give the optimal solution. Kenneth Baker (1989) provides a comprehensive testing of these heuristics, as well as a number of others, and finds that, in general, the Silver–Meal heuristic dominates both the LUC and part-period balancing heuristics.

It is possible to compute true optimal lot sizes. Optimal in this context means the policy that minimizes the total holding and setup cost over the planning horizon. Appendix 9–A provides the **Wagner–Whitin algorithm** that guarantees an optimal solution to the problem of production planning with time-varying demands. While tedious to solve by hand, the Wagner–Whitin algorithm can be implemented easily on a computer and solved quickly and efficiently.

Given modern computing power we would recommend the use of the Wagner-Whitin algorithm generally. However, the Silver–Meal heuristic is more easily explained to management, which can sometimes be an advantage, and is usually close to optimal.

Incorporating Lot-Sizing Algorithms into the Explosion Calculus

EXAMPLE 9.4

Returning to Example 9.1 concerning the Harmon Music Company, consider the impact of lot sizing on the explosion calculus. Consider first the valve casing assembly. The time-phased net requirements for the valve casing assembly are

Week	4	5	6	7	8	9	10	11	12	13
Time-phased net requirements	42	42	32	12	26	112	45	14	76	38

The setup cost for the valve casing assembly is \$132, and the holding cost is $h = \$0.60$ per assembly per week. We will determine the lot sizing by the Silver–Meal heuristic.

Starting in week 4:

$$C(1) = 132,$$

$$C(2) = \frac{132 + 0.6(42)}{2} = 78.6,$$

$$C(3) = \frac{132 + 0.6[42 + (2)(32)]}{3} = 65.2,$$

$$C(4) = \frac{132 + 0.6[42 + (2)(32) + (3)(12)]}{4} = 54.3,$$

$$C(5) = \frac{132 + 0.6[42 + (2)(32) + (3)(12) + (4)(26)]}{5} = 55.92.$$

Since $C(5) > C(4)$, we terminate computations and set $y_4 = 42 + 42 + 32 + 12 = 128$.
Starting in week 8:

$$C(1) = 132,$$
$$C(2) = \frac{132 + 0.6(112)}{2} = 99.6,$$
$$C(3) = \frac{132 + 0.6[112 + (2)(45)]}{3} = 84.4,$$
$$C(4) = \frac{132 + 0.6[112 + (2)(45) + (3)(14)]}{4} = 69.6,$$
$$C(5) = \frac{132 + 0.6\,[112 + (2)(45) + (3)(14) + (4)(76)]}{5} = 92.16.$$

Hence, $y_8 = 26 + 112 + 45 + 14 = 197$.
Production occurs next in week 12. It is easy to show that $y_{12} = 76 + 38 = 114$.
A summary of the MRP calculations using the Silver–Meal (S–M) heuristic to determine lot sizes for the valve casing assembly follows.

Week	4	5	6	7	8	9	10	11	12	13	14	15	16	17
Net requirements					42	42	32	12	26	112	45	14	76	38
Time-phased net requirements	42	42	32	12	26	112	45	14	76	38				
Planned order release (S–M)	128	0	0	0	197	0	0	0	114	0				
Planned deliveries					128	0	0	0	197	0	0	0	114	0
Ending inventory					86	44	12	0	171	59	14	0	38	0

It is interesting to compare the holding and setup cost of the policy using the Silver–Meal heuristic with the previous solutions using lot for lot and the EOQ formula. There are exactly three setups in our solution, resulting in a total setup cost of $(132)(3) = \$396$. The sum of the ending inventories each week is $86 + 44 + \cdots + 38 + 0 = 424$. The total holding cost is $(0.6)(424) = \$254.40$. Hence, the total cost of the Silver–Meal solution for this assembly amounts to \$650.40. Compare this to the costs of \$1,320 using lot for lot and \$919.80 using the EOQ solution.

It is also interesting to note what would have been the result if we had employed the Wagner–Whitin algorithm to find the true optimal solution. The optimal solution for this problem turns out to be $y_4 = 154$, $y_9 = 171$, and $y_{12} = 114$, with a total cost of \$610.20, which is only a slight improvement over the Silver–Meal heuristic.

We will now consider how the planned order release for the valve casing assembly affects the scheduling for lower-level components. In particular, if we consider the MRP calculations for the valves and assume that the lot sizing for the valves is determined by the Silver–Meal heuristic as well, we obtain

Week	1	2	3	4	5	6	7	8	9	10	11	12	13
Gross requirements				384	0	0	0	591	0	0	0	342	0
Scheduled receipts					96								
On-hand inventory			186	0	96	96	96	0					
Net requirements				198	0	0	0	495	0	0	0	342	0
Time-phased net requirements	198	0	0	0	495	0	0	0	342	0			
Planned order release (S–M)	198	0	0	0	495	0	0	0	342	0			

Note in this calculation summary that the scheduled receipts in period 5 must be held until period 8 before they can be used to offset demand. This results from the zero gross requirements in periods 5 through 8.

The calculation of the planned order release is based on a setup cost of $80 and a holding cost of $0.07 per valve per week. It is interesting to note that the Silver–Meal heuristic resulted in a lot-for-lot production rule in this case. This results from the lumpy demand pattern caused by the lot sizing applied at a higher level of the product structure. Both Silver–Meal and Wagner–Whitin give the same results for this example.

PROBLEMS FOR SECTION 9.4

10. Perform the MRP calculations for the valves in the example of this section, using the gross requirements schedule that results from EOQ lot sizing for the valve casing assemblies. Use $K = \$150$ and $h = 0.4$.
11. a. Determine the planned order release for the motherboards in problem 4 assuming that one uses the EOQ formula to schedule production. Use $K = \$180$ and $h = 0.40$.
    b. Using the results from part (a), determine the gross requirements schedule for the DRAM chips, which are ordered from an outside supplier. The order cost is $25.00 and the holding cost is $0.01 per chip per week. What order schedule with the vendor results if the EOQ formula is used to determine the lot size?
    c. Repeat the calculation of part (b) for the add-in boards. Use the same value of the setup cost and a holding cost of 28 cents per board per week.
12. a. Discuss why the EOQ formula may give poor results for determining planned order releases.
    b. If the forecasted demand for the end item is the same each period, will the EOQ formula result in optimal lot sizing at each level of the product structure? Explain.
13. The problem of lot sizing for the valve casing assembly described for Harmon Music Company in Section 9.3 was solved using the EOQ formula. Determine the lot sizing for the 10 periods using the following methods.
    a. Silver–Meal.
    b. Least unit cost (see Appendix 9–A).
    c. Part-period balancing (see Appendix 9–A).
    d. Which lot-sizing method resulted in the lowest cost for the 10 periods?
14. A single inventory item is ordered from an outside supplier. The anticipated demand for this item over the next 12 months is 6, 12, 4, 8, 15, 25, 20, 5, 10, 20, 5, 12. Current inventory of this item is 4, and ending inventory should be 8. Assume a holding cost of $1 per period and a setup cost of $40. Determine the order policy for this item based on
    a. Silver–Meal.
    b. Least unit cost (see Appendix 9–A).
    c. Part-period balancing (see Appendix 9–A).
    d. Which lot-sizing method resulted in the lowest cost for the 12 periods?
15. For the counterexample (Example 9.3), which shows that the Silver–Meal heuristic may give a suboptimal solution, do either the least unit cost or the part-period balancing heuristics give the optimal solution?
16. Discuss the advantages and disadvantages of the following lot-sizing methods in the context of an MRP scheduling system: lot for lot, EOQ, Silver–Meal, least unit cost, and part-period balancing.

17. The time-phased net requirements for the base assembly in a table lamp over the next six weeks are listed below.

Week	1	2	3	4	5	6
Requirements	335	200	140	440	300	200

The setup cost for the construction of the base assembly is \$200, and the holding cost is \$0.30 per assembly per week.
a. What lot sizing do you obtain from the EOQ formula?
b. Determine the lot sizes using the Silver–Meal heuristic.
c. Determine the lot sizes using the least unit cost heuristic (see Appendix 9–A).
d. Determine the lot sizes using part-period balancing (see Appendix 9–A).
e. Compare the holding and setup costs obtained over the six periods using the policies found in parts (a) through (d) with the cost of a lot-for-lot policy.

18. For the example presented in problem 6, assume that the setup cost for both components A and B is \$100 and that the holding costs are respectively \$0.15 and \$0.25 per component per week. Using the Silver–Meal algorithm, determine the planned order releases for both components A and B and the resulting gross requirements schedules for components C, D, and E.

Problems 19–25 are based on the material appearing in Appendix 9–A.

19. Anticipated demands for a four-period planning horizon are 23, 86, 40, and 12. The setup cost is \$300 and the holding cost is $h = \$3$ per unit per period.
a. Enumerate all the exact requirements policies, compute the holding and setup costs for each, and find the optimal production plan.
b. Solve the problem by backward dynamic programming.

20. Anticipated demands for a five-period planning horizon are 14, 3, 0, 26, 15. Current starting inventory is four units, and the inventory manager would like to have eight units on hand at the end of the planning horizon. Assume that $h = 1$ and $K = 30$. Find the optimal production schedule. (Hint: Modify the first and the last period's demands to account for starting and ending inventories.)

21. A small manufacturing firm that produces a line of office furniture requires casters at a fairly constant rate of 75 per week. The MRP system assumes a six-week planning horizon. Assume that it costs \$266 to set up for production of the casters and the holding cost amounts to \$1 per caster per week.
a. Compute the EOQ and determine the number of periods of demand to which this corresponds by forming the ratio (EOQ)/(demand per period). Let T be this ratio rounded to the nearest integer. Determine the policy that produces casters once every T periods.
b. Using backward dynamic programming with $N = 6$ and $\mathbf{r} = (75, 75, \ldots, 75)$, find the optimal solution. Does your answer agree with what you obtained in part (a)?

22. a. Based on the results of problem 21, suggest an approximate lot-sizing technique. Under what circumstances would you expect this method to give good results?
b. Use this method to solve Example 9.9 (see Appendix 9–A). By what percentage does the resulting solution differ from the optimal?

23. Solve problem 17 using the Wagner–Whitin algorithm.

24. If we were to solve Example 9.4 using the Wagner–Whitin algorithm, we would obtain (154, 0, 0, 0, 0, 171, 0, 0, 114, 0) as the planned order release for the valve casing assembly. What are the resulting planned order releases for the valves?

25. Consider the example of Noname Computer Company presented in problem 4. Suppose that the setup cost for the production of the motherboards is \$180 and the holding cost is h = \$0.40 per motherboard per week. Using part-period balancing, determine the planned order release for the motherboards and the resulting gross requirements schedule for the DRAM chips. (Hint: Use the net demand for computers after accounting for starting inventory and returns.)

9.5 LOT SIZING WITH CAPACITY CONSTRAINTS

Assume that in addition to known requirements ($r_1, \dots, r_n$) in each period, there are also production capacities ($c_1, \dots, c_n$). Hence, we now wish to find the optimal production quantities ($y_1, \dots, y_n$) subject to the constraints $y_i \le c_i$, for $1 \le i \le n$.

The introduction of capacity constraints clearly makes the problem far more realistic. As lot-sizing algorithms can be incorporated into an MRP planning system, production capacities will be an important part of any realizable solution. However, they also make the problem more complex. The rather neat result that optimal policies always order exact requirements is no longer valid. Determining true optimal policies is difficult and time-consuming and is probably not practical for most real problems.

Even finding a feasible solution may not be obvious. Consider the simple four-period scheduling in Example 9.10 (see Appendix 9–A) with vector $\mathbf{r}$ = (52, 87, 23, 56), but now suppose that the production capacity in each period is $\mathbf{c}$ = (60, 60, 60, 60). First we must determine if the problem is feasible; that is, whether at least one solution exists. On the surface the problem looks solvable, as the total requirement over the four periods is 218 and the total capacity is 240. But this problem is infeasible; the most that can be produced in the first two periods is 120, but the requirements for those periods sum to 139.

We have the following feasibility condition:

$$\sum_{i=1}^{j} c_i \ge \sum_{i=1}^{j} r_i \qquad \text{for } j = 1, \dots, n.$$

Even when the feasibility condition is satisfied, it is not obvious how to find a feasible solution. Consider the following example.

EXAMPLE 9.5

$$\mathbf{r} = (20, 40, 100, 35, 80, 75, 25),$$
$$\mathbf{c} = (60, 60, 60, 60, 60, 60, 60).$$

Checking for feasibility, we have that:

$r_1 = 20,$	$c_1 = 60;$
$r_1 + r_2 = 60,$	$c_1 + c_2 = 120;$
$r_1 + r_2 + r_3 = 160,$	$c_1 + c_2 + c_3 = 180;$
$r_1 + r_2 + r_3 + r_4 = 195,$	$c_1 + c_2 + c_3 + c_4 = 240;$
$r_1 + r_2 + r_3 + r_4 + r_5 = 275,$	$c_1 + c_2 + c_3 + c_4 + c_5 = 300;$
$r_1 + r_2 + r_3 + r_4 + r_5 + r_6 = 350,$	$c_1 + c_2 + c_3 + c_4 + c_5 + c_6 = 360;$
$r_1 + r_2 + r_3 + r_4 + r_5 + r_6 + r_7 = 375,$	$c_1 + c_2 + c_3 + c_4 + c_5 + c_6 + c_7 = 420.$

The feasibility test is satisfied, so we know at least that a feasible solution exists. However, it is far from obvious how we should go about finding one. Scheduling on a lot-for-lot basis is not going to work because of the capacity constraints in periods 3, 5, and 6.

We will present an approximate lot-shifting technique to obtain an initial feasible solution. The method is to back-shift demand from periods in which demand exceeds capacity to prior periods in which there is excess capacity. This process is repeated for each period in which demand exceeds capacity until we construct a new requirements schedule in which lot for lot is feasible. In the example, the first period in which demand exceeds capacity is period 3. We replace r_3 with c_3. The difference of 40 units must now be redistributed back to periods 1 and 2. We consider the first prior period, which is period 2. There are 20 units of excess capacity in period 2, which we absorb. We still have 20 units of demand from period 3 that are not yet accounted for; this is added to the requirement for period 1. Summarizing the results up until this point, we have

$$\begin{array}{rcllllllll} & & \ \ 40 & 60 & 60 & & & & \\ \mathbf{r}' & = & (\cancel{20} & \cancel{40} & \cancel{100} & 35 & 80 & 75 & 25), \\ \mathbf{c} & = & (60 & 60 & 60 & 60 & 60 & 60 & 60). \end{array}$$

The next period in which demand exceeds capacity is period 5. The excess demand of 20 units can be back-shifted to period 4. Finally, the 15 units of excess demand in period 6 can be back-shifted to periods 4 (5 units) and 1 (10 units). The feasibility condition guarantees that this process leads to a feasible solution.

This leads to

$$\begin{array}{rcllllllll} & & \ \ 50 & & & 60 & & & \\ & & \ \ \cancel{40} & 60 & 60 & \cancel{55} & 60 & 60 & \\ \mathbf{r}' & = & (\cancel{20} & \cancel{40} & \cancel{100} & \cancel{35} & \cancel{80} & \cancel{75} & 25), \\ \mathbf{c} & = & (60 & 60 & 60 & 60 & 60 & 60 & 60). \end{array}$$

Hence, the modified requirements schedule obtained is

$$\mathbf{r}' = (50 \quad 60 \quad 60 \quad 60 \quad 60 \quad 60 \quad 25).$$

Setting $\mathbf{y} = \mathbf{r}'$ gives a feasible solution to the original problem.

We have that lot for lot for the modified requirements schedule $\mathbf{r}'$ is feasible for the original problem. However, we would like to see if we can discover an improvement—that is, another feasible policy that has lower cost. There are a variety of reasonable improvement rules that one can use. Appendix 9–A continues this problem with a standard improvement rule.

PROBLEMS FOR SECTION 9.5

26. Consider Example 9.1 presented in Section 9.3 of scheduling the production of the valve casing assembly.
    a. Suppose that the production capacity in any week is 100 valve casings. Using the algorithm presented in this section, determine the planned order release for the valve casings.
    b. What gross requirements schedule for the valves does the lot sizing you obtained in part (a) give?

c. Suppose that the production capacity for the valves is 200 valves per week. Is the gross requirements schedule from part (b) feasible? If not, suggest a modification in the planned order release computed in part (a) that would result in a feasible gross requirements schedule for the valves.

27. Solve problem 14 assuming a maximum order size of 20 per month.
28. a. Solve problem 17 assuming the following production capacities.

Week	1	2	3	4	5	6
Capacity	600	600	600	400	200	200

b. On a percentage basis, how much larger are the total holding and setup costs in the capacitated case than in the solutions obtained from parts (b), (c), and (d) of problem 17?

29. The method of rescheduling the production of a lot to one or more prior periods if the increase in the holding cost is less than the cost of a setup also can be used when no capacity constraints exist. This method is an alternative heuristic lot scheduling technique for the uncapacitated problem. For problem 14, start with a lot-for-lot policy and consider shifting lots backward as we have done in this section, starting with the final period and ending with the first period. Compare the total cost of the policy that you obtain with the policies derived in problem 14.

9.6 SHORTCOMINGS OF MRP

MRP is a closed production system with two major inputs: (1) the master production schedule for the end item and (2) the relationships between the various components, modules, and subassemblies composing the production process for that end item. The method is logical and seemingly sensible for scheduling production lot sizes. However, many of the assumptions made are unrealistic. In this section, we will discuss some of these assumptions, the problems that arise as a result of them, and the means for dealing with these problems.

Uncertainty

Underlying MRP is the assumption that all required information is known with certainty. However, uncertainties do exist. The two key sources of uncertainty are the forecasts for future sales of the end item and the estimation of the production lead times from one level to another. Forecast uncertainty usually means that the realization of demand is likely to be different from the forecast of that demand. In the production planning context, it also could mean that updated forecasts of future demands are different from earlier forecasts of those demands. Forecasts must be revised when new orders are accepted, prior orders are canceled, or new information about the marketplace becomes available. That has two implications in the MRP system. One is that *all* the lot-sizing decisions that were determined in the last run of the system could be incorrect, and, even more problematic, former decisions that are currently being implemented in the production process may be incorrect.

The analysis of stochastic inventory models in Chapter 5 showed that an optimal policy included safety stock to protect against the uncertainty of demand. That is, we would order to a level exceeding expected demand. The same logic can be applied to MRP systems. The manner in which uncertainty transmits itself through a complex multilevel productive system is not well understood. For that reason, it is not recommended to include independent safety stock at all levels of the system. Rather, by using the methods in Chapter 5, suitable safety levels can be built into the forecasts for the end item. These will be transmitted automatically down through the system to the lower levels through the explosion calculus.

EXAMPLE 9.6

Consider Example 9.1 on the Harmon Music Company. Suppose that the firm wishes to incorporate uncertainty into the demand forecasts for weeks 8 through 17. Based on historical records of trumpet sales maintained by the firm, an analyst finds that the ratio of the standard deviation of the forecast error to the mean demand each week is near 0.3. [In symbols, this is written σ/μ and is known as the coefficient of variation (see Chapter 6).] Furthermore, weekly demand is closely approximated by a normal distribution. Harmon has decided that it would like to produce enough trumpets to meet all the demand each week with probability .90. (In the terminology used in Chapter 5, this means that they are using a Type 1 service level of 90 percent for the trumpets.)

The safety stock is of the form σ_z where z is the appropriate cut-off point from the normal table. Here $z = 1.28$. Incorporating safety stock into the demand forecasts, we obtain

Week	8	9	10	11	12	13	14	15	16	17
Predicted demand (μ)	77	42	38	21	26	112	45	14	76	38
Standard deviation (σ)	23.1	12.6	11.4	6.3	7.8	33.6	13.5	4.2	22.8	11.4
Mean demand plus safety stock ($\mu + \sigma_z$)	107	58	53	29	36	155	62	19	105	53

Of course, this is not the only way to compute safety stock. Alternatives are to employ a Type 2 service criterion or to use a stock-out cost model instead of a service level model. The next step is to net out the scheduled receipts and anticipated on-hand inventory to arrive at a revised MPS for the trumpets. The explosion calculus would now proceed as before, except that the safety stock that is included in the revised MPS would automatically be transmitted to the lower-level assemblies.

Safety lead times are used to compensate for the uncertainty of production lead times in MRP systems. Simply put, this means that the estimates for the time required to complete a production batch at one level and transport it to the next level would be multiplied by some safety factor. If a safety factor of 1.5 were used at Harmon Music Company, the lead times for the components would be revised as follows: bell assembly, 3 weeks; valve casing assembly, 6 weeks; slide assemblies, 3 weeks; valves, 4.5 weeks. Conceptually, safety lead times make sense if the essential uncertainty is in the production times from one level to the next, and safety stocks make sense if the essential uncertainty is in the forecast of the demand for the end item. In practice, both sources of uncertainty are generally present, and some mixture of both safety stocks and safety lead times is used.

Capacity Planning

Another important issue that is not treated explicitly by MRP is the capacity of the production facility. The type of capacitated lot-sizing method we discussed earlier will deal with production capacities at one level of the system but will not solve the overall capacity problem. The problem is that even if lot sizes at some level do not exceed the production capacities, there is no guarantee that when these lot sizes are translated to gross requirements at a lower level, these requirements also can be satisfied with the existing capacity. That is, a feasible production schedule at one level may result in an infeasible requirements schedule at a lower level.

Capacity requirements planning (CRP) is the process by which the capacity requirements placed on a work center or group of work centers are computed by using the output of the MRP planned order releases. If the planned order releases result in an infeasible requirements schedule, there are several possible corrective actions. One is to schedule overtime at the bottleneck locations. Another is to revise the MPS so that the planned order releases at lower levels can be achieved with the current system capacity. This is clearly a cumbersome way to solve the problem, requiring an iterative trial-and-error process between the CRP and the MRP.

As an example of CRP, consider the manufacture of trumpets discussed in Example 9.1 and throughout the rest of this chapter. Suppose that the valves are produced in three work centers: 100, 200, and 300. At work center 100, the molten brass is poured into the form used to shape the valve. At work center 200, the holes are drilled in the appropriate positions in the valves (there are three hole configurations, depending upon whether the valve is number 1, 2, or 3). Finally, at work center 300, the valve is polished and the surface appropriately graded to ensure that the valve does not stick in operation. A summary of the appropriate information for the work centers is given in the following table.

Work Center	Worker Time Required to Produce One Unit (hours/unit)	Machine Throughput (units/day)
100	0.1	120
200	0.25	100
300	0.15	160

According to this information, there would be a total of six minutes (0.1 hour) of worker time required to produce a single valve at work center 100, and the existing equipment can support a maximum throughput of 120 valves per day. Consider the planned order releases obtained for the valves resulting from the Silver–Meal lot scheduling rule given in Section 9.4.

Week	2	3	4	5	6	7	8	9	10	11
Planned order release (S–M)	198	0	0	0	495	0	0	0	342	0

This planned order release translates to the following capacity requirements at the three work centers.

Week	2	3	4	5	6	7	8	9	10	11
Labor time requirements (hours):										
Work center 100	19.8	0	0	0	49.5	0	0	0	34.2	0
Work center 200	49.5	0	0	0	123.75	0	0	0	85.5	0
Work center 300	29.7	0	0	0	72.25	0	0	0	51.3	0
Machine time requirements (days):										
Work center 100	1.65	0	0	0	4.125	0	0	0	2.85	0
Work center 200	1.98	0	0	0	4.95	0	0	0	3.42	0
Work center 300	1.24	0	0	0	3.09	0	0	0	2.14	0

The capacity requirements show whether the planned order release obtained from the MRP is feasible. For example, suppose that the requirement of 123.75 labor hours in week 6 at work center 200 exceeds the capacity of this work center. This means that the current lot sizing is infeasible and some corrective action is required. One possibility is to split the lot scheduled for week 6 by producing some part of it in a prior week. Another is to adjust the lot sizing for the valve casing assembly at the next higher level of the product structure to accommodate the capacity constraints at the current level. In either case, substantial changes in the initial production plan may be required.

This example suggests an interesting speculation. Would it not perhaps make more sense to determine where the bottlenecks occur *before* attempting to explode the MPS through the various levels of the system? In this way, a feasible production plan could be found that would meet capacity constraints. Additional refinements could then be considered.

Rolling Horizons and System Nervousness

Thus far, our view of MRP is that it is a static system. Given known requirements for the end items over a specified planning horizon, one determines both the timing and the sizes of the production lot sizes for all the lower-level components. In practice, however, the production planning environment is dynamic. The MRP system may have to be rerun each period and the production decisions reevaluated. Often it is the case that only the lot-sizing decisions for the current planning period need to be implemented. We use the term **rolling horizons** to refer to the situation in which only the first-period decision of an N-period problem is implemented. The full N-period problem is rerun each period to determine a new first-period decision.

When using rolling horizons, the planning horizon should be long enough to guarantee that the first-period decision does not change. Unfortunately, certain demand patterns are such that even for long planning horizons, the first-period decision does not remain constant. Consider the following simple example (from Carlson, Beckman, and Kropp, 1982).

EXAMPLE 9.7

Suppose that the demand follows the cyclic pattern 190, 210, 190, 210, 190, … . For a five-period planning horizon, the requirements schedule for periods 1 to 5 is

$$\mathbf{r} = (190, 210, 190, 210, 190).$$

Furthermore, suppose that $h = 1$ and $K = 400$. The optimal solution for this problem obtained from the Wagner–Whitin algorithm is

$$\mathbf{y} = (190, 400, 0, 400, 0).$$

However, suppose that the planning horizon is chosen to be six periods instead of five periods. The requirements schedule for a six-period planning horizon is

$$\mathbf{r} = (190, 210, 190, 210, 190, 210).$$

The optimal solution in this case is

$$\mathbf{y} = (400, 0, 400, 0, 400, 0).$$

That is, the first-period production quantity has changed from 190 to 400. If we go to a seven-period planning horizon, then y_1 will be 190. With an eight-period planning horizon, y_1 again becomes 400. One might think that this cycling of the value of y_1 would continue indefinitely. It turns out that this is *not* the case, however. For planning horizons of $n \geq 21$ periods, the value of y_1 remains fixed at 190. [We are grateful to

Lawrence Robinson of Cornell University for pointing out that this anomaly does not continue for all values of n as was claimed in earlier editions and by Carlson et al., 1982.] However, even when there is eventual convergence of y_1, as in this example, the cycling for the first 20 periods could be troublesome when using rolling planning horizons.

Another common problem that results when using MRP is "**nervousness.**" The term was coined by Daniel Steele (1973), who used it to refer to the changes that can occur in a schedule when the horizon is moved forward one period. Some of the causes of nervousness include unanticipated changes in the MPS because of updated forecasts, late deliveries of raw materials, failure of key equipment, absenteeism of key personnel, and unpredictable yields.

There has been some analytical work on the nervousness problem. Some use the term **system nervousness** specifically to mean that a revised schedule requires a setup in a period in which the prior schedule did not (Carlson, Jucker, and Kropp, 1979; 1983). There is an interesting technique to reduce this particular type of nervousness: Let $(\hat{y}_1, \hat{y}_2, \ldots, \hat{y}_N)$ be the existing production schedule and $(y_1, y_2, \ldots, y_N)$ be a revised schedule based on new demand information. Suppose that besides the usual costs of holding and setup, there is an additional cost of v if the new schedule $\mathbf{y}$ calls for a setup in a period that the old schedule $\hat{y}$ did not. This means that there is an additional setup cost associated with the new schedule in those periods in which no setup was called for in the old schedule. This method increases the setup cost from K to $K + v$ if $\hat{y}_k = 0$ prior to determining the new schedule $\mathbf{y}$. Re-solving the problem with the modified setup costs using any of the lot-sizing algorithms previously discussed will result in fewer setup revisions. The revision cost v reflects the relative importance of the cost of nervousness.

Additional Considerations

Although MRP would seem to be the most logical way to schedule production in batch-type operations, the basic method has some very serious shortcomings as we saw above. Listed below are other difficulties.

Lead Times Dependent on Lot Sizes

The MRP calculus assumes that the production lead time from one level to the next is a fixed constant independent of the size of the lot. In many contexts this assumption is clearly unreasonable. One would expect the lead time to increase if the lot size increases. Including a dependence between the production lead time and the size of the production run into the explosion calculus seems to be extremely difficult.

MRP II: Manufacturing Resource Planning

As we have noted, MRP is a closed production planning system that converts an MPS into planned order releases. **Manufacturing resource planning (MRP II)** is a philosophy that attempts to incorporate the other relevant activities of the firm into the production planning process. In particular, the financial, accounting, and marketing functions of the firm are tied to the operations function. As an example of the difference between the perspectives offered by MRP and MRP II, consider the role of the master production schedule. In MRP, the MPS is treated as input information. In MRP II, the MPS would be considered a part of the system and, as such, would be considered a decision variable as well. Hence, the production control manager would work with the marketing manager to determine when the production schedule should be altered to incorporate revisions in the forecast and new order commitments. Ultimately, all divisions of the company would work together to find a production schedule consistent with the overall business plan and long-term financial strategy of the firm. MRP II is consistent with an S&OP process as described in Chapter 3.

Another important aspect of MRP II is the incorporation of capacity resource planning (CRP). Capacity considerations are not explicitly accounted for in MRP. MRP II is a closed-loop cycle in which lot sizing and the associated shop floor schedules are compared to capacities and recalculated to meet capacity restrictions. However, capacity issues continue to be an important issue in both MRP and MRP II operating systems.

Obviously, such a global approach to the production scheduling problem is quite ambitious. However, today's enterprise systems can facilitate such robust processes when used with an understanding of the core ideas behind production scheduling.

Imperfect Production Processes

An implicit assumption made in MRP is that there are no defective items produced. Because requirements for components and subassemblies are computed to exactly satisfy end-item demand forecasts, losses due to defects can seriously upset the balance of the production plan. In some industries, such as semiconductors, yield rates can be as low as 10 to 20 percent for new products. As long as yields are stable, incorporating yield losses into the MRP calculus is not difficult. One computes net demands and lot sizes in the same way, and in the final step divides the planned order release figures by the average yield. For example, if a particular process has a yield rate of 78 percent, one would multiply the planned order releases by $1/0.78 = 1.28$. The problem is much more complex if yields are random and variances are too large to be ignored. Using mean yields would result in substantial stock-outs. One would have to develop a kind of newsvendor model that balanced the cost of producing too many and too few and determine an appropriate safety factor. Because of the dependencies of successive levels, this would be a difficult problem to model mathematically. Monte Carlo computer simulation would be a good alternative. There has been some work on modeling random yields in the context of MRP systems (e.g., Inderfurth, 2009) but the methods are not trivial. Random yields more generally are discussed by several researchers. For example, Steven Nahmias and Kamran Moinzadeh (1997) consider a mathematical model of random yields in a single-level lot-sizing problem. Candace Yano and Hau Lee (1995) provide a survey of lot-sizing problems with random yields.

Data Integrity

An MRP system can function effectively only if the numbers representing inventory levels are accurate. It is easy for incorrect data to make their way into the scheduling system. This can occur if a shipment is not recorded or is recorded incorrectly at some level, items entering inventory for rework are not included, scrap rates are higher than anticipated, and so on. In order to ensure the integrity of the data used to determine the size and the timing of lot sizes, physical stock-taking may be required at regular intervals. An alternative to complex physical stock-taking is a technique known as **cycle counting**. Cycle counting simply means directly verifying the on-hand levels of the various inventories comprising the MRP system. For example, are the 45 units of part A557J indicated on the current record the actual count of this part number?

Efficient cycle counting can be achieved in a variety of ways. Stockrooms may have containers that only hold a fixed number or weight of items. Coded shelving systems could be used to more easily identify items with part numbers. Certain areas could be made accessible only to specific personnel. Cycle counting systems can be based on number or on weight. Furthermore, an error in the inventory level must be considered in relative terms. Based on the importance of the item, different percentage errors may be considered acceptable. Different error tolerances should be applied to weigh-counted items versus hand-counted items. If MRP is to have a positive impact on the overall production scheduling problem, the inventory records must be an accurate reflection of the actual state of the system.

Order Pegging

In some complex systems, a single component may be used in more than one item at the next higher level of the system. For example, a company producing many models of toy cars may use the same-sized axle in each of the cars. Gross requirements for axles would be the sum of the gross requirements generated by the MPS for each model of car. Hence, when one component is used in several items, the gross requirements schedule for this component comes from several sources. If a shortage of this component occurs, it is useful for the firm to be able to identify the particular items higher in the tree structure that would be affected. In order to do this, the gross requirements schedule is broken down by the items that generate it and each requirement is "**pegged**" with the part number of the source of the requirement. Pegging adds considerable complexity to the information storage requirements of the system and should only be considered when the additional information is important in the decision-making process.

PROBLEMS FOR SECTION 9.6

30. MRP systems have been used with varying degrees of success. Describe under what circumstances MRP might be successful and under what circumstances it would not be successful.
31. Discuss the advantages and disadvantages of including safety stock in MRP lot-sizing calculations. Do you think that a production control manager would be reluctant to build to safety stock if he or she is behind schedule?
32. For what reason is the capacitated lot-sizing method discussed in Section 9.5 not adequate for solving the overall capacity problem?
33. Planned order releases (POR) for three components, A, B, and C, are given below. Suppose that the yields for these components are respectively 84 percent, 92 percent, and 70 percent. Assuming lot-for-lot scheduling, how should these planned order releases be adjusted to account for the fact that the yields are less than 100 percent?

Week	6	7	8	9	10	11	12	13	14	15	16	17
POR(A)				200	200	80	80	200	400	400	400	
POR(B)			100	100	40	40	100	200	200	200		
POR(C)	200	400	280	100	280	600	800	800	400			

34. Define the terms *rolling horizons* and *system nervousness* in the context of MRP systems.

9.7 PULL SCHEDULING FUNDAMENTALS

Pull-scheduling, kanban control, just-in-time, Toyota production system (TPS), lean production, and zero inventories are all terms that have been used to describe essentially the same thing: a system of moving material through a plant that requires a minimum of inventory. (Although, as noted earlier, the Toyota production system and lean production are actually much broader philosophies than just production control techniques.)

Some have speculated that the roots of the system go back to the situation in postwar Japan. Devastated by the war, the Japanese firms were cash poor and did not have the luxury of investing cash in excess inventories. Thus, lean production was born from necessity. However, as Japanese cars started gaining in popularity in the United States, it quickly became clear that they were far superior to American- and European-made cars in terms of quality, value, efficiency, and reliability. We know today that lean production and the quality initiatives of the 1950s played important roles in this success.

Two developments were key to the success of this new approach to mass production: the **kanban** system and **SMED** (which stands for single minute exchange of dies). As mentioned earlier, kanban is a Japanese word for *card* or *ticket.* It is a manual information system developed and used by Toyota for implementing pull production. The approach is also facilitated by the use of **heijunka**, or demand smoothing. The ideas behind kanban, SMED, and heijunka will be covered in this section. The larger strategy behind these ideas was introduced in Chapter 1. In Chapter 10 we will discuss the quality management aspects of lean production, particularly as they relate to eliminating waste.

The Mechanics of Kanban

There are a variety of different types of kanban tickets, but two are the most prevalent. These are withdrawal kanbans and production ordering kanbans. A withdrawal kanban is a request for parts to a work center from the prior level of the system. A production ordering kanban is a signal for a work center to produce additional lots. The manner in which these two kanban tickets are used to control the flow of production is depicted in Figure 9–8.

The process is as follows: Parts are produced at work center 1, stored in an intermediate location (known as the store), and subsequently transported to work center 2. Parts are transported in small batches represented by the circles in the figure. Production flows from left to right in the diagram. The detailed steps in the process are as follows (the numbers below appear in the appropriate locations in Figure 9–8).

1. When the number of tickets on the withdrawal kanban reaches a predetermined level, a worker takes these tickets to the store location.
2. If there are enough canisters available at the store, the worker compares the part number on the production ordering kanbans at the store with the part number on the withdrawal kanbans.
3. If the part numbers match, the worker removes the production ordering kanbans from the containers, places them on the production ordering kanban post, and places the withdrawal kanbans in the containers.
4. When a specified number of production ordering kanbans have accumulated, work center 1 proceeds with production.
5. The worker transports parts picked up at the store to work center 2 and places them in a holding area until they are required for production.

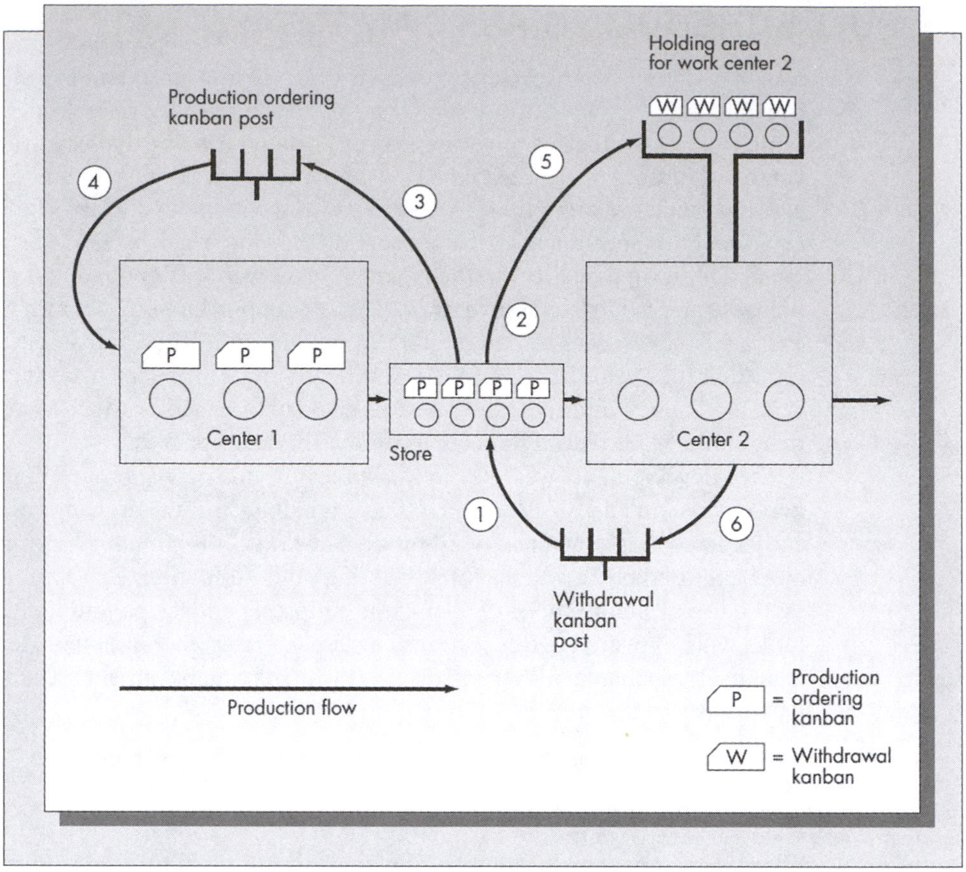

Figure 9–8 Kanban system for two production centers

6. When the parts enter production at work center 2, the worker removes the withdrawal kanbans and places them on the withdrawal kanban post. (Note that production ordering kanbans for work center 2 are then attached to the parts produced at that work center. These kanban tickets are not shown in Figure 9–8.)

One computes the number of kanban tickets in the system in advance. Toyota uses the following formula (Monden, 1981):

$$y = \frac{\bar{D}L + w}{a},$$

where

y = Number of kanbans.
$\bar{D}$ = Expected demand per unit of time.
L = Lead time (processing time + waiting time between processes + conveyance time).
w = Policy variable specifying the level of buffer stock, generally around 10 percent of $\bar{D}L$.
a = Container capacity (usually no more than 10 percent of daily demand).

This formula implies that the maximum level of inventory is given by $ay = \bar{D}L + w$. The ideal value of w is zero. However, it is difficult to balance a system so perfectly that buffer stock is eliminated entirely.

As mentioned earlier, the kanban system is a manual information system for implementing pull production. Pull systems also can be implemented in other ways, which may be more efficient than the kanban method. More will be said about this later in the section.

Single Minute Exchange of Dies

One of the key components of the success of Toyota's production system was the concept of **single minute exchange of dies (SMED)**, championed by Shigeo Shingo. Shingo is generally credited with developing and implementing SMED at Toyota in 1970, which has become an important part of the overall Toyota production system. The basic theory given in Chapter 4 tells us that small lot sizes will be optimal only if fixed costs are small (as K decreases in the EOQ formula, so does the value of Q). The most significant component of the cost of setting up for a new operation in a plant is the time required to change over the machinery for that operation, since the production line must be frozen during the changeover operation. This involves changing some set of tools and/or dies required in the process, hence the origin of the term SMED. (A die is a tool used for shaping or impressing an object or material.)

Die-changing operations are required in automotive plants when switching over the production line from one car model to another. These operations typically took about four hours. The idea behind SMED is that a significant portion of the die-changing operation can be done off-line while the previous dies are still in place and the line continues to operate. According to Shingo (1981), this is accomplished by dividing the die-changing operation into two components: inside exchange of die (IED) and outside exchange of die (OED). The OED operation is to be performed while the line is running in advance of the actual exchange. The goal is to structure the die change so that there are as many steps as possible in the OED portion of the operation.

While this idea sounds simple, it has led to dramatic improvements in throughput rates in many circumstances, both within and outside the automotive industry. Shingo (1981) describes some of the successes.

- At Toyota, an operation that required eight hours for exchange of dies and tools for a bolt maker was reduced over a year's time to only 58 seconds.
- At Mitsubishi Heavy Industry, a tool-changing operation for a boring machine was reduced from 24 hours to only 2 minutes and 40 seconds.
- At H. Weidmann Company of Switzerland, changing the dies for a plastic molding machine was reduced from 2.5 hours to 6 minutes and 35 seconds.
- At Federal-Mogul Company of the United States, the time required to exchange the tools for a milling machine was reduced from 2 hours to 2 minutes.

Of course, SMED cannot be applied in all manufacturing contexts. Even in contexts where it can be applied, the benefits of the die-changing time reduction can be realized only if the process is integrated into a carefully designed and implemented overall manufacturing control system.

Heijunka

Another key principle behind the TPS is known as **heijunka**, the Japanese word for production leveling or smoothing. This does not mean building a single product on the line. Instead, the production schedule maintains a stable mix of products and firm monthly schedules. So, for example, if one quarter of cars have sunroofs then a car with a sunroof will go down the production line exactly every four cars. As sunroofs are more time consuming to install, this will allow the workers to take a little more time every four production cycles, making this up in the other three cycles.

One might ask whether this sort of planned production is contradictory to pull production. In some sense it is. However, it is also what allows pull production to work within Toyota. By smoothing out all variation in end demand (as well as by using SMED), the product can "flow like water" throughout the production process with minimum inventories. Indeed, much of what makes TPS, and lean more generally, work so well is its pragmatic approach. Since it is not possible to smooth out the variation in end customer demand, that demand must be buffered. Once the demand is buffered, even more variation can be eliminated through using heijunka.

Relationship with Suppliers

For pull control to work properly, it must be coordinated with the purchasing system and with purchasing strategies. One complaint about lean production is that it merely pushes system uncertainty and higher inventories onto the supplier. This has certainly been the case for companies that follow pull production control but do not perform heijunka (i.e., the smoothing out of end-customer demand variability). In this case, greater flexibility on the part of the suppliers is necessary; they must be able to react quickly and provide sufficiently reliable parts to relieve the manufacturer of the necessity to inspect all incoming lots. Furthermore, multiple sourcing becomes difficult under such a system. That is, the firm may be forced to deal with a single supplier in order to develop the close relationship that the system requires. Single sourcing presents risks for both suppliers and manufacturers. The manufacturer faces the risk that the supplier will be unable to supply parts when they are needed, and the supplier faces the risk that the manufacturer will suffer reverses and demand will drop.

James Freeland (1991) provided a list of characteristics contrasting conventional and JIT purchasing behaviors. [We have largely moved away from using the term JIT in this chapter; Freeland used the term, which was popular in the 1990s.]

Conventional Purchasing	JIT Purchasing
1. Large, infrequent deliveries.	1. Small, frequent deliveries.
2. Multiple suppliers for each part.	2. Few suppliers; single sourcing.
3. Short-term purchasing agreements.	3. Long-term agreements.
4. Minimal exchange of information.	4. Frequent information exchange.
5. Prices established by suppliers.	5. Prices negotiated.
6. Geographical proximity unimportant.	6. Geographical proximity important.

In his study, Freeland noted that the industries that seemed to benefit most from JIT purchasing were those that typically had large inventories. Companies without JIT purchasing tended to be more job-shop oriented or make-to-order oriented. Vendors that entered into

JIT purchasing agreements tended to carry more safety stock, suggesting manufacturers are reducing inventories at the expense of the vendors. The JIT deliveries were somewhat more frequent, but the differences were not as large as one might have expected. Geographical separation of vendors and purchasers was a serious impediment to successful implementation of JIT purchasing. The automotive industry was one that reported substantial benefit from JIT purchasing arrangements. In other industries, such as computers, the responses were mixed; some companies reported substantial benefits and some reported few benefits.

Advantages and Disadvantages of Pull Production

Champions of lean and the Toyota production system would have one believe that all other production planning systems are now obsolete. The zeal with which they promote the method is reminiscent of the enthusiasm that heralded MRP in the early 1970s. At that time, some claimed that classical inventory analysis was no longer valid. However, each new production method should be viewed as an addition to, rather than a replacement for, what is already known. Pull production can be a useful tool in the right circumstances, but it is far from a panacea for the problems facing every industry.

Pull production and EOQ are not mutually exclusive. Setup time and, consequently, setup cost reduction result in smaller lot sizes. Smaller lot sizes require increased efficiency and reliability of the production process but considerably less investment in raw materials, work-in-process inventories, and finished-goods inventories.

Kanban is a manual information system used to support pull production. However, pull and kanban are not necessarily wedded. Pull production can be implemented through electronic boards rather than physical kanban cards. A shortcoming of kanban is the time required to transmit new information through the system. Figure 9–9 considers a schematic of a serial production process with six levels. With the kanban system, the direction of the flow of information is opposite to the direction of the flow of the production. Consider the consequences of a sudden change in the requirements at level 6. This change is transmitted first to level 5, then to level 4, and so on. There could be a substantial time lag from the instant that the change occurs at level 6 until the information reaches level 1.

A centralized information processing system will help to alleviate this problem. If there are sudden changes in the requirements at one end of the system resulting from unplanned changes in demand or breakdowns of key equipment, these changes will be instantly transmitted to the entire system. However, in a strict pull system, even if implemented electronically, the actual response to the change event will still have the same time lag. This is because production is in response to actual production so it will take a while for the change event to propagate up the chain.

MRP has an important advantage over kanban in this regard. One of the strengths of MRP is its ability to react to forecasted changes in the pattern of demand. The MRP system recomputes production quantities based on these changes and makes this information available to all levels simultaneously. MRP allows planning to take place at all levels in a way that pull production, and especially kanban, will not. As Harlan Meal (1984) points out:

> [Pull] production works well when the overall production rate is constant, but it is unsatisfactory for communicating basic changes in production rate to earlier stages in the process… . On the other hand using the HPP [hierarchical production planning] approach, plant managers do not rely on their short term signals to establish their early stage production rates. (p. 110)

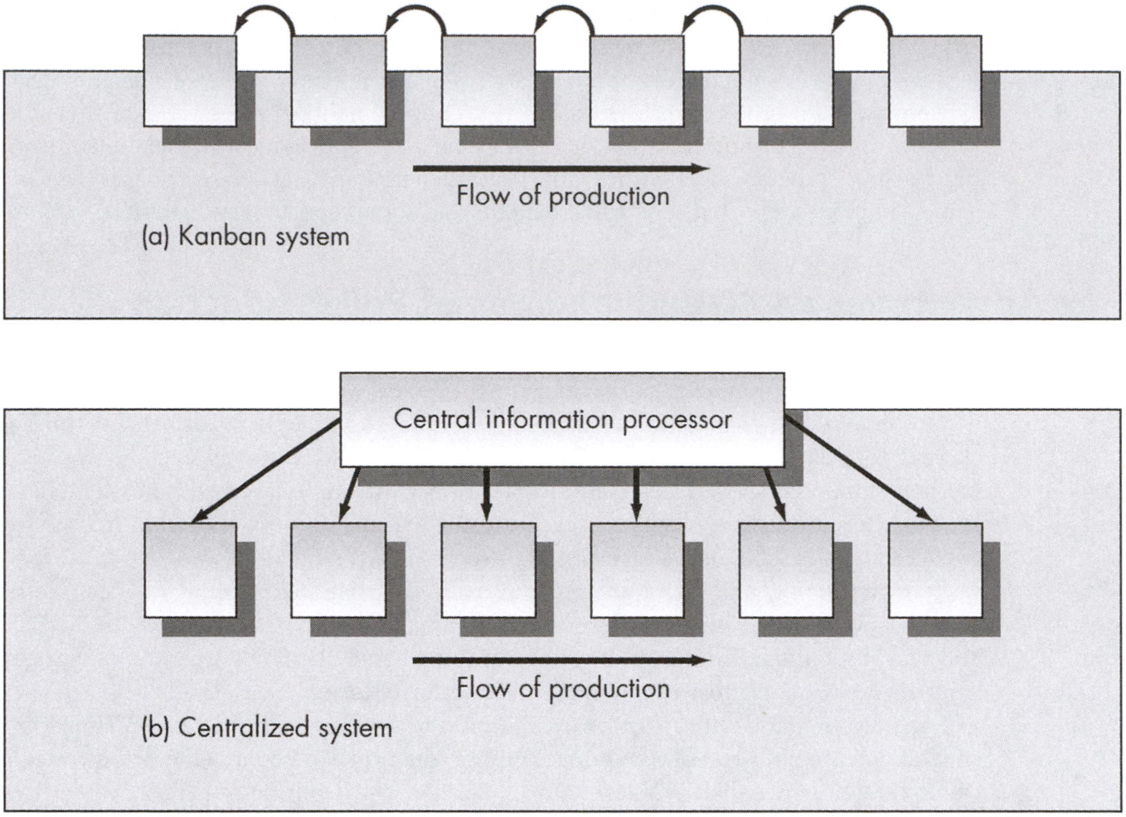

Figure 9–9 Kanban information system versus centralized information system

Pull production is most efficient when the pattern of demand is stable and predictable. As mentioned earlier, Toyota makes sure this is the case by applying heijunka. Changes in the demand may result from predictable causes or random fluctuations or both. MRP makes use of forecasts of anticipated changes in demand and transmits this information to all parts of the productive system. However, neither MRP nor pull production is designed to protect against random fluctuations of demand. Both methods could be unstable in the face of high demand variance.

Another potential shortcoming of pull production is the idle time that may result when unscheduled breakdowns occur. Part of this is deliberate. If a breakdown occurs, then workers' attention can be focused on the problem immediately. If workers are only familiar with their own operation (which is not the case in Toyota), there will be significant worker idle time when a breakdown occurs. This is consistent with the trade-off curve presented in Figure 11–4 of Chapter 11 on shop floor control and sequence scheduling. Buffer inventories between successive operations provide a means of smoothing production processes. However, buffer inventories also have disadvantages. They can mask underlying problems. A popular analogy is to compare a production process with a river and the level of inventory with the water level in the river. When the water level is high, the water will cover the rocks. Likewise, when inventory levels are high, problems are masked. However, when the water level (inventory) is low, the rocks (problems) are evident (see Figure 9–10).

Seen in this light, lean production can be incorporated easily into an overall quality control strategy. Six-sigma quality management, discussed in Chapter 10, and lean can work together

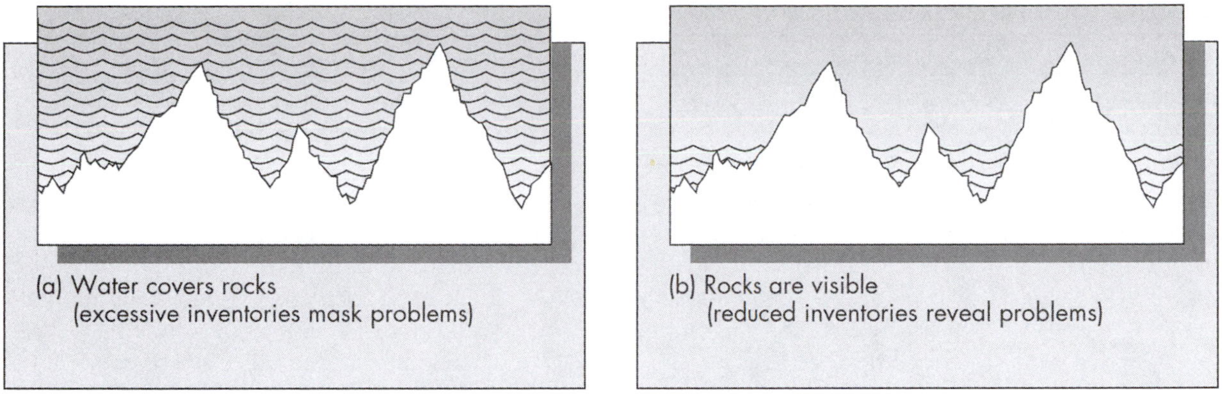
(a) Water covers rocks
(excessive inventories mask problems)
(b) Rocks are visible
(reduced inventories reveal problems)

Figure 9–10 River inventory analogy illustrating the advantages of lean inventories

not only to reduce inventory costs but also to bring about significant improvements in product quality. Indeed, "lean-six-sigma" programs have become very popular in recent years.

Although reducing excess work-in-process inventory can have many benefits, low inventory is not necessarily the answer for all manufacturing situations. According to Robert Stasey and Carol McNair (1990):

> Inventory in a typical plant is like insurance, insurance that a problem in one area of a plant won't affect work performed in another. When problems creating the need for insurance are solved, then inventories disappear from the plant floor. (p. 13)

The implication is that we merely eliminate all sources of uncertainty in the plant and the need for inventories disappears. The problem is that there are some sources of variation that can never be eliminated. One is variation in consumer demand. Pull production is effective only if final demand is regular or if it has been smoothed out by adjusting finished goods inventories. Another may be sources of variation inherent in the production process or in the equipment. Can one simply legislate away all sources of uncertainty in the manufacturing environment? Of course not. Hence, although the underlying principles of pull production are sound, it is not a cure-all and will not necessarily be the right method for every production situation.

Table 9–1 briefly summarizes the primary advantages and disadvantages discussed in this section.

PROBLEMS FOR SECTION 9.7

35. Discuss the concepts of push versus pull and how they relate to kanban control.
36. What is the difference between a pull system and a kanban system? Can pull be implemented without kanban?
37. A regional manufacturer of table lamps plans on using a manual kanban information system. On average the firm produces 1,200 lamps monthly. Production lead time is 18 days, and the firm plans to have 15 percent buffer stock. Assume 20 working days per month.
    a. If each container holds 15 lamps, what will be the total number of kanban tickets required? (Use the formula given in this section.)
    b. What is the maximum WIP inventory the company can expect to have using this system?

Table 9–1 Summary of Advantages and Disadvantages of Pull and Kanban

Feature	Advantages	Disadvantages
Small work-in-process inventories	1. Decreases inventory costs. 2. Improves production efficiency. 3. Points out quality problems quickly.	1. May result in increased worker idle time. 2. May decrease the production rate.
Kanban information flow system	1. Provides for efficient lot tracking. 2. Inexpensive means of implementing pull. 3. Allows for predetermined level of WIP inventory by presenting number of kanban tickets. 4. Allows for an easy on-site visualization of the state of the system.	1. Slow to react to changes in demand. 2. Ignores known information about future demand patterns. 3. Does not provide off-site visualization of the state of the system (although computer-based pull systems can).
Coordinated inventory and purchasing	1. Inventory reduction. 2. Improved coordination of different systems. 3. Improved relationships with vendors.	1. Decreased opportunity for multiple sourcing. 2. Suppliers must react more quickly. 3. Improved reliability required of suppliers

c. Suppose that each lamp costs the firm $30 to produce. If carrying costs are based on a 20 percent annual interest rate, what annual carrying cost for WIP inventory is the company incurring? (You may wish to review the discussion of holding costs in Chapter 4.)

38. What is SMED? In what way can die-changing operations be reduced by orders of magnitude? Why is it an advantage to do so?
39. Explain how the dual-card kanban system operates.
40. Discuss the advantages and disadvantages of each of the following features of pull control.
    a. Small lot sizes.
    b. Integrated purchasing and inventory control.
    c. Kanban information system.
    d. Heijunka production smoothing.

9.8 A COMPARISON OF MRP, PULL, AND HYBRID SYSTEMS

MRP and pull are fundamentally different systems for manufacturing control. As we noted earlier, MRP is a push system, while pull is a reactive system. If a problem develops and the line is shut down, pull reacts immediately, since requests for new material are discontinued. In this way, one might say that pull reacts to uncertainties and MRP does not. However, pull is clearly not going to work well when it is known that demands will vary significantly over time (e.g., when demand is highly seasonal). MRP builds this information into the planning structure while pull does not.

For most manufacturing environments, implementing a pure pull system is simply not feasible. Suppliers may not be located in close enough proximity to allow inputs to be delivered according to a rigid schedule. Demands for products may be highly variable, making it impractical to ignore this information in the planning process. It may be difficult to move products in small batches. Implementing SMED may not be possible in some environments. When setup costs are very high, it makes economic sense to produce in large lots and store items rather than change over production processes frequently. With that said, however, major reductions in WIP inventories can be achieved in the vast majority of traditional manufacturing plants. Plants that run lean run better.

Toyota's enormous success in reducing inventory-related costs while producing high-quality products with high throughput rates has formed the basis for much of the support for pull-based systems. However, it is unclear if pull is primarily responsible for Toyota's success. Toyota's way of doing business differs from that of U.S. automakers in many dimensions. Is its success a direct result of kanban-control methods, or can it be attributed to other factors that might be more difficult to emulate? In order to determine under what circumstances pull would be most advantageous, researchers developed a large-scale simulator to compare pull, MRP, and ROP (reorder point) manufacturing environments (Krajewski, King, Ritzman, and Wong, 1987). The comparison included 36 distinct factors aggregated into eight major categories, with each factor varied from one to five levels. The eight categories considered and a brief summary of the factors in these categories are listed below.

1. *Customer influence.* Demand forecast error.
2. *Vendor influence.* Vendor reliability (order size received compared to order size requested, and time received compared to time requested).
3. *Buffer mechanisms.* Capacity buffer and safety stock and safety lead times.
4. *Product structure.* Pyramid (few end items) or inverted pyramid (many end items) structures.
5. *Facility design.* Routing patterns, shop emphasis, and length of routings.
6. *Process.* Scrap rates, equipment failures, worker flexibility, and capacity imbalances.
7. *Inventory.* Reporting accuracy, lot-sizing rules, and setup times.
8. *Other factors.* Number of items, number of workstations, and several other factors.

The results obtained were very interesting. Pull worked well only in favorable manufacturing environments. This means little or no demand variability, reliable vendors, and small setup times for production. Pull showed poor performance when one or more of these factors became unfavorable. In fact, in a favorable environment both ROP and MRP gave good results. This suggests that the primary benefit comes from creating a favorable manufacturing environment, as opposed to simply implementing a pull system. Perhaps greater benefit would result from carefully evaluating and (where possible) rectifying the key problems affecting manufacturing than by blindly implementing a new production planning system. Alternatively, perhaps it is the *process* of implementing a pull system that offers the greatest benefit.

New approaches to manufacturing control are quickly tagged with three-letter acronyms and proselytized with the zealousness of a religion. JIT was no exception. Paul Zipkin (1991) offers a very thoughtful piece on the JIT "revolution" (p. 4). His main point is that JIT can have real benefits, but if the pragmatism is not distinguished from the romance, the consequence could be disaster. He cites an example of a manager of a computer assembly plant who was ordered to reduce his WIP inventory to zero in six months. This was the result, apparently, of the experiences of the CFO, who had attended a JIT seminar that inspired him to promote massive inventory reductions. It is likely stories such as these that have led to the decreased use of the term "JIT."

At its best, lean production (or its predecessor JIT) is a set of methods that should be implemented on a continuous improvement basis for reducing inventories at every level of the supply chain. At its worst, lean is a romantic idea that when applied blindly can be very damaging to worker morale, relationships with suppliers, and ultimately the bottom line. Inventory savings can often be an illusion, as illustrated by Jitendra Chhikara and ElliottWeiss (1995). They show in three case studies that if inventory reductions are not tied to accounting systems and cash flows, inventory reductions do not translate to reduced inventory costs. Pull production for its own sake does not make sense. It must be carefully integrated into the entire supply and manufacturing chain in order to realize its benefits.

Finally, as noted by Uday Karmarkar (1989), the issue is not to make a choice between MRP and pull but to make the best use of both techniques. Pull reacts very slowly to sudden changes in demand, while MRP incorporates demand forecasts into the plan. Does that mean that Toyota, famous for their pull system, ignores demand forecasts in their manufacturing planning and control? Very unlikely. As discussed earlier, Toyota runs a make-to-stock process for its completed cars, using heijunka to smooth the demand pattern. Understanding what different methodologies offer and their limitations leads to a manufacturing planning and control system that is well designed and efficient. Improvements should be incremental, not revolutionary.

Pull and MRP are certainly not the only ways to approach manufacturing control. In their book, Wallace Hopp and Mark Spearman (1996) explore a production planning system based on fixing WIP inventory (based on Spearman, Hopp, and Woodruff 1989). If reduction of work-in-process is a goal, then why not design a system from scratch that forces a desired level of WIP? They have designed just such a system, which they call **CONWIP** (for *CONstant Work-In-Process*). In a CONWIP system, each time an item is completed (exits the production line), initiation of production of a new item is begun at the start of the line. In this way the WIP stays fixed. One can adjust the size of the WIP based on variability and cost considerations. CONWIP is similar in principle to kanban control in that both are pull systems. However, with CONWIP, the information flow is not from a stage in the process to the previous stage but from the end of the process to the beginning. The method sounds intriguing but does not appear to have been widely adopted.

Another variant on kanban is **POLCA**, which stands for "Paired-cell Overlapping Loops of Cards with Authorization" (see Suri, 2018). It has been designed for systems with a large variety of customized orders where standard kanban control does not work well. Rajan Suri, the inventor of the system, states that POLCA is particularly suited to companies manufacturing high-mix, low-volume and customized products. However, POLCA is another system yet to reach a wide adoption point.

Flow Manufacturing seeks to integrate MRP and pull production. In such a system, planning (e.g., components forecasting) is done based on MRP, while shop-floor execution is performed using kanban cards. Although this idea has often been used as an advertisement tool for an enterprise system, it does appear to be a sensible approach.

The final production control variant we mention here is **synchronous production** or **drum-buffer-rope** production. Introduced by Eliyahu Goldratt in his novel *The Goal*, the key idea behind this technique is to synchronize flow through the bottleneck (defined as the machine with the slowest processing rate). The "rope" is a pull-control mechanism used to release raw material based on the bottleneck, and the "buffer" makes sure the bottleneck is never idle. The bottleneck is the "drum" beating out the pace for the system. Like many of the other good ideas described above, this production control mechanism does not appear to have achieved wide implementation. Anecdotally, Goldratt wrote *The Goal* as a means of selling his production control software. Since then, the software has become obsolete, but the book has become a runaway success. The appeal of the book is more in its big ideas (particularly around the so-called **theory of constraints**) than its specific production control ideas. However, it does provide the following "rules" for production scheduling.

1. Do not balance capacity—balance flow.
2. The level of utilization of a non-bottleneck resource is not determined by its own potential but by some other constraint in the system.
3. Utilization and activation of a resource are not the same.
4. An hour lost at a bottleneck is an hour lost for the entire system.
5. An hour saved at a non-bottleneck is a mirage.
6. Bottlenecks govern both throughput and inventory in the system.
7. Transfer batch may not and many times should not be equal to the process batch.
8. Process batch should be variable along its route and over time.
9. Priorities can be set only by examining the system's constraints. Lead time is a derivative of the schedule.

In summary, there is no one ideal system for production control. All systems have both advantages and disadvantages. In order to decide which system to implement, you must understand your operations and supply chain strategy and how the proposed system meshes with it.

9.9 FLEXIBLE MANUFACTURING SYSTEMS

Flexible manufacturing systems (**FMS**s) were introduced in Chapter 1. They allow many parts to be produced at the same machine and also for the same part to be produced at different machines. These two points make production control of FMS a more complex task than that for systems where parts have a single fixed flow through the plant. The core MRP assumption that lead times are known and fixed is particularly problematic for FMS because lead time is a decision that depends on the production control decisions.

Queueing models, as introduced in Chapter 8, can be useful in aiding with FMS system design and control. In an FMS, each machine corresponds to a separate queue. The jobs correspond to the customers, and the queue is the stack of jobs waiting for processing at a machine. Because an FMS is a collection of machines connected by a materials handling system, the entering stream of customers for one machine is the result of the accumulation of departing streams from other machines. Hence, an FMS is a special kind of queueing network.

Such queueing networks can be used to compute a variety of measures relating to machine utilization and system performance. An important issue that arises during the initial design phase is system capacity. The system capacity is the maximum production rate that the FMS can sustain. It depends on both the configuration of the FMS and the assumptions one makes about the nature of the arrival stream of jobs. Paul Schweitzer's (1977) analysis showed that the output rate of a flexible manufacturing system is simply the capacity of the slowest machine in the system.

To be more precise, suppose that machine i can process jobs at a rate μ_i. If the probabilities that an entering job visits machines in a given order are known, one can find e_i, the expected number of visits of a newly arriving job to machine i, using queueing theory methods. It follows that $1/e_i$ is the rate that jobs arrive to machine i, so μ_i/e_i is the output rate of machine i. The "slowest" machine, where slow is used in a relative sense, is that machine whose output rate is least. Hence, the output rate of the system is $\min_{1 \le i \le n} (\mu_i/e_i)$. Note that this analysis is consistent with Goldratt's claim that the output rate of the system is the production rate of the bottleneck, i.e., slowest machine (Goldratt and Cox, 2014).

Such results can be very valuable when considering the design of the system and its potential utility to the firm. However, in some circumstances queueing models are only of limited value. In particular, in order to obtain explicit results for complex queueing networks, one often must assume that the arrival and the service processes are purely random. That is, both the time between the arrival of successive jobs and the time required for a machine to complete its task are exponential random variables. (The properties of the exponential distribution are discussed in detail in Chapters 2 and 8.) It is unlikely that both interarrival times and job performance times in flexible manufacturing systems are accurately described by exponential distributions. There are circumstances under which exponential queueing formulas do give reasonable approximations to more complex cases, but simulation should be used to validate analytical formulas. Queueing networks were discussed in Section 8.5 and Supplement 2.5. Even with their limitations, queueing models provide a powerful means for analyzing the performance characteristics and capacity limitations of flexible manufacturing systems.

Advantages of Flexible Manufacturing Systems

When used correctly, these systems can provide substantial advantages over more rigid designs, including the following.

1. *Reduced work-in-process inventory.* The design of the system limits the number of pallets available for moving parts through the system. Hence, the WIP inventory never exceeds a predetermined level. In this sense, the FMS is similar to the just-in-time system, in which the level of WIP inventory is a decision variable whose value may be chosen in advance.
2. *Increased machine utilization.* Numerically controlled machines often have a utilization rate of 50 percent or less. However, an efficient FMS may have a utilization rate as high as 80 percent. The improved utilization is a result of both the reduction of changeover time of machine settings and tooling and the ability to better balance the system workload.
3. *Reduced manufacturing lead time.* Without an FMS, parts might have to be processed through several different work centers. As a result, there could be substantial transportation time between work centers and substantial queueing time at work centers. Because an FMS reduces transportation, setup, and changeover time, it results in significant reduction in lead time for production.
4. *Ability to handle a variety of different part configurations.* As noted earlier in this section, the FMS is more flexible than a fixed transfer line but not as flexible as a standalone numerically controlled machine. Depending on the tooling available for the machines, parts may be launched into the system with little or no setup time required. Also, the FMS can process part configurations simultaneously.
5. *Reduced labor costs.* The number of workers required to manage an FMS can be as much as a factor of 10 fewer than the number required in a traditional job shop. Even when numerically controlled machines are used on a stand-alone basis, at least one worker is required per machine, and workers are required to transport the parts between machines. The automated materials handling capability of the FMS leads to significant reductions in labor requirements.

Disadvantages of Flexible Manufacturing Systems

Although there are many potential advantages of FMS, there is one factor that has delayed installation of these systems. That factor is cost. Most FMSs are very costly (although their cost has come down in recent years).

In addition to the direct costs of the equipment and the space, several indirect costs also are incurred. In order to manage the flow of materials, a sophisticated software system is required. Effective software can be extremely expensive, may require customization, often has bugs, and generally requires worker training. The cost of equipment such as feeders, conveyors, and transfer devices may not be part of the initial purchase cost. Other indirect costs include site preparation, spare parts to support the machinery, and disruptions that might result during the installation period. Furthermore, any company that purchases an FMS must anticipate a decline in productivity that accompanies the introduction of new technology.

It is no wonder that so many U.S. companies have trouble justifying the investment in FMS. However, Robert Kaplan (1986) argues that traditional cost accounting methods may ignore some important considerations. One is that evaluation of alternative investments based on discounted cash flow assumes a status quo. That is, NPV analysis assumes that factors such as market share, price, labor costs, and the company's competitive position in the marketplace will remain constant. However, the values of some of these variables will change, and are more likely to degenerate if the company retains outmoded production methods.

Furthermore, most firms prefer to invest in a variety of small projects rather than make a major capital outlay for a single project. Such a philosophy is safer in the short run but could be suboptimal in the long run. An example of an industry that failed to invest in new technology is the railroad industry. Because of outmoded equipment and facilities, many firms in this industry have been unable to stay competitive and profitable.

Cost is not the only problem. The FMS may experience downtime for a variety of reasons. Planned downtime could be the result of scheduled maintenance and scheduled tool changeovers. Unplanned downtime could result from mechanical failures of the machines or electrical failures. If a single machine goes down, the system can continue to function, but if either the materials handling system or the central computer fails, the entire FMS is crippled.

Two final issues with FMSs are quality and speed. Although the quality of parts produced by an FMS is usually very good, it is often not quite as high as that of a system designed specifically for only one part. Similarly, a system designed for only one part will have a very high processing speed for that part while an FMS will be fast in general but slower than a purpose-build system.

The decisions that need to be made when using FMSs and the corresponding mathematical modeling methods for these decisions are detailed in Table 9–2.

9.10 HISTORICAL NOTES

The specific term *MRP* appears to date from the 1960s, although the concept of materials planning based on predicted demand and product structures was around even longer. In fact, Alfred Sloan (1964) referred to a purely technical calculation by analysts at General Motors of the quantity of materials needed for production of a given number of cars in 1921. The term *BOM* (bill of materials) *explosion* was commonly used to describe what is now called MRP.

The books by Joseph Orlicky (1975) and Colin New (1974) did a great deal to legitimize MRP as a valid and identifiable technique (although the term seems to have been around since the mid-1960s). The well-known practitioners George Plossl and Oliver Wight (1971) should also be credited for popularizing the method. The first computerized MRP systems were implemented in about 1970 (Anderson, Schroeder, Tupy, and White,1982). The number of installed systems has increased at an exponential rate since that time.

Table 9–2 Levels of Decision Making in Flexible Manufacturing Systems

Time Horizon	Major Issues	Modeling Methods
Long term (months–years)	Design of the system Parts mix changes	Queueing networks Simulation
	System modification and/or expansion	Optimization
Medium term (days–weeks)	Production batching Maximizing of machine utilization	Balancing methods Network simulation
	Planning for fluctuations in demand and/or availability of resources	
Short term (minutes–hours)	Work order scheduling and dispatching	Tool operation and allocation program
	Tool management	Work order dispatching algorithm
	Reaction to system failures	Simulation

Interestingly, much of the work on optimal and suboptimal lot-sizing methods predates the formal recognition of MRP. Harvey Wagner and Thomson Whitin (1958) first recognized the optimality of an exact requirements policy for periodic-review inventory control systems with time-varying demand and developed the dynamic programming algorithm described in this chapter. Edward Silver and Harvey Meal (1973), as mentioned earlier, addressed opportunities for replenishment at the start of periods with their Silver–Meal heuristic. John DeMatteis (1968) is generally credited with the part-period balancing approach. However, an article at the time refers to both part-period balancing (called the least total cost approach) and least unit cost methods, suggesting that these methods already were well known (Gorham, 1968). It is likely that both methods were developed by practitioners before 1968 but were not reported in the literature.

The lot-shifting algorithm for the capacitated problem outlined in Section 9.5 is very similar to one developed by Reuven Karni (1981). Paul Dixon and Edward Silver (1981) explored similar ideas. Lot sizing is not always used by practitioners in operational MRP systems because small errors at a high level of the product structure are telescoped into large errors at a lower level. Furthermore, lot-sizing algorithms require estimation of the holding and setup costs and more calculations than the simple lot-for-lot policy.

The historical references to JIT are contained within the chapter. It is unclear who coined the term JIT, but the concept is clearly derived from Toyota's kanban system. SMED (single minute exchange of dies) has played an important role in the success of the Japanese lean production methods. Shigeo Shingo is generally credited with its development.

9.11 SUMMARY

Materials requirements planning (MRP) is a set of procedures for converting forecast demand for a manufactured product into a requirements schedule for the components, subassemblies, and raw materials comprising that product. A closely related concept is that of the master production schedule (MPS), which is a specification of the projected needs of the end product by time period. The explosion calculus represents the set of rules and procedures for converting the MPS into the requirements at lower levels. The information required to do the explosion calculus is contained in the product structure diagram and the indented bill-of-materials list. The two key pieces of information contained in the product structure diagram are the production lead times needed to produce the specific component and the multiplier giving the number of units of the component required to produce one item at the next higher level of the product structure.

Many MRP systems are based on a lot-for-lot production schedule. That is, the number of units of a component produced in a period is the same as the requirements for that component in that period. However, if setup and holding costs can be estimated accurately, it is possible to find other lot-sizing rules that are more economical. The optimal lot-sizing procedure is the Wagner–Whitin algorithm. However, its calculations are relatively complex. We explored three heuristic methods that require fewer calculations than the Wagner–Whitin algorithm, although none of these methods will necessarily give an optimal solution. They are the Silver–Meal heuristic, the least unit cost heuristic, and part-period balancing.

We also treated the dynamic lot-sizing problem when capacity constraints exist. One of the limitations of MRP is that capacities are ignored. This is especially important if lot sizing is incorporated into the system. Finding optimal solutions to a capacity-constrained inventory system subject to time-varying demand is an extremely difficult problem. (For those of you familiar with the term, the problem is said to be NP complete, which is a reference to the level of difficulty.) We presented a straightforward heuristic method for obtaining a solution to the capacitated lot-sizing problem. However, incorporating such a method into the MRP system will, in and of itself, *not* solve the complete capacitated MRP problem. Even if a particular lot-sizing schedule is feasible at one level, there is no guarantee that it will result in a feasible requirements schedule at a lower level.

Truly optimal lot-sizing solutions for an MRP system would require formulating the problem as an integer program in order to determine the optimal decisions for all levels simultaneously. For real assembly systems, which can be as many as 10 to 15 levels deep, this would result in an enormous mathematical programming problem. In view of the host of other issues concerned with implementing MRP, the marginal benefit one might achieve from multilevel optimization would probably not justify the effort involved. Today's enterprise systems, which embed MRP within a larger information system, appear to do the job reasonably efficiently; the algorithms they use are typically proprietary.

System nervousness is one problem that arises when implementing an MRP system. The term refers to the unanticipated changes in a schedule that result when the planning horizon is rolled forward by one period. Another difficulty is that in many circumstances, production lead times depend on the lot sizes; MRP assumes that production lead times are fixed. Still another problem is that the yields at various levels of the process may not be perfect. If the yield rates can be accurately estimated in advance, these rates can be factored into the calculations in a straightforward manner. However, in many industries the yield rates may be difficult to estimate in advance.

Manufacturing resource planning (MRP II) attempts to deal with some of the problems of implementing MRP by integrating the financial, accounting, and marketing functions into the production-planning function. Both MRP and MRP II are predecessors of enterprise resource planning (ERP) systems, which are now more commonly called enterprise systems.

In this chapter we also discussed pull scheduling or just-in-time production, which stands in contrast to MRP (a push system). One key goal of pull systems is to reduce work-in-process to a bare minimum. The concepts are based on the production control systems used by Toyota. Parts are transferred from one level to the next only when requested. Pull is often implemented using kanban cards. In addition to reduced inventories, this approach allows workers to quickly locate quality control problems. Because parts are moved through the system in small batches, defects can be identified quickly. By the nature of the system, stopping production at one location automatically stops production on the entire line so that the source of the problem can be identified and corrected before defective inventories build up.

Pull production has several disadvantages vis-à-vis MRP as well as several advantages. Pull systems will react more quickly if a problem develops. However, they are very slow to react to changes in the pattern of demand. MRP, on the other hand, builds forecasts directly into the explosion calculus.

The chapter ended with a discussion of flexible manufacturing systems (FMSs). An FMS is a collection of machines linked by an automated materials handling system, which is generally controlled by a central computer. FMSs are used primarily in the metal-working industries and are an appropriate choice when there is medium to large volume and a moderate variety of part types required. The downside to these systems is cost, speed, and quality. A dedicated system can usually produce the same part faster and with a higher quality and will often cost less than a full flexible system. The upside is, of course, the flexibility they provide. However, this flexibility is only of value when combined with smart production control systems.

ADDITIONAL PROBLEMS FOR PUSH AND PULL PRODUCTION

41. CalcIt produces a line of calculators. One model, IT53, is a solar-powered scientific model with a liquid crystal display (LCD). The calculator is pictured in Figure 9–11.

    Each calculator requires four solar cells, 40 buttons, one LCD display, and one main processor. All parts are ordered from outside suppliers, but final assembly is done by CalcIt. The processors must be in stock three weeks before the anticipated completion date of a batch of calculators to allow enough time to set the processor in the casing, connect the appropriate wiring, and allow the setting paste to dry. The buttons must be in stock two weeks in advance and are set by hand into the calculators. The LCD displays and the solar cells are ordered from the same supplier and need to be in stock one week in advance.

    Based on firm orders that CalcIt has obtained, the master production schedule for IT53 for a 10-week period starting at week 8 is given by

Week	8	9	10	11	12	13	14	15	16	17
MPS	1,200	1,200	800	1,000	1,000	300	2,200	1,400	1,800	600

    Determine the gross requirements schedule for the solar cells, the buttons, the LCD display, and the main processor chips.

42. Consider the example of the CalcIt Company for problem 41. Suppose that the buttons used in the calculators cost \$0.02 each, and the company estimates a fixed cost of \$12 for placing and receiving orders of the buttons from an outside supplier. Assume that holding costs are based on a 24 percent annual interest rate and that there are 48 weeks to a year. Using the gross requirements schedule for the buttons determined in problem 41, what order policy does the Silver–Meal heuristic recommend for the buttons? (Hint: Express h as a cost per 10,000 units and divide each demand by 10,000.)

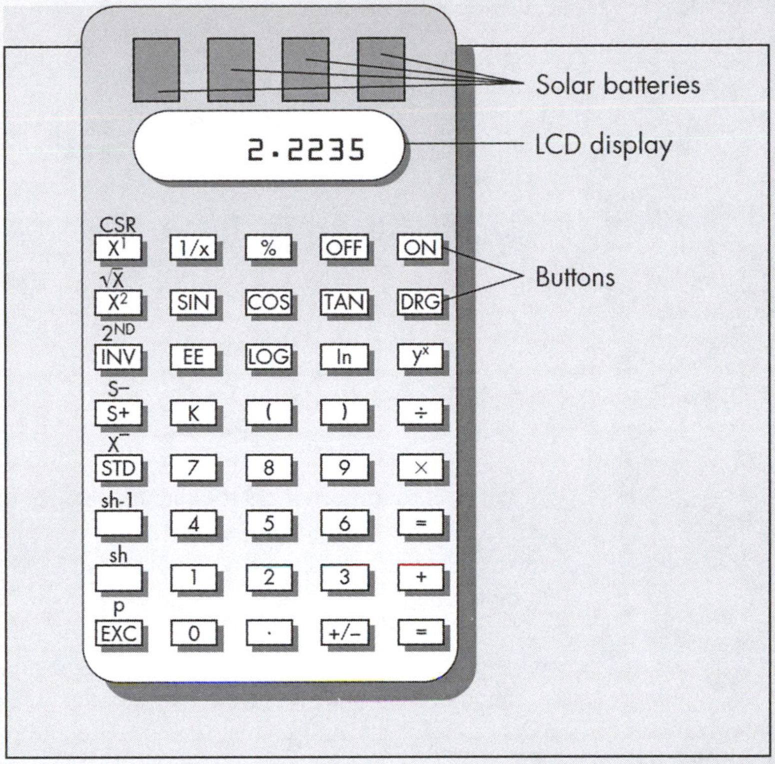

Figure 9–11 CalcIt Model IT53 Scientific Calculator (problem 41)

43. Solve problem 42 using part period balancing and least unit cost. Compare the costs of the resulting solutions to the cost of the solution obtained by using the Silver–Meal heuristic.
44. *Work-in-process* (WIP) *inventory* is a term that refers to the inventory of components and subassemblies in a manufacturing process. Assuming lot-for-lot scheduling at all levels, theoretically the WIP inventory should be zero. Do you think that this is likely to be the case in a real manufacturing environment? Discuss the possible reasons for large WIP inventories occurring even when an MRP system is used.
45. Vivian Lowe is planning a surprise party for her husband's 50th birthday. She has decided to serve shish kabob. The recipe that she is using calls for two pineapple chunks for each shrimp. She plans to size the kabobs so that each has three shrimp. She estimates that a single pineapple will yield about 50 chunks, but from past experience, about 1 out of every 10 pineapples is bad and has to be thrown out. She has invited 200 people and expects that about half will show up. Each person generally eats about 2 kabobs.
    a. How many pineapples should she buy?
    b. Suppose that the number of guests is a random variable having the normal distribution with mean 100 and variance 1,680. If she wants to make enough kabobs to feed all the guests with probability 95 percent, how many pineapples should she buy?
46. In this chapter, we assumed that the "time bucket" was one week. This implies that forecasts are reevaluated and the MRP system rerun on a weekly basis.
    a. Discuss the potential advantages of using a shorter time bucket, such as a day.
    b. Discuss the potential advantages of using a longer time bucket, such as two weeks or a month.

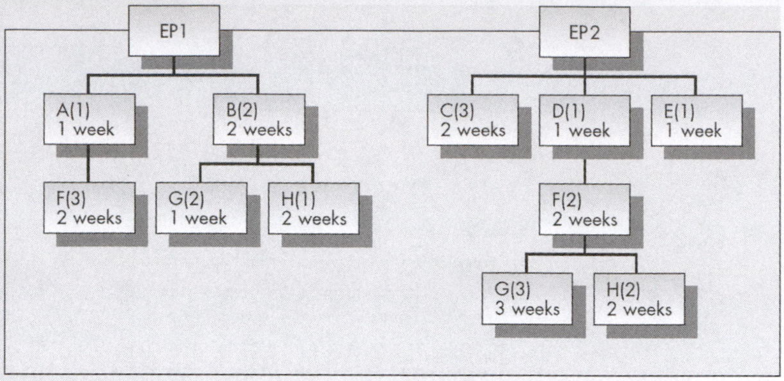

Figure 9–12 Product structure diagrams (for problem 48)

47. Develop a spreadsheet that reproduces the calculations for Example 9.1. As in the example, the spreadsheet should include the net predicted demand for trumpets. The columns should correspond to weeks and should be labeled 1 to 18. Below the net predicted demand for trumpets should be the calculations for the valve casing assembly, and below that the calculations for the valves. For each component, include rows for the following information: (1) gross requirements, (2) scheduled receipts, (3) on-hand inventory, (4) time-phased net requirements, and (5) lot-for-lot planned order release. Your spreadsheet should automatically update all calculations if the net predicted demand for trumpets changes.
48. Two end products, EP1 and EP2, are produced in the Raleigh, North Carolina, plant of a large manufacturer of furniture products located in the Southeast. The product structure diagrams for these products appear in Figure 9–12.

    Suppose that the master production schedules for these two products are as follows.

Week	18	19	20	21	22	23	24
EP1	120	112	76	22	56	90	210
EP2	62	68	90	77	26	30	54

    Assuming lot-for-lot production, determine the planned order releases for components F, G, and H.
49. A component used in a manufacturing facility is ordered from an outside supplier. Because the component is used in a variety of end products, the demand is high. Estimated demand (in thousands) over the next 10 weeks is listed below.

Week	1	2	3	4	5	6	7	8	9	10
Demand	22	34	32	12	8	44	54	16	76	30

    The components cost 65 cents each and the interest rate used to compute the holding cost is 0.5 percent per week. The fixed order cost is estimated to be $200. (Hint: Express h as the holding cost per thousand units.)
    a. What ordering policy is recommended by the Silver–Meal heuristic?
    b. What ordering policy is recommended by the part-period balancing heuristic?
    c. What ordering policy is recommended by the least unit cost heuristic?
    d. Which method resulted in the lowest-cost policy for this problem?

50. A popular heuristic lot-sizing method is known as period order quantity (POQ). The method requires determining the average number of periods spanned by the EOQ and choosing the lot size to equal this fixed period supply. Let A be the total demand over an N-period planning horizon $[\lambda = (\Sigma r_i/n)]$ and assume that EOQ is computed as described in Section 9.4. Then $P = \text{EOQ}/\lambda$ rounded to the nearest integer. For the example in Section 9.4, $P = 139/43.9 = 3.17$, which is rounded to 3. The POQ would call for setting the lot size equal to three periods of demand. For the example in Section 9.4, the resulting planned order release would be 116, 0, 0, 150, 0, 0, 135, 0, 0, 38.
   a. Compare the cost of the policy obtained by this method for the example in Section 9.4 to that obtained using the EOQ.
   b. What are the advantages of this approach over EOQ?
   c. Do you think that this method will be more cost effective in general than the heuristic methods discussed in Section 9.4?
   d. Solve problem 17 using this method and compare the total holding and setup cost with that obtained by the other methods.
   e. Solve problem 49 using this method and compare the total holding and setup cost with that obtained by the other methods.
51. The campus store at a large Midwestern university sells legal pads to students, faculty, and staff. They sell more pads near exam time. During a typical 10-week quarter, the pattern of sales is 2,280, 1,120, 360, 3,340, 1,230, 860, 675, 1,350, 4,600, 1,210. The pads cost the store $1.20 each, and holding costs are based on a 30 percent annual interest rate. The cost of employee time, paperwork, and handling amounts to $30 per order. Assume 50 weeks per year.
   a. What is the optimal order policy during the 10-week quarter based on the Silver–Meal heuristic? Using this policy, what are the total holding and ordering costs incurred over the 10-week period?
   b. The bookstore manager has decided that it would be more economical if the demand were the same each week. In order to even out the demand, he limits weekly sales (to the annoyance of his clientele). Hence, assume that the total demand for the 10-week quarter is still the same, but the sales are constant from week to week. Determine the optimal order policy in this case and compare the total holding and ordering cost over the 10 weeks to the answer you obtained in part (a). (You may assume continuous time for the purposes of your calculations, so that the optimal lot size is the EOQ.)
   c. Based on the results of parts (a) and (b), do you think that it is more economical in general to face a smooth or a spiky demand pattern?
52. Develop a spreadsheet template for finding Silver–Meal solutions for general lot-sizing problems. Store the holding and setup cost parameters in separate cell locations so that they can be inputted and changed easily. Allow for 30 periods of demand to be inputted in column 2 of the spreadsheet. List the period numbers 1, 2, … , 30 in column 1. Work out the logic that gives $C(j)$ in column 3.

   One would use such a spreadsheet in the following way. Input requirements period by period until an increase is observed in column 3.This identifies the first forecast horizon. Now replace entries in column 2 with zeros and input requirements starting at the current period. Continue until the next forecast horizon is identified. Repeat this process until the end of the planning horizon. Use this method to find the Silver–Meal solution for the following production planning problems.

a. Solve problem 14 in this manner.
b. Weekly demands for 2-inch rivets at a division of an aircraft company are predicted to be (in gross).

Week	1	2	3	4	5	6	7	8	9	10	11	12
Demand	240	280	370	880	950	120	135	450	875	500	400	200
Week	13	14	15	16	17	18	19	20	21	22	23	24
Demand	600	650	1,250	250	800	700	750	200	100	900	400	700

Setup costs for ordering the rivets are estimated to be \$200, and holding costs amount to 10 cents per gross per week. Find the lot sizing given by the Silver–Meal method.

53. Along the lines described in problem 52, construct a spreadsheet for finding the least unit cost lot-sizing rule.
   a. Solve both parts (a) and (b) of problem 52.
   b. Which method, least unit cost or Silver–Meal, gave the more cost-effective solution?

APPENDIX 9–A

LOT SIZING FOR TIME-VARYING DEMAND

The Silver–Meal heuristic considered in Section 9.4 is easy to use and gives lot sizes with costs that are generally near the true optimal. However, there are another two heuristics that are popular, namely the **least unit cost** (**LUC**) heuristic and the **part-period balancing** heuristic. In addition, it is possible to compute true optimal lot sizes. Optimal in this context means the policy that minimizes the total holding and setup cost over the planning horizon. This appendix outlines the two aforementioned heuristics and shows how optimal policies can be determined by casting the problem as a shortest-path problem. It also shows how dynamic programming can be used to find the shortest path. Finally, it gives an improvement heuristic for use in lot sizing with capacity constraints.

Least Unit Cost

The least unit cost (LUC) heuristic is similar to the Silver–Meal method except that instead of dividing the cost over j periods by the number of periods, j, we divide it by the total number of units demanded through period j, $r_1 + r_2 + \ldots + r_j$. We choose the order horizon that minimizes the cost per unit of demand rather than the cost per period.

Define $C(T)$ as the average holding and setup cost per unit for a T-period order horizon. Then,

$$C(1) = K/r_1,$$
$$C(2) = (K + hr_2)/(r_1 + r_2),$$
$$C(j) = [K + hr_2 + 2hr_3 + \ldots + (j - 1)\, hr_j] \,/\, (r_1 + r_2 + \ldots + r_j).$$

As with the Silver–Meal heuristic, this computation is stopped when $C(j) > C(j-1)$, and the production level is set equal to $r_1 + r_2 + \ldots + r_{j-1}$. The process is then repeated, starting at period j and continuing until the end of the planning horizon is reached.

EXAMPLE 9.8

Assume the same requirements schedule and costs as given in Example 9.2. Starting in period 1:

$$C(1) = 80/18 = 4.44,$$
$$C(2) = [80 + (2)(30)] / (18 + 30) = 2.92,$$
$$C(3) = [80 + (2)(30) + (2)(2)(42)] / (18 + 30 + 42) = 3.42.$$

Because $C(3) > C(2)$, we stop and set $y_1 = r_1 + r_2 = 48$.
Starting in period 3:

$$C(1) = 80 / 42 = 1.90,$$
$$C(2) = [80 + (2)(5)] / (42 + 5) = 1.92.$$

Because $C(2) > C(1)$, stop and set $y_3 = r_3 = 42$.
Starting in period 4:

$$C(1) = 80 / 5 = 16,$$
$$C(2) = [80 + (2)(20)] / (5 + 20) = 4.8.$$

As we have reached the end of the horizon, we set $y_4 = r_4 + r_5 = 5 + 20 = 25$. The solution obtained by the LUC heuristic is $\mathbf{y} = (48, 0, 42, 25, 0)$. It is interesting to note that the policy obtained by this method is different from that for the Silver–Meal heuristic. It turns out that the Silver–Meal method gives the optimal policy, with cost \$310, whereas the LUC gives a suboptimal policy, with cost \$340.

Part-Period Balancing

Another approximate method for solving the lot-sizing problem is part-period balancing. Although the Silver–Meal technique seems to give better results in a greater number of cases, part-period balancing seems to be more popular in practice.

The method is to set the order horizon equal to the number of periods that most closely matches the total holding cost with the setup cost over that period. The order horizon that exactly equates holding and setup costs will rarely be an integer number of periods (hence the origin of the name of the method).

EXAMPLE 9.9

Again consider Example 9.2. Starting in period 1, we find

Order Horizon	Total Holding Cost
1	0
2	60
3	228

Because 228 exceeds the setup cost of 80, we stop. As 80 is closer to 60 than to 228, the first order horizon is two periods. That is, $y_1 = r_1 + r_2 = 18 + 30 = 48$.

We start the process again in period 3.

Order Horizon	Total Holding Cost
1	0
2	10
3	90

We have exceeded the setup costs of 80, so we stop. Because 90 is closer to 80 than is 10, the order horizon is three periods. Hence $y_3 = r_3 + r_4 + r_5 = 67$. The complete part period balancing solution is $\mathbf{y} = (48, 0, 67, 0, 0)$, which is different from both the Silver–Meal and LUC solutions. This solution is optimal, as it also has a total cost of $310.

The Wagner–Whitin Algorithm

Assume that:

1. Forecasted demands over the next n periods are known and given by the vector $\mathbf{r} = (r_1, \ldots, r_n)$.
2. Costs are charged against holding at $\$h$ per unit per period and $\$K$ per setup. We will assume that the holding cost is charged against ending inventory each period.

In order to get some idea of the potential difficulty of this problem, consider the following simple example.

EXAMPLE 9.10

The forecast demand for an electronic assembly produced at Hi-Tech, a local semiconductor fabrication shop, over the next four weeks is 52, 87, 23, 56. There is only one setup each week for production of these assemblies, and there is no back-ordering of excess demand. Assume that the shop has the capacity to produce any number of the assemblies in a week.

Consider the total number of feasible production policies for Hi-Tech over the four-week period. Let $y_1, \ldots, y_4$ be the order quantities in each of the four weeks. Clearly $y_1 \geq 52$ in order to guarantee that we do not stock out in period 1. If we assume that ending inventory in period 4 is zero (which will be easy to show is optimal), then $y_1 \leq 218$, the sum of all the demands. Hence y_1 can take any one of 167 possible values. Consider y_2. The number of feasible values of y_2 depends upon the value of y_1. As no stock-out is permitted to occur in period 2, $y_1 + y_2 \geq 52 + 87 = 139$. If

$$y_1 \leq 139, \qquad \text{then } 139 - y_1 \leq y_2 \leq 218 - y_1,$$

and if

$$y_1 > 139, \qquad \text{then } 0 \leq y_2 \leq 218 - y_1.$$

With a little effort, one can show that this results in a total of 10,200 different values of just the pair (y_1, y_2). It is thus clear that for even moderately sized problems the number of feasible solutions is enormous.

Searching all the feasible policies is unreasonable. However, an important discovery by Wagner and Whitin reduces considerably the number of policies one must consider as candidates for optimality.

The Wagner–Whitin algorithm is based on the following observation.

Result. An optimal policy has the property that each value of **y** is exactly the sum of a set of future demands. (We will call this an exact requirements policy.) That is,

$$
\begin{array}{lll}
y_1 = r_1, & \text{or } y_1 = r_1 + r_2, \ldots, & \text{or } y_1 = r_1 + r_2 + \ldots + r_n, \\
y_2 = 0, & \text{or } y_2 = r_2, & \text{or } y_2 = r_2 + r_3, \ldots, \\
 & \text{or } y_2 = r_2 + r_3 + \ldots + r_n, & \\
y_n = 0, & \text{or } y_n = r_n. &
\end{array}
$$

An exact requirements policy is completely specified by designating the periods in which ordering is to take place. The number of exact requirements policies is much smaller than the total number of feasible policies.

EXAMPLE 9.10 (CONTINUED)

We continue with the four-period scheduling problem. Because y_1 must satisfy exact requirements, we see that it can assume values of 52, 139, 162, or 218 only—that is, only four distinct values. Ignoring the value of y_1, y_2 can assume values 0, 87, 110, 166. It is easy to see that every exact requirements policy is completely determined by specifying in what periods ordering should take place. That is, each such policy is the form $(i_1, \ldots, i_n)$, where the values of i_j are either 0 or 1. If $i_j = 1$, then production takes place in period j. Note that $i_1 = 1$ because we must produce in period 1 to avoid stocking out, whereas $i_2, \ldots, i_n$ will each be either 0 or 1. For example, the policy (1, 0, 1, 0) means that production occurs in periods 1 and 3 only. It follows that $\mathbf{y} = (139, 0, 79, 0)$. For this example, there are exactly $2^3 = 8$ distinct exact requirements policies.

A convenient way to look at the problem is as a one-way network with the number of nodes equal to exactly one more than the number of periods. Every path through the network corresponds to a specific exact requirements policy. The network for the four-period problem appears in Figure 9–13.

For any pair (i, j) with $i < j$, if the arc (i, j) is on the path, it means that ordering takes place in period i and the order size is equal to the sum of the requirements in periods $i, i + 1, \ldots, j - 1$. Period j is the next period of ordering. Note that all paths end at period $n + 1$. The policy of ordering in periods 1 and 3 only would correspond to the path 1–3–5. The path 1–2–4–5 means that ordering is to take place in periods 1, 2, and 4.

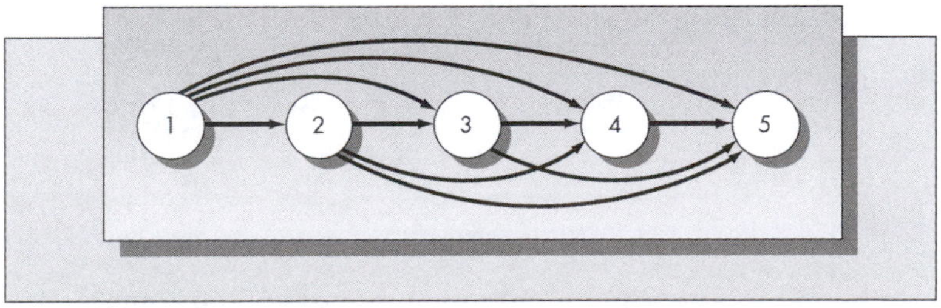

Figure 9–13 Network representation for lot scheduling (Example 9.10)

The next step is to assign a value to each arc in the network. The value or "length" of the arc (i, j), called c_{ij}, is defined as the setup and holding cost of ordering in period i to meet requirements through period $j - 1$. For example, c_{15} = the cost of ordering in period 1 to satisfy the demands in periods 1 through 4.

Finally, we would like to determine the minimum-cost production schedule, or shortest path through the network. As we will see, **dynamic programming** is one method of solving this problem. However, for a small problem, the optimal policy can be found by simply enumerating the paths through the network and choosing one with minimum cost.

EXAMPLE 9.11

We will solve Example 9.10 using path enumeration. Recall that $\mathbf{r} = (52, 87, 23, 56)$. In addition, assume that there is a cost of holding of $h = \$1$ per unit per period and a cost of $K = \$75$ per setup.

The first step is to compute c_{ij} for $1 \le i \le 4$ and $i + 1 \le j \le 5$.

$$c_{12} = 75 \text{ (setup cost only)},$$
$$c_{13} = 75 + 87 = 162,$$
$$c_{14} = 75 + (23 \times 2) + 87 = 208,$$
$$c_{15} = 75 + (56 \times 3) + (23 \times 2) + 87 = 376,$$
$$c_{23} = 75,$$
$$c_{24} = 75 + 23 = 98,$$
$$c_{25} = 75 + 23 + (56 \times 2) = 210,$$
$$c_{34} = 75,$$
$$c_{35} = 75 + 56 = 131,$$
$$c_{45} = 75.$$

Summarizing these costs in matrix form gives the following.

i	*j* 1	2	3	4	5
1		75	162	208	376
2			75	98	210
3				75	131
4					75

As there are only eight exact requirements policies, we can solve this problem by enumerating the policies and comparing the costs.

Path	Cost
1–2–3–4–5	$300
1–2–4–5	248
1–2–5	285
1–2–3–5	281
1–3–4–5	312
1–3–5	293
1–4–5	283
1–5	376

It follows that the optimal path is 1–2–4–5 at a cost of \$248. This corresponds to ordering in periods 1, 2, and 4 only. The optimal ordering policy is $y_1 = 52$, $y_2 = 110$, $y_3 = 0$, $y_4 = 56$.

Solution by Dynamic Programming

The total number of exact requirements policies for a problem of n periods is 2^{n-1}. As n gets large, total enumeration is not efficient. Dynamic programming is a recursive solution technique that can significantly reduce the number of computations required, although it too can be quite cumbersome.

Dynamic programming is based on the **principle of optimality**. One version of this principle is that if a problem consists of exactly n stages and there are $r < n$ stages remaining, the optimal policy for the remaining stages is independent of policy adopted in the previous stages. Because dynamic programming is not used anywhere else in this text, we will not present a detailed discussion of it (for a brief overview, see Hillier and Lieberman, 2015; for a more in-depth treatment at a mathematical level consistent with ours, see Nemhauser, 1966).

Define f_k as the minimum cost starting at node k, assuming that an order is placed in period k. The principle of optimality for this problem results in the following system of equations:

$$f_k = \min_{j>k}(c_{kj} + f_j) \qquad \text{for } k = 1, \ldots, n.$$

The initial condition is $f_{n+1} = 0$.

EXAMPLE 9.12

We will solve Example 9.10 by dynamic programming in order to illustrate the technique. One starts with the initial condition and works backward from period $n + 1$ to period 1. [The original Wagner–Whitin algorithm is based on a forward dynamic programming formulation; although the forward formulation has some advantages for planning horizon analysis, we feel that backward recursion is more natural and more intuitive.] In each period one determines the value of j that achieves the minimum.

$$f_5 = 0.$$

$$f_4 = \min_{j>4}(c_{4j} + f_j) = 75 \text{ at } j = 5 \text{ (the only possible value of } j).$$

$$f_3 = \min_{j>3}(c_{3j} + f_j) = \min\begin{Bmatrix} c_{34} + f_4 \\ c_{35} + f_5 \end{Bmatrix} = \min\begin{Bmatrix} 75 + 75 \\ 131 + 0 \end{Bmatrix} = \min\begin{Bmatrix} 150 \\ 131 \end{Bmatrix} = 131 \text{ at } j = 5.$$

$$f_2 = \min_{j>2}(c_{2j} + f_j) = \min\begin{Bmatrix} c_{23} + f_3 \\ c_{24} + f_4 \\ c_{25} + f_5 \end{Bmatrix} = \min\begin{Bmatrix} 75 + 131 \\ 98 + 75 \\ 210 + 0 \end{Bmatrix} = \min\begin{Bmatrix} 206 \\ 173 \\ 210 \end{Bmatrix} = 173 \text{ at } j = 4.$$

Finally,

$$f_1 = \min_{j>1}(c_{1j} + f_j) = \min\begin{Bmatrix} c_{12} + f_2 \\ c_{13} + f_3 \\ c_{14} + f_4 \\ c_{15} + f_5 \end{Bmatrix} = \min\begin{Bmatrix} 75 + 173 \\ 162 + 131 \\ 208 + 75 \\ 376 + 0 \end{Bmatrix} = \min\begin{Bmatrix} 248 \\ 293 \\ 283 \\ 376 \end{Bmatrix} = 248 \text{ at } j = 2.$$

To determine the optimal order policy, we retrace the solution back from the beginning. In period 1 the optimal value of j is $j = 2$. This means that the production level in period 1 is equal to the demand in period 1, so that $y_1 = r_1 = 52$. The next order period is period 2. The optimal value of j in period 2 is $j = 4$, which implies that the production quantity in period 2 is equal to the sum of the demands in periods 2 and 3, or $y_2 = r_2 + r_3 = 110$. The next period of ordering is period 4. The optimal value of j in period 4 is $j = 5$. This gives $y_4 = r_4 = 56$. Hence, the optimal order policy is $\mathbf{y} = (52, 110, 0, 56)$.

The Improvement Step for Lot Sizing with Capacity Constraints

Section 9.5 considered lot sizing with capacity constraints. While the resulting schedule was feasible, it is easy to improve as follows.

For each lot that is scheduled, starting from the last and working backward to the beginning, determine whether it is cheaper to produce the units composing that lot by shifting production to prior periods of excess capacity. By eliminating a lot, one reduces setup cost in that period to zero, but shifting production to prior periods increases the holding cost. The shift is made only if the additional holding cost is less than the setup cost. We illustrate the process with an example.

EXAMPLE 9.13

Assume that $K = \$450$ and $h = \$2$.

$$\mathbf{r} = (100, 79, 230, 105, 3, 10, 99, 126, 40),$$

$$\mathbf{c} = (120, 200, 200, 400, 300, 50, 120, 50, 30).$$

Computing the cumulative sum of requirements and capacities for each period makes it easy to see the problem as feasible. However, because requirements exceed capacities in some periods, lot for lot is not feasible. We back-shift excess demand to prior periods in order to obtain the modified requirements schedule $\mathbf{r}' = (100, 109, 200, 105, 28, 50, 120, 50, 30)$.

Lot for lot for the modified requirements schedule $\mathbf{r}'$ is feasible for the original problem. The initial feasible solution requires nine setups at a total setup cost of $9 \times 450 = \$4,050$. The holding cost of the initial policy is $2(0 + 30 + 0 + 0 + 25 + 65 + 86 + 10) = \$432$.

In order to do the improvement step, it is convenient to arrange the data in a table.

	1	2	3	4	5	6	7	8	9
$\mathbf{r}'$	100	109	200	105	28	50	120	50	30
$\mathbf{c}$	120	200	200	400	300	50	120	50	30
$\mathbf{y}$	100	109	200	105	28	50	120	50	30
Excess capacity	20	91	0	295	272	0	0	0	0

Starting from the last period, consider the final lot of 30 units. There is enough excess capacity in prior periods to consider shifting this lot. The latest period that this lot can be scheduled for is period 5. The extra holding cost incurred by making the shift is $2 \times 30 \times 4 = \$240$. As this is cheaper than the setup cost of \$450, we make the shift and increase y_5 from 28 to 58 and reduce the excess capacity in period 5 from 272 to 242.

Now consider the lot of 50 units scheduled in period 8. This lot can also be shifted to period 5 with a resulting additional holding cost of $2 \times 50 \times 3 = \$300$. Again this is cheaper than the setup cost, so we make the shift. At this point we have $y_5 = 108$, and the excess capacity in period 5 is reduced to 192.

The calculations are summarized on our table in the following way.

	1	**2**	**3**	**4**	**5**	**6**	**7**	**8**	**9**
r′	100	109	200	105	28	50	120	50	30
c	120	200	200	400	300	50	120	50	30
					108				
					~~58~~			0	0
y	100	109	200	105	~~28~~	50	120	~~50~~	~~30~~
					192				
					~~242~~				
Excess capacity	20	91	0	295	~~272~~	0	0	0	0

Next consider the lot of 120 units scheduled in period 7. At this point, we still have 192 units of excess capacity in period 5. The additional holding cost of shifting the lot of 120 from period 7 to period 5 is $2 \times 120 \times 2 = \$480$. This exceeds the setup cost of \$450, so we do not make the shift.

It is clearly advantageous to shift the lot of 50 units in period 6 to period 5, thus reducing the excess capacity in period 5 to 142 and increasing the lot size in period 5 from 108 to 158. Doing so results in the following.

	1	**2**	**3**	**4**	**5**	**6**	**7**	**8**	**9**
r′	100	109	200	105	28	50	120	50	30
c	120	200	200	400	300	50	120	50	30
					158				
					~~108~~				
					~~58~~			0	0
y	100	109	200	105	~~28~~	0	120	~~50~~	~~30~~
					142				
					~~192~~				
					~~242~~				
Excess capacity	20	91	0	295	~~272~~	0	0	0	0

At this point it may seem that we are done. However, there is enough capacity in period 4 to shift the entire lot of 158 units from period 5 to period 4. The additional holding cost of this shift is $2 \times 158 = \$316$. Because this is cheaper than the setup cost, we make the shift.

Summarizing these calculations on the table gives

	1	2	3	4	5	6	7	8	9
r′	100	109	200	105	28	50	120	50	30
c	120	200	200	400	300	50	120	50	30
					0				
					~~158~~				
					~~108~~				
				263	~~58~~	0		0	0
y	100	109	200	~~105~~	~~28~~	~~50~~	120	~~50~~	~~30~~
					300				
					142				
					192				
				137	242	50		50	30
Excess capacity	20	91	0	~~295~~	~~272~~	~~0~~	0	~~0~~	~~0~~

At this point, no additional lot shifting is possible. The solution we have obtained is

$$\mathbf{y} = (100, 109, 200, 263, 0, 0, 120, 0, 0)$$

for the original requirements schedule

$$\mathbf{r} = (100, 79, 230, 105, 3, 10, 99, 126, 40).$$

We will compute the cost of this solution and compare it with that of our initial feasible solution. There are five setups at a total setup cost of $5 \times 450 = \$2,250$. The holding cost is $2(0 + 30 + 0 + 158 + 155 + 145 + 166 + 40 + 0) = 2 \times 694 = \$1,388$. The total cost of this policy is $3,638, compared with $4,482 for our initial feasible policy. For this example the improvement step resulted in a cost reduction of close to 20 percent.

APPENDIX 9–B

Glossary of Notation for Chapter 9

$C(T)$ = Average holding and setup cost per period (for Silver–Meal heuristic) or per unit (LUC heuristic) if the current order spans T periods.

c_i = Production capacity in period i.

c_{ij} = Cost associated with arc (i, j) in network representation of lot scheduling problem used for Wagner–Whitin algorithm.

f_j = Minimum cost from period i to the end of the horizon (refer to the dynamic programming algorithm for Wagner–Whitin).

h = Holding cost per unit per time period.

K = Setup cost for initiating an order.

r_i = Requirement for period i.

y_i = Production lot size in period i.

BIBLIOGRAPHY

Anderson, J. C., R. G. Schroeder, S. E. Tupy, and E. M. White. "Material Requirements Planning Systems: The State of the Art." *Production and Inventory Management* 23 (1982), pp. 51–66.

Baker, K. R. "Lot-sizing Procedures and a Standard Data Set: A Reconciliation of the Literature." *Journal of Manufacturing and Operations Management*, 2 (1989), pp. 199-221.

Carlson, R. C., S. L. Beckman, and D. H. Kropp. "The Effectiveness of Extending the Horizon in Rolling Production Scheduling." *Decision Sciences* 13 (1982), pp. 129–46.

Carlson, R. C., J. V. Jucker, and D. H. Kropp. "Less Nervous MRP Systems: A Dynamic Economic Lot-Sizing Approach." *Management Science* 25 (1979), pp. 754–61.

Carlson, R. C., J. V. Jucker, and D. H. Kropp. "Heuristic Lot Sizing Approaches for Dealing with MRP System Nervousness." *Decision Sciences* 14 (1983), pp. 156–69.

Chhikara, J., and E. N. Weiss. "JIT Savings—Myth or Reality?" *Business Horizons* 38 (May–June 1995), pp. 73–78.

Davenport, T. H. "Putting the Enterprise into the Enterprise System." *Harvard Business Review,* July (1998).

DeMatteis, J. J. "An Economic Lot Sizing Technique: The Part-Period Algorithm." *IBM Systems Journal* 7 (1968), pp. 30–38.

Dixon, P. S., and E. A. Silver. "A Heuristic Solution Procedure for the Multi-Item, Single-Level, Limited Capacity Lot Sizing Problem." *Journal of Operations Management* 2 (1981), pp. 23–39.

Freeland, J. R. "A Survey of Just-in-Time Purchasing Practices in the United States." *Production and Inventory Management Journal*, Second Quarter 1991, pp. 43–50.

Gartner. "Gartner Says Worldwide Supply Chain Management Software Revenue Grew 13.9 Percent in 2017." Stamford, CT: Author, July 18, 2018, https://www.gartner.com/en/newsroom/press-releases/2018-07-18-gartner-says-worldwide-supply-chain-management-software-revenue-grew-13-percent-in-2017

Goldratt, E. M., and J. Cox. *The Goal: A Process of Ongoing Improvement*. 30th Anniversary Edition (2014).

Gorham, T. "Dynamic Order Quantities." *Production and Inventory Management* 9 (1968), pp. 75–81.

Hillier, F. S., and G. J. Lieberman. *Introduction to Operations Research*, 10th ed. McGraw Hill, 2015.

Hopp, W., and M. Spearman. *Factory Physics*. Long Grove IL: Waveland Press, 1996.

Inderfurth, K. "How to Protect Against Demand and Yield Risks In MRP Systems." *International Journal of Production Economics*, 121 (2009), pp.474-481.

Kaplan, R. S. "Must CIM Be Justified by Faith Alone?" *Harvard Business Review* 64 (1986), pp. 69–76.

Karmarkar, U. "Getting Control of Just-In-Time." *Harvard Business Review* 67 (September–October 1989), pp. 122–31.

Karni, R. "Maximum Part Period Gain (MPG)—A Lot Sizing Procedure for Unconstrained and Constrained Requirements Planning Systems." *Production and Inventory Management* 22 (1981), pp. 91–98.

Koch, C. "Nike Rebounds: How (and Why) Nike Recovered from Its Supply Chain Disaster." CIO, June (2004)

Krajewski, L. J., B. E. King, L. P. Ritzman, and D. S. Wong. "Kanban, MRP, and Shaping the Manufacturing Environment." *Management Science* 33 (1987), pp. 39–57.

Mabert, V. A., A. Soni, and M. A. Venkataramanan. "Enterprise Resource Planning: Common Myths Versus Evolving Reality." *Business Horizons*, 44 (3), May (2001), pp. 69-76.

Meal, H. "Putting Production Decisions Where They Belong." *Harvard Business Review* 62 (1984), pp. 102–11.

Monden, Y. "Adaptable Kanban System Helps Toyota Maintain Just-in-Time Production." *Industrial Engineering* 13, no. 5 (1981), pp. 28–46.

Nahmias, S., and K. Moinzadeh. "Lot Sizing with Randomly Graded Yields." *Operations Research* 46, no. 6 (1997), pp. 974–86.

Nemhauser, G. L. *Introduction to Dynamic Programming.* New York: John Wiley & Sons, 1966.

New, C. *Requirements Planning.* Essex, England: Gower Press, 1974.

Nicholas, J., and A. Soni. *The Portal to Lean Production.* Boca Raton, Auerbach Publications, 2006.

Orlicky, J. *Materials Requirements Planning.* New York: McGraw-Hill, 1975.

Plossl, G., and O. Wight. *Materials Requirements Planning by Computer.* Washington, DC: American Production and Inventory Control Society, 1971.

Schweitzer, P. J. "Maximum Throughput in Finite Capacity Open Queueing Networks with Product-Form Solutions." *Management Science* 24 (1977), pp. 217–23.

Shingo, S. *Study of "Toyota" Production System from Industrial Engineering Viewpoint.* Tokyo: Japan Management Association, 1981.

Silver, E. A., and H. C. Meal. "A Heuristic for Selecting Lot Size Quantities for the Case of a Deterministic Time-Varying Demand Rate and Discrete Opportunities for Replenishment." *Production and Inventory Management* 14 (1973), pp. 64–74.

Silver, E. A., D. F. Pyke, and D. J. Thomas. *Inventory and Production Management in Supply Chains.* 4th ed. CRC Press, Taylor and Francis Group, FL, 2017.

Sloan, A. *My Years with General Motors.* Garden City, NY: Doubleday, 1964.

Spearman, M. L., W. J. Hopp, and D. L. Woodruff "A Hierarchical Control Architecture for Constant Work-in-Process (CONWIP) Production Systems." *Journal of Manufacturing and Operations Management* 2 (1989), pp. 147-171.

Spearman, M. L., D. L. Woodruff, and W. J. Hopp. "CONWIP: a pull alternative to kanban." *International Journal of Production Research* 28 (1990), pp. 879-894.

Stasey, R., and C. J. McNair. *Crossroads: A JIT Success Story.* New York: McGraw-Hill/Irwin, 1990.

Steele, D. C. "The Nervous MRP System: How to Do Battle." *Production and Inventory Management* 16 (1973), pp. 83–89.

Suri, R. *The Practitioner's Guide to POLCA - The Production Control System for High-mix, Low-volume and Custom Products.* Productivity Press (2018).

Wagner, H. M., and T. M. Whitin. "Dynamic Version of the Economic Lot Size Model." *Management Science* 5 (1958), pp. 89–96.

Yano, C. A., and H. L. Lee. "Lot Sizing with Random Yields: A Review." *Operations Research* 43 (Mar. - Apr. 1995), pp. 311-334.

Zipkin, P. "Does Manufacturing Need a JIT Revolution?" *Harvard Business Review* 69 (January–February 1991), pp. 40–50.

CHAPTER

10

Quality and Assurance

"Quality in a service or product is not what you put into it. It is what the client or customer gets out of it."

—Peter Drucker

CHAPTER OVERVIEW

Purpose

To understand what quality means in the operations context, how it can be measured, and how it can be improved.

KEY POINTS

1. *What is quality?* While we all have a sense of what we mean by quality, defining it precisely as a measurable quantity is not easy. A useful definition is conformance to specifications. This is something that can be measured and quantified. If it can be quantified, it can be improved. However, this definition falls short of capturing all the aspects of what we mean by quality and how it is perceived by the customer.
2. *Quality movements.* Several agencies worldwide promote quality in their respective countries through formal recognition. This process was started in Japan with the Deming Prize, established and funded by quality guru W. Edwards Deming. The United States recognizes outstanding quality with the Malcolm Baldrige National Quality Award. The International Standards Organization's certification ISO 9000 requires firms to clearly document their policies and procedures. While the certification process can be costly in both time and money, it is often required to do business in many countries. Lean and Six Sigma are two other important quality-related movements.
3. *Statistical process control and the $\overline{X}$ chart.* Statistical methods can assist with the task of monitoring quality in the context of manufacturing. The underlying basis of statistical control charts is the normal distribution. The normal distribution (bell-shaped distribution) has the property that the mean plus and minus two standard deviations ($\mu \pm 2\sigma$) contains about 95 percent of the population, and the mean plus and minus three standard deviations ($\mu \pm 3\sigma$) contains more than 99 per cent of the population. It is these properties that form the basis for statistical control charts. Consider a manufacturing process producing an item with a measurable quantity that must conform to a given specification. One averages the measurements of this quantity in subgroups (typically of size four or five). The central limit theorem guarantees that the distribution of the average measurement will be approximately normally distributed. If the average of a subgroup lies outside two or three sigma limits of the normal distribution, it is unlikely that this deviation is due to chance.

This signals an out-of-control situation, which might require intervention into the process. This is the basis for the $\bar{X}$ chart.

4. *The R chart.* While the $\bar{X}$ chart is a valuable way to test for a shift in the underlying mean of a process, it does not signal shifts in the process variation. To monitor process variation, one computes the range of subgroup measurements (that is, the largest value minus the smallest value in the subgroup). Since the range of a sample is proportional to the standard deviation of a sample, this statistic can be used to monitor process variation. This is the purpose of the *R* chart. The *R* chart establishes upper and lower control limits on the average range of subgroups and signals when the process variation has gone out of control.
5. *The* p *and* c *charts.* The $\bar{X}$ and *R* charts are useful when measuring quality along a single scalar dimension such as length or weight. In other cases, one might be interested in whether the item functions or not. Under these circumstances, the *p* chart is appropriate. The *p* chart is based on the binomial distribution. Either an item has the appropriate attribute or it doesn't. When the observed value of *p* (the proportion of good items) undergoes a sudden shift, it signals a possible out-of-control situation.

   The *c* chart is based on the Poisson distribution. The Poisson distribution describes events that occur completely at random over time or space. In the statistical quality control context, consider a situation where a certain number of defects are acceptable, such as minor dents on an automobile, but too many are considered unacceptable. In this case, the *c* chart would be an appropriate means of monitoring the process. The parameter *c* is the average rate of occurrence of flaws, and an out-of-control signal is tripped when the observed value of *c* is too high. Note that both the *p* and *c* charts are typically implemented with a normal distribution, since, under the right circumstances, the normal distribution provides a good approximation to both the binomial and Poisson distributions.
6. *Process capability.* A process that is in control is reliably producing within a given range (typically the three-sigma control limits). However, this does not determine whether the process is capable of reliably meeting the required specifications. In order to determine if a process is capable, the specification limits must be compared with the control limits. If the control limits fall inside the specification limits, then the process is said to be capable. The capability ratio compares the spread of the specification limits with the variability of the process. The capability index extends the capability ratio to also account for processes that are not centered between the specification limits.
7. *Acceptance sampling.* Acceptance sampling occurs after a lot of items is produced, rather than during the manufacturing process. It can be performed by the manufacturer or by the consumer. In many cases, 100 percent inspection of items is impractical, impossible, or too costly. For these reasons, a more common approach is to sample a subset of the lot and choose to accept or reject the lot based on the results of the sampling. The most common sampling plans are (1) single sampling, (2) double sampling, and (3) sequential sampling. In the case of single sampling, one samples *n* items from a lot of *N* items (where $n < N$) and rejects the lot if the number of defects exceeds a specified level.
8. *Design for Quality.* Listening to the customer is important when designing quality products. This process includes customer surveys and focus groups to find out what the customer wants, distilling this information, prioritizing customer needs, and linking those needs to the design of the product. One means of accomplishing the last item on the list is quality function deployment (QFD). By putting a greater investment up front in sound product design, the consumer will be rewarded with superior products, and the firm will be rewarded with customer loyalty.

What is quality? Consider the following hypothetical remarks.

1. "That hairdryer was a real disappointment. It really didn't dry my hair as well as I expected."
2. "I was thrilled with my last car. I sold it with 150,000 miles and hardly had any repairs."
3. "I love buying from that company's website. I always get what I order within two days."
4. "The refrigerator works fine, but I think the shelves could have been laid out better."
5. "That park had great rides, but the lines were a mess."
6. "Our quality is great. We've got less than six defectives per one million parts produced."

In each case, the speaker is referring to a different aspect of quality. In the first case, the product simply didn't perform the task it was designed to do. That is, its function was substandard. The repair record of an automobile is really an issue of reliability rather than quality, per se. In the third case, it is delivery speed that translates to quality service for that customer. The fourth case refers to a product that does what it is supposed to do, but the consumer is disappointed with the product design. The product quality (the rides) at the amusement park were fine, but the logistics of the park management were a disappointment. The final case refers to the statistical aspects of quality control. Clearly there are many dimensions to quality.

Given these dimensions, how should we define quality? Traditional thinking would say that quality is conformance to specifications; that is, does the product do what it was designed to do? Some feel that this definition is the only meaningful definition of quality, because conformance is something that can be measured. According to Philip Crosby (1979):

> That is precisely the reason we must define quality as "conformance to requirements" if we are to manage it. Thus, those who want to talk about quality of life must talk about that life in specific terms, such as desirable income, health, pollution control, political programs, and other items that can each be measured. (p. 15)

Crosby makes a good point. By defining quality in terms of conformance, we avoid making unreasonable comparisons. Is a Rolls-Royce a better-quality product than a Toyota Corolla? Not necessarily. The Toyota may be a higher-quality product relative to *what it was designed to do.*

This does not tell the entire story, however. Just as beauty is in the eye of the beholder, so is quality in the mind of the customer. If the customer is not happy with the product, it is not high quality. Viewed in this way, quality is a measure of the conformance of the product to the *customer's needs.*

Why is this different? Conformance to specifications assumes a given design and the specifications resulting from that design. Conformance to customer needs means that the design of the product is part of the evaluation. Given two washing machines with comparable repair records, what determines which one the consumer will buy? The answer is a combination of aesthetics, features, and design. Viewing quality in this broader way is both good and bad. It is good in that it gets at the heart of the issue; quality is what the customer thinks it is. It is bad in that it makes it difficult to measure quality and thus difficult to improve it.

Poor quality can be very costly. There are **internal failure costs** (that occur prior to transfer of ownership of the product to customers) that include scrap, rework, retesting, downtime, purchasing errors, and excessive overtime. Then there are **external failure costs** (that occur after transfer of ownership of the product to customers) that include customer complaints, warranties, product recall, returned material, loss of customer goodwill, and loss of reputation. However, there are also process costs associated with avoiding failures. **Appraisal costs** include incoming goods inspection, quality audits, field performance testing, and inspection and testing of intermediate and final products. Finally, **prevention costs** consist of quality planning, process control, quality training, and quality improvement process implementation.

Management must grapple with the difficult problems of knowing how much to invest in quality and determining the best way to go about making that investment. Often this is a strategic decision, as discussed in Chapter 1. In theory, there is an optimal trade-off between the cost of poor quality (internal plus external failure costs) and the investment required in the process to improve the quality (appraisal and prevention costs), as represented in Figure 10–1. However, there are two issues associated with this view of quality.

The first issue with Figure 10–1 is that, although such curves can be drawn in principle, evaluating the costs of poor quality is difficult. Direct costs, such as those resulting from scrap, rework, and inspection, are relatively easy to determine. But how does one factor in the costs of lost consumer loyalty? No one would deny that marketing is essential, but has the emphasis on marketing in some firms been at the expense of manufacturing? We must acknowledge that the modern consumer is more educated and more discriminating than ever before. Clever advertisements will not sell second-rate products, or at least not for long. Firms ignore issues of quality at their peril.

The second issue with Figure 10–1 is that many people believe that the process cost associated with avoiding defects is not fixed but is a function of the quality-orientation of the firm. That is, as a firm improves its processes generally, the conformance cost of quality will decrease. This is in part why, these days, quality management is often linked with lean production. Lean Six Sigma programs have become common as a way to perform process improvements that both drive out waste (i.e., lean) and decrease variance (i.e., Six Sigma quality). [We will capitalize "six sigma" when we are referring to the specific quality initiative founded

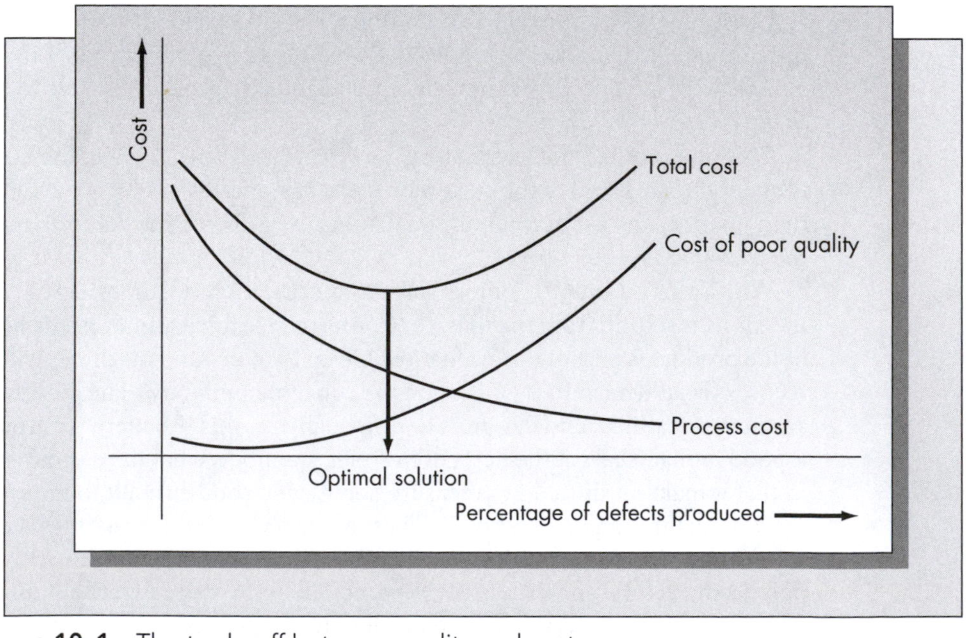

Figure 10–1 The trade-off between quality and cost

by Motorola and will leave it uncapitalized when referring to six standard deviations.] This chapter will explore these ideas more fully.

Statistical quality control dates back to the 1930s. Its roots lie in the work of Walter Shewhart, a scientist employed by Bell Telephone Laboratories. W. Edwards Deming, the man credited with bringing the quality control message to the Japanese, was a student of Shewhart. Deming has stressed in his teachings that understanding the concept of statistical variation of processes is a key step in designing an effective quality control program. One needs to understand process variation in order to know how to produce products that conform to specifications. Deming has become a demigod in Japan, where his teachings ignited the Japanese quality revolution.

This chapter begins with a broad discussion of key trends in quality management including some of Deming's ideas. Then the bulk of the chapter is aimed at providing an understanding of the essentials underlying statistical quality control. Statistical quality control constitutes a set of techniques based on the theories of probability and statistical sampling for monitoring process variation and for determining if manufactured lots meet desired quality levels. However, delivering quality to the customer is a far broader problem than is addressed by statistical issues alone. We therefore also consider quality from the management perspective. Two fundamental pillars of statistical quality control are control charts and acceptance sampling. Briefly, a control chart is a graphical means for determining if the underlying distribution of some measurable variable seems to have undergone a shift. Acceptance sampling is the set of procedures for inferring characteristics of a lot from the characteristics of a sample of items from that lot. Whereas "zero defects" thinking might suggest that these approaches are obsolete, we disagree. Process monitoring and statistical sampling continue to be used. It is important to understand the underlying theory behind these methods to know when and how they should be applied. The chapter concludes with a discussion of designing for quality.

10.1 QUALITY BASICS

Deming's view on quality was that "Quality is a predictable degree of uniformity and dependability at low cost and suited to the market." This definition matches well with our statement in the introduction that "quality is a measure of the conformance of the product to the customer's needs." David Garvin (1988) expands on these ideas and suggests that quality be considered along eight basic dimensions.

- Performance.
- Features.
- Reliability.
- Conformance.
- Durability.
- Serviceability.
- Aesthetics.
- Perceived quality.

We could lump the first five dimensions together under the general heading of conformance to requirements (the definition suggested by Crosby, 1979) and the last three under the heading of customer satisfaction (as suggested by Feigenbaum, 2004). By further breaking down these two categories, Garvin gives a better appreciation for the complexity of the quality issue.

Another of Deming's key principles was that quality should be built into the process; firms should concentrate on improving processes rather than on targets or quotas. He also made the claim that management is responsible for 85 percent of quality problems; hence, they must give the workers the tools and skills to meet the required goals. He developed his ideas into 14 key points as follows.

1. Constancy of purpose.
2. New philosophy.
3. End mass inspection.
4. End ordering on price alone.
5. Improve the process of production.
6. Train on the job with a purpose.
7. Supervise for information.
8. Drive out fear.
9. Break down barriers between departments.
10. Eliminate numerical goals for workforce.
11. Eliminate work standards and quotas.
12. Reduce barriers to pride in work.
13. Retrain for future.
14. Create a structure for top management to push the above.

The interested reader is referred to Deming (2000) to learn more about Deming's ideas of quality management.

The Deming Prize and the Baldrige Award

Using the royalties from a book based on his 1950 lectures, Deming established a national quality prize in Japan. The funding for the prize continues to be primarily from Deming's royalties, but it is supplemented by a Japanese newspaper company and private donations. The **Deming Prize** is awarded to (1) those who have achieved excellence in research in the theory or application of statistical quality control, (2) those who have made remarkable contributions to the dissemination of statistical quality control methods, and (3) those corporations that have attained commendable results in the practice of statistical quality control.

The prize has been awarded almost every year in several categories since 1951. Recipients include such well-known firms as Toyota Motor Co., Hitachi, Fuji Film, Nippon Electric, and NEC. Prizes also have been awarded to specific divisions of large firms and specific plants within a division. It is a highly sought-after national honor in Japan (Aguayo, 1990).

As mentioned in Chapter 1, the U.S. Department of Commerce established the **Malcolm Baldrige National Quality Award** largely as a response to the success of the Deming Prize in Japan. The award was named for the late secretary of commerce, who died in an accident in 1987, the same year that Congress and President Ronald Reagan enacted the Malcolm Baldrige National Quality Improvement Act. The award is made each year in six categories: (1) manufacturing companies or subsidiaries, (2) service companies or subsidiaries, (3) small businesses, (4) education, (5) healthcare, and (6) nonprofit. Applying for the Baldrige Award is an involved process requiring a sincere commitment on the part of the firm seeking the award. A board of overseers appointed by the secretary of commerce has final say. They base their evaluation on the following nine-point value system.

1. Total quality management.
2. Human resource utilization.
3. Performance.
4. Measurables.
5. Customer satisfaction.
6. World-class quality.
7. Quality early in the process.
8. Innovation.
9. External leadership.

The evaluation process utilizes an elaborate scoring system and includes site visits from the evaluation committee. Funds to support the examining process are donated by individuals and firms. The Baldrige Award is not an end in itself but rather a means to an end. The process of preparing an application forces the firm to take a good hard look at its quality efforts and provides a blueprint for self-examination. The Baldrige application is an excellent means for ferreting out problems and suggesting where improvements need to be made.

ISO 9000

The International Organization for Standardization (ISO), based in Switzerland, established guidelines for quality management, known as the series of standards **ISO 9000**. This was the first attempt at developing a true uniform global standard for quality. First published in 1987, ISO standards were revised in 1994, 2000, 2008, and in 2015. For a firm to obtain ISO 9000 certification, it must carefully document its systems and procedures. Purchasing, materials handling, manufacturing, and distribution are all subject to ISO documentation and certification. ISO certification is very different from the Baldrige process. ISO certification is not an award, nor even necessarily a judgment about the quality of products. It is a certification of the manufacturing and business processes used by the firm.

With the establishment of the ISO standard came a host of consulting organizations specializing in taking a company through the certification process. Certification is not cheap. Aside from the direct cost of consultants, there are substantial indirect costs associated with developing the necessary documentation. When he was president of Reynolds Aluminum Supply Company, John Rudin voiced this opinion in an article about ISO 9000.

> From an out-of-pocket standpoint, just to get one facility registered gets to be somewhere in the $25,000 to $30,000 range. For RASCO as a whole, you're talking a million or so, if we were to do all the facilities. Right now we don't see the value in having a plaque to hang on the wall for a million dollars, but we do see the value in having a good, back-to-basics, well-documented process, which is what ISO is really all about anyway. (Kuster, 1995, p. 5)

The firm seeking certification must document quality-related procedures, system and contract-related policies, and record keeping. According to Jay Velury (1995), a firm should expect to document about 20 procedures for each section required for certification. Given the expense in time and money, why should anyone bother? There are several advantages of achieving certification. One is from the process itself. Careful documentation of quality practices reveals where those practices fall short. The process of continuous improvement starts with knowing where one stands. Furthermore, many firms now require suppliers to obtain certification. It could be a prerequisite to doing business in many circumstances. Thomas Boldlund, then president of Oregon Steel, was quoted about the advantages of certification.

> ISO 9000 allows us to participate in markets that 10 years ago we could not serve even if we wanted to. When we meet a customer, the first thing they ask is "Can you meet the quality standards of your competition?" and second: "Can you produce the grades the Europeans and Japanese produce today?" With ISO 9000, the answer is yes. (Kuster,1995, p. 5)

The ISO continues to develop international standards. In 1996 the organization announced the ISO 14000 series of environmental standards (Alexander, 1996). ISO 14000 is a series of international standards for environmental management systems. The motivation for the ISO in adopting a uniform environmental standard is the disparity of these standards in different countries. It is hoped that this new standard will not only spur international trade but, more importantly, improve quality of life by improving the environment. Both ISO 9000 and ISO 14000 guidelines are important first steps in the movement toward a global uniform standard.

Six Sigma Quality

As discussed in Chapter 1 and the Snapshot Application for this section, Motorola has become famous for instituting its "**Six Sigma**" quality thrust. Sigma, in this context, is the Greek letter, commonly (and throughout this text) used for standard deviation. An entire training industry has now developed around Six Sigma certification. Yellow, green, black, and master black belts are offered in Six Sigma. While General Electric and Motorola both offer Six Sigma certification, so do a host of other institutions. Indeed, there is no standard certification body, and the belts mean different things depending on who is doing the certifying. For example, some green belt certifications require experience leading a quality improvement project, while others do not.

SNAPSHOT APPLICATION

Motorola Leads the Way with Six-sigma Quality Programs

The Motorola Corporation has compiled an impressive record of defining and implementing new quality initiatives and translating those initiatives into profits. As we see in this chapter, traditional quality methods assume that an out-of-control condition corresponds to an observation falling outside of $\pm 2\sigma$ or $\pm 3\sigma$. Motorola decided that this standard was too loose, giving too many defects. In the 1980s, they moved toward a 6σ standard. To achieve this goal, Motorola established the practice of quality system reviews (QSRs) and placed a great deal of emphasis on classical statistical process control (SPC) techniques. Eventually, their program was trademarked as Six Sigma quality.

To achieve the 6σ standard, Motorola infused the total quality management philosophy into its entire organization. Quality programs were not assigned to and policed by a single group but became part of everyone's job. Motorola's approach was based on the following key ingredients.

- Overriding objective of total customer satisfaction.
- Uniform quality metrics for all parts of the business.
- Consistent improvement expectations throughout the firm.
- Goal-directed incentive plans for management and employees.
- Coordinated training programs.

Motorola carefully documented the road map to follow to its quality goals. The first step in the process was a detailed audit of several key parts of the business, including: control of new product development, control of suppliers (both internal and external), monitoring of processes and equipment, human resource considerations, and assessing customer satisfaction. Top management took part in many of these audits by paying regular visits to customers, chairing meetings of the operating policy committees, and recognizing executives who made outstanding contributions to the company's quality initiatives.

Sanjoy Kumar and Yash Gupta (1993) reported on their experience with a quality management program put in place in Motorola's Austin, Texas, assembly plant. In May of 1988, management began the process of implementing an SPC program at Austin. The process began by bringing in an outside consultant to design the program and assigning an internal coordinator to ultimately take over the duties of the consultant. To be sure that employees bought into the initiative, management organized participative problem-solving teams. Each team included a manufacturing manager, a group leader, operators from the two shifts, a representative from the Quality Assurance (QA) Department, and an engineer. Austin had a total of six teams. To further ensure a buy-in from all employees, management initiated a plantwide training program in SPC. Training was tailored to job function.

The plant's QA Department instituted a certification program at Austin for vendors. As a result, about 60 percent of the vendors supplying the plant were certified. Within the plant, traditional SPC methods were employed: attribute data were collected and charted, and machines were shut down when out-of-control situations were detected. Members of the QA team employed design of experiments techniques to identify causes of problems.

What was the bottom line? Over the first two years of the initiative, the Austin plant reported a decrease in scrap rates of 56 percent. By developing a clear-cut and coordinated strategy, Motorola was able to achieve major improvements in traditional quality measures in this facility. Motorola's overall success is a testament to the fact that this was not an isolated example. It demonstrates that American companies can compete effectively with overseas competitors when the quality effort is a true companywide initiative.

In 1988, Motorola was a winner of the first Malcolm Baldrige National Quality Award. In 2002 their Commercial, Government, and Industrial Solutions Sector (CGISS) won the award. Of the 102 award recipients since 1988, Motorola is one of 4 Fortune 500 companies to win the award. (National Institute of Standards and Technology, 2019). Subunits of large corporations have won the award (as was the case with Motorola in 2002). Motorola continues to be considered a leader in quality management and continues to offer Six Sigma certification.

Six-sigma quality is defined as defect rates of less than 3.4 parts per million. The extremely acute reader will notice that six standard deviations away from the mean of a normal distribution would actually result in a much lower defect rate than this (already small) rate. That is because the quality target of 3.4 parts per million also allows the original mean of the distribution to shift by plus or minus one and a half standard deviations. This is explained further in Section 10.6.

The methodologies within Six Sigma are based on earlier quality initiatives of both Deming and the **Total Quality Management** movement. Deming's **Plan-Do-Study-Act** cycle has been evolved into **DMAIC** (Define, Measure, Analyze, Improve, and Control) for existing processes and **DMADV** (Define, Measure, Analyze, Define, and Verify) for new products or processes. One of the keys to Six Sigma's success has been its focus on measurable outcomes. That is, projects can justify their costs based on the savings they realize.

Six Sigma's focus on process improvement has a large overlap with lean production ideas. This is not surprising given Deming's influence on both Toyota (the birthplace of lean) and Total Quality Management (the predecessor to Six Sigma). Rather than trying to differentiate the two, many companies embrace **Lean Six Sigma** quality and process improvement. Belt certification is now being offered in Lean Six Sigma. The idea behind Lean Six Sigma appears to have been developed by Barbara Wheat, Chuck Mills, and Mike Carnell (2001).

Lean Quality

The Toyota production system embraces quality control; hence, solid quality management is a core part of any lean production system. Key aspects of lean quality systems include the following.

1. *Quality at the source.* Quality should be built into the process rather than found through later inspection.
2. *Automatic inspection/100 percent inspection.* In order to achieve quality at the source, there should be automatic inspection of all parts as they progress through production.
3. *Quality first, output second.* Output is always considered secondary to quality production. That is, quality should never be sacrificed for the sake of meeting production quotas. **Andon lights** are used to signal quality issues, and team leaders are tasked with helping workers fix the problem. Under-capacity scheduling allows for planning, problem solving, and maintenance.
4. *Worker responsibility and authority.* Workers are tasked with implementing quality at the source and are instructed to stop the production line (in an assembly line environment) if the problem cannot be fixed in real time. The Japanese word for this is **jidoka**.
5. *Measurement tools.* Lean production makes strong use of statistical methods for quality and process control. Control charts are heavily used to make sure processes remain stable.
6. *Fail-safe methods.* Lean quality methods emphasize **visual control**. If something is out of place or not correct, this should be immediately obvious. The Japanese word for this is **poka-yoke**.
7. *Housekeeping.* Workers are expected to keep their work areas orderly using the principles of the 5Ss. While the original S words are Japanese, they have been translated into English as sort, straighten, sweep, standardize, and self-discipline or sort, set in order, shine, standardize, and sustain. Often a 6th S-word "safety" is added to the list.
8. *Continuous improvement.* The Japanese term for **continuous improvement** is **kaizen**. Like quality at the source, it requires total employee involvement. The essence of kaizan is willingness of workers to spot quality problems; halt production when necessary; generate ideas for improvement; analyze problems; and perform different functions.

The goal of continuous improvement is to eliminate waste, or **muda** as the Japanese call it. The standard implementation of lean has seven types of waste (e.g., Costantino, 1998, p. 376).

1. *Overproduction.* Producing more than is needed before it is needed.
2. *Waiting.* Any nonwork time waiting for tools, supplies, etc.

3. *Conveyance.* Wasted effort to transport materials, parts, or finished goods into or out of storage or between processes.
4. *Processing.* Providing higher quality than is necessary, extra operations, etc.
5. *Inventory.* Maintaining excess inventory of raw materials, parts in process, or finished goods.
6. *Motion.* Any wasted motion to pick up parts or stack parts. Also wasted walking.
7. *Correction.* Repair or reword.

Lean Six Sigma programs focus on eliminating these wastes using process improvement techniques and measuring the effects using statistical methods. It also emphasizes decreasing sources of variability, so that product can "flow like water" throughout the process.

10.2 STATISTICAL BASIS OF CONTROL CHARTS

A process is **in control** if a stable system of chance causes is operating. That is, the underlying probability distribution generating observations is not changing with time. One of Deming's key contributions was explaining that a system in control will still have **natural variation**; in fact, it may have a lot of variation. If a system is in control, then only management has the power to reduce defects because only management has the power to change the process. Pleading with workers or setting quotas will have no effect other than demoralizing the workforce.

It therefore becomes important to determine whether a variation from specifications (i.e., a defect) is **normal** or from a **common cause** or whether there is **special** or **assignable cause** for the variation. Common causes relate to the design of the system. If only common causes exist, then one expects a bell shaped or normal distribution from the average of a sample. Further, this distribution should not change over time, and the process is said to be in control.

Control charts provide a simple graphical means of monitoring whether a process is in control in real time. Although easy to construct and easy to use, control charts are based on rigorous statistical principles. They have gained wide acceptance in industry and are preferred to more conventional statistical methods.

A control chart maps the output of a production process over time and signals when a change in the probability distribution generating observations seems to have occurred. To construct a control chart, one uses information about the probability distribution of process variation and fundamental results from probability theory. A result that forms the basis for a class of control charts is known as the **central limit theorem**. Roughly, the central limit theorem says that the distribution of sums of independent and identically distributed random variables approaches the normal distribution as the number of terms in the sum increases. (The central limit theorem is also used in Chapter 12 to justify the use of the normal distribution to describe project completion time in PERT networks.) Generally, the distribution of the sum converges very quickly to a normal distribution. In order to illustrate the central limit theorem, suppose that X is a random variable with the uniform distribution on the interval (0, 1). The probability density function of X is pictured in Figure 10–2.

The density of X bears little resemblance to a normal density. Assume that the three random variables X_1, X_2, and X_3 are independent random variables, each of which has the uniform distribution on the interval (0, 1). Consider the random variable $W = X_1 + X_2 + X_3$. One can derive the distribution of W by convoluting the distributions of X_1, X_2, and X_3. We will not present the details here. (The interested reader should refer to a graduate-level text in probability such as DeGroot and Schervish, 2012.) The density function of W appears

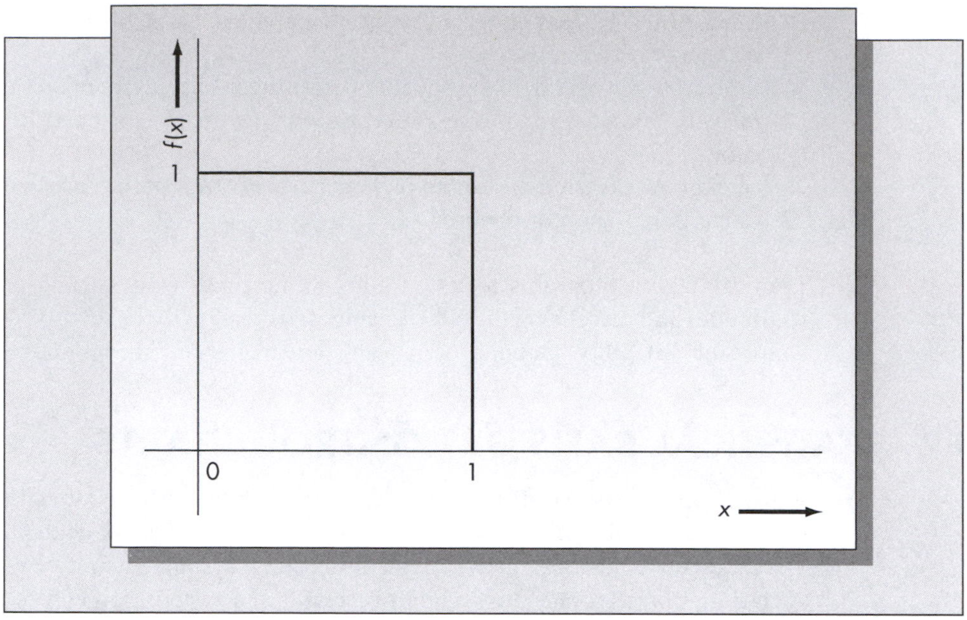

Figure 10–2 Probability density of a uniform variate on (0, 1)

in Figure 10–3. The resemblance to a normal density is now quite striking. In Figure 10–4 we have graphed the probability density function of W and the associated normal approximation. Notice how closely the two curves agree. Were we to continue to add independent uniform random variables, the agreement would be even closer.

In quality control, the central limit theorem justifies the assumption that the distribution of $\bar{X}$, the sample mean, is approximately normally distributed. Recall the definition of the sample mean: If $(X_1, X_2, \ldots, X_n)$ is a random sample, then the sample mean $\bar{X}$ is defined as

$$\bar{X} = \frac{1}{n}\sum_{i=1}^{n} X_i.$$

Suppose that a variable Z has the standard normal distribution. Then, from Table A–1 at the back of this book or from the Excel function NORM.S.DIST,

$$P\{-3 \le Z \le 3\} = .9974.$$

In words, this means that the likelihood of obtaining a value of Z either larger than 3 or less than −3 is .0026, or roughly 3 chances in 1,000. This is the basis of the so-called three-sigma limits that have become the de facto standard in quality control. Now consider the sample mean $\bar{X}$, which the central limit theorem tells us is approximately normally distributed. Suppose that the mean of each sample value is μ and the standard deviation of each sample value is σ. Then it is well known that the mean of $\bar{X}$ is also μ and the standard deviation of $\bar{X}$ is $\sigma/\sqrt{n}$. Therefore, the standardized variate

$$Z = \frac{\bar{X} - \mu}{\sigma\sqrt{n}}$$

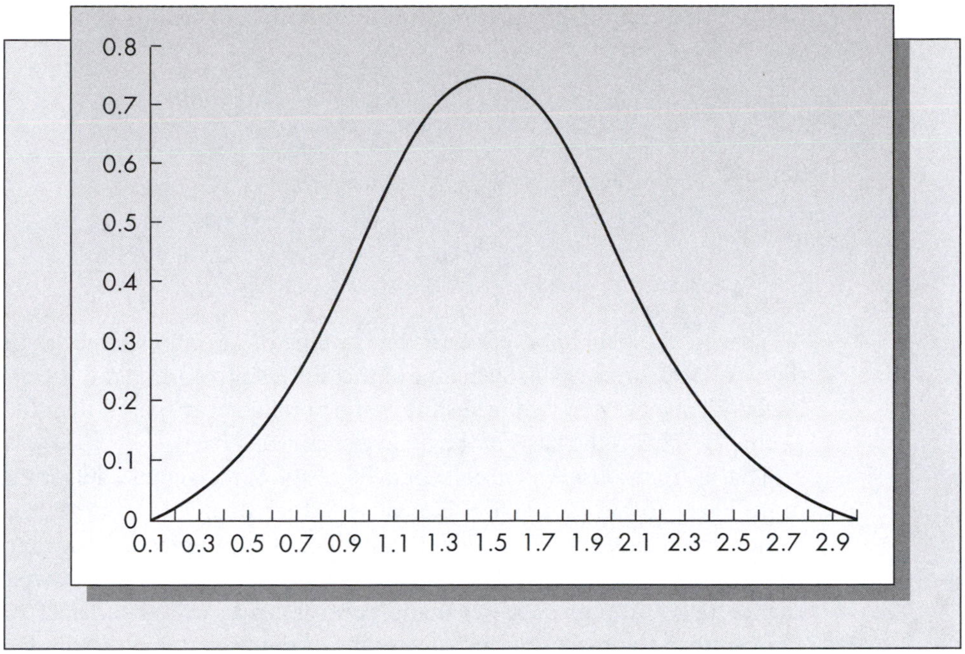

Figure 10–3 Density of the sum of three uniform random variables

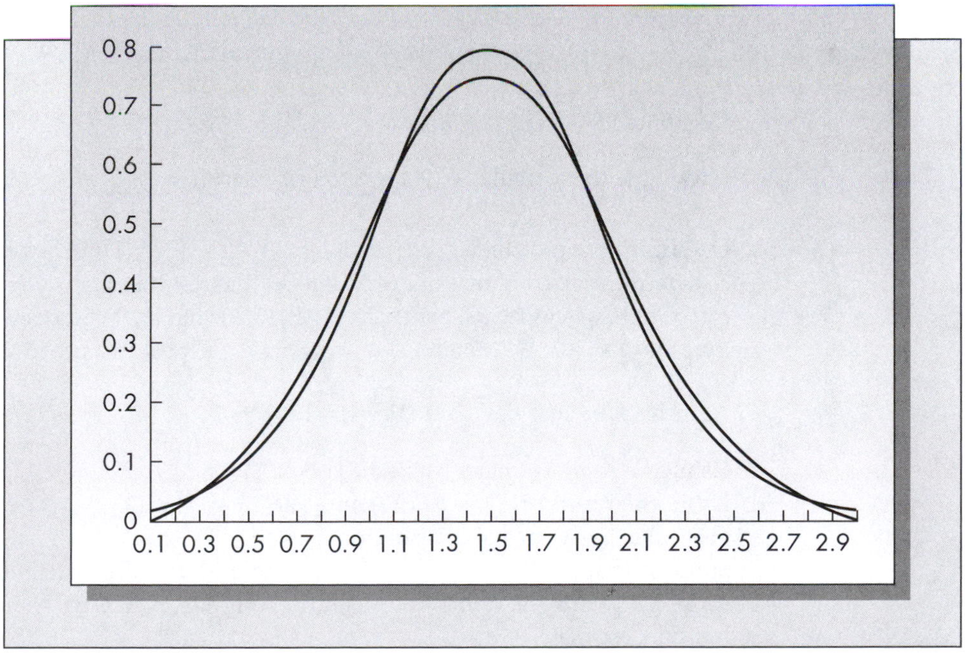

Figure 10–4 Density of the sum of three uniform random variables and the normal approximation

has (approximately) the normal distribution with zero mean and unit variance. It follows that

$$P\left\{-3 \le \frac{\bar{X}-\mu}{\sigma/\sqrt{n}} \le 3\right\} = .9974,$$

which is equivalent to

$$P\left\{\mu - \frac{3\sigma}{\sqrt{n}} \le \bar{X} \le \mu + \frac{3\sigma}{\sqrt{n}}\right\} = .9974.$$

That is, the likelihood of observing a value of $\bar{X}$ either larger than $\mu + 3\sigma/\sqrt{n}$ or less than $\mu - 3\sigma/\sqrt{n}$ is .0026. Such an event is sufficiently rare that if it were to occur, it is more likely to have been caused by a shift in the population mean, μ, than to have been the result of chance. This is the basis of the theory of control charts.

That is, if an observation falls outside of the three-sigma control limits, then it is much more likely to have an assignable cause, rather than just coming from natural variation. Hence, that cause should be investigated. Possible assignable causes include operator error, a bad batch of raw material, an environment change, etc. Further, over time a process may change (e.g., through tool wear) and control charts will also detect such a shift. Changes can occur in the mean, variance, or shape of the underlying distribution of the production process. The process is said to be **out of control** if such changes occur.

PROBLEMS FOR SECTION 10.2

1. Suppose that X_1 and X_2 are independent random variables having the uniform distribution on (0, 1, 2, 3, 4, 5). That is,

$$f(j) = P\{X_i = j\} = \tfrac{1}{6} \qquad \text{for } i = 1, 2, \text{ and } 0 \le j \le 5.$$

   a. Compute $E(X_1)$ and $\text{Var}(X_1)$.

   [Hint: Use the formulas $E(X) = \sum_{j=0}^{5} jf(j)$, $\text{Var}(X) = \sum_{j=0}^{5} j^2 f(j) - (E(X))^2$.]

   b. Determine the probability distribution for $Y = X_1 + X_2$. [Hint: For each possible value of Y, determine all combinations of X_1 and X_2 that result in that value. For example, $Y = 3$ can be obtained by $(X_1, X_2) = (0, 3), (1, 2), (2, 1)$, and $(3, 0)$. Since each pair has probability $\left(\tfrac{1}{6}\right)\left(\tfrac{1}{6}\right) = \left(\tfrac{1}{36}\right)$, we obtain $P\{Y = 3\} = \tfrac{4}{36} = \tfrac{1}{9}$. Repeat this process for all values of Y.]

   As a check be sure that $\sum_y P\{Y = y\} = 1.0$.

   c. Using the results of part (b), find $P\{1.5 < Y < 6.5\}$.

   d. Using the results $E(Y) = 2E(X_1)$ and $\text{Var}(Y) = 2\text{Var}(X_1)$, approximate the answer to part (c) using a normal distribution.

   e. Suppose that $X_1, X_2, \ldots, X_{20}$ are independent identically distributed discrete random variables having the uniform distribution on (0, 1, 2, 3, 4, 5). Using a normal approximation, estimate

$$P\left\{\sum_{i=1}^{20} X_i \le 75\right\}.$$

f. Do you think that the approximation computed in part (d) or part (e) is more accurate? Why?

2. The following data represent the observed number of defective chips produced each hour based on observing the system for 30 successive hours.

0	3	5	2	6	8	3	5	4	6	6	9	5	1	2
1	2	5	3	3	0	1	0	7	1	7	5	4	4	3

a. Plot a frequency histogram for the numbers of failures per hour.

b. Compute the mean and the variance of the sample. Use the formulas

$$\bar{X} = \frac{1}{n}\sum_{i=1}^{n} X_i, \quad s^2 = \frac{1}{n-1}\left(\sum_{i=1}^{n} X_i^2 - n\bar{X}^2\right).$$

(Note: You can streamline the calculations by grouping the data first, as there are only 10 distinct values.)

c. If the number of defective chips produced hourly is normally distributed with mean and variance computed in part (b), determine the probability that fewer than five defectives are observed in any particular hour.

d. If each chip costs the company $5 to produce and defective chips are discarded, how long (in an expected-value sense) would it take to pay off a new piece of equipment costing $45,000 that would reduce defectives to half of the current level? Assume 40-hour production weeks and 48 weeks per year.

3. The tensile strength of a heavy-duty plastic bag used in trash compactors is normally distributed with mean 150 pounds per square inch and standard deviation 12 pounds per square inch. An independent landscape contractor uses them to haul refuse that requires 120-pounds-per-square-inch tensile strength. What proportion of the compactor bags will not meet the requirements?

4. A credit rating company recommends granting of credit cards based on several criteria. One is annual income. If the annual income of applicants is normally distributed with mean $66,000 and standard deviation $12,800 and the company recommends no applicant unless his or her income exceeds $45,000, what fraction of the applicants are denied on this basis?

5. a. What is the probability that a normal variable exceeds two-sigma limits? (That is, what is the probability of observing a value of the random variable larger than $\mu + 2\sigma$ or less than $\mu - 2\sigma$?)

   b. If "deciding that the process is out of control" means observing a realization of the sample mean exceeding k sigma limits, discuss the advantages and disadvantages of using $k = 2$ versus $k = 3$.

6. The members of a golf club have handicaps that are normally distributed with mean 15 and standard deviation 3.5. In a particular event, foursomes are chosen by grouping four players chosen at random from the club. The handicap of the foursome is the arithmetic average of the handicaps of the four players comprising the foursome. In what proportion of the foursomes will the handicap of the foursome be less than 10 or more than 20? (Hint: The standard deviation of the average of four independent identically distributed random variables is exactly half the standard deviation of one of them.)

10.3 CONTROL CHARTS FOR VARIABLES: THE $\bar{X}$ AND R CHARTS

As discussed in the previous section, when the observed value of the sample mean of a group of observations falls outside the appropriate three-sigma limits, it is likely that there has been a change in the probability distribution generating observations. To illustrate how one develops and interprets control charts, consider the following example.

EXAMPLE 10.1

Wonderdisk produces a line of plug-compatible disks for IBM equipment. Building 35 is responsible for production of the read/write arms for the model A55C disk. The arms are approximately 2.875 inches in length. The design engineers have established a tolerance of ± 0.025 inch for the arm lengths and advertise this figure in the published specifications.

The company usually produces 40 arms per day. On 30 consecutive production days, five arms are sampled randomly from each day's production and measured. The resulting measurements appear in Table 10–1.

These observations show that there is some variation in the length of the arms. However, there appears to be no discernible pattern to this variation. Define the random variable X to be the length of an arm selected at random. We may then interpret Table 10–1 as 150 independent observations on the random variable X.

Howard Hamilton, an industrial engineer working for Wonderdisk, is given the job of analyzing these data. The first thing that Howard notices is that the established tolerances of ± 0.025 inch were often exceeded. This becomes most evident by computing the range of daily observations. The range of a sample is the maximum of the observations minus the minimum of the observations. For the 30 days of data, Howard observes that the range exceeds 0.05 in four cases. In order to obtain a clearer idea of what proportion of the population lies outside the specified tolerances, Howard develops a frequency histogram of the 150 measurements. This histogram appears in Figure 10–5 (p. 588). The histogram suggests that the measurements are normally distributed. Howard used a goodness-of-fit test to verify the normality of the observations. Figure 10–6 shows the theoretical normal curve.

One determines the theoretical normal curve in the following way. Because the normal distribution depends upon two parameters, μ and σ, we must estimate these values from the sample data. From the theory of statistics, we know that the "best" estimators of the population mean and variance are the sample mean, $\bar{X}$, and the sample variance, s^2, given by

$$\bar{X} = \frac{1}{n}\sum_{i=1}^{n} X_i,$$

$$s^2 = \frac{1}{n-1}\sum_{i=1}^{n}(X_i - \bar{X})^2 = \frac{1}{n-1}\left(\sum_{i=1}^{n} X_i^2 - n\bar{X}^2\right).$$

Interpret $X_1, X_2, \ldots, X_n$ as the sample values (the random sample). The sample mean based on all 150 observations is 2.875, and the sample variance is 0.0002434. The estimate of the population standard deviation is the square root of the sample variance, which is 0.0156.

Howard can now estimate the fraction of the arms produced that fall outside the advertised tolerances. An arm will exceed the tolerance if it is longer than 2.90 inches or shorter than 2.85 inches. The area of the crosshatched region in Figure 10–6 represents the probability that this will occur, which is evidently not negligible. In fact,

Table 10–1 Tracking Arm Data

Sample Number	Measurements of the Length of a Tracking Arm					Average	Range
1	2.8971	2.8477	2.8624	2.8606	2.8971	2.8730	.0494
2	2.8863	2.8541	2.8677	2.8838	2.8854	2.8755	.0322
3	2.8772	2.8708	2.8920	2.8892	2.8840	2.8826	.0212
4	2.8808	2.8650	2.8686	2.8874	2.8804	2.8764	.0224
5	2.8633	2.8993	2.8650	2.8909	2.9131	2.8863	.0497
6	2.8743	2.8571	2.8863	2.8473	2.8739	2.8678	.0390
7	2.8820	2.8612	2.8805	2.8737	2.8933	2.8781	.0322
8	2.8847	2.8630	2.8846	2.8969	2.8916	2.8842	.0339
9	2.8569	2.8934	2.8926	2.8585	2.8721	2.8747	.0365
10	2.8784	2.8795	2.8794	2.8608	2.8672	2.8731	.0187
11	2.8821	2.8544	2.9053	2.8495	2.8670	2.8717	.0558
12	2.8643	2.8533	2.8718	2.8565	2.8724	2.8637	.0191
13	2.8675	2.8578	2.8971	2.8709	2.8908	2.8768	.0394
14	2.8495	2.8701	2.8741	2.8699	2.8766	2.8680	.0271
15	2.8822	2.8731	2.8551	2.8782	2.8687	2.8714	.0271
16	2.8731	2.8675	2.8743	2.8520	2.8900	2.8714	.0379
17	2.9054	2.9190	2.8752	2.8477	2.8639	2.8822	.0713
18	2.8759	2.8832	2.8660	2.8667	2.8674	2.8718	.0172
19	2.8676	2.8775	2.8793	2.8943	2.9048	2.8847	.0373
20	2.8765	2.8613	2.8737	2.8524	2.8767	2.8681	.0243
21	2.9052	2.8851	2.8895	2.8904	2.8723	2.8885	.0328
22	2.8606	2.8837	2.9017	2.8628	2.8455	2.8709	.0562
23	2.8752	2.8722	2.8618	2.8637	2.8725	2.8691	.0133
24	2.8566	2.8929	2.9035	2.9109	2.8594	2.8847	.0543
25	2.8495	2.8749	2.8873	2.8557	2.8673	2.8669	.0378
26	2.8736	2.8606	2.8797	2.8522	2.8802	2.8693	.0280
27	2.8449	2.8908	2.8851	2.8798	2.8610	2.8723	.0459
28	2.8589	2.8800	2.9025	2.8974	2.8606	2.8799	.0437
29	2.8910	2.8546	2.8744	2.8775	2.8634	2.8722	.0364
30	2.8607	2.8769	2.8771	2.8934	2.8706	2.8757	.0326

$$
\begin{aligned}
P\{X > 2.90 \text{ or } X < 2.85\} &= P\{|X - \mu| > 0.025\} \\
&= P\left\{\left|\frac{X-\mu}{\sigma}\right| > 1.602\right\} \\
&= P\{|Z| > 1.602\} = .11.
\end{aligned}
$$

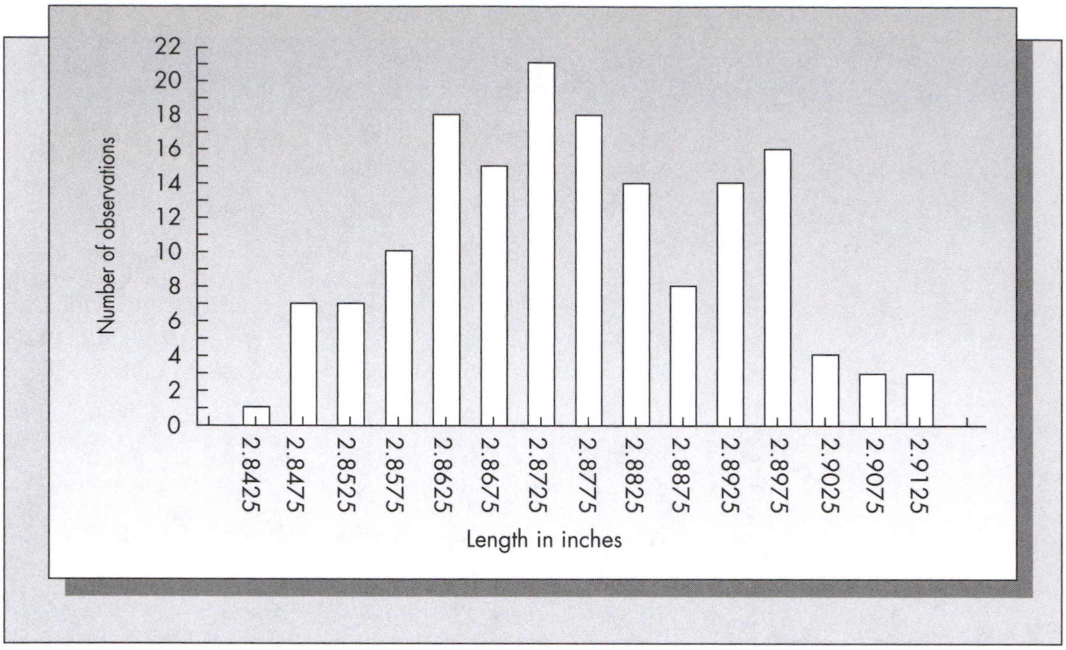

Figure 10–5 Frequency histogram of 150 measurements

We can find the probability, .11, in a table of the normal distribution (Table A–1 at the back of this book) or with Excel. Hence, about 11 percent of the arms produced over the last 30 days do not meet the company's published specifications. However, failure rates for the disks due to incompatibility of the arms has been extremely low (less than 1 percent). Howard presented these results to the director of manufacturing, who discussed

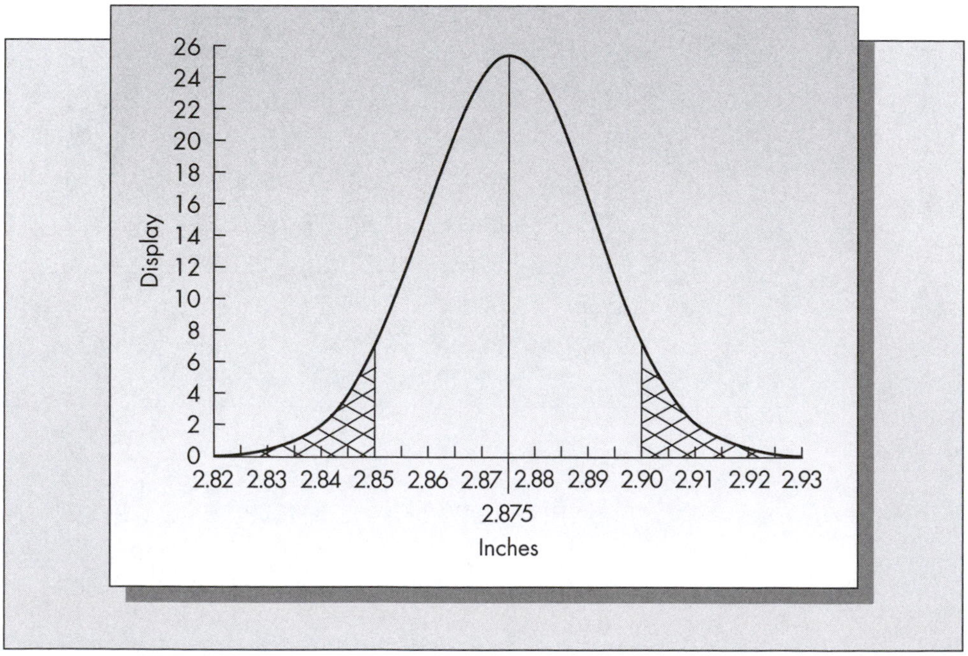

Figure 10–6 Theoretical normal curve of arm length

the problem with the company president and the director of engineering. After some additional investigation, Howard concluded that the original tolerances were not consistent with design requirements of the disk. It was found that a tolerance of ± 0.05 inch would be sufficient. The tighter tolerances were based on an earlier design, and the department simply forgot to revise its figures. Later testing showed that a tolerance of ± 0.05 inch was much more realistic and consistent with the operation of the disk. Howard confirmed that the revised tolerances would include more than 99 percent of the population.

This example raises an important point in the application of statistical principles to quality control. If control charts are to be used to compare the characteristics of manufactured items with preset design specifications, then the desired tolerances and the observed statistical variation in the sample must be consistent. If the tolerances are much tighter than the variation observed in the sample, as in Example 10.1, then they often will be exceeded even when the process is in control. The opposite situation also can occur: the observed tolerances may be much *wider* than the observed variation in the population. In this case an observation may be out of control relative to other sample values but may fall within desired tolerances. Whether the process is in control would yield little information about whether parts are meeting specifications. Section 10.6 on process capability explores these ideas more fully.

$\bar{X}$ Charts

Consider Example 10.1. Let us say that Howard decides to construct an $\bar{X}$ **chart** for the data summarized in Table 10–1. An $\bar{X}$ chart requires that the data be broken down into subgroups of fixed size. The size of the subgroups for the example is $n = 5$. The subgroup size should be at least 4 for the central limit theorem to apply.

To construct an $\bar{X}$ chart, it is necessary to estimate the sample mean and the sample variance of the population. This can be done using the given formulas. However, it generally is not recommended that one use the sample standard deviation as an estimator of σ when constructing an $\bar{X}$ chart. For s to be an accurate estimator for σ, it is necessary that the underlying mean of the sample be constant. Because the purpose of an $\bar{X}$ chart is to determine whether a shift in the mean has occurred, we should not assume a priori that the mean is constant when estimating σ. An alternative method for estimating the sample variation that remains accurate when the population mean changes uses data ranges. Even if the process mean shifts, the ranges will be stable as long as the process variation is stable. There is a relationship between the standard deviation of the population and the range of the subgroups of a given size that depends on the subgroup size. That is, there exists a constant d_2 such that

$$\hat{\sigma} = \frac{\bar{R}}{d_2},$$

where $\bar{R}$ is the average of the observed ranges and $\hat{\sigma}$ is an estimate of the population standard deviation. The constants d_2 for various subgroup sizes appear in Table A–5 at the back of this book. For the data presented in Table 10–1, the average of the 30 ranges turns out to be 0.035756. The value of d_2 for subgroups of size 5 is 2.326. Hence the estimator for σ based on this data is

$$\hat{\sigma} = 0.035756 / 2.326 = 0.01537,$$

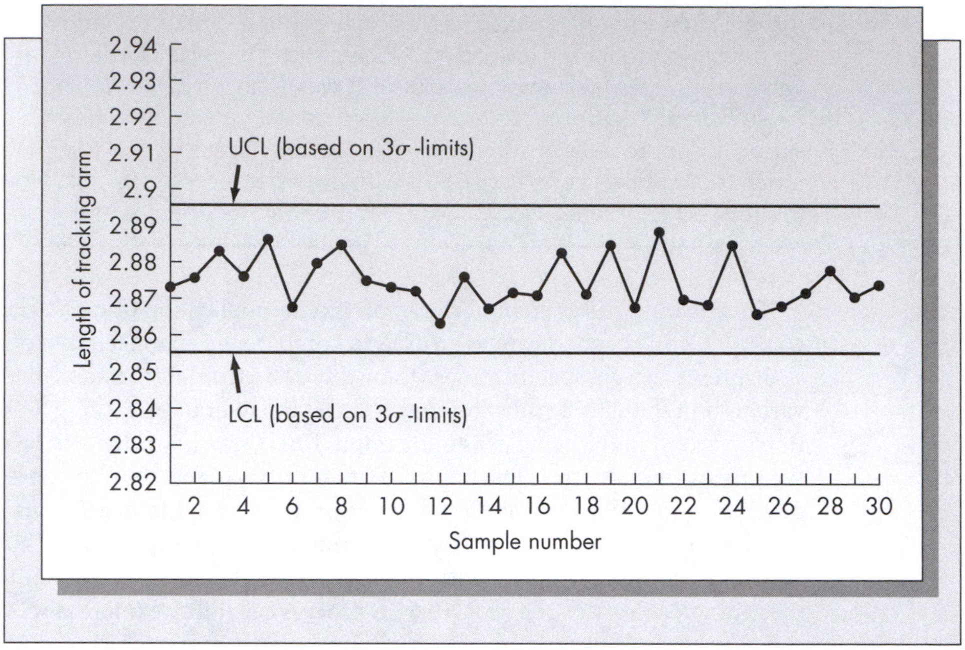

Figure 10–7 $\bar{X}$ chart for tracking arm data

which is quite close to the estimate of the standard deviation using the sample standard deviation *s*.

Given estimators for the mean and the standard deviation of the group average, the control charts are constructed in the following way. Lines are drawn for the upper and the lower control limits at $\bar{X} \pm 3\sigma / \sqrt{n}$ .The group averages are graphed on a daily basis. The process is said to be out of control if an observation falls outside of the control limits. The $\bar{X}$ chart for the sample of 30 days for the tracking arm appears in Figure 10–7. Notice that the process appears to be in control, as all observations fall within the 3σ limits.

Relationship to Classical Statistics

Here we consider statistical control charts in the context of classical statistical hypothesis testing. The null hypothesis is that the underlying process is in control. That is, we have the hypotheses

H_0: Process is in control.

H_1: Process is out of control.

We interpret the word **control** as meaning that the underlying chance mechanism generating observations over time is stable. For $\bar{X}$ charts, we test whether the process mean has undergone a shift. There are two ways that we can come to the wrong conclusion: reject the null hypothesis when it is true (conclude that the process is out of control when it is in control) and reject the alternative hypothesis when it is true (conclude the process is in control when it is out of control). These are called, respectively, Type 1 and Type 2 errors. We use the symbol α to represent the probability of a Type 1 error and β to represent the probability of a Type 2 error. A test is a rule that indicates when to reject H_0 based on the sample values. A test requires specification of an acceptable value of α. Conceptually, we are doing the same thing when we use control charts.

The hypothesis that the process is in control is rejected if an observed value of $\bar{X}$ falls outside the control limits. We can set the values of the upper control limit (UCL) and lower control limit (LCL) based on the specification of any value of α.

$$\begin{aligned}
\alpha &= P\{\text{Type 1 error}\} \\
&= P\{\text{Out-of-control signal is observed} \mid \text{Process is in control}\} \\
&= P\{\bar{X} < \text{LCL or } \bar{X} > \text{UCL} \mid \text{True mean is } \mu\} \\
&= P\{\bar{X} < \text{LCL} \mid \mu\} + P\{\bar{X} > \text{UCL} \mid \mu\} \\
&= P\left\{\frac{\bar{X}-\mu}{\sigma/\sqrt{n}} < \frac{\text{LCL}-\mu}{\sigma/\sqrt{n}}\right\} + P\left\{\frac{\bar{X}-\mu}{\sigma/\sqrt{n}} > \frac{\text{UCL}-\mu}{\sigma/\sqrt{n}}\right\} \\
&= P\left\{Z < \frac{\text{LCL}-\mu}{\sigma/\sqrt{n}}\right\} + P\left\{Z > \frac{\text{UCL}-\mu}{\sigma/\sqrt{n}}\right\}.
\end{aligned}$$

Because the normal distribution is symmetric, we set

$$\frac{\text{LCL}-\mu}{\sigma/\sqrt{n}} = -z_{\alpha/2}, \qquad \frac{\text{UCL}-\mu}{\sigma/\sqrt{n}} = z_{\alpha/2},$$

which gives

$$\text{LCL} = \mu - \frac{\sigma z_{\alpha/2}}{\sqrt{n}}, \qquad \text{UCL} = \mu + \frac{\sigma z_{\alpha/2}}{\sqrt{n}}.$$

Setting $z_{\alpha/2} = 3$, we obtain the popular three-sigma control limits. This is equivalent to choosing a value of $\alpha = .0026$. This particular value of α is the one that is traditionally used; it is not necessarily the only one that makes sense. In some applications, one might wish to increase the likelihood of recognizing when the process goes out of control. One would then use a larger value of α, which would result in tighter control limits. For example, a value of α of .05 would result in two-sigma rather than three-sigma limits. This increases the probability of a Type 1 error (but decreases the probability of a Type 2 error).

R Charts

The $\bar{X}$ chart is used to test for a shift in the mean value of a process. In many instances we are also interested in testing for a shift in the variance of the process. Process variation can be monitored by examining the sample variances of the subgroup observations. However, the ranges of the subgroups give roughly the same information and are much easier to compute. The theory behind the ***R* chart** is that when the underlying population is normal, there is a relationship between the range of the sample and the standard deviation of the sample that depends on the sample size. If $\bar{R}$ is the average of the ranges of all the subgroups of size n, then we have from earlier in this section

$$\hat{\sigma} = \bar{R} / d_2,$$

where d_2, which depends on n, appears in Table A–5 at the back of this book.

Normally, one would develop an R chart before an $\overline{X}$ chart in order to obtain a reliable estimator of the variance. The estimator $\hat{\sigma}$ is less sensitive to changes in the process mean than is the estimator s. The purpose of the R chart is to determine if the process variation is stable. The upper and lower limits for this chart are given by the formulas

$$\text{LCL} = d_3\overline{R},$$
$$\text{UCL} = d_4\overline{R}.$$

The values of the constants d_3 and d_4 appear in Table A–6 at the back of this book. The values given for these constants assume three-sigma limits for the range process.

EXAMPLE 10.1 (CONTINUED)

Again consider the data for the tracking arm in Table 10–1. The ranges of the samples of size $n = 5$ appear in the final column of the table. As stated earlier, the average of these 30 ranges is 0.035756. This is the value of $\overline{R}$ and becomes the center point for the R chart. The upper and lower control limits for R are computed using the given formulas and Table A–6. For the case of $n = 5$, we have $d_3 = 0$ and $d_4 = 2.11$, thus resulting in the following control limits for Wonderdisk's R chart:

$$\text{LCL} = (0)(0.035756) = 0,$$
$$\text{UCL} = (2.11)(0.035756) = 0.07545.$$

These are three-sigma limits.

Wonderdisk's R chart appears in Figure 10–8. Because all observed values of R fall within the control limits, the process variation is in control.

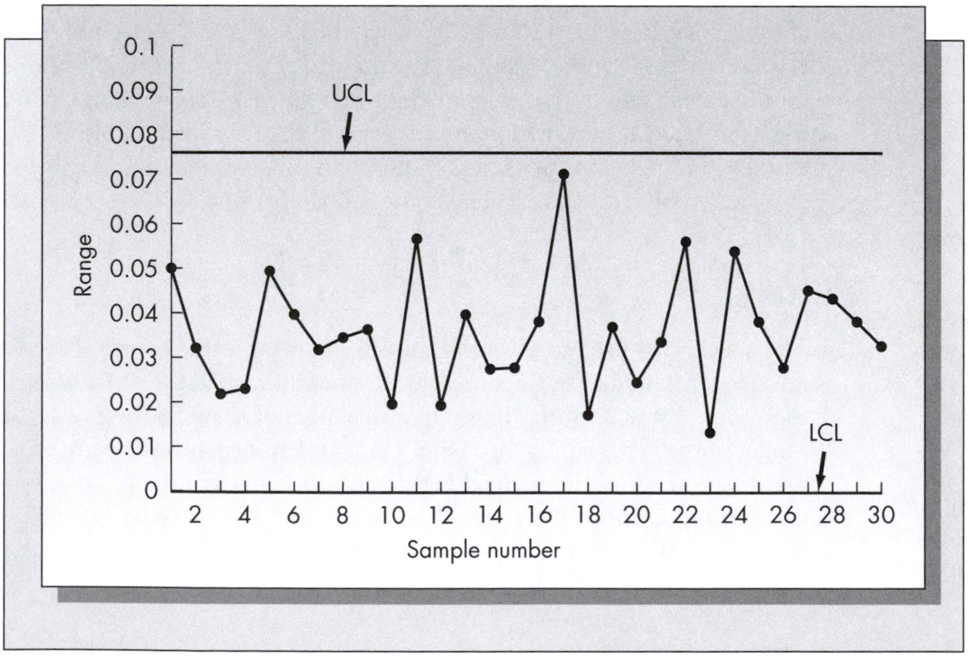

Figure 10–8 R chart for tracking arm data

R charts are used for testing whether there is a shift in the process variation. As discussed earlier, for the values of the estimators used to construct the $\overline{X}$ chart to be correct, the process variance should be constant. That is, it is recommended that the R chart be used *before* the $\overline{X}$ chart, since an $\overline{X}$ chart assumes that the process variation is stable.

R charts are not the only means for testing the stability of the process variation. One could also use a σ chart. One plots the sample standard deviations of subgroups over time to determine when and if a statistically significant shift in these values occurs. Sigma charts are rarely used in practice for two reasons.

1. It is more work to compute the sample standard deviations for each subgroup than it is to compute the ranges.
2. R charts and σ charts will almost always give the same results.

For these reasons we will not discuss σ charts in this text.

PROBLEMS FOR SECTION 10.3

7. The quality control group of a manufacturing company is planning to use control charts to monitor the production of a certain part. The specifications for the part require that each unit weigh between 13.0 and 15.5 ounces with a target value of 14.25. A sample of 75 observations results in the following:

$$\sum_{i=1}^{75} X_i = 1{,}065, \qquad \sum_{i=1}^{75} X_i^2 = 15{,}165.$$

   a. Are the specifications consistent with the statistical variation in the sample? (Hint: Use the computing formula for the sample variance: $s^2 = \dfrac{1}{n-1}\left[\sum_{i=1}^{n} X_i^2 - n\overline{X}^2\right]$.)
   b. What problems do you anticipate if the group attempts to use $\overline{X}$ and R charts for this process?
   c. If the 75 observations are statistically stable, what percentage of the manufactured items will fall outside the tolerances?

8. Control charts for $\overline{X}$ and R are maintained on the shear strength of spot welds. One hundred observations divided into subgroups of size five are used as a baseline to construct the charts, and estimates of μ and σ are computed from these observations. Assume that the 100 observations are $X_1, X_2, \ldots, X_{100}$ and the ranges of the 20 subgroups are $R_1, R_2, \ldots, R_{20}$. From these baseline data the following quantities are computed:

$$\sum_{i=1}^{100} X_i = 97{,}500, \qquad \sum_{j=1}^{20} R_j = 1{,}042.$$

   Using this information, compute the values of three-sigma limits for both charts.

9. For problem 8, suppose that the probability of concluding that the process is out of control when it is actually in control is set at .02. Find the upper and the lower control limits for the resulting $\overline{X}$ chart.

10. Suppose that the process mean shifts to 1,011 in problem 8.
    a. What is the probability that the $\bar{X}$ chart will *not* indicate an out-of-control condition from a single subgroup sampling? (This is precisely the Type 2 error.)
    b. What is the probability that the $\bar{X}$ chart will not indicate an out-of-control condition after sampling 20 subgroups?
11. A printing service monitors the quality of the produced colors with light-sensitive equipment. The accuracy measure is a number with a target value of zero. Suppose that an $\bar{X}$ chart with subgroups of size five is used to monitor the process and the control limits are UCL = 1.5 and LCL = –1.5. Assume that the estimate for the process mean is zero and for the process standard deviation is 1.30.
    a. What is the value of a for this control chart?
    b. Find the UCL and LCL based on three-sigma limits.
    c. Suppose that the process mean shifts to 1. What is the probability that the shift is detected on the first subgroup after the shift occurs?
12. An R chart is used to monitor the variation in the weights of packages of chocolate chip cookies produced by a large national producer of baked goods. An analyst has collected a baseline of 200 observations to construct the chart. Suppose the computed value of $\bar{R}$ is 3.825.
    a. If subgroups of size six are to be used, compute the value of three-sigma limits for this chart.
    b. If an $\bar{X}$ chart based on three-sigma limits is used, what is the difference between the UCL and LCL?
13. A process is monitored using an $\bar{X}$ chart with UCL = 13.8 and LCL = 8.2. The process standard deviation is estimated to be 6.6. If the $\bar{X}$ chart is based on three-sigma limits,
    a. What is the estimate of the process mean?
    b. What is the size of each of the sampling subgroups?

10.4 CONTROL CHARTS FOR ATTRIBUTES: THE *p* CHART

$\bar{X}$ and R charts are valuable tools for process control when the output of the process can be expressed as a single real variable. This is appropriate when there is a single quality dimension such as length, width, or hardness. In two circumstances control charts for variables are not appropriate: (1) when one's concern is whether an item has a particular attribute (for example, the issue might be whether the item functions) and (2) when there are many different quality variables. In case (2) it is not practical or cost-effective to maintain separate control charts for each variable. Either the item has the desired attributes or it does not.

When using control charts for attributes, each sample value is either a 1 or a 0. A 1 means that the item is acceptable, and a 0 means that it is not. Let n be the size of the sampling subgroup and define the random variable X as the total number of defectives in the subgroup. We will assume that each subgroup represents a sampling from one day's production. The theory would be exactly the same whether the sampling interval is one hour, one day, or one month. We also assume that defects occur independently of each other. Because X counts the number of independent defectives in a fixed sample size, the underlying distribution of X is binomial with parameters n and p. Interpret p as the proportion of defectives produced and n as the number of items sampled in each group (typically, n is the number of items sampled each day). A ***p* chart** would be used to determine if there is a significant shift in the true value of p.

Although one could construct p charts based on the exact binomial distribution, it is more common to use a normal approximation. Also, as our interest is in estimating the value of p, we track the random variable X/n, whose expectation is p, rather than $\overline{X}$ itself. It is easy to show that

$$E(X/n) = p,$$
$$\mathrm{Var}(X/n) = p(1-p)/n.$$

For large n, the central limit theorem tells us that X/n is approximately normally distributed with parameters $\mu = p$ and $\sigma = \sqrt{p(1-p)/n}$. Using a normal approximation, the traditional three-sigma limits are

$$\mathrm{UCL} = p + 3\sqrt{\frac{p(1-p)}{n}},$$
$$\mathrm{LCL} = p - 3\sqrt{\frac{p(1-p)}{n}}.$$

The estimate for p, the true proportion of defectives in the population, is $\overline{p}$, the average fraction of defectives observed over some reasonable baseline period. The process is said to be in control as long as the observed fraction defective for each subgroup remains within the upper and the lower control limits.

EXAMPLE 10.2

Xezet, a maker of smart phones, inspects a sample of 50 phones from each day's output. Based on a variety of attributes, the quality inspector classifies each phone as acceptable or not. The experience over a typical 20-day period is summarized in Table 10–2.

To construct a control chart for the fraction defective, it is necessary to have an accurate estimate of the true fraction defective in the entire population, p. Based on the data in Table 10–2 we construct a preliminary control chart to determine if the baseline data are in control. The total number of defectives observed during the 20 days is 176.

Table 10–2 Number of Rejected Phones

Date	Number Rejected	Date	Number Rejected
3/18	3	4/1	0
3/19	10	4/2	4
3/20	13	4/3	9
3/21	4	4/4	22
3/22	12	4/5	7
3/25	14	4/8	6
3/26	8	4/9	18
3/27	7	4/10	3
3/28	19	4/11	9
3/29	1	4/12	7

The total production over the same period of time is 1,000 phones. Hence, the current estimate of the proportion of defectives in the population is 176/1,000 = .176. The current estimator for σ is

$$\hat{\sigma} = \sqrt{\frac{\bar{p}(1-\bar{p})}{n}} = \sqrt{\frac{(.176)(.824)}{50}} = 0.54.$$

Based on three-sigma limits we obtain

$$\text{UCL} = 0.176 + (3)(0.054) = 0.338,$$
$$\text{LCL} = 0.176 - (3)(0.054) = 0.014.$$

Figure 10–9 is the preliminary control chart for the fraction defective. Notice that four points are out of control. These correspond to days 9 (3/28), 11 (4/1), 14 (4/4), and 17 (4/9). We need not worry about a point that falls below the lower control limit, because that shows a *better*-than-expected rate of defectives. The production manager considers the three remaining out-of-control points and realizes that they correspond to three days when a key employee was absent from work for personal reasons. The employee's job requires some very complex smart-screen installation equipment. The high rate of defectives on these days was apparently the result of a temporary employee's lack of experience with equipment.

Because the out-of-control points were explained by an assignable cause, we eliminate these points from our sample. The baseline data now consist of the data listed in Table 10–2 with the three days corresponding to the out-of-control points eliminated. That is, our database now consists of a total of 17 days. Our estimate of p is now recomputed based on the revised data. We obtain

$$\bar{p} = 117/(17)(50) = .138,$$

and

$$\hat{\sigma} = \sqrt{(.138)(.862)/50} = 0.049.$$

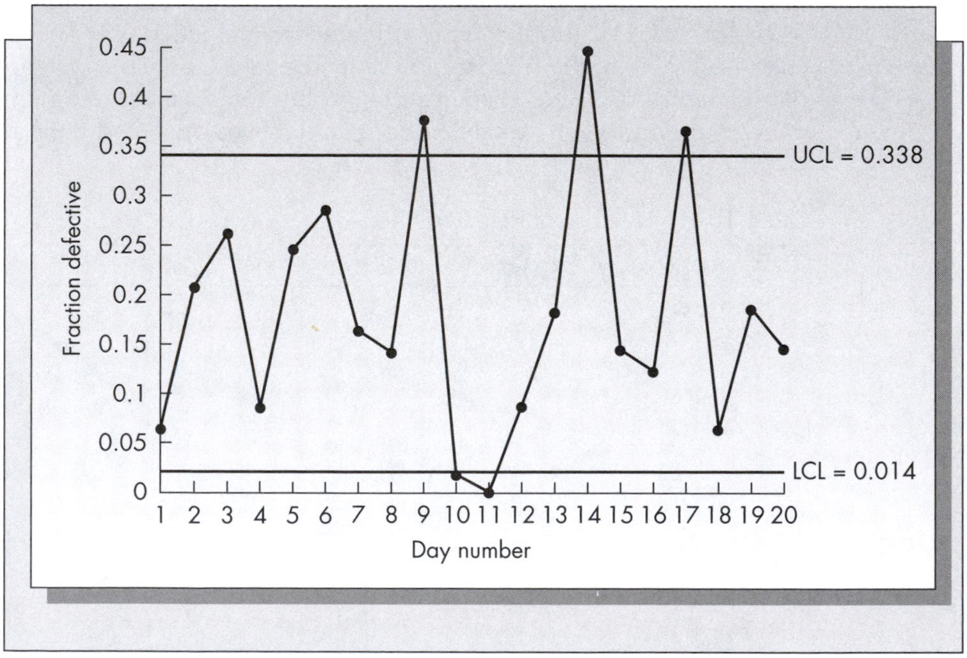

Figure 10–9 Preliminary p chart for Xezet phone data (Example 10.2)

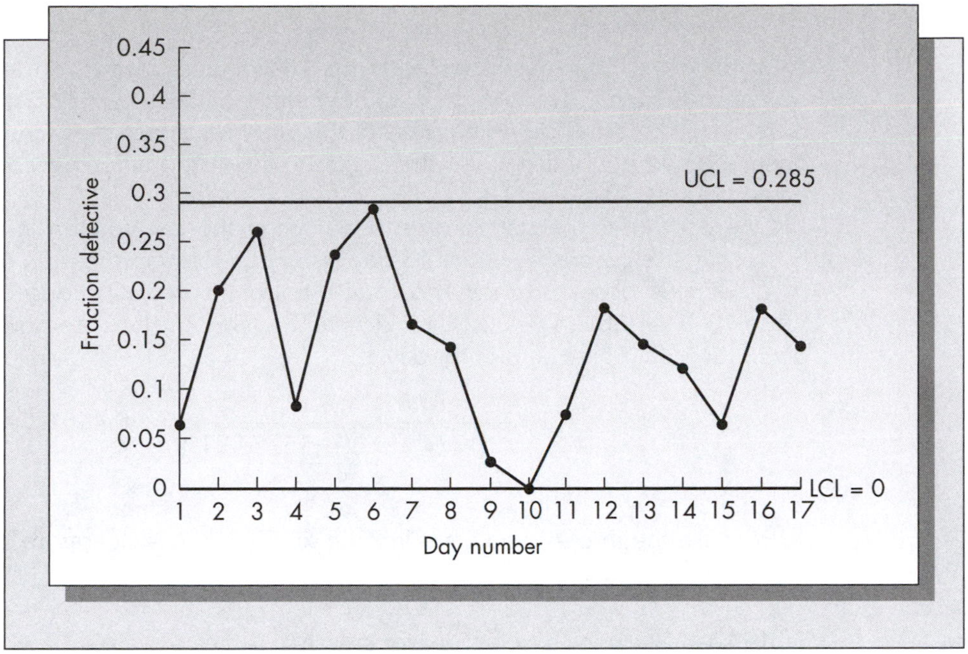

Figure 10–10 Revised *p* chart for Xezet phone data (Example 10.2)

Based on these new estimators for *p* and σ we obtain

$$UCL = 0.138 + (3)(0.049) = 0.285,$$
$$LCL = 0.138 - (3)(0.049) = -0.009.$$

If LCL < 0, we set LCL = 0, because it is impossible to observe a value of *p* that is negative. In Figure 10–10 we graph the revised *p* chart for the phone data. Notice that with the three out-of-control points eliminated, all remaining points now fall within the control limits. This will not always be the case, as the upper and the lower control limits are closer in Figure 10–10 than they were in Figure 10–9. The revised control limits would now be used to monitor future observations of the fraction defective for this process.

p Charts for Varying Subgroup Sizes

Previously, we assumed that the number of items inspected in each subgroup was the same. This assumption is reasonable when the subgroups are sampled periodically from large lots. However, in many circumstances few items are produced each day and are consequently subject to 100 percent inspection. If daily production varies, the subgroup size also will vary. Here we base the analysis on the standardized variate *Z*:

$$Z = \frac{p - \overline{p}}{\sqrt{\overline{p}(1-\overline{p})/n}},$$

which is approximately standard normal independent of *n*. The lower and the upper control limits would be set at −3 and +3, respectively, to obtain three-sigma limits, and the control chart would monitor successive values of the standardized variate *Z*.

EXAMPLE 10.3

A producer of industrial lathes performs 100 percent inspection and testing of each day's production. The lathe production varies from day to day based on anticipated orders and the production schedule for the other machines the company produces. The observed number of defective lathes and the daily production over the past 15 days are given in Table 10–3.

One computes the standardized Z values in the last column in the following way. First, estimate p from the entire 15-day history by forming the ratio of the total number of defectives observed over the total number of items produced. For the data in Table 10–3 we obtain $\bar{p} = 137/1{,}211 = .1131$. The Z values are now computed by the given formula. For example, for day 1,

$$Z = \frac{8/82 - 137/1{,}211}{\sqrt{\dfrac{(137/1{,}211)(1 - 137/1{,}211)}{82}}} = -0.4451.$$

Because the three-sigma limits in Z are simply ± 3, it is clear from Table 10–3 that this process is in control.

Table 10–3 Observed Defectives for Industrial Lathes (Example 10.3)

Day Number	Production	Number of Defectives	Standardized Z Value
1	82	8	−0.4451
2	76	12	1.2320
3	85	6	−1.2382
4	53	5	−0.4319
5	30	3	−0.2270
6	121	14	0.0893
7	63	11	1.5404
8	80	9	−0.0178
9	88	7	−0.9946
10	97	8	−0.9532
11	91	13	0.8953
12	77	14	1.9029
13	71	6	−0.7614
14	95	8	−0.8899
15	102	13	0.4566

PROBLEMS FOR SECTION 10.4

14. A bakery produces an average of 12,000 bread rolls of various types each day. An inspector samples 25 from each day's production and tests them for taste, color, and overall quality. During 23 consecutive production days, the inspector rejected the following numbers of rolls.

Day	Number Rejected	Day	Number Rejected
1	2	13	1
2	3	14	1
3	2	15	0
4	1	16	0
5	3	17	0
6	4	18	2
7	7	19	3
8	1	20	1
9	3	21	2
10	3	22	2
11	0	23	1
12	0		

a. Based on these observations, compute three-sigma limits for a p chart.
b. Do any of the points fall outside the control limits? If so, recompute the three-sigma limits after eliminating the out-of-control points.

15. Applied Machines produces large test equipment for integrated circuits. The machines are made to order, so the production rate varies from month to month. Before shipping, each machine is subject to extensive testing. Based on the tests the machine is either passed or sent back for rework. During the past 20 months the firm has had to rework the following numbers of machines.

Month	Number Produced	Number Reworked	Month	Number Produced	Number Reworked
1	23	3	11	17	3
2	28	3	12	4	0
3	16	1	13	14	2
4	6	0	14	0	0
5	41	2	15	18	6
6	32	4	16	0	0
7	29	5	17	33	4
8	19	2	18	46	5
9	12	1	19	21	7
10	7	1	20	29	7

Determine if the process was in control for the 20-month period using a standardized version of the p chart. Assume three-sigma limits for the control chart. (Hint: Do you actually have 20 months of data?)

16. Consider the example of Applied Machines presented in problem 15. Based on the estimate of the probability that a machine is sent back for rework computed from the 20 months of data, determine the following.

a. If the company produces 35 machines in one particular month, how many, on average, require rework?
b. Out of 100 machines produced, what is the probability that more than 20 percent of them require rework? (Use the normal approximation to the binomial for your calculations. It is discussed in Appendix 10–A.)

17. Over a period of 12 consecutive production hours, samples of size 50 resulted in the following proportions of defective items.

Sample Number	Proportion Defective	Sample Number	Proportion Defective
1	.04	7	.10
2	.02	8	.10
3	.06	9	.06
4	.08	10	.08
5	.08	11	.04
6	.04	12	.04

a. What are the three-sigma control limits for this process?
b. Do any of the sample points fall outside of the control limits?
c. The company claims a defect rate of 3 percent for these items. Are the observed proportions consistent with a target value of 3 percent defectives? What difficulty would arise if the control limits were based on a target value of 0.03? In view of the company's claims, what difficulty would arise if the control limits computed in part (a) were used?

10.5 THE *c* CHART

There are control charts other than $\overline{X}$, R, and p charts. Although the form of the distribution of the appropriate random variable depends on the application, the basic approach is the same. In general, one must determine the probability distribution of the random variable of interest and find upper and lower control limits that contain the universe of observations with a desired level of confidence. Usually one sets the probability of falling outside the control limits to be less than .01.

The p chart is appropriate when classifying an item as either good or bad. However, often we are concerned with the number of defects in an item or collection of items. An item is acceptable if the number of defects is not too large. For example, a refrigerator that has a few scratches might be considered acceptable, but one that has too many scratches might be considered unacceptable. As another example, for a textile mill that manufactures cloth, both the manufacturer and the consumer would be concerned with the number of defects per yard of cloth.

The ***c* chart** is based on the observation that if the defects are occurring completely at random, then the probability distribution of the number of defects per unit of production has the Poisson distribution. If c represents the true mean number of defects in a unit of production, then the likelihood that there are k defects in a unit is

$$P\{\text{Number of defects in one unit} = k\} = \frac{e^{-c}c^k}{k!} \qquad \text{for } k = 0, 1, 2, \ldots.$$

In using a control chart for the number of defects, the sample size should be the same at each inspection. One estimates the value of c from baseline data by computing the sample mean of the observed number of defects per unit of production. When $c \geq 20$, the normal distribution provides a reasonable approximation to the Poisson. Because the mean and the variance of the Poisson are both equal to c, it follows that for large c,

$$Z = \frac{X - c}{\sqrt{c}}$$

is approximately standard normal. Using the traditional three-sigma limits, the upper and the lower control limits for the c chart are

$$\text{LCL} = c - 3\sqrt{c},$$
$$\text{UCL} = c + 3\sqrt{c}.$$

One develops and uses a c chart in the same way as $\overline{X}$, R, and p charts.

EXAMPLE 10.4

Leatherworks produces various leather goods in its plant in Montpelier, Vermont. Inspections of the past 20 units of a leather portfolio revealed the following numbers of defects.

Unit Number	Number of Defects Observed	Unit Number	Number of Defects Observed
1	4	11	2
2	3	12	3
3	3	13	6
4	0	14	1
5	2	15	5
6	5	16	4
7	4	17	1
8	2	18	1
9	3	19	2
10	3	20	2

Most defects are the result of natural marks in the leather, but even so, the firm does not want to ship products with too many defects in the leather. Using these 20 data points as a baseline, determine upper and lower control limits that include the universe of observations with probability .95. What control limits result from using a normal approximation of the Poisson?

Solution

To estimate c we compute the sample mean of the data, which is found by adding the total number of observed defects and dividing by the number of observations. This gives $c = 56/20 = 2.8$. To be certain that a c chart is appropriate here, we should do a goodness-of-fit test of these data for a Poisson distribution with parameter 2.8. We will leave it to the reader to check that the data do indeed fit a Poisson distribution. (Goodness-of-fit tests are described in almost every statistics text. The most common is the chi-square test.)

To determine exact control limits, we use Table A–3 in the back of the book. Because the Poisson is a discrete distribution, it is very unlikely that we will be able to find control limits that contain exactly 95 percent of the probability. From the table we see that the probability that the number of defects is less than or equal to zero is 1.0000 – .9392 = .0608, which is too large. Hence, we will set the lower control limit to zero. By symmetry, the upper limit should correspond to a right tail of about .025, which occurs at $k = 7$. Hence we would recommend control limits of LCL = 0 and UCL = 7.

For a normal distribution, approximately two standard deviations from the mean include 95 percent of the probability. Hence, the control limits based on a normal approximation of the Poisson are

$$\text{LCL} = c - 2\sqrt{c} = 2.8 - (2)\sqrt{2.8} = -0.55 \qquad \text{(set to zero)},$$
$$\text{UCL} = c + 2\sqrt{c} = 2.8 + (2)\sqrt{2.8} = 6.2.$$

The normal approximation is not very accurate in this case because c is too low.

PROBLEMS FOR SECTION 10.5

18. Amertron produces electrical wiring in 100-foot rolls. The quality inspection process involves selecting rolls of wire at random and counting the number of defects on each roll. The last 20 rolls examined revealed the following numbers of defects.

Roll	Number of Defects	Roll	Number of Defects
1	4	11	2
2	6	12	5
3	2	13	5
4	4	14	7
5	1	15	4
6	9	16	8
7	5	17	6
8	5	18	4
9	3	19	6
10	3	20	4

a. If the number of defects per 100-foot roll of wire follows a Poisson distribution, what is the estimate of c obtained from these observations?
b. Using a normal approximation to the Poisson, what are the three-sigma control limits that you would use to monitor this process?
c. Are all 20 observations within the control limits?

19. Amertron, discussed in problem 18, has established a policy of passing rolls of wire having five or fewer defects.

a. Based on the exact Poisson distribution, what is the proportion of the rolls that pass inspection? (See Table A–3 in the back of the book.)
b. Estimate the answer to part (a) using the normal approximation to the Poisson.

20. A large national producer of cookies and baked goods uses a *c* chart to monitor the number of chocolate chips in its chocolate chip cookies. The company would like to have an average of six chips per cookie. One cookie is sampled each hour. The results of the last 12 hours appear below.

Hour	Number of Chips per Cookie	Hour	Number of Chips per Cookie
1	7	7	3
2	4	8	6
3	3	9	3
4	3	10	2
5	5	11	4
6	4	12	4

a. Assuming a target value of $c = 6$, what are the upper and the lower control limits for a *c* chart?
b. Are the 12 observations consistent with a target value of $c = 6$? If those 12 observations constitute a baseline, what upper and lower control limits result? (Use the normal approximation for your calculations.)

21. For the company mentioned in problem 20, a purchaser of a bag of chocolate chip cookies discovers a cookie that has no chips in it and charges the company with fraudulent advertising. Suppose the company produces 300,000 cookies per year. If the expected number of chips per cookie is six, how many cookies baked each year would have no chips? See Table A–3 in the back of the book.

10.6 PROCESS CAPABILITY

A process is said to be **capable** if it can reliably meet specifications (i.e., produce items that meet the required tolerances). In order to be capable, the process must both be in control and the three-sigma limits should fall within the specification limits. It is very important to realize that control limits simply show what the current process is producing; they do not reflect goals or targets for the process. In addition to upper and lower control limits, it is often helpful to plot the upper and lower specification or tolerance limits for the process.

The **process capability ratio,** C_p, compares the specifications range with the natural variation of the process as follows:

$$C_p = \frac{\text{USL} - \text{LSL}}{6\sigma},$$

where LSL and USL are the **lower** and **upper specification limits**, respectively. That is, it compares what the customer desires with what the process is reliably able to produce. A process is not capable if $C_p < 1$. If $C_p = 1$, then the process is only capable if the mean of the process is centered within the specification limits.

The **process capability index,** C_{pk}, is used to account for both the **spread** and **setting** of the process as follows:

$$C_{pk} = \min\left(\frac{\bar{X}-LSL}{3\sigma}, \frac{USL-\bar{X}}{3\sigma}\right).$$

Unlike C_p, C_{pk} can be zero or negative. A $C_{pk} = 0$ implies that half the items are defective, and a negative C_{pk} implies that more than half are defective. A capable process should have a C_{pk} of at least 1, and higher is always better.

EXAMPLE 10.5

Reconsider the tracking arm production process of Example 10.1. The mean and standard deviation of the process were found to be 2.875 and 0.0156 respectively. The lower and upper specification limits were 2.85 and 2.9, respectively. In this case

$$C_p = \frac{2.9-2.85}{6\times 0.0156} = 0.534.$$

Because this is less than one, as discussed in the example, the process is not capable. When the specification limits are changed to 2.825 and 2.925, respectively, $C_p = 1.068$, making the process capable.

Now suppose that the engineers are comfortable with the new USL of 2.925 but believe the original LSL of 2.85 should stand. Because the process is no longer centered on the specification limits, we must use the process capability index as follows.

$$C_{pk} = \min\left(\frac{\bar{X}-LSL}{3\sigma}, \frac{USL-\bar{X}}{3\sigma}\right) = \min\left(\frac{2.875-2.85}{3\times 0.0156}, \frac{2.925-2.875}{3\times 0.0156}\right)$$
$$= \min(0.534, 1.068) = 0.534.$$

The process is not capable. Note that it is not a coincidence that the values in this calculation match the earlier calculated values.

A six-sigma process is one with a process capability index of 2. That is, the specifications range is double what the process can reliably produce. In other words, the process spread is half what is needed according to the specifications and therefore very unlikely to be violated. The standard definition of Six Sigma quality, namely 3.4 defects per million parts, allows for a shift in the mean by 1.5σ. In particular, suppose that X is normally distributed with mean $\mu = \frac{LSL+USL}{2} + 1.5\sigma$ and standard deviation σ. Further, suppose $USL - LSL = 12\sigma$ (6σ on either side of the midpoint). Then

$$P(LSL \le X \le USL) = P\left(\frac{LSL-USL}{2} - 1.5\sigma \le X-\mu \le \frac{USL-USL}{2} - 1.5\sigma\right)$$
$$= P(-6\sigma - 1.5\sigma \le X-\mu \le 6\sigma - 1.5\sigma) = P(-7.5 \le Z \le 4.5)$$

where Z is the standard normal random variable. Putting this into Excel (using the NORM.S.DIST function) gives $P(LSL \le X \le USL) = 0.999996602$ or 3.398 defects per

SNAPSHOT APPLICATION

Navistar Scores with Six-sigma Quality Program

Navistar International is a major U.S. manufacturer of trucks, buses, and engines with several plants around the world. In 1985 Navistar's worldwide workforce numbered over 110,000. Because of a crippling United Auto Workers (UAW) strike and a recession, the company had to severely trim the workforce to survive, reducing its numbers to around 12,000. To combat cost and quality problems it was experiencing at the time, Navistar decided to launch a Six Sigma quality program in the mid-1990s.

As noted in this section, Six Sigma quality means defect rates of 3.4 parts per million or less. While Six Sigma programs rarely achieve such low defect rates, the goal is clear: Do what needs to be done in the organization to effect a fundamental change in both management's and labor's attitudes about quality. Quality programs do not come free, however. Navistar paid a consulting company more than $6 million to implement its program. One immediate result was that Navistar's stock price grew over 400 percent in the 14 months following implementation of the program. (The price of a company's stock is influenced by many factors, so it isn't clear what role the Six Sigma program played.)

Six Sigma programs have their own culture. Specially trained employees are dubbed black belts after one month's training and master black belts after additional training. The black belts are assigned specific projects and have the power to go directly to top management with proposed solutions. For such an approach to work, management as well as employees must be firmly committed to the program. Does everyone believe in the value of Six Sigma programs? Evidently not. For example, Charles Holland, president of a consulting company based in Knoxville that specializes in statistical quality control methods, dubs the Six Sigma program a "silver bullet" sold at "outrageous prices."(-Franklin, 1999, p. 8).

If this is true, what motivated Navistar to plunk down $6 million for this program? According to John Horne, the company's chief executive in 1995, the company needed an antidote to the slide it was experiencing: "We didn't have a strategy; most companies don't." The strategy that Horne adopted was to go after the company's problems at the plant level. Quality control problems had been dogging Navistar's plants for years. The target of the Six Sigma program was the massive 4,000 square foot plant in Springfield, Ohio. (Navistar did not implement Six Sigma in all its plants for various reasons. For example, union opposition prevented implementation in the Canadian plant located in Chatham, Ontario.)

What was the result in Springfield? The effort was credited with $1 million of savings the first year and greater savings in subsequent years. The total savings in the Springfield plant were projected to be $26 million, well above the $6 million cost of the program. Sometimes kaizen (continuous improvement) is simply not enough to fix a troubled system. While expensive, Six Sigma can provide the jump start needed to turn things around, as it did with Navistar.

As of this writing, the Navistar Springfield plant is still going strong. They have invested heavily in new equipment in the last few years. In 2017 they announced that they were at "full-run-rate production" on a new van production line (Navera, 2017).

million parts. If the process was not shifted, that is the mean fell halfway between the specification limits, then a six-sigma process would only produce 2 defects per billion.

PROBLEMS FOR SECTION 10.6

22. Show that if the mean is decreased by 15σ (instead of increased in the text above) then the defect rate is also 3.4 defects per million parts.

23. Calculate the process capability ratio and process capability index for the process in problem 7.
24. Suppose that the shear strength of spot welds in problem 8 have a specification range of 950 to 1025. Calculate the process capability ratio and process capability index for this process. Is the process capable? Why or why not?
25. For the printing service in problem 11, suppose that the colors are acceptable within a range of plus or minus 1. Calculate the process capability ratio and process capability index for this process. Is the process capable? Why or why not?

10.7 OVERVIEW OF ACCEPTANCE SAMPLING

Control charts provide a convenient way to monitor a process in real time to determine if a shift in the process parameters appears to have occurred. Another important aspect of quality control is to determine the quality of manufactured goods *after* they have been produced. In many cases 100 percent inspection is either impossible or impractical. Hence, a sample of items is inspected and quality parameters of large lots of items are estimated based on the results of the sampling.

To be more specific, **acceptance sampling** addresses the following problem. If a sample is drawn from a large lot of items and the sample is subject to 100 percent inspection, what inferences can we draw about the quality of the lot based on the quality of the sample? Statistical analysis provides a means for extrapolating the characteristics of a sample to the characteristics of the lot and a means for determining the probability of coming to the wrong conclusion.

Obviously, 100 percent inspection of all items in the lot will reduce the probability of an incorrect conclusion to zero. However, there are several reasons that 100 percent inspection is either not feasible or not desirable; some of the reasons are listed below.

1. In most cases 100 percent inspection is too costly. It is virtually impossible for high-volume transfer lines and continuous production processes.
2. In some cases 100 percent inspection may be impossible, such as when inspection involves destructive testing of the item. For example, determining the lifetime of a light bulb requires burning the bulb until it fails.
3. If the inspection is done by the consumer rather than the producer, 100 percent inspection by the consumer provides little incentive to the producer to improve quality. It is cheaper for the producer to repair or replace the items returned by the consumer than it is to improve the quality of the production process. However, if the consumer returns the entire lot based on the results of sampling, it provides a much greater motivation to the producer to improve the quality of outgoing lots.

In this chapter we discuss the following three sampling plans. The first (single sampling) is given in the next section while the remaining two may be found in Appendices 10–D and 10–E.

1. *Single sampling plans*. Single sampling plans are by far the most popular and easiest to use of the plans we will discuss. Two numbers, n and c, determine a single sampling plan. If there are more than c defectives in a sample of size n, the lot is rejected; otherwise it is accepted.

2. *Double sampling plans.* In a double sampling plan, we first select a sample of size n_1. If the number of defectives in the sample is less than or equal to c_1, the lot is accepted. If the number of defectives is greater than c_2, then the lot is rejected. However, if the number of defectives is larger than c_1 and less than or equal to c_2, a second sample of size n_2 is drawn. The lot is now accepted if the cumulative number of defectives in both samples is less than or equal to a third number, c_3. (Often $c_3 = c_2$.)
3. *Sequential sampling.* A double sampling plan can obviously be extended to a triple sampling plan, which can be extended to a quadruple sampling plan, and so on. A sequential sampling plan is the logical conclusion of this process. Items are sampled one at a time, and the cumulative number of defectives is recorded at each stage of the process. Based on the value of the cumulative number of defectives, there are three possible decisions at each stage.
   a. Reject the lot.
   b. Accept the lot.
   c. Continue sampling.

A complex sampling plan may have desirable statistical properties, but the acceptance and rejection regions could be difficult to calculate and the plan difficult to implement. The right sampling plan for a particular environment may not be the most mathematically sophisticated. As with any analytical tool, the potential benefits must be weighed against the potential costs.

Peter Kolesar (1993) makes the point that with improving quality standards, the value of acceptance sampling diminishes. When defect rates are low (e.g., for six-sigma quality), acceptance sampling becomes very inefficient. For example, suppose the defect rate in a six-sigma capable process increased by a factor of 10. In that case, the probability of finding a defect in a sample of, say, 1,000 units would be only .034. We wouldn't expect to see a single defect until we have sampled at least 29 lots on average! However, we should keep in mind that six-sigma quality is not possible in all industries. Acceptance sampling is a valuable tool for immature or cutting-edge processes with high defect rates.

We will use the following notation throughout the remainder of this chapter.

N = Number of pieces in a given lot or batch.
n = Number of pieces in the sample ($n < N$).
M = Number of defectives in the lot.
β = Consumer's risk—the probability of accepting bad lots.
α = Producer's risk—the probability of rejecting good lots.
c = Rejection level.
X = Number of defectives in the sample.
p = Proportion of defectives in the lot.
p_0 = Acceptable quality level (AQL).
p_1 = Lot tolerance percent defective (LTPD).

Assume that N is a known constant. If N is very large relative to the sample size, n, it can be assumed to be infinite. In that case it will not enter into the calculations. Although M is a constant as well, its value is *not* known in advance. In fact, only 100 percent inspection will reveal the true value of M. Often we are interested in analyzing the behavior of the sampling plan for various values of M. The consumer's risk and the producer's risk depend upon the sampling plan. Finally, X, the number of defectives in the sample, is a random variable. This means that were we to repeat the sampling experiment with a different random sample of size n, we would not necessarily observe the same number of defectives. Based on statistical properties of the population as a whole, we can determine the form of the probability distribution of X.

The acceptable quality level, p_0, is the desired or target level of the proportion of defectives in the lot. If the true proportion of defectives in the lot is less than or equal to p_0, the lot is considered to be acceptable. The lot tolerance percent defective, p_1, is an unacceptable proportion of defectives in the lot. The lot is considered unacceptable if the proportion of defectives exceeds p_1. Because of the imprecision of statistical sampling, we allow a gray area between p_0 and p_1. When the AQL and LTPD are equal, large sample sizes may be required to achieve acceptable values of α and β (the Type 1 and Type 2 error probabilities).

10.8 SINGLE SAMPLING FOR ATTRIBUTES

The goal of all sampling procedures is to estimate the properties of a population from the properties of the sample. In particular, we wish to test two hypotheses.

$$\mathrm{H}_0\text{: Lot is of acceptable quality } (p \le p_0).$$
$$\mathrm{H}_1\text{: Lot is of unacceptable quality } (p \ge p_1).$$

The test is of the form: Reject H_0 if $X > c$. The value of c depends on the choice of the Type 1 error probability α. The Type 1 error probability is the probability of rejecting H_0 when it is true. In the context of the quality control problem, this is the probability of rejecting the lot when it is acceptable. This is also known as the producer's risk. In equation form,

$$\alpha = P\{\text{Reject H}_0 \mid \text{H}_0 \text{ true}\} = P\{\text{Reject lot} \mid \text{Lot is good}\} = P\{X > c \mid p = p_0\}.$$

The exact distribution of X is *hypergeometric* with parameters n, N, and M. That is,

$$P\{X = m\} = \frac{\binom{M}{m}\binom{N-M}{n-m}}{\binom{N}{n}} \qquad \text{for } 0 \le m \le \min(M, n),$$

where

$$\binom{N}{n} = \frac{N!}{n!(N-n)!}.$$

In most applications, N is much larger than n, so the binomial approximation to the hypergeometric is satisfactory. In that case

$$P\{X = m\} = \binom{n}{m} p^m (1-p)^{n-m} \qquad \text{for } 0 \le m \le n,$$

where $p = M/N$ is the true proportion of defectives in the lot.

Using the binomial approximation, the producer's risk and the consumer's risk are given by

$$\alpha = P\{X > c \mid p = p_0\} = \sum_{m=c+1}^{n} \binom{n}{m} p_0^m (1-p_0)^{n-m},$$

$$\beta = P\{X \le c \mid p = p_1\} = \sum_{m=0}^{c} \binom{n}{m} p_1^m (1-p_1)^{n-m}.$$

Most statistical tests require specification of the probability of Type 1 error, α. Values of α, n, and p_0 will determine a unique value of c, which can be found from tables of the cumulative binomial distribution. However, because the binomial is a discrete distribution, it may not be possible to find c to match exactly the desired value of α. When p is small and n is moderately large ($n > 25$ and $np < 5$), the Poisson distribution provides an adequate approximation to the binomial. For very large values of n such that $np(1 - p) > 5$, the normal distribution provides an adequate approximation to the binomial. Refer to Appendix 10–A for a detailed discussion of these approximations.

EXAMPLE 10.6

Spire Records is a small West Coast retail chain of stores specializing in traditional vinyl records. One of Spire's suppliers is B&G, which ships refurbished records to Spire in 100-record lots. After some negotiation, Spire and B&G have agreed that a 10 percent rate of defectives is acceptable and a 30 percent rate of defectives is unacceptable. From each lot of 100 records, Spire has established the following sampling plan: 10 records are sampled, and if more than 2 are found to be warped, scratched, or defective in some other way, the lot is rejected. Consider the consumer's and the producer's risk associated with this sampling plan.

From the given information, we have that $p_0 = .1$, $p_1 = .3$, $n = 10$, and $c = 2$. Hence,

$$\alpha = P\{X > c \mid p = p_0\} = P\{X > 2 \mid p = .1\} = 1 - P\{X \le 2 \mid p = .1\}$$

$$= 1 - \sum_{k=0}^{2} \binom{10}{k} (.1)^k (.9)^{10-k} = 1 - .9298 = .0702.$$

$$\beta = P\{X \le c \mid p = p_1\} = P\{X \le 2 \mid p = .3\}$$

$$= \sum_{k=0}^{2} \binom{10}{k} (.3)^k (.7)^{10-k} = .3828.$$

Note that the parameter values of $n = 10$, $p = .1$, and $n = 10$, $p = .3$ imply that neither the normal nor the Poisson approximation is accurate. (The reader may wish to check, using Table A–3 at the back of this book, that using the Poisson distribution with $\lambda = np$ gives α and β the approximate values of .0803 and .4216, respectively.)

Derivation of the OC Curve

The **operating characteristic** (OC) **curve** measures the effectiveness of a test to screen lots of varying quality. The OC curve is a function of p, the true proportion of defectives in the lot, and is given by

$$\text{OC}(p) = P\{\text{Accepting the lot} \mid \text{True proportion of defectives} = p\}.$$

We will now derive the form of the OC curve for the particular case of a single sampling plan with sample size n and rejection level c. In that case,

$$\begin{aligned}\text{OC}(p) &= P\{X \le c \mid \text{Proportion of defectives in lot} = p\}\\ &= \sum_{k=0}^{c} \binom{n}{k} p^k (1-p)^{n-k}.\end{aligned}$$

Ideally, the sampling procedure would be able to distinguish perfectly between good and bad lots. Figure 10–11 shows the ideal OC curve.

EXAMPLE 10.6 (CONTINUED)

Consider again Example 10.6 of Spire Records. The OC curve for its single sampling plan is given by

$$\text{OC}(p) = \sum_{k=0}^{2} \binom{10}{k} p^k (1-p)^{10-k}.$$

The graph of Spire's OC curve appears in Figure 10–12. An examination of the figure shows that this particular sampling plan is more advantageous for the supplier, B&G,

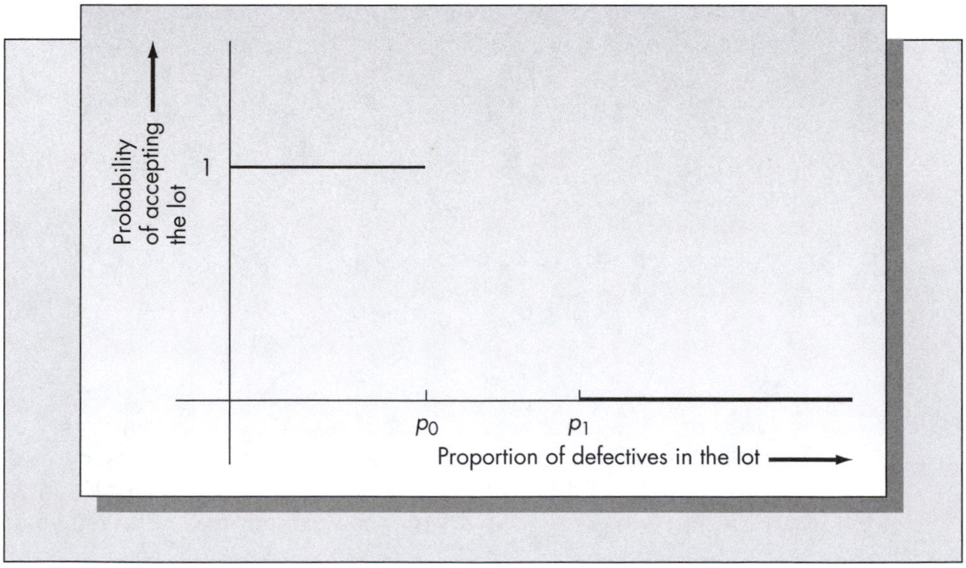

Figure 10–11 The ideal OC curve

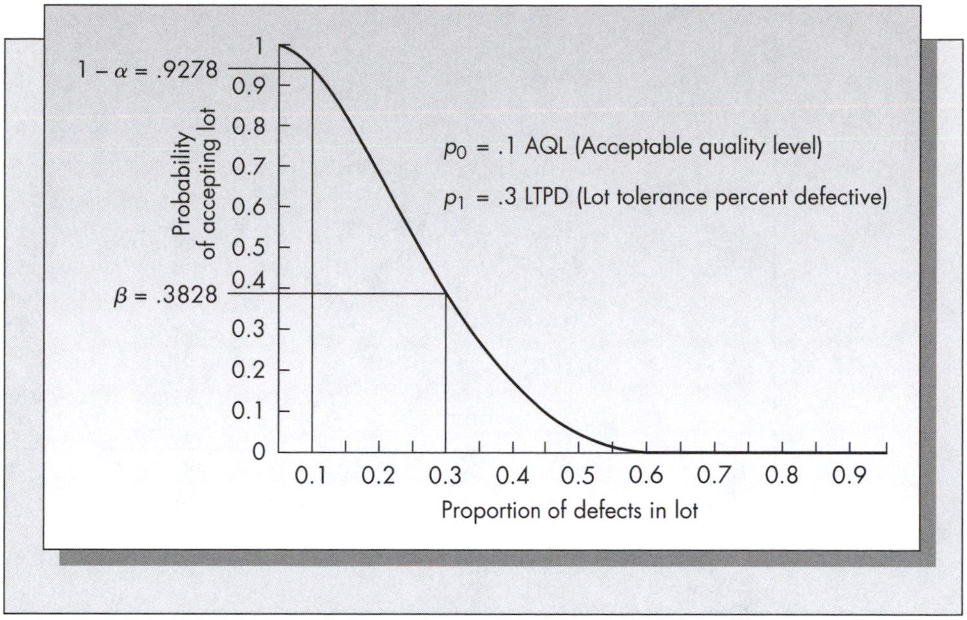

FIGURE 10–12 OC curve for Spire Records ($n = 10$)

than it is for Spire. The value of $\beta = .3828$ means that Spire is passing almost 40 percent of the lots that contain 30 percent defectives. Furthermore, the probability of accepting lots with proportions of defectives as high as 40 percent and even 50 percent is not negligible. This accounts for Spire's experience that there seemed to be many customer returns of B&G label records.

Herman Sondle, an employee of Spire enrolled in a local Masters program, was asked to look into the problem with B&G records. He discovered the cause of the trouble by analysis of the OC curve pictured in Figure 10–12. In order to decrease the chances that Spire receives bad lots from B&G, he suggested that the sampling plan be modified by setting $c = 0$. The resulting consumer's risk is

$$\beta = P\{X \leq 0 \mid p = .3\} = (.3)^0(.7)^{10} = .028,$$

or approximately 3 percent. This seemed to be an acceptable level of risk, so the firm instituted this policy. Unfortunately, the proportion of rejected batches *increased* dramatically. The resulting value of the producer's risk, α, is

$$\alpha = P\{X > 0 \mid p = .1\} = 1 - P\{X = 0 \mid p = .1\} = 1 - (.9)^{10} = .6513.$$

That is, about 65 percent of the good batches were being rejected by Spire under the new plan. B&G threatened to discontinue shipments to Spire unless it returned to its original sampling plan.

Spire management didn't know what to do. If it returned to the original plan, it faced the risk of losing customers who would go elsewhere to purchase higher-quality refurbished records. If it continued with the current plan, it risked losing B&G as a supplier. Fortunately Sondle, who had been studying quality control methods, was able to propose a solution. If the sample size were increased, the power of the test would improve. Eventually, a test could be devised that would have acceptable levels of both the consumer's and the producer's risk. Because B&G insisted on no more than a 10 percent probability of rejecting good lots, Spire also wanted no more than a 10 percent probability of accepting bad lots.

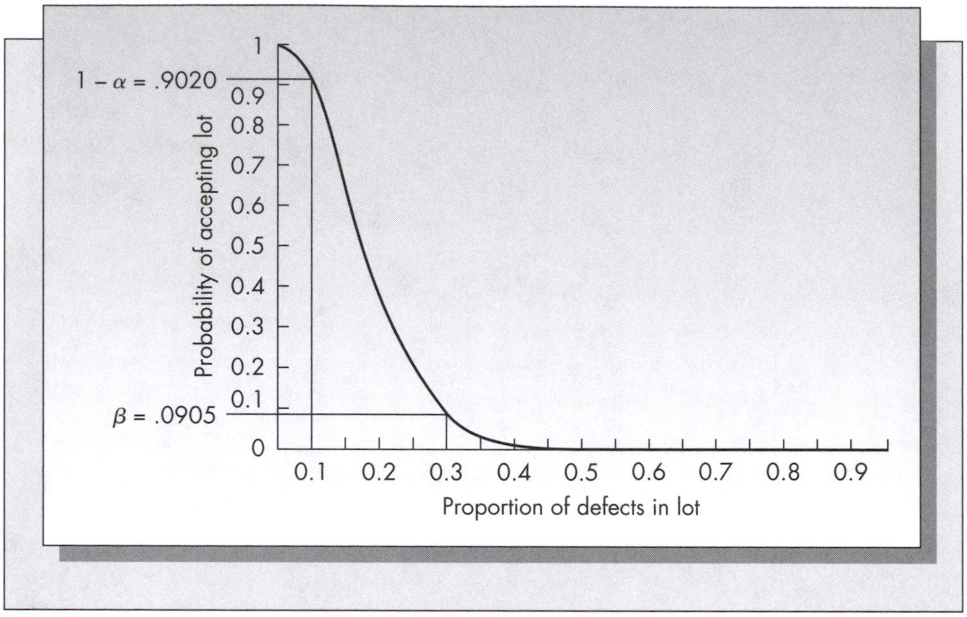

FIGURE 10–13 Revised OC curve for Spire Records (n = 25)

After some experimentation, Herman found that a sample size of $n = 25$ with a rejection level of $c = 4$ seemed to meet the requirements of both B&G and Spire. The exact values of α and β for this test are

$$\alpha = P\{X > 4 \mid p = .1, n = 25\} = .0980.$$
$$\beta = P\{X \leq 4 \mid p = .3, n = 25\} = .0905.$$

Of course, the improved efficiency of this plan did not come without some cost. The employee time required to inspect B&G records increased by two and a half times. B&G and Spire agreed to share the additional cost of the inspection. The OC curve for the sampling plan with $n = 25$ and $c = 4$ appears in Figure 10–13. Notice how much more closely this approximates the ideal curve than does the OC curve for the original plan pictured in Figure 10–12.

PROBLEMS FOR SECTION 10.8

26. Samples of size 20 are drawn from lots of 100 items, and the lots are rejected if the number of defectives in the sample exceeds 2. If the true proportion of defectives in the lot is 5 percent, determine the probability that a lot is accepted using
    a. The exact hypergeometric distribution.
    b. The binomial approximation to the hypergeometric.
    c. The Poisson approximation to the binomial.
    d. The normal approximation to the binomial.
27. A producer of electronic toys purchases the main processor chips in lots of 1,000. The producer would like to have a 1 percent rate of defectives but will normally not refuse a lot unless it has 4 percent or more defectives. Samples of 50 are drawn from each lot, and the lot is rejected if more than two defectives are found.
    a. What are p_0, p_1, n, and c for this problem?
    b. Compute α and β. Use the Poisson approximation for your calculations.

28. A company employs the following sampling plan. It draws a sample of 10 percent of the lot being inspected. If 1 percent or less of the sample is defective, the lot is accepted. Otherwise the lot is rejected.
    a. If a lot contains 500 items of which 10 are defective, what is the probability that the lot is accepted?
    b. If a lot contains 1,000 items of which 20 are defective, what is the probability that the lot is accepted?
    c. If a lot contains 10,000 items of which 200 are defective, what is the probability that the lot is accepted?
29. Hemispherical Conductor produces the 80J84 microprocessor, which the Sayle Company plans to use in a heart–lung machine. Because of the sensitivity of the application, Sayle has established a value of AQL of .001 and LTPD of .005. Sayle purchases the microprocessors in lots of 500 and tests 100 from each lot. The testing requires destruction of the microprocessor. The lot is rejected if any defectives are found.
    a. What are the values of p_0, p_1, n, and c used?
    b. Compute α and β.
    c. In view of your answer to part (b), what problem could Sayle run into?
30. Determine a sampling plan for Spire Records that results in $\alpha = .05$, $\beta = .05$, AQL $= .10$, and LTPD $= .30$. Discuss the advantages and the disadvantages of the plan you obtain as compared to the current plan of $n = 25$ and $c = 4$. (Refer to Example 10.6.)

10.9 DESIGNING QUALITY INTO THE PRODUCT

Traditional quality control methods focus on sampling and inspection. Sampling plans and inspection policies have the ultimate goal of producing an acceptable percentage of defects. Viewing quality in terms of the entire product cycle, including design, production, and consumption, shows that it is economical to design quality into the product. Applying statistical design of experiments to the problem of product design, has been developed by the Japanese statistician and consultant Genichi Taguchi. Taguchi's initial contributions were in **off-line** quality control methods. His later work incorporated economic issues as well.

Quality has significant economic value to the consumer, and product design plays an important role in the product's quality. What does it mean to design for quality? It means that the number of parts that fail easily, or those that significantly complicate the manufacturing process, should be minimized. In particular, how can the design be simplified to eliminate small parts such as screws and latches that are difficult to assemble and may be likely trouble spots down the road?

Further, the most creative design in the world is useless if it can't be manufactured economically into a reliable product. Successful linking of the design and the manufacturing processes is a hallmark of Japan's success in consumer products. The **design for manufacturability** (DFM) movement in the United States had its roots in Japanese manufacturing methods. Further, as mentioned earlier, Six Sigma programs have a design methodology, DMADV (Define, Measure, Analyze, Define, and Verify), which is also sometimes known as **Design for Six Sigma**.

The Design Cycle

The **design cycle** is an important part of the quality chain. Taguchi, Elsayed, and Hsiang (1989) recommend the following three steps in the engineering design cycle.

1. *System design.* This is the basic prototype design that meets performance and tolerance specifications of the product. It includes selection of materials, parts, components, and system assembly.
2. *Parameter design.* After the system design is developed, the next step is optimization of the system parameters. Given a system design, there are generally several system parameters whose values need to be determined. A typical design parameter might be the gain for a transistor that is part of a circuit. One needs to find a functional relationship between the parameter and the measure of performance of the system to determine an optimal value of the parameter. In the example, the measure of performance might be the voltage output of the circuit. The goal is to find the parameter value that optimizes the performance measure. The Taguchi method considers this issue.
3. *Tolerance design.* The purpose of this step is to determine allowable ranges for the parameters whose values are optimized in step 2. Achieving the optimal value of a parameter may be very expensive, whereas a suboptimal value could give the desired quality at lower cost. The tolerance design step requires explicit evaluation of the costs associated with the system parameter values.

The same concepts can be applied to the design of the production process once the product design has been completed. The system design phase corresponds to the design of the actual manufacturing process. In the parameter design phase, one identifies parameters that affect the manufacturing process. Typical examples are temperature variation, raw material variation, and input voltage variation. In the tolerance design phase, one determines acceptable ranges for the parameters identified in phase 2.

The area of off-line quality control (as opposed to the subject of this chapter, which might be referred to as on-line quality control) involves techniques for achieving these three design objectives. Taguchi methods, based on the theory of design of experiments, give new approaches to solving these problems.

Optimizing a parameter value may not always be the overall optimal solution. In some cases, redesigning the product to be less sensitive to the parameter in question might be more economical. In this spirit, Raghu Kackar (1985) described a Japanese tile manufacturer who solved the problem of sensitivity to temperature.

> A Japanese ceramic tile manufacturer knew in 1953 that it is more costly to control causes of manufacturing variations than to make a process insensitive to these variations. The Ina Tile Company knew that an uneven temperature distribution in the kiln caused variation in the size of tiles. Since uneven temperature distribution was an assignable cause of variation, a process quality control approach would have been to devise methods for controlling the temperature distribution. This approach would have increased manufacturing cost. The company wanted to reduce the size variation without increasing cost. Therefore, instead of controlling temperature distribution they tried to find a tile formulation that reduced the effect of uneven temperature distribution on the uniformity of tiles. Through a designed experiment, the Ina Tile Company found a cost-effective method for reducing tile size variation caused by uneven temperature distribution in the kiln. The company found that increasing the content of lime in the tile formulation from 1 percent to 5 percent reduced the tile size variation by a factor of 10. This discovery was a breakthrough for the ceramic tile industry. (p. 176)

Taguchi's method is based on assuming a loss function, say $L(y)$, where y is the value of some functional characteristic and $L(y)$ is the quality loss measured in dollars. In keeping with much of the classical theory of statistics and control, Taguchi recommends a quadratic loss function. The quadratic form is the result of using the first two terms of a Taylor series expansion. Given an explicit form for the loss function, one can address such questions as the benefits of tightening tolerances and the value of various inspection policies. [See Taguchi et al.1989) and Logothetis and Wynn (1989) for a discussion of the general theory. Applications to specific industries are treated by Dehnad (1989).]

Ease of Manufacturability

Geoffrey Boothroyd and Peter Dewhurst were among the first to develop an effective scoring system for designs in terms of their ease of manufacturability. With Winston Knight, they have published three editions of a book summarizing the methodology they developed over a period of several years (Boothroyd et al., 2011). The original focus was on assembly efficiency, which is the ratio of the theoretical minimum assembly time over an estimate of the actual assembly time based on the current product design. Later editions address more general DFM issues including numbers of parts, types of parts, and types of fasteners. These rules, which are very detailed, recommend that simpler designs with fewer parts are preferred. Such designs lead to products that are easier and less expensive to manufacture and are less likely to fail in use.

Karl Ulrich and Steven Eppinger (1995) recommend that designers keep track of design complexity via a scorecard approach. (See Example 10.7.) A scorecard provides a way to compare different designs objectively and a means of keeping track of the complexity of the manufacturing process for every product design.

EXAMPLE 10.7

Scorecard of Manufacturing Complexity Example

Complexity Drivers	Revision 1	Revision 2
Number of new parts introduced	6	5
Number of new vendors introduced	3	2
Number of custom parts introduced	2	3
Number of new "major tools" introduced	2	2
Number of new production processes introduced	0	0
Total	13	12

Source: Ulrich and Eppinger, 1995.

This example is meant to be illustrative only. In practice, the team would have to decide on the relative importance of the drivers and apply suitable weights. Scorecards such as this force the design team to take a good hard look at the manufacturing consequences of their decisions.

A classic example of a successful DFM effort was the IBM Proprinter. The Proprinter was a dot matrix printer aimed at the ever-expanding PC printer market dominated by the Japanese in the early 1980s. IBM developed a video showing someone assembling the printer by hand in a matter of minutes. In designing the Proprinter, IBM followed classic DFM methodology.

The Proprinter had very few separate parts and virtually no screws and fasteners, without any compromise in functionality. The result was that IBM was able to assemble the Proprinter in the United States and remain cost competitive with Japanese rivals (Epson, in particular) that dominated the market at that time. The Proprinter was a very successful product for IBM.

The number of parts in a product is not the only measure of manufacturability. Exactly how parts are designed and put together also plays an important role (Boothroyd et al., 2011). Listed below are ideal characteristics of a part.

- Part should be inserted into the top of the assembly.
- Part is self-aligning.
- Part does not need to be oriented.
- Part requires only one hand for assembly.
- Part requires no tools.
- Part is assembled in a single, linear motion.
- Part is secured immediately upon insertion.

While there are some clear successes in applying DFM, the methodology has yet to gain universal acceptance. The following reasons are the most common for not implementing DFM in the design phase (Boothroyd et al., 2011).

1. *No time.* Designers are pushed to finish their designs quickly to minimize the design-to-manufacture time for a new product. The DFM approach is time intensive. Designing to reduce assembly costs and product complexity cannot be done haphazardly.
2. *Not invented here.* New ideas are always resisted. It would be better if the impetus for DFM came from the designers themselves, but more often it comes from management. Designers resent having a new approach thrust upon them by outsiders (as does anyone).
3. *Low assembly costs.* Since assembly costs often account for a small portion of total manufacturing costs, one might argue that there is little point to doing a design for assembly (DFA) analysis. However, savings often can be greater than one might think.
4. *Low volume.* One might argue that DFM analysis is not worthwhile for low-volume items, but Boothroyd and colleagues argue that the opposite is true. When volumes are low, redesign is unlikely once production begins. This means that doing it right the first time is even more important.
5. *We already do it.* Many firms have used some simple rules of thumb for design (such as limiting the number of bends in a sheet metal part). While such rules make sense in isolation, they are unlikely to lead to the best overall design for the product.
6. *DFM leads to products that are difficult to service.* This is not likely to be true. Products that are easier to assemble are easier to disassemble, and thus easier to service.

Paul Dvorak (1994) offered other reasons for the slow acceptance of DFM. Classical accounting systems may not be able to recognize the cost savings from new designs. A design that reduces fixed setup costs would not be viewed as cost effective since in many accounting systems fixed costs are considered part of overhead. An activity-based accounting system would not have this problem. However, most agree that the greatest obstacle to the acceptance of DFM is resistance to change.

Finally, we should not forget that product design extends far beyond issues of manufacturability only. Aesthetic issues are important as well and may be the dominant factor for some products. Another important issue is the process of narrowing down the field of choices to a final design. Since the interest in this book is in manufacturing-related issues, we will not discuss these broader design issues but refer the interested reader to Stuart Pugh (1991).

Listening to the Customer

Another important aspect of the process of designing quality products is giving people what they want. A perfectly designed and built coffee maker sold in a place where no one drinks coffee is, by definition, a failure. Hence, part of the process of delivering quality to the customer is knowing what the customer wants.

While listening to the customer is an important part of the manufacturing/design cycle link, it is generally more closely associated with marketing than with operations. Still, the boundaries separating the functional areas of business are becoming fuzzier. Manufacturing cannot operate in a vacuum; it must be part of the link with the customer.

Finding out what the customer wants and incorporating those wants into product design and manufacture is a multistep process outlined below.

- Obtaining the data.
- Characterizing customer needs.
- Prioritizing customer needs.
- Linking needs to design.

There are several means for obtaining the raw data. Traditionally, customer opinion is solicited through interviews and surveys. There are many issues to be aware of when considering interviews with customers or potential customers. How many customer responses are enough? The right answer depends on several factors. How many market segments are there for the product? How many attributes are important? What methods will be used to interpret the results? Next, there is the question of how to solicit the information from the customer. Should one conduct interviews or surveys? The answer is unclear. Both have advantages. Interviews allow more open-ended responses, but the biases of the interviewer could slant the results. Both surveys and interviews depend on how questions are worded. For example, suppose Mr. Coffee is considering a new design for a coffee maker. A question like "What should the capacity of an automatic coffee maker be?" automatically assumes that the customer is concerned about capacity. The question "Do you prefer an 8- or a 12-cup coffee maker?" imposes even more assumptions (Dahan, 1995).

Focus groups are another popular technique for soliciting customer feedback. The focus group format has the advantage of being open-ended; the specific wording of questions is not as important as it is with surveys or interviews. However, focus groups have disadvantages. The moderator can affect the flow of the discussion. Also, participants with strong personalities are likely to dominate the group.

Once the database is developed, the customer needs and desires must be prioritized and grouped. Several methods are available for this. One that has received a great deal of attention in the marketing literature is conjoint analysis (Green and Rao, 1971). Conjoint analysis is a statistically based technique to estimate utilities for product attributes based on customer preference data.

Once attributes are determined and grouped, one needs to link those attributes to the design and manufacturing processes. This can be done with **quality function deployment** (QFD). With QFD, customer needs are related to the product attributes and/or aspects of the production process through a matrix. The user provides estimates of the correlation between

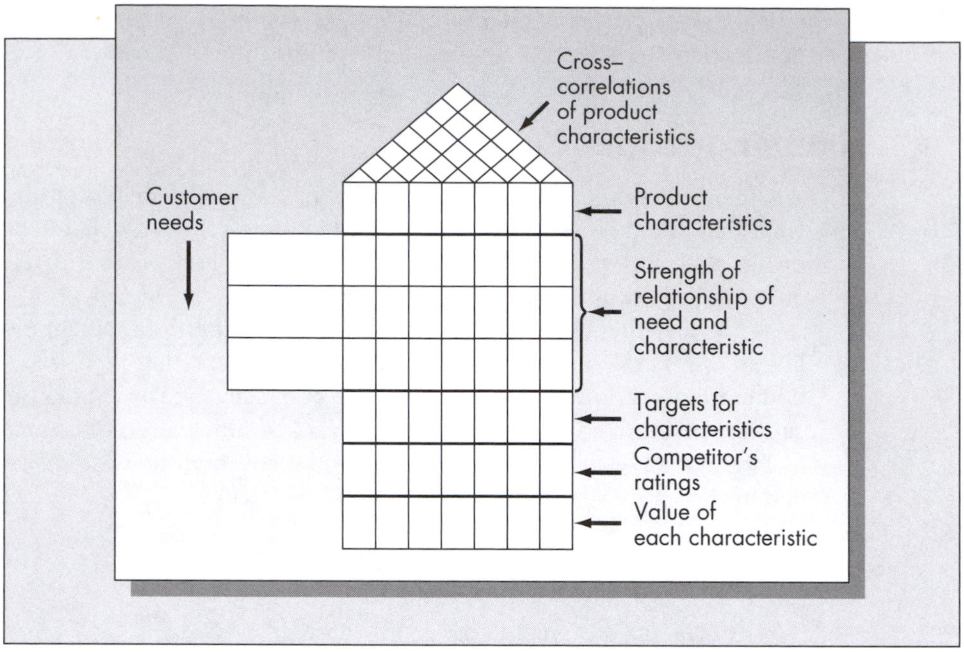

FIGURE 10–14 The house of quality: The QFD planning matrix

attributes and needs in a "roof" portion of the matrix. The resulting figure looks a little like a house, hence the term **house of quality** to describe the resulting matrix (see Figure 10–14). QFD is used in conjunction with traditional methods such as surveys and focus groups. The strength of the correlations between customer needs and product attributes or design characteristics shows where the emphases should be placed when considering design of new products or design changes in existing products. The interested reader is referred to Louis Cohen (2010) for a comprehensive discussion of QFD methods.

This section only touched the surface of the issue of soliciting customer opinion and integrating that opinion into the process of product design and manufacture. However, one must be careful not to put *too* much emphasis on the voice of the customer. Real innovations do not come from the customer but from visionaries. By letting the customer be king, innovation could be stifled.

> Still, more and more companies are learning that sometimes your customers can actually lead you astray. The danger lies in becoming a feedback fanatic, slavishly devoted to customers, constantly trying to get in even better touch through more focus groups, more surveys. Not only can that distract you from the real work at hand, but it may also cause you to create new offerings that are safe and bland. (Martin, 1995, p. 121)

Home run new products like Post-it^{TM} notes developed by 3M were not the result of marketing surveys. Many products that looked like winners from the market research flopped in the marketplace. One example is New Coke, which won hands down in taste tests but was never accepted by consumers. Of course, there are probably more examples of flops that resulted from not listening to the customer. The moral here is that while the customer is an important part of the equation, customer surveys and the like should not be used to the exclusion of ingenuity and vision. Recently, **design thinking** has been promoted as a technique to balance innovation with the voice of the customer. This technique has been promoted by IDEO, which is a consulting company headquartered in California.

10.10 HISTORICAL NOTES

The desire to maintain quality of manufactured goods is far from new. The prehistoric man whose weapons and tools did not function did not survive. However, the statistical quality control methods discussed in this chapter were devised only in the last 90 years or so. Walter Shewhart (1931), who was an employee of the Bell Telephone Laboratories, conceived the idea of the control chart and published a monograph summarizing economic control of quality in manufacturing. Harold Dodge and Harry Romig, also employees of Bell Labs, are generally credited with laying the foundations of the theory of acceptance sampling.

As with most new methodology, U.S. industry was slow to adopt statistical quality control techniques. However, as part of the war effort, the U.S. government decided to adopt sampling inspection methods for army ordnance in 1942. As a result of this and other wartime governmental activities, knowledge and acceptance of the techniques discussed in this chapter became widespread after the war. Many of the acceptance sampling techniques discussed in this chapter are part of Military Standard 105. In fact, sampling plans are often described in terms of their military designation. Military Standard 105 and its various revisions and additions form the basis for most of the acceptance sampling plans in use today.

Abraham Wald was responsible for developing the theory of sequential analysis, which forms the basis for the formulas that appear in Appendix 10-E. Wald's work proved to be a major milestone in the advance of the theory of acceptance sampling. Wald was part of a research team organized at Columbia University in 1942. The U.S. government considered his work to be so significant that they withheld publication until June of 1945.

W. Edwards Deming is given much of the credit for transferring statistical quality control technology to Japan. However, the three leaders of the quality movement in the United States during the 1950s, W. Edwards Deming, Joseph M. Juran, and Armand V. Feigenbaum, each contributed to the Japanese quality movement, although Deming is certainly the name that comes to mind first. Deming's success in Japan was the result of a seminar he presented on statistical quality control in 1950. He repeated that seminar several times, became active in the Japanese quality movement, and ultimately became a national hero in Japan. Deming recommended both the application of statistical methods and a systematic approach to solving quality problems. His approach later became known as the Plan, Do, Check, Act (PDCA) method.

Juran stressed the managerial aspects of quality rather than the statistical aspects. He also presented several seminars to the Japanese targeted at middle to upper management. The Juran Institute, located in Wilton, Connecticut, was founded in 1979 and continues to provide consulting and training in the quality area (including lean manufacturing management training and Six Sigma certification). [However, Philip Crosby Associates, founded the same year and now a subsidiary of The Capability Group, Inc., is arguably more successful.]

Prior to World War II, the quality of manufactured goods in Japan was poor. At that time the Japanese sought to compete on price rather than on quality. However, after the war, with the help of the quality gurus above, the focus changed dramatically. Nowhere is this seen more clearly than in the automobile industry.

In the 1950s virtually all the automobiles sold in the United States were made in the United States. In 1955 the big three automakers (Ford, G.M., and Chrysler) accounted for 95 percent of the U.S. sales; the majority of the remaining 5 percent were sales made by other (now defunct) U.S. nameplates. By 1961 U.S. automakers accounted for close to 50 percent of the world market in passenger vehicles, with Japan accounting for about 2 percent. Today, the U.S. manufacturers account for about 15 percent of the world market in passenger automobiles, while Japan's market share is over 20 percent. Meanwhile Korean automobile manufacturer Hyundai occupies 8.3 percent market share (third only to Toyota and Volkswagen), and Chinese automobile companies are ascending.

At first U.S. manufacturers were slow to recognize the higher quality of their Japanese competitors, but eventually the difference became too large to ignore. In 1980 Japan passed the United States as the world's largest auto-producing nation, and in 2007 Toyota became the world's largest car company. In 1977 Detroit recalled more cars than they sold (Leonard, 1995)! Eventually, U.S. manufacturers woke up to the need to change. Lean and quality programs are now standard practice. However, the quality gap has still not closed entirely. In the 2018 What Car? Reliability Survey, the top 3 most reliable car companies were Japanese, and no U.S. car companies were in the top 10 (Daily Mail, 2018). Tesla was at the bottom of the list.

Most of the methods discussed in this chapter are in widespread use throughout industry in the United States and overseas. Optimization models for designing control charts have not enjoyed the same level of acceptance, however. The model outlined in Appendix 10–C is from Kenneth Baker (1971), although more complex and comprehensive economic models for design of $\bar{X}$ charts were developed earlier (see, for example, Duncan, 1956). For a comprehensive discussion of the history of the early quality movement in the United States, see Kolesar (1993).

Total quality management (TQM) is a term that came into vogue but fell out of favor as international standards for quality management appeared. The concept of total quality control seems to have been first introduced by Feigenbaum in an article in 1946. Feigenbaum wrote *Quality Control: Principles, Practice, and Administration* while a graduate student at MIT. Subsequently, he published four editions of *Total Quality Control,* which have been translated into more than twenty languages. He provided the following definition (Feigenbaum, 1961).

> Total quality control is an effective system for integrating the quality-development, quality-maintenance, and quality-improvement efforts of the various groups in an organization so as to enable marketing, engineering, production, and service at the most economical levels which allow for full customer satisfaction. (p. 1)

Feigenbaum's approach was to define quality in terms of the customer. As we noted in the introduction of this chapter, most definitions of quality concern either conformance to specifications or customer satisfaction.

A program that received a great deal of attention in the early 1980s was **quality circles** (QCs). Quality circles were an attempt to emulate the Japanese organizational structure. A quality circle is a small group of roughly 6 to 12 employees who meet on a regular basis for the purpose of identifying, analyzing, and suggesting solutions to problems affecting work. Typically, such a group would meet for about four hours per month on company time. The group might be given special training on statistical quality control methods, group dynamics, and general problem-solving techniques.

There was a dramatic increase in interest in quality circles in the United States starting around 1980. Edward Lawler and Susan Mohrman (1985) reported that 44 percent of the companies in the United States with more than 500 employees had some type of quality circle program in place in 1982 and that 75 percent of these programs were started after 1980. However, they suggested that Quality Circles was inherently an unstable organizational structure. With no decision-making power, the only role of the QC group was to serve as a formal "suggestion box." It was natural that employees would eventually lose interest. Quality circles have largely been replaced by Six Sigma and similar initiatives.

10.11 SUMMARY

This chapter covered quality and assurance. Quality has many definitions. We argued that at its heart, it should measure conformance of the product to the customer's needs. We discussed Garvin's eight dimensions of quality: (1) performance, (2) features, (3) reliability, (4) conformance, (5) durability, (6) serviceability, (7) aesthetics, and (8) perceived quality. Assurance is the avoidance of quality problems. Lean has become an important tool for developing quality assured processes. ISO 9000 certification is also largely about quality assurance.

Two highly sought-after national prizes are the Deming Prize in Japan and the Baldrige Award in the United States. These awards recognize exceptional industry efforts in implementing quality. An early winner of the Baldrige award was Motorola, who was one of three winners in the first year of the award. In part they won because of the Six Sigma Quality program they developed.

In this chapter we outlined the techniques for control chart design and acceptance sampling. The basis for the methodology of this chapter is classical probability and statistics. The underpinnings of control charts include fundamental results from probability theory such as the law of large numbers and the central limit theorem.

We discussed control charts in the first half of this chapter. A control chart is a graphical device that is used to determine when a shift in the value of a parameter of the underlying distribution of some measurable quantity has occurred. When this happens, the process is said to have gone out of control. The design of the control chart depends on the particular parameter that is being monitored.

The most common control chart is the $\bar{X}$ chart. An $\bar{X}$ chart is designed to monitor a single measurable variable, such as weight or length. Subgroups of size n are sampled on a regular basis, and the sample mean of the subgroup is computed and placed on a graph. Because the sample mean is approximately normally distributed independently of the form of the distribution of the population, the likelihood that a single observation falls outside three-sigma limits is sufficiently small that when such an event occurs it is unlikely to be due to chance. Rather, it is more likely to be the result of a shift in the true mean of the process.

The second type of control chart treated in this chapter is the R chart. An R chart is designed to determine when a shift in the process variation has occurred. The symbol R stands for range. The range is the difference between the largest and the smallest observations in a subgroup. Because a close relationship exists between the value of the range of the sample and the underlying population variance, R charts are used to measure the stability of the variance of a process.

Often control charts for variables are inappropriate. The p chart is a control chart for attributes. When using p charts, items are classified as either acceptable or not. The p chart utilizes the normal approximation of the binomial distribution and may be used when subgroups are of equal or of varying sizes.

The last control chart presented was the c chart. The c chart is used to monitor the number of defectives in a unit of production. The parameter c is the average number or rate of defects per unit of production. The c chart is based on the Poisson distribution, and the control limits are derived using the normal approximation to the Poisson.

The basis for control charts is the idea that all processes have a degree of natural variation that should be measured (and, ideally, reduced using lean process improvement techniques). A system that is in control is not necessarily also capable. Process capability is the ability to reliably produce items that are within specification. The capability ratio and capability index compare the spread of the specification limits with the natural variability of the process.

The chapter also considered acceptance sampling. The purpose of an acceptance sampling scheme is to determine if the proportion of defectives in a large lot of items is acceptable based on the results of sampling a relatively small number of items from the lot. The simplest sampling plan is single sampling. A single sampling plan is specified by two numbers: n and c. Interpret n as the size of the sample and c as the acceptance level. A double sampling plan requires specification of five numbers: n_1, n_2, c_1, c_2, and c_3, although often $c_2 = c_3$. Based on the results of an initial sample size of n_1, the lot is accepted or rejected, or an additional sample of size n_2 is drawn. The logical extension of double sampling is sequential sampling, in which items are sampled one at a time and a decision is made after each item is sampled about whether to accept the lot, reject the lot, or continue sampling.

Methods for eliciting the voice of the customer such as conjoint analysis and quality function deployment (QFD) were discussed as well. Off-line quality methods are directed at the problem of designing quality into the product. Taguchi methods, largely based on the theory of design of experiments, are an important development in this area. The Taguchi methods identify important process and design parameters and attempt to find overall optimum values of these parameters, relative to some measure of performance of the system. The chapter concluded with a discussion of design for manufacturability and the contributions of Boothroyd, Dewhurst, and Knight to this area.

ADDITIONAL PROBLEMS ON QUALITY AND ASSURANCE

31. What is an operational definition of quality? Is it possible for a 13-inch TV selling for \$100 to be of superior quality to a 50-inch smart TV selling for \$800?
32. In what ways could each of the factors listed below contribute to poor quality?
    a. Management
    b. Labor
    c. Equipment maintenance
    d. Equipment design
    e. Control and monitoring
    f. Product design
33. Figure 10–1 presents a conceptual picture of the trade-off between process cost and the costs of losses due to poor quality. What are the costs of poor quality and what difficulties might arise when attempting to measure these costs?
34. $\bar{X}$ and R charts are maintained on a single quality dimension. A sudden shift in the process occurs, causing the process mean to increase by 2σ, where σ is the true process standard deviation. No shift in the process variation occurs. Assuming that the $\bar{X}$ chart is based on 3σ, limits and subgroups of size $n = 6$, what proportion of the points on the $\bar{X}$ chart would you expect to fall outside the limits after the shift occurs?
35. XYZ produces bearings for bicycle wheels and monitors the process with an $\bar{X}$ chart for the diameter of the bearings. The $\bar{X}$ chart is based on subgroups of size 4. The target value is 0.37 inch, and the upper and lower limits are 0.35 and 0.39 inch, respectively (assume

that these are based on three-sigma limits). Wheeler, which purchases the bearings from XYZ to construct the wheels, requires tolerances of 0.39 ± 0.035 inch. Oversized bearings are ground down and undersized bearings are scrapped. What proportion of the bearings does Wheeler have to grind, and what proportion must it scrap? Can you suggest what Wheeler should do to reduce the proportion of bearings that it must scrap?

36. A process that is in statistical control has an estimated mean value of 180 and an estimated standard deviation of 26.
    a. Based on subgroups of size 4, what are the control limits for the $\bar{X}$ and R charts?
    b. Suppose that a shift in the mean occurs so that the new value of the mean is 162. What is the probability that the shift is detected in the first subgroup after the shift occurs?
    c. On average, how many subgroups would need to be sampled after the shift occurred before it was detected?
    d. If the specification limits for the process are [100, 260], what are the capability ratio and capability index for the original mean (180)? What are they for the shifted process (mean 162)?
37. Consider the data presented in Table 10–1 for the tracking arm example. Suppose that an R chart is constructed based on sample numbers 1 to 15 only.
    a. What is the estimate of σ obtained from these 15 observations only?
    b. What are the values of the UCL and LCL for an R chart based on these observations only?
38. Discuss the advantages and disadvantages of the following strategies in control chart design. In particular, what are the economic trade-offs attendant to each strategy?
    a. Choosing a very small value of α.
    b. Choosing a very small value of β.
    c. Choosing a large value of n.
    d. Choosing a small value for the sampling interval.
39. The construction of the p chart described in this chapter requires specification of both UCL and LCL levels. Is the LCL meaningful in this context? In particular, what does it mean when an observed value of p is less than the LCL? If only the UCL is used to signal an out-of-control condition, should the calculation of the UCL be modified in any way? (Hint: The definition of the Type 1 error probability, α, will be different. The hypothesis test for determining if the process is in control is one-sided rather than two-sided.)
40. A manufacturer of large appliances maintains a p chart for the production of washing machines. The machines may be rejected because of cosmetic or functional defectives. Based on sampling 30 machines each day for 50 consecutive days, the current estimate of p is .0855.
    a. What are the control limits for a p chart? (Assume 3σ limits.)
    b. Suppose that the percentage of defective washing machines increases to 20 percent. What is the probability that this shift is detected on the first day after it occurs?
    c. On average, how many days would be required to detect the shift?
41. A maker of personal computers, Noname, purchases DRAM chips from two different manufacturers, A and B. Noname uses the following sampling plan: A sample of 10 percent of the chips is drawn, and the lot is rejected if two or more defective chips are discovered. The two manufacturers supply the chips in lots of 100 and 1,000, respectively.
    a. For each manufacturer, determine the true proportion of defectives in the lot that would result in 90 percent of the lots being accepted. You may use the Poisson approximation for your calculations.
    b. Would you say that this plan is fair?

42. Graph the AOQ curves for manufacturers A and B mentioned in problem 41. Estimate the values of the AOQL in each case.
43. Consider the sampling plan discussed in problem 41. Would a fairer plan be to reject the lot if more than 10 percent of the chips in a sample are defective? Which of the two manufacturers mentioned would be at an advantage if this plan were adopted?
44. Assuming AQL = 5 percent and LTPD = 10 percent, determine the values of α and β for the plan described in problem 41 for manufacturers A and B and for the plan described in problem 43 for manufacturers A and B.
45. Graph the OC curves for the sampling plan described in problem 41 for both manufacturers A and B.
46. $\bar{X}$ control charts are used to maintain control of the manufacture of the cases used to house a generic brand of personal computer. Separate charts are maintained for length, width, and height. The length chart has UCL = 20.5 inches, LCL = 19.5 inches, and a target value of 20 inches. This chart is based on using subgroups of size 4 and three-sigma limits. However, the customer's specifications require that the target length should be 19.75 inches with a tolerance of ± 0.75 inch. What percentage of the cases shipped will fall outside the customer's specifications?
47. A p chart is used to monitor the fraction defective of an integrated circuit to be used in a commercial pacemaker. A sample of 15 circuits is taken from each day's production for 30 consecutive working days. A total of 17 defectives are discovered during this period.
    a. Determine the three-sigma control limits for this process.
    b. Suppose that α, the probability of Type 1 error (that is, the probability of drawing the conclusion that the process is out of control when it is in control), is set to be .05. What control limits do you now obtain? (Use a normal approximation for your calculations.)
48. A single sampling plan is used to determine the acceptability of shipments of a bearing assembly used in the manufacture of skateboards. For lots of 500 bearings, samples of $n = 20$ are taken. The lot is rejected if any defectives are found in the sample.
    a. Suppose that AQL = .01 and LTPD = .10. Find α and β.
    b. Is this plan more advantageous for the consumer or the producer?
49. A double sampling plan is constructed as follows. From a lot of 200 items, a sample of 10 items is drawn. If there are zero defectives, the lot is accepted. If there are two or more defectives, the lot is rejected. If there is exactly one defective, a second sample of 10 items is drawn. If the combined number of defectives in both samples is two or less, the lot is accepted; otherwise it is rejected. If the lot has 10 percent defectives, what is the probability that it is accepted?
50. Hammerhead produces heavy-duty nails, which are purchased by Modulo, a maker of prefabricated housing. Modulo buys the nails in lots of 10,000 and subjects a sample to destructive testing to determine the acceptability of the lot. Modulo has established an AQL of 1 percent and an LTPD of 10 percent.
    a. Assuming a single sampling plan with $n = 100$ and $c = 2$, find α and β.
    b. Derive the sequential sampling plan that achieves the same values of α and β as the single sampling plan derived in part (a).
    c. By estimating the ASN curve, find the maximum value of the expected sample size Modulo will require if it uses the sequential plan derived in part (b).
    d. Suppose that the sequential sampling plan derived in part (b) is used. One hundred nails are tested with the following result: the first 80 are acceptable, the 81st is defective, and the remaining 19 are acceptable. By graphing the acceptance and the rejection

regions, determine whether the sequential sampling plan derived in part (b) would recommend acceptance or rejection on or before testing the 100th nail.

51. For the single sampling plan derived in part (a) of problem 50, suppose lots that are not passed are returned to Hammerhead.
   a. Estimate the graph of the AOQ curve by computing AOQ(p) for various values of p.
   b. Using the results of part (a), estimate the maximum proportion of defective nails that Modulo will be using in its construction.

52. Twenty sets of four measurements of the diameters in inches of Hot Shot golf balls were

Sample				
1	2.13	2.18	2.05	1.96
2	2.08	2.10	2.02	2.20
3	1.93	1.98	2.03	2.06
4	2.01	1.94	1.91	1.99
5	2.00	1.90	2.14	2.04
6	1.92	1.95	2.02	2.05
7	2.00	1.94	2.00	1.90
8	1.93	2.02	2.04	2.09
9	1.87	2.13	1.90	1.92
10	1.89	2.14	2.16	2.10
11	1.93	1.87	1.94	1.99
12	1.86	1.89	2.07	2.06
13	2.04	2.09	2.03	2.09
14	2.15	2.02	2.11	2.04
15	1.96	1.99	1.94	1.98
16	2.03	2.06	2.09	2.02
17	1.95	1.99	1.87	1.92
18	2.05	2.03	2.06	2.04
19	2.12	2.02	1.97	1.95
20	2.03	2.01	2.04	2.02

   a. Enter the data into a spreadsheet and compute the means and the ranges for each sample.
   b. Using the results of part (a), develop $\bar{X}$ and R charts similar to Figures 10–7 and 10–8. Assume three-sigma limits for the X chart.
   c. Develop a histogram based on the 80 observations. Assume class intervals (1) 1.80–1.849, (2) 1.85–1.899, (3) 1.90–1.949, (4) 1.95–1.999, (5) 2.0–2.049, (6) 2.05–2.099, (7) 2.10–2.149, (8) 2.15–2.20. Based on your histogram, what distribution might accurately describe the diameter of a golf ball selected at random?

53. A p chart is used to monitor the number of riding lawn mowers produced. The numbers that are sent back for rework because they did not pass inspection are listed on the next page.

Day	Number Produced	Number Rejected
1	400	23
2	480	18
3	475	24
4	525	34
5	455	17
6	385	17
7	372	12
8	358	19
9	395	24
10	405	29
11	385	16
12	376	19
13	395	23
14	405	14
15	415	25
16	440	34
17	380	26
18	318	19

Enter the data into a spreadsheet and compute standardized Z values for a p chart. Graph the Z values. Is this process in control?

54. Samples of size 50 are drawn from lots of 1,000 items. The lot is rejected if there are more than two defectives in the sample. Using a binomial approximation, graph the OC curve as a function of p, the proportion of defectives in the lot. For an AQL of .01 and an LTPD of .10, find α and β.
    a. Graph the OC curve and identify the Type 1 and Type 2 error probabilities (that is, develop a graph similar to Figure 10–12).
    b. Graph the AOQ curve and identify the value of the AOQL.
55. (Appendix 10–B) Consider the data presented in problem 14. Problem 14 required testing whether the process was in control using a p chart. Test the hypothesis that the value of p is the same each day using classical statistical methods. That is, test the hypothesis

$$H_0: p_1 = p_2 = \ldots = p_k$$

versus

$$H_1: \text{Not all the } p_i \text{ are equal,}$$

where k is the number of days in the data set ($k = 23$ in this case) and p_i is the true proportion of defectives for day i. Define x_i as the number of rolls rejected in day i and p' as the estimate of p obtained from the data [p' was computed in problem 14(a)]. The test statistic is given by the formula

$$\chi^2 = \sum_{i=1}^{k} \frac{(x_i - np')^2}{np'(1-p')},$$

where n is the number of items sampled each day ($n = 25$ in this case). The test is to reject H_0 if $\chi^2 > \chi^2_{\alpha,k-1}$, where $\chi^2_{\alpha,k-1}$ is a number obtained from a table of the χ^2 distribution. For $\alpha = .01$ (which is larger than the α value that yields 3σ limits on a p chart), $\chi^2_{01,22} = 40.289$. Based on this value of α, does the χ^2 test indicate that this process is out of control? If the answer you obtained is different from that of problem 14(b), how do you account for the discrepancy? Which method is probably better suited for this application?

56. (Appendix 10–C) A quality control engineer is considering the optimal design of an $\bar{X}$ chart. Based on his experience with the production process, there is a probability of .03 that the process shifts from an in-control to an out-of-control state in any period. When the process shifts out of control, it can be attributed to a single assignable cause; the magnitude of the shift is 2σ. Samples of n items are made hourly, and each sampling costs \$0.50 per unit. The cost of searching for the assignable cause is \$25, and the cost of operating the process in an out-of-control state is \$300 per hour.
    a. Determine the hourly cost of operating the system when $n = 6$ and $k = 2.5$.
    b. Estimate the optimal value of k for the case $n = 6$. If you are doing the calculations by hand, use $k = 0.5$, 1, 1.5, 2, 2.5, and 3.0. If you are using a computer, use $k = 0.1$, 0.2, … , 2.9, 3.0.
    c. Determine the optimal control chart design that minimizes average annual costs.
57. (Appendix 10–C) Consider the application of the economic design of $\bar{X}$ charts for Wonderdisk presented in this chapter. Without actually performing the calculations, discuss what the effect on the optimal values of n and k is likely to be if
    a. δ increased from 1 to 2.
    b. π increased from .05 to .10.
    c. a_1 decreased to 1.
    d. a_2 increased to 150.
58. (Appendix 10–C) Under what circumstances would the following assumptions not be accurate?
    a. The assumption that the probability law describing the number of periods until the process goes out of control follows the geometric distribution.
    b. The assumption that an out-of-control condition corresponds to a shift of the mean equal to $\delta\sigma$.
    c. The assumption that the search cost is a fixed constant, a_2.
59. (Appendix 10–C) Discuss the following pro and con positions on using optimization models to design control charts.

    Con: "These models are useless to me because I don't feel I can accurately estimate the values of the required inputs."

    Pro: "The choice of specific values of n and k in the construction of X bar charts means that you are assuming values for the various system costs and parameters. You might as well take the bull by the horns and obtain the best estimates you can and use those to design the X bar chart."

60. (Appendix 10–C) Suppose that "the process going out of control" corresponds to a shift of the mean from μ to $\mu + \sigma$ with probability .25, $\mu + 2\sigma$ with probability .25, $\mu - \sigma$ with probability .25, and $\mu - 2\sigma$ with probability .25. What modifications in the model are required? In particular, if we write the shift in the form $\mu \pm \delta\sigma$, show how to compute β_1 and β_2 that would correspond to values of $\delta = 1$ and $\delta = 2$, respectively. If the out-of-control costs were now represented by a_3 when $\delta = 1$ and a_4 when $\delta = 2$, determine an expression for the average annual operating costs. (Assume all other costs and system parameters remain the same.)
61. (Appendix 10–C) A local contractor manufactures the speakers used in telephones. The phone company requires the speakers to ring at a specified noise level (in decibels). An $\bar{X}$ chart is being designed to monitor this variable. The process of sampling speakers from the line requires hitting the speakers with a fixed-force clapper and measuring the decibel level on a meter designed for that purpose. The cost of sampling is \$1.25 per speaker. When the process goes out of control, the thickness of the speakers is incorrect. The cost of searching for an assignable cause is estimated to be \$50. The cost of operating the process in an out-of-control state is estimated to be \$180 per hour. Out of control corresponds to a shift of 2σ in the decibel level, and the probability that the process shifts out of control in any hour is .03.
    a. The company uses an $\bar{X}$ chart based on 3σ limits and subgroups of size 4. What is its hourly cost?
    b. What are the optimal values of n and k for this process and the associated optimal cost?
62. (Appendix 10–D) Consider the double sampling plan for Spire Records presented in this chapter.
    a. Suppose that the true proportion of defectives in the lot is 10 percent. On average, how many records will have to be sampled before the lot is either accepted or rejected?
    b. Suppose that the true proportion of defectives in the lot is 30 percent. On average, how many records will have to be sampled before the lot is either accepted or rejected?
63. (Appendix 10–D) For the double sampling plan for Spire Records presented in this chapter, what is the probability that a lot is rejected on the first sample? Perform the computation for both $p = p_0$ and $p = p_1$.
64. (Appendix 10–D) Consider the double sampling plan for Spire Records described in this chapter. Over a period of one year, 3,860 records are subject to inspection using this plan. If 60 percent of these batches are "good" (that is, in 60 percent of the batches the proportion of defectives is exactly 10 percent) and 40 percent are "bad" (that is, in 40 percent of the batches the proportion of defectives is exactly 30 percent), then what is the expected number of batches
    a. Accepted?
    b. Rejected?
    c. Accepted on the first sample?
    d. Accepted on the second sample?
    e. Rejected on the first sample?
    f. Rejected on the second sample?
65. (Appendix 10–D) Graph the OC curve for the double sampling plan with $n_1 = 20$, $n_2 = 10$, $c_1 = 3$, and $c_2 = c_3 = 5$, as described in this chapter. If you are doing this by hand, evaluate the curve at $p = 0, .2, .4, .6, .8$, and 1 only. (Hint: The OC curve for this sampling plan has the form $\mathrm{OC}(p) = P\{X \le 3 \mid p, n = 20\}$ +

$$P\{X = 4 \mid p, n = 20\}\, P\{Y \le 1 \mid p, n = 10\} + P\{X = 5 \mid p, n = 20\}\, P\{Y = 0 \mid p, n = 10\}.)$$

66. (Appendix 10–D) By trial and error devise a double sampling plan for Spire Records that achieves $\alpha \approx .10$ and $\beta \approx .10$.

67. (Appendix 10–D) Consider the following double sampling plan. First select a sample of 5 from a lot of 100. If there are four or more defectives in the sample, reject the lot. If there is only one or fewer, accept the lot. If there are two or three defectives, sample an additional five items and reject the lot if the combined number of defectives in both samples is five or more. If the lot has 10 defectives, what is the probability that a lot passes the inspection?
68. (Appendix 10–D) For the double sampling plan described in problem 67, determine the following.
    a. The probability that the lot is rejected based on the first sample.
    b. The probability that the lot is rejected based on the second sample.
    c. The expected number of items sampled before the lot is accepted or rejected.
69. (Appendix 10–E) Develop a spreadsheet from which one may obtain a graph such as Figure 10–18 for sequential sampling. Store the values of p_0, p_1, α, and β in cell locations so that these can be altered at will. Print a graph for $p_0 = .05$, $p_1 = .20$, $\alpha = .05$, and $\beta = .10$. Allow for $n \leq 100$.
70. (Appendix 10–E) Sequential sampling resulted in the following: the first 40 items were good, item 41 was defective, item 68 was defective, and items 86 and 87 were defective. Place these results on the graph you obtained in problem 69. Is the lot accepted, rejected, or neither one before testing the 87th item?
71. (Appendix 10–E) A manufacturer of aircraft engines uses a sequential sampling plan to accept or reject incoming lots of microprocessors used in the engines. Assume an AQL of 1 percent and an LTPD of 5 percent. Determine a sequential sampling plan assuming $\alpha = .05$, $\beta = .10$. Graph the acceptance and rejection regions.
72. (Appendix 10–E) Consider the sequential sampling plan described in problem 71. Suppose that a lot of 1,000 microprocessors is inspected. Suppose that the 31st, 89th, 121st, and 122nd chips tested are found defective. Assuming the sequential sampling plan derived in problem 71, will the lot be accepted, rejected, or neither by the time the 122nd chip has been tested?
73. (Appendix 10–E) Estimate the ASN curve for the plan derived in problem 71. According to your curve, what is the expected number of microprocessors that must be tested when the true proportion of defectives in the lot is
    a. 0.1 percent?
    b. 1.0 percent?
    c. 10 percent?
74. (Appendix 10–E) Consider the example of Hemispherical Conductor and the Sayle Company discussed in problem 29. Devise a sequential sampling plan for Sayle that results in $\alpha = .05$ and $\beta = .20$. What are the advantages and disadvantages of this plan over the Sayle sampling plan derived in problem 29?
75. (Appendix 10–E) Estimate the ASN curve for the sampling plan derived in problem 74. On average, how many of the microprocessors would have to be tested if
    a. $p = .001$?
    b. $p = .005$?
    c. $p = .01$?
76. (Appendix 10–F) If defective items are replaced and $N >> n$, show by differential calculus that the value of p for which AOQ(p) achieves its maximum value satisfies

$$\frac{d\text{OC}(p)}{dp} = -\frac{\text{OC}(p)}{p}.$$

77. (Appendix 10–F) Consider the single sampling plan with $n = 10$ and $c = 0$.
    a. Derive an analytical expression for the OC curve as a function of p.
    b. Using the results of problem 76, determine the value of p at which the AOQ curve is a maximum.
    c. Using the results of parts (a) and (b), determine the maximum value of the average outgoing quality.
78. (Appendix 10–F) Consider the single sampling plan discussed in problem 27. If defective items are replaced and $N >> n$, graph the AOQ curve and determine the value of the AOQL.

APPENDIX 10–A

APPROXIMATING DISTRIBUTIONS

Several probability approximations were used in this chapter. This appendix will discuss the motivation and justification for these approximations.

The complexity of a probability distribution depends upon the number of parameters that are required to specify it. The distributions considered in this chapter in descending order of complexity are

Distribution	Parameters
Hypergeometric	n, N, M
Binomial	n, p
Poisson	λ
Normal	μ, σ

It is not clear from this chart why the normal distribution should be simpler than the Poisson, since the normal is a two-parameter distribution and the Poisson one. The reason is that all normal probabilities can be obtained from a single table of the standard normal distribution, and the Poisson distribution must be tabled separately for distinct values of λ.

The binomial approximation to the hypergeometric. The **hypergeometric distribution** (whose formula appears in Section 10.8) is the probability that if n items are drawn from a lot of N items of which M are defective, then there are exactly m defectives in the sample. The experiment that gives rise to the hypergeometric may be thought of as sampling the items one by one without replacement. If N is much larger than n, the probability that any item sampled is defective is very close to M/N. In that case the hypergeometric probability would be close to the binomial probability with $p = M/N$ and $n = n$. Note that the binomial distribution corresponds to sampling *with* replacement. If $N > 10n$, the binomial should provide an adequate approximation to the hypergeometric.

The Poisson approximation to the binomial. The **Poisson distribution** can be derived as the limit of the binomial as $n \to \infty$ and $p \to 0$, but with the product np remaining constant. Write $\lambda = np$. Then for large n and small p,

$$P\{X = m\} \approx \frac{e^{-\lambda}\lambda^m}{m!} \qquad \text{for } m = 0, 1, 2, \ldots.$$

It is not obvious under what circumstances this approximation is adequate. In general, $p < .1$ and $n > 25$ should hold, but if p is very small, then smaller values of n are acceptable, and if n is very large, then larger values of p are acceptable. For example, for $n = 10$ and $p = .01$, the binomial probability that $X = 1$ is .0914 and the Poisson probability is .0905. Values of p close to 1, such as $p = .99$, also would be acceptable because the binomial distribution with $p = .01$ is a mirror image of a binomial distribution with $p = .99$.

Normal approximations. The central limit theorem says (roughly) that the distribution of a sum of n independent identically distributed random variables approaches the normal distribution as n grows large. Because the binomial distribution is derived as the sum of n independent identically distributed Bernoulli random variables, when n is large the normal gives a good approximation to the binomial. As the normal approximation is more accurate when p is near .5, a good rule of thumb is that the approximation should be used only if $np(1 - p) > 5$.

Whenever the normal distribution is used to approximate any other distribution, it is necessary to express μ and σ in terms of the original parameters. In the binomial case, $\mu = np$ and $\sigma = \sqrt{np(1-p)}$.

Because the normal random variable is continuous and the binomial random variable is discrete, the approximation can be improved by using the "**continuity correction**." In the binomial case, the events $\{X > 2\}$ and $\{X \geq 3\}$ are identical, but in the normal case they are not. The continuity correction would suggest approximating either of these cases by $\{X > 2.5\}$. The general rule is to express the original event in terms of both $>$ and $\geq$ (or $<$ and $\leq$) and to use a cutoff number halfway between the two.

For example, suppose that $n = 25$, $p = .40$, and we wish to determine $P\{X \leq 10\}$. Since $\{X \leq 10\} = \{X < 11\}$, the continuity correction cutoff is at 10.5. The exact binomial probability is .5858. The normal approximation at 10 gives

$$P\{X \leq 10\} \approx P\left\{Z < \frac{10-(25)(.40)}{\sqrt{(25)(.40)(.60)}}\right\} = P\{Z < 0\} = 0.5,$$

and with the continuity correction

$$P\{X \leq 10\} \approx P\left\{Z < \frac{10.5-(25)(.40)}{\sqrt{(5)(.40)(.60)}}\right\} = P\{Z < .2041\} = 0.5948.$$

The normal distribution also may be used to approximate the Poisson when λ is large ($\lambda > 10$). In that case, use $\mu = \lambda$, $\sigma = \sqrt{\lambda}$, and the continuity correction as described. For example, suppose that we wish to use a normal approximation of the probability that a Poisson random variable with parameter $\lambda = 15$ exceeds 8. Since $\{X > 8\} = \{X \geq 9\}$, the continuity correction cutoff falls at 8.5. Hence,

$$P\{X > 8\} \approx P\left\{Z > \frac{8.5-15}{\sqrt{15}}\right\} = P\{Z > -1.68\} = .9535.$$

The exact Poisson probability is .9626.

APPENDIX 10–B

CLASSICAL STATISTICAL METHODS AND CONTROL CHARTS

Control charts signal nonrepresentative observations in a sample. The hypothesis that a shift in the process has occurred also can be tested by classical statistical methods. For example, consider the p chart. In constructing the p chart, we are testing the hypothesis that a shift has occurred in the underlying value of p, the true proportion of defectives in the lot. A $2 \times n$ contingency table also can be used to test if p has changed. One variable is time, and the other variable is the proportion of defectives observed. The χ^2 test would be used to test whether or not there exists a relationship between the two variables; that is, whether the proportion of defectives changes with time.

It is not necessarily true that the χ^2 test will give the same results as a p chart. In general, the χ^2 test will recommend rejection of the hypothesis that the data are homogeneous based on the average of the departures from the estimated mean, and the control chart will recommend rejection of the hypothesis that the process is in control based on a large deviation of a single observation. It is important to understand this difference to determine which would be a more appropriate procedure. It is probably true that in the context of manufacturing one is more concerned with extreme deviations of a few observations than the average of many deviations, thus providing one reason for the preference among practitioners for control chart methodology. Another reason for the preference for control charts is that they are easy to use and understand. Quality control managers are more familiar with control charts than they are with classical statistical methods.

APPENDIX 10–C

ECONOMIC DESIGN OF $\bar{X}$ CHARTS

The design of an $\bar{X}$ chart requires the determination of various parameters. These include the amount of time that elapses between sampling, the size of the sample drawn in each interval, and the upper and lower control limits. The penalties associated with the upper and lower control limits are reflected in the Type 1 and Type 2 errors. This section will incorporate explicit costs of these errors into the analysis as well as costs of sampling and considers the problem of designing an $\bar{X}$ chart based on cost minimization.

The model treated here does not include the sampling interval as a decision variable. In many circumstances the sampling interval is determined from considerations other than cost. There are convenient or natural times to sample based on the nature of the process, the items being produced, or personnel constraints.

We will consider the following three costs.

1. Sampling cost.
2. Search cost.
3. Cost of operating out of control.

1. *The sampling cost.* We assume that exactly n items are sampled each period. In most cases, sampling requires workers' time, so personnel costs are incurred. There also may be costs associated with the equipment required for the sampling. Furthermore, sampling may

require destructive testing, adding the cost of the item itself. We will assume that for each item sampled, there is a cost of a_1. It follows that the sampling cost incurred each period is $a_1 n$.

2. *The search cost.* When an out-of-control condition is signaled, the presumption is that there is an assignable cause for the condition. The search for the assignable cause generally will require that the process be shut down. When an out-of-control signal occurs, there are two possibilities: either the process is truly out of control, or the signal is a false alarm. In either case, we will assume that there is a cost a_2 incurred each time a search is required for an assignable cause of the out-of-control condition. The search cost could include the costs of shutting down the facility, labor time required to identify the cause of the signal, time required to determine if the out-of-control signal was a false alarm, and the cost of testing and possibly adjusting the equipment. Note that the search cost is probably a random variable. It might not be possible to predict the degree of effort required to search for an assignable cause of the out-of-control signal. When that is the case, interpret a_2 as the *expected* search cost.
3. *The cost of operating out of control.* The third and final cost that we will consider is the cost of operating the process after it has gone out of control. There is a greater likelihood that defective items are produced if the process is out of control. If defectives are discovered during inspection, they would either be scrapped or repaired at a future time. An even more serious consequence is that a defective item becomes part of a larger subassembly, which must be disassembled or scrapped. Finally, defective items can make their way into the marketplace, resulting in possible costs of warranty claims, liability suits, and overall customer dissatisfaction. Assume that there is a cost of a_3 each period that the process is operated in an out-of-control condition.

We consider the economic design of $\bar{X}$ charts only. Assume that the process mean is μ and the process standard deviation is σ. A sufficient history of observations is assumed to exist so that μ and σ can be estimated accurately. We also assume that an out-of-control condition means that the underlying mean undergoes a shift from μ to $\mu + \delta\sigma$ or to $\mu - \delta\sigma$. Hence, out of control means that the mean shifts by δ standard deviations.

Define a cycle as the time interval from the start of production just after an adjustment to detection and elimination of the assignable cause of the next out-of-control condition. A cycle consists of two parts. Define T as the number of periods that the process remains in control directly following an adjustment and S as the number of periods that the process remains out of control until a detection is made. A cycle is the sum $T + S$. Successive cycles are pictured in Figure 10–15. Note that both T and S are random variables, so the length of each cycle is a random variable as well. The probability distribution of T is given subsequently.

The $\bar{X}$ chart is assumed to be constructed using the following control limits:

$$\text{UCL} = \mu + \frac{k\sigma}{\sqrt{n}}.$$
$$\text{LCL} = \mu - \frac{k\sigma}{\sqrt{n}}.$$

Throughout this chapter we have assumed that $k = 3$, but this may not always be optimal. The goal of the analysis of this section is to determine the economically optimal values of

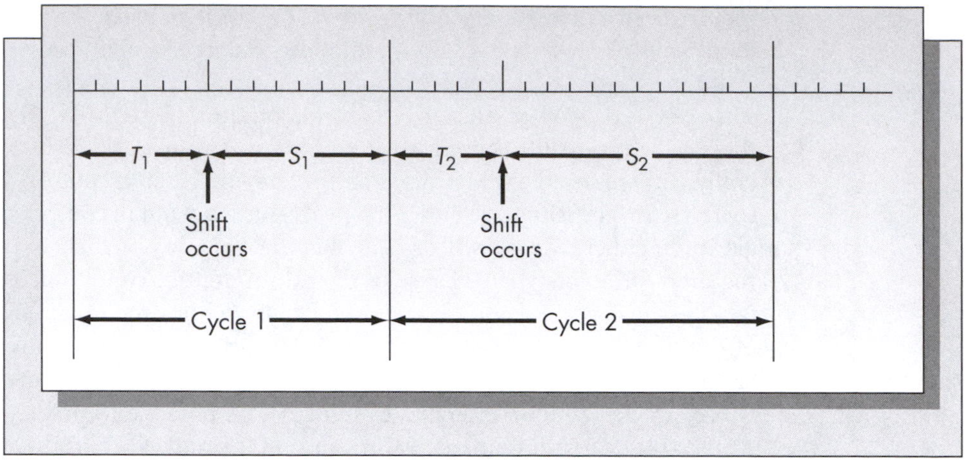

FIGURE 10–15 Successive cycles in process monitoring

both k and n. The method of analysis is to determine an expression for the total cost incurred in one cycle and an expression for the expected length of each cycle. In the spirit of regenerative process (see Ross, 1996, for example), we have the result that

$$E(\text{Cost per unit time}) = \frac{E(\text{Cost per cycle})}{E(\text{Length of cycle})}.$$

After determining an expression for the expected cost per unit time, we will find the optimal values of n and k that minimize this cost. (Similar ideas were used in inventory control models in Chapters 4 and 5 and will be used to analyze age replacement models in Section 13.6.)

Assume that T, the number of periods that the system remains in control following an adjustment, is a discrete random variable having the geometric distribution. That is,

$$P\{T = t\} = \pi(1 - \pi)^t \qquad \text{for } t = 0, 1, 2, 3, \ldots.$$

The geometric model arises in the following fashion. Suppose that in any period the process is in control. Then π is the conditional probability that the process will shift out of control in the next period. The geometric distribution is the discrete analog of the exponential distribution. Like the exponential distribution, the geometric distribution also has the memoryless property. (A detailed discussion of the memoryless property of the exponential distribution is given in Section 13.1.) In the present context, the memoryless property implies that there is no aging or decay in the production process. That is, the process is equally likely to shift out of control just after an assignable cause has been found and corrected as it is many periods later. This assumption will be accurate when process shifts are due to random causes or when the process is recalibrated on an ongoing basis.

An out-of-control signal is indicated when

$$|\bar{X} - \mu| > \frac{k\sigma}{\sqrt{n}}.$$

As in earlier sections of this chapter, let α be the probability of Type 1 error. Type 1 error occurs when an out-of-control signal is observed but the process is in control. It follows that

$$\alpha = P\left\{\left|\bar{X}-\mu\right| > \frac{k\sigma}{\sqrt{n}} \middle| E\left(\bar{X}\right)=\mu\right\}$$
$$= P\left\{\left|\frac{\bar{X}-\mu}{\sigma/\sqrt{n}}\right| > k \middle| E\left(\bar{X}\right)=\mu\right\} = P\left\{\left|Z\right|>k\right\} = 2\Phi\left(-k\right).$$

where Φ is the cumulative standard normal distribution function.

The Type 2 error probability, β, is the probability of not detecting an out-of-control condition. Here we assume that an out-of-control condition means that the process mean has shifted to $\mu + \delta\sigma$ or to $\mu - \delta\sigma$. Suppose that we condition on the event that the mean has shifted from μ to $\mu + \delta\sigma$. The probability that the shift is not detected after observing a sample of n observations is

$$\beta = P\left\{\left|\bar{X}-\mu\right| \le \frac{k\sigma}{\sqrt{n}} \middle| E\left(\bar{X}\right)=\mu+\delta\sigma\right\}$$
$$= P\left\{\frac{-k\sigma}{\sqrt{n}} \le \bar{X}-\mu \le \frac{k\sigma}{\sqrt{n}} \middle| E\left(\bar{X}\right)=\mu+\delta\sigma\right\}$$
$$= P\left\{-k-\delta\sqrt{n} \le \frac{\bar{X}-\mu-\delta\sigma}{\sigma/\sqrt{n}} \le k-\delta\sqrt{n} \middle| E\left(\bar{X}\right)=\mu+\delta\sigma\right\}$$
$$= P\left\{-k-\delta\sqrt{n} \le Z \le k-\delta\sqrt{n}\right\}$$
$$= \Phi\left(k-\delta\sqrt{n}\right) - \Phi\left(-k-\delta\sqrt{n}\right).$$

If we had conditioned on $E(\bar{X}) = \mu - \delta\sigma$, we would have obtained

$$\beta = \Phi\left(k+\delta\sqrt{n}\right) - \Phi\left(-k+\delta\sqrt{n}\right).$$

Using the symmetry of the normal distribution [specifically, that $\Phi(t) = 1 - \Phi(-t)$ for any t], it is easy to show that these two expressions for β are the same.

Consider the random variables T and S. We assume that T is a geometric random variable assuming values 0, 1, 2, … . One can show that

$$E\left(T\right) = \frac{1-\pi}{\pi}$$

(see, for example, DeGroot-Schervish, 2012). The random variable S is the number of periods that the process remains out of control after a shift occurs. The probability that a shift is not detected when the process is out of control is exactly β. It follows that S is also a geometric random variable except that it assumes only the values 1, 2, 3, … . That is,

$$P\{S = s\} = (1 - \beta)\,\beta^{s-1} \qquad \text{for } s = 1, 2, 3, \ldots .$$

Because S is defined on the set 1, 2, 3, … , the expected value of S is $E(S) = 1/(1 - \beta)$. It follows that the expected cycle length, say C, is given by

$$E(C) = E(T+S) = E(T) + E(S) = \frac{1-\pi}{\pi} + \frac{1}{1-\beta}.$$

Consider the expected sampling cost incurred in a cycle. Each period there are n items sampled. As there are, on the average, $E(C)$ periods per cycle, it follows that the sampling cost per cycle is $a_1 nE(C)$.

We now compute the expected search cost. The process is shut down each time an out-of-control signal is observed. One or more of these signals could be a false alarm. Suppose that there are exactly M false alarms in a cycle. The random variable M has the binomial distribution with probability of "success" (i.e., a false alarm) of α for a total of T trials. It follows that $E(M) = \alpha E(T)$. The expected number of searches per cycle is exactly $1 + E(M)$, as the final search is assumed to discover and correct the assignable cause. Hence, the total search cost in a cycle is

$$a_2[1 + \alpha E(T)] = a_2[1 + \alpha(1 - \pi)/\pi].$$

We also assume that there is a cost of a_3 for each period that the process is operated in an out-of-control condition. The process is out of control for exactly S periods. Hence, the expected out-of-control cost is $a_3 E(S) = a_3/(1 - \beta)$.

It follows that the expected cost per cycle is

$$a_1 nE(C) + a_2[1 + \alpha(1 - \pi)/\pi] + a_3/(1 - \beta).$$

Dividing by the expected length of a cycle, $E(C)$, gives the average cost per unit time as

$$a_1 n + \frac{a_2\left[1 + \alpha\frac{1-\pi}{\pi}\right] + \frac{a_3}{1-\beta}}{\frac{1-\pi}{\pi} + \frac{1}{1-\beta}}$$

$$= a_1 n + \frac{a_2\left[1 + \alpha\frac{1-\pi}{\pi}\right] + \frac{a_3}{1-\beta}}{\frac{1-\beta(1-\pi)}{(1-\beta)\pi}}$$

$$= a_1 n + \frac{a_2(1-\beta)\left[\pi + \alpha(1-\pi)\right] + a_3\pi}{1-\beta(1-\pi)}.$$

We will write this as $G(n, k)$ to indicate that the optimization requires searching for the best n and k, where $n = 1, 2, 3, \ldots$ and $k > 0$. Note that α depends on k and β depends on both n and k. The goal is to find the values of n and k that minimize $G(n, k)$. This is a complex optimization problem because both a and β require evaluation of the cumulative normal distribution function.

EXAMPLE 10.8

Consider Example 10.1 of Wonderdisk, introduced in Section 10.3. Howard Hamilton would like to design an $\bar{X}$ chart in an economically optimal fashion. Based on his experience with the process and an analysis of the past history of failures, he decides that the geometric distribution accurately describes changes in the process state.

In order to use the model described in this section, he must estimate various costs and system parameters. The first is the sampling cost. Here sampling requires measuring the length of a tracking arm. This requires moving the arm to a different location, mounting it on a special brace to protect it, and measuring the length with calipers designed for the purpose. The process requires about 12 minutes of a technician's time. The technician is paid \$15 per hour, so the sampling cost is \$15/5 = \$3 per item sampled.

The second cost to estimate is the search cost. The time spent searching for an assignable cause of an out-of-control signal is usually about 30 minutes. If a problem is not discovered within that time, it is generally assumed that the out-of-control signal was a false alarm. The arms generate a revenue for the company of about \$1,200 daily. Assuming an eight-hour workday, the cost of shutting down production comes to about \$1,200/8 = \$150 per hour. Hence, the search cost is \$75.

The third cost required by the model is the cost of operating the process in an out-of-control condition. If the process is out of control, the proportion of defective arms produced increases. Most of the defective arms show up in the final testing phase of the drives. If a drive has a defective arm, the drive is disassembled and the arm replaced. Some defective arms pass inspection and are shipped to the customer with the disk drive. Wonderdisk provides purchasers a 14-month warranty, and it is likely that a problem with the drive will develop during the warranty period if the arm is defective. Howard estimates that the cost of operating the process out of control is about \$300 per hour, but he is not very confident about this estimate.

The model also requires estimates of π and δ. Recall that π represents the probability that the process will shift from an in-control state to an out-of-control state during one period. In the past, out-of-control signals have occurred at a rate of about one for every 10 hours of operation. As half of these have been false alarms, a reasonable estimate of the proportion of periods in which a shift has occurred is about one out of 20, or $\pi = .05$. The constant δ represents the degree of the shift as measured in multiples of the process standard deviation. In the past, the shifts have averaged about one standard deviation, so the estimate of δ is 1.

In order to simplify the calculations, Howard decides to use an approximation to the standard normal cumulative distribution function. The one he uses is the following:

$$\Phi(z) = 0.500232 - 0.212159z^{2.08388} + 0.5170198z^{1.068529} + 0.041111z^{2.82894}.$$

This approximation (Herron, 1985) is accurate to within 0.5 percent for $0 < z < 3$. Howard Hamilton decides that this is accurate enough for his purposes. The optimization scheme he adopts is the following. Because n is a discrete variable and represents the number of items sampled in each subgroup, it is unlikely that n would exceed 10. Furthermore, because k is the number of standard deviations of $\bar{X}$ used in the control chart, it is unlikely that k would exceed 3. Howard writes a computer program to evaluate $G(n, k)$ for $k = 0, 0.1, 0.2, \ldots, 2.8, 2.9, 3.0$ and $n = 1, 2, 3, \ldots, 10$. (These calculations were actually done using a popular spreadsheet program.) For each fixed value of n, the function $G(n, k)$ appears to be convex in the variable k (convex functions were discussed in Chapters 4 and 5). For the given parameter values and $n = 4$, Figure 10–16

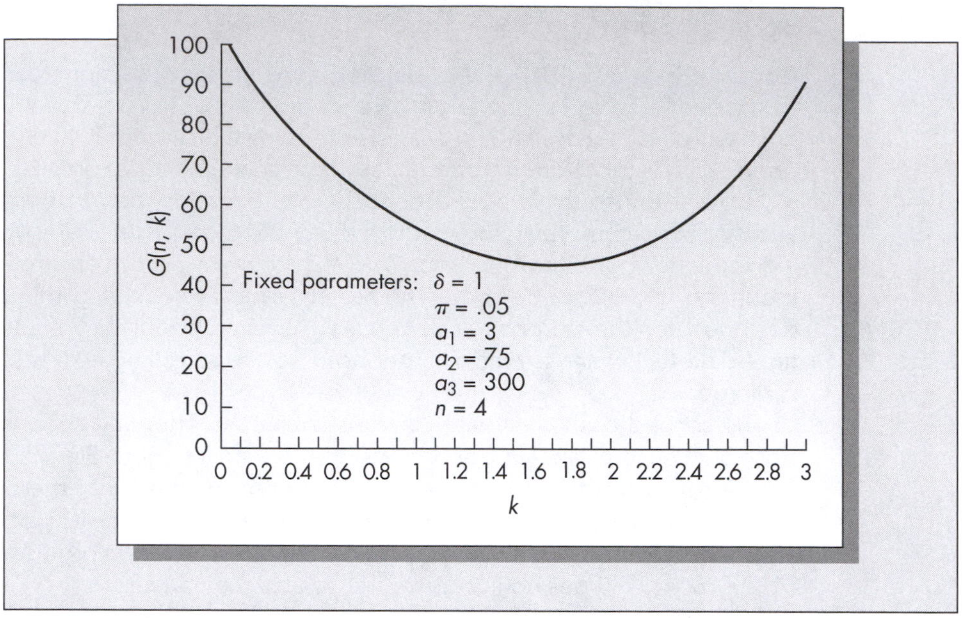

FIGURE 10–16 The behavior of G(*n*, *k*) as a function of *k* (Example 10.8)

shows the function $G(n, k)$ as a function of k. The graph shows that the minimum cost occurs at about $k = 1.7$ and equals about \$45 per hour.

Table 10–4 gives the results of the calculations for these parameter settings. According to the table, the optimal subgroup size is 4 and the optimal k is 1.74. These results were reasonably close to the current policy and give Hamilton confidence that his estimates were at least in the right ballpark. When he presented the results to his

Table 10–4 Optimal Values of *k* for Various *n*

	Fixed parameters are:		$\delta = 1$	
			$\pi = 0.05$	
			$a_1 = 3$	
			$a_2 = 75$	
			$a_3 = 300$	
n	**Optimal *k***	α	β	**Cost**
1	1.13	.25	.54	\$54.4
2	1.52	.14	.50	48.8
3	1.60	.11	.44	46.1
4	1.74	.08	.39	45.2
5	1.86	.07	.34	45.3
6	1.97	.05	.30	46.2
7	2.06	.04	.26	47.6
8	2.14	.03	.23	49.3
9	2.21	.02	.21	51.3
10	2.27	.02	.18	53.4

Table 10–5 Optimal Values of k for Various n (Revised) Fixed Parameters: Same as in Table 10–4 except a_3 = 1,000

n	Optimal k	α	β	Cost
1	0.67	.48	.30	$111.6
2	0.93	.33	.29	102.8
3	1.13	.25	.26	96.9
4	1.35	.18	.25	93.3
5	1.43	.15	.21	91.3
6	1.59	.11	.19	90.5
7	1.70	.09	.16	90.4
8	1.79	.07	.14	90.9
9	1.86	.06	.12	91.8
10	1.94	.05	.11	93.3

boss, however, his boss expressed concern about the large value of β, the probability of Type 2 error.

> You mean to tell me that if we operate the system optimally, then there is almost a 40 percent chance that we won't be able to detect when the system has gone out of control? That sounds pretty high. Considering the push that is on all over the company for an improvement in quality, I find that figure very disturbing. How does that turn out to be optimal?

The reason that the Type 2 error was so large was the assumption that the cost to the company was $300 for each hour that the process was operated out of control. After giving the matter some thought, Howard's boss decided that a value of $1,000 was probably closer to the mark and was more consistent with the corporate goal of improving quality. Howard repeated his calculations substituting a value of a_3 = 1,000. The results are presented in Table 10–5.

With the revised value of a_3 = 1,000, the optimal value of the sample size n increased to 7 with a corresponding value of k = 1.70. The cost per hour at the optimal solution increased to $90.40. Although the limits on the control chart remained the same, the effect of increasing the sample size from 4 to 7 resulted in a dramatic decrease in the value of β from .39 to .16. Howard's boss was far happier with the values of α and β that resulted from the new design.

APPENDIX 10–D

DOUBLE SAMPLING PLANS FOR ATTRIBUTES

Five numbers define a double sampling plan: n_1, n_2, c_1, c_2, and c_3. The plan is implemented in the following way. One draws an initial sample of size n_1 and determines the number of defectives in the sample. If the number of defectives in the sample is less than or equal to c_1, the lot is accepted. If the number of defectives in the sample is larger than c_2, the lot is rejected.

However, if the number of defectives is larger than c_1 but less than or equal to c_2, another sample of size n_2 is drawn. If the number of defectives in the combined samples is less than or equal to c_3, the lot is accepted. If not, the lot is rejected. Most double sampling plans assume that $c_3 = c_2$. We will make that assumption as well from this point on.

A double sampling plan is obviously more difficult to construct and more difficult to implement than a single sampling plan. However, it does have some advantages over single plans. First, a double sampling plan may give similar levels of the consumer's and the producer's risks but require less sampling in the long run than a single plan. Also, there is the psychological advantage in double sampling plans of providing a second chance before rejecting a lot.

EXAMPLE 10.9

Consider again Example 10.6 concerning Spire Records. Herman Sondle decides to experiment with a few double sampling plans to see if he can achieve similar levels of efficiency with less sampling. Unfortunately, because the plan depends on four different numbers, considerable trial-and-error experimentation is necessary. Let us consider the computation of the consumer's and the producer's risks for the following sampling plan:

$$n_1 = 20 \qquad c_1 = 3,$$
$$n_2 = 10 \qquad c_2 = 5.$$

Define

$X =$ Number of defectives observed in the first sample.

$Y =$ Number of defectives observed in the second sample.

$Z =$ Number of defectives observed in the combined samples ($Z = X + Y$).

The OC curve is

$$\begin{aligned} OC(p) = P\{\text{Lot is accepted} \mid p\} &= P\{\text{Lot is accepted on first sample} \mid p\} \\ &+ P\{\text{Lot is accepted on second sample} \mid p\} \end{aligned}$$

where

$$P\{\text{Lot is accepted on the first sample} \mid p\} = P\{X \le 3 \mid p\}$$

and

$$\begin{aligned} &P\{\text{Lot is accepted on the second sample} \mid p\} \\ &= P\left\{\begin{matrix}\text{Lot is neither accepted nor rejected on the} \\ \text{first sample and the lot is accepted on the} \\ \text{second sample}\end{matrix}\middle| p\right\} \\ &= P\{3 < X \le 5, \quad Z \le 5 | p\}. \end{aligned}$$

Computation of this joint probability must be done carefully, as X and Z are *dependent* random variables.

Consider $p = \text{AQL} = .1$.

$$\begin{aligned} &P\{\text{Lot is accepted on the first sample} \mid p = .1\} \\ &= P\{X \le 3 \mid p = .1, n = 20\} = .8670, \end{aligned}$$

$P\{\text{Lot is accepted on the second sample} \mid p = .1\}$

$$\begin{aligned} &= P\{X = 4 \mid p = .1, n = 20\}\, P\{Y \le 1 \mid p = .1, n = 10\} \\ &+ P\{X = 5 \mid p = .1, n = 20\}\, P\{Y \le 0 \mid p = .1, n = 10\} \\ &= (.0898)(.7361) + (.0319)(.3487) = .0772. \end{aligned}$$

Summing:

$$P\{\text{Lot is accepted} \mid p = .1\} = .8670 + .0772 = .9442.$$

Repeating similar calculations with $p = .3$ gives

$$\begin{aligned} &P\{\text{Lot is accepted} \mid p = .3\} \\ &= .1071 + (.1304)(.1493) + (.1789)(.0282) \\ &= .1316. \end{aligned}$$

Hence, it follows that for this case we obtain

$$\alpha = 1 - .9442 = .0558,$$
$$\beta = .1316.$$

Experimentation with other values of n_1, n_2, c_1, and c_2 can lead to double sampling plans that more closely match the desired values of α and β. Tables are available for optimizing double sampling plans. (See, for example, Duncan, 1986, pp. 232–33.)

APPENDIX 10–E

SEQUENTIAL SAMPLING PLANS

Double sampling plans may be extended to triple sampling plans, which also may be extended to higher-order plans. The logical conclusion of this process is the sequential sampling plan. In a sequential plan, items are sampled one at a time. After each sampling, two numbers are recorded: the number of items sampled and the cumulative number of defectives observed. Based on these numbers, one of three decisions is made: (1) accept the lot, (2) reject the lot, or (3) continue sampling. Unlike single and double sampling plans, there will always exist a sequential sampling plan that will give specific values of p_0, p_1, α, and β. Sequential sampling plans are defined by three regions: the acceptance region, the rejection region, and the sampling region. The three regions are separated by straight lines. The lines have the forms

$$L_1 = -h_1 + sn,$$
$$L_2 = h_2 + sn,$$

where n is the number of items sampled. Note that L_1 and L_2 are both linear functions of the variable n. The y intercepts are respectively $-h_1$ and h_2, and the slope of each line is s. As the lines have the same slope, they are parallel. The sequential sampling plan is implemented in the following manner. The cumulative number of defectives is graphed, together with the lines for L_1 and L_2. When the cumulative number of defectives exceeds L_2, the lot is rejected, and when the cumulative number of defectives falls below L_1, the lot is accepted. As long as the cumulative number of defectives lies between L_1 and L_2, sampling continues.

Figure 10–17 shows two examples of the results of sampling for the same sequential sampling plan. In Case A the sampling led to acceptance of the lot, and in Case B it led to rejection of the lot.

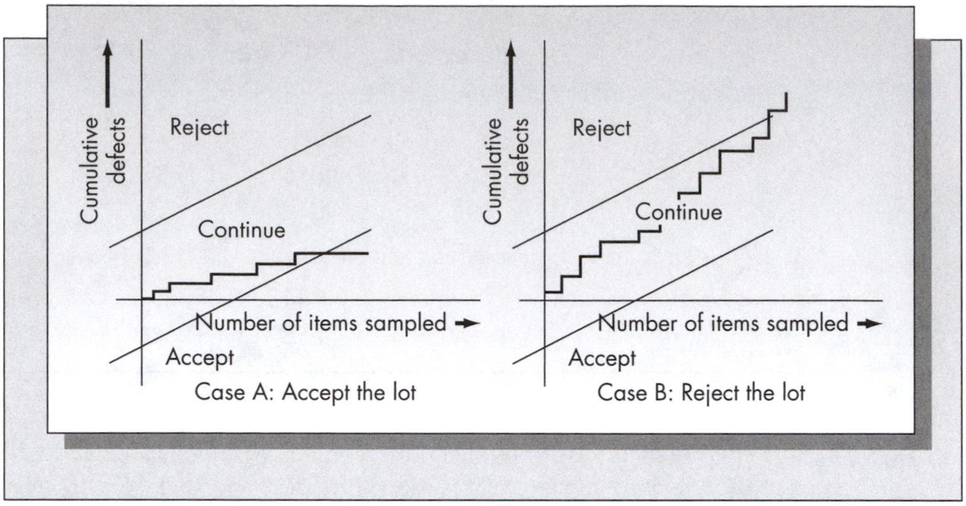

FIGURE 10–17 Two realizations of a sequential sampling plan

The equations for h_1, h_2, and s are

$$h_1 = \frac{\log \dfrac{1-\alpha}{\beta}}{\log \dfrac{p_1(1-p_0)}{p_0(1-p_1)}},$$

$$h_2 = \frac{\log \dfrac{1-\beta}{\alpha}}{\log \dfrac{p_1(1-p_0)}{p_0(1-p_1)}},$$

$$s = \frac{\log \dfrac{1-p_0}{1-p_1}}{\log \dfrac{p_1(1-p_0)}{p_0(1-p_1)}}.$$

EXAMPLE 10.10

Consider again Example 10.6 of Spire Records. Herman Sondle has experimented with various sampling plans to achieve the desired levels of the consumer's and the producer's risks. He decides to construct a sequential sampling plan to see how it compares with the single and the double plans that were presented earlier. Spire and B&G agreed on the following values of the AQL, LTPD, consumer's risk, and producer's risk:

$$p_0 = .1, \qquad \alpha = .1,$$
$$p_1 = .3, \qquad \beta = .1.$$

Notice that the denominators in the expressions for h_1, h_2, and s are the same. We compute the denominator first. (We will use log to the base 10 in our calculations.

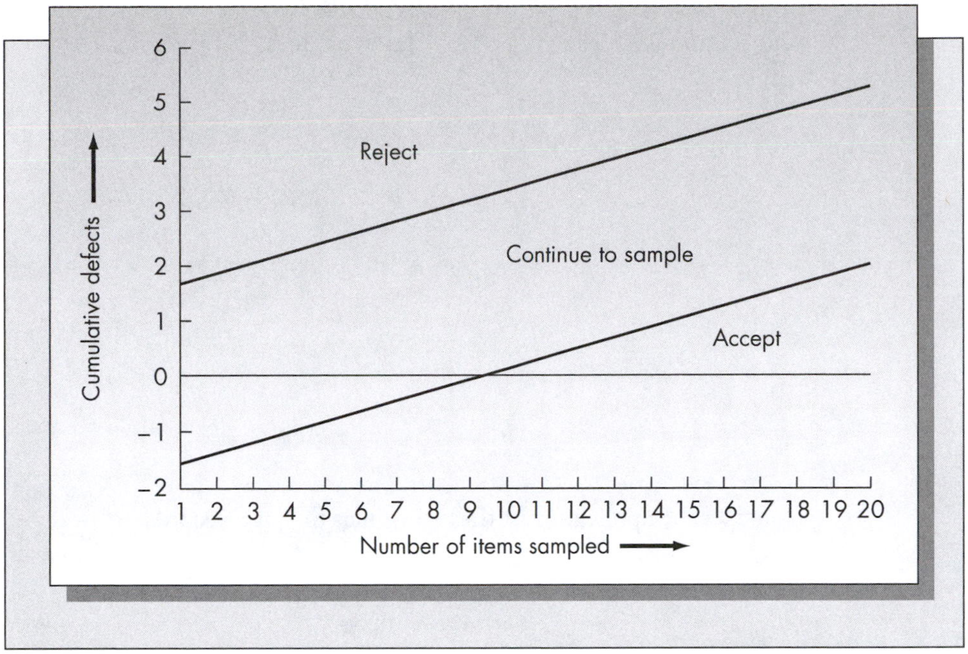

FIGURE 10–18 Sequential sampling plan for Spire Records (Example 10.10)

Because all formulas involve the ratio of logarithms, the results will be the same whether one uses base 10 or base e.)

$$\log[(.3)(.9)/(.1)(.7)] = 0.58626.$$

Hence,

$$h_1 = \log(.9/.1)/0.58626 = 0.9542/0.58626 = 1.6277,$$
$$h_2 = h_1 = 1.6277 \qquad (\text{since } \alpha = \beta \text{ for this case}),$$
$$s = \log(.9/.7)/0.58626 = 0.10914/0.58626 = 0.18617.$$

Figure 10–18 shows the decision regions for Spire's sequential sampling plan.

When Herman suggested that Spire Records implement the sequential sampling plan shown in Figure 10–18, he met with considerable resistance from some of his coworkers responsible for the inspection of incoming stock. With a single sampling plan with $n = 25$, they argued that at least they would know in advance how many records they would have to check. With the sequential plan, however, they argued that they might have to sample the entire lot without the plan recommending either acceptance or rejection. Although Herman had heard that sequential plans were more efficient, he had difficulty convincing his coworkers to try the plan.

The coworkers in the example were correct in seeing that the number of items sampled when using sequential sampling is a random variable. The *expected* sample size that results from a sequential sampling plan depends on the proportion of defectives in the lot, p. The average sample number (ASN) curve gives the expected sample size for a sequential sampling plan as a function of p. We will estimate the ASN curve by obtaining its value at five specific points: when $p = 0$, $p = p_0$, $p = s$, $p = p_1$, and $p = 1$. It is easy to find the ASN curve at these points. In most cases one can obtain an adequate approximation to the ASN curve knowing only its value at these five points. The five values are

$$\text{At } p = 0, \qquad \text{ASN} = \frac{h_1}{s}.$$

$$\text{At } p = p_0, \qquad \text{ASN} = \frac{(1-\alpha)h_1 - \alpha h_2}{s - p_0}.$$

$$\text{At } p = s, \qquad \text{ASN} = \frac{h_1 h_2}{s(1-s)}.$$

$$\text{At } p = p_1, \qquad \text{ASN} = \frac{(1-\beta)h_2 - \beta h_1}{p_1 - s}.$$

$$\text{At } p = 1, \qquad \text{ASN} = \frac{h_2}{1-s}.$$

Consider the application of these formulas to Spire Records:

$$\text{ASN}(0) = \frac{1.6277}{0.18617} = 8.74,$$

$$\text{ASN}(.1) = \frac{(.9)(1.6277) - (.1)(1.6277)}{0.18617 - .1} = 15.11,$$

$$\text{ASN}(.18617) = \frac{(1.6277)(1.6277)}{(0.18617)(1 - 0.18617)} - 17.49,$$

$$\text{ASN}(.3) = \frac{(.9)(1.6277) - (.1)(1.6277)}{.3 - 0.18617} = 11.44,$$

$$\text{ASN}(1) = \frac{1.6277}{1 - 0.18617} = 2.0.$$

The ASN curve is a unimodal curve whose maximum value lies between p_0 and p_1. Based on this and the five points computed, we obtain the estimated ASN curve shown in Figure 10–19. We see from the figure that the expected sample size required for the sequential sampling plan for Spire Records will be at most 18 items. This is clearly an improvement over the single sampling plan with $n = 25$ and $c = 4$, which resulted in similar values of α and β. It is important to keep in mind, however, that the actual sample size required in the sequential plan for the inspection of any particular batch of records is a random variable. The ASN curve gives only the *expected* value of this random variable for any specified value of p. Thus, it is possible that in specific instances, the actual sample size could be larger than 18, or even larger than 25.

APPENDIX 10–F

AVERAGE OUTGOING QUALITY

The purpose of a sampling plan is to screen out lots of unacceptable quality. However, because sampling is a statistical process, it is possible that bad lots will be passed and good lots will be rejected. A fundamental issue related to the effectiveness of any sampling plan is to determine the quality of product that results *after* the inspection process is completed.

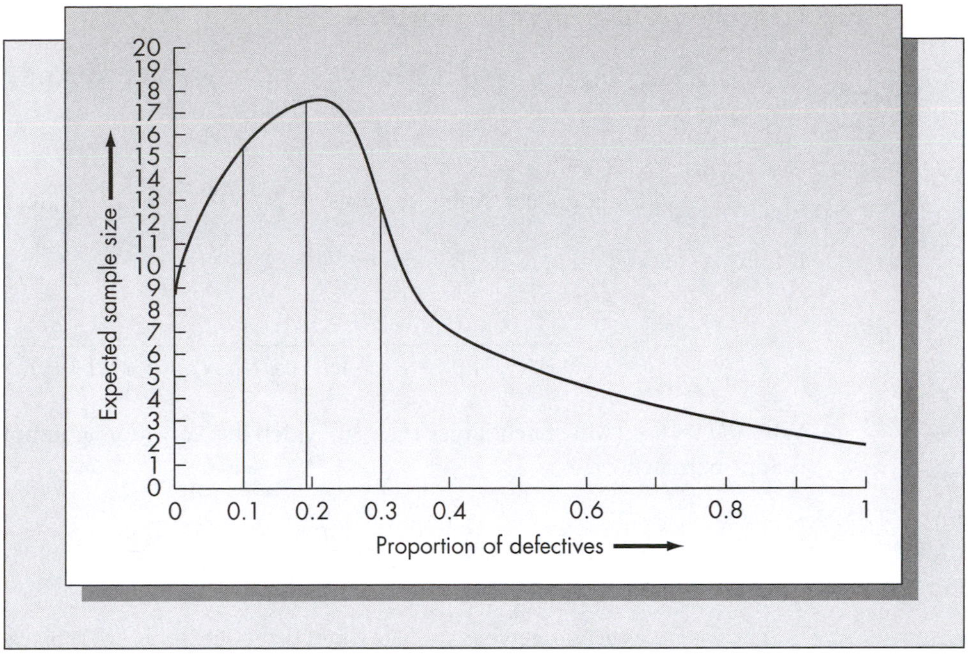

FIGURE 10–19 ASN curve for Spire Records (estimated)

The calculation of the **average outgoing quality** of an inspection process depends on the assumption that one makes about lots that do not pass inspection and the assumption that one makes about defective items. Assume that rejected lots are subject to 100 percent inspection. We derive the average outgoing quality curve under two conditions: (1) defective items in samples and in rejected lots are not replaced and (2) defective items in samples and in rejected lots are replaced.

The average outgoing quality (AOQ) is the long-run ratio of the expected number of defectives and the expected number of items successfully passing inspection. That is,

$$\text{AOQ} = \frac{E\{\text{outgoing number of defectives}\}}{E\{\text{outgoing number of items}\}}.$$

The OC curve is the probability that a lot is accepted as a function of p. That is,

$$\text{OC}(p) = P\{\text{lot is accepted} \mid p\}.$$

For convenience we will refer to this term as P_a.

Case 1: Defective items are not replaced. Suppose that lots are of size N and samples are of size n. Then the expected number of defectives and the expected number of items shipped are

		Number of Defectives	Number of Items
Accept lot	P_a	$(N - n)p$	$N - np$
Reject lot	$1 - P_a$	0	$N(1 - p)$

From this tree diagram we see that

$$E\{\text{outgoing number of defectives}\} = P_a(N - n)p + (1 - P_a)(0) = P_a(N - n)p$$

and

$$E\{\text{outgoing number of items}\} = P_a(N - np) + (1 - P_a)N(1 - p).$$

It follows that the ratio, AOQ, is given by

$$\text{AOQ} = \frac{P_a(N-n)p}{P_a(N-np)+(1-P_a)N(1-p)} = \frac{P_a(N-n)p}{N-np-p(1-P_a)(N-n)}.$$

When $N >> n$ (N is much larger than n), which is a common assumption, this expression is approximately

$$\text{AOQ} \approx \frac{P_a p}{P_a + (1-P_a)(1-p)} = \frac{P_a p}{1-p(1-P_a)}.$$

The formulas are somewhat simpler when defective items are replaced with good items. *Case 2: Defective items are replaced.* In this case the tree diagram becomes

		Number of Defectives	Number of Items
Accept lot	P_a	$(N - n)p$	N
Reject lot	$1-P_a$	0	N

The AOQ is given by

$$\text{AOQ} = \frac{P_a(N-n)p}{N},$$

which is approximately

$$\text{AOQ} = P_a p = \text{OC}(p)p$$

when $N >> n$.

This last formula is the one most commonly used in practice, primarily because of its simplicity. An important measure of the effectiveness of a sampling plan is the **average outgoing quality limit** (AOQL), which is defined as the maximum value of the AOQ curve.

EXAMPLE 10.11

Consider the case of Spire Records. In Spire's case, we can assume that lots are large compared to samples and that all defectives are replaced. In that case AOQ $\approx$ OC(p) p. In Figure 10–20 we have generated the AOQ curves for Spire's single sampling plans

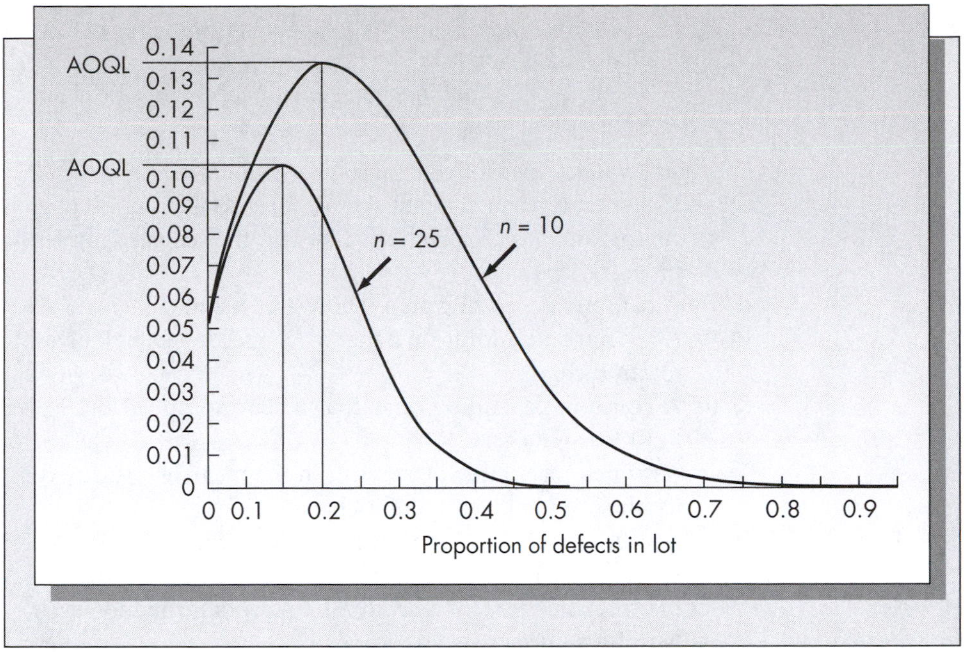

FIGURE 10–20 AOQ curves for Spire Records (Example 10.11)

with $n = 10$ and $n = 25$. Note how the larger sample size significantly improves the average outgoing quality. From Figure 10–20 we see that if Spire uses a single sampling plan with $n = 25$ and $c = 4$, the store can expect that the proportion of defective B&G records on its shelves will be no more than about 10.2 percent.

APPENDIX 10–G

Glossary of Notation for Chapter 10 on Quality and Assurance

Note: This chapter uses accepted notation for control charts and acceptance sampling. As a result, the same symbol may have one meaning in the context of control charts and another in the context of acceptance sampling.

a_1 = Cost of sampling one item.

a_2 = Cost of searching for an assignable cause.

a_3 = Cost of operating the process in an out-of-control state.

AOQ = Average outgoing quality.

AOQL = Average outgoing quality limit. The maximum value of the AOQ curve.

AQL = Acceptable quality level.

α = $P\{\text{Type 1 error}\}$.

Control chart usage: α represents the probability of obtaining an out-of-control signal when the process is in control. *Acceptance sampling usage:* α is the probability of rejecting good lots.

β = $P\{\text{Type 2 error}\}$.

Control chart usage: β represents the probability of not obtaining an out-of-control signal when the process has gone out of control.*Acceptance sampling usage:* β is the probability of accepting bad lots.

$c = \begin{cases} \textit{Control chart usage}\text{: The expected number of defects per unit of production.} \\ \textit{Acceptance sampling usage}\text{: The acceptance level for a single sampling plan.} \end{cases}$

c_1 = Acceptance level for first sample for a double sampling plan.
c_2 = Rejection level for the first sample for a double sampling plan.
c_3 = Acceptance level for both the first and the second samples for a double sampling plan. (Often $c_2 = c_3$.)
d_2 = A constant depending on n that relates $\bar{R}$ and σ.
d_3 = A constant depending on n that when multiplied by $\bar{R}$ gives the lower control limit for an R chart.
d_4 = A constant depending on n that when multiplied by $\bar{R}$ gives the upper control limit for an R chart.
δ = Assumed magnitude of the shift in the mean as measured in standard deviations for the economic design of $\bar{X}$ charts.
LCL = Lower control limit for a control chart.
LSL = Lower specification limit for an item.
LTPD = Lot tolerance percent defective. Unacceptable quality level.
M = Number of defectives in a lot.
μ = Population mean.
N = Number of items in lot. Used in acceptance sampling.

$n = \begin{cases} \textit{Control chart usage}\text{: Size of each subgroup for an } \bar{X} \text{ chart.} \\ \textit{Acceptance sampling usage}\text{: Number of items sampled from a lot for a single sampling plan.} \end{cases}$

n_1 = Size of the first sample for a double sampling plan.
n_2 = Size of the second sample in a double sampling plan.
OC(p) = Operating characteristic curve.

$p = \begin{cases} \textit{Control chart usage}\text{: True proportion of defective items produced.} \\ \textit{Acceptance sampling usage}\text{: True proportion of defectives in a lot.} \end{cases}$

p_0 = AQL
p_1 = LTPD
π = Probability that the process goes out of control in a single period.
R = Range of a sample. The difference between the largest and the smallest values in the sample.
s = Sample standard deviation of a random sample.
σ = Population standard deviation.
$\hat{\sigma}$ = Estimator for σ.
UCL = Upper control limit for a control chart.
USL = Upper specification limit for an item.

$X = \begin{cases} \textit{Control chart usage}\text{: Value of a single measurement from the population.} \\ \textit{Acceptance sampling usage}\text{: Number of defectives observed in a sample of } n \text{ items.} \end{cases}$

$\bar{X}$ = Arithmetic average of a random sample of n independent measurements.
Z = Standard normal variate.
$z_{\alpha/2}$ = The number such that the probability of observing a value of Z that exceeds $z_{\alpha/2}$ is $\alpha/2$.

BIBLIOGRAPHY

Aguayo, R. *Dr. Deming: The American Who Taught the Japanese about Quality*. New York: Lyle Stuart, 1990.

Alexander, F. "ISO 14001: What Does It Mean for IEs?" *IIE Solutions* 2 (January 1996), pp. 15–18.

Baker, K. R. "Two Process Models in the Economic Design of an Chart." *AIIE Transactions* 13 (1971), pp. 257–63.

Boothroyd, G.; P. Dewhurst; and W. A. Knight. *Product Design for Manufacture and Assembly,* 3rd ed. Boca Raton, FL: CRC Press, 2011.

Cohen, L. *Quality Function Deployment and Six Sigma: A QFD Handbook*, 2nd ed. Boston: Pearson, 2010.

Costantino, B. "Cedar Works: Making the Transition to Lean. In J. K. Liker (Ed.), *Becoming Lean: Inside Stories of U.S. Manufacturers* (pp. 303–388). Productivity Press, Portland, OR, 1998.

Crosby, P. B. *Quality Is Free*. New York: McGraw-Hill, 1979.

Dahan, E. "Note on Listening to the Customer, Part I." Teaching note, Graduate School of Business, Stanford University, Stanford, CA, 1995.

Daily Mail. Suzuki and Lexus top new reliability survey, Tesla on the bottom. 06/09/2018. https://www.driven.co.nz/news/news/suzuki-and-lexus-top-new-reliability-survey-tesla-on-the-bottom/

DeGroot, M. H. and M. Schervish, *Probability and Statistics*. 4th ed. Boston: Addison Wesley, 2012.

Dehnad, K. *Quality Control, Robust Design, and the Taguchi Method*. Pacific Grove, CA: Wadsworth Cole, 1989.

Deming, W. Edwards. *Out of the Crisis*. The MIT Press Reprint Edition, Cambridge, MA, 2000.

Duncan, A. J. "The Economic Design of $\bar{X}$ Charts Used to Maintain Current Control of a Process." *Journal of the American Statistical Association* 51 (1956), pp. 228–42.

Duncan, A. J. *Quality Control and Industrial Statistics*. 5th ed. New York: McGraw-Hill/Irwin, 1986.

Dvorak, P. "Manufacturing Puts a New Spin on Design." *Machine Design* 67 (August 22, 1994), pp. 67–74.

Feigenbaum, A. V. *Total Quality Control*. New York: McGraw-Hill, 1961.

Feigenbaum, A. V. *Total Quality Control*. 4th ed. New York: McGraw-Hill, 2004.

Franklin, S., "In Pursuit of Perfection," *Chicago Tribune,* Sunday April 4, 1999, section 5, pp. 7–8.

Garvin, D. A. *Managing Quality*. New York: Free Press, 1988.

Green, P., and V. R. Rao. "Conjoint Measurement for Quantifying Judgmental Data." *Journal of Marketing Research* 8 (1971), pp. 355–63.

Herron, D. A. Private communication, 1985.

Kackar, R. N. "Off-Line Quality Control, Parameter Design, and the Taguchi Method." *Journal of Quality Technology* 17 (1985), pp. 176–88.

Kolesar, P. "Scientific Quality Management and Management Science." In *Handbooks in Operations Research and Management Science*, vol. 4, *Logistics of Production and Inventory*, ed. S. Graves, A. H. G. Rinnooy Kan; and P. Zipkin. Chapter 13. Amsterdam: North Holland, 1993.

Kumar, S., and Y. Gupta. "Statistical Process Control at Motorola's Austin Assembly Plant." *Interfaces* 23, no. 2 (March–April 1993), pp. 84–92.

Kuster, T. "ISO 9000: A 500-lb Gorilla?" *Metal Center News* 35, no. 10 (September 1995), pp. 5–6.

Lawler, E. E., and S. A. Mohrman. "Quality Circles after the Fad." *Harvard Business Review* 63 (1985), pp 65–71.

Leonard, J. See the USA. *New York Magazine*, 5th June 1995.

Logothetis, N., and H. P. Wynn. *Quality through Design*. Oxford: Clarendon Press, 1989.

Martin, J. "Ignore Your Customer." *Fortune*, May 1, 1995, pp. 121–26.

National Institute of Standards and Technology. "Baldridge FAQS: Baldridge Award Recipients." Gaithersburg, MD: U.S. Department of Commerce, 2019.

Navera, T. "New Navistar Line Hits Full Production in Springfield," Dayton Business Journal, June 7, 2017.

Pugh, S. *Total Design*. Workingham, England: Addison Wesley, 1991.

Ross, S. M. *Stochastic Processes*. 2nd Edition. USA: John Wiley & Sons, 1996.

Shewhart, W. A. *Economic Control of the Quality of Manufactured Product*. New York: D. Van Nostrand, 1931.

Taguchi, G.; A. E. Elsayed; and T. Hsiang. *Quality Engineering in Production Systems*. New York: McGraw-Hill, 1989.

Ulrich, K. T., and S. D. Eppinger. *Product Design and Development*. New York: McGraw-Hill, 1995.

Velury, J. "Integrating ISO 9000 into the Big Picture." *IIE Solutions* 1 (October 1995), pp. 26–29.

Wheat, B., C. Mills, and M. Carnell. *Leaning into Six Sigma: The Path to Integration of Lean Enterprise and Six Sigma*. New York: McGraw-Hill, 2001.

CHAPTER

11

Operations Scheduling

"When someone tells you they are too 'busy'… it's not a reflection of their schedule;
it's a reflection of YOUR spot on their schedule."

—Steve Maraboli

CHAPTER OVERVIEW

Purpose

To gain an understanding of the key methods and results for sequence scheduling in a job shop environment.

KEY POINTS

1. *The job shop scheduling problem.* A job shop is a set of machines and workers who use the machines. Jobs may arrive all at once or randomly throughout the day. Different jobs require different equipment and possibly different personnel. The shop foreman must determine the sequence in which to schedule the jobs in the shop to make the most efficient use of the resources (both human and machine) available. The relevant characteristics of the sequencing problem include
   - The pattern of arrivals.
   - Number and variety of machines.
   - Number and types of workers.
   - Patterns of job flow in the shop.
   - Objectives for evaluating alternative sequencing rules.
2. *Sequencing rules.*
   - *First come first served (FCFS).* Schedule jobs in the order they arrive to the shop.
   - *Shortest processing time (SPT) first.* Schedule the next job with the shortest processing time.
   - *Earliest due date (EDD).* Schedule the jobs that have the earliest due date first.
   - *Critical ratio (CR) scheduling.* The critical ratio is (due date − current time)/processing time. Schedule the job with the smallest CR value next.
3. *Sequencing results.* A common criterion for evaluating the effectiveness of sequencing rules is the mean flow time. The flow time of any job is the amount of time that elapses from the point that the job arrives in the shop to the point that the job is completed. The mean flow time is the average of all the flow times for all the jobs. SPT scheduling minimizes the mean flow time. If the objective is to minimize the maximum lateness, then the jobs should be scheduled by EDD.

All the preceding results apply to a single machine or single facility. When scheduling jobs on multiple machines, the problem is much more complex with few known results. Consider the case of n jobs that must be scheduled on two machines. The main result discovered in this case is that the optimal solution is to sequence the jobs in the same order on both machines (known as a permutation schedule). This means that there are a possible $n!$ feasible solutions. This can, of course, be a very large number. However, there is a procedure that efficiently computes the optimal sequence for n jobs on two machines (Johnson, 1954). Essentially the same algorithm can be applied to three machines under very special circumstances. The problem of scheduling two jobs on m machines can be solved efficiently by a graphical procedure.

4. *Sequence scheduling in a stochastic environment.* The problems alluded to previously assume all information is known with certainty. Real problems are more complex in that there is generally some type of uncertainty present. One source of uncertainty could be the job times. In that case, the job times, say $t_1, t_2, \ldots, t_n$, are assumed to be independent random variables with a known distribution function. The optimal sequence for a single machine in this case is very much like scheduling the jobs in SPT order based on expected processing times.

   When scheduling jobs with uncertain processing times on multiple machines, one must assume that the distribution of job times follows an exponential distribution. The exponential distribution is the only one possessing the so-called memoryless property, which turns out to be crucial in the analysis. When the objective is to minimize the expected makespan (that is, the total time to complete all jobs), it turns out that the longest expected processing time (LEPT) first rule is optimal.

   Another source of uncertainty in a job shop is the order in which jobs arrive to the shop. In a factory setting, jobs are likely to arrive at random times during the day. Queueing theory can shed some light on how much time elapses from the point a job arrives until its completion.
5. *Line balancing.* Another problem that arises in the factory setting is that of balancing an assembly line. While line balancing is not a sequence scheduling problem found in a job shop environment, it is certainly a scheduling problem arising within the plant. Assume we have an item flowing down an assembly line and that a total of n tasks must be completed on the item. The problem is to determine which tasks should be placed where on the line. Typically, an assembly line is broken down into stations and some subset of tasks is assigned to each station. The goal is to balance the time required at each station while taking into account the precedence relationships existing among the individual tasks.

   Optimal line balances are difficult to find. We consider one heuristic method, which gives reasonable results in most circumstances.

Scheduling is an important aspect of operations control in both manufacturing and service industries. With increased emphasis on time to market and time to volume as well as improved customer satisfaction, efficient scheduling will gain increasing emphasis in the operations function in the coming years.

In some sense, much of what has been discussed so far in this text can be considered a subset of production scheduling. Aggregate planning, treated in Chapter 3, is aimed at macroscheduling of workforce levels and overall production levels for the firm. Detailed inventory control, discussed in Chapters 4 and 5, concerns methods of scheduling production at the individual item level; Chapter 7 treated vehicle scheduling; and MRP, discussed in Chapter 9, provides production schedules for end items and subassemblies in the product structure.

There are many different types of scheduling problems faced by the firm. A partial list follows.

1. *Job shop scheduling.* Known more commonly in practice as shop floor control, job shop scheduling is the set of activities in the shop that transform inputs (a set of requirements) to outputs (products to meet those requirements). Much of this chapter will be concerned with sequencing issues on the shop floor, and more will be said about this problem in Section 11.1.
2. *Personnel scheduling.* Scheduling personnel is an important problem for both manufacturing and service industries. Although shift scheduling on the factory floor may be considered one of the functions of shop floor control, personnel scheduling is a much larger problem. Scheduling health professionals in hospitals and other health facilities is one example. Determining whether to meet peak demand with overtime shifts, night shifts, or subcontracting is another example of a personnel scheduling problem.
3. *Facilities scheduling.* This problem is particularly important when facilities become a bottleneck resource. Scheduling operating rooms at hospitals is one example. As the need for health care increases, some hospitals and health maintenance organizations (HMOs) find that facilities are strained. A similar problem occurs in colleges and universities in which enrollments have increased without commensurate increases in the size of the physical plant.
4. *Vehicle scheduling.* Manufacturing firms must distribute their products in a cost-efficient and timely manner. Some service operations, such as dial-a-ride systems, involve pick-ups and deliveries of goods and/or people. Vehicle routing is a problem that arises in many contexts. Problems such as scheduling snow removal equipment, postal and bank deliveries, and shipments to customers with varying requirements at different locations are some examples. Vehicle scheduling is discussed in Section 7.9.
5. *Vendor scheduling.* For firms with lean systems, scheduling deliveries from vendors is an important logistics issue. Purchasing must be coordinated with the entire product delivery system to ensure that lean production systems function efficiently. (Lean is discussed in Chapters 1, 9 and 10 of this book.)
6. *Project scheduling.* A project may be broken down into a set of interrelated tasks. Although some tasks can be done concurrently, many tasks cannot be started until others are completed. Complex projects may involve thousands of individual tasks that must be coordinated for the project to be completed on time and within budget. Project scheduling is an important component of the planning function, which we treat in detail in Chapter 12.
7. *Dynamic versus static scheduling.* Most scheduling theory that we review in this chapter views the scheduling problem as a static one. Numerous jobs arrive simultaneously to be processed on a set of machines. In practice, many scheduling problems are dynamic in the sense that jobs arrive continuously over time. One example is the problem faced by an air traffic controller who must schedule runways for arriving planes. The problem is a dynamic one in that planes arrive randomly and runways are freed up and committed randomly over time. Dynamic scheduling problems, treated in Section 11.9, are analyzed using the tools of queueing theory (discussed in detail in Chapter 8 and Supplement 2).

Scheduling is a complex but extremely important operations function. The purpose of this chapter is to give the reader a sampling of the kinds of results one can obtain using analytical models and to show how these models can be used to solve certain classes of scheduling problems Our focus is primarily on job shop scheduling, but we consider several other scheduling problems as well.

11.1 PRODUCTION SCHEDULING AND THE HIERARCHY OF PRODUCTION DECISIONS

Crucial to controlling production operations is the detailed scheduling of various aspects of the production function. We may view the production function in a company as a hierarchical process. First, the firm must forecast demand for aggregate sales over some predetermined planning horizon. These forecasts provide the input for determining the sales and operations planning function discussed in Chapter 3. The production plan then must be translated into the master production schedule (MPS). The MPS results in specific production goals by product and time period.

Materials requirements planning (MRP), treated in detail in Chapter 9, is one method for meeting specific production goals of finished-goods inventory generated by the MPS. The MRP system "explodes" the production levels one obtains from the MPS analysis back in time to obtain production targets at each level of assembly by time period. The result of the MRP analysis is specific planned order releases for final products, subassemblies, and components.

Finally, the planned order releases must be translated into a set of tasks and the due dates associated with those tasks. This level of detailed planning results in the shop floor schedule. Because the MRP or other lot scheduling system usually recommends revisions in the planned order releases, shop floor schedules change frequently. The hierarchy of production planning decisions is shown schematically in Figure 11–1.

Shop floor control means scheduling personnel and equipment in a work center to meet the due dates for a collection of jobs. Often, jobs must be processed through the machines in the work center in a unique order or sequence. Figure 11–2 shows the layout of a typical job shop.

Both jobs and machines are treated as indivisible. Jobs must wait, or queue up, for processing when machines are busy. This is referred to as discrete processing. Production scheduling in continuous-process industries, such as sugar or oil refining, has a very different character.

Although there are many problems associated with operations scheduling, our concern in this chapter will be job sequencing. Given a collection of jobs remaining to be processed on a collection of machines, the problem is how to sequence these jobs to optimize some specified criterion. Properly choosing the sequencing rule can effect dramatic improvements in the throughput rate of the job shop.

11.2 IMPORTANT CHARACTERISTICS OF JOB SHOP SCHEDULING PROBLEMS

Listed below are significant issues for determining optimal or approximately optimal scheduling rules.

1. *The job arrival pattern.* We often view the job shop problem as a static one in which we take a "snapshot" of the system at a point in time and proceed to solve the problem based on the value of the current state. However, the number of jobs waiting to be processed

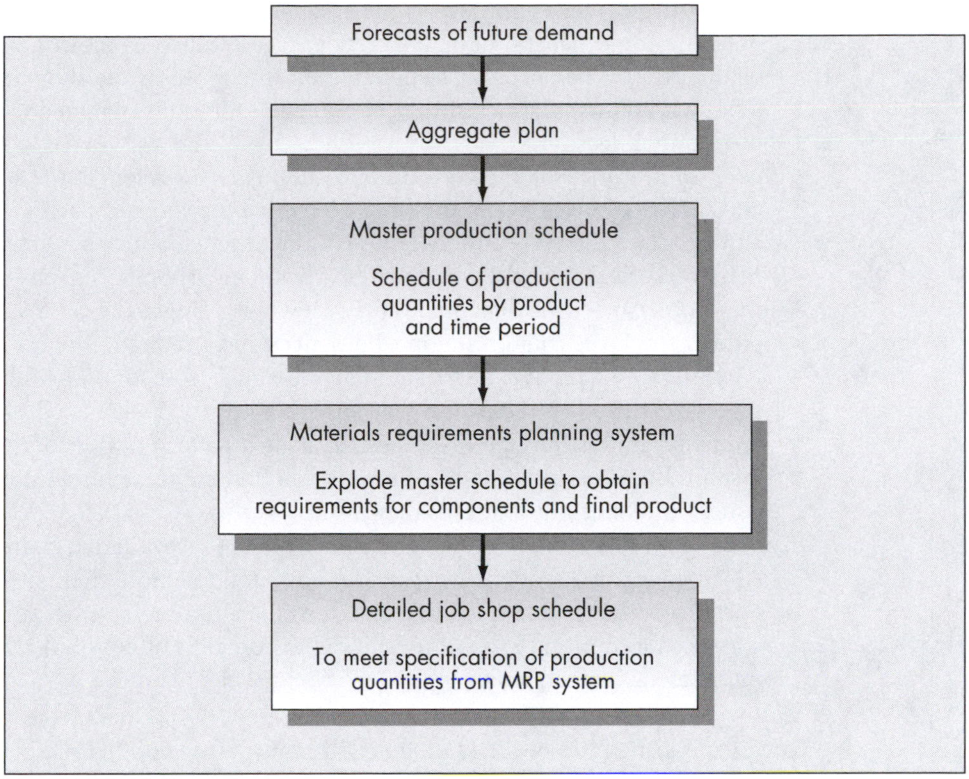

Figure 11–1 Hierarchy of production decisions

is constantly changing. Hence, although many of the solution algorithms we consider view the problem as being static, most practical shop scheduling problems are dynamic in nature.

2. *Number and variety of machines in the shop.* A particular job shop may have unique features that could make implementing a solution obtained from a scheduling algorithm difficult. For example, it is generally assumed that all machines of a given type

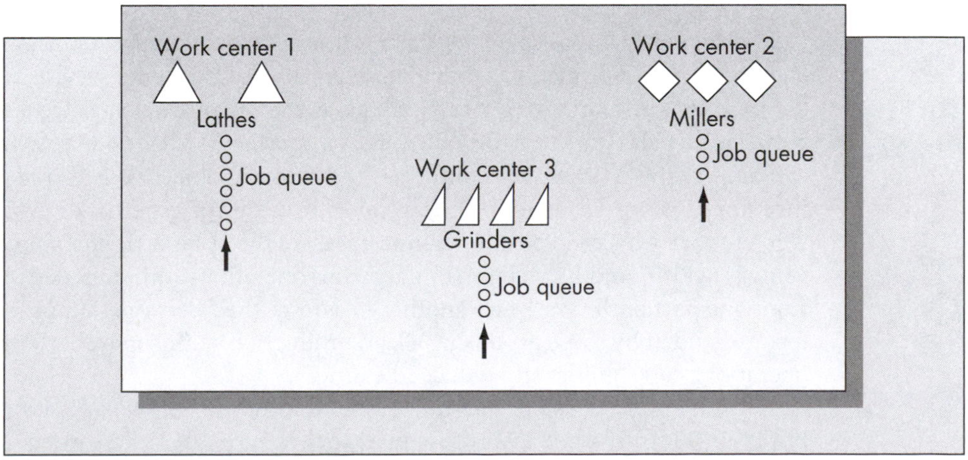

Figure 11–2 Typical job shop layout

are identical. This is not always the case, however. The throughput rate of a particular machine could depend upon a variety of factors, such as the condition of the machine or the skill of the operator. Depending on the layout of the shop and the nature of the jobs, constraints might exist that would make solutions obtained from an "all-purpose" procedure infeasible.

3. *Number of workers in the shop*. Both the number of workers in the shop and the number and variety of machines in the shop determine the shop's capacity. Capacity planning is an important aspect of production planning. Many control systems, such as traditional MRP discussed in Chapter 9, do not explicitly incorporate capacity considerations. Furthermore, capacity is dynamic. A breakdown of a single critical machine or the loss of a critical employee could result in a bottleneck and a reduction in the shop's capacity.
4. *Particular flow patterns*. The solutions obtained from the scheduling algorithms presented in this chapter require that jobs be completed in a fixed order. However, each sequence of jobs through machines results in a pattern of flow of materials through the system. Because materials-handling issues often are treated separately from scheduling issues, infeasible flow patterns may result.
5. *Evaluation of alternative rules*. The choice of objective will determine the suitability and effectiveness of a sequencing rule. It is common for more than one objective to be important, so it may be impossible to determine a unique optimal rule. For example, one may wish to minimize the time required to complete all jobs but also to limit the maximum lateness of any single job.

One of the difficulties of scheduling is that many, often conflicting, objectives are present. The goals of different parts of the firm are not always the same. Some of the most common objectives are

1. Meet due dates.
2. Minimize work-in-process (WIP) inventory.
3. Minimize the average flow time through the system.
4. Provide for high machine/worker time utilization (minimize machine/worker idle time).
5. Provide for accurate job status information.
6. Reduce setup times.
7. Minimize production and worker costs.

It is obviously impossible to optimize all seven objectives simultaneously. In particular, (1) and (3) are aimed primarily at providing a high level of customer service, while (2), (4), (6), and (7) are aimed primarily at providing a high level of plant efficiency. Determining the trade-off between cost and quality is one of the most important strategic issues facing firms today.

Some of these objectives conflict. If the primary objective is to reduce work-in-process inventory (as, for example, with lean inventory control systems, discussed in Chapter 9), it is likely that worker idle time will increase. As the system tightens up by reducing the inventory within and between manufacturing operations, differences in the throughput rate from one part of the system to another may force the faster operations to wait. Although not recommended by those espousing a lean production philosophy, buffer inventories between operations can significantly reduce idle time.

As an example, consider the simple system composed of two operations in series, pictured in Figure 11–3. If work-in-process inventory is zero, then the throughput of the system at any point in time is governed by the smaller of the throughputs of the two operations. If

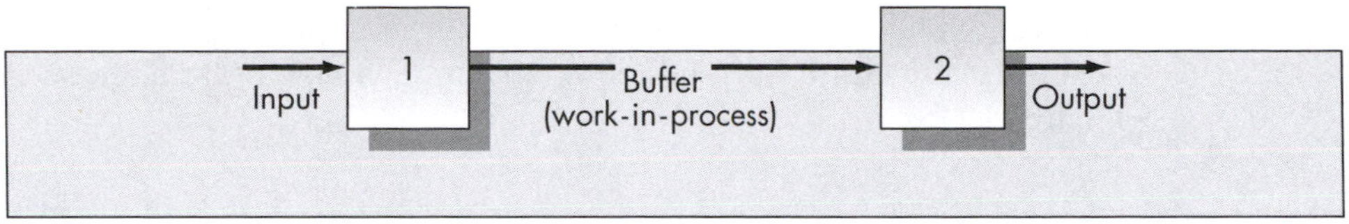

Figure 11–3 A process composed of two operations in series

operation 1 is temporarily halted by a machine failure, then operation 2 also must remain idle. However, if there is a buffer inventory placed between the operations, then 2 can continue to operate while 1 is undergoing repair or recalibration.

Finding the proper mix between WIP inventory and worker idle time is equivalent to choosing a point on the trade-off curve of these conflicting objectives. (Trade-off, or exchange, curves were discussed in Chapter 5 in the context of multi-item inventory control.) Such a curve is pictured in Figure 11–4a. A movement from one point to another along such a curve does not necessarily imply that the system has improved, but rather that different weights

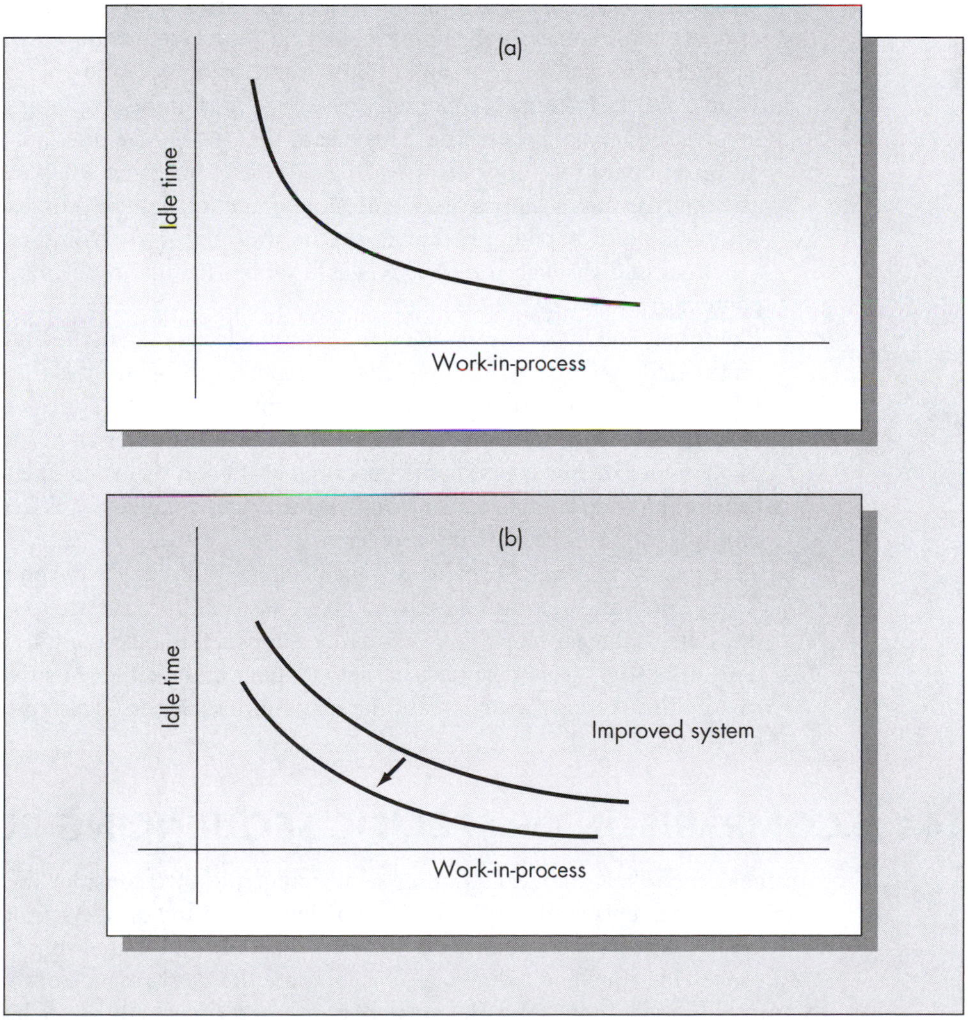

Figure 11–4 Conflicting objectives in job shop management

are being applied to the two objectives. A true improvement in the overall system would mean that the entire trade-off curve undergoes a downward shift, such as that pictured in Figure 11–4b.

11.3 JOB SHOP SCHEDULING TERMINOLOGY

In general, a **job shop scheduling problem** is one in which n jobs must be processed through m machines. The complexity of the problem depends upon a variety of factors, such as what job sequences are permissible and what optimization criteria are chosen. In this section we define some of the terms that will be used throughout this chapter.

1. *Flow shop*. In a **flow shop** each of the n jobs must be processed through the m machines in the same order, and each job is processed exactly once on each machine. This is what we typically think of as an assembly line.
2. *Job shop*. A general **job shop** differs from a flow shop in that not all jobs are assumed to require exactly m operations, and some jobs may require multiple operations on a single machine. Furthermore, in a job shop each job may have a different required sequencing of operations. General job shop problems are extremely complex. All-purpose solution algorithms for solving general job shop problems do not exist.
3. *Parallel processing versus sequential processing*. Most of the problems that we will consider involve **sequential processing**. This means that the m machines are distinguishable, and different operations are performed by different machines. In **parallel processing** we assume that the machines are identical, and any job can be processed on any machine. An example of parallel processing occurs in a phone switching center, in which calls are processed through the next available server. Parallel processing is discussed in the context of stochastic scheduling in Section 11.8.
4. *Flow time*. The **flow time** of job i is the time that elapses from the initiation of the first job on the first machine to the completion of job i. Equivalently, it is the amount of time that job i spends in the system. The **mean flow time**, which is a common measure of system performance, is the arithmetic average of the flow times for all n jobs.
5. *Makespan*. The **makespan** is the flow time of the job that is completed last. It is also the time required to complete all n jobs. Minimizing the makespan is a common objective in multiple-machine sequencing problems.
6. *Tardiness and lateness*. **Tardiness** is the positive difference between the completion time (flow time) and the due date of a job. A tardy job is one that is completed after its due date. **Lateness** refers to the difference between the job completion time and its due date and differs from tardiness in that lateness can be either positive or negative. Minimizing the average tardiness and the maximum tardiness is also a common scheduling objective.

11.4 A COMPARISON OF SPECIFIC SEQUENCING RULES

In the comparison and evaluation of sequencing rules, we consider the job shop at a fixed point in time. This section will focus on a single machine only. Assume that there is a collection of jobs that must be processed on the machine and that each job has associated with it a processing time and a due date. We compare the performance of four sequencing rules commonly used in practice. The purpose of this section is to illustrate how these sequencing rules affect various measures of system performance.

1. **First-come, first-served (FCFS).** Jobs are processed in the sequence in which they entered the shop.
2. **Shortest processing time (SPT).** Jobs are sequenced in increasing order of their processing times. The job with the shortest processing time is first, the job with the next shortest processing time is second, and so on.
3. **Earliest due date (EDD).** Jobs are sequenced in increasing order of their due dates. The job with the earliest due date is first, the job with the next earliest due date is second, and so on.
4. **Critical ratio (CR).** Critical ratio scheduling requires forming the ratio of the remaining time until the due date divided by the processing time of the job and scheduling the job with the smallest ratio next.

We compare the performance of these four rules for a specific case based on mean flow time, average tardiness, and number of tardy jobs. The purpose of the next example is to help the reader develop a familiarity with the mechanics of scheduling before presenting formal results.

EXAMPLE 11.1

A machining center in a job shop for a local fabrication company has five unprocessed jobs remaining at a particular point in time. The jobs are labeled 1, 2, 3, 4, and 5 in the order that they entered the shop. The respective processing times and due dates are given in the following table.

Job Number	Processing Time	Due Date
1	11	61
2	29	45
3	31	31
4	1	33
5	2	32

First-Come, First-Served

Because the jobs are assumed to have entered the shop in the sequence that they are numbered, FCFS scheduling means that the jobs are scheduled in the order 1, 2, 3, 4, 5.

Sequence	Completion Time	Due Date	Tardiness
1	11	61	0
2	40	45	0
3	71	31	40
4	72	33	39
5	74	32	42
Totals	268		121

$$\text{Mean flow time} = 268/5 = 53.6.$$
$$\text{Average tardiness} = 121/5 = 24.2.$$
$$\text{Number of tardy jobs} = 3.$$

The tardiness of a job is equal to zero if the job is completed prior to its due date and is equal to the number of days late if the job is completed after its due date.

Shortest Processing Time

Here jobs are sequenced in order of increasing processing time.

Job	Processing Time	Completion Time	Due Date	Tardiness
4	1	1	33	0
5	2	3	32	0
1	11	14	61	0
2	29	43	45	0
3	31	74	31	43
Totals		135		43

$$\text{Mean flow time} = 135/5 = 27.0.$$
$$\text{Average tardiness} = 43/5 = 8.6.$$
$$\text{Number of tardy jobs} = 1.$$

Earliest Due Date

Here jobs are completed in the order of their due dates.

Job	Processing Time	Completion Time	Due Date	Tardiness
3	31	31	31	0
5	2	33	32	1
4	1	34	33	1
2	29	63	45	18
1	11	74	61	13
Totals		235		33

$$\text{Mean flow time} = 235/5 = 47.0.$$
$$\text{Average tardiness} = 33/5 = 6.6.$$
$$\text{Number of tardy jobs} = 4.$$

Critical Ratio Scheduling

After each job has been processed, we compute

$$\frac{\text{Due date} - \text{Current time}}{\text{Processing time}},$$

which is known as the critical ratio, and schedule the next job in order to minimize the value of the critical ratio. The idea behind critical ratio scheduling is to provide a balance between processing times and due dates. The ratio will grow smaller as the current time approaches the due date, and more priority will be given to those jobs with longer processing times. One disadvantage of the method is that the critical ratios need to be recalculated each time a job is scheduled.

It is possible that the numerator will be negative for some or all of the remaining jobs. When that occurs, it means that the job is late, and we will assume that late jobs are automatically scheduled next. If there is more than one late job, then the late jobs are scheduled in SPT sequence.

First we compute the critical ratios starting at time $t = 0$.

Current time: t = 0

Job	Processing Time	Due Date	Critical Ratio	
1	11	61	61/11	(5.545)
2	29	45	45/29	(1.552)
3	31	31	31/31	(1.000)
4	1	33	33/1	(33.00)
5	2	32	32/2	(16.00)

The minimum value corresponds to job 3, so job 3 is performed first. As job 3 requires 31 units of time to process, we must update all the critical ratios in order to determine the next job to process. We move the clock to time $t = 31$ and recompute the critical ratios.

Current time: t = 31

Job	Processing Time	Due Date–Current Time	Critical Ratio	
1	11	30	30/11	(2.727)
2	29	14	14/29	(0.483)
4	1	2	2/1	(2.000)
5	2	1	1/2	(0.500)

The minimum is 0.483, which corresponds to job 2. Hence, job 2 is scheduled next. Since job 2 has a processing time of 29, we update the clock to time $t = 31 + 29 = 60$.

Current time: t = 60

Job	Processing Time	Due Date–Current Time	Critical Ratio
1	11	1	1/11 (.0909)
4	1	–27	$-27/1 < 0$
5	2	–28	$-28/2 < 0$

Jobs 4 and 5 are now late, so they are given priority and scheduled next. Since they are scheduled in SPT order, they are done in the sequence job 4, then job 5. Finally, job 1 is scheduled last.

Summary of the Results for Critical Ratio Scheduling			
Job	Processing Time	Completion Time	Tardiness
3	31	31	0
2	29	60	15
4	1	61	28
5	2	63	31
1	11	74	13
Totals		289	87

$$\text{Mean flow time} = 289/5 = 57.8.$$

$$\text{Average tardiness} = 87/5 = 17.4.$$

$$\text{Number of tardy jobs} = 4.$$

Below is the summary for all four scheduling rules.

Summary of the Results of Four Scheduling Rules			
Rule	Mean Flow Time	Average Tardiness	Number of Tardy Jobs
FCFS	53.6	24.2	3
SPT	27.0	8.6	1
EDD	47.0	6.6	4
CR	57.8	17.4	4

11.5 OBJECTIVES IN JOB SHOP MANAGEMENT: AN EXAMPLE

EXAMPLE 11.2

An air traffic controller is faced with the problem of scheduling the landing of five aircraft. Based on the position and runway requirements of each plane, she estimates the landing times as follows.

Plane:	1	2	3	4	5
Time (in minutes):	26	11	19	16	23

Only one plane may land at a time. The problem is essentially the same as that of scheduling five jobs on a single machine, with the planes corresponding to the jobs, the landing times to the processing times, and the runway to the machine.

1. With the given information, two reasonable objectives would be to minimize the total time required to land all planes (i.e., the makespan) or the average time required to land the planes (the mean flow time). The makespan for any sequence is clearly 95 minutes, the sum of the landing times. However, as we saw in Example 11.1, the mean flow time is not sequence independent, and the shortest-processing-time

rule minimizes the mean flow time. We will show in Section 11.6 that SPT is the optimal sequencing rule for minimizing mean flow time for a single machine in general.

2. An alternative objective might be to land as many people as quickly as possible. In this case we also would need to know the number of passengers on each plane. Suppose the following numbers.

Plane	1	2	3	4	5
Landing time	26	11	19	16	23
Number of passengers	180	12	45	75	252

   The appropriate objective in this case might be to minimize the weighted makespan or the weighted sum of the completion times, where the weights would correspond to the number of passengers in each plane. Notice that the objective function would now be in units of passenger-minutes.

3. An issue that we have not yet addressed is the time that each plane is scheduled to arrive.

Plane	1	2	3	4	5
Landing time	26	11	19	16	23
Scheduled arrival time	5:30	5:45	5:15	6:00	5:40

   Sequencing rules that ignore due dates could give very poor results in terms of meeting the arrival times. Some possible objectives related to due dates include minimizing the average tardiness and minimizing the maximum tardiness.

4. Thus far we have ignored special conditions that favor some planes over others. Suppose that plane number 4 has a critically low fuel level. This would probably result in plane 4 taking precedence. Priority constraints could arise in other ways as well. Planes that are scheduled for continuing flights or planes carrying precious or perishable cargo also might be given priority.

The purpose of this section was to demonstrate the difficulties of choosing an objective function for job sequencing problems. The optimal sequence is highly sensitive to the choice of the objective, and the appropriate objective is not always obvious.

PROBLEMS FOR SECTIONS 11.1–11.5

1. Discuss each of the following objectives listed and the relationship each has with job shop performance.
   a. Reduce WIP inventory.
   b. Provide a high level of customer service.
   c. Reduce worker idle time.
   d. Improve factory efficiency.
2. In problem 1, why are (a) and (c) conflicting objectives, and why are (b) and (d) conflicting objectives?
3. Define the following terms.
   a. Flow shop.
   b. Job shop.

c. Sequential versus parallel processing.
d. Makespan.
e. Tardiness.

4. Four trucks, 1, 2, 3, and 4, are waiting on a loading dock at XYZ Company that has only a single service bay. The trucks are labeled in the order that they arrived at the dock. Assume the current time is 1:00 PM The times required to unload each truck and the times that the goods they contain are due in the plant are given in the following table.

Truck	Unloading Time (minutes)	Time Material Is Due
1	20	1:25 PM
2	14	1:45 PM
3	35	1:50 PM
4	10	1:30 PM

Determine the schedules that result for each of the rules FCFS, SPT, EDD, and CR. In each case compute the mean flow time, average tardiness, and number of tardy jobs.

5. Five jobs must be scheduled for batch processing on a mainframe computer system. The processing times and the promised times for each of the jobs are listed here.

Job	1	2	3	4	5
Processing time	40 min	2.5 hr	20 min	4 hr	1.5 hr
Promised time	11:00 AM	2:00 PM	2:00 PM	1:00 PM	4:00 PM

Assume that the current time is 10:00 AM
a. If the jobs are scheduled according to SPT, find the tardiness of each job and the mean tardiness of all jobs.
b. Repeat the calculation in part (a) for EDD scheduling.

11.6 AN INTRODUCTION TO SEQUENCING THEORY FOR A SINGLE MACHINE

Assume that n jobs are to be processed through one machine. For each job i, define the following quantities.

t_i = Processing time for job i
d_i = Due date for job i
W_i = Waiting time for job i
F_i = Flow time for job i
L_i = Lateness of job i
T_i = Tardiness of job i
E_i = Earliness of job i

The processing time and the due date are constants that are attached to the description of each job. The waiting time for a job is the amount of time that the job must wait before its processing can begin. For the cases that we consider, it is also the sum of the processing times for all the preceding jobs. The flow time is simply the waiting time plus the job processing time ($F_i = W_i + t_i$).The flow time of job i and the completion time of job i are the same. We will define the lateness of job i as $L_i = F_i - d_i$ and assume that lateness can be either a positive or a negative quantity. Tardiness is the positive part of lateness ($T_i = \max[L_i, 0]$), and earliness is the negative part of lateness ($E_i = \max[-L_i, 0]$).

Other related quantities are maximum tardiness T_{max}, given by the formula

$$T_{max} = \max\{T_1, T_2, \ldots, T_n\}$$

and the mean flow time F', given by the formula

$$F' = \frac{1}{n}\sum_{i=1}^{n} F_i.$$

As we are considering only a single machine, every schedule can be represented by a permutation (that is, ordering) of the integers 1, 2, … , n. There are exactly $n!$ different permutation schedules [$n! = n(n - 1) \ldots (2)(1)$].

Shortest-Processing-Time Scheduling

We have the following result.

THEOREM 1.1

The scheduling rule that minimizes the mean flow time F′ is SPT.

Theorem 11.1 is easy to prove. Let [1], [2], … , [n] be any permutation of the integers 1, 2, 3, … , n. The flow time of the job that is scheduled in position k is given by

$$F_{[k]} = \sum_{i=1}^{n} t_{[i]}.$$

It follows that the mean flow time is given by

$$F' = \frac{1}{n}\sum_{k=1}^{n} F_{[k]} = \frac{1}{n}\sum_{k=1}^{n}\sum_{i=1}^{k} t_{[i]}.$$

The double summation term may be written in a different form. Expanding the double summation, we obtain

$$\begin{aligned} &k = 1 : t_{[1]} \\ &k = 2 : t_{[1]} + t_{[2]} \\ &\vdots \\ &k = n : t_{[1]} + t_{[2]} + \cdots + t_{[n]}. \end{aligned}$$

By summing down the column rather than across the row, we may rewrite F' in the form

$$nt_{[1]} + (n - 1)t_{[2]} + \ldots + t_{[n]},$$

which is clearly minimized by setting

$$t_{[1]} \leq t_{[2]} \leq \ldots t_{[n]},$$

which is exactly the SPT sequencing rule.

We have the following corollary to Theorem 11.1.

COROLLARY 11.1

The following measures are equivalent:

1. *mean flow time*
2. *mean waiting time*
3. *mean lateness.*

Taken together, Corollary 11.1 and Theorem 11.1 establish that SPT minimizes mean flow time, mean waiting time, and mean lateness for single-machine sequencing.

Earliest-Due-Date Scheduling

If the objective is to minimize the maximum lateness, then the jobs should be sequenced according to their due dates. That is, $d_{[1]} \leq d_{[2]} \leq \ldots \leq d_{[n]}$. We will not present a proof of this result. The idea behind the proof is to choose some schedule that does not sequence the jobs in order of their due dates; that implies that there is some value of k such that $d_{[k]} > d_{[k+1]}$. One shows that by interchanging the positions of jobs k and $k + 1$, the maximum lateness is reduced.

Minimizing the Number of Tardy Jobs

There are many instances in which the penalty for a late (tardy) job remains the same no matter how late it is. For example, any delay in the completion of all tasks required for preparation of a space launch would cause the launch to be aborted, independent of the length of the delay.

We will describe an algorithm from J. Michael Moore (1968) that minimizes the number of tardy jobs for the single machine problem.

Step 1. Sequence the jobs according to the earliest due date to obtain the initial solution. That is $d_{[1]} > d_{[2]} \leq \ldots \leq d_{[n]}$.

Step 2. Find the first tardy job in the current sequence, say job $[i]$. If none exists, go to step 4.

Step 3. Consider jobs $[1], [2], \ldots, [i]$. Reject the job with the largest processing time. Return to step 2.

Step 4. Form an optimal sequence by taking the current sequence and appending to it the rejected jobs. The jobs appended to the current sequence may be scheduled in any order because they constitute the tardy jobs.

EXAMPLE 11.3

A machine shop processes custom orders from a variety of clients. One of the machines, a grinder, has six jobs remaining to be processed. The processing times and promised due dates (both in hours) for the six jobs are given here.

Job	1	2	3	4	5	6
Due date	15	6	9	23	20	30
Processing time	10	3	4	8	10	6

The first step is to sequence the jobs according to the EDD rule.

Job	2	3	1	5	4	6
Due date	6	9	15	20	23	30
Processing time	3	4	10	10	8	6
Completion time	3	7	17	27	35	41

We see that the first tardy job is job 1, and there are a total of four tardy jobs. We now consider jobs 2, 3, and 1 and reject the job with the longest processing time. This is clearly job 1. At this point, the new current sequence is

Job	2	3	5	4	6
Due date	6	9	20	23	30
Processing time	3	4	10	8	6
Completion time	3	7	17	25	31

The first tardy job in the current sequence is now job 4. We consider the sequence 2, 3, 5, 4 and reject the job with the longest processing time, which is job 5. The current sequence is now

Job	2	3	4	6
Due date	6	9	23	30
Processing time	3	4	8	6
Completion time	3	7	15	21

Clearly there are no tardy jobs at this stage. The optimal sequence is 2, 3, 4, 6, 5, 1 or 2, 3, 4, 6, 1, 5. In either case the number of tardy jobs is exactly 2.

Precedence Constraints: Lawler's Algorithm

Eugene Lawler's (1973) algorithm is a powerful technique for solving a variety of constrained scheduling problems. The objective function is assumed to be of the form

$$\min \max_{1 \le i \le n} g_i(F_i)$$

where g_i is any nondecreasing function of the flow time F_i. Furthermore, the algorithm handles *any* **precedence constraints**. Precedence constraints occur when certain jobs must be completed before other jobs can begin; they are quite common in scheduling problems. Some examples of functions g_i that one might consider are $g_i(F_i) = F_i - d_i = L_i$, which corresponds to minimizing maximum lateness, or $g_i(F_i) = \max(F_i - d_i, 0)$, which corresponds to minimizing maximum tardiness.

Lawler's algorithm first schedules the job to be completed last, then the job to be completed next to last, and so on. At each stage one determines the set of jobs not required to precede any other. Call this set V. Among the set V, choose the job k that satisfies

$$g_k(\tau) = \min_{i \in V}(g_i(\tau)),$$

where $\tau = \sum_{i=1}^{n} t_i$ and corresponds to the processing time of the current sequence.

Job k is now scheduled last. Consider the remaining jobs and again determine the set of jobs that are not required to precede any other remaining job. After scheduling job k, this set may have changed. The value of τ is reduced by t_k and the job scheduled next to last is now determined. The process is continued until all jobs are scheduled. Note that as jobs are scheduled, some of the precedence constraints may be relaxed, so the set V is likely to change at each iteration.

SNAPSHOT APPLICATION

Millions Saved with Scheduling System for Fractional Aircraft Operators

Celebrities, corporate executives, and sports professionals are a large part of the group that uses private planes for travel. For many of these people, it doesn't make economic sense to purchase planes. An attractive alternative is fractional ownership, especially for those who have only occasional need of a plane. Fractional ownership of private planes provides owners with the flexibility to fly to over 5,000 destinations (as opposed to about 500 for the commercial airlines). Other advantages include privacy, personalized service, fewer delays, and the ability to conduct business on the plane.

The concept of a fractional aircraft program is similar to that of a time-share condominium, except that the aircraft owners are guaranteed access at any time with as little as four hours' notice. The fees are based on the number of flight hours the owner will require: one-eighth share owners are allotted 100 hours of annual flying time, one-quarter share owners 200 hours, and so forth. The entire system is coordinated by a fractional management company (FMC). Clearly, the problem of scheduling the planes and crews can become quite complex.

When scheduling planes and crews, the FMC must determine schedules that (1) meet customer requests on time, (2) satisfy maintenance and crew restrictions, and (3) allow for specific aircraft trip assignments and requests. The profitability of the FMC will depend upon how efficiently they perform these tasks. A group of consultants attacked this problem and developed a scheduling system known as ScheduleMiser (Martin, Jones, and Keskinocak, 2003). The inputs to this system are trip requests, aircraft availability, and aircraft restrictions over a specified planning horizon. Note that even though owners are guaranteed service with only four hours' notice, the vast majority of trips are booked at least three days or more in advance. This gives the FMC a reliable profile of demand over a two- to three-day planning horizon. Note that aircraft schedules must be coordinated with crew schedules, as crew work rules cannot be violated.

ScheduleMiser is the underlying engine that drives the larger planning system known as Flight Ops. ScheduleMiser is based on a mixed-integer mathematical formulation of the problem. The objective function consists of five terms delineating the various costs in the system. Several sets of constraints are included to ensure that demands are filled, crews are properly scheduled, and planes are not overbooked. This system was adopted and implemented by Raytheon Travel Air in November of 2000 (now Flight Options) for scheduling their fleet of over 100 aircraft. Raytheon reported a savings of over \$4.4 million in the first year of implementation of this system. This is only one example of many mathematical-based scheduling systems that have been implemented in the airline industry.

EXAMPLE 11 .4

Tony D'Amato runs a local body shop that does automotive painting and repairs. On a particular Monday morning he has six cars waiting for repair. Three (1, 2, and 3) are from a car rental company, and he has agreed to finish these cars in the order of the dates that they were promised. Cars 4, 5, and 6 are from a retail dealer who has requested that car 4 be completed first because a customer is waiting for it. The resulting precedence constraints can be represented as two disconnected networks, as pictured in Figure 11–5.

The times required to repair each of the cars (in days) and the associated promised completion dates are

Job	1	2	3	4	5	6
Processing time	2	3	4	3	2	1
Due date	3	6	9	7	11	7

Determine how the repair of the cars should be scheduled through the shop in order to minimize the maximum tardiness.

Solution

1. First, we find the job scheduled last (sixth). Among the candidates for the last position are those jobs that are not predecessors of other jobs. These are 3, 5, and 6. The total processing time of all jobs is 2 + 3 + 4 + 3 + 2 + 1 = 15. (This is the current value of r.) As the objective is to minimize the maximum tardiness, we compare the tardiness of these three jobs and pick the one with the smallest value. We obtain min{15 – 9, 15 – 11, 15 – 7} = min{6, 4, 8} = 4, corresponding to job 5. Hence job 5 is scheduled last (position 6).
2. Next we find the job scheduled fifth. The candidates are jobs 3 and 6 only. At this point the value of τ is 15 – 2 =13. Hence, we find min{13 – 9, 13 – 7}= min{4, 6} = 4, which corresponds to job 3. Hence, job 3 is scheduled in the fifth position.
3. Find the job scheduled fourth. Because job 3 is no longer on the list, job 2 now becomes a candidate. The current value of τ = 13 – 4 = 9. Hence, we compare

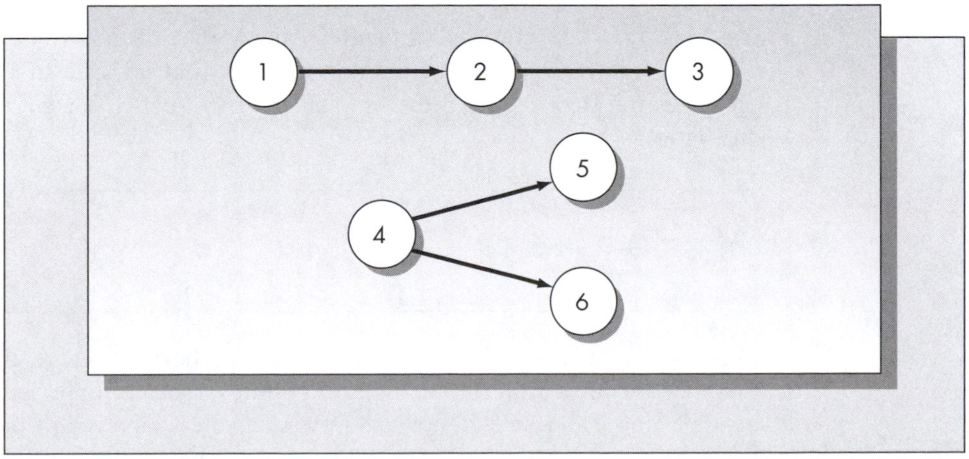

Figure 11–5 Precedence constraints for Example 11.4

min{9 – 6, 9 – 7} = min{3, 2} = 2, which corresponds to job 6. Schedule job 6 in the fourth position.

4. Find the job scheduled third. Job 6 has been scheduled, so job 4 now becomes a candidate along with job 2, and $\tau = 9 - 1 = 8$. Hence, we look for min{8 – 6, 8 – 7} = min{2, 1} = 1, which occurs at job 4.
5. At this point we would find the job scheduled second. However, we are left with only jobs 1 and 2, which, because of the precedence constraints, must be scheduled in the order 1–2.

Summarizing the results, the optimal sequence to repair the cars is 1–2–4–6–3–5.

In order to determine the value of the objective function, the maximum tardiness, we compute the flow time for each job and compare it to the due date. We have

Job	Processing Time	Flow Time	Due Date	Tardiness
1	2	2	3	0
2	3	5	6	0
4	3	8	7	1
6	1	9	7	2
3	4	13	9	4
5	2	15	11	4

Hence, the maximum tardiness is four days. The reader should verify that any other sequence results in a maximum tardiness of at least four days.

PROBLEMS FOR SECTION 11.6

6. Consider the information given in problem 4. Determine the sequence that the trucks should be unloaded in order to minimize
   a. mean flow time.
   b. maximum lateness.
   c. number of tardy jobs.
7. On May 1, a lazy student suddenly realizes that he has done nothing on seven different homework assignments and projects that are due in various courses. He estimates the time required to complete each project (in days) and also notes their due dates.

Project	1	2	3	4	5	6	7
Time (days)	4	8	10	4	3	7	14
Due date	4/20	5/17	5/28	5/28	5/12	5/7	5/15

Because projects 1, 3, and 5 are from the same class, he decides to do those in the sequence that they are due. Furthermore, project 7 requires results from projects 2 and 3, so 7 must be done after 2 and 3 are completed. Determine the sequence in which he should do the projects in order to minimize the maximum lateness.

8. Eight jobs are to be processed through a single machine. The processing times and due dates are given here.

Job	1	2	3	4	5	6	7	8
Processing time	2	3	2	1	4	3	2	2
Due date	5	4	13	6	12	10	15	19

Furthermore, assume that the following precedence relationships must be satisfied:

$$2 \rightarrow 6 \rightarrow 3.$$

$$1 \rightarrow 4 \rightarrow 7 \rightarrow 8.$$

Determine the sequence in which these jobs should be done in order to minimize the maximum lateness subject to the precedence restrictions.

9. Jane Reed bakes breads and cakes in her home for parties and other affairs on a contract basis. Jane has only one oven for baking. One particular Monday morning she finds that she has agreed to complete five jobs for that day. Her husband John will make the deliveries, which require about 15 minutes each. Suppose that she begins baking at 8:00 AM.

Job	Time Required	Promised Time
1	1.2 hr	11:30 AM
2	40 min	10:00 AM
3	2.2 hr	11:00 AM
4	30 min	1:00 PM
5	3.1 hr	12:00 NOON
6	25 min	2:00 PM

Determine the sequence in which she should perform the jobs in order to minimize
a. mean flow time.
b. number of tardy jobs.
c. maximum lateness.

10. Seven jobs are to be processed through a single machine. The processing times and due dates are given here.

Job	1	2	3	4	5	6	7
Processing time	3	6	8	4	2	1	7
Due date	4	8	12	15	11	25	21

Determine the sequence of the jobs in order to minimize
a. mean flow time.
b. number of tardy jobs.
c. maximum lateness.
d. What is the makespan for any sequence?

11.7 SEQUENCING ALGORITHMS FOR MULTIPLE MACHINES

We now extend the analysis of Section 11.6 to the case in which several jobs must be processed on more than one machine. Assume that n jobs are to be processed through m machines. The number of possible schedules is staggering, even for moderate values of both n and m. For each machine, there are $n!$ different orderings of the jobs. If the jobs may be processed on the machines in any order, it follows that there are a total of $(n!)^m$ possible schedules. For example, for $n = 5$, $m = 5$, there are $24{,}883 \times 10^{10}$ (about 25 billion) possible schedules. Even with the availability of inexpensive computing today, enumerating all feasible schedules for even moderate-sized problems is impossible or, at best, impractical.

In this section we will present some known results for scheduling jobs on more than one machine. A convenient way to represent a schedule is via a **Gantt chart**. As an example, suppose that two jobs, I and J, are to be scheduled on two machines, 1 and 2. The processing times are

	Machine 1	Machine 2
Job I	4	1
Job J	1	4

Assume that both jobs must be processed first on machine 1 and then on machine 2. The possible schedules appear in four Gantt charts in Figure 11–6. The first two schedules are known as permutation schedules. That means that the jobs are processed in the same sequence on both machines. Clearly, for this example, the permutation schedules provide better system performance in terms of both total and average flow time.

Recall that the total flow time (or makespan) is the total elapsed time from the initiation of the first job on the first machine until the completion of the last job on the last machine. For the given schedules, the makespans (total flow time) are respectively 9, 6, 10, and 10.

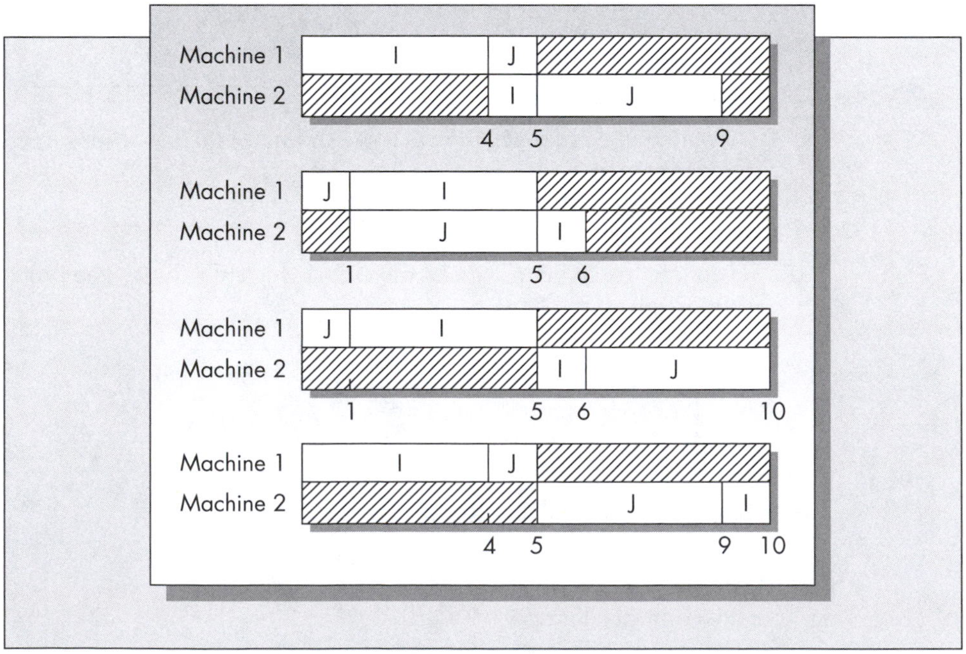

Figure 11–6 All possible schedules for two jobs on two machines

The mean flow time is also used as a measure of system performance. For the first schedule in the example, the mean flow time is $(5 + 9)/2 = 7$. For the second schedule, it is $(5 + 6)/2 = 5.5$, and so on.

A third possible objective is minimization of the mean idle time in the system. The mean idle time is the arithmetic average of the idle times for each machine. In schedule 1, we see that machine 1 is idle for 4 units of time (between times 5 and 9) and machine 2 is idle for 4 units of time as well (between times 0 and 4). Hence the mean idle time for schedule 1 is 4. In schedule 2, both machines 1 and 2 are idle for 1 unit of time, giving a mean idle time of 1. The mean idle times for schedules 3 and 4 are 5 units of time.

Scheduling *n* Jobs on Two Machines

Assume that n jobs must be processed through two machines and that each job must be processed in the order machine 1 then machine 2. Furthermore, assume that the optimization criterion is to minimize the makespan. The problem of scheduling on two machines turns out to have a relatively simple solution.

THEOREM 11.2

The optimal solution for scheduling n jobs on two machines is always a permutation schedule.

Theorem 11.2 means that one can restrict attention to schedules in which the sequence of jobs is the same on both machines. This result can be demonstrated as follows. Consider a schedule for n jobs on two machines in which the sequencing of the jobs on the two machines is different.

Machine 1	. . .	I	. . .	J				
Machine 2					. . .	J	. . .	I

By reversing the position of these jobs on either machine, the flow time decreases. By scheduling the jobs in the order I–J on machine 2 the pair (I, J) on machine 2 may begin after I is completed on machine 1, rather than having to wait until J is completed on machine 1.

Because the total number of permutation schedules is exactly $n!$, determining optimal schedules for two machines is roughly of the same level of difficulty as determining optimal schedules for one machine.

A very efficient algorithm for solving the two-machine problem was discovered by Selmer Johnson (1954). Following Johnson's notation, denote the machines by A and B. It is assumed that the jobs must be processed first on machine A and then on machine B. Suppose that the jobs are labeled i, for $1 \le i \le n$, and define

$$A_i = \text{Processing time of job } i \text{ on machine A.}$$
$$B_i = \text{Processing time of job } i \text{ on machine B.}$$

Johnson's result is that the following rule is optimal for determining an order in which to process the jobs on the two machines.

Rule: Job i precedes job $i + 1$ if $\min(A_i, B_{i+1}) < \min(A_{i+1}, B_i)$.

The following is an easy procedure to implement this rule.

1. List the values of A_i and B_i in two columns.
2. Find the smallest remaining element in the two columns. If it appears in column A, then schedule that job next. If it appears in column B, then schedule that job last.
3. Cross off the jobs as they are scheduled. Stop when all jobs have been scheduled.

EXAMPLE 11.5

Five jobs are to be scheduled on two machines. The processing times are

Job	Machine A	Machine B
1	5	2
2	1	6
3	9	7
4	3	8
5	10	4

The first step is to identify the minimum job time. It is 1, for job 2 on machine A. Because it appears in column A, job 2 is scheduled first and row 2 is crossed out. The next smallest processing time is 2, for job 1 on machine B. This appears in the B column, so job 1 is scheduled last. The next smallest processing time is 3, corresponding to job 4 in column A, so that job 4 is scheduled next. Continuing in this fashion, we obtain the optimal sequence

2–4–3–5–1.

The Gantt chart for the optimal schedule is pictured in Figure 11–7. Note that there is no idle time between jobs on machine A. This is a feature of all optimal schedules.

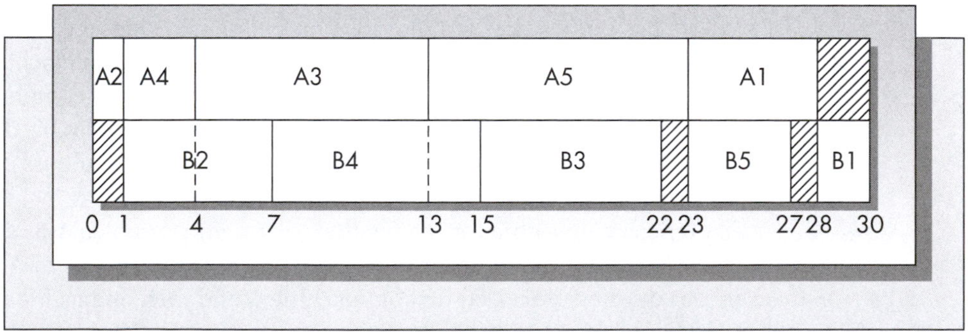

Figure 11–7 Gantt chart for the optimal schedule for Example 11.5

Extension to Three Machines

The problem of scheduling jobs on three machines is considerably more complex. If we restrict attention to total flow time only, it is still true that a permutation schedule is optimal (this is not necessarily the case for average flow time). Label the machines A, B, and C.

The three-machine problem can be reduced to (essentially) a two-machine problem if the following condition is satisfied:

$$\min A_i \geq \max B_i \qquad \text{or} \qquad \min C_i \geq \max B_i.$$

It is only necessary that *either one* of these conditions be satisfied. If that is the case, then the problem is reduced to a two-machine problem in the following way.

Define $A_i' = A_i + B_i$, and define $B_i' = B_i + C_i$. Now solve the problem using the rules described for two machines, treating A_i' and B_i' as the processing times. The resulting permutation schedule will be optimal for the three-machine problem.

EXAMPLE 11.6

Consider the following job times for a three-machine problem. Assume that the jobs are processed in the sequence A–B–C.

	Machine		
Job	**A**	**B**	**C**
1	4	5	8
2	9	6	10
3	8	2	6
4	6	3	7
5	5	4	11

Checking the conditions, we find

$$\min A_i = 4,$$
$$\max B_i = 6,$$
$$\min C_i = 6,$$

so that the required condition is satisfied. We now form the two columns A′ and B′.

	Machine	
Job	**A′**	**B′**
1	9	13
2	15	16
3	10	8
4	9	10
5	9	15

The problem is now solved using the two-machine algorithm. The optimal solution is

1–4–5–2–3.

Note that because of ties in column A, the optimal solution is not unique.

If the conditions for reducing a three-machine problem to a two-machine problem are not satisfied, this method will usually give reasonable, but possibly suboptimal, results. As long as the objective is to minimize the makespan or total flow time, a permutation schedule is optimal for scheduling on three machines. (It is not necessarily true, however, that a permutation schedule is optimal for three machines when using an average flow time criterion.)

Note that we assume that the machines are different and the processing proceeds sequentially: all jobs are assumed to be processed first on machine 1, then on machine 2. For example, machine 1 might be a drill press and machine 2 a lathe. A related problem that we discuss in the context of stochastic scheduling is that of parallel processing on identical machines. In this case the machines are assumed to perform the same function, and any job may be assigned to any machine. For example, a collection of 10 jobs might require processing on either one of two drill presses. The results for parallel processing suggest that SPT is an effective rule for minimizing mean flow time, but longest processing time first (LPT) is often more effective for minimizing total flow time or makespan. We will discuss parallel processing in the context of random job times in Section 11.8.

The Two-Job Flow Shop Problem

Assume that two jobs are to be processed through m machines. Each job must be processed by the machines in a particular order, but the sequences for the two jobs need not be the same. We present a graphical procedure for solving this problem developed by Sheldon Akers (1956).

1. Draw a Cartesian coordinate system with the processing times corresponding to the first job on the horizontal axis and the processing times corresponding to the second job on the vertical axis. On each axis, mark off the operation times in the order in which the operations must be performed for that job.
2. Block out areas corresponding to each machine at the intersection of the intervals marked for that machine on the two axes.
3. Determine a path from the origin to the end of the final block that does not intersect any of the blocks and that minimizes the vertical movement. Movement is allowed only in three directions: horizontal, vertical, and 45-degree diagonal. The path with minimum vertical distance will indicate the optimal solution. Note that this will be the same as the path with minimum horizontal distance.

This procedure is best illustrated by an example.

EXAMPLE 11.7

A regional manufacturing firm produces a variety of household products. One is a wooden desk lamp. Prior to packing, the lamps must be sanded, lacquered, and polished. Each operation requires a different machine. There are currently shipments of two models awaiting processing. The times required for the three operations for each of the two shipments are listed on p. 677.

Job 1		Job 2	
Operation	**Time**	**Operation**	**Time**
Sanding (A)	3	A	2
Lacquering (B)	4	B	5
Polishing (C)	5	C	3

The first step is to block out the job times on each of the axes. Refer to Figure 11–8 for this step. Every feasible schedule is represented by a line connecting the origin to the tip of block C, with the condition that the line not go through a block. Only three types of movement are allowed: horizontal, vertical, and 45-degree diagonal. Horizontal movement implies that only job 1 is being processed, vertical movement implies that only job 2 is being processed, and diagonal movement implies that both jobs are being processed. Minimizing the flow time is the same as maximizing the time that both jobs are being processed. This is equivalent to finding the path from the origin to the end of block C that maximizes the diagonal movement and therefore minimizes either the horizontal or the vertical movement.

Two feasible schedules for this problem are represented in Figure 11–8. The total time required by any feasible schedule can be obtained in two ways: it is either the total time represented on the horizontal axis (12 in this case) plus the total vertical movement (4 and 3 respectively), or the total time on the vertical axis (10 in this case) plus the total horizontal movement (6 and 5 respectively). Schedule 1 has total time 16, and schedule 2 has total time 15. Schedule 2 turns out to be optimal for this problem.

The Gantt chart for the optimal schedule appears in Figure 11–9.

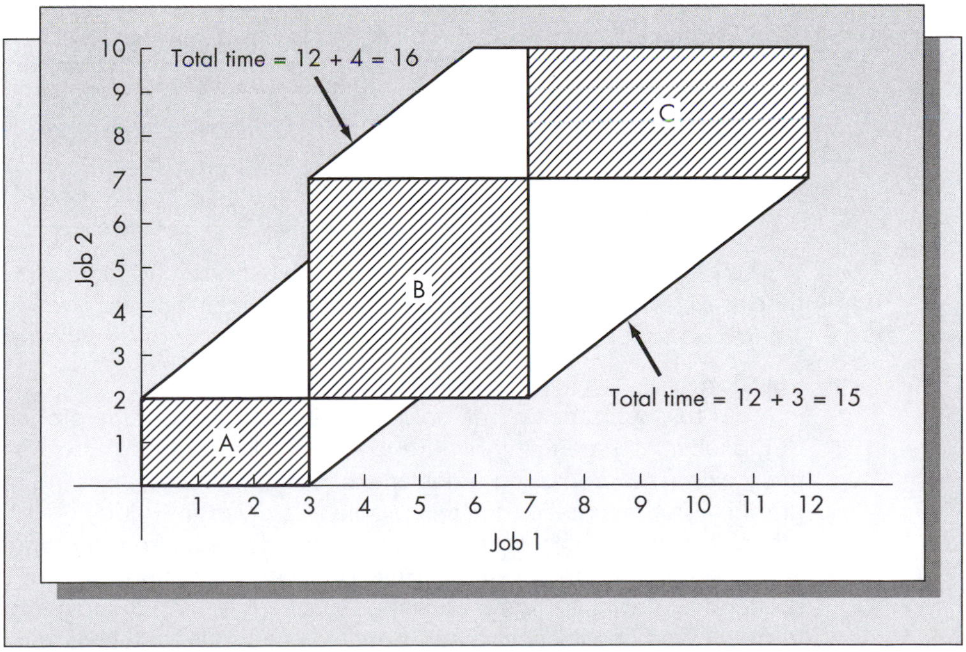

Figure 11–8 Graphical solution of Example 11.7

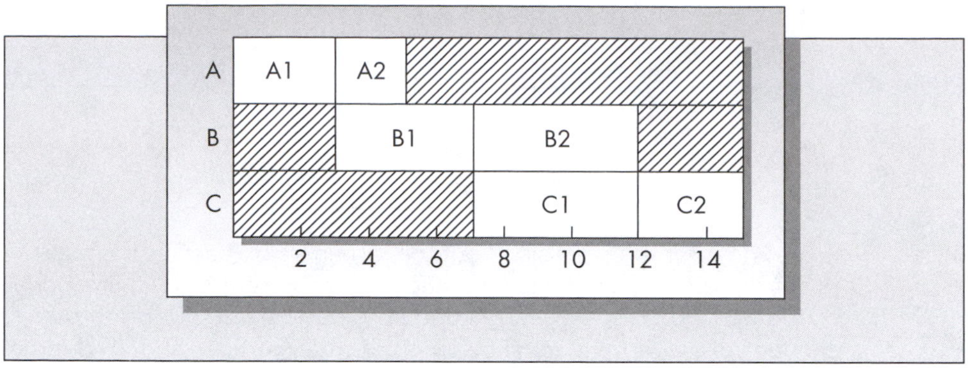

Figure 11–9 Gantt chart for optimal solution to Example 11.7

We should note that this method does *not* require the two jobs to be processed in the same sequence on the machines. We present another example to illustrate the case in which the sequence of the jobs is different.

EXAMPLE 11.8

Reggie Sigal and Bob Robinson are roommates who enjoy spending Sunday mornings reading the Sunday newspaper. Reggie likes to read the main section first, followed by the sports section, then the comics, and finally the classifieds. Bob also starts with the main section, but then goes directly to the classifieds, followed by the sports section and finally the comics. The times required (in tenths of an hour) for each to read the various sections are

Reggie		Bob	
Required Sequence	**Time**	**Required Sequence**	**Time**
Main section (A)	6	Main section (A)	4
Sports (B)	1	Classifieds (D)	3
Comics (C)	5	Sports (B)	2
Classifieds (D)	4	Comics (C)	5

The goal is to determine the order of the sections read for each to minimize the total time required to complete reading the paper. In this problem we identify Reggie as job 1 and Bob as job 2. The sections of the paper correspond to the machines.

Solution

In order to obtain the optimal solution, we first block out the processing times for each of the jobs. Assume that job 1 (Reggie) is blocked out on the x axis and job 2 (Bob) on the y axis. The processing times are sequenced on each axis in the order stated. The graphical representation for this problem is given in Figure 11–10. In the figure, two different feasible schedules are represented as paths from the origin to the point (16, 14). The top path represents a schedule that calls for Bob to begin reading the main section first (job 2 is processed first on machine A), and the lower path calls for Reggie to begin first. The lower path turns out to be optimal for this problem with a processing time of 20.

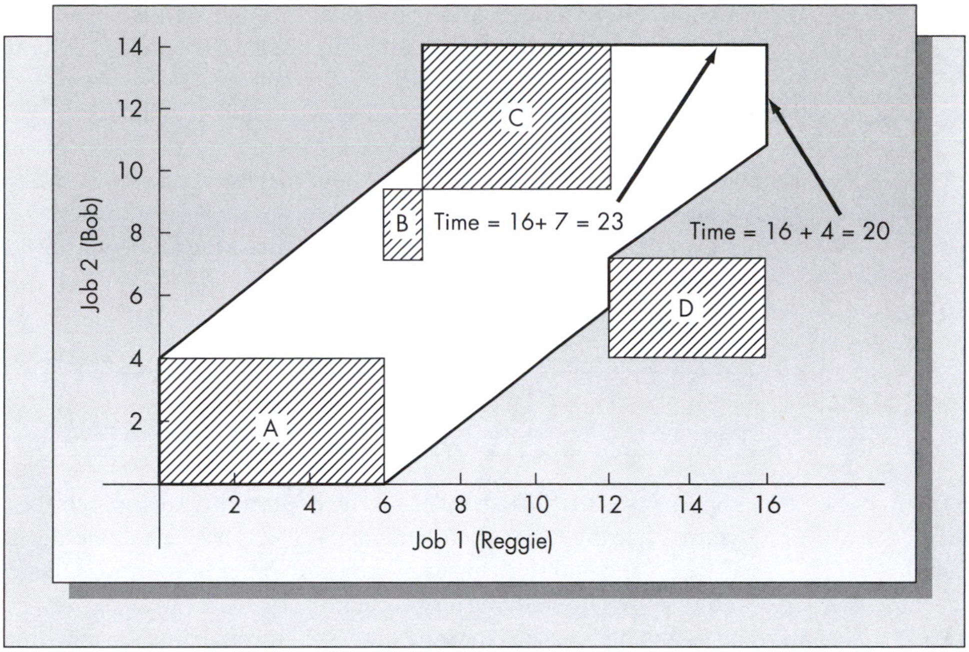

Figure 11–10 Graphical solution of Example 11.8

The optimal processing time for this problem is 20. As the time units are in tenths of an hour, this is exactly two hours. One converts the lower path in Figure 11–10 to a Gantt chart in the following way. From time 0 to 6, Reggie reads A and Bob is idle. Between times 6 and 12, the 45-degree line indicates that both Bob and Reggie are reading. Reggie reads B for 1 time unit and C for 5 time units, and Bob reads A for 4 time units and D for 3 time units. Reggie is now idle for 1 time unit because the path is vertical at this point, and begins reading D at time 13. When Reggie completes D, he is done. Starting at time 15, Bob reads C and completes his reading at time 20. Figure 11–11 shows the Gantt chart indicating the optimal solution.

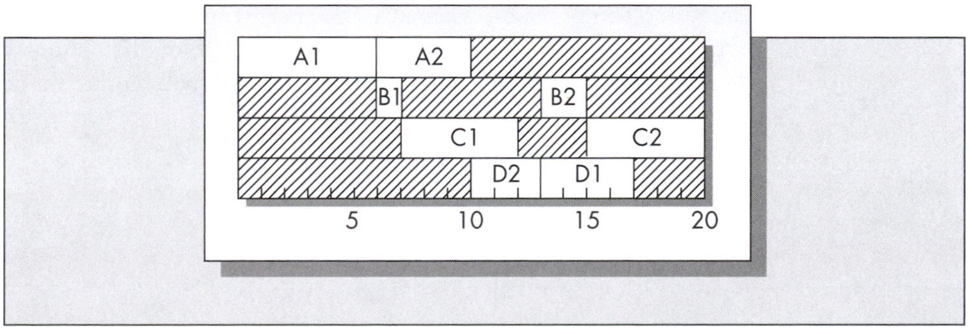

Figure 11–11 Gantt chart for optimal solution to Example 11.8

PROBLEMS FOR SECTION 11.7

11. Consider Example 11.6, illustrating the use of Johnson's algorithm for three machines. List all optimal solutions for this example.

12. Suppose that 12 jobs must be processed through six machines. If the jobs may be processed in any order, how many different possible schedules are there? If you were to run a computer program that could evaluate 100 schedules every second, how much time would the program require to evaluate all feasible schedules?
13. Two law students, John and Marsha, are planning an all-nighter to prepare for their law boards the following day. Between them they have one set of materials in the following five subjects: contracts, torts, civil law, corporate law, and patents. Based on their previous experience, they estimate that they will need the following amount of time (in hours) with each set of materials.

	Contracts	Torts	Civil	Corporate	Patents
John	1.2	2.2	0.7	0.5	1.5
Marsha	1.8	0.8	3.1	1.1	2.3

They agree that Marsha will get the opportunity to see each set of notes before John. Assume that they start their studying at 8:00 PM Determine the exact times that each will begin and end studying each subject in order to minimize the total time required for both to complete studying all five subjects.
14. The following four jobs must be processed through a three-machine flow shop.

	Machine		
Job	A	B	C
1	4	2	6
2	2	3	7
3	6	5	6
4	3	4	8

Find the optimal sequencing of the jobs in order to minimize the makespan. What is the makespan at the optimal solution? Draw a Gantt chart illustrating your solution.
15. Ashley and Samantha Brown are two sisters currently attending university together. Each requires advising in five subjects: history, English, mathematics, science, and religion. They estimate that the time (in minutes) that each will require for advising is

	Ashley	Samantha
Math	40	20
History	15	30
English	25	10
Science	15	35
Religion	20	25

They think that the five advisers will be available all day. Ashley would like to visit the advisers in the order given in the table, and Samantha would prefer to see them in the order math, religion, English, science, and history. At what times should each plan to see the advisers in order to minimize the total time for both to complete their advising?

16. Two jobs must be processed through four machines in the same order. The processing times in the required sequence are

Job 1		**Job 2**	
Machine	**Time**	**Machine**	**Time**
A	5	A	2
B	4	B	4
C	6	C	3
D	3	D	5

Determine how the two jobs should be scheduled in order to minimize the total makespan; draw the Gantt chart indicating the optimal schedule.

17. Peter Minn is planning to go to the Department of Motor Vehicles to have his driver's license renewed. Because he is over 75 he will need to take both the written and driving tests. His granddaughter, Patricia, who is accompanying him, is applying for a new license. In both cases there are five steps that are required: (A) having a picture taken, (B) signing a signature verification form, (C) passing a written test, (D) passing an eye test, and (E) passing a driving test.

For renewals the steps are performed in the order A, B, C, D, and E with average times required, respectively, of 0.2, 0.1, 0.3, 0.2, and 0.6 hour. In the case of new applications, the steps are performed in the sequence D, B, C, E, and A, with average times required of 0.3, 0.2, 0.7, 1.1, and 0.2 hour, respectively. Peter and Pat go on a day when the department is essentially empty. How should they plan their schedule in order to minimize the time required for both to complete all five steps?

11.8 STOCHASTIC SCHEDULING: STATIC ANALYSIS

Single Machine

An issue we have not yet addressed is uncertainty of the processing times. In practice it is possible and even likely that the exact completion time of one or more jobs may not be predictable. This section looks at optimal sequencing rules when processing times are uncertain. We assume that processing times are independent of one another.

In the case of processing on a single machine, most of the results are quite similar to those discussed earlier for the deterministic case. Suppose that n jobs are to be processed through a single machine. Assume that the job times, $t_1, t_2, \ldots, t_n$, are random variables with known distribution functions. The goal is to minimize the *expected* average weighted flow time; that is,

$$\text{minimize } E\left(\frac{1}{n}\sum_{i=1}^{n} u_i F_i\right),$$

where u_i are the weights and F_i is the (random) flow time of job i.

Michael Rothkopf (1966) has shown that the optimal solution is to sequence the jobs so that job i precedes job $i + 1$ if

$$\frac{E(t_i)}{u_i} < \frac{E(t_{i+1})}{u_{i+1}}.$$

Notice that if we set all the weights $u_i = 1$, then this rule is simply to order the jobs according to the minimum expected processing time; that is, it is essentially the same as SPT.

In the case of due date scheduling with random processing times, results are also similar to the deterministic case. If the objective is to minimize the maximum over all jobs of the probability that a job is late, then the optimal schedule is to order the jobs according to earliest due date or according to earliest expected due date when due dates are themselves random (Banerjee, 1965).

Multiple Machines

Somewhat more interesting results exist for scheduling jobs with random job times on multiple machines. In fact, there are results available for this case that are *not* true for the deterministic problem.

An assumption that is usually made for the multiple-machine problem is that the distribution of job times is exponential. This assumption is needed because the exponential distribution is the only one having the *memoryless property*. (This property is discussed in detail in Chapter 13 on reliability and maintenance.) The requirement that job times be exponentially distributed is severe in the context of scheduling. In most job shops, if processing times cannot be predicted accurately in advance, it is unlikely that they would be exponentially distributed. Why? The memoryless property tells us that the probability that a job is completed in the next instant of time is independent of the length of time already elapsed in processing the job. Certain applications, such as telephone systems and shared computer applications, may be accurately modeled in this manner, but by and large the exponential law would not accurately describe processing times for most manufacturing job shops.

With these caveats in mind, we present several results for scheduling jobs on multiple machines with random job times. Consider the following problem: n jobs are to be processed through two identical parallel machines. Each job needs to be processed only once on either machine. The objective is to minimize the expected time that elapses from time zero until the last job has completed processing. This is known as the expected makespan. We assume that the n jobs have processing times $t_1, t_2, \ldots, t_n$, which are exponential random variables with rates $\mu_1, \mu_2, \ldots, \mu_n$. This means that the expected time required to complete job i, $E(t_i)$, is $1/\mu_i$.

Parallel processing is different from flow shop processing. In flow shop processing, the jobs are processed first on machine 1 and then on machine 2. In parallel processing, the jobs need to be processed on only one machine, and any job can be processed on either machine. Assume that at time $t = 0$ machine 1 is occupied with a prior job, job 0, and the remaining processing time of job 0 is t_0, which could be a random or deterministic variable. The remaining jobs are processed as follows: let [1], [2], ... , [n] be a permutation of the n jobs. Job [1] is scheduled on the vacant machine. Job [2] follows either job 0 on machine 1 or job [1] on machine 2, depending on which is completed first. Each successive job is then scheduled on the next available machine.

Let $T_0 \leq T_1 \leq \ldots \leq T_n$ be the completion times of the successive jobs. The makespan is the time of completion of the last job, which is T_n. The expected value of the makespan is minimized by using the longest-expected-processing-time-first rule (LEPT). [However, if we consider flow time, then SEPT (shortest expected processing time first) minimizes the expected flow time on two machines.] Note that this is exactly the opposite of the SPT rule for a single machine. The optimality of scheduling the jobs in decreasing order of their expected size rather than in increasing order (as the SPT rule does) is more likely a result of parallel processing than of randomness of the job times.

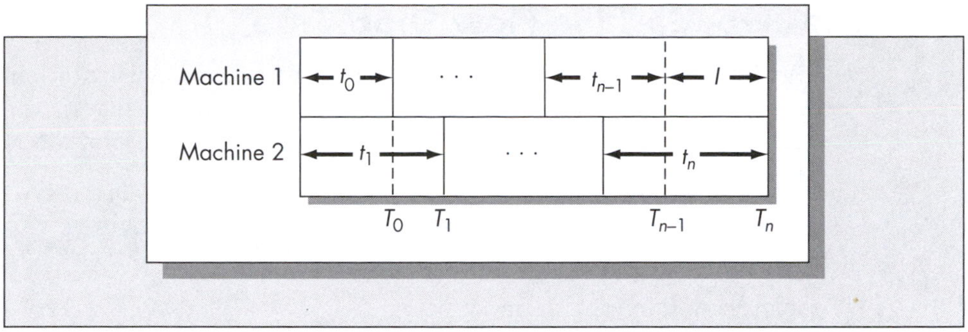

Figure 11–12 Realization of parallel processing on two machines with random job times

We intuitively show the optimality of LEPT for this problem as follows: Consider the schematic diagram in Figure 11–12, which gives a particular realization of the processing times for an arbitrary sequencing of the jobs. In Figure 11–12 the random variable I corresponds to the idle time of the machine that does not process the job completed last. Intuitively, we would like to make I as small as possible in order to minimize the expected makespan. This can be shown more rigorously as follows.

From the picture, it is clear that

$$T_n + T_{n-1} = \sum_{i=0}^{n} t_i$$

and

$$T_n = T_{n-1} - I.$$

Solving for T_{n-1} in the second equation and substituting into the first gives

$$T_n + T_n - I = \sum_{i=0}^{n} t_i$$

or

$$2T_n = \sum_{i=0}^{n} t_i + I.$$

As Σt_i is fixed independent of the processing sequence, it follows that minimizing $E(T_n)$ is equivalent to minimizing $E(I)$. Because I is minimized when the processing time of the last job is minimized, we schedule the jobs in order of decreasing expected processing time. Note that this result does *not* necessarily carry over to the case of parallel processing on two machines with deterministic processing times. However, in the deterministic case, scheduling the longest job first will generally give good results when minimizing total makespan. SPT is superior for minimizing mean flow time.

A class of problems we will not discuss, but for which there are several interesting results, are problems in which jobs are to be processed through m nonidentical processors and the processing time does not depend on the job. See Rhonda Righter (1988) for the characterization of the optimal scheduling rule for this problem.

The Two-Machine Flow Shop Case

An interesting question is whether there is a stochastic analog to Johnson's algorithm for scheduling n jobs on two machines in a flow shop setting—that is, where each job must be processed through machine 1 first, then through machine 2. Johnson's algorithm tells us that job i precedes job $i + 1$ if

$$\min(A_i, B_{i+1}) < \min(A_{i+1}, B_i)$$

in order to minimize the makespan.

Now suppose that $A_1, A_2, \ldots, A_n$ and $B_1, B_2, \ldots, B_n$ are exponential random variables with respective rates $a_1, a_2, \ldots, a_n$ and $b_1, b_2, \ldots, b_n$. We now wish to minimize the expected value of the makespan. As the minimum of two exponential random variables has a rate equal to the sum of the rates, it follows that

$$E[\min(A_i, B_{i+1})] = \frac{1}{a_i + b_{i+1}}.$$

$$E[\min(A_{i+1}, B_i)] = \frac{1}{a_{i+1} + b_i}.$$

It follows that Johnson's condition translates in the stochastic case to the condition that

$$a_i - b_i \le a_{i+1} - b_{i+1},$$

so that the jobs should be scheduled in the order of decreasing values of the difference in the rates.

EXAMPLE 11.9

Consider Example 11.5, used to illustrate Johnson's algorithm, but assume that the job times are random variables having the exponential distribution with mean times given in the example. Hence, we have the following.

	Expected Times		Rates		
Job	A	B	A	B	Differences
1	5	2	0.20	0.500	–0.30
2	1	6	1.00	0.170	0.83
3	9	7	0.11	0.140	–0.03
4	3	8	0.33	0.125	0.21
5	10	4	0.10	0.250	–0.15

Ordering the jobs according to decreasing values of the differences in the final column results in the sequence

2–4–3–5–1,

which is exactly the same sequence we found in the deterministic case using Johnson's algorithm.

This section considered several solution procedures when job times are random variables. Even when job times are known with certainty, randomness resulting from other sources may still be present. For example, when considering scheduling as a dynamic problem, one must determine the pattern of arrivals to the system. It is common for jobs to arrive according to some random process and queue up for service. Queueing theory and simulation are useful tools for dealing with randomness of this type. (See Conway, Maxwell, and Miller, 1967, for the application of both simulation and queueing to operations scheduling problems.)

PROBLEMS FOR SECTION 11.8

18. Consider Example 11.2 in Section 11.5 on determining the optimal sequence to land the planes. Suppose that the landing times are random variables with standard deviation equal to one-third of the mean in each case.
    a. In what sequence should the planes be landed in order to minimize the expected average weighted flow time, if the weights to be used are the reciprocals of the number of passengers on each plane?
    b. For the sequence you found in part (a), what is the probability that all planes are landed within 100 minutes? Assume that the landing times are independent normally distributed random variables. Will your answer change if the planes are landed in a different sequence?
19. A computer center has two identical computers for batch processing. The computers are used as parallel processors. Job times are estimated by the user, but experience has shown that an exponential distribution gives an accurate description of the actual job times. Suppose that at a point in time there are eight jobs remaining to be processed with the following expected job times (expressed in minutes).

Job	1	2	3	4	5	6	7	8
Expected time	4	8	1	50	1	30	20	6

    a. In what sequence should the jobs be processed in order to minimize the expected completion time of all eight jobs (i.e., the makespan)?
    b. Assume that computer A is occupied with a job that has exactly two minutes of processing time remaining and computer B is idle. If job times are deterministic, show the start and end times of each job on each computer using the sequence derived in part (a).
20. Six ships are docked in a harbor awaiting unloading. The times required to unload the ships are random variables with respective means of 0.6, 1.2, 2.5, 3.5, 0.4, and 1.8 hours. The ships are given a priority weighting based on tonnage. The respective tonnages are 12, 18, 9, 14, 4, and 10. In what sequence should the ships be unloaded in order to minimize the expected weighted time?
21. Solve problem 13 assuming that the times required by John and Marsha are exponentially distributed random variables with expected times given in problem 13.
22. Five sorority sisters plan to attend a social function. Each requires hair styling and fitting for a gown. Assume that the times required are independent exponentially distributed random variables with the mean times for the fittings of 0.6, 1.2, 1.5, 0.8, and 1.1 hours, respectively, and mean times for the stylings of 0.8, 1.6, 1.0, 0.7, and 1.3 hours, respectively. Assume that the fittings are done before the stylings and that there is only a single hair stylist and a single seamstress available. In what sequence should they be scheduled in order to minimize the total expected time required for fittings and stylings?

11.9 STOCHASTIC SCHEDULING: DYNAMIC ANALYSIS

The scheduling algorithms discussed thus far in this chapter are based on the assumption that all jobs arrive for processing simultaneously. In practice, however, scheduling jobs on machines is a **dynamic** problem. We use the term *dynamic* here to mean that jobs are arriving randomly over time, and decisions must be made on an ongoing basis as to how to schedule those jobs.

Queueing theory provides a means of modeling some dynamic scheduling problems. Chapter 8 and Supplement 2 provide a review of basic queueing theory. In this section familiarity with the results presented in Supplement 2 is assumed.

Consider the following problem. Jobs arrive completely at random to a single machine. This means that the arrival process is a *Poisson* process. Assume that the mean arrival rate is λ. We initially will assume that processing times are exponentially distributed with mean $1/\mu$. This means that the average processing rate is μ and processing times are independent identically distributed exponential random variables. Finally, we assume that jobs are processed on a first-come, first-served (FCFS) basis. In queueing terminology, we are assuming an M/M/1/FCFS queue. Other processing sequences also will be considered.

Basic queueing theory answers several questions about the performance characteristics of this scheduling problem. First, the probability distribution of the number of jobs in the system (the number waiting to be processed plus the number being processed) in the steady state is known to be geometric with parameter $\rho = \lambda/\mu$. That is, if L is the number of jobs in the system in steady state, then

$$P\{L = i\} = \rho^i(1 - \rho) \qquad \text{for } i = 0, 1, 2, 3, \ldots.$$

The expected number of jobs in the system is $\rho/(1 - \rho)$. This implies that a solution exists only for $\rho < 1$. Intuitively this makes sense: the rate at which jobs arrive in the system must be less than the rate at which they are processed to guarantee that the queue does not grow without bound. [What is not obvious, however, is what happens at the boundary when $\lambda = \mu$ or $\rho = 1$. It turns out that when the processing and the arrival rates are equal (and interarrival times and job processing times are random), the queue still grows without bound. Essentially, after a long time all states become equally likely including the ones with a large number of jobs.]

Minimizing mean flow time is a common objective not only in static scheduling but also in dynamic scheduling. The flow time of a job begins the instant the job joins the queue of unprocessed jobs and continues until its processing is completed. For the dynamic scheduling problem, the flow time of a job is a random variable; it depends on the realization of the processing times of preceding jobs as well as its own processing time. The queueing term for the flow time of a job is the waiting time in the system and is denoted by the symbol W. Supplement 2 shows that the distribution of the flow time for the M/M/1/FCFS queue is exponential with parameter $\mu - \lambda$. That is,

$$P\{W > t\} = e^{-(\mu-\lambda)t} \qquad \text{for all } t > 0.$$

Also derived are the distribution for the waiting time in the queue and the expected number of jobs in the queue waiting to be processed. We can see the application of these formulas to the dynamic scheduling problem in the following example.

EXAMPLE 11.10

A student computer laboratory has a single printer. Jobs queue up on the printer from the network server in the lab and are completed on a first-come, first-served basis. The average printing job requires four minutes, but the times vary considerably. Experience has shown that the distribution of times closely follows an exponential distribution. At peak times, about 12 students per hour require use of the printer, but the arrival of jobs to the printer can be assumed to occur completely at random.

Assuming a peak traffic period, determine the following.

a. The average number of jobs in the printer queue.
b. The average flow time of a job.
c. The probability that a job will wait more than 30 minutes before it begins processing.
d. The probability that there are more than six jobs in the system.

Solution

First, we must determine the service and arrival rates. As each printing job requires an average of 4 minutes, it follows that the service *rate* is $\mu = 1/4$ per minute or 15 per hour. The arrival rate is given as $\lambda = 12$ per hour. The traffic intensity $\rho = 12/15 = .8$.

a. The average number of jobs in the queue is l_q, which is given by (refer to Supplement 2)

$$l_q = \frac{\rho^2}{1-\rho} = \frac{.64}{.2} = 3.2.$$

b. The average flow time of a job is the same as the waiting time in the system. From Chapter 8,

$$w = \frac{\rho}{\lambda(1-\rho)} = \frac{.8}{(12)(.2)} = 0.3333 \text{ hour (20 minutes)}.$$

c. Here we are interested in $P\{W_q > 0.5\}$. The distribution of W_q is essentially exponential with parameter $\mu - \lambda$, but with positive mass at zero.

$$P\{W_q > t\} = \rho e^{-(\mu-\lambda)t} = .8e^{-(3)(0.5)} = .1785.$$

d. Here we wish to determine $P\{L > 6\}$. From the solution of Example 8.5 in Chapter 8, we showed that

$$P\{L > k\} = \rho^{k+1} = .8^7 = .2097.$$

Selection Disciplines Independent of Job Processing Times

Although our sense of fair play says that the service discipline should be FCFS, there are many occasions when jobs are processed in other sequences. For example, consider a manufacturing process in which parts are stacked as they are completed. The next stage in the process may simply take the parts from the top of the stack, resulting in a last-come, first-served (LCFS) discipline. Similarly, one could envision a situation in which completed parts are thrown into a bin and taken out at random, resulting in service occurring in a random order.

Queueing theory tells us that as long as the selection discipline *does not depend on the processing times*, the mean flow times (and hence mean numbers in the system and mean queue lengths) are the same. [Many of the results quoted in this section are discussed more fully in Chapter 9 of Conway et al. (1967). In fact, even though this book is more than 50 years old, it still provides the most comprehensive treatment of the application of queueing theory to scheduling problems.] However, the **variance** of the flow times does depend on the selection discipline. The flow time variance is greatest with LCFS and least with FCFS among the three disciplines FCFS, LCFS, and random. [This is strictly true only if we eliminate the possibility of preemption. Preemption means that a newly arriving job is allowed to interrupt the service of a job already in progress. We will not treat preemptive disciplines here.] The second moments of the flow time are given by

$$E_{\text{LCFS}}(W^2) = \frac{1}{1-\rho} E_{\text{FCFS}}(W^2),$$
$$E_{\text{RANDOM}}(W^2) = \frac{1}{1-\rho/2} E_{\text{FCFS}}(W^2).$$

(The second equation has been proved for exponential processing times only, but it has been conjectured that it holds in general.) Recall that the variance of a random variable is the second moment minus the mean squared, so we can obtain the variance of W directly from these formulas.

EXAMPLE 11.10 (*CONTINUED*)

Determine the variance of the flow times for the printer during peak hours assuming (1) FCFS, (2) LCFS, and (3) random selection disciplines.

Solution

Because under FCFS the flow time is exponentially distributed with parameter $\mu - \lambda$, it follows that the mean flow time is $1/(\mu - \lambda)$ and the variance of the flow time is $1/(\mu - \lambda)^2$.

$$E_{\text{FCFS}}(W) = \frac{1}{\mu - \lambda} = \frac{1}{15-12} = \frac{1}{3} \text{ hour.}$$
$$\text{Var}_{\text{FCFS}}(W) = \frac{1}{(3)^2} = \frac{1}{9}.$$

Because $\text{Var}(W) = E(W)^2 - (E(W))^2$, it follows that

$$E_{\text{FCFS}}(W)^2 = \text{Var}_{\text{FCFS}}(W) + (E_{\text{FCFS}}(W))^2 = \frac{1}{9} + \left(\frac{1}{3}\right)^2 = \frac{2}{9}.$$

From these results, we obtain

$$E_{\text{LCFS}}(W)^2 = \left(\frac{1}{1-.8}\right)\left(\frac{2}{9}\right) = \frac{10}{9}, \quad \text{giving}$$
$$\text{Var}_{\text{LCFS}}(W) = \frac{10}{9} - \left(\frac{1}{3}\right)^2 = 1.0.$$

Similarly,

$$E_{\text{RANDOM}}(W)^2 = \left(\frac{1}{1-.4}\right)\left(\frac{2}{9}\right) = 0.3704, \quad \text{giving}$$
$$\text{Var}_{\text{RANDOM}}(W) = 0.3704 - \left(\frac{1}{3}\right)^2 = 0.2593.$$

Hence, we see that if the jobs are processed on the printer on an LCFS basis, the variance of the flow time is 1.0, as compared to 2/9 for FCFS. Processing jobs in a random order also increases the variance of the flow time over FCFS, but by a much smaller degree. Intuitively, the variance is larger for LCFS than for FCFS, because when jobs are processed in the opposite order of their arrival, it is likely that a job that has been in the queue for a while will continue to be "bumped" by newly arriving jobs. The result will be a very long flow time. A similar phenomenon occurs in the random case, but the effect is not as severe.

Selection Disciplines Dependent on Job Processing Times

One of the goals of research into dynamic queueing models is to discover optimal selection disciplines. Previously we stated that the average measures of performance (w, w_q , l, and l_q) are independent of the selection discipline as long as the selection discipline does not depend on the job processing times. However, it was shown in Chapter 8 that selection based on shortest processing times can make a large difference to average waiting time. Furthermore, consider the case in which job processing times are realized at the instant the job joins the queue. This assumption is reasonable for most industrial scheduling applications. For machine processing problems, the work content is likely to be a multiple of the number of parts that have to be processed, so the processing time is known at the instant a job joins the queue. An example in which this does not hold would be a bank. It is generally not possible to tell how long a customer will require for service until he or she enters service and reveals the nature of the transaction.

When job processing times are realized when a job joins the queue, it is possible to use a selection discipline that is dependent on job times. One such discipline, discussed at length earlier in this chapter, is the SPT rule: The next job processed is the one with the shortest processing time. It turns out that the SPT rule is effective for the dynamic problem as well as the static problem.

SPT scheduling can significantly reduce the size of the queue in the dynamic scheduling problem. In Figure 11–13 we show just how dramatic this effect can be. This figure assumes an M/M/1 queue. Define the relative flow time as the ratio

$$\frac{E_{\text{SPT}}\ (\text{Flow Time})}{E_{\text{FCFS}}\ (\text{Flow Time})}.$$

We see that as the traffic intensity increases, the advantage of SPT at reducing flow time (and hence the number in the system and the queue length) improves. The queue is reduced by "cleaning out" the shorter jobs. For values of ρ near 1, this ratio could be as low as 0.2. Because of Little's formula (see Chapter 8), the expected number in the system and the flow time are proportional, so this curve also represents the ratio of the expected numbers in the system for the respective selection disciplines.

An interesting question is what happens to the variance of the performance measures under each selection discipline. Again assuming exponential service times, Table 11–1 gives the variances of the flow times for SPT and FCFS as a function of the traffic intensity, ρ.

What we see from this table is that for low values of the traffic intensity (less than about .7), SPT has slightly lower variance than FCFS. However, as the traffic intensity approaches 1, the variance under SPT increases dramatically. For $\rho = .99$, the variance of the flow time

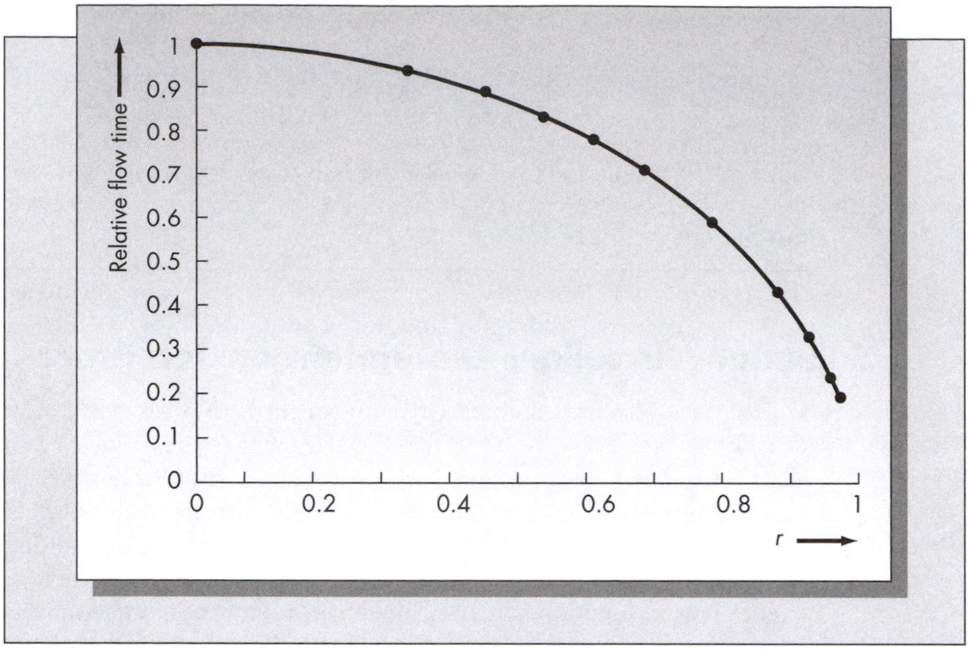

Figure 11–13 Relative flow times: SPT versus FCFS

Source: Adapted from tabled results in Conway et al., 1967, p. 184.

under SPT is more than 16 times that for FCFS. Hence, the reduction in the mean flow time achieved by SPT comes at the cost of possibly increasing the variance.

A large portion of the research in dynamic scheduling considers priority disciplines, which means that incoming jobs are classified into groups and priority is given to certain groups over others. Priority scheduling is common in hospital emergency rooms. It also occurs in scheduling batch jobs on a mainframe computer, as certain users may be given priority over others.

Table 11–1 Variance of the Flow Time under FCFS and SPT

ρ	FCFS	SPT
.1	1.2345	1.179
.2	1.5625	1.482
.3	2.0408	1.896
.4	2.6666	2.563
.5	4.0000	3.601
.6	6.2500	5.713
.7	11.1111	12.297
.8	25	32.316
.9	100	222.2
.95	400	1,596.5
.98	2,500	22,096
.99	10,000	161,874

Source: Conway et al., 1967, p. 189.

Priorities may be preemptive or nonpreemptive. Preemptive priority means that a newly arriving job with a higher priority than the job currently being processed is permitted to interrupt the service of the job in process. The interrupted job may continue where it left off at a later time, or it may have to start processing again from scratch. We will not pursue this complex area but note that SPT is quite robust; it is optimal for a large class of priority scheduling problems.

The $c\mu$ Rule

Consider the following scheduling problem. Jobs arrive randomly to a single machine with exponential processing times. We allow the jobs in the queue to have different service rates μ_i. That is, at any moment, suppose that there are n jobs waiting to be processed. We index the jobs 1, 2, 3, … , n and assume that the time required to complete job i has the exponential distribution with mean $1/\mu_i$. In addition, suppose that there is a return of c_i if job i is completed by some fixed time t. The issue is, what is the best choice for the next job to be processed if the objective is to maximize the total expected earnings?

The optimal policy is to choose the job with the largest value of $c_i\mu_i$ (Derman, Lieberman, and Ross, 1978). Notice that if the weights are set equal to 1, the $c\mu$ rule is exactly the same as SPT in expectation. Hence, this can be considered to be a type of weighted SPT scheduling rule. It turns out that the $c\mu$ rule is optimal for several other versions of the stochastic scheduling problem. We refer the interested reader to Michael Pinedo (1983) and the references listed there.

PROBLEMS FOR SECTION 11.9

23. A computer system queues up batch jobs and processes them on an FCFS basis. Between 2 and 5 PM, jobs arrive at an average rate of 30 per hour and require an average of 1.2 minutes of computer time. Assume the arrival process is Poisson and the processing times are exponentially distributed.
    a. What is the expected number of jobs in the system and in the queue in the steady state?
    b. What are the expected flow time and the time in the queue in the steady state?
    c. What is the probability that the system is empty?
    d. What is the probability that the queue is empty?
    e. What is the probability that the flow time of a job exceeds 10 minutes?
24. Consider the computer system of problem 23.
    a. Compute the variance of the flow times assuming FCFS, LCFS, and random selection disciplines.
    b. Using a normal approximation of the flow time distribution for the LCFS and random cases, estimate the probability that the flow time in the system exceeds 10 minutes in each case.
25. A critical resource in a manufacturing operation experiences a very high traffic intensity during the shop's busiest periods. During these periods, the arrival rate is approximately 57 jobs per hour. Job processing times are approximately exponentially distributed with mean 1 minute.
    a. Compute the expected flow time in the system assuming an FCFS processing discipline and the expected flow time under SPT using Figure 11–13.
    b. Compute the probability that a job waits more than 30 minutes for processing under FCFS.
    c. Using a normal approximation, estimate the probability that a job waits more than 30 minutes for processing under LCFS.

11.10 ASSEMBLY LINE BALANCING

The problem of balancing an assembly line is a classic industrial engineering problem. Even though much of the work in the area goes back to the mid-1950s and early 1960s, the basic structure of the problem is relevant to the design of production systems today, even in automated plants. The problem is characterized by a set of n distinct tasks that must be completed on each item. The time required to complete task i is a known constant t_i. The goal is to organize the tasks into groups, with each group of tasks being performed at a single workstation. In most cases, the amount of time allotted to each workstation is determined in advance, based on the desired rate of production of the assembly line. This is known as the **cycle time** and is denoted by C. A schematic of a typical assembly line is given in Figure 11–14. In the figure, circles represent tasks to be done at the corresponding stations.

Assembly line balancing is traditionally thought of as a facilities design and layout problem. Assigning tasks to workstations has traditionally been a one-shot decision made at the time the plant is constructed and the line is set up. However, the nature of the modern factory is changing. New plants are being designed with flexibility in mind, allowing new lines to be brought up and old ones restructured on a continuous basis. In such an environment, line balancing is more like a dynamic scheduling problem than a one-shot facilities layout problem.

There are a variety of factors that contribute to the difficulty of the problem. First, there are precedence constraints; some tasks may have to be completed in a particular sequence. Another problem is that some tasks cannot be performed at the same workstation. For example, it might not be possible to work on the front end and the back end of a large object such as an automobile at the same workstation. This is known as a **zoning restriction**. Still other complications might arise. For example, certain tasks may have to be completed at the same workstation, and other tasks may require more than one worker.

Finding the optimal balance for an assembly line is a difficult combinatorial problem even when the problems previously described are not present. Several relatively simple heuristics have been suggested for determining an approximate balance. Many of these methods require few calculations and make it possible to solve large problems by hand.

Let $t_1, t_2, \ldots, t_n$ be the time required to complete the respective tasks. The total work content associated with the production of an item, say T, is given by

$$T = \sum_{i=1}^{n} t_i.$$

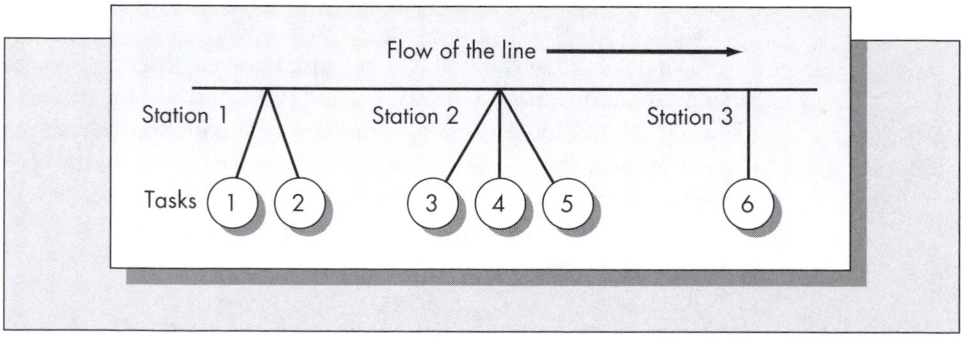

Figure 11–14 Schematic of a typical assembly line

SNAPSHOT APPLICATION

Manufacturing Divisions Realize Savings with Scheduling Software

Motorola Schedules Engine Controllers

Motorola Corporation adopted software to schedule the remanufacture of 1,000 engine controllers each month in the Automotive and Industrial Electronics Group in its Seguin, Texas, plant. Utilization of used engine controllers improved from 35 percent to over 70 percent and the time required for scheduling was reduced from 100 to only 8 hours per month as a consequence of implementing the software. The problem of scheduling cores remanufacture is a very complex one, requiring five or six employees when it was done manually. According to Eileen Svoboda, a process improvement manager:

> Remanufacturing in itself is a unique and complex process because the raw material—the core—arrives randomly through time, and the material required to remanufacture it is unknown until after the core is inspected. Right now in our business we have 7,000 Motorola core model numbers that can be remanufactured into approximately 800 different customer models. To keep all this straight in the old manual system we relied on paper trails, as well as the memory and expertise of a few key individuals in Texas, Michigan, and Illinois.

The software allowed the group to schedule one month's worth of production in detail, as well as some additional planning functions for a 26-week time horizon. Motorola considered four different scheduling systems and chose the one considered best able to both handle complex material assignment needs and easily run what-if simulations. Although the simulation feature played an important part in the decision, the core of the scheduling software was an optimization package, and, according to Berclain, Motorola's primary goal was optimization of materials utilization.

The system's database incorporated manufacturing data, including sequence of operations or routings, bills of materials, and setup sequencing information. It distinguished raw materials, subassembly, and finished goods inventory levels. It exchanged data with other inventory control systems, and it detailed production orders for all work centers.

HP Implements Materials Scheduling

The Hewlett-Packard Corporation of Palo Alto, California, experienced enormous growth in the highly competitive PC business. Part of this success was attributable to careful management of the materials in the PC product plants.

HP's scheduling system was PC-based used in conjunction with its MRP II system. According to Dr. Lee Seaton, a senior industry analyst with HP:

> The materials content of a personal computer constitutes over 90 percent of its cost. Unfortunately, we frequently had too much of one thing and too little of something else. And we knew it was only going to get worse, since product life cycles continue to accelerate, recently down from 15 months to under one year. We had to reduce product write-offs and have a smoother cut-off for a given product set. Too often, we'd launch a successful product, but either not have enough to sell or at the end of the cycle have excessive material write-offs.

While material control was normally the domain of MRP II, HP was not having the desired level of success with MRP II. Dr. Seaton claimed that 30 percent of the MRP runs were considered useless. Updates of inventory levels or bills of materials were not entered into the system in a timely fashion, making the results out of date. The new scheduling tools allowed HP employees to verify the validity of MRP runs with desktop systems. Also, the software provided the capability for determining the consequences of business decisions in a spreadsheet or graphical form. Engineering changes could more easily be incorporated into the materials plan, and excess material could be utilized more effectively by constructing "matched sets." The key to HP's success according to Seaton was "getting information into the hands of the decision makers."

Source: These two applications are discussed in Parker (1995).

For a cycle time of C, the minimum number of workstations possible is $[T/C]$, where the brackets indicate that the value of T/C is to be rounded to the next larger integer. Because of the discrete and indivisible nature of the tasks and the precedence constraints, it is often true that more stations are required than this ideal minimum value. If there is leeway in the choice of the cycle time, it is advisable to experiment with different values of C to see if a more efficient balance can be obtained.

We will present one heuristic method known as the **ranked positional weight technique** (Helgeson and Birnie, 1961). The method places a weight on each task based on the total time required by all the succeeding tasks. Tasks are assigned sequentially to stations based on these weights. We illustrate the method by example.

EXAMPLE 11.11

The final assembly of Noname personal computers, a generic mail-order PC clone, requires a total of 12 tasks. The assembly is done at the Lubbock, Texas, plant using various components imported from the Far East. The tasks required for the assembly operations are

1. Drill holes in the metal casing and mount the brackets to hold disk drives.
2. Attach the motherboard to the casing.
3. Mount the power supply and attach it to the motherboard.
4. Place the main processor and memory chips on the motherboard.
5. Plug in the graphics card.
6. Attach the hard disk controller and the power supply.
7. Mount the hard disk drive. Attach the hard disk controller and the power supply to the hard drive.
8. Set switch settings on the motherboard for the specific configuration of the system.
9. Attach the monitor to the graphics board prior to running system diagnostics.
10. Run the system diagnostics.
11. Seal the casing.
12. Attach the company logo and pack the system for shipping.

The holes must be drilled and the motherboard attached to the casing prior to any other operations. Once the motherboard has been mounted, the power supply, memory, processor chips, graphics card, and disk controller can be installed. Based on the memory configuration and the choice of graphics adapter, the switch settings on the motherboard are determined and set. The monitor must be attached to the graphics board so that the results of the diagnostic tests can be read. Finally, after all other tasks are completed, the diagnostics are run and the system is packed for shipping. The job times and precedence relationships for this problem are summarized in the following table. The network representation of this particular problem is given in Figure 11–15.

Task	Immediate Predecessors	Time
1	—	12
2	1	6
3	2	6
4	2	2
5	2	2
6	2	12

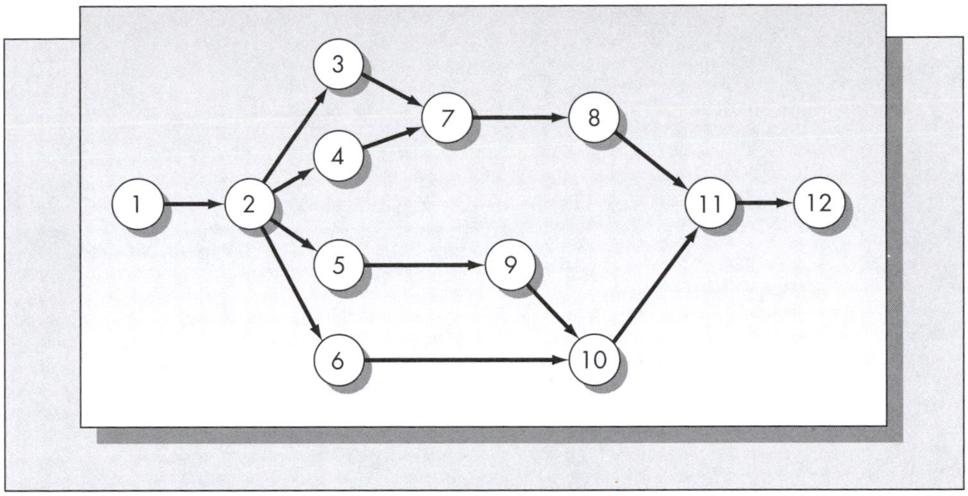

Figure 11–15 Precedence constraints for Noname computer (Example 11.11)

7	3, 4	7
8	7	5
9	5	1
10	9, 6	4
11	8, 10	6
12	11	7

Suppose that the company is willing to hire enough workers to produce one assembled machine every 15 minutes. The sum of the task times is 70, which means that the minimum number of workstations is the ratio 70/15 = 4.67 rounded to the next larger integer, which is 5. This does not mean that a five-station balance necessarily exists.

The solution procedure requires determining the positional weight of each task. The positional weight of task i is defined as the time required to perform task *i* plus the times required to perform all tasks having task *i* as a predecessor. As task 1 must precede all other tasks, its positional weight is simply the sum of the task times, which is 70. Task 2 has positional weight 58. From Figure 11–15 we see that task 3 must precede tasks 7, 8, 11, and 12, so that the positional weight of task 3 is $t_3 + t_7 + t_8 + t_{11} + t_{12} = 31$. The other positional weights are computed similarly. The positional weights are listed in Table 11–2 (p. 696).

The next step is to rank the tasks in the order of decreasing positional weight. The ranking for this case is 1, 2, 3, 6, 4, 7, 5, 8, 9, 10, 11, 12. Finally, the tasks are assigned sequentially to stations in the order of the ranking, and assignments are made only as long as the precedence constraints are not violated.

Let us now consider the balance obtained using this technique assuming a cycle time of 15 minutes. Task 1 is assigned to station 1. That leaves a slack of three minutes at this station. However, because task 2 must be assigned next, in order not to violate the precedence constraints, and the sum $t_1 + t_2$ exceeds 15, we close station 1. Tasks 2, 3, and 4 are then assigned to station 2, resulting in an idle time of only one minute at this station. Continuing in this manner, we obtain the following balance for this problem.

Station	1	2	3	4	5	6
Tasks	1	2, 3, 4	5, 6, 9	7, 8	10,11	12
Idle time	3	1	0	3	5	8

Table 11–2 Positional Weights for Example 11.11

Task	Positional Weight
1	70
2	58
3	31
4	27
5	20
6	29
7	25
8	18
9	18
10	17
11	13
12	7

Notice that although the minimum possible number of stations for this problem is five, the ranked positional weight technique results in a six-station balance. As the method is only a heuristic, it is possible that there is a solution with five stations. In this case, however, the optimal balance requires six stations when $C = 15$ minutes.

The head of the firm assembling Noname computers is interested in determining the minimum cycle time that would result in a five-station balance. If we increase the cycle time from $C = 15$ to $C = 16$, then the balance obtained is

Station	1	2	3	4	5
Tasks	1	2, 3, 4, 5	6, 9	7, 8, 10	11, 12
Idle time	4	0	3	0	3

This is clearly a much more efficient balance. The total idle time has been cut from 20 minutes per unit to only 10 minutes per unit. The number of stations decreases by about 16 percent, while the cycle time increases by only about 7 percent. Assuming that a production day is seven hours, a value of $C = 15$ minutes would result in a daily production level of 28 units per assembly operation and a value of $C = 16$ minutes would result in a daily production level of 26.25 units per assembly operation. Management would have to determine whether the decline in the production rate of 1.75 units per day per operation is justified by the savings realized with five rather than six stations.

An alternative choice is to stay with the six stations but see if a six-station balance can be obtained with a cycle time less than 15 minutes. It turns out that for values of the cycle time of both 14 minutes and 13 minutes, the ranked positional weight method will give six-station balances. The $C = 13$ solution is

Station	1	2	3	4	5	6
Tasks	1	2, 3	6	4, 5, 7, 9	8, 10	11, 12
Idle time	1	1	1	1	4	0

Thirteen minutes appear to be the minimum cycle time with six stations. The total idle time of eight minutes resulting from the balance above is two minutes less than that achieved with five stations when $C = 16$. The production rate with six stations and $C = 13$ would be 32.3 units per day per operation. Increasing the number of stations from five to six results in a substantial improvement in the throughput rate.

In this section we presented the ranked positional weight heuristic for solving the assembly line balancing problem. Other heuristic methods exist as well. One is COMSOAL, a computer-based heuristic developed by Albert Arcus (1966). The method is efficient for large problems involving many tasks and workstations.

There are optimal procedures for solving the line balancing problem, but the calculations are complex and time-consuming, requiring either dynamic programming (Held, Karp, and Shareshian, 1963) or integer programming (Thangavelu and Shetty, 1971). Other interest in the line balancing problem has focused on issues relating to uncertainty in the performance times for the individual tasks (see Hillier and Boling, 1986).

Virtually all assembly line balancing procedures assume that the objective is to minimize the total idle time at all workstations. However, as we saw in this section, an optimal balance for a fixed cycle time may not be optimal in a global sense. Meir Rosenblatt and Robert Carlson (1985) suggest that most assembly line balancing procedures are based on an incorrect objective. The authors claim that maximizing profit (rather than minimizing idle time) would give a different solution to most assembly line balancing problems, and they present several models in which both numbers of stations and cycle time are decision variables.

PROBLEMS FOR SECTION 11.10

26. Consider the example of Noname computers presented in this section.
    a. What is the minimum cycle time that is possible? What is the minimum number of stations that would theoretically be required to achieve this cycle time?
    b. Based on the ranked positional weight technique, how many stations are actually required for the cycle time indicated in part (a)?
    c. Suppose that the owner of the company that sells Noname computers finds that he is receiving orders for approximately 100 computers per day. How many separate assembly lines are required assuming (*i*) the best five-station balance, (*ii*) the best six-station balance (both determined in the text), and (*iii*) the balance you obtained in part (b)? Discuss the trade-offs involved with each choice.
27. A production facility assembles inexpensive telephones on a production line. The assembly requires 15 tasks with precedence relationships and activity times as

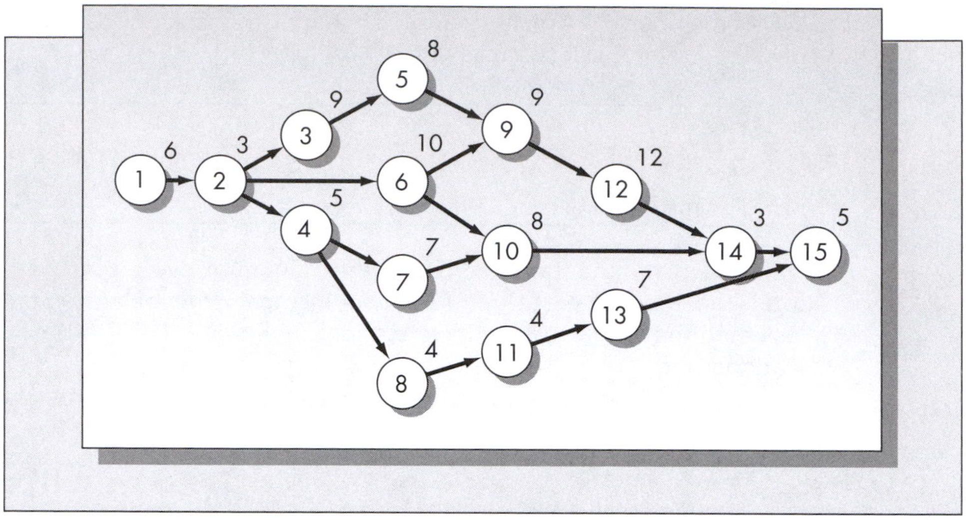

Figure 11–16 (For problem 27)

shown in Figure 11–16. The activity times appear next to the node numbers in the network.

a. Determine positional weights for each of the activities.
b. For a cycle time of 30 units, what is the minimum number of stations that could be achieved? Find the $C = 30$ balance obtained using the ranked positional weight technique.
c. Is there a solution with the same number of stations you found in part (b) but with a lower cycle time? In particular, what appears to be the minimum cycle time that gives a balance with the same number of stations you found in part (b)?

28. For the data given in problem 27, determine by experimentation the minimum cycle time for a three-station balance.
29. Consider the assembly line balancing problem represented by the network in Figure 11–17. The performance times are shown above the nodes.
    a. Determine a balance for C = 15.
    b. Determine a balance for $C = 20$.
    c. What appears to be the minimum cycle time that can be achieved using the number of stations you obtained in part (a)?
    d. What appears to be the minimum cycle time that can be achieved using the number of stations you obtained in part (b)?

11.11 HISTORICAL NOTES

One of the earliest monographs in the open literature that considered sequencing issues was written by Melvin Salveson (1952). The first significant published work to present analytical results for optimal sequencing strategies was authored by Selmer Johnson (1954). Richard Bellman's (1956) review article discusses sequence scheduling (among other topics) and presents an interesting proof of the optimality of Johnson's algorithm based on dynamic programming arguments. Bellman appears to be the first to have recognized the optimality of a permutation schedule for scheduling n jobs on three machines.

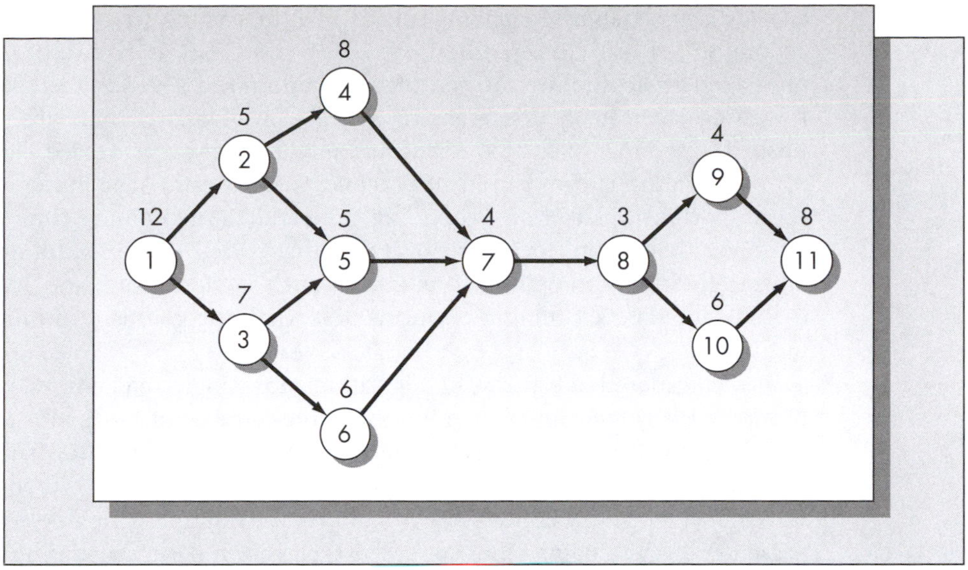

Figure 11–17 (For problem 29)

Johnson's work spawned considerable interest in scheduling. The excellent monograph by Richard Conway, William Maxwell, and Louis Miller (1967), for example, lists over 200 publications up until 1967. It is likely that considerably more than 200 papers have been published since that time. Much of the research in sequence scheduling has focused on stochastic scheduling. Gideon Weiss (1982) provides an interesting synopsis of work in this area. That jobs should be scheduled in order of decreasing expected processing time when scheduling is on two parallel processors and the objective is to minimize the expected makespan appears to have been discovered at about the same time by John Bruno and Peter Downey (1977) and Michael Pinedo and Weiss (1979).

Conway and colleagues (1967) also cover much of the work on dynamic scheduling under uncertainty. The rule, apparently discovered by Cyrus Derman, G. L. Lieberman, and Sheldon Ross (1978) in an entirely different context, led to several extensions. See, for example, Rhonda Righter and J. George Shanthikumar (1989).

11.12 SUMMARY

Scheduling problems arise in many contexts in managing operations. One of the areas explored in this chapter was optimal sequencing rules for scheduling jobs on machines, a problem that arises in *shop floor control*. We also considered dynamic scheduling problems, which are more typical of the kinds of scheduling problems that one encounters in management of service systems.

Much of the emphasis of the chapter was on determining efficient *sequencing rules*. The form of the optimal sequencing rule depends on several factors, including the pattern of arrivals of jobs, the configuration of the job shop, constraints, and the optimization objectives. Typical objectives in job shop management include meeting due dates, minimizing work-in-process, minimizing the average flow time, and minimizing machine idle time.

We discussed four sequencing rules: FCFS (first-come, first-served), SPT (shortest processing time first), EDD (earliest due date), and CR (critical ratio). For sequencing jobs on a single machine, we showed that SPT optimized several objectives, including mean flow time. However, SPT sequencing could be problematic. Long jobs would be constantly pushed to the rear of the job queue and might never be processed. For that reason, pure SPT scheduling is rarely used in practice. Critical ratio scheduling attempts to balance the importance placed on processing time and the remaining time until the due date. However, there is little evidence to suggest that critical ratio scheduling performs well relative to the common optimization criteria such as mean flow time. As one would expect, earliest-due-date scheduling performs best when the goal is to minimize the maximum tardiness.

We considered a variety of algorithms for single-machine sequencing, including Moore's (1968) algorithm for minimizing the number of tardy jobs and Lawler's (1973) algorithm for minimizing any nondecreasing function of the flow subject to precedence constraints. An excellent summary of algorithms for sequence scheduling can be found in Simon French (1982). Several techniques for multiple-machine scheduling also were presented in this chapter. Johnson's (1954) classic work on scheduling n jobs through two machines is discussed, as well as Akers's (1956) graphical method for two jobs through m machines.

Stochastic scheduling problems were treated in two contexts: static and dynamic. Static problems are stochastic counterparts to the problems treated earlier in the chapter, except that job completion times are random variables. Most of the simple results for static stochastic scheduling problems require the assumption that job times have the exponential distribution. Because of the memoryless property of the exponential distribution, this assumption is probably not satisfied in most manufacturing job shops. The dynamic case occurs when jobs arrive randomly over time. Queueing theory, discussed in detail in Chapter 8 and Supplement 2, is the means for analyzing such problems.

The assembly line balancing problem is one in which a collection of tasks must be performed on each item. Furthermore, the tasks must be performed in a specified sequence. The problem is to allot the tasks to workstations on an assembly line. The quality of the balance is measured by the idle time remaining at each station. Determining the best mix of cycle time (amount of time allotted to each station) and number of stations is an extremely difficult analytical problem. We discussed one very simple heuristic solution method, known as the ranked positional weight technique, which provides an efficient balance quickly.

While most of this chapter concerns analytical methods for developing schedules, it is important to keep in mind that many real scheduling problems are too complex to be modeled mathematically. In such cases, simulation can be a very valuable tool. A computer-based simulation is a model of a real process expressed as a computer program. By running simulations under different scenarios, consequences of alternative strategies can be evaluated easily. Simulations are particularly effective when significant randomness or variation exists.

There are a number of excellent texts on scheduling. For further reading in this area, we would suggest the book by Michael Pinedo (2016) and the book by Kenneth Baker and Dan Trietsch (2019).

ADDITIONAL PROBLEMS ON SCHEDULING

30. Mike's Auto Body Shop has five cars waiting to be repaired. The shop is quite small, so only one car can be repaired at a time. The number of days required to repair each car and the promised date for each are given in the following table.

Cars	Repair Time (days)	Promised Date
1	3	5
2	2	6
3	1	9
4	4	11
5	5	8

Mike has agreed to provide a rental car to each customer whose car is not repaired on time Compare the performance of the four sequencing rules FCFS, SPT, EDD, and CR relative to minimizing average tardiness.

31. For each of the problems listed, indicate precisely what or who would correspond to jobs and who or what would correspond to machines. In each case discuss what objectives might be appropriate and special priorities that might exist.
    a. Treating patients in a hospital emergency room.
    b. Unloading cargo from ships at port.
    c. Serving users on a time-shared computer system.
    d. Transferring long-distance phone calls from one city to another.
32. Six patients are waiting in a hospital emergency room to receive care for various complaints, ranging from a sprained ankle to a gunshot wound. The patients are numbered in the sequence in which they arrived and are given a priority weighting based on the severity of the problem. There is only one doctor available.

Patient	1	2	3	4	5	6
Time required	20 min	2 hr	30 min	10 min	40 min	1 hr
Priority	1	10	3	5	2	2

    a. Suppose that the patients are scheduled on an FCFS basis. Compute the mean flow time and the *weighted* mean flow time, where the weight is the priority number associated with each patient.
    b. Perform the calculations in part (a) for SPT sequencing.
    c. Determine a sequence different from those in parts (a) and (b) that achieves a lower value of the weighted mean flow time.
33. Consider Mike's Auto Body Shop mentioned in problem 30. What sequencing of the jobs minimizes the
    a. mean flow time?
    b. maximum lateness?
    c. number of tardy jobs?
34. Consider the situation of the emergency room mentioned in problem 32. Determine the sequence in which patients should be treated in order to minimize the *weighted* value of the mean flow time.

35. Barbara and Jenny's Ice Cream Company produces four different flavors of ice cream: vanilla, chocolate, strawberry, and peanut fudge. Each batch of ice cream is produced in the same large vat, which must be cleaned prior to switching flavors. One day is required for the cleaning process.

At the current time they have the following outstanding orders of ice cream.

Flavor	Order Size (gallons)	Due Date
Vanilla	385	3
Chocolate	440	8
Strawberry	200	6
Peanut fudge	180	12

It takes one day to produce the ice cream, and a maximum of 100 gallons can be produced at one time. Cleaning is required only when flavors are switched. The production for one flavor will be completed prior to beginning production for another flavor. Cleaning begins at the start of a day.

Treating each ice cream flavor as a different job, find the following.

a. The sequence in which the ice cream flavors should be produced in order to minimize the mean flow time for all of the flavors.

b. The optimal sequence to produce the flavors in order to minimize the number of flavors that are late.

36. Consider Barbara and Jenny's Ice Cream Company, mentioned in problem 35. Suppose that if vanilla or strawberry is produced after chocolate or peanut fudge, an extra day of cleaning is required. For that reason, they decide that the vanilla and strawberry will be produced before the chocolate and peanut fudge.

a. Find the optimal sequencing of the flavors to minimize the maximum lateness using Lawler's algorithm.

b. Enumerate all the feasible sequences and determine the sequence that minimizes the maximum lateness by evaluating and comparing the objective function value for each case.

37. Irving Bonner, an independent computer programming consultant, has contracted to complete eight computer programming jobs. Some jobs must be completed in a certain sequence because they involve program modules that will be linked.

Job	Time Required (days)	Due Date
1	4	June 8
2	10	June 15
3	2	June 10
4	1	June 12
5	8	July 1
6	3	July 6
7	2	June 25
8	6	June 29

Precedence restrictions:

$$1 \rightarrow 2 \rightarrow 5 \rightarrow 6.$$
$$4 \rightarrow 7 \rightarrow 8.$$

Assume that the current date is Monday, June 1, and that Bonner does not work on weekends. Using Lawler's algorithm, find the sequence in which he should be performing the jobs in order to minimize maximum lateness subject to the precedence constraints.

38. William Beebe owns a small shoe store. He has 10 pairs of shoes that require resoling and polishing. He has a machine that can resole one pair of shoes at a time, and the time required for the operation varies with the type and condition of the shoe and the type of sole that is used. Shoes are polished on a machine dedicated to this purpose as well, and polishing is always done after resoling. His assistant generally does the polishing while Mr. Beebe does the resoling. The resoling and polishing times (in minutes) are as follows.

Shoes	Resoling Time	Polishing Time
1	14	3
2	28	1
3	12	2
4	6	5
5	10	10
6	14	6
7	4	12
8	25	8
9	15	5
10	10	5

In what order should the shoes be repaired in order to minimize the total makespan for these 10 jobs?

39. A leatherworks factory has two punch presses for punching holes in the various leather goods produced in the factory prior to sewing. Suppose that 12 different jobs must be processed on one or the other punch press (i.e., parallel processing). The processing times (in minutes) for these 12 jobs are given in the following table.

Job	1	2	3	4	5	6	7	8	9	10	11	12
Time	26	12	8	42	35	30	29	21	25	15	4	75

Assume that both presses are initially idle. Compare the performance of SPT and LPT (longest processing time) rules for this example. (In parallel processing the next job is simply scheduled on the next available machine.)

40. An independent accountant is planning to prepare tax returns for six of her clients. Prior to her actually preparing each return, her secretary checks the client's file to be sure all the necessary documentation is there and obtains all the tax forms needed for the preparation of the return. Based on past experience with the clients, her secretary estimates that the following times (in hours) are required for preparation of the return and for the accountant to complete the necessary paperwork prior to filing each return.

Client	Secretary Time	Accountant Time
1	1.2	2.5
2	1.6	4.5
3	2.0	2.0
4	1.5	6.0
5	3.1	5.0
6	0.5	1.5

In what order should the work be completed in order to minimize the total time required for all six clients?

41. Five Hong Kong tailors—Simon, Pat, Choon, Paul, and Wu—must complete alterations on a suit for the duke and a dress for the duchess as quickly as possible. On the dress, Choon must first spend 45 minutes cutting the fabric, then Pat will spend 75 minutes sewing the bodice, Simon will need 30 minutes stitching the sleeves, Paul 2 hours lacing the hem, and finally Wu will need 80 minutes for finishing touches. For the suit, Pat begins by shortening the sleeves, which requires 100 minutes. Paul then sews in the lining in 1.75 hours; Wu next spends 90 minutes sewing on the buttons and narrowing the lapels; finally, Choon presses and cleans the suit in 30 minutes. Determine precisely when each tailor should be performing each task in order to minimize the total time required to complete the dress and the suit. Assume that the tailors start working at 9 AM and take no breaks. Draw the Gantt chart indicating your solution.
42. The assembly of a transistorized clock radio requires a total of 11 tasks. The task times and predecessor relationships are given in the following table.

Task	Time (seconds)	Immediate Predecessors
1	4	
2	38	
3	45	
4	12	1, 2
5	10	2
6	8	4
7	12	5
8	10	6
9	2	7
10	10	8, 9
11	34	3, 10

a. Develop a network for this assembly operation.
b. What is the minimum cycle time that could be considered for this operation? What is the minimum number of stations that could be used with this cycle time?
c. Using the ranked positional weight technique, determine the resulting balance using a cycle time of 45 seconds.

d. Determine by experimentation the minimum cycle time that results in a four-station balance.
e. What is the daily production rate for this product if the company adopts the balance you determined in part (c)? (Assume a six-hour day for your calculations.) What would have to be done if the company wanted a higher rate of production?

43. Suppose in problem 42 that additional constraints arise from the fact that certain tasks cannot be performed at the same station. In particular suppose that the tasks are zoned as follows.

Zone 1	Tasks 2, 3, 1, 4, 6
Zone 2	Tasks 5, 8, 7, 9
Zone 3	Tasks 10, 11

Assuming that only tasks in the same zone category can be performed at the same station, determine the resulting line balance for problem 42 based on a 45-second cycle time.

44. The Southeastern Sports Company produces golf clubs on an assembly line in its plant in Marietta, Georgia. The final assembly of the woods requires the eight operations given in the following table.

Task	Time Required (min.)	Immediate Predecessors
Polish shaft	12	
Grind the shaft end	14	
Polish club head	6	
Imprint number	4	3
Connect wood to shaft	6	1, 2, 4
Place and secure connecting pin	3	5
Place glue on other end of shaft	3	1
Set in grips and balance	12	6, 7

a. Draw a network to represent the assembly operation.
b. What is the minimum cycle time that can be considered? Determine the balance that results from the ranked positional weight technique for this cycle time.
c. By experimentation, determine the minimum cycle time that can be achieved with a three-station balance.

45. Develop a template that computes several measures of performance for first-come, first-served job sequencing. Allow for up to 20 jobs so that column 1 holds the numbers 1, 2, … , 20. Column 2 should be the processing times to be inputted by the user, and column 3 the due dates also inputted by the user. Column 4 should be the tardiness and column 5 the flow time. Develop the logic to compute the mean flow time, the average tardiness, and the number of tardy jobs. (When computing the average of a column, be sure that your spreadsheet does not treat blanks as zeros.)
a. Use your template to find the mean flow time, average tardiness, and number of tardy jobs for problem 7 (Section 11.6), assuming FCFS sequencing.

b. Find the mean flow time, average tardiness, and number of tardy jobs for problem 8, assuming FCFS sequencing.

c. Find the mean flow time, average tardiness, and number of tardy jobs assuming FCFS sequencing for the following 20-job problem.

Job	Processing Time	Due Date	Job	Processing Time	Due Date
1	10	34	11	17	140
2	24	38	12	8	120
3	16	60	13	23	110
4	8	52	14	25	160
5	14	25	15	40	180
6	19	95	16	19	140
7	26	92	17	6	130
8	24	61	18	23	190
9	4	42	19	25	220
10	12	170	20	14	110

46. a. Solve problem 45(a) assuming SPT sequencing. To do this you may use the spreadsheet developed in problem 45 and simply sort the data in the first three columns, using column 2 (the processing time) as a sort key.

b. Solve problem 45(b) assuming SPT sequencing.

c. Solve problem 45(c) assuming SPT sequencing.

47. a. Solve problem 45(a) assuming EDD sequencing. In this case one sorts on the due date column.

b. Solve problem 45(b) assuming EDD sequencing.

c. Solve problem 45(c) assuming EDD sequencing.

BIBLIOGRAPHY

Akers, S. B. "A Graphical Approach to Production Scheduling Problems." *Operations Research* 4 (1956), pp. 244–45.

Arcus, A. L. "COMSOAL: A Computer Method for Sequencing Operations for Assembly Lines." *International Journal of Production Research* 4 (1966), pp. 259–77.

Baker, K. R., and D. Triestch. *Introduction to Sequencing and Scheduling,* 2nd ed. New York: John Wiley & Sons, 2019.

Banerjee, B. P. "Single Facility Sequencing with Random Execution Times." *Operations Research* 13 (1965), pp. 358–64.

Bellman, R. E. "Mathematical Aspects of Scheduling Theory." *SIAM Journal of Applied Mathematics* 4 (1956), pp. 168–205.

Bruno, J., and P. Downey. "Sequencing Tasks with Exponential Service Times on Two Machines." Technical Report, Department of Electrical Engineering and Computer Science, University of California at Santa Barbara, 1977.

Conway, R. W., W. L. Maxwell, and L. W. Miller. *Theory of Scheduling*. Reading, MA: Addison Wesley, 1967.

Derman, C., G. J. Lieberman, and S. M. Ross. "A Renewal Decision Problem." *Management Science* 24 (1978), pp. 554–61.

French, S. *Sequencing and Scheduling: An Introduction to the Mathematics of the Job Shop*. Chichester, England: Ellis Horwood Limited, 1982.

Held, M., R. M. Karp, and R. Shareshian. "Assembly Line Balancing—Dynamic Programming with Precedence Constraints." *Operations Research* 11 (1963), pp. 442–59.

Helgeson, W. P., and D. P. Birnie. "Assembly Line Balancing Using the Ranked Positional Weight Technique." *Journal of Industrial Engineering* 12 (1961), pp. 394–98.

Hillier, F. S., and R. W. Boling. "On the Optimal Allocation of Work in Symmetrically Unbalanced Production Line Systems with Variable Operation Times." *Management Science* 25 (1986), pp. 721–28.

Johnson, S. M. "Optimal Two and Three Stage Production Schedules with Setup Times Included." *Naval Research Logistics Quarterly* 1 (1954), pp. 61–68.

Kilbridge, M. D., and L. Wester. "A Heuristic Method of Line Balancing." *Journal of Industrial Engineering* 12 (1961), pp. 292–98.

Lawler, E. L. "Optimal Seque ncing of a Single Machine Subject to Precedence Constraints." *Management Science* 19 (1973), pp. 544–46.

Martin, C., D. Jones, and P. Keskinocak. "Optimizing On-Demand Aircraft Schedules for Fractional Aircraft Operators," *Interfaces*, 33, no. 5, September–October 2003, pp. 22–35.

Moore, J. M. "An n-job, One Machine Sequencing Algorithm for Minimizing the Number of Late Jobs." *Management Science* 15 (1968), pp. 102–109.

Parker, K. "Dynamism and Decision Support." *Manufacturing Systems* 13, no. 4 (April 1995), pp. 12–24.

Pinedo, M. "Stochastic Scheduling with Release Dates and Due Dates." *Operations Research* 31 (1983), pp. 554–72.

Pinedo, M. *Scheduling, Theory, Algorithms and Systems,* 5th ed. New York: Springer Science + Business Media, 2016.

Pinedo, M., and G. Weiss. "Scheduling Stochastic Tasks on Two Parallel Processors." *Naval Research Logistics Quarterly* 26 (1979), pp. 527–36.

Righter, R. "Job Scheduling to Minimize Weighted Flowtime on Uniform Processors." *Systems and Control Letters* 10 (1988), pp. 211–16.

Righter, R., and J. G. Shanthikumar. "Scheduling Multiclass Single-Server Queueing Systems to Stochastically Maximize the Number of Successful Departures." *Probability in the Engineering and Informational Sciences* 3 (1989), pp. 323–33.

Rosenblatt, M. and R. Carlson. "Designing a Production Line to Maximize Profit." *HE Transactions* 17 (1985), pp. 117–22.

Rothkopf, M. H. "Scheduling with Random Service Times." *Management Science* 12 (1966), pp. 707–13.

Salveson, M. E. "Production Planning and Scheduling." *Econometrica* 20 (1952), pp. 554-90.

Thangavelu, S. R., and C. M. Shetty. "Assembly Line Balancing by Zero One Integer Programming." *AIIE Transactions* 3 (1971), pp. 61–68.

Weiss, G. "Multiserver Stochastic Scheduling." *Deterministic and Stochastic Scheduling* 84 (1982), pp. 157–179)

CHAPTER

12 Project Scheduling

"Man does not plan to fail, he just fails to plan."

—Anonymous

CHAPTER OVERVIEW

Purpose

To understand how mathematical and graphical techniques are used to assist with the task of scheduling complex projects in an organization.

KEY POINTS

1. *Project representation.* There are two convenient graphical techniques for representing a project. One is a Gantt chart and the other is with a network. The Gantt chart was used in Chapter 11 to represent sequence schedules on multiple machines. However, representing a project as a Gantt chart has one significant drawback. Precedence relationships (that is, specifying which activities must precede other activities) are not displayed. To overcome this inadequacy, we often represent a project as a network rather than a Gantt chart.
2. *Network representation and critical path identification.* In the network representation of a project, the goal is to identify the critical, or longest, path. A network is a set of nodes and directed arcs. Nodes correspond to milestones in the project (completion of some subset of activities) and arcs to specific activities. In the spirit of "a chain is only as strong as its weakest link," a project cannot be completed until all the activities along the critical path are completed. The length of the critical path gives the earliest completion time of the project. Activities not along the critical path (noncritical activities) have slack time—that is, they can be delayed without necessarily delaying the project. In Section 12.2, we present an algorithm for identifying the critical path in a network. (This is only one of several solution methods.)
3. *Time costing methods.* Consider a construction project. Each additional day that elapses results in higher costs. These costs include direct labor costs for the personnel involved in the project, costs associated with equipment and material usage, and overhead costs. Suppose one has the option of decreasing the time of selected activities at some cost. As the times required for activities along the critical path are decreased, the expediting costs increase but the costs proportional to the project time decrease. Hence, there is some optimal time for the project that balances these two competing costs. The problem of cost-optimizing the time of a project can be solved manually or via linear programming.

4. *Project scheduling with uncertain activity times.* In some projects, such as construction projects, the time required to do specific tasks can be predicted accurately in advance. In most cases, past experience can be used as an accurate guide, even for novel projects. However, this is not the case with research projects. When undertaking the solution of an unsolved problem, or designing an entirely new piece of equipment, it is difficult, if not impossible, to predict activity times accurately in advance. A more reasonable assumption is that activity times are random variables with some specified distribution. These distributions can then be used for the PERT scheduling method or in a Monte Carlo simulation.
5. *PERT scheduling.* A method that explicitly allows for uncertain activity times is the project evaluation and review technique (PERT). This technique was developed by the Navy to assist with planning the Polaris submarine project in 1958. Using the PERT approach, planners estimate a minimum time, a, a maximum time, b, and a most likely time, m, for each activity. These estimates are then used to construct a beta distribution for the time of each activity. The PERT assumption is that the critical path will be the path with the longest expected completion time (which is not necessarily the case), and the total project time will be the sum of the times along the critical path. Assuming activity times are independent random variables, one computes the mean and variance along the critical path by summing the means and variances of the activity times. The central limit theorem is then used to justify the assumption that the project completion time has a normal distribution with mean and variance computed as previously described. Note that this is only an approximation, since there is no guarantee that the path with the longest expected completion time will turn out to be the critical path. Determining the true distribution of project completion time appears to be a very difficult problem in general. However, PERT provides a reasonable approximation and is certainly an improvement over the deterministic critical path method (CPM).
6. *Resource considerations.* Consider a department within a firm in which several projects are simultaneously ongoing. Suppose that each member of the department is working on more than one project at a time. Since the time of each worker is limited, each project manager is competing for a limited resource, namely, the time of the workers. One could imagine other cases where the limited resource might be a piece of equipment, such as a single supercomputer in a company. In these cases, incorporating resource constraints into the project planning function can be quite a challenge. We present an example of balancing resources but know of no general-purpose method for solving this problem.

Rome wasn't built in a day, and neither were the pyramids of Egypt, the Empire State Building, the Golden Gate Bridge, or the Eiffel Tower. These were all complex projects that required careful planning and coordinating. How does one organize and monitor such massive projects? Effective project management could make or break a project. While many different skills are required to be an effective project manager, quantitative techniques–the subject of this chapter—can be an enormous help.

What are the consequences of poor project management? One is cost overruns. How often have we heard members of Congress express dismay at the cost overruns in military projects? In some cases, these overruns could not be avoided. Unforeseen obstacles arose or technological problems could not be solved as easily as anticipated. In many cases, however, these problems were a consequence of poor project scheduling and management.

Large complex projects involving governments in partnership with business are perhaps the most vulnerable to delays and overruns. A case in point was the Trans-Alaska Pipeline, designed to transport large quantities of oil from Prudhoe Bay on Alaska's north slope to Port Valdez on the Gulf of Alaska. It was an extremely complex and massive design and construction project (Goodman, 1988). Political roadblocks, environmental concerns, and contract disputes plagued the project from the start. The list of important players changed several times, resulting in some firms (Bechtel, in particular) not having enough time to do an adequate job of project planning. The Alyeska Pipeline Service Company, which was responsible for much of the actual building of the pipeline, incurred excessive cost overruns, partly due to poor project management. In retrospect, it is clear that many of the problems with the project were a consequence of the fact that there was never a single project team to oversee the entire integrated project cycle.

Project scheduling and project management methods have been an important part of doing business for many firms. For example, Lockheed Martin, headquartered in Bethesda, Maryland, is a global security and aerospace company. The majority of Lockheed Martin's business is with the U.S. Department of Defense and U.S. federal government agencies. The company uses project scheduling methods not only for monitoring and control of projects but also for the process of preparing bids and developing proposals. In fact, Lockheed was part of the team that developed PERT, a technique considered in detail in this chapter.

This chapter reviews analytical techniques for project management. Effective people management is an important factor in getting projects done on time and within budget. The firm must create a structure and environment conducive to properly motivating employees. Poor organizational design and incentive structures can be just as serious a problem as poor project planning.

The project management methods considered in this chapter have been used to plan long-term projects such as launching new products, organizing research projects, and building new production facilities. The methods also have been used for smaller projects such as building residential homes. Two techniques that are treated in detail are the **critical path method (CPM)** and the **project evaluation and review technique (PERT).** Both methods were developed almost simultaneously for solving very different project management problems in the late 1950s. Although the two labels are used interchangeably today, we will retain the terminology consistent with the original intent of the methods. That is, CPM deals with purely deterministic problems, whereas PERT allows randomness in the activity times.

The following are primary elements of critical path analysis.

1. *Project definition.* This is a clear statement of the project, the goals of the project, and the resources and personnel that the project requires.
2. *Activity definitions.* The project must be broken down into a set of indivisible tasks or activities. The project planner must specify the work content of each activity and estimate the activity times. Often the most difficult part of project planning is finding the best way to break down the project into a collection of distinct activities.
3. *Activity relationships.* An important part of the project planning methodology is a specification of the interrelationships among the activities. Known as **precedence constraints,** these describe the logical sequence to complete the activities comprising the project.
4. *Project scheduling.* A **project schedule** is a specification of the starting and ending times of *all* activities comprising the project. Using the techniques of this chapter, we will show how specification of the activity times and precedence constraints yields an efficient schedule.

5. *Project monitoring.* Once the activities have been suitably defined and a schedule determined, the proper controls must be put in place to ensure that project milestones are met. The project manager must be prepared to revise existing schedules if unforeseen problems arise.

Because of the level of detail and precision required, critical path analysis is most effective when the project can be easily expressed as a well-defined group of activities. Construction projects fall into this category. Precedence constraints are straightforward, and activity times are not difficult to estimate. For this reason, CPM has found wide acceptance in the construction industry. However, there are case studies reported in the literature of successful applications in a variety of environments, including military, government, and nonprofit.

12.1 REPRESENTING A PROJECT AS A NETWORK

As we saw in Chapter 11 on shop floor scheduling and control, a **Gantt chart** is a convenient graphical means of picturing a schedule. A Gantt chart is a horizontal bar graph on which each activity corresponds to a separate bar. Consider the following simple example.

EXAMPLE 12.1

Suppose that a project consists of five activities, A, B, C, D, and E. Figure 12–1 shows a schedule for completing this project, in the form of a Gantt chart. According to the figure, task A starts at 12:00 and finishes at 1:30. Tasks B and C start at 1:30, B finishes at 2:30, C finishes at 3:30, and so on. The project is finally completed at 6:00.

Notice that Figure 12–1 gives no information about the relationships among the activities. For example, the figure implies that E cannot start until D is completed. However, suppose that E is permitted to start any time after A, B, and C finish. Then E can start at 3:30, and the project can be completed at 5:00 rather than at 6:00.

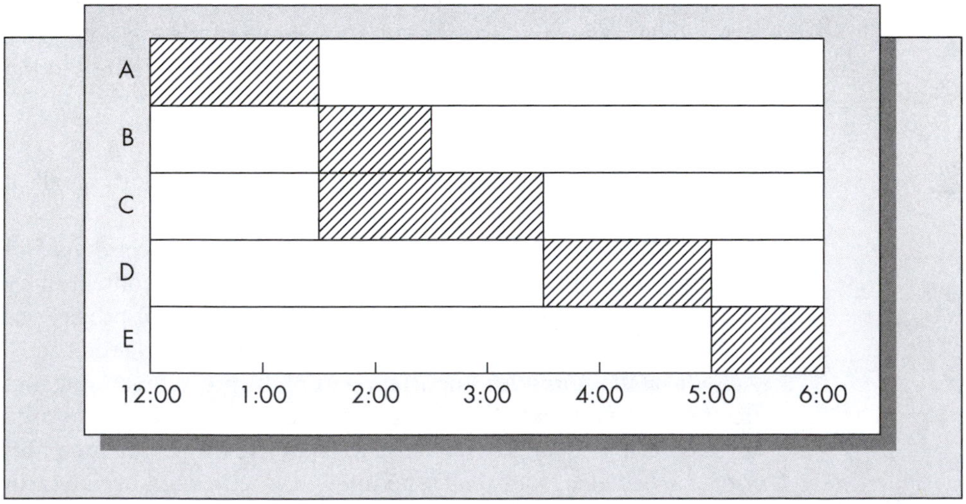

Figure 12–1 Gantt chart for five activities (Example 12.1)

Although the Gantt chart is a useful means of representing the schedule, it is not a very useful planning tool because it does not show the precedence constraints. For this reason, one graphically represents the project as a network. Networks, unlike Gantt charts, explicitly show the precedence constraints.

A **network** is a collection of nodes and directed arcs. In the traditional network representation of a project, a node corresponds to an event and an arc to an activity or task. Events may be (1) the start of a project, (2) the completion of a project, or (3) the completion of some collection of activities. This method of representation is known as **activity-on-arrow** and is historically the most common means of representing a project as a network. An alternative method of representation is **activity-on-node**. Although the latter method has some advantages, it is rarely used in practice. The activity-on-node method will be illustrated in Section 12.2.

EXAMPLE 12.2

Suppose a project consists of the five activities A, B, C, D, and E that satisfy the following precedence relationships.

1. Neither A nor B has any immediate predecessors.
2. A is an immediate predecessor of C.
3. B is an immediate predecessor of D.
4. C and D are immediate predecessors of E.

The network for this project appears in Figure 12–2. Node 1 is always the initiation of the project. All activities having no immediate predecessors emanate from node 1. As A is an immediate predecessor of C, we must have a node representing the completion of A. Similarly, as B is an immediate predecessor of D, there must be a node representing the completion of B. Notice that although both A and B must also be completed before E, they are not *immediate* predecessors of E.

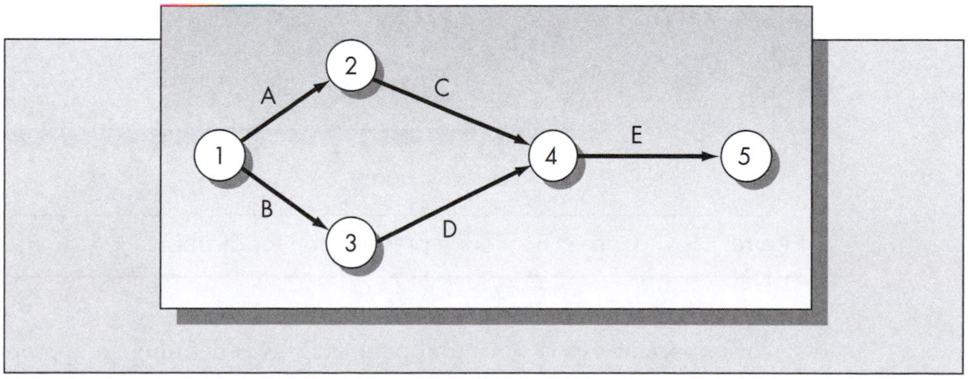

Figure 12–2 Project network for Example 12.2

Representing a project by a network is not always this straightforward.

EXAMPLE 12.3

Consider Example 12.2, except now replace (3) with (3′) and the precedence relationships below.

3′. A and B are immediate predecessors of D.

Try to find a network representation of this project. One might think that the correct representation is the one in Figure 12–3, as this network implies that D must wait for both A and B. However, this representation is incorrect, because it also shows that C must wait for both A and B as well.

The set of precedent relations (1), (2), (3′), and (4) require that we introduce a pseudo activity between nodes 2 and 3. The correct representation of the system is given in Figure 12–4. The pseudo activity, labeled P, is a directed arc from node 2 to node 3 and is represented by a broken line. Note that the direction of the arrow is important. With the pseudo activity, node 3 corresponds to the completion of both activities A and B, whereas node 2 still corresponds to the completion of only activity A.

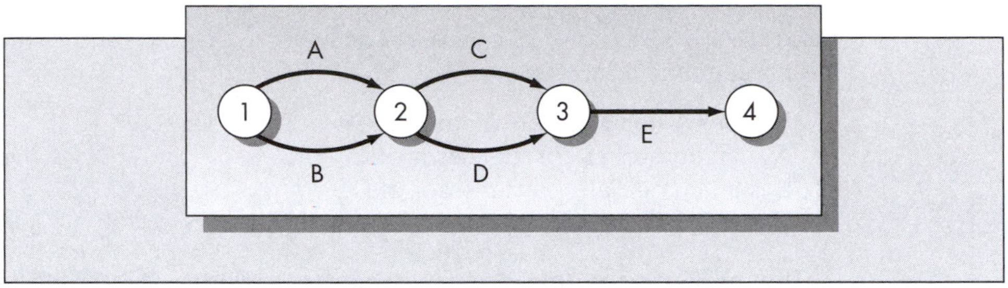

Figure 12–3 Incorrect network representation for Example 12.3

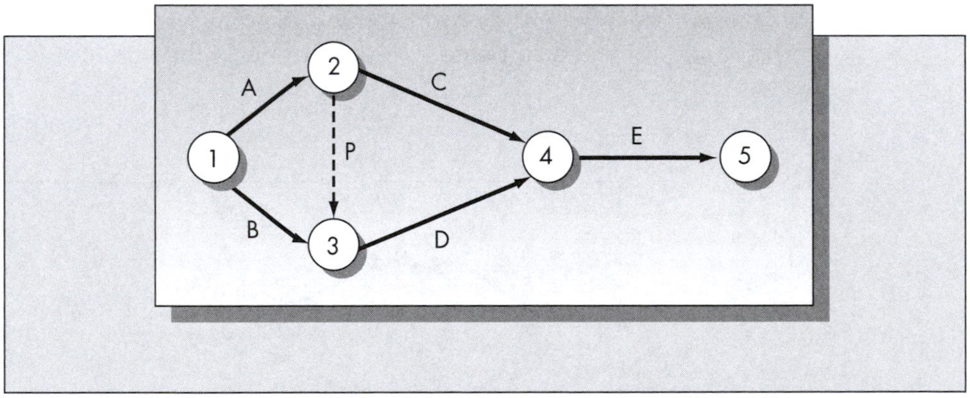

Figure 12–4 Correct network representation for Example 12.3

An important aspect of critical path analysis is defining an appropriate set of tasks. A problem that may arise is that one defines the tasks too broadly. Suppose in Example 12.3 that E could start when only half of C was completed. This means that C would have to be further broken down into two other activities, one representing the first half of C and one representing the second half of C. Conversely, it is possible to define activities too narrowly. This can result in an overly complicated network representation, with portions of the network resembling Figure 12–3.

12.2 CRITICAL PATH ANALYSIS

Once having represented the project as a network, we can begin to explore the methods available for answering a variety of questions. For example,

1. What is the minimum time required to complete the project?
2. What are the starting and ending times for each of the activities?
3. Which activities can be delayed without delaying the project?

A path through the network is a sequence of activities from node 1 to the final node. One labels a path by the sequence of nodes visited or by the sequence of arcs (activities) traversed. In the network pictured in Figure 12–4 there are exactly three distinct paths: 1–2–4–5 (A–C–E), 1–2–3–4–5 (A–P–D–E), and 1–3–4–5 (B–D–E). Each path is a sequence of activities satisfying the precedence constraints. Because the project is completed when all activities are completed, it follows that *all* paths from the initial node to the final node must be traversed. Hence, the minimum time to complete the project must be the same as the length of the *longest* path in the network.

Consider Example 12.3 and suppose that activity times are the same as those indicated in Figure 12–1. The activity times and the precedence constraints follow.

Activity	Time (hours)	Immediate Predecessors
A	1.5	
B	1.0	
C	2.0	A
D	1.5	A and B
E	1.0	C and D

The lengths of the three paths are listed below.

Path	Time Required to Complete (hours)
A–C–E	4.5
A–P–D–E	4.0
B–D–E	3.5

To complete the project, all three paths must be completed. The longest path for this example is obviously A–C–E. This is known as the **critical path**. The **length** of the critical path is the *minimum* completion time of the project. The activities that lie along the critical path are known as the critical activities. A delay in a critical activity results in a delay in the project. However, activities that do not lie along the critical path have some slack. **Slack** is the amount of time that an activity can be delayed without delaying the project. A delay in a noncritical activity does not necessarily delay the project.

Enumerating all paths is, in general, not an efficient way to find the critical path. Later, we consider a general procedure for identifying the critical path and the start and finish times for all activities comprising the project that does *not* require path enumeration. First, we consider a programming project in Example 12.4.

EXAMPLE 12.4

Simon North and Irving Bonner, computer consultants, are considering embarking on a joint project that will involve development of a relatively small commercial software package for personal computers. The program involves scientific calculations for a specialized portion of the engineering market. North and Bonner have broken down the project into nine tasks.

The first task is to undertake a market survey in order to determine exactly what the potential clientele will require and what features of the software are likely to be the most attractive. Once this stage is completed, the actual development of the program can begin. The programming requirements fall into two broad categories: graphics and source code. Because the system will be interactive and icon driven, the first task is to identify and design the icons. After the programmers have completed the icon designs, they can proceed with the second part of the graphics development, design of the input/output screens. These include the various menus and report generators required in the system.

The second part of the project is coding the modules that do the scientific calculations. The first step is to develop a detailed flowchart of the system. After they complete the flowchart, the programmers can begin work on the modules. There are a total of four modules. The work on modules 1 and 2 can begin immediately after completion of the flowchart. Module 3 requires parts of module 1, so the work on module 3 cannot begin until module 1 is finished. The programming of module 4 cannot start until both modules 1 and 2 are completed. Once the graphics portion of the project and the modules are completed, the two separate phases of the system must be merged and the entire system tested and debugged.

North has managed to obtain some funding for the project, but his source requires that the program be completed and ready to market in 25 weeks. In order to determine whether this is feasible, the two programmers have divided the project into nine indivisible tasks and have estimated the time required for each task. The list of these tasks, the times required, and the precedence relationships are given below.

Task	Time Required (weeks)	Immediate Predecessors
A. Perform market survey	3	
B. Design graphic icons	4	A
C. Develop flowchart	2	A
D. Design input/output screens	6	B, C
E. Module 1 coding	5	C
F. Module 2 coding	3	C
G. Module 3 coding	7	E
H. Module 4 coding	5	E, F
I. Merge modules and graphics and test program	8	D, G, H

The total of the task times is 43 weeks. Based on this, the programmers conclude that it would be impossible to complete their project in the required 25 weeks. Fortunately, they see the error in their thinking before breaking off relations with their

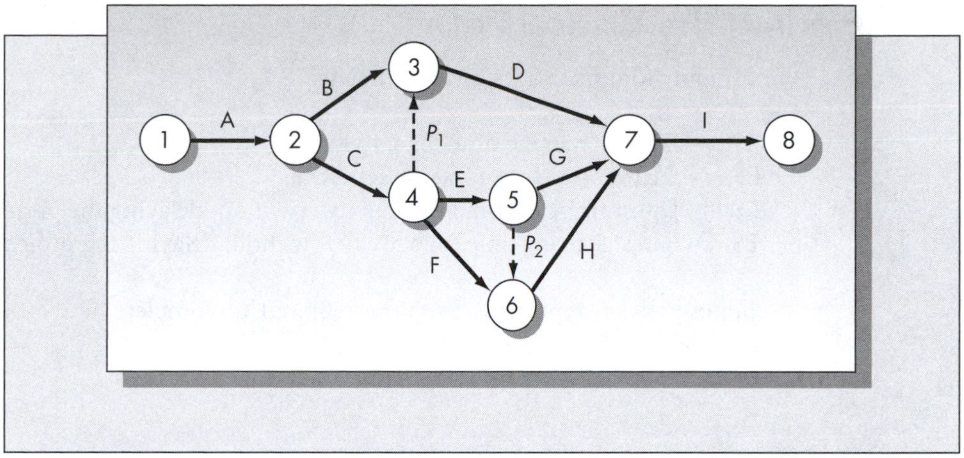

Figure 12–5 Network for Example 12.4

funding source. Since some of the activities can be done concurrently, the project should take fewer than 43 weeks. The network representation for this project is given in Figure 12–5. In order to satisfy the precedence constraints we require two pseudo activities, P_1 and P_2. We will determine the critical path *without* enumerating all paths through the network. The critical path calculations yield both the critical path *and* the allowable slack for each activity as well.

For illustrative purposes, we represent the project network using the activity-on-node format in Figure 12–6. One advantage of this format is that pseudo activities are not required, although for certain networks dummy starting and ending nodes may be needed. Given a list of activities and immediate predecessors, it should be easy to find the network representation in either format.

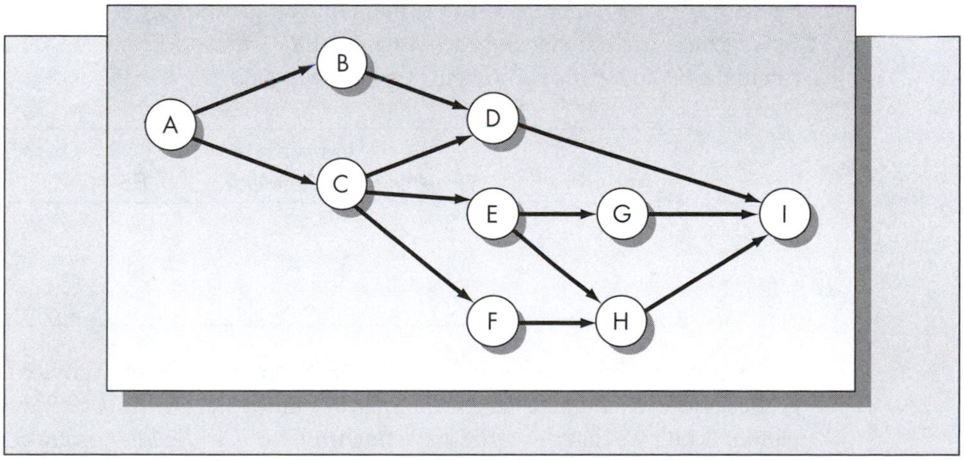

Figure 12–6 Network for Example 12.4: Activity-on-node representation

Practitioners prefer the activity-on-arrow method, believing that it is a more intuitive way to represent the project. We will use it exclusively for the remainder of the chapter.

Finding the Critical Path

We compute four quantities for each activity:

ES_i = Earliest starting time for activity *i*,
EF_i = Earliest finishing time for activity *i*,
LF_i = Latest finishing time for activity *i* (without delaying the project),
LS_i = Latest starting time for activity *i* (without delaying the project).

Suppose that t_i represents the time required to complete activity *i*. Then it is easy to see that

$$\mathrm{EF}_i = \mathrm{ES}_i + t_i$$

and

$$\mathrm{LS}_i = \mathrm{LF}_i - t_i.$$

There are two steps in the process.

1. *Compute the earliest times for each activity.* One computes the earliest times by a forward pass through the network; that is, the computations proceed from node 1 to the final node.
2. *Compute the latest times for each activity.* One computes the latest times by a backward pass through the network; that is, the computations proceed from the final node back to node 1.

We will illustrate the computations using Example 12.4. The first step is to set the earliest starting times for all activities emanating from node 1 to zero and add the activity times to obtain the earliest finishing times. In this case, the only activity emanating from node 1 is activity A. Because B has A as an immediate predecessor, it follows that $\mathrm{ES}_\mathrm{B} = \mathrm{EF}_\mathrm{A}$. Similarly for C, $\mathrm{ES}_\mathrm{C} = \mathrm{EF}_\mathrm{A}$. Compute the earliest finishing times for B and C by adding the activity times to the earliest starting times: $\mathrm{EF}_\mathrm{B} = \mathrm{ES}_\mathrm{B} + t_\mathrm{B}$, and $\mathrm{EF}_\mathrm{C} = \mathrm{ES}_\mathrm{C} + t_\mathrm{C}$. Summarizing the calculations up to this point yields the following.

Activity	Time	Immediate Predecessor	ES	EF
A	3	—	0	3
B	4	A	3	7
C	2	A	3	5

Calculating the earliest starting time for D is more difficult. Activity D has two immediate predecessors, B and C. That means that D cannot start until *both* B and C have been completed. It follows that the earliest starting time for D is the *later* of the earliest finishing times for B and C. That is,

$$\mathrm{ES}_\mathrm{D} = \max(\mathrm{EF}_\mathrm{B}, \mathrm{EF}_\mathrm{C}) = \max(7, 5) = 7.$$

General Rule: The earliest starting time of an activity is the *maximum* of the earliest finishing times of its immediate predecessors.

Continuing in this manner, we obtain the following earliest times for all remaining activities.

Activity	Time	Immediate Predecessor	ES	EF
A	3	—	0	3
B	4	A	3	7
C	2	A	3	5
D	6	B, C	7	13
E	5	C	5	10
F	3	C	5	8
G	7	E	10	17
H	5	E, F	10	15
I	8	D, G, H	17	25

At this point we have actually determined the length of the critical path. It is the maximum of the earliest finish times, or 25 weeks in this case. However, we must find the latest times before we can identify the critical activities.

The computation of the latest times proceeds by working backward through the network and applying essentially the dual of the procedure for the earliest times. The first step is to set the latest finishing time of all the activities that enter the final node to the maximum value of the earliest finishing times. In this case there is only a single ending activity, so we set

$$LF_I = 25.$$

The latest starting time is obtained by subtracting the activity time, so

$$LS_I = 25 - 8 = 17.$$

Next we must determine the latest finishing time for all the activities that enter node 7, which are D, G, and H. Because these activities end when I begins, we have

$$LF_D = LF_G = LF_H = LS_I = 17.$$

One finds the latest starting times for D, G, and H by simply subtracting the activity times. The following summarizes the calculations up to this point.

Activity	Time	Immediate Predecessor	ES	EF	LS	LF
A	3	—	0	3		
B	4	A	3	7		
C	2	A	3	5		
D	6	B, C	7	13	11	17
E	5	C	5	10		
F	3	C	5	8		
G	7	E	10	17	10	17
H	5	E, F	10	15	12	17
I	8	D, G, H	17	25	17	25

Because F ends when H begins, $LF_F = LS_H = 12$, and $LS_F = 12 - 3 = 9$. Now consider activity E. From Figure 12–5, E has both G and H as immediate successors. This means that E must end prior to the time that both G and H start. Hence the latest finishing time for E is the *earlier* of the latest start times for G and H. That is, $LF_E = \min(LS_G, LS_H) = \min(10, 12) = 10$.

General Rule: The latest finishing time of an activity is the *minimum* of the latest start times of its immediate successors.

According to the network diagram of Figure 12–5, C has both E and F as immediate successors. Hence $LF_C = \min(LS_E, LS_F) = \min(5, 9) = 5$, and $LS_C = 5 - 2 = 3$. Because B has only D as an immediate successor, $LF_B = LS_D = 11$, and $LS_B = 11 - 4 = 7$. Finally, A has both B and C as immediate successors, so $LF_A = \min(LS_B, LS_C) = \min(7, 3) = 3$, and $LS_A = 3 - 3 = 0$.

The complete summary of the calculations for this example follows.

Activity	Time	Immediate Predecessors	ES	EF	LS	LF	Slack
A	3	—	0	3	0	3	0
B	4	A	3	7	7	11	4
C	2	A	3	5	3	5	0
D	6	B, C	7	13	11	17	4
E	5	C	5	10	5	10	0
F	3	C	5	8	9	12	4
G	7	E	10	17	10	17	0
H	5	E, F	10	15	12	17	2
I	8	D, G, H	17	25	17	25	0

We have added a column labeled "slack." This is the difference between the columns LS and ES (it is also the difference between the columns LF and EF). The activities with zero slack—A, C, E, G, and I—are the critical activities and constitute the critical path.

Figure 12–7 is a Gantt chart that shows the starting and the ending times for each activity. The noncritical activities are shown starting at the earliest times, although they can be scheduled at any time within the period marked by the slack.

Example 12.4 (continued)

Suppose that the two programmers begin the project on June 1. Consider the following questions.

a. On what date should the merging of the graphics and the program modules begin in order to guarantee that the programmers complete the project on time?
b. What is the latest time that the screen development can be completed without delaying the project?
c. Suppose that North discovers a bug in a coding of module 2, so that the time required to complete module 2 is longer than anticipated. Will this necessarily delay the project?

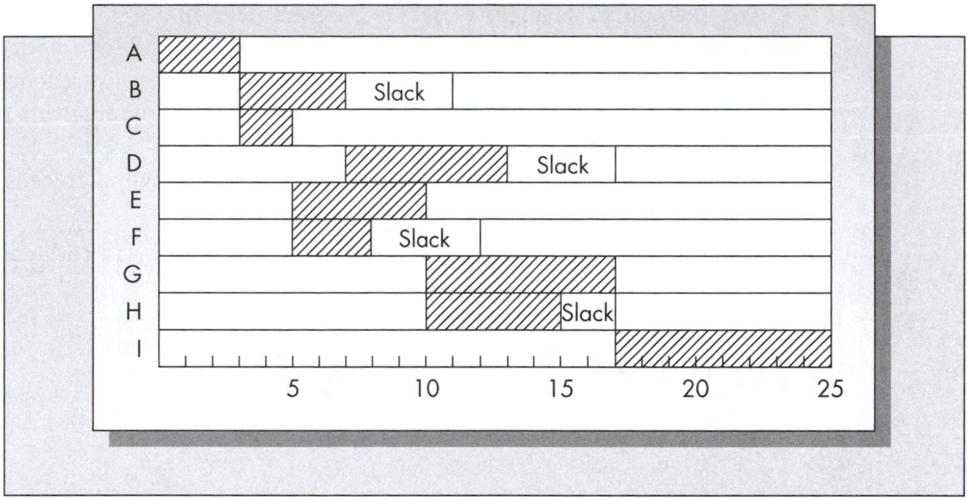

Figure 12–7 Gantt chart for software development project of Example 12.4

d. Suppose that a similar problem is discovered in module 1. Will the project necessarily be delayed in this case?
e. If Bonner is responsible for the coding of the modules and North is responsible for the screen development, and all time estimates are accurate, will the programmers be able to complete the project within 25 weeks?

Solution

a. The merging of the graphics and the program modules is activity I. This is a critical activity, so it must be started no later than its earliest start time, which is week 17. The calendar date would be September 14.
b. The screen development is activity D. This is not a critical activity and has a latest finishing time of 17. The calendar date is again September 14.
c. Module 2 is activity F. Because F is not critical, a delay of up to 4 weeks is permitted.
d. Module 1 is activity E. This is a critical activity, so a delay will necessarily delay the project.
e. The answer is no. The problem is that, assuming that Bonner can work only on a single module at one time, the current schedule requires concurrent programming of modules 1 and 2 and modules 3 and 4. The programmers would have to take in another partner or subcontract some of the module development in order to complete the project within 25 weeks.

PROBLEMS FOR SECTIONS 12.1 AND 12.2

1. Answer the following questions for Example 12.4.
   a. What event is represented by node 6 in Figure 12–5?
   b. What group of activities will have to be completed by week 16 in order to guarantee that the project will not be delayed?
   c. Suppose that module 4 is coded by a subcontractor who delivers the module after 7 weeks. How much will it delay the project if the subcontractor does not start until week 11?

2. For Example 12.4, suppose that the programmers choose not to obtain additional help to complete the project; that is, Bonner will code the modules without outside assistance.
   a. In what way will this alter the project network and the precedence constraints?
   b. Find the critical path of the project network you obtained in part (a). How much longer is it than the one determined in Example 12.4?
3. A project consisting of eight activities satisfies the following precedence constraints.

Activity	Time (weeks)	Immediate Predecessors
A	3	
B	5	A
C	1	A
D	4	B, C
E	3	B
F	3	E, D
G	2	F
H	4	E, D

   a. Construct a network for this project. (You should need only one pseudo activity.)
   b. Compute the earliest and the latest starting and finishing times for each activity and identify the critical path.
   c. Draw a Gantt chart of the schedule for this project based on earliest starting times.
4. Examine the project network pictured in Figure 12–8.
   a. List the immediate predecessors of each activity.
   b. Try to determine the critical path by enumerating all paths from node 1 to node 10.
   c. Compute the earliest starting and finishing times for all activities and identify the critical activities.
5. Consider the network pictured in Figure 12–9.
   a. Determine the immediate predecessors of each activity from the network representation. (Hint: Be certain that you consider only *immediate* predecessors.)
   b. Redraw the network based on the results of part (a) with only two pseudo activities.

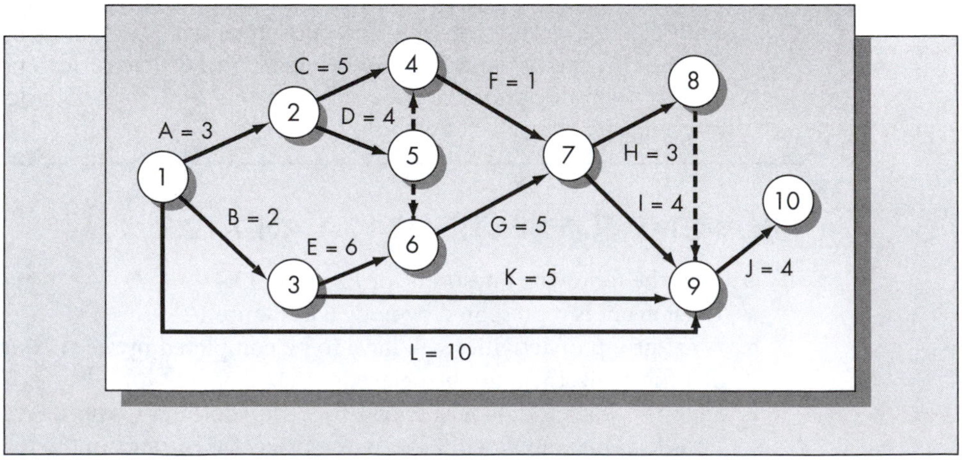

Figure 12–8 Project network for problem 4

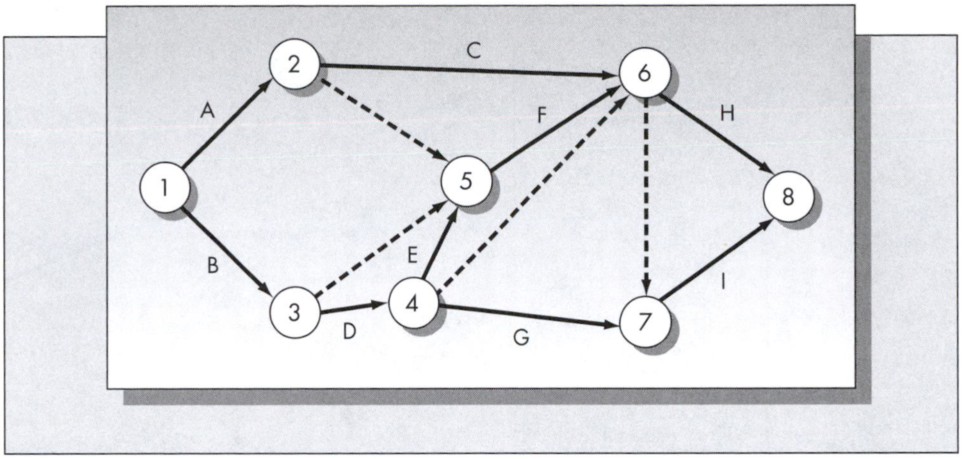

Figure 12–9 Network for problem 5

6. A project consists of seven activities. The precedence relationships are given in the following table.

Activity	Time (days)	Immediate Predecessors
A	32	
B	21	
C	30	
D	45	A
E	26	A, B
F	28	C
G	20	E, F

a. Construct a network for this project.
b. Compute the earliest and the latest starting and finishing times for each activity and identify the critical path.
c. Draw a Gantt chart of the schedule for this project based on earliest starting times.
d. What group of activities will have to be completed by day 60 in order to guarantee that the project will not be delayed?

12.3 TIME COSTING METHODS

Besides assisting with the scheduling of large projects, CPM is a useful tool for project costing and comparing alternative schedules based on cost. This section will consider the costs of expediting activities and how one incorporates expediting costs into the project management framework.

In Section 12.2, we assumed that the time required to complete each activity is known and fixed. In this section we will assume that activity times can be reduced at additional cost. Assume that the time specified for completing an activity is the **normal time**. The minimum possible time required is defined as the **expedited time**. Furthermore, assume that the costs of completing the activity in each of these times are known. Then the CPM assumption is

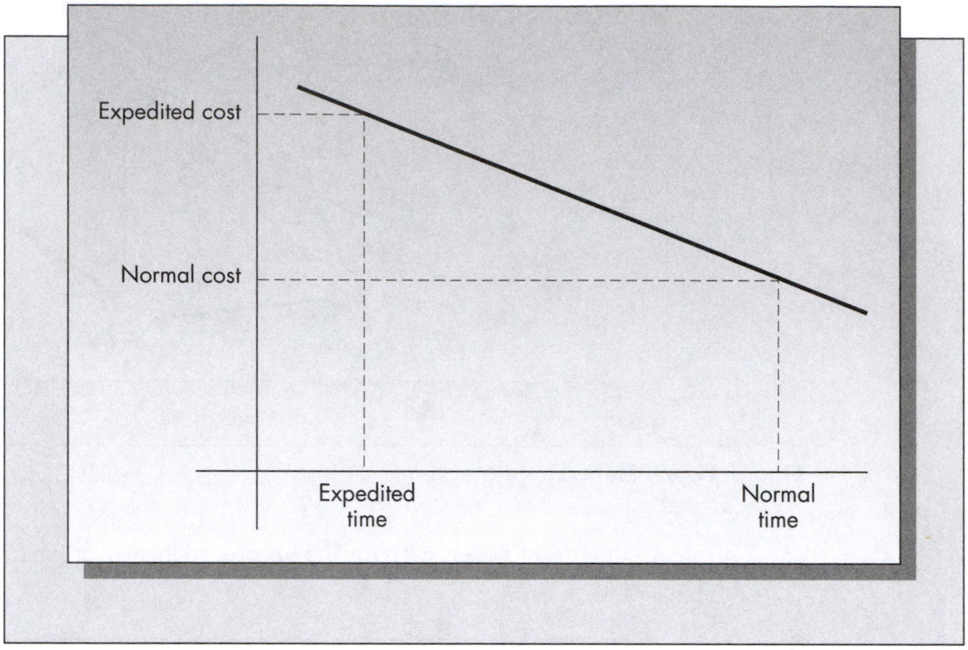

Figure 12–10 The CPM cost–time linear model

that the costs of completing an activity at times between the normal and the expedited times lie along a straight line, as pictured in Figure 12–10. Assuming that the expediting cost function is linear should be reasonable in most circumstances.

There are two types of costs in most projects: direct and indirect costs. Direct costs include costs of labor, material, equipment, and so on. Indirect costs include costs of overhead such as rents, interest, utilities, and any other costs that increase with the length of the project.

Indirect costs and direct costs are respectively increasing and decreasing functions of the project completion time. When these functions are convex, the total cost function, which is their sum, also will be convex. This means that there will be a value of the project time between the normal and the expedited times that is optimal in the sense of minimizing the total cost. Convex cost functions are pictured in Figure 12–11. Figure 12–10 shows the direct cost as a function of time for a given activity.

The general approach is to reduce the project time by one week (or in whatever unit of time activities are measured) until no further reductions are possible or until an optimal solution is identified. At each reduction, one computes the resulting additional direct cost. One continues this process until the minimum total cost solution is identified. A difficulty that arises is that as particular activities are expedited, it is possible that new paths will become critical.

EXAMPLE 12.5

We saw that it is possible for the two computer consultants in Example 12.4 (with some additional help) to complete their project in 25 weeks. Once they complete the project and place the program on the market, the consultants anticipate that they will receive an average of $1,500 per week for the first three years that the product is available. By completing the project earlier, they hope to realize this income sooner.

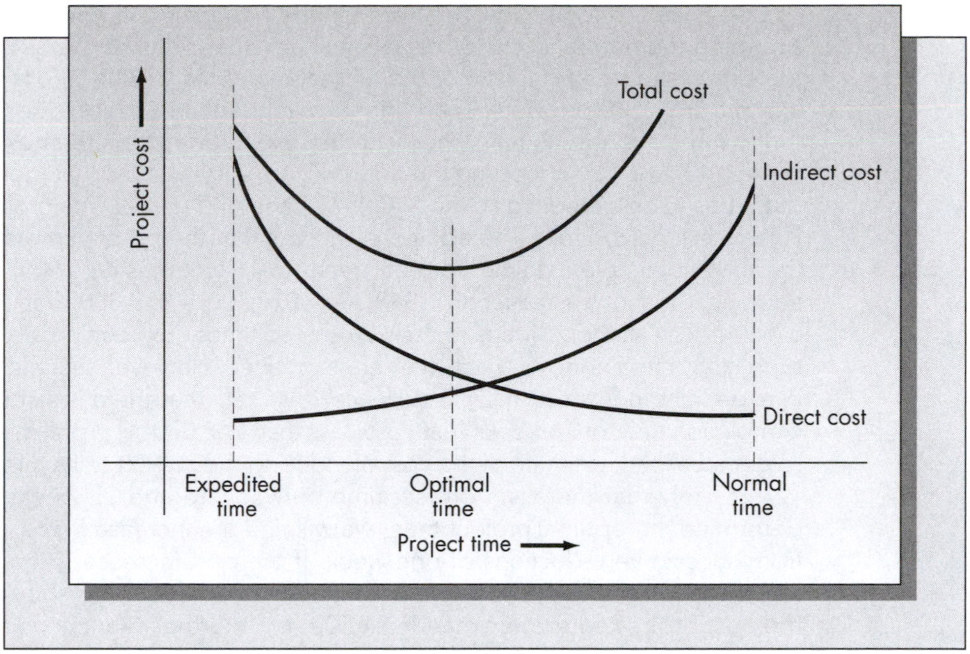

Figure 12–11 Optimal project completion time

They carefully consider each activity and the possibility of reducing the activity time and the associated costs. They estimate that the normal costs are $500 per week for activities that the consultants themselves do, and either more or less for activities that they contract out. Most of the activities can be expedited by subcontracting parts of the programming. Expediting costs vary based on the nature of the task.

They obtain the following estimates.

Activity	Normal Time (weeks)	Expedited Time (weeks)	Normal Cost	Expedited Cost	Cost per Week
A	3	1	$1,000	$3,000	$1,000
B	4	3	4,000	6,000	2,000
C	2	2	2,000	2,000	
D	6	4	3,000	6,000	1,500
E	5	4	2,500	3,800	1,300
F	3	2	1,500	3,000	1,500
G	7	4	4,500	8,100	1,200
H	5	4	3,000	3,600	600
I	8	5	8,000	12,800	1,600

The final column, cost per week, shows the slope of the cost curve pictured in Figure 12–10. One computes the slope from the formula

$$\text{Cost per week} = \frac{\text{Expedited cost} - \text{Normal cost}}{\text{Normal time} - \text{Expedited time}}.$$

One finds the total cost of performing the project in the normal time of 25 weeks by summing the normal costs for each of the activities. This sum is $29,500. If all activities are expedited, the total cost of the project increases to $48,300. Is this additional cost worth the $1,500 per week in additional revenue they expect to realize?

If we replace the normal times with the expedited times, then using the methods of Section 12.2 (or just inspecting the network), it is easy to show that there will be two critical paths based on expedited times: A–C–E–G–I and A–C–E–H–I. The project completion time is reduced to 16 weeks. The additional income that the programmers realize by reducing the project completion time from 25 weeks to 16 weeks is (9)(1,500) = $13,500, but the additional cost of the reduction is $48,300 – $29,500 = $18,800. Hence, it is not economical to reduce all the activities to their expedited times. (Expediting all activities means that both critical and noncritical activities are expedited. However, there is clearly no economy in expediting noncritical activities. Hence, it is likely that there is a solution with a project completion time of 16 weeks that costs less than $48,300. The preliminary analysis is used to get a ballpark estimate of the cost of reducing the project to its minimum time.)

It is likely that there is a project time between 16 and 25 weeks that is optimal. To determine the optimal project time, we will find the increase in the direct cost resulting from successive reductions of one week. If the cost increase is less than $1,500, then the reduction is clearly economical, and additional reductions should be considered. If the cost of the reduction exceeds $1,500, then further reductions are not economical and the process should be terminated.

The key point is that in order to reduce the time required to complete the project by one week, it is necessary to reduce the time of an activity *along the current critical path,* or activities along the current critical paths if more than one path is critical. Reducing the time of a noncritical activity will not reduce the project time.

The initial results are listed below.

Project Time	Critical Path(s)	Critical Activities	Current Time	Expedited Time	Cost to Reduce by One Week
25	A–C–E–G–I	A	3	1	$1,000
		C	2	2	
		E	5	4	1,300
		G	7	4	1,200
		I	8	5	1,600

The least expensive activity to reduce is A. We can reduce activity A to 1 week without introducing any new critical paths. Because the cost of each weekly reduction is less than $1,500, it is economical to reduce A to its minimum time, which is one week. At that point we have the following.

Project Time	Critical Path(s)	Critical Activities	Current Time	Expedited Time	Cost to Reduce by One Week
23	A–C–E–G–I	A	1	1	
		C	2	2	
		E	5	4	$1,300
		G	7	4	1,200
		I	8	5	1,600

The next cheapest activity to reduce is G. The critical path will remain the same until G is reduced to five weeks. (Consider reducing G by one week at a time to be certain that no additional paths become critical.) When G is reduced to five weeks, both paths A–C–E–G–I and A–C–E–H–I are critical. Reducing G from seven weeks to five weeks results in the following.

Project Time	Critical Path(s)	Critical Activities	Current Time	Expedited Time	Cost to Reduce by One Week
21	A–C–E–G–I				
	A–C–E–H–I	A	1	1	
		C	2	2	
		E	5	4	$1,300
		G	5	4	1,200
		H	5	4	600
		I	8	5	1,600

In order to further reduce the project time to 20 weeks, it is necessary to be certain that we make the reduction along both critical paths. At first it would seem that we should reduce H because its marginal cost is least. However, this is not the case. If we reduce H to four weeks, then only the critical path A–C–E–H–I is reduced and not the path A–C–E–G–I. If we reduce both H and G, the increase in the direct cost is $1,800, which is not economical. Does this necessarily mean that it is not worth reducing the project time any further? The answer is no.

Note that the activities A, C, E, and I lie simultaneously along both critical paths. Hence, a reduction in the activity time of any one of these four activities will result in a reduction of the project time. Among these activities, E can be reduced from five to four weeks for under $1,500. Making this reduction obtains the following results.

Project Time	Critical Path(s)	Critical Activities	Current Time	Expedited Time	Cost to Reduce by One Week
20	A–C–E–G–I				
	A–C–E–H–I	A	1	1	
		C	2	2	
		E	4	4	
		G	5	4	$1,200
		H	5	4	600
		I	8	5	1,600

At this point the cost of reducing the project by an additional week exceeds $1,500, so we have reached the optimal solution. The reduction from 25 weeks to 20 weeks costs a total of (1,000)(2) + (1,200)(2) + 1,300 = $5,700 in additional direct costs and results in a return of (1,500)(5) = $7,500 in additional revenue. If all cost and time estimates are correct, the programmers have realized a savings of $1,800 by taking costing into account.

PROBLEMS FOR SECTION 12.3

7. Consider Example 12.5 in which the two programmers are to develop a commercial software package.
   a. What is the minimum project time and the total direct cost of completing the project in that time?
   b. Suppose that all the activity times are reduced to their minimum values. Explain why the total direct cost obtained in Example 12.5 is different from the cost you obtained in part (a).
8. Consider the project described in problem 3 with the normal and the expedited costs and times given in the following table.

Activity	Immediate Predecessor	Normal Time	Expedited Time	Normal Cost	Expedited Cost
A	—	3	2	$200	$250
B	A	5	3	600	850
C	A	1	1	100	100
D	B, C	4	2	650	900
E	B	3	2	450	500
F	E, D	3	2	500	620
G	F	2	1	500	600
H	E, D	4	2	600	900

   a. Consider successive reductions in the project time of one week and find the direct cost of the project after each reduction.
   b. Suppose that indirect costs are $150 per week. Find the optimal project completion time and the optimal total project cost.
9. Discuss the assumption that the cost–time curve is linear. What shape might be more realistic in practice?
10. Consider problem 6. Suppose that the normal and the expedited costs and times are as given in the following table.

Activity	Normal Time	Expedited Time	Normal Cost	Expedited Cost
A	32	26	$200	$500
B	21	20	300	375
C	30	30	200	200
D	45	40	500	800
E	26	20	700	1,360
F	28	24	1,000	1,160
G	20	18	400	550

If indirect costs amount to $100 per day, determine the optimal time to complete the project and the optimal project completion cost.

12.4 SOLVING CRITICAL PATH PROBLEMS WITH LINEAR PROGRAMMING

Finding critical paths in project networks and finding minimum cost schedules can be accomplished by using either dedicated project scheduling software or linear programming. If one expects to solve project scheduling problems on a continuing basis, it is worth the investment in a dedicated software product. However, linear programming is a useful tool for solving a moderately sized problem on an occasional basis. Many linear programming packages, such as Excel's Solver, are widely available and easy to learn and use. This section shows how to formulate and solve critical path problems as linear programs. (This chapter assumes that the reader is familiar with formulating problems as linear programs and interpreting computer output. Supplement 1provides a discussion of linear programming and how linear programs can be solved with Excel.)

Based on the choice of objective functions, the linear programming solution will give either earliest or latest start times for each node in the project network. We first formulate the earliest start time problem. Assume that the network representation of the project consists of nodes 1 through m, with node 1 representing the starting time of the project and node m representing the ending time of the project.

Let

x_i = Earliest start time for node i.

t_{ij} = Time required to complete activity (i, j).

[Note: We used letters to represent activities in the earlier sections, but most computer-based systems represent activities in the form (i, j), where i is the origination node and j is the destination node.]

Then the minimum project completion time is the solution to the following linear program:

$$\min \sum_{i=1}^{m} x_i$$

subject to $x_j - x_i \geq t_{ij}$ for all pairs of nodes corresponding to activity (i, j).

$$x_i \geq 0 \qquad \text{for } 1 \leq i \leq m.$$

The constraints guarantee that there is sufficient time separating each node to account for the activity times represented by the arc between the nodes. As the objective function minimizes the x_i values, the linear programming solution will give the earliest start times. The latest start times can be found by replacing the objective function with

$$\min \left\{ m x_m - \sum_{i=1}^{m-1} x_i \right\}.$$

As x_m is the project completion time, we still wish to find the smallest allowable value of x_m, so its sign must remain positive. Because each $x_i \leq x_m$, the multiplier m for the term x_m guarantees that the objective function is positive to ensure that we obtain a bounded solution. We reverse the sign of the remaining node variables so that the minimization will seek their largest values but will still seek the minimum value for x_m. This means that the variable values will be the latest start times for the activities emanating from each of the nodes. Once the earliest and the latest start times for all the nodes have been determined, it is easy to translate this into earliest and latest start times for the activities using the network representation of the project. The resulting value of x_m, the minimum project completion time, is the same for both formulations.

EXAMPLE 12.4 (REVISITED)

We will solve Example 12.4 using linear programming. From the network representation of the project given in Figure 12–5, we see that there are a total of eight nodes. The formulation of the problem that gives the earliest start times is

$$\min \sum_{i=1}^{8} x_i$$

subject to the following.

$x_2 - x_1 \geq 3$	(A)
$x_3 - x_2 \geq 4$	(B)
$x_4 - x_2 \geq 2$	(C)
$x_3 - x_3 \geq 0$	(P_1)
$x_5 - x_3 \geq 5$	(E)
$x_6 - x_4 \geq 3$	(F)
$x_6 - x_5 \geq 0$	(P_2)
$x_7 - x_3 \geq 6$	(D)
$x_7 - x_6 \geq 5$	(H)
$x_7 - x_5 \geq 7$	(G)
$x_8 - x_7 \geq 8$	(I)
$xi \geq 0$	for $1 \leq i \leq 8$.

The activity giving rise to each constraint is shown in parentheses. Constraints for the pseudo activities must be included as well. Also, the constraint corresponding to P_1 is $x_3 - x_4 \geq 0$, and not vice versa, because P_1 corresponds to the directed arc from node 4 to node 3. The relevant portion of the Excel output is given here.

Objective (Min)				
	Cell	Name	Original Value	Final Value
	\$M\$7	Min Value	0	77
Variable Cells				
	Cell	Name	Original Value	Final Value
	\$B\$5	x1	0	0
	\$C\$5	x2	0	3

$D$5	x3	0	7
$E$5	x4	0	5
$F$5	x5	0	10
$G$5	x6	0	10
$H$5	x7	0	17
$I$5	x8	0	25

The minimum project completion time is the value of x_8, which is 25 weeks. The earliest times correspond to nodes rather than activities, so these times must be converted to activity times. This is easy to do by referring to the network representation in Figure 12–5. For example, because both B and C emanate from node 2, they would have earliest start times of 3 (the value of x_2). Verify that the earliest start times for the other activities obtained in this way agree with the solution we obtained in Section 12.2.

Changing the objective function to $8x_8 - \Sigma x_i$ and rerunning Excel gives the following output.

Objective (Min)			
Cell	Name	Original Value	Final Value
$M$7	Min Value	148	142
Variable Cells			
Cell	Name	Original Value	Final Value
$B$5	x1	0	0
$C$5	x2	3	3
$D$5	x3	7	11
$E$5	x4	5	5
$F$5	x5	10	10
$G$5	x6	10	12
$H$5	x7	17	17
$I$5	x8	25	25

This now gives the latest start times for all the activities. The noncritical activities are those with the latest start times that differ from the earliest start times; the magnitude of the difference is the slack time. Again, you should assure yourself that these results agree with those obtained in Section 12.2.

Linear Programming Formulation of the Cost–Time Problem

Linear programming also can be used to find the optimal completion time when expediting costs are included. The formulation can result in a large linear program even for moderately sized problems. We again will refer to activities by the pair (i, j), where i is the origination node and j is the destination node. Define M_{ij} as the expedited time for activity (i, j) and N_{ij}

as the normal time for activity (i, j). Suppose that the cost–time function (Figure 12–10) has the representation $a_{ij} - b_{ij} = t_{ij}$, where a_{ij} is the y intercept, b_{ij} the slope, and t_{ij} the activity time. Let C be the indirect cost per day. Then the linear programming formulation of the problem of finding the optimal project completion time is

$$\min \sum_{\text{all }(i,j)} [a_{ij} - b_{ij}t_{ij}] + Cx_m$$

subject to

$$\begin{aligned} x_j - x_i &\geq t_{ij} && \text{for all activities } (i, j), \\ t_{ij} &\leq N_{ij}, && \\ t_{ij} &\geq M_{ij}, && \\ x_i &\geq 0 && \text{for } 1 \leq i \leq m, \\ t_{ij} &\geq 0 && \text{for all activities } (i, j). \end{aligned}$$

Note that the a_{ij} terms in the objective function are constants and can be eliminated without altering the solution. These constants can be added later to find the optimal value of the objective function.

The activity times, t_{ij}, are now problem variables rather than given constants. This linear programming problem is considerably larger than the one in Section 12.3. There is now a variable for each node in the network *and* a variable for each activity. However, the added burden is still less work than finding the minimum cost solution manually.

EXAMPLE 12.5 (CONTINUED)

We will solve Example 12.5 using linear programming. The relevant data for developing the linear programming formulation are listed below.

Activity	Node Representation	a_{ij}	b_{ij}	M_{ij}	N_{ij}
A	(1, 2)	4,000	1,000	1	3
B	(2, 3)	12,000	2,000	3	4
C	(2, 4)	—	—	2	2
D	(3, 7)	12,000	1,500	4	6
E	(4, 5)	9,000	1,300	4	5
F	(4, 6)	6,000	1,500	2	3
G	(5, 7)	12,900	1,200	4	7
H	(6, 7)	6,000	600	4	5
I	(7, 8)	20,800	1,600	5	8

The resulting linear program is

$$\text{Min}\{-1{,}000t_{12} - 2{,}000t_{23} - 1{,}500t_{37} - 1{,}300t_{45} - 1{,}500t_{46} - 1{,}200t_{57} - 600t_{67} - 1{,}600t_{78} + 1{,}500x_8\}$$

subject to

$$
\begin{aligned}
x_2 - x_1 - t_{12} &\geq 0,\\
x_3 - x_2 - t_{23} &\geq 0,\\
x_4 - x_2 - t_{24} &\geq 0,\\
x_3 - x_4 &\geq 0,\\
x_5 - x_4 - t_{45} &\geq 0,\\
x_6 - x_4 - t_{46} &\geq 0,\\
x_6 - x_5 &\geq 0,\\
x_7 - x_3 - t_{37} &\geq 0,\\
x_7 - x_6 - t_{67} &\geq 0,\\
x_7 - x_5 - t_{57} &\geq 0,\\
x_8 - x_7 - t_{78} &\geq 0,\\
1 \leq t_{12} &\leq 3,\\
3 \leq t_{23} &\leq 4,\\
4 \leq t_{45} &\leq 5,\\
2 \leq t_{46} &\leq 3,\\
4 \leq t_{57} &\leq 7,\\
4 \leq t_{67} &\leq 5,\\
5 \leq t_{78} &\leq 8,\\
4 \leq t_{37} &\leq 6,\\
t_{24} &\leq 2,
\end{aligned}
$$

$$
\begin{aligned}
x_i &\geq 0 \quad \text{for } i \leq 1 \leq 8,\\
t_{ij} &\geq 0 \quad \text{for all activities } (i, j).
\end{aligned}
$$

The upper and the lower bound constraints on the t_{ij} variables would have to be entered as two separate constraints into the computer, giving a total of 28 constraints. The Excel output for this problem follows.

Objective (Min)

Cell	Name	Original Value	Final Value
\$M\$7	Min Value	0	77

Variable Cells

Cell	Name	Original Value	Final Value
\$B\$5	x1	0	0
\$C\$5	x2	0	1
\$D\$5	x3	0	6
\$E\$5	x4	0	3
\$F\$5	x5	0	7
\$G\$5	x6	0	7
\$H\$5	x7	0	12
\$I\$5	x8	0	20
\$J\$5	t12	0	1
\$K\$5	t23	0	4
\$L\$5	t24	0	2

Cell	Name	Original Value	Final Value
$M$5	t37	0	6
$N$5	t45	0	4
$O$5	t46	0	3
$P$5	t57	0	5
$QS5	t67	0	5
$R$5	t78	0	8

The optimal length of the project is given by the value of $x_8 = 20$, and the optimal activity times are the values of t_{ij}. Notice that these times agree with the activity times we found to be optimal in Section 12.3. The total cost of this solution is found by adding Σa_j, which is $82,700, and the cost for activity *C*, which was not included in the objective function. (We could have included the cost of activity *C* in the objective function by adding the term $1{,}000t_{24}$, as t_{24} is fixed at 2.) The resulting value for the total cost of the project at 20 days is $82,700 + $2,000 – $19,500 = $65,200.

In general, linear programming is not an efficient way to solve large network problems. Algorithms that exploit the network structure of the problem are far more efficient. Many commercial software products based on such algorithms are available for project scheduling. Even PC-based software products are capable of solving large projects. Linear programming is a useful tool for solving moderately sized problems. However, for those with a large project scheduling problem, or with an ongoing need for project scheduling, we recommend a dedicated project scheduling program. See Section 12.9 for a discussion of project management software.

PROBLEMS FOR SECTION 12.4

11. Solve problem 3 by linear programming.
12. Solve problem 4 by linear programming.
13. Formulate problem 5 as a linear program.
14. Solve problem 6 by linear programming.
15. Solve problem 8 by linear programming.
16. Solve problem 10 by linear programming.

12.5 PERT: PROJECT EVALUATION AND REVIEW TECHNIQUE

PERT is a generalization of CPM to allow uncertainty in the activity times. When activity times are difficult to predict, PERT can provide estimates of the effect of this uncertainty on the project completion time. However, for reasons that will be given in detail, the results of the analysis are only approximate. Let T_i be the time required to complete activity *i*. In this section we will assume that T_i is a random variable. Furthermore, we assume that the collection of random variables $T_1, T_2, \ldots, T_n$ is mutually independent. The first issue addressed is the appropriate form of the distribution of these random variables. Define the following quantities:

a = Minimum activity time,
b = Maximum activity time,
m = Most likely activity time.

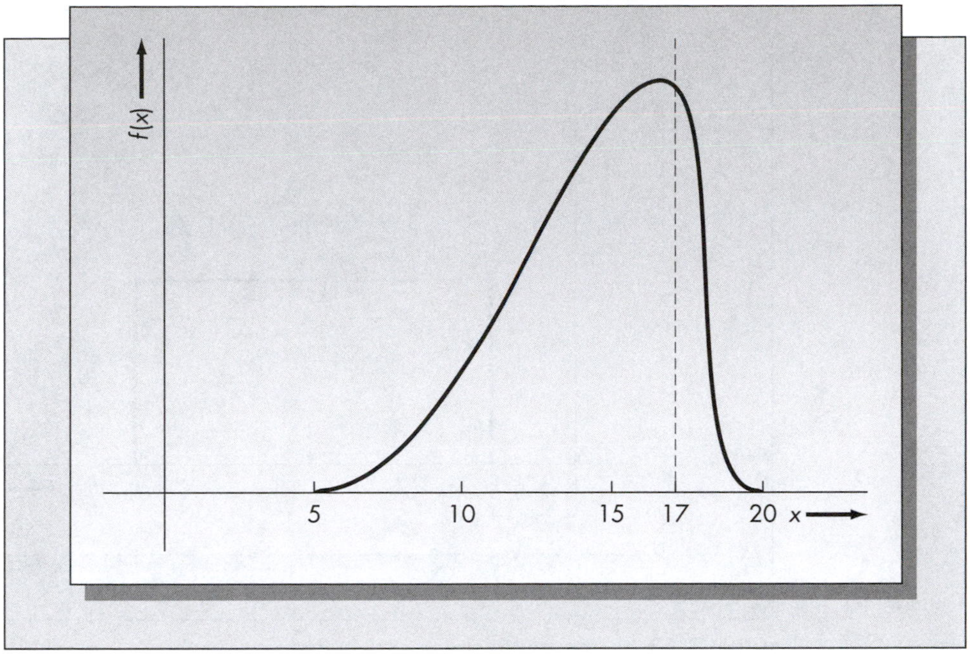

Figure 12–12 Probability density of activity time

As an example, suppose that $a = 5$ days, $b = 20$ days, and $m = 17$ days. Then the probability density function of the activity time is as pictured in Figure 12–12. The density function should be zero for values less than 5 and more than 20 and should be a maximum at $t = 17$ days. The point where the density is a maximum is known as the **mode** in probability theory. The **beta distribution** is a type of probability distribution defined on a finite interval that may have its modal value anywhere on the interval. For this reason, the beta distribution is usually used to describe the distribution of individual activity times. The assumption of beta-distributed activity times is used to justify simple approximation formulas for the mean and the variance, but it is rarely used to make probabilistic statements concerning individual activities.

The beta distribution assumption is used to justify the approximations of the mean μ and the standard deviation σ of each activity time. The traditional PERT method is to estimate μ and σ from a, b, and m using the following formulas:

$$\mu = \frac{a + 4m + b}{6}, \qquad \sigma = \frac{b - a}{6}.$$

The approximation for the standard deviation is based on the following property of the normal distribution: limits at distance 3σ to either side of the mean for a normal variate include all the population with probability exceeding .99. In view of this property, it is assumed that there are six standard deviations from a to b. The approximation for the mean is obtained by assuming the variance approximation as well as the beta distribution for the activity time. Given the variance, a, b, and m, the mean is determined by solving a cubic equation. Calculating μ for various values of the other parameters and developing the best fit by linear regression results in the one-four-one weighting scheme. (See Archibald and Villoria, 1967, p. 449.)

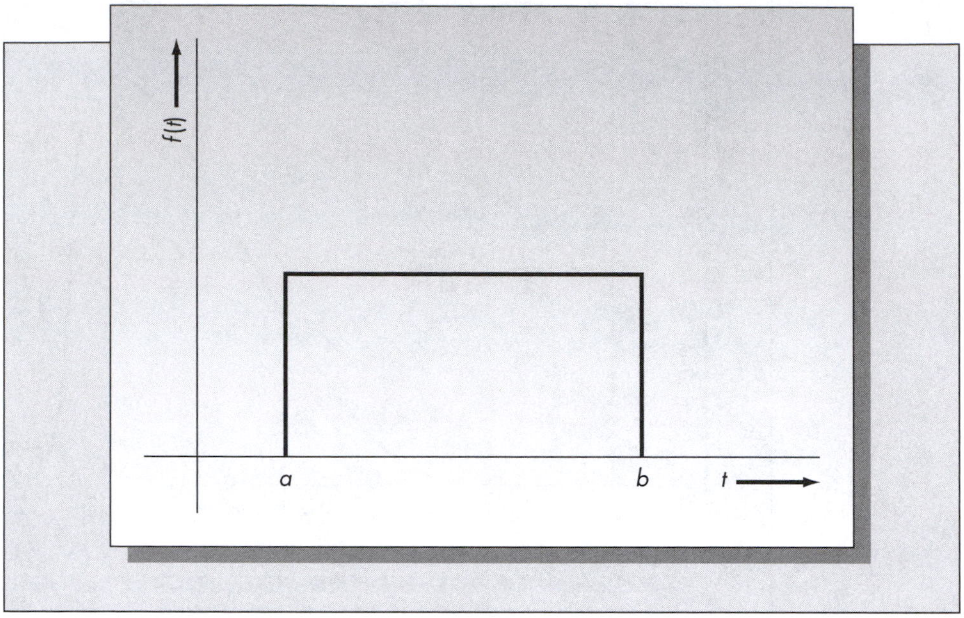

Figure 12–13 Uniform density of activity times

Squaring both sides in the formula for σ gives

$$\sigma^2 = \frac{(b-a)^2}{36}$$

where σ^2 is the variance.

The uniform distribution is a special case of the beta distribution. If the job time has a uniform distribution from a to b, the density function would be rectangular, as pictured in Figure 12–13. The variance of a uniform variate is $(b - a)^2/12$. Because we would expect the variance to be less for a peaked distribution such as the one pictured in Figure 12–12, the approximation for σ^2 recommended earlier seems reasonable. Note that the one-four-one weighting scheme gives the correct value of the mean, $(a + b)/2$, if one substitutes $m = (a + b)/2$ as the mode.

In PERT, one assumes that the distribution of the total project time is normal. The **central limit theorem** is used to justify this assumption. Roughly, the central limit theorem says that the distribution of the sum of independent random variables is approximately normal as the number of terms in the sum grows large. In most cases, convergence occurs quickly. Hence, because the total project time is the sum of the times of the activities along the critical path, it should be approximately normally distributed as long as activity times are independent.

Suppose that the critical activity times are $T_1, T_2, \ldots, T_k$. Then the total project time T is

$$T = T_1 + T_2 + \ldots + T_k.$$

It follows that the mean project time, $E(T)$, and the variance of the project time, $\mathrm{Var}(T)$, are given by

$$E(T) = \mu_1 + \mu_2 + \cdots + \mu_k,$$
$$Var(T) = \sigma_1^2 + \sigma_2^2 + \cdots + \sigma_k^2.$$

These formulas are based on the following facts from probability theory: the expected value of the sum of *any* set of random variables is the sum of the expected values, and the variance of a sum of *independent* random variables is the sum of the variances. Independence of activity times is required in order to easily obtain the variance of the project time and to justify the application of the central limit theorem. One could modify the variance formula to incorporate correlations among the activities, but the assumption of normality of the project completion time might no longer be accurate. For these reasons, explicit treatment of the dependencies among activity times is rare in practice.

Summarizing the PERT method.

1. For each activity obtain estimates of a, b, and m. These estimates should be supplied by the project manager or by someone familiar with similar projects.
2. Using these estimates, compute the mean and the variance of each of the activity times from the given formulas.
3. Based on the mean activity times, use the methods of Section 12.4 to determine the critical path.
4. Once the critical activities are identified, add the means and the variances of the critical activities to find the mean and the variance of the total project time.
5. The total project time is assumed to be normally distributed with the mean and the variance determined in step 4.

Using the assumption that the project time is normally distributed, we can address a variety of issues. An example will illustrate the method.

EXAMPLE 12.6

Consider the case in example 12.4 of the two computer consultants developing a software project. Before embarking on the project, they decide that it is important to consider the uncertainty of the times required for certain tasks. As with any software project, unanticipated bugs can surface and cause significant delays. Based on their past experience, the programmers decide that the values of a, b, and m are as follows.

Activity	Min (a)	Most Likely (m)	Max (b)	$\mu = \frac{a+4m+b}{6}$	$\sigma^2 = \frac{(b-a)^2}{36}$
A	2	3	4	3	0.11
B	2	4	10	4.67	1.78
C	2	2	2	2	0
D	4	6	12	6.67	1.78
E	2	5	8	5	1.00
F	2	3	8	3.67	1.00
G	3	7	10	6.83	1.36
H	3	5	9	5.33	1.00
I	5	8	18	9.17	4.69

Using these values, we have computed the means and the variances of each of the activity times. Activity C, the design of the flowchart, requires precisely two days. Hence, the variance of this activity time is zero.

The traditional PERT method is to compute the critical path based on the mean activity times. In this example, the introduction of uncertainty does not alter the critical path. It is still A–C–E–G–I. (However, it is possible that in other cases the critical path based on the mean times will *not* be the same as the critical path based on the most likely times.) The expected project completion time, $E(T)$, is simply the sum of the mean activity times along the critical path. In this example,

$$E(T) = 3 + 2 + 5 + 6.83 + 9.17 = 26 \text{ weeks.}$$

Similarly, the variance of the project completion time, $\mathrm{Var}(T)$, is the sum of the variances of the activities along the critical path. In this case,

$$\mathrm{Var}(T) = 0.11 + 0 + 1.0 + 1.36 + 4.69 = 7.16.$$

The assumption made is that the total project completion time T is a normal random variable with mean $\mu = 26$ and standard deviation $\sigma = \sqrt{7.16} = 2.68$. We can now answer a variety of specific questions concerning this project.

EXAMPLE 12.7

Answer the following questions about the project scheduling problem described in Example 12.6.

1. What is the probability that the project can be completed in under 22 weeks?
2. What is the probability that the project requires more than 28 weeks?
3. Find the number of weeks required to complete the project with probability .90.

Solution

1. We wish to compute $P\{T < 22\}$.

$$P\{T < 22\} = P\left\{\frac{T-\mu}{\sigma} < \frac{22-\mu}{\sigma}\right\} = P\left\{Z < \frac{22-26}{2.68}\right\}$$
$$= P\{Z < -1.5\} = .0668.$$

Z is the standard normal variate. The probability is from Table A–1 of the standard normal distribution in Appendix A. The solution is pictured in Figure 12–14.

2. $$P\{T > 28\} = P\left\{\frac{T-\mu}{\sigma} > \frac{28-\mu}{\sigma}\right\} = P\left\{Z > \frac{28-26}{2.68}\right\}$$
$$= P\{Z > .75\} = .2266.$$

The solution to this problem is pictured in Figure 12–15.

3. Here we wish to find the value of t such that $P\{T \le t\} = .90$:

$$.90 = P\{T \le t\} = P\left\{Z < \frac{t-\mu}{\sigma}\right\}.$$

It follows that

$$\frac{t-\mu}{\sigma} = z_{90}.$$

or $t = \mu + \sigma z_{.90} = 26 + (2.68)(1.28) = 29.43$ weeks.

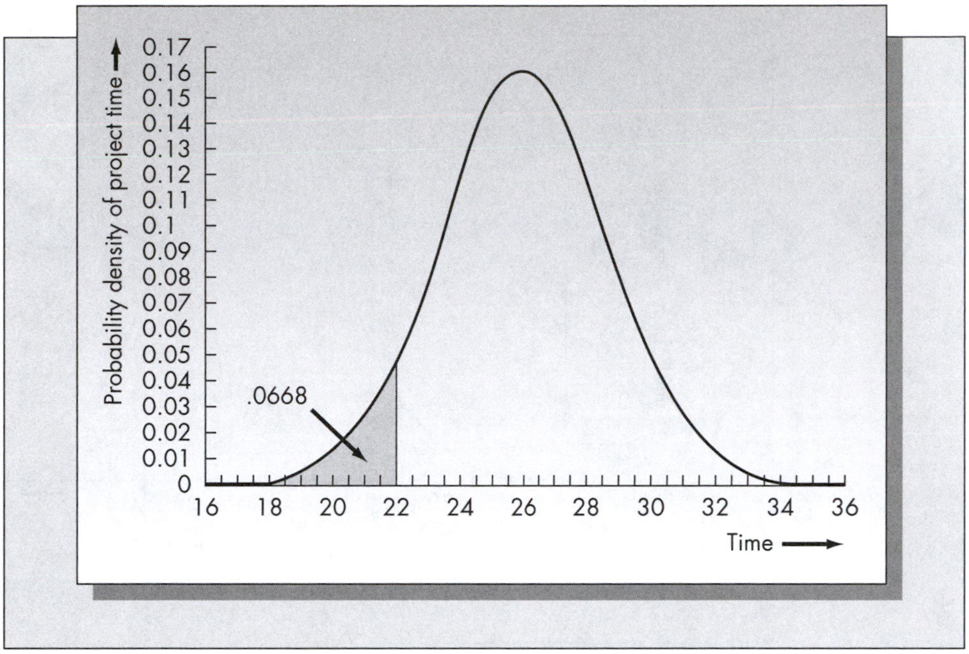

Figure 12–14 Answer to Example 12.7, part 1

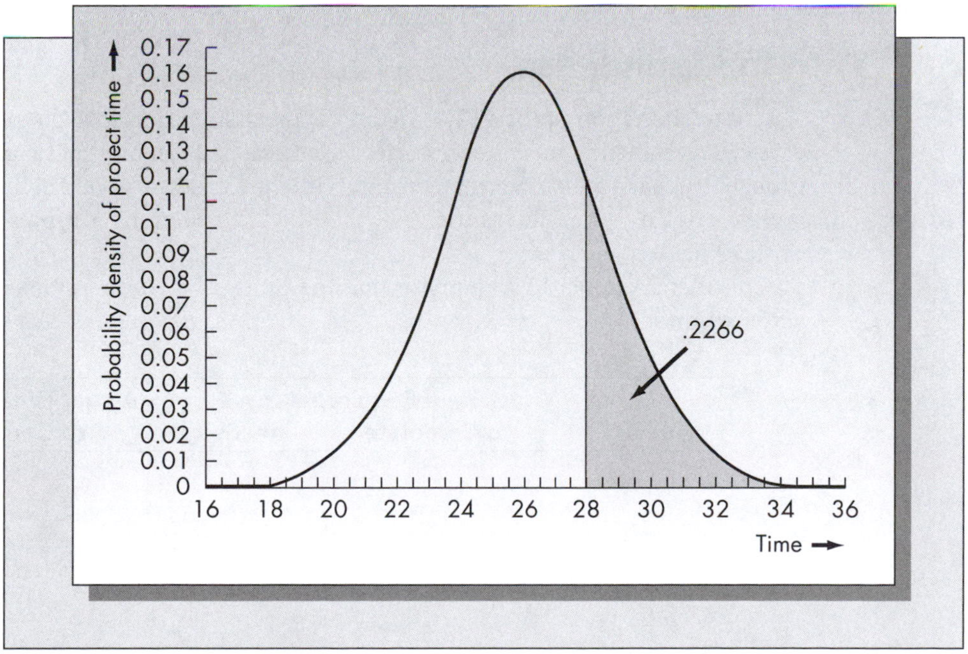

Figure 12–15 Answer to Example 12.7, part 2

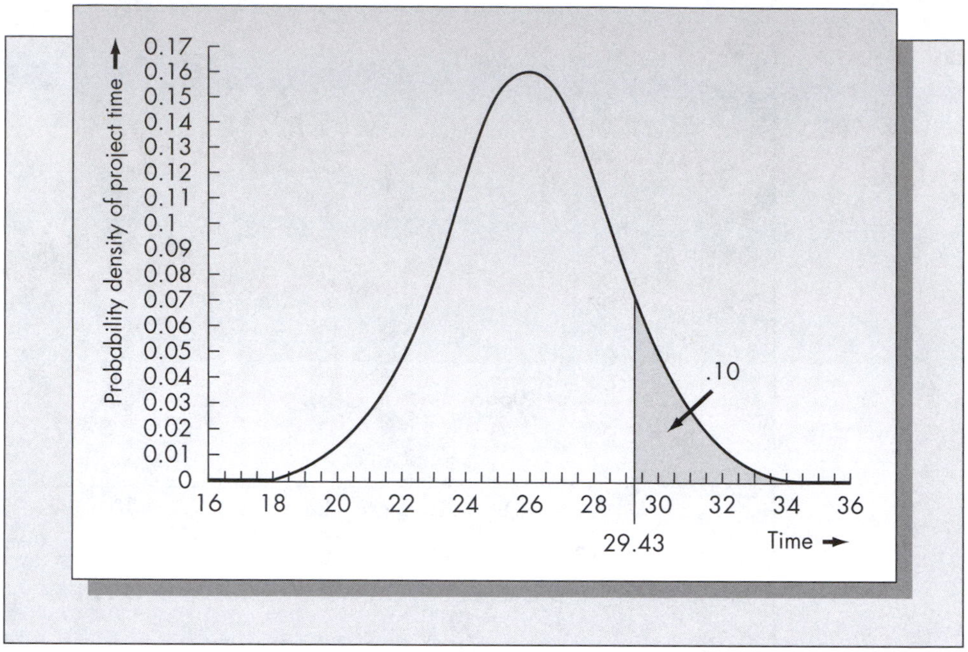

Figure 12–16 Answer to Example 12.7, part 3

The notation $z_{.90}$ stands for the 90th percentile of the standard normal distribution. That is, $P\{Z \leq z_{.90}\} = .90$. Its value, 1.28, is found from Table A–1 in Appendix A at the back of the book or through Excel. The solution is pictured in Figure 12–16.

Path Independence

A serious limitation of the PERT method is the assumption that the path with the longest expected completion time is necessarily the critical path. In most networks there is a positive probability that a path other than the one with the longest expected completion time, in fact, will be critical. When this is the case, the PERT calculations just presented could be very misleading.

Consider Example 12.7. Suppose that the project has been completed in the following activity times.

Activity	Actual Time Required to Complete	Activity	Actual Time Required to Complete
A	3.4	F	5.0
B	4.0	G	6.2
C	2.0	H	7.2
D	7.0	I	13.0
E	3.5		

With these activity times, the critical path is no longer A–C–E–G–I; it is now A–C–F–H–I. Hence, the PERT assumption that a fixed path is critical is not accurate.

Unfortunately, determining the exact distribution of the critical path is difficult. In general, this distribution is not known. The difficulty is that, because different paths include the same activities, the times required to complete different paths are *dependent* random variables.

Depending on the particular configuration of the project network, assuming independence of two or more paths may be more accurate than assuming a single critical path. Consider the following example.

EXAMPLE 12.8

Consider the project pictured in Figure 12–17. In the figure we have included the means and the variances of the activity times. Assume that these times are expressed in weeks. There are exactly two paths from node 1 to node 9: A–C–E–G–I and B–D–F–H–I. The expected times of these paths are 41 and 40 weeks, respectively. Using PERT, we would assume that the critical path is A–C–E–G–I. However, there is almost an equal likelihood that path B–D–F–H–I will turn out to be critical after the activity times are realized.

In this example, the two paths have only activity I in common. Hence, the paths are almost statistically independent. Performing the calculations assuming two independent paths could give very different results from assuming a unique critical path.

Suppose that we wish to determine the probability that the project is completed within 43 weeks. Let T_1 be the time required to complete path A–C–E–G–I and T_2 the time required to complete path B–D–F–H–I. Using the methods of Section 12.4, we conclude that T_1 and T_2 are approximately normally distributed, with

$$E(T_1) = 41,$$
$$\text{Var}(T_1) = 0.13 + 1 + 1.15 + 4.4 + 1 = 7.68,$$

and

$$E(T_2) = 40,$$
$$\text{Var}(T_2) = 0.11 + 3.1 + 0.86 + 3.0 + 1 = 8.07.$$

Let T be the total project time. Clearly, $T = \max(T_1, T_2)$. It follows that

$$P\{T < 43\} = P\{\max(T_1, T_2) < 43\} = P\{T_1 < 43, T_2 < 43\}$$

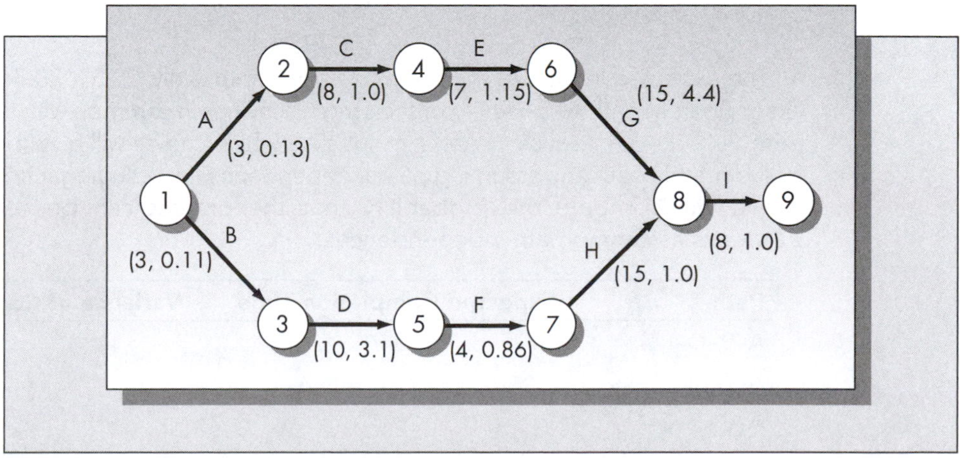

Figure 12–17 Network for PERT Example 12.8

(the last equality follows from the fact that if the maximum of two quantities is less than a constant, then both quantities must be less than that constant)

$$= P\{T_1 < 43\}\, P\{T_2 < 43\}$$

(which follows from the assumption of path independence)

$$= P\left\{Z < \frac{43-41}{\sqrt{7.68}}\right\} P\left\{Z < \frac{43-40}{\sqrt{8.07}}\right\}$$

$$= P\{Z < 0.72\} P\{Z < 1.05\}$$

$$= (.7642)(.8413) = .6429.$$

The methods of Section 12.4 would have given the estimate of completing the project within 43 weeks as .7642. For this network, .6429 is far more accurate.

Certain calculations are more complex if we assume path independence. For example, suppose that we wanted to know the number of weeks required to complete the project with probability .90. Then we wish to find t to satisfy

$$P\{T_1 < t\} P\{T_2 < t\} = .90$$

or

$$P\left\{Z < \frac{t-41}{2.77}\right\} P\left\{Z < \frac{t-40}{2.84}\right\} = .90.$$

One calculates t by trial and error. Because the probabilities are likely to be close, a good starting guess for the value of each probability is $\sqrt{.90} = .95$, which gives $t = 45.6$ for the first term and $t = 44.7$ for the second term. The correct value is approximately $t = 45.2$ weeks, which results in a value of .904 for the product of the two probabilities.

Because the two paths for this project have only one activity in common, the assumption of path independence is quite reasonable and the answers obtained in this manner far more accurate than those found by assuming a unique critical path. In most networks, however, paths may have many activities in common, and the assumption of path independence may be inaccurate. Consider the network for Example 12.4 pictured in Figure 12–5. In this example, there are a total of five paths.

A–B–D–I

A–C–P_1–D–I

A–C–E–G–I

A–C–E–P_2–H–I

A–C–F–H–I

The expected lengths of the five paths are, respectively, 23.51, 20.84, 26, 23, and 23.17. Unfortunately, path A–C–F–H–I contains three activities in common with the longest expected path, A–C–E–G–I. In such a case, it is not clear which choice will give more accurate results: including this path and assuming path independence or excluding it from consideration.

We will compute the probability that the project can be completed in under 22 weeks assuming path independence.

Path	Expected Completion Time	Variance of Completion Time
A–B–D–I	23.5	8.36
A–C–P_1–D–I	20.8	6.58
A–C–E–G–I	26.0	7.16
A–C–E–P_2–H–I	23.0	6.80
A–C–F–H–I	23.2	6.80

If T is the project completion time and $T_1, \ldots, T_5$ are the times required to complete each of the five paths listed in the table, above, then

$$T = \max(T_1, \ldots, T_5).$$

It follows that

$$\begin{aligned} P\{T < 22\} &= P\{T_1 < 22, T_2 < 22, T_3 < 22, T_4 < 22, T_5 < 22\} \\ &\approx P\{T_1 < 22\}\, P\{T_2 < 22\}\, P\{T_3 < 22\}\, P\{T_4 < 22\}\, P\{T_5 < 22\}. \end{aligned}$$

Again, assuming a normal distribution for each of the path completion times, we have

$$P\{T_1 < 22\} = P\left\{Z < \frac{22 - 23.5}{\sqrt{8.36}}\right\} = P\{Z < -0.52\} = .3015,$$

$$P\{T_2 < 22\} = P\left\{Z < \frac{22 - 20.8}{\sqrt{6.58}}\right\} = P\{Z < -0.47\} = .6808,$$

$$P\{T_3 < 22\} = .0668 \text{ (from Example 9.7, Part 1).}$$

$$P\{T_4 < 22\} = P\left\{Z < \frac{22 - 23}{\sqrt{6.8}}\right\} = P\{Z < -0.38\} = .3520,$$

$$P\{T_5 < 22\} = P\left\{Z < \frac{22 - 23.2}{\sqrt{6.8}}\right\} = P\{Z < -0.46\} = .3228.$$

Hence, it follows that

$$P\{T < 22\} \approx (.3015)(.6808)(.0668)(.3520)(.3228) = .0016.$$

The true value of the probability will fall somewhere between .0016 and the probability computed by traditional PERT methods, .0668. Assuming path independence can have a very significant effect on the probabilities. For this example, it is safe to say that the likelihood that the consultants complete the project in less than 22 weeks is far less than 6.68 percent and is probably well under 1 percent. Monte Carlo simulation (see Chapter 8) is another tool to estimate these probabilities with reasonable accuracy.

PROBLEMS FOR SECTION 12.5

17. Referring to Example 12.4 and Figure 12–5, what is the probability that node 6 (i.e., the completion of activities A, C, E, and F) is reached before 12 weeks have elapsed?
18. With reference to Example 12.4 and Figure 12–5, what is the conditional probability that the project is completed by the end of week 25, given that activities A through H (node 7) are completed at the end of week 15? Assume that the time required to complete activity I is normally distributed for your calculation.
19. Consider the following PERT time estimates.

Activity	Immediate Predecessors	*A*	*m*	*b*
A	—	2	5	9
B	A	1	6	8
C	A	3	5	12
D	B	2	4	12

Activity	Immediate Predecessors		*A*	*m*	*b*
	E	B, C	4	6	8
	F	B	6	7	8
	G	D, E	1	2	6
	H	F, G	4	6	16

a. Draw a network for this project and determine the critical path based on the *most likely times* by inspection.
b. Assuming that the critical path is the one you identified in part (a), what is the probability that the project will be completed before 28 weeks? Before 32 weeks?
c. Assuming that the critical path is the one you identified in part (a), how many weeks are required to complete the project with probability .95?

20. Consider the project network pictured in Figure 12–18. Assume that the times attached to each node are the means and the variances of the project completion times, respectively.
a. Identify all paths from nodes 1 to 6.
b. Which path is critical based on the expected completion times?
c. Determine the probability that the project is completed within 20 weeks assuming that the path identified in part (b) is critical.
d. Using independence of paths A–C–E and B–F–G only, recompute the answer to part (c).
e. Recompute the answer to part (c) assuming independence of the paths identified in part (a).
f. Which answer, (c), (d), or (e), is probably the most accurate?

21. Consider the project described in problem 3. Suppose that the activity times are random variables with a constant **coefficient of variation** of 0.2. (The coefficient of variation is the ratio σ/μ.) Assume that the times given in problem 3 are the mean activity times.
a. Compute the standard deviation of each of the activity times.
b. Find the mean and the variance of the path with the longest expected completion time.
c. Using the results of part (b), estimate the probability that the project is completed within 20 weeks.
d. Using the results of part (b), estimate the number of weeks required to complete the project with probability .85.

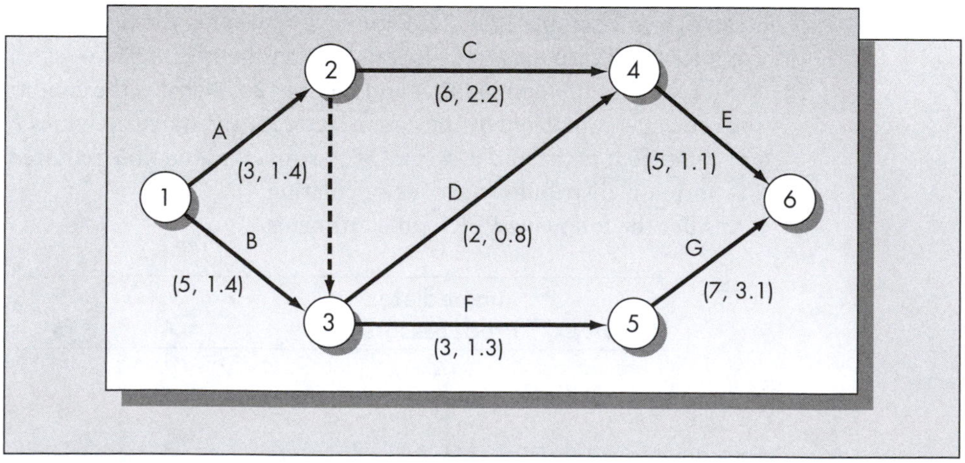

Figure 12–18 Project network (for problem 20)

SNAPSHOT APPLICATION

Warner Robins Streamlines Aircraft Maintenance with CCPM Project Management

Annual spending on aircraft maintenance and repair amounts to close to $100 billion annually worldwide, about half of which is accounted for by the military. Of this amount, about $20 billion is spent by the U.S. military alone. Warner Robins Air Logistics Center near Macon, Georgia, provides maintenance, repair, and overhaul for its customer, the U.S. Air Force. Warner Robins supports several aircraft including the C-5 Galaxy, C-17 Globemaster, C-130 Hercules transport, and the F-15 Eagle fighter jet.

A team of researchers considered the task of scheduling operations for the C-5. The C-5 line has 24 frontline supervisors and about 460 mechanics organized into several skill groups. The line operates in two shifts. Scheduling repair is a challenge for several reasons. Repairs due to damage are quite unpredictable, and even scheduled maintenance may have unforeseen problems crop up. Typical program depot maintenance may require as much as 40,000 to 50,000 worker-hours, but this figure could be as much as 10,000 hours more than anticipated. These long lead times and inherent uncertainties make due-date scheduling extremely difficult.

To help ameliorate these problems, the research team implemented a new approach to project scheduling. The approach is based on concepts developed by Elihu Goldratt (the developer of OPT, a defunct manufacturing scheduling tool, and author of *The Goal*). Goldratt labeled his method critical chain project management (CCPM), an alternative to PERT for scheduling projects with uncertain activity times. In PERT, one considers the distribution of each activity time, and effectively builds in a buffer for each activity. CCPM considers a buffer for the entire project only. One then identifies those activities that consume the buffer after the fact. As an example, in the case of the C-5 line, the managers were able to identify floorboard replacement as an activity that consistently consumed the buffer. By focusing attention on this activity, management was able to reduce the time for this activity by 45 percent, thus resulting in substantial improvements in overall project completion times. Other activities were also identified as being trouble spots, and the overall efficiency of the C-5 schedule significantly improved. The research team estimated overall revenue savings of almost $50 million annually as a result of improved scheduling of the C-5 maintenance, repair, and overhaul operation. Similar scheduling methods were also being considered for the other aircraft programs at the base.

Source: Srinivasan, Best, and Chandrasekaran, 2007.

12.6 RESOURCE CONSIDERATIONS

Resource Constraints for Single-Project Scheduling

An implicit assumption made throughout this chapter is that sufficient resources are available and only the **technological constraints** (precedence relationships) are important for setting schedules. In most environments, however, resource constraints cannot be ignored. Examples of limited resources that would affect project schedules are workers, raw materials, and equipment. Because traditional CPM ignores resource considerations, one manager described it as a "feasible procedure for producing a nonfeasible schedule."

Determining optimal schedules for complex project networks subject to resource limitations is an extremely difficult combinatorial problem. A single project may require a variety of different resources, or many different projects may compete for one or more resources. Heuristic (approximate) methods are generally used to modify schedules obtained by more conventional means.

Allocation of scarce resources among competing activities has been an area of considerable interest. This section considers the allocation process using the example of the programming project discussed earlier in the chapter.

EXAMPLE 12.9

Consider Example 12.4 about the two programmers developing a software package. An analysis based only on precedence considerations showed that the project could be completed in 25 weeks without expediting any activities. This analysis, however, is based on the assumption of unlimited resources. In fact, as we noted earlier, it is not possible for the programmers to complete the project within 25 weeks without obtaining additional help. What is the minimum time required for them to complete the project if they do not obtain additional assistance?

Assume that North is responsible for activities B (design of the graphic icons) and D (design of the input/output screens), and Bonner has the expertise required for the development of modules 1 and 2 (activities E and F). Furthermore, assume that either of the programmers can perform activities G and H (modules 3 and 4), but both must work on the final merging and testing. Consider the Gantt chart pictured in Figure 12–7. It shows an infeasible schedule because activities D, E, and F must be done simultaneously between weeks 7 and 8, and activities D, G, and H must be done simultaneously between weeks 10 and 13.

If only Bonner and North are to do the programming, they must reorder the activities so as not to schedule more than two activities simultaneously. Furthermore, they must be certain that these are not two activities that can only be done by one of them. First, they try to find a feasible schedule without delaying the project by rescheduling the noncritical activities within the allowable slack. From Figure 12–7 it is clear that no matter how one rearranges the noncritical activities within the available slack, there is no way to avoid scheduling three activities simultaneously at some point.

This means that they cannot complete the project within 25 weeks without additional help. We determine a feasible schedule by considering the sequence of activities performed by each programmer. At week 3 North begins work on B and Bonner on C. Note that activity A (the market survey) doesn't require either North or Bonner. At week 5 Bonner begins work on E, and at week 7 North begins work on D. Because F is module 2 coding, which must be done by Bonner, F must now follow E, so that F starts at week 10. Both G and H can begin at week 13 when both programmers are free. Finally, I starts at week 20 and ends at week 28. Hence, the project time has increased from 25 weeks to 28 weeks as a result of resource considerations.

Figure 12–19 shows the Gantt chart for the modified schedule. This schedule is feasible because there are never more than two activities scheduled at any point in time. North begins work at week 3 and works on the activities B, D, G, and I, while Bonner, who also begins working at week 3, performs activities C, E, F, H, and I. Note that activities G and H are interchangeable.

Including resource considerations has a number of interesting consequences. For one, the critical path is no longer the same. In the previous example, from a resource perspective, Activities B, D, and F no longer have any slack, so they are now critical. For example, activity B is shown as having slack in Figure 12.19 (using the standard definition of slack); however, because North needs to start activity D at week 7, there is no actual slack for activity B in the schedule. Furthermore, there is considerably less slack time overall.

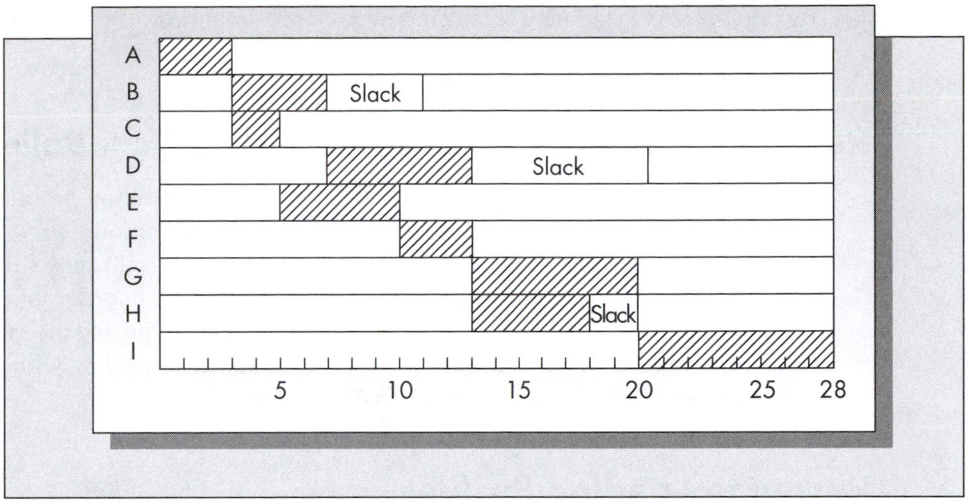

Figure 12–19 Gantt chart for modified schedule for Example 12.9

In general, the inclusion of resource constraints has the following effects.

1. The total amount of scheduled slack is reduced.
2. The critical path may be altered. Furthermore, the zero-slack activities may not necessarily lie along one or more critical paths.
3. Earliest- and latest-start schedules may not be unique. They depend on the particular rules that are used to resolve resource limitations.

Several heuristic methods for solving this problem exist. Most involve ranking the activities according to some criterion and resolving resource conflicts according to the sequence of the ranking. Some of the ranking rules that have been suggested include the following.

1. *Minimum job slack.* Priority is given to activities with the smallest slack.
2. *Latest finishing times.* When resource conflicts exist, this rule assigns priority to the activity with the minimum latest finishing time.
3. *Greatest resource demand.* This rule assigns priority on the basis of the total resource requirements of all types, giving highest priority to the activities having greatest resource demand. The rationale behind this method is to give priority to potential bottleneck activities.
4. *Greatest resource utilization.* This rule gives priority to that combination of activities that results in the maximum resource utilization (minimum idle time) in any scheduling interval.

These rules (among others) were compared by Edward Davis and James Patterson (1975). Their results indicated that the first two methods on the list tended to be the best performers.

Optimal solution methods for project scheduling under resource constraints exist as well. The formulation may be an integer program requiring some type of branch-and-bound procedure, or it may be solved by a technique known as implicit enumeration. For large networks these methods require substantial computer time but can provide efficient solutions for moderately sized networks with few resource limitations. Patterson (1984) compares and contrasts three optimal solution methods.

Resource Constraints for Multiproject Scheduling

Dealing with resource constraints in single-product scheduling problems can be difficult, but the difficulties are magnified significantly when a common pool of resources is shared by a number of otherwise independent projects. Figure 12–20 shows this type of problem. Two projects require resources A and B. Delaying activities in order to resolve resource conflicts can have far-reaching consequences for all projects requiring the same resources. Commercial computer systems exist that have the capability of dealing with tens of projects and resource types, and possibly thousands of activities.

Resource Loading Profiles

Project planning is a useful means for generating schedules of interrelated activities. One significant kind of planning result involves the loading profiles of the required resources. A loading profile is a representation over time of the resources needed. As long as the requirements associated with each activity are known, one can easily obtain the resulting loading profiles for all required resources.

EXAMPLE 12.10

We illustrate the procedure with the case first introduced in Example 12.4. Suppose that the two programmers are developing the program on a single multiuser computer system. The system can segment the random access memory (RAM) between the two users and can segment the permanent storage on the hard disk as well. Both RAM and permanent storage are measured in gigabytes (GB). The requirements for RAM

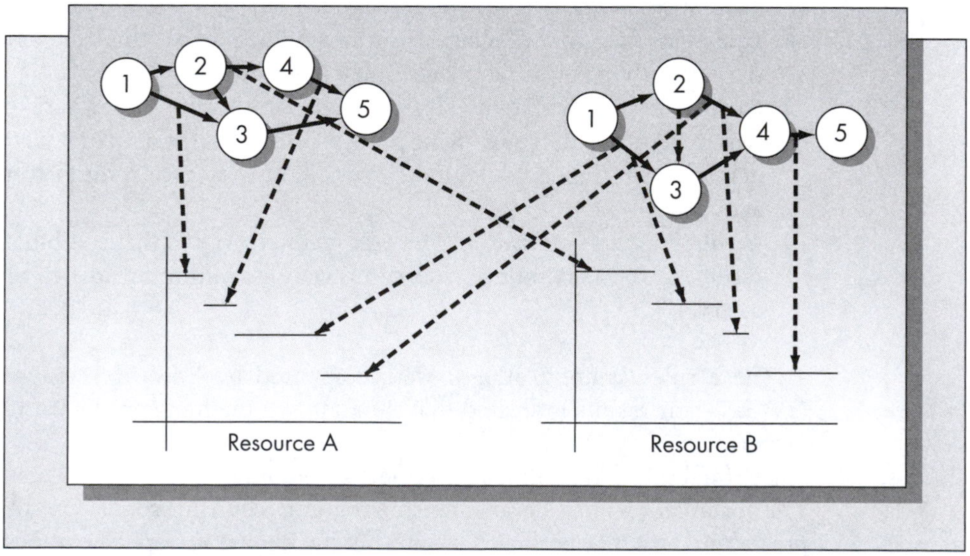

Figure 12–20 Two projects sharing two resources

memory and permanent memory for each of the activities comprising the project are given in the following table.

Activity	RAM Required (GB)	Permanent Storage Required (GB)
A	0	0
B	1	60
C	0.5	5
D	2	30
E	1	10
F	3	5
G	1.5	15
H	2	10
I	4	80

Assume that the activities comprising the project follow the starting and ending times pictured in the Gantt chart of Figure 12–19. Figure 12–21 shows the resulting load profiles of RAM and permanent memory. According to these profiles, the system will require at least 5 GB of RAM and 80 GB of permanent storage. Notice that we are assuming that these requirements are *not* cumulative. Once a portion of the project is completed, the results can be stored offline and retrieved at a later time.

Resources are either consumable or nonconsumable. In Example 12.10 the resources were nonconsumable. The workforce is another example of a nonconsumable resource. Typical consumable resources are cash or fuels. An issue that arises with consumable resources is the cumulative amount of the resource consumed at each point in time. We will explore load profiles for consumable resources in the problems at the end of this section.

A desirable feature of load profiles is that they be as smooth as possible. Large variations in resource requirements make planning difficult and may result in exceeding resource availability at some point in time. The idea behind **resource leveling** is to reschedule noncritical activities within the available slack in order to smooth out the pattern of resource usage. Often it is possible to do this rescheduling by inspection. For larger networks a systematic method is desirable.

In summary, resource loading profiles provide an important means of determining the requirements imposed by any particular schedule. Given a project schedule, one can usually construct the loading profiles without the aid of a computer. Rather than a strict forecast of requirements, the profiles are often more useful as rough planning guides when significant variations in activity time are anticipated. They are probably more widely used than any other resource analysis technique (Moder, Phillips, and Davis, 1983).

PROBLEMS FOR SECTION 12.6

22. For Example 12.10, suppose the following requirements for RAM and permanent storage.

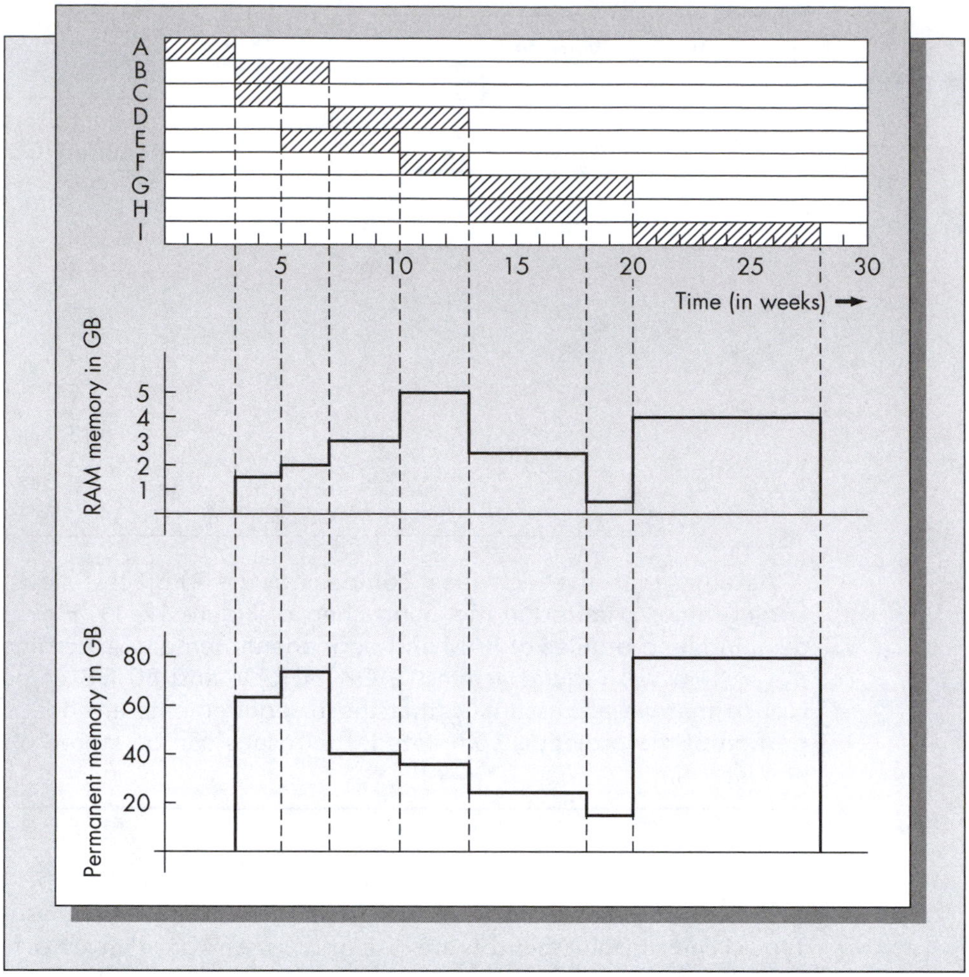

Figure 12–21 Load profiles for RAM and permanent memory (refer to Example 12.10)

Activity	RAM Required (GB)	Permanent Storage Required (GB)
A	0	0
B	1.5	30
C	2.5	20
D	0.5	10
E	2.0	40
F	1.5	15
G	2.0	25
H	1.5	20
I	4.0	50

a. Determine the load profiles for RAM and permanent storage assuming that the activities are scheduled according to the Gantt chart of Figure 12–7.

b. Determine the load profiles for RAM and permanent storage assuming that the activities are scheduled according to the Gantt chart of Figure 12–19.

23. Consider the project described in problem 3. Three machines (M1, M2, and M3) are required to complete the project. The requirements for each activity follow.

Activity	Machines Required
A	M1, M2
B	M1, M3
C	M2
D	M1, M2
E	M2
F	M1, M3
G	M2, M3
H	M1

Determine the minimum time required to complete the project if there is only one machine of each type available. How many weeks are added to the project time when resource requirements are considered?

24. Consider the project described in problem 6. The tasks require both welders and pipe fitters, with the following requirements.

Activity	Number of Welders Required	Number of Pipe Fitters Required
A	6	10
B	3	15
C	8	8
D	0	20
E	10	6
F	10	9
G	4	14

a. Determine the load profiles for welders and pipe fitters assuming an early-start schedule.

b. Determine the load profiles assuming a late-start schedule.

25. Consider the project network pictured in Figure 12–22 (p. 752). Activity times, measured in days, are shown in parentheses next to each activity label.

a. Determine the earliest and the latest starting and finishing times for all activities. Draw a Gantt chart based on earliest starting times but indicate activity slack where appropriate. How many days are required to complete the project?

b. A single critical piece of equipment is required in order to complete the following activities: A, B, C, D, G, and H. Determine a feasible schedule for the project assuming that none of these activities can be done simultaneously.

c. Two resources, R1 and R2, are used for each activity. Assume that these are both consumable resources with the following daily requirements.

Activity	Daily Requirement of R1	Daily Requirement of R2
A	4	0
B	8	6
C	10	9
D	18	4
E	12	3
F	5	12
G	3	2
H	0	6

Determine resource loading profiles based on the schedule found in part (b).

d. Based on the results of part (c), determine the *cumulative* amounts of resources R1 and R2 consumed if the schedule found in part (b) is used.

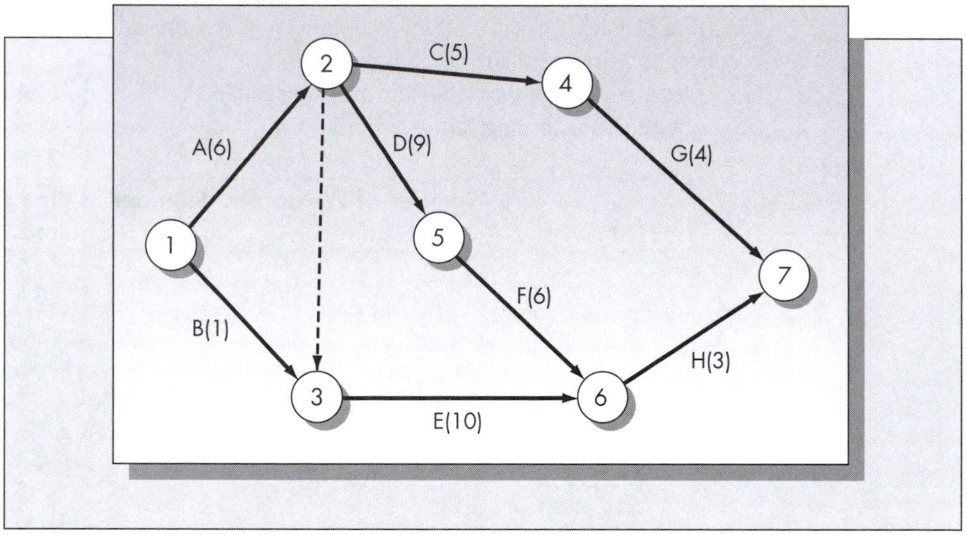

Figure 12–22 Project network (for problem 25)

12.7 ORGANIZATIONAL ISSUES IN PROJECT MANAGEMENT

This chapter has been concerned with reviewing techniques for assisting with the project management function. Successful project management also depends on effective people management. How the organization is structured can be an important factor in whether or not a project succeeds.

The classic structure of an organization is a line organization. That means that there is a clear pyramid structure. Vice presidents report to the president, directors report to vice presidents, middle managers report to directors, and so forth. In a line organization, usually one person at the bottom is assigned to coordinate several employees who may be in other departments at the same level of the organization. The individual responsible will be given the title of project leader or project officer. The line organization is probably the weakest organizational structure for interdepartmental project management.

At the other end of the organizational spectrum is the divisional project organization. In this setting, employees are assigned to the project organization for the life of the project. They report only to the project leader. This is the strongest organizational structure for successful project completion. However, the divisional project organization requires frequent shifting of employees among projects, which can be disruptive for the employees and expensive for the company.

More recently, firms have been experimenting with the matrix organization. In the matrix organization, the firm is organized both horizontally and vertically. The vertical structure is the same as in the traditional line organization. The horizontal direction corresponds to individual projects that may span several functional departments. Each employee reports vertically to his or her functional superior and horizontally to his or her project leader.

The matrix organization is a compromise between the pure line organization and the project organization. However, when project teams span functional departmental boundaries, problems arise. Typically, employees' first loyalty will be to their direct functional superior. Dual subordination causes conflicts when demands are made on an employee from two directions at once. Hence, in order for the matrix concept to work, the project leader must be empowered to set priorities and provide incentives for outstanding performance from project members. In addition, there must be a shared sense of responsibility for the project among project team members. Managing split loyalties among team members is no easy task, however. Texas Instruments Corporation, for example, long advocated the matrix organization but found that ambiguous lines of authority were causing problems.

The structure of the project team varies with the application. For example, a project team assigned the task of implementing a manufacturing planning and control system might consist of five to eight employees. It would ideally be comprised of one representative from each of marketing, engineering, production planning, line manufacturing, and management information systems departments. A team member from finance also might be desirable. The project team should be freed from other responsibilities and physically isolated from their functional departments during the project's course—a project organization (Jacobs, Berry, Whybark, and Vollman, 2010). The best team is comprised of experienced employees from within the company who are familiar with the business. The project team leader should be someone who will ultimately be a user of the new system, such as a professional in production planning or manufacturing.

12.8 HISTORICAL NOTES

It is generally recognized that the two network-based scheduling systems discussed in this chapter, CPM and PERT, were developed almost simultaneously in the United States in the late 1950s. CPM was a result of a joint effort by the Du Pont Company and Remington Rand (Walker and Sayer, 1959). The technique resulted from a study aimed at reducing the time required to perform plant overhaul, maintenance, and construction. The CPM methodology outlined in this report included the cost–time trade-off of indirect and direct costs discussed in Section 12.3, as well as the methods for developing the project network and identifying the critical path. That linear programming could be used to solve project scheduling problems appears to have been first discovered by Abraham Charnes and William Cooper (1962).

PERT was developed as a result of the Polaris Weapons System program undertaken by the United States Navy in 1958. PERT was the result of a joint effort by Lockheed Aircraft Corporation, the Navy Special Projects Office, and the consulting firm of Booz, Allen, and Hamilton. Although the PERT system shared many features with the CPM system, the PERT project focused on activity time uncertainty rather than on project costs. An interesting issue related to the PERT approach is the justification for the approximation formulas for the mean and variance. Both Frank Grubbs (1962) and M. W. Sasieni (1986) raised this issue. It appears that the formula for the variance is assumed (based on properties of the normal distribution), and the formula for the mean was obtained as a consequence of the assumption of the beta distribution and the formula assumed for the variance (see Archibald and Villoria, 1967, p. 449).

Joseph Moder, Cecil Phillips, and Edward Davis (1983, pp. 10–12) discuss Harmonygraph, a little-known planning device that precedes the development of PERT and CPM by almost 30 years and may have been the true genesis of project planning methodology. In 1931 a Polish scientist named Karol Adamiecki (1931) developed and published the Harmonygraph, which is basically a vertical Gantt chart modified to include immediate predecessors. The technique requires the use of sliding tabs for each activity.

Research into project networks continues. One area of continuing interest is resource-constrained networks (Patterson, 1984). Another is networks with random activity times. As we indicated in Section 12.5, the PERT methodology gives only an approximation of the distribution of the project completion time. Interest has also focused on developing efficient simulations (Sullivan, Hayya, and Schaul, 1982) or determining the exact distribution of the project completion time assuming other than a beta distribution for the time required for each activity (Kulkarni and Adlakha, 1986).

12.9 PROJECT MANAGEMENT SOFTWARE FOR THE PC

Project management software was available shortly after mainframe computers were sold to businesses. Most of the major early computer manufacturers (IBM, Honeywell, Control Data, among others) marketed some type of project management software as early as the 1950s. With the spread of personal computers to the business community in the 1980s, software providers realized that there was a significant market for PC-based project management software. Project management tools, along with word processing and spreadsheets, became an important component of the software suite available for personal computers.

Many open source and proprietary software products are available. Microsoft® Project continues to be a popular seller. There is now a Monte Carlo simulation add-in available for Microsoft® Project which, while less popular, allows for a rigorous treatment of project uncertainty. Cloud-based project management platforms are also available. There are project management packages dedicated to specific applications. Several are available for helping legal departments manage their workloads and packages are available for construction/building applications. These are only two examples. The demand for both special-purpose and general-purpose project management software continues to grow.

Project Management Helps United Stay on Schedule

United Airlines has been a long-time user of project management methods and software to help keep its business on track. United owns 788 aircraft and makes 4,900 flights every day. For an operation of this magnitude to work properly takes both a commitment from its employees and quality project management tools. United began using Project Workbench 2.0 software in 1994 to integrate more than 80 mainframe applications in diverse areas such as architecture development and technical support under one operational due date (Ouelette, 1994). In addition, the software assisted United with managing its worldwide fleet maintenance program.

The software allowed United to link dependencies among an unlimited number of distinct projects and to calculate schedules for these projects. The software was more expensive than most of the general-purpose software aimed at the mass market (see the discussion of project management software in Section 12.9). However, United believed that the additional cost was more than offset by the flexibility in scheduling multiple projects with dependent activity links.

Thomas Brothers Plans Staffing with Project Management Software

Thomas Brothers Maps, Inc. supplied street guides and maps that covered many West Coast cities and counties. Before being sold to Rand McNally in 1998, they had 230 workers and annual revenues exceeding $20 million. The president of the firm began using client–server project management software to plan a schedule for hiring and training cartographers. PlanView software handled planning for the 250 projects the company undertook in 1995 and allowed the company to track exactly how each worker was deployed.

One of the features of PlanView was the ability of the software to track resources as well as schedules. PlanView looked at the labor pool in terms of resource overload rather than as a succession of tasks and potential bottlenecks. The firm could track its progress on a large number of multiple projects by maintaining a running schedule of the workload for each cartographer. Another feature of PlanView was that the package used a standard SQL database and interfaced easily with other database products (Oracle 7 for Thomas Brothers). The project management information supplied by the system also tracked of other business functions, such as the cost of sales.

Florida Power and Light Takes Project Management Seriously

Florida Power and Light (FPL), a major utility company, is responsible for managing two nuclear power generating facilities at Turkey Point and St. Lucie. To handle the management of these facilities as well as several other functions within the firm, FPL established an independent project management department. At its peak, the department had 17 project control personnel to support more than 600 engineers and analysts. Separate dedicated groups were established for each plant.

To deal effectively with contractors, the group established project control and execution reporting requirements for all major contracted work. This process required FPL to identify their major contractors and begin negotiations with those contractors to implement the system. It turned out that four contractors were responsible for 80 percent of the workload. Within a year, each of these contractors had implemented a satisfactory project control system.

To make the system more user friendly, FPL abandoned its traditional mainframe software and replaced it with Welcom's PC-based software. While the PC-based system was not as powerful as the mainframe package for handling resource modeling issues, the feeling was that local ownership of the process afforded by PCs would compensate for the new software's limitations. Since that time, FLP has transitioned to Primavera 3 project management software, which allows the best of both worlds through its provision of local project dashboards connected to a central enterprise system.

Self-assessment of FPL's project management function showed that customers were very satisfied with the project management function and scheduling support afforded by the project management group. In fact, FPL was awarded the Deming Prize several years ago for its commitment to continuous improvement and quality of service. The success of its project management initiatives played an important role in this achievement (Cooprider, 1994). More recently, it has won the Nuclear Energy Institute's 2016 top innovation award for its continued excellence in continuous improvement.

12.10 SUMMARY

When large projects consist of many interrelated activities that must be completed in a specific sequence, project management techniques provide useful tools for preparing and administering schedules for these projects. Project planning takes place at many levels of an organization, with projects lasting from a few days to months or even years. This chapter focused on the critical path method (CPM) and its extensions. Networks are a convenient means of representing a project. There are two ways of using networks to represent projects: activity-on-arrow and activity-on-node. Using the activity-on-arrow method, the nodes of the network correspond to completion of some subset of activities. When nodes rather than arrows are used to represent activities, pseudo activities are not required. However, activity-on-arrow is far more common in practice and, in general, more intuitively appealing.

The critical path is the longest path or chain through the network. The length of the critical path is the minimum project completion time, and the activities that lie along the critical path are known as the critical activities. Delay in a critical activity delays the project. Noncritical activities have slack, which means that they can be delayed without necessarily delaying the project. Although for small networks the critical path can be identified readily by inspection, for larger networks an algorithmic procedure is required. This chapter presented a method, involving both forward and backward passes through the network, that specified the earliest and the latest starting and ending times for all activities.

One of the goals of the early development work on CPM was to consider the effect of project costing. We assume that costs are either direct or indirect. Direct costs include labor, material, and equipment; these costs increase if the project time decreases. Indirect costs include costs of rents, interest, and utilities; these costs increase if the project time increases. The goal of the analysis is to determine the optimal time to perform the project that minimizes the sum of indirect and direct costs.

Linear programming is one means of solving project scheduling problems. The linear programming formulations considered here solve both the CPM problem and the CPM problem with cost–time trade-offs. Although not treated in this chapter, linear programming also can be used to solve some cost–time problems when the direct costs are nonlinear functions of the activity duration times.

PERT is an extension of critical path analysis to incorporate uncertainty in the activity times. For each activity, three time estimates are required: (1) the minimum activity time (called a), (2) the maximum activity time (called b), and (3) the most likely activity time (called m). Based on these estimates, one approximates the mean and the standard deviation of the activity time. The project time is assumed to follow a normal distribution with mean equal to the sum of the means along the path with the longest expected completion time and the variance equal to the sum of the variances along this same path. Depending on the configuration of the network, assuming path independence of two or more paths could give more accurate results. Although the PERT approach is only approximate, it does provide a measure of the effect of uncertainty of activity times on the total project completion time.

Resource considerations also were considered. Traditional CPM and PERT methods ignore the fact that schedules may be infeasible because of insufficient resources. Typical examples of scarce resources that might give rise to an infeasible project schedule are workforce, raw materials, and equipment. When schedules are infeasible, noncritical activities should be rescheduled within the available slack if possible. If not, critical activities may have to be delayed and the project completion date moved ahead. Resource loading profiles are a useful tool for determining the requirements placed on resources by any schedule. These profiles can be used as rough planning guides as major projects evolve.

We also discussed organizational design for effective project management. From a project management perspective, the traditional line organization is the weakest organizational structure. On the other end of the spectrum is the project organization. Here, employees are freed from their usual responsibilities for the life of the project. Some firms have experimented with matrix organizations, which is a hybrid of the two designs. Most companies, however, have retained the traditional line structure.

The chapter concluded with a brief overview of the software available for project management. The explosion of the personal computer has been accompanied by an explosion of software. Project management is no exception. There is an entire range of software products available. Although many of the PC-based software can handle multiple projects and resources, very large systems require more powerful tools. There are programs available for cloud-based and other client–server systems that allow for multiple users, projects, and resources. These packages can interface with large databases and other parts of the firm's financial and production systems.

ADDITIONAL PROBLEMS ON PROJECT SCHEDULING

26. Two brothers have purchased a small lot, in the center of town, where they intend to build a gas station. The station will have two pumps, a service area for water and tire maintenance, and a main building with restrooms, office, and cash register area. Before they begin excavating the site, the local authorities must approve the location for a gasoline station and be certain that the placement of the storage tanks will not interfere with water, gas, and electric lines that are already in place.

    Once the site has been approved, the excavation can begin. After excavation, the three primary parts of the construction start: laying in the gasoline tanks, building the water and tire service area (including installation of the air compressor), and constructing the main building. The surfacing can begin after all building is completed. After surfacing, the site must be cleaned and the station's signs erected. However, before the station can open for business, the air compressor must be inspected, tested, and approved.

    The activities and the time required for each of them are listed below.

Activity	Time Required (weeks)
A: Obtain site approval	4
B: Begin site excavation	2
C: Place and secure gasoline tanks	3
D: Install gasoline pumps	1
E: Connect and test gasoline pumps	1
F: Construct service area	2
G: Install and connect water and air compressor	3
H: Test compressor	1
I: Construct main building including the restrooms, the office, and the cash register area	5

Activity	Time Required (weeks)
J: Install plumbing and electrical connections in the main building	3
K: Cover tanks and surface the area	4
L: Clean site	2
M: Erect station signs	1

a. Based on the description of the project, determine the activity precedence relationships and develop a network for the project.
b. Determine the critical path and the earliest and the latest starting and finishing times for each activity.
c. Draw the Gantt chart for this project based on earliest times.
d. Suppose that the air compressor fails to function correctly and must be replaced. It takes two weeks to obtain another compressor and test it. Will the project necessarily be delayed as a result?
e. List the activities that must be completed by the end of the 15th week in order to guarantee that the project is not delayed.
f. Solve this problem using linear programming.

27. A scene is being shot for a film. A total of 11 distinct activities have been identified for the filming. First, the script must be verified for continuity, the set erected and decorated, and the makeup applied to the actors. After the set is completed, the lighting is set in place. After the makeup is applied, the actors get into costume. When these five activities are completed, the first rehearsal of the scene commences, which is followed by the second scene rehearsal with the cameraperson. While the rehearsals are going on, verifications of the audio and the video equipment are made. After both rehearsals and verifications are completed, the scene is shot. Afterward, it is viewed by the director to determine if it needs to be reshot.

Listed below are the activities and the time required for each activity.

Activity	Time Required (days)
A. Check script for story continuity	2.0
B. Decorate set; place necessary props	4.5
C. Check lighting of scene	1.0
D. Apply makeup to actors	0.5
E. Costumes for actors	1.5
F. First rehearsal (actors only)	2.5
G. Video verification	2.0
H. Sound verification	2.0
I. Second rehearsal (with camera and lights)	2.0
J. Shoot scene	3.5
K. Director's OK of scene	1.5

a. Develop a network for the filming of the scene.
b. Compute the earliest and the latest finishing and starting times for each of the activities and identify the critical path.
c. Draw the Gantt chart for this project. Assume that activities with slack are scheduled so that there is equal slack before and after the activities.
d. Suppose that the video verification (G) shows that the equipment is faulty. Four additional days are required to obtain and test new equipment. How much of a delay in the total time required to film the scene will result?
e. One of the costumes is damaged as it is being fitted (activity E). How much extra time is available for repair without delaying the project?
f. What kind of delays can you envision as a result of the uncertainty in the time of activity K?
g. Solve this problem by linear programming.

28. A guidance and detection system is being built as part of a large defense project. The detection portion consists of radar and sonar subsystems. Separate equipment is required for each of the subsystems. In each case, the equipment must be calibrated prior to production. After production, each subsystem is tested independently. The radar and the sonar are combined to form the detection system, which also must be tested prior to integration with the guidance system. The final test of the entire system requires complex equipment. These activities and the times required are listed below.

Activity	Time Required (days)
A. Calibrate machine 1 (for radar)	2.0
B. Calibrate machine 2 (for sonar)	3.5
C. Calibrate machine 3 (for guidance)	1.5
D. Assemble and prepare final test gear	7.0
E. Make radar subsystem	4.5
F. Make sonar subsystem	5.0
G. Make guidance subsystem	4.5
H. Test radar subsystem	2.0
I. Test sonar subsystem	3.0
J. Test guidance subsystem	2.0
K. Assemble detection subsystem (radar and sonar)	1.5
L. Test detection subsystem	2.5
M. Final assembly of three systems	2.5
N. Testing of final assembly	3.5

a. Construct a network for this project.
b. Determine the earliest and the latest starting and finishing times for all the activities and identify the critical path.
c. Draw a Gantt chart for this project based on the earliest times.
d. How much time is available for assembling and calibrating the final test gear without delaying the project?

e. What are the activities that must be completed by the end of 10 days to guarantee that the project is not delayed?
f. What are the activities that must be started by the end of 10 days to guarantee that the project is not delayed?
g. Solve this problem using linear programming.

29. Consider the filming of the scene described in problem 27. Based on past experience, the director is not very confident about the time estimates for some of the activities. The director estimates the following minimum, most likely, and maximum times.

Activity	*a*	*m*	*b*
F	2	3	12
I	1	2	8
J	3	4	10
K	1	2	7

a. Including these PERT time estimates, how long is the filming expected to take?
b. What is the probability that the number of days required to complete the filming of the scene is at least 30 percent larger than the answer you found in part (a)?
c. For how many days should the director plan in order to be 95 percent confident that the filming of the scene is completed?

30. Consider the following project time and cost data.

Activity	Immediate Predecessors	Normal Time	Expedited Time	Normal Cost	Expedited Cost
A	—	6	6	$200	$200
B	A	10	4	600	1,000
C	A	12	9	625	1,000
D	B	6	5	700	800
E	B	9	7	200	500
F	C, D	9	5	400	840
G	E	14	10	1,000	1,440
H	E, F	10	8	1,100	1,460

a. Develop a network for this project.
b. Compute the earliest and the latest starting and finishing times for each of the activities. Find the slack time for each activity and identify the critical path.
c. Suppose that the indirect costs of the project amount to $200 per day. Find the optimal number of days to perform the project by expediting one day at a time. What is the total project cost at the optimal solution? What savings have been realized by expediting the project?
d. Solve this problem using linear programming.

APPENDIX 12–A

Glossary of Notation for Chapter 12

a = Minimum activity time for PERT.
b = Maximum activity time for PERT.
EF_i = Earliest finishing time for activity i.
ES_i = Earliest starting time for activity i.
LF_i = Latest finishing time for activity i.
LS_i = Latest starting time for activity i.
m = Most likely activity time for PERT.
M_{ij} = Expedited time for activity (ij).
μ = Expected activity time estimate for PERT.
N_{ij} = Normal time for activity (ij).
σ = Estimate of the standard deviation of the activity time for PERT.
T = Project completion time for PERT; T is a random variable.
t_{ij} = Time required for activity (ij); t_{ij} is a constant in the standard formulation and a variable in the cost–time formulation.
x_i = Earliest start time for node i (linear programming formulation).

BIBLIOGRAPHY

Adamiecki, Karol. "Harmonygraph." *Polish Journal of Organizational Review,* 1931. (In Polish.)

Archibald, R. D., and R. L. Villoria. *Network-Based Management Systems (PERT/CPM)*. New York: John Wiley & Sons. 1967.

Charnes, A., and W. W. Cooper. "A Network Interpretation and a Directed Subdual Algorithm for Critical Path Scheduling." *Journal of Industrial Engineering* 13 (1962), pp. 213–19.

Cooprider, D. H. "Overview of Implementing a Project Control System in the Nuclear Utility Industry." *Cost Engineering* 36, no. 3 (March 1994), pp. 21–24.

Davis, E. W., and J. H. Patterson. "A Comparison of Heuristic and Optimum Solutions in Resource-Constrained Project Scheduling." *Management Science* 21 (1975), pp. 944–55.

Goodman, L. J. *Project Planning and Management*. New York: Van Nostrand Reinhold, 1988.

Grubbs, F. E. "Attempts to Validate Certain PERT Statistics, or 'Picking on Pert.'" *Operations Research* 10 (1962), pp. 912–15.

Jacobs, F. R., W. L. Berry, D. C. Whybark, and T. E. Vollmann, T. E. *Manufacturing Planning and Control Systems,* 6th ed. New York: McGraw-Hill, 2010.

Kulkarni, V. G., and V. G. Adlakha. "Markov and Markov Regenerative PERT Networks." *Naval Research Logistics Quarterly* 34 (1986), pp. 769–81.

Moder, J. J., C. R. Phillips, and E. W. Davis. *Project Management with CPM, PERT, and Precedence Diagramming*. 3rd ed. New York: Van Nostrand Reinhold, 1983.

Ouelette, T. "Project Management Helps Airline Stick to Schedule." *Computerworld* 28, no. 45 (November 7, 1994), p. 79.

Patterson, J. H. "A Comparison of Exact Approaches for Solving the Multiple Constrained Resource Project Scheduling Problem." *Management Science* 30 (1984), pp. 854–67.

Sasieni, M. "A Note on PERT Times." *Management Science* 32 (1986), pp. 1652–53.

Srinivasan, M. M., W. D. Best, and S. Chandrasekaran. "Warner Robins Air Logistics Center Streamlines Aircraft Repair and Overhaul," *Interfaces* 37 (2007), pp. 7–21.

Sullivan, R. S., J. C. Hayya, and R. Schaul. "Efficiency of the Antithetic Variate Method for Simulating Stochastic Networks." *Management Science* 28 (1982), pp. 563–72.

United States Navy, Special Projects Office, Bureau of Ordnance. "PERT: Program Evaluation Research Task." Phase I Summary Report. Washington, D.C., July 1958.

Walker, M. R., and J. S. Sayer. "Project Planning and Scheduling." Report 6959. Wilmington, DE: E. I. du Pont de Nemours & Co., Inc., March 1959.

CHAPTER 13

Reliability and Maintainability

"Simplicity is prerequisite for reliability."

—Edsger W. Dijkstra

CHAPTER OVERVIEW

Purpose

To gain an appreciation of the importance of reliability, to understand the mechanisms by which products fail, and to acquire an understanding of the mathematics underlying these processes.

KEY POINTS

1. *Preparation.* The topics in this chapter (reliability theory, warranties, and age replacement) are rarely treated in texts on operations. They are included here because of their importance and relevance to the quality movement. However, the mathematics of reliability is complex. One must have a basic understanding of random variables, probability density and distribution functions, and elementary stochastic processes. Several of these methods were also used in Chapter 5, Chapter 8, and in Supplement 2 on queueing. We suggest the reader carefully review those discussions of the exponential distribution.
2. *Reliability of a single component.* Consider a single item whose time of failure cannot be predicted in advance—that is, it is a random variable, T. We assume that we know both the distribution function and density functions of T: $F(t)$ and $f(t)$, respectively. Several important quantities associated with T include the survival function $R(t) = 1 - F(t)$, which is the probability that the item survives beyond t, and the failure rate function, defined as $r(t) = f(t)/R(t)$.

   An important case occurs when the failure rate function is a constant independent of $\boldsymbol{t}$. This results in the failure time distribution having the exponential distribution. The exponential distribution is the only one possessing the memoryless property. In this context, it means that the item is neither getting better nor getting worse with age. Decreasing and increasing failure rate functions, respectively, represent the cases where the reliability of an item is improving or declining with age. The Weibull distribution is a popular choice for representing both increasing and decreasing failure rate functions.
3. *The Poisson process in reliability modeling.* The Poisson process is perhaps the most important stochastic process for applications. When interfailure times are

independent and identically distributed (IID) exponential random variables, one can show that the total number of failures up to any point t follows a Poisson distribution, and the time for n failures follows an Erlang distribution. Because the exponential distribution is memoryless, this process accurately describes events that occur completely at random over time.

4. *Reliability of complex equipment.* Items prone to failure are generally constructed of more than a single component. In a series system, the system fails when any one of the components fails. In a parallel system, the system fails only when all components fail. A third possibility is a K out of N system. Here the system functions as long as at least K components function. Section 13.4 shows how to derive the time to failure distributions for these systems based on the time to failure distributions of the components comprising the systems.
5. *Maintenance models.* Preventive maintenance means replacing an item before it fails. Clearly, this only makes sense for items that are more likely to fail as they age. By replacing items on a regular basis before they fail, one can avoid the disruptions that result from unplanned failures. Based on knowledge of the items' failure mechanisms and costs of planned and unplanned replacements, one can derive optimal replacement strategies. The simplest case gives a formula for optimal replacement times, which is very similar to the EOQ formula derived in Chapter 4.
6. *Warranties.* A warranty is an agreement between the buyer and seller of an item in which the seller agrees to provide restitution to the buyer in the event the item fails within the warranty period. Warranties are common for almost all consumer goods, and extended warranties are a big business. Section 13.8 examines two kinds of warranties: the free replacement warranty and the pro rata warranty. The free replacement warranty is just as it sounds: the seller agrees to replace the item when it fails during the warranty period. In the case of the pro rata warranty, the amount of restitution depends on the remaining time of the warranty. (Pro rata warranties are common for tires, for example, where the return depends on the remaining tread on the tire.)
7. *Software reliability.* Software is playing an increasingly important role in our lives. With the explosive growth of personal computers, the market for personal computer software has become enormous. Microsoft took advantage of this growth to become one of the world's major corporations within a decade of its founding. There is a lot more to the software industry than personal computers, however. Large databases, such as those managed by the IRS or your state Department of Motor Vehicles require massive information retrieval systems. Software failures can be just as catastrophic as hardware failures, causing major systems to fail.

Our critical systems continue to grow and become more complex. As complexity grows, reliability is threatened. During the week of August 11, 2003, a downed power line near Cleveland, Ohio, triggered one of the worst power outages in U.S. history. Virtually the entire East Coast of the United States and parts of Canada were affected, with several deaths attributed to the blackout. How could such a thing have occurred? The answer is that our electrical grid is linked all over the country and can be brought down even by minor problems. Such systems need to be designed with more attention to their reliability. As our population grows and our basic systems become more complex, such catastrophes will become more common. We depend on the reliability of our infrastructure every day.

In operations management, quality has been a key issue. The dramatic success of Japanese manufacturing has been attributed to a large extent to the quality of their goods. Quality is multidimensional (see Section 10.1), but reliability is certainly a key component. In Table 1–3, we reported on competitive priorities in Europe, Japan, and the United States. Product reliability ranked number one for the group of Japanese firms surveyed. Japanese automakers have had a steadily growing market share in the United States. Perceived product quality is probably the key reason that so many people choose to purchase Japanese automobiles. But what dimension of quality is most important? A likely answer is product reliability. Annual surveys conducted by the Consumer's Union attest to the continued exceptional reliability of Japanese-made automobiles.

Reliability as a field separated from the mainstream of statistical quality control in the 1950s with the postwar growth of the aerospace and the electronics industries in the United States. The Department of Defense took a keen interest in reliability studies when it became painfully apparent that there was a serious problem with the reliability of military components and systems. David Garvin (1988) reported that in 1950 only one-third of the Navy's electronic devices were working properly at any given time; for every vacuum tube the military had in an operational state, there were nine tubes in warehouses or on order. The yearly cost of maintaining some military systems in an operable state was as high as 10 times the original cost of the equipment (Amstadter, 1971).

What is the difference between statistical quality control and reliability? Statistical quality control is concerned with monitoring processes to ensure that the manufactured product conforms to specifications. The random variables of interest are numbers of defects and degree of conformance variation. Reliability considers the performance of a product over time. The random variables of interest concern the amount of elapsed time between failures after the product is placed into service. A definition of reliability that emphasizes its close association with quality has been suggested by Patrick O'Connor and Andre Kleyner (2012): reliability is a time-based concept of quality. Alternatively, reliability is the probability that a component or system will perform a required function for a given period of time when used under stated operating conditions (Ebeling, 2019).

When the first edition of this book was published, a reviewer commented that from the student's point of view, reliability was "a narrow engineering issue which almost never makes it to the front page of *The Wall Street Journal*." This couldn't be further from the truth. Reliability failures cause significant disasters.

The Ford Pinto was recalled in 1978, seven years after it was introduced, for modifications to the fuel tank to reduce leakage and fires resulting from rear-end collisions (Ebeling, 2019). The Three Mile Island disaster in Pennsylvania in 1979 resulted in a partial meltdown of the nuclear reactor (see Snapshot Application). The 1986 explosion of the space shuttle Challenger was the result of the failure of rubber O-rings. The largest recall in automotive history—about 100 million vehicles worldwide—began in 2008 and continued through 2020. Takata airbag inflators used by some 34 automobile brands were recalled because an explosion in the inflator could cause metal fragments to pass through the airbag and enter the interior of the vehicle at high speed, resulting in injury or death. Cracks in the spillway for the Oroville Dam in California in 2017 caused the evacuation of 188,000 people.

Reliability concerns each of us every day. Our lives depend on the reliability of cars, commuter trains, and airplanes. Our livelihoods depend on the reliability of power generation, telephones, and computers. Our health depends on the reliability of pollution control systems, heating and air-conditioning systems, and emergency medical care systems.

Reliability and risk are closely related. Risks of poor reliability are of concern to both the producer and the consumer. Listed below are some aspects of risk from the producer's point of view.

1. *Competition.* Product reliability is an important component of perceived quality by the consumer. Highly unreliable products do not gain customer loyalty and eventually disappear.
2. *Customer requirements.* The U.S. government required weapons systems with clearly specified reliability levels when it found that maintenance costs for these systems were becoming prohibitive. Today, reliability requirements established by the buyer are common.
3. *Warranty and service costs.* Warranties, which will be treated in the second part of this chapter, are a significant financial burden to the manufacturer when products are unreliable. Automobile firms commonly provide extended warranties as an incentive to the consumer to boost sales. Often they are priced as an add-on to the base price, thus providing a further income source for the manufacturer (so long as their price exceeds the cost to the firm from defects).
4. *Liability costs.* Largely as an outgrowth of the efforts of Ralph Nader, the U.S. Congress has enacted legislation that makes a manufacturer liable for the consequences of failures in product performance resulting from faulty design or manufacture. Liability losses have had the effect of shifting some of the costs of poor reliability from the consumer to the manufacturer.

Below are some of the risks of poor reliability borne by the consumer.

1. *Safety.* There is no doubt that equipment failure results in human death. Approximately 35,000 Americans die in automobile accidents on the nation's roads each year. Undoubtedly, some portion of these are attributable to mechanical failure. Travelers die in airplane accidents, many of which are the result of equipment failure, such as the 2019 crashes resulting from design issues with Boeing 737 aircrafts. The failure of the nuclear plants at Three Mile Island (1979), Chernobyl (1986), and Fukishima (2011) resulted in human death and injury from radiation exposure. Safety and reliability are closely linked.
2. *Inconvenience.* Even though many failures do not result in death, they can be a source of frustration and delay. Delays at airports are common because some piece of equipment on the plane fails to operate properly. Automobile breakdowns may leave motorists stranded for hours. Failure of communication equipment, computer equipment, or power generation plants can cripple businesses.
3. *Cost.* Poor reliability costs everyone in the end. For this reason, consumers are willing to pay a premium for products with higher reliability. As discussed in Chapter 1, product reliability is a key competitive priority for many brands.

Why study reliability? We need to understand the probability laws governing failure patterns in order to better design processes to build reliable systems. Incorrect analysis can lead to disastrous consequences. The U.S. Nuclear Regulatory Commission (1975) predicted that it would be hundreds of years before we could expect a major accident in a nuclear plant. Given that we have since observed three major accidents, the analysis in this report is clearly flawed.

Reliability is an issue of concern for operations management from two perspectives. First, to implement effective quality management, we must understand how and why products fail. Reliability will continue to be a key component of quality. Designing an effective quality delivery system will require understanding the randomness of failure patterns. Second, we need to understand the failure patterns of the equipment used in the manufacturing process. In this way, we can develop effective maintenance policies for that equipment.

13.1 RELIABILITY OF A SINGLE COMPONENT

Introduction to Reliability Concepts

In order for readers to better understand and appreciate some of the definitions and concepts introduced in this chapter, this section starts with an example.

EXAMPLE 13.1

In 1970 the U.S. Army purchased 1,000 identical capacitors for use in short-distance radio transmitters. The army maintained detailed records on the failure pattern of the capacitors with the following results.

Number of years of operation	1	2	3	4	5	6	7	8	9	10	>10
Number of failures	220	158	121	96	80	68	47	40	35	25	110

Based on these data, the army wanted to estimate the probability distribution associated with the failure of a capacitor chosen at random.

Define the random variable T as the time that the capacitor will operate before failure. We can estimate the cumulative distribution function of T, $F(t)$, using the data provided in the table. In symbols, $F(t) = P\{T \leq t\}$, and in words, $F(t)$ is the probability that a component chosen at random fails at or before time t.

In order to estimate $F(t)$ from the given data, we find the cumulative number of failures and the proportion of the total this number represents each year.

Number of years of service	1	2	3	4	5	6	7	8	9	10	>10
Cumulative failures	220	378	499	595	675	743	790	830	865	890	1,000
Proportion of total	.220	.378	.499	.595	.675	.743	.790	.830	.865	.890	1.0

The proportions are estimates of $F(t)$ for $t = 1, 2, \ldots$. These probabilities may be used directly to compute various quantities of interest by treating T as a discrete random variable, or they may be used to estimate the parameters of a continuous distribution. We will use the discrete version to answer the following questions about the lifetime of the capacitors.

a. The probability that a capacitor chosen at random lasts more than five years.
b. The proportion of the original 1,000 capacitors put into operation that fail in year 6.
c. The proportion of components that survive for at least five years that fail in year 6.
d. The proportion of components that survive at least eight years that fail in year 9.

Solution

a. $P\{T > 5\} = 1 - P\{T \le 5\} = 1 - .675 = .325.$

b. $P\{T = 6\} = P\{T \le 6\} - P\{T \le 5\} = .743 - .675 = .068.$

c. At first glance, this question might appear to be the same as part (b). However, there is an important difference. Part (b) asks for the proportion of the original set of capacitors failing in year 6, whereas part (c) asks for the proportion of components lasting at *least five years* that fail in year 6. This is a conditional probability.

$$P\{T = 6 \mid T > 5\} = \frac{P\{T = 6, T > 5\}}{P\{T > 5\}} = \frac{P\{T = 6\}}{P\{T > 5\}} = \frac{.068}{.325} = 0.209.$$

Notice that the events $\{T = 6, T > 5\}$ and $\{T = 6\}$ are equivalent since $\{T = 6\} \subset \{T > 5\}$.

d. $P\{T = 9 \mid T \ge 8\} = \dfrac{P\{T = 9\}}{P\{T \ge 8\}} = \dfrac{.035}{.170} = 0.206,$

which is almost precisely the same as the proportion of components surviving more than five years that fail in year 6. In fact, the proportion of components that survive n years and fail in year $n + 1$ is very close to .20 for $n = 0, 1, 2, \ldots, 9$. As we will later see, this results in a failure distribution with some unusual properties.

Preliminary Notation and Definitions

Many new concepts were introduced in Example 13.1. We will now formalize the definitions to be used throughout this chapter.

As before, define the random variable T as the lifetime of the component. We assume that T has cumulative distribution function $F(t)$ given by

$$F(t) = P\{T \le t\}.$$

In what follows we will treat $F(t)$ as a differentiable function of t, so that the probability density function $f(t)$, given by the equation

$$f(t) = \frac{dF(t)}{dt},$$

will exist.

In addition to the distribution and density functions of the random variable T, we will be interested in related functions. One is the **reliability function** (also known as **the survival function**). The reliability function of the component, which we call $R(t)$, is given by

$$R(t) = P\{T > t\} = 1 - F(t).$$

In words, $R(t)$ is the probability that a new component will survive past time t. Notice that this implies that $F(t)$ is the probability that a new component will not survive past time t.

Consider the following conditional probability:

$$P\{t < T \le t + s \mid T > t\}.$$

This is the conditional probability that a new component will fail between t and $t + s$ given that it lasts beyond t. We may think of this conditional probability in the following way. Interpret t as now and s as an increment of time into the future. The event $\{T > t\}$ means that the component has survived until the present—in other words, it is still working. The conditional event $\{t < T \le t + s \mid T > t\}$ means that the component is working now but will fail before an additional s units of time have passed.

Recall from elementary probability theory that for any events A and B

$$P\{A \mid B\} = \frac{P\{A \cap B\}}{P\{B\}}.$$

In the special case where $A \subset B$, $A \cap B = A$, so

$$P\{A \mid B\} = \frac{P\{A\}}{P\{B\}} \qquad \text{when } A \subset B.$$

Identify the event $A = \{t < T \le t + s\}$ and $B = \{T > t\}$. A little reflection shows that $A \subset B$ in this particular case, so

$$P\{t < T \le t + s \mid T > t\} = \frac{P\{t < T \le t + s\}}{P\{T > t\}} = \frac{F(t + s) - F(t)}{R(t)}.$$

We will divide by s and let s approach zero.

$$\lim_{s \to 0} \frac{1}{s} \frac{F(t + s) - F(t)}{R(t)} = \frac{f(t)}{R(t)}.$$

This ratio turns out to be a fundamental quantity in reliability theory.

Define

$$r(t) = \frac{f(t)}{R(t)}.$$

We call $r(t)$ the **failure rate function**. Its derivation is the best way to understand what the failure rate function means. For positive s, the conditional probability used to derive $r(t)$ is the probability that a component that has survived up until time t fails between times t and $t + s$. Dividing by s and letting s go to zero is the same way one derives a first derivative. Hence, the failure rate function is the rate of change of the conditional probability of failure at time t. It can be considered a measure of the likelihood that a component that has survived up until time t fails in the next instant of time.

The failure rate function is a fundamental quantity in reliability theory but, like the probability density function, does not have a direct physical interpretation. However, for values of Δt sufficiently small, the term $r(t)\Delta t$ is the probability that an item that survives to time t fails between t and $t + \Delta t$. How does one determine if Δt is sufficiently small? In general, Δt should be small relative to the lifetime of a typical component. However, the only way to be certain as to whether $r(t)\Delta t$ is a good approximation to this conditional probability is to compute $P\{t < T \le t + \Delta t \mid T > t\}$ directly.

EXAMPLE 13.2

The length of time that a particular piece of equipment operates before failure is a random variable with cumulative distribution function

$$F(t) = 1 - e^{-0.043t^{2.6}}.$$

Consider the following.

a. The failure rate function.
b. The probability that the equipment operates for more than five years without experiencing failure.
c. Suppose that 100 pieces of the equipment are placed into service in year 0. What fraction of the units surviving four years fail in year 5? Can one accurately estimate this proportion using only the failure rate function?
d. What fraction of units surviving four years fail in the first month of year 5? Can this be accurately estimated using the failure rate function?

Solution

a. $f(t) = \dfrac{dF(t)}{dt} = -e^{-0.043t^{2.6}} \dfrac{d}{dt}(-0.043t^{2.6})$

$$= (0.043)(2.6)t^{1.6}e^{-0.043t^{2.6}}$$

$$= 0.1118t^{1.6}e^{-0.043t^{2.6}}.$$

Since $R(t) = 1 - F(t) = e^{-0.043t^{2.6}}$, it follows that

$$r(t) = \frac{f(t)}{R(t)} = 0.1118t^{1.6}.$$

b. $P\{T > 5\} = R(5) = e^{-0.043(5)^{2.6}} = e^{-2.8235} = 0.0594.$

c. We will compute this directly and compare the result with $r(t)\Delta t$. The proportion of units surviving four years that fail in year 5 is

$$P\{4 < T \le 5 \mid T > 4\} = \frac{F(5) - F(4)}{R(4)} = \frac{R(4) - R(5)}{R(4)}$$

$$= \frac{e^{-0.043(4)^{2.6}} - e^{-0.043(5)^{2.6}}}{e^{-0.043(4)^{2.6}}}$$

$$= \frac{0.2059 - 0.0594}{0.2059}$$

$$= 0.7115.$$

This means that about 71 percent of the machines surviving four years will fail in the fifth year. As about 94 percent of the units fail in the first five years of operation [$F(5) = .9406$], a value of $\Delta t = 1$ year is probably too large for $r(t)\Delta t$ to give a good approximation. In fact, we see that $r(4)(1) = (0.1118)(4)^{1.6} = 1.0274$.

d. Because one month corresponds to $\frac{1}{12} = 0.0833$ year, we wish to compute

$$P\{4 < T \le 4.0833 \mid T > 4\} = \frac{F(4.0833) - F(4)}{R(4)} = \frac{0.2059 - 0.1887}{0.2059} = 0.0836.$$

Here $\Delta t = \frac{1}{12}$ should be sufficiently small to use the failure rate function to estimate this probability. We obtain $r(4)\,\frac{1}{12} = 0.0856$, which is quite accurate.

The Exponential Failure Law

The **exponential distribution** plays a fundamental role in reliability theory and practice because it accurately describes the failure characteristics of many types of operating equipment. The exponential law can be derived in several ways. We will consider a derivation that utilizes the failure rate function.

We know that the failure rate function $r(t)$ is given by the formula

$$r(t) = f(t)/R(t)$$

and is a measure of the likelihood that a unit that has been operating for t units of time fails in the next instant. Consider the following case: $r(t) = \lambda$, for some constant $\lambda > 0$. This means that the likelihood that a working unit fails in the next instant of time is independent of how long it has been operating. This implies that the unit does not exhibit any signs of aging. It is equally likely to fail in the next instant whether it is new or old. We will derive the probability distribution of the lifetime T that corresponds to a constant failure rate function.

We can determine a solution to the equation $r(t) = \lambda$ by noting that since $R(t) = 1 - F(t)$,

$$f(t) = \frac{dR(t)}{dt} = -R'(t).$$

Hence, the equation $r(t) = \lambda$ may be written in the form

$$\frac{-R'(t)}{R(t)} = \lambda$$

or

$$R'(t) = -\lambda R(t).$$

This is the simplest first-order linear differential equation. Its solution is

$$R(t) = e^{-\lambda t} + c,$$

where c is a constant that is easily shown to be zero in this case. It follows that the distribution function $F(t)$ is given by

$$F(t) = 1 - e^{-\lambda t}$$

and the density function $f(t)$ is given by

$$f(t) = \lambda e^{-\lambda t}.$$

This is known as the exponential distribution. It depends on the single parameter λ, which represents a rate of occurrence. If T has the exponential distribution with parameter λ, then T corresponds to the lifetime of a component that exhibits no aging over time; that is, a component that has survived up until time t_1 is equally likely to fail in the next instant of time as one that has survived up until time t_2 for any times t_1 and t_2. The expected failure time is $1/\lambda$. The standard deviation of the failure time is also $1/\lambda$. The exponential density and distribution functions appear in Figure 13–1.

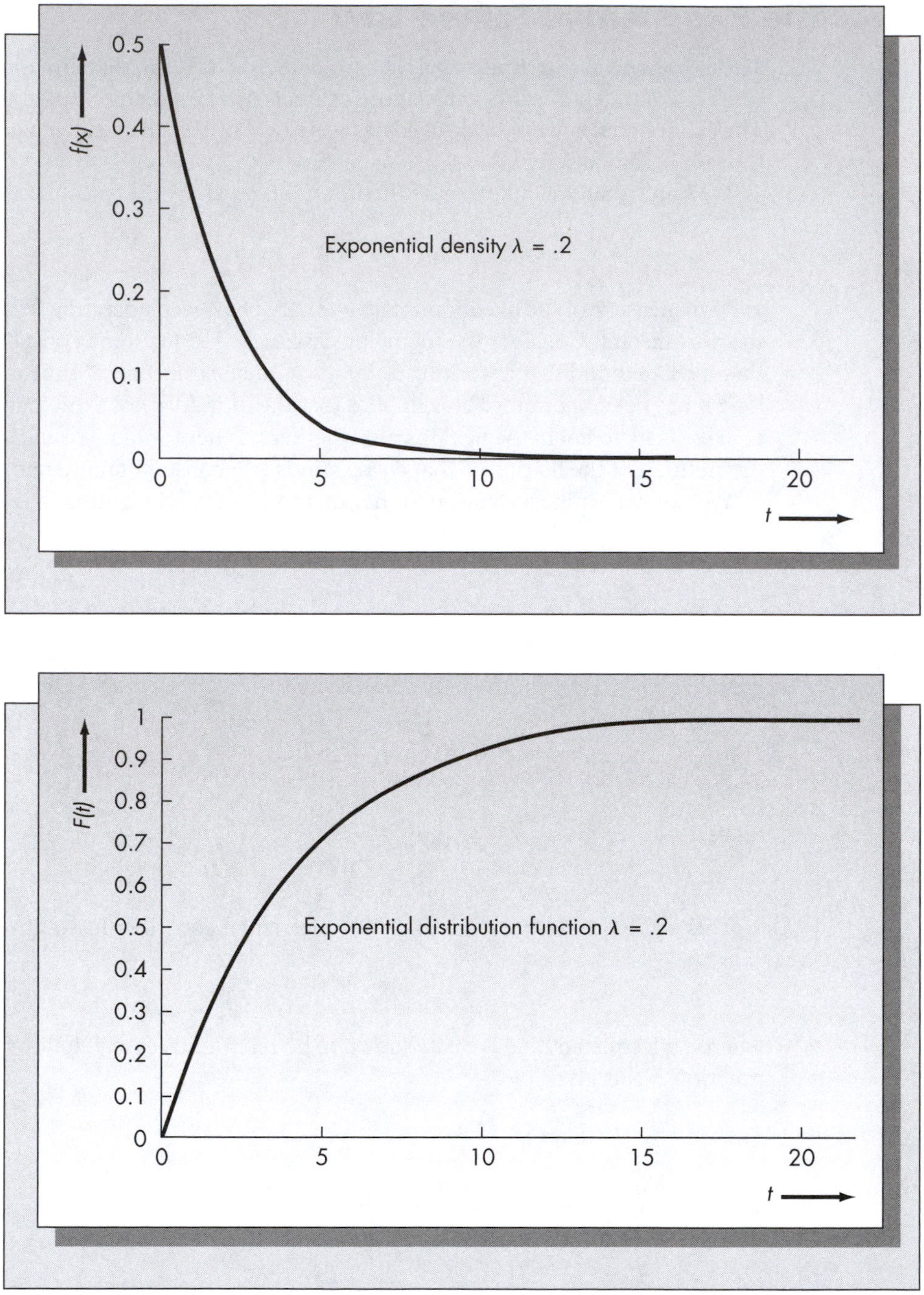

Figure 13–1 The exponential density and distribution functions

EXAMPLE 13.3

Because the exponential distribution has a constant failure rate function, it is likely that the failure law for the capacitor described in Example 13.1 is exponential. This follows because we observed that the proportion of the capacitors having survived for n years that failed in year $n + 1$ was the same for $n = 0, 1, \ldots$, which was approximately

20 percent. In order to estimate the value of λ, note that the proportion failing in the first year is also 20 percent, which gives $F(1) = 0.20 = 1 - e^{-\lambda}$. Solving for λ results in $e^{-\lambda} = 0.8$, or $\lambda = -\ln(0.8) = 0.223$.

We can compare the expected number of capacitors that would fail if the true lifetime distribution were exponential versus the actual number that failed to see if the exponential distribution provides a reasonable fit of the data.

Number of years of operation	1	2	3	4	5	6	7	8	9	10	>10
Number of failures	220	158	121	96	80	68	47	40	35	25	110
Expected number of failures under exponential law with $\lambda = 0.223$	200	160	128	102	82	66	52	42	34	27	107

The expected number of failures for each year is obtained by multiplying the probability of failure for a given year, assuming the exponential law, by the total number of units. For example, one finds the expected number of failures for year 3 by multiplying 1,000 by $F(3) - F(2) = e^{-2\lambda} - e^{-3\lambda} = 0.1280$. Clearly there is a close agreement between the actual number of failures and the expected number, indicating that the exponential distribution provides a good fit for the observed historical data. (It is easy to verify by a formal goodness-of-fit test that the exponential distribution fits these data very closely.)

We can now answer the following inquiries about the capacitors using the exponential distribution directly.

a. The probability that a capacitor chosen at random lasts more than eight years.
b. The proportion of capacitors that survive three years that also survive at least three additional years.

Solution

a. $P\{T > 8\} = e^{-\lambda t} = e^{-(.223)(8)} = e^{-1.784} = .1680.$
b. We wish to compare $P\{T > 6 \mid T > 3\}$. Using the laws of conditional probability, we have

$$P\{T > 6 \mid T > 3\} = \frac{P\{T > 6, T > 3\}}{P\{T > 3\}}$$

$$= \frac{P\{T > 6\}}{P\{T > 3\}} = \frac{.2624}{.5122} = .5122.$$

It is not a coincidence that this is the same as the unconditional probability that a new capacitor lasts more than three years. This property is known as the memoryless property of the exponential distribution.

The **memoryless property** (see Chapter 8 for additional discussion) of the exponential distribution relates to the following conditional probability:

$$P\{T > t + s \mid T > t\}.$$

This is the probability that the component survives past time $t + s$ given that it has survived until time t. If we think of time t as now, then this is the probability that a component that is currently functioning continues to function for at least another s units of time. If T follows an exponential failure law, then

$$
\begin{aligned}
P\{T>t+s \mid T>t\} &= \frac{P\{T>t+s, T>t\}}{P\{T>t\}} \\
&= \frac{P\{T>t+s\}}{P\{T>t\}} \\
&= \frac{e^{-\lambda(t+s)}}{e^{-\lambda t}} \\
&= e^{-\lambda s} \\
&= P\{T>s\}.
\end{aligned}
$$

Note that as $\{T>t+s\} \subset \{T>t\}$, the events $\{T>t+s, T>t\}$ and $\{T>t+s\}$ are equivalent. The last expression, $P\{T>s\}$, is the **unconditional** probability that a new component will last at least s units of time. That is, we have demonstrated that if the component has been operating for t units of time without failure, then the probability that it continues to operate for at least another s units of time is the same as the probability that a new component operates for at least s units of time. This means that there is no aging. The likelihood of failure is independent of how long the component has been operating. However, we required that the lifetime distribution be exponential. In fact, the exponential distribution is the only continuous distribution possessing the memoryless property; that is, it is the only one for which $P\{T>t+s \mid T>t\} = P\{T>s\}$.

When we say that an item fails completely at random, we mean that the failure law for the item is exponential. Events that occur completely at random over time follow a Poisson process. The Poisson process is discussed in Section 13.3 as well as Chapter 8 and Supplement 2.

PROBLEMS FOR SECTION 13.1

1. Three hundred identical fuses placed into service simultaneously on January 1, 2006, experienced the following numbers of failures through December 31, 2018.

Year	2013	2014	2015	2016	2017	2018
Number of failures	13	19	16	34	21	38

Assume that there were no failures before 2013.

a. Based on these data, estimate the cumulative distribution function (CDF) of a fuse chosen at random.

Using the results of part (a), estimate the following probabilities for a fuse chosen at random.

b. Lasts more than 5 years.
c. Lasts more than 10 years.
d. Lasts more than 12 years.
e. That a fuse that has survived for 10 years fails in the 11th year of operation.

2. Suppose that the cumulative distribution function of the lifetime of a piece of operating equipment is given by $F(t) = 1 - e^{-0.6t} - 0.6te^{-0.6t}$, where t is measured in years of continuous operation.

a. Determine the reliability function.
b. Determine the failure rate function.
c. What is the probability that this piece of equipment fails in the first year of operation?
d. What is the probability that this piece of equipment fails in the fifth year of operation?
e. What proportion of the equipment surviving four years fails in the fifth year? (Calculate without using the failure rate function.)
f. Does $r(4)$ closely approximate the answer to part (e)? Why or why not?
g. What proportion of the equipment surviving four years fails in the first month of the fifth year? (Calculate using the failure rate function.)

3. A large number of identical items are placed into service at time 0. The items have a failure rate function given by $r(t) = 1.105 + 0.30t$, where t is measured in years of operation.
   a. Derive $R(t)$ and $F(t)$.
   b. If 300 items are still operating at time $t = 1$ year, approximately how many items would you expect to fail between year 1 and year 2?
   c. Does the value of $r(1)$ yield a good approximation to the conditional probability computed in part (b)? Why or why not?
   d. Repeat the calculation of part (b) but determine the expected number of items that fail between $t = 1$ year and $t = 1$ year plus 1 week. Does $r(t)\Delta t$ provide a reasonable approximation to the conditional probability in this case? Why or why not?
4. A microprocessor that controls the tuner in color TVs fails completely at random (that is, according to the exponential distribution). Suppose that the likelihood that a microprocessor that has survived for k years fails in year $k + 1$ is .0036. What is the cumulative distribution function of the time until failure of the microprocessor?
5. A pressure transducer regulates a climate control system in a factory. The transducer fails according to an exponential distribution with rate one failure every five years on average.
   a. What is the cumulative distribution function of the time until failure?
   b. What is the probability that a transducer chosen at random functions for eight years without failure?
   c. What is the probability that a transducer that has functioned for eight years continues to function for another eight years?
6. For the pressure transducer mentioned in problem 5, use the failure rate function to estimate the likelihood that a transducer that has been operating for six years fails in the seventh year. How close is the approximation to the exact answer?

13.2 INCREASING AND DECREASING FAILURE RATES

Although the constant failure rate function that leads to the exponential law is significant in reliability theory, there are other important failure laws as well. Most of us are more familiar with items that possess increasing failure rate functions. That is, they are more likely to fail as they get older. Decreasing failure rate functions also occur frequently. New products often have a high failure rate because of the "burn-in" phase, in which the defective items in the population are weeded out.

An important class of failure rate functions that includes both increasing and decreasing failure rate functions is of the form

$$r(t) = \alpha\beta t^{\beta-1} \qquad \text{where } \alpha \text{ and } \beta > 0.$$

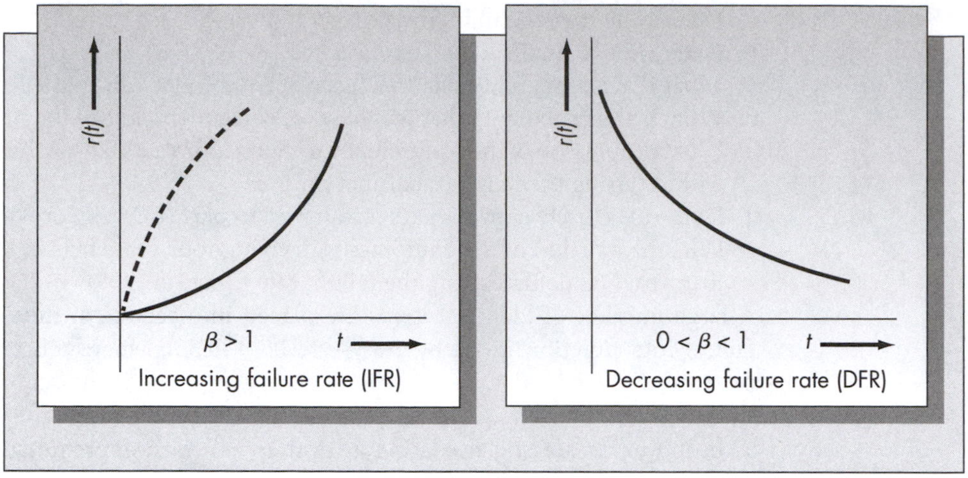

Figure 13–2 Failure rate functions for the Weibull lifetime distribution

Here $r(t)$ is a polynomial function in the variable t that depends on the two parameters α and β. When $\beta > 1$, $r(t)$ is increasing, and when $0 < \beta < 1$, $r(t)$ is decreasing. Typical failure rate functions for these cases are pictured in Figure 13–2.

This form of $r(t)$ will yield another differential equation in $R(t)$. It can be shown that the solution in this case will be

$$R(t) = e^{-\alpha t^{\beta}} \qquad \text{for all } t \geq 0$$

or

$$F(t) = 1 - e^{-\alpha t^{\beta}} \qquad \text{for all } t \geq 0.$$

This distribution is known as the **Weibull distribution**. It depends on the two parameters α and β, and as we saw earlier, when $0 < \beta < 1$, it corresponds to the lifetime of an item with a decreasing failure rate, and when $\beta > 1$, it corresponds to the lifetime of an item with an increasing failure rate. Because it is often true that empirical failure rate functions (i.e., those that are observed from test data) are closely approximated by polynomials, the Weibull distribution is an accurate description of the failure law of many types of operating equipment. Note that when $\beta = 1$ the Weibull reduces to the exponential. Various Weibull densities appear in Figure 13–3.

EXAMPLE 13.4

A local manufacturer of copying equipment includes a repair warranty with each copier. Virtually all of his equipment exhibit an increasing failure rate. Based on historical repair data, the failure rate for Model 25cc7 is accurately described by the function $r(t) = 2.7786t^{1.3}$, where t is measured in months of continuous operation. What is the probability that the time between two successive failures of this equipment exceeds two months of operation?

Solution

Because $r(t)$ is a polynomial in t, the distribution of the time until failure is the Weibull distribution. It is necessary to identify the values of α and β. We have that

$$2.7786t^{1.3} = \alpha\beta t^{\beta-1}.$$

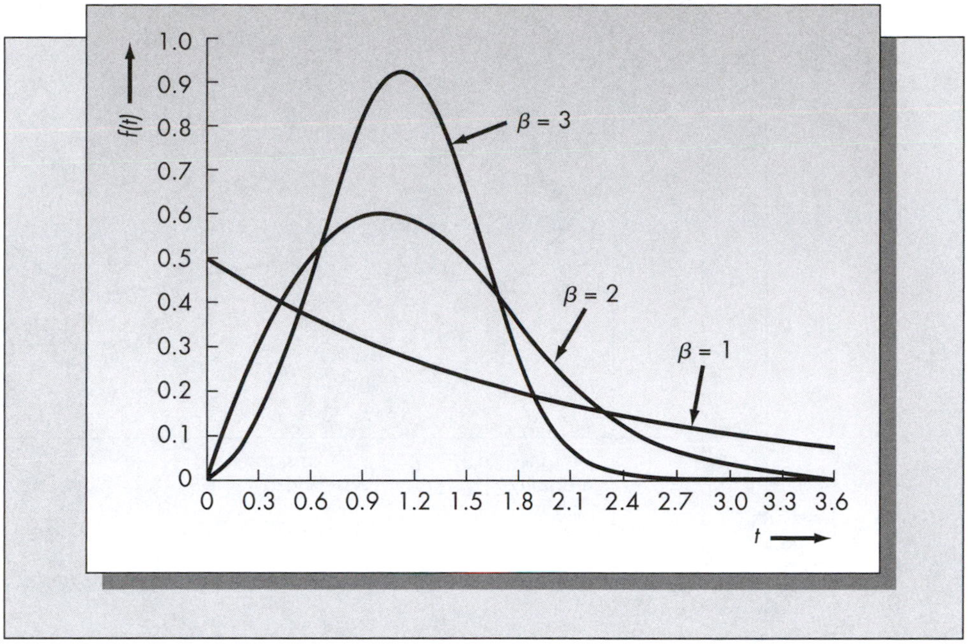

Figure 13–3 Weibull densities for various values of β ($\alpha = 0.5$)

It follows that $\beta - 1 = 1.3$, or $\beta = 2.3$. Since $\alpha\beta = 2.7786$, we obtain

$$\alpha = 2.7786/\beta = 2.7786/2.3 = 1.208.$$

We are required to compute $P\{T > 2\} = R(2)$. Substituting $t = 2$ into the equation for $R(t)$ we obtain

$$R(2) = e^{-1.208 \times (2)^{2.3}} = 4.977 \times 10^{-4}.$$

Sometimes neither increasing nor decreasing failure rate functions accurately describe the failure characteristics of particular equipment. A typical case in point is the "bathtub" failure rate function pictured in Figure 13–4. In the early phases of the product life, the failure rate is decreasing. This follows because defective components fail quickly, causing failure rates to be high initially. This is commonly known as the infant mortality stage. Once bad components are weeded out, the failure rate remains constant until aging begins. At that time we enter the wear-out phase, and the failure rate starts to increase.

If $r(t)$ is an arbitrary failure rate function, then it can be shown that the reliability function $R(t)$ is given by

$$R(t) = \exp\left(-\int_0^t r(u)du\right).$$

It is easy to verify that both the exponential and the Weibull cases satisfy this relationship. Sometimes, such as with the bathtub failure rate function, finding an explicit representation for the function $r(t)$ is difficult. In such cases it is possible to approximate $r(t)$ by step functions. We will not illustrate the procedure here.

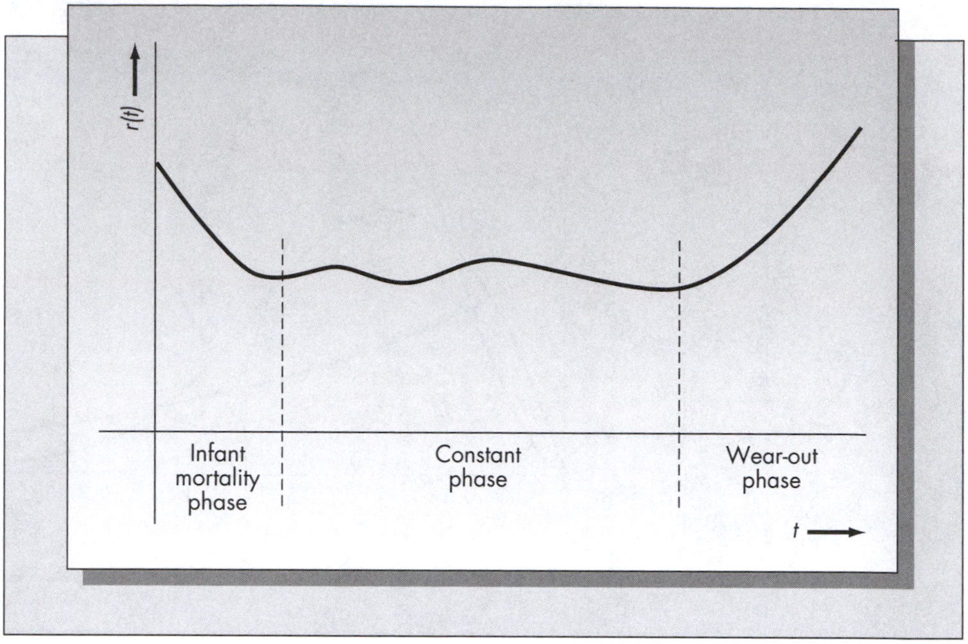

Figure 13–4 "Bathtub" failure rate function

PROBLEMS FOR SECTION 13.2

7. A piece of equipment has a lifetime T (measured in years) that is a continuous random variable with cumulative distribution function

$$F(t) = 1 - e^{-t/10} - (t/10)\, e^{-t/10} \qquad \text{for all } t \geq 0.$$

   a. What is the probability density function of T?
   b. What is the probability that a piece of equipment survives more than 20 years?
   c. What is the probability that a piece of equipment survives more than 10 years but fewer than 20 years?
   d. What is the probability that a piece of equipment survives more than 20 years given that it has survived for 10 years?

8. For the equipment mentioned in problem 7,
   a. Derive the failure rate function $r(t)$ and draw a graph of the function.
   b. Without using the failure rate function, determine the probability that a piece of equipment that has survived 20 years of operation fails in the 21st year.
   c. Does $r(20)$ accurately estimate your answer to part (b)? Why or why not?

9. The Air Force maintains enormous amounts of data on engine failure times. A particular engine has experienced a failure pattern whose failure rate function is closely approximated by $r(t) = 0.000355e^{2.2t}$, where t is in flying hours.
   a. What are the reliability and the cumulative distribution functions of the time until failure?
   b. Determine the value of t such that the likelihood that an engine fails before t is the same as the likelihood that an engine fails after t.

10. A sample of high-capacity resistors is tested until failure and the results fitted to a Weibull probability model. Based on these tests, the reliability function of a resistor is estimated to be

$$R(t) = e^{-0.0013 t^{1.83}}.$$

a. What is the failure rate function for these resistors?
b. Is this resistor more likely to fail as it ages?
c. What is the probability that a resistor will function for more than 30 hours without failure?
d. Suppose that a resistor has been operating for 50 hours. What is the probability that it fails in the 51st hour? [Use the results of part (a) for your calculations.]

13.3 THE POISSON PROCESS IN RELIABILITY MODELING

[Note: The Poisson process was also discussed in Chapter 8 as a model of random arrivals.]

Consider a single piece of operating equipment that fails completely at random. As we saw in Section 13.2, that means that the time until failure follows the exponential distribution. Suppose that when the item fails, it is immediately repaired or that the repair time is sufficiently small compared with the interfailure time that it can be ignored. We then have a process in which events (failures) occur over time, and the times between successive failures are independent identically distributed exponential random variables. Let $T_1, T_2, \ldots$ be random variables corresponding to the times between successive failures. Each of the random variables has distribution function $F(t)$ and reliability function $R(t)$ given by

$$F(t) = 1 - e^{-\lambda t},$$
$$R(t) = e^{-\lambda t},$$

where λ is the rate at which failures occur. We also define a related sequence of random variables $W_1, W_2, \ldots$, which are defined by the equations

$$W_1 = T_1,$$
$$W_2 = T_1 + T_2,$$
$$W_3 = T_1 + T_2 + T_3,$$

and so on.

Interpret W_n as the time of the nth failure. Finally, we introduce the process $N(t)$. Define $N(t)$ as the number of failures that occur up until time t. $N(t)$ is a stochastic process because for each fixed value of t, $N(t)$ is a random variable. Clearly, $N(t)$ is closely related to both the times between failures, $T_1, T_2, \ldots$, and the times of failures $W_1, W_2, \ldots$. When $T_1, T_2, \ldots$ are independent exponentially distributed random variables, $N(t)$ is a Poisson process. A realization of a Poisson process is pictured in Figure 13–5.

We will proceed with the analysis of the Poisson process in the following way. We start with the knowledge of the distribution of the interfailure times, $T_1, T_2, \ldots$. From that we can derive the distribution of the times until failure $W_1, W_2, \ldots$. We obtain the distribution of $N(t)$ by using the following equivalence of events:

$$\{N(t) < n\} = \{W_n > t\}.$$

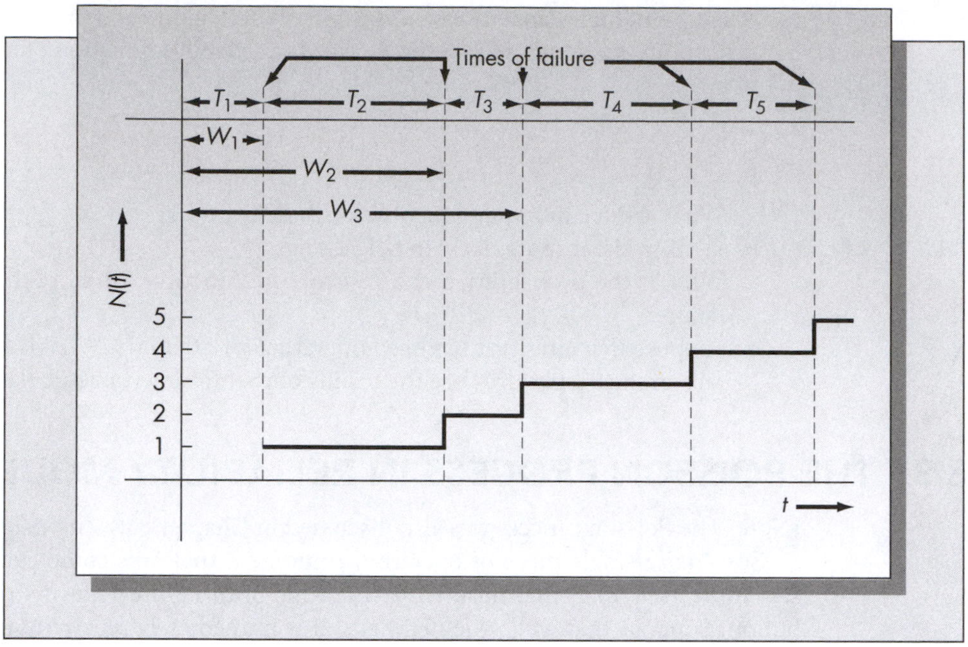

Figure 13–5 Realization of a Poisson process

A little reflection should convince you of the truth of this identity. In order for the left-hand side to be true, the number of failures up until time t must be fewer than n. If that is the case, then the time of the nth failure must be after t, which gives the right-hand side. Similarly, if the nth failure occurs after t, then the number of failures up until time t must be fewer than n.

The analysis requires obtaining the cumulative distribution function of the times until failure. Since

$$W_n = T_1 + T_2 + \ldots + T_n,$$

the distribution of W_n can be obtained by forming the n-fold convolution of the exponential distribution, which leads to the **Erlang** distribution:

$$P\{W_n > t\} = \sum_{k=0}^{n-1} \frac{e^{-\lambda t}(\lambda t)^k}{k!}.$$

The Erlang distribution is named for Agner Krarup Erlang in recognition of his pioneering work in the area of queueing (see Chapter 8). (That W_n has an Erlang distribution is shown in Hillier and Lieberman, 2015.) We can now derive the distribution of the number of failures up until time t, $N(t)$.

We have that

$$\begin{aligned} P\{N(t) = n\} &= P\{N(t) < n+1\} - P\{N(t) < n\} \\ &= \sum_{k=0}^{n} \frac{e^{-\lambda t}(\lambda t)^k}{k!} - \sum_{k=0}^{n-1} \frac{e^{-\lambda t}(\lambda t)^k}{k!} \\ &= \frac{e^{-\lambda t}(\lambda t)^n}{n!}. \end{aligned}$$

Table 13–1 Summary of Results for the Poisson Process

Random Variable	Distribution	Parameter(s)	Mean	Variance
Time between failure, T_n	Exponential	λ	$1/\lambda$	$1/\lambda^2$
Time of *n*th failure, W_n	Erlang	λ, n	n/λ	n/λ^2
Number of failures until time t, $N(t)$	Poisson	λt	λt	λt

This is exactly the Poisson distribution with parameter λt. The Poisson distribution is a discrete distribution that assumes values only on the nonnegative integers 0, 1, 2, ... It has mean and variance given by λt. The process by which (1) events occur completely at random over time, (2) the interevent times are exponential random variables, (3) the times of events are Erlang random variables, and (4) the number of events up until any time t is a Poisson random variable is called the Poisson process. It is a fundamental stochastic process that crops up in many fields besides reliability. The single parameter λ, the rate at which events occur, defines the process. Table 13–1 summarizes the important points about the Poisson process.

As W_n is the sum of independent identically distributed random variables, when n is reasonably large, the central limit theorem implies that the normal distribution may be used to approximate the Erlang. Also note that the reliability function of the Erlang is of precisely the same form as the cumulative Poisson distribution. Hence, a table of the Poisson distribution (Table A–3 at the back of this book) may be used to obtain exact Erlang probabilities. Furthermore, note that for large values of λt, the normal distribution is an adequate approximation of the Poisson distribution. (However, the normal should not be used to approximate the exponential.)

EXAMPLE 13.5

A local military base maintains a sensitive radar device that signals incursion of enemy planes into U.S. airspace. Breakdowns of the device occur completely at random at an average rate of three per year. The equipment is generally repaired the same day that it fails. Determine the following probabilities.

a. The probability that the time between two successive failures is less than one month.
b. The probability that there are exactly five breakdowns in any given year.
c. The probability that there are more than 15 failures in a four-year period.
d. The average time for 100 failures to occur.
e. The probability that the 25th failure occurs after 10 years of operation.

Solution

As with all probability problems, it is a good idea to have a ballpark estimate of the answer before beginning formal calculations. This will serve as a verification of your calculations.

a. Because there are three failures per year on average, there will be on average four months between failures. Hence, the probability that the time between two successive failures is less than a month should be less than .5.

   Let T be the time between any two successive failures. Then we know that T has the exponential distribution with parameter $\lambda = 3$ per year. We must be sure to express all units of time in terms of years as we solve this problem. We compute

$$P\{T < 1/12\} = 1 - \exp(-\lambda t) = 1 - \exp(-3/12) = .22.$$

b. Here we count the number of breakdowns in a given year, so that the appropriate distribution is Poisson with parameter $\lambda t = 3 \times 1 = 3$.

$$P\{N(1) = 5\} = \frac{e^{-3}3^5}{5!} = .1008.$$

c. We wish to compute $P\{N(4) > 15\}$, where $N(4)$ has the Poisson distribution with parameter $\lambda t = 3 \times 4 = 12$. From Table A–3, we obtain

$$P\{N(4) > 15\} = P\{N(4) \geq 16\} = .1556.$$

This probability also can be approximated using the normal distribution. In order to use a normal approximation, the standardized variate Z is constructed by subtracting the mean and dividing by the standard deviation. Hence,

$$P\{N(4) > 15\} \approx P\{Z > (15 - 12)/\sqrt{12}\} = P\{Z > 0.8660\} = .1922.$$

This approximation can be improved by using the continuity correction. The continuity correction is appropriate when approximating a discrete random variable with a continuous random variable. Since $\{N(4) > 15\} = \{N(4) \geq 16\}$, we go halfway between and use 15.5. (The continuity correction is discussed in detail in Appendix 10–A of Chapter 10.) Hence, $P\{N(4) > 15\} \approx P\{Z > (15.5 - 12)/\sqrt{12}\} = P\{Z > 1.01\} = .1562$. This is very close to the exact answer, .1556.

d. Here we are interested in $E(W_{100})$. From Table 13–1, $E(W_{100}) = 100/\lambda = 100/3 = 33.33$ years.

e. The time of the 25th failure is the random variable W_{25}. Again we will use the normal approximation, but we do not require the continuity correction because both the Erlang and the normal distributions are continuous. From the table we have that $E(W_{25}) = 25/\lambda = 25/3 = 8.33$, and $\mathrm{Var}(W_{25}) = 25/\lambda^2 = 25/9$. Hence, $\sigma = 5/3 = 1.67$. It follows that

$$P\{W_{25} > 10\} \approx P\left\{Z > \frac{10 - 8.33}{1.67}\right\} = P\{Z > 1\} = .1587.$$

Many applications involve monitoring many pieces rather than a single piece of equipment. For example, a repairer is responsible for maintaining all equipment in his or her location, and the airlines must maintain an entire fleet of planes. Failures of collections of equipment are treated below.

Series Systems Subject to Purely Random Failures

Consider a bank of items labeled 1, 2, …, N and assume that each of the items fails completely at random, that is, according to an exponential failure law. Furthermore, we assume that the items fail independently. A series system implies that the bank fails when the first item in the bank fails. Let $T_1, T_2, \ldots, T_N$ be the failure times associated with each piece of equipment. Then

$$P\{T_i > t\} = \exp(-\lambda_i t) \qquad \text{for } 1 \leq i \leq N.$$

Define the random variable $T = \min(T_1, T_2, \ldots, T_N)$. Then T represents the time that the next component fails. It is also the time the bank fails.

$$P\{T > t\} = P\{\min(T_1, T_2, \ldots T_N) > t\}$$
$$= P\{T_1 > t, T_2 > t, \ldots T_N > t\}$$

(this follows because if the minimum of a group of numbers exceeds a fixed number, then it must be true that all members of the group exceed it as well)

$$= P\{T_1 > t\} \times P\{T_2 > t\} \times \cdots \times P\{T_N > t\}$$

(this follows from the independence of the individual failure times)

$$= e^{-\lambda_1 t} e^{-\lambda_2 t} \cdots e^{-\lambda_N t}$$
$$= \exp\left(-\sum_{i=1}^{N} \lambda_i t \right),$$

which is exactly the exponential failure law with $\lambda = \Sigma \lambda_i$. If we assume that units that fail are repaired quickly, then the number of failures of the bank up until any time t, say $N(t)$, will be a Poisson process with rate λ.

PROBLEMS FOR SECTION 13.3

11. Automobiles arrive at a tollbooth on a highway completely at random according to a Poisson process with rate $\lambda = 4$ cars per hour. Determine the following probabilities.
    a. The probability that the time between any two successive arrivals exceeds 20 minutes.
    b. The probability that exactly four cars arrive in any given hour.
    c. The probability that more than five cars arrive in any given hour.
    d. The probability that more than 10 cars arrive in any given two-hour period.
    e. A person working for the transportation department begins counting arrivals at the tollbooth at 8 AM. What is the probability that he counts 20 arrivals before 12 noon? (Use a normal approximation for your calculations.)
12. Herman's Hardware uses a neon light in its store window that is left burning continuously. The light has an average lifetime of 1,250 hours and fails completely at random. Lights that burn out are replaced instantly.
    a. On average, how many neon lights does Herman's use in one year?
    b. Suppose that the lights cost \$37.50 each and the store owner has budgeted \$300 annually for them. What is the probability that Herman exceeds his annual budget in any given year?
    c. What is the probability that two bulbs will be used within the same month? (Assume that one month equals 30 days for your calculation.)
13. An electronic module used by the Navy in a sonar device requires replacement on average once every 16 months and fails according to a Poisson process. Suppose that the Navy places these sonar devices into service on the same date in eight different aircraft carriers. If the modules are replaced immediately after failure and the budget allows for exactly 40 spares over five years, what is the probability that the budget is exceeded? (Hint: Use a normal approximation to the Poisson.)

14. Determine the following probabilities for problem 13.
    a. The probability that a single carrier sent on a six-month mission will not require replacement of the module during that time.
    b. The probability that the time of the fifth failure is more than one year after the devices are placed into service.
    c. The expected time to use all 40 spares.

13.4 FAILURES OF COMPLEX EQUIPMENT

Many applications of **reliability theory** involve predicting the failure patterns of equipment from knowledge of the failure patterns of the components comprising that equipment. For example, a well-known study published by the U.S. Nuclear Regulatory Commission (1975) claimed that nuclear plants are safe. In the study, reliability theory was used to predict the likelihood of a major problem occurring in a nuclear plant by analyzing the failure rate of the various components comprising the plant. Unfortunately, incorrect assumptions about the independence of these components led to the conclusion that major nuclear accidents were virtually impossible. That is obviously not the case.

Section 13.3 showed that a bank of items connected in series, each of which has the exponential failure law, will also have the exponential failure law. That result is a special case of one that will be derived in this section.

Components in Series

A series system will function only if every component functions. A schematic diagram of a series system appears in Figure 13–6(a).

Define T_i as the time until failure of the ith component and let T_S be the time of failure of the entire series system. As before, we have that $T_S = \min(T_1, T_2, \ldots, T_N)$. Recall the

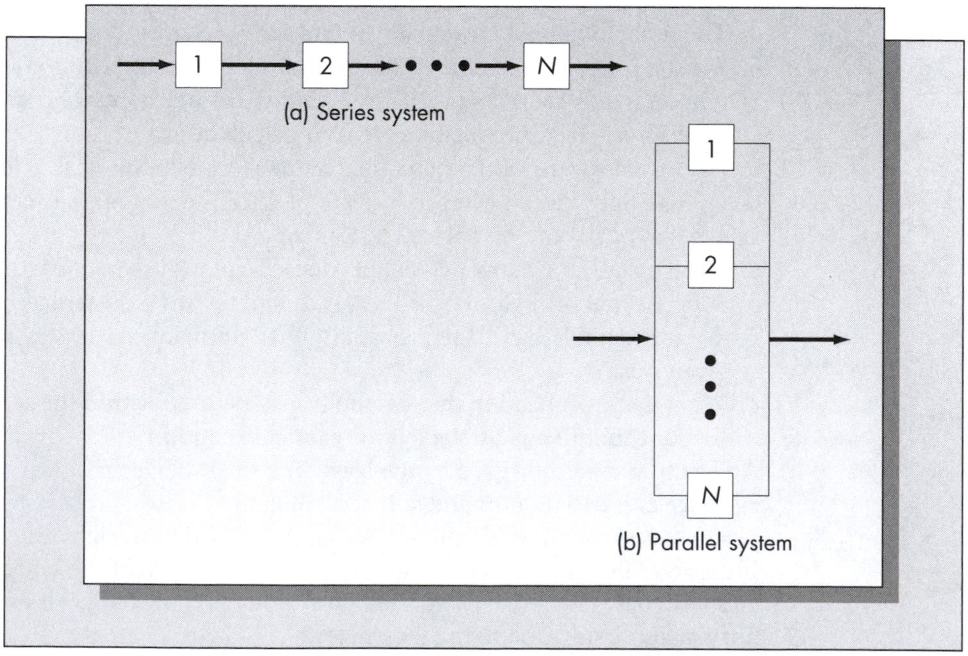

Figure 13–6 Systems of components in both series and parallel

definition of the reliability function, $R(t) = P\{T > t\}$. We will derive $R_S(t)$, the reliability function of the system in terms of the reliability functions of each of the components, $R_i(t)$. Using essentially the same arguments as in Section 13.3, we have

$$\begin{aligned} R_S(t) &= P\{T_S > t\} = P\{\min(T_1, T_2, \ldots, T_N) > t\} \\ &= P\{T_1 > t, T_2 > t, \ldots T_N > t\} \\ &= P\{T_1 > t\} \times P\{T_2 > t\} \times \cdots \times P\{T_N > t\} \\ &= R_1(t) \times R_2(t) \times \cdots \times R_N(t). \end{aligned}$$

For N identical components each having reliability function $R(t)$, this becomes simply $[R(t)]^N$.

To express the cumulative distribution function of the series system in terms of the distribution function of each of the individual components, substitute $F(t) = 1 - R(t)$. For N identical components in series we obtain

$$F_S(t) = 1 - [1 - F(t)]^N.$$

Components in Parallel

Figure 13–6(b) shows a parallel system of components. A parallel system functions if any one of the components functions. Parallel systems occur when redundancy is included to increase reliability. Define T_P as the time of failure of a parallel system of identical components. It is clear that $T_P = \max(T_1, T_2, \ldots, T_N)$. For a parallel system, it is more convenient to determine the distribution function of the time until failure rather than that of the reliability function. We have that

$$\begin{aligned} F_P(t) &= P\{\max(T_1, T_2, \ldots T_N) \le t\} \\ &= P\{T_1 \le t, T_2 \le t, \ldots T_N \le t\} \end{aligned}$$

(because if the largest member of a group is less than a given number, then all the members of the group are also less than that number)

$$= F_1(t) \times F_2(t) \times \ldots \times F_N(t),$$

which reduces to $[F(t)]^N$ in the case of N identical components. The reliability function of a parallel system with N identical components is $R_P(t) = 1 - [1 - R(t)]^N$.

Expected Value Calculations

A useful result from probability theory that may simplify calculating expected values is the following.

THEOREM

If T is a nonnegative random variable with cumulative distribution function F(t) and probability density function f(t), then one can compute the expected value in two ways:

$$E(T) = \int_0^\infty t f(t)\,dt = \int_0^\infty [1 - F(t)]\,dt.$$

The second equation can streamline calculations of the expected time until failure. We will use it to compute the expected time until failure of a parallel system of N identical components, each of which has the exponential failure law with parameter λ.

We have that

$$E(T_P)=\int_0^\infty [1-(1-e^{-\lambda t})^N]dt.$$

In order to perform the integration, we make the change of variable $v = 1 - e^{-\lambda t}$, which gives $dv = \lambda e^{-\lambda t}\, dt$, or

$$dt=\frac{1}{\lambda}\frac{1}{e^{-\lambda t}}dv=\frac{1}{\lambda}\frac{1}{1-v}dv.$$

Hence,

$$E(T_P)=\frac{1}{\lambda}\int_0^1 \frac{1-v^N}{1-v}dv.$$

The expression in the integrand is just the finite geometric series $1 + v + v^2 +\ldots+ v^{N-1}$. Hence, it follows that

$$\begin{aligned}E(T_P)&=\frac{1}{\lambda}\int_0^1 (1+v+v^2+\cdots+v^{N-1})dv\\&=\frac{1}{\lambda}(1+\tfrac{1}{2}+\tfrac{1}{3}+\cdots+1/n).\end{aligned}$$

This implies that for a system with k components in parallel, the expected lifetime of the system is increased by $1/(k + 1)\lambda$ by adding one additional component.

EXAMPLE 13.6

Wizard, a popular brand of electric garage door opener, includes two 40-watt bulbs that go on when the garage door is opened. A bulb will generally last about one year in normal operation. Three neighbors, James, Smith, and Walker have Wizard openers in their garages. Each time a bulb burns out, James replaces both bulbs. Smith, on the other hand, replaces only the bulb that has burned out, and Walker replaces both bulbs only after both have burned out. Assume that light bulbs fail according to an exponential law.

a. Over a 10-year period, how many bulbs, on average, will each neighbor require?
b. What percentage of the time will Walker have only one bulb burning?
c. Is there any advantage of James's strategy over Smith's?

Solution

a. Both James and Smith treat the system as a pure series system of two components; the system fails when one of the bulbs fails. If each bulb has failure rate $\lambda_i = 1$, then the system has failure rate $\lambda = \lambda_1 + \lambda_2 = 2$. Because James replaces two bulbs at each occurrence of a failure, he uses an average of 4 bulbs per year, or 40 bulbs in 10 years. Smith, on the other hand, only requires one bulb each time the system fails, so he uses an average of 2 bulbs per year, or 20 bulbs over 10 years.

Walker's policy of replacing both bulbs only after both have failed is equivalent to a pure parallel system of two components. The expected lifetime of a parallel system of two components with failure rate 1 is $1 + \frac{1}{2} = 1.5$ years. Hence, every 1.5 years he requires two bulbs, thus resulting in an average of 6.67 replacements over the 10 years, which amounts to a total of 13.33 bulbs.

b. One might think that half the time he would have one bulb operating and half the time he would have two bulbs operating. However, this turns out not to be the case. Because the failure rate of the series system is $\lambda = 2$, the first bulb will fail on average after six months. As the parallel system has, on average, a 1.5-year lifetime, the remaining bulb will last an average of one year. Hence, he will be operating his garage door an average of 66.67 percent of the time with only one bulb.

c. Because the failure law is exponential, there is absolutely no advantage of James's strategy over Smith's. They will both have to make replacements equally often (twice a year), but James will use twice as many light bulbs. However, if the failure law is not exponential, it is possible that James's method will result in fewer occasions in which a replacement must be made.

This problem raises an interesting point. In installations in which many lights are used, such as a Las Vegas hotel sign, it is a common strategy to periodically replace all bulbs at once. However, as we saw in the problem, if the bulbs follow an exponential failure law, this strategy will result in no fewer unplanned replacements but will lead to using far more bulbs.

K Out of *N* Systems

Assume that a system consists of N components. A K out of N system is one in which the system functions only if at least K of the components function, where $1 \le K \le N$. A typical example of a K out of N system is a four-engine airplane that can fly as long as at least two of its engines are operating.

To analyze a K out of N system, we use a binomial framework. Think of each component as a separate Bernoulli trial. Identify a success with a functioning component and a failure with a nonfunctioning component. We assume that all components are identical, so that each has the same reliability function $R(t)$ and the same distribution function $F(t)$. Fix a point in time, t. Let $p = P\{\text{a component functions at time } t\} = R(t)$. The probability that the system functions at time t, $R_K(t)$, is the probability that there are at least K successes in N trials of a binomial experiment with $p = P\{\text{success}\}$ at each trial.

Hence,

$$R_K(t) = \sum_{j=K}^{N} \binom{N}{j} p^j (1-p)^{N-j}$$

$$= \sum_{j=K}^{N} \binom{N}{j} R(t)^j F(t)^{N-j}.$$

Note that both series and parallel systems are special cases of K out of N systems. A series system is an N out of N system, and a parallel system is a 1 out of N system. A series system of N identical components will always have lower reliability than a single component, whereas a parallel system of N identical components will always have higher reliability than a single component.

However, whether a K out of N system is more reliable than a single component depends upon the reliability of the individual components. Figure 13–7 graphs the reliability of 2 out

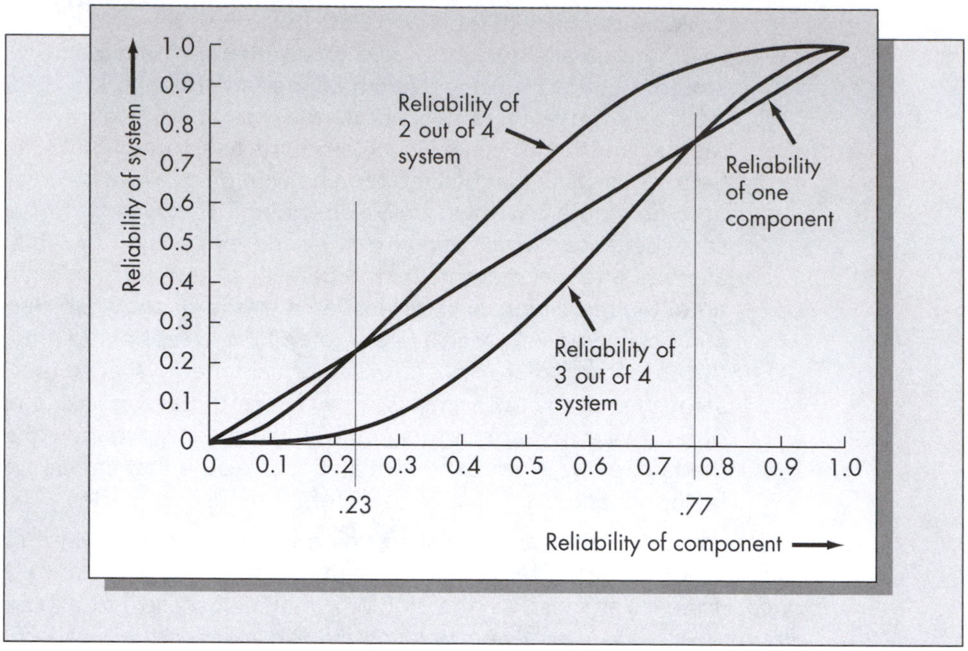

Figure 13–7 Reliability of 2 out of 4 and 3 out of 4 systems

of 4 and 3 out of 4 systems as a function of the reliability of each component. The 45-degree line represents the reliability of a single component. The reliability curve for the 2 out of 4 system crosses the 45-degree line at about $p = .23$ and for the 3 out of 4 system at about $p = .77$. This means that a 2 out of 4 system is preferred to a single component only if the reliability of the components comprising the system exceeds .23 and a 3 out of 4 system is preferred to a single component only if the component reliability exceeds .77.

PROBLEMS FOR SECTION 13.4

15. A subassembly in an industrial robot consists of 12 components in series, each of which fails completely at random at a rate of once every 50 years.
    a. What is the mean time between failures for this part of the robot?
    b. What is the probability that this subassembly does not fail within eight years of operation?
16. Show that the time until failure of a series system of n identical components, each of which has the Weibull lifetime distribution, also has a distribution of the Weibull type.
17. Consider the following three systems: (a) a single component with failure rate one per year, (b) two components in series, each of which has failure rate one every two years, and (c) two components in parallel, each of which has failure rate two per year. Compare the reliability of these three systems assuming an exponential failure law.
18. A design engineer is considering the number of levels of redundancy to build into a particular circuit. The circuit will be part of a sensitive piece of equipment with cost estimated at \$500 per failure. Each additional level of redundancy costs \$100. If each component fails at random at a rate of one failure every five years, what level of redundancy most closely equates the cost of the design with the expected failure cost of the equipment over its 10-year life cycle?

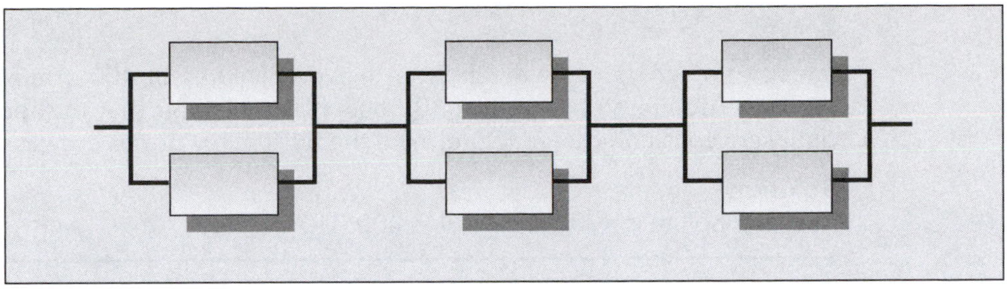

Figure 13–8 System of components (for problem 19)

19. Consider the system of six identical components pictured in Figure 13–8. If each component has constant failure rate λ, derive the distribution function of the time until failure of the system.
20. An aircraft engine fails with probability p. Assume that for an aircraft to successfully complete a flight, at least half the engines must operate. Show that for $0 < p < \frac{1}{3}$, a four-engine plane is preferred to a two-engine plane and for $\frac{1}{3} < p < 1$, a two-engine plane is preferred to a four-engine plane.

13.5 INTRODUCTION TO MAINTENANCE MODELS

The maintenance of complex equipment often can account for a large portion of the costs associated with that equipment. It has been estimated, for example, that maintenance costs in the military comprise almost one-third of all the operating costs incurred. Clearly, the issues of reliability and maintenance are closely connected.

This section introduces the following standard maintenance terminology.

1. MTBF = **Mean time between failures**. This corresponds to the expected time between failures in our previous notation and equals $1/\lambda$.
2. MTTR = **Mean time to repair**. This is the expected value of the repair time R.
3. **Availability** = Average fraction of time the equipment operates. It is given by the formula

$$\text{Availability} = \frac{E(T_i)}{E(T_i) + E(R_i)} = \frac{\text{MTBF}}{\text{MTBF} + \text{MTTR}}.$$

We may think of a single piece of equipment that has successive failure times T_1, T_2, … and successive repair times R_1, R_2, ….A schematic diagram of such a system appears in Figure 13–9.

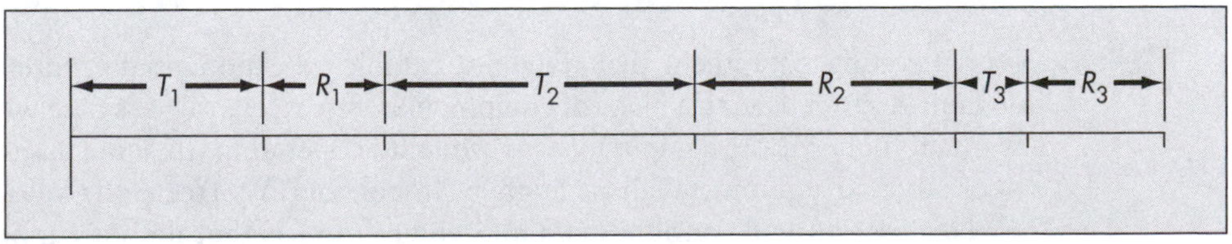

Figure 13–9 Realization of failure and repair times

EXAMPLE 13.7

A copier machine has a mean time between failures of 400 operating hours. Repairs typically require an average of 10 hours from the time that the repair call is received until service is completed. Determine the availability of this copier.

Solution
The availability is 400/(400 + 10) = 400/410 = .9756.

Define a repair cycle as the time between two successive repairs. Often we must determine the distribution of a repair cycle rather than just its expectation. A single repair cycle is the sum of the failure time T_i and the repair time R_i. The exact distribution of the sum of two random variables is the **convolution** of the individual distributions. (See DeGroot and Schervish, 2012, for example.)

If the interfailure and the repair times can be reasonably approximated by the normal distribution, then the sum of the two also will be approximately normal.

EXAMPLE 13.8

Suppose that the time between failures, T_i, is approximately normally distributed with mean 400 hours and variance 10,000. The repair time of the equipment is also approximately normally distributed with mean 10 hours and variance 11.6. Find the probability that there are more than six repair cycles within a one-year period. Assume that one year corresponds to 2,000 hours of operation.

Solution
We have that

$$E(T_i + R_i) = 400 + 10 = 410,$$
$$\text{Var}(T_i + R_i) = 10{,}000 + 11.6 = 10{,}011.6.$$

It follows that

$$E\left[\sum_{i=1}^{6}(T_i + R_i)\right] = 6 \times 410 = 2{,}460,$$

$$\text{Var}\left[\sum_{i=1}^{6}(T_i + R_i)\right] = 6 \times 10{,}011.6 = 60{,}069.6,$$

$$P\left\{\sum_{i=1}^{6}(T_i + R_i) \le 2{,}000\right\} = P\left\{Z \le \frac{2{,}000 - 2{,}460}{\sqrt{60{,}069.6}}\right\}$$
$$= P\{Z \le -1.88\} = .03.$$

13.6 DETERMINISTIC AGE REPLACEMENT STRATEGIES

For operating equipment that does not exhibit an exponential failure law, there are often advantages to replacing a piece of equipment before it fails. This is true when the cost of repair is much higher if the equipment fails while it is operating. In some cases, such as in military operations, an equipment failure might be impossible to correct and could result in loss of lives.

This section will consider age replacement models that do not explicitly account for the uncertainty of the failure process. Rather, the aging mechanism is subsumed in the cost structure; in particular, it is assumed that the cost of maintaining the equipment increases as the equipment ages. Section 13.7 will consider models of planned replacement that explicitly include the uncertainty of the failure process.

The models we consider are appropriate for both continuously operating equipment, such as radar or power generating units, and intermittently operating equipment, such as automobiles. In the latter case, we would keep track of operating time rather than clock time.

Assume that the replacement cost of the item is K. Also assume that the instantaneous cost rate of operating an item of age u is $C(u)$. We will consider various forms for $C(u)$ but assume initially that $C(u) = au$.

Based on the values of the various costs, there will be some optimal point at which to replace the item in order to minimize the total cost per unit time. The total cost function may be thought of as the sum of two components: maintenance and replacement. The marginal cost of maintenance increases over time, and the marginal cost of replacement decreases over time. The optimal replacement age minimizes the average cost function. In the cases we consider, the average cost function is convex, thus making the optimal solution easy to find.

The Optimal Policy in the Basic Case

We make the following assumptions.

1. The equipment used is operating continuously.
2. We ignore downtime for repair and maintenance.
3. The planning horizon is infinite.
4. Every new piece of equipment has identical characteristics.
5. Only maintenance and replacement costs are considered.
6. The objective is to minimize the long-run costs of replacement and maintenance.
7. The cost rate of maintaining an item of age u is au, and the replacement cost of the item is K. There is no salvage value.

The decision variable is the amount of time that elapses from the point that a piece of equipment is purchased until it is replaced with a new item. Figure 13–10 shows successive replacement cycles.

The object of the analysis is to determine the value of t that minimizes the total cost of maintenance and replacement over an infinite horizon. A replacement cycle is the time between successive replacements. Because all replacement cycles are identical, we may restrict attention only to the costs incurred in a single cycle.

$$\text{Total replacement cost per cycle} = K,$$

$$\text{Total maintenance costs per cycle } = \int_0^t C(u)du = \int_0^t au\ du = \frac{at^2}{2}.$$

The average cost per unit time is just the total cycle cost divided by the length of the cycle. Let $G(t)$ be the average cost per unit time if the replacement time is t. Then

$$G(t) = \frac{1}{t}\left(K + \frac{at^2}{2}\right) = \frac{K}{t} + \frac{at}{2}.$$

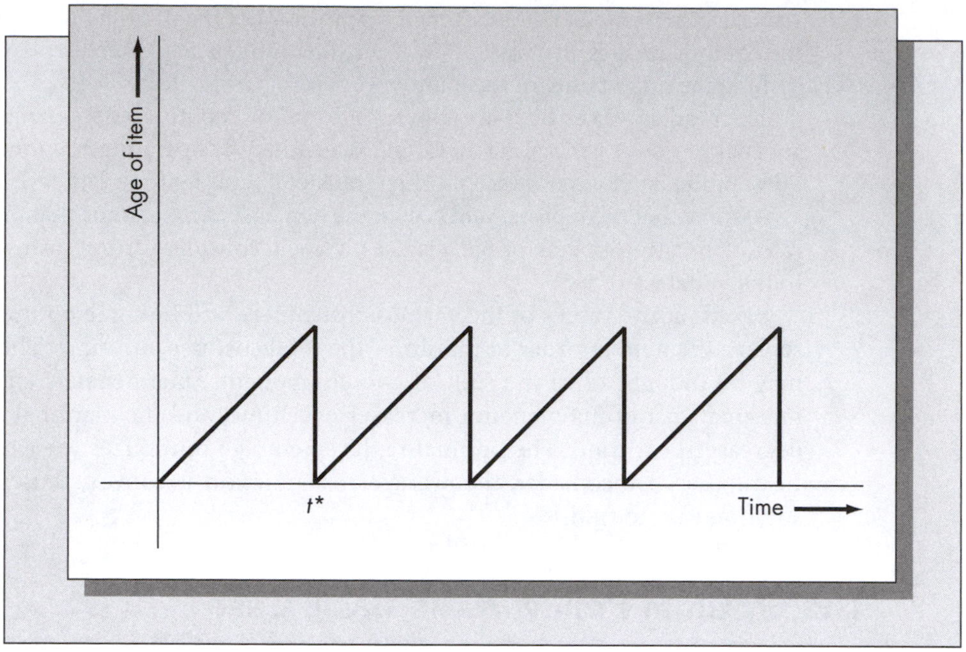

Figure 13–10 Optimal age replacement strategy

As $G''(t) = K/t^3 > 0$, it follows that $G(t)$ is a convex function of the single variable t. The function $G(t)$ is pictured in Figure 13–11. The goal is to find the optimal value of t, say t^*, that minimizes $G(t)$. Since $G(t)$ is convex, it follows that the optimal solution satisfies

$$G'(t) = \frac{-K}{t^2} + \frac{a}{2} = 0,$$

which results in

$$t^* = \sqrt{\frac{2K}{a}},$$

which is analogous to the EOQ formula from Chapter 4.

EXAMPLE 13.9

We will use the simple version of the age replacement model to estimate the number of years that one should keep a car. Although the model does not exactly describe the car replacement problem, we can use it as an approximation. Assume that the maintenance cost rate of a car u years old is $400u$ dollars. This means that the maintenance cost during the first year is \$200, during the second year is \$600, during the third year is \$1,000, and so on. (These numbers are obtained by computing $at^2/2$ for $t = 1, 2,$ and 3 and subtracting the costs of the previous years.) This is probably rising a bit faster than our actual maintenance costs would. Assume that a new car costs

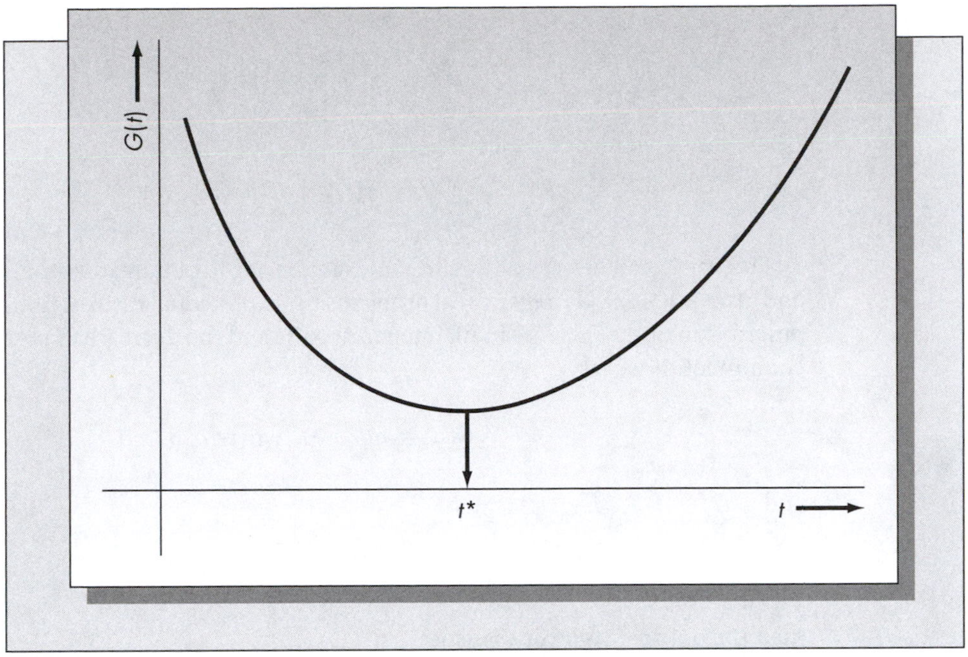

Figure 13–11 Computation of the optimal replacement age

\$10,000. According to the formula, the optimal number of years that we should hold the car is

$$t^* = \sqrt{\frac{2K}{a}} = \sqrt{\frac{(2)(10{,}000)}{400}} = 7.07 \text{ years.}$$

A General Age Replacement Model

When we include a salvage value and allow for more general maintenance cost functions, the model becomes considerably more complex. As before, suppose that $C(u)$ is an arbitrary function representing the maintenance cost rate of an item of age u. Define $S(u)$ as the salvage value of an item of age u. Then the total cost incurred in the cycle is

$$K + \int_0^t C(u)du - S(t).$$

The average cost per unit time is

$$G(t) = \frac{K}{t} + \frac{1}{t}\int_0^t C(u)du - \frac{S(t)}{t}.$$

The optimal value of t, t^*, is the solution to

$$G'(t) = \frac{-K}{t^2} - \frac{H(t)}{t^2} + \frac{C(t)}{t} + \frac{S(t)}{t^2} - \frac{S'(t)}{t} = 0,$$

or

$$tC(t) + S(t) = K + H(t) + tS'(t),$$

where for convenience we let

$$H(t) = \int_0^t C(u)du.$$

Finding t^* can be very difficult. [For example, try to obtain a solution assuming $C(u) = au$ and $S(u) = K - bu$.] For many real problems the exponential distribution provides an accurate description of the increase in maintenance costs and the decrease in resale value of operating equipment. If we let

$$C(u) = ae^{bu}, \qquad \text{where } a, b > 0,$$

and

$$S(u) = ce^{-du}, \qquad \text{where } c, d > 0,$$

then the optimal value of t satisfies

$$tae^{bt} + ce^{-dt} = K + \int_0^t ae^{bu}du + t\frac{d}{dt}(ce^{-dt}).$$

It is easy to show that

$$H(t) = \int_0^t ae^{bu}du = \frac{a}{b}(e^{bt} - 1)$$

and

$$\frac{d}{dt}(ce^{-dt}) = -cde^{-dt},$$

so the equation defining an optimal solution is

$$tae^{bt} + ce^{-dt} = K + \frac{a}{b}(e^{bt} - 1) - tcde^{-dt}.$$

Rearranging terms gives

$$ae^{bt}\left(t - \frac{1}{b}\right) + ce^{-dt}(1 + dt) + \frac{a}{b} = K.$$

The goal is to find the value of t that makes the left-hand side of the equation as close to K, the replacement cost, as possible. This is a difficult equation to solve for t because it involves both exponentials and constants (a transcendental equation). Spreadsheets provide a convenient method of obtaining a solution. One simply computes the left-hand side of the equation for various values of t and graphically determines the point at which this function crosses the value of K or uses the Goal Seek function built into Excel.

EXAMPLE 13.10

Consider again finding the optimal time to replace an automobile. Exponential functions are more realistic than linear functions and should give an accurate estimate of the true optimal time for replacement. As in Example 13.9, assume that the replacement cost of the automobile is \$10,000. Furthermore, assume that the car loses 15 percent of its value each year. This is probably a reasonable estimate of the decline in the resale value of most new cars. This means that the car is worth (0.85)(10,000) = \$8,500 after one year, (0.85)(0.85)(10,000) = \$7,225 after two years, and so on.

We wish to determine c and d so that $S(t) = ce^{-dt}$ agrees with these values. Because the salvage value at time $t = 0$ is exactly the replacement cost, we have $S(0) = 10{,}000$. Substituting, we obtain

$$S(0) = ce^{-d(0)} = c = 10{,}000.$$

The value of the car after one year is (0.85)(10,000), which corresponds to $S(1)$. Hence,

$$S(1) = (0.85)(10{,}000) = ce^{-d(1)} = ce^{-d}.$$

Since $c = 10{,}000$, we obtain

$$e^{-d} = (0.85),$$

$$d = -\ln(0.85) = 0.1625.$$

Hence, it follows that

$$S(t) = 10{,}000e^{-0.1625t}.$$

Now consider the maintenance costs. Assume, as in Example 13.9, that maintenance costs for the first year of operation amount to \$200. This is equivalent to

$$H(1) = 200$$

or

$$(a/b)(e^{b} - 1) = 200.$$

Furthermore, suppose that the maintenance costs increase at a rate of 40 percent per year. This means that

$$\frac{C(t)}{C(t-1)} = 1.4.$$

Substituting for $C(t)$ gives

$$\frac{ae^{bt}}{ae^{b(t-1)}} = e^{b} = 1.40,$$

or $b = \ln(1.4) = 0.3365$. It follows that

$$a = \frac{(200)(b)}{e^{b} - 1} = \frac{(200)(0.3365)}{0.4} = 168.25.$$

Combining these results, it follows that the optimal time to replace the automobile, t^*, is the value of t that solves

$$168.25e^{0.3365t}(t - 2.972) + 10{,}000e^{-0.1625t}(1 + 0.1625t) + 500 = 10{,}000.$$

One method of estimating the solution to this equation is to graph the function represented by the left-hand side of the equation. We have done so in Figure 13–12. The minimum-cost replacement age is about 4.5 years. That means that for an automobile

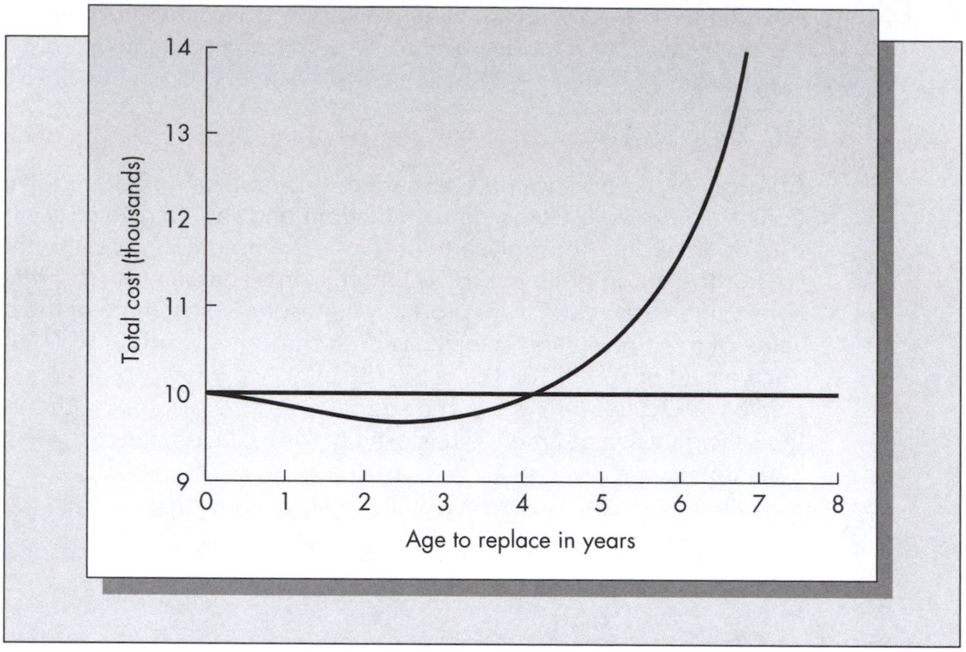

Figure 13–12 Optimal number of years to replace auto (M = .4)

costing \$10,000 that declines in value at the rate of 15 percent per year, and for which the maintenance cost is \$200 the first year and increases at the rate of 40 percent per year, the optimal strategy, which minimizes average costs of replenishment less salvage plus maintenance, is to replace the car about once every four and a half years.

To see the effect of the maintenance cost, we have re-solved the example with a value of 20 percent rather than 40 percent for the rate of increase of maintenance costs per year. This is probably more accurate for most cars. The solution is represented graphically in Figure 13–13. Here, the optimal replacement time is approximately nine years.

Let M represent the rate at which maintenance costs increase each year expressed as a decimal ($M = .20$ in Figure 13–13). Let I_0 be the maintenance cost in the first year, and D the yearly rate of depreciation, also expressed as a fraction ($D = 0.15$ in the example). Then it can be shown that

$$b = \ln(1 + M),$$
$$a = I_0 b/M,$$
$$c = K,$$
$$d = \ln(1 - D).$$

The models presented in this section do not consider the effects of inflation. If inflation effects were included, the resulting solutions would not change appreciably because the inflation is applied to both the replacement and the maintenance costs. The method is to compute the present value of all future discounted costs and determine the replacement strategy that minimizes the resulting function. We will not consider discounted cash flows in this section.

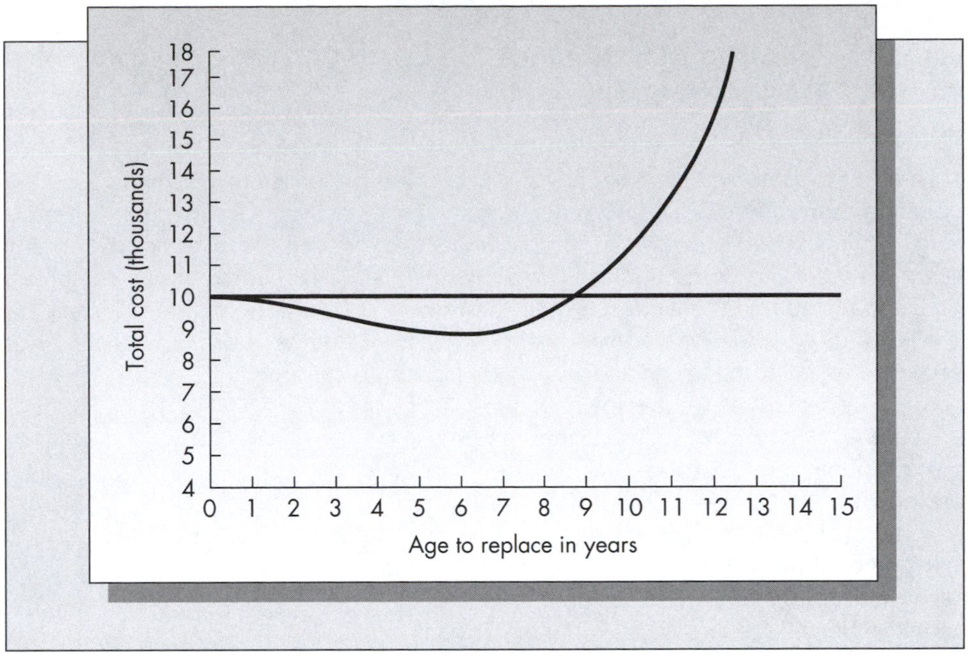

Figure 13–13 Optimal number of years to replace auto (M = .2)

PROBLEMS FOR SECTION 13.6

21. For the basic age replacement model, consider a piece of equipment that costs \$18,000 to replace. The *total* maintenance costs for five years of operation are estimated to be \$2,400. Assuming a linear maintenance cost rate, find the value of a and the optimal age at which the equipment should be replaced.
22. For the basic age replacement model, derive the optimal replacement age when the maintenance cost rate $C(u)$ has the form $C(u) = a\sqrt{u}$ for some constant $a > 0$.
23. Suppose for the simple age replacement model that the maintenance cost is the same every year [that is, $C(u) = a$ for all $u \geq 0$]. What is the optimal replacement age? Why is this so?
24. The army is attempting to determine the optimal replacement age for a piece of field equipment. The equipment costs \$280,000 to replace. The manufacturer will supply a rebate toward the next purchase that declines at a rate of 20 percent per year. Maintenance costs for the first year are estimated to be \$1,000, and they increase roughly at the rate of 18 percent per year. Estimate the number of years that the army should hold the equipment before making a replacement.
25. Try to determine the optimal replacement age when $C(u) = au$ and $S(u) = K - bu$. What difficulty do you encounter?

13.7 PLANNED REPLACEMENT UNDER UNCERTAINTY

The purpose of preventive maintenance is to decrease the likelihood that an item will require replacement because of failure. At the heart of such a policy is the assumption that it costs more to make a repair or replacement at the time of failure than at some

SNAPSHOT APPLICATION

Reliability-Centered Maintenance Improves Operations at Three Mile Island Nuclear Plant

The Three Mile Island nuclear facility located on the Susquehanna River about 10 miles from Harrisburg, Pennsylvania, is notorious in one respect. It was the site of the worst nuclear power generating plant accident in the United States. In March of 1979, Unit 2 underwent a core meltdown as safety systems failed to lift nuclear fuel rods from the core. The facility was shut down as a result of the accident for the next six and one-half years, finally reopening in October 1985. The plant, operated by GPU Nuclear Corporation, has compiled one of the most impressive records in the industry since it has reopened. On the basis of its capacity factor (proportion of up-time), the plant was ranked best in the world in 1989 (Fox, Snyder, and Smith, 1994).

In 1987, GPU began to consider the benefits of a reliability-centered maintenance (RCM) approach to preventive maintenance. They identified 28 out of a total of 134 systems as viable candidates for RCM. These 28 systems included the main turbine, the cooling water system, the main generator, and circulating water. The RCM process relied on the following four basic principles.

- Preserve system functions.
- Identify equipment failures that defeat those functions.
- Prioritize failure modes.
- Define preventive maintenance tasks for high-priority failure modes.

The RCM project spanned the period of September 1988 to June 1994, A total of 3,778 components in the 28 subsystems came under consideration. By the end of the program, preventive maintenance policies included more than 5,400 tasks for these components. The cost of implementing RCM was substantial: about $30,000 per system. However, these costs were more than offset by the benefits. Over the period 1990 to 1994, records show a significant decline in plant equipment failures. Listed below are additional benefits from a reliability-based maintenance program.

- Increased plant availability.
- Optimized spare parts inventories.
- Identification of component failure modes.
- Discovery of new plant failure scenarios.
- Training for engineering personnel.
- Identification of components that benefit from revised preventive maintenance strategies.
- Identification of potential design improvements.
- Improved documentation.

GPU learned several lessons from its RCM experience (Fox et al., 1994). One is that it is better for the internal maintenance organization, rather than an outside agency, to direct the process. This avoids the "we versus they" syndrome. Successful implementation is also more likely in this case. A cost analysis checklist was developed to screen failure modes. Finally, the team evolved an efficient multiuser relational database software system to facilitate RCM evaluations. This system reduced the time required to perform the necessary analyses by 50 percent.

The lesson learned from this case is that a carefully designed and implemented reliability-based preventive maintenance program can have big payoffs for high stakes systems.

predetermined time. For example, preventive maintenance can be accomplished at a convenient time when the system is not operating versus a failure that occurs during production when the line must be stopped to determine the cause of the failure and repair the problem. Leading to the cost of planned replacements being less than the cost of unplanned replacements.

Because of the memoryless property of the exponential distribution, if an item or group of items obeys an exponential failure law, then there is no advantage to replacing prior to failure. In the exponential case, the likelihood that failure will occur in a time Δt is the same just after a planned replacement as it is for an item that has been operating for an arbitrary amount of time. Hence, planned replacement strategies can have value only if the items exhibit aging, that is, have an increasing failure rate function.

Planned Replacement for a Single Item

Consider a single piece of continuously operating equipment whose lifetime is a random variable T with known cumulative distribution function $F(t)$. We assume that T is a continuous random variable. Suppose that it costs c_1 to replace the item when it fails and $c_2 < c_1$ to replace the item prior to failure. We assume that planned replacements are made exactly t units of time after the last replacement. The goal is to find the optimal value of t to minimize the average cost per unit time of both planned and unplanned replacements.

A cycle is the time between successive replacements. Because the process "restarts" itself after each replacement, irrespective of whether the replacement was planned or unplanned, we may use the renewal method to obtain an expression for the expected cost per unit time. That is,

$$E\left(\text{cost per unit time}\right)=\frac{E(\text{cost per cycle})}{E(\text{length of a cycle})}.$$

This approach also was used in a variety of other places, including Section 13.6 on age replacement and in much of Chapters 4 and 5 on inventory modeling. Successive replacement cycles are pictured in Figure 13–14.

We have that

$$E\left(\text{cost per cycle}\right)= c_1P\left\{\text{replacement is result of failure}\right\}+ c_2P\left\{\text{replacement is planned}\right\}.$$

Notice that $P\{\text{replacement is result of failure}\} = P\{T \le t\} = F(t)$, and $P\{\text{replacement is planned}\} = P\{T > t\} = 1 - F(t)$, where T is the lifetime of the item placed into service at the end of the previous cycle. It follows that

$$E(\text{cost per cycle}) = c_1F(t) + c_2[1 - F(t)].$$

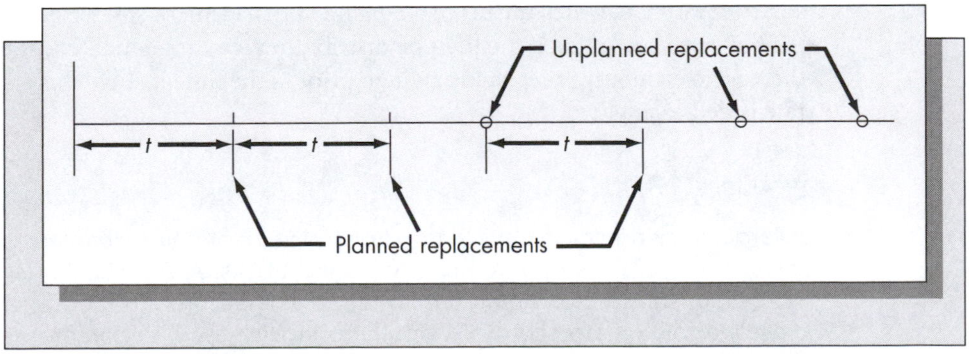

Figure 13–14 Successive cycles for planned replacement of a single item

Let T be the time of failure of the item placed into service at the end of the previous cycle. Then, clearly, the next replacement will occur at $\min(T, t)$. Hence,

$$E(\text{length of cycle}) = E[\min(T,t)] = \int_0^\infty \min(x,t) f(x)dx$$
$$= \int_0^t xf(x)dx + t\int_0^\infty f(x)dx$$
$$= \int_0^t xf(x)dx + t[1 - F(t)].$$

It follows that the expected cost per unit time, say $G(t)$, is given by

$$G(t) = \frac{c_1 F(t) + c_2[1 - F(t)]}{\int_0^t xf(x)dx + t[1 - F(t)]}.$$

The goal is to find t to minimize $G(t)$. The optimization may be cumbersome depending upon the form of the lifetime distribution $F(t)$.

We will now show that there is no advantage to planned replacement when the lifetime distribution is exponential. Suppose that $F(t) = 1 - e^{-\lambda t}$. Then the expected length of each cycle is

$$\int_0^t xe^{-\lambda x}dx + te^{-\lambda t} = \frac{1}{\lambda}[1 - e^{-\lambda t}(1 + \lambda t)] + te^{-\lambda t}$$
$$= \frac{1}{\lambda}(1 - e^{-\lambda t}).$$

(The expression for $\int_0^t xe^{-\lambda x}dx$ can be obtained by integration by parts or can be found in a table of integrals.) It follows that

$$G(t) = \frac{c_1(1 - e^{-\lambda t}) + c_2 e^{-\lambda t}}{\frac{1}{\lambda}(1 - e^{-\lambda t})} = \frac{c_1 - (c_1 - c_2)e^{-\lambda t}}{\frac{1}{\lambda} - \frac{1}{\lambda}e^{-\lambda t}}.$$

As $t \to \infty$, the term $e^{-\lambda t} \to 0$, so $G(\infty) = \lambda c_1$. Furthermore, substituting $t = 0$ results in $G(0) = \infty$. It can be shown by either calculus or direct computation that the function $G(t)$ is monotonically decreasing; a typical case is pictured in Figure 13–15. Hence, the optimal solution is $t = \infty$, which means that a planned replacement should never be made.

We have shown that if the lifetime distribution is exponential (constant failure rate), then there is no economy in replacing an item prior to the time it fails. This also holds if the failure rate is decreasing.

EXAMPLE 13.11

A large trucking company, Harley Brown, Inc., maintains detailed records on the mortality of the tires used on company-owned trucks. A statistical analysis of the data on tire failure shows that the lifetime of a tire as measured in thousands of miles of use is closely approximated by the Weibull probability law with parameters $\alpha = 0.00235$ and $\beta = 2.3$. The company estimates a cost of \$450 if a tire fails during use. This is a result of the lost time and the potential liability of an accident. Tires replaced before failure cost \$220 each. The company would like to find the optimal timing of tire replacement.

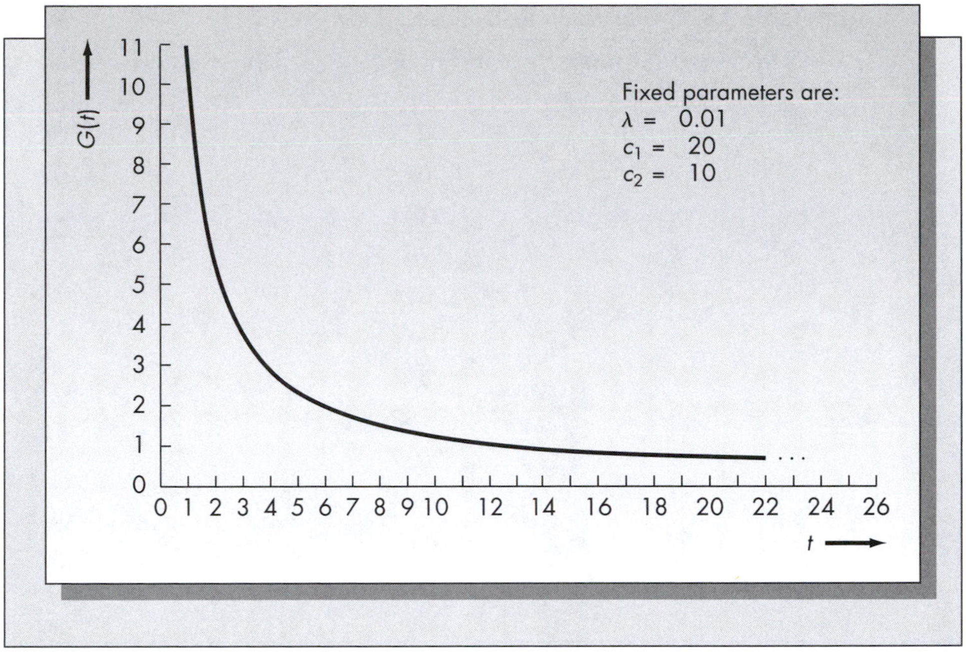

Figure 13–15 The function G(t)

The difficulty with calculating the optimal solution for this problem is determining an expression for the term $\int_0^t xf(x)dx$ that appears in the denominator of $G(t)$.

To circumvent the problem of finding an analytical expression for this integral, we will obtain a discrete approximation to the failure law and perform the calculation as if the lifetime distribution were discrete rather than continuous. The probability of failure within the first 1,000 miles is

$$P\{T \leq 1\} = 1 - F(1) = .0023.$$

The probability of failure after 1,000 miles but before 2,000 miles of wear is

$$P\{T \leq 2\} - P\{T \leq 1\} = R(1) - R(2) = .0092.$$

The remainder of the failure probabilities are computed in a similar fashion and appear in Table 13–2.

Using these discrete probabilities we may now compute $G(t)$ directly. The partial expectation term

$$\int_0^t xf(x)dx \approx \sum_{k=1}^{t} kp_k,$$

where the probabilities p_k are as given in Table 13–2. Note that this approximation assumes that all failures occur at multiples of 1,000 miles.

The remainder of the terms comprising $G(t)$ can be obtained directly from the Weibull reliability function

$$R(t) = e^{-\alpha t^{\beta}}.$$

Hence, the approximate form of $G(t)$ may be written

$$G(t) = \frac{c_1 - (c_1 - c_2)R(t)}{\sum_{k=1}^{t} kp_k + tR(t)}.$$

Table 13–2 Failure Probabilities for Example 13.11

Lifetime (thousands of miles)	Probability of Failure	Lifetime (thousands of miles)	Probability of Failure
1	.0023	14	.0625
2	.0092	15	.0580
3	.0175	16	.0527
4	.0264	17	.0469
5	.0354	18	.0408
6	.0440	19	.0348
7	.0517	20	.0291
8	.0582	21	.0238
9	.0632	22	.0191
10	.0664	23	.0150
11	.0679	24	.0116
12	.0677	25	.0087
13	.0658		

$G(t)$ appears in Table 13–3 for the parameter values $\alpha = 0.00235$, $\beta = 2.3$, $c_1 = 450$, and $c_2 = \$220$. The function appears to be convex and is minimized at $t = 13$. Hence, the optimal policy calls for replacing the tires after about 13,000 miles of wear. The value of the objective function at the optimal solution is $G(13) = 33.21$. This means that at the optimal solution, the replacement cost is \$33.21 per thousand miles of use per tire.

Table 13–3 The Function $G(t)$ for Example 13 .11

t	$G(t)$	t	$G(t)$
1	220.54	14	33.24
2	111.45	15	33.35
3	75.91	16	33.53
4	58.82	17	33.72
5	49.14	18	33.93
6	43.20	19	34.13
7	39.38	20	34.32
8	36.89	21	34.48
9	35.27	22	34.62
10	34.25	23	34.74
11	33.64	24	34.84
12	33.33	25	34.91
13	33.21		

Block Replacement for a Group of Items

In certain circumstances it is more economical to replace groups of items at the same time rather than one by one. Example 13.11 showed that the optimal policy was to replace a truck tire after about 13,000 miles of use. Depending upon the time and the expense involved in changing truck tires, it could be more economical to replace all the tires on a truck when a replacement is made. The costs of transporting the truck to a service area, placing the truck on a lift, and paying a technician to mount and balance the tires could be comparable to the cost of the tire itself. If all the tires were replaced simultaneously, this cost would be incurred less often than if the tires were individually replaced.

This section will consider a model to determine the optimal time to replace an entire group of items. In order to avoid intricate mathematics that are beyond the scope of this book, we will assume that the lifetime of each operating unit is a discrete random variable with a known distribution. That is, suppose that p_k is the probability that an item fails in period k assuming the item was placed into service at period 0. These probabilities may be estimated directly from historical data or computed from a continuous distribution as in Example 13.11.

Assume that n_0 items are placed into service at time 0. Suppose there is no block replacement and all items that fail in a period are replaced at the end of that period. We also will assume for simplicity that p_k is the actual proportion of units k periods old that fail. Then the number of failures occurring in period 1 is $n_1 = n_0 p_1$.

In period 2 the proportion of the original group of items that fail is $n_0 p_2$, and the proportion of the items placed into service in period 1 that fail is $n_1 p_1$. Hence, the expected number of failures in period 2 is $n_2 = n_0 p_2 + n_1 p_1$. Continuing with this argument, we obtain

$$n_k = n_0 p_k + n_1 p_{k-1} + \ldots + n_{k-1} p_1.$$

Now suppose that individual replacements cost a_1 each and the entire block of n_0 can be replaced for a_2. If all n_0 items were replaced at the end of each period, the cost each period would be $a_2 + a_1 n_1$. If all n_0 items were replaced at the end of every other period, the cost incurred every two periods would be $a_2 + a_1(n_1 + n_2)$ or an average per period cost of $[a_2 + a_1(n_1 + n_2)] / 2$. Similarly, the average per period cost of replacing all n_0 items after k periods is

$$G(k) = \frac{a_2 + a_1 \sum_{j=1}^{k} n_j}{k}.$$

The optimal number of periods to replace all n_0 items is the value of k that minimizes $G(k)$. The minimum value of $G(k)$ should be compared to the expected cost per period assuming that items are replaced as they fail. Let

$$E(T) = \sum_{k=1}^{\infty} k p_k$$

represent the expected lifetime of a single item. Then $\lambda = 1/T$ is the failure rate of a single item. It follows that the cost of making replacements to items on a one-at-a-time basis is $a_1 \lambda$ per item or $n_0 a_1 \lambda$ for the entire block of items. This should be compared to the optimal value of $G(k)$ to determine if a block replacement strategy is economical.

EXAMPLE 13.12

A large sign is lit by 8,000 bulbs. The bulbs cost $2 each to replace as they fail but can be replaced for 30 cents each when they are replaced all at once. Based on past experience, bulbs fail according to the following probabilities.

Months of Service	Probability of Failure
1	.02
2	.03
3	.03
4	.05
5	.08
6	.09
7	.07
8	.10
9	.11
10	.13
11	.15
12	.14

The first step is to compute n_k. We have

$$n_0 = 8{,}000,$$
$$n_1 = n_0 p_1 = (8{,}000)(.02) = 160,$$
$$n_2 = n_0 p_2 + n_1 p_1 = (8{,}000)(.03) = (160)(.02) = 243,$$

and so on.

These values are used to compute $G(k)$. The results of the calculation appear in Table 13–4. Using values of $a_2 = (0.30)(8{,}000) = \$2{,}400$ and $a_1 = 2$, we see that the optimal time to replace the block of bulbs is after four months with an expected monthly cost of $1,135.

It is interesting to compare block replacement with a policy of replacing the bulbs only if they fail. Each bulb has an expected lifetime of

$$E(T) = \sum_{k=1}^{12} k p_k = 8.22 \text{ months.}$$

Hence, the failure rate is 1/8.22 = 0.12165 failure per month per bulb. For 8,000 bulbs, this amounts to an average number of failures of 973.24 per month. The resulting replacement cost is $1,946.47 monthly, which is considerably more than the cost of replacing bulbs as a block every four months that has an expected monthly cost of $1,135.

PROBLEMS FOR SECTION 13.7

26. Consider Example 13.11 of Harley Brown, Inc. Without performing the calculations, discuss what the effect on the optimal replacement policy would likely be if the parameter values were $c_1 = \$800$, $c_2 = \$300$, and $\beta = 1$.

Table 13–4 *G(k)* for Example 13.12

k	P_k	n_k	G_k
1	.02	160.0	2,720.0
2	.03	243.2	1,603.2
3	.03	249.2	1,235.2
4	.05	417.1	1,135.0
5	.08	671.1	1,176.4
6	.09	778.4	1,239.8
7	.07	654.6	1,249.7
8	.10	930.5	1,326.1
9	.11	1,064.0	1,415.2
10	.13	1,298.4	1,533.4
11	.15	1,542.9	1,674.5
12	.14	1,562.4	1,795.4

27. Repeat the calculations for the Harley Brown trucking company using the following parameter values: $\alpha = 0.0156$, $\beta = 1.8$, $c_1 = 1{,}000$, $c_2 = 600$.
28. An expensive piece of equipment is used in the masking operation for semiconductor manufacture. A capacitor in the equipment fails randomly. The capacitor costs \$7.50, but if it burns out while the machine is in use, the production process must be halted. Here the replacement cost is estimated to be \$150. Based on past experience, the lifetime distribution of the capacitor is as follows.

Number of Months of Service	Probability of Failure
1	.08
2	.12
3	.16
4	.26
5	.22
6	.16

How often should the capacitors be replaced in order to minimize the expected monthly cost of planned and unplanned replacement?
29. A large electronic pipe organ contains 100 fuses. Because of the power demands of the organ, the fuses burn out at a fairly regular rate, but newer fuses last longer than older ones. The probability distribution of the lifetime of a fuse is closely approximated by the Weibull law with $\alpha = 0.0204$ and $\beta = 1.8$. Assume that t is hours of playing time. The fuses cost \$1.35 each when replaced as a block but \$12 each when replaced just after a failure.
    a. Express the lifetime distribution as a discrete distribution assuming t is measured in hours. (Follow the procedure used in Example 13.11 for the Harley Brown trucking company.)

b. Determine the optimal time to replace all 100 fuses and the average hourly cost of that policy.

c. Compare the answer you obtained in part (b) with the cost of replacing the fuses as they fail. Which policy would you recommend?

30. Tires that fail in service result in significantly higher replacement costs than those replaced before failure. For an 18-wheeler (that is, a truck with 18 tires), failure on the road costs $300, whereas all 18 tires can be replaced prior to failure at a cost of $75 per tire. The following table gives the probability of failure.

Number of Miles	Probability of Failure
0–5,000	.05
5,001–10,000	.15
10,001–15,000	.20
15,001–20,000	.40
20,001–25,000	.20

If tires are replaced at multiples of 5,000 miles only, what is the optimal age replacement policy?

31. The Navy uses a certain type of vacuum tube in a sonar scanning device. Based on past experience, the vacuum tube exhibits the following failure pattern.

Number of Months of Operation	Probability of Failure
1	.1
2	.1
3	.2
4	.1
5	.3
6	.2

Failures during operation cost $200 each, but the tube can be replaced before failure for $50. Find the optimal replacement strategy.

32. A local newsletter is printed on a printer with a cartridge that may break or run out of ink during operation. The cartridges can be replaced for $7.50, but if they fail when the newsletter is being printed, the cost is estimated to be $25 because of the delay in publication. The failure distribution of the cartridges follows.

Weeks of Use	Probability of Failure
1	.1
2	.2
3	.3
4	.4

Determine the optimal time to replace the cartridges.

33. Mactronics produces industrial robots. Each robot contains a part with a lifetime of at least one year and a maximum of five years. Lifetimes between one and five years are equally likely. If the part fails during operation, replacement costs are estimated to be \$400, whereas the part can be replaced before failure for \$50. When should the part be replaced? [Hint: The lifetime distribution is uniform on $\{1, 2, 3, 4, 5\}$. Assuming discrete variables, this means that $f(x) = \frac{1}{5}$ for $x = 1, 2, \ldots, 5$.]
34. A firm has purchased 30 Mactronics robots. If replaced as a block, the parts cost \$20 each but \$400 each when replaced after failure.
    a. Find the optimal block replacement strategy.
    b. Find the cost of replacing the items as they fail.
    c. Compare the cost of the block replacement policy obtained in part (a) with the solution obtained in part (b). Is the block replacement strategy preferred to an individual replacement strategy?

13.8 ANALYSIS OF WARRANTY POLICIES

An important issue related to the reliability of operating equipment is the protection afforded to the consumer who experiences failure of the equipment prior to its intended lifetime. Buyers and sellers perceive warranties differently. From the seller's point of view, the warranty is a means of limiting liability by specifying consumer responsibilities. These responsibilities include proper use of the product and heeding the warnings. From a marketing perspective, the warranty also can serve as an inducement to purchase the product. From the buyer's point of view, the warranty is a means of reducing or eliminating the economic penalty if the product fails to operate properly for a reasonable period of time. Warranties are particularly important to the consumer for products that are likely to experience high failure rates early in the product lifetime.

This section will present mathematical models for determining the economic value of a warranty. Such models could be used to find the portion of the cost of an item that could reasonably be attributed to the costs of satisfying a warranty commitment. In the models, we consider both the structure of the warranty and the reliability of the product to find the value of a warranty.

We must distinguish between repairable and nonrepairable items. Nonrepairable items include most electronic components, items in which failure corresponds to destruction of the item (burning out of a bulb or blowout of a tire, for example), or items typically not repaired but replaced (such as batteries that fail to hold a charge). Most major appliances, such as washing machines and fridges, fall into the category of repairable items. Repairable items that fail during the warranty period are typically repaired rather than replaced. The mathematical models presented in this section assume nonrepairable items.

Warranties for nonrepairable consumer goods generally take one of two forms. One is the **free replacement warranty**; if a failure occurs during the warranty period, a new item is supplied without charge. The second type of warranty is the **pro rata warranty**. Here, the consumer is given a rebate proportional to the amount of time remaining in the warranty period. The rebate is used to reduce the cost of a replacement item.

The Free Replacement Warranty

Assume that a single piece of operating equipment is placed into service and fails completely at random (that is, according to an exponential failure law) with known failure rate λ. The item is assumed to operate continuously. For intermittently operating equipment, clock time could be measured in operating hours rather than elapsed hours.

We will use the following notation:

T = Lifetime of an item chosen at random.
λ = Failure rate of an item chosen at random.
$F(t)$ = Cumulative distribution function of the random variable t.
C_1 = Cost of purchasing a new item with free replacement warranty.
K = Cost of purchasing a new item without any warranty.
W_1 = Time that the free replacement warranty is in effect after purchase.

If a failure occurs during the warranty period, the item is replaced free of charge. Assume that the consumer purchases a new item when a failure occurs after the expiration of the warranty. The new item has an identical free replacement warranty. Let Y be a random variable representing the time between successive purchases by the consumer. From Figure 13–16, we see that

$$Y = W_1 + \text{Time until the first failure after the warranty expires.}$$

It can be shown that

$$E(Y) = W_1 + 1/\lambda.$$

This expression for $E(Y)$ is valid only when the lifetime distribution is exponential. It results from the property of the exponential failure law that the time of the first failure after a fixed time (known as the forward recurrence time in probability theory) has the same distribution as the time between two successive failures. This result is true *only* if the failure law is exponential.

We will now compute the cost per unit time for an item that is replaced infinitely many times and has a free replacement warranty.

Each time that an item is purchased after the warranty expires constitutes the start of a new cycle. As the cost per cycle is C_1, it follows that the average cost per unit time under the free replacement warranty is

$$\frac{C_1}{E(Y)} = \frac{C_1}{W_1 + 1/\lambda} = \frac{\lambda C_1}{\lambda W_1 + 1}.$$

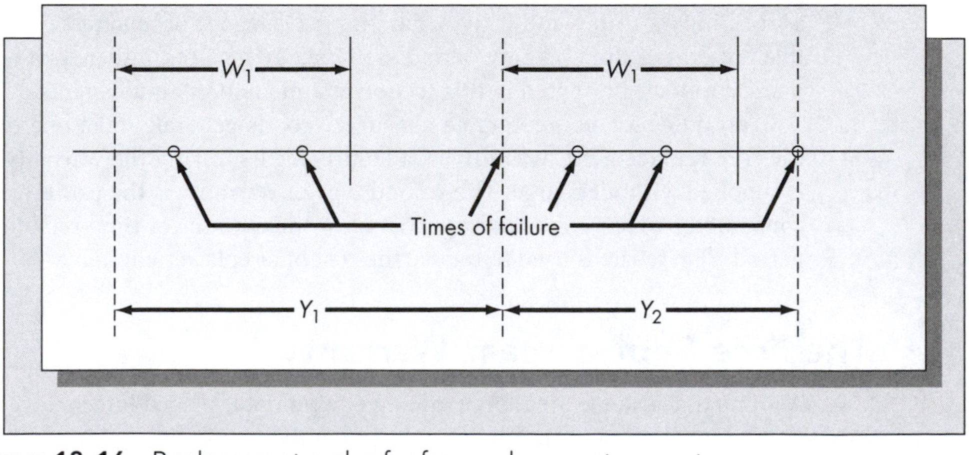

Figure 13–16 Replacement cycles for free replacement warranty

Without the warranty, the cost of replacement per unit time is simply λK. Let C_1^* be the cost of an item with a free replacement warranty that is indifferent to the cost of an item without the warranty, K. Then C_1^* solves

$$\lambda K = \frac{\lambda C_1^*}{\lambda W_1 + 1}$$

or

$$C_1^* = (\lambda W_1 + 1) K.$$

By definition $C_1^* - K$ is the economic value of the warranty. If $C_1 < C_1^*$, then the warranty should be purchased.

The Pro Rata Warranty

Assume all notation previously presented in this section. We also define

C_2 = Cost of a new item with a pro rata warranty.
W_2 = Effective warranty period with a pro rata warranty.

Consider an item purchased under a pro rata warranty that fails at a random time T. There are two cases.

Case 1: $T < W_2$. In this case, the fraction of the warranty period that has expired is T/W_2. The cost of the replacement item is then $C_2(T/W_2)$.
Case 2: $T \geq W_2$. Here, the warranty has expired, and the replacement cost is C_2.

Both of these cases can be represented mathematically by the expression

$$\frac{C_2}{W_2} \min(W_2, T).$$

To determine the expected life-cycle cost, we need to find an expression for $E[\min(W_2, T)]$. We have that

$$E[\min(W_2,T)] = \int_0^{\infty} \min(W_2,t) \lambda e^{-\lambda t} dt$$
$$= \int_0^{W_2} t\lambda e^{-\lambda t} dt + W_2 \int_{W_2}^{\infty} \lambda e^{-\lambda t} dt,$$

where $\lambda e^{-\lambda t}$ is the probability density of the time until failure, T.

The first integral requires integration by parts, and the second is the reliability function evaluated at W_2. It is easy to show that

$$\int_0^{W_2} t\lambda e^{-\lambda t} dt = \frac{1}{\lambda}\left[1 - e^{-\lambda W_2}(1 + \lambda W_2)\right]$$

and

$$W_2 \int_{W_2}^{\infty} \lambda e^{-\lambda t} dt = W_2 e^{-\lambda W_2},$$

Combining terms, it follows that

$$E[\min(W_2, T)] = \frac{1}{\lambda}(1 - e^{-\lambda W_2}).$$

The pro rata warranty starts anew with each purchase. Hence, each purchase begins a new cycle. It follows that the expected cost per unit time following a pro rata warranty is

$$\frac{C_2}{W_2} \frac{E[\min(W_2, T)]}{E(T)} = \frac{C_2}{W_2} \frac{(1/\lambda)(1 - e^{-\lambda W_2})}{1/\lambda} = \frac{C_2(1 - e^{-\lambda W_2})}{W_2}.$$

The cost of the pro rata warranty that is indifferent to the cost of the item without the warranty, C_2^*, solves

$$\frac{C_2^*(1 - e^{-\lambda W_2})}{W_2} = \lambda K,$$

or

$$C_2^* = \frac{\lambda K W_2}{1 - e^{-\lambda W_2}}.$$

The value of the pro rata warranty is $C_2^* - K$. If $C_2 > C_2^*$, then the pro rata warranty should not be purchased, and if $C_2 < C_2^*$, the pro rata warranty should be purchased. This assumes that the consumer is willing to base his or her decision on expected values. If the consumer is risk averse, the indifference value will be slightly higher than C_2^*.

EXAMPLE 13.13

You are considering purchasing a battery for your automobile. You find the same battery offered at three different stores. Store A sells the battery for \$21 and offers no warranty or guarantee. Store B sells the battery for \$40 and offers a free replacement if the battery fails to hold a charge for the first two years of operation. Store C sells the battery for \$40 as well but offers a pro rata warranty for the anticipated lifetime of the battery, which is advertised to be five years. The failure rate of the battery depends on the usage and the conditions, but from past experience you estimate that the time between failure is about once every three years.

We will determine the values of C_1^* and C_2^* for both warranties. For the full replacement warranty, we have that

$$K = 21,$$
$$C_1 = 40,$$
$$\lambda = \frac{1}{3},$$
$$W_1 = 2.$$

It follows that

$$C_1^* = (\lambda W_1 + 1)K = [(\tfrac{1}{3})(2) + 1](21) = \$35.$$

This means that the value of the free replacement warranty is $35 – $21 = $14, which is less than the $19 difference in the prices. On the basis of expected costs, the battery with no warranty (store A) is preferred to the one with the free replacement warranty (store B).

For the case of the pro rata warranty offered by store C, $W_2 = 5$ and the remaining parameters are as defined before. In that case we obtain

$$C_2^* = \frac{\lambda K W_2}{1 - e^{-\lambda W_2}} = \frac{\frac{1}{3}(21)(5)}{1 - e^{-5/3}} = \$43.15.$$

The value of the pro rata warranty is $43.15 – $21.00 = $22.15, which exceeds the difference between the price of the battery with and without the warranty and exceeds the value of the free replacement warranty. Hence, on the basis of this analysis, the pro rata warranty is the preferred choice.

Extensions and Criticisms

A criticism of Example 13.13 is the assumption that the failure law for the batteries is exponential. This means that a new battery has the same probability of failing in its first year of operation as a four-year-old battery has of failing in its fifth year of operation. In the case of batteries, it would seem that a failure law incorporating aging, such as the Weibull, would be more accurate. The models discussed can be extended to include more general types of failure patterns, but the calculations required to determine an optimal policy are very complex. In particular, one must determine the renewal function, which is generally a difficult computation. Problem 41 considers a case in which the failure law follows an Erlang distribution.

The two types of warranties discussed in this section, the free replacement warranty and the pro rata warranty, account for the majority of consumer warranties for nonrepairable items. In the military context, another type of warranty is very common. This is known as the **reliability improvement warranty** (RIW). In this case, the supplier agrees to repair or replace items that fail within a specified warranty period and provides a pool or pools of spares and perhaps one or more repair facilities. This type of warranty is intended to provide an incentive to the supplier to make initial reliability high and to provide improvements in the reliability of existing units if possible.

For repairable items, warranties usually cover all or some of the cost of effecting repairs during the warranty period. Analysis of warranties for repairable items is more complex than that for nonrepairable items. Different levels of repair are possible, and the likelihood of failure during the warranty period may depend upon usage that can vary significantly from one user to another.

For repairable items, an issue closely related to warranties is service contracts. The primary difference between a warranty and a service contract is that the cost of the warranty is usually assumed to be included in the purchase price of the item, whereas a service contract is an additional item whose purchase is at the option of the buyer. A service contract can be thought of as an extended warranty beyond the normal warranty period. The analytical approach discussed here also can be applied to determining the economic value of a service contract. However, complex failure laws and different costs associated with various levels of repair should be allowed. In addition, the pricing of consumer service contracts relies heavily on the assumption that the consumer is risk averse. That is, a typical consumer would prefer to pay more than a service contract is worth in an expected value sense to reduce the risk of incurring an expensive repair. In that sense, the service contract serves the role of an insurance policy. Perhaps it is the tendency of consumers to be risk averse that justifies the rather high prices that are charged for service contracts for many consumer products.

PROBLEMS FOR SECTION 13.8

35. For Example 13.13, what value of the warranty period equates the full replacement warranty with no warranty? (That is, for what value of W_1 is the consumer indifferent to purchasing from store A or store B?)
36. For Example 13.13, what value of the warranty period for the full replacement warranty equates the full replacement and the pro rata warranties? (For what value of W_1 is the consumer indifferent to purchasing from store B or store C, assuming $W_2 = 5$?)
37. A producer of calculators estimates that the calculators fail at a rate of one every five years. The calculators are sold for \$25 each with a one-year free replacement warranty but can be purchased from an unregistered mail-order source for \$18.50 without the warranty. Is it worth purchasing the calculator with the warranty?
38. For problem 37, what length of period of the warranty equates the replacement costs of the calculator with and without the warranty?
39. Zemansky's sells tires with a pro rata warranty. The tires are warranted to deliver 50,000 miles with the warranty payment based on the remaining tread on the tire. The tires fail on the average after 35,000 miles of wear. Suppose the tires sell for \$50 each with the warranty. If failures occur completely at random, what would be a consistent price for the tires if no warranty were offered?
40. Habard's, a chain of hardware stores, sells a variety of tools and home repair items. One of their best wrenches sells for \$5.50. Habard's will include a three-year free replacement warranty for an additional \$1.50. The wrench is expected to be subject to heavy use and, based on past experience, will fail randomly at a rate of one every eight years. Is it worth purchasing the warranty?
41. Consider the case in which the failure mechanism for the product does not obey the exponential law. In that case, the cost under the free replacement warranty that is indifferent to the cost of buying the item without a warranty is given by

$$C_1^* = K[M(W_1) + 1],$$

where $M(t)$ is known as the renewal function.
If the time between failures, T, follows an Erlang law with parameters λ and 2, then

$$M(t) = \frac{\lambda t}{2} - 0.25 + 0.25e^{-2\lambda t} \qquad \text{for all } t \geq 0.$$

(See, for example, Barlow and Proschan, 1965, p. 57.)
a. For Example 13.13, determine the indifference value of the item with a free replacement warranty when the failure law follows an Erlang distribution. Assume that $\lambda = \frac{2}{3}$ to give the same value of $E(T)$ as in the example.
b. Is the value of the warranty larger or smaller than in the corresponding exponential case? Explain the result intuitively.

13.9 SOFTWARE RELIABILITY

Software reliability is a problem with characteristics different from hardware reliability problems. Typically, new software possesses a few "bugs," or errors. Ideally, one would like to remove all the bugs from the software before its release, but that may be impossible. It is more reasonable to release the software when the number of bugs has been reduced to an acceptable level. Predicting the number of remaining bugs is, however, a difficult problem.

The importance of software reliability cannot be overemphasized.

> The tiniest software bug can fell the mightiest machine—often with disastrous consequences. During the past five years, software defects have killed sailors, maimed patients, wounded corporations and threatened to cause the government-securities market to collapse. Such problems are likely to grow as industry and the military increasingly rely on software to run systems of phenomenal complexity. (Davis, 1987)

Several models exist for estimating software reliability. However, we will not present these models in detail. Z. Jelinski and Paul Moranda (1972) have suggested the following approach. Let N be the total initial error content (i.e., the number of bugs) in the software. As the software undergoes testing, the number of bugs is reduced. They assume that the failure rate (that is, the likelihood of detecting a bug) is proportional to the number of bugs remaining in the program, where φ is the proportionality constant. That is, the time until detection of the first bug has the exponential distribution with parameter $N\varphi$; the time between detection of the first and the second bugs has exponential distribution with parameter $(N - 1)\varphi$, and so on.

Hence, as bugs are removed from the program, the amount of time required to detect the next bug increases. After n bugs have been removed, one will have observed the values of T_1, T_2, …, T_n representing the time between successive detections. These observations are used to estimate φ and N using the maximum likelihood principle. Based on these estimates, one could predict exactly how much testing would be required in order to achieve a certain level of reliability in the software.

Martin Shooman (1972) suggests using a normalized error rate to measure the error content in the program. He defines

$$p(t) = \text{Errors per total number of instructions per month of debugging time}$$

and develops a reliability model based on first principles. He demonstrates how this model can be used to build a functional relationship between the amount of time devoted to debugging and the reliability of the program.

13.10 HISTORICAL NOTES

Much of the theory of reliability, life testing, and maintenance strategies has its roots in actuarial theory developed by the insurance industry. Sophisticated mathematical models for predicting survival probabilities date back to the turn of the previous century. Alfred Lotka (1939) discusses some of the connections between equipment replacement models and actuarial studies. The work of Waloddi Weibull (1939 and 1951) laid the foundations for the subject of fatigue life in materials.

Interest in reliability problems became considerably more widespread during World War II when attempts were made to understand the failure laws governing complex military systems. During the 1950s, problems concerning life testing and missile reliability began to receive serious attention. In 1952 the Department of Defense established the Advisory Group on Reliability of Electronic Equipment, which published its first report on reliability in June of 1957.

The origins of the specific age replacement models presented in this chapter are unclear. However, sophisticated age replacement models date back as far as the early 1920s (see Taylor, 1923, and Hotelling, 1925). The stochastic planned replacement models presented in Section 13.7 form the basis for much of the research in replacement theory, but the origins of these models are unclear as well.

Section 13.8 on warranties is based on the article by Wallace Blischke and Ernest Scheuer (1975). Extensions and corrections of their work can be found in John Mamer (1982). Readers interested in pursuing further reading should refer to the excellent texts by Richard Barlow and Frank Proschan (1965 and 1975) on reliability models and Ilya Gertsbakh's (2010) book on maintenance strategies. Efraim Turban (1967) and Lawrence Mann (1976) discuss issues concerning the application of maintenance models.

13.11 SUMMARY

The purpose of this chapter was to review the terminology and the methodology of the theory and application of reliability and maintenance models. Reliability theory is an area of study that has received considerable attention from mathematicians. However, the mathematics is of interest not only for its own sake. These models are extremely useful in an operational setting in considering such issues as failure characteristics of operating equipment, economically sound maintenance strategies, and the value of product warranties and service contracts.

The complexity of the analysis depends upon the assumptions made about the random variable T, which represents the lifetime of a single item or piece of operating equipment. The distribution function of T, $F(t)$, is the probability that the item fails at or before time t ($P\{T \leq t\}$), whereas the reliability function of T, $R(t)$, is the probability that the item fails after time t ($P\{T > t\}$). An important quantity related to these functions is the failure rate function $r(t)$, which is the ratio $f(t)/R(t)$ of the probability density function and the reliability function. If Δt is sufficiently small, the term $r(t)\Delta t$ can be interpreted as the conditional probability that the item will fail in the next Δt units of time given that it has survived up until time t.

The failure rate function provides considerable information about the aging characteristics of operating equipment. In a manufacturing environment, we would expect that most operating equipment would have an increasing failure rate function. That means it would be more likely to fail as it ages. A decreasing failure rate function can arise when the likelihood of early failure is high due to defectives in the population. The Weibull probability law can be used to describe the failure characteristics of equipment having either an increasing or a decreasing failure rate function.

Of interest is the case in which the failure rate function is constant. This case gives rise to the exponential distribution for the lifetime of a single component. The exponential distribution is the only continuous distribution possessing the memoryless property. This means that the conditional probability that an item that has been operating up until time t fails in the next s units of time is independent of t.

The Poisson process describes the situation in which a single piece of operating equipment fails according to the exponential distribution and is replaced immediately upon failure. When this occurs, the number of failures in a given time has the Poisson distribution, the time between successive failures has the exponential distribution, and the time for n failures to occur has the Erlang distribution.

The chapter considered the reliability functions of complex systems of components. It showed how to obtain the reliability functions for components in series and parallel from the reliability functions of the individual components. The chapter also considered K out of N systems, which function only if at least K components function.

Reliability issues form the basis of the maintenance models discussed in the latter half of the chapter. An important measure of a system's performance is the availability, which is the proportion of the time that the equipment operates. We treated both deterministic age replacement models, which do not explicitly include the likelihood of equipment failure, and stochastic age replacement models, which do. The stochastic models allow for replacing the equipment before failure. This is of interest when items have an increasing failure rate function and unplanned failures are more costly than planned failures.

We discussed the economic value of warranties. A warranty is a promise supplied by the seller to the buyer to either replace the item with a new one if it fails during the warranty period (free replacement warranty) or provide a discount on the purchase of a new item proportional to the remaining amount of time (or wear) in the warranty period (pro rata warranty). The issues surrounding warranties and service contracts are similar, but service contract models are considerably more complex, owing to the need to include multiple levels of repair. Finally, we concluded the chapter with a discussion of software reliability.

ADDITIONAL PROBLEMS ON RELIABILITY AND MAINTAINABLITY

42. A large national producer of appliances has traced customer experience with a popular toaster oven. A survey of 5,000 customers who purchased the oven early in 2010 revealed the following.

Year	Number of Breakdowns
2010	188
2011	58
2012	63
2013	72
2014	54
2015	71

a. Using these data, estimate p_k = the probability that a toaster oven fails in its kth year of operation, for $k = 1, \ldots, 6$.
b. What is the likelihood that a toaster oven will last at least six years without failure based on these data?
c. The discrete failure rate function has the form $r_k = p_k/R_{k-1}$, where R_k is the probability that a unit survives through period k. Determine the failure rate function for the first five years of operation from the given data.

d. Suppose that you purchased a toaster oven at the beginning of 2017 and it is still operating at the end of 2020. If the reliability has not changed appreciably from 2010 to 2020, use the results of part (c) to obtain the probability that it will fail during the first two months of calendar year 2021.

43. Six thousand light bulbs light a large hotel and casino marquee. Each bulb fails completely at random, and each has an average lifetime of 3,280 hours. Assuming that the marquee stays lit continuously and bulbs that burn out are replaced immediately, how many replacements must be made each year on the average?

44. The owner of the hotel mentioned in problem 43 has decided that in order to decrease the number of burned-out bulbs, she will replace all 6,000 bulbs at the start of each year in addition to replacing the bulbs as they burn out. Comment on the effectiveness of this strategy.

45. The owner of the hotel in problem 43 falls on hard times and dispenses with replacement of the bulbs. She notices that more than half of the bulbs have burned out before the advertised average lifetime of 3,280 hours and decides to sue the light bulb manufacturer for false advertising. Do you think she has a case? (Hint: What fraction of the bulbs would be expected to fail prior to the mean lifetime?)

46. Continuing with the example of problem 43, determine the following.
   a. The proportion of bulbs lasting more than two years.
   b. The probability that a bulb chosen at random fails in the first three months of operation.
   c. The probability that a bulb that has lasted for 10 years fails in the next three months of operation.

47. Assume that the bulbs in problem 43 are not replaced as they fail.
   a. What fraction of the 6,000 bulbs are expected to fail in the first year?
   b. What fraction of the bulbs surviving the first year are expected to fail in the second year?
   c. What fraction of the bulbs surviving the nth year are expected to fail in year $n + 1$ for any value of $n = 1, 2 \ldots$?
   d. Using the results from part (c), how many of the original 6,000 bulbs would be expected to fail in the fourth year of operation?

48. The mean value of a Weibull random variable is given by the formula

$$\mu = a^{-1/\beta}\,\Gamma(1 + 1/\beta),$$

where Γ represents the gamma function. The gamma function has the property that $\Gamma(k) = (k - 1)\Gamma(k - 1)$ for any value of $k > 1$ and $\Gamma(1) = 1$. Notice that if k is an integer, this results in $\Gamma(k) = (k - 1)!$. If k is not an integer, one must use the recursive definition for $\Gamma(k)$; the following table gives the values of $\Gamma(k)$ for $1 \le k \le 2$.

k	**Γ(k)**	**k**	**Γ(k)**
1.00	1.0000	1.55	.8889
1.05	.9735	1.60	.8935
1.10	.9514	1.65	.9001
1.15	.9330	1.70	.9086
1.20	.9182	1.75	.9191
1.25	.9064	1.80	.9314

k	Γ(*k*)	*k*	Γ(*k*)
1.30	.8975	1.85	.9456
1.35	.8912	1.90	.9612
1.40	.8873	1.95	.9799
1.45	.8857	2.00	1.0000
1.50	.8862		

For example, this table would be used as follows: $\Gamma(3.6) = (2.6)\Gamma(2.6) = (2.6)(1.6)\Gamma(1.6) = (2.6)(1.6)(.8935) = 3.717$.

a. Compute the expected failure time for Example 13.4 regarding copier equipment.
b. Compute the expected failure time for a piece of operating equipment whose failure law is given in Example 13.2.
c. Determine the mean failure time for $\alpha = 1.35$ and $\beta = 0.20$.
d. Determine the mean failure time for $\alpha = 0.90$ and $\beta = 0.45$.

49. Suppose that a particular light bulb is advertised as having an average lifetime of 2,000 hours and is known to satisfy an exponential failure law. For simplicity, suppose that the bulb is used continuously. Find the probabilities of the bulb lasting as indicated below.
a. More than 3,000 hours.
b. Less than 1,500 hours.
c. Between 2,000 and 2,500 hours.

50. Applicational Materials sells several pieces of equipment used in the manufacture of silicon-based microprocessors. In 2020 the company filled 130 orders for model a55212. Suppose that the machines fail according to a Weibull law. In particular, the cumulative distribution function $F(t)$ of the time until failure of any machine is given by

$$F(t) = 1 - e^{-0.0475t^{1.2}} \qquad \text{for all } t \geq 0,$$

where t is in years.
a. What is the failure rate function for this piece of equipment?
b. Of the original 130 machines sold in 2020, how many would one expect *not* to experience a breakdown before January 2024? Assume for the sake of simplicity that all the machines were sold on January 1, 2020.
c. Using the results of part (a), estimate the fraction of machines that have survived 10 years of use that will break down during the 11th year of operation [or you may compute this directly if you did not get the answer to part (a)].

51. A local cab company maintains a fleet of 10 cabs. Each time a cab breaks down, it is repaired the same day. Assume that breakdowns of individual cabs occur completely at random at a rate of two per year.
a. What is the probability that any particular cab will run for a full year without suffering a breakdown?
b. What is the probability that the entire fleet will run for one month without a breakdown?
c. On average, how many breakdowns would one expect for the fleet in a typical three-month period?
d. What is the probability that there are more than five breakdowns between Thanksgiving Day (November 28) and New Year's Day (January 1)?
e. For what reason might your answer in part (d) be too low?

52. A collection of 30 Christmas tree lights are arranged in a pure series circuit; that is, if one of the lights burns out, then the entire string goes out. Suppose that each light fails completely at random at a rate of one failure every year. What is the probability that the lights will burn from the beginning of Christmas Eve (December 24) to the end of New Year's Day (January 1) without failure?
53. A piece of industrial machinery costs $48,000 to replace and has essentially no salvage value. Over the first five years of operation, maintenance costs amounted to $8,000. If the maintenance cost rate is a linear function of time, what is the optimal age at which to replace the machinery?
54. For an automobile that you own or would like to own, estimate the correct values of the replacement cost, the rate of depreciation, the initial maintenance cost, and the rate at which the maintenance cost increases. Based on these estimates, determine the optimal number of years that you should wait before replacing your car.

APPENDIX 13–A

Glossary of Notation on Reliability and Maintainability

a = Maintenance cost rate per unit time for simple age replacement model. Also used as a parameter of the exponential maintenance cost function for the exponential age replacement model.

a_1 = Cost of replacing an item when it fails. Used in the model of block replacement.

a_2 = Cost of replacing the entire block of n_0 items. Used in the model of block replacement.

α = A parameter of the Weibull probability function.

b = A parameter of the exponential maintenance cost function for the exponential age replacement model.

β = A parameter of the Weibull probability function.

c = A parameter of the exponential salvage value function for the exponential age replacement model.

c_1 = Cost of replacing an item when it fails. Used in the model of planned replacement under uncertainty for a single item.

c_2 = Cost of replacing an item before it fails. Used in the model of planned replacement under uncertainty for a single item.

C_1 = Cost of purchasing a new item with a free replacement warranty.

C_2 = Cost of purchasing a new item with a pro rata warranty.

d = A parameter of the exponential salvage value function for the exponential age replacement model.

$f(t)$ = Density function of the random variable T.

$F(t)$ = $P\{T \le t\}$ = Cumulative distribution function of T.

$F_K(t)$ = Cumulative distribution function of the random variable T_K, the lifetime of a K out of N system.

$F_P(t)$ = Cumulative distribution function of the random variable T_P, the lifetime of a parallel system.

$F_S(t)$ = Cumulative distribution function of the random variable T_S, the lifetime of a series system.

$G(t)$ = Average cost per unit time for age replacement and warranty models.
K = Fixed replacement cost for age replacement models. Also used to denote the replacement cost of a new item without a warranty in warranty models.
λ = The parameter of the exponential failure law; the expected number of failures per unit time when failures occur at random.
MTBF = Mean time between failures.
MTTR = Mean time to repair.
n_0 = Total number of items in a block. Used in the block replacement model.
$N(t)$ = Number of failures occurring in the interval $(0, t)$ when failures follow a Poisson process. $N(t)$ has the Poisson distribution with parameter λt.
$r(t)$ = Failure rate function of T.
$R(t)$ = $P\{T > t\}$ = Reliability function of T.
$R_K(t)$ = Reliability function of a K out of N system of components.
$R_P(t)$ = Reliability function of a parallel system of components.
$R_S(t)$ = Reliability function of a series system of components.
T = Random variable corresponding to the lifetime of an operating unit.
T_K = Lifetime of a K out of N system of components.
T_P = Lifetime of a parallel system of components.
T_S = Lifetime of a series system of components.
W_1 = Warranty period for a free replacement warranty.
W_2 = Warranty period for a pro rata warranty.
W_n = Time for n failures to occur when failures follow a Poisson process. W_n has the Erlang distribution with parameters λ and n.

BIBLIOGRAPHY

Amstadter, B. L. *Reliability Mathematics*. New York: McGraw-Hill, 1971.

Barlow, R. E., and F. Proschan. *Mathematical Theory of Reliability*. New York: John Wiley & Sons, 1965.

Barlow, R. E., and F. Proschan. *Statistical Theory of Reliability and Life Testing*. New York: Holt, Rinehart & Winston, 1975.

Blischke, W. R., and E. M. Scheuer. "Calculation of the Cost of Warranty Policies as a Function of Estimated Life Distributions." *Naval Research Logistics Quarterly* 22 (1975), pp. 681–96.

Davis, R. "As Complexity Rises, Tiny Flaws in Software Pose a Growing Threat." *The Wall Street Journal*, January 28, 1987, p. 1.

DeGroot, M. H., and M. Schervish. *Probability and Statistics*. 4th ed. New York: Pearson, 2012.

Ebeling, C. E. *An Introduction to Reliability and Maintainability Engineering*, 3rd ed. Long Grove, IL: Waveland Press, 2019.

Fox, B. H., M. G. Snyder, and A. M. Smith. "Reliability-Centered Maintenance Improves Operations at TMI Nuclear Plant." *Power Engineering*, November 1994, pp. 75–78.

Garvin, D. A. *Managing Quality*. New York: The Free Press, 1988.

Gertsbakh, I. B. *Reliability Theory with Applications to Preventive Maintenance*. Heidelberg: Spring-Verlag, 2010.

Hillier, F. S., and G. J. Lieberman. *Introduction to Operations Research*. 10th ed. New York: McGraw-Hill, 2015.

Hotelling, H. "A General Mathematical Theory of Depreciation." *Journal of the American Statistical Association* 20 (September 1925), pp. 340–53.

Jelinski, Z., and P. Moranda. "Software Reliability Research." In *Statistical Computer Performance Evaluation*, ed. W. Freiberger. New York: Academic Press, 1972.

Lotka, A. J. "A Contribution to the Theory of Self-Renewing Aggregates with Special Reference to Industrial Replacement." *Annals of Mathematical Statistics* 10 (1939), pp. 1–25.

Mamer, J. "Cost Analysis of Pro Rata and Free Replacement Warranties." *Naval Research Logistics Quarterly* 29 (1982), pp. 345–56.

Mann, L. L. *Maintenance Management*. Lexington, MA: Lexington Books, 1976.

O'Connor, P. D. T, and A. Kleyner. *Practical Reliability Engineering*. 5th ed. New York: John Wiley & Sons, 2012.

Shooman, M. L. "Probabilistic Models for Software Reliability Prediction." In *Statistical Computer Performance Evaluation*, ed. W. Freiberger, pp. 485–502. New York: Academic Press, 1972.

Taylor, J. S. "A Statistical Theory of Depreciation." *Journal of the American Statistical Association*, December 1923, pp. 1010–23.

Turban, E. "The Use of Mathematical Models in Plant Maintenance Decision Making." *Management Science* 13 (1967), pp. 20–27.

U.S. Nuclear Regulatory Commission. *Reactor Safety Study*. WASH–1400, NU REG 75–01, 1975.

Weibull, W. "A Statistical Theory of the Strength of Materials." *Ing. Vetenskaps Akad Handl.*, no. 151 (1939).

Weibull, W. "A Statistical Distribution Function of Wide Applicability." *Journal of Applied Mechanics* 18 (1951), pp. 293–97.

Appendix Tables

TABLE A–1 Areas under the Normal Curve

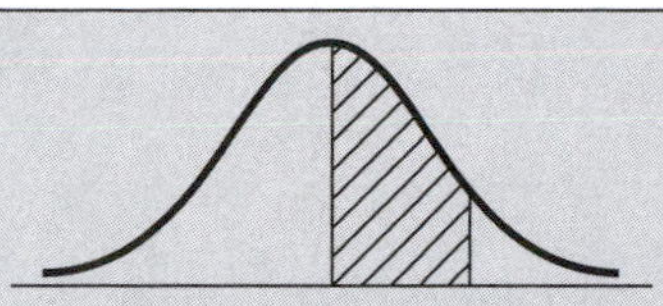

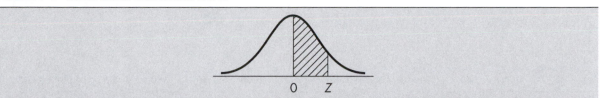

Z	.00	.01	.02	.03	.04	.05	.06	.07	.08	.09
.00	.0000	.0040	.0080	.0120	.0160	.0199	.0239	.0279	.0319	.0359
.10	.0398	.0438	.0478	.0517	.0557	.0596	.0636	.0675	.0714	.0753
.20	.0793	.0832	.0871	.0910	.0948	.0987	.1026	.1064	.1103	.1141
.30	.1179	.1217	.1255	.1293	.1331	.1368	.1406	.1443	.1480	.1517
.40	.1554	.1591	.1628	.1664	.1700	.1736	.1772	.1808	.1844	.1879
.50	.1915	.1950	.1985	.2019	.2054	.2088	.2123	.2157	.2190	.2224
.60	.2257	.2291	.2324	.2357	.2389	.2422	.2454	.2486	.2517	.2549
.70	.2580	.2611	.2642	.2673	.2703	.2734	.2764	.2793	.2823	.2852
.80	.2881	.2910	.2939	.2967	.2995	.3023	.3051	.3078	.3106	.3133
.90	.3159	.3186	.3212	.3238	.3264	.3289	.3315	.3340	.3365	.3389
1.00	.3413	.3438	.3461	.3485	.3508	.3531	.3554	.3577	.3599	.3621
1.10	.3643	.3665	.3686	.3708	.3729	.3749	.3770	.3790	.3810	.3830
1.20	.3849	.3869	.3888	.3907	.3925	.3944	.3962	.3980	.3997	.4015
1.30	.4032	.4049	.4066	.4082	.4099	.4115	.4131	.4147	.4162	.4177
1.40	.4192	.4207	.4222	.4236	.4251	.4265	.4279	.4292	.4306	.4319
1.50	.4332	.4345	.4357	.4370	.4382	.4394	.4406	.4418	.4429	.4441
1.60	.4452	.4463	.4474	.4484	.4495	.4505	.4515	.4525	.4535	.4545
1.70	.4554	.4564	.4573	.4582	.4591	.4599	.4608	.4616	.4625	.4633
1.80	.4641	.4649	.4656	.4664	.4671	.4678	.4686	.4693	.4699	.4706
1.90	.4713	.4719	.4726	.4732	.4738	.4744	.4750	.4756	.4761	.4767
2.00	.4772	.4778	.4783	.4788	.4793	.4798	.4803	.4808	.4812	.4817
2.10	.4821	.4826	.4830	.4834	.4838	.4842	.4846	.4850	.4854	.4857
2.20	.4861	.4864	.4868	.4871	.4875	.4878	.4881	.4884	.4887	.4890
2.30	.4893	.4896	.4898	.4901	.4904	.4906	.4909	.4911	.4913	.4916
2.40	.4918	.4920	.4922	.4925	.4927	.4929	.4931	.4932	.4934	.4936
2.50	.4938	.4940	.4941	.4943	.4945	.4946	.4948	.4949	.4951	.4952
2.60	.4953	.4955	.4956	.4957	.4959	.4960	.4961	.4962	.4963	.4964
2.70	.4965	.4966	.4967	.4968	.4969	.4970	.4971	.4972	.4973	.4974
2.80	.4974	.4975	.4976	.4977	.4977	.4978	.4979	.4979	.4980	.4981
2.90	.4981	.4982	.4982	.4983	.4984	.4984	.4985	.4985	.4986	.4986
3.00	.4987	.4987	.4987	.4988	.4988	.4989	.4989	.4989	.4990	.4990
3.10	.4990	.4991	.4991	.4991	.4992	.4992	.4992	.4992	.4993	.4993
3.20	.4993	.4993	.4994	.4994	.4994	.4994	.4994	.4995	.4995	.4995
3.30	.4995	.4995	.4995	.4996	.4996	.4996	.4996	.4996	.4996	.4997
3.40	.4997	.4997	.4997	.4997	.4997	.4997	.4997	.4997	.4997	.4998
3.50	.4998	.4998	.4998	.4998	.4998	.4998	.4998	.4998	.4998	.4998
3.60	.4998	.4998	.4999	.4999	.4999	.4999	.4999	.4999	.4999	.4999
3.70	.4999	.4999	.4999	.4999	.4999	.4999	.4999	.4999	.4999	.4999
3.80	.4999	.4999	.4999	.4999	.4999	.4999	.4999	.4999	.4999	.4999

The values in the body of the table are the areas between the mean and the value of Z. Example: If we want to find the area under the standard normal curve between $Z = 0$ and $Z = 1.96$, we find the $Z = 1.90$ row and .06 column (for $Z = 1.90 + .06 = 1.96$) and read .4750 at the intersection. For convenience, this information also appears in Table A–4.

Source: P. Billingsley, D. J. Croft, D. V. Huntsberger, and C. J. Watson, *Statistical Inference for Management and Economics* (Boston: Allyn & Bacon, 1986). Reprinted with permission.

TABLE A–2 Cumulative Binomial Probabilities

$$P(X \le r) = \sum_{k=0}^{r} \binom{n}{k} p^k (1-p)^{n-k}$$

where X is the number of successes in n trials

$n = 5$

r	.01	.05	.10	.20	.30	.40	*p* .50	.60	.70	.80	.90	.95	.99	r
0	0.9510	0.7738	0.5905	0.3277	0.1681	0.0778	0.0313	0.0102	0.0024	0.0003	0.0000	0.0000	0.0000	0
1	0.9990	0.9774	0.9185	0.7373	0.5282	0.3370	0.1875	0.0870	0.0308	0.0067	0.0005	0.0000	0.0000	1
2	1.0000	0.9988	0.9914	0.9421	0.8369	0.6826	0.5000	0.3174	0.1631	0.0579	0.0086	0.0012	0.0000	2
3	1.0000	1.0000	0.9995	0.9933	0.9692	0.9130	0.8125	0.6630	0.4718	0.2627	0.0815	0.0226	0.0010	3
4	1.0000	1.0000	1.0000	0.9997	0.9976	0.9898	0.9688	0.9222	0.8319	0.6723	0.4095	0.2262	0.0490	4

$n = 10$

r	.01	.05	.10	.20	.30	.40	*p* .50	.60	.70	.80	.90	.95	.99	r
0	0.9044	0.5987	0.3487	0.1074	0.0282	0.0060	0.0010	0.0001	0.0000	0.0000	0.0000	0.0000	0.0000	0
1	0.9957	0.9139	0.7361	0.3758	0.1493	0.0464	0.0107	0.0017	0.0001	0.0000	0.0000	0.0000	0.0000	1
2	0.9999	0.9885	0.9298	0.6778	0.3828	0.1673	0.0547	0.0123	0.0016	0.0001	0.0000	0.0000	0.0000	2
3	1.0000	0.9990	0.9872	0.8791	0.6496	0.3823	0.1719	0.0548	0.0106	0.0009	0.0000	0.0000	0.0000	3
4	1.0000	0.9999	0.9984	0.9672	0.8497	0.6331	0.3770	0.1662	0.0473	0.0064	0.0001	0.0000	0.0000	4
5	1.0000	1.0000	0.9999	0.9936	0.9527	0.8338	0.6230	0.3669	0.1503	0.0328	0.0016	0.0001	0.0000	5
6	1.0000	1.0000	1.0000	0.9991	0.9894	0.9452	0.8281	0.6177	0.3504	0.1209	0.0128	0.0010	0.0000	6
7	1.0000	1.0000	1.0000	0.9999	0.9984	0.9877	0.9453	0.8327	0.6172	0.3222	0.0702	0.0115	0.0001	7
8	1.0000	1.0000	1.0000	1.0000	0.9999	0.9983	0.9893	0.9536	0.8507	0.6242	0.2639	0.0861	0.0043	8
9	1.0000	1.0000	1.0000	1.0000	1.0000	0.9999	0.9990	0.9940	0.9718	0.8926	0.6513	0.4013	0.0956	9

$n = 15$

r	.01	.05	.10	.20	.30	.40	*p* .50	.60	.70	.80	.90	.95	.99	r
0	0.8601	0.4633	0.2059	0.0352	0.0047	0.0005	0.0000	0.0000	0.0000	0.0000	0.0000	0.0000	0.0000	0
1	0.9904	0.8290	0.5490	0.1671	0.0353	0.0052	0.0005	0.0000	0.0000	0.0000	0.0000	0.0000	0.0000	1
2	0.9996	0.9638	0.8159	0.3980	0.1268	0.0271	0.0037	0.0003	0.0000	0.0000	0.0000	0.0000	0.0000	2
3	1.0000	0.9945	0.9444	0.6482	0.2969	0.0905	0.0176	0.0019	0.0001	0.0000	0.0000	0.0000	0.0000	3
4	1.0000	0.9994	0.9873	0.8358	0.5155	0.2173	0.0592	0.0093	0.0007	0.0000	0.0000	0.0000	0.0000	4
5	1.0000	0.9999	0.9978	0.9389	0.7216	0.4032	0.1509	0.0338	0.0037	0.0001	0.0000	0.0000	0.0000	5
6	1.0000	1.0000	0.9997	0.9819	0.8689	0.6098	0.3036	0.0950	0.0152	0.0008	0.0000	0.0000	0.0000	6
7	1.0000	1.0000	1.0000	0.9958	0.9500	0.7869	0.5000	0.2131	0.0500	0.0042	0.0000	0.0000	0.0000	7
8	1.0000	1.0000	1.0000	0.9992	0.9848	0.9050	0.6964	0.3902	0.1311	0.0181	0.0003	0.0000	0.0000	8
9	1.0000	1.0000	1.0000	0.9999	0.9963	0.9662	0.8491	0.5968	0.2784	0.0611	0.0022	0.0001	0.0000	9
10	1.0000	1.0000	1.0000	1.0000	0.9993	0.9907	0.9408	0.7827	0.4845	0.1642	0.0127	0.0006	0.0000	10
11	1.0000	1.0000	1.0000	1.0000	0.9999	0.9981	0.9824	0.9095	0.7031	0.3518	0.0556	0.0055	0.0000	11
12	1.0000	1.0000	1.0000	1.0000	1.0000	0.9997	0.9963	0.9729	0.8732	0.6020	0.1841	0.0362	0.0004	12
13	1.0000	1.0000	1.0000	1.0000	1.0000	1.0000	0.9995	0.9948	0.9647	0.8329	0.4510	0.1710	0.0096	13
14	1.0000	1.0000	1.0000	1.0000	1.0000	1.0000	1.0000	0.9995	0.9953	0.9648	0.7941	0.5367	0.1399	14

TABLE A–2 (*continued*)

n = 20

r	.01	.05	.10	.20	.30	.40	*p* .50	.60	.70	.80	.90	.95	.99	*r*
0	0.8179	0.3585	0.1216	0.0115	0.0008	0.0000	0.0000	0.0000	0.0000	0.0000	0.0000	0.0000	0.0000	0
1	0.9831	0.7358	0.3917	0.0692	0.0076	0.0005	0.0000	0.0000	0.0000	0.0000	0.0000	0.0000	0.0000	1
2	0.9990	0.9245	0.6769	0.2061	0.0355	0.0036	0.0002	0.0000	0.0000	0.0000	0.0000	0.0000	0.0000	2
3	1.0000	0.9841	0.8670	0.4114	0.1071	0.0160	0.0013	0.0000	0.0000	0.0000	0.0000	0.0000	0.0000	3
4	1.0000	0.9974	0.9568	0.6296	0.2375	0.0510	0.0059	0.0003	0.0000	0.0000	0.0000	0.0000	0.0000	4
5	1.0000	0.9997	0.9887	0.8042	0.4164	0.1256	0.0207	0.0016	0.0000	0.0000	0.0000	0.0000	0.0000	5
6	1.0000	1.0000	0.9976	0.9133	0.6080	0.2500	0.0577	0.0065	0.0003	0.0000	0.0000	0.0000	0.0000	6
7	1.0000	1.0000	0.9996	0.9679	0.7723	0.4159	0.1316	0.0210	0.0013	0.0000	0.0000	0.0000	0.0000	7
8	1.0000	1.0000	0.9999	0.9900	0.8867	0.5956	0.2517	0.0565	0.0051	0.0001	0.0000	0.0000	0.0000	8
9	1.0000	1.0000	1.0000	0.9974	0.9520	0.7553	0.4119	0.1275	0.0171	0.0006	0.0000	0.0000	0.0000	9
10	1.0000	1.0000	1.0000	0.9994	0.9829	0.8725	0.5881	0.2447	0.0480	0.0026	0.0000	0.0000	0.0000	10
11	1.0000	1.0000	1.0000	0.9999	0.9949	0.9435	0.7483	0.4044	0.1133	0.0100	0.0001	0.0000	0.0000	11
12	1.0000	1.0000	1.0000	1.0000	0.9987	0.9790	0.8684	0.5841	0.2277	0.0321	0.0004	0.0000	0.0000	12
13	1.0000	1.0000	1.0000	1.0000	0.9997	0.9935	0.9423	0.7500	0.3920	0.0867	0.0024	0.0000	0.0000	13
14	1.0000	1.0000	1.0000	1.0000	1.0000	0.9984	0.9793	0.8744	0.5836	0.1958	0.0113	0.0003	0.0000	14
15	1.0000	1.0000	1.0000	1.0000	1.0000	0.9997	0.9941	0.9490	0.7625	0.3704	0.0432	0.0026	0.0000	15
16	1.0000	1.0000	1.0000	1.0000	1.0000	1.0000	0.9987	0.9840	0.8929	0.5886	0.1330	0.0159	0.0000	16
17	1.0000	1.0000	1.0000	1.0000	1.0000	1.0000	0.9998	0.9964	0.9645	0.7939	0.3231	0.0755	0.0010	17
18	1.0000	1.0000	1.0000	1.0000	1.0000	1.0000	1.0000	0.9995	0.9924	0.9308	0.6083	0.2642	0.0169	18
19	1.0000	1.0000	1.0000	1.0000	1.0000	1.0000	1.0000	1.0000	0.9992	0.9885	0.8784	0.6415	0.1821	19

n = 25

r	.01	.05	.10	.20	.30	.40	*p* .50	.60	.70	.80	.90	.95	.99	*r*
0	0.7778	0.2774	0.0718	0.0038	0.0001	0.0000	0.0000	0.0000	0.0000	0.0000	0.0000	0.0000	0.0000	0
1	0.9742	0.6424	0.2712	0.0274	0.0016	0.0001	0.0000	0.0000	0.0000	0.0000	0.0000	0.0000	0.0000	1
2	0.9980	0.8729	0.5371	0.0982	0.0090	0.0004	0.0000	0.0000	0.0000	0.0000	0.0000	0.0000	0.0000	2
3	0.9999	0.9659	0.7636	0.2340	0.0332	0.0024	0.0001	0.0000	0.0000	0.0000	0.0000	0.0000	0.0000	3
4	1.0000	0.9928	0.9020	0.4207	0.0905	0.0095	0.0005	0.0000	0.0000	0.0000	0.0000	0.0000	0.0000	4
5	1.0000	0.9988	0.9666	0.6167	0.1935	0.0294	0.0020	0.0001	0.0000	0.0000	0.0000	0.0000	0.0000	5
6	1.0000	0.9998	0.9905	0.7800	0.3407	0.0736	0.0073	0.0003	0.0000	0.0000	0.0000	0.0000	0.0000	6
7	1.0000	1.0000	0.9977	0.8909	0.5118	0.1536	0.0216	0.0012	0.0000	0.0000	0.0000	0.0000	0.0000	7
8	1.0000	1.0000	0.9995	0.9532	0.6769	0.2735	0.0539	0.0043	0.0001	0.0000	0.0000	0.0000	0.0000	8
9	1.0000	1.0000	0.9999	0.9827	0.8106	0.4246	0.1148	0.0132	0.0005	0.0000	0.0000	0.0000	0.0000	9
10	1.0000	1.0000	1.0000	0.9944	0.9022	0.5858	0.2122	0.0344	0.0018	0.0000	0.0000	0.0000	0.0000	10
11	1.0000	1.0000	1.0000	0.9985	0.9558	0.7323	0.3450	0.0778	0.0060	0.0001	0.0000	0.0000	0.0000	11
12	1.0000	1.0000	1.0000	0.9996	0.9825	0.8462	0.5000	0.1538	0.0175	0.0004	0.0000	0.0000	0.0000	12
13	1.0000	1.0000	1.0000	0.9999	0.9940	0.9222	0.6550	0.2677	0.0442	0.0015	0.0000	0.0000	0.0000	13
14	1.0000	1.0000	1.0000	1.0000	0.9982	0.9656	0.7878	0.4142	0.0978	0.0056	0.0000	0.0000	0.0000	14
15	1.0000	1.0000	1.0000	1.0000	0.9995	0.9868	0.8852	0.5754	0.1894	0.0173	0.0001	0.0000	0.0000	15
16	1.0000	1.0000	1.0000	1.0000	0.9999	0.9957	0.9461	0.7265	0.3231	0.0468	0.0005	0.0000	0.0000	16
17	1.0000	1.0000	1.0000	1.0000	1.0000	0.9988	0.9784	0.8464	0.4882	0.1091	0.0023	0.0000	0.0000	17
18	1.0000	1.0000	1.0000	1.0000	1.0000	0.9997	0.9927	0.9264	0.6593	0.2200	0.0095	0.0002	0.0000	18
19	1.0000	1.0000	1.0000	1.0000	1.0000	0.9999	0.9980	0.9706	0.8065	0.3833	0.0334	0.0012	0.0000	19
20	1.0000	1.0000	1.0000	1.0000	1.0000	1.0000	0.9995	0.9905	0.9095	0.5793	0.0980	0.0072	0.0000	20
21	1.0000	1.0000	1.0000	1.0000	1.0000	1.0000	0.9999	0.9976	0.9668	0.7660	0.2364	0.0341	0.0001	21
22	1.0000	1.0000	1.0000	1.0000	1.0000	1.0000	1.0000	0.9996	0.9910	0.9018	0.4629	0.1271	0.0020	22
23	1.0000	1.0000	1.0000	1.0000	1.0000	1.0000	1.0000	0.9999	0.9984	0.9726	0.7288	0.3576	0.0258	23
24	1.0000	1.0000	1.0000	1.0000	1.0000	1.0000	1.0000	1.0000	0.9999	0.9962	0.9282	0.7226	0.2222	24

Source: P. Billingsley, D. J. Croft, D. V. Huntsberger, and C. J. Watson, *Statistical Inference for Management and Economics* (Boston: Allyn & Bacon, 1986). Reprinted with permission.

TABLE A–3 Cumulative Poisson Probabilities

$$\sum_{x=x'}^{\infty} \frac{e^{-m}m^x}{x!}$$

x'	m = 0.1	0.2	0.3	0.4	0.5	0.6	0.7	0.8	0.9	1.0
0	1.0000	1.0000	1.0000	1.0000	1.0000	1.0000	1.0000	1.0000	1.0000	1.0000
1	.0952	.1813	.2592	.3297	.3935	.4512	.5034	.5507	.5934	.6321
2	.0047	.0175	.0369	.0616	.0902	.1219	.1558	.1912	.2275	.2642
3	.0002	.0011	.0036	.0079	.0144	.0231	.0341	.0474	.0629	.0803
4	.0000	.0001	.0003	.0008	.0018	.0034	.0058	.0091	.0135	.0190
5	.0000	.0000	.0000	.0001	.0002	.0004	.0008	.0014	.0023	.0037
6	.0000	.0000	.0000	.0000	.0000	.0000	.0001	.0002	.0003	.0006
7	.0000	.0000	.0000	.0000	.0000	.0000	.0000	.0000	.0000	.0001

x'	m = 1.1	1.2	1.3	1.4	1.5	1.6	1.7	1.8	1.9	2.0
0	1.0000	1.0000	1.0000	1.0000	1.0000	1.0000	1.0000	1.0000	1.0000	1.0000
1	.6671	.6988	.7275	.7534	.7769	.7981	.8173	.8347	.8504	.8647
2	.3010	.3374	.3732	.4082	.4422	.4751	.5068	.5372	.5663	.5940
3	.0996	.1205	.1429	.1665	.1912	.2166	.2428	.2694	.2963	.3233
4	.0257	.0338	.0431	.0537	.0656	.0788	.0932	.1087	.1253	.1429
5	.0054	.0077	.0107	.0143	.0186	.0237	.0296	.0364	.0441	.0527
6	.0010	.0015	.0022	.0032	.0045	.0060	.0080	.0104	.0132	.0166
7	.0001	.0003	.0004	.0006	.0009	.0013	.0019	.0026	.0034	.0045
8	.0000	.0000	.0001	.0001	.0002	.0003	.0004	.0006	.0008	.0011
9	.0000	.0000	.0000	.0000	.0000	.0000	.0001	.0001	.0002	.0002

x'	m = 2.1	2.2	2.3	2.4	2.5	2.6	2.7	2.8	2.9	3.0
0	1.0000	1.0000	1.0000	1.0000	1.0000	1.0000	1.0000	1.0000	1.0000	1.0000
1	.8775	.8892	.8997	.9093	.9179	.9257	.9328	.9392	.9450	.9502
2	.6204	.6454	.6691	.6916	.7127	.7326	.7513	.7689	.7854	.8009
3	.3504	.3773	.4040	.4303	.4562	.4816	.5064	.5305	.5540	.5768
4	.1614	.1806	.2007	.2213	.2424	.2640	.2859	.3081	.3304	.3528
5	.0621	.0725	.0838	.0959	.1088	.1226	.1371	.1523	.1682	.1847
6	.0204	.0249	.0300	.0357	.0420	.0490	.0567	.0651	.0742	.0839
7	.0059	.0075	.0094	.0116	.0142	.0172	.0206	.0244	.0287	.0335
8	.0015	.0020	.0026	.0033	.0042	.0053	.0066	.0081	.0099	.0119
9	.0003	.0005	.0006	.0009	.0011	.0015	.0019	.0024	.0031	.0038
10	.0001	.0001	.0001	.0002	.0003	.0004	.0005	.0007	.0009	.0011
11	.0000	.0000	.0000	.0000	.0001	.0001	.0001	.0002	.0002	.0003
12	.0000	.0000	.0000	.0000	.0000	.0000	.0000	.0000	.0001	.0001

TABLE A–3 ***(continued)***

					m					
x′	3.1	3.2	3.3	3.4	3.5	3.6	3.7	3.8	3.9	4.0
0	1.0000	1.0000	1.0000	1.0000	1.0000	1.0000	1.0000	1.0000	1.0000	1.0000
1	.9550	.9592	.9631	.9666	.9698	.9727	.9753	.9776	.9798	.9817
2	.8153	.8288	.8414	.8532	.8641	.8743	.8838	.8926	.9008	.9084
3	.5988	.6201	.6406	.6603	.6792	.6973	.7146	.7311	.7469	.7619
4	.3752	.3975	.4197	.4416	.4634	.4848	.5058	.5265	.5468	.5665
5	.2018	.2194	.2374	.2558	.2746	.2936	.3128	.3322	.3516	.3712
6	.0943	.1054	.1171	.1295	.1424	.1559	.1699	.1844	.1994	.2149
7	.0388	.0446	.0510	.0579	.0653	.0733	.0818	.0909	.1005	.1107
8	.0142	.0168	.0198	.0231	.0267	.0308	.0352	.0401	.0454	.0511
9	.0047	.0057	.0069	.0083	.0099	.0117	.0137	.0160	.0185	.0214
10	.0014	.0018	.0022	.0027	.0033	.0040	.0048	.0058	.0069	.0081
11	.0004	.0005	.0006	.0008	.0010	.0013	.0016	.0019	.0023	.0028
12	.0001	.0001	.0002	.0002	.0003	.0004	.0005	.0006	.0007	.0009
13	.0000	.0000	.0000	.0001	.0001	.0001	.0001	.0002	.0002	.0003
14	.0000	.0000	.0000	.0000	.0000	.0000	.0000	.0000	.0001	.0001

					m					
x′	4.1	4.2	4.3	4.4	4.5	4.6	4.7	4.8	4.9	5.0
0	1.0000	1.0000	1.0000	1.0000	1.0000	1.0000	1.0000	1.0000	1.0000	1.0000
1	.9834	.9850	.9864	.9877	.9889	.9899	.9909	.9918	.9926	.9933
2	.9155	.9220	.9281	.9337	.9389	.9437	.9482	.9523	.9561	.9596
3	.7762	.7898	.8026	.8149	.8264	.8374	.8477	.8575	.8667	.8753
4	.5858	.6046	.6228	.6406	.6577	.6743	.6903	.7058	.7207	.7350
5	.3907	.4102	.4296	.4488	.4679	.4868	.5054	.5237	.5418	.5595
6	.2307	.2469	.2633	.2801	.2971	.3142	.3316	.3490	.3665	.3840
7	.1214	.1325	.1442	.1564	.1689	.1820	.1954	.2092	.2233	.2378
8	.0573	.0639	.0710	.0786	.0866	.0951	.1040	.1133	.1231	.1334
9	.0245	.0279	.0317	.0358	.0403	.0451	.0503	.0558	.0618	.0681
10	.0095	.0111	.0129	.0149	.0171	.0195	.0222	.0251	.0283	.0318
11	.0034	.0041	.0048	.0057	.0067	.0078	.0090	.0104	.0120	.0137
12	.0011	.0014	.0017	.0020	.0024	.0029	.0034	.0040	.0047	.0055
13	.0003	.0004	.0005	.0007	.0008	.0010	.0012	.0014	.0017	.0020
14	.0001	.0001	.0002	.0002	.0003	.0003	.0004	.0005	.0006	.0007
15	.0000	.0000	.0000	.0001	.0001	.0001	.0001	.0001	.0002	.0002
16	.0000	.0000	.0000	.0000	.0000	.0000	.0000	.0000	.0001	.0001

					m					
x′	5.1	5.2	5.3	5.4	5.5	5.6	5.7	5.8	5.9	6.0
0	1.0000	1.0000	1.0000	1.0000	1.0000	1.0000	1.0000	1.0000	1.0000	1.0000
1	.9939	.9945	.9950	.9955	.9959	.9963	.9967	.9970	.9973	.9975
2	.9628	.9658	.9686	.9711	.9734	.9756	.9776	.9794	.9811	.9826
3	.8835	.8912	.8984	.9052	.9116	.9176	.9232	.9285	.9334	.9380
4	.7487	.7619	.7746	.7867	.7983	.8094	.8200	.8300	.8396	.8488
5	.5769	.5939	.6105	.6267	.6425	.6579	.6728	.6873	.7013	.7149
6	.4016	.4191	.4365	.4539	.4711	.4881	.5050	.5217	.5381	.5543
7	.2526	.2676	.2829	.2983	.3140	.3297	.3456	.3616	.3776	.3937
8	.1440	.1551	.1665	.1783	.1905	.2030	.2159	.2290	.2424	.2560
9	.0748	.0819	.0894	.0974	.1056	.1143	.1234	.1328	.1426	.1528

(continued)

TABLE A–3 (*continued*)

x′	m = 5.1	5.2	5.3	5.4	5.5	5.6	5.7	5.8	5.9	6.0
10	.0356	.0397	.0441	.0488	.0538	.0591	.0648	.0708	.0772	.0839
11	.0156	.0177	.0200	.0225	.0253	.0282	.0314	.0349	.0386	.0426
12	.0063	.0073	.0084	.0096	.0110	.0125	.0141	.0160	.0179	.0201
13	.0024	.0028	.0033	.0038	.0045	.0051	.0059	.0068	.0078	.0088
14	.0008	.0010	.0012	.0014	.0017	.0020	.0023	.0027	.0031	.0036
15	.0003	.0003	.0004	.0005	.0006	.0007	.0009	.0010	.0012	.0014
16	.0001	.0001	.0001	.0002	.0002	.0002	.0003	.0004	.0004	.0005
17	.0000	.0000	.0000	.0001	.0001	.0001	.0001	.0001	.0001	.0002
18	.0000	.0000	.0000	.0000	.0000	.0000	.0000	.0000	.0000	.0001

x′	m = 6.1	6.2	6.3	6.4	6.5	6.6	6.7	6.8	6.9	7.0
0	1.0000	1.0000	1.0000	1.0000	1.0000	1.0000	1.0000	1.0000	1.0000	1.0000
1	.9978	.9980	.9982	.9983	.9985	.9986	.9988	.9989	.9990	.9991
2	.9841	.9854	.9866	.9877	.9887	.9897	.9905	.9913	.9920	.9927
3	.9423	.9464	.9502	.9537	.9570	.9600	.9629	.9656	.9680	.9704
4	.8575	.8658	.8736	.8811	.8882	.8948	.9012	.9072	.9129	.9182
5	.7281	.7408	.7531	.7649	.7763	.7873	.7978	.8080	.8177	.8270
6	.5702	.5859	.6012	.6163	.6310	.6453	.6594	.6730	.6863	.6993
7	.4098	.4258	.4418	.4577	.4735	.4892	.5047	.5201	.5353	.5503
8	.2699	.2840	.2983	.3127	.3272	.3419	.3567	.3715	.3864	.4013
9	.1633	.1741	.1852	.1967	.2084	.2204	.2327	.2452	.2580	.2709
10	.0910	.0984	.1061	.1142	.1226	.1314	.1404	.1498	.1505	.1695
11	.0469	.0514	.0563	.0614	.0668	.0726	.0786	.0849	.0916	.0985
12	.0224	.0250	.0277	.0307	.0339	.0373	.0409	.0448	.0490	.0534
13	.0100	.0113	.0127	.0143	.0160	.0179	.0199	.0221	.0245	.0270
14	.0042	.0048	.0055	.0063	.0071	.0080	.0091	.0102	.0115	.0128
15	.0016	.0019	.0022	.0026	.0030	.0034	.0039	.0044	.0050	.0057
16	.0006	.0007	.0008	.0010	.0012	.0014	.0016	.0018	.0021	.0024
17	.0002	.0003	.0003	.0004	.0004	.0005	.0006	.0007	.0008	.0010
18	.0001	.0001	.0001	.0001	.0002	.0002	.0002	.0003	.0003	.0004
19	.0000	.0000	.0000	.0000	.0001	.0001	.0001	.0001	.0001	.0001

x′	m = 7.1	7.2	7.3	7.4	7.5	7.6	7.7	7.8	7.9	8.0
0	1.0000	1.0000	1.0000	1.0000	1.0000	1.0000	1.0000	1.0000	1.0000	1.0000
1	.9992	.9993	.9993	.9994	.9994	.9995	.9995	.9996	.9996	.9997
2	.9933	.9939	.9944	.9949	.9953	.9957	.9961	.9964	.9967	.9970
3	.9725	.9745	.9764	.0781	.9797	.9812	.9826	.9839	.9851	.9862
4	.9233	.9281	.9326	.9368	.9409	.9446	.9482	.9515	.9547	.9576
5	.8359	.8445	.8527	.8605	.8679	.8751	.8819	.8883	.8945	.9004
6	.7119	.7241	.7360	.7474	.7586	.7693	.7797	.7897	.7994	.8088
7	.5651	.5796	.5940	.6080	.6218	.6354	.6486	.6616	.6743	.6866
8	.4162	.4311	.4459	.4607	.4754	.4900	.5044	.5188	.5330	.5470
9	.2840	.2973	.3108	.3243	.3380	.3518	.3657	.3796	.3935	.4075

TABLE A–3 (*continued*)

					m					
x′	**7.1**	**7.2**	**7.3**	**7.4**	**7.5**	**7.6**	**7.7**	**7.8**	**7.9**	**8.0**
10	.1798	.1904	.2012	.2123	.2236	.2351	.2469	.2589	.2710	.2834
11	.1058	.1133	.1212	.1293	.1378	.1465	.1555	.1648	.1743	.1841
12	.0580	.0629	.0681	.0735	.0792	.0852	.0915	.0980	.1048	.1119
13	.0297	.0327	.0358	.0391	.0427	.0464	.0504	.0546	.0591	.0638
14	.0143	.0159	.0176	.0195	.0216	.0238	.0261	.0286	.0313	.0342
15	.0065	.0073	.0082	.0092	.0103	.0114	.0127	.0141	.0156	.0173
16	.0028	.0031	.0036	.0041	.0046	.0052	.0059	.0066	.0074	.0082
17	.0011	.0013	.0015	.0017	.0020	.0022	.0026	.0029	.0033	.0037
18	.0004	.0005	.0006	.0007	.0008	.0009	.0011	.0012	.0014	.0016
19	.0002	.0002	.0002	.0003	.0003	.0004	.0004	.0005	.0006	.0006
20	.0001	.0001	.0001	.0001	.0001	.0001	.0002	.0002	.0002	.0003
21	.0000	.0000	.0000	.0000	.0000	.0000	.0001	.0001	.0001	.0001

					m					
x′	**8.1**	**8.2**	**8.3**	**8.4**	**8.5**	**8.6**	**8.7**	**8.8**	**8.9**	**9.0**
0	1.0000	1.0000	1.0000	1.0000	1.0000	1.0000	1.0000	1.0000	1.0000	1.0000
1	.9997	.9997	.9998	.9998	.9998	.9998	.9998	.9998	.9999	.9999
2	.9972	.9975	.9977	.9979	.9981	.9982	.9984	.9985	.9987	.9988
3	.9873	.9882	.9891	.9900	.9907	.9914	.9921	.9927	.9932	.9938
4	.9604	.9630	.9654	.9677	.9699	.9719	.9738	.9756	.9772	.9788
5	.9060	.9113	.9163	.9211	.9256	.9299	.9340	.9379	.9416	.9450
6	.8178	.8264	.8347	.8427	.8504	.8578	.8648	.8716	.8781	.8843
7	.6987	.7104	.7219	.7330	.7438	.7543	.7645	.7744	.7840	.7932
8	.5609	.5746	.5881	.6013	.6144	.6272	.6398	.6522	.6643	.6761
9	.4214	.4353	.4493	.4631	.4769	.4906	.5042	.5177	.5311	.5443
10	.2959	.3085	.3212	.3341	.3470	.3600	.3731	.3863	.3994	.4126
11	.1942	.2045	.2150	.2257	.2366	.2478	.2591	.2706	.2822	.2940
12	.1193	.1269	.1348	.1429	.1513	.1600	.1689	.1780	.1874	.1970
13	.0687	.0739	.0793	.0850	.0909	.0971	.1035	.1102	.1171	.1242
14	.0372	.0405	.0439	.0476	.0514	.0555	.0597	.0642	.0689	.0739
15	.0190	.0209	.0229	.0251	.0274	.0299	.0325	.0353	.0383	.0415
16	.0092	.0102	.0113	.0125	.0138	.0152	.0168	.0184	.0202	.0220
17	.0042	.0047	.0053	.0059	.0066	.0074	.0082	.0091	.0101	.0111
18	.0018	.0021	.0023	.0027	.0030	.0034	.0038	.0043	.0048	.0053
19	.0008	.0009	.0010	.0011	.0013	.0015	.0017	.0019	.0022	.0024
20	.0003	.0003	.0004	.0005	.0005	.0006	.0007	.0008	.0009	.0011
21	.0001	.0001	.0002	.0002	.0002	.0002	.0003	.0003	.0004	.0004
22	.0000	.0000	.0001	.0001	.0001	.0001	.0001	.0001	.0002	.0002
23	.0000	.0000	.0000	.0000	.0000	.0000	.0000	.0000	.0001	.0001

					m					
x′	**9.1**	**9.2**	**9.3**	**9.4**	**9.5**	**9.6**	**9.7**	**9.8**	**9.9**	**10**
0	1.0000	1.0000	1.0000	1.0000	1.0000	1.0000	1.0000	1.0000	1.0000	1.0000
1	.9999	.9999	.9999	.9999	.9999	.9999	.9999	.9999	1.0000	1.0000
2	.9989	.9990	.9991	.9991	.9992	.9993	.9993	.9994	.9995	.9995
3	.9942	.9947	.9951	.9955	.9958	.9962	.9965	.9967	.9970	.9972
4	.9802	.9816	.9828	.9840	.9851	.9862	.9871	.9880	.9889	.9897

(*continued*)

TABLE A–3 ***(continued)***

	m									
x′	**9.1**	**9.2**	**9.3**	**9.4**	**9.5**	**9.6**	**9.7**	**9.8**	**9.9**	**10**
5	.9483	.9514	.9544	.9571	.9597	.9622	.9645	.9667	.9688	.9707
6	.8902	.8959	.9014	.9065	.9115	.9162	.9207	.9250	.9290	.9329
7	.8022	.8108	.8192	.8273	.8351	.8426	.8498	.8567	.8634	.8699
8	.6877	.6990	.7101	.7208	.7313	.7416	.7515	.7612	.7706	.7798
9	.5574	.5704	.5832	.5958	.6082	.6204	.6324	.6442	.6558	.6672
10	.4258	.4389	.4521	.4651	.4782	.4911	.5040	.5168	.5295	.5421
11	.3059	.3180	.3301	.3424	.3547	.3671	.3795	.3920	.4045	.4170
12	.2068	.2168	.2270	.2374	.2480	.2588	.2697	.2807	.2919	.3032
13	.1316	.1393	.1471	.1552	.1636	.1721	.1809	.1899	.1991	.2084
14	.0790	.0844	.0900	.0958	.1019	.1081	.1147	.1214	.1284	.1355
15	.0448	.0483	.0520	.0559	.0600	.0643	.0688	.0735	.0784	.0835
16	.0240	.0262	.0285	.0309	.0335	.0362	.0391	.0421	.0454	.0487
17	.0122	.0135	.0148	.0162	.0177	.0194	.0211	.0230	.0249	.0270
18	.0059	.0066	.0073	.0081	.0089	.0098	.0108	.0119	.0130	.0143
19	.0027	.0031	.0034	.0038	.0043	.0048	.0053	.0059	.0065	.0072
20	.0012	.0014	.0015	.0017	.0020	.0022	.0025	.0028	.0031	.0035
21	.0005	.0006	.0007	.0008	.0009	.0010	.0011	.0013	.0014	.0016
22	.0002	.0002	.0003	.0003	.0004	.0004	.0005	.0005	.0006	.0007
23	.0001	.0001	.0001	.0001	.0001	.0002	.0002	.0002	.0003	.0003
24	.0000	.0000	.0000	.0000	.0001	.0001	.0001	.0001	.0001	.0001

	m									
x′	**11**	**12**	**13**	**14**	**15**	**16**	**17**	**18**	**19**	**20**
0	1.0000	1.0000	1.0000	1.0000	1.0000	1.0000	1.0000	1.0000	1.0000	1.0000
1	1.0000	1.0000	1.0000	1.0000	1.0000	1.0000	1.0000	1.0000	1.0000	1.0000
2	.9998	.9999	1.0000	1.0000	1.0000	1.0000	1.0000	1.0000	1.0000	1.0000
3	.9988	.9995	.9998	.9999	1.0000	1.0000	1.0000	1.0000	1.0000	1.0000
4	.9951	.9977	.9990	.9995	.9998	.9999	1.0000	1.0000	1.0000	1.0000
5	.9849	.9924	.9963	.9982	.9991	.9996	.9998	.9999	1.0000	1.0000
6	.9625	.9797	.9893	.9945	.9972	.9986	.9993	.9997	.9998	.9999
7	.9214	.9542	.9741	.9858	.9924	.9960	.9979	.9990	.9995	.9997
8	.8568	.9105	.9460	.9684	.9820	.9900	.9946	.9971	.9985	.9992
9	.7680	.8450	.9002	.9379	.9626	.9780	.9874	.9929	.9961	.9979
10	.6595	.7576	.8342	.8906	.9301	.9567	.9739	.9846	.9911	.9950
11	.5401	.6528	.7483	.8243	.8815	.9226	.9509	.9696	.9817	.9892
12	.4207	.5384	.6468	.7400	.8152	.8730	.9153	.9451	.9653	.9786
13	.3113	.4240	.5369	.6415	.7324	.8069	.8650	.9083	.9394	.9610
14	.2187	.3185	.4270	.5356	.6368	.7255	.7991	.8574	.9016	.9339
15	.1460	.2280	.3249	.4296	.5343	.6325	.7192	.7919	.8503	.8951
16	.0926	.1556	.2364	.3306	.4319	.5333	.6285	.7133	.7852	.8435
17	.0559	.1013	.1645	.2441	.3359	.4340	.5323	.6250	.7080	.7789
18	.0322	.0630	.1095	.1728	.2511	.3407	.4360	.5314	.6216	.7030
19	.0177	.0374	.0698	.1174	.1805	.2577	.3450	.4378	.5305	.6186

TABLE A–3 *(concluded)*

	m									
x′	**11**	**12**	**13**	**14**	**15**	**16**	**17**	**18**	**19**	**20**
20	.0093	.0213	.0427	.0765	.1248	.1878	.2637	.3491	.4394	.5297
21	.0047	.0116	.0250	.0479	.0830	.1318	.1945	.2693	.3528	.4409
22	.0023	.0061	.0141	.0288	.0531	.0892	.1385	.2009	.2745	.3563
23	.0010	.0030	.0076	.0167	.0327	.0582	.0953	.1449	.2069	.2794
24	.0005	.0015	.0040	.0093	.0195	.0367	.0633	.1011	.1510	.2125
25	.0002	.0007	.0020	.0050	.0112	.0223	.0406	.0683	.1067	.1568
26	.0001	.0003	.0010	.0026	.0062	.0131	.0252	.0446	.0731	.1122
27	.0000	.0001	.0005	.0013	.0033	.0075	.0152	.0282	.0486	.0779
28	.0000	.0001	.0002	.0006	.0017	.0041	.0088	.0173	.0313	.0525
29	.0000	.0000	.0001	.0003	.0009	.0022	.0050	.0103	.0195	.0343
30	.0000	.0000	.0000	.0001	.0004	.0011	.0027	.0059	.0118	.0218
31	.0000	.0000	.0000	.0001	.0002	.0006	.0014	.0033	.0070	.0135
32	.0000	.0000	.0000	.0000	.0001	.0003	.0007	.0018	.0040	.0081
33	.0000	.0000	.0000	.0000	.0000	.0001	.0004	.0010	.0022	.0047
34	.0000	.0000	.0000	.0000	.0000	.0001	.0002	.0005	.0012	.0027
35	.0000	.0000	.0000	.0000	.0000	.0000	.0001	.0002	.0006	.0015
36	.0000	.0000	.0000	.0000	.0000	.0000	.0000	.0001	.0003	.0008
37	.0000	.0000	.0000	.0000	.0000	.0000	.0000	.0001	.0002	.0004
38	.0000	.0000	.0000	.0000	.0000	.0000	.0000	.0000	.0001	.0002
39	.0000	.0000	.0000	.0000	.0000	.0000	.0000	.0000	.0000	.0001
40	.0000	.0000	.0000	.0000	.0000	.0000	.0000	.0000	.0000	.0001

Source: *CRC Standard Mathematical Tables,* 16th ed., The Chemical Rubber Company, 1968.

TABLE A–4
Normal Probability Distribution and Partial Expectations

Standardized Variate	Probabilities		Partial Expectations	
z	$F(z)$	$1 - F(z)$	$L(z)$	$L(-z)$
.00	.5000	.5000	.3989	.3989
.01	.5040	.4960	.3940	.4040
.02	.5080	.4920	.3890	.4090
.03	.5120	.4880	.3841	.4141
.04	.5160	.4840	.3793	.4193
.05	.5200	.4800	.3744	.4244
.06	.5239	.4761	.3697	.4297
.07	.5279	.4721	.3649	.4349
.08	.5319	.4681	.3602	.4402
.09	.5359	.4641	.3556	.4456
.10	.5398	.4602	.3509	.4509
.11	.5438	.4562	.3464	.4564
.12	.5478	.4522	.3418	.4618
.13	.5517	.4483	.3373	.4673
.14	.5557	.4443	.3328	.4728
.15	.5596	.4404	.3284	.4784
.16	.5636	.4364	.3240	.4840
.17	.5685	.4325	.3197	.4897
.18	.5714	.4286	.3154	.4954
.19	.5753	.4247	.3111	.5011
.20	.5793	.4207	.3069	.5069
.21	.5832	.4168	.3027	.5127
.22	.5871	.4129	.3027	.5186
.23	.5910	.4090	.2944	.5244
.24	.5948	.4052	.2904	.5304
.25	.5987	.4013	.2863	.5363
.26	.6026	.3974	.2824	.5424
.27	.6064	.3936	.2784	.5484
.28	.6103	.3897	.2745	.5545
.29	.6141	.3859	.2706	.5606
.30	.6179	.3821	.2668	.5668
.31	.6217	.3783	.2630	.5730
.32	.6255	.3745	.2592	.5792
.33	.6293	.3707	.2555	.5855
.34	.6331	.3669	.2518	.5918
.35	.6368	.3632	.2481	.5981
.36	.6406	.3594	.2445	.6045
.37	.6443	.3557	.2409	.6109
.38	.6480	.3520	.2374	.6174
.39	.6517	.3483	.2339	.6239
.40	.6554	.3446	.2304	.6304
.41	.6591	.3409	.2270	.6370
.42	.6628	.3372	.2236	.6436
.43	.6664	.3336	.2203	.6503
.44	.6700	.3300	.2169	.6569
.45	.6736	.3264	.2137	.6637
.46	.6772	.3228	.2104	.6704
.47	.6808	.3192	.2072	.6772
.48	.6844	.3156	.2040	.6840
.49	.6879	.3121	.2009	.6909
.50	.6915	.3085	.1978	.6978

TABLE A–4
(continued)

Standardized Variate z	Probabilities $F(z)$	$1 - F(z)$	Partial Expectations $L(z)$	$L(-z)$
.51	.6950	.3050	.1947	.7047
.52	.6985	.3015	.1917	.7117
.53	.7019	.2981	.1887	.7187
.54	.7054	.2946	.1857	.7257
.55	.7088	.2912	.1828	.7328
.56	.7123	.2877	.1799	.7399
.57	.7157	.2843	.1771	.7471
.58	.7190	.2810	.1742	.7542
.59	.7224	.2776	.1714	.7614
.60	.7257	.2743	.1687	.7687
.61	.7291	.2709	.1659	.7759
.62	.7324	.2676	.1633	.7833
.63	.7357	.2643	.1606	.7906
.64	.7389	.2611	.1580	.7980
.65	.7422	.2578	.1554	.8054
.66	.7454	.2546	.1528	.8128
.67	.7486	.2514	.1503	.8203
.68	.7517	.2483	.1478	.8278
.69	.7549	.2451	.1453	.8353
.70	.7580	.2420	.1429	.8429
.71	.7611	.2389	.1405	.8505
.72	.7642	.2358	.1381	.8581
.73	.7673	.2327	.1358	.8658
.74	.7703	.2297	.1334	.8734
.75	.7733	.2267	.1312	.8812
.76	.7764	.2236	.1289	.8889
.77	.7793	.2207	.1267	.8967
.78	.7823	.2177	.1245	.9045
.79	.7852	.2148	.1223	.9123
.80	.7881	.2119	.1202	.9202
.81	.7910	.2090	.1181	.9281
.82	.7939	.2061	.1160	.9360
.83	.7967	.2033	.1140	.9440
.84	.7996	.2004	.1120	.9520
.85	.8023	.1977	.1100	.9600
.86	.8051	.1949	.1080	.9680
.87	.8067	.1922	.1061	.9761
.88	.8106	.1894	.1042	.9842
.89	.8133	.1867	.1023	.9923
.90	.8159	.1841	.1004	1.0004
.91	.8186	.1814	.0986	1.0086
.92	.8212	.1788	.0968	1.0168
.93	.8238	.1762	.0955	1.0250
.94	.8264	.1736	.0953	1.0330
.95	.8289	.1711	.0916	1.0416
.96	.8315	.1685	.0899	1.0499
.97	.8340	.1660	.0882	1.0582
.98	.8365	.1635	.0865	1.0665
.99	.8389	.1611	.0849	1.0749
1.00	.8413	.1587	.0833	1.0833
1.01	.8438	.1562	.0817	1.0917
1.02	.8461	.1539	.0802	1.1002
1.03	.8485	.1515	.0787	1.1087
1.04	.8508	.1492	.0772	1.1172
1.05	.8531	.1469	.0757	1.1257

(continued)

TABLE A–4
(continued)

Standardized Variate	Probabilities		Partial Expectations	
z	$F(z)$	$1 - F(z)$	$L(z)$	$L(-z)$
1.06	.8554	.1446	.0742	1.1342
1.07	.8577	.1423	.0728	1.1428
1.08	.8599	.1401	.0714	1.1514
1.09	.8621	.1379	.0700	1.1600
1.10	.8643	.1357	.0686	1.1686
1.11	.8665	.1335	.0673	1.1773
1.12	.8686	.1314	.0659	1.1859
1.13	.8708	.1292	.0646	1.1946
1.14	.8729	.1271	.0634	1.2034
1.15	.8749	.1251	.0621	1.2121
1.16	.8770	.1230	.0609	1.2209
1.17	.8790	.1210	.0596	1.2296
1.18	.8810	.1190	.0584	1.2384
1.19	.8830	.1170	.0573	1.2473
1.20	.8849	.1151	.0561	1.2561
1.21	.8869	.1131	.0550	1.2650
1.22	.8888	.1112	.0538	1.2738
1.23	.8907	.1093	.0527	1.2827
1.24	.8925	.1075	.0517	1.2917
1.25	.8943	.1057	.0506	1.3006
1.26	.8962	.1038	.0495	1.3095
1.27	.8980	.1020	.0485	1.3185
1.28	.8997	.1003	.0475	1.3275
1.29	.9015	.0985	.0465	1.3365
1.30	.9032	.0968	.0455	1.3455
1.31	.9049	.0951	.0446	1.3446
1.32	.9066	.0934	.0436	1.3636
1.33	.9082	.0918	.0427	1.3727
1.34	.9099	.0901	.0418	1.3818
1.35	.9115	.0885	.0409	1.3909
1.36	.9131	.0869	.0400	1.4000
1.37	.9147	.0853	.0392	1.4092
1.38	.9162	.0838	.0383	1.4183
1.39	.9177	.0823	.0375	1.4275
1.40	.9192	.0808	.0367	1.4367
1.41	.9207	.0793	.0359	1.4459
1.42	.9222	.0778	.0351	1.4551
1.43	.9236	.0764	.0343	1.4643
1.44	.9251	.0749	.0336	1.4736
1.45	.9265	.0735	.0328	1.4828
1.46	.9279	.0721	.0321	1.4921
1.47	.9292	.0708	.0314	1.5014
1.48	.9306	.0694	.0307	1.5107
1.49	.9319	.0681	.0300	1.5200
1.50	.9332	.0668	.0293	1.5293
1.51	.9345	.0655	.0286	1.5386
1.52	.9357	.0643	.0280	1.5480
1.53	.9370	.0630	.0274	1.5574
1.54	.9382	.0618	.0267	1.5667
1.55	.9394	.0606	.0261	1.5761
1.56	.9406	.0594	.0255	1.5855
1.57	.9418	.0582	.0249	1.5949
1.58	.9429	.0571	.0244	1.6044
1.59	.9441	.0559	.0238	1.6138

TABLE A–4
(continued)

Standardized Variate	Probabilities		Partial Expectations	
z	$F(z)$	$1 - F(z)$	$L(z)$	$L(-z)$
1.60	.9460	.0540	.0232	1.6232
1.61	.9463	.0537	.0227	1.6327
1.62	.9474	.0526	.0222	1.6422
1.63	.9484	.0516	.0216	1.6516
1.64	.9495	.0505	.0211	1.6611
1.65	.9505	.0495	.0206	1.6706
1.66	.9515	.0485	.0201	1.6801
1.67	.9525	.0475	.0197	1.6897
1.68	.9535	.0465	.0192	1.6992
1.69	.9545	.0455	.0187	1.7087
1.70	.9554	.0446	.0183	1.7183
1.71	.9564	.0436	.0178	1.7278
1.72	.9573	.0427	.0174	1.7374
1.73	.9582	.0418	.0170	1.7470
1.74	.9591	.0409	.0166	1.7566
1.75	.9599	.0401	.0162	1.7662
1.76	.9608	.0392	.0158	1.7558
1.77	.9616	.0384	.0154	1.7854
1.78	.9625	.0375	.0150	1.7950
1.79	.9633	.0367	.0146	1.8046
1.80	.9641	.0359	.0143	1.8143
1.81	.9649	.0351	.0139	1.8239
1.82	.9656	.0344	.0136	1.8436
1.83	.9664	.0336	.0132	1.8432
1.84	.9671	.0329	.0129	1.8529
1.85	.9678	.0322	.0126	1.8626
1.86	.9685	.0314	.0123	1.8723
1.87	.9693	.0307	.0119	1.8819
1.88	.9699	.0301	.0116	1.8916
1.89	.9706	.0294	.0113	1.9013
1.90	.9713	.0287	.0111	1.9111
1.91	.9719	.0281	.0108	1.9208
1.92	.9726	.0274	.0105	1.9305
1.93	.9732	.0268	.0102	1.9402
1.94	.9738	.0262	.0100	1.9500
1.95	.9744	.0256	.0097	1.9597
1.96	.9750	.0250	.0094	1.9694
1.97	.9756	.0244	.0092	1.9792
1.98	.9761	.0239	.0090	1.9890
1.99	.9767	.0233	.0087	1.9987
2.00	.9772	.0228	.0085	2.0085
2.01	.9778	.0222	.0083	2.0183
2.02	.9783	.0217	.0080	2.0280
2.03	.9788	.0212	.0078	2.0378
2.04	.9793	.0207	.0076	2.0476
2.05	.9798	.0202	.0074	2.0574
2.06	.9803	.0197	.0072	2.0672
2.07	.9808	.0192	.0072	2.0770
2.08	.9812	.0188	.0068	2.0868
2.09	.9817	.0183	.0066	2.0966
2.10	.9821	.0179	.0065	2.1065
2.11	.9826	.0174	.0063	2.1163
2.12	.9830	.0170	.0061	2.1261
2.13	.9834	.0166	.0060	2.1360
2.14	.9838	.0162	.0058	2.1458

(continued)

TABLE A–4
(continued)

Standardized Variate	Probabilities		Partial Expectations	
z	$F(z)$	$1 - F(z)$	$L(z)$	$L(-z)$
2.15	.9842	.0158	.0056	2.1556
2.16	.9846	.0154	.0055	2.1655
2.17	.9850	.0150	.0053	2.1753
2.18	.9854	.0146	.0052	2.1852
2.19	.9857	.0143	.0050	2.1950
2.20	.9861	.0139	.0049	2.2049
2.21	.9864	.0136	.0048	2.2148
2.22	.9868	.0132	.0046	2.2246
2.23	.9871	.0129	.0045	2.2345
2.24	.9875	.0125	.0044	2.2444
2.25	.9878	.0122	.0042	2.2542
2.26	.9881	.0119	.0041	2.2641
2.27	.9884	.0116	.0040	2.2740
2.28	.9887	.0113	.0039	2.2839
2.29	.9890	.0110	.0038	2.2938
2.30	.9893	.0107	.0037	2.3037
2.31	.9896	.0104	.0036	2.3136
2.32	.9898	.0102	.0035	2.3235
2.33	.9901	.0099	.0034	2.3334
2.34	.9904	.0096	.0033	2.3433
2.35	.9906	.0094	.0032	2.3532
2.36	.9909	.0091	.0031	2.3631
2.37	.9911	.0089	.0030	2.3730
2.38	.9913	.0087	.0029	2.3829
2.39	.9916	.0084	.0028	2.3928
2.40	.9918	.0082	.0027	2.4027
2.41	.9920	.0080	.0026	2.4126
2.42	.9922	.0078	.0026	2.4226
2.43	.9925	.0075	.0025	2.4325
2.44	.9927	.0073	.0024	2.4424
2.45	.9929	.0071	.0023	2.4523
2.46	.9931	.0069	.0023	2.4623
2.47	.9932	.0068	.0022	2.4722
2.48	.9934	.0066	.0021	2.4821
2.49	.9936	.0064	.0021	2.4921
2.50	.9938	.0062	.0020	2.5020
2.51	.9940	.0060	.0019	2.5119
2.52	.9941	.0059	.0019	2.5219
2.53	.9943	.0057	.0018	2.5318
2.54	.9945	.0055	.0018	2.5418
2.55	.9946	.0054	.0017	2.5517
2.56	.9948	.0052	.0017	2.5617
2.57	.9949	.0051	.0016	2.5716
2.58	.9951	.0049	.0016	2.5816
2.59	.9952	.0048	.0015	2.5915
2.60	.9953	.0047	.0015	2.6015
2.61	.9955	.0045	.0014	2.6114
2.62	.9956	.0044	.0014	2.6214
2.63	.9957	.0043	.0013	2.6313
2.64	.9959	.0041	.0013	2.6413
2.65	.9960	.0040	.0012	2.6512
2.66	.9961	.0039	.0012	2.6612
2.67	.9962	.0038	.0012	2.6712
2.68	.9963	.0037	.0011	2.6811
2.69	.9964	.0036	.0011	2.6911
2.70	.9965	.0035	.0011	2.7011

TABLE A–4
(concluded)

Standardized Variate	Probabilities		Partial Expectations	
z	$F(z)$	$1 - F(z)$	$L(z)$	$L(-z)$
2.71	.9966	.0034	.0010	2.7110
2.72	.9967	.0033	.0010	2.7210
2.73	.9968	.0032	.0010	2.7310
2.74	.9969	.0031	.0009	2.7409
2.75	.9970	.0030	.0009	2.7509
2.76	.9971	.0029	.0009	2.7609
2.77	.9972	.0028	.0008	2.7708
2.78	.9973	.0027	.0008	2.7808
2.79	.9974	.0026	.0008	2.7908
2.80	.9974	.0026	.0008	2.8008
2.81	.9975	.0025	.0007	2.8107
2.82	.9976	.0024	.0007	2.8207
2.83	.9977	.0023	.0007	2.8307
2.84	.9977	.0023	.0007	2.8407
2.85	.9978	.0022	.0006	2.8506
2.86	.9979	.0021	.0006	2.8606
2.87	.9979	.0021	.0006	2.8706
2.88	.9980	.0020	.0006	2.8806
2.89	.9981	.0019	.0006	2.8906
2.90	.9981	.0019	.0005	2.9005
2.91	.9982	.0018	.0005	2.9105
2.92	.9982	.0018	.0005	2.9205
2.93	.9983	.0017	.0005	2.9305
2.94	.9984	.0016	.0005	2.9405
2.95	.9984	.0016	.0005	2.9505
2.96	.9985	.0015	.0004	2.9604
2.97	.9985	.0015	.0004	2.9704
2.98	.9986	.0014	.0004	2.9804
2.99	.9986	.0014	.0004	2.9904
3.00	.9986	.0014	.0004	3.0004

Source: R. G. Brown, *Decision Rules for Inventory Management* (Hinsdale, IL.: Dryden, 1967). Adapted from Table VI, pp. 95–103.

TABLE A–5
Factors d_2 for R Charts

Number of Observations in Subgroup, n	Factor d_2, $d_2 = \frac{\bar{R}}{\sigma}$
2	1.128
3	1.693
4	2.059
5	2.326
6	2.534
7	2.704
8	2.847
9	2.970
10	3.078
11	3.173
12	3.258
13	3.336
14	3.407
15	3.472
16	3.532
17	3.588
18	3.640
19	3.689
20	3.735
21	3.778
22	3.819
23	3.858
24	3.895
25	3.931
30	4.086
35	4.213
40	4.322
45	4.415
50	4.498
55	4.572
60	4.639
65	4.699
70	4.755
75	4.806
80	4.854
85	4.898
90	4.939
95	4.978
100	5.015

Note: These factors assume sampling from a normal population.

Source: E. L. Grant and R. S. Leavenworth, *Statistical Quality Control,* 6th ed. (New York: McGraw-Hill, 1988).

TABLE A–6
Factors d_3 and d_4 for *R* Charts

Number of Observations in Subgroup, n	Factors for *R* Chart	
	Lower Control Limit d_3	Upper Control Limit d_4
2	0	3.27
3	0	2.57
4	0	2.28
5	0	2.11
6	0	2.00
7	.08	1.92
8	.14	1.86
9	.18	1.82
10	.22	1.78
11	.26	1.74
12	.28	1.72
13	.31	1.69
14	.33	1.67
15	.35	1.65
16	.36	1.64
17	.38	1.62
18	.39	1.61
19	.40	1.60
20	.41	1.59

Note: These factors assume sampling from a normal population. They give the upper and the lower control limits for an *R* chart as follows:

$$\text{LCL} = d_3 R$$

$$\text{UCL} = d_4 R$$

Source: E. L. Grant and R. S. Leavenworth, *Statistical Quality Control,* 6th ed. (New York: McGraw-Hill, 1988).

INDEX

D

E

F

N

O

Q

R

T

U

V

W

X

Z